AF680974

Contemporary Physicists

Gian-Carlo Rota
David Sharp

Editors

Birkhäuser
Boston · Basel · Stuttgart

James Glimm
Arthur Jaffe

Collected Papers
Volume 1

Quantum Field Theory and
Statistical Mechanics

Expositions

Birkhäuser
Boston · Basel · Stuttgart
1985

James Glimm
Courant Institute
New York, N.Y.
USA

Arthur Jaffe
Harvard University
Cambridge, Mass.
USA

Library of Congress Cataloging in Publication Data

Glimm, James.
Quantum field theory and statistical mechanics.

(Contemporary physicists)
Reprint of articles originally published 1969–1977.
Includes bibliographies.
1. Quantum field theory—Addresses, essays, lectures.
2. Statistical mechanics—Addresses, essays, lectures.
I. Jaffe, Arthur, 1937– . II. Title. III. Series.
QC174.46.G585 1985b 530.1'43 84–21524
ISBN 3-7643-3271-9 (Basel)
ISBN 0-8176-3271-9 (Boston)

CIP-Kurztitelaufnahme der Deutschen Bibliothek

Glimm, James:
Collected papers / James Glimm ; Arthur Jaffe. –
Boston ; Basel ; Stuttgart : Birkhäuser
ISBN 3-7643-3273-5
(Contemporary physicists)

NE: Jaffe, Arthur:; Glimm, James: [Sammlung];
Jaffe, Arthur: [Sammlung]

Vol. 1. Quantum field theory
and statistical mechanics : expositions. – 1985.
ISBN 3-7643-3271-9

All rights reserved.
No part of this publication may be reproduced, stored in a retrieval system, or transmitted in any form or by any means, electronic, mechanical, photocopying or otherwise, without the prior permission of the copyright owner.

© 1985 Birkhäuser Boston Inc.
Printed in Switzerland
ISBN 0-8176-3273-5 (complete set) 3-7643-3273-5 (complete set)
ISBN 0-8176-3271-9 (Vol. 1) 3-7643-3271-9 (Vol. 1)
ISBN 0-8176-3272-7 (Vol. 2) 3-7643-3272-7 (Vol. 2)

James Glimm/Arthur Jaffe

Collected Papers – Volume 1

Collected Papers – Volume 2

I Construction of $P(\varphi)_2$

II Phase Cell Localization and ${\varphi_3}^4$ Stability

III Phase Transitions Exist

IV Phase Transitions and Critical Behavior

V Particle Structure

VI Bounds on Coupling Constants

VII Confinement and Instantons

VIII Reflection Positivity

Introduction

This volume contains a selection of expository articles on quantum field theory and statistical mechanics by James Glimm and Arthur Jaffe. They include a solution of the original interacting quantum field equations and a description of the physics which these equations contain. Quantum fields were proposed in the late 1920s as the natural framework which combines quantum theory with relativity. They have survived ever since.

The mathematical description for quantum theory starts with a Hilbert space H of state vectors. Quantum fields are linear operators on this space, which satisfy nonlinear wave equations of fundamental physics, including coupled Dirac, Maxwell and Yang-Mills equations. The field operators are restricted to satisfy a "locality" requirement that they commute (or anti-commute in the case of fermions) at space-like separated points. This condition is compatible with finite propagation speed, and hence with special relativity. Asymptotically, these fields converge for large time to linear fields describing free particles. Using these ideas a scattering theory had been developed, based on the existence of local quantum fields.

Whenever physicists had attempted to find solutions to such equations for quantum fields, they ran into formal difficulties. The requirement of Lorentz invariance forced the quantum fields to exhibit local singularities which appeared incompatible with the possibility to interpret nonlinear functions of the fields and therefore with nonlinear equations of motion. In practical terms, calculations led to infinite answers, which physicists learned to avoid only by adhering to a set of rules known as "renormalization." One interpretation of renormalization is that certain constants in the quantum mechanical Hamiltonian must be infinite, as is demonstrated in the paper **An infinite renormalization of the Hamiltonian is necessary.**

The rules to perform calculations were developed in the 1940s for the coupled equations of Maxwell and Dirac, i.e., for electrodynamics, by Bethe, Dyson, Feynman, Schwinger, Tomonaga, Weisskopf, and others. They gave a procedure for calculating two important observable deviations from Dirac's equations: the Lamb shift in the hydrogen spectrum and the "anomalous" magnetic moment of the electron and hence led physicists to conclude that quantum field theory was a correct interpretation of nature. The tenth of one percent correction to the magnetic moment predicted by Dirac can now be measured to one part in 10^7 and provides the most accurate comparison in physics between experiment and calculation. For this reason, physicists accept quantum field theory and apply these ideas to understand weak and strong forces as well as electromagnetic and gravitational ones. The quest for appropriate equations to describe elementary particles led to a unified framework for weak and electromagnetic interactions, namely a quantum theory with $SU(2) \times U(1)$ gauge fields. One speculates that SU(5) may also unify the strong forces and even further that a quantum theory including gravity can be formulated.

In spite of the success of renormalization in electrodynamics, and its extension to other quantum field interactions, there was no picture of how to fit quantum fields into a traditional mathematical framework. In the mid 1950s, a number of physicists including Haag, Jost, Lehmann, Symanzik, Wightman, and Zimmermann set up a general scheme into which a putative field theory could fit, while Hepp, Ruelle, and others investigated consequences of their proposal. Within this structure one could derive general relations such as the connection between spin and statistics, the PCT theorem, and scattering theory. These features are independent of the detailed equations of motion. However no non-trivial examples were known. This gap led to an effort over the last twenty years to demonstrate that quantum fields do exist and to analyze their structure starting from classical Lagrangians. This program has become known as *Constructive Quantum Field Theory,* or CQFT.

Ultimately CQFT has opened new directions both in physics and in mathematics. The two sets of lecture notes, **Field Theory Models and Boson Quantum Field Models,** develop the basic theme of CQFT from the Hamiltonian point of view. They appeared shortly after the first success of constructing two dimensional quantum field models and were intended as a first draft of a comprehensive monograph explaining the subject to both physicists and mathematicians; as such, they contain a comprehensive exposition and review of the literature. They were presented at summer schools in Les Houches (1970) and in London (1971) respectively, but the monograph was never completed.

The basic idea which led to understanding CQFT was *phase cell localization.* Phase space is the product of position and momentum spacer. The uncertainty principle (for instance in the form that a function and its Fourier transform cannot simultaneously have bounded support) places limits on allowed phase cell localization. One looked at parts of quantum fields, or degrees of freedom, approximately localized in a bounded region of phase space, namely in phase cell volumes of size O(1). One generically finds a finite number of degrees of freedom per unit volume. This guiding principle of analyzing degrees of freedom in geometrically increasing or geometrically decreasing momentum ranges, but O(1) phase space volume, was implemented throughout CQFT. It led to expansion methods for the study both of infrared problems and of ultraviolet problems. Related ideas underlie the renormalization group approach.

In 1972 the subject turned in a different direction by incorporating certain decade-old ideas of K. Symanzik relating Euclidean field theory to quantum mechanics via function space integration. While function space integration had been the major tool in the proof of estimates for the Hamiltonian approach above, and entered through the Feynman-Kac representation of the heat kerne!, the explicit Euclidean symmetry in Symanzik's approach came to play a central role. Work focused on studying the existence and properties of Euclidean field; using the Osterwalder-Schrader reconstruction theorem one could infer the analytic continuation from Euclidean fields to quantum fields and consequences for the study of the Hamiltonian and its spectrum. Work from this point of view was developed by Nelson, Guerra, Rosen, Simon, and Fröhlich, as well as the authors.

The 1973 Erice lectures, **The particle structure of the weakly coupled $P\phi_2$ model and other applications of high temperature expansions,** reflect the point of view adopted by the authors that a combination of both methods can provide deep insights. **Part 1: physics of quantum field models** deals with physics which can be studied in the laboratory provided by mathematically exact quantum field models, as well as how to extend the constructive approach to dimension three. Questions of particle structure – e.e., the spectrum of the quantum mechanical Hamiltonian H, its bound states and scattering, and questions of phase transitions, symmetry breaking, and critical phenomena had begun to complement renormalization as central issues of interest. Later developments led to work by Spencer, Zirilli, Dimock, Eckmann and others on two particle scattering via an exact Bethe-Salpeter equation. In **Particles and bound states and progress toward unitarity and scaling,** the authors propose a program to study n-particle scattering using kernels they define which are n-particle irreducible in a given channel. Their application to the spectral properties of scattering has only been partially understood.

The second portion of the Erice lectures, **The cluster expansion,** presents one version of a technical tool to investigate some of these problems. This expansion technique has come to be known as the "cluster expansion," because it invokes a decoupling strategy similar to high temperature expansions in statistical physics also known by that name. The cluster expansion also adopts in detail the phase cell localization ideas discussed above. Originally, the cluster expansion was developed to study large-distance effects in correlations, so called infrared-phenomena associated with the spectrum of H. It now turns out that these ideas can be used in the analysis of with certain short-distance questions of local regularity, or ultraviolet phenomena.

It is within the Euclidean framework that the relation between statistical mechanics and quantum theory became evident. The study of Euclidean fields boils down to the study of the statistical mechanics of classical field theory (classical Grassmanian fields in the case of Fermions). The probability distribution is a Gibbs distribution with the exponential weight exp(-S), where the functional of the fields S is the classical (Euclidean) action functional. For this reason, it is not surprising that the phase transitions and critical phenomena of statistical physics appear in quantum field theory. This connection between field theory and statistical mechanics led to the first existence proof for degenerate ground states in a continuum quantum field theory (i.e. the existence of multiple vacua and symmetry breaking), through an analysis of the equilibrium states of the corresponding statistical mechanics system in Euclidean field theory. One aspect of this joint work of the authors and Thomas Spencer is described in **Existence of phase transitions for ϕ_2^4 quantum fields,** a paper presented at a June 1975 conference in Marseilles. The detailed proof, as well as the unification of these ideals with the cluster expansion methods, is published in journal articles. Once again the theme of phase cell localization enters, with a decomposition of the field into high and low momentum parts. The low momentum degrees of freedom in each bounded space-time volume are analyzed according to a Peierls-type argument from statistical physics, while the high momentum degrees of freedom are shown to be fluctuations which have only a small influence on the structure of the phase diagram. Extension of these ideas and methods have led to other developments in the theory of phase transitions in statistical physics as well as in quantum field theory.

The physics of the particle structure and its connection with phase transitions through the approach to the critical point is discussed further in the articles, **Critical problems in quantum field theory, Critical exponents and renormalization in the ϕ^4 scaling limit,** and **A tutorial course in constructive field theory.**

In summary it is fitting to publish together here expositions of the two alternative approaches to CQFT: the Hamiltonian approach based on Hilbert space methods and the Euclidean approach based on functional integrals and probability theory. The text *Quantum Physics* by the authors concentrates on the latter. In many ways the technical estimates for scalar fields are more straightforward within the probabilistic framework. In the case of fermion and gauge theories, the question is still debatable. Physicists and mathematicians now are interested again in understanding the operator formulation of quantum field theory, as well as the Euclidean formulation, since it may have advantages for the study of certain open problems. Thus the authors hope that making these papers accessible will provide a valuable working reference.

Bibliography

An infinite renormalization of the Hamiltonian is necessary, J. Math. Phys. 10, 2213–2214 (1969).

Field theory models, in *Statistical Mechanics and Quantum Field Theory,* C. De Witt and R. Stora, Editors, Gordon and Breach, New York 1971.

Boson quantum field models, in *Mathematics of Contemporary Physics,* R. Streater, Editor, Academic Press, London, 1972. (References at the end of the volume.)

The particle structure of the weakly coupled $P(\phi)_2$ model and other applications of high temperature expansions, Part I: physics of quantum field models, and Part II: the cluster expansion, with T. Spencer, in *Constructive Quantum Field Theory,* G. Velo and A. S. Wightman, Editors, Lecture Notes in Physics 25, Springer, New York 1973.

Particles and bound states and and progress toward unitarity and scaling, in *International Symposium on Mathematical Problems in Theoretical Physics,* H. Araki, Editor, Lecture Notes in Physics 39, Springer, New York 1975.

Critical problems in quantum fields, in *Les Méthods Mathématiques de la Théorie Quantique des Champs,* F. Guerra, D. Robinson, and R. Stora, Editors, CNRS, Paris 1976.

Existence of phase transitions for ϕ_2^4 quantum fields, with T. Spencer, in *Les Méthods Mathématiques de la Théorie Quantique des Champs,* F. Guerra, D. Robinson, and R. Stora, Editors, CNRS, Paris 1976.

Critical exponents and renormalization in the ϕ^4 scaling limit, in *Quantum Dynamics: Models and Mathematics,* L. Streit, Editor, Springer, Vienna 1976.

A tutorial course in constructive field theory, in *New Developments in Quantum Field Theory and Statistical Mechanics, Cargèse 1976,* P. Mitter and M. Levy, Editors, Plenum, New York 1977.

I Infinite Renormalization of the Hamiltonian Is Necessary

Infinite Renormalization of the Hamiltonian Is Necessary

JAMES GLIMM*
Courant Institute of Mathematical Sciences, New York University, New York, New York

AND

ARTHUR JAFFE†
Lyman Laboratory of Physics, Harvard University, Cambridge, Massachusetts

(Received 28 April 1969)

We show that the unrenormalized Hamiltonian in quantum field theory is unbounded from below whenever lowest-order perturbation theory indicates that this is true. We conclude that perturbation theory is an accurate guide to the divergence of the vacuum energy in quantum field theory.

In this paper we show that the unrenormalized Hamiltonian is unbounded from below when lowest-order perturbation theory predicts that this is true.

The proof is a simple calculation. The unrenormalized Hamiltonian H is a densely defined bilinear form on Fock space.[1] We choose a sequence Ω_κ of unit vectors in the domain of H. As $\kappa \to \infty$, the expectation values of the Hamiltonian in the vectors Ω_κ tends to $-\infty$. The ground state of H given by first-order perturbation theory motivates our choice of the vectors Ω_κ.

We concentrate on the boson self-interaction Φ^{2n} in $(s + 1)$-dimensional space–time. The second-order vacuum energy has a momentum divergence for $s \geq 2$, $n \geq 2$ and for $s \geq 3$, $n = 1$, and these cases will be treated first. Afterwards, the remaining cases (which have a volume divergence) will be treated.

The Hamiltonian that we study is

$$H = H_0 + \lambda \int :\Phi(x)^{2n}: dx = H_0 + \lambda H_I. \qquad (1)$$

The methods and the results hold equally for the spatially cut-off Hamiltonian:

$$H(g) = H_0 + \lambda \int :\Phi(x)^{2n}: g(x)\, dx, \quad 0 \leq g(x) \leq 1. \qquad (2)$$

We work on the Fock space; H_0 is the free-field Hamiltonian for mass $m > 0$, and Φ is the standard boson field.

We study first the cases $s \geq 2$, $n \geq 2$ and $s \geq 4$, $n \geq 1$. The vectors Ω_κ are defined by

$$\Omega_\kappa = c(\psi_0 - \lambda \psi_{2n}),$$

where ψ_0 is the Fock no-particle vector and ψ_{2n} is a $2n$-particle vector. The constant c is chosen so that $\|\Omega_\kappa\| = 1$, and

$$\psi_{2n}(k_1, \cdots, k_{2n}) = \left(\sum_{i=1}^{2n} \mu(k_i)\right)^{-1} ((2n)!)^{\frac{1}{2}} h\left(\sum_{i=1}^{2n} k_i\right) \prod_{i=1}^{2n} [\mu(k_i)^{\tau - \frac{1}{2}} \chi_\kappa(k_i)]. \qquad (3)$$

Here $\mu(k) = (k^2 + m^2)^{\frac{1}{2}}$ and h is a smooth, positive, rapidly decreasing function. The function $\chi_\kappa(k)$ equals unity if $|k| \leq \kappa$; it equals zero otherwise. We choose

$$\tau = -(s - 1)\left(1 - \frac{1}{2n}\right) + \frac{\epsilon + 2}{2n},$$

---

* Supported by the New York Science and Technology Foundation, Grant SSF-(8)-8.

† Alfred P. Sloan Foundation Fellow. Supported in part by the Air Force Office of Scientific Research, AF49(638)-1380.

[1] A. Galindo Proc. Natl. Acad. Sci. (U.S.) **48**, 1128 (1962).

where ϵ is in the interval $0 < \epsilon < \frac{1}{2}$. We remark that the choices $h = \delta$, $\kappa = \infty$, $\tau = 0$ would give the ground state in first-order perturbation theory. With the above restrictions on n, s, and ϵ, we have $\tau < -\frac{1}{8}$. Furthermore,

$$\epsilon = 2n\tau + (2n - 1)s - (2n + 1).$$

Theorem 1:

$$\lim_{\kappa \to \infty} (\Omega_\kappa, H\Omega_\kappa) = -\infty.$$

Proof: We compute the inner product as

$$(\Omega_\kappa, H\Omega_\kappa) = c^2\{P_0 + \lambda P_1 + \lambda^2 P_2 + \lambda^3 P_3\}.$$

In this expansion it is easy to see that

$$P_0 = P_1 = 0$$

and that

$$P_2 = (\psi_{2n}, H_0\psi_{2n}) - (\psi_0, H_I\psi_{2n}) - (\psi_{2n}, H_I\psi_0).$$

The proof is completed by showing that, for large κ,

$$c^{-2} = 1 + \lambda^2 \|\psi_{2n}\|^2 = O(1),$$
$$(\psi_{2n}, H_0\psi_{2n}) = O(1),$$
$$P_3 = O(1),$$

and that for some positive constant D,

$$D\kappa^\epsilon \le (\psi_0, H_I\psi_{2n}) = (\psi_{2n}, H_I\psi_0).$$

These orders of growth are established by standard power-counting arguments.[2] This completes the proof. The remaining case $n = 1, s = 3$ is handled by similar methods, modified to deal with a logarithmic divergence.

In the cases not covered by Theorem 1, perturbation theory predicts no momentum divergence. Thus, when g has compact support, perturbation theory predicts that $H(g)$ is bounded from below. This lower bound has been proved rigorously.[3] Perturbation theory predicts that H has a vacuum-energy divergence which is linear in the volume, and thus it predicts a lower bound for $H(g)$ which is linear in the volume. (The "volume" here is the area of the support of g.)

It is known that the true bounds are no worse than this prediction. Thus, for $s = 1$ or $s = 2$, $n = 1$, the lower bound diverges no faster than a constant times the volume.[3] We now show that, for $s = 1$, the H defined in Eq. (1) is unbounded from below. The same proof shows that the lower bound on $H(g)$ in Eq. (2) tends to $-\infty$ as $g \to 1$, and similar results hold for the case $s = 2$, $n = 1$.

Let

$$h_V(k) = Vh(kV).$$

In the definition (3) for ψ_{2n}, we substitute h_V for h and set $\tau = 0$, $\kappa = \infty$. We define

$$\Omega_V = \psi_0 - V^{-\frac{1}{2}}\lambda\psi_{2n}.$$

As before, one proves the following theorem:

Theorem 2:

$$\lim_{V \to \infty} (\Omega_V, H\Omega_V) = -\infty,$$

$$\limsup_{V \to \infty} \|\Omega_V\|^2 < \infty.$$

We conclude that perturbation theory is an accurate guide to the divergence of the vacuum energy in quantum field theory.

[2] S. Weinberg, Phys. Rev. **118**, 839 (1960).

[3] For the case $s = 1$, any n, see E. Nelson in *Mathematical Theory of Elementary Particles* (M.I.T. Press, Cambridge, Mass., 1966); J. Glimm, Commun. Math. Phys. **8**, 12 (1968); J. Glimm and A. Jaffe (to be published). The case $s = 2$, $n = 1$ can be computed explicitly or estimated.

II Quantum Field Theory Models: Part I. The φ_2^{2n} Model

* Supported in part by the U.S. Air Force Office of Scientific Research, Contract AF-49(638)-1719.

† Alfred P. Sloan Foundation Fellow. Supported in part by the U.S. Air Force Office of Scientific Research, Contract F 44 620-70-C-0030.

Introduction

Quantum fields are believed to provide a correct description of the interactions between particles. In the case of quantum electrodynamics, calculations based on perturbation theory provide exact agreement between theory and experiment within the limits of experimental accuracy. These calculations rely on infinite renormalizations and therefore reveal that quantum fields are highly singular. In these lectures, we will be concerned with the qualitative structure of quantum fields. Given a classical Lagrangian, we seek a mathematically complete construction of the corresponding quantum fields. Because of the difficulty of this problem, we have restricted ourselves to two space time dimensions. With this restriction, the program has been successful. Quantum fields have been constructed for the ϕ^{2n} and Yukawa interactions in two dimensions. Many properties of the fields have been verified for these models, and the ϕ^{2n} theory obeys all the Haag-Kastler axioms. Two major unanswered questions are:

a) Can the existence of local quantum fields be extended to three space time dimensions?

b) For the two dimensional models, can detailed properties of the fields be established in order to describe elementary particles and their bound states? Is standard perturbation theory asymptotic to the exact solution of the models? Do standard dispersion relations hold for the cross sections?

In quantum field theory there are four types of divergences, the infra red, ultraviolet, infinite volume and particle number divergences. The infra red divergence is associated with the divergence of integrals such as

$$\int_{|k|\leq 1} |k|^{-1}\, dk .$$

This divergence occurs at low frequencies (i.e. small momenta) and it occurs only when at least one particle has zero mass. This divergence is more singular in lower dimensions. Thus there is no necessity to analyse the infra red difficulties in two dimensions where they are worse than in the physical world of four dimensions, and we take all masses to be positive.

The ultraviolet divergences are associated with integrals such as

$$\int (k^2 + m^2)^{-1/2}\, dk$$

and these integrals become more divergent in higher dimensions. The ultraviolet divergences arise in the definition of the nonlinear terms in the field equations. For the ϕ^{2n} model, the classical field equation is

$$\phi_{tt} - \phi_{xx} + m_0^2\phi + 2n\phi^{2n-1} = 0.$$

The ultraviolet divergences have a rather trivial nature in the corresponding quantum field equation, because of the restriction to two dimensions. The Yukawa model is more singular and there is an infinite shift in the boson mass (and in the vacuum energy) due to the ultraviolet divergences. These divergences are predicted by second order perturbation theory. The mass divergence is cancelled by a renormalization of the field equation. We substitute

$$m^2 - \delta(m^2) = m_0^2$$

for m_0^2 in the field equation. Here m_0 and $-\delta m^2$ are infinite for the Yukawa model but finite for the ϕ^{2n} model. The infinite part of δm^2 is chosen to cancel other infinities in the theory. This leaves δm^2 uniquely fixed, modulo some finite contribution. According to conventional ideas, the latter is determined by a requirement placed on the solution ϕ of the field equations. Namely, ϕ should describe an elementary particle of (physical) mass $m = m_{\text{phys.}}$. Since we do not have satisfactory control over the finite part of the mass renormalization in our models, we do not discuss this point further.

The infinite volume divergence arises from the translation invariance of the theory, i.e. from conservation of momentum. In calculations, this divergence occurs as integrals of the form

$$\int_R dx = \int_{\hat{R}} \delta(k)^2 \, dk.$$

In a similar but less obvious fashion, the ultraviolet divergences are forced by the invariance of the theory under Lorentz rotations.

The divergence for large particle numbers has a different character from the ultraviolet or volume divergences. This divergence is not cancelled by some other infinity. Rather, it remains in the theory and reflects the fact that the Taylor's expansion of the S-matrix,

$$S = S(\lambda) \sim \sum \frac{\lambda^n}{n!} \left[\left(\frac{d}{d\lambda} \right)^n S(\lambda) \bigg|_{\lambda=0} \right],$$

appears to diverge. Here λ is the coupling constant, and is inserted into the field equation as a coefficient multiplying the nonlinear term. The Taylor's series is simply the perturbation expansion, and because it presumably diverges, we must use nonperturbative methods in an essential manner.

We now discuss the methods used to control the ultraviolet and volume divergences; we then discuss the particle number divergence separately. We first introduce an approximation into the problem to eliminate the divergences. For the ultraviolet divergences we use a momentum cutoff.

For example we may replace ϕ by $\phi_\varkappa = h_\varkappa * \phi$, where $h_\varkappa(x) = \varkappa h(\varkappa x)$ is a suitable function with total integral one. For the infinite volume divergence, we may use either (or occasionally both) of two approximations. We may destroy translation invariance by introducing a space cutoff into the nonlinear terms of the field equations, e.g.

$$\phi^{2n-1}(x, t) \to g(x)\, \phi^{2n-1}(x, t)$$

where the cutoff function g is nonnegative and has compact support. Alternatively, we may replace R by a circle (quantization in a box with periodic boundary conditions). For the cutoff theories there are no infinite divergences, but there are cancellations between large finite quantities. We perform the cancellations first, and then estimate the remainders in order to take the limits $\varkappa \to \infty$, $V \to \infty$ and $g \to 1$. In this program, the major difficulty is to obtain estimates which are valid uniformly as the cutoffs are removed.

In order to avoid the particle number divergence, we use nonperturbative methods. We use a Hamiltonian, which serves to reduce the nonlinear field equation to an equivalent linear problem. Let H be the total energy operator and let $H(g)$ and $H(g, \varkappa)$ be the total energy operators associated with the cutoff fields and cutoff field equations. The main point is to realize H (or $H(g)$) as a self adjoint operator, because a simple calculation shows that

$$\phi(x, t) = e^{itH(g)}\, \phi(x, 0)\, e^{-itH(g)}$$

solves the g-cutoff field equations. Furthermore, using finite propagation speed (i.e. the hyperbolic character of the field equations), we can piece together different g-cutoff fields for different g's to obtain a single field ϕ independent of g. The study of $H(g)$ is the linear problem which must be approached by nonperturbative methods. In fact,

$$H(g) = \lim_{\varkappa \to \infty} H(g, \varkappa) \quad \text{and} \quad H = \lim_{g \to 1} H(g)$$

are defined as limits, and most of the difficulty in our proofs occurs in the estimates on operators that we need to control these limits.

The Cauchy data $\phi(x, 0)$ and $\dot{\phi}(x, 0) = \pi(x, 0)$ in the equation above are time zero free fields. We begin our construction with $\phi(x, 0)$ given in the standard fashion, acting on Fock space. However, a complete treatment of the limit $g \to 1$ and the corresponding infinite volume renormalization will force us to modify this hypothesis. We must choose a non-Fock representation of $\phi(x, 0)$ and $\pi(x, 0)$, and the particular representation is related intimately to the interaction and the field equations we are solving.

In summary, our construction of the ϕ_2^{2n} model follows the outline of Figure 1. The Yukawa$_2$ model follows the same outline, but is at present completed only through the construction of the field operators and the verification of the field equations. A similar outline should hold for models in three space time dimensions.

$H(g, \varkappa) = H(g, \varkappa)^*$

↓

$H(g) = H(g)^*$ ⟶ Local Lorentz generator

↓ ↓

Finite propagation speed,
Bounded functions of fields

↓

Field operators,
Field equations ⟶ Lorentz covariance automorphisms

↓

Physical Hilbert space, $\mathscr{F}_{\text{ren}}$,
Hamiltonian H, Vacuum Ω,
Non Fock representation
of the field operators

Figure 1. Construction of the ϕ_2^{2n} quantum field theory.

Our final field theory should satisfy the usual requirements of physics: For the ϕ_2^{2n} model we show that the Hamiltonian is positive, $0 \leqq H$, and that H has a ground state Ω, $H\Omega = 0$. On the physical Hilbert space $\mathscr{F}_{\text{ren}}$, $\phi(x, t) = e^{itH} \phi(x, 0) e^{-itH}$. The field is local, so that a boson field satisfies $[\phi(x, t), \phi(x', t')] = 0$ if $|x - x'| > |t - t'|$. The theory is Lorentz covariant; the Lorentz transformations should be unitarily implemented, and so we expect that

$$U(a, \Lambda)\, \phi(x, t)\, U(a, \Lambda)^* = \phi(\{a, \Lambda\}\, (x, t)).$$

Physical particles should occur in the theory as discrete eigenvectors for the operator $(H^2 - P^2)^{1/2}$, where P is the momentum operator; see Glimm-Jaffe[9] for a more detailed discussion. The scattering of particles, in principle, is completely described by the fields, see Jost[1].

Part I The ϕ_2^{2n} Model

The ϕ_2^4 theory is the least singular and most highly developed of our quantum field theory models. The $P(\phi)_2$ theory is essentially as complete, with P a positive polynomial. Wightman[2] proposed the construction of local field theories as a limit of cutoff models. Two early results were the proof of semiboundedness of the Hamiltonian by Nelson[3] and the construction of space and momentum cutoff models by Jaffe[2,3]. Subsequent development of the theory was made by Glimm[3], by Glimm and Jaffe[2,3,4,5] and by Cannon and Jaffe[1]. These results were extended to the $P(\phi)_2$ theory by Rosen[1,2,3]. Simplifications of some of the proofs have been given by Segal[2,3] and by Hoegh-Krohn and Simon[1]. Segal[2] announced results for $P(\phi)_2$ which are independent of (and somewhat weaker than) Rosen's[1]. Other contributions to the $P(\phi)_2$ program are due to Lanford (see Wightman[1], Glimm and Jaffe[2]), Guenin[1], Segal[1], Jaffe and Powers[1], Federbush[1], Glimm and Jaffe[6,10-11], Jaffe, Lanford and Wightman[1], Hoegh-Krohn[1,2] and Simon[1]. For a further discussion of bibliographical points, see the cited articles.

1 Fock Space

1.1 Annihilation-creation forms

The Fock space $\mathscr{F}$ for our $P(\phi)_2$ model is the Hilbert space completion of the symmetric tensor algebra over $L_2(R)$,

$$\mathscr{F} = \mathfrak{S}(L_2(R)) = \bigoplus_{n=0}^{\infty} \mathscr{F}_n, \tag{1.1.1}$$

where $\mathscr{F}_n$ is the space of n noninteracting particles:

$$\mathscr{F}_n = L_2(R) \underset{s}{\otimes} L_2(R) \underset{s}{\otimes} \dots \underset{s}{\otimes} L_2(R) \qquad (n \text{ factors}). \tag{1.1.2}$$

The variable $k \in R$ denotes momentum. For $\psi = \{\psi_0, \psi_1, \dots\} \in \mathscr{F} = \mathscr{F}_0 \oplus \mathscr{F}_1 \oplus \dots$, we have $\|\psi\|^2 = \sum_{n=0}^{\infty} \|\psi_n\|^2$. The no particle space $\mathscr{F}_0$ is the complex numbers and $\Omega_0 = \{1, 0, \dots\} \in \mathscr{F}$ is the (bare) vacuum or (bare) no particle vector. We define the fractional order energy operator N_τ by

$$(N_\tau \psi)_n (k_1, \dots, k_n) = \sum_{j=1}^{n} \mu(k_j)^\tau \psi_n(k_1, \dots, k_n) \tag{1.1.3}$$

where $\mu(k) = (k^2 + m^2)^{1/2}$. Thus $N_0 = N$ is the number operator and $N_1 = H_0$ is the free energy operator. The annihilation operator $a(k)$ is defined by the formula

$$(a(k)\psi)_{n-1} (k_1, \dots, k_{n-1}) = n^{1/2} \psi_n(k, k_1, \dots, k_{n-1}). \tag{1.1.4}$$

We take as a domain for $a(k)$ the set $\mathscr{D}$ of vectors $\psi = \{\psi_n\}$ with a finite number of particles and with wave functions ψ_n in the Schwartz space $\mathscr{S}(R^n)$. Any monomial $A(k)^* A(k')$, where

$$A(k') = a(k_1') \dots a(k_s'),$$

$$A(k)^* = a(k_1)^* \dots a(k_r)^*,$$

is a bilinear form on $\mathscr{D} \times \mathscr{D}$, and for $\theta \in \mathscr{D}$,

$$\langle \theta, A(k)^* A(k') \theta \rangle \in \mathscr{S}(R^{r+s}).$$

Thus we define the weak integral of bilinear forms

$$W = \int A^*(k) \, w(k, k') \, A(k') \, dk \, dk' \tag{1.1.5}$$

for any distribution w in $\mathscr{S}'(R^{r+s})$. On $\mathscr{D} \times \mathscr{D}$, we compute that if $w(k, k') = \mu(k)^\tau \, \delta(k - k')$,

$$\begin{aligned} N_\tau &= \int a^*(k) \, \mu(k)^\tau \, \delta(k - k') \, a(k') \, dk \, dk' \\ &= \int a^*(k) \, \mu(k)^\tau \, a(k) \, dk. \end{aligned} \tag{1.1.6}$$

For the Y_2 model of Part II, our Fock space will be

$$\mathscr{F} = \mathscr{F}_b \otimes \mathscr{F}_f = \bigoplus_0^\infty \mathscr{F}_{n_b n_f} \tag{1.1.7}$$

where $\mathscr{F}_b$ is given by (1.1.1) and $\mathscr{F}_f$ is the Hilbert space completion of the antisymmetric tensor algebra

$$\begin{aligned} \mathscr{F}_f &= \mathfrak{G}(L_2(R) \oplus L_2(R)) \\ &= \bigoplus_{n=0}^\infty (L_2(R) \oplus L_2(R)) \otimes_a \ldots \otimes_a (L_2(R) \oplus L_2(R)). \end{aligned} \tag{1.1.8}$$

We have fermion particle and antiparticle annihilation operators

$$b(p) = b(p, 1), \qquad b'(p) = b(p, -1)$$

satisfying the commutation relations

$$b(p_1, \varepsilon_1) \, b(p_2, \varepsilon_2)^* + b(p_2, \varepsilon_2)^* \, b(p_1, \varepsilon_1) = \delta(p_1 - p_2) \, \delta_{\varepsilon_1, \varepsilon_2}$$

and also

$$a(k_1) \, a(k_2)^* - a(k_2)^* \, a(k_1) = \delta(k_1 - k_2).$$

The notation

$$a(k) = b(k, 0)$$

will also be convenient. On the Fock space (1.1.7),

$$N_\tau = \sum_{\varepsilon = 0, \pm 1} \int b(p, \varepsilon)^* \, \mu(p, \varepsilon)^\tau \, b(p, \varepsilon) \, dp$$

where $\mu(p, \varepsilon) = (p^2 + m(\varepsilon)^2)^{1/2}$ and $m(\varepsilon)$ is the rest mass of the corresponding particle. We take the domain $\mathscr{D}$ as before, and then W defined by

$$W = W_{\varepsilon, \varepsilon'} = \int A(k, \varepsilon)^* \, w(k, k', \varepsilon, \varepsilon') \, A(k', \varepsilon') \, dk \, dk' \tag{1.1.9}$$

is a bilinear form on $\mathscr{D} \times \mathscr{D}$ and a weak integral of bilinear forms where

$$A(k, \varepsilon)^* = b(k_1, \varepsilon_1)^* \ldots b(k_r, \varepsilon_r)^*.$$

We find it convenient to introduce a periodic approximation in configuration space. Under this approximation, the Fourier transform or momentum space variable $k \in R$ is replaced by a discrete variable $k \in Z_V$, where $Z_V = (2\pi/V)\,Z$ is the lattice of multiples (by $2\pi/V$) of integers. Thus we define $\mathscr{F}_V$, the boson Fock space for volume V, as

$$\mathscr{F}_V = \mathfrak{S}(l_2(Z_V)) = C \oplus l_2(Z_V) \oplus \left[l_2(Z_V) \underset{s}{\otimes} l_2(Z_V)\right] \oplus \dots. \tag{1.1.10}$$

We identify $\mathscr{F}_V$ with the subspace of $\mathscr{F}$ consisting of piecewise constant functions which are constant on each cube of volume $(2\pi/V)^j$ centered about a lattice point

$$\{k_1, \dots, k_j\} \in Z_V \times \dots \times Z_V = Z_V^j.$$

The periodic annihilation and creation operators $a_V(k)$ and $a_V^*(k)$ can be extended from $\mathscr{F}_V$ to $\mathscr{F}$ through the formulas

$$a_V^\#(k) = \left(\frac{V}{2\pi}\right)^{1/2} \int_{-\pi/V}^{\pi/V} a^\#(k + l)\, dl, \tag{1.1.11}$$

where $a^\#(k) = a(k)^*$ or $a(-k)$. We define

$$H_{0,V} = \int a(k)^* \,\mu(k_V)\, a(k)\, dk \tag{1.1.12}$$

with k_V a lattice point close to k,

$$k_V \in Z_V, \qquad |k - k_V| \leqq \pi/V. \tag{1.1.13}$$

On $\mathscr{F}_V$, this definition agrees with the standard formula

$$\sum_{k \in Z_V} a_V^*(k)\, \mu(k)\, a_V(k).$$

There is a similar periodic approximation to the Yukawa Fock space (1.1.7) which we will use without further comment. On (1.1.7) we have the periodic approximation

$$N_{\tau,V} = \sum_{\varepsilon = 0, \pm 1} \int b(k, \varepsilon)^* \,\mu(k_V, \varepsilon)^\tau\, b(k, \varepsilon)\, dk \tag{1.1.14}$$

to N_τ. On the boson Fock space, only the term $\varepsilon = 0$ occurs in N_τ or $N_{\tau,V}$.

1.2 N_τ estimates

The important operators of field theory have simple expressions in terms of the annihilation creation forms (1.1.5) and (1.1.9). These forms are dominated by powers of the operator N_τ. We call such bounds N_τ estimates.

Vectors in the tensor product $L_2(R^s) \otimes \mathscr{F}$ describe s labeled particles and an indefinite number of unlabeled particles, just as vectors in $\mathscr{F}$ describe an indefinite number of particles. Corresponding to the smooth domain $\mathscr{D} \subset \mathscr{F}$ defined in Section 1.1, we introduce the domain

$$\mathscr{S}(R^s) \otimes \mathscr{D} \subset L_2(R^s) \otimes \mathscr{F}$$

consisting of vectors with a finite number of particles (at least s particles; the first s variables are not symmetrized) and wave functions in $\mathscr{S}$. The annihilation operator $A(k, \varepsilon) = b(k_1, \varepsilon_1) \ldots b(k_s, \varepsilon_s)$ maps $\mathscr{D}$ into $\mathscr{D}$, and depends on the parameter k. If we regard $k \in R^s$ as a variable and not a parameter, then

$$A(., \varepsilon) : \mathscr{D} \to \mathscr{S}(R^s) \otimes \mathscr{D} \subset L_2(R^s) \otimes \mathscr{F}. \tag{1.2.1}$$

The operator

$$B(., \varepsilon) = A(., \varepsilon) \prod_{i=1}^{s} (\mu(k_i, \varepsilon_i)^{\tau_i/2} N_{\tau_i}^{-1/2}) \tag{1.2.2}$$

also maps $\mathscr{D}$ into $\mathscr{S}(R^s) \otimes \mathscr{D}$. Here $N_\tau^{-1/2}$ denotes the operator which gives zero on Ω_0 and is the inverse to $N_\tau^{1/2}$ on $\Omega_0^\perp$.

PROPOSITION 1.2.1 $\quad \|B(., \varepsilon)\| \leqq 1.$

Proof We express B as a product of operators mapping $\mathscr{S}(R^j) \otimes \mathscr{D} \to \mathscr{S}(R^{j+1}) \otimes \mathscr{D}$. It is sufficient to bound each of these operators; this amounts to proving the proposition in the case $s = 1$. To rearrange B into such a product we note that

$$b(k, \varepsilon)\, N_\tau^{-1/2} = (N_\tau + \mu(k, \varepsilon)^\tau)^{-1/2}\, b(k, \varepsilon).$$

Let $s = 1$, $\theta \in \mathscr{D}$. Then

$$\begin{aligned} \|\mu(., \varepsilon)^{\tau/2}\, b(., \varepsilon)\, \theta\|^2 &= \int \langle \theta, b(k, \varepsilon)^* \mu(k, \varepsilon)^\tau b(k, \varepsilon)\, \theta \rangle\, dk \\ &\leqq \langle \theta, N_\tau \theta \rangle = \|N_\tau^{1/2}\theta\|^2. \end{aligned}$$

If $\theta \perp \Omega_0$, we substitute $\psi = (N_\tau + aI)^{1/2}\, \theta$, $0 \leqq a$, to complete the proof.

Let $\prod_{i=1}^{r+s} \mu(k_i)^{-\tau_i/2}\, w$ be a bounded operator from $L_2(R^s) \otimes \mathscr{F} \to L_2(R^r) \otimes \mathscr{F}$. Then the integral (1.1.9) defines W as a bilinear form on

$$\mathscr{D}\left(\prod_{i=1}^{r} N_{\tau_i}^{1/2}\right) \times \mathscr{D}\left(\prod_{i=r+1}^{s} N_{\tau_i}^{1/2}\right)$$

and we have

PROPOSITION 1.2.2

$$\left\|\left(\prod_{i=1}^{r} N_{\tau_i}^{-1/2}\right) W \prod_{i=r+1}^{s} N_{\tau_i}^{-1/2}\right\| \leqq \left\|\prod_{i=1}^{r+s} \mu(k_i)^{-\tau_i/2}\, w\right\|.$$

Proof Both norms are operator norms. We apply Proposition 1.2.1.

We see that $N_\tau^{1/2}$ is sufficient to dominate each annihilation or creation operator. In the case of fermions, fewer powers of N_τ are needed provided the kernel $w(k)$ of w is sufficiently well behaved and is a scalar function. Let

$$\{1, \ldots, r + s\} = A \cup B \cup C$$

be a disjoint union with

$$C \subset \{1, \ldots, r\}, \qquad A \subset \{r + 1, \ldots, r + s\}$$

and let $|A|$, $|B|$ and $|C|$ denote the number of elements of A, B, C.

We now let $w(k, k') \in \mathscr{S}'(R^{r+s})$, so that w is a densely defined bilinear form on $\mathscr{S}(R^r) \times \mathscr{S}(R^s)$. We denote by $\|w\|$ the norm of the operator w from $L_2(R^s)$ to $L^2(R^r)$ given by the kernel $w(k, k')$, $k = k_1, \ldots, k_r$; $k' = k_{r+1}, \ldots, k_{r+s}$. Let W be given by (1.1.9).

PROPOSITION 1.2.3 (*a*) *There is a constant less than* $0(|r - s| + 1|)^{|B|/2}$ *such that*

$$\|N_\tau^{-|C|/2} W N_\tau^{-|A|/2} (N + I)^{-|B|/2}\| \leqq \text{const.} \left\| \prod_{l \in A \cup C} \mu(k_l)^{-\tau/2} w \right\|.$$

(*b*) *Suppose that B contains at least one fermion variable.*

$$\|N_\tau^{-|C|/2} W N_\tau^{-|A|/2} (N + I)^{-(|B|-1)/2}\| \leqq \text{const.} \left\| \prod_{l \in A \cup C} \mu(k_l)^{-\tau/2} w \right\|_2.$$

(*c*) *Let W contain exactly j boson variables. Then*

$$\| W(N + I)^{-j/2} \| \leqq |w|_s$$

where $|w|_s$ *is an appropriate Schwartz space norm.*

(*d*) *Let* $D \subset B$ *be a set of fermion variables* $p_1, \ldots, p_j$. *We write* $w = w(k, p)$ *to distinguish the variables p in D, and we assume* $w(k, p) = w(k, p_V)$. *Then*

$$\|N_\tau^{-|C|/2} W N_\tau^{-|A|/2} (N + I)^{-(|B|-|D|)/2}\| \leqq |w|_{1,2},$$

where

$$|w|_{1,2} = 0(V^{|D|/2}) \left\| \prod_{l \in A \cup C} \mu(k_l)^{-\tau/2} \int |w(., p)| \, dp \right\|_2.$$

Proof (a) We use Proposition 1.2.2 with $\tau_l = \tau$ or $\tau_l = 0$, but replace factors $N^{-1/2}$ by factors $(N + I)^{-1/2}$, using $\|(N + I)^{1/2} N^{-1/2}\| \leqq 2$. These factors can be placed on either side of W because w is a scalar kernel and

$$W(N + I)^{-1/2} = (N + I)^{-1/2} W(N + (r - s + 1) I)^{1/2} (N + I)^{-1/2}.$$

Since $\|(N + (r - s + 1) I)^{1/2} (N + I)^{-1/2}\| \leqq (|r - s| + 1)^{1/2}$, part (*a*) is established.

(b) Let k_j be a fermion variable in B. The cannonical anticommutation relations yield

$$\{b(f, \varepsilon), b(f, \varepsilon)^*\} = \|f\|^2$$

and $\|b(f, \varepsilon)\| = \|f\|$, since bb^* and b^*b have orthogonal ranges. Hence

$$\left\| \int w(k, k')\, b(k_j, \varepsilon_j)\, dk_j \right\|^2 = \int |w|^2\, dk_j$$

and the same bound holds if k_j is a creating variable. We apply Proposition 1.2.2 to the operator valued kernel

$$v(k_1, \ldots, k_{j-1}, k_{j+1}, \ldots, k_{r+s}) = \int dk_j\, w(k, k')\, b(k_j, \varepsilon_j)$$

with $\tau_i = \tau$ or 0. The operator norm of v satisfies

$$\begin{aligned}\left\| \prod_{i \in A \cup C} \mu(k_i)^{-\tau/2}\, v \right\| &\leqq \left(\int \prod_{i \in A \cup C} \mu(k_i)^{-\tau}\, \|v(k)\|^2\, dk_1 \ldots dk_{j-1}\, dk_{j+1} \ldots dk_{r+s} \right)^{1/2} \\ &= \left(\int \prod_{i \in A \cup C} \mu(k_i)^{-\tau}\, |w(k, k')|^2\, dk\, dk' \right)^{1/2} \\ &= \left\| \prod_{i \in A \cup C} \mu(k_i)^{-\tau/2}\, w \right\|_2 .\end{aligned}$$

Since w is a scalar kernel, the proof of part (a) above yields (b).

(c) We treat the fermion variables first. As in the proof of (b), the boson variables are handled by an application of Proposition 1.2.2, with an operator valued kernel. Thus we may suppose $j = 0$. Let $\{e_i\}$ be an orthonormal basis for $L_2(R)$ consisting of eigenfunctions of the Hermite operator

$$-\frac{d^2}{dk^2} + k^2 = H(k)$$

with eigenvalues $\lambda_i = 2i + 1$. We may suppose $w \in L_2$, and then

$$w = \Sigma \alpha_{i_1 \ldots i_{r+s}}\, e_{i_1} \otimes \ldots \otimes e_{i_{r+s}}$$

with $\Sigma |\alpha|^2 = \|w\|_2^2 < \infty$. Since

$$\left\| \int e_{i_\nu}(k_\nu)\, b(k_\nu, \varepsilon_\nu)\, dk_\nu \right\| = 1,$$

it is sufficient to show that $\Sigma |\alpha| \leqq |w|_s$ for some Schwartz norm $|.|_s$. However,

$$\begin{aligned}\Sigma |\alpha| &= \Sigma |\alpha_{i_1 \ldots i_{r+s}}| \prod_\nu (2i_\nu + 1) \prod_\nu (2i_\nu + 1)^{-1} \\ &\leqq \left(\Sigma_{i_1 \ldots i_{r+s}} \prod_\nu (2i_\nu + 1)^{-2} \right)^{1/2} \left(\Sigma_{i_1 \ldots i_{r+s}} |\alpha|^2 \prod_\nu (2i_\nu + 1)^2 \right)^{1/2} \\ &= \text{const.} \left\| \prod_\nu H(k_\nu)\, w \right\|_2 = |w|_s .\end{aligned}$$

We remark that for $j = 0$, the norm $|w|_s$ may be replaced by

$$\inf \Sigma|\alpha|,$$

where the infinum ranges over the choice of an orthonormal basis for $L^2(R)$.

(d) We write W in the form

$$W = \int A(k_1)^* A(p_1)^* w(k, p) A(p_2) A(k_2) dk\, dp$$

where $\{p\} = \{p_1, p_2\}$, $\{k\} = \{k_1, k_2\}$. We first estimate

$$X(k) = \int A(p_1)^* w(k, p) A(p_2) dp$$

$$= \left(\frac{2\pi}{V}\right)^{|D|/2} \sum_{p \in Z_V \times \ldots \times Z_V} A_V(p_1)^* w(k, p) A_V(p_2).$$

Since $\|b_V(k, \varepsilon)\| = 1$ for $\varepsilon = \pm 1$, we have

$$\|X(k)\| \leqq \left(\frac{2\pi}{V}\right)^{|D|/2} \sum_{p \in Z_V \times \ldots \times Z_V} |w(k, p)|$$

$$\leqq 0(|V|^{|D|/2}) \int |w(k, p)| dp.$$

By applying Proposition 1.2.2 to $\int A(k_1)^* X(k) A(k_2) dk$, we obtain, by the methods of part (a),

$$\|N_\tau^{-|C|/2} W N_\tau^{-|A|/2} (N + I)^{-(|B|-|D|)/2}\|$$

$$\leqq \text{const.} \left\| \prod_{i \in A \cup C} \mu(k_i)^{-\tau/2} \|X(k)\| \right\|_2$$

$$\leqq |w|_{1,2}.$$

To simplify the discussion of the Yukawa$_2$ interaction in part II, we give the following corollary to Proposition 1.2.3.

COROLLARY 1.2.4 *Let W_i have scalar kernels w_i and equal*

$$W_1 = \int a(k)^* b(p_1)^* b'(p_2)^* w_1(k, p_1, p_2) dk\, dp_1\, dp_2,$$

$$W_2 = \int b(p_1)^* b'(p_2)^* a(-k) w_2(k, p_1, p_2) dk\, dp_1\, dp_2,$$

$$W_3 = \int a(k)^* b(p_1, \varepsilon)^* b(-p_2, \varepsilon) w_3(k, p_1, p_2) dk\, dp_1\, dp_2, \quad \varepsilon = \pm 1,$$

$$W_4 = \int b(p_1) b(p_2) w_4(p_1, p_2) dp_1\, dp_2.$$

Then

$$\|(N_\tau + I)^{-1/2} W_i (N_\tau + I)^{-1/2}\| \leqq \text{const. } M_i,$$

and

$$\| W_4(N_\tau + I)^{-1/2} \| \leqq \text{const.}\, M_4,$$

where

$$M_1 = \|(\mu + \omega_1 + \omega_2)^{-\tau/2}\, w_1\|_2, \quad M_2 = \|(\omega_1 + \omega_2)^{-\tau/2}\, \mu^{-\tau/2}\, w_2\|_2$$

$$M_3 = \|\mu^{-\tau/2}\, \omega_2^{-\tau/2}\, w_3\|_2, \quad M_4 = \|(\omega_1 + \omega_2)^{-\tau/2}\, w_4\|_2 .$$

Proof We prove the estimate on W_1; the other cases are similar. Write $W_1 = R_1 + R_2 + R_3$ where the kernel $r_j = wE_j$ and E_j is the characteristic function of the subset of R^3 in which the jth variable is largest in magnitude. Apply Proposition 1.2.3 (b) to each R_j. Since $\mu^{-\tau/2}E_1 \leqq \text{const.}\,(\mu + \omega_1 + \omega_2)^{-\tau/2}\, E_1$, $\omega_1^{-\tau/2}E_2 \leqq \text{const.}\,(\mu + \omega_1 + \omega_2)^{-\tau/2}\, E_2$, $w_2^{-\tau/2}E_3 \leqq \text{const.}\,(\mu + \omega_1 + \omega_2)^{-\tau/2}\, E_3$, the bound

$$\|(N_\tau + I)^{-1/2}\, R_j(N_\tau + I)^{-1/2}\| \leqq \text{const.}\, \|(\mu + \omega_1 + \omega_2)^{-\tau/2}\, r_j\|_2$$

follows.

We now apply the N_τ estimates to estimate Wick polynomials in the boson field operators. Since

$$[\mu(k)/\mu(k_V)]^{\pm 1} \leqq \text{const.}$$

we have

$$[N_\tau/N_{\tau,V}]^{\pm 1} \leqq \text{const.}$$

and so the estimates could be given in terms of the operators $N_{\tau,V}$ instead.

Our time zero fields coincide with free fields,

$$\phi(x) = (4\pi)^{-1/2} \int e^{-ikx}\, \mu(k)^{-1/2}\, \{a^*(k) + a(-k)\}\, dk \tag{1.2.3}$$

$$\pi(x) = i(4\pi)^{-1/2} \int e^{-ikx}\, \mu(k)^{1/2}\, \{a^*(k) - a(-k)\}\, dk. \tag{1.2.4}$$

Similarly we define periodic fields $\phi_V(x)$ and $\pi_V(x)$ by replacing $\mu(k)$ above by $\mu(k_V)$. The canonical commutation relations

$$[\phi(x), \pi(y)] = i\, \delta(x - y) \tag{1.2.5}$$

hold in exponentiated form. The Wick powers of ϕ are given by the formula

$$:\phi(x)^n: = \sum_{j=0}^{n} \binom{n}{j} w_x(k)\, a^*(k_1) \ldots a^*(k_j)\, a(-k_{j+1}) \ldots a(-k_n) \tag{1.2.6}$$

where

$$w_x(k) = (4\pi)^{-n/2}\, \{\mu(k_1) \ldots \mu(k_n)\}^{-1/2}\, e^{-i(\Sigma_l k_l)x}. \tag{1.2.7}$$

The space averaged Wick powers,

$$:\phi^n(g): = \int :\phi(x)^n:\, g(x)\, dx, \tag{1.2.8}$$

$$:\phi_V^n: = \int_{-V/2}^{V/2} :\phi_V(x)^n:\, dx, \tag{1.2.9}$$

have kernels

$$w_g(k) = (2\pi)^{1/2} (4\pi)^{-n/2} \{\mu(k_1) \dots \mu(k_n)\}^{-1/2} \tilde{g}(\Sigma k_i) \qquad (1.2.10)$$

$$w_V(k) = V(4\pi)^{-n/2} \{\mu(k_{1,V}) \dots \mu(k_{n,V})\}^{-1/2} \delta_{0,\Sigma k_{i,V}} \qquad (1.2.11)$$

PROPOSITION 1.2.5 *Let $g \in L_2$ and let $0 < \varepsilon$. Then*

$$\sup_i \mu(k_i)^{-\varepsilon+1/2} w_g \in L_2(R^n)$$

$$\sup_i \mu(k_i)^{-\varepsilon+1/2} w_V \in l_2(Z_V^n)$$

and the L_2 norms are dominated by const. $\|g\|_2$.

We omit the elementary proof. Let $P = \Sigma c_j \xi^j$ be a real polynomial, and (henceforth) let g be real. Then the operators

$$:P(\phi, g): = \Sigma c_j \int :\phi(x)^j: g(x)\, dx,$$

$$:P(\phi_V): = \Sigma c_j \int_{-V/2}^{V/2} :\phi_V(x)^j: dx$$

are symmetric and defined on the domain $\mathscr{D}(N^{n/2})$, where n is the degree of P, by Propositions 1.2.3-4. These operators are also essentially self adjoint on this domain, as we will see later.

PROPOSITION 1.2.6 *Let $(-\Delta + 1)^{-1/4} f \in L_2$ and let $(-\Delta + 1)^{1/4} h \in L_2$ with f and h real. Then $\phi(f)$ and $\pi(h)$ are essentially self adjoint on $\mathscr{D}$.*

Proof For $\theta \in \mathscr{D}$, we compute

$$\|\phi(f)^n \theta\| \leqq 0(c^n n!^{1/2}),$$

using Proposition 1.2.3 (a). Hence θ is an analytic vector for $\phi(f)$ and since $\mathscr{D}$ is dense, $\phi(f)$ is essentially self adjoint on $\mathscr{D}$, by a theorem of Nelson[1]. The same proof works for $\pi(h)$; the essential point is that the kernel lies in L_2.

The introduction of Q space in Chapter 2 is based on the following proposition. Let $\mathfrak{M}$ be the strongly closed operator algebra (von Neumann algebra) generated by the spectral projections of the time zero fields $\phi(f)$ $f \in C_0^\infty$.

PROPOSITION 1.2.7 *$\mathfrak{M}\Omega_0$ is dense in $\mathscr{F}$.*

Proof Let $\theta \perp \mathfrak{M}\Omega_0$. Since Ω_0 is an analytic vector for $\phi(f)$, the Taylor series for $\exp\left(i \sum_{\mu=1}^{\nu} t_\mu \phi(f_\mu)\right) \Omega_0$ converges strongly and we have

$$0 = \partial_{t_1} \dots \partial_{t_\nu} \left\langle \theta, \exp\left(i \sum_{\mu=1}^{\nu} t_\mu \phi(f_\mu)\right) \Omega_0 \right\rangle \Bigg|_{t=0}$$

$$= i^\nu \langle \theta, \phi(f_1) \dots \phi(f_\nu) \Omega_0 \rangle.$$

Let θ_j be the j particle component of θ. We prove by induction that $\theta_j = 0$. If $\nu = 0$, we have from the above equation $\theta_0 = 0$. Suppose $\theta_i = 0$ for all i less than j. Then

$$0 = \langle \theta, \phi(f_1) \dots \phi(f_j) \Omega_0 \rangle = \langle \theta_j, a^*(f_1) \dots a^*(f_j) \Omega_0 \rangle$$

and $\theta_j = 0$ also, since the vectors $a^*(f_1) \dots a^*(f_j) \Omega_0$ are total in $\mathscr{F}_j$. Hence $\theta = 0$ and the proof is complete.

The Hamiltonians for bosons with a $P(\phi)$ interaction are

$$H(g) = H_0 + :P(\phi, g):, \tag{1.2.12}$$

$$H(V) = H_0 + :P(\phi_V):. \tag{1.2.13}$$

Later we will add a constant to renormalize these operators, and much later we will take the limit $g \to 1$. We also consider

$$H(g, V) = H_{0,V} + :P(\phi_V, g): \tag{1.2.14}$$

and the momentum cutoff Hamiltonians

$$H(g, \varkappa), \quad H(V, \varkappa) \quad \text{and} \quad H(g, V, \varkappa).$$

In these operators, the field ϕ or ϕ_V in the interaction term P is replaced by the momentum cutoff field $\phi_\varkappa$ or $\phi_{V,\varkappa}$:

$$\phi_\varkappa = (4\pi)^{-1/2} \int_{|k| \leq \varkappa} e^{-ikx} \mu(k)^{-1/2} \{a(k)^* + a(-k)\} \, dk$$

and for $\phi_{V,\varkappa}$ we replace $\mu(k)$ by $\mu(k_V)$. We define

$$Z_{V,\varkappa} = Z_V \cap \{k : |k| \leqq \varkappa\}.$$

1.3 Direct sums and tensor products

The operators H_0 and $H_0 + H_I$ are partial differential operators in a continuum of variables. The only way we are able to make use of this fact is to approximate these operators by partial differential operators in a finite number of variables. This section is devoted to the Fock space aspects of this approximation. Let $\mathscr{H}$ be a closed linear subspace of $\mathscr{F}_1 = L_2(R)$. We will be interested in the case $\mathscr{H} = l_2(Z_{V,\varkappa}) \subset L_2(R)$. Let

$$\mathscr{H}^\perp = \mathscr{F}_1 \ominus \mathscr{H} = L_2(R) \ominus \mathscr{H}.$$

The Fock spaces $\mathfrak{S}(\mathscr{H})$ and $\mathfrak{S}(\mathscr{H}^\perp)$ are both subspaces of $\mathscr{F}$. Let S be the symmetrization projection which maps each space $L_2(R^n)$ onto $\mathscr{F}_n$ and let $\mathfrak{S}_i(\mathscr{H})$ be the i particle subspace of $\mathfrak{S}(\mathscr{H})$.

LEMMA 1.3.1 *The mapping*

$$S\binom{i+j}{i}^{1/2} : \mathfrak{S}_i(\mathscr{H}) \otimes \mathfrak{S}_j(\mathscr{H}^\perp) \to \mathscr{F}_{i+j} \tag{1.3.1}$$

is isometric.

Proof The symmetrization projection onto $\mathscr{F}_n$ is a sum over the permutation group, acting by interchange of variables, divided by $n!$. Acting on a function $f \in \mathfrak{S}_i(\mathscr{H}) \otimes \mathfrak{S}_j(\mathscr{H}^\perp)$, the $(i+j)!$ permutations yield $(i+j)!/i!\,j! = \binom{i+j}{i}$ distinct (and orthogonal) permuted functions, and each such function occurs $i!\,j!$ times. It follows that

$$\|Sf\|^2 = \binom{i+j}{i} \frac{i!^2\, j!^2}{(i+j)!^2} \|f\|^2 = \binom{i+j}{i}^{-1} \|f\|^2$$

and the proof is complete.

Let $N(\mathscr{H})$ be the number operator on $\mathfrak{S}(\mathscr{H})$. We can write the isometry above as

$$S\left(\frac{N(\mathscr{H}) \otimes I + I \otimes N(\mathscr{H}^\perp)}{N(\mathscr{H}) \otimes I}\right)^{1/2} = U. \tag{1.3.2}$$

PROPOSITION 1.3.2 *(1.3.2) defines a unitary mapping of* $\mathfrak{S}(\mathscr{H}) \otimes \mathfrak{S}(\mathscr{H}^\perp)$ *onto* $\mathscr{F}$.

Proof For distinct i and j, the ranges of the maps (1.3.1) are orthogonal. This is easily seen for functions

$$f = \{S(f_1 \otimes \dots \otimes f_i)\} \otimes \{S(f_{i+1} \otimes \dots \otimes f_{i+j})\} \tag{1.3.3}$$

which factor, and such functions span the spaces $\mathfrak{S}_i(\mathscr{H}) \otimes \mathfrak{S}_j(\mathscr{H}^\perp)$. To complete the proof we show that the range of (1.3.2), restricted to

$$\sum_{i+j=n} \mathfrak{S}_i(\mathscr{H}) \otimes \mathfrak{S}_j(\mathscr{H}^\perp)$$

equals $\mathscr{F}_n$. If $h \in \mathscr{F}_n$ is orthogonal to the range, then $h \perp f_1 \otimes \dots \otimes f_n$ where each $f_l \in \mathscr{H}$ or $\mathscr{H}^\perp$, using the fact that h is symmetric. However, the $f = f_1 \otimes \dots \otimes f_n$ which factor span $L_2(R^n)$ and so $h = 0$. This completes the proof.

For $h \in L_2(R)$, we define $a(h) = \int a(k)\, h(k)^-\, dk$, and if $h \in \mathscr{H}$ or $\mathscr{H}^\perp$, we define the annihilation operator $a_1(h)$ or $a_2(h)$ acting on $\mathfrak{S}(\mathscr{H})$ or $\mathfrak{S}(\mathscr{H}^\perp)$.

PROPOSITION 1.3.3 *For* $h \in \mathscr{H}$,

$$a(h)\, U = U(a_1(h) \otimes I).$$

Proof It is enough to establish the identity for vectors (1.3.3) in $\mathfrak{S}_i(\mathscr{H}) \otimes \mathfrak{S}_j(\mathscr{H}^\perp)$ which factor. For such f the left and right members of this identity are

$$\binom{i+j}{i}^{1/2} (i+j)^{-1/2} \sum_{l=1}^{i+j} \left(\int \bar{h} f\, dk\right) S(f_1 \otimes \dots \otimes f_{l-1} \otimes f_{l+1} \otimes \dots \otimes f_{i+j})$$

and

$$\binom{i+j-1}{i-1}^{1/2} i^{-1/2} \sum_{l=1}^{i} \left(\int \bar{h} f_l\, dk\right) S(f_1 \otimes \dots \otimes f_{l-1} \otimes f_{l+1} \otimes \dots \otimes f_{i+j}).$$

Since $h \perp f_l$ for $l > i$, the summations in the two expressions above are equal. The factorials are also equal and the proposition is proved.

THEOREM 1.3.4 *Let* $\mathscr{H} = l_2(Z_{V,\varkappa})$, *regarded as a subspace of* $L_2(R)$ *by conventions of section 1.1. The unitary operator* U *transforms* $H_{0,V}$ *and* $H_I(g, V, \varkappa)$ *as follows*:

$$H_{0,V} \to (H_{0,V} \restriction \mathfrak{S}(\mathscr{H})) \otimes I + I \otimes (H_{0,V} \restriction \mathfrak{S}(\mathscr{H})^{\perp}), \tag{1.3.3}$$

$$H_I(g, V, \varkappa) \to (H_I(g, V, \varkappa) \restriction \mathfrak{S}(\mathscr{H})) \otimes I \tag{1.3.4}$$

Proof (1.3.4) follows from Proposition 1.3.3 and the fact that the kernels of $H_I(g, V, \varkappa)$ belong to $\mathscr{H} \otimes \ldots \otimes \mathscr{H}$. (1.3.3) is most easily seen by a direct calculation, and is based on the fact that $\mathscr{H}$ is an invariant subspace for the multiplication operator $\mu(k_V)$, acting on $L_2(R)$.

2 *Q* space

To say that $\phi(x)$ is a field operator means that $\langle \theta, \phi(x)\, \theta \rangle$ is the expected value of the field strength at the space point x if the field is in the state θ. We diagonalize the $\phi(x)$'s simultaneously so that $\mathscr{F}$ is represented as an L_2 space,

$$\mathscr{F} \approx L_2(Q)$$

and under this isomorphism, each $\phi(f)$ goes into a multiplication operator. For each point $q \in Q$, the field strength at x (or averaged near x) takes on a definite value and so Q may be regarded as the configuration space of the classical field. The state $\theta(q) \in L_2(Q)$ determines a probability distribution, $|\theta|^2\, dq$, which gives the probability that the quantum field will be in some region in Q space. The interaction Hamiltonian $:P(\phi, g):$ is a multiplication operator in the Q space representation; this is the principal merit of Q space. Fock space was set up in order to diagonalize H_0. On $L_2(Q)$, H_0 is a Hermite operator and the particle structure (1.1.1) in $\mathscr{F}$ results from an expansion of $L_2(Q)$ in Hermite functions.

2.1 $:\phi^n:$ *as a multiplication operator*

PROPOSITION 2.1.1 *There is a measure space* Q *with a measure* dq *of total mass 1, and a unitary operator* $W: \mathscr{F} \to L_2(Q)$ *such that*

$$W \mathfrak{M} W^* = L_\infty(Q), \tag{2.1.1}$$

$$W \Omega_0 = 1. \tag{2.1.2}$$

Q is the spectrum of $\mathfrak{M}$ or of a weakly dense subalgebra of $\mathfrak{M}$.

The existence of Q and W follows from one form of the spectral theorem for a family of commuting normal operators (see for instance Kunze and Segal[1]) and from the fact that $\mathfrak{M}$ has a cyclic vector Ω_0, Proposition 1.2.7.

PROPOSITION 2.1.2 *Let* $g = \bar{g} \in L_2$. *Then* $:P(\phi, g):$ *is essentially self adjoint on the domain* $\mathscr{D}$, *and* $W :P(\phi, g):^- W^* \in L_p(Q)$ *for any* $p < \infty$.

Proof Let $T = (:P(\phi, g): \upharpoonright \mathscr{D})^-$. We assert that $\mathfrak{M}\Omega_0 \in \mathscr{D}(T)$ and that T commutes with $\mathfrak{M}$ on the domain $\mathfrak{M}\Omega_0$. We have

$$\lim_n \sum_{j \leqq n} :P(\phi, g): (i\phi(f))^j \Omega_0/j!$$

$$= \lim_n \sum_{i \leqq n} (i\phi(f))^j :P(\phi, g): \Omega_0/j!$$

$$= e^{i\varphi(f)} T\Omega_0 .$$

Thus $\exp(i\phi(f))\Omega_0 \in \mathscr{D}(T)$ and

$$T \exp(i\phi(f))\Omega_0 = \exp(i\phi(f)) T\Omega_0 .$$

The same argument applies to linear combinations in the exponentials $\exp(i\phi(f))$, and to polynomials in the exponentials (since polynomials in the exponentials can be written as linear combinations). An arbitrary element M of $\mathfrak{M}$ is the strong limit of polynomials M_j in the operators $\exp(i\phi(f))$. Hence $M\Omega_0 = \lim M_j\Omega_0$ and

$$\lim TM_j\Omega_0 = \lim M_jT\Omega_0 = MT\Omega_0 ,$$

which proves our assertion.

On Q space, our assertion states that $L_\infty \subset \mathscr{D}(WTW^*)$ and that WTW^* commutes with bounded multiplication operators. Let $t = WT\Omega_0 \in L_2(Q)$. Then WTW^*, on the domain L_∞, is multiplication by t. Since multiplication by a real L_2 function is essentially self adjoint on the domain L_∞, WTW^* and T are extensions of self adjoint operators. Since they are symmetric, they are self adjoint.

On Fock space, Proposition 1.2.3 (a) shows that $\Omega_0 \in \mathscr{D}(T^j)$ for any integer j. Thus 1 is in the domain of multiplication by t^j and $t \in L_p$ for any $p < \infty$. This completes the proof. In the following we let $:P(\phi, g): = T$.

Let B be an open subset of R and let $\mathfrak{M}(B)$ be the von Neumann algebra generated by the operators of the form $\exp(i\phi(f))$, where $f = \bar{f} \in C_0^\infty$ and $\operatorname{supp} f \subset B$.

PROPOSITION 2.1.3 *Let* $g \in L_2$ *and let* $\operatorname{supp} g \subset B^-$. *Then* $\exp(it :P(\phi, g):) \in \mathfrak{M}(B)$ *for all real* t.

Proof If $T_\varkappa \to T$ strongly on $\mathscr{D}$ and if each $T_\varkappa$ is self adjoint, then $\exp(iT_\varkappa t) \underset{s}{\to} \exp(iTt)$ since $\mathscr{D}$ is a core for T. See Kato [1, page 502]. We approximate T by $T_\varkappa = :P(\phi_\varkappa, g_\varkappa):$. Here the $g_\varkappa$ are chosen so that $g_\varkappa \to g$ in L_2, and the distance from $\operatorname{supp} g_\varkappa$ to $\sim B$ is greater than $\varkappa^{-1}$. We let $\phi_\varkappa(x) = (h_\varkappa * \phi)(x)$, where $h_\varkappa(x) = \varkappa h(\varkappa x)$, $\int h(x)\,dx = 1$ and $h \in C_0^\infty$ with support in the interval $[-1/2, 1/2]$. We note that the Fourier

2*

transform of $h_\varkappa$ is $\tilde{h}(k/\varkappa)$, which converges pointwise to 1 as $\varkappa \to \infty$. By Proposition 1.2.5, the kernels of $T_\varkappa$ converge in L_2 to the kernels of T, so by Proposition 1.2.3 (a)

$$\|(T_\varkappa - T)(N + I)^{-n/2}\| \to 0.$$

We note that by Wick's theorem (as in the proof of Lemma 2.1.5 below) there is a polynominal P_1 such that

$$:P(\phi_\varkappa, g_\varkappa): = P_1(\phi_\varkappa, g_\varkappa).$$

By definition, the bounded functions of $\phi_\varkappa(x)$, $x \in \operatorname{supp} g_\varkappa$, belong to $\mathfrak{M}(B)$, and hence so do the bounded functions of $P_1(\phi_\varkappa, g_\varkappa)$. By limits, the same is true for $:P(\phi, g):$ and the proof is complete.

We saw that $:P(\phi, g):$, as a function on Q, is in L_p for any $p < \infty$; it is thus "almost bounded". When $0 \leqq P, 0 \leqq g$, we have $:P(\phi, g):$ almost bounded from below in a much more restricted sense.

THEOREM 2.1.4 *Let* $0 \leqq P$ *and* $0 \leqq g$. *Then as a function on* Q, $\exp(-:P(\phi, g):) \in L_p$ *for all* $p < \infty$.

We write

$$:P(\phi, g): = :P(\phi_\varkappa, g): + (:P(\phi, g): - :P(\phi_\varkappa, g):) \qquad (2.1.3)$$
$$= P_{g,\varkappa} + P'_{g,\varkappa}$$

and we bound the two terms above separately in the following lemmas.

LEMMA 2.1.5 *Let P have degree n. Then*

$$-\int g(x)\, dx\, 0 (\log \varkappa)^{n/2} \leqq P_{g,\varkappa}.$$

Proof Let $c_\varkappa = \|\phi_\varkappa(x)\Omega_0\|^2$. By (1.2.15), $c_\varkappa = 0(\log \varkappa)$. Wick's theorem asserts that

$$:\phi_\varkappa(x)^j: = \sum_{i=0}^{[j/2]} (-1)^i \frac{j!}{(j-2i)!\, i!} c_\varkappa^i \phi_\varkappa(x)^{j-2i}.$$

This formula follows from the commutation relations and a combinatorial argument. The factorials give the number of ways of selecting i pairs from j objects. Since $0 \leqq P$, it follows that

$$-0(c_\varkappa^{n/2}) \leqq :P(\phi_\varkappa(x)):,$$

which yields the lemma.

LEMMA 2.1.6 *Regard* $P'_{g,\varkappa}$ *as a function on* Q. *Let* $\varepsilon > 0$. *With a constant independent of* $\varkappa$ *and* j *(but depending on* g*)*

$$\int |P'_{g,\varkappa}(q)|^{2j}\, dq \leqq j!^n\, (\text{const. } \varkappa^{-1/2+\varepsilon})^{2j}$$

Proof We use the identity

$$\int P'_{g,\varkappa}(q)|^{2j}\, dq = \|(P'_{g,\varkappa})^j \Omega_0\|^2$$

to estimate the integral. If $P'_{g,\varkappa}$ is written as a sum of monomials (1.1.5) in the annihilation creation operators, then the kernels $w_\varkappa$ which occur satisfy the bound

$$\|w_\varkappa\|_2 \leqq 0(\varkappa^{-1/2+\varepsilon})$$

by Proposition 1.2.5. Since $(P'_{g,\varkappa})^l \Omega_0$ is a state with at most ln particles, we have

$$\|(P'_{g,\varkappa})^l \Omega_0\| \leqq \|P'_{g,\varkappa}(N+I)^{-n/2}\| \, \|(N+I)^{n/2}(P'_{g,\varkappa})^{l-1} \Omega_0\|$$
$$0(\varkappa^{-(1/2)+\varepsilon})\,(nl+1)^{n/2}\, \|(P'_{g,\varkappa})^{l-1} \Omega_0\|,$$

using Proposition 1.2.3(a). An induction completes the proof.

Proof of Theorem 2.1.4 Let $\varkappa_\nu = 2^\nu$ and let

$$Q_\nu = \{q : |P'_{g,\varkappa_\nu}(q)| \leqq 1,\ |P'_{g,\varkappa_{\nu-1}}(q)| > 1\}$$

for $\nu = 2, 3, \ldots,$

$$Q_1 = \{q : |P'_{g\varkappa_1}(q)| \leqq 1\}.$$

On Q_ν, $-0(\log \varkappa_\nu)^{n/2} \leqq :P(\phi, g):$ by Lemma 2.1.5.
Thus

$$\int \exp(-t :P(\phi, g):)\, dq \leqq \sum_{\nu=1}^{\infty} \exp(0(\log \varkappa_\nu)^{n/2}) \int_{Q_\nu} dq$$

and the proof is completed by the estimate

$$\int_{Q_\nu} dq \leqq \exp(-0(\varkappa_\nu{}^{-(1-2\varepsilon)/n})). \tag{2.1.4}$$

To derive this bound on the measure of Q_ν, we use Lemma 2.1.6, being careful to make an optimal choice for j. For $\nu \geqq 2$,

$$\int_{Q_\nu} dq \leqq \int |P'_{g,\varkappa_{\nu-1}}|^{2j}\, dq$$
$$\leqq j!^n\, 0(\varkappa_\nu^{-1/2+\varepsilon})^{2j}.$$

We use Sterling's formula to bound $j!$; the choice $j = \varkappa_\nu^{-(1-2\varepsilon)/n}$ then completes the proof.

The Theorem 2.1.4 is sufficient for the study of $H(g)$, but to control the limit $g \to 1$, we need an estimate independent of g, given in the following theorem. Since the proof is lengthy, we only indicate the main ideas of the proof. For details, see Glimm-Jaffe[(5)], where a similar result is proved.

THEOREM 2.1.7 *Let* $0 \leqq P$, $0 \leqq g$ *and let* $|g(x)| \leqq M$. *With a constant independent of* g,

$$\int \exp(-t{:}P(\phi, g){:})\, dq \leqq e^{\text{const.}\, V},$$

where V *is the length of the set of points within distance one of suppt* g.

Proof We decompose g,

$$g = \sum_{\alpha=-\infty}^{\infty} g_\alpha$$

so that suppt $g_\alpha \subset [\alpha - \frac{1}{2}, \alpha + \frac{1}{2}]$. There are V nonzero terms in the sum. Thus

$${:}P(\phi, g){:} = \Sigma_\alpha P(\phi, g_\alpha) = \Sigma P_\alpha$$

also decomposes, and as in (2.1.3), each P_α is split into a low momentum portion and a tail. The idea behind the proof is that the functions P_α and P_β are nearly independent, for $|\alpha - \beta|$ large. If they were exactly independent for $\alpha \neq \beta$, then we would have

$$\int \exp(-t{:}P(\phi, g){:})dq = \Pi_\alpha \int \exp(-t\, P_\alpha)\, dq = e^{\text{const.}\, V},$$

using Theorem 2.1.4 to bound each factor. Although the integral does not factor as above, we obtain bounds for the integral which do factor and this gives the upper bound $\Pi_\alpha\, e^{\text{const.}} = e^{\text{const.}\, V}$ for the integral.

Let $\nu = \{\nu(\alpha)\}$ be a multi-index and let

$$Q_\nu = \{q : |P'_{g_\alpha, \varkappa_{\nu(\alpha)}}(q)| \leqq 1 \text{ for all } \alpha \text{ and } |P'_{g_\alpha, \varkappa_{\nu(\alpha)-1}}(q)| > 1 \text{ if } \nu(\alpha) > 1\}.$$

Then

$$\int \exp(-t{:}P(\phi, g){:})\, dq$$
$$\leqq \Sigma_\nu \exp(0\, \Sigma_\alpha (\log \varkappa_\alpha)^{n/2}) \int_{Q_\nu} dq.$$

The next lemma completes the proof; on comparison with (2.1.4) we see that this is the bound which factors. The proof is as in Glimm-Jaffe[5].

LEMMA 2.1.8 *With the above notation,*

$$\int_{Q_\nu} dq \leqq \exp(-0(\Sigma_\alpha \varkappa_{\nu(\alpha)}^{-(1-2\varepsilon)/n})).$$

Remark on the proof Let $j = \{j(\alpha)\}$ be a multi-index. Then

$$\int_{Q_\nu} dq \leqq \int \Pi_\alpha |P'_{g_\alpha, \varkappa_{\nu(\alpha)-1}}|^{2j(\alpha)}\, dq;$$

we omit from the product α's with $g_\alpha = 0$ or with $\varkappa_{\nu(\alpha)} = 1$. This integral is equal to

$$\|\Pi_\alpha(P'_{g_\alpha,\varkappa_{\nu(\alpha)-1}})^{j(\alpha)}\,\Omega_0\|^2.$$

The detailed estimates take advantage of the fact that P_α and P_β or $P'_{g_\alpha,\varkappa_{\nu(\alpha)-1}}$ and $P'_{g_\beta,\varkappa_{\nu(\beta)-1}}$ create and annihilate particles with nearly orthogonal wave functions for $|\alpha - \beta|$ large.

2.2 H_0 as a Hermite operator

The semigroup $\exp(-tH_0)$ is smoothing; it maps L_p into L_q for $1 < p$, $q < \infty$ and t sufficiently large.

PROPOSITION 2.2.1 *Let f_1 and f_2 be nonnegative functions in $L_2(Q)$. Then $0 \leqq \langle f_1, \exp(-tH_0)\,f_2\rangle$, i.e. $\exp(-tH_0)$ has a nonnegative kernel.*

Proof We approximate H_0 strongly on $\mathscr{D}$ by operators of the form $A = A_1 + \ldots + A_j$ where

$$A_i = \mu(k_i)\,a^*(e_i)\,a(e_i),$$

$$a^\#(e_i) = a_V^\#(k_i), \qquad k_i \in Z_V.$$

Since $\mathscr{D}$ is a core for H_0, strong convergence of the operators A implies strong convergence of the resolvents (see Kato [1, page 429]) and strong convergence of the semigroups, Kato [1, page 502]. Strong convergence is simple to verify, so it is sufficient to prove that e^{-tA} has a nonnegative kernel.

Let $\mathscr{H}$ be the span of the e_i and let $\mathfrak{M}(\mathscr{H})$ be the von Neumann algebra generated by the spectral projections of the $\phi(e)$, $e \in \mathscr{H}$. Let $\mathfrak{M}^+$ and $\mathfrak{M}(\mathscr{H})^+$ be the positive cones in $\mathfrak{M}$ and $\mathfrak{M}(\mathscr{H})$. We postpone the proof of the following lemma.

LEMMA 2.2.2

$$\cup_{\mathscr{K}}\{\mathfrak{M}(\mathscr{K})^+\,\Omega_0 : \mathscr{H} \subset \mathscr{K} \subset L_2(R), \dim \mathscr{K} < \infty\}$$

is dense in $\mathfrak{M}^+\Omega_0$.

Using the lemma, we may suppose that f_1 and f_2 are in $\mathfrak{M}(\mathscr{K})^+$. Thus we are reduced to a problem with a finite number of degrees of freedom.
Let

$$\mathfrak{S}(\mathscr{K}) = C \oplus \mathscr{K} \oplus (\mathscr{K} \otimes_s \mathscr{K}) \oplus \ldots \subset \mathscr{F}$$

be the Fock space over $\mathscr{K}$. Let $e_{j+1}, \ldots, e_l$ be an orthonormal basis for $\mathscr{K} \ominus \mathscr{H}$ and let

$$q_i = 2^{-1/2}(a^*(e_i) + a(e_i)),$$

$$p_i = 2^{-1/2}\,i(a^*(e_i) - a(e_i)).$$

Then

$$A = \sum_{i=1}^{j} \frac{\mu_i}{2}(p_i^2 + q_i^2 - 1)$$

and $[q_\alpha, p_\beta] = i\,\delta_{\alpha\beta}$. By the von Neumann uniqueness theorem for irreducible representations of the commutation relations (in a finite number of degrees of freedom) or by explicit calculations using Hermite functions (see Bargmann[1]) we have a unitary equivalence of $\mathfrak{S}(\mathscr{K})$ with $L_2(R^l, \pi^{-l/2}\, e^{-|q|^2}\, dq)$. Furthermore this unitary equivalence carries q_α into multiplication by the αth coordinate, q_α, and it carries p_α into

$$-ie^{-|q|^2/2}\frac{\partial}{\partial q_\alpha}e^{|q|^2/2}.$$

The equivalence carries Ω_0 into the function 1 and $\mathfrak{M}(\mathscr{K})^+\,\Omega_0$ into positive functions, and so it is sufficient to show that e^{-tA} is carried into an operator with a positive kernel. In fact, this kernel is given by the formula

$$\prod_{\alpha=1}^{j}(1 - e^{-2\mu_\alpha t})^{-1/2}\exp\left[-\frac{(q'_\alpha - e^{-\mu_\alpha t}q_\alpha)^2}{1 - e^{-2\mu_\alpha t}} + (q'_\alpha)^2\right] \tag{2.2.1}$$
$$\times \prod_{\beta=j+1}^{l}\delta(q_\beta - q'_\beta).$$

The quickest derivation of this formula is to check that the kernel and (2.2.1) satisfy the same parabolic equation with the same Cauchy data, because then by a uniqueness theorem, they must coincide.

Proof of Lemma 2.2.2 Let $h \in \mathfrak{M}^+$. Since

$$\cup\{\mathfrak{S}(\mathscr{K}) : \mathscr{H} \subset \mathscr{K} \subset L_2(R),\ \dim\mathscr{K} < \infty\}$$

is dense in $\mathscr{F}$, we may approximate $h\Omega_0$ by vectors $h_{\mathscr{K}}\Omega_0$, $h_{\mathscr{K}} \in \mathfrak{M}(\mathscr{K})$. Thus as functions on Q, the $\mathfrak{M}(\mathscr{K})$ are dense in $\mathfrak{M}$ in the L_2 norm. By the triangle inequality,

$$-|h - h_{\mathscr{K}}| \leqq |h| - |h_{\mathscr{K}}| \leqq |h - h_{\mathscr{K}}|,$$

and it follows that the $\mathfrak{M}(\mathscr{K})^+$ are dense in $\mathfrak{M}^+$ in the L_2 norm, which proves the lemma.

PROPOSITION 2.2.3 (*Glimm*[3]) *e^{-tH_0} is a contraction on $L_1(Q)$ for $t \geqq 0$ and for sufficiently large t, e^{-tH_0} is a constraction from $L_2(Q)$ to $L_4(Q)$.*

Remark A semigroup e^{-tH_0} acting on a measure space Q, dq, with a positive self adjoint generator H_0 is called *hypercontractive* (Hoegh Krohn-Simon[1]) if it is a contraction on L_1 for all t and a bounded operator from L_2 to L_4 for some $t > 0$. As we will see, these hypotheses give strong conclusions, by a simple application of the Riesz-Thorin and Stein convexity theorems.

Proof $\|e^{-tH_0} f\|_1 \leqq \|e^{-tH_0}|f|\|_1$

$$= \langle \Omega_0, e^{-tH_0}|f|\,\Omega_0\rangle = \langle \Omega_0, |f|\,\Omega_0\rangle = \|f\|_1,$$

and so e^{-tH_0} is a contraction on L_1.

To prove that e^{-tH_0} is a contraction from L_2 to L_4, we may replace the semigroup by strongly approximating semigroups, namely by $e^{-tH_{0,V}}$. Furthermore, we may replace $f \in L_2$ by vectors in a dense subset of L_2. We choose the approximating f's to lie in subspaces of the form $\mathfrak{S}(\mathcal{H})$ where $\mathcal{H}$ is a finite dimensional subspace of $L_2(R)$ with the orthonormal basis $e_1, \dots, e_j$ given above. For such an f, $e^{-tH_0} f = e^{-tA} f$, with A as in Proposition 2.2.1. Thus we are reduced to the case of a finite number of degrees of freedom, and the kernel of e^{-tA} is given by (2.2.1) with $j = l$. The following lemma reduces the problem to a single degree of freedom.

LEMMA 2.2.4 *Let K_1 and K_2 be operators on σ finite measure spaces Q_1 and Q_2, acting as contractions from $L_2(Q_i)$ to $L_4(Q_i)$. Then $K_1 \otimes K_2$ is a contraction from $L_2(Q_1 \times Q_2)$ to $L_4(Q_1 \times Q_2)$.*

Proof We identify K_1 and $K_1 \otimes I$, K_2 and $I \otimes K_2$. Then

$$\begin{aligned}
\|K_1 \otimes K_2 f\|_4^4 &= \iint |K_1(K_2 f)(q_1, q_2)|^4 \, dq_1 \, dq_2 \\
&\leqq \int \Big(\int |(K_2 f)(q_1, q_2)|^2 \, dq_1 \Big)^2 dq_2 \\
&= \iint |(K_2 f)(q_1, q_2)|^2 \, dq_1 \int |(K_2 f)(p_1, q_2)|^2 \, dp_1 \, dq_2 \\
&= \iiint |(K_2 f)(q_1, q_2)|^2 \, |(K_2 f)(p_1, q_2)|^2 \, dq_2 \, dp_1 \, dq_1 \\
&\leqq \iint \Big(\int |(K_2 f)(q_1, q_2)|^4 \, dq_2 \Big)^{1/2} \\
&\qquad \times \Big(\int |(K_2 f)(p_1, p_2)|^4 \, dp_2 \Big)^{1/2} dp_1 \, dq_1 \\
&\leqq \iint \Big(\int |f(q_1, q_2)|^2 \, dq_2 \Big) \Big(\int |f(p_1, p_2)|^2 \, dp_2 \Big) dp_1 \, dq_1 \\
&= \|f\|_2^4.
\end{aligned}$$

We now take $A = A_1$ and $j = 1$ in (2.2.1). Let

$$\begin{aligned}
a^2(q) &= \pi^{-1} \int (1 - e^{-2\mu t})^{-1} \exp\Big(-2 \frac{(q' - e^{-\mu t} q)^2}{1 - e^{-2\mu t}} + 2(q')^2 \Big) e^{-(q')^2} \, dq' \\
&= \pi^{-1} \int (1 - e^{-2\mu t})^{-1} \exp\Big(-2 \frac{(q' - e^{-\mu t} q)^2}{1 - e^{-2\mu t}} + (q')^2 \Big) dq'.
\end{aligned}$$

By the Schwarz inequality

$$|(e^{-tA} f)(q)| \leqq a(q) \, \|f\|_2,$$

and so the norm of e^{-tA}, as a mapping from L_2 to L_4, is bounded by $\|a\|_4$. The exponent in the integrand defining a^2 is proportional to

$$-2(q' - e^{-\mu t} q)^2 + (1 - e^{-2\mu t})(q')^2$$
$$= -\left[(1 + e^{-2\mu t})^{1/2} q' - \frac{e^{-\mu t}}{(1 + e^{-2\mu t})^{1/2}} q\right]^2 + \left[\frac{e^{-\mu t}}{1 + e^{-2\mu t}} - 2 e^{-2\mu t}\right] q^2.$$

The first term above makes the integral defining a^2 converge. Thus

$$a^2(q) \leqq \text{const. } \exp\left(\left[\frac{e^{-\mu t}}{1 + e^{-2\mu t}} - 2 e^{-2\mu t}\right] q^2\right)$$

and since the coefficient of q^2 is small for large t, $a^2 \in L_4$ for large t, and e^{-tA} is a bounded map from L_2 to L_4.

Let $\mathscr{H}' = 1^{\perp}$ in $L_2(R, \pi^{-1/2} e^{-q^2} dq)$. Since the norm of $e^{-tA} : \mathscr{H}' \to \mathscr{H}'$ tends to zero as $t \to \infty$, so does the norm of $e^{-tA} : \mathscr{H}' \to L_4$. Let $f' \in \mathscr{H}'$ and let $f = 1 + f'$. Then

$$(e^{-tA} f)^4 = \sum \binom{4}{j} (e^{-tA} f')^j$$

and for large t,

$$\|e^{-tA} f\|_4^4 \leqq \sum_{j=0,2,3,4} \binom{4}{j} \|e^{-tA} f'\|_j^j$$
$$\leqq \sum_{j=0,2,3,4} \binom{4}{j} \|e^{-tA} f'\|_4^j$$
$$\leqq 1 + 0(1)(\|e^{-tA} f'\|_4^2 + \cdots + \|e^{-tA} f'\|_4^4)$$
$$\leqq 1 + o(1)(\|f'\|_2^2 + \|f'\|_2^4) \leqq \|f\|.$$

The $j = 1$ term drops out because $\int e^{-tA} f' = \langle 1, e^{-tA} f' \rangle = \langle 1, f' \rangle = 0$. We used the inequalities $\|h\|_2 \leqq \|h\|_3 \leqq \|h\|_4$, valid because the total measure is 1.

THEOREM 2.2.5 *For some $M > 0$, e^{-tH_0} is a contraction from L_p to L_q if*

$$\frac{1}{p} \leqq (1 + Mt) \frac{1}{q}$$

and

$$\left(1 - \frac{1}{q}\right) \leqq (1 + Mt)\left(1 - \frac{1}{p}\right).$$

Proof By duality ($H_0 = H_0^*$) the $L_p \to L_q$ norm equals the $L_{q'} \to L_{p'}$ norm if $L_{p'} = L_p^*, L_{q'} = L_q^*$. By the Riesz-Thorin convexity theorem (Dunford Schwartz[(1)]), the $L_p \to L_q$ norm is bounded by one in a convex region in the square $0 \leqq 1/p \leqq 1$, $0 \leqq 1/q \leqq 1$. Combining these facts with

Proposition 2.2.3, we see that e^{-tH_0} is a contraction from L_p to L_p for $0 \leqq t$ and $1 \leqq p \leqq \infty$. Since the identity operator is a contraction from L_p to L_q for $q > p$, so is e^{-tH_0}. For large t, we again use duality, convexity and Proposition 2.2.3. We find that e^{-tH_0} is a contraction from L_4 to L_8, and from L_{2^n} to $L_{2^{n+1}}$. Thus for $t = Mn$ it is a contraction from L_2 to L_{2^n} and (by duality) from $L_{1/(1-2^{-n})}$ to L_2. For large t, the theorem follows.

For small t, we use the Stein interpolation theorem, Stein[1,2], following Segal[3]. Since $0 \leqq H_0$, e^{-tH_0} is complex analytic in the strip $0 \leqq \operatorname{Re} t \leqq T$. On the boundary $\operatorname{Re} t = 0$, it is a contraction from L_2 to L_2 and on the boundary $\operatorname{Re} t = T$ it is a contraction from L_2 to L_4. Thus it is a contraction from L_2 to L_q, where $1/q$ is a convex combination of $\frac{1}{2}$ and $\frac{1}{4}$, namely,

$$\frac{(1 - \operatorname{Re} t/T)}{2} + \frac{(\operatorname{Re} t/T)}{4} = \frac{1}{q}.$$

The theorem now follows from duality and the Riesz-Thorin convexity theorem.

We complete this section by analysing the Hamiltonian $H(g, V, \varkappa)$. After a trivial reduction, the Hamiltonian has only a finite number of degrees of freedom, and is transformed into the operator $-\Delta$ + potential.

THEOREM 2.2.6 *(Jaffe[3,4])* $H(g, V, \varkappa)^n$ *is essentially self adjoint on* $C^\infty(H_0) = \bigcap_j \mathscr{D}(H_0^j)$.

Proof We use Theorem 1.3.4 to transform the Hamiltonian and write (1.3.3-4) as

$$H_0 \to A_1 \otimes I + I \otimes A_2 = A_1^\sim + A_2^\sim, \tag{2.2.2}$$

$$H_I(g, V, \varkappa) \to B \otimes I = B^\sim. \tag{2.2.3}$$

LEMMA 2.2.7 *Let* A_1 *and* A_2 *be self adjoint operators in Hilbert spaces* $\mathscr{H}_1$ *and* $\mathscr{H}_2$ *respectively. If for some* $n \geqq 1$, A_1^n *and* A_2^n *are essentially self adjoint on domains* $\mathscr{D}_1$ *and* $\mathscr{D}_2$ *respectively, then* $C_j = (A_1 \otimes I + I \otimes A_2)^j$ *is essentially self adjoint on* $\mathscr{D}_1 \otimes \mathscr{D}_2$ *for* $1 \leqq j \leqq n$. *Here* $\mathscr{D}_1 \otimes \mathscr{D}_2$ *is the algebraic tensor product.*

Proof We first note that

$$\mathscr{D}(A_1^n) \otimes \mathscr{D}(A_2^n) \subset \mathscr{D}(C_j^-),$$

which is a consequence of the inequality

$$\|A_i^j x_i\| \leqq \|A_i^n x_i\| + \|x_i\|$$

for $x_i \in \mathscr{D}_i$, $i = 1,2$. Hence if $y_\nu = (x_{1\nu} \otimes x_{2\nu}) \to y$ and $A_i^n x_{i\nu} \to A_i^n x_i$, $i = 1,2$, we have

$$C_j y_\nu \to C_j^- y, \qquad j = 1, \ldots, n.$$

Furthermore

$$C_j^-(x_1 \otimes x_2) = (A_1 \otimes I + I \otimes A_2)^j (x_1 \otimes x_2).$$

Let x_i have compact support with respect to the spectral measure of A_i. Then x_i is an analytic vector for A_i^n, and $x_1 \otimes x_2$ is an analytic vector for C_j^-. The linear span of such vectors is dense, and so C_j^- is self adjoint by Nelson's theorem.

We apply the lemma to $A_1 + B$ and A_2 of (2.2.2-3). A_2^n is essentially self adjoint on $C^\infty(A_2)$. We are now reduced to considering $(A_1 + B)^n$, a problem with a finite number of degrees of freedom. We use a Schrödinger representation, similar to that of Lemma 2.2.2,

$$\mathfrak{S}(\mathscr{H}) \approx L_2(R^l, dq),$$

$$A_1 + B \to H = -\Delta + W$$

where W is a semibounded polynomial. (The lower bound on B, used here and above follows from the proof of Lemma 2.1.5.) The above isomorphism carries $C^\infty(A_1)$ onto the Schwartz space $\mathscr{S}$. By a Kato perturbation argument, we may without loss of generality add a constant to W, so that $1 \leqq W$, and we must then prove the essential self adjointness of H^n on $\mathscr{S}$.

LEMMA 2.2.8 *Let $\chi \in \mathscr{S}'$ be a distribution solution of the equation $\Delta\chi = W\chi$. Then $\chi \in \mathscr{O}_M$, i.e. χ and all its derivatives are continuous and polynomially bounded.*

Proof The fundamental solution to the Laplace equation is const. r^{2-l}. Let $\phi \in C_0^\infty$ be chosen so that ϕ is a constant in a neighborhood of the origin and

$$\Delta E = \delta - \omega, \quad \omega \in C_0^\infty,$$

where $E = r^{2-l}\phi$. Note that E and $\partial_j E$ are integrable and have compact support. Thus convolution by E or $\partial_j E$ maps $\mathscr{S}$ into $\mathscr{S}$ and $\mathscr{S}'$ into $\mathscr{S}'$ and convolution by E takes a C^m polynomially bounded function into a C^{m+1} polynomially bounded function. Let $F\chi = E * (W\chi)$. Then

$$\chi - \omega * \chi = \Delta E * \chi = E * \Delta\chi = E * (W\chi) = F\chi$$

and iterating n times, we get

$$\chi = \sum_{j=0}^{n-1} F^j(\omega * \chi) + F^n\chi.$$

Since $\chi \in \mathscr{S}'$ and $\omega \in C_0^\infty \subset \mathscr{S}$, it follows that $\omega * \chi \in \mathscr{O}_M$, Schwartz[(1)], and so $F^j(\omega * \chi) \in \mathscr{O}_M$. Thus we need only study $F^n\chi$. We write $\chi = \partial^\nu\theta$ where ν is a multi-index and θ is continuous and polynomially bounded. We integrate the derivatives in $F^n\chi$ by parts to move them from θ to the W's or the E's, and for n sufficiently large a given E is differentiated at most once. There are at least $n - |\nu|$ undifferentiated E's and so $F^n\chi$ (and χ) is a $C^{n-|\nu|}$ polynomially bounded function. Thus $\chi \in \mathscr{O}_M$.

The n-fold application of the next lemma completes the proof of Theorem 2.2.6.

LEMMA 2.2.9 *Let $\mathscr{S}_0$ be a dense subspace of $\mathscr{S}$. Then $H\mathscr{S}_0 = (-\Delta + W)\mathscr{S}_0$ is dense in $\mathscr{S}$.*

Proof H maps $\mathscr{S}$ continuously into $\mathscr{S}$. If $H\mathscr{S}_0$ is not dense in $\mathscr{S}$, there is a $\chi \in \mathscr{S}'$ orthogonal to $H\mathscr{S}_0$ (and to $H\mathscr{S}$) and then χ is a distribution solution of the equation $\Delta\chi = W\chi$. By Lemma 2.2.8, $\chi \in \mathcal{O}_M$. Since W is real, we can suppose χ is real also. By Green's theorem and the fact that $1 \leqq W$, we have

$$\int_{r^2 \leqq a^2} \chi^2 r^{l-1}\, dr\, d\omega \leqq 2^{-1} r^{l-1} \int_{r=a} (\partial_r \chi)^2\, d\omega$$

where $r^{l-1}\, dr\, d\omega = dq$. If we define $F(a) = \int_{r=a} \chi^2\, d\Omega$ then F has at most polynomial growth, since $\chi \in \mathcal{O}_M$. But by Green's theorem above,

$$0 \leqq \int_0^a r^{l-1}\, F(r)\, dr \leqq a^{l-1}\, F'(a).$$

Thus $F(a)$ is monotone increasing and for $1 < a$,

$$\left(\frac{a-1}{a}\right)^{l-1} F(a-1) \leqq \int_{a-1}^{a} \left(\frac{r}{a}\right)^{l-1} F(r)\, dr$$

$$\leqq \int_0^a \left(\frac{r}{a}\right)^{l-1} F(r)\, dr \leqq F'(a).$$

Integrating from a to $a + 1$,

$$((a-1)/a)^{l-1}\, F(a-1) \leqq F(a+1) - F(a),$$

$$(1 + ((a-1)/a)^{l})\, F(a-1) \leqq F(a+1)$$

and F grows exponentially, a contradiction. Thus $F = 0$ and the proof is complete.

3 The Hamiltonian $H(g)$

3.1 Positivity of the energy and higher order estimates

Let

$$E(g, \varkappa) = \inf \text{spectrum}\, (H_0 + H_I(g, \varkappa)), \tag{3.1.1}$$

$$H(g, \varkappa) = H_0 + H_I(g, \varkappa) - E(g, \varkappa). \tag{3.1.2}$$

These equations change the definition of $H(g, \varkappa)$ by including the finite vacuum energy renormalization. We prove that $E(g) = \lim_{\varkappa \to \infty} E(g, \varkappa)$ is finite, so that after the finite renormalization, $0 \leqq H(g) = H_0 + H_I(g) - E(g)$. The finiteness of the vacuum energy is a special feature of the space cutoff ϕ_2^{2n} model (Glimm-Jaffe[6]). However, a fundamental property, which should be true for all models, is that the shift in the vacuum energy under suitable perturbations A is finite. With $H(g, \varkappa)$ renormalized by (3.1.1-2), let

$$\delta E(g, \varkappa, A) = \inf \text{spectrum} \, (H(g, \varkappa) + A). \tag{3.1.3}$$

If $\delta E(\pm A)$ remains finite as $\varkappa \to \infty$ or $g \to 1$ then we have uniform estimates

$$\pm A \leqq \text{const.} \, (H(g, \varkappa) + I) \tag{3.1.4}$$

and

$$\|R^{1/2} A R^{1/2}\| \leqq \text{const.} \tag{3.1.5}$$

where $R = R(\zeta) = (H(g, V, \varkappa) - \zeta)^{-1}$. If $0 \leqq A$, we also have $\|A^{1/2}R^{1/2}\| \leqq \text{const.}$

The higher order estimates have the form

$$\pm A \leqq \text{const.} \, (H(g, \varkappa) + I)^n$$

and lead to the bound $\|R^{n/2} A R^{n/2}\| \leqq \text{const.}$ The A's which can be so dominated depend upon the model. They can be N_τ or related operators, or, for higher order estimates, N_τ^j. Combining these estimates with the N_τ estimates of Sec. 1.2, we can estimate $\|R^{n/2} A R^{m/2}\|$ where A is an annihilation creation form whose kernel has a finite L_2 or operator type norm.

THEOREM 3.1.1 *(Glimm-Jaffe[5]) With the hypothesis and assumptions of Theorem 2.1.7,*

$$0 \leqq -E(g, \varkappa) \leqq \text{const. diam. suppt.}\, g.$$

The constant is independent of g and $\varkappa$, *for* $\varkappa \leqq \infty$, $g \to 1$.

Remark Nelson[3] derived a bound uniform in $\varkappa$ but depending on a periodic volume V. The box was replaced by a fixed g cutoff by Glimm[3]. The passage from separate estimates on H_0 and H_I (cf. Theorem 2.1.3 or 2.1.6 and 2.2.5) to the bound on E has been simplified by Segal[2] and is valid in the hypercontractive setting, with Theorem 2.1.3 as hypothesis on the perturbing potential. See also Glimm-Jaffe[12].

Proof $E(g, \varkappa) \leqq 0$ since $\langle \Omega_0, (H_0 + H_I)\Omega_0 \rangle = 0$. Our proof of the lower bound on E requires essential self adjointness of the sum $H_0 + H_I$ on the domain $\mathscr{D}(H_0) \cap \mathscr{D}(H_I)$. Thus for the present we estimate only $E(g, V, \varkappa)$ (cf. Theorem 2.2.6) and E_j, the vacuum energy of

$$H_j = H_0 + H_I^{(j)} = H_0 + F_j(H_I(g)),$$

where

$$F_j(\lambda) = \begin{cases} \lambda & \text{for} \quad |\lambda| \leqq j \\ 0 & \text{for} \quad |\lambda| > j. \end{cases}$$

Since $H_I^{(j)}$ is bounded, $H_0 + H_I^{(j)}$ is self adjoint. The proof for $E(g, \varkappa)$ will follow from essential self adjointness proved in section 3.2.

Let c = const. diameter support g. For t fixed but sufficiently large, and H_I equal to either $H_I(g, V, \varkappa)$ or $H_I^{(j)}$,

$$\|e^{-tH_0/2}\, e^{-tH_I}\, e^{-tH_0/2}\| \leqq e^c,$$

since Theorems 2.1.7 and 2.2.5 bound the three factors above (from right to left) as operators from $L_2 \to L_4$, $L_4 \to L_2$ and $L_2 \to L_2$. In fact, by Hölder's inequality

$$\|e^{-tH_I}\psi\|_2 \leqq \|e^{-tH_I}\|_4\, \|\psi\|_4$$

so $e^c = \|e^{-tH_I}\|_4$. Thus $e^{-tH_I} \leqq e^c\, e^{tH_0}$, and by monotonicity of the square root, see Kato [1, p. 292],

$$e^{-(t/n)H_I} \leqq e^{c/n}\, e^{(t/n)H_0},$$

or

$$e^{-(t/2n)H_0}\, e^{-(t/n)H_I}\, e^{-(t/2n)H_0} \leqq e^{c/n}.$$

We use the Trotter product formula

$$e^{tC} = \operatorname*{st\,lim}_{n\to\infty} (e^{tA/n}\, e^{tB/n})^n,$$

valid when A, B and $C = (A + B)^-$ generate contraction semigroups, see Nelson[4]. It follows that

$$\|e^{-t(H_0+H_I)}\| \leqq e^c = \|e^{-tH_I}\|_4.$$

Hence $-(c/t) \leqq (H_0 + H_I)$, and the proof for $E(g, V, \varkappa)$ or E_j is complete.

COROLLARY 3.1.2 *Let P_1 be a polynomial with degree less than or equal to the degree of P. Then with*

$$A = H_0/(\text{diam. suppt. } g)$$

or with $|f| \leqq$ const. g *and*

$$A = \int :P_1(\phi_\varkappa(x)):\, f(x)\, dx/(\text{diam. suppt. } g)$$

the estimates (3.1.4-5),

$$\pm A \leqq \text{const.}\,(H(g, V, \varkappa) + I),$$

$$\|R^{1/2} A R^{1/2}\| \leqq \text{const.},$$

are valid uniformly as $\varkappa \to \infty$ and $g \to 1$.

Proof $0 \leqq 2^{-1}H_0 + H_I(g, V, \varkappa) - 2^{-1}E(2g, V, \varkappa)$ and so

$$H_0 \leqq 2(2^{-1}H_0 + 2^{-1}H_0 + H_I(g, V, \varkappa) - 2^{-1}\, E(2g, V, \varkappa))$$

$$= 2(H_0 + H_I(g, V, \varkappa) - E(g, V, \varkappa)) + 2E(g, V, \varkappa) - E(2g, V, \varkappa).$$

The bound on E from Theorem 3.1.1 completes this case and the estimate in the second case is similar.

Remark The estimates (3.1.4-5), uniform in $\varkappa$ and g, for $A = H_0^{\text{loc}}$ or N_τ^{loc} would lead to estimates on the vacuum expectation values in the limit $\varkappa \to \infty$, $g \to 1$. (See chapter 4.)

We now turn to higher order estimates. These estimates are used in the study of the field equations, in the proof of Lorentz covariance and in the construction of in and out fields for $H(g)$. First we state the results.

THEOREM 3.1.3 (*Rosen*[2]) *Let i be a positive integer. For $\varepsilon > 0$, $i \geqq 3$ and $j = j(i, \deg. P, \varepsilon)$ sufficiently large, there is a constant such that*

$$H_0^{3-\varepsilon} N^{i+\varepsilon-3} \leqq \text{const.}\,(H(g) + \text{const.})^j.$$

If $\varepsilon > 2$, we may take $j = i$.

If degree $P \leqq 4$, we have (Glimm-Jaffe[2]*)*

$$H_0^2 \leqq \text{const.}\,(H(g)\ \text{const.})^2.$$

The theorem is proved first for $H(g, V, \varkappa)$ with constants independent of V and $\varkappa$ and then transferred to the limit, $H(g)$. (See Sec. 3.2.) The proof is based on the pull through formula

$$a(k)\,R(\zeta) = R(\zeta - \mu(k))\,a(k) - R(\zeta - \mu(k))\,[a(k), H_I]\,R(\zeta). \quad (3.1.6)$$

This formula has also been used in standard field theory calculations, see for instance Schweber [1, p. 359].

PROPOSITION 3.1.4 *Let R be the resolvent of $H(g, V, \varkappa)$. With a constant independent on V and $\varkappa$,*

$$\|R^{1/2}[a(k), H_I]\,R^{1/2}\| \leqq \text{const.}\,\mu(k)^{-1/2}.$$

Proof We apply Corollary 3.1.2 with $P_1 = \mu(k)^{1/2}\,[a(k), H_I]$. P_1 has k and x dependent coefficients but they are bounded uniformly and $\deg P_1 = \deg P - 1$. The factor $\mu^{1/2}$ is permitted because of the factor $\mu^{-1/2}$ in the field occuring in H_I.

To illustrate the method, we give a formal derivation of the weak second order estimate

$$H_0 N_\tau \leqq \text{const.}\,(H(g, V, \varkappa) + \text{const.})^2$$

for $\tau < 0$. We have

$$H_0 N_\tau = N_{\tau+1} + \int a^*(k)\,H_0 \mu(k)^\tau\,a(k)\,dk,$$

and the first term is dominated by the first order estimate, Corollary 3.1.2, with $A = N_{\tau+1} \leqq \text{const.}\,H_0$. To bound the second term, we show

$$\int \|H_0^{1/2}\,a(k)\,R(\zeta)\,\theta\|^2\,\mu(k)^\tau\,dk \leqq \text{const.}\,\|\theta\|^2. \quad (3.1.8)$$

By (3.1.6) and the triangle inequality, the left member is bounded by

$$2 \int \|H_0^{1/2} R(\zeta - \mu(k))\, a(k)\, \theta\|^2\, \mu(k)^\tau\, dk$$

$$+ 2 \int \|H_0^{1/2}\, R(\zeta - \mu(k))\, [a(k), H_I]\, R(\zeta)\, \theta\|^2\, \mu(k)^\tau\, dk. \qquad (3.1.9)$$

We use Corollary 3.1.2 to bound $\|H_0 R^{1/2}\|$. The first term in (3.1.9) is bounded by

$$\text{const.} \int \|R(\zeta - \mu(k))^{1/2}\, a(k)\, \theta\|^2\, \mu(k)^\tau\, dk$$

$$\leqq \text{const.} \int \|(N_\tau + I + \mu(k)^\tau)^{-1/2}\, a(k)\, \theta\|^2\, \mu(k)^\tau\, dk$$

$$= \text{const.} \int \|a(k)\, (N_\tau + I)^{-1/2}\, \theta\|^2\, \mu(k)^\tau\, dk$$

$$= \text{const.}\ \|N_\tau^{1/2} (N_\tau + I)^{-1/2}\, \theta\|^2 \leqq \text{const.}\ \|\theta\|^2$$

and the second term is bounded by

$$\text{const.} \int \|R(\zeta - \mu(k))^{1/2}\, [a(k), H_I]\, R(\zeta)\, \theta\|^2\, \mu(k)^\tau\, dk$$

$$\leqq \text{const.} \int \mu(k)^{-1+\tau}\, dk\, \|\theta\|^2 \leqq \text{const.}\ \|\theta\|^2.$$

3.2 Domains of essential self adjointness

THEOREM 3.2.1 *The Hamiltonian $H(g)$ is essentially self adjoint on $\mathscr{D}(H_0) \cap \mathscr{D}(H_I(g))$ and on the subdomains*

$$C^\infty(H_0) = \bigcap_n \mathscr{D}(H_0^n), \qquad (3.2.1)$$

$$\mathscr{D}(H_0) \cap L_p(Q), \qquad 2 < p < \infty. \qquad (3.2.2)$$

The two domains result from distinct proofs of essential self adjointness. The domain (3.2.1) is associated with the Fock space proof while (3.2.2) comes from the Q space proof. For the ϕ^4 theory, $H(g)$ is self adjoint on $\mathscr{D}(H_0) \cap \mathscr{D}(H_I(g))$.

The Fock space proof uses N_τ estimates and Rosen's higher order estimates, and it generalizes the original ϕ_2^4 proof of Glimm and Jaffe[(2)]. The first ϕ_2^{2n} proof of essential self adjointness was given on path space (Rosen[(1)]). This path space proof was simplified and placed in Q space by Segal[(3)]; see also Høegh-Krohn and Simon[(1)].

Proof of (3.2.1) in Fock space We will show that the resolvents

$$R(g, V, \varkappa) = (H_{0,V} + H_I(g, V, \varkappa) + \text{const.})^{-1}$$

converge in norm as $V \to \infty$, $\varkappa \to \infty$. Furthermore, the Hamiltonians converge strongly on $\mathscr{D}(H_0 N^i)$, provided deg $P \leqq 2(i+1)$, by a calculation based on Proposition 1.2.3:

$$\|\{(H_0 + H_I(g)) - (H_{0,V} + H_I(g, V, \varkappa))\} (N+I)^{-i} (H_0 + I)^{-1}\| \to 0.$$

Hence $H_{0,V} + H_I(g, V, \varkappa)$ has a strong graph limit (see Sec. 6.2) and it follows that

$$R(g) = \lim_{V,\varkappa} R(g, V, \varkappa)$$

is the resolvent of a self adjoint operator T, and

$$\{H_0 + H_I(g) + \text{const.}\} \restriction D(H_0 N^i) \subset T.$$

The operator T is essentially self adjoint on the range of $R(g)^j$, for any j. We next show that

$$\text{Range } R(g)^j \subset D(H_0 N^i), \tag{3.2.3}$$

if j is large enough. The essential self adjointness on $\mathscr{D}(H_0 N^i)$ follows, and hence the essential self adjointness on (3.2.1) follows from an N_τ estimate.

Let A be a self adjoint operator and $\psi_j \in \mathscr{D}(A)$. If

$$\psi_n \to \psi, \qquad \|A\psi_n\| \leqq \text{const.}, \tag{3.2.4}$$

then $\psi \in \mathscr{D}(A)$. To prove this we note that for $\psi \in \mathscr{D}(A)$,

$$|\langle A\chi, \psi\rangle| = \left|\lim_n \langle A\chi, \psi_n\rangle\right| = \left|\lim_n \langle \chi, A\psi_n\rangle\right| \leqq \|\chi\| \, \|A\psi_n\|$$
$$\leqq \text{const. } \|\chi\|.$$

Hence $\psi \in \mathscr{D}(A^*) = \mathscr{D}(A)$.

By Theorem 3.1.3, for $\psi_{\varkappa,V} = R(g, V, \varkappa)^j \theta$, $\theta \in \mathscr{F}$, we have

$$\|H_0 N^i \psi_{\varkappa,V}\| \leqq \text{const.},$$

uniformly in $\varkappa$, V. Hence resolvent convergence and (3.2.4) prove (3.2.3).

To prove norm convergence of the resolvents, it is sufficient to prove the norm convergence of their j^{th} power. If $\delta R = R_1 - R_2$ and $\delta H = H_1 - H_2$ denote differences in the resolvents and the Hamiltonians arising from different cutoffs, then calculating as bilinear forms on the domain $C^\infty(H_1) \times C^\infty(H_2)$ we have by Theorem 3.1.3,

$$\begin{aligned}
\|R_1^j - R_2^j\| &\leqq \sum_{\alpha=0}^{j-1} \|R_1^\alpha \, \delta R \, R_2^{j-\alpha-1}\| \\
&\leqq \sum_{\alpha=0}^{j-1} \|R_1^{\alpha+1} \, \delta H \, R_2^{j-\alpha}\| \\
&\leqq \text{const.} \sum_{\alpha=0}^{j-1} \|(N+I)^{-(\alpha+1)} \, \delta H (N+I)^{-(j-\alpha)}\|.
\end{aligned}$$

Hence, we choose $j \geqq \frac{1}{2}$ deg. P.

Each term in the sum over α above converges to zero, by Proposition 1.2.3. This completes the proof, and furthermore shows that Theorem 3.1.3 is valid for $H(g)$, in addition to $H_{0,V} + H_I(g, V, \varkappa)$ as indicated in Sec. 3.1.

The Q space proof of (3.2.1) extends to the hypercontractive setting under the hypothesis that $H_I(g)$ and $\exp(-H_I(g))$ belong to L_p for all $p < \infty$, as in Proposition 2.1.2 and Theorem 2.1.4. Let H_j be the Hamiltonian (3.1.1).

LEMMA 3.2.2 *Let* $1 < p, q < \infty$. *The semigroups* $e^{-tH_j}: L_p \to L_p$ *are strongly continuous in* $t \geqq 0$ *and exponentially bounded with a bound uniform in* j. *If* t *is sufficiently large, the same is true for* $e^{-tH_j}: L_p \to L_q$.

Proof We use the Trotter product formula,

$$e^{-tH_j} f = L_2 \lim_n (e^{-tH_0/n} e^{-tH_I(j)/n})^n f,$$

valid since $H_I^{(j)}$ is a bounded perturbation of the positive and self adjoint operator H_0. Let $f \in L_\infty$. Each factor $e^{-tH_0/n}$ maps $L_r \to L_{r+Mt/n}$ with norm 1, by Theorem 2.2.5 and each factor $\exp(-tH_I^{(j)}/n)$ maps $L_r \to L_{r-Mt/2n}$ with norm $e^{O(t/n)}$ by Theorem 2.1.4 and the Hölder inequality, for r bounded away from 1 and ∞. Here $t > 0$. Thus the Trotter approximants to $e^{-tH_j} f$ have the required exponential bounds. The approximants converge strongly in L_2 and they are weakly compact in L_p (and in L_q for large t). Thus they converge weakly in L_p or L_q and $e^{-tH_j} f$ has the required exponential bound:

$$\|e^{-tH_j} f\|_p \leqq e^{O(t)} \|f\|_p, \qquad 0 \leqq t$$

$$\|e^{-tH_j} f\|_q \leqq e^{O(t)} \|f\|_p, \qquad \text{large } t.$$

For $p = 2$, strong continuity results from self adjointness. For $1 < p < 2$, it results from the denseness of L_2 in L_p and the continuity for $p = 2$. For $2 < p < \infty$, we use duality and reduce to the case $1 < p < 2$.

LEMMA 3.2.3 *Let* $1 < p, q < \infty$. *The semigroups* $e^{-tH_j}: L_p \to L_p$ *converge strongly as* $j \to \infty$ *to continuous exponentially bounded semigroups, and for large* t, *the maps* $e^{-tH_j}: L_p \to L_q$ *converge in norm.*

Proof In the Duhamel formula

$$e^{-tH_i} f - e^{-tH_j} f = \int_0^t e^{-(t-u)H_i} (H_j - H_i) e^{-uH_j} f\, du$$

we take $f \in L_\infty$. Then $e^{-uH_j} f$ is bounded in L_r with a bound uniform j and u, $0 \leqq u \leqq t$. We have

$$H_j - H_i = H_I^{(j)} - H_I^{(i)} \to 0 \text{ in } L_q,$$

for all $q \to \infty$. Strong convergence follows from Hölder's inequality; Lemma 3.2.2 controls the factor $e^{-(t-u)H_i}$. The convergence is uniform in bounded t, and consequently the limiting semigroups are continuous. For

3*

large t, we take $f \in L_p$ and we decompose the integration into two regions: $\int_0^t = \int_0^{t/2} + \int_{t/2}^t$. For $0 \leqq u \leqq t/2$, $e^{-uH_j}: L_p \to L_p$ and $e^{-(t-u)H_i}: L_{p-\varepsilon} \to L_q$ with bounds uniform in i and j. Also $H_i - H_j: L_p \to L_{p-\varepsilon}$ converges to zero as $i, j \to \infty$ by Hölder's inequality. Reversing the role of the exponential factors for $u > t/2$, the proof is complete.

Let T be the self adjoint generator of the limiting semigroup on L_2.

Proof of (3.2.2) Consider the domain $\mathscr{D}_t = e^{-tT} L_\infty(Q)$. Then $\mathscr{D}_t \subset L_p$, for all $p < \infty$. Let $f = e^{-tT} g \in \mathscr{D}_t, g \in L_\infty$, and let $f_j = e^{-tH_j} g$. Then $f_j \to f$ in L_p and $H_j f_j \to Tf$ in L_2 by the functional calculus for self adjoint operators. Moreover $f \in L_p \subset \mathscr{D}(H_I(g))$ and

$$H_I^{(j)} f_j \to H_I(g) f$$

in L_2 by Hölder's inequality. By subtraction,

$$H_0 f_j = H_j f_j - H_I^{(j)} f_j$$

converges in L_2. This implies that $f \in \mathscr{D}(H_0)$ and

$$H_0 f = \lim_j H_0 f_j = Tf - H_I(g) f.$$

Thus

$$H_0 + H_I(g) \restriction \mathscr{D}_t = T \restriction \mathscr{D}_t.$$

Since $H_0 + H_I(g)$ is symmetric on $\mathscr{D}_t$ and T is essentially self adjoint on $\mathscr{D}_t$, the theorem follows.

Remark The semigroups and resolvents, as operators on L_2, are norm continuous in the cutoff $\varkappa$ in the potential $H_I(g, \varkappa)$ and they are norm continuous in V and $\varkappa$ in $H_{0,V} + H_I(g, V, \varkappa)$. Using the formulas

$$\inf \text{spectrum } A = t^{-1} \log \|e^{-tA}\| = \|(A + b)^{-1}\|^{-1} + b$$

we see that the vacuum energies $E(g, \varkappa)$ and $E(g, V, \varkappa)$ converge as $\varkappa, V \to \infty$. Henceforth, we renormalize all Hamiltonians as in (3.1.2), so that $H(g)$ now denotes

$$H(g) = H_0 + H_I(g) - E(g), \tag{3.2.5}$$

and

$$0 = \inf \text{spectrum } H(g). \tag{3.2.6}$$

Since the E's converge, the semigroups and resolvents of the renormalized Hamiltonians still converge as before.

3.3 *The spectrum of* $H(g)$

Let m be the mass in H_0. After the vacuum energy renormalization (3.2.5) the spectrum of $H(g)$ has only separated discrete eigenvalues of finite multiplicity in the spectral interval $[0, m)$ (Glimm and Jaffe[4]). Høegh-Krohn[2] proved the existence of continuous spectrum for $H(g)$ in the interval $[m, \infty)$. The discrete eigenvalues in $(0, m)$ may result from a downward mass shift: particles localized in suppt g may have a physical mass less than m. We propose that such eigenvalues should be eliminated by the addition to $H(g)$ of a finite mass renormalization counter term,

$$-1/2\,\delta m^2 \int :\phi^2(x): g(x)^2\,dx.$$

For the box cutoff Hamiltonian $H(V)$ we lose the continuum, but since $H(V)$ and the periodic momentum operator P_V commute, we may discuss their joint spectrum.

THEOREM 3.3.1 *The joint spectrum of $H(V)$ and P_V is purely discrete and is contained in the forward light cone. The same is true for P_V and the momentum cutoff Hamiltonian $H(V, \varkappa)$.*

Proof We only show that $H(V)$ has discrete spectrum; see Glimm-Jaffe[10] for the remainder. We use the representation (2.2.2-3) for

$$H(V) = A_1^{\sim} + B^{\sim} + A_2^{\sim} - E(V) = (A_1 + B) \otimes I + I \otimes A_2 - E(V),$$

with $\varkappa = \infty$, $g = 1$, and then it is enough to show that $H_{0,V} + H_I(V)$, acting on $\mathfrak{S}(\mathscr{H}) = \mathfrak{S}(l_2(Z_V))$, has pure discrete spectrum. However $H_{0,V}$ has a compact resolvent on $\mathfrak{S}(\mathscr{H})$ and the first order estimate

$$H_{0,V} \leqq \text{const.}\,(H_{0,V} + H_I(V) + \text{const.})$$

implies that $H_{0,V} + H_I(V)$ has a compact resolvent on $\mathfrak{S}(\mathscr{H})$ also, so the proof is complete.

Remark The same proof shows that $H(g, V, \varkappa)$, as an operator on $\mathfrak{S}(l_2(Z_{V,\varkappa})) = \mathscr{F}_{\varkappa,V}$, has a compact resolvent. We now use this fact.

THEOREM 3.3.2 *In the spectral interval $[0, m)$, $H(g)$ has pure discrete spectrum with finite multiplicity and no accumulation points.*

Proof First we prove this result for $H(g, V, \varkappa)$, using the representation (1.3.3-4). Since

$$m \leqq A_2 = I \otimes (H_{0,V} \restriction \mathfrak{S}(\mathscr{H}^{\perp})),$$

the spectrum and multiplicity of $H(g, V, \varkappa)$ in the interval $[0, m)$ coincides with that of $H(g, V, \varkappa) \restriction \mathfrak{S}(l_2(Z_{V,\varkappa}))$. By the remark above, this proves the theorem for $H(g, \varkappa, V)$. The proof for $H(g)$ follows from the norm convergence of the resolvents and the following proposition.

PROPOSITION 3.3.3 *Let A_j be a sequence of self adjoint operators, and let the resolvents $R_j(\zeta) = (A_j - \zeta)^{-1}$ converge in norm to the resolvents of a self adjoint operator A. If each A_j has pure discrete spectrum with finite multiplicity and no accumulation points in some fixed spectral interval (a, b) then the same is true of A.*

Proof For any continuous function F, vanishing at infinity, $F(R_j(\zeta)) \to F(R(\zeta))$ in norm. If suppt $F \subset ((b-\zeta)^{-1}, (a-\zeta)^{-1})$ then $F(R_j(\zeta))$ is compact. Thus $F(R(\zeta))$ is compact and the proof is complete.

THEOREM 3.3.4 *$H(g)$ has a unique vacuum. That is, 0 is a simple eigenvalue for $H(g)$. The same statement is valid for the Hamiltonians $H(g, \varkappa)$, $H(g, V, \varkappa)$, $H(V)$ and $H(V) + \theta P_V$, $|\theta| < 1$.*

Let Ω_g be the vacuum for $H(g)$. Then $H(g)\,\Omega_g = 0$ and Ω_g is a limit of the $\Omega_{g,\varkappa}$ or $\Omega_{g,V,\varkappa}$. The existence of the vacuum follows from (3.2.6) and Theorem 3.3.2. The uniqueness is based on a general proposition.

DEFINITION 3.3.5 *Let A be a bounded operator on a measure space $L_2(Q, dq)$. A has a positive ergodic kernel provided $0 \leqq \langle f, Ah\rangle$ whenever f and h are nonnegative functions in L_2 and*

$$0 < \langle f, A^j h\rangle \tag{3.3.1}$$

whenever $0 \leqq f, h$, $0 \neq f, h$, for some j depending on f and h.

For positive kernels the ergodic hypothesis is trivially seen to be equivalent to the irreducibility of A and $L_\infty(Q)$ as operators on $L_2(Q)$.

PROPOSITION 3.3.6 *Let A have a positive ergodic kernel, and suppose that $\|A\|$ is an eigenvalue of A. Then $\|A\|$ is a simple eigenvalue and the corresponding eigenvector can be chosen to be a strictly positive function.*

Proof Since A maps positive functions into positive functions, it also maps real functions into real functions. If $f \in L_2$ satisfies $Af = \|A\|\, f$, then so do Re f and Im f. Therefore we may take f to be real. Since $\|A^j\| = \|A\|^j$ and $A^j f = \|A\|^j f$, we infer that

$$\|A^j\|\;\|f\|^2 = \langle f, A^j f\rangle \leqq \langle |f|, A^j|f|\rangle \leqq \|A^j\|\;\|f\|^2$$

and so $\langle f, A^j f\rangle = \langle |f|, A^j|f|\rangle$. Writing $f = f^+ - f^-$ where $f^\pm$ are the positive and negative parts of f,

$$\begin{aligned}&\langle f^+, A^j f^+\rangle - \langle f^+, A^j f^-\rangle - \langle f^-, A^j f^+\rangle + \langle f^-, A^j f^-\rangle \\ &= \langle f^+, A^j f^+\rangle + \langle f^+, A^j f^-\rangle + \langle f^-, A^j f^+\rangle + \langle f^-, A^j f^-\rangle,\end{aligned}$$

or

$$\langle f^+, A^j f^-\rangle + \langle f^-, A^j f^+\rangle = 0.$$

Unless $f^+ = 0$ or $f^- = 0$, each of the two terms above is strictly positive, by a suitable choice of j. Thus $f^+ = 0$ or $f^- = 0$ and we may choose f to be nonnegative. If $0 \leqq h$, $0 \neq h$, then for some j, $0 < \langle h, A^j f\rangle = \|A\|^j \times \langle h, f\rangle$. Thus $hf \neq 0$ and f is positive almost everywhere. Finally if h is a second eigenvector for the eigenvalue $\|A\|$ then a scalar multiple of h is also positive almost everywhere, and h and f are not orthogonal. Hence $\|A\|$ must be a simple eigenvalue.

Proof of Theorem 3.3.4 Let $A = e^{-H(g)}$ in Proposition 3.3.6. The Trotter approximation to A shows that A has a positive kernel, by Proposition 2.2.1. The ergodic property (3.3.1) may be established directly (Glimm-Jaffe[4]) or the irreducibility of A and $L_\infty(Q)$ may be established (Høegh-Krohn and Simon[1]). Let B be a bounded operator commuting with A and $L_\infty(Q)$. We prove $B = \lambda I$. Since $H_0 + H_I(g) - H_I^{(j)}$ converges to H_0 on a core for H_0, we have

$$e^{-tH_0} = \text{st.} \lim_{j\to\infty} \lim_{n\to\infty} (e^{-(t/n)H(g)} e^{(t/n)H_I(j)})^n.$$

Hence B commutes with e^{-tH_0}, and $e^{-tH_0} B\Omega_0 = B e^{-tH_0}\Omega_0 = B\Omega_0$. Since zero is a simple eigenvalue of H_0, $B\Omega_0 = \lambda\Omega_0$, and since $\mathfrak{M}\Omega_0$ is dense, $B - \lambda I = 0$. This completes the proof.

Høegh-Krohn[2] has constructed the in and out annihilation and creation operators $a_\pm$ for the Hamiltonian $H(g)$. Let

$$a^\#(h) = \int a(k)\, h(k)^-\, dk \quad \text{or} \quad \int a^*(k)\, h(k)\, dk,$$

$h \in L_2(R)$ and let

$$a_t^\#(h) = e^{-itH(g)} e^{itH_0} a^\#(h)\, e^{-itH_0} e^{itH(g)}.$$

PROPOSITION 3.3.7 *On the domain $H(g)^{1/2}$, the following strong limits exist:*

$$a_\pm^\#(h) = \lim_{t\to\pm\infty} a_t^\#(h).$$

Sketch of proof For large j, one can evaluate the weak derivative

$$\frac{d}{dt} a_t^\#(h) = -e^{-itH(g)} [iH_I(g), a^\#(h_t)]\, e^{itH(g)}$$

on the domain $\mathscr{D}(H(g)^j) \times \mathscr{D}(H(g)^j)$, where $h_t = e^{it\mu(k)} h$. The proof uses the higher order estimate, Theorem 3.1.3. Thus with $n = \deg P$ and j large,

$$\|(a_{t_1}^\#(h) - a_{t_2}^\#(h))\, (H(g) + I)^{-j}\|$$

$$\leq \int_{t_1}^{t_2} \|[H_I(g), a^\#(h_s)]\, (H(g) + I)^{-j}\|\, ds$$

$$\int_{t_1}^{t_2} \|[H_I(g), a^\#(h_s)]\, (N + I)^{-n/2}\|\, ds.$$

The integrand is estimated by N_τ estimates of section 1.1.2 with $\tau = 0$. For $h \in C_0^\infty$, h vanishing in a neighborhood of the origin, the kernels in the commutator have L_2 norms bounded by $|s|^{-r}$ for any r, as $s \to \pm\infty$. This gives convergence for a restricted class of h's on a smaller domain, and the full result follows from the uniform estimate

$$\|a_\tau^{\#}(h)\,(H(g) + I)^{-1/2}\| \leqq \text{const.}\ \|h\|_2,$$

which is a consequence of Corollary 3.1.2 and an N_τ estimate.

Using the $a_\pm^{\#}$, we can realize $\mathscr{F}$ as a tensor product, $\mathscr{F} \approx \mathscr{F}_\pm \otimes \mathscr{V}_\pm$, where $\mathscr{V}_\pm$ is the space of simultaneous null vectors for the $a_\pm(h)$ and $F_\pm$ is the closure of the range of polynomials in the $a_\pm^*(h)$ applied to Ω_g.

THEOREM 3.3.8 (Høegh-Krohn[2]) *With the above tensor product decomposition, $H(g)$ is transformed into*

$$(H(g) \restriction \mathscr{F}_\pm) \otimes I + I \otimes (H(g) \restriction \mathscr{V}_\pm).$$

Moreover $H(g) \restriction \mathscr{F}_\pm$ is unitarily equivalent to H_0. $H(g) \restriction \mathscr{V}_\pm$ is a positive operator and in any interval $[0, m - \varepsilon]$, $\varepsilon > 0$, the spectrum is finite dimensional.

4 Removing the space cutoff

Having studied the cutoff Hamiltonian $H(g)$, we are now prepared to construct our ϕ_2^{2n} field theory model, by taking the limit $g \to 1$. The experimentally observed phenomena in quantum field theory are the particles and the scattering cross sections, which specify the measured interactions between the particles. According to conventional ideas in physics, particles are labeled by their masses, which are discrete eigenvalues of the mass operator $M = (H^2 - P^2)^{1/2}$. Here $H = \lim H(g)$ and P is the generator of space translations. The scattering cross sections are expressed in terms of the S matrix and via the Lehmann–Symanzik–Zimmermann reduction formulas (Schweber[1]), the S matrix is expressed in terms of the vacuum expectation values

$$\langle \Omega, \phi(x_1, t_1) \ldots \phi(x_n, t_n)\, \Omega \rangle \tag{4.1}$$

of the field operators. In this chapter we take the limit $g \to 1$ in $H(g)$, in its vacuum vector Ω_g and in the fields

$$\phi_g(x, t) = e^{itH(g)}\, \phi(x, 0)\, e^{-itH(g)}, \tag{4.2}$$

$$\pi_g(x, t) = e^{itH(g)}\, \pi(x, 0)\, e^{-itH(g)}. \tag{4.3}$$

Here $\phi(x, 0) = \phi(x)$ and $\pi(x, 0)$ are the time zero free fields (1.2.3-4). While we also expect that the limit $g \to 1$ exists for the vacuum expectation values (4.1), this has not been proved.

Field theory may also be expressed in terms of bounded operators and operator algebras, and we do this by verifying the Haag-Kastler axioms in Chapters 4 and 5. With sufficient control on the spectrum of M, the existence of the S matrix would follow from the Haag-Ruelle theory, see for instance Jost[1].

4.1 The limit $g \to 1$ on Fock space

We begin by studying bounded operators and we remove the space cutoff g in the Heisenberg picture dynamics,

$$A \to A(t) = \sigma_t(A) = e^{itH(g)} A\, e^{-itH(g)}. \tag{4.1.1}$$

Using the finite propagation speed, we prove that σ_t is independent of g. Let B be a bounded open region of space and let $\mathfrak{A}(B)$ be the von Neumann algebra (weakly closed *-algebra) generated by the operators

$$e^{i\phi(h)}, \quad e^{i\pi(h)} \tag{4.1.2}$$

where suppt $h \subset B$ and $\phi(h) = \int \phi(h)\, h(x)\, dx$ and $\pi(h)$ are time zero free fields. Let $\mathfrak{A}$ be the C^*-algebra (norm closed *-algebra) generated by the $\mathfrak{A}(B)$'s. $\mathfrak{A}$ is called the algebra of quasilocal observables, and an operator A in some $\mathfrak{A}(B)$ is a local observable. The topologies (weak vs. norm) in the definitions of $\mathfrak{A}(B)$ and $\mathfrak{A}$ are determined by the renormalizations required for ϕ_2^{2n}. When one studies distinct representations of a given algebra of operators, it is a general rule that von Neumann algebras and weak closures can be used whenever the represention (or quasi-equivalence class of the representation) does not change; otherwise C^*-algebras and norm closures are necessary. Thus the locally Fock property, that we establish for the algebras $\mathfrak{A}(B)$, justifies the use of the weak closure for $\mathfrak{A}(B)$ but not for $\mathfrak{A}$, see Sections 4.2-4.3. For the Yukawa$_2$ theory, we have a field algebra $\mathfrak{A}_f$, an algebra of observables $\mathfrak{A} \subset \mathfrak{A}_f$ and the corresponding local subalgebras $\mathfrak{A}_f(B)$ and $\mathfrak{A}(B)$. $\mathfrak{A}_f(B)$ is the von Neumann algebra generated by the operators (4.1.2) and the time zero fermion fields $\psi^\alpha(h)$, suppt $h \subset B$, while $\mathfrak{A}(B)$ is generated by (4.1.2) and the fermion currents $\psi^\alpha(h_1)^* \psi^\beta(h_2)$, $\alpha, \beta = 1, 2$, denoting components of the fields. The quasilocal algebras $\mathfrak{A}$ and $\mathfrak{A}_f$ are defined as the C^*-algebras generated by the corresponding local subalgebras $\mathfrak{A}(B)$ or $\mathfrak{A}_f(B)$. For B and C disjoint the commutation relations yield

$$[\mathfrak{A}(B), \quad \mathfrak{A}(C)] = 0 \tag{4.1.3}$$

and in the fermion case

$$[\mathfrak{A}_f(B), \mathfrak{A}(C)] = 0. \tag{4.1.4}$$

Let B_t be the set of points within distance $|t|$ of B.

DEFINITION *Let σ be a one parameter automorphism group of A. The propagation speed $s(\sigma)$ is the infimum of the numbers v such that*

$$\sigma_t(\mathfrak{A}(B)) \subset \mathfrak{A}(B_{vt}) \tag{4.1.5}$$

for all t and all B. Let $\sigma_t^{(1)}$ and $\sigma_t^{(2)}$ be automorphism groups of $\mathfrak{A}$ having finite propagation speed. We define the product $\sigma^{(1)} \times \sigma^{(2)}$ of $\sigma^{(1)}$ and $\sigma^{(2)}$ if for each $A \in \cup_B \mathfrak{A}(B)$, the sequence $(\sigma_{t/n}^{(1)} \cdot \sigma_{t/n}^{(2)})^n (A)$ of Trotter approximants converges in the strong operator topology, as $n \to \infty$. The limit defines an automorphism group of $\mathfrak{A}$, which we denote by $\sigma^{(1)} \times \sigma^{(2)} = \sigma^{(3)}$.

For example, if $\sigma_t^{(j)}$, $j = 1, 2, 3$ are implemented by unitary groups with self-adjoint generators $H^{(j)}$, respectively,

$$\sigma_t^{(j)}(A) = \exp(iH^{(j)}t)\, A \exp(-iH^{(j)}t),$$

and if $H^{(3)} = (H^{(1)} + H^{(2)})^-$ then $\sigma^{(3)} = \sigma^{(1)} \times \sigma^{(2)}$ by Trotter's theorem.

PROPOSITION 4.1.1 *The propagation speed is subadditive: If* $\sigma^{(3)} = \sigma^{(1)} \times \sigma^{(2)}$,

$$s(\sigma^{(3)}) \leqq s(\sigma^{(1)}) + s(\sigma^{(2)}).$$

Proof Since the local algebras $\mathfrak{A}(B)$ are dense in $\mathfrak{A}$, the Trotter limit

$$\sigma_t^{(3)}(A) = \text{s.} \lim_{n\to\infty} (\sigma_{t/n}^{(1)} \sigma_{t/n}^{(2)})^n (A)$$

for $A \in \mathfrak{A}(B)$ proves subadditivity. (This proof is abstracted from Segal[1].)

THEOREM 4.1.2 *In (4.1.1) let* $A \in \mathfrak{A}(B)$ and let $g \equiv 1$ *on* B_t. *Then* $\sigma_t(A)$ *is independent of g and σ extends to a one parameter automorphism group of* $\mathfrak{A}$. *The propagation speed of σ equals one.*

Proof In the notation of Proposition 4.1.1, let $H^{(1)} = H_0$, $H^{(2)} = H_I(g)$. Then $s(\sigma^{(1)}) = 1$, as follows from Lorentz covariance of the free field, for example, or from explicit solutions of the classical wave equation. Moreover $s(\sigma^{(2)}) = 0$. In fact for $A \in \mathfrak{A}(C)$, let $g = g_1 + g_2$ where $g_1 = 0$ on $\sim C$ and $g_2 = 0$ on C. Then

$$\sigma_t^{(2)}(A) = \exp(itH_I(g_1))\, A \exp(-itH_I(g_1)) \in \mathfrak{A}(C)$$

since $\exp(itH_I(g_2))$ commutes with $\mathfrak{A}(C)$ by (4.1.3) and Proposition 2.1.3. The same reasoning shows that the Trotter approximants in (4.1.7) are independent of g. By Theorem 3.2.1, $H^{(1)} + H^{(2)} = H(g)$ is essentially self adjoint, and by Proposition 4.1.1, $\sigma^{(3)}$ is an automorphism of $\mathfrak{A}$ with propagation speed at most one. Furthermore $\sigma_t^{(3)}(A)$ is independent of g for $A \in \mathfrak{A}(B_t)$ provided $g \equiv 1$ on B_t. The rest of the theorem follows from $\sigma_t^{(1)} = \sigma_t^{(3)} \times \sigma_{-t}^{(2)}$, showing $1 \leqq s(\sigma^{(3)})$.

Space translations can be introduced in a simple fashion. With $P = \int a^*(k)\, ka(k)\, dk$, we have

$$e^{-iPy}\, \phi(x, t)\, e^{+iPy} = \phi(x + y, t) \tag{4.1.8}$$

and for bounded observables $A \in \mathfrak{A}$, we have the space translation automorphism $A \to e^{-iPy} A\, e^{+iPy}$. If $a = (y, t) \in R^2$, we define

$$\sigma_a(A) = e^{-iPy}\, \sigma_t(A)\, e^{+iPy}. \tag{4.1.9}$$

One checks that $a \to \sigma_a$ is a representation of R^2 by automorphisms of $\mathfrak{A}$.

Our next goal is to remove the space cutoff in the field ϕ.

LEMMA 4.1.3 *The fields* ϕ_g *and* π_g *of (4.2-3) are bilinear forms on the domain* $C^\infty(H(g)) \times C^\infty(H(g))$, *and as such, they are continuous in x and t.*

Remark (Rosen[2]) The fields are C^2 in x and t. For ϕ_2^4 (and presumably for ϕ_2^{2n}) they are C^∞ in x and t.

Proof Let $b_x = e^{ikx}\mu(k)^{\pm 1/2}$. Then $\phi_g(x, 0) = \phi(x)$ and $\pi_g(x, 0)$ are linear combinations of the bilinear forms $a^{\#}(b_x) = \int b_x(k)\, a^{\#}(k)\, dk$. By an N_τ estimate,

$$(H_0 + I)^{-1-\varepsilon}\, a^{\#}(b_x)\, (H_0 + I)^{-1-\varepsilon}$$

is a bounded operator, bounded by $\|\mu(k)^{-1-\varepsilon}\, b_x\|_2$, and depending continuously on x. Thus by the higher order estimates, Theorem 3.1.3,

$$(H(g) + I)^{-j}\, e^{itH(g)}\, a^{\#}(b_x)\, e^{-itH(g)}\, (H(g) + I)^{-j}$$

is a bounded operator, depending continuously on x and t. The lemma follows from this fact.

We now consider the space time averaged fields

$$\phi_g(f) = \int f(x, t)\, \phi_g(x, t)\, dx\, dt, \quad \pi_g(f) = \int f(x, t)\, \pi_g(x, t)\, dx\, dt \tag{4.1.10}$$

and the sharp time fields

$$\phi_g(f(\cdot, t)) = \int f(x, t)\, \phi_g(x, t)\, dx, \quad \pi_g(f(\cdot, t)) = \int f(x, t)\, \pi_g(x, t)\, dx. \tag{4.1.11}$$

LEMMA 4.1.4 *Let f be a real function in $\mathscr{S}(R^2)$. The fields (4.1.10-11) are symmetric operators on the domain $C^\infty(H(g))$.*

Proof The kernel $b_t(k) = \mu(k)^{\pm 1/2} \int f(x, t)\, e^{ikx}\, dx$ for

$$A_t = \int f(x, t)\, \phi_g(x, 0)\, dx \quad \text{or} \quad \int f(x, t)\, \pi_g(x, 0)\, dx$$

is in L_2. Thus by a N_τ estimate and Cor. 3.1.2

$$\|A_t(H(g) + I)^{-1/2}\| \leqq \text{const.}\ \|A_t(N + I)^{-1/2}\| \leqq \text{const.}\ \|b_t\|_2. \tag{4.1.12}$$

Hence

$$e^{itH(g)}\, A_t\, e^{-itH(g)}\, (H(g) + I)^{-1/2}$$

is a bounded operator, as is its integral over t. This completes the proof.

Remark It is easy to see that the sharp time fields are essentially self adjoint on $C^\infty(H(g))$. By a unitary equivalence, it is sufficient to show this at $t = 0$. By (4.1.12), we may use the domain

$$\mathscr{D}(H(g)) \supset \mathscr{D}(H_0) \cap \mathscr{D}(N^n)$$

and then Proposition 1.2.5 gives essential self adjointness. The space time averaged fields are also self adjoint (see Glimm-Jaffe[4]), but the proof is more difficult.

THEOREM 4.1.5 *Let B be an open region in R^2, let $f \in C_0^\infty(B)$ and let $g \equiv 1$ on $\{y : |y - x| \leqq t, x, t \in B\}$. Then $\phi_g(f)^- = \phi(f)$ and $\pi_g(f)^- = \pi(f)$ are independent of g.*

Proof The spectral projections of the sharp time fields are independent of g, by Theorem 4.1.2, and thus so are the (closed) sharp time fields themselves. Hence $\phi_g(f) \upharpoonright C^\infty(H_0)$ is independent of g. By Theorem 3.2.1, $H(g)$ is essentially self adjoint on $C^\infty(H_0) \subset \mathscr{D}(H(g))$. Thus by (4.1.12), integrated over t,

$$(\phi_g(f) \upharpoonright C^\infty(H_0))^- \supset \phi_g(f) \upharpoonright C^\infty(H(g)) = \phi_g(f)$$

and the proof is complete.

It is a special feature of the $P(\phi)_2$ model that Wick polynomials in the field $\phi(x, t)$ are defined. We start with the momentum cutoff field

$$\phi_\varkappa(x, t) = \int \delta_\varkappa(x - y)\, \phi(y, t)\, dy$$

where $\delta_\varkappa \to \delta$ as $\varkappa \to \infty$. To fix the arbitrary constants in the Wick powers we use the vacuum Ω_g or the vacuum Ω of Sec. 4.3, and we define

$$:\phi_\varkappa^n(x, t): = \phi_\varkappa^n(x, t) + \sum_{j=0}^{n-1} a_{j,n}\, \phi_\varkappa^j(x, t) \tag{4.1.13}$$

where the constants $a_{j,n}$ are determined inductively by the equations

$$[:\phi_\varkappa^n(x, t):, \pi(y, t)] = in\, \delta_\varkappa(x - y) :\phi_\varkappa^{n-1}(x, t): \tag{4.1.14}$$

and

$$\langle \Omega_g, :\phi_\varkappa^n(x, t): \Omega_g \rangle = 0. \tag{4.1.15g}$$

One can show that these equations determine the $a_{j,n}$, that the limit

$$:\phi^n(x, t): = \lim_{\varkappa \to \infty} :\phi_\varkappa^n(x, t): \tag{4.1.16}$$

exists and that

$$e^{itH(g)} :\phi^n(x, 0): e^{-itH(g)} = :\phi^n(x, t):. \tag{4.1.17g}$$

For polynomials P_1 whose degree is bounded by the degree of P, we could presumably apply Cor. 3.1.2 to define Wick ordering in the physical vacuum Ω of Sec. 4.3, replacing (4.1.15g) by

$$\langle \Omega, :\phi_\varkappa^n(x, t): \Omega \rangle = 0, \quad 0 < n \leqq \deg P \tag{4.1.15}$$

and (4.1.17g) by

$$e^{itH} :P_1(\phi(x, 0)): e^{-itH} = :P_1(\phi(x, t)):. \tag{4.1.17}$$

We have $:P(\phi(x, 0)): = \,\vdots Q(\phi(x, 0))\vdots$ for some polynomial Q determined by the $a_{j,n}$ in (4.1.13) and $\deg P = \deg Q$. We note that the Wick dots : : are determined by the Eqs. (4.1.14-5g) with $g = 0$.

The verification of the field equation

$$\phi_{tt}(x, t) - \phi_{xx}(x, t) + m^2\phi(x, t) + j(x, t) = 0 \tag{4.1.18}$$

is straightforward, using higher order estimates, N_τ estimates and the commutation relations, see Glimm and Jaffe[4]. The nonlinear term j is

$$\begin{aligned} j(x, t) &= \vdots Q'(\phi(x, t)) \vdots \\ &= e^{itH(g)} : P'(\phi(x, 0)) : e^{-itH(g)}. \end{aligned}$$

If we remove the linear term from Q' and absorb it into m^2, thereby redefining m^2 in (4.1.18), a formal argument due to Källen and Lehmann (see Schweber[1]) indicates that $m_{\text{phys}} \leqq m$. In particular the new m^2 is then nonnegative.

4.2 The physical vacuum is locally Fock

According to formal perturbation theory, the vacuum Ω_g of the cutoff theory tends weakly to zero as $g \to 1$. Thus we consider the limit of states rather than vectors. We define ω_g as the linear functional

$$\omega_g(A) = \langle \Omega_g, A\,\Omega_g \rangle, \quad A \in \mathfrak{A} \tag{4.2.1}$$

and obtain a physical vacuum ω by a limiting process as $g \to 1$. Let g and h be fixed nonnegative C_0^∞ functions, let

$$\left.\begin{aligned} \operatorname{suppt} h \subset [-1, 1], \quad \int h(x)\,dx = 1 \\ g \equiv 1 \quad \text{on} \quad [-3, +3] \end{aligned}\right\} \tag{4.2.2}$$

and define

$$g_n(x) = g(n^{-1}x)$$

$$\omega_n(A) = n^{-1} \int \omega_{g_n(\cdot + y)}(A)\, h(n^{-1}y)\, dy. \tag{4.2.3}$$

We choose ω as a limit point of the sequence ω_n. (We believe that the generalized sequence $\{\omega_g\}$ of (4.2.1) converges as $g \to 1$, but this seems difficult to prove. If it converges, it necessarily converges to ω.) Then ω is a state on the C^*-algebra $\mathfrak{A}$. Using ω to define an inner product, we construct the Hilbert space $\mathfrak{F}_{\text{ren}}$ of physical states in Sec. 4.3.

The existence of the limit point ω is a trivial consequence of general w^* compactness theorems. This section is devoted to regularity properties of ω, as will be needed in Sec. 4.3 to show that the unitary translation group acts continuously on $\mathfrak{F}_{\text{ren}}$ and that the unbounded field operators ϕ act on $\mathfrak{F}_{\text{ren}}$. Our main result (Glimm-Jaffe[5]) is

THEOREM 4.2.1 *Let B be a bounded open region in space. The sequence* $\omega_n \restriction \mathfrak{A}(B)$ *lies in a norm compact subset of the dual Banach space* $\mathfrak{A}(B)^*$. *A limit point* $\omega \restriction \mathfrak{A}(B)$ *is a normal state of* $\mathfrak{A}(B)$. ω *(without restriction to a subalgebra) is a limit of a subsequence of the* ω_n.

The proof of this theorem is lengthy, and we will only indicate the main ideas. Our starting point is the uniform estimate from Cor. 3.1.2,

$$\omega_{g_n}(H_0) \leqq \text{const. } n. \tag{4.2.4}$$

In order to remove the factor n on the right, we use a localization in configuration space. Let ζ be a nonnegative C_0^∞ function and let ζ act as a multiplication operator on L_2.

Then $\zeta(-\Delta + m^2)^{\tau/2}\,\zeta$ is a localization of $(-\Delta + m^2)^{\tau/2}$. In momentum space this operator has the kernel

$$w(k_1, k_2) = \int \zeta^{\sim}(k_1 - l)\,\mu^{\tau}(l)\,\zeta^{\sim}(l - k_2)\,dl$$

and using w, we introduced the localized fractional energy operator

$$N_{\tau,\zeta} = \int a^*(k_1)\,w(k_1, k_2)\,a(k_2)\,dk.$$

Similarly for $\tau < \frac{1}{4}$, we can replace ζ by the characteristic function χ_B of an interval B, to obtain a sharply localized operator $N_{\tau,B}$.

LEMMA 4.2.2 *For* $\tau < \frac{1}{4}$, $\omega_n(N_{\tau,B}) \leqq$ const. *The constant depends on the diameter of B, but is otherwise independent of B, and it is independent of n.*

Sketch of proof Using classical methods of Fourier analysis, one can show that

$$\chi_B(-\Delta + m^2)^{\tau}\,\chi_B \leqq \text{const. } \zeta(-\Delta + m^2)^{2\tau+\varepsilon}\,\zeta$$

if $\tau < \frac{1}{2}$ and if $\zeta \equiv 1$ on a neighborhood of B^-. Furthermore if $\zeta_j(x) = \zeta(x + j)$ then

$$\Sigma_j \zeta_j(-\Delta + m^2)^{\tau}\,\zeta_j \leqq \text{const. } (-\Delta + m^2)^{\tau}.$$

These inequalities on the single particle space $L_2(R)$ lead immediately to the estimate

$$\Sigma_j N_{\tau,B+j} \leqq \text{const. } H_0$$

where $B + j = \{x + j : x \in B\}$.

Let $U(y)$ be the unitary operator on $\mathscr{F}$ which implements translation by y, as in (4.1.8). We have

$$U(y)\,H(g_n)\,U(y)^* = H(g_n(\cdot + y))$$

and by the uniqueness of the vacuum Ω_g,

$$U(y)\,\Omega_{g_n} = \Omega_{g_n(\cdot + y)}.$$

Thus

$$\omega_n(N_{\tau,B}) = n^{-1} \int \langle \Omega_{g_n(.+y)}, N_{\tau,B}\Omega_{g_n(.+y)}\rangle\, h(n^{-1}y)\, dy$$

$$= n^{-1} \int \langle \Omega_{g_n}, U(y)^* N_{\tau,B} U(y)\, \Omega_{g_n}\rangle\, h(n^{-1}y)\, dy$$

$$\leqq n^{-1}\, \|h\|_\infty \int \omega_{g_n}(N_{\tau,\,B+y})\, dy$$

$$= n^{-1}\, \|h\|_\infty \int_0^1 \omega_{g_n}(\Sigma_j N_{\tau,\,B+\theta+j})\, d\theta$$

$$\leqq \text{const.}\ n^{-1}\, \omega_{g_n}(H_0)$$

$$\leqq \text{const.}$$

We use the configuration space decomposition

$$L_2(R) = L_2(B) \oplus L_2(\sim B)$$

to obtain the factorization

$$\mathscr{F} \approx \mathscr{F}_B \otimes \mathscr{F}_{\sim B} = \mathfrak{S}(L_2(B)) \otimes \mathfrak{S}(L_2(\sim B))$$

as in Sec. 1.3. In this factorization,

$$N_{\tau,B} = (N_{\tau,B} \upharpoonright \mathscr{F}_B) \otimes I$$

and for $0 < \tau < \frac{1}{4}$, one can also show that $N_{\tau,B} \upharpoonright \mathscr{F}_B$ has a compact resolvent.

Now we return to the sequence ω_n and the proof of Theorem 4.2.1. Each state ω_n can be expressed in terms of a density matrix as a convex linear combination of vector states,

$$\omega_n(A) = \Sigma \alpha_j \langle \theta_j, A\theta_j\rangle, \qquad \theta_j \in \mathscr{F}.$$

Using the uniform estimate of Lemma 4.2.2, we can find a "regularized" state ω_n^ε which approximates ω_n on the full algebra $\mathfrak{A}$:

$$\|\omega_n - \omega_n^\varepsilon\| \leqq \varepsilon.$$

The state ω_n^ε is also a linear combination of certain vector states $A \rightarrow \langle \theta_j^\varepsilon, A\theta_j^\varepsilon\rangle$, and the regularization of ω_n^ε consists in a bound (uniform in n) on the number of particles which occur in the vectors θ_j. We are not able to bound the total number of particles—such a bound would make the limit state ω a Fock state—but using Lemma 4.2.2 we may require that the total number of particles localized in the space interval $[-a, a]$ grows no faster than, say, $0(a^3)$. Furthermore, because $\tau > 0$ and because $N_{\tau,B} \upharpoonright \mathscr{F}_B$ has a compact inverse, we may require that the wave functions of particles localized in $[-a, a]$ lie in a compact subset of L_2, as $n \rightarrow \infty$. In fact there is a bound, uniform in n, on the L_2 norm of the fractional derivatives of order $\tau/2$ for these wave functions.

Since $\omega_n - \omega_n^\varepsilon$ is small in norm, it is sufficient to prove norm compactness with ω_n replaced by ω_n^ε. In fact, if $\{\omega_n^\varepsilon\}$ lies in a norm compact subset of $\mathfrak{A}(B)^*$, there is a norm convergent subsequence $\{\omega_{n_j}^\varepsilon\}$ by a three ε argument,

$$\begin{aligned}\|(\omega_{n_i} - \omega_{n_j}) \restriction \mathfrak{A}(B)\| &\leqq 2\varepsilon + \|(\omega_{n_i}^\varepsilon - \omega_{n_j}^\varepsilon) \restriction \mathfrak{A}(B)\| \\ &\leqq 2\varepsilon + o(1),\end{aligned}$$

and hence $\{\omega_n\}$ lies in a norm compact subset of $\mathfrak{A}(B)^*$.

We now restrict our attention to the ω_n^ε.

The next step in the proof is to regularize the operators A in which the state ω_n^ε is evaluated. The regularization begins with a general argument based on Kaplansky's density theorem. Let $\mathfrak{A}_0(B)$ be a strongly dense *-subalgebra of $\mathfrak{A}(B)$. By Kaplansky's theorem, the unit ball,

$$\{A : A \in \mathfrak{A}_0(B),\ \|A\| \leqq 1\}$$

of $\mathfrak{A}_0(B)$ is strongly dense in the unit ball of $\mathfrak{A}(B)$, see Dixmier[1]. As a corollary,

$$\|(\omega_n^\varepsilon - \omega_m^\varepsilon) \restriction \mathfrak{A}(B)\| = \|(\omega_n^\varepsilon - \omega_m^\varepsilon) \restriction \mathfrak{A}_0(B)\|.$$

Thus it is sufficient to consider operators $A \in \mathfrak{A}_0(B)$. We choose $\mathfrak{A}_0(B)$ to be the linear span generated by the operators

$$e^{i(\phi(f_1) + \pi(f_2))}$$

$f_i \in C^\infty$, suppt $f_i \subset B$. For $A \in \mathfrak{A}_0(B)$ the Taylor series expansion of the above exponentials converges in the expression $\omega_n^\varepsilon(A)$. Let A^δ be the (unbounded) operator which results from a truncation of the Taylor series. The most difficult part of the proof is to show that the truncation can be done (independently of n) so that

$$\|\omega_n^\varepsilon(A - A^\delta)\| \leqq \delta\|A\|, \tag{4.2.5}$$

and so that for a subsequence n_i (chosen independently of A)

$$\|(\omega_{n_i}^\varepsilon - \omega_{n_j}^\varepsilon)(A^\delta)\| \leqq o(1)\,\|A\|, \tag{4.2.6}$$

as $i, j \to \infty$. Given (4.2.5)–(4.2.6), norm compactness follows.

The idea behind these estimates is very simple. The operator $A \in \mathfrak{A}(B)$ observes mainly vectors localized near B, and ω_n^ε has only vectors with a finite number of particles localized near B. The notions of localization for vectors in Fock space and for operators in $\mathfrak{A}(B)$ are distinct and differ on account of the factors $\mu(k)^{\pm 1/2}$ that occur in the definition of $\phi(f)$ and $\pi(f)$, e.g. for real f,

$$\phi(f) = a(\mu^{-1/2}\tilde{f}) + a(\mu^{-1/2}\tilde{f})^*.$$

The difference between the two notions of localization arises from the tail of the kernel $k_\pm(x, y)$ of the operator $(-\Delta + m^2)^{\pm 1/4}$. As $|x - y| \to \infty$,

$k_{\pm} = 0(\exp(-m|x - y|))$, so that the probability of the operator A observing vectors localized far from B is exponentially small, but nonzero. We choose a sufficiently large interval B_1 containing B, and let $\mathscr{F} = \mathscr{F}_{B_1} \otimes \mathscr{F}_{\sim B_1}$. Since our two notions of localization differ only by an exponential tail, we introduce a negligible error by retaining in A^δ only the part acting on on $\mathscr{F}_{B_1}$.

A basic tool in the proof of (4.2.6) is the estimate

$$\omega_n^\varepsilon(N_{\tau, B_1}) \leqq \text{const.}, \quad 0 < \tau < \tfrac{1}{4}, \tag{4.2.7}$$

which follows from Lemma 4.2.2 and the definition of ω_n^ε. Since the operator N_{τ,B_1} has a compact resolvent on $\mathscr{F}_{B_1}$, we use (4.2.7) to establish that a subsequence of the states ω_n^ε is norm convergent on the algebra of all bounded operators on $\mathscr{F}_{B_1}$. Using the fact that the bounded functions of A^δ are bounded operators on $\mathscr{F}_{B_1}$, we are able to establish (4.2.6).

The difficulty in proving (4.2.5) is to obtain an estimate that is uniform in the norm of A. There is no convenient estimate for the norm of an operator in terms of its kernel.

We now complete the proof of the theorem. Normal states of a von Neumann algebra $\mathfrak{A}$ are states which can be expressed as an infinite convex linear combination of vector states. Such states can be characterized in terms of the order property:

$$\omega(\sup A_\alpha) = \sup \omega(A_\alpha)$$

for each monotonically increasing family $\{A_\alpha\}$ which is bounded from above, Dixmier[1]. Using the order definition, a $3 - \varepsilon$ argument proves that the norm limit of normal states is normal. Any limit point of the ω_n can (when restricted to $\mathfrak{A}(B)$) be obtained as a norm limit of a subsequence. By the diagonal process, we can choose one subsequence of the ω_n that converges in norm on each $\mathfrak{A}(B)$. Hence this subsequence converges w^* on $\mathfrak{A}$.

4.3 The Hilbert space of physical states

Let ω be a limit point of the sequence $\{\omega_n\}$ of approximate vacuums. We use ω to define the inner product

$$A, B \to \omega(A^*B) \tag{4.3.1}$$

on $\mathfrak{A} \times \mathfrak{A}$. We introduce the corresponding (pseudo) norm

$$\|A\| = \omega(A^*A)^{1/2}. \tag{4.3.2}$$

The set $\mathfrak{N}$ of null vectors with respect to this norm is a left ideal in $\mathfrak{A}$. We divide by this ideal and complete the quotient $\mathfrak{A}/\mathfrak{N}$ in the norm (4.3.2). We obtain a new Hilbert space, which we call $\mathscr{F}_{\text{ren}}$; (4.3.1) defines the inner produce in $\mathscr{F}_{\text{ren}}$. The map

$$A : B + \mathfrak{N} \to AB + \mathfrak{N} = \pi(A)(B + \mathfrak{N}) \tag{4.3.3}$$

defines a representation $A \to \pi(A)$ of $\mathfrak{A}$ as operators on $\mathscr{F}_{\text{ren}}$. The above is the GNS construction, see Dixmier[(2)].

$\mathscr{F}_{\text{ren}}$ is the Hilbert space of physical states. $\Omega = I + \mathfrak{N} \in \mathscr{F}_{\text{ren}}$ is the physical vacuum, and

$$\omega(A) = \langle \Omega, \pi(A)\, \Omega \rangle \tag{4.3.4}$$

as one sees from (4.3.1) and the definition of π. In some weak sense, one should regard Ω as a limit point of the vacuums Ω_g. We use the local Fock property of ω, Theorem 4.2.1, to show that π is also locally Fock.

THEOREM 4.3.1 *Let B be a bounded region of space at time zero. There is a unitary operator $U_B: \mathscr{F} \to \mathscr{F}_{\text{ren}}$ such that for $A \in \mathfrak{A}(B)$,*

$$\pi(A) = U_B A U_B^*.$$

Remark As a corollary, we extend the representation π to the unbounded field operators $\phi(f)$, $f \in C_0^\infty$, and preserve local properties (such as the canonical commutation relations and the field equation) satisfied by ϕ. In order to do this, the normality of the representation π is sufficient. Normality of π does not depend upon the theorems of Araki and of Griffin.

Proof The theorem is a simple consequence of Theorem 4.2.1 and theorems of Araki[(2)] and Griffin[(1)]. Araki showed that $\mathfrak{A}(B)$ is a factor of type III in the sense of Murray and von Neumann and Griffin showed that isomorphisms between separable factors of type III are unitarily implementable. In order to apply Griffin's theorem we need only the ultraweak continuity (normality) of $\pi \restriction \mathfrak{A}(B)$. In fact on the one hand if $\pi \restriction \mathfrak{A}(B)$ is continuous it must be an isomorphism because there are no weakly closed ideals in a factor, and on the other hand the continuous image of a factor of type III is a factor of type III. (See Dixmier[(1)].) The continuity of $\pi \restriction \mathfrak{A}(B)$ which we need is equivalent to a bound of the form

$$|\langle \theta, \pi(A)\, \theta \rangle| \leqq \Sigma\, |\langle \psi_j, A\psi_j \rangle|$$

where $\theta \in \mathscr{F}_{\text{ren}}$, $A \in \mathfrak{A}(B)$ and the ψ_j in $\mathscr{F}$ depend on B and the choice of θ. However, this continuity results from the continuity of $\omega \restriction \mathfrak{A}(B_1)$ for B_1 bounded because

$$|\omega(A)| \leqq \Sigma |\langle \psi_j, A\psi_j \rangle|$$

by Theorem 4.2.1 and so

$$\begin{aligned} |\langle \pi(C)\, \Omega, \pi(A)\, \pi(C)\, \Omega \rangle| &= |\omega(C^* A C)| \\ &\leqq \Sigma\, |\langle C\psi_j, AC\psi_j \rangle| \end{aligned}$$

where $C \in \mathfrak{A}(B_1)$, $B \subset B_1$. As B_1 varies, vectors of the form $\theta = \pi(C)\, \Omega$ are dense in $\mathscr{F}_{\text{ren}}$, and the proof is complete.

THEOREM 4.3.2 *$\mathscr{F}_{\text{ren}}$ is separable and π is an isomorphism. There is a continuous unitary representation U of R^2 such that*

$$U(a)\, \pi(A)\, U(a)^* = \pi(\sigma_a(A)), \tag{4.3.5}$$

$$U(a)\, \Omega = \Omega. \tag{4.3.6}$$

The spectrum of the generator H of time translations is contained in the interval $[0, \infty)$.

Proof The separability results from the facts that $\pi \upharpoonright \mathfrak{A}(B)$ is ultraweakly continuous and that each $\mathfrak{A}(B)$ is ultraweakly separable. π is an isomorphism because $\mathfrak{A}$ is simple as a C^*-algebra. In fact for any representation π of $\mathfrak{A}$, $\pi \upharpoonright \mathfrak{A}(B)$ is an isomorphism because $\mathfrak{A}(B)$, as a separable type III factor, is a simple C^*-algebra. Hence $\pi \upharpoonright \mathfrak{A}(B)$ is isometric, and since this is true for every B, π is isometric, and consequently an isomorphism. We define

$$U(a)\,\pi(A)\,\Omega = \pi(\sigma_a(A))\,\Omega\,.$$

Then $U(\cdot)$ is a unitary representation of R^2, provided $\omega = \omega \circ \sigma$, i.e. provided ω is a fixed point under the space time translation automorphisms, as one can see by a short calculation. To show that ω is a fixed point for space translations, we must utilise the space integration in the definition of ω_n and to show the invariance under time translations we must choose $H(g_n(\cdot + y))$ to implement the time automorphism. The restriction on the supports of g and h were introduced so that this choice of the cutoff Hamiltonian is permitted for any y in the support of h. We omit further details. (See Glimm-Jaffe[5].)

To show that H is positive, we introduce the spectral resolution $H = \int \mu \, dE(\mu)$ and the function

$$F(t) = \omega(A^* \sigma_{0,t}(A)) = \langle \pi(A)\,\Omega, e^{iHt}\,\pi(A)\,\Omega \rangle$$
$$= \int e^{i\mu t}\, d\|E(\mu)\,\pi(A)\,\Omega\|^2,$$

and we must show that only the interval $[0, \infty)$ contributes to the above integral. In other words we must show that F has a Fourier transform with support in $[0, \infty)$. However, this property is easily established for the approximating functions

$$F_n(t) = \omega_n(A^* \sigma_{0,t}(A))$$

(using the positivity of $H(g_n(\cdot + y))$) and it is preserved under limits. This completes the proof. For further results, see Glimm-Jaffe[11].

5 Lorentz covariance and the Haag-Kastler axioms

5.1 Lorentz covariance

We introduce the algebras of local observables $\mathfrak{A}(B)$, where $B \subset R^2$ is a bounded open region of space time. Self adjoint elements A in $\mathfrak{A}(B)$ are interpreted as observables which can be observed by measurements performed in the region B. We start with the self adjoint field operator

$$\phi(f) = \int \phi(x, t)\, f(x, t)\, dx\, dt$$

4*

where f is a real, C_0^∞ function. This operator measures the field strength, averaged over space time points with the weight function f. We define $\mathfrak{A}(B)$ as the von Neumann algebra generated by bounded functions of the fields localized in B,

$$\mathfrak{A}(B) = \{\exp(i\phi(f)) : \operatorname{supp} f \subset B\}''.$$

Let $\mathfrak{A}$ be the C^* algebra generated by $\cup_B \mathfrak{A}(B)$. $\mathfrak{A}$ coincides with the algebra $\mathfrak{A}$ introduced in Sec. 4.1, but B and $\mathfrak{A}(B)$ are distinct from the time zero regions and algebras of Sec. 4.1.

The Lorentz group $\mathscr{P} = \{a, \Lambda\}$ is the semidirect product of R^2 with R^1,

$$\{a_1, \Lambda_1\}\{a_2, \Lambda_2\} = \{a_1 + \Lambda_1 a_2, \Lambda_1\Lambda_2\}. \tag{5.1.1}$$

Here $a \in R^2$ is a space time translation and Λ is the Lorentz rotation:

$$\Lambda_\beta : (x, t) \to (x \cosh\beta + t \sinh\beta, x \sinh\beta + t \cosh\beta).$$

We formulate Lorentz covariance in terms of automorphisms. Our main result is the following: There exists a representation $\sigma_{\{a,\Lambda\}}$ of $\mathscr{P}$ by *-automorphisms of $\mathfrak{A}$, such that

$$\sigma_{\{a,\Lambda\}}(\mathfrak{A}(B)) = \mathfrak{A}(\{a, \Lambda\} B) \tag{5.1.2}$$

for all bounded open sets B and all $\{a, \Lambda\} \in \mathscr{P}$.

The Lorentz group composition law gives

$$\sigma_{\{a,\Lambda\}} = \sigma_{\{a,I\}}\,\sigma_{\{0,\Lambda\}}. \tag{5.1.3}$$

The space time covariance is an easy consequence of the results of Chapter 4:

$$\sigma_a(\mathfrak{A}(B)) = \sigma_{\{a,I\}}(\mathfrak{A}(B)) = \mathfrak{A}(\{a, I\} B) = \mathfrak{A}(B + a). \tag{5.1.4}$$

To see this we make use of the fact that $\mathfrak{A}(B)$ coincides with the von Neumann algebra generated by sharp time fields,

$$\mathfrak{A}(B) = \left\{\exp\left(i \int \phi(x, t) f(x, t)\, dx\right) : \operatorname{supp} f \subset B\right\}'',$$

see Glimm-Jaffe[(4)]. In (5.1.4), the time translation is implemented locally by the unitary operators $\exp(it\, H(g))$, and the space translation is implemented by $\exp(-iPx)$. Since

$$\exp(-iPx)\, H(g) \exp(iPx) = H(g_x)$$

$$g_x(y) = g(y - x),$$

the space and time automorphisms commute. Returning to (5.1.3), we conclude that the existence of the automorphism representation $\sigma_{\{a,\Lambda\}}$ follows easily from the construction of the pure Lorentz transformation $\sigma_{\{0,\Lambda\}} = \sigma_\Lambda$. We obtain σ_Λ by constructing locally correct infinitesimal generators.

We now sketch the formal ideas in the proof of Cannon-Jaffe[1] of Lorentz covariance for the $(\phi^4)_2$ model. The extension to the $P(\phi)_2$ model is given by Rosen[3].

Formally, the operator,

$$M = M_0 + M_I(g)$$
$$= \tfrac{1}{2}\int \{:\pi(x)^2: + :(\nabla\phi(x))^2: + m^2 :\phi(x)^2:\}\, x\, dx + H_I(xg) \qquad (5.1.5)$$

is an infinitesimal generator of Lorentz transformations in a region B if the cutoff function g equals one on a sufficiently large interval. Unfortunately the methods of Sec. 2 and 3 can not be applied to the M of (5.1.5), since M is not formally positive. This problem does not affect the study of M_0, since M_0 acts in a simple fashion. It is easy to prove from the explicit form of M_0 acting on n particle states that M_0 is essentially self adjoint on the domain $\mathscr{D}$. It is not known, however, whether M of (5.1.5) is self adjoint.

To avoid this trouble we restrict our attention to regions B_1 contained in the quadrant $x > |t| + 1$. For such regions B_1, we may replace (5.1.5) by

$$M = \int H(x)\, xg(x)\, dx \qquad (5.1.6)$$

where $xg(x)$ is nonnegative.

Here $H(x)$ is the energy density (usually denoted $T_{00}(x)$),

$$H(x) = \tfrac{1}{2}:\{\pi(x)^2 + (\nabla\phi)^2\,(x) + m^2(\phi^2)\,(x)\}: + H_I(x)$$
$$\equiv H_0(x) + H_I(x),$$

and is formally positive. Thus (5.1.6) is formally positive. In fact it is technically convenient to use different spatial cutoffs in the free and the interaction part of M. Our final formula for M is

$$M = M(g_0, g) = \alpha H_0 + H_0(xg_0) + H_I(xg), \qquad (5.1.7)$$

where $0 < \alpha$ and $0 \leqq xg_0(x), xg(x)$. In order that (5.1.7) be formally correct, we require that

$$\alpha + xg_0 = x = xg \quad \text{on} \quad [1, R]$$

with R sufficiently large. For technical reasons we require that

$$\alpha + xg_0 = x \quad \text{on} \quad \operatorname{supp} g. \qquad (5.1.8)$$

By our above restrictions on g_0 and g we have

$$\operatorname{supp} g_0, \quad \operatorname{supp} g \subset \{x : \alpha \leqq x\}.$$

We show that the operator M of (5.1.7) is essentially self adjoint on $\mathscr{D}$, and it generates Lorentz rotations in an algebra $\mathfrak{A}(B_1)$,

$$\exp(i\beta M)\,\mathfrak{A}(B_1)\exp(-i\beta M) \subset \mathfrak{A}(\{0, \Lambda_\beta\}\, B_1) \qquad (5.1.9)$$

provided B_1 and $\{0, \Lambda_\beta\}$ B_1 are contained in the region

$$\{x, t: 1 + |t| < x < R - |t|\} \tag{5.1.10}$$

where M is formally correct. These results permit us to define the Lorentz rotation automorphism σ_Λ on an arbitrary local algebra $\mathfrak{A}(B)$. Using a space time translation σ_a, we can translate B into a region

$$B + a = B_1 \subset \{x, t: |t| + 1 < x\}.$$

For R large enough, B_1 and $\{0, \Lambda_\beta\}$ B_1 are contained in (5.1.10), and we define σ_{Λ_β} by the formula

$$\sigma_{\Lambda_\beta} \restriction \mathfrak{A}(B) = \sigma_{\{-\Lambda a, I\}}\sigma_{\{0,\Lambda_\beta\}}\sigma_{\{a,I\}} \restriction \mathfrak{A}(B). \tag{5.1.11}$$

THEOREM 5.1.1 *Let $M(g_0, g)$ be given by (5.1.7), with α, g_0 and g restricted as above. Then M is essentially self adjoint on $\mathscr{D}$, and for the ϕ_2^4 theory M is self adjoint.*

By Theorems 3.1.1 and 3.2.1, $\alpha H_0 + H_I(xg)$ is bounded from below and essentially self adjoint on $\mathscr{D}$; for the ϕ_2^4 theory it is self adjoint on $\mathscr{D}(H_0) \cap \mathscr{D}(H_I(xg))$. For the ϕ_2^4 theory, the second order estimate

$$(H_0(xg_0))^2 \leqq a(\alpha H_0 + \gamma H_0(xg_0) + H_I(xg) + bI)^2$$

is valid uniformly in γ, $0 \leqq \gamma \leqq 1$. In particular, this estimate shows that $H_0(xg_0)$ is a densely defined operator. In fact there is a cancellation between the pair creation terms of $:\pi^2(xg_0):$ and $:(\nabla\phi)^2 (xg_0):$ and similarly between the pair annihilation terms to give the unbounded operator $H_0(xg_0)$ defined on $\mathscr{D}(H_0)$. More important, this estimate exhibits $M(g_0, g)$ as the final term in a finite sequence of Kato perturbations, namely $\alpha H_0 + \gamma_j H_0(xg_0) + H_I(xg)$.

In order to prove (5.1.9), we prove the corresponding identity for the field operators.

THEOREM 5.1.2 *Let B_1 and $\{0, \Lambda_\beta\}$ B_1 be contained in the set (5.1.10). We have the identity between self adjoint operators:*

$$\exp(i\beta M)\,\phi(f)\exp(-i\beta M) = \phi(f_{\Lambda_\beta})$$

$$= \int \phi(\Lambda_\beta(x, t))\, f(x, t)\, dx\, dt,$$

provided supp $f \subset B_1$.

The field $\phi(x, t)$ is also transformed correctly, in the sense of bilinear forms. Since the transformations

$$f \to f_{\{a,\Lambda\}}(\cdot) = f(\{a, \Lambda\}^{-1}\cdot)$$

and

$$\phi(f) \to \phi(f_{\{a,\Lambda\}})$$

obviously satisfy the group law in $\mathscr{P}$, and since $\mathfrak{A}$ is generated by the field operators, Theorem 5.1.2 implies that the automorphisms $\sigma_{\{a, \Lambda\}}$ defined by (5.1.3), (5.1.9) and (5.1.11) also satisfy the group law in $\mathscr{P}$. In particular (5.1.11) is independent of the choice of a.

The proof of Theorem 5.1.2 is reduced to the verification of the equation

$$\left\{x\frac{\partial}{\partial t} + t\frac{\partial}{\partial x}\right\}\phi(x, t) = [iM, \phi(x, t)] \tag{5.1.12}$$

as an equation for bilinear forms on a suitable domain. Since M is self adjoint, we can integrate (5.1.12) to yield the theorem.

The formal proof of (5.1.12) can be understood as follows: We compute for $H = H_0 + H_I(g)$,

$$\begin{aligned}[iM, \phi(x, t)] &= [iM, e^{itH}\,\phi(x, 0)\,e^{-itH}] \\ &= e^{itH}[iM(-t), \phi(x, 0)]\,e^{-itH},\end{aligned} \tag{5.1.13}$$

where

$$M(-t) = e^{-itH}\,M\,e^{itH}.$$

Again formally

$$M(-t) = \sum_{n=0}^{\infty} \frac{(-t)^n}{n!}\,ad^n(iH)\,(M).$$

If M and H were the correct global Lorentz generator and Hamiltonian they would satisfy

$$\begin{aligned}[iH, M] &= ad(iH)\,(M) = P, \\ [iH, [iH, M]] &= 0,\end{aligned} \tag{5.1.14}$$

and

$$M(-t) = M - Pt.$$

Thus by (5.1.13)

$$[iM, \phi(x, 0)] = [iM_0, \phi(x, 0)] = x\,\pi(x, 0)$$

and

$$[iP, \phi(x, 0)] = -(\nabla\phi)\,(x, 0),$$

we have (5.1.12). The difficulty with this formal argument is that H and M do not obey (5.1.14), since they are correct only in B_1. We have instead the equations

$$\begin{aligned}[iH, M] &= P^{\text{loc}} \\ [iH, [iH, M]] &= R^{\text{loc}},\end{aligned} \tag{5.1.15}$$

where P^{loc} acts like the momentum operator in B_1, i.e.

$$[P^{\text{loc}}, \phi(x, t)] = [P, \phi(x, t)], \quad x, t \in B_1.$$

Hence $[iH, P^{\text{loc}}] = R^{\text{loc}}$ is not identically zero, but commutes with $\mathfrak{A}(B_1)$. Formally, further commutators of R^{loc} with H are localized outside B_1, and (5.1.12) follows formally even for our approximate, but locally correct H and M.

In order to convert this formal argument into a mathematical theorem, more work is necessary. We make a Taylor series expansion for

$$F(-t) = \langle \Omega, [iM(-t), \phi(x, 0)]\, \Omega \rangle, \tag{5.1.16}$$

with $\Omega \in C^{\infty}(H)$, namely

$$F(-t) = F(0) - tF'(0) + \tfrac{1}{2} t^2 F''(s)$$

where $|s| \leqq |t|$. By (5.1.15), we have

$$F''(-s) = \langle e^{isH}\Omega, [iR^{\mathrm{loc}}, \phi(x, s)]\, e^{isH}\Omega \rangle .$$

We note that $(x, t) \in B_1$, so that with $|s| \leqq |t|$, $(x, s) \in B_1$. Hence $[R^{\mathrm{loc}}, \phi(x,s)] = 0$, which is an operator identity after integration over x with a C_0^{∞} function. Thus

$$F''(s) = 0, \qquad |s| \leqq |t|$$

and

$$\begin{aligned} F(-t) &= F(0) - tF'(0) \\ &= \langle \Omega, \{[iM, \phi(x, 0)] - t[iP^{\mathrm{loc}}, \phi(x, 0)]\}\, \Omega \rangle \\ &= \langle \Omega, \{x\pi(x, 0) + t(\nabla\phi)\,(x, 0)\}\, \Omega \rangle . \end{aligned}$$

We conclude that

$$[iM(-t), \phi(x, 0)] = x\pi(x, 0) + t\, \nabla\phi(x, 0)$$

and inserting this relation in (5.1.13) yields (5.1.12). This completes the formal outline of the proof of Lorentz covariance.

5.2 *The Haag-Kastler axioms*

In the physics literature, one encounters axiom schemes for quantum field theory. These axioms formulate the physical ideas of space time localization, Lorentz covariance and locality. Wightman's axioms[1] deal directly with the field operators, acting on the Hilbert space of physical states. The Haag-Kastler[1] axioms are independent of the choice of Hilbert space and the representation of the operators; hence these axioms involve less structure and in principle they are more general. The Haag-Kastler framework avoids certain technical problems by working with bounded functions of the fields, i.e. bounded observables. On the other hand there is no direct way to formulate the field equations in terms of the bounded observables. Rather than comparing these axiom schemes, we are interested in verifying them in our models.

We describe the Haag-Kastler axioms as follows: To each bounded open region B, we associate a von Neumann algebra $\mathfrak{A}(B)$ and we let $\mathfrak{A}$ be the C^* algebra generated by the union of the algebras $\mathfrak{A}(B)$. The axioms are

1) Isotony: If $B_1 \subset B_2$ then $\mathfrak{A}(B_1) \subset \mathfrak{A}(B_2)$.

2) Locality: If B_1 and B_2 are spacelike separated then $\mathfrak{A}(B_1)$ commutes with $\mathfrak{A}(B_2)$.

3) Lorentz covariance: There is a representation $\{a, \Lambda\} \to \sigma_{\{a,\Lambda\}}$ of the Lorentz group by *-automorphism of $\mathfrak{A}$ with the property

$$\sigma_{\{a,\Lambda\}} \mathfrak{A}(B) = \mathfrak{A}(\{a, \Lambda\} B).$$

4) $\mathfrak{A}$ has a faithful irreducible representation, i.e. $\mathfrak{A}$ is primitive. (The role of this axiom is to exclude classical field theory.)

The Haag-Kastler axioms are valid for the $P(\phi)_2$ quantum field theory, by results of Chapter 4 and Sec. 5.1.

The Wightman axioms are more detailed.

(a) There is a Hilbert space H and a domain $\mathscr{D}$ which is dense in H and invariant under the space time averaged field operators:

$$\phi(f)\,\mathscr{D} \subset \mathscr{D}.$$

(b) There is a strongly continuous unitary representation $\{a, \Lambda\} \to U(a, \Lambda)$ of the Lorentz group on $\mathscr{H}$, and

$$U(a, \Lambda)\,\phi(f)\,U(a, \Lambda)^* = \phi(f_{\{a,\Lambda\}}).$$

(c) For $a = (x, t) \in R^2$, $U(a, I) = \exp(itH - ixP)$ has a unique fixed vector Ω which lies in $\mathscr{D}$ and is cyclic with respect to polynomials in the fields. The joint spectrum of P and H lies in the forward light cone.

(d) The fields $\phi(f)$ and $\phi(g)$ commute if f and g have space like separated supports.

In the $P(\phi)_2$ quantum field theory, parts of the Wightman axioms have been established; presumably they are all valid. In (a) we would expect to take $\mathscr{D}$ as either $C^\infty(H)$ or else the cyclic space generated by polynomials in the fields, applied to the vacuum Ω. Bounds on derivatives of the field hold see Glimm-Jaffe[(10-11)]. The estimate

$$\pm\phi(f) \leqq |f|\,(H(g) + I),$$

uniform in g, would be sufficient to establish (a). Here $|f|$ is some Schwartz space norm, independent of g. The stronger estimate,

$$N_\tau^{\rm loc} \leqq \text{const.}\,(H(g) + I)$$

should be valid, uniformly in g. Here $N_\tau^{\rm loc}$ denotes one of the local energy operators introduced in Sec. 4.2.

Axiom (b), Lorentz covariance, is known for the translation subgroup, but not for the full Lorentz group. It is an open problem whether the vacuum ω of Sec. 4.2 is invariant under the Lorentz rotation σ_Λ (or whether a Lorentz invariant average of the family $\{\omega \circ \sigma_\Lambda\}$ of states is locally Fock and so usable as a vacuum state). Given a Lorentz invariant locally Fock state on the C^* algebra $\mathfrak{A}$, the general theory assures the existence of the representation U required by (b).

In axiom (c), the spectral condition was proved in Glimm-Jaffe(10–11). The existence of Ω is known, but not its uniqueness. From a mass gap in the spectrum of $H(g)$, uniform as $g \to 1$, we would conclude the uniqueness of ω as a ground state for H and thus the uniqueness of Ω.

Axiom (d) has been established, see Chapter 4 and Glimm-Jaffe(4,5).

II Quantum Field Theory Models: Part II. The Yukawa Model

Part II The Yukawa$_2$ Model

The scalar Yukawa$_2$ coupling involves a fermion field ψ and a boson field ϕ. The classical interaction Lagrangian density $-\lambda\bar{\psi}\psi\phi$ leads to the coupled equations

$$(\gamma_0\partial_t + \gamma_1\partial_x + M)\psi - \lambda\psi\phi = 0,$$

$$(\partial_t^2 - \partial_x^2 + m^2)\phi + \lambda\bar{\psi}\psi = 0.$$

The corresponding quantum equations are more singular, since the interaction produces an infinite shift in the mass of a single boson at rest. Hence the renormalized boson field equation has the form

$$(\partial_t^2 - \partial_x^2 + m^2)\phi + j_{\text{ren}} = 0,$$

where

$$j_{\text{ren}} = \lambda\bar{\psi}\psi - \delta m^2\phi$$

and $\delta m^2 = -\infty$. In the equation above, only the difference j_{ren} has a meaning, and it is defined as a limit of a cutoff expression. In the cutoff expression approximating j_{ren}, the corresponding two terms are well defined and the difference is taken in the ordinary sense. Since the locally correct Hamiltonian $H(g)$ contains terms with infinite coefficients, the Yukawa$_2$ theory is more singular than the $P(\phi)_2$ models of Part I. The Hamiltonian $H(g)$ is defined as a limit of self adjoint approximate Hamiltonians $H(g, \varkappa)$.

The renormalization cancellations which occur agree with those predicted by formal perturbation theory. Thus here (and elsewhere in the theory) we find that simple calculations in low order perturbation theory provide excellent predictions. In addition to predicting correctly the renormalization cancellations, perturbation theory has provided conjectures (subsequently verified) about the domains of operators and the validity of uniform estimates expressing the domination of one operator by another.

Work on the doubly cutoff Yukawa theory was done by Y. Kato and Mugibayashi[1] and by Lanford[1] who proved the existence of the theory with $\varkappa, g$ or $\varkappa, V$ cutoffs. The first results on the limit of $H(g, \varkappa)$ as $\varkappa \to \infty$ were obtained by Glimm[1-3] and extended by Hepp[1-2]. They show that $H(g)$ is bounded from below and is an operator on an explicitly given domain. The theory was developed by Glimm and Jaffe[1,3,8,9], who proved that $H(g) = H(g)^*$ and that the propagation speed is finite. Refinements and extensions are due to Federbush[2], Eckmann[1] Dimock[1] and Schrader[1-2]. Osterwalder[1] has renormalized the more singular (but unphysical) ϕ_4^3 interaction.

6 Preliminaries

6.1 The Yukawa$_2$ Hamiltonian

We consider the Fock space $\mathscr{F} = \mathscr{F}_b \otimes \mathscr{F}_f$ of Chapter 1 for a bose particle of mass $m > 0$ and fermions of mass $M > 0$. For convenience, we assume the stability condition $m < 2M$, although this is not necessary for our results. The free fermion field is $\psi_0 = (\psi_0^{(1)}, \psi_0^{(2)})$,

$$\psi_0(x, t) = \begin{Bmatrix} \psi_0^{(1)}(x, t) \\ \psi_0^{(2)}(x, t) \end{Bmatrix} = (4\pi)^{-1/2} \int e^{-ipx}(e^{i\omega t} b'(p)^* \begin{Bmatrix} \nu(-p) \\ \nu(p) \end{Bmatrix}$$

$$+ e^{-i\omega t} b(-p) \begin{Bmatrix} \nu(p) \\ -\nu(-p) \end{Bmatrix} \omega(p)^{-1/2}\, dp. \tag{6.2.1}$$

Here $\omega(p) = (p^2 + M^2)^{1/2}$, $\nu(p) = (\omega(p) + p)^{1/2}$ and $\psi_0(x, t)$ satisfies the free Dirac equation

$$\left(\gamma_0 \frac{\partial}{\partial t} + \gamma_1 \frac{\partial}{\partial x} + M\right) \psi_0(x, t) = 0,$$

$$\gamma_0 = \begin{Bmatrix} 0 & i \\ i & 0 \end{Bmatrix}, \qquad \gamma_1 = \begin{Bmatrix} 0 & i \\ -i & 0 \end{Bmatrix},$$

with the canonical anticommutation relations

$$\{\psi_0^{(i)}(x, t), \psi_0^{(j)}(y, t)^*\} = \delta(x - y)\, \delta_{ij}.$$

At time zero, we set the interacting field $\psi(x, t)$ equal to the free field, $\psi(x) = \psi_0(x, 0)$. The conjugate field $\bar{\psi}$ is defined by

$$\bar{\psi}(x) = (\psi^{(2)}(x)^*, \psi^{(1)}(x)^*) = -i\psi(x)^* \gamma_0,$$

and the scalar current density at time zero is defined by

$$\begin{aligned} j(x) &= :\bar{\psi}(x)\, \psi(x): \\ &= :\psi^{(2)}(x)^*\, \psi^{(1)}(x) + \psi^{(1)}(x)^*\, \psi^{(2)}(x): \end{aligned}$$

The Wick dots : : indicate that in each monomial of creation and annihilation operators the creation operators are permuted to the left of the annihilation operators. In addition, a factor -1 is introduced for each permutation of a pair of adjacent fermion operators.

We use the scalar boson field of Sec. 1.2. The unrenormalized Hamiltonian with a spatial cutoff g is

$$H_{un}(g) = H_0 + \lambda \int j(x)\, \phi(x)\, g(x)\, dx.$$
$$= H_0 + H_I(g).$$

By the N_τ estimates of Proposition 1.2.2, the Hamiltonian $H_{un}(g)$ is a bilinear form on $\mathscr{D} \times \mathscr{D}$, where $\mathscr{D}$ is the domain of vectors with a finite number of particles and wave functions in the Schwartz space $\mathscr{S}(R^m)$. We write

$$H_I(g) = W^C + W + W^A$$

where the pair creation term W^C equals

$$W^C = \int \{a(k)^* + a(-k)\}\, b(p_1)^*\, b'(p_2)^*\, w^C(k, p_1, p_2)\, dk\, dp_1\, dp_2,$$

and the kernel w^C equals

$$w^C(k, p_1, p_2)$$
$$= -\frac{\lambda}{4\pi}\tilde{g}(k + p_1 + p_2)\, (\mu\omega_1\omega_2)^{-1/2}\, (\omega_1\omega_2 - p_1p_2 - M^2)^{1/2} \operatorname{sgn}(p_1 - p_2)$$
$$= -\bar{w}^C(-k, -p_1, -p_2) = -w^C(k, p_2, p_1).$$

The pair annihilation terms are $W^A = (W^C)^*$.

$$W^A = \int w^C(k, p_1, p_2)\, b'(-p_1)\, b(-p_2)\, \{a(k)^* + a(-k)\}\, dk\, dp_1\, dp_2.$$

Note that in W^A we have adopted the arbitrary convention that the particle annihilator is placed to the right of the antiparticle annihilator. In W^C, we follow the adjoint of this convention. This gives subsequent formulas a more natural form, by suitably fixing an arbitrary sign in the kernels.

The boson emission and absorption terms W equal

$$W = \int w(k, p_1, p_2)\, \{a(k)^* + a(-k)\}\, \{b(p_1)^*\, b(-p_2) + b'(p_1)^*\, b'(-p_2)\}$$
$$\times\, dk\, dp_1\, dp_2,$$

with

$$w(k, p_1, p_2) = -\frac{\lambda}{4\pi}\tilde{g}(k + p_1 + p_2)\, (\mu\omega_1\omega_2)^{-1/2}\, (\omega_1\omega_2 + p_1p_2 + M^2)^{1/2}.$$

It is convenient to express the interaction in terms of diagrams introduced by Friedrichs[1],

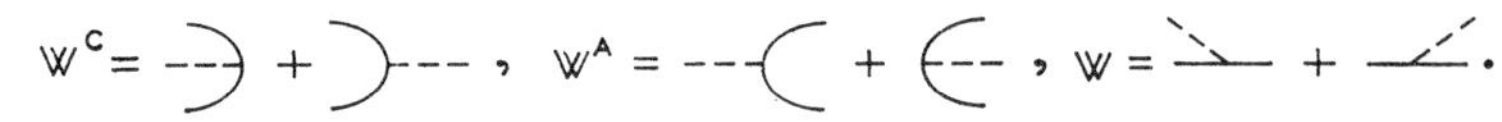

Each diagram represents a Wick monomial W defined in (1.1.9), or possibly a sum of similar monomials. A line pointing to the right stands for an annihilation operator $b(k, \varepsilon)$ while a line pointing to the left stands for a creation form $b(k, \varepsilon)^*$. A solid line denotes a fermion, $\varepsilon = \pm 1$, and a dotted line denotes a boson, $\varepsilon = 0$. A kernel is assigned to each vertex and the diagram is the integral over the momenta of the product of the kernel and the corresponding creation annihilation forms. For clarity, we might designate a variable to be associated with each line in the diagram. For instance, one contribution to W^C is

$$= \int a(k)^* \, b(p_1)^* \, b(p_2)^* \, w^C(k, p_1, p_2) \, dk \, dp_1 \, dp_2,$$

and one contribution to W is

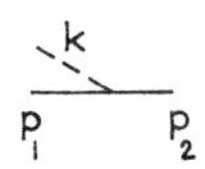

$$= \int a(k)^* \, \{b(p_1)^* \, b(-p_2) + b'(p_1)^* \, b'(-p_2)\} \, w(k, p_1, p_2) \times dk \, dp_1 \, dp_2 .$$

The use of these Friedrichs diagrams provides a convenient shorthand notation for certain equations. The Friedrichs diagrams differ from Feynman diagrams in the asymmetry between creators and annihilators, depending on the orientation of lines. This asymmetry is useful for the discussion of operators defined on some dense domain. For instance, both W^A and $W^C = (W^A)^*$ are bilinear forms on $\mathscr{D} \times \mathscr{D}$, but only W^A is an operator on this domain.

The product $W_1 W_2$ of two Wick monomials W_1 and W_2 can be expressed as a sum of Wick monomials,

$$W_1 W_2 = \sum_{\substack{\text{finite} \\ \text{sum}}} W_j .$$

Such a representation is obtained by using the commutation relations to permute the creation forms to the left of the annihilation forms. The sum occurs because each use of the commutation relations introduces an extra term $\{b(p_1, \varepsilon), b(p_2, \varepsilon')^*\} = \delta_{\varepsilon\varepsilon'} \, \delta(p_1 - p_2)$ and $[a(k), a(k')^*] = \delta(k - k')$. These commutation relations lower the number of creation and annihilation operators by two, and yield an integration over the kernels that remain. Such a term is said to be *contracted* and is indicated in the Friedrichs diagram by joining an annihilation leg (pointing right) to a creation leg on its right (the creation leg points left). In fact, Wick's theorem says that $W_1 W_2$ equals the sum over all possible contractions, with the uncontracted legs in Wick order.

Let us take as a particular example, the product of two diagrams in $H_I(g)^2$.

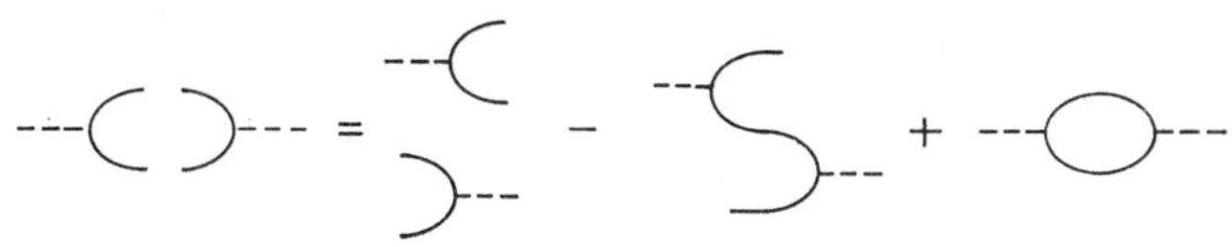

In this example,

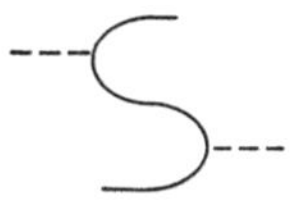

$$= \int dk\, dk'\, dp\, dp' \{b(p)^* \, b(-p') + b'(p)^* \, b'(-p')\} \, a(k)^* \, a(-k') \times \int dp_1 \, w^C(k, p_1, p) \, w^C(k', p', -p_1),$$

$$= \int dk\, dk'\, a(k)^* \, a(-k') \times \int dp_1 \, dp_2 \, w^C(k, p_1, p_2) \, w^C(k', -p_2, -p_1).$$

The kernel of the last term is not defined since for all k, k' the integral

$$\int dp_1 \, dp_2 \, w^C(k, p_1, p_2) \, w^C(k', -p_2, -p_1),$$

diverges. This divergence is one reason that $H_I(g)$ is a bilinear form on $\mathscr{D} \times \mathscr{D}$ but is not an operator.

The divergences are removed from $H_I(g)$ by the introduction of a cutoff function $\chi_\varkappa(k, p_1, p_2)$. We replace $w^C(k, p_1, p_2)$ and $w(k, p_1, p_2)$ by

$$w_\varkappa^C(k, p_1, p_2) = w^C(k, p_1, p_2) \, \chi_\varkappa(k, p_1, p_2)$$

and

$$w_\varkappa(k, p_1, p_2) = w(k, p_1, p_2) \, \chi_\varkappa(k, p_1, p_2).$$

We choose $\chi_\varkappa$ so that $w_\varkappa^C$ and $w_\varkappa$ are L_2 functions; by Proposition 1.2.3b the corresponding cutoff interaction Hamiltonian $H_I(g, \varkappa)$ is an operator on $\mathscr{D}(N)$. For instance, if $\chi_\varkappa(k, p_1, p_2) = \chi(k/\varkappa) \, \chi(p_1/\varkappa) \, \chi(p_2/\varkappa)$ and $\chi(k)$ has compact support, we say that $\chi_\varkappa$ is a sharp cutoff in momentum space. If the Fourier transform of χ has compact support, we say that $\chi_\varkappa$ is a sharp cutoff in position space. More details will be given in Definition 6.3.2.

We now describe the counterterms $c(g, \varkappa)$ suggested by perturbation theory. The renormalized Hamiltonian

$$H_{\text{ren}}(g, \varkappa) = H_0 + H_I(g, \varkappa) + c(g, \varkappa)$$

is required to be bounded from below, uniformly in $\varkappa$. In addition, a proper choice of counterterms is required to yield an operator in the limit as $\varkappa \to \infty$ rather than a bilinear form.

The no-particle vector is the ground state of H_0 and satisfies $H_0\Omega_0 = 0$. In second order perturbation theory the ground state energy of $H_{un}(g, \varkappa) = H_0 + H_I(g, \varkappa)$ is

$$\begin{aligned} E_2(g, \varkappa) &= -\|H_0^{-1/2} H_I(g, \varkappa)\,\Omega_0\|^2 \\ &= -\int |w_\varkappa^C(k, p_1, p_2)|^2 (\mu + \omega_1 + \omega_2)^{-1}\, dk\, dp_1\, dp_2 . \end{aligned}$$

Since $E_2(g, \varkappa)$ diverges as $\varkappa \to \infty$, and the corresponding first order vectors $\Omega_0 - H_0^{-1} H_I(g, \varkappa)\,\Omega_0$ converge as $\varkappa \to \infty$, the unrenormalized Hamiltonian $H_{un}(g)$ is unbounded from below as a bilinear form on $\mathscr{D} \times \mathscr{D}$. In all orders of perturbation theory greater than two, the additive contribution to the vacuum energy of $H_{un}(g)$ is finite, except for modifications to the mass explained below. Hence our theory is superrenormalizable. In other words the divergences of the renormalization constants become less severe in each higher order of perturbation theory.

In order to obtain a theory with positive energy, we subtract $E_2(g, \varkappa)$ from $H_{un}(g, \varkappa)$. However, there is still an infinite shift in second order perturbation theory for the energy of the one-boson, zero-momentum state δ, namely

$$\begin{aligned} &-\langle \delta, \{H_I(g, \varkappa)\,(H_0 - m)^{-1} H_I(g, \varkappa) - E_2(g, \varkappa)\}\,\delta\rangle \\ &= -\left(\frac{\lambda}{4\pi}\right)^2 \int \tilde{g}|(p_1 + p_2)|^2 \frac{(\omega_1\omega_2 - p_1 p_2 - M^2)}{m\omega_1\omega_2} \\ &\quad \times \left\{\frac{1}{\omega_1 + \omega_2 - m} + \frac{1}{\omega_1 + \omega_2 + m}\right\} \\ &\quad \times |\chi_\varkappa(0, p_1, p_2)|^2\, dp_1\, dp_2 . \end{aligned}$$

Here we neglect the normalization of the continuum state δ. Since

$$\langle \delta, :\phi^2(g^2):\, \delta\rangle = \frac{\|g\|_2^2}{2\pi m},$$

we have for the second order mass shift

$$\begin{aligned} \delta m_2^2(g, \varkappa) = &-\frac{\lambda^2}{4\pi\|g\|_2^2} \int |\tilde{g}(p_1 + p_2)|^2 \left(\frac{\omega_1\omega_2 - p_1 p_2 - M^2}{\omega_1\omega_2}\right) \\ &\times \left(\frac{1}{\omega_1 + \omega_2 + m} + \frac{1}{\omega_1 + \omega_2 - m}\right) |\chi_\varkappa(0, p_1, p_2)|^2\, dp_1\, dp_2 . \end{aligned}$$

We modify this expression by a g dependent constant and define the g independent mass shift

$$\delta m^2(\varkappa) = -\frac{\lambda^2}{2\pi}\int\left|\chi_\varkappa\left(0,\frac{\xi}{2},\frac{-\xi}{2}\right)\right|^2\omega(\xi)^{-1}\,d\xi + \text{const.} + o(1), \qquad (6.1.1)$$

where the constant is independent of g, $\varkappa$ and $\chi_\varkappa$, and $o(1)$ vanishes as $\varkappa \to \infty$. (See Lemma 6.3.3.) We define

$$E(g,\varkappa) = E_2(g,\varkappa) + M(g) + o(1), \qquad (6.1.2)$$

where the constant $M(g)$ is independent of $\varkappa$ and $\chi_\varkappa$, and $o(1)$ vanishes as $\varkappa \to \infty$.

The renormalized Hamiltonian $H(g,\varkappa)$ is defined by

$$\begin{aligned} H(g,\varkappa) &= H_0 + H_I(g,\varkappa) + c(g,\varkappa) \\ &= H_0 + H_I(g,\varkappa) - \tfrac{1}{2}\,\delta m^2(\varkappa)\int :\phi(x)^2:\,g(x)^2\,dx - E(g,\varkappa). \end{aligned} \qquad (6.1.3)$$

In the following for simplicity of notation, we suppress the constants and the term $o(1)$ in (6.1.1) and (6.1.2). Their possible inclusion is elementary, and we use them in Chapter 8.

We also introduce a Hamiltonian $H(g,\varkappa,\varrho)$ with a lower cutoff ϱ. For $\varrho = 0$, $H(g,\varkappa,0) = H(g,\varkappa)$, while $H(g,\varkappa,\infty) = H_0$. To obtain the lower cutoff interaction Hamiltonian $H_I(g,\varkappa,\varrho)$, we replace the kernels $w_\varkappa$ and $w_\varkappa^C$ by

$$w_{\varkappa,\varrho}(k,p_1,p_2) = w_\varkappa(k,p_1,p_2)\,\{1-\theta_\varrho(p_1-p_2)\}$$

and

$$w_{\varkappa,\varrho}^C(k,p_1,p_2) = w_\varkappa^C(k,p_1,p_2)\,\{1-\theta_\varrho(p_1-p_2)\}$$

where $\theta_\varrho(p)$ is the characteristic function of the interval $|p| \leqq \varrho$. The counterterms corresponding to the lower cutoff ϱ are

$$c(g,\varkappa,\varrho) = -\tfrac{1}{2}\,\delta m^2(\varkappa,\varrho)\int :\phi(x)^2:\,g(x)^2\,dx - E(g,\varkappa,\varrho), \qquad (6.1.4)$$

$$\delta m^2(\varkappa,\varrho) = -\frac{\lambda^2}{2\pi}\int_{|\xi|>\varrho}\left|\chi_\varkappa\left(0,\frac{\xi}{2},\frac{-\xi}{2}\right)\right|^2\omega(\xi)^{-1}\,d\xi, \qquad (6.1.5)$$

$$E(g,\varkappa,\varrho) = \int |w_{\varkappa,\varrho}^C(k,p_1,p_2)|^2\,(\mu+\omega_1+\omega_2)^{-1}\,dk\,dp_1\,dp_2. \qquad (6.1.6)$$

Then we define

$$H(g,\varkappa,\varrho) = H_0 + H_I(g,\varkappa,\varrho) + c(g,\varkappa,\varrho). \qquad (6.1.7)$$

6.2 Limits of self adjoint operators

We often encounter a sequence H_n of self adjoint operators. For H_n, we might choose a sequence of approximate Hamiltonians, or approximate field operators. In this section, we study notions of convergence that yield a self adjoint limit as $n \to \infty$. We choose types of convergence that we are able to verify directly in our models. The results generalize Glimm and Jaffe[(3)].

One type of convergence is strong convergence of the resolvents $R_n(\zeta) = (H_n - \zeta)^{-1} \to R$ and their adjoints. However, the limit R need not have an inverse (e.g. $H_n = nI$). Whenever R^{-1} is densely defined, R is the resolvent of a self adjoint operator H. In order to assure the invertibility of R, we introduce the notion of *graph convergence* (and later *dense boundedness*) of the sequence of operators H_n. We prove in Theorem 6.2.6 that convergence of the resolvents plus dense boundedness of $\{H_n\}$ yields a self adjoint limit. From functional analysis, one obtains the equivalence of the standard notions of convergence, for example the convergence of the resolvents $R_n(\zeta)$ *to the resolvent of a self adjoint operator* or the strong convergence of the unitary groups $U_n(t) = \exp(itH_n)$. In the following sections we prove dense boundedness and resolvent convergence for approximate Hamiltonians in the Yukawa$_2$ field theory. Analogous results for the $P(\phi)_2$ theory were proved in Part I.

Let H_n be a sequence of densely defined operators on a Hilbert space $\mathscr{H}$. The graph $\mathscr{G}(H_n) \subset \mathscr{H} \oplus \mathscr{H}$ is the set of pairs $\{\psi, H_n\psi : \psi \in \mathscr{D}(H_n)\}$. Let

$$\mathscr{G}_\infty = \{\psi, \chi : \psi = \lim \psi_n, \psi_n \in \mathscr{D}(H_n), \quad \chi = \text{weak}\lim H_n\psi_n\},$$

$$\mathscr{D}_\infty = \{\psi : \{\psi, \chi\} \in \mathscr{G}_\infty \quad \text{for some } \chi\}.$$

In general, $\mathscr{G}_\infty$ will not be the graph of an operator. If $\mathscr{G}_\infty$ is the graph of an operator H, then $\mathscr{D}_\infty$ is the domain of H.

DEFINITION 6.2.1 *A sequence H_n of densely defined operators has a weak graph limit H, if $\mathscr{G}_\infty$ is the graph of a densely defined operator H.*

PROPOSITION 6.2.2 *Let $H_n = H_n^*$. The weak graph limit H of $\{H_n\}$ exists if and only if $\mathscr{D}_\infty$ is dense. In that case H is symmetric.*

Proof If the weak graph limit exists, $\mathscr{D}(H) = \mathscr{D}_\infty$ is dense by definition. Conversely, suppose that $\mathscr{D}_\infty$ is dense and $\{0, \chi\} \in \mathscr{G}_\infty$. We prove $\chi = 0$, so that $\mathscr{G}_\infty$ is the graph of an operator H.

Let $w.\lim \phi_n = \phi$, st. $\lim \psi_n = \psi$. We show that $\langle \psi_n, \phi_n \rangle \to \langle \psi, \phi \rangle$. By the principle of uniform boundedness, $\|\phi_n\| \leq$ const. Hence

$$\langle \psi_n, \phi_n \rangle - \langle \psi, \phi \rangle = \langle \psi, \phi_n - \phi \rangle + \langle \psi_n - \psi, \phi_n \rangle \to 0.$$

Let $\psi_n \in \mathscr{D}(H_n)$, $\lim \psi_n = \psi = 0$, $w.\lim H_n\psi_n = \chi$. For $\theta \in \mathscr{D}_\infty$, there is a sequence $\theta_n \in \mathscr{D}(H_n)$ with $\theta = \lim \theta_n$ and $\Lambda = w.\lim H_n\theta_n$. Thus we have $0 = \langle \psi, \Lambda \rangle = \lim \langle \psi_n, H_n\theta_n \rangle = \lim \langle H_n\psi_n, \theta_n \rangle = \langle \chi, \theta \rangle$. Since $\mathscr{D}_\infty$ is dense, $\chi = 0$ and $\mathscr{G}_\infty = \mathscr{G}(H)$.

To prove that H is symmetric, we note for $\theta \in \mathscr{D}_\infty$,

$$\langle H\theta, \theta \rangle = \lim \langle H_n\theta_n, \theta_n \rangle = \lim \langle \theta_n, H_n\theta_n \rangle = \langle \theta, H\theta \rangle.$$

While the graph limit of a sequence of self adjoint operators is symmetric, it is not necessarily self adjoint. We combine graph convergence of the H_n with convergence of the resolvents $R_n(\zeta) = (H_n - \zeta)^{-1}$. Our main results are Theorems 6.2.3 and 6.2.6.

THEOREM 6.2.3 *Let H be the weak graph limit of a sequence $\{H_n\}$ of self adjoint operators and for some ζ, let the resolvents and their adjoints converge strongly,*

$$R_n(\zeta) \to R \quad \text{and} \quad R_n(\zeta)^* = R_n(\bar{\zeta}) \to R^*.$$

Then H is self adjoint and $R = (H - \zeta)^{-1}$.

Proof We show that R has a densely defined inverse, namely Null $R = 0 = (\text{Range } R)^{\perp}$. We use a simplification of Kleinstein[(1)]. Suppose $\chi \in \text{Null } R$, $\theta \in \mathscr{D}(H)$. Using $\theta = \lim \theta_n$, $H\theta = w.\lim H_n\theta_n$, we have

$$0 = \langle (H - \bar{\zeta})\,\theta, R\chi\rangle = \lim \langle (H_n - \bar{\zeta})\,\theta_n, R_n\chi\rangle = \lim \langle \theta_n, \chi\rangle = \langle \theta, \chi\rangle.$$

Since $\mathscr{D}(H)$ is dense, $\chi = 0$ and Null $R = 0$.

We now prove $\mathscr{D}(H) \subset \text{Range } R$, so $(\text{Range } R)^{\perp} = 0$. Let $\chi \in \mathscr{H}$, $\theta \in \mathscr{D}(H)$.

$$\begin{aligned}\langle \chi, R(H - \zeta)\,\theta\rangle &= \langle R^*\chi, (H - \zeta)\,\theta\rangle = \lim \langle R_n(\bar{\zeta})\,\chi, (H_n - \zeta)\,\theta_n\rangle \\ &= \lim \langle \chi, \theta_n\rangle = \langle \chi, \theta\rangle.\end{aligned}$$

Thus $R(H - \zeta)\,\theta = \theta$, $\mathscr{D}(H) \subset \text{Range } R$ and R^{-1} is densely defined.

On the other hand, for all χ,

$$R_n(\zeta)\,\chi \to R\chi \quad \text{and} \quad (H_n - \zeta)\,R_n(\zeta)\,\chi = \chi.$$

Thus $R\chi \in \mathscr{D}(H)$ and $(H - \zeta)\,R\chi = \chi$. Therefore $\mathscr{D}(H) = \text{Range } R$ and $R = (H - \zeta)^{-1}$. Likewise $R^* = (H - \bar{\zeta})^{-1}$. Since

$$\text{Range}\,(H - \zeta) = \mathscr{D}(R) = \mathscr{H} = \mathscr{D}(R^*) = \text{Range}\,(H - \bar{\zeta}),$$

H is self adjoint.

LEMMA 6.2.4 *Let $H_n = H_n^*$, and let $\mathscr{D}$ be a dense set of vectors with the property that for each $\psi \in \mathscr{D}$, there exists a sequence ψ_n with*

$$\psi_n \in \mathscr{D}(H_n), \qquad \psi_n \to \psi, \qquad \|H_n\psi_n\| \leqq \text{const}.$$

Suppose furthermore that for the sequence $\{\chi_n\}$ approximating a χ in $\mathscr{D}$ as above, the inner products $\langle \chi_n, H_n\psi_n\rangle$ converge as $n \to \infty$. Then $\{H_n\}$ has a weak graph limit H, and $\mathscr{D} \subset \mathscr{D}(H)$.

Proof Given $\varepsilon > 0$ and $\theta \in \mathscr{H}$, choose $\bar{\theta} \in \mathscr{D}$ such that $\|\theta - \bar{\theta}\| < \varepsilon$. Thus for $\bar{\theta}_n \to \bar{\theta}$, $\|H_n\bar{\theta}_n\| \leqq \text{const.}$,

$$\langle \theta, H_n\psi_n\rangle = \langle \theta - \bar{\theta}, H_n\psi_n\rangle + \langle \bar{\theta} - \bar{\theta}_n, H_n\psi_n\rangle + \langle \bar{\theta}_n, H_n\psi_n\rangle.$$

Hence

$$|\langle\theta, H_n\psi_n - H_m\psi_m\rangle| \leqq \|\theta - \bar\theta\| \{\|H_n\psi_n\| + \|H_m\psi_m\|\}$$
$$+ \|\bar\theta - \bar\theta_n\| \|H_n\psi_n\| + \|\bar\theta - \bar\theta_m\| \|H_m\psi_m\|$$
$$+ |\langle\bar\theta_n, H_n\psi_n\rangle - \langle\bar\theta_m, H_m\psi_m\rangle|$$
$$\leqq o(1)$$

since $\|\theta - \bar\theta\| < \varepsilon$, $\|\bar\theta - \bar\theta_n\| \leqq o(1)$ and the last term is small by assumption. Hence $H_n\psi_n$ converges weakly, and the proposition follows by the density of $\mathscr{D}$ and Proposition 6.2.2.

DEFINITION 6.2.5 *The sequence* $\{H_n\}$ *is densely bounded if there is a dense set* $\mathscr{D} \subset \mathscr{H}$, *such that for each* $\psi \in \mathscr{D}$ *there exists a sequence* $\psi_n \in \mathscr{D}(H_n)$ *with*

$$\lim_{n\to\infty} \psi_n = \psi, \qquad \langle\psi_n, |H_n| \psi_n\rangle \leqq \text{const.}$$

Our main result is the following:

THEOREM 6.2.6 *Let* $\mathscr{H}$ *be separable and let* $H_n = H_n^*$. *Assume that the sequence* $\{H_n\}$ *is densely bounded and that for some complex* ζ, *the strong limits* $R = \lim R_n(\zeta)$ *and* $R^* = \lim R_n(\zeta)^*$ *exist. Then* $R = (H - \zeta)^{-1}$ *and*

$$H = H^* = \text{graph} \lim_{n\to\infty} H_n.$$

Proof We assume at first that $0 \leqq H_n$. In this case we may take ζ negative by a simple Neumann series argument (Kato [1, p. 427]) so that

$$I \leqq (H_n - \zeta).$$

We choose in $\mathscr{D}$ a countable dense subset, $\mathscr{D}_1$. For $\psi_n \to \psi \in \mathscr{D}_1$ and $\chi_n \to \chi \in \mathscr{D}_1$ as in Definition 6.2.5, the inner products

$$M_n = \langle\chi_n, (H_n - \zeta)^{1/2} \psi_n\rangle$$

are bound in magnitude, uniformly in n, by

$$\|\chi_n\| \, \|(H_n - \zeta)^{1/2} \psi_n\| \leqq \text{const.}$$

Thus there exists a convergent subsequence M_{n_j} of the M_n, and by the diagonal process, one subsequence n_j converging for every pair of limits $\psi, \chi \in \mathscr{D}_1$.

Let $B_j = (H_{n_j} - \zeta)^{1/2}$. By Lemma 6.2.4, $B =$ weak graph limit B_j exists. Also $R_n(\zeta)^{1/2} \to R^{1/2}$, so by Theorem 6.2.3 $B = B^* = R(\zeta)^{-1/2}$. The square of a self adjoint operator is self adjoint so $R(\zeta)^{-1} = B^2$ is self adjoint. Since the full sequence $R_n(\zeta)$ converges strongly to $R(\zeta)$, and since Range $R(\zeta) = \mathscr{D}(B^2)$ is dense, the graph limit H of the H_n exists, and $R(\zeta) = (H - \zeta)^{-1}$.

5*

We have used convergence of $R_n(\zeta)^{1/2}$. Since $R_n(\zeta)$ is positive, $R_n(\eta)$ converges strongly for all $\eta < \zeta$. Hence the representation

$$R_n(\zeta)^{1/2} = \pi^{-1} \int_0^\infty \lambda^{-1/2} R_n(\zeta - \lambda)\, d\lambda$$

proves the convergence of the square root.

To prove the theorem in the general case, we use the operators $|H_n| = (H_n^2)^{1/2} = H_n'$, and

$$\begin{aligned} R_n'(\pm i) &= (H_n' \mp i)^{-1} \\ &= |\mathrm{Re}\, R_n(\pm i)| + i\, \mathrm{Im}\, R_n(\pm i). \end{aligned}$$

The resolvents $R_n'(\pm i)$ converge strongly, because by the Neumann series argument refered to above, we may take $\zeta = \pm i$ without loss of generality. Convergence of $R_n(\pm i)$ yields convergence of $|\mathrm{Re}\, R_n(\pm i)|$, etc, and convergence of $R_n'(\pm i)$. By the theorem as proved so far,

$$R'(\pm i) = |\mathrm{Re}\, R(\pm i)| + i\, \mathrm{Im}\, R(\pm i)$$

is the resolvent of a self adjoint operator. Thus

$$0 = \mathrm{Null}\, R'(\pm i) = \mathrm{Null}\, R(\pm i)$$

and so

$$0 = \{\mathrm{Range}\, R(\mp i)\}^{\perp}.$$

As a consequence of the strong convergence of $R_n(\mp i)$ to an operator $R(\mp i)$ with dense range, it follows that

$$H = \mathrm{graph} \lim_{n \to \infty} H_n$$

exists. By Theorem 6.2.3, $H = H^* = R(i)^{-1} + i$.

We now define the strong graph limit. Let

$$\mathscr{G}_\infty^s = \{\psi, \chi : \psi = \lim \psi_n,\ \psi_n \in \mathscr{D}(H_n),\ \chi = \lim H_n \psi_n\}.$$

$$\mathscr{D}_\infty^s = \{\psi : \{\psi, \chi\} \in \mathscr{G}_\infty^s \quad \text{for some } \chi\}.$$

Definition 6.2.7 *The sequence H_n has a strong graph limit H if $\mathscr{G}_\infty^s$ is the graph of a densely defined operator H.*

The existence of a strong graph limit of a sequence H_n implies the existence of the weak graph limit.

Lemma 6.2.8 *The strong graph limit of a sequence H_n exists if and only if $\mathscr{D}_\infty^s$ is dense. If $H =$ st. graph $\lim H_n = H_n^*$, then H is closed and symmetric.*

Proof We follow the proof of Proposition 6.2.2 except to establish that H is closed. Let $\{\psi_m, \chi_m\} \to \{\psi, \chi\}$ be a Cauchy sequence in $\mathscr{G}(H)$.

Let π_n be the projection in $\mathscr{H} \oplus \mathscr{H}$ onto $\mathscr{G}(H_n)$. To show $\{\psi, \chi\} \in \mathscr{G}(H)$, we prove $\|\{\psi, \chi\} - \pi_n\{\psi, \chi\}\| \to 0$.

$$\begin{aligned}\|\{\psi, \chi\} - \pi_n\{\psi, \chi\}\| &\leqq \|\{\psi, \chi\} - \{\psi_m, \chi_m\}\| + \|\{\psi_m, \chi_m\} - \pi_n\{\psi_m, \chi_m\}\| \\ &\quad + \| \pi_n\{\psi_m, \chi_m\} - \pi_n\{\psi, \chi\}\| \\ &\leqq 2\|\{\psi, \chi\} - \{\psi_m, \chi_m\}\| + \|\{\psi_m, \chi_m\} - \pi_n\{\psi_m, \chi_m\}\|.\end{aligned}$$

The first term is $o(1)$ for m large, by assumption. The second term is the distance from $\{\psi_m, \chi_m\}$ to $\mathscr{G}(H_n)$. Since $\{\psi_m, \chi_m\} \in \mathscr{G}(H)$, there is a sequence $\{\psi_{mn}, \chi_{mn}\}$ in $\mathscr{G}(H_n)$ converging to $\{\psi_m, \chi_m\}$ for m fixed. Thus

$$\begin{aligned}\|\{\psi_m, \chi_m\} - \pi_n\{\psi_m, \chi_m\}\| &\leqq \|\{\psi_m, \chi_m\} - \{\psi_{mn}, \chi_{mn}\}\| \\ &\leqq o(1)\end{aligned}$$

for m fixed. Hence $\mathscr{G}(H)$ is closed.

Question Is the weak graph limit of a sequence H_n closed?

Remark If Theorem 6.2.3 or Theorem 6.2.6 is valid, then the strong graph limit of the sequence H_n exists and

$$w.\text{ graph lim } H_n = \text{st. graph lim } H_n = H = H^*.$$

The assumption that the sequence $\{H_n\}$ is densely bounded may be replaced by the assumption that $\{|H_n|^\varepsilon\}$ is densely bounded, for some $\varepsilon > 0$.

6.3 Properties of the cutoffs

In this section we discuss elementary properties of the allowed momentum cutoff functions $\chi_\varkappa$ and the corresponding cutoff operators $H_I(g, \varkappa, \varrho)$ and $c(g, \varkappa, \varrho)$. We start with a function $\chi(p)$ in $\mathscr{S}(R^1)$, for which $\chi(0) = 1$, $\chi(p) = \chi(-p)$, and we define

$$\chi_\varkappa(k, p_1, p_2) = \chi\left(\frac{k}{\varkappa}\right)\chi\left(\frac{p_1}{\varkappa}\right)\chi\left(\frac{p_2}{\varkappa}\right). \tag{6.3.1}$$

LEMMA 6.3.1 *The cutoff (6.3.1) satisfies*

$$|\chi_\varkappa| \leqq 0(1). \tag{a}$$

$$|\chi_\varkappa| \leqq 0(\varkappa)(\mu + \omega_1 + \omega_2)^{-1} \tag{b}$$

$$|\chi_\varkappa - 1| \leqq 0(\varkappa^{-1})(\mu + \omega_1 + \omega_2). \tag{c$_1$}$$

Let $\eta = p_1 + p_2$, $\xi = p_1 - p_2$. For some $\varepsilon > 0$, (c$_2$)

$$\int_{|\xi| > \varrho} \left|\chi_\varkappa(k, p_1, p_2) - \chi_\varkappa\left(0, \frac{\xi}{2}, \frac{-\xi}{2}\right)\right| \omega(\xi)^{-1}\, d\xi \leqq 0(\varrho^{-\varepsilon}\varkappa^{-\varepsilon})(\omega(\eta) + \mu(k))^{2\varepsilon}.$$

$$\chi_\varkappa(k, p_1, p_2) = \chi_\varkappa(k, p_2, p_1) = \chi_\varkappa(-k, -p_1, -p_2). \tag{d}$$

Proof Since $\chi \in \mathscr{S}(R^1)$, properties (a) and (b) follow. Property (c_1) results from

$$\left|\chi\left(\frac{p}{\varkappa}\right)-1\right| \leqq \int_0^{p\varkappa^{-1}} |\chi'(t)|\,dt \leqq 0(\varkappa^{-1})\,\mu(p).$$

To prove (c_2), we note that the first derivatives of $\chi_\varkappa(k, p_1, p_2)$ are $0(\varkappa^{-1})$. Thus

$$\left|\chi_\varkappa(k, p_1, p_2) - \chi_\varkappa\left(0, \frac{\xi}{2}, \frac{-\xi}{2}\right)\right| \leqq 0(\varkappa^{-1})\,(\mu(k) + \omega(\eta)).$$

Using (a) and (b)

$$\left|\chi_\varkappa(k, p_1, p_2) - \chi_\varkappa\left(0, \frac{\xi}{2}, \frac{-\xi}{2}\right)\right| \leqq 0(\varkappa^{-2\varepsilon})\,(\mu + \omega(\eta))^{2\varepsilon}\,0(\varkappa^{\varepsilon})\,\omega(\xi)^{-\varepsilon}$$

$$= 0(\varkappa^{-\varepsilon})\,\omega(\xi)^{-\varepsilon}\,(\mu + \omega(\eta))^{2\varepsilon},$$

from which (c_2) follows. Property (d) is obvious.

For each cutoff $\chi_\varkappa$ of the form (6.3.1), we introduce a second cutoff that is sharp in momentum space. Let $\theta(x)$ be the characteristic function of the interval $|x| \leqq 1$, and let

$$\chi'_\varkappa(k, p_1, p_2) = \chi_\varkappa(k, p_1, p_2)\,\theta(k\varkappa^{-1+\alpha})\,\theta(p_1\varkappa^{-1+\alpha})\,\theta(p_2\varkappa^{-1+\alpha}), \tag{6.3.2}$$

for some $0 < \alpha < 1$. Thus the cutoff $\chi'_\varkappa$ is sharp at momenta $0(\varkappa^{1-\alpha})$. The cutoff $\chi'_\varkappa$ of (6.3.2) satisfies the conditions (a)–(d) of Lemma 6.3.1, except that (c_1) must be replaced by

$$(c_1')\ |\chi'_\varkappa - 1| \leqq 0(\varkappa^{-1})\,(\mu + \omega_1 + \omega_2) \quad \text{on} \quad S_{\varkappa^{1-\alpha}},$$

$$S_r = \{k, p_1, p_2 : |k| \leqq r, |p_1| \leqq r, |p_2| \leqq r\}.$$

We do not prove these elementary estimates that can be found in Glimm-Jaffe [8, Sec. 4.1].

DEFINITION 6.3.2 *A function $\chi_\varkappa(p, k_1, k_2)$ is a general momentum cutoff if $\chi_\varkappa$ satisfies the conditions of Lemma 6.3.1, and is piecewise continuous with at most a finite number of jump discontinuities across cubes S_r.*

Our proofs hold for general cutoffs $\chi_\varkappa$, since the proofs involve only (a)–(d) and the piecewise continuity of $\chi_\varkappa$. Condition (b) ensures that the kernels $w^C_\varkappa$ and $w_\varkappa$ are L_2 functions. Condition (d) makes $H_I(g, \varkappa)$ a symmetric operator. Condition (c_2) is used in estimating the mass renormalization cancellation.

We next show that $\delta m^2(\varkappa)$ of (6.1.1) differs from the second order mass shift $\delta m_2^2(g, \varkappa)$ by a constant plus $o(1)$.

LEMMA 6.3.3 *We have*

$$\delta m^2(\varkappa) = \delta m_2^2(g, \varkappa) + M(g) + o(1)$$

where the constant $M(g)$ is independent of $\varkappa$ and the choice of cutoff $\chi_\varkappa$, and where $o(1)$ vanishes as $\varkappa \to \infty$.

Proof We write the integrands defining $\delta m^2(\varkappa)$ and $\delta m_2^2(g, \varkappa)$ as products. Let

$$\begin{aligned}
\alpha_1 &= \frac{1}{2}\left(\frac{\omega_1\omega_2 - p_1 p_2 - M^2}{\omega_1\omega_2}\right), & \beta_1 &= 1 \\
\alpha_2 &= \frac{1}{2}\left(\frac{1}{\omega_1 + \omega_2 - m} + \frac{1}{\omega_1 + \omega_2 + m}\right), & \beta_2 &= \omega(\xi)^{-1} \\
\alpha_3 &= |\chi_\varkappa(0, p_1, p_2)|^2, & \beta_3 &= \left|\chi_\varkappa\left(0, \frac{\xi}{2}, -\frac{\xi}{2}\right)\right|^2.
\end{aligned} \tag{6.3.3}$$

We have the elementary bounds for $0 \leqq \varepsilon \leqq 1$,

$$\begin{aligned}
|\alpha_1 - \beta_1| = \frac{1}{2}\left|\frac{\omega_1\omega_2 + p_1 p_2 + M^2}{\omega_1\omega_2}\right| &\leqq \text{const.}\ \omega(\eta)^{2\varepsilon}\,\omega(\xi)^{-\varepsilon} \\
|\alpha_2 - \beta_2| &\leqq \text{const.}\ (\omega_1 + \omega_2)^{-1}\,\omega(\xi)^{-1}\,\omega(\eta).
\end{aligned} \tag{6.3.4}$$

Here we use the stability condition $m < 2M$. Using (a) and (c_2) of Lemma 6.3.1, for some $\varepsilon > 0$,

$$\int |\alpha_3 - \beta_3|\,\omega(\xi)^{-1}\,d\xi \leqq 0(\varkappa^{-\varepsilon})\,\omega(\eta)^{2\varepsilon}.$$

Note that

$$\begin{aligned}
\beta_1\beta_2\beta_3 - \alpha_1\alpha_2\alpha_3 = (\beta_1\beta_2 - \alpha_1\alpha_2) &+ (\beta_1\beta_2 - \alpha_1\alpha_2)(\beta_3 - 1) \\
&+ \alpha_1\alpha_2(\beta_3 - \alpha_3);
\end{aligned}$$

$$\delta m^2(\varkappa) - \delta m_2^2(g, \varkappa) = -\frac{\lambda^2}{2\pi}\|g\|^{-2}\int |\tilde{g}(\eta)|^2(\beta_1\beta_2\beta_3 - \alpha_1\alpha_2\alpha_3)\,d\xi\,d\eta.$$

Thus

$$\begin{aligned}
\delta m^2(\varkappa) - \delta m_2^2(g, \varkappa) = &-\frac{\lambda^2}{2\pi}\|g\|^{-2}\int |\tilde{g}(\eta)|^2(\beta_1\beta_2 - \alpha_1\alpha_2)\,d\xi\,d\eta \\
&-\frac{\lambda^2}{2\pi}\|g\|^{-2}\int |\tilde{g}(\eta)|^2(\beta_1\beta_2 - \alpha_1\alpha_2)(\beta_3 - 1)\,d\xi\,d\eta \\
&+ 0(\varkappa^{-\varepsilon}).
\end{aligned} \tag{6.3.5}$$

By (6.3.4), $|\beta_1\beta_2 - \alpha_1\alpha_2| \leqq \text{const. } \omega(\eta)^{2\varepsilon}\omega(\xi)^{-1-\varepsilon}$, the integrals on the right of (6.3.5) converge. Hence by conditions (a) and (c_1),

$$\delta m^2(\varkappa) = \delta m_2^2(g, \varkappa) - \frac{\lambda^2}{2\pi}\|g\|^{-2}\int |\tilde{g}(\eta)|^2 (\beta_1\beta_2 - \alpha_1\alpha_2)\, d\xi\, d\eta + 0(\varkappa^{-\varepsilon}),$$

and since $(\beta_1\beta_2 - \alpha_1\alpha_2)$ is independent of $\chi_\varkappa$, the lemma is proved.

LEMMA 6.3.4 *The Hamiltonian $H(g, \varkappa, \varrho)$ is self adjoint on $\mathscr{D}(H_0)$.*

Proof Since $0 < -\delta m^2(\varkappa)$, Proposition 3.2.1 implies that $H_0 + c(g, \varkappa)$ is essentially self adjoint on $\mathscr{D}(H_0)$. This operator is self adjoint on $\mathscr{D}(H_0)$, since Theorem 3.1.3 with $\deg P = 2$ yields

$$\|H_0\theta\| \leqq \text{const. } \|\{H_0 + c(g, \varkappa)\}\,\theta\|.$$

We now show that $H_I(g, \varkappa, \varrho)$ is infinitesimally small in the sense of Kato [1, p. 287] with respect to H_0, completing the proof of the lemma. We approximate the kernel $w^C_{\varkappa,\varrho}$ of $W^C_{\varkappa,\varrho}$ by a function $r \in \mathscr{S}(R^3)$ so that $\|w^C_{\varkappa,\varrho} - r\|_2 < \varepsilon/2$. Hence by Proposition 1.2.3(b–c), for each $\varkappa$, ϱ we have

$$\begin{aligned}\|W^C_{\varkappa,\varrho}\theta\| &\leqq \|w^C_{\varkappa,\varrho} - r\|\;\|(N + I)\,\theta\| + |r|_s\;\|(N + I)^{1/2}\,\theta\|\\ &\leqq \varepsilon\|N\theta\| + \text{const. } \|\theta\|.\end{aligned}$$

The bounds on $W^A_{\varkappa,\varrho}$ and $W_{\varkappa,\varrho}$ are similar.

For use in chapters 7 and 8, we prove related estimates.

LEMMA 6.3.5 *Let $\varepsilon > 0, \tau > 0$. Then*

$$\begin{aligned}&\pm\{H_I(g, \varkappa, \varrho) - H_I(g, \varkappa, \varrho_1)\} \leqq \varepsilon N_\tau + \text{const.},\\ &\pm\{c(g, \varkappa, \varrho) - c(g, \varkappa, \varrho_1)\} \leqq \text{const. } (N_\tau + I)\end{aligned} \tag{6.3.6}$$

with constants uniform in $\varkappa$ and ϱ, $\varrho \leqq \varrho_1$.
For fixed $\varkappa < \infty$, there exists $\delta > 0$ such that

$$\pm H_I(g, \varkappa, \varrho) \pm c(g, \varkappa, \varrho) \leqq 0(\varrho^{-\delta})\,(N + I). \tag{6.3.7}$$

Proof Let

$$H_I(g, \varkappa, \varrho) - H_I(g, \varkappa, \varrho_1) = \delta H_I = \delta W^C + \delta W + \delta W^A.$$

We approximate the kernel $\delta w_{\varkappa,\varrho}$ by $\delta w_{\varkappa,\varrho,V} = \delta w_{\varkappa,\varrho}(k, p_{1V}, p_{2V})$, using the notation of (1.1.3). Since $\gamma = (\omega_1\omega_2 + p_1p_2 + M^2)^{1/2}(\omega_1\omega_2)^{-1/2}$ is bounded and has uniformly bounded derivatives, $|\gamma - \gamma_V| \leqq 0(V^{-1})$. Also $|\tilde{g}(k + \eta) - \tilde{g}(k + \eta_V)| \leqq 0(V)^{-1} h(k + \eta)$, where $h(\cdot)$ is rapidly decreasing and $|\chi_\varkappa - \chi_{\varkappa,V}| \leqq 0(V^{-1})$. Also, $\theta_\varrho(\xi) - \theta_\varrho(\xi_V)$ is nonzero only in a ξ interval of length $0(V^{-1})$. Thus

$$\begin{aligned}|\delta w_{\varkappa\varrho} - \delta w_{\varkappa\varrho V}| \leqq{}& \{0(V^{-1})\,\theta_{\varrho_1}(\xi) + |\theta_\varrho(\xi) - \theta_\varrho(\xi_V)| + |\theta_{\varrho_1}(\xi) - \theta_{\varrho_1}(\xi_V)|\}\\ &\times \mu^{-1/2}\, h(k + \eta)\end{aligned}$$

and

$$\|\mu^{-\tau/2}(\delta w_{\varkappa\varrho} - \delta w_{\varkappa\varrho V})\|_2 \leqq 0(V^{-1/2}).$$

Hence by Proposition 1.2.3(b),

$$\pm\{\delta W_{\varkappa,\varrho} - \delta W_{\varkappa,\varrho,V}\} \leqq 0(V^{-1/2})(N_\tau + I). \tag{6.3.8}$$

By Proposition 1.2.3(d), we can neglect two fermions in $\delta W_{\varkappa,\varrho,V}$ so

$$\begin{aligned}|\langle\theta, \delta W_{\varkappa,\varrho,V}\theta\rangle| &\leqq \|N_\tau^{1/2}\theta\|\ \|\theta\|\ |\mu^{-\tau/2}\delta w_{\varkappa,\varrho,V}|_{1,2}\\ &\leqq \varepsilon\langle\theta, N_\tau\theta\rangle + 0(V)\ \|\theta\|^2.\end{aligned} \tag{6.3.9}$$

By choosing V large enough to make (6.3.8) small, we obtain by (6.3.9),

$$\pm\delta W \leqq \varepsilon N_\tau + \text{const.},$$

uniformly in $\varkappa$ and $\varrho \leqq \varrho_1$. Identical bounds hold on δW^C and δW^A, so the bound (6.3.6) on δH_I is proved.

For fixed $\varkappa$, $w^C_{\varkappa,\varrho}$ and $w_{\varkappa,\varrho}$ are L_2 functions with L_2 norms equal $O(\varrho^{-\delta})$. Hence the bound (6.3.7) on H_I follows from Corollary 1.2.4.

We now study the counterterms. Since

$$\begin{aligned}-\delta E &= -E(g, \varkappa, \varrho) + E(g, \varkappa, \varrho_1)\\ &= \|(\mu + \omega_1 + \omega_2)^{-1/2}\, w^C_\varkappa\{\theta_\varrho(\xi) - \theta_{\varrho_1}(\xi)\}\|_2^2\\ &\leqq \|(\mu + \omega_1 + \omega_2)^{-1/2}\, w^C_\varkappa\theta_{\varrho_1}(\xi)\|_2^2,\end{aligned}$$

and since for $0 < \delta \leqq 1$,

$$|w^C_\varkappa| \leqq \text{const.}\ |\tilde{g}(k + \eta)|\ \mu^{-1/2}\ \omega(\xi)^\sigma\ \omega(\eta)^{-\delta/2},$$

we have $|\delta E| \leqq$ const., uniformly in $\varkappa$. Also

$$|\delta m^2(\varkappa, \varrho) - \delta m^2(\varkappa, \varrho_1)| \leqq |\delta m^2(\varkappa, \varrho_1)| \leqq \text{const.},$$

so by Proposition 1.2.3 we have a $\varkappa$ and ϱ independent estimate

$$\pm\delta c \leqq \text{const.}\ (N + I).$$

For fixed $\varkappa$, $|\delta m^2(\varkappa, \varrho)| \leqq 0(\varrho^{-\delta})$, and $|E(g, \varkappa, \varrho)| \leqq 0(\varrho^{-\delta})$. The remaining bounds of the lemma follow.

7 First and Second Order Estimates

We prove estimates for the Yukawa$_2$ Hamiltonian of the form

$$N_\tau \leqq \text{const.}\ \big(H(g, \varkappa) + \text{const}\big)., \qquad \tau < 1 \tag{7.1}$$

and

$$N_\tau^2 = \text{const.}\ \big(H(g, \varkappa) + \text{const.}\big)^2, \qquad \tau < 1/2, \tag{7.2}$$

with constants independent of $\varkappa$. We call these inequalities first and second order estimates respectively, since they involve first and second powers of the Hamiltonian. The results of Sec. 6.3 yield $\varkappa$ dependent estimates for $\tau \leqq 1$. However, according to perturbation theory, the estimates (7.1) and (7.2) are the best possible estimates uniform in $\varkappa$.

We now verify in perturbation theory for which values of τ the ground state of $H(g, \varkappa)$ gives expectation values of N_τ or N_τ^2 that are uniformly bounded in $\varkappa$.

To first order perturbation theory, the ground state of $H(g,\varkappa)$ is $\Omega^{(1)} = \Omega_0 - H_0^{-1} H_I(g, \varkappa)\, \Omega_0$. Thus

$$\begin{aligned}\langle \Omega^{(1)}, N_\tau \Omega^{(1)} \rangle &= \| N_\tau^{1/2} H_0^{-1} H_I(g, \varkappa)\, \Omega_0 \|^2 \\ &= \| (\mu^\tau + \omega_1^\tau + \omega_2^\tau)^{1/2} (\mu + \omega_1 + \omega_2)^{-1} w_\varkappa^C \|_2^2 .\end{aligned}$$

The integral diverges logarithmically for $\tau = 1$, since the integral then equals the second order vacuum energy $E^{(2)}(g, \varkappa)$. Thus $\langle \Omega^{(1)}, N_\tau \Omega^{(1)} \rangle$ is bounded uniformly in $\varkappa$ for $\tau < 1$, and we expect a uniform first order estimate to be valid for $\tau < 1$.

Likewise, we compute

$$\begin{aligned}\langle \Omega^{(1)}, N_\tau^2 \Omega^{(1)} \rangle &= \| N_\tau H_0^{-1} H_I(g, \varkappa)\, \Omega_0 \|^2 \\ &= \| (\mu^\tau + \omega_1^\tau + \omega_2^\tau) (\mu + \omega_1 + \omega_2)^{-1} w_\varkappa^C \|_2^2 ,\end{aligned}$$

which is bounded uniformly in $\varkappa$ for $\tau < \frac{1}{2}$. Hence we expect a uniform second order estimate to hold for $\tau < \frac{1}{2}$. In fact, uniform first order estimates do not hold for $\tau = 1$ and uniform second order estimates do not hold for $\tau = \frac{1}{2}$.

7.1 Positivity of the Yukawa$_2$ Hamiltonian

THEOREM 7.1.1 *Glimm*[(2)]. *Let $\tau < 1$, g be fixed. There are constants a, b independent of $\varkappa$ and ϱ such that*

$$N_\tau \leqq a\big(H(g, \varkappa, \varrho) + bI\big). \tag{7.1.1}$$

Formal motivation Before starting the proof of (7.1.1), we motivate the ideas involved. We use an approximate diagonalization of $H(g, \varkappa, \varrho)$ that exhibits the Hamiltonian as a positive operator plus a small error. Formal perturbation theory suggests that there is a unitary operator T, the wave operator, for which

$$H(g, \varkappa, \varrho) = T H_0 T^* . \tag{7.1.2}$$

Since H_0 is positive, the positivity of H would follow.

The wave operator $T_\pm$ is by definition the limit of operators $T(t) = \exp(-iHt) \exp(iH_0 t)$ if the limits $t \to \infty$ or as $t \to -\infty$ exist. For well behaved problems in potential scattering such limits exist. Under further restrictions (repulsive forces) the limit is unitary. Assuming further that $\frac{d}{dt} T(t) \to 0$, we have

$$\frac{d}{dt} T(t) = -i(HT(t) - T(t) H_0) \to HT - TH_0 = 0.$$

While such a unitary operator T does not exist for relativistic field theory, the formal perturbation expansion for T provides a useful tool.

Since

$$H_0 = \sum_{\varepsilon = 0, \pm 1} b(k, \varepsilon)^* \mu(k, \varepsilon) b(k, \varepsilon) dk,$$

the relation (7.1.2) would yield

$$H(g, \varkappa, \varrho) = \sum_{\varepsilon = 0, \pm 1} \int \hat{b}(k, \varepsilon)^* \mu(k, \varepsilon) \hat{b}(k, \varepsilon) dk \tag{7.1.3}$$

where $\hat{b}(k, \varepsilon) = Tb(k, \varepsilon) T^*$ are the annihilation operators of the incoming or outgoing asymptotic particles.

Divergences occur in perturbation theory for (7.1.3) only in terms involving $\hat{b}$ to first order. Hence we only need to calculate T to first order in order to generate the renormalization counterterms. We estimate the error introduced in (7.1.3) by retaining only first order terms in T.

We write $T = I + X$, and compute X to first order. Since T is assumed unitary, $TT^* = I$, and to first order $X^* = -X$. Thus to first order, we have by (7.1.2), and the fact that $c(g, \varkappa, \varrho)$ is second order,

$$H_0 + H_I = TH_0T^* = (I + X) H_0(1 - X)$$

$$= H_0 + [X, H_0]$$

or $T = I - \Gamma H_I$, where $[H_0, \Gamma H_I] = H_I$. Hence to first order

$$\hat{b}(k, \varepsilon) = b(k, \varepsilon) + [b(k, \varepsilon), \Gamma H_I(g, \varkappa, \varrho)].$$

We now give a proof based on these formal ideas.

Proof of Theorem 7.1.1 We first prove the theorem for ϱ sufficiently large. It is no loss of generality to take $\tau < 1$ close to one, since $N_\sigma \leqq$ const. $\times N_\tau$ for $\sigma \leqq \tau$. We use an approximate form of (7.1.2-3). Only the pair creation terms W^C yield divergent contributions to (7.1.2-3), so we only retain this term in the definition of $\hat{b}$. To simplify computations, we define ΓW^C so that $[H_0, \Gamma W^C] = W^C +$ error. In particular,

$$\Gamma W^C_{\varkappa,\varrho} = \int w^C_{\varkappa,\varrho}\{a(k)^* (\mu + \omega_1 + \omega_2)^{-1} + a(-k)(\omega_1 + \omega_2)^{-1}\}$$

$$\times\ b(p_1)^* b'(p_2)^* dk\, dp_1\, dp_2.$$

Diagramatically, we write

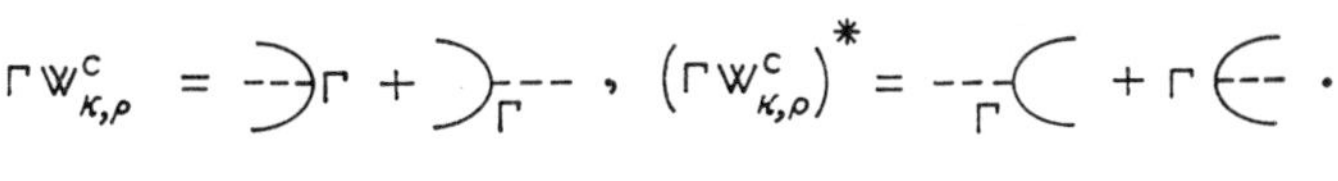

We have

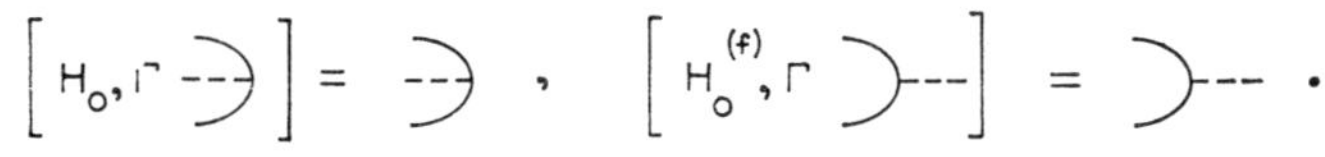

We define

$$\hat{b}(p,\varepsilon) = b(p,\varepsilon) + [b(p,\varepsilon), \Gamma W^C_{\varkappa,\varrho}]. \tag{7.1.4}$$

In particular

$$\hat{b}(p,+1) = b(p,1) + \int w^C(k,p,p_2)\, b'(p_2)^* \times \left\{ a(k)^* \frac{1}{\mu+\omega+\omega_2} + a(-k)\frac{1}{\omega+\omega_2} \right\} dk\, dp_2,$$

$$\hat{b}(p,0) = b(p,0) + \int w^C(p,p_1,p_2)\,(\mu+\omega_1+\omega_2)^{-1}\, b(p_1)^*\, b'(p_2)^*\, dp, dp_2,$$

$$\hat{b}(p,-1) = b(p,-1) - \int w^C(k,p_1,p)\, b(p_1)^* \times \left\{ a(k)^* \frac{1}{\mu+\omega+\omega_1} + a(-k)\frac{1}{\omega+\omega_1} \right\} dk\, dp_1.$$

Let $1-\tau$ and c be sufficiently small so that

$$0 \leqq \bar{\mu}(k,\varepsilon) \equiv \mu(k,\varepsilon) - c\mu(k,\varepsilon)^\tau, \qquad \varepsilon = 0, \pm 1.$$

We define a positive approximation to $H(g,\varkappa,\varrho)$ by

$$\hat{H} = \sum_{\varepsilon=0,\pm 1} \int \hat{b}(k,\varepsilon)^*\, \bar{\mu}(k,\varepsilon)\, \hat{b}(k,\varepsilon)\, dk. \tag{7.1.5}$$

Expanding (7.1.5) with (7.1.4) yields many terms. We write

$$0 \leqq \hat{H} = -cN_\tau + H_1 + H_2. \tag{7.1.6}$$

We choose H_1 to be the terms in (7.1.5) that resemble $H(g,\varkappa,\varrho)$, and we will prove for $\varepsilon > 0$ and $\varrho > \varrho_1$ sufficiently large,

$$\pm\{H_1 - H(g,\varkappa,\varrho)\} \leqq \varepsilon(N_\tau + I). \tag{7.1.7}$$

We also estimate the error term H_2 by

$$H_2 \leqq \varepsilon(N_\tau + I). \tag{7.1.8}$$

Adding (7.1.7-8) and using (7.1.6) we obtain the desired result,

$$(c - 2\varepsilon)\, N_\tau \leqq H_1 + H_2 - 2\varepsilon N_\tau \leqq H(g,\varkappa,\varrho) + I.$$

In terms of diagrams we define

$$H_1 = H_0 + W^C_{\varkappa,\varrho} + W^A_{\varkappa,\varrho}$$

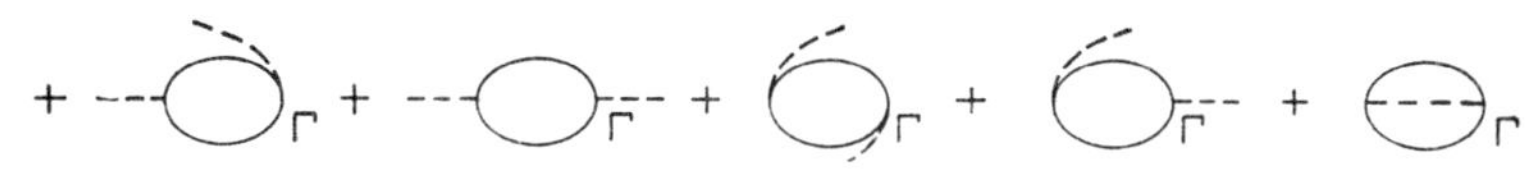

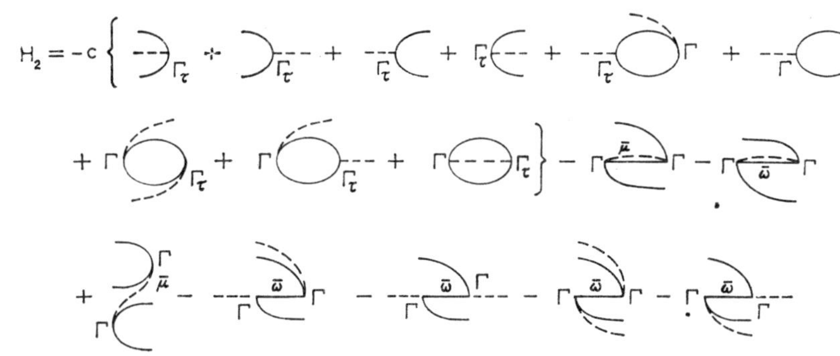

$$= Z_1 + Z_2 + \cdots + Z_{16}.$$

We now explain the notation in the above equations. The vertices that appear in H_1 and H_2 are the vertices in $W^C_{\varkappa,\varrho}, \Gamma W^C_{\varkappa,\varrho}, \Gamma_\tau W^C_{\varkappa,\varrho}$ and their adjoints-with the exception of Z_7 as explained below. The Γ_τ vertices are defined by the equations

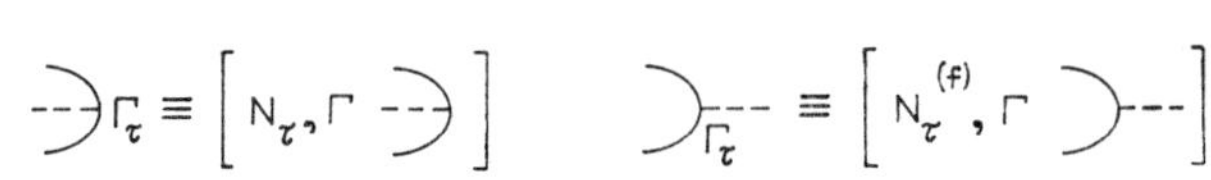

In Z_{10} and Z_{12} the boson line has an extra factor $\bar{\mu}$. In $Z_{11}, Z_{13}, \ldots, Z_{16}$ the contracted fermion (or antifermion) line has an extra factor $\bar{\omega} = \bar{\mu}(\cdot, 1)$. The derivation of these expansions for H_1 and H_2, where $\hat{H} = -cN_\tau + H_1 + H_2$, is elementary. We insert the definition (7.1.4) into (7.1.5) and use relations such as

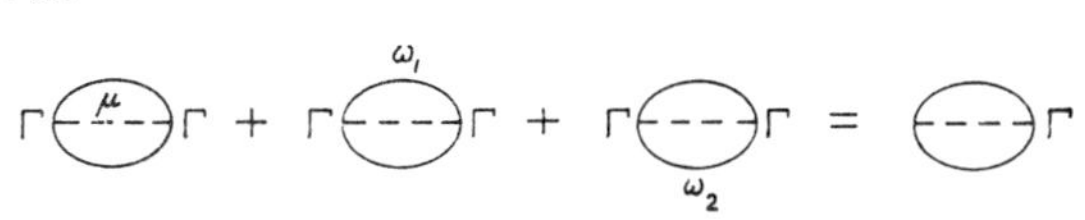

and

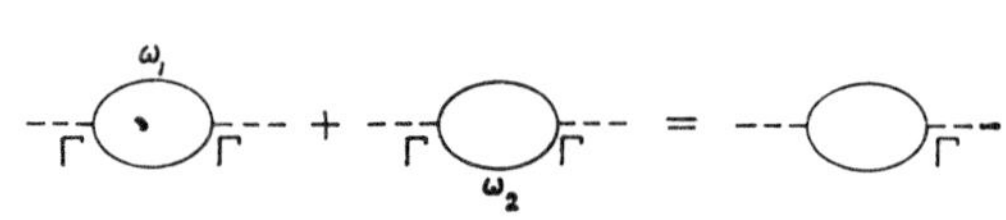

Finally, the kernel of Z_7 is

$$\int \frac{w^C_{\varkappa,\varrho}(k, p_1, p_2)\, w^C_{\varkappa,\varrho}(k', -p_1, -p_2)}{(\mu + \omega_1 + \omega_2)\,(\mu' + \omega_1 + \omega_2)} \{\omega_1^\tau + \omega_2^\tau + c^{-1}\mu\}\, dp_1\, dp_2.$$

LEMMA 7.1.2 *The inequalities (7.1.7-8) are valid.*

Proof To prove (7.1.7) we estimate the difference

$$H_1 - H(g, \varkappa, \varrho) = -W_{\varkappa,\varrho} + \tfrac{1}{2}\delta m^2(\varkappa, \varrho)\int :\phi(x)^2: g(x)^2 + E_2(g, \varkappa, \varrho)$$

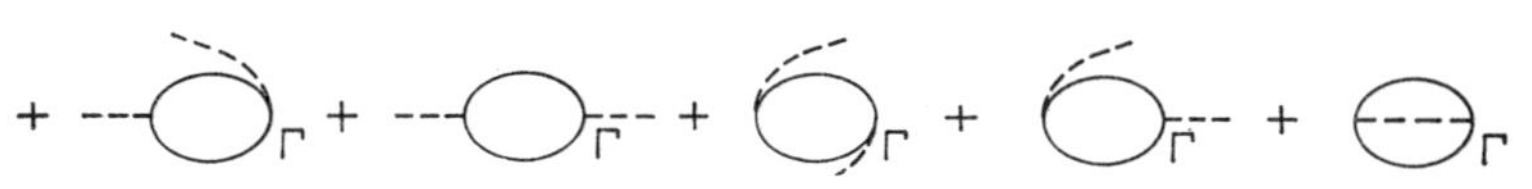

The kernel w of W satisfies for $0 < \delta \leqq \frac{1}{2}$,

$$|w| \leqq \text{const.}\, \mu^{-1/2}\, |\tilde{g}(k + \eta)|\, \omega(\eta)^{2\delta}\, \omega(\xi)^{-\delta}. \tag{7.1.9}$$

For τ sufficiently close to 1, we conclude that $\|(\mu\omega_2)^{-\tau/2}\, w\|_2 < \infty$, and for $0 < \varepsilon$ sufficiently small,

$$\|(\mu\omega_2)^{-\tau/2}\, w_{\varkappa,\varrho}\|_2 \leqq 0(\varrho^{-\varepsilon}).$$

Thus by Proposition 1.2.2,

$$\pm W_{\varkappa,\varrho} \leqq 0(\varrho^{-\varepsilon})\,(N_\tau + I) \tag{7.1.10}$$

where $0(\varrho^{-\varepsilon})$ is independent of $\varkappa$.

We now bound the counterterms in (7.17.). We note that the vacuum energy $E_2(g, \varkappa, \varrho)$ cancels exactly. The four mass counterterms in $c(g, \varkappa, \varrho)$ have the form

$$\int \beta(k, k') :a(k)^{\#}\, a(k')^{\#}: dk\, dk'$$

where $\beta(k, k') = -\frac{1}{2}\delta m^2(\varkappa, \varrho)\,(4\pi)^{-1}\,(\mu\mu')^{-1/2}\,(\tilde{g} * \tilde{g})\,(k + k')$. The corresponding term in H_1 only differs by the kernel $\gamma_i(k, k')$ replacing $\beta(k, k')$, $1 \leqq i \leqq 4$. Here

$$\gamma_i(k, k') = \int w^{C}_{\varkappa,\varrho}(k, p_1, p_2)\, w^{C}_{\varkappa,\varrho}(k', -p_2, -p_1)\, \delta_i\, dp_1\, dp_2,$$

where

$$\delta_1 = (\omega_1 + \omega_2 + \mu')^{-1}, \qquad \delta_2 = (\omega_1 + \omega_2)^{-1},$$

$$\delta_3 = (\omega_1 + \omega_2 + \mu)^{-1}, \qquad \delta_4 = (\omega_1 + \omega_2 + \mu')^{-1}.$$

We show that $\|\beta - \gamma_i\|_2 \leqq 0(\varrho^{-\varepsilon})$, independent of $\varkappa$. Hence by Proposition 1.2.2,

$$\pm \int (\gamma_i - \beta)\,(k, k') : a(k)^{\#}\, a(k')^{\#} : dk\, dk' \leqq 0(\varrho^{-\varepsilon})\,(N_\tau + I). \tag{7.1.11}$$

The bounds (7.1.10-11) prove (7.1.7).

We now estimate $\|\beta - \gamma_1\|_2$. We write with the notation of (6.3.3),

$$\beta(k, k') = \frac{\lambda^2}{4\pi}(\mu\mu')^{-1/2}\int \beta_1\beta_2\beta_3\, \tilde{g}(k + \eta)\, \tilde{g}(k' - \eta)\, \{1 - \theta_\varrho(\xi)\}\, d\xi\, d\eta,$$

and

$$\gamma_1(k, k') = \frac{\lambda^2}{4\pi}(\mu\mu')^{-1/2}\int \beta_1\bar{\beta}_2\bar{\beta}_3\, \tilde{g}(k + \eta)\, \tilde{g}(k' - \eta)\, \{1 - \theta_\varrho(\xi)\}\, d\xi\, d\eta,$$

where $\tilde{\beta}_1 = \alpha_1, \tilde{\beta}_2 = \delta_1, \tilde{\beta}_3 = \chi_\varkappa(k, p_1, p_2)\chi_\varkappa(k', -p_1, -p_2)$. We use the bounds (6.3.4) and

$$|\beta_2 - \tilde{\beta}_2| \leqq \text{const.}\,(\omega_1 + \omega_2 + \mu')^{-1}\,\omega(\xi)^{-1}\,(\omega(\eta) + \mu')$$
$$\leqq \text{const.}\,\omega(\xi)^{-1-2\varepsilon}\,(\omega(\eta) + \mu')^{2\varepsilon},$$
$$\int |\beta_3 - \tilde{\beta}_3|\,\omega(\xi)^{-1}\,\{1 - \theta_\varrho(\xi)\}\,d\xi \leqq 0(\varrho^{-\varepsilon})\,(\omega(\eta) + \mu + \mu')^{3\varepsilon},$$

the latter following by condition (c_2) of Lemma 6.1.1. Using these estimates we obtain from the identity

$$\beta_1\beta_2\beta_3 - \tilde{\beta}_1\tilde{\beta}_2\tilde{\beta}_3 = (\beta_1 - \tilde{\beta}_1)\,\tilde{\beta}_2\tilde{\beta}_3 + \beta_1(\beta_2 - \tilde{\beta}_2)\,\tilde{\beta}_3 + \beta_1\beta_2(\beta_3 - \tilde{\beta}_3)$$

the estimate

$$|(\beta - \gamma_1)\,(k, k')| \leqq 0(\varrho^{-\varepsilon})\,(\mu\mu')^{-1/2+3\varepsilon}\,h(k + k'),$$

where $h(\cdot)$ is rapidly decreasing. Taking L_2 norms gives $\|\beta - \gamma_1\|_2 \leqq 0(\varrho^{-\varepsilon})$. The proof that $\|\beta - \gamma_i\|_2 \leqq 0(\varrho^{-\varepsilon})$ for $i = 2, 3, 4$ is similar, and (7.1.7) is proved.

We next prove (7.1.8). The terms $Z_1, \ldots, Z_6, Z_8, Z_9$ are the same as $-c(H_1 - H_0)$, except that each diagram also has one Γ_τ vertex. For $0 < \delta = 1 - \tau$, we estimate $Z_1, \ldots, Z_4$ by

$$\left\|(\mu + \omega_1 + \omega_2)^{-\tau/2}\left(\frac{\mu^\tau + \omega_1^\tau + \omega_2^\tau}{\mu + \omega_1 + \omega_2}\right) w^C_{\varkappa,\varrho}\right\|_2$$
$$\leqq \text{const.}\,\|(\mu + \omega_1 + \omega_2)^{-(1+\delta)/2}\,w^C_{\varkappa,\varrho}\|_2 \leqq 0(\varrho^{-\delta/4}),$$

uniformly in $\varkappa$. Hence for ϱ sufficiently large,

$$\pm Z_i \leqq \varepsilon(N_\tau + I), \qquad i = 1, 2, 3, 4 \tag{7.1.12}$$

by Corollary 1.2.4.

The extra Γ_τ factor yields L_2 kernels for $Z_5, \ldots, Z_9$. These bounds are uniform in $\varkappa$. Using the lower cutoff ϱ, we obtain norms that are $0(\varrho^{-\varepsilon})$. Hence (7.1.12) is also valid for $Z_5, \ldots, Z_9$. The estimates on Z_{10} and Z_{11} are similar, since the kernel of Z_{10} or Z_{11} is dominated by

$$|z_{10}(p_1, p_2)| + |z_{11}(p_1, p_2)| \leqq$$
$$\leqq 0(\varrho^{-\varepsilon})\int \mu^{-1}(\mu + \omega + \omega_1)^{-1+\varepsilon}\,|\tilde{g}(k + p_1 + p)\,\tilde{g}(-k - p + p_2)|\,dp\,dk$$
$$\leqq 0(\varrho^{-\varepsilon})\,\omega_1^{-1+2\varepsilon}\,h(p_1 + p_2),$$

where $h(\cdot)$ is rapidly decreasing. Thus $\|z_{10}\|_2 + \|z_{11}\|_2 \leqq 0(\varrho^{-\varepsilon})$.

We now estimate the singly contracted terms, $Z_{12}, \ldots, Z_{16}$. For vectors $\theta \in \mathscr{D}$,

$$\langle\theta, Z_{12}\theta\rangle \int = \|Z_{17}(k)\,\theta\|^2\,dk$$

where

$$Z_{17}(k) = \bar{\mu}(k)^{1/2} \int b'(-p_1)\, b(-p_2)\, w^C_{\varkappa,\varrho}(k, p_1, p_2)\, (\omega_1 + \omega_2)^{-1}\, dp_1\, dp_2 .$$

By Corollary 1.2.4,

$$\begin{aligned}\|Z_{17}(k)\,(N_\tau + I)^{-1/2}\| &\leqq \text{const.}\ \mu^{1/2} \|\,(\omega_1 + \omega_2)^{-1-\tau/2}\, w^C_{\varkappa,\varrho}\|_2 \\ &\leqq 0(\varrho^{-\varepsilon})\, \mu^{-(1+\tau-2\varepsilon)/2} \in L_2 .\end{aligned}$$

Hence the bound (7.1.12) is valid for Z_{12}.

Thus the terms Z_i, $i = 1, 2, \ldots, 12$ satisfy (7.1.12). The terms $Z_{13} + \cdots + Z_{16}$ are negative, aside from a small error satisfying (7.1.12). We now prove this. Let

$$\begin{aligned} Z_{18}(p) = \int b(-p_1)\, w^C_{\varkappa,\varrho}(k, -p, p_1) \bar{\omega}(p)^{1/2} \Big\{ a(k)^* \frac{1}{\omega(p) + \omega(p_1)} \\ + a(-k) \frac{1}{\mu(k) + \omega(p) + \omega(p_1)} \, dp_1\, dk \end{aligned}$$

and let $Z'_{18}(p)$ have $b'(p_1)$ in place of $b(p_1)$. Then

$$\begin{aligned} Z_{13} + \cdots + Z_{16} = &- \int Z_{18}(p)^* Z_{18}(p)\, dp \\ &+ \int b(p_1)^*\, b(-p_2)\, K(p_1, p_2)\, dp_1\, dp_2 \\ &- \int Z'_{18}(p)^* Z'_{18}(p)\, dp \\ &+ \int b'(p_1)^*\, b'(-p_2)\, K(p_1, p_2)\, dp_1\, dp_2 . \end{aligned}$$

Here

$$\begin{aligned} &|K(p_1, p_2)| \\ &= \left| \int dk\, dp\, w^C_{\varkappa,\varrho}(-k, p_1, -p)\, w^C_{\varkappa,\varrho}(k, p_2, p) \bar{\omega}(p) (\omega + \omega_1 + \mu)^{-1} (\omega + \omega_2 + \mu)^{-1} \right| \\ &\leqq 0(\varrho^{-\delta})\, (\omega_1 \omega_2)^{-1/2 + 3\delta}\, h(p_1 + p_2), \end{aligned}$$

where $h(\cdot)$ is rapidly decreasing. Since $K(p_1, p_2) \in L_2$, it follows that for small δ,

$$Z_{13} + \cdots + Z_{16} \leqq 0(\varrho^{-\delta})\, (N_\tau + I).$$

and the proof of (7.1.8) is complete. This pıoves the lemma and the theorem for ϱ sufficiently large.

To complete the proof of the theorem, we estimate $H(g, \varkappa, \varrho)$ for $\varrho \leqq \varrho_1$, with the theorem valid for ϱ_1. Hence

$$\begin{aligned} H(g, \varkappa, \varrho) &= H(g, \varkappa, \varrho_1) + \{H(g, \varkappa, \varrho) - H(g, \varkappa, \varrho_1)\} \\ &\geqq \text{const.}\ N_\tau - I + \{H(g, \varkappa, \varrho) - H(g, \varkappa, \varrho_1)\} . \end{aligned}$$

The next lemma completes the proof of the theorem.

LEMMA 7.1.3 *Let $\varepsilon > 0$. Then*

$$0 \leqq \varepsilon N_\tau + \{H(g, \varkappa, \varrho) - H(g, \varkappa, \varrho_1)\} + \text{const.},$$

with a constant uniform in $\varkappa$ and $\varrho \leqq \varrho_1$.

Proof Since

$$0 \leqq -\tfrac{1}{2}\{\delta m^2(\varkappa, \varrho) - \delta m^2(\varkappa, \varrho_1)\} = M,$$

we have by Theorem 3.1.1

$$0 \leqq \varepsilon N + M \int :\phi(x)^2: g(x)^2\, dx + \text{const.}$$

The lemma now follows by Lemma 6.3.5.

7.2. The second order estimate

THEOREM 7.2.1 *Glimm and Jaffe*[8]*, Dimock*[1]*. Let $\tau < 1$ and let g be fixed. There are constants a and b independent of $\varkappa$ and ϱ such that*

$$(N_{\tau/2})^2 \leqq a(H(g, \varkappa, \varrho) + bI)^2. \tag{7.2.1}$$

We introduce the resolvent $R(\zeta) = R_{\varkappa,\varrho}(\zeta) = (H(g, \varkappa, \varrho) - \zeta)^{-1}$ and we prove that $\|N_{\tau/2} R(\zeta)\| \leqq$ const., for ζ sufficiently negative. Our two basic tools are the first order estimate of Theorem 7.1.1 yielding $\|N_\tau^{1/2} R(\zeta)^{1/2}\| \leqq$ const., and the pull-through formulas,

$$b(k, \varepsilon)\, R(\zeta) = R(\zeta - \mu(k, \varepsilon))\, b(k, \varepsilon) - R(\zeta - \mu(k, \varepsilon))\, X(k, \varepsilon)\, R(\zeta), \tag{7.2.2}$$

where

$$X(k, \varepsilon) = [b(k, \varepsilon), H_I(g, \varkappa, \varrho) + c(g, \varkappa, \varrho)]. \tag{7.2.3}$$

A complete treatment of the domains of these equations is given in Glimm and Jaffe[8].

LEMMA 7.2.2 *Let $\delta > 0$. Then*

$$\|(N_\tau + I)^{-1/2}\, X(p, \pm 1)\, (N_\tau + I)^{-1/2}\| \leqq 0(\omega(p)^{-\tau/2+\delta}).$$

Proof We apply Corollary 1.2.4 to the various terms in $X(p, \pm 1)$. For instance, the pure creation term has a kernel $w^C_{\varkappa,\varrho}(k, p_1, p_2)$ satisfying

$$\|\omega_2^{-\tau/2}\, w^C_{\varkappa,\varrho}(\cdot, p, \cdot)\|_2^2 \leqq \text{const.} \int dp_2\, \omega(p + p_2)^{-1-\delta/2}\, \omega(p_2)^{-\tau+\delta/2}$$
$$\leqq \text{const.}\, \omega(p)^{-\tau+\delta},$$

and this yields a bound of the desired form. The analogous bound does not hold for $\varepsilon = 0$, and the $\varepsilon = 0$ term requires a renormalization cancellation.

We now see how Lemma 7.2.2 is useful in the proof in the theorem. With $N_\tau = \sum_\varepsilon N_\tau^{(\varepsilon)}$, it is sufficient to bound each $(N_\tau^{(\varepsilon)})^2$. Since

$$(N_{\tau/2}^{(\varepsilon)})^2 = N_\tau^{(\varepsilon)} + \int b(k, \varepsilon)^*\, N_{\tau/2}^{(\varepsilon)}\, b(k, \varepsilon)\, \mu(k, \varepsilon)^{\tau/2}\, dk,$$

and since $R(\zeta)\, N_\tau R(\zeta)$ is uniformly bounded by Theorem 7.1.1, it is sufficient to prove that for $\theta \in \mathscr{D}$, and for $\varepsilon = 0, \pm 1$,

$$M_\varepsilon = \int \|(N_{\tau/2}^{(\varepsilon)})^{1/2}\, b(k, \varepsilon)\, R(\zeta)\, \theta\|^2\, \mu(k, \varepsilon)^{\tau/2}\, dk \leqq \text{const. } \|\theta\|^2.$$

Using (7.2.2) we obtain

$$M_\varepsilon \leqq 2 \int \|(N_{\tau/2}^{(\varepsilon)})^{1/2}\, R(\zeta - \mu(k, \varepsilon))\, b(k, \varepsilon)\, \theta\|^2\, \mu(k, \varepsilon)^{\tau/2}\, dk$$

$$+ 2 \int \|(N_{\tau/2}^{(\varepsilon)})^{1/2}\, R(\zeta - \mu(k, \varepsilon))\, X(k, \varepsilon)\, R(\zeta)\, \theta\|^2\, \mu(k, \varepsilon)^{\tau/2}\, dk.$$

Since

$$\|(N_{\tau/2}^{(\varepsilon)})^{1/2}\, R(\zeta - \mu(k, \varepsilon))\, (N_{\tau/2} + \mu(k, \varepsilon)^{\tau/2})^{1/2}\| \leqq \text{const.},$$

and

$$\int \|(N_{\tau/2} + \mu(k, \varepsilon)^{\tau/2})^{-1/2}\, b(k, \varepsilon)\, \theta\|^2\, \mu^{\tau/2}\, dk$$

$$= \int \|b(k, \varepsilon)\, N_{\tau/2}^{-1/2}\, \theta\|^2\, \mu^{\tau/2}\, dk \leqq \|\theta\|^2,$$

we have

$$M_\varepsilon \leqq \text{const. } \|\theta\|^2 + \int \|(N_{\tau/2}^{(\varepsilon)})^{1/2}\, R(\zeta - \mu(k, \varepsilon))\, X(k, \varepsilon)\, R(\zeta)\, \theta\|^2\, \mu(k, \varepsilon)^{\tau/2}\, dk. \tag{7.2.4}$$

For $\varepsilon = \pm 1$, we use (7.2.2), Lemma 7.2.2 and the relation $\{b(k, \varepsilon), X(k', \varepsilon)\} = 0$ to obtain the bound

$$\int \|(N_{\tau/2}^{(\varepsilon)})^{1/2}\, R(\zeta - \mu(k, \varepsilon))\, X(k, \varepsilon)\, R(\zeta)\, \theta\|^2\, \mu(k, \varepsilon)^{\tau/2}\, dk$$

$$= \int \|b(k', \varepsilon)\, R(\zeta - \mu(k, \varepsilon))\, X(k, \varepsilon)\, R(\zeta)\, \theta\|^2\, \mu(k', \varepsilon)^{\tau/2}\, \mu(k, \varepsilon)^{\tau/2}\, dk\, dk'$$

$$\leqq 2 \int \|R(\zeta - \mu - \mu')\, X(k, \varepsilon)\, R(\zeta - \mu')\, b(k', \varepsilon)\, \theta\|^2\, (\mu\mu')^{\tau/2}\, dk\, dk'$$

$$+ 4 \int \|R(\zeta - \mu - \mu')\, X(k, \varepsilon)\, R(\zeta - \mu')\, X(k', \varepsilon)\, R(\zeta)\, \theta\|^2\, (\mu\mu')^{\tau/2}\, dk dk'$$

$$\leqq \text{const. } \|\theta\|^2.$$

Hence $M_{\pm 1} \leqq$ const. $\|\theta\|^2$.

The proof of the theorem is now reduced to the case $\varepsilon = 0$. We remark that it is easy to prove a divergent estimate

$$\|(N_\tau + I)^{-1/2}\, X(k, 0)\, (N_\tau + I)^{-1/2}\| \leqq 0(\varkappa^\delta \mu^{-1/2}), \quad \delta > 0, \tag{7.2.5}$$

yielding $M_0 \leqq 0(\varkappa^{2\delta})\, \|\theta\|^2$, and therefore by the above bounds

$$(N_{\tau/2})^2 \leqq 0(\varkappa^{2\delta})\, (H(g, \varkappa, \varrho) + bI)^2. \tag{7.2.6}$$

To prove (7.2.5), we note that

$$X(k, 0) = [a(k), H_I(g, \varkappa, \varrho)] - \delta m^2(\varkappa, \varrho)\, (4\pi\mu)^{-1/2} \int e^{-ikx}\, \phi(x)\, g(x)^2\, dx. \tag{7.2.7}$$

By conditions (a), (b) of Lemma 6.1.1, for $1 - \tau < \delta$,

$$\|(\omega_1 + \omega_2)^{-\tau/2} \left(|w^C_{\varkappa,\varrho}(k, \cdot, \cdot)| + |w_{\varkappa,\varrho}(k, \cdot, \cdot)|\right)\|_2 \leqq 0(\varkappa^\delta \mu^{-1/2}).$$

Thus by Corollary 1.2.4,

$$\|(N_\tau + I)^{-1/2} [a(k), H_I(g, \varkappa, \varrho)] (N_\tau + I)^{-1/2}\| \leqq 0(\varkappa^\delta \mu^{-1/2}).$$

Since $\delta m^2(\varkappa, \varrho) \leqq 0(\log \varkappa)$, the bounds (7.2.5-6) hold. A proof of resolvent convergence and self adjointness can be based on this estimate. The proof of (7.2.1) follows by isolating and cancelling the divergence in (7.2.7).

LEMMA 7.2.3 *Dimock*[1]. *For* $\tau < 1$, $0 < \delta$,

$$\|(N^{(0)}_{\tau/2})^{1/2} R(\zeta - \mu) X(k, 0) R(\zeta)\| \leqq 0(\mu^{-1+\delta}). \tag{7.2.8}$$

Inserting (7.2.8) into (7.2.4) yields the theorem. We do not give the lengthy proof of (7.2.8). In order to illustrate the ideas involved, we estimate one particular renormalization cancellation. We write

$$X(k, 0) = X_1(k) + X_2(k) + X_1(-k)^*$$

where

$$X_1(k) = \int w^C_{\varkappa,\varrho}(k, p_1, p_2)\, b'(-p_1)\, b(-p_2)\, dp_1\, dp_2$$
$$- \frac{\delta m^2(\varkappa, \varrho)}{8\pi} \int dk' (\mu\mu')^{-1/2} (\tilde{g} * \tilde{g}) (k + k') \{a(k')^* + a(-k')\},$$

$$X_2(k) = \int w_{\varkappa,\varrho}(k, p_1, p_2) \sum_{\varepsilon = \pm 1} b(p_1, \varepsilon)^* b(-p_2, \varepsilon)\, dp_1\, dp_2 .$$

The renormalization cancellations occur in X_1 and X_1^*, while X_2 can be estimated by Proposition 1.2.2,

$$\|(N_\tau + I)^{-1/2} X_2(k) (N_\tau + I)^{-1/2}\| \leqq 0(\mu^{-(1+\tau)/2}),$$

which is sufficient for (7.2.8). We now prove the lemma for X_1 replacing X. We simplify $X_1 R$ by using the pull through formula

$$\begin{aligned} b'(-p_1)\, b(-p_2)\, R &= b'(-p_1)\, R(\zeta - \omega_2)\, b(-p_2) \\ &+ R(\zeta - \omega_1 - \omega_2)\, X(-p_2, +1)\, R(\zeta - \omega_1)\, b'(-p_1) \\ &- R(\zeta - \omega_1 - \omega_2)\, X(-p_2, +1)\, R(\zeta - \omega_1) \times X(-p_1, -1)\, R(\zeta) \\ &+ R(\zeta - \omega_1 - \omega_2)\, X(-p_1, -1)\, R(\zeta - \omega_2) \times X(-p_2, +1)\, R(\zeta) \\ &- R(\zeta - \omega_1 - \omega_2)\, V(-p_1, -p_2)\, R(\zeta), \\ &= B_1(-p_1, -p_2) + \cdots + B_5(-p_1, -p_2), \end{aligned} \tag{7.2.9}$$

where

$$V(p_1, p_2) = \int w^C_{\varkappa,\varrho}(k', p_2, p_1) \{a(k')^* + a(-k')\} dk'$$
$$= \{b'(p_1), [b(p_2), H_I(g, \varkappa, \varrho) + c(g, \varkappa, \varrho)]\}.$$

Each B_i, except, B_5, gives a bounded contribution to $R^{1/2}X_1(k) R$, namely for $i = 1, 2, 3, 4$,

$$\left\| R(\zeta)^{1/2} \int B_i(-p_1, -p_2) w^C_{\varkappa,\varrho}(k, p_1, p_2) dp_1 dp_2 \right\| = 0(\mu^{-1+\delta}).$$

This bound is established by the use of N_τ estimates and Lemma 7.2.2. The bound for $i = 5$ involves a renormalization cancellation. We combine B_5 with the counterterm in X_1. Thus we estimate

$$R^{1/2} \int B_5(p_1, p_2) w^C_{\varkappa,\varrho}(k, -p_1, -p_2) dp_1 dp_2 R - R^{1/2}\left(\frac{1}{8\pi}\right)\delta m^2(\varkappa, \varrho)$$
$$\times \int dk'(\mu\mu')^{-1/2} (\tilde{g} * \tilde{g}) (k + k') \{a(k')^* + a(-k')\} R$$
$$= R^{1/2} \int \alpha(k, k') \{a(k')^* + a(-k')\} dk'R,$$

where

$$\int \alpha(k, k') \{a(k')^* + a(-k')\} dk' = -\int w^C_{\varkappa,\varrho}(k, -p_1, -p_2) R(\zeta - \omega_1 - \omega_2)$$
$$\times V(p_1, p_2) dp_1 dp_2 - \frac{1}{8\pi} \delta m^2(\varkappa, \varrho) \int dk'(\mu\mu')^{-1/2} (\tilde{g} * \tilde{g}) (k + k')$$
$$\times \{a(k')^* + a(-k')\}.$$

By Proposition 1.2.3(a), $\|V(p_1, p_2) R^{1/2}\| \leqq 0(\omega(p_1 + p_2)^{-1/2})$. Hence we can replace $R(\zeta - \omega_1 - \omega_2)$ by $(\omega_1 + \omega_2)^{-1}$ yielding α' for α above. The error is estimated by

$$\int \left\| R^{1/2} \int w^C_{\varkappa,\varrho}(k, -p_1, -p_2) R(\zeta - \omega_1 - \omega_2) (\omega_1 + \omega_2)^{-1} (H(g, \varkappa, \varrho) - \zeta) \right.$$
$$\left. \times V(p_1, p_2) R \right\| dp_1 dp_2$$
$$\leqq 0(\mu^{-1/2}) \int |\tilde{g}(k - p_1 - p_2)| (\omega_1 + \omega_2)^{-3/2} \omega(p_1 + p_2)^{-1/2} dp_1 dp_2$$
$$\leqq 0(\mu^{-3/2+\delta}).$$

Thus we need to estimate the renormalization cancellation,

$$R^{1/2} \int \alpha'(k, k') \{a(k')^* + a(-k')\} dk' R,$$

where

$$\alpha'(k, k') = -w^C_{\varkappa,\varrho}(k, p_1, p_2)\, w^C_{\varkappa,\varrho}(k', -p_2, -p_1)\, (\omega_1 + \omega_2)^{-1}\, dp_1\, dp_2$$

$$- \frac{1}{8\pi} \delta m^2(\varkappa, \varrho)\, (\mu\mu')^{-1/2}\, (\tilde{g} * \tilde{g})\, (k + k').$$

In Sec. 7.1, in the proof of (7.1.11), we established the inequality $|\alpha'(k, k')| \leqq \text{const.}\, (\mu\mu')^{-1/2+\delta/2}\, h(k + k')$, where h is rapidly decreasing. Hence

$$\left\| R^{1/2} \int \alpha'(k, k') \{a(k')^* + a(-k')\}\, dk'\, R \right\| \leqq 0(\mu^{-1+\delta}). \qquad (7.2.10)$$

Combining (7.2.10) with the estimates on $B_1, \ldots, B_4$ proves

$$\|R^{1/2}\, X_1(k)\, R\| \leqq 0(\mu^{-1+\delta}),$$

proving (7.2.8) for X_1 in the place of X. The X_1^* term is more difficult, but the methods are the same.

Using these methods, it is also possible to prove that for $\zeta, \zeta_1 < 0$ and $\delta > 0$,

$$\|R(\zeta + \zeta_1)\, X(k, 0)\, R(\zeta)\| \leqq 0((|\zeta_1|\, \mu)^{-1/2+\delta}). \qquad (7.2.11)$$

8 Resolvent Convergence and Self Adjointness

8.1 Resolvent convergence for the Hamiltonians

We prove norm convergence of the resolvents $R_{\varkappa,\varrho}(\zeta) = (H(g, \varkappa, \varrho) - \zeta)^{-1}$ of the cutoff Hamiltonians $H(g, \varkappa, \varrho)$. Let ζ be real and sufficiently negative.

THEOREM 8.1.1 *There exist bounded self adjoint operators $R_\varrho(\zeta)$ independent of $\chi_\varkappa$ such that*

$$\|R_{\varkappa,\varrho}(\zeta) - R_\varrho(\zeta)\| \to 0 \quad \text{as} \quad \varkappa \to \infty \qquad (8.1.1)$$

uniformly in ϱ and

$$\|R_\varrho(\zeta) - (H_0 - \zeta)^{-1}\| \to 0 \quad \text{as} \quad \varrho \to \infty. \qquad (8.1.2)$$

Remarks The proof of resolvent convergence follows Glimm-Jaffe[8]. We inspect the difference $R_{\varkappa_1,\varrho} - R_{\varkappa_2,\varrho} = -R_{\varkappa_1,\varrho}\, \delta H\, R_{\varkappa_2,\varrho}$, where $\delta H = H(g, \varkappa_1, \varrho) - H(g, \varkappa_2, \varrho)$. The first step is to perform explicit cancellations between the divergent counterterms δc and certain parts of the interaction Hamiltonian δH_I. We then estimate the remainder, and show that it converges to zero in norm (uniformly in ϱ). The main tool in the renormalization cancellations is the pull through formula. We use this formula to generate a perturbation series with a finite number of terms. The leading terms exhibit the renormalization cancellations, while the remaining terms can be estimated directly by combining the first and second order energy estimates of Chapter 7 with the N_τ estimates of Chapter 1. We thereby show that $R_{\varkappa,\varrho}$ is a Cauchy sequence (in norm) as $\varkappa \to \infty$, and therefore has a limit R_ϱ.

In order to prove the cutoff independence of the limit R_ϱ, we apply the same methods to show that $\|R_{1,\varkappa,\varrho} - R_{2,\varkappa,\varrho}\| \to 0$, where the subscripts 1 and 2 refer to distinct cutoff functions $\chi_{1,\varkappa}$ and $\chi_{2,\varkappa}$ satisfying the conditions of Lemma 6.3.1. This proves that the limit R_ϱ is independent of the family $\{\chi_\varkappa\}$ of cutoff functions used to define the approximate Hamiltonians $H(g, \varkappa)$.

To establish (8.1.2), we use the fact that the convergence (8.1.1) is uniform in ϱ, and write

$$\|R_\varrho(\zeta) - (H_0 - \zeta)^{-1}\| \leqq \|R_\varrho(\zeta) - R_{\varkappa,\varrho}(\zeta)\| + \|R_{\varkappa,\varrho}(\zeta) - (H_0 - \zeta)^{-1}\|.$$

For $\varkappa$ large, the first term is small uniformly in ϱ. With $\varkappa$ fixed, the second term is seen to converge from Lemma 6.3.5:

$$\|R_{\varkappa,\varrho}(\zeta) - (H_0 - \zeta)^{-1}\| \leqq \| R_{\varkappa,\varrho}(\zeta)\,(H_I(g, \varkappa, \varrho) + c(\varkappa, \varrho))\,(H_0 - \zeta)^{-1}\|$$

$$\leqq \text{const.}\ \|(N + I)^{-1}\,(H_I(g, \varkappa, \varrho) + c(\varkappa, \varrho))\,(N + I)^{-1}\|$$

$$\leqq 0(\varrho^{-\delta}).$$

In order to study the convergence of $R_{\varkappa 1,\varrho}\,\delta H R_{\varkappa 2,\varrho}$, we write

$$\delta H = \delta H_I + \delta c = \delta W^C + \delta W + \delta W^A + \delta c,$$

following the notation of Sec. 6.1. In perturbation theory we find that $H_0 + W^C$, W and $W^A + c$ are individually defined as bilinear forms on the domain of $H(g, \varkappa, \varrho)$. Thus we expect that $R_{\varkappa 1}\,\delta W^C R_{\varkappa 2}$ and $R_{\varkappa 1}\,\delta W R_{\varkappa 2}$ are well behaved as $\varkappa_1, \varkappa_2 \to \infty$ so long as $\varkappa_1 \leqq \varkappa_2$. On the other hand, both $R_{\varkappa 1}\,\delta W^A\,R_{\varkappa 2}$ and $R_{\varkappa 1}\,\delta c\,R_{\varkappa 2}$ diverge as $\varkappa_1, \varkappa_2 \to \infty$, but their sum converges to zero. We must isolate the renormalization cancellations that occur in this sum, and estimate the remainder after cancellation. We note that writing $R_{\varkappa 2} - R_{\varkappa 1} = -R_{\varkappa 2}\,\delta H R_{\varkappa 1}, \varkappa_1 \leqq \varkappa_2$, would reverse the role of δW^C and δW^A in the proof.

We choose the first cutoff $\chi_{1,\varkappa}$ to be sharp in momentum space, and to have the form (6.3.2),

$$\chi'_\varkappa = \chi_{1,\varkappa} = \chi_\varkappa h_\varkappa,$$

where

$$h_\varkappa(k, p_1, p_2) = \theta\left(\frac{k}{\varkappa^{1-\alpha}}\right)\theta\left(\frac{p_1}{\varkappa^{1-\alpha}}\right)\theta\left(\frac{p_2}{\varkappa^{1-\alpha}}\right).$$

We define $H'(g, \varkappa, \varrho)$ to be the Hamiltonian corresponding to the cutoff $\chi'_\varkappa = \chi_\varkappa h_\varkappa$, and we let $R'_{\varkappa,\varrho}(\zeta)$ be the resolvent of $H'(g, \varkappa, \varrho)$. We now study the difference between $H'_1(g, \varkappa_1, \varrho)$ (given by a cutoff $\chi'_{\varkappa 1} = \chi_{1,\varkappa 1} h_{\varkappa 1}$) and $H_2(g, \varkappa_2, \varrho)$ (given by some other cutoff $\chi_{2,\varkappa 2}$). Let $\delta H = H'_1(g, \varkappa_1, \varrho)$

$- H_2(g, \varkappa_2, \varrho)$. To simplify the notation, we set $\varkappa_1 = \varkappa, \chi_1 = \chi$ and $R_{2,\varkappa_2} = R_{\varkappa_2}$. We take ζ real and sufficiently negative. We state our three basic lemmas.

LEMMA 8.1.2 *Let $\varkappa \leqq \varkappa_2$. Then for some $\delta > 0$,*

$$\|R'_{\varkappa,\varrho}\, \delta W\, R_{\varkappa_2,\varrho}\| \leqq 0(\varkappa^{-\delta}\varrho^{-\delta}).$$

LEMMA 8.1.3 *Let $\varkappa \leqq \varkappa_2$. Then for some $\delta > 0$,*

$$\|R'_{\varkappa,\varrho}\, \delta W^C\, R_{\varkappa_2,\varrho}\| \leqq 0(\varkappa^{-\delta}\varrho^{-\delta}).$$

LEMMA 8.1.4 *Let $\varkappa \leqq \varkappa_2$. Then for some $\delta > 0$,*

$$\|R'_{\varkappa,\varrho}\{\delta W^A + \delta c\}\, R_{\varkappa_2,\varrho}\| \leqq 0(\varkappa^{-\delta}\varrho^{-\delta}).$$

Proof of Theorem 8.1.1 We use the three lemmas to prove the theorem. The first step is to prove convergence of $R'_{\varkappa,\varrho}$ as $\varkappa \to \infty$. Hence we pick $\chi_{2,\varkappa} = \chi'_\varkappa$. Thus $H_2(g, \varkappa, \varrho) = H'_1(g, \varkappa, \varrho)$. By Lemmas 8.1.2–8.1.4.

$$\begin{aligned} \|R'_{\varkappa,\varrho} - R'_{\varkappa_2,\varrho}\| &\leqq \|R'_{\varkappa,\varrho}\, \delta W^C\, R'_{\varkappa_2,\varrho}\| \\ &\quad + \|R'_{\varkappa,\varrho}\, \delta W\, R'_{\varkappa_2,\varrho}\| \\ &\quad + \|R'_{\varkappa,\varrho}\{\delta W^A + \delta c\}\, R'_{\varkappa_2,\varrho}\| \\ &\leqq 0(\varkappa^{-\delta}\varrho^{-\delta}). \end{aligned}$$

Thus for a particular cutoff $\chi'_\varkappa$,

$$R_\varrho(\zeta) = \underset{\varkappa\to\infty}{\text{norm limit}}\, R'_{\varkappa,\varrho}$$

exists, and the limit is uniform in ϱ.

Next we let $\chi_{2,\varkappa}$ be any permissable cutoff, as in Definition 6.3.2. Again, we have by the lemmas

$$\begin{aligned} \|R_\varrho(\zeta) - R_{\varkappa,\varrho}(\zeta)\| &\leqq \|R_\varrho - R'_{\varkappa,\varrho}\| + \|R'_{\varkappa,\varrho} - R_{\varkappa,\varrho}\| \\ &\leqq 0(\varkappa^{-\delta}\varrho^{-\delta}). \end{aligned}$$

Hence $R_{\varkappa,\varrho}(\zeta)$ converges to $R_\varrho(\zeta)$, and $R_\varrho(\zeta)$ is independent of the choice of cutoff function $\chi_\varkappa = \chi_{1,\varkappa}$. We therefore have proved (8.1.1). The proof of (8.1.2) was given above.

We now prove Lemmas 8.1.2–8.1.3 and outline the main ideas in the proof of Lemma 8.1.4.

Proof of Lemma 8.1.2 The kernel δw of δW is bounded by

$$\begin{aligned} |\delta w| &\leqq \text{const.}\, \mu^{-1/2}\, |\tilde{g}(k+\eta)|\, \omega(\eta)^{6\delta}\, \omega(\xi)^{-3\delta} |\, \chi'_\varkappa - \chi_{2,\varkappa_2}|\, \{1 - \theta_\varrho(\xi)\} \\ &\leqq 0(\varkappa^{-\delta(1-\alpha)}\varrho^{-\delta})\, (\mu + \omega_1 + \omega_2)^\delta\, \omega(\xi)^\delta\, \mu^{-1/2}\, |\tilde{g}(k+\eta)|\, \omega(\eta)^{6\delta}\, \omega(\xi)^{-3\delta}. \end{aligned}$$

Here we use condition (c_1) and (c_1') of Lemma 6.3.1 to estimate the change of cutoffs in the region $S_{\varkappa^{1-\alpha}}$, and the bound $(\mu + \omega_1 + \omega_2)^{-\delta} \leqq 0(\varkappa^{-\delta(1-\alpha)})$ on the complement of $S_{\varkappa^{1-\alpha}}$. Hence for $\tau > 1$, sufficiently close to one, and for a new δ,

$$\|\mu^{-\tau/2}\, \omega_2^{-\tau/2}\, \delta w\|_2 \leqq 0(\varkappa^{-\delta}\varrho^{-\delta}).$$

By Theorem 7.1.1 and Proposition 1.2.3(b),

$$\|R'_{\varkappa,\varrho}\, \delta W\, R_{\varkappa_2,\varrho}\| \leqq \text{const.}\ \|(N_\tau + I)^{-1/2}\, \delta W (N_\tau + I)^{-1/2}\| \leqq 0(\varkappa^{-\delta}\varrho^{-\delta}).$$

Proof of Lemma 8.1.3 In this lemma we use the special properties of $\chi'_\varkappa$. We first modify $H'(g, \varkappa, \varrho)$ by using a cutoff mass counterterm

$$-\tfrac{1}{2}\, \delta m^2(\varkappa, \varrho) \int :\phi_\varkappa(x)^2 :\, g(x)^2\, dx,$$

where $\phi_\varkappa$ has a sharp momentum cutoff at $\varkappa^{1-\alpha}$,

$$\phi_\varkappa(x) = (4\pi)^{-\frac{1}{2}} \int e^{-ikx}\{a(k)^* + a(-k)\}\mu^{-\frac{1}{2}}\theta(k\varkappa^{-1+\alpha})dk.$$

We call the corresponding Hamiltonian $\hat{H}(g, \varkappa, \varrho)$ and let $\hat{R}$ be the resolvent of $\hat{H}$. This approximation has the advantage that $\hat{R}_{\varkappa,\varrho}(\zeta)$ commutes with the high momentum part of N_τ,

$$\hat{N}_{\tau,\varkappa} = \sum_{\varepsilon = \pm 1, 0} \int b(k, \varepsilon)^*\, \hat{\mu}(k, \varepsilon)\, b(k, \varepsilon)\, dk,$$

where

$$\hat{\mu}(k, \varepsilon) = \begin{cases} \mu(k, \varepsilon) & \text{for } \varkappa^{1-\alpha} \leqq |k| \\ 0 & \text{otherwise.} \end{cases}$$

Thus for $\tau < 1$,

$$\|\hat{N}_\tau^{1/2}\, \hat{R}_{\varkappa,\varrho} N_\tau^{1/2}\| = \|\hat{R}_{\varkappa,\varrho}\hat{N}_\tau^{1/2} N_\tau^{1/2}\| \leqq \text{const.},$$

and

$$\|\hat{R}_{\varkappa,\varrho}\, \delta W^C R_{\varkappa_2,\varrho}\| \leqq \text{const.}\ \|(N_\tau + I)^{-1/2}\, (\hat{N}_\tau + I)^{-1/2}\, \delta W^C (N_\tau + I)^{1/2}\|. \tag{8.1.3}$$

We estimate (8.1.3) by N_τ estimates. Such estimates for $\hat{N}_\tau$ involve an energy factor $(1 + \hat{\mu})^{-\tau/2}$ in place of $\mu^{-\tau/2}$. We write $\delta W^C = A_1 + A_2 + B$ with kernels α_1, α_2 and β which are nonzero in disjoint regions. We suppose α_1 has support in the region where $|p_1| \geqq \varkappa^{1-\alpha}$, α_2 has support in the region where $|p_2| \geqq \varkappa^{1-\alpha}$, and β has support in $|p_1| \leqq \varkappa^{1-\alpha}$, $|p_2| \leqq \varkappa^{1-\alpha}$. We note that for δ sufficiently small and τ sufficiently close to 1,

$$\|(\hat{\omega}_1\omega_2)^{-\tau/2}\, \alpha_1\|^2 + \|(\omega_1\hat{\omega}_2)^{-\tau/2}\, \alpha_2\|^2 = \|(\omega_1\omega_2)^{-\tau/2}(\alpha_1 + \alpha_2)\|^2 \leqq 0(\varkappa^{-\delta}\varrho^{-\delta}),$$

since by condition (c_1) and (c_1') on the cutoffs $\chi_\varkappa, \chi'_\varkappa$,

$$\begin{aligned} |\alpha_1| + |\alpha_2| &\leqq \text{const}\, \mu^{-1/2}\, |\tilde{g}(k + \eta)|\, |\chi'_\varkappa - \chi_{2,\varkappa_2}|\, \{1 - \theta_\varrho(\xi)\} \\ &\leqq 0(\varkappa^{-\delta(1-\alpha)}\varrho^{-\delta})\, (\mu + \omega_1 + \omega_2)^{2\delta}\, \mu^{-1/2}\, |\tilde{g}(k + \eta)|. \end{aligned} \tag{8.1.4}$$

Thus by Proposition 1.2.3(a), with one $\hat{\omega}$ variable, and a new δ,

$$\|(N_\tau + I)^{-1/2}\, (\hat{N}_\tau + I)^{-1/2}\, (A_1 + A_2)\, (N_\tau + I)^{-1/2}\| \leqq 0(\varkappa^{-\delta}\varrho^{-\delta}). \tag{8.1.5}$$

We next bound the contribution B to δW^C. On the support of β, both $|p_1|$ and $|p_2|$ are less than $\varkappa^{1-\alpha}$. We write $B = B_1 + B_2$ with kernels β, and β_2, where $\beta_1 = \beta\theta\left(\frac{k}{\varkappa^{1-\alpha}}\right)$ and $\beta_2 = \beta\left(1 - \theta\left(\frac{k}{\varkappa^{1-\alpha}}\right)\right)$. Since $\hat{\mu}\beta_2 = \mu\beta_2$, we have

$$\begin{aligned}\|(N_\tau + I)^{-1/2} (\hat{N}_\tau + I)^{-1/2} B_2(N_\tau + I)^{-1/2}\| \\ &\leqq \text{const. } \|\mu^{-\tau/2} (\omega_1 + \omega_2)^{-\tau/2} \beta_2\|_2 \\ &\leqq 0(\varkappa^{-\delta}\varrho^{-\delta})\end{aligned}$$

as in (8.1.4)–(8.1.5). Hence the $\hat{N}_\tau$ factor is useful for estimating the high momentum parts of δW^C, namely A_1, A_2 and B_2.

For the low momentum contribution B_1 we use the standard N_τ estimate

$$\|(N_\tau + I)^{-1/2} B_1(N_\tau + I)^{-1/2}\| \leqq \text{const. } \|(\mu + \omega_1 + \omega_2)^{-\tau/2} \beta_1\|_2. \quad (8.1.6)$$

On $S_{\varkappa^{1-\alpha}}$ we have

$$\begin{aligned}|\delta\chi| = |\chi'_\varkappa - \chi_{2,\varkappa_2}| = |\chi_\varkappa - \chi_{2,\varkappa_2}| &\leqq 0((\mu + \omega_1 + \omega_2)\varkappa^{-1}). \\ &\leqq 0(\varkappa^{-\alpha}).\end{aligned}$$

Hence by (8.1.4),

$$\begin{aligned}\|(\mu + \omega_1 + \omega_2)^{-\tau/2} \beta_1\|_2^2 &\leqq 0(\varkappa^{-2\alpha}) \int_{S_{\varkappa^{1-\alpha}}} (\mu + \omega_1 + \omega_2)^{-\tau} |\beta_1|^2 \, dk \, dp_1 \, dp_2 \\ &\leqq 0(\varrho^{-\delta}\varkappa^{-2\alpha}) \varkappa^{2\delta+(1-\tau)} \\ &\leqq 0(\varkappa^{-\delta}\varrho^{-\delta}),\end{aligned}$$

for δ and $1 - \tau$ sufficiently small. Combining these estimates,

$$\|\hat{R}_{\varkappa,\varrho} \delta W^C R_{\varkappa_2,\varrho}\| \leqq 0(\varkappa^{-\delta}\varrho^{-\delta}).$$

To complete the proof of the lemma, we must estimate the error introduced by replacing R' with $\hat{R}$.

$$\begin{aligned}\|R'_{\varkappa,\varrho} \delta W^C R_{\varkappa_2,\varrho}\| &\leqq \|\hat{R}_{\varkappa,\varrho} \delta W^C R_{\varkappa_2,\varrho}\| + \|\hat{R}_{\varkappa,\varrho} \delta\hat{H} R'_{\varkappa,\varrho} \delta W^C R_{\varkappa_2\varrho}\|, \\ &\leqq 0(\varkappa^{-\delta}\varrho^{-\delta}) + \|\hat{R}_{\varkappa,\varrho} \delta\hat{H} R'_{\varkappa,\varrho} \delta W^C R_{\varkappa_2,\varrho}\|. \quad (8.1.8)\end{aligned}$$

Here $\delta\hat{H} = \hat{H} - H' = -\frac{1}{2}\delta m^2(\varkappa, \varrho) \int :\phi_\varkappa^2 - \phi^2: g(x)^2 \, dx$.

We now use standard estimates to prove that (8.1.8) is small.

$$-\frac{1}{2}\delta m^2(\varkappa, \varrho) = \int_{|\xi|>\varrho} \left|\chi'_\varkappa\left(0, \frac{\xi}{2}, \frac{-\xi}{2}\right)\right|^2 \omega(\xi)^{-1} \, d\xi \leqq 0(\varkappa^\delta\varrho^{-\delta}),$$

$$\|(N_\tau + I)^{-1/2} W^C(N_\tau + I)^{-1/2}\| \leqq 0(\varkappa^\delta),$$

$$\left\|(N + I)^{-1/2} \int :\phi_\varkappa^2 - \phi^2: g(x)^2 \, dx (N_\tau + I)^{-1/2}\right\| \leqq 0(\varkappa^{-(1-\alpha)/2}).$$

Hence for δ sufficiently small

$$\|\hat{R}_{\varkappa,\varrho}\,\delta H R'_{\varkappa,\varrho}\, W^C R_{\varkappa_2,\varrho}\| \leqq 0(\varkappa^{-(1-\alpha)/2+2\delta}\varrho^{-\delta})$$
$$\leqq 0(\varkappa^{-\delta}\varrho^{-\delta}),$$

and the lemma follows by (8.1.8).

We do not give the lengthy proof of Lemma 8.2.4, see Glimm-Jaffe [8, Sec. 4], but we illustrate the type of cancellations that occur, and how some of the convergent terms are estimated. There are two parts in W^A,

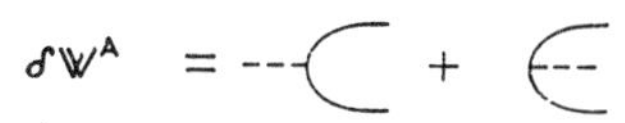

The first of these diagrams cancels two of the mass counterterms. The second diagram cancels the remaining two mass counterterms and the vacuum energy $-\delta E$. We use the expansion (7.2.9) for $\delta W^A R_{\varkappa_2,\varrho}$, and we consider only the term corresponding to B_5 in (7.2.9),

$$Z = -\int \{a(-k) + a(k)^*\}\, R_{\varkappa_2,\varrho}(\zeta - \omega_1 - \omega_2)\, V(-p_1, -p_2)\, R_{\varkappa_2,\varrho}(\zeta)$$
$$\times\ \delta w^C(k, p_1, p_2)\, dk\, dp_1\, dp_2$$

because all the cancellations occur between the counterterms and $R'Z$. We study the $a(-k)$ contribution to $R'Z$. It contains the vacuum energy cancellation and part of the mass renormalization cancellation. Let

$$C = -R'_{\varkappa,\varrho}\int a(-k)\, R_{\varkappa_2,\varrho}(\zeta - \omega_1 - \omega_2)\, a(k')^*\, R_{\varkappa_2,\varrho}(\zeta)$$
$$\times\ w^C_{\varkappa_2,\varrho}(k', -p_1, -p_2)\, \delta w^C(k, p_1, p_2)\, dk\, dk'\, dp_1\, dp_2 .$$

Using the pull through formula, we have

$$C = -R'_{\varkappa,\varrho}\int R_{\varkappa_2,\varrho}(\zeta - \omega_1 - \omega_2 - \mu)\, |w^C_\varrho(k, p_1, p_2)|^2\, \delta\chi(k, p_1, p_2)$$
$$\times\ \chi_{\varkappa_2}(-k, -p_1, -p_2)\, dk\, dp_1\, dp_2\, R_{\varkappa_2,\varrho}(\zeta),$$
$$-R'_{\varkappa,\varrho}\int a(k')^*\, a(-k)\, R_{\varkappa_2,\varrho}(\zeta - \omega_1 - \omega_2 - \mu')\, R_{\varkappa_2,\varrho}(\zeta)$$
$$\times\ w^C_{\varkappa_2,\varrho}(k', -p_1, -p_2)\, \delta w^C(k, p_1, p_2)\, dk\, dk'\, dp_1\, dp_2,$$
$$+R'_{\varkappa,\varrho}\int a(-k)\, R_{\varkappa_2,\varrho}(\zeta - \omega_1 - \omega_2)\, X(k', 0)$$
$$\times\ R_{\varkappa_2,\varrho}(\zeta - \omega_1 - \omega_2 - \mu')\, R_{\varkappa_2,\varrho}(\zeta)$$
$$w^C_{\varkappa_2,\varrho}(k', -p_1, -p_2)\, \delta w^C(k, p_1, p_2)\, dk\, dk'\, dp_1\, dp_2$$
$$= C_1 + C_2 + C_3 .$$

In this expression, $\delta\chi(k, p_1, p_2) = \chi'_\varkappa(k, p_1, p_2) - \chi_{2,\varkappa_2}(k, p_1, p_2)$. The term C_1 cancels the vacuum energy counterterm. The term C_2 cancels the two mass terms

$$R'_{\varkappa,\varrho}\{-\tfrac{1}{2}\delta m_1^2(\varkappa) + \tfrac{1}{2}\delta m_2^2(\varkappa_2)\}$$
$$\times \int a(k')^* a(-k)(4\pi)^{-1}(\mu\mu')^{-1/2}(\tilde{g} * \tilde{g})(k + k')\, dk\, dk'\, R_{\varkappa_2,\varrho},$$

while C_3 converges to zero in norm.

$$\left\| R'_{\varkappa,\varrho} \int a(-k)\, \delta w^C_{\varkappa,\varrho}(k, p_1, p_2)\, dk \right\| \leqq \text{const.}\, \|\mu^{-\tau/2}\, \delta w^C(\cdot, p_1, p_2)\|_2$$
$$\leqq 0(\varkappa^{-\delta}\varrho^{-\delta})\, \omega(\eta)^{-1+3\delta}(\omega_1\omega_2)^\delta.$$

Hence by (7.2.11) we have

$$\|C_3\| \leqq 0(\varkappa^{-\delta}\varrho^{-\delta}) \int \omega(\eta)^{-1+3\delta}(\omega_1\omega_2)^\delta(\omega_1 + \omega_2)^{-(3/2)+\delta}$$
$$\times (\mu')^{-(1/2)+\delta}|\tilde{g}(k' + \eta)|\, dk'\, dp_1\, dp_2$$
$$\leqq 0(\varkappa^{-\delta}\varrho^{-\delta}).$$

We now show how C_1 cancels. We prove

$$\|C_1 - R'_{\varkappa,\varrho}\, \delta E\, R_{\varkappa_2,\varrho}\| \leqq 0(\varkappa^{-\delta}\varrho^{-\delta}).$$

We note that $-\delta E = -E'_1(g, \varkappa, \varrho) + E_2(g, \varkappa_2, \varrho)$,

$$-\delta E = \int |w^C_\varrho(k, p_1, p_2)|^2(\mu + \omega_1 + \omega_2)^{-1}(|\chi'_\varkappa|^2 - |\chi_{2,\varkappa_2}|^2)\, dk\, dp_1\, dp_2,$$

$$|\chi_{2,\varkappa_2}|^2 - |\chi'_\varkappa|^2 = (\delta\chi)\chi_{2,\varkappa_2}(-k, -p_1, -p_2) + \delta\chi(-k, -p_1, -p_2)\chi'_\varkappa.$$

The term $(\delta\chi)\chi_{2,\varkappa_2}(-k, -p_1 - p_2)$ will cancel with C_1, while the low momentum part $(\delta\chi)(-k, -p_1, -p_2)\chi'_\varkappa$ will converge to zero.

$$C_1 - R'_{\varkappa,\varrho}\, \delta E\, R_{\varkappa_2,\varrho} = R'_{\varkappa,\varrho} \int |w^C_\varrho|^2\, \delta\chi(-k, -p_1, -p_2)$$
$$\times \chi'_\varkappa(\mu + \omega_1 + \omega_2)^{-1}\, dk\, dp_1\, dp_2\, R_{\varkappa_2,\varrho}$$
$$+ R'_{\varkappa,\varrho} \int |w^C_\varrho|^2 R_{\varkappa_2,\varrho}(\zeta - \mu - \omega_1 - \omega_2)$$
$$\times (\omega_1 + \omega_2 + \mu)^{-1}(\delta\chi)\chi_{2,\varkappa_2}(-k, -p_1, -p_2)$$
$$\times dk\, dp_1\, dp_2. \tag{8.1.9}$$

The first term in (8.1.9) vanishes because $|\chi'_\varkappa| \leqq 0((\mu + \omega_1 + \omega_2)^{-\delta}\varkappa^{(1-\alpha)\delta})$, and $\chi'_\varkappa$ has support on $S_{\varkappa^{1-\alpha}}$. On that set $|\delta\chi| \leqq 0((\mu + \omega_1 + \omega_2)\varkappa^{-1})$, so

$$\int_{S_{\varkappa^{1-\alpha}}} |w^C_\varrho|^2 |\delta\chi(-k, -p_1, -p_2)|(\mu + \omega_1 + \omega_2)^{-1-\delta}\, dk\, dp_1\, dp_2 \leqq 0(\varkappa^{-\delta+\varepsilon}\varrho^{-\varepsilon/2}),$$

where we can choose $\varepsilon < \alpha\,\delta/2$. Hence the first term in (8.1.10) is $0(\varkappa^{-\alpha\delta/2}\varrho^{-\varepsilon/2})$. The second term in (8.1.9) is also $0(\varkappa^{-\delta}\varrho^{-\delta})$, but in this case the kernel has degree of convergence 1. The cancellation of C_2 with the mass counterterm and the other contributions to $R'\{Z + \delta c\}\,R$ are dealt with in a similar fashion.

8.2 Self adjointness and graph convergence

THEOREM 8.2.1 *The limit*

$$R(\zeta) = \text{norm}\lim_{\varkappa\to\infty} (H(g, \varkappa) - \zeta)^{-1}$$

is the resolvent of a self adjoint operator $H(g)$, independent of the cutoff function $\chi_\varkappa$, and

$$H(g) = \text{strong graph}\lim_{\varkappa\to\infty} H(g, \varkappa).$$

Proof In Theorem 8.1.1 we established norm convergence and $\chi_\varkappa$ independence for the resolvents. Self adjointness and graph convergence follow from Theorem 6.2.6. In order to verify the hypothesis of Theorem 6.2.6, we need only to show that the family $H(g, \varkappa)$ is densely bounded as $\varkappa \to \infty$. For some ζ, we let

$$\mathscr{D} \bigcup_{\varrho<\infty} \text{Range } R_\varrho(\zeta).$$

Since Range $(H_0 - \zeta)^{-1} = \mathscr{D}(H_0)$ is dense, and since by Theorem 8.1.1

$$\lim_{\varrho\to\infty} R_\varrho(\zeta)\,\theta = (H_0 - \zeta)^{-1}\,\theta,$$

it follows that $\mathscr{D}$ is dense. For $\chi = R_\varrho(\zeta)\,\theta \in \mathscr{D}$, we define $\chi_\varkappa = R_{\varkappa,\varrho}(\zeta)\,\theta$. Hence $\chi_\varkappa \in \mathscr{D}(H(g, \varkappa, \varrho)) = \mathscr{D}(H_0) = \mathscr{D}(H(g, \varkappa))$ and $\lim \chi_\varkappa = \chi$. To prove dense boundedness we estimate

$$\langle\chi_\varkappa, H(g, \varkappa)\,\chi_\varkappa\rangle = \langle\chi_\varkappa, \theta\rangle + \zeta\|\chi_\varkappa\|^2 + \langle\chi_\varkappa, \{H(g, \varkappa) - H(g, \varkappa, \varrho)\}\,\chi_\varkappa\rangle. \tag{8.2.1}$$

Let

$$\delta H = H(g, \varkappa) - H(g, \varkappa, \varrho) = \delta H_I + \delta c\,.$$

We have for fixed ϱ, θ,

$$|\langle\chi_\varkappa, \delta H\,\chi_\varkappa\rangle| = |\langle\theta, R_{\varkappa,\varrho}\,\delta H\,R_{\varkappa,\varrho}\,\theta\rangle|$$

$$\leqq \text{const.}\ \|(N_\tau + I)^{-1/2}\,(\delta H_I + \delta c)\,(N_\tau + I)^{-1/2}\| \leqq \text{const.},$$

uniformly in $\varkappa$ by Lemma 6.3.5. We conclude from (8.2.1) that $H(g, \varkappa)$ is densely bounded, and the proof is complete.

The norm convergence of the resolvents $H(g, \varkappa)$ has immediate consequences in terms of the spectrum of $H(g)$. We now introduce a new definition of $E(g, \varkappa)$. Let

$$E(g, \varkappa) = \inf \text{spectrum}\left\{H_0 + H_I(g, \varkappa) - \tfrac{1}{2}\,\delta m^2\,(\varkappa) \int :\phi(x)^2\!: g(x)^2\,dx\right\}. \tag{8.2.1}$$

Thus we renormalize $H(g, \varkappa)$ by adding a constant to the vacuum energy, chosen so that

$$\inf \text{spectrum } H(g, \varkappa) = 0.$$

We let $E_2(g, \varkappa)$ be the second order vacuum energy counterterm used so far, namely

$$E_2(g, \varkappa) = -\|(\mu + \omega_1 + \omega_2)^{-1/2} w_\varkappa^C\|^2.$$

COROLLARY 8.2.2 *The limit*

$$\lim_{\varkappa \to \infty} \{E(g, \varkappa) - E_2(g, \varkappa)\}$$

exists and is finite. Hence with our new renormalization,

$$0 = \inf \text{spectrum } H(g).$$

This definition of $E(g, \varkappa)$ is consistent with the definition of Sec. 6.1,

$$E(g, \varkappa) = E_2(g, \varkappa) + \text{const.} + o(1).$$

COROLLARY 8.2.3 *There is a vacuum vector Ω_g for $H(g)$, namely $H(g)\,\Omega_g = 0$. In the spectral interval* $[0, \min\{m, M\})$, *the operator $H(g)$ is compact.*

The proofs of these corollaries follow Theorem 3.3.2. In particular, the space of vacuum vectors is finite dimensional. We note that the convergence of $R_\varkappa(\zeta)$ as $\varkappa \to \infty$ is uniform in λ for bounded intervals of λ. Since each $R_\varkappa(\zeta)$ is continuous in λ, for λ real, the limit $R(\zeta)$ is continuous in λ. Hence for λ sufficiently close to zero, the eigenvalue zero of $H(g)$ is simple, see Kato [1, p. 213]. It is an interesting problem whether the vacuum is unique for all values of λ, or for some bounded, but g independent, interval in λ at least after a suitable finite mass renormalization. Do the vacuum state(s) lie in the charge zero sector?

A dense domain for the operator $H(g)$ has been constructed explicitly in terms of a truncated power series expansions for the wave operator, Glimm(1,4), Hepp(1,2). We finish this section by outlining the construction in closed form of a dense domain for $H(g)^{1/2}$. Let

$$T_\varkappa = \exp(-\Gamma W_{\varkappa,\varrho}^C). \tag{8.2.2}$$

Thus $T_\varkappa$ is the truncation of the perturbation expansion for the wave operator for the Hamiltonian $H(g, \varkappa, \varrho)$.

For ϱ sufficiently large, we prove that $T_\varkappa$ maps $\mathscr{D}$ into vectors $\chi_\varkappa = T_\varkappa\theta$ in $\mathscr{D}(H(g, \varkappa))$, such that

$$\chi = \lim_{\varkappa \to \infty} \chi_\varkappa = \exp(-\Gamma W_\varrho^C)\,\theta \in \mathscr{D}(H(g))^{1/2}.$$

For the vectors $\chi_\varkappa = T_\varkappa\theta$, we prove

$$\langle \chi_\varkappa, H(g, \varkappa)\,\chi_\varkappa \rangle \to \langle \chi, H(g)\,\chi \rangle.$$

We first establish that $T_\varkappa$ is defined on $\mathscr{D}$ for all $\varkappa \leqq \infty$. For $\tau > \frac{1}{2}$ the kernel and $(\omega_1 + \omega_2)^{-\tau} w^C_{\varkappa,\varrho}$ are in L_2, and for $\varkappa < \varkappa_1$,

$$\|(\omega_1 + \omega_2)^{-\tau} w^C_{\varkappa,\varrho}\|_2 \leqq 0(\varrho^{-\delta}),$$

$$\|(\omega_1 + \omega_2)^{-\tau} (w^C_{\varkappa,\varrho} - w^C_{\varkappa_1,\varrho})\|_2 \leqq 0(\varkappa^{-\delta}\varrho^{-\delta}). \tag{8.2.3}$$

Hence by Proposition 1.2.3(b),

$$\|\Gamma W^C_{\varkappa,\varrho}(N + I)^{-1}\| \leqq M\varrho^{-\delta}, \quad \|\Gamma(W^C_{\varkappa,\varrho} - W^C_{\varkappa_1,\varrho})(N + I)^{-1}\| \leqq M\varkappa^{-\delta}\varrho^{-\delta}$$

for a constant M independent of $\varkappa$. For θ a vector with fewer than n particles, $\Gamma W^C_{\varkappa,\varrho}$ has fewer than $(n + 3)$ particles. Thus

$$\begin{aligned}\|(\Gamma W^C_{\varkappa,\varrho})^j \theta\| &= \|(\Gamma W^C_{\varkappa,\varrho}(N + I)^{-1} (N + I))^j \theta\| \\ &\leqq (M\varrho^{-\delta})^j \max\{(2n)^j, (6j)_j\},\end{aligned}$$

and for ϱ sufficiently large the exponential series (8.2.2) converges on $\mathscr{D}$, for $\varkappa \leqq \infty$. Furthermore, by (8.2.3), for $\theta \in \mathscr{D}$, and for ϱ sufficiently large,

$$\lim_{\varkappa\to\infty} T_\varkappa\theta = T\theta, \quad \lim_{\varkappa\to\infty} N^j T_\varkappa\theta = N^j T\theta, \tag{8.2.4}$$

where $T\theta = T_\infty\theta$, and j is any integer. Also, for $\tau < \frac{1}{2}$,

$$N_\tau T_\varkappa\theta \to N_\tau T\theta.$$

In order to study $(H_0 + W^C)$, we need to compute $[H_0, T_\varkappa]$, and this involves

$B_{\kappa,\rho} = [W^C, \Gamma W^C] =$ 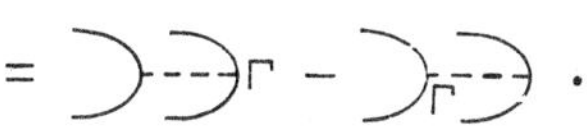

The kernel $b_{\varkappa,\varrho}$ of $B_{\varkappa,\varrho}$ is

$$\int dk\, w^C_{\varkappa,\varrho}(k, p_1, p_2)\, w^C_{\varkappa,\varrho}(-k, p_1', p_2')\, \{(\mu + \omega_1' + \omega_2')^{-1} - (\omega_1 + \omega_2)^{-1}\}. \tag{8.2.5}$$

The norm of $b_{\varkappa,\varrho}$ can be estimated using the relation

$$\begin{aligned}&|(\mu + \omega_1' + \omega_2')^{-1} - (\omega_1 + \omega_2)^{-1}| \\ &\qquad = (\mu + \omega_1' + \omega_2')^{-1} (\omega_1 + \omega_2)^{-1} |(\mu + \omega_1' + \omega_2' - \omega_1 - \omega_2)|.\end{aligned}$$

We conclude that for $\varkappa \leqq \varkappa_1$, and $0 < \tau < \frac{1}{2}$,

$$\|(\omega_1 + \omega_2 + \omega_1' + \omega_2')^{-\tau/2} (b_{\varkappa,\varrho} - b_{\varkappa_1,\varrho})\|_2 < 0(\varkappa^{-\delta}).$$

Thus $(N_\tau + I)^{-1} B_{\varkappa,\varrho}(N + I)^{-1}$ is bounded, and norm convergent as $\varkappa \to \infty$.

On the domain $\mathscr{D}$,

$$H_0 T_\varkappa = T_\varkappa H_0 - W^C_{\varkappa,\varrho} T_\varkappa - \tfrac{1}{2} B_{\varkappa,\varrho} T_\varkappa.$$

Thus for $\theta \in \mathscr{D}$,

$$(H_0 + W_\varkappa^C)\,\theta = T_\varkappa H_0\theta + (W_\varkappa^C - W_{\varkappa,\varrho}^C)\,T_\varkappa\theta - \tfrac{1}{2}B_{\varkappa,\varrho}T_\varkappa\theta,$$

and the first two terms on the right converge as $\varkappa \to \infty$, since

$$\|(W_\varkappa^C - W_{\varkappa,\varrho}^C)\,\theta\| \leqq \text{const. } \|w_\varkappa^C\theta_\varrho(\xi)\|_2\ \|(N + I)\,\theta\|$$
$$\leqq 0(\varkappa^{-\delta}).$$

Also

$$\langle T_\varkappa\theta, B_{\varkappa,\varrho}T_\varkappa\theta\rangle \to \langle T\theta, B_\varrho T\theta\rangle.$$

We prove that $W_\varkappa^A + c(g,\varkappa)$ has a strong graph limit on $T_\varkappa\mathscr{D}$. In fact

$$W_\varkappa^A T_\varkappa = T_\varkappa W_\varkappa^A + D_{\varkappa,\varrho}T_\varkappa,$$

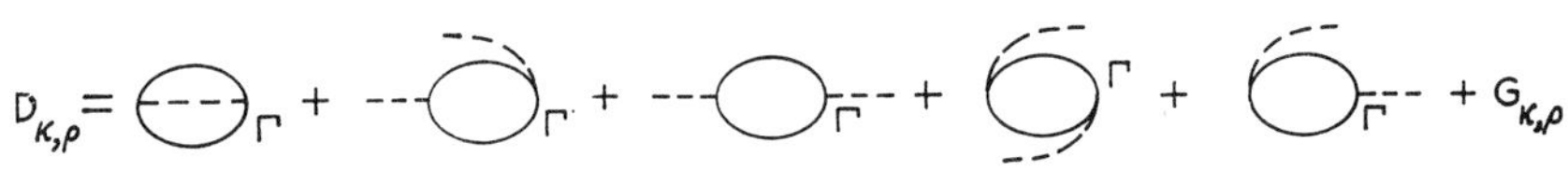

$$= F_{\varkappa,\varrho} + G_{\varkappa,\varrho}$$

where

$$\{F_{\varkappa,\varrho} + c(g,\varkappa)\}\,(N + I)^{-1}$$

is bounded and norm convergent as $\varkappa \to \infty$. The error $G_{\varkappa,\varrho}$ is a finite sum of diagrams $G_{\varkappa,\varrho,j}$, where $G_{\varkappa,\varrho,j}(N + I)^{-i}$ is bounded, for $i = i(j)$ sufficiently large, and norm convergent as $\varkappa \to \infty$. Hence for $\theta \in \mathscr{D}$,

$$W_\varkappa^A T_\varkappa\theta = T_\varkappa W_\varkappa^A(H_0 + I)^{-2}\,(H_0 + I)^2\,\theta$$
$$+ \{D_{\varkappa,\varrho} + c(g,\varkappa)\}\,(N + I)^{-i}\,(N + I)^i\,T_\varkappa\theta$$

and the two terms on the right converge as $\varkappa \to \infty$.

By similar methods, one can prove that for $\theta \in \mathscr{D}$,

$$\langle T_\varkappa\theta, W_\varkappa T_\varkappa\theta\rangle \to \langle T\theta, WT\theta\rangle.$$

Hence $\bigcup_\varrho T\mathscr{D}$ is a dense subset of $\mathscr{D}(H(g)^{1/2})$.

It is possible to use a better dressing transformation and obtain strong graph convergence of $H(g,\varkappa)$ on an explicit domain $T_\varkappa\mathscr{D}$. The transformation

$$T_\varkappa = \exp(-\Gamma V_{\varkappa,\varrho}), \tag{8.2.6}$$

where

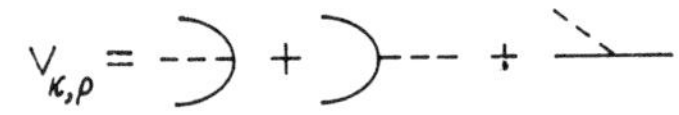

and

$$= \sum_{\varepsilon = \pm 1} \int b(p_1,\varepsilon)^*\, b(-p_2,\varepsilon)\, a(k)^*\, w(k,p_1,p_2) \frac{1}{\mu(k) + \omega_1}\, dk\, dp_1\, dp_2$$

seems to be correct, but the details have not been checked. A modification of (8.2.6) is used in Hepp[(2)].

9 The Heisenberg Picture

9.1 Finite propagation speed

We introduce the algebras $\mathfrak{A}(B)$ of time zero observables for the Yukawa$_2$ theory. We study the field algebras $\mathfrak{A}_f(B)$ generated by $\psi(f)$ and $\exp\{i(\phi(f) + \pi(g))\}$, $\operatorname{supp} f, g \subset B$, and also the algebras of observables $\mathfrak{A}(B)$, satisfying $[\mathfrak{A}(B_1), \mathfrak{A}(B_2)] = 0$ if $B_1 \cap B_2 = \emptyset$. Let $\mathfrak{A}$ be the C^* algebra obtained as the norm closure of $\bigcup_B \mathfrak{A}(B)$ as in Sec. 4.1.

We prove that the Hamiltonians $H(g)$ induce a g independent automorphism group σ of A, given locally by the equation

$$\sigma_t(A) = e^{itH(g)} A \, e^{-itH(g)} = A(t). \tag{9.1.1}$$

For the $P(\phi)_2$ theory of Part I, the properties of σ_t followed as a consequence of the essential self adjointness of $H(g)$ and the Trotter product formula. For the Yukawa$_2$ theory, this proof does not apply since the Hamiltonian $H(g)$ is defined only as a limit of the Hamiltonians $H(g, \varkappa)$ with a momentum cutoff $\varkappa$, and it has no mathematical meaning when written as the sum

$$H(g) = H_0 + H_I(g) + c(g). \tag{9.1.2}$$

On the other hand, the approximate Hamiltonians $H(g, \varkappa)$ have infinite propogation speed, so we are faced with proving that a limit of approximate nonlocal theories is local.

THEOREM 9.1.1 *Let A belong to $\mathfrak{A}(B)$ or $\mathfrak{A}_f(B)$, and let g equal one on B_t. Then (9.1.1) defines a g independent automorphism on $\mathfrak{A}$ or $\mathfrak{A}_f$, and the propagation speed of σ is at most one*:

$$\sigma_t : \mathfrak{A}(B) \to \mathfrak{A}(B_t),$$

$$\sigma_t : \mathfrak{A}_f(B) \to \mathfrak{A}_f(B_t).$$

We outline the main steps of the proof. See Glimm and Jaffe[9] for more details. We introduce a new approximate theory, which has the required finite propagation speed. We define a g independent transformation $A \to A_\Lambda(t)$, where $A \in \mathfrak{A}(B)$, $A_\Lambda(t) \in \mathfrak{A}(B_t)$. However, $A_\Lambda(t)$ is not given by the dynamics of $H(g)$, but rather by an approximate dynamics. We then prove that weak $\lim_{\Lambda \to \infty} A_\Lambda(t) = A(t)$, where $A(t)$ is defined in (9.1.1), provided g equals one on B_t. Since the algebras $\mathfrak{A}(B)$ are weakly closed, $A(t) \in \mathfrak{A}(B_t)$ and (9.1.1) does define a g independent automorphism of $\mathfrak{A}$.

In carrying out this procedure, we choose an approximation $A_\Lambda(t)$ which obviously possesses the desired property of finite propagation speed. Thus the difficult feature of the proof to establish the convergence of $A_\Lambda(t)$ to $A(t)$.

In order to minimize the loss of locality in $H(g, \varkappa)$, we choose a sharp cutoff in configuration space.

$$H_I(g, \varkappa) = \lambda \int \phi_\varkappa(x) :\psi_\varkappa(x)\, \psi_\varkappa(x):\, g(x)\, dx,$$

where $\phi_\varkappa(x) = \delta_\varkappa^* \phi$, $\psi_\varkappa(x) = \delta_\varkappa^* \phi$, and

$$\delta_\varkappa(x) = \varkappa\tilde{\chi}(\varkappa x) \in C_0^\infty, \quad \int \delta_\varkappa(x)\, dx = 1.$$

With this cutoff, $\phi_\varkappa(x)$ is localized (in the sense of fields) in an interval of diameter $0(\varkappa^{-1})$ centered at x. In other words, $\exp(i\phi_\varkappa(x)) \in \mathfrak{A}(B)$, where $B = (x - 0(\varkappa^{-1}), x + 0(\varkappa^{-1}))$. For definiteness, let us suppose that diam. supp. $\delta_\varkappa \leqq (2\varkappa)^{-1}$.

If we attempt to apply Proposition 4.1.1 to the Hamiltonian $H(g, \varkappa) = H_0 + H_I(g, \varkappa) + c(g, \varkappa)$, we wish to write $H(g, \varkappa) = H_1 + H_2$ where H_1 has propagation speed $s(\sigma^{(1)}) \leqq 1$ and H_2 has propagation speed $s(\sigma^{(2)}) = 0$. Since the ultraviolet cutoff enters the counterterms only in the coefficients $\delta m^2(\varkappa)$ and $E(g, \varkappa)$, Proposition 4.1.1 shows that $H_1 = H_0 + c(g, \varkappa)$ generates an automorphism with propagation speed $s(\sigma^{(1)}) \leqq 1$. However, the interaction term $H_2 = H_I(g, \varkappa)$ has an infinite propagation speed. If for each s it were possible to write

$$H_I(g, \varkappa) = A_1 + A_2$$

where

$$\exp(iA_1 t) \in \mathfrak{A}(B_s) \quad \text{and} \quad \exp(iA_2 t) \in \mathfrak{A}(\sim B_s),$$

then H_2 would generate an automorphism with propagation speed zero. The difficulty with writing such a decomposition for H_I arises from the contribution

$$\delta H_I(s) = \lambda \int_{I_s} \phi_\varkappa(x) :\psi_\varkappa(x)\, \psi_\varkappa(x):\, g(x)\, dx, \tag{9.1.2}$$

where I_s is the set of points within the distance $2\varkappa^{-1}$ from the boundary of B_s. Following these ideas, we replace $H_I(g, \varkappa)$ by the (time dependent) interaction Hamiltonian

$$H_I(s) = H_I(g, \varkappa) - \delta H_I(s)$$

and we replace $H(g, \varkappa)$ by the time dependent Hamiltonian $H(s) = H(g, \varkappa) - \delta H_I(s)$ as a step in defining the approximate but local dynamics $A \to A_\Lambda(t)$, see Figure 2.

We construct a unitary propagator $W_\varkappa(t)$ generated by a sequence of Hamiltonians $H(t_j)$. Each Hamiltonian $H(t_j)$ is used to propagate for a time interval of length $0(\varkappa^{-1})$ and the $H(t_j)$ dynamics carries $\mathfrak{A}(B_{t_j})$ into $\mathfrak{A}(B_{t_j+\tau})$ in an elapsed time τ, $|\tau| \leq 0(\varkappa^{-1})$. The propagator $W_\varkappa(t)$ implements a g independent map of $A \in \mathfrak{A}(B)$ into $\hat{A}(t) \in \mathfrak{A}(B_t)$,

$$\hat{A}(t) = W_\varkappa^*(t)\, A\, W_\varkappa(t). \tag{9.1.3}$$

We let n equal the integer part of $|\varkappa t|$, that is $n = [|\varkappa t|]$, and we define

$$W_\varkappa(t) = e^{-i\frac{t}{n}H(0)}\, e^{-i\frac{t}{n}H\left(\frac{t}{n}\right)} e^{-i\frac{t}{n}H\left(\frac{2t}{n}\right)} \ldots e^{-i\frac{t}{n}H\left(\frac{(n-1)}{n}t\right)}. \qquad (9.1.4)$$

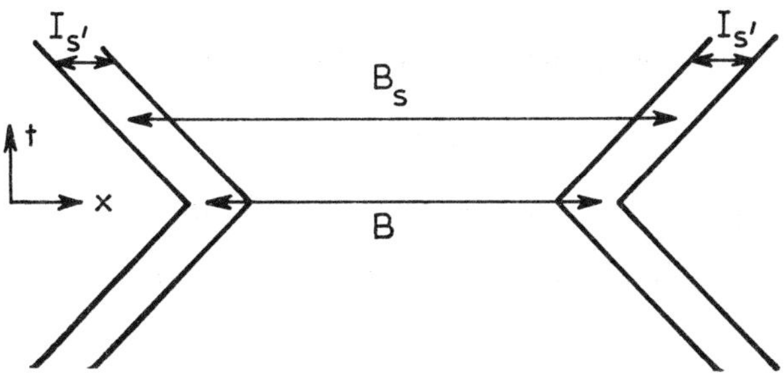

Figure 2. Localization of the time dependent Hamiltonian $H(s)$. $H(s)$ is localized outside I_s and $\delta H(s)$ is localized inside I_s.

There are $n = 0(\varkappa)$ factors in this product. We note that the propagator $W_\varkappa(t)$ is defined separately for each bounded open set B. The most general open subset of the real line is a countable disjoint union of open intervals. To simplify the proof, we take B to be the interval $(-1, 1)$. The same proof applies to any finite union of intervals, and the general case is covered by a limiting argument. In fact if B is a countable disjoint union of of open intervals and if supp $f \subset B$ then supp f is contained in a finite union of open intervals contained in B. Thus the proof we give shows that σ_t maps a set of generators for $\mathfrak{A}(B)$ into $\mathfrak{A}(B_t)$ and consequently $\sigma_t\mathfrak{A}(B) \subset \mathfrak{A}(B_t)$ for a general open set B.

LEMMA 9.1.2 *Let $A \in \mathfrak{A}(B)$ and let $\hat{A}(t)$ be defined by (9.1.3-4). Then $\hat{A}(t) \in \mathfrak{A}(B_t)$ and $\hat{A}(t)$ is independent of g.*

Proof As in Proposition 4.1.1, the Hamiltonian $H(s)$ induces a g independent mapping

$$A \to \exp(i\tau H(s))\, A \exp(-i\tau H(s))$$

of $\mathfrak{A}(B_s)$ into $\mathfrak{A}(B_{|s|+|\tau|})$, for $|\tau| \leqq tn^{-1}$. Composing n such maps, we obtain the lemma.

For technical reasons, to aid the proof of convergence as $\varkappa \to \infty$, we smooth out the approximation $\hat{A}(t)$ of (9.1.3-4). Let $0 \leqq x \leqq 1$. We scale the length of time propagation by x and define

$$W_\varkappa(t; x) = e^{-i\frac{tx}{n}H(0)}\, e^{-i\frac{tx}{n}H\left(\frac{t}{n}\right)}\, e^{-i\frac{tx}{n}H\left(\frac{2t}{n}\right)} \ldots e^{-i\frac{tx}{n}H\left(\frac{n-1}{n}t\right)}. \qquad (9.1.5)$$

For $A \in \mathfrak{A}(B)$, the operator

$$W_\varkappa^*(t; x)\, A\, W_\varkappa(t; x)$$

also belongs to $\mathfrak{A}(B_t)$ and is g independent. This is easily understood, since we use each Hamiltonian $H(j\,t/n)$, $j = 0, 1, \ldots, n-1$, for a time interval of length $|x\,t/n| \leqq |t/n|$. (Note that $x > 1$ can not be used.) We can average over $0 \leqq x \leqq 1$ to obtain our final approximation

$$A_\Lambda(t) = \int W_\varkappa^*(t; x)\, A\, W_\varkappa(t; x)\, f_\Lambda(x)\, dx, \tag{9.1.6}$$

where f_Λ is a positive continuous function such that

$$\int f_\Lambda(x)\, dx = 1, \quad \text{suppt}\, f_\Lambda(x) \subset [1 - \Lambda^{-1}, 1].$$

In (9.1.6) we will let $\varkappa = \varkappa(\Lambda)$ depend on Λ. Theorem 9.1.1 follows from

THEOREM 9.1.3 *For a suitable choice of the function* $\varkappa(\Lambda)$,

$$\operatorname*{weak\,lim}_{\Lambda\to\infty} A_\Lambda(t) = A(t).$$

The proof of Theorem 9.1.3 is divided into several main steps:

Step 1 We show that $W_\varkappa(t; x)$ is the boundary value of an analytic function $W_\varkappa(t; z)$, analytic in the half plane $\operatorname{Im} tz < 0$ and bounded uniformly in $\varkappa$ and z.

Step 2 We prove that for $\operatorname{Im} tz < 0$,

$$\operatorname*{weak\,lim}_{\varkappa\to\infty} W_\varkappa(t; z) = W(t; z)$$

exists and equals $e^{-itzH(g)}$.

Step 3 We use step 2 to prove weak convergence of $W_\varkappa(t; x)$ as measures (with respect to the variable x) to $e^{-itxH(g)} = U(tx)$.

Step 4 We prove weak convergence of

$$A_\Lambda(t) - A(t)$$

to zero.

The first two steps depend crucially on properties of $\delta H(s)$. In particular, we use first and second order estimates on $\delta H(s)$ to show that $\delta H(s)$ is small as $\varkappa \to \infty$. The first order estimate is

$$\pm\delta H(s) \leqq 0(\varkappa^{-\tau/2+\delta})\,(N_\tau + I). \tag{9.1.7}$$

The second order estimate is divergent. For any positive ε, there is a constant M such that for all large $\varkappa$

$$(\delta H(s))^2 \leqq \varepsilon H^2 + M \log \varkappa. \tag{9.1.8}$$

We also use the convergence of $R_\varkappa(\zeta) = (H(g, \varkappa) - \zeta)^{-1}$ to $(H(g) - \zeta)^{-1}$ proved in Sec. 8. After proving steps one and two, steps three and four follow by general methods.

We prove the first order estimate (9.1.7) by using an N_τ estimate. We have

$$H_I(s) = H_I(g, \varkappa) - H_I(\delta g, \varkappa) = H_I(g, \varkappa) - \delta H_I(s),$$

7*

where

$$\delta g = g(x)\,\theta_{2/\varkappa}(|x| - |s| - 1).$$

Hence

$$|(\delta g)^{\sim}| \leqq \|\delta g\|_1 \leqq 2\|\theta_{2/\varkappa}\|_1 = 0(\varkappa^{-1}).$$

Since $\delta H_I(s)$ is localized in a region of space of size $0(\varkappa^{-1})$, we lose momentum conservation as $\varkappa \to \infty$, but we have gained a factor $\varkappa^{-1}$ from the size $0(\varkappa^{-1})$ of the region I_s. By N_τ estimates of Proposition 1.2.3,

$$\begin{aligned}\|(N_\tau + I)^{-1/2}\,\delta H_I(s)\,(N_\tau + I)^{-1/2}\| &\leqq \text{const. } \|\omega_1^{-\tau/2}(|\delta w| + |\delta w^C|)\|_2 \\ &\leqq \text{const. } \left(\int \omega_1^{-\tau}\mu^{-1}\,0(\varkappa^{-2})\,|\chi_\varkappa|^2\,dk\,dp_1\,dp_2\right)^{1/2} \\ &= 0(\varkappa^{-\tau/2+\delta}),\end{aligned}$$

which proves (9.1.7).

To prove (9.1.8), we expand $\delta(H_I(s))^2$ as a sum of Wick monomials,

$$(\delta H_I(s))^2 = $$ 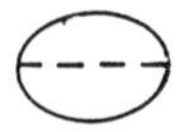 $$+ \text{ other terms.}$$

The first term above is a divergent constant,

$$0 \leqq$$ 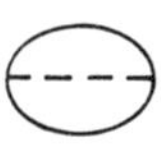 $$\leqq \int \mu^{-1}\varkappa^{-2}|\chi_\varkappa|^2\,dk\,dp_1\,dp_2 \leqq M\log\varkappa,$$

while the other terms are convergent and can be dominated by εH^2 + const., using the first and second order estimates of Sec. 7. (See Glimm-Jaffe[(9)] for details.) We remark that (9.1.8) can be avoided in the proof of self adjointness of $H(s)$, but it will be needed in Lemma 9.1.5 below.

LEMMA 9.1.4. *The Hamiltonian $H(s)$ is self adjoint, and bounded from below uniformly in $\varkappa$ and s.*

Proof By (9.1.8), for $\varepsilon < 1$, the perturbation $\delta H(s)$ is small in the sense of Kato with respect to $H(g, \varkappa)$. The self adjointness of $H(s)$ follows. The uniform lower bound on

$$H(s) = H(g, \varkappa) - \varepsilon N_\tau + \varepsilon N_\tau - \delta H(s)$$

follows from Theorem 7.1.1 and the first order estimate (9.1.7) on $\delta H(s)$.

Since $H(s)$ is bounded uniformly from below, we add a constant so that $I \leqq H(s)$. Similarly we let $I \leqq H(g, \varkappa)$. For convenience of notation, we let $R = H(g, \varkappa)^{-1}$, $H = H(g, \varkappa)$. With this normalization $\|W_\varkappa(t, z)\| \leqq 1$ for $\operatorname{Im} tz < 0$ and step 1 of the proof is completed.

As preparation for step 2 of Theorem 9.1.3, i.e. weak convergence of the propagators for $\operatorname{Im} tz < 0$, we prove three lemmas.

LEMMA 9.1.5 *Let $t \geqq 0$. Then*

$$\|H^{1/2}e^{-tH(s)}R^{1/2}\| \leqq (1 + Mt\log\varkappa).$$

Proof We prove that

$$F(t) = \|H^{1/2}\, e^{-tH(s)}\, R^{1/2}\theta\|^2$$

satisfies the inequality $\dfrac{dF}{dt} \leqq M \log \varkappa \|\theta\|^2$, which yields the lemma on integration. This bound on dF/dt follows from

$$\frac{dF}{dt} = -\langle e^{-tH(s)}R^{1/2}\theta,\ \{HH(s) + H(s)\,H\}\, e^{-tH(s)}\, R^{1/2}\theta\rangle$$

and

$$\begin{aligned} -(HH(s) + H(s)\,H) &= -2H^2 + H\,\delta H_I(s) + \delta H_I(s)\,H \\ &\leqq -H^2 + (\delta H_I(s))^2 \leqq M \log \varkappa. \end{aligned}$$

The final inequality uses the second order estimate (9.1.8).

LEMMA 9.1.6 *Let* $0 \leqq t \leqq T$ *and let* $\tau < 1$. *Then*

$$\|R^{1/2}\{e^{-tH} - e^{-tH(s)}\}\, R^{1/2}\| \leqq 0(t\varkappa^{-\tau/2}).$$

Proof *Let* $0 < \varepsilon, \tau + \varepsilon < 1$. Then

$$\begin{aligned} \|R^{1/2}\{e^{-tH} - e^{-tH(s)}\}\, R^{1/2}\| &\leqq \int_0^t \|e^{-(t-r)H}\, R^{1/2}\, \delta H_I(s)\, R^{1/2} H^{1/2}\, e^{-rH(s)}\, R^{1/2}\|\, dr \\ &\leqq \int_0^t 0(\varkappa^{-(\tau+\varepsilon)/2})\,(1 + Mr \log \varkappa)\, dr \\ &\leqq 0(\varkappa^{-\tau/2} t), \end{aligned}$$

by (9.1.7) and Lemma 9.1.5.

Let $U_\varkappa(t) = \exp(-itH(g, \varkappa))$, $U(t) = \exp(-itH(g))$.

LEMMA 9.1.7 *For* ty *positive but sufficiently small, there is a positive* δ *such that*

$$\|R^{1/2}\{W_\varkappa(t; -iy) - U_\varkappa(-iyt)\}\, R^{1/2}\| \leqq 0(\varkappa^{-\delta}).$$

Proof The difference $R^{1/2}\{W_\varkappa(t; -iy) - U_\varkappa(-iyt)\}\, R^{1/2}$ is a sum of $n = 0(\varkappa)$ terms of the form

$$\prod_{j=0}^{r-1} \left(R^{1/2}\, e^{-y\left(\frac{t}{n}\right)H(jt/n)}\, H^{1/2}\right) R^{1/2}\{e^{-ytH/n} - e^{-ytH(rt/n)/n}\}\, R^{1/2}\, U(-iyt/n)^{n-r-1}.$$

Using Lemmas 9.1.5 and 9.1.6, we obtain the bound

$$\begin{aligned} \|R^{1/2}\{W_\varkappa(t; -iy) - U_\varkappa(-iyt)\}\, R^{1/2}\| &\leqq 0(n)\left(1 + M\frac{ty}{n}\log\varkappa\right)^n 0(\varkappa^{-\tau/2})\left(\frac{ty}{n}\right) \\ &\leqq 0(\varkappa^{-\tau/2 + Mty}), \end{aligned}$$

which is $0(\varkappa^{-\delta})$ for y sufficiently small.

THEOREM 9.1.8 *Let* Im $tz < 0$. *Then*

$$\lim_{\varkappa \to \infty} \langle \theta, \{W_\varkappa(t; z) - U_\varkappa(tz)\} \theta \rangle = 0,$$

and convergence is uniform on compact sets of z.

Proof Since $W_\varkappa(t; z)$ and $U_\varkappa(tz)$ are bounded uniformly in $\varkappa$ convergence on a dense set is sufficient. For $\theta \in D(H(g))$, there is a sequence of vectors $\theta_\varkappa \in D(H(g, \varkappa))$ such that $\theta_\varkappa \to \theta$ and $H(g, \varkappa)\, \theta(\varkappa) \to H(g)\, \theta$. Let

$$F_\varkappa(z) = \langle \theta, \{W_\varkappa(t; z) - U_\varkappa(tz)\} \theta \rangle .$$

For $z = -iy$ and $0 < ty$, we have

$$\begin{aligned} |F_\varkappa(z)| &\leqq \text{const.}\ \|\theta - \theta_\varkappa\| + |\langle H_\varkappa \theta_\varkappa, R_\varkappa \{W_\varkappa(t; z) - U_\varkappa(tz)\} R_\varkappa H_\varkappa \theta_\varkappa \rangle| \\ &\leqq \text{const.}\ \|\theta - \theta_\varkappa\| + 0(\varkappa^{-\tau/2 + Mty}) \\ &\to 0 . \end{aligned}$$

For fixed θ, the functions $F_\varkappa(z)$ are bounded uniformly in the half plane Im $tz < 0$, and are analytic in z. Hence convergence for small $y = iz$ implies convergence for all z in the half plane, by Vitali's convergence theorem for normal families of analytic functions. This completes the second step of the proof of Theorem 9.1.3.

COROLLARY 9.1.9 *For some* $\varkappa(\Lambda)$ *such that* $\varkappa(\Lambda) \to \infty$,

$$\operatorname*{weak\,lim}_{\Lambda \to \infty} \int \{W_{\varkappa(\Lambda)}(t; x) - U_{\varkappa(\Lambda)}(tx)\} f_\Lambda(x)\, dx = 0 .$$

Proof The corollary is a consequence of the convergence of the boundary values of $F_\varkappa(z)$ as measures, as $\varkappa \to \infty$, namely for fixed $f \in C([0, 1])$ and fixed θ,

$$\int F_\varkappa(x) f(x)\, dx \to 0 \quad \text{as} \quad \varkappa \to \infty . \tag{9.1.10}$$

It follows that for some choice $\varkappa = \varkappa(\Lambda)$, $\varkappa(\Lambda) \to \infty$ as $\Lambda \to \infty$ and

$$\int F_{\varkappa(\Lambda)}(x) f_\Lambda(x)\, dx \to 0 . \tag{9.1.11}$$

By the diagonal process, we can pick one function $\varkappa(\Lambda)$ yielding convergence for a countable dense set of vectors θ_i. Since $|F_\varkappa(x)| \leqq 2\|\theta\|^2$, this proves that (9.1.11) is valid for all θ, completing the proof of the corollary.

To establish (9.1.10), we approximate f by a function h analytic in a complex neighborhood of [0, 1], and such that

$$\int_0^1 |f(x) - h(x)|\, dx \equiv \|f - h\|_1 \leqq \varepsilon .$$

Thus by the Cauchy integral theorem,

$$\left| \int F_\varkappa(x) f(x)\, dx \right| \leqq \|f - h\|_1 + \left| \int_C F_\varkappa(z)\, h(z)\, dz \right|,$$

where C is a path from 0 to 1, lying in the half space $\operatorname{Im} tz > 0$ and in the domain of analyticity of h. By Theorem 9.1.8, $F_\varkappa(z)$ converges to zero, pointwise on C. Thus (9.1.10) is valid and the proof is complete.

We now take up the fourth step, completing the proof of Theorem 9.1.3.

Proof of Theorem 9.1.3 It is sufficient to prove that $A_\Lambda(t) - U^*_{\varkappa(\Lambda)}(t) \times A U_{\varkappa(\Lambda)}(t)$ converges weakly to zero, and since $\{U_{\varkappa(\Lambda)}(tx) - U_{\varkappa(\Lambda)}(t)\}$ converges strongly to zero for $x \in \operatorname{supp} f_\Lambda$, it is sufficient to prove

$$\begin{aligned} 0 &= \operatorname*{weak\,lim}_{\Lambda\to\infty} \int dx\, f_\Lambda(x)\, \{W^*_{\varkappa(\Lambda)}(t; x)\, A\, W_{\varkappa(\Lambda)}(t; x) - U^*_{\varkappa(\Lambda)}(tx)\, A\, U_{\varkappa(\Lambda)}(tx)\} \\ &= \operatorname*{weak\,lim}_{\Lambda\to\infty} \{A_\Lambda(t) - \bar{A}_\Lambda(t)\} \end{aligned}$$

for any $A \in \mathfrak{A}(B)$. The equation above defines $\bar{A}_\Lambda(t)$. By the Schwarz inequality,

$$\begin{aligned} \langle \Omega, \{A_\Lambda(t) - \bar{A}_\Lambda(t)\}\, \Omega\rangle| &\leqq 2\|A\|\, \|\Omega\| \int dx\, f_\Lambda(x)\, \|\{W_\varkappa(t; x) - U_\varkappa(tx)\}\Omega\| \\ &\leqq 2\|A\|\, \|\Omega\| \left(\int dx\, f_\Lambda(x)\, \|\{W_\varkappa(t; x) - U_\varkappa(tx)\}\,\Omega\|^2\right)^{1/2}. \end{aligned}$$

Since

$$\|(W - U)\,\Omega\|^2 = 2 \operatorname{Re} \langle (U - W)\,\Omega, U\Omega\rangle,$$

the theorem follows from Corollary 9.1.9.

9.2 The field operators

We use the preceeding results to construct local quantum fields $\phi(x, t)$ and $\psi(x, t)$ that are independent of the spatial cutoff function g. These fields obey equations of motion defined as limits (as $\varkappa \to \infty$) of cutoff equations. When averaged with C_0^∞ functions, the field bilinear forms uniquely determine operators $\phi(f)$ and $\psi(f)$.

We define the sharp time fields

$$\phi(f, t) = e^{itH(g)}\, \phi(f, 0)\, e^{-itH(g)},$$

$$\psi(f, t) = e^{itH(g)}\, \psi(f, 0)\, e^{-itH(g)}.$$

THEOREM 9.2.1 (*Glimm and Jaffe*[9]). *The sharp time fields $\phi(f, t)$, $\psi(f, t)$ are g independent if g equals 1 on a sufficiently large set. The domain of $H(g)$ is a core for $\phi(f, t)$. The fields obey canonical equal time commutation and anticommutation relations.*

THEOREM 9.2.2 *The bilinear form $\phi(f, t)$ is a C^∞ function of t on $C^\infty(H(g)) \times C^\infty(H(g))$. The weak limits of*

$$\left(\frac{d}{dt}\right)^n R_\varkappa(\zeta)^{n+1} e^{itH(g,\varkappa)} \phi(f, 0) e^{-itH(g,\varkappa)} R_\varkappa(\zeta)^{n+1}$$

exist as $\varkappa \to \infty$. The corresponding strong limits exist for the fermion fields.

The field equations can be defined as the limits, as $\varkappa \to \infty$, of the space averaged, cutoff equations. In fact the cutoff equations

$$\left(\frac{\partial^2}{\partial t^2} - \frac{\partial^2}{\partial x^2} + m^2\right)\phi_{g,\varkappa}(x, t) + j_{g,\varkappa}(x, t) = 0,$$

$$\left(\gamma_0 \frac{\partial}{\partial t} + \gamma_1 \frac{\partial}{\partial x} + M\right)\psi_{g,\varkappa}(x, t) + J_{g,\varkappa}(x, t) = 0$$

after averaging with a C_0^∞ function have weak limits as $\varkappa \to \infty$ on $C^\infty(H(g)) \times C^\infty(H(g))$. See Glimm-Jaffe[9]. See Dimock[1] for further properties of the fields and currents.

We also define the space time averaged fields $\phi(f)$, where $f = \bar{f} \in C_0^\infty(R^2)$.

$$\phi(f) = \int \phi(x, t) f(x, t)\, dx\, dt.$$

Following the proof for the $P(\phi)_2$ theory (Glimm and Jaffe [4, Sec. 3.2]) we conclude that $\phi(f)$ is essentially self adjoint on $\mathscr{D}(H(g))$ provided g equals 1 on a sufficiently large set.

THEOREM 9.2.3 *Schrader*[1]. *Let f be a real C_0^∞ function. The self adjoint operator $\phi(f)$ is independent of g.*

Proof The idea of the proof is to approximate $\phi(f)$ by bounded self adjoint operators C_n that are independent of g. For this approximation, and for $\theta \in \mathscr{D}(H(g))$,

$$\lim_{n\to\infty} C_n\theta = \phi(f)\,\theta.$$

Since $\mathscr{D}(H(g))$ is a core for $\phi(f)$, the resolvents $R_n(\zeta) = (C_n - \zeta)^{-1}$ converge strongly to $(\phi(f) - \zeta)^{-1}$. Since $R_n(\zeta)$ is independent of g, so is the limit.

To choose the approximation C_n, let

$$A(t) = \left(\int \phi(x, 0) f(x, t)\, dx\right)^- = \int \lambda\, dE(\lambda, t).$$

By Theorem 9.2.1,

$$e^{itH(g)} A(t)\, e^{-itH(g)}$$

is independent of g, and hence so is the spectral measure

$$E^t(\lambda, t) = e^{itH(g)} E(\lambda, t)\, e^{-itH(g)}.$$

We define the bounded g-independent operator C_n by the equation

$$C_n = \int dt \int_{|\lambda| \leqq n} \lambda \, dE^t(\lambda, t),$$

and for $\theta \in \mathscr{D}(H(g))$

$$\{C_n - \phi(f)\} \theta = - \int dt \int_{|\lambda| \geqq n} \lambda \, dE^t(\lambda, t) \theta .$$

This yields the estimate

$$\begin{aligned} \| \{C_n - \phi(f)\} \theta \| \leqq \int dt \Big\| \int_{|\lambda| \geqq n} \lambda \, dE(\lambda, t) \, (A(t)^2 + I)^{-1}, \\ \times \big(A(t)^2 + I\big)\big(H(g) + I\big)^{-1} \big(H(g) + I\big) e^{-itH(g)} \theta \Big\| \\ \leqq \text{const.} \, \|(H(g) + I) \theta\| \int dt \Big\| \int_{|\lambda| \geqq n} \lambda(\lambda^2 + 1)^{-1} \, dE(\lambda, t) \Big\| \end{aligned}$$

where we use the second order estimate, Theorem 7.2.1, and the N_τ estimate,

$$\|\big(A(t)^2 + I\big) (N + I)^{-1}\| \leqq \text{const.}$$

Hence $\| \{C_n - \phi(f)\} \theta \| \to 0$, and the theorem is proved.

Bibliography

H. Araki, [1] Von Neumann algebras of local observables for the free scalar field, *J. Math. Phys.*, **5**, 1–13 (1964).

[2] The type of von Neumann algebra associated with the free field, *Prog. Theor. Phys.*, **32**, 956–965 (1964).

[3] On the algebra of all local observables, *Prog. Theor. Phys.* **32**, 844–854 (1964).

V. Bargmann, [1] On a Hilbert space of analytic functions and an associated integral transform. Part I, *Comm. Pure Appl. Math.*, **14**, 187–214 (1961).

J. Bjorken and S. Drell, [1] *Relativistic quantum fields*, McGraw Hill, New York (1965).

N. Bogoliubov and D. Shirkov, [1] *Introduction to the theory of quantized fields*, Interscience, New York (1959).

J. Cannon, [1] Quantum field theoretic properties of a model of Nelson: Domain and eigenvector stability for perturbed linear operators, *J. Funct. Analy.*, to appear.

[2] Convergence criteria for a sequence of semigroups, to appear.

J. Cannon and A. Jaffe, [1] Lorentz covariance of the $(\phi^4)_2$ quantum field theory, *Commun. Math. Phys.*, **17**, 261–321 (1970).

J. Dixmier, [1] *Les algèbres d'opérateurs dans l'espace Hilbertien (algèbres de von Neumann)*, Gauthier-Villars, Paris (1957).

[2] *Les C*-algèbras et leurs représentations*, Gauthier-Villars, Paris (1964).

J. Dimock, [1] Estimates, renormalized currents and field equations for the Yukawa$_2$ field theory, Harvard University thesis (1971).

N. Dunford and J. T. Schwartz, [1] *Linear operators, Part I: general theory*, Interscience, New York (1958).

J.-P. Eckmann, [1] Estimates on graphs (unpublished). J.–P. Eckmann and K. Osterwalder, [1] On the uniqueness of the Hamiltonian and of the representation of the CCR for the quartic boson interaction in three dimensions, *Helo. Phys. Acta*, to appear.

H. Epstein, V. Glaser, and A. Jaffe, [1] Nonpositivity of the energy density in quantized field theories, *Nuovo Cimento*, **36**, 1016–1022 (1965).

J. Fabrey, [1] Exponentical representations of the canonical commutation relations *Commun. Math. Phys.*, **19**, 1–30 (1970).

P. Federbush, [1] A partially alternate derivation of a result of Nelson, *J. Math. Phys.*, **10**, 50–52 (1969).

[2] Renormalization of some one-space dimensional quantum field theories by unitary transformations, preprint.

[3] Unitary renormalization of $(\phi^4)_{2+1}$, to appear.

[4] A convergent expansion for the resolvent of $:\phi^4:_{1+1}$, to appear.

P. Federbush and B. Gidas, [1] Renormalization of the one-space dimensional Yukawa model by unitary transformations, preprint.

J. Fell, [1] The dual spaces of C^*-algebras, *Trans. Amer. Math. Soc.*, **94**, 365–403 (1960).

K. Friedrichs, [1] *Perturbation of spectra in Hilbert space*, American Mathematical Society, Providence (1965).

J. Glimm, [1] Yukawa coupling of quantum fields in two dimensions, I. *Commun. Math. Phys.*, **5**, 343–386 (1967).

[2] Yukawa coupling of quantum fields in two dimensions, II. *Commun. Math. Phys.*, **6**, 120–127 (1967).

[3] Boson fields with nonlinear self interaction in two dimensions, *Commun. Math. Phys.*, **8**, 12–25 (1968).

[4] Boson fields with the $:\phi^4:$ interaction in three dimensions, *Commun. Math. Phys.*, **10**, 1–47 (1968).

[5] The foundations of quantum field theory, *Advances in Math.*, **3**, 101–125 (1969).

[6] Models for quantum field theory, in *Rendiconti di Fisica Internazionale de Fisica "E. Fermi" XLV Corso*, Academic Press, New York (1969).

J. Glimm and A. Jaffe, [1] A Yukawa interaction in infinite volume, *Commun. Math. Phys.*, **11**, 9–18 (1968).

[2] A $\lambda\phi^4$ quantum field theory without cutoffs. I., *Phys. Rev.*, **176**, 1945–1951 (1968).

[3] Singular perturbations of self adjoint operators, *Comm. Pure Appl. Math.*, **22**, 401–414 (1969).

[4] The $\lambda(\phi^4)_2$ quantum field theory without cutoffs. II. The field operators and the approximate vacuum, *Ann. Math.*, **91**, 362–401 (1970).

[5] The $\lambda(\phi^4)_2$ quantum field theory without cutoffs. III. The physical vacuum, *Acta. Math.*, **125**, 203–261 (1970).

[6] Infinite renormalization of the Hamiltonian is necessary, *Jour. Math. Phys.*, **10**, 2213–2214 (1969).

[7] Rigorous quantum field theory models, *Bull. A.M.S.*, **76**, 407–410 (1970).

[8] Self-adjointness of the Yukawa_2 Hamiltonian, *Ann. of Phys.*, **60**, 321–383 (1970).

[9] The Yukawa_2 quantum field theory without cutoffs, *Jour. Funct. Analysis*, **7**, 323–357 (1971).

[10] The energy-momentum spectrum and vacuum expectation values in quantum field theory, *Jour. Math. Phys.*, **11**, 3335–3338 (1970).

[11] The energy momentum spectrum and vacuum expectation values in quantum field theory. II, *Commun. Math. Phys.* **22**, 1–22 (1971).

[12] Positivity and self adjointness of the $P(\phi)_2$ Hamiltonian, *Commun. Math. Phys.*, to appear.

E. Griffin, [1] Some contributions to the theory of rings of operators II, *Trans. Amer. Math. Soc.*, **79**, 389-400 (1955).

L. Gross, [1] Existence and uniqueness of physical ground states, to appear.

M. Guenin, [1] On the interaction picture, *Comm. Math. Phys.*, **3**, 120–132 (1966).

R. Haag and D Kastler, [1] An algebraic approach to quantum field theory, *Jour. Math. Phys.*, **5**, 848–861 (1964).

D. Hall and A. Wightman, [1] A theorem on invariant functions with applications to relativistic quantum field theory, *Mat.-Fys. Medd. Kong. Danske Videns. Skelskab*, **31**, No. 5 (1957).

K. Hepp, [1] Renormalized Hamiltonians for a class of quantum fields with infinite mass and charge renormalizations. In *Anniversary volume for N. N. Bogoliubov*, Nauka, Moscow (1969).

[2] *Théorie de la renormalisation*, Springer-Verlag, Heidelberg (1969).

R. Høegh-Krohn, [1] A general class of quantum fields without cut-offs in two space-time dimensions, *Commun. Math. Phys.*, **21**, 244–255 (1971).

[2] On the spectrum of the space cutoff $:P(\phi):$ Hamiltonian in two space-time dimensions, *Commun. Math. Phys.*, **21**, 256–260 (1971).

R. Høegh-Krohn and B. Simon, [1] Hypercontractive semi-groups and two-dimension self-coupled bose fields, *I. Funct. Analysis*, to appear.

A. Jaffe, [1] Divergence of perturbation theory for bosons, *Commun. Math. Phys.*, **1**, 127–149 (1965).

[2] Existence theorems for a cut-off $\lambda\phi^4$ field theory, in *Mathematical theory of elementary particles*, ed. R. Goodman and I. Segal, M.I.T. Press, Cambridge (1966).

[3] *The dynamics of a cut-off $\lambda\phi^4$ field theory*, Princeton University thesis (1965).

[4] *Constructive quantum field theory*, E.T.H. Lecture Notes, Zürich (1968).

[5] Constructing the $\lambda(\phi^4)_2$ theory, in *Rendiconti di Fisica Internazionale de Fisica "E. Fermi"—XLV Corso*, Academic Press, New York, 120–151 (1969).

[6] Whither axiomatic field theory?, *Reviews of Modern Physics*, **41**, 576–580 (1969).

A. Jaffe, O. Lanford, and A. Wightman, [1] A general class of cut-off model field theories, *Commun. Math. Phys.*, **15**, 47–68 (1969).

A. Jaffe and R. Powers, [1] Infinite volume limit of a $\lambda\phi^4$ field theory, *Commun. Math. Phys.*, **7**, 218–222 (1968).

R. Jost, [1] *General theory of quantum fields*, American Mathematical Society, Providence (1965).

M. Kac, [1] *Probability and related topics in physical sciences*, Interscience Publications, New York (1959).

T. Kato, [1] *Perturbation theory for linear operators*, Springer-Verlag, New York (1966).

Y. Kato and N. Mugibayasi, [1] Regular perturbation and asymptotic limits of operators in quantum field theory, *Prog. Theor. Phys.*, **30**, 103–133 (1963).

[2] Asymptotic fields in the $\lambda(\phi^4)_2$ quantum field theory, *Prog. Theor. Phys.* **45**, 628-639 (1971).

A. Kleinstein, [1] The Hamiltonian in a class of models for a quantum field theory, Columbia University Thesis (1970).

J. Konrady, [1] Almost positive perturbations of positive self adjoint operators, to appear,

R. Kunze and I. Segal, [1] *Integrals and operators*, McGraw-Hill, New York (1968).

O. Lanford, [1] Construction of quantum fields interacting by a cut-off Yukawa coupling, Princeton University Thesis (1966).

D. Masson, [1] Essential self adjointness of semi bounded operators: An extension of the Kato-Rellich theorem, to appear.

D. Masson and W. Mc Clary, [1] On the essential self adjointness of the $(g(x)\phi^4)_2$ Hamiltonian, *Commun. Math. Phys.*, **21**, 71-74 (1971).

M. Naimark, [1] *Normed rings*, P. Noordhoff, Groningen (1964).

E. Nelson, [1] Analytic vectors, *Ann. of Math.*, **70**, 572–615 (1959).

[2] Interaction of nonrelativistic particles with a quantized scalar field, *J. Math. Phys.*, **5**, 1190–1197 (1964).

[3] A quartic interaction in two dimensions, in *Mathematical theory of elementary particles*, ed. R. Goodman and I. Segal, M.I.T. Press, Cambridge (1966).

[4] *Topics in dynamics: I flows*, Princeton University Press, Princeton (1970).

K. Osterwalder, [1] Cubic boson theories in two, three and four dimensions, Fortschr. d. Phys., to appear.

L. Rosen, [1] A $\lambda\phi^{2n}$ field theory without cutoffs, *Commun. Math. Phys.*, **16**, 157–183 (1970).
[2] The $(\phi^{2n})_2$ quantum field theory: higher order estimates, Comm. Pure appl. Math., to appear.
[3] The $(\phi^{2n})_2$ quantum field theory: Lorentz covariance, preprint.

L. Rosen and B. Simon, [1] The $(\phi^{2n})_2$ field Hamiltonian for complex coupling constant, to appear.

L. Schwartz, [1] Théorie des distributions, Hermann, Paris (1966).

S. Schweber, [1] *An introduction to relativistic quantum field theory*, Row, Peterson and Co., New York (1961).

R. Schrader, [1] A remark on Yukawa plus boson selfinteraction in two space dimensions, *Commun. Math. Phys.* **21**, 164–170 (1971).
[2] Yukawa quantum field theory in two space time dimensions without cutoffs, to appear.

I. Segal, [1] Notes toward the construction of nonlinear relativistic quantum fields I; The Hamiltonian in two space-time dimensions as the generator of a C^*-automorphism group, *P.N.A.S.*, **57**, 1178–1183 (1967).
[2] Notes toward the construction of nonlinear relativistic quantum fields III: Properties of the C^*-dynamics for a certain class of interactions, *Bull. A.M.S.*, **75**, 1390–1395 (1969).
[3] Construction of nonlinear local quantum processes: I, *Ann. Math.* **92**, 462–481 (1970).

B. Simon, [1] Borel summability of the ground state energy in spatially cutoff $(\phi^4)_2$, *Phys. Rev. Letters*, **25**, 1583–1586 (1970).

E. Stein, [1] Interpolation of linear operators, *Trans. A.M.S.*, **83**, 482–492 (1956).
[2] *Paley-Littlewood theory and related topics,Ann. Math. Studies.* (1970).

R. Streater and A. Wightman, *PCT, spin* and *statistics, and all that*, W. A. Benjamin, New York (1964).

S. Weinberg, [1] High energy behavior in quantum field theory, *Phys. Rev.*, **118**, 838–849 (1960).

G.-C. Wick, [1] The evaluation of the collision matrix, *Phys. Rev.*, **80**, 268–272 (1950).

A. S. Wightman, [1] Quantum field theory in terms of vacuum expectation values, *Phys. Rev.*, **101**, 860–866 (1956).
[2] Introduction to some aspects of the relativistic dynamics of quantized fields, in *1964 Cargèse Summer School Lectures*, Ed. by M. Lévy, Gordon and Breach, New York (1967), pp. 171–291.

A. S. Wightman and L. Gårding, [1] Fields as operator valued distributions in relativistic quantum theory, *Arch. för Physik*, **28**, 129–184 (1964).

III Boson Quantum Field Models: Part I. General Results

Boson Quantum Field Models

J. GLIMM*

Courant Institute of Mathematical Sciences, New York University, New York, USA

A. JAFFE†

Lyman Laboratory of Physics, Harvard University, Cambridge, Massachussetts, USA

PART I. GENERAL RESULTS

1. INTRODUCTION

Quantum fields, from a mathematical point of view, are highly singular. These fields are believed to describe the interactions of elementary particles. For the interaction of electrons with light (photons), the quantum field description is exact within the limits of experimental accuracy (5 significant figures). For these reasons, i.e. the mathematical difficulties and the importance to physics, the problem of formulating the mathematical foundations of quantum field theory has attracted the attention of both mathematicians and physicists over a period of several decades. On the side of the physicists, the most striking achievements were the calculation in the late 1940's and early 1950's of the Lamb shift and the anomalous magnetic moment of the electron together with the development of the renormalization method on which these calculations were based. Of the mathematicians, J. von Neumann was the first to realize that new mathematical theories would be required to formulate quantum field theory correctly and this realization was one of the motives for developing the theory of operator algebras.

* Supported in part by the National Science Foundation, NSF-GP-24003.

† Supported in part by the Air Force Office of Scientific Research, Contract F44620-70-C-0030, and by the National Science Foundation, Grant GP-31239X.

The calculations referred to above are based on a perturbation method, and they depend essentially on the fact that a numerical coefficient, multiplying the interaction term in the field equation, is small ($\sim 137^{-\frac{1}{2}}$). For strong interactions (protons, neutrons, mesons, etc.), the numerical coefficient is about 15, and perturbation calculations have had only limited success. To obtain a deeper understanding of these problems, Wightman, Haag and Kastler and others have formulated precisely the basic principles (axioms) of quantum field theory, and have sought to derive rigorously consequences of these axioms. High points of this program include the PCT theorem, the spin and statistics theorem and the Haag–Ruelle and the Lehmann, Symanzik and Zimmermann scattering theories.

However the axiomatic program leaves open the question of whether quantum fields with conventional interactions satisfy the axioms, or indeed whether nontrivial fields can be shown to exist in any reasonable sense. In the past five years there has been considerable progress on the existence problem. We have worked in two- and three-dimensional space–time. This is a simplifying assumption, which reduces but does not eliminate the basic singularities of quantum field theory. The results for two dimensions are summarized as follows.

THEOREM *In two space time dimensions, quantum fields exist. These fields are known to satisfy all or most of the Haag–Kastler axioms and most of the Wightman axioms.*

The fields in this theorem have conventional Yukawa or polynomial Boson interactions. In three space time dimensions, the situation looks promising, although the results to date are of a preliminary and technical nature, and apply only to the ϕ^4 boson interaction. The methods, in their present form, do not seem to apply to four dimensions.

We will consider here only the polynomial boson interactions, because they are simpler than the Yukawa interaction. As physics background, we want to assume only a familiarity with ordinary quantum mechanics. In ordinary quantum mechanics, the time evolution

$$\psi(q, t) = (e^{-itH}\psi)(q, 0) \tag{1.1}$$

is governed by the Schrödinger equation

$$i\frac{\partial}{\partial t}\psi = H\psi, \tag{1.2}$$

where H is the Hamiltonian or energy operator. For example

$$H = -\frac{1}{2}\sum_{l=1}^{n} m_l^{-1}\frac{\partial^2}{\partial q_l^2} + V(q) \tag{1.3}$$

acts on $L_2(\mathbf{R}^n)$, and $q \in \mathbf{R}^n$. The quantum field differs from the above in having an infinite number of degrees of freedom, i.e. $n = \infty$. The singularities of quantum field theory can all be traced to the fact that $n = \infty$.

For $n = \infty$, we have no satisfactory analogue of Lebesgue measure. It turns out, however, that Gaussian measures, which do generalize to the infinite dimensional case, are more suitable to our purposes. We use a Gaussian measure d_Bq defined by a quadratic form B on the infinite dimensional space $Q = \mathscr{S}'(\mathbf{R})$ of configurations of the classical field. Formally d_Bq has the density $\exp(-B(q,q))\,dq$. Then $L_2(Q, d_Bq)$ replaces $L_2(\mathbf{R}^n)$ above. $\mathscr{S}'$ denotes the space of tempered distributions.

The Hamiltonian H is the second most important operator in quantum field theory. To study H, we will use both the theory of a single self-adjoint operator and the theory of operator algebras. H is self-adjoint, and the use of self-adjoint operator techniques is hardly surprising. In the main, we are concerned with resolvents $R(\zeta) = (H - \zeta)^{-1}$ and semigroups e^{-tH}. We use criteria for the self adjointness of a limit $H = \lim_\kappa H(\kappa)$ of self-adjoint operators, expressed in terms of the convergence of the associated resolvents and semigroups.

The use of operator algebra techniques to study a self-adjoint operator is less standard, and results from the fact that $n = \infty$. Gaussian measures d_Bq on an infinite dimensional space are very sensitive to changes in the quadratic form B. If the Jacobian $\det B_1^{\frac{1}{4}}B_2^{-\frac{1}{4}}$ does not exist (e.g. if $\det B_1^{\frac{1}{4}}B_2^{-\frac{1}{4}}$ equals zero or infinity) then the measures $d_{B_1}q$ and $d_{B_2}q$ are mutually singular. Moreover the field operators act in a natural fashion on $L_2(Q, d_{B_i}q)$, $i = 1, 2$, and these two representations of the field operators are then unitarily inequivalent. Thus in many natural approximation or limiting procedures we find operators acting on one Hilbert space, in one representation, converging to operators acting on another Hilbert space, in a unitarily inequivalent representation. Plainly we will have to think of operators independently of the Hilbert spaces on which they act, and this is exactly what operator algebras accomplish. Non-Gaussian measures seem to be equally sensitive to changes in the parameters of the problem. Typically unitary inequivalence of representations is characterized by the divergence of some integral or infinite series.

In addition to the general theories of Gaussian measures, self-adjoint operators and operator algebras, we need some information which is special to the interactions and problems we are considering. For polynomial Boson interactions in one space dimension, we have approximate Hamiltonians

$$H(g) = H_0 + H_I(g) \tag{1.4}$$

where

$$0 \leqslant g \in L_1(\mathbf{R}) \cap L_2(\mathbf{R}), \tag{1.5}$$

and the approximation is removed by the limit $g \to 1$. We show that $H(g)$ is essentially self-adjoint on

$$\mathscr{D}(H_0) \cap \mathscr{D}(H_I(g)). \tag{1.6}$$

$H_I(g)$ is a multiplication operator on $L_2(Q, d_B q)$, and as such, $H_I(g) \in L_p(Q, d_B q)$ for all $p < \infty$. H_0 acts on $L_2(Q, d_B q)$ as a Hermite operator, and it is conveniently studied in terms of a Hermite function expansion for $L_2(Q, d_B q)$. This Hermite expansion is called Fock space.

We require that the polynomial P in the interaction be positive. Let $p = \deg P$. Then $H_I(g)$ is "nearly lower bounded" in the sense that for any positive number κ, we can write

$$H_I(g) = H_I(g, \kappa) + H_I(g, \kappa)' \tag{1.7}$$

where

$$-O(\ln \kappa)^{p/2} \leqslant H_I(g, \kappa) \tag{1.8}$$

and $H_I(g, \kappa)'$ is a "small" unbounded operator, with magnitude

$$|H_I(g, \kappa)'| \sim O(\kappa^{-\frac{1}{4}+\varepsilon}). \tag{1.9}$$

In general, estimates involving positivity, such as (1.8) are performed in the Schrödinger representation, i.e. on $L_2(Q, d_B q)$, while estimates on small but indefinite remainders, such as (1.9), are performed on Fock space.

2. HERMITE OPERATORS

Before passing to the limit $n = \infty$, we study the Hermite operator with a finite number of degrees of freedom. We begin with one degree of freedom, and then the Hermite operator

$$H_0(\mu) = \frac{1}{2}\left[-\left(\frac{d}{dq}\right)^2 + \mu^2 q^2 - \mu \right], \tag{2.1}$$

acting on $L_2(R)$, is the Hamiltonian for the quantized harmonic oscillator. Here μ is an arbitrary positive normalization factor.

We introduce the annihilation and creation operators

$$b = 2^{-\frac{1}{2}}(\mu^{\frac{1}{2}}q + i\mu^{-\frac{1}{2}}p)$$
$$b^* = 2^{-\frac{1}{2}}(\mu^{\frac{1}{2}}q - i\mu^{-\frac{1}{2}}p). \tag{2.2}$$

where $p = -id/dq$. The importance of these operators derives from the representation

$$H_0(\mu) = \mu b^* b. \tag{2.3}$$

As a convenient domain for b and b^*, let

$$\mathscr{D} = \{P(q)\,e^{-\mu q^2/2} : P \text{ is a polynomial}\}. \tag{2.4}$$

Then

$$b\mathscr{D} \subset \mathscr{D}, \qquad b^*\mathscr{D} \subset \mathscr{D} \tag{2.5}$$

and so $\mathscr{D}$ is an invariant domain for b, b^*, p and q. On $\mathscr{D}$, the commutation relations

$$[b, b^*] = I \tag{2.6a}$$

$$[q, p] = iI \tag{2.6b}$$

are valid. We set $e_0(q) = (\mu/\pi)^{1/4} \exp(-\mu q^2/2)$. Then $\|e_0\|_2 = 1$ and

$$be_0 = 0. \tag{2.7}$$

Let

$$e_j = \|b^{*j}e_0\|^{-1}\, b^{*j}e_0.$$

To justify this definition, we compute

$$\begin{aligned} \langle b^{*j}e_0, b^{*l}e_0\rangle &= \langle bb^{*j}e_0, b^{*(l-1)}e_0\rangle \\ &= \langle [b, b^{*j}]e_0, b^{*(l-1)}e_0\rangle \\ &= j < b^{*(j-1)}e_0, b^{*(l-1)}e_0\rangle \\ &= j!\,\delta_{jl} \end{aligned} \tag{2.8}$$

and see that $\|b^{*j}e_0\| \neq 0$. This calculation also shows that the e_j's form an orthogonal family. e_j is the jth Hermite function. From (2.8) we have

$$\left.\begin{aligned} b^*e_j &= (j+1)^{\frac{1}{2}}\, e_{j+1} \\ b\,e_j &= j^{\frac{1}{2}}\, e_{j-1}, \qquad j > 0. \end{aligned}\right\} \tag{2.9}$$

It follows from (2.3) and (2.9) that

$$H_0(\mu)\, e_j = j\mu e_j.$$

Thus $H_0(\mu)$ has eigenvalues $0, \mu, 2\mu, \ldots$ and eigenfunctions $e_0, e_1, \ldots$. Since $e_j \in \mathscr{D}$,

$$e_j(q) = P_j(q)\, e_0(q)$$

for some polynomial P_j. P_j is the jth Hermite polynomial. From (2.2) and (2.9) we see that P_j has degree j and that the coefficient of q^j in P_j is positive.

To complete our analysis of $H_0(\mu)$, we must show that the eigenfunctions e_j are complete in L_2. The linear span of the eigenfunctions is exactly $\mathscr{D}$. $\mathscr{D}^-$ contains the functions $\exp(i\lambda q)\exp(-\mu q^2/2)$, since the Taylor series in powers of λ for such functions converge in L_2. To establish this statement, we invert the relation between b, b^* and q, p, obtaining

$$\begin{aligned} q &= (2\mu)^{-\frac{1}{2}}(b^* + b) \\ p &= i2^{-\frac{1}{2}}\mu^{\frac{1}{2}}(b^* - b). \end{aligned} \tag{2.10}$$

Then

$$\begin{aligned} (\mu/\pi)^{\frac{1}{2}} \int q^j e^{-\mu q^2}\, dq &= (2\mu)^{-j/2} \langle (b^* + b)^j e_0, e_0 \rangle \\ &\leqslant (\text{const.})^j\, O(j\,!)^{\frac{1}{2}}, \end{aligned}$$

and the required L_2 convergence follows. With $\exp(i\lambda q)\exp(-\mu q^2/2) \in \mathscr{D}^-$ for all λ, we have

$$(2\pi)^{-\frac{1}{2}} \int f(\lambda)^\sim e^{i\lambda q} e^{-\mu q^2/2}\, d\lambda = f(q)\, e^{-\mu q^2/2}$$

in $\mathscr{D}^-$ for all $f \in \mathscr{S}$, and so $\mathscr{D}^- = L_2$.

The operators b and b^* are uniquely characterized by the relations (2.5), (2.6a) and (2.7). This uniqueness theorem is valid without any restriction on the number of degrees of freedom. There is another uniqueness theorem, the von Neumann uniqueness theorem, which assumes the commutation relations (2.6) in integrated form

$$e^{isq} e^{itp} = e^{itp} e^{isq} e^{-ist}, \tag{2.11}$$

see the lectures of B. Simon. For a finite number of degrees of freedom, (2.11) determines the action of p and q on a Hilbert space $\mathscr{H}$ uniquely, up to unitary equivalence and multiplicity. However for an infinite number of degrees of freedom, (2.11) does not lead to uniqueness. The failure of the von Neumann uniqueness theorem for an infinite number of degrees of freedom indicates again that operator algebras will play a larger role in field theory than in ordinary quantum mechanics.

Definition 2.1. Let $\mathscr{E}$ be a real prehilbert space (= incomplete Hilbert space). A representation of the canonical commutation relations over $\mathscr{E}$ is a pair of linear maps

$$f \to b(f), \qquad g \to b^*(g)$$

from $\mathscr{E}$ to operators $b(f)$ and $b^*(g)$ defined on a dense domain $\mathscr{D}$ in a (complex) Hilbert space $\mathscr{H}$ such that

$$b(f)\mathscr{D} \subset \mathscr{D}, \qquad b^*(g)\mathscr{D} \subset \mathscr{D}$$

$$[b(f), b^*(g)]\theta = \langle f, g\rangle\theta$$

$$[b(f), b(g)]\theta = 0 = [b^*(f), b^*(g)]\theta$$

and

$$\langle \theta_1, b(f)\theta_2\rangle = \langle b^*(f)\theta_1, \theta_2\rangle$$

for all θ, θ_1 and θ_2 in $\mathscr{D}$ and all f and g in $\mathscr{E}$.

Definition 2.2. A representation of the canonical commutation relations is a Fock representation if there is a unit vector $\Omega \in \mathscr{D}$ such that

$$b(f)\Omega = 0$$

for all $f \in \mathscr{E}$ and such that $\mathscr{D}$ is spanned algebraically by vectors of the form

$$\{b^*(g_1) \dots b^*(g_m)\Omega : g_l \in \mathscr{E};\ m = 0, 1, \dots\}.$$

Here Ω is called the Fock vacuum vector.

Example. Let $\mathscr{H} = L_2(\mathbf{R})$, $\mathscr{E} = \mathbf{R}$, $\mathscr{D}$ as above, $\Omega = e_0$, $b(\lambda) = \lambda b$ and $b^*(\lambda) = \lambda b^*$. This defines a Fock representation of the canonical commutation relations.

THEOREM 2.3. *The Fock representation of the canonical commutation relations over $\mathscr{E}$ is unique up to unitary equivalence. If $\{b_i, b_i^*\}$, $i = 1, 2$, are two Fock representations over $\mathscr{E}$ with vacuums Ω_i, then the unitary equivalence operator U is uniquely determined, if we require*

$$U\Omega_1 = \Omega_2.$$

Proof. Let

$$\theta_i = b_i^*(f_1) \dots b_i^*(f_n)\Omega_i$$

$$\psi_i = b_i^*(g_1) \dots b_i^*(g_m)\Omega_i.$$

We compute $\langle \psi_i, \theta_i \rangle$, and obtain the same value for $i = 1$ and $i = 2$. Then U, defined by the equations $U\theta_1 = \theta_2$, extends by linearity to a unitary operator from $\mathscr{H}_1$ to $\mathscr{H}_2$ giving the required equivalence. Furthermore if U' is any unitary operator from $\mathscr{H}_1$ onto $\mathscr{H}_2$ and if

$$U'\Omega_1 = \Omega_2$$
$$U'b_1{}^*(f)U'^* = b_2{}^*(f)$$

for all $f \in \mathscr{E}$, then $U'\theta_1 = \theta_2$ and so U is uniquely determined.

From the commutation relations and the fact that $b_i\Omega_i = 0$, we have

$$\langle \psi_i, \theta_i \rangle = \sum_{l=1}^{m} \langle f_1, g_l \rangle$$
$$\times \langle b_i{}^*(g_1) \ldots b_i{}^*(g_{l-1}) b_i{}^*(g_{l+1}) \ldots b_i{}^*(g_m)\Omega_i, b_i{}^*(f_2) \ldots b_i{}^*(f_n)\Omega_i \rangle.$$

Continuing by induction, we have

$$\langle \psi_i, \theta_i \rangle = \delta_{m,n} \sum_{\sigma} \prod_{l=1}^{n} \langle f_l, g_{\sigma(l)} \rangle$$

where the sum ranges over all permutations σ on n elements. This completes the proof. ■

For n degrees of freedom, the oscillator Hamiltonian is a sum

$$H_0 = \sum_{l=1}^{n} H_0(\mu_l) \tag{2.12}$$

of n one-dimensional Hermite operators, each acting on one of the n variables q_l of $Q = \mathbf{R}^n$. The eigenfunctions of H_0 are tensor products

$$e(q) = \prod_{l=1}^{n} e_{j(l)}(q_l)$$

of the one-dimensional Hermite functions. H_0 acts on the Hilbert space $\mathscr{H} = L_2(Q)$ and $\mathscr{H}$ is a tensor product,

$$\mathscr{H} = L_2(Q) = L_2(\mathbf{R}^n) = \bigotimes_{l=1}^{n} L_2(Q_l), \tag{2.13}$$

where Q_l is a copy of the real numbers, $\mathbf{R}$.

In Section 3, we take the limit $n \to \infty$. In order to construct a measure on the resulting infinite tensor product space Q, it is essential to have a measure on each factor whose total mass is one. Furthermore, we want H_0 to be a positive self-adjoint operator on $\mathscr{H}$ in the limit $n \to \infty$, and so it is natural to require that the ground state e_0 of H_0 lie in $\mathscr{H}$. To ensure that $e_0 \in \mathscr{H}$ in the limit $n \to \infty$, we replace p_l, q_l and H_0 by unitarily equivalent operators, before taking the limit. We write $L_2(Q, dq)$ to indicate the use of Lebesgue measure. The multiplication operator

$$Uf = e_0 f$$

is a unitary operator from $L_2(Q, e_0{}^2 dq)$ onto $L_2(Q, dq)$, and $U^* g = e_0{}^{-1} g$. Then

$$U^* q_l U = q_l \tag{2.14}$$

$$U^* p_l U = -i \frac{\partial}{\partial q_l} + i \mu_l q_l \tag{2.15}$$

and $U^* H_0 U$ act on $L_2(Q, e_0{}^2 dq)$. The eigenfunctions of $U^* H_0 U$ are tensor products

$$\prod_{l=1}^{n} P_{j(l)}(q_l)$$

of the one-dimensional Hermite polynomials and the ground state of $U^* H_0 U$ is

$$\prod_{l=1}^{n} P_0(q_l) \equiv 1.$$

Certainly $1 \in L_2(Q, d\nu(q))$ for any measure $d\nu$ of total mass one. For comparison with later chapters, we write

$$e_0{}^2\, dq = \text{const}\, e^{-B(q,q)}\, dq$$

where

$$B(q, q) = \sum_{l=1}^{n} \mu_l q_l{}^2.$$

The finite constant μ, subtracted from $H_0(\mu)$ in (2.1), is a renormalization. With this renormalization, the bottom of the spectrum of $H_0(\mu)$, H_0 and $U^* H_0 U$ are all zero. This renormalization is optional for finite n, but if

$$\sum_{l=1}^{\infty} \mu_l = \infty,$$

as is generally the case, this renormalization is crucial in order to take the limit $n \to \infty$.

THEOREM 2.4. *The kernel of* $\exp(-tU^*H_0U)$ *is positive.*

Proof. Because H_0 is a sum of terms, each acting on a distinct factor in (2.13), the above kernel factors also, and so it is sufficient to consider the case of one degree of freedom. We show that in this case the kernel is

$$k(q,q') = 2^{-\frac{1}{4}}(1-e^{-2\mu t})^{-\frac{1}{2}} \exp\left[-\frac{\mu(q'-e^{-\mu t}q)^2}{2(1-e^{-\mu t})} + \mu q'^2\right] \tag{2.16}$$

relative to the measure $e_0{}^2\,dq$ on Q. The theorem follows from (2.16). ∎

We prove equality of Fourier transforms. Let

$$Kf(q) = \int k(q,q')f(q')\,e_0(q')^2\,dq'.$$

We show that $Ke^{i\lambda q} = U^*e^{-tH_0}Ue^{i\lambda q}$, and since the vectors $e^{i\lambda q}$ are dense in $L_2(Q, e_0{}^2dq)$, this completes the proof. An elementary calculation gives

$$Ke^{i\lambda q} = \exp\left[-\lambda^2(1-e^{-2t\mu})/2\mu\right]\exp\left(i\lambda e^{-t\mu}q\right).$$

However since

$$\begin{aligned} e^{i\lambda q}e_0 &= e^{-\lambda^2/2\mu}\exp\left[i\lambda(2\mu)^{-\frac{1}{2}}b^*\right]e_0 \\ &= e^{-\lambda^2/2\mu}\sum_{n=0}^{\infty}\left(\frac{i\lambda}{\sqrt{2\mu}}\right)^n (n!)^{-\frac{1}{2}}\,e_n, \end{aligned}$$

we have

$$\begin{aligned} e^{-tH_0}e^{i\lambda q}e_0 &= e^{-\lambda^2/2\mu}\sum_{n=0}^{\infty}\left(\frac{i\lambda}{\sqrt{2\mu}}e^{-t\mu}\right)^n (n!)^{-\frac{1}{2}}\,e_0 \\ &= \exp\left[-\lambda^2/2\mu\right]\exp\left[i\lambda(2\mu)^{-\frac{1}{2}}e^{-t\mu}b^*\right]e_0 \\ &= \exp\left[-\lambda^2(1-e^{-2t\mu})/2\mu\right]\exp\left[i\lambda e^{-t\mu}q\right]e_0. \end{aligned}$$

Hence

$$U^*e^{-tH_0}Ue^{i\lambda q} = \exp\left[-\lambda^2(1-e^{-2t\mu})/2\mu\right]\exp\left[i\lambda e^{-t\mu}q\right]$$

and the proof is complete. ∎

3. GAUSSIAN MEASURES AND THE SCHRÖDINGER REPRESENTATION

In this section we introduce the quantum field operators ϕ and π. They are analogous to the operators q_l and p_l of Section 2, and we go somewhat out of our way to make the analogy evident, so that the definitions should seem natural to mathematicians familiar with ordinary quantum mechanics. For simplicity we confine our attention to the quantization of the Klein–Gordon equation

$$\phi_{tt} - \phi_{xx} + m^2\phi = 0, \tag{3.1}$$

although the methods of this chapter apply to more general linear equations. Let $\pi = \partial\phi/\partial t$. The energy of a solution ϕ of (3.1) is

$$H_0(\phi, \pi) = \tfrac{1}{2}\int (\pi(x,t)^2 + \nabla\phi(x,t)^2 + m^2\phi(x,t)^2)\, dx \tag{3.2}$$

and using (3.1), it is easy to see that H is independent of the time, t. We note that (3.1) is a Hamiltonian system:

$$\dot{\phi}(x,t) = \delta H_0/\delta\pi(x,t)$$

$$\dot{\pi}(x,t) = -\,\delta H_0/\delta\phi(x,t)$$

and that $\pi(x,t)$ is the momentum variable conjugate to $\phi(x,t)$. Configuration space Q is a linear space of functions $\phi(x) = \phi(x, 0)$ and phase-space is a linear space of pairs of functions $\{\phi(x), \pi(x)\}$. It is convenient to take Q to be the real† valued Schwartz space, $Q = \mathcal{S}_{\mathbf{R}}'(\mathbf{R})$. Then H_0 is real valued and finite on a dense subspace of phase space.

$H_0(\phi, \pi)$ is a quadratic form. For comparison with Section 2, we want to diagonalize H_0, so we take Fourier transforms. With

$$\pi(x) = (2\pi)^{-\frac{1}{2}}\int e^{-ikx}\pi^{\sim}(k)\, dk, \tag{3.3}$$

one sees that

$$H_0(\phi,\pi) = \tfrac{1}{2}\int \left[(\operatorname{Re}\pi^{\sim}(k))^2 + (\operatorname{Im}\pi^{\sim}(k))^2\right] dk$$

$$+ \tfrac{1}{2}\int \mu(k)^2 \left[(\operatorname{Re}\phi^{\sim}(k))^2 + (\operatorname{Im}\phi^{\sim}(k))^2\right] dk, \tag{3.4}$$

where

$$\mu(k) = (k^2 + m^2)^{\frac{1}{2}}.$$

Let

$$\mu_x = (-(d/dx)^2 + m^2)^{\frac{1}{2}}. \tag{3.4}$$

We want to quantize the free field Hamiltonian H_0 of (3.2), (3.4) according to the prescription of Section 2. Thus we seek a Gaussian

† See Note 1 on page 143.

measure $d_B q$ on Q which formally has a density proportional to $\exp(-B(q,q))$, where

$$B(f,g) = \int \mu(k)[\operatorname{Re} f(k)^{\sim} \operatorname{Re} g(k)^{\sim} + \operatorname{Im} f(k)^{\sim} \operatorname{Im} g(k)^{\sim}]\, dk = \langle f, \mu_x g\rangle. \tag{3.5}$$

For each finite dimensional set F of linear coordinate functions on Q, we define a finite dimensional Gaussian measure. These measures satisfy a consistency condition when $F_1 \subset F_2$ and they determine the measure $d_B q$. This measure is countably additive† on Q. The countable additivity is not obvious and it depends on the choice $Q = \mathscr{S}_{\mathbf{R}}'(\mathbf{R})$. Smaller spaces than $\mathscr{S}_{\mathbf{R}}'(\mathbf{R})$ could be chosen, but it seems that Q cannot be a space of functions.

The coordinate functions on Q are just the elements of $Q' = \mathscr{S}_{\mathbf{R}}(\mathbf{R})$. For $f \in Q'$, the coordinate function is

$$Q \ni q \to q(f) = \int q(x) f(x)\, dx. \tag{3.6}$$

A polynomial on Q is just a polynomial in the coordinate functions on Q. If $f_1, \ldots, f_n \in Q$ and if P is a polynomial in n variables, then

$$\Phi(q) = P(q(f_1), \ldots, q(f_n)) \tag{3.7}$$

is a polynomial on Q. If $f_1, \ldots, f_n$ lie in a closed subspace F of Q', then we say that Φ in (3.7) is based on F. If Φ is based on F then the function $\Phi(\mu_x^{-\frac{1}{2}} q)$ is based on $\mu_x^{-\frac{1}{2}} F$. We use this change of variables, $\Phi(q) \to \Phi(\mu_x^{-\frac{1}{2}} q)$ in the definition of $d_B q$. Formally, this change of variables maps a density proportional to $\exp(-B(q,q))$ onto a density proportional to $\exp(-\|q\|_2^2)$, since formally

$$\int \Phi(q) e^{-B(q,q)}\, dq = \text{const} \int \Phi(\mu_x^{-\frac{1}{2}} q) \exp(-\|q\|_2^2)\, dq.$$

Let F be a finite dimensional subspace of Q' and let $\{f_1, \ldots f_n\}$ be a basis for F chosen so that $\{\mu_x^{-\frac{1}{2}} f_1, \ldots, \mu_x^{-\frac{1}{2}} f_n\}$ is an orthonormal basis for $\mu_x^{-\frac{1}{2}} F$. If Φ and P are related by (3.7), we define

$$\int \Phi(q)\, d_B q = \pi^{-n/2} \int P(\lambda_1, \ldots, \lambda_n) \exp[-\Sigma \lambda_i^2]\, d\lambda. \tag{3.8}$$

LEMMA 3.1. *The integral* (3.8) *is independent of the finite dimensional space F on which Φ is based.*

Proof. If Φ is based on F and G, it is also based on $F + G$, and so we may suppose that $F \subset G$. We choose a basis $\{g_1, \ldots, g_n\}$ for G so that

† See Note 2 on page 143.

$\{g_1, \ldots, g_m\}$ is a basis for F and so that the set $\{\mu_x^{-\frac{1}{2}} g_l\}$ is orthonormal. There are polynomials P_F and P_G such that

$$\Phi(q) = P_F(q(g_1), \ldots, q(g_m)) = P_G(q(g_1), \ldots, q(g_n)).$$

Thus P_G is independent of the final $n - m$ variables and equals P_F as a function of the first m variables. Hence integrating over the final $n - m$ variables, we have

$$\pi^{-n/2} \int P_G(\lambda_1, \ldots, \lambda_n) \exp\left(- \sum_{l=1}^{n} \lambda_l^2 \right) d\lambda$$

$$= \pi^{-m/2} \int P_F(\lambda_1, \ldots, \lambda_m) \exp\left(- \sum_{l=1}^{m} \lambda_l^2 \right) d\lambda$$

and the lemma is proved. ∎

With F held fixed, the integral (3.8) is countably additive. Moreover we can replace P by a bounded Borel function and the consistency condition, Lemma 3.1, and its proof remain valid. If Y_F is a Borel set in R^n, we define

$$Y = \{q : q(f_1), \ldots, q(f_n) \in Y_F\}. \tag{3.9}$$

Such a set is called a Borel cylinder set, based on F. The Borel cylinder sets form a ring, in the sense of measure theory, and the Borel sets in Q are the elements of the generated σ-ring. The integral defines a measure ν on cylinder sets. If $\{\mu_x^{-\frac{1}{2}} f_j\}$ is orthonormal and if Y and Y_F are related as above, then

$$\begin{aligned} \nu(Y) &= \int \chi_Y(q)\, d_B q = \pi^{-n/2} \int \chi_{Y_F}(\lambda) \exp(-\Sigma \lambda_l^2\, d\lambda) \\ &= \pi^{-n/2} \int_{Y_F} \exp(-\Sigma \lambda_l^2)\, d\lambda, \end{aligned} \tag{3.10}$$

where χ_Y is the characteristic function of Y.

PROPOSITION 3.2. *The measure ν, defined on the ring of Borel cylinder sets, is countably additive.*

We postpone the proof to the end of this Section. As an immediate corollary of Proposition 3.2 and general results of measure theory, we have

THEOREM 3.3. *The measure ν extends uniquely to define a countably additive measure $\nu = d_B q$ on the σ-ring of Borel sets.*

Let $\mathscr{H} = L_2(Q, d_B q)$. We call $\mathscr{H}$ Schrödinger space, or more precisely, the Schrödinger space of the free field because the (free) field operators $\phi(f)$ defined below are represented on $\mathscr{H}$ as multiplication operators, as in the Schrödinger representation of ordinary quantum mechanics. Let $\mathscr{D} \subset \mathscr{H}$ be the set of complex polynomial functions on Q.

PROPOSITION 3.4. *$\mathscr{D}$ is dense in $\mathscr{H}$.*

Proof. Let $\mathscr{D}_F$ be the set of polynomials based on F and let $\mathscr{H}_F$ be the set of L_2 functions based on F. Then $\mathscr{D}_F$ is dense in $\mathscr{H}_F$ since the Hermite functions are complete in L_2. (See Chapter 2.) By general measure theory, $\bigcup_F \mathscr{H}_F$ is dense in $\mathscr{H}$, and this completes the proof. ∎

We define $\phi(f)$ to be the multiplication operator

$$\phi(f) : \Phi(q) \to q(f)\Phi(q). \tag{3.11}$$

Then $\phi(f)$ contains $\mathscr{D}$ in its domain and $\phi(f)\mathscr{D} \subset \mathscr{D}$. $\phi(f)$ is interpreted as a position operator, where position refers not to the location of a point x in physical space, R, but rather to the location of the classical time-zero field $\phi(x)$ in its configuration space $Q = \mathscr{S}_{\mathbf{R}}'(\mathbf{R})$. In fact $\phi(f)$ is the position operator which measures the distance from the origin of Q in the coordinate direction f. It is not hard to show that $\phi(f) \in L_p(Q, d_B q)$ for all $p < \infty$, and so $\phi(f)$ defines a self-adjoint operator on $L_2(Q, d_B q)$.

On the domain $\mathscr{D}$, we define

$$\pi(f) = -i\frac{\partial}{\partial\phi(f)} + i\,\phi(\mu_x f). \tag{3.12}$$

The second term is required to make $\pi(f)$ symmetric, and occurs because of our use of a Gaussian, rather than Euclidean measure (cf. (2.15)). Let $\Omega_0 \in \mathscr{D}$ be the function identically equal to one. On a vector

$$\theta = \phi(f_1) \ldots \phi(f_n)\Omega_0 \in \mathscr{D} \tag{3.13}$$

we have

$$\pi(f)\theta = -i\sum_{l=1}^{n} \langle f, f_l\rangle \phi(f_1) \ldots \phi(f_{l-1})\phi(f_{l+1}) \ldots \phi(f_n)\Omega_0 + i\,\phi(\mu_x f)\theta.$$

The inner product $\langle\, ,\rangle$ is the Euclidean inner product,

$$\langle f, f_l\rangle = \int f(x) f_l(x)\, dx.$$

The commutation relations

$$[\phi(f), \pi(g)] = i\langle f, g\rangle I \tag{3.14}$$

$$[\phi(f), \phi(g)] = 0 = [\pi(f), \pi(g)] \tag{3.15}$$

may be verified directly on $\mathscr{D}$.

In analogy with (2.2) we introduce the operators

$$b(f) = 2^{-\frac{1}{2}}(\phi(\mu_x{}^{\frac{1}{2}}f) + i\pi(\mu_x{}^{-\frac{1}{2}}f)) \tag{3.16}$$

$$b^*(f) = 2^{-\frac{1}{2}}(\phi(\mu_x{}^{\frac{1}{2}}f) - i\pi(\mu_x{}^{-\frac{1}{2}}f)). \tag{3.17}$$

THEOREM 3.5. *The representation of b and b^* defined by* (3.16–17) *is the Fock representation for the canonical commutation relations over Q'.*

Proof. The commutation relations for b and b^* follow from (3.14–17). Since ϕ and π are defined on $\mathscr{D}$ and map $\mathscr{D}$ into $\mathscr{D}$, the same is true for b and b^*. Since $\Omega_0 \equiv 1$, we have

$$-i\frac{\partial}{\partial\phi(f)}\Omega_0 = 0$$

$$i\pi(\mu_x{}^{-\frac{1}{2}}f)\Omega_0 = -\phi(\mu_x{}^{\frac{1}{2}}f)\Omega_0$$

and

$$b(f)\Omega_0 = 0.$$

Since $\mathscr{D}$ is spanned by polynomials in ϕ, acting on Ω_0, it is also spanned by polynomials in b^*, acting on Ω_0. The fact that $b^*(f)$ is contained in the adjoint of $b(f)$ follows from (3.16–17) and the fact that $\pi(f)$ is symmetric, proved below.

LEMMA 3.6. *$\pi(f)$ is a symmetric operator on the domain $\mathscr{D}$.*

Proof. It is sufficient to show that

$$\langle\pi(f)\theta_1, \theta_2\rangle = \langle\theta_1, \pi(f)\theta_2\rangle$$

where θ_1 and θ_2 are vectors of the form (3.13). We reason by induction on the number n_1 of factors $\phi(f)$ in the definition of θ_1. Assuming the identity for n_1, we have

$$\begin{aligned}\langle\pi(f)\phi(g)\theta_1, \theta_2\rangle &= i\langle f, g\rangle\langle\theta_1, \theta_2\rangle + \langle\phi(g)\pi(f)\theta_1, \theta_2\rangle \\ &= i\langle f, g\rangle\langle\theta_1, \theta_2\rangle + \langle\theta_1, \pi(f)\phi(g)\theta_2\rangle \\ &= \langle\theta_1, \phi(g)\pi(f)\theta_2\rangle \\ &= \langle\phi(g)\theta_1, \pi(f)\theta_2\rangle\end{aligned}$$

and so the identity holds for $n_1 + 1$. Thus it is sufficient to assume $n_1 = 0$ and establish

$$\langle \pi(f)\Omega_0, \theta \rangle = \langle \Omega_0, \pi(f)\theta \rangle.$$

This is equivalent to the identity

$$\langle \Omega_0, \phi(f)\phi(f_1) \ldots \phi(f_n)\Omega_0 \rangle$$

$$= \frac{1}{2} \sum_{l=1}^{n} \langle \mu_x^{-1} f, f_l \rangle \langle \Omega_0, \phi(f_1) \ldots \phi(f_{l-1})\phi(f_{l+1}) \ldots \phi(f_n)\Omega_0 \rangle \quad (3.18)$$

once we substitute $\mu_x^{-1} f$ for f.

To establish (3.18), we choose a finite dimensional subspace F of Q' containing all the f's, and we choose a basis $\{g_1, \ldots, g_m\}$ for F with the set $\{\mu_x^{-\frac{1}{2}} g_1, \ldots, \mu_x^{-\frac{1}{2}} g_m\}$ orthonormal. Since (3.18) is linear in the f's, we may suppose that each of the f's is one of the basis elements, and set $f = g_1$. Then (3.18) is equivalent to the identity

$$\int_{R^m} \lambda_1 \lambda_{l_1} \ldots \lambda_{l_n} \exp\left(- \sum_{i=1}^{m} \lambda_i^2 \right) d\lambda$$

$$= \frac{1}{2} \sum_{i=1}^{n} \delta_{1 l_i} \int_{R^m} \lambda_{l_1} \ldots \lambda_{l_{i-1}} \lambda_{l_{i+1}} \ldots \lambda_{l_n} \exp\left(- \sum_{i=1}^{m} \lambda_i^2 \right) d\lambda.$$

The integral over $\lambda_2, \ldots, \lambda_m$ contributes the same factor to each side of the identity, and so we are reduced to proving

$$\int_R \lambda^j e^{-\lambda^2} d\lambda = \frac{j-1}{2} \int_R \lambda^{j-2} e^{-\lambda^2} d\lambda.$$

The latter is proved by integration by parts. This completes the proof of the lemma and the theorem. ∎

Inverting the relations between b, b^* and ϕ, π, we have

$$\left.\begin{aligned} \phi(f) &= 2^{-\frac{1}{2}}(b^*(\mu_x^{-\frac{1}{2}} f) + b(\mu_x^{-\frac{1}{2}} f)) \\ \pi(f) &= i2^{-\frac{1}{2}}(b^*(\mu_x^{\frac{1}{2}} f) - b(\mu_x^{\frac{1}{2}} f)), \end{aligned}\right\} \quad (3.19)$$

which generalizes (2.10).

We now turn to the proof of Proposition 3.2. The measure is finitely additive and it is regular in the sense that

$$\nu(Y) = \inf_{Y' \supset Y} \nu(Y')$$

where Y' ranges over open cylinder sets and Y is a Borel cylinder set. Let

$$h = -(d/dx)^2 + x^2$$

be the Hermite operator as studied in Chapter 2, and let

$$\|q\|_j = \|h^{j/2}q\|_{L_2}.$$

The norms $\|\cdot\|_j$ determine the topology in Q and Q'. Let Q_j be the subspace of Q on which $\|\cdot\|_j$ is finite. Then

$$\ldots \supset Q_j \supset Q_{j+1} \supset \ldots$$

and

$$Q = \bigcup_j Q_j, \qquad Q' = \bigcap_j Q_j.$$

Let

$$S(r,j) = \{q : \|q\|_j \leqslant r\}.$$

PROPOSITION 3.7. *Let ν be a finitely additive regular measure defined on Borel cylinder sets in Q. Suppose that for any $\varepsilon > 0$ there is a $j = j(\varepsilon)$ and an $r = r(\varepsilon)$ such that for any cylinder set Y disjoint from $S(r,j)$, we have $\nu(Y) \leqslant \varepsilon$. Then ν is countably additive.*

Remark. The condition on j and r can be reformulated by saying that the inner measure of $\sim S(r,j)$ is bounded by ε. The converse to the proposition is easily established, i.e. if ν is countably additive then such $j(\varepsilon)$ and $r(\varepsilon)$ exist.

Proof. Let $Y = \bigcup_{k=1}^{\infty} Y_k$ be a Borel cylinder set, expressed as a disjoint union of Borel cylinder sets. Let $Y_0 = Q \sim Y$, and then we must show that

$$\sum_{k=0}^{\infty} \nu(Y_k) = 1.$$

By finite additivity,

$$\sum_{k=0}^{\infty} \nu(Y_k) = \lim_{L\to\infty} \sum_{k=0}^{L} \nu(Y_k) = \lim_{L\to\infty} \nu\left(\bigcup_{k=0}^{L} Y_k\right) \leqslant 1.$$

Because ν is regular, the proof is completed by showing that

$$\sum_{k=0}^{\infty} \nu(Z_k) \geqslant 1$$

whenever Z_k is an open cylinder set containing Y_k.

Let $\varepsilon > 0$ be given. We use the fact that $S(r, j)$ is a ball in a Hilbert space, and so is weakly compact. By weak compactness, there is a finite number of these sets, $Z_0, \ldots, Z_l$, which form a cover for $S(r, j)$. Let

$$Z = Q \sim \bigcup_{k=0}^{l} Z_k.$$

Then Z is a cylinder set, disjoint from $S(r, j)$ and so

$$\varepsilon \geqslant \nu(Z) = \nu\left(Q \sim \bigcup_{k=0}^{l} Z_k\right) \geqslant 1 - \sum_{k=0}^{l} \nu(Z_k) \geqslant 1 - \sum_{k=0}^{\infty} \nu(Z_k).$$

Hence

$$\sum_{k=0}^{\infty} \nu(Z_k) \geqslant 1 - \varepsilon,$$

and since this is true for all $\varepsilon > 0$, the proof is complete. ■

Proof of Proposition 3.2. We choose j sufficiently negative, so that the operator

$$C = h^{j/2} \mu_x^{-\frac{1}{2}}$$

is Hilbert Schmidt on L_2, and we apply the criterion of Proposition 3.7. We choose

$$r = r(\varepsilon) = \left[\text{Trace}\, C^*C\, \varepsilon^{-1} \pi^{-\frac{1}{2}} \int_R \lambda^2 e^{-\lambda^2} d\lambda\right]^{\frac{1}{2}}.$$

Let Z be a Borel cylinder set disjoint from $S(r, j)$ and let Z be based on a finite dimensional subspace F of Q'. Let $G = \mu_x^{-\frac{1}{2}} F$ and let P_G be the orthogonal projection in L_2 onto G. With

$$D = h^{j/2} \mu_x^{-\frac{1}{2}} P_G,$$

we have

$$\text{Trace}\, D^*D \leqslant \text{Trace}\, C^*C.$$

We choose a basis $\{f_1, \ldots, f_n\}$ for F so that the vectors $g_i = \mu_x^{-\frac{1}{2}} f_i$ are eigenvectors for D^*D. Furthermore we can require that the set $\{g_i\}$ is orthonormal. Let δ_i be the corresponding eigenvalue.

Using these coordinates, we define

$$Z_F = \{\lambda \in \mathbf{R}^n : \lambda_i = q(f_i),\ q \in Z\}$$

$$S(r, j)_F = \{\lambda \in \mathbf{R}^n : \lambda_i = q(f_i),\ q \in S(r, j)\}.$$

Then

$$\nu(Z) = \pi^{-n/2} \int_{Z_F} \exp\left(-\Sigma \lambda_i^2\right) d\lambda.$$

Since Z is a cylinder set, and since Z and $S(r,j)$ are disjoint, the projections Z_F and $S(r,j)_F$ are disjoint also. Hence

$$\nu(Z) \leqslant \pi^{-n/2} \int_{\sim S(r,j)_F} \exp(-\Sigma\lambda_i^2)\, d\lambda.$$

However by a change of variables, $q \to \mu_x^{-\frac{1}{2}} q$, we see that

$$\begin{aligned} S(r,j)_F &= \{\lambda \in \mathbf{R}^n : \lambda_i = q(g_i),\ \|Cq\|_{L_2} \leqslant r,\ q \in Q\} \\ \supset S'(r,j)_F &= \{\lambda \in \mathbf{R}^n : \lambda_i = q(g_i),\ \|Dq\|_{L_2} \leqslant r,\ q \in G\} \\ &= \{\lambda \in \mathbf{R}^n : \textstyle\sum_i \delta_i \lambda_i^2 \leqslant r^2\}. \end{aligned}$$

Thus

$$\begin{aligned} \nu(Z) &\leqslant \pi^{-n/2} \int_{\Sigma_i \delta_i \lambda_i^2 \geqslant r^2} \exp(-\Sigma\lambda_i^2)\, d\lambda \\ &\leqslant \pi^{-n/2} \int_{R^n} \frac{\sum_i \delta_i \lambda_i^2}{r^2} \exp(-\textstyle\sum_i \lambda_i^2)\, d\lambda \\ &\leqslant \operatorname{Trace} C^*C\, r^{-2} \pi^{-n/2} \sup_k \int_{R^n} \lambda_k^2 \exp(-\textstyle\sum_i \lambda_i^2)\, d\lambda \\ &\leqslant \varepsilon \end{aligned}$$

and the proof is complete. ∎

4. HERMITE EXPANSIONS AND FOCK SPACE

In Fock space the free Hamiltonian H_0 is diagonalized. This fact is one of the merits of Fock space and the other is that some calculations or estimates are easier to do in Fock space than in Schrödinger space. In order to diagonalize H_0, we must utilize a momentum space (rather than a configuration space) representation for Fock space. First we define Fock space directly and then we identify it as the Hermite expansion of the Schrödinger space $L_2(Q, d_B q)$ using Theorem 2.3.

Let $\mathscr{F}_n$ be the space of symmetric L_2 functions defined on R^n; $\mathscr{F}_0$ is the complex numbers. Let

$$\mathscr{F} = \sum_{n=0}^{\infty} \mathscr{F}_n \tag{4.1}$$

$$\Omega_0 = 1 \in \mathscr{F}_0 \subset \mathscr{F}. \tag{4.2}$$

$\mathscr{F}_n$ is called the n particle subspace of $\mathscr{F}$ and $\mathscr{F}$ is Fock space. With $\theta = \{\theta_0, \theta_1, \ldots\} \in \mathscr{F}$, θ_n is the n particle component of θ. The number (of particles) operator N is defined by the formula

$$N\theta = \{0, \theta_1, 2\theta_2, \ldots, n\theta_n, \ldots\}. \tag{4.3}$$

Ω_0 is the vacuum or no particle state. These definitions refer to particles whose dynamics is governed by the free Hamiltonian H_0 (cf. Corollary 4.5). These particles have a provisional role in our construction and they are not the physical particles, which emerge from the theory as $|t| \to \infty$.

Let S_n be the projection of $L_2(\mathbf{R}^n)$ onto $\mathscr{F}_n$ and let $\mathscr{D}$ be the dense domain in $\mathscr{F}$ spanned algebraically by Ω_0 and vectors of the form

$$S_n f_1(k_1) \ldots f_n(k_n) \tag{4.4}$$

where $f_i \in \mathscr{S}_C(R)$ and $n = 1, 2, \ldots$.

The annihilation operator $a(k)$, $k \in \mathbf{R}$, is defined on $\mathscr{D}$ by the formula

$$(a(k)\theta)_n(k_1, \ldots, k_n) = (n+1)^{\frac{1}{2}}\theta_{n+1}(k, k_1, \ldots, k_n).$$

Then $a(k)\mathscr{D} \subset \mathscr{D}$, and so for $k = k_1, \ldots, k_n \in \mathbf{R}^n$, we define the product

$$a(k) = a(k_1) \ldots a(k_n), \tag{4.5}$$

also mapping $\mathscr{D}$ into $\mathscr{D}$. $a(k)$ is not closable and its adjoint, considered as an operator, has domain $\{0\}$. Nonetheless $a^*(k)$ is a bilinear form on $\mathscr{D} \times \mathscr{D}$. Similarly $a^*(k)a(k') = a^*$ is a bilinear form on $\mathscr{D} \times \mathscr{D}$.

For $\theta_1, \theta_2 \in \mathscr{D}, k \in \mathbf{R}^m, k' \in \mathbf{R}^n$

$$\langle \theta_1, a^*(k)a(k')\theta_2 \rangle \in \mathscr{S}_C(\mathbf{R}^{m+n})$$

and so for any distribution $w \in \mathscr{S}_C'(\mathbf{R}^{m+n})$, the weak integral

$$W = \int_{\mathbf{R}^{m+n}} a^*(k) w(k; k') a(k)\, dk\, dk' = a^{\#}(w) \tag{4.6}$$

is also a bilinear form on $\mathscr{D} \times \mathscr{D}$. W is called a Wick monomial of degree m, n, and any linear combination of Wick monomials is a Wick polynomial. Let $\mathscr{W}(\mathscr{S}')$ be the class of Wick polynomials and for any subspace $\mathscr{X} \subset \mathscr{S}'$, let $\mathscr{W}(\mathscr{X})$ be the class of Wick polynomials whose kernels w are restricted to lie in $\mathscr{X}$. $\mathscr{W}$ is a general class of forms. The important operators of field theory have natural expressions in terms of Wick monomials. Moreover any bounded operator on $\mathscr{F}$ has an expansion in Wick monomials. The expansion converges in terms of bilinear forms on $\mathscr{D} \times \mathscr{D}$.

We also introduce the more restricted class $\mathscr{V}(\mathscr{S}')$ of bilinear forms on $\mathscr{D} \times \mathscr{D}$ expressible as a linear combination of the forms

$$V = \sum_{j=0}^{n} \binom{n}{j} \int_{R^n} v(k) a^*(k_1) \ldots a^*(k_j) a(-k_{j+1}) \ldots a(-k_n)\, dk \tag{4.7}$$

with symmetric kernels $v \in \mathscr{S}'$ having real Fourier transforms. As before, $\mathscr{V}(\mathscr{X})$ is the subclass of $\mathscr{V}(\mathscr{S}')$ obtained by restricting the kernels v to lie in $\mathscr{X}$. As we will see later, $\mathscr{V}(\mathscr{S}')$ is a "maximal abelian" class of forms, and it is a general class of forms expressible as functions of the ϕ's.

We now define the configuration space annihilation and creation operators, using the Fourier transform with the sign convention (3.3). For real test functions $f \in \mathscr{S}_{\mathbf{R}}(\mathbf{R})$, let

$$\begin{aligned} b^*(f) &= \int a^*(-k) f(k)^\sim \, dk \\ b(f) &= \int a(k) f(k)^\sim \, dk. \end{aligned} \tag{4.8}$$

The commutation relations

$$[b(f), b^*(g)] = \int f(g) g(x)\, dx = \int f(k)^\sim g(-k)^\sim \, dk$$

follow from the corresponding commutation relations for a and a^*,

$$[a(k_1), a^*(k_2)] = \delta(k_1 - k_2), \tag{4.9}$$

and (4.9) can be verified directly.

THEOREM 4.1. *With the above definitions, b and b^* are the Fock representation for the canonical commutation relations over $Q' = \mathscr{S}_{\mathbf{R}}(\mathbf{R})$.*

The proof is routine, and will be omitted. By Theorem 2.3, we identify $\mathscr{F}$ and $\mathscr{H} = L_2(Q, d_B q)$. We now define

$$\begin{aligned} \phi(x) &= (4\pi)^{-\frac{1}{2}} \int e^{-ikx} (a^*(k) + a(-k)) \mu(k)^{-\frac{1}{2}} \, dk \\ \pi(x) &= i (4\pi)^{-\frac{1}{2}} \int e^{-ikx} (a^*(k) - a(-k)) \mu(k)^{\frac{1}{2}} \, dk. \end{aligned} \tag{4.10}$$

Since the kernels in (4.10) belong to $\mathscr{S}_C'$, $\phi(x)$ and $\pi(x)$ are bilinear forms on $\mathscr{D} \times \mathscr{D}$. Combining (3.19), (4.8) and (4.10), we have

$$\begin{aligned} \phi(f) &= 2^{-\frac{1}{2}} \int f(k)^\sim \mu(k)^{-\frac{1}{2}} (a^*(-k) + a(k))\, dk \\ &= (4\pi)^{-\frac{1}{2}} \iint e^{-ikx} (a^*(k) + a(-k)) \mu(k)^{-\frac{1}{2}} f(x)\, dk\, dx \\ &= \int \phi(x) f(x)\, dx \end{aligned}$$

and similarly $\pi(f) = \int \pi(x) f(x)\, dx$. This calculation justifies the definition (4.10). For the interchange of k and x integration, see the proof of Theorem 4.4 below.

We resume our analysis of the Wick monomial W in (4.6).

THEOREM 4.2. *Let w be the kernel of a bounded operator from $S_n L_2(\mathbf{R}^n)$ to $S_m L_2(\mathbf{R}^m)$, with norm $\|w\|$. Then*

$$(N + I)^{-m/2} W (N + I)^{-n/2} \tag{4.11}$$

is also bounded, with norm at most $\|w\|$.

Proof. Let A be the operator (4.11). Then $A : \mathscr{F}_{r+n} \to \mathscr{F}_{r+m}$ and

$$\|A\| = \sup_r \|A \restriction \mathscr{F}_{r+n}\|.$$

For $\theta_1 \in \mathscr{F}_{r+m} \cap \mathscr{D}$ and $\theta_2 \in \mathscr{F}_{r+n} \cap \mathscr{D}$, we have

$$\langle \theta_1, A\theta_2 \rangle = (r + m + 1)^{-m/2} (r + n + 1)^{-n/2} \langle \theta_1, W\theta_2 \rangle$$

and

$$\langle \theta_1, W\theta_2 \rangle = \left(\frac{(r+m)!}{r!} \frac{(r+n)!}{r!} \right)^{\frac{1}{2}} \int \theta_1(k, p)^{-} w(k; k') \theta_2(k', p)\, dk\, dk'\, dp,$$

where $p \in \mathbf{R}^r$. Thus

$$\begin{aligned} |\langle \theta_1, A\theta_2 \rangle| &\leqslant \|w\| \int \|\theta_1(\cdot, p)\|_2 \, \|\theta_2(\cdot, p)\|_2 \, dp \\ &\leqslant \|w\| \left(\int \|\theta_1(\cdot, p)\|_2^{\,2} \, dp \right)^{\frac{1}{2}} \left(\int \|\theta_2(\cdot, p)\|_2^{\,2} \, dp \right)^{\frac{1}{2}} \\ &= \|w\| \, \|\theta_1\| \, \|\theta_2\| \end{aligned}$$

and the proof is complete. ∎

COROLLARY 4.3. *Let $a + b \geqslant m + n$. Then*

$$\|(N + I)^{-a/2} W (N + I)^{-b/2}\| \leqslant (1 + |m - n|)^{|a-m|/2} \|w\|.$$

Proof. We transfer $|a - m|$ factors of $(N + I)^{-\frac{1}{2}}$ from one side of W to the other. We use the identity

$$(N + (\alpha + 1)I)^{-\frac{1}{2}} W = W(N + (\alpha + m - n + 1)I)^{-\frac{1}{2}}$$

with $\alpha = 0$ if $m > n$ and $\alpha = n - m$ if $m \leqslant n$, and we use

$$\left\| (N + (\alpha + 1)I)^{\frac{1}{2}} (N + I)^{-\frac{1}{2}} \right\| \leqslant (1 + \alpha)^{\frac{1}{2}}$$

for $\alpha = |m - n|$. ∎

Remark. It follows that the bilinear form defines uniquely an operator on the domain $\mathscr{D}(N^{(m+n)/2})$.

Now let W_1 and W_2 be two Wick monomials, with degrees i, j and m, n respectively. Assuming that the kernels w_1 and w_2 are sufficiently regular (e.g. in $\mathscr{S}$), the product $W_1 W_2$ may be defined. $W_1 W_2$ is not a Wick monomial because the creation operators in W_2 precede the annihilation operators in W_1. However using the commutation relations (4.9), the a's and a^*'s can be ordered as in (4.6). Because of the δ function on the right side of (4.9), the repeated use of (4.9) expresses $W_1 W_2$ as a sum of terms. Letting U_r be the contribution to the sum formed by all terms with r δ functions (r contractions) we have

$$W_1 W_2 = \sum_{r=0}^{\min\{j,m\}} U_r \tag{4.12}$$

and

$$U_r = \int a^*(k_1) \dots a^*(k_{i+m-r}) u_r(k; k') a(k_1') \dots a(k'_{j+n-r})\, dk\, dk'. \tag{4.13}$$

If w_1 and w_2 are each symmetric in their creating variables ($k_1, \dots, k_i$ or $k_1, \dots, k_m$, resp.) and also in their annihilating variables, then

$$\begin{aligned} u_r &= u_r(k, k'; k'', k''') \\ &= r! \binom{j}{r} \binom{m}{r} \int w_1(k; p, k'_{r+1}, \dots, k_j') w_2(p, k''_{r+1}, \dots, k_m''; k''')\, dp \end{aligned} \tag{4.14}$$

where $k \in \mathbf{R}^i$, $p \in \mathbf{R}^r$ and $k''' \in \mathbf{R}^n$. The term U_0 is called the Wick product and is denoted $:W_1 W_2:$. The kernel of $:W_1 W_2:$ is just $w_1 \otimes w_2$, that is the product of w_1 and w_2, considered as functions of distinct variables. For any two tempered distributions w_1 and w_2, $w_1 \otimes w_2$ is again a tempered distribution, and so $:W_1 W_2:$ is always defined, even if the product $W_1 W_2$ is not. Similarly we define the Wick product $:U \dots V W:$ of an arbitrary number of Wick monomials. If P is a noncommutative polynomial in l indeterminants and if $W_1, \dots, W_l$ are Wick monomials, then $P(W_1, \dots, W_l)$ is called a formal expression in the $a^{\#}$'s. We define $:P(W_1, \dots, W_l):$ by linearity. As bilinear forms we have the identity

$$W_1 W_2 = :W_1 W_2: + :U_1: + :U_2: + \dots.$$

Since U_l is in general not zero, the Wick product on formal expressions cannot be regarded as a product on bilinear forms.

As a special case of the above discussion, we have $:\phi^n(x):\in\mathscr{V}(\mathscr{S}')$ with the kernel

$$v(k, x) = (4\pi)^{-n/2} \exp(-ix\textstyle\sum_l k_l) \displaystyle\prod_{i=1}^{n} \mu(k_l)^{-\frac{1}{2}}. \tag{4.15}$$

The kernel belongs to $\mathscr{S}_C'(\mathbf{R}^n)$ as a function of k, and x is a parameter. For $n = 2$, one can check that

$$\phi^2(x) - :\phi^2(x): = (4\pi)^{-1} \int \mu(k)^{-1}\, dk. \tag{4.16}$$

The integral on the right is logarithmically divergent and since $:\phi^2(x):$ exists, we conclude that $\phi^2(x)$ does not exist. The facts that $\phi^2(x)$ does not exist (or is identically infinite), that $\phi(x)$ is a bilinear form but not an operator and that the Gaussian measure $d_B q$ is concentrated on distributions but not on functions all are different facets of the same phenomenon.

We are now ready to define the free Hamiltonian H_0. According to the correspondence principle, we obtain H_0 by substituting the quantized fields ϕ and π in the classical free field Hamiltonian (3.2) at time $t = 0$. We have already seen that $\phi^2(x)$ is identically equal to $+\infty$, and the same is true for π^2 and $\nabla\phi^2$. However we are also allowed to subtract a constant, as was done in (2.1) in the case of one degree of freedom. Thus we define

$$H_0 = \int \tfrac{1}{2} :(\pi^2(x) + \nabla\phi^2(x) + m^2\phi^2(x)):\, dx. \tag{4.17}$$

This formidable looking expression is actually very simple, thanks to our special choice of the bilinear form B. In analogy with (2.3), we have

THEOREM 4.4. $H_0 = \int \mu(k) a^*(k) a(k)\, dk.$

Proof. This is a routine calculation, and so we omit some details. The pure creation part of H_0 is

$$2^{-1}(4\pi)^{-1} \int a^*(k_1)a^*(k_2)\, e^{-ix(k_1+k_2)}(-\mu(k_1)\mu(k_2) + (-k_1k_2 + m^2)) \times \mu(k_1)^{-\frac{1}{2}}\mu(k_2)^{-\frac{1}{2}}\, dk_1\, dk_2\, dx,$$

and the point is to justify the interchange of k and x integration. For θ_1, $\theta_2 \in \mathscr{D}$, let

$$f(k_1, k_2) = \langle \theta_1, a^*(k_1)a^*(k_2)\theta_2 \rangle.$$

Then $f \in \mathscr{S}_C(R^2)$ and the above contribution to $\langle \theta_1, H_0 \theta_2 \rangle$ is

$$2^{-1}(4\pi)^{-1} \int f(k_1, k_2)\, e^{-ix(k_1+k_2)}(-\mu(k_1)\mu(k_2) - k_1 k_2 + m^2) \times \mu(k_1)^{-\frac{1}{2}}\mu(k_2)^{-\frac{1}{2}}\, dk_1\, dk_2\, dx$$
$$= 4^{-1} \int f(k_1, k_2)\delta(k_1 + k_2)(-\mu(k_1)\mu(k_2) - k_1 k_2 + m^2) \times \mu(k_1)^{-\frac{1}{2}}\mu(k_2)^{-\frac{1}{2}}\, dk_1\, dk_2 = 0.$$

Similarly the pure annihilation part of H_0 is zero, and the remaining contribution to H_0 gives

$$(4\pi)^{-1} \int a^*(k_1)a(-k_2)\, e^{-ix(k_1+k_2)}(\mu(k_1)\mu(k_2) - k_1 k_2 + m^2)) \times \mu(k_1)^{-\frac{1}{2}}\mu(k_2)^{-\frac{1}{2}}\, dk_1\, dk_2\, dx.$$

As above we interchange the x and k integrations to obtain

$$2^{-1} \int a^*(k_1)a(-k_2)\delta(k_1 + k_2)(\mu(k_1)\mu(k_2) - k_1 k_2 + m^2) \times \mu(k_1)^{-\frac{1}{2}}\mu(k_2)^{-\frac{1}{2}}\, dk_1\, dk_2$$
$$= \int \mu(k)a^*(k)a(k)\, dk,$$

as bilinear forms on the domain $\mathscr{D} \times \mathscr{D}$. ∎

COROLLARY 4.5. *H_0 leaves each subspace $\mathscr{F}_n$ invariant, and on $\mathscr{F}_n$, H_0 is the multiplication operator*

$$\sum_{l=1}^{n} \mu(k_l).$$

COROLLARY 4.6. *H_0 is essentially self-adjoint as an operator on the domain $\mathscr{D}$.*

The proofs are elementary. From now on, we let H_0 denote the self-adjoint operator, with

$$\mathscr{D}(H_0) = \left\{ \theta : \sum_{n=1}^{\infty} \left\| \sum_{l=1}^{n} \mu(k_l)\theta_n \right\|_2 < \infty \right\}. \tag{4.18}$$

III Boson Quantum Field Models: Part II. The Solution of Two-Dimensional Boson Models

PART II. THE SOLUTION OF TWO-DIMENSIONAL BOSON MODELS

5. THE INTERACTION HAMILTONIAN

Two-dimensional boson quantum field models employ three cut-offs for treatment of the Hamiltonian operator. The cut-offs are approximations which simplify the Hamiltonian. After obtaining properties of the cut-off

Hamiltonians, the cut-offs are removed by some limit procedure, and one obtains properties of the full Hamiltonian. The first cut-off is a space cut-off, indexed by a non-negative function $g \in L_\infty(\mathbf{R})$ of compact support. g limits the interaction of particles to the space region, support g. The second cut-off is a momentum cut-off κ, and κ indicates roughly the largest magnitude of the momentum occurring in the interaction Hamiltonian. Finally we have a third auxiliary cut-off, which makes evident the fact that the sum of the free and the cut-off interaction Hamiltonians are essentially self-adjoint.

We fix a positive polynomial P of degree p. The interaction Hamiltonian is

$$H_I = \int :P(\phi(x)): dx \in \mathscr{V}(\mathscr{S}'). \tag{5.1}$$

The kernel of the term of degree n in H_I is proportional to

$$\delta\left(\sum_{l=1}^{n} k_l\right) \prod_{l=1}^{n} \mu(k_l)^{-\frac{1}{2}} \tag{5.2}$$

after an interchange of x and k integration, as in the proof of Theorem 4.4. The kernel belongs to $\mathscr{S}_C'(\mathbf{R}^n)$ but not to L_2, because of the δ-function. The δ-function results from conservation of momentum, or what is the same, the translation invariance of H_I. Because of the δ-function, the sum $H_0 + H_I$ is exceedingly singular. After adding an infinite constant to $H_0 + H_I$, we will realize it as a self-adjoint operator on a new Hilbert space, associated with a (presumably) non-Fock representation of the canonical commutation relations. The space cut-off Hamiltonian $H(g)$ is much less singular. We define

$$H_I(g) = \int :P(\phi(x)): g(x)\, dx, \tag{5.3}$$

and here the kernel is proportional to

$$v = g\left(-\sum_{l=1}^{n} k_l\right)^{\sim} \prod_{l=1}^{n} \mu(k_l)^{-\frac{1}{2}}. \tag{5.4}$$

The momentum cut-off field $\phi_\kappa(x)$ is defined by the formula

$$\phi_\kappa(x) = (4\pi)^{-\frac{1}{2}} \int \xi(k\kappa^{-1})\, e^{-ixk}\, \mu(k)^{-\frac{1}{2}}\big(a^*(k) + a(-k)\big)\, dk \tag{5.5}$$

where ξ is some fixed element of $\mathscr{S}_C(\mathbf{R})$ satisfying $\xi(0) = 1$ and $\xi(k) = \xi(-k)^-$, so that ξ is the Fourier transform of a real function. We define

$$H_I(g, \kappa) = \int :P(\phi_\kappa(x)): g(x)\, dx \tag{5.6}$$

and

$$H(g, \kappa) = H_0 + H_I(g, \kappa). \tag{5.7}$$

The kernel of $H_I(g, \kappa)$ is proportional to

$$v_\kappa = g\left(-\sum_{l=1}^{n} k_l\right)^{\sim} \prod_{l=1}^{n} (\mu(k_l)^{-\frac{1}{2}}\xi(k_l\kappa^{-1})). \tag{5.8}$$

The g cut-off is introduced in order to bring the kernel v into L_2. Although $v_\kappa \in \mathscr{S}$ and $H_I(g, \kappa) \in \mathscr{V}(\mathscr{S})$, the extra regularity does not help. Rather, the importance of the κ cut-off is that $H_I(g, \kappa)$ is semi-bounded, while $H_I(g)$ is not. To prove this statement, we need Wick's theorem.

Let A_j be a set of disjoint unordered pairs, selected from $\{1, \ldots, n\}$, let $|A_j|$ be the subset of $\{1, \ldots, n\}$ covered by the pairs in A_j and let $\mathscr{A}_j$ be the set of all A_j's.

THEOREM 5.1. *The relations*

$$\prod_{l=1}^{n} [a^*(k_l) + a(-k_l)]$$

$$= \sum_{j=0}^{[n/2]} \sum_{A_j \in \mathscr{A}_j} \prod_{(\alpha,\beta) \in A_j} \delta(k_\alpha + k_\beta) : \prod_{l \notin |A_j|} (a^*(k_l) + a(-k_l)): \tag{5.9}$$

and

$$:\prod_{l=1}^{n} [a^*(k_l) + a(-k_l)]:$$

$$= \sum_{j=0}^{[n/2]} (-1)^j \sum_{A_j \in \mathscr{A}_j} \prod_{(\alpha,\beta) \in A_j} \delta(k_\alpha + k_\beta) \prod_{l \notin |A_j|} (a^*(k_l) + a(-k_l)) \tag{5.10}$$

are valid in $\mathscr{W}(\mathscr{S})$, after multiplying by smooth test functions and integrating over k. The identity is then an operator identity on the domain $\mathscr{D}$.

Proof. We use (4.9) to reorder the a^* and a operators on the left side. For each pair k_α, k_β of variables, there is one interchange of order between an a^* and an a operator, and so there are two terms, one with a δ-function $\delta(k_\alpha + k_\beta)$ and the other with the a^* and a in reversed order. Thus the expansion of the left side contains one term for each way of selecting disjoint unordered pairs from $\{1, \ldots, n\}$. This term contains δ-functions for the pairs selected and reordered a^* and a operators for the variables not lying in such pairs. The minus sign in (5.10) comes from the direction in which the reordering takes place, i.e. $a^*a = aa^* - \delta$ in (5.10). This completes the proof. ∎

Let

$$c_\kappa = (4\pi)^{-1} \int \xi(k\kappa^{-1})\mu(k)^{-1}\, dk. \tag{5.11}$$

COROLLARY 5.2. *As an operator identity on the domain* $\mathscr{D}$, *we have*

$$:\phi_\kappa{}^n(x): = \sum_{j=0}^{[n/2]} (-1)^j c_\kappa{}^j \frac{n!}{(n-2j)!j!2^j} \phi_\kappa{}^{n-2j}(x).$$

Proof. The number of elements of $\mathscr{A}_j$ is just $n!/(n-2j)!j!2^j$. ∎

COROLLARY 5.3. *Let* $g \in L_1$. *As an operator inequality on* $\mathscr{D} \times \mathscr{D}$, *we have*

$$-O(\ln \kappa)^{p/2} \leqslant H_I(g, \kappa).$$

Proof. $c_\kappa = O(\ln \kappa)$ in (5.11). Since P is a positive polynomial, the inequality follows from Corollary 5.2. ∎

The inequality in Corollary 5.3 extends by closure to an inequality for self-adjoint operators, see Theorem 5.6.

LEMMA 5.4. *Let* $f_1, f_2 \in \mathscr{S}_R(\mathbf{R})$. *Then* $\phi(f_1) + \pi(f_2)$ *is essentially self-adjoint on* $\mathscr{D}$.

Proof. We establish the inequality

$$\|(\phi(f_1) + \pi(f_2))^n\theta\| \leqslant O(n!)^{\frac{1}{2}}(\|\mu_x{}^{-\frac{1}{2}}f_1\|_2 + \|\mu_x{}^{\frac{1}{2}}f_2\|_2)^n \tag{5.12}$$

for $\theta \in \mathscr{D}$. Since $\mathscr{D}$ is also invariant, θ is then an analytic vector for $[\phi(f_1) + \pi(f_2)] \restriction \mathscr{D}$. Since $\mathscr{D}$ is dense, the lemma follows from Nelson's theorem. (See Simon's chapter.)

To prove (5.12), we expand the nth power on the left as a sum of 4^n terms, each of the form

$$\|a^{\#}(h_1) \ldots a^{\#}(h_j)\theta\| \tag{5.13}$$

where $a^{\#} = a$ or a^* and h_j is proportional to $\mu_x^{\pm\frac{1}{2}}f_i$. If M is the maximum number of particles in θ, (5.13) is dominated by

$$\prod_{j=1}^{n} \|a^{\#}(h_i)(N+I)^{-\frac{1}{2}}\| \left\| \prod_{j=1}^{n} (N + (j+1)I)^{\frac{1}{2}}\theta \right\|$$

$$\leqslant (M+1)^n n!^{\frac{1}{2}} \prod_{j=1}^{n} \|h_i\|_2 \|\theta\|$$

as in the proof of Corollary 4.3. This completes the proof. ∎

We let $\phi(f)$ now denote the self-adjoint closure, $\phi(f) = (\phi(f) \restriction \mathscr{D})^-$. Let $\mathfrak{M}$ be the von Neumann algebra generated by the operators $\{e^{i\phi(f)} : f \in \mathscr{S}\}$.

By definition, a von Neumann algebra is a *-algebra of bounded operators, closed under strong limits. It is easy to see that $\mathfrak{M}$ is also generated by the joint spectral projections of finite families $\phi(f_1), \ldots, \phi(f_n)$, $f_i \in \mathscr{S}$. Such a spectral projection is a multiplication operator on $L_2(Q, d_B q)$. In fact it is multiplication by the characteristic function of a Borel cylinder set. Conversely, any such a multiplication operator is a spectral projection, and so these multiplication operators generate $\mathfrak{M}$. We now prove a simple density theorem.

PROPOSITION 5.5. $\mathfrak{M} = L_\infty(Q)$.

Proof. $\mathfrak{M} \subset L_\infty(Q)$ by construction. To prove the converse, let $\mathscr{R} = \{Y : \chi_Y \in \mathfrak{M}\}$. Then $\mathscr{R}$ is a ring of subsets of Q, containing the Borel cylinder sets. Since a monotone sequence of projections has a strong limit, $\mathscr{R}$ is closed under sequential monotone limits. Hence $\mathscr{R}$ contains all Borel sets, and the proposition follows from this fact. ∎

THEOREM 5.6. *Let $V \in \mathscr{V}(L_2)$. Then V is essentially self-adjoint on $\mathscr{D}$.*

Proof. Let V be defined as an operator on the domain $\mathscr{D}$. We assert that $\mathfrak{M}\Omega_0 \subset \mathscr{D}(V^-)$ and that $\mathfrak{M}$ commutes with V^-. Since $V\Omega_0 = \theta \in \mathscr{H} = L_2(Q)$, it follows that $V^- \restriction \mathfrak{M}\Omega_0$ is multiplication by a real L_2 function. Since any real L_2 multiplication operator is essentially self-adjoint on $L_\infty(Q) = \mathfrak{M}\Omega_0$, $V^- \restriction \mathfrak{M}\Omega_0$ is essentially self-adjoint and V^- is self-adjoint. To prove the assertion, we compute

$$\lim_n \sum_{j \leqslant n} V(i\phi(f))^j \Omega_0/j! = \lim_n \sum_{j \leqslant n} (i\phi(f))^j V\Omega_0/j! = e^{i\phi(f)} V\Omega_0.$$

The series above converges, as in (5.10), since $V\Omega_0$ has a finite number of particles. Thus

$$\exp(i\phi(f))\Omega_0 \in \mathscr{D}(V^-) \qquad \text{and} \qquad V^- \exp(i\phi(f))\Omega_0 = \exp(i\phi(f))V^-\Omega_0.$$

The same argument applies to linear combinations in the exponentials, and to polynomials in the exponentials (since polynomials in the exponentials can be written as linear combinations). An arbitrary element M of $\mathfrak{M}$ is the strong limit of such polynomials, M_j. Hence $M\Omega_0 = \lim M_j\Omega_0$ and

$$\lim V^- M_j \Omega_0 = \lim M_j V^- \Omega_0 = MV^-\Omega_0,$$

which proves our assertion. ∎

THEOREM 5.7. *Let $V \in \mathscr{V}(L_2)$. Then $V \in L_r(Q, d_B q)$ for all $r < \infty$, and if v is the kernel of V,*

$$\|V\|_r \leqslant \text{const } \|v\|_2.$$

Proof. For $r = 2j$ an even integer, we have

$$\|V\|_r^r = \int_Q V^{2j}\, d_B q = \langle \Omega_0, V^{2j}\Omega_0 \rangle$$

$$= \|V^j \Omega_0\|^2 \leqslant \prod_{l=1}^{j} \|(N+I)^{n(l-1)/2} V (N+I)^{-nl/2}\|^2$$

$$\leqslant \text{const}\, \|v\|_2^{2j}$$

by Corollary 4.3. This proves the theorem for $r = 2j$, and since the total measure of Q is finite, the theorem is also valid for smaller values of r, and hence for all $r < \infty$. ∎

PROPOSITION 5.8. *For $g \in L_2$, v and v_κ belong to $L_2(\mathbf{R}^n)$ and for any $\varepsilon > 0$,*

$$\|v - v_\kappa\| \leqslant O(\kappa^{-\frac{1}{2}+\varepsilon}).$$

For g_1 and g_2 in L_2, the associated kernels v_1 and v_2 of (5.4) *satisfy*

$$\|v_1 - v_2\| \leqslant O(\|g_1 - g_2\|_2).$$

Proof. Let Σ be the sector

$$\mu(k_1) \leqslant \ldots \leqslant \mu(k_n).$$

Since v and v_κ are symmetric functions,

$$\|v - v_\kappa\|_2^2 = n! \int_\Sigma |v - v_\kappa|^2\, dk$$

$$= \text{const} \int_\Sigma \left|1 - \prod_i \xi(k_i\kappa^{-1})\right|^2 \prod_l \mu(k_l)^{-1} \left|g^\sim\left(-\sum_i k_i\right)\right|^2 dk.$$

Now

$$\left|1 - \prod_i \xi(k_i\kappa^{-1})\right|^2 \leqslant \text{const}\, \sup_i |1 - \xi(k_i\kappa^{-1})|.$$

Since $|1 - \xi(k_i\kappa^{-1})|\, \mu(k_n)^{-1}$ is bounded from above by const κ^{-1} and also by const $\mu(k_n)^{-1}$ on the sector Σ, we have

$$|1 - \xi(k_i\kappa^{-1})|\, \mu(k_n)^{-1} \leqslant \text{const}\, \kappa^{-1+\varepsilon} \mu(k_n)^{-\varepsilon}$$

on Σ. Thus

$$\|v - v_\kappa\|_2^{\,2} \leqslant \text{const}\, \kappa^{-1+\varepsilon} \int_\Sigma \prod_{l<n} \mu(k_l)^{-1} \mu(k_n)^{-\varepsilon} \left| g^{\sim}\left(- \sum_i k_i \right) \right|^2 dk$$

$$\leqslant \text{const}\, \kappa^{-1+\varepsilon} \int_\Sigma \prod_{l<n} \mu(k_l)^{-1-\varepsilon/n-1} \left| g^{\sim}\left(- \sum_i k_i \right) \right|^2 dk$$

$$\leqslant O(\kappa^{-1+\varepsilon})\, O(\|g\|_2^{\,2}),$$

where we integrate first over k_n. This proves the first estimate, and it also proves the second, since $v_1 - v_2$ is the kernel associated with the cut-off function $g_1 - g_2$. We note that g is not required to be positive in this proposition. ■

Remark. It follows that $H_I(g) \in \mathscr{V}(L_2)$ and so it is a self-adjoint operator. Corollary 5.3 and Proposition 5.8 express the fact that $H_I(g)$ consists of a semibounded part and a small remainder.

Let B be a bounded open region of space and let $\mathfrak{M}(B)$ be the von Neumann algebra generated by the field operators $e^{i\phi(f)}$ where $f \in \mathscr{S}_{\mathbf{R}}(\mathbf{R})$ and where suppt $f \subset B$.

THEOREM 5.9. *Let* $g(x) \equiv 0$ *on* $\sim B$. *Then* $\exp\left(itH_I(g)\right) \in \mathfrak{M}(B)$ *for all* $t \in \mathbf{R}$.

Proof. The multiplication operators $V \in \mathscr{V}(L_2)$ are each essentially self-adjoint on the dense domain L_∞. If a sequence V_j converges, as L_2 functions, to a limit V, then the sequence also converges strongly as operators on the domain L_∞, and in this case the unitary groups $\exp\,(itV_j)$ converge strongly also. By Theorem 5.7, the convergence of the kernels v_j in L_2 implies the convergence of V_j in $L_2(Q, d_B q)$. Since $\mathfrak{M}(B)$ is closed under strong limits, it is sufficient to prove the theorem for kernels v_j approximating the kernels v of $H_I(g)$ in L_2.

As our first approximation, we choose

$$g_j(x) = \begin{cases} g(x) & \text{if} \quad \text{dist}\,(x, \text{Bdry}\, B) > j^{-1} \\ 0 & \text{otherwise.} \end{cases}$$

Then $g_j \to g$ in L_2, and the kernels v_j converge in L_2 by Proposition 5.8. Next we approximate $H_I(g_j)$ by $H_I(g_j, \kappa_l)$, $\kappa_l \to \infty$. Again the kernels converge in L_2 by Proposition 5.8. As a further assumption, we choose the momentum cut-off function ξ to be the Fourier transform of a function of compact support. Then

$$\phi_\kappa(x) = 2\pi\kappa^{-1} \int \eta(\kappa(x-y))\phi(y)\,dy$$

and for $x \in \operatorname{supp} g_j$ and l sufficiently large,

$$\operatorname{supp} \eta(\kappa_l(x - \cdot)) \subset B.$$

Thus $\exp(it\phi_\kappa(x)) \in \mathfrak{M}(B)$ by definition and so the spectral projections of $\phi_\kappa(x)$ belong to $\mathfrak{M}(B)$. The same is true for $:P(\phi_\kappa(x)):$, since it is a polynomial in $\phi_\kappa(x)$ (Corollary 5.2) and the same is true for $H_I(g,\kappa)$ since the integration over x converges strongly on the core $L_\infty(Q)$ of the operator $H_I(g,\kappa)$ (Theorem 5.7). This completes the proof. ∎

6. THE FREE HAMILTONIAN

We now develop properties of the operator H_0. For the proof of essential self-adjointness of $H(g)$, the key property of H_0 is the fact that e^{-tH_0} has a nonnegative kernel. This fact results from H_0 being a second order elliptic operator with real coefficients, and it implies the existence of a positive measure on the associated path space. For removal of the space cut-off g, the key property of H_0 is its finite propagation speed. The finite propagation speed follows from the same property for the classical field, and is a consequence of the hyperbolic character of the equation (3.1).

Definition 6.1. Let X be a measure space and let dx be a measure on X. Let A be a bounded operator on $L_2(X, dx)$. Then A has a nonnegative kernel if for any two nonnegative functions θ and ψ in $L_2(X, dx)$,

$$0 \leqslant \langle \theta, A\psi \rangle. \tag{6.1}$$

If B is another bounded operator on $L_2(X, dx)$, and if $B - A$ has a nonnegative kernel, then we say that the kernel of B is greater than the kernel of A.

THEOREM 6.2. *e^{-tH_0} has a nonnegative kernel.*

Proof. We take three reductions and then appeal to the finite dimensional case, Theorem 2.4. The first reduction is to replace e^{-tH_0} by operators converging strongly to it. We choose the approximating operators to have the form e^{-tH_n} where

$$H_n = \sum_{l=1}^{j(n)} \mu_{l,n} a^*(e_{l,n}) a(e_{l,n}), \qquad 0 \leqslant \mu_{l,n}, \qquad e_{l,n} \in \mathscr{S}(R)$$

and where the family $\{e_{l,n}\}$ for fixed n is orthonormal. Furthermore since the kernel $w(k;p) = \mu(k)\delta(k-p)$ of H_0 satisfies the configuration reality condition, $w(k;p) = w(-k;-p)^-$, we can require that $e_{l,n}(k) = e_{l,n}(-k)^-$ and then $e_{l,n}$ is the Fourier transform of a real function $g_{l,n}$. Hence by (4.8),

$$H_n = \sum_{l=1}^{j(n)} \mu_{l,n} b^*(g_l) b(g_l).$$

The second reduction is to replace θ and ψ by strongly approximating vectors $|\theta_\nu| \geqslant 0$ and $|\psi_\nu| \geqslant 0$. We want θ_ν and ψ_ν to depend on a finite number of degrees of freedom, and we accomplish this by requiring that θ_ν and ψ_ν belong to $\mathscr{D}$. $\mathscr{D}$ is dense in $L_2(Q)$ and by the triangle inequality

$$-|\theta - \theta_\nu| \leqslant |\theta| - |\theta_\nu| \leqslant |\theta - \theta_\nu|$$

we see that absolute values of elements of $\mathscr{D}$ are dense in the positive elements of $L_2(Q)$.

Let G be a finite dimensional subspace of $\mathscr{S}_{\mathbf{R}}(\mathbf{R})$ such that $g_l \in G$ and such that θ_ν and ψ_ν are polynomials based on $F = \mu_x^{\frac{1}{2}} G$. Since $\mu_l = 0$ is permitted, we may suppose that the set $\{g_l\}$ is a basis for G.

The third reduction is to appeal to the unitary equivalence of Theorem 2.3. Let $\mathscr{D}_G$ be the linear subspace of $\mathscr{D}$ spanned algebraically by polynomials based on $\mu_x^{\frac{1}{2}} G$ and let $\mathscr{H}_G = \mathscr{D}_G^- \subset L_2(Q)$. Then $\mathscr{H}_G$ is also the cyclic subspace generated by the operators $b^*(g_l)$ applied to Ω_0, and we check easily that

$$G \ni g \to b^*(g), b(g)$$

is the Fock representation for the canonical commutation relations over G. Since

$$e^{-tH_n} b^*(g_l) = b^*(\exp(-t\mu_{l,n}) g_l)\, e^{-tH_n},$$

e^{-tH_n} leaves $\mathscr{H}_G$ invariant, and so the unitary equivalence U between $\mathscr{H}_G$ and $L_2(\mathbf{R}^{2j(n)})$ given by Theorem 2.3 preserves the inner product (6.1). Let $\mathfrak{M}(F)$ be the von Neumann algebra generated by $\{e^{i\phi(f)} : f \in F\}$. Then $\mathfrak{M}(F)$ also leaves $\mathscr{H}_G$ invariant, and one checks that $(\mathfrak{M}(F)\Omega_0)^- = \mathscr{H}_G$. For $M \in \mathfrak{M}(F)$, $UM \cdot 1 = M' \in L_\infty(\mathbf{R}^{2j(n)})$. The map $M \to M'$ preserves products, hence square roots and hence absolute values. Thus U preserves absolute values on $\mathscr{H}_G$. The theorem now follows from Theorem 2.4. ∎

COROLLARY 6.3. *Let $V(q) \in L_r(Q, d_B q)$ for some $r \geqslant 1$. Suppose that $H_0 + V$ is essentially self-adjoint on the domain*

$$\mathscr{D}(H_0) \cap \mathscr{D}(V)$$

and suppose that for some constant M,

$$-M \leqslant V,$$

Then

$$0 \leqslant \text{kernel}\, e^{-t(H_0+V)} \leqslant \text{kernel}\, e^{tM}\, e^{-tH_0}.$$

Proof. We use the Trotter formula to approximate $\exp(-t(H_0+V))$. Let θ and ψ be nonnegative, as elements of $L_2(Q, d_B q)$, and let

$$\theta_j = e^{-(jt/n)H_0}\,\theta$$
$$\psi_j = (e^{-(t/n)H_0}\, e^{-(t/n)V})^{n-j}\,\psi.$$

Then θ_j and ψ_j are nonnegative by Theorem 6.2 and

$$\langle\theta, (e^{-(t/n)H_0}\, e^{-(t/n)V})^n\psi\rangle = \langle\theta_0, \psi_0\rangle = \langle\theta_1, e^{-(t/n)V}\psi_1\rangle$$
$$\leqslant e^{(t/n)M}\langle\theta_1, \psi_1\rangle \leqslant \dots \leqslant e^{tM}\langle\theta_n, \psi_n\rangle = \langle\theta, e^{tM}\, e^{-tH_0}\psi\rangle.$$

The proof is completed by the limit $n \to \infty$. ∎

Let

$$\begin{aligned} \phi_0(x,t) &= e^{itH_0}\,\phi(x)\,e^{-itH_0} \\ \pi_0(x,t) &= e^{itH_0}\pi(x)\,e^{-itH_0} \end{aligned} \tag{6.2}$$

$$\begin{aligned} \phi_0(f,t) &= \textstyle\int \phi_0(x,t) f(x)\,dx \\ \pi_0(f,t) &= \textstyle\int \pi_0(x,t) f(x)\,dx. \end{aligned} \tag{6.3}$$

Because of the commutation relations

$$\begin{aligned} e^{itH_0}a^*(f) &= a^*(e^{it\mu(\cdot)}f)\,e^{itH_0} \\ [H_0, a^*(f)] &= a^*(\mu(\cdot)f) \\ [H_0, a(f)] &= -\,a(\mu(\cdot)f), \end{aligned} \tag{6.4}$$

H_0 and e^{itH_0} map $\mathscr{D}$ into $\mathscr{D}$. Thus $\phi_0(x,t)$ and $\pi_0(x,t)$ are bilinear forms defined on $\mathscr{D} \times \mathscr{D}$.

THEOREM 6.4. *Let* θ *and* ψ *be in* $\mathscr{D}$. *Then*

$$F(x,t) = \langle\theta, \phi_0(x,t)\psi\rangle$$

is a solution of the Klein–Gordon equation (3.1), *with Cauchy data*

$$F(x,0) = \langle\theta, \phi(x)\psi\rangle$$
$$(d/dt)\,F(x,0) = \langle\theta, \pi(x)\psi\rangle.$$

Proof. The formal calculation

$$\frac{d}{dt} F(x, t) = (4\pi)^{-1} \int e^{-ikx} \mu(k)^{-\frac{1}{2}}$$
$$\times \langle e^{-itH_0}\theta, [iH_0, a^*(k) + a(-k)]\, e^{-itH_0}\psi\rangle\, dk$$

is valid on $\mathscr{D} \times \mathscr{D}$, since H_0 and e^{itH_0} map $\mathscr{D}$ into $\mathscr{D}$. By (6.4),

$$\frac{d}{dt} F(x, t) = (4\pi)^{-1}\, i \int e^{-ikx} \mu(k)^{\frac{1}{2}}$$
$$\times \langle e^{-itH_0}\theta, \big(a^*(k) - a(-k)\big)\, e^{-itH_0}\psi\rangle\, dk$$
$$= \langle \theta, \pi_0(x, t)\psi\rangle.$$

Similarly,

$$\frac{d^2}{dt^2} F(x, t) = -\,(4\pi)^{-1} \int e^{-ikx} \mu(k)^{3/2}$$
$$\times \langle e^{-itH_0}\theta, \big(a^*(k) + a(-k)\big)\, e^{-itH_0}\psi\rangle\, dk,$$

and the proof is completed by writing

$$\mu(k)^{3/2} = \mu(k)^2 \mu(k)^{-\frac{1}{2}} = (k^2 + m^2)\mu(k)^{-\frac{1}{2}}. \blacksquare$$

The fundamental solution E for the Klein–Gordon equation with data

$$E(x, 0) = 0, \qquad E_t(x, 0) = \delta(x)$$

is denoted $E = \Delta(x, t)$ in the physics literature. As is well known, $\Delta(x, t)$ has support in the double light-cone $|x| \leqslant |t|$.

COROLLARY 6.5. *As bilinear forms on* $\mathscr{D} \times \mathscr{D}$,

$$\phi_0(x, t) = \int \Delta(x - y, t)\pi(y)\, dy + \int \Delta_t(x - y, t)\phi(y)\, dy$$
$$\pi_0(x, t) = \int \Delta_t(x - y, t)\pi(y)\, dy + \int \Delta_{tt}(x - y, t)\phi(y)\, dy.$$

THEOREM 6.6. *Let f_1 and f_2 belong to $\mathscr{S}_{\mathbf{R}}(\mathbf{R})$. The operator*

$$\phi_0(f_1, t) + \pi_0(f_2, t)$$

is essentially self-adjoint on the domain $\mathscr{D}$.

Proof. e^{itH_0} is a unitary operator which leaves $\mathscr{D}$ invariant and transforms $\phi_0(f_1, t) + \pi_0(f_2, t)$ onto $\phi(f_1) + \pi(f_2)$, which is essentially self-adjoint by Lemma 5.4. ∎

Let B be a bounded open region of space and let $\mathfrak{A}(B)$ be the von Neumann algebra generated by the operators

$$\exp\left(i(\phi(f_1) + \pi(f_2))\right)$$

where $f_1, f_2 \in \mathscr{S}_{\mathbf{R}}(\mathbf{R})$ and suppt f_1, suppt $f_2 \subset B$. For any real number t, let B_t be the set of points with distance less than $|t|$ to B.

THEOREM 6.7. $e^{itH_0}\mathfrak{A}(B)\, e^{-itH_0} \subset \mathfrak{A}(B_t)$.

Remark. We reformulate the theorem by saying that H_0 has propagation speed at most one.

Proof. By Corollary 6.5,

$$\phi_0(f_1, t) + \pi_0(f_2, t) = \int_{y \in B_t} h_1(y)\phi(y) + h_2(y)\pi(y)\, dy$$

where $h_1 = f_1 * \Delta_t + f_2 * \Delta_{tt}$ and $h_2 = f_1 * \Delta + f_2 * \Delta_t$. The identity is valid as bilinear forms on $\mathscr{D} \times \mathscr{D}$, but by Theorem 6.6, it extends to an identity between essentially self-adjoint operators. Since the unitary group generated by the right side belongs to $\mathfrak{A}(B_t)$ by definition, the unitary group generated by the left side also belongs to $\mathfrak{A}(B_t)$, and the proof is complete. ∎

THEOREM 6.8. *If B and C are disjoint bounded open regions of space then the operators in $\mathfrak{A}(B)$ commute with the operators in $\mathfrak{A}(C)$.*

Proof. Let $f_{i,B} \in \mathscr{S}_{\mathbf{R}}(\mathbf{R})$ be supported in B and let $f_{i,C} \in \mathscr{S}_{\mathbf{R}}(\mathbf{R})$ be supported in C, $i = 1, 2$, and let

$$B = \exp\left(i[\phi(f_{1,B}) + \pi(f_{2,B})]\right)$$
$$C = \exp\left(i[\phi(f_{1,C}) + \pi(f_{2,C})]\right).$$

From the commutation relations (3.13), valid on the domain $\mathscr{D}$, and the fact that vectors in $\mathscr{D}$ are analytic for operators of the form $\phi(f_1) + \pi(f_2)$, we see that B and C commute. Since such operators generate $\mathfrak{A}(B)$ and $\mathfrak{A}(C)$ respectively, the theorem follows. ∎

COROLLARY 6.9. *Let $g \in L_2$, and let $g(x) \equiv 0$ on B. Then* $\exp\left(itH_I(g)\right) \in \mathfrak{A}(B)'$ *for all $t \in \mathbf{R}$.*

Proof. As in the proof of Theorem 5.9, we approximate $H_I(g)$ by polynomials in the field $\phi(x)$, with x localized in bounded open sets $C_j \subset \sim B$.

7. SELF-ADJOINTNESS OF $H(g)$

PROPOSITION 7.1. *Let H_j be a sequence of self-adjoint operators. The resolvents $R_j(\zeta) = (H_j - \zeta)^{-1}$ converge strongly to the resolvent of a self-adjoint operator H provided the following three conditions are satisfied:*

(1) *The operators H_j are bounded from below, uniformly in j.*

(2) *The operators H_j converge strongly on some dense domain $\mathscr{D}$.*

(3) *The operators e^{-tH_j} converge strongly, uniformly in t, for t bounded away from zero and infinity.*

Proof. By (1) and (3), the resolvents

$$R_j(\zeta) = \int_0^\infty e^{-t(H_j-\zeta)}\,dt$$

converge for ζ sufficiently negative. From Kato, (1966b p.428, Theorem 1.3), $R(\zeta) = \lim_j R_j(\zeta)$ is the resolvent of a closed operator H once we show that the null space of $R(\zeta)$ is zero. Moreover for negative ζ, $R(\zeta)$ is self-adjoint, and thus so is H.

Let $R(\zeta)\theta = 0$. Then for $\psi \in \mathscr{D}$,

$$\langle\psi, \theta\rangle = \langle(H_j - \zeta)\psi,\ R_j(\zeta)\theta\rangle \to \langle(H - \zeta)\psi,\ R(\zeta)\theta\rangle = 0$$

using (2), and so $\theta = 0$ as required. This completes the proof. ■

Remark. The conclusion of the proposition remains valid if we replace (1) and (3) by the assumption that the sequences $R_j(\zeta)$ and $R_j(\zeta)^*$ converge strongly for some ζ.

PROPOSITION 7.2. *Let $H_j = (H_0 + V_j)^-$ where H_0 and V_j are self-adjoint. Assume the hypothesis of Proposition 7.1 and the conditions*

(4) *V_j converges to a self-adjoint limit V on a core $\mathscr{D}_0$ for V*

(5) *Range $e^{-tH_j} \subset \mathscr{D}(H_0) \cap \mathscr{D}(V_j)$*

(6) *The sequence $V_j e^{-tH_j}$ converges strongly, $0 < t$. Then $H = (H_0 + V)^-$ is essentially self-adjoint on $\mathscr{D}(H_0) \cap \mathscr{D}(V)$ and for $t > 0$*

$$\text{Range}\, e^{-tH} \subset \mathscr{D}(H_0) \cap \mathscr{D}(V).$$

Remark. $H_0 + V$ is taken with domain $\mathscr{D}(H_0) \cap \mathscr{D}(V)$.

Proof. From the functional calculus and Proposition 7.1, $H_j e^{-tH_j}$ converges to He^{-tH} for $t > 0$. By (6),

$$H_0\, e^{-tH_j}\,\theta = H_j\, e^{-tH_j}\,\theta - V_j\, e^{-tH_j}\,\theta$$

converges for any vector θ. Since H_0 is closed, $e^{-tH}\theta \in \mathscr{D}(H_0)$ and

$$H_0\, e^{-tH}\theta = H\, e^{-tH}\theta - \lim V_j\, e^{-tH_j}\theta.$$

Let $\psi \in \mathscr{D}_0$. By (4), $V_j\psi \to V\psi$, so

$$\left\langle \psi, \lim_j V_j e^{-tH_j}\theta \right\rangle = \lim_j \langle V_j\psi,\ e^{-tH_j}\theta\rangle = \langle V\psi, e^{-tH}\theta\rangle$$
$$= \langle \psi, (V \restriction \mathscr{D}_0)^*\, e^{-tH}\theta\rangle.$$

Since $\mathscr{D}_0$ is a core for V and since V is self-adjoint, $(V \restriction \mathscr{D}_0)^* = V$ and so Range $e^{-tH} \subset \mathscr{D}(V)$. Thus

$$H \restriction (\text{Range}\, e^{-tH}) = (H_0 + V) \restriction (\text{Range}\, e^{-tH}).$$

Since H is essentially self-adjoint on the range of e^{-tH},

$$H = [(H_0 + V) \restriction \text{Range}\, e^{-tH}]^- \subset (H_0 + V)^-.$$

Thus $(H_0 + V)^-$ is a symmetric extension of the self-adjoint operator H, and so $H = (H_0 + V)^-$. ∎

THEOREM 7.3. *Let $V \in \mathscr{V}(L_2)$ be bounded from below. Then $H = (H_0 + V)^-$ is essentially self-adjoint on $\mathscr{D}(H_0) \cap \mathscr{D}(V)$ and for $t > 0$ and $V' \in \mathscr{V}(L_2)$,*

$$\text{Range}\, e^{-tH} \subset \mathscr{D}(H_0) \cap \mathscr{D}(V').$$

Proof. Let

$$F_j(\lambda) = \begin{cases} \lambda & \text{for} \quad |\lambda| \leq j \\ 0 & \text{otherwise} \end{cases}$$

and let $V_j = F_j(V)$. Then $H_j = H_0 + V_j$ is self-adjoint and bounded from below, uniformly in j. $V_j \to V$ on $L_r(Q)$ by the Lebesgue bounded convergence theorem. We choose the domain $\mathscr{D}$ as in (4.4). Since $\mathscr{D} \subset L_r(Q)$ also, for all $r < \infty$, V_j and H_j converge on $\mathscr{D}$. This verifies (1) and (2). In (4) we take $\mathscr{D}_0 = L_\infty(Q)$, while (5) follows from the inclusions

$$\text{Range}\, e^{-tH_j} \subset \mathscr{D}(H_0) = \mathscr{D}(H_0) \cap \mathscr{D}(V_j).$$

LEMMA 7.4. *The sequence $F_j(V')e^{-tH_j}$ converges in norm, for $t > 0$ and $V' \in \mathscr{V}(L_2)$.*

The lemma yields (3) and (6) and moreover for $\psi \in L_\infty(Q)$

$$\left\langle \psi, \lim_j F_j(V')e^{-tH_j}\theta \right\rangle = \langle \psi, (V' \restriction \mathscr{D}_0)^* e^{-tH}\theta\rangle = \langle \psi, V'e^{-tH}\theta\rangle$$

as in the proof of Proposition 7.2. Thus Range e^{-tH} is contained in $\mathscr{D}(V')$ and the theorem is proved. ∎

Proof of Lemma 7.4. The Duhamel formula states that

$$e^{-tH_j} - e^{-tH_l} = -\int_0^t e^{-sH_j}\delta V e^{-(t-s)H_l}\, ds \tag{7.1}$$

where

$$\delta V = H_j - H_l = V_j - V_l.$$

Let $j \leqslant l$. For vectors θ and $\psi \in L_2(Q, d_B q)$,

$$\begin{aligned}|\langle \theta, (e^{-tH_j} - e^{-tH_l})\psi\rangle| &\leqslant t \sup_{0 \leqslant s \leqslant t} \langle e^{-sH_j}|\theta|, |\delta V| e^{-(t-s)H_l}|\psi|\rangle \\ &\leqslant \text{const} \sup_s \langle e^{-sH_0}|\theta|, |\delta V| e^{-(t-s)H_0}|\psi|\rangle \\ &\leqslant j^{-1} \text{const} \sup_s \langle |\theta|, e^{-sH_0} V^2 e^{-(t-s)H_0}|\psi|\rangle \\ &\leqslant j^{-1} \text{const} \sup_s \|\theta\| \, \|\psi\| \, \|e^{-sH_0} V^2 e^{-(t-s)H_0}\|,\end{aligned}$$

using Corollary 6.3 and the inequality $|\delta V| \leqslant j^{-1}V^2$. Thus

$$\|e^{-tH_j} - e^{-tH_l}\| \leqslant j^{-1} \text{const} \sup_s \|e^{-sH_0} V^2 e^{-(t-s)H_0}\|.$$

Similarly

$$\begin{aligned}&\|F_j(V')e^{-tH_j} - F_l(V')e^{-tH_l}\| \\ &\qquad \leqslant j^{-1} \text{const} \sup_s \|(1 + V'^2)e^{-sH_0} V^2 e^{-(t-s)H_0}\| \\ &\qquad\quad + j^{-1} \text{const} \|(V')^2 e^{-tH_0}\|.\end{aligned}$$

N and H_0 are commuting self-adjoint operators and $N \leqslant m^{-1}H_0$ where m is the mass in (3.1). If $\deg V$, $\deg V' \leqslant n$,

$$\|(V')^2 e^{-tH_0}\| \leqslant \|(V')^2(N + I)^{-n}\| \, \|(N + I)^n e^{-tH_0}\| < \infty$$

by these properties of N and H_0 and by Theorem 4.2. If $t - s > t/2$, then

$$\begin{aligned} &\|(1 + V'^2)e^{-sH_0}V^2 e^{-(t-s)H_0}\| \\ &\qquad \leqslant \|(1 + V'^2)(N + I)^{-n}\| \, \|e^{-sH_0}\| \\ &\qquad\quad \times \|(N + I)^n V^2 (N + I)^{-2n}\| \, \|(N + I)^{2n} e^{-tH_0/2}\| < \infty \end{aligned}$$

by Corollary 4.3. If $s > t/2$, we use the factor e^{-sH_0} to dominate the powers of $N + I$, and so the proof is complete. ∎

THEOREM 7.4. *$H(g) = (H_0 + H_I(g))^-$ is essentially self-adjoint on $\mathscr{D}(H_0) \cap \mathscr{D}(H_I(g))$. $H(g)$ is bounded from below and for $t > 0$ and $V' \in \mathscr{V}(L_2)$, we have*

$$\text{Range}\, e^{-tH} \subset \mathscr{D}(H_0) \cap \mathscr{D}(V').$$

Proof. Let $j = \kappa$, $H_j = H(g, \kappa)$. We verify conditions (1)–(6) of Propositions 7.1 and 7.2. Conditions (2), (4) and (5) are trivial. We prove (1). The modifications required for (3) and (6) are routine, and similarly for the convergence of $V'e^{-tH_j}$. As before, these facts complete the proof.

Our proof of (1) is a specialization to two dimensions of methods developed for the three-dimensional ϕ^4 model. We use the nth order Duhamel expansion, and to simplify the formulas, we use time ordered products. For example the integrand in (7.1) is expressed as

$$\left(\exp\left[-\int_0^t H(\sigma, s)\, d\sigma\right] \delta V(s)\right)_+$$

where

$$H(\sigma, s) = \begin{cases} H_j & \text{for} \quad \sigma \leqslant s \\ H_l & \text{for} \quad \sigma \geqslant s \end{cases}$$

and $\delta V(s)$ inserts δV at the time s. The subscript $+$ denotes antitime ordering (earlier times to the left in the product).

Let $\kappa_\nu = \exp(\nu^{2/p})$ where p is the degree of the polynomial P in H_I. Let

$$h_\nu = \begin{cases} H(g, \kappa_\nu) & \text{if} \quad \kappa_\nu \leqslant \kappa \\ H(g, \kappa) & \text{if} \quad \kappa \leqslant \kappa_\nu \end{cases}$$

and let $\delta h_\nu = H(g, \kappa) - H(g, \kappa_\nu)$. Iterating (7.1) we obtain

$$\exp(-tH(g, \kappa)) = \sum_{n=0}^{\infty} (-1)^n \int \left(\exp\left[-\int_0^t H(\sigma, s)\, d\sigma\right] \prod_{\nu=1}^{n} \delta h_\nu(s_\nu)\right)_+ ds \tag{7.2}$$

where the time integration ds extends over the domain

$$0 \leqslant s_1 \leqslant \ldots \leqslant s_n \leqslant t$$

and $H(\sigma, s) = h_\nu$ if $s_{\nu-1} \leqslant \sigma \leqslant s_\nu$. The series terminates for $\kappa_n \geqslant \kappa$.

To dominate $|\delta h_\nu|$, we use the upper bound

$$|\delta h_\nu| \leqslant \kappa_\nu^{-\frac{1}{4}} + \kappa_\nu^{\frac{1}{4}}(\delta h_\nu)^2 \equiv f_\nu. \tag{7.3}$$

Then

$$\left|\left\langle \theta, \left(\exp\left[-\int_0^t H(\sigma, s)\, d\sigma\right] \prod_{\nu=1}^n \delta h_\nu(s_\nu)\right)_+ \psi \right\rangle\right|$$

$$\leqslant \|\theta\|\, \|\psi\|\, e^{O(n)t} \left\| \left(\exp\left[-\int_0^t H_0\, d\sigma\right] \prod_{\nu=1}^n f_\nu(s_\nu)\right)_+ \right\| \tag{7.4}$$

by Corollary 6.3, since $-O(n) \leqslant H_I(g, \kappa_\nu)$ for $\nu \leqslant n$.

Each f_ν is a sum of a bounded number of Wick monomials,

$$f_\nu = \sum_{l=1}^{L} d_{\nu,l}.$$

The L_2 norm of the kernel of $d_{\nu,l}$ can be bounded in terms of the L_2 norms of the kernels of δh_ν, by the definition (7.3) and the Schwarz inequality. By Proposition 5.8 we get the bound $O(\kappa_\nu^{-1/4})$ as the bound on the kernel of $d_{\nu,l}$, since the kernel of $(\delta h_\nu)^2$ in (7.3) has an L_2 norm bounded by $O(\kappa_\nu^{-1+\varepsilon}) \leqslant O(\kappa_\nu^{-\frac{1}{2}})$. Thus by Corollary 4.3

$$\|d_{\nu,l}(N + I)^{-p}\| \leqslant O(\kappa_\nu^{-1/4}) = \exp\left(-\tfrac{1}{4}\nu^{2/p}\right). \tag{7.5}$$

The maximum length $s_j - s_{j-1}$ of the time intervals in the time ordered product is $t/(n+1)$ at least, since there are $(n+1)$ intervals of total length t. We insert the product

$$(N + I)^p(N + (1 + 2p)I)^p \ldots \big(N + (1 + (n-1)2p)I\big)^p$$
$$\leqslant (2p)^{np}(n-1)!\,(N + I)^{np}$$

and its inverse in the maximum time interval in (7.4). We use $(N + \alpha I)^{-p}W = W\big(N + (\alpha + m - n)I\big)^{-p}$ to move the inverse powers and obtain a factor $\big(N + (1 + \alpha)I\big)^{-p}$, $\alpha \geqslant 0$, in each time interval. This yields the bound

$$\left\| \left(\exp\left[-\int_0^t H_0 d\sigma\right] \prod_{\nu=1}^n f_\nu(s_\nu)\right)_+ \right\|$$

$$\leqslant (4np)^{np}(n!)^p \prod_{\nu=1}^n \kappa_\nu^{-\frac{1}{4}} \|N^{pn} e^{-tH_0/n+1}\|$$

$$\leqslant M^n(n!)^{3p} \prod_{\nu=1}^n \kappa_\nu^{-\frac{1}{4}},$$

for some constant M. The product

$$\prod_{\nu=1}^{n} \kappa_\nu^{-\frac{1}{4}} = \exp\left((-\tfrac{1}{4}) \sum_{\nu=1}^{n} \nu^{2/p}\right) = \exp\left(-O(n^{1+2/p})\right)$$

is rapidly converging to zero, and it dominates the other factors above. Substituting in (7.4) and (7.2), we obtain

$$\|e^{-tH(g,\kappa)}\| \leqslant \sum_{n=0}^{\infty} e^{O(tn)} \exp\left(-O(n^{1+2/p})\right),$$

which is bounded uniformly in κ. This completes the proof.

8. THE LOCAL ALGEBRAS AND THE LORENTZ GROUP AUTOMORPHISMS

Haag and Kastler have formulated the following axioms for quantum field theory. Their axioms are independent of the underlying representation of the field operators and so they can be verified at the present stage of the construction, with the field operators acting on $\mathscr{F} = L_2(Q, d_B q)$, the Fock space of the bare particles, rather than the Hilbert space of physical states (i.e. the Fock space of the physical, or in and out particles, see Section 10). The axioms are:

(a) To each bounded open region B of space time there is associated a C^*-algebra $\mathfrak{A}(B)$ containing the identity.

(b) Isotony: If $B_1 \subset B_2$ then $\mathfrak{A}(B_1) \subset \mathfrak{A}(B_2)$.

(c) Locality: If B_1 and B_2 are space like separated then $\mathfrak{A}(B_1)$ commutes with $\mathfrak{A}(B_2)$.

(d) The algebra $\mathfrak{A}$ of local observables is defined as the norm closure of $\bigcup_B \mathfrak{A}(B)$. $\mathfrak{A}$ is *primitive*; in other words $\mathfrak{A}$ has a faithful irreducible representation.

(e) Poincaré covariance: Let $\{a, \Lambda\}$ be an element of the inhomogeneous Lorentz group G. There is a representation $\sigma_{\{a,\Lambda\}}$ of G by a group of * automorphisms of $\mathfrak{A}$ such that

$$\sigma_{\{a,\Lambda\}}(\mathfrak{A}(B)) = \mathfrak{A}(\{a,\Lambda\}B). \tag{8.1}$$

In the two-dimensional boson models we are considering the time translations (i.e. the dynamics) and the Lorentz rotations are the only non-trivial axioms. The use of automorphisms is equivalent to working in the Heisenberg picture, in which the observables, represented by Hilbert space operators, change with time, while the states, represented by Hilbert space vectors, remain fixed. In a quantum system with a Hamiltonian H, the Heisenberg picture dynamics is given by the formula

$$A(t) = e^{itH} A(0)\, e^{-itH}$$

and then $A(t)$ is the observable at time t corresponding to the time zero observable $A(0)$. In our model we have locally correct Hamiltonians $H(g)$ but no global Hamiltonian, and we construct the Heisenberg picture dynamics nonetheless. We do this by restricting the observables to lie in the local algebras $\mathfrak{A}(B)$ and by using the finite propagation speed implicit in (8.1).

THEOREM 8.1. *Let B be a bounded open region of space, let $t \in R$, let g be a nonnegative function in $L_1 \cap L_2$ and let g be incidentally equal to one on B_t. For $A \in \mathfrak{A}(B)$,*

$$\sigma_t(A) \equiv e^{itH(g)} A\, e^{-itH(g)}$$

is independent of g and $\sigma_t(A) \in \mathfrak{A}(B_t)$.

Proof. Let

$$\sigma_t^{(0)}(A) = e^{itH_0} A\, e^{-itH_0}$$
$$\sigma_t^I(A) = e^{itH_I} A\, e^{-itH_I}.$$

Trotter's product formula is valid for the unitary group $\exp\big(it(H_0 + H_I(g)\big)$. This fact yields the product formula for the associated automorphism group:

$$\sigma_t(A) = \lim_{n\to\infty} (\sigma_{t/n}^{(0)} \sigma_{t/n}^I)^n (A). \tag{8.2}$$

Each interaction automorphism σ^I maps each $\mathfrak{A}(B_s)$ into itself and is independent of g on $\mathfrak{A}(B_s)$ for $|s| \leqslant |t|$. To see this, let $\chi(B_s)$ be the characteristic function of B_s. We assert that

$$\sigma_{t/n}^I(C) = \exp\big(i(t/n)H_I(\chi(B_s))\big) C \exp\big(-i(t/n)H_I(\chi(B_s))\big) \tag{8.3}$$

for $C \in \mathfrak{A}(B_s)$ and that $\sigma_{t/n}^I(C) \in \mathfrak{A}(B_s)$ also. In other words the interaction automorphism has propagation speed zero and is independent of g on $\mathfrak{A}(B_s)$ for $|s| \leqslant |t|$. The theorem follows from (8.2), (8.3) and Theorem 6.7. To prove (8.3), we write

$$H_I(g) = H_I(\chi(B_s)) + H_I(g[1 - \chi(B_s)]).$$

as a sum of commuting self-adjoint operators. By Theorem 5.6, $\exp\big(itH_I(\chi(B_s))\big) \in \mathfrak{A}(B_s)$ and so the right side of (8.3) belongs to $\mathfrak{A}(B_s)$. By Corollary 6.8,

$$\exp\big(itH_I(g[1 - \chi(B_s)])\big) \in \mathfrak{A}(B_s)'$$

and (8.3) follows. ∎

Definition 8.2. Let B be a bounded open region of space time and for any time t, let $B(t) = \{x : x, t \in B\}$ be the time t time slice of B. We define $\mathfrak{A}(B)$ to be the von Neumann algebra generated by

$$\bigcup_s \sigma_s\Big(\mathfrak{A}(B(s))\Big). \tag{8.4}$$

THEOREM 8.3. *The Haag–Kastler axioms* (a)–(e) *are valid for these local algebras* $\mathfrak{A}(B)$.

Proof. (except Lorentz rotations). The axioms (a) and (b) are obvious, while (c) follows easily from the finite propagation speed, Theorem 8.1, together with the time zero locality, Theorem 6.8. (d) is a well known property of the free field (Proposition 9.8). Because the time zero fields coincide with the time zero free fields, and because the time zero fields generate $\mathfrak{A}$ by Theorem 8.1 and the definition of the local algebras, the free field result carries over to our model with interaction $H_I \neq 0$.

In the Poincaré covariance axiom (e), the time translation is given by σ_t. Let $B + t$ be the time translate of the space time region B. Then $(B + t)(s) = B(s - t)$ and so

$$\begin{aligned}\sigma_t \bigcup_s \sigma_s\Big(\mathfrak{A}(B(s))\Big) &= \bigcup_s \sigma_{s+t}\Big(\mathfrak{A}(B(s))\Big) \\ &= \bigcup_s \sigma_s\Big(\mathfrak{A}(B(s - t))\Big) = \bigcup_s \sigma_s\Big(\mathfrak{A}((B + t)(s))\Big).\end{aligned}$$

Thus $\sigma_t(\mathfrak{A}(B)) = \mathfrak{A}(B + t)$, and (8.1) is verified for time translations. Since the local algebras are norm dense in $\mathfrak{A}$ and since automorphisms of C^*-algebras preserve the norm, σ_t extends to an automorphism of $\mathfrak{A}$.

To define the space translation automorphism σ_x, let

$$P = \int k\, a^*(k) a(k)\, dk$$

and define

$$\sigma_x(A) = e^{-ixP} A\, e^{ixP}. \tag{8.5}$$

Then

$$e^{-ixP}\phi(y)\, e^{ixP} = \phi(x + y)$$

$$e^{-ixP}\pi(y)\, e^{ixP} = \pi(x + y).$$

We leave as an exercise the proof of the following lemma, which completes the proof of Theorem 8.5 except for Lorentz rotations. ∎

LEMMA 8.4. $\sigma_x(\mathfrak{A}(B)) = \mathfrak{A}(B + x)$, σ_x *extends to a* $*$ *automorphism of* $\mathfrak{A}$, *and* $x, t \to \sigma_x\sigma_t = \sigma_t\sigma_x$ *defines a two parameter abelian automorphism group of* $\mathfrak{A}$.

To construct automorphisms for the full Lorentz group and to complete the proof of Theorem 8.1, there are four separate steps. The first is to construct a self-adjoint locally correct generator for Lorentz rotations, along the lines of Theorem 7.4. This generator then defines a locally correct unitary group and automorphism group. The second step is to prove this statement for the fields, by showing that the field $\phi(x, t)$, considered as a bilinear form on a suitable domain, is transformed locally correctly by our unitary group. The third step is to show that the local algebras $\mathfrak{A}(B)$ are also transformed correctly. The final step is to reconstruct the Lorentz group automorphisms from the locally correct pieces given by the first three steps. This final step is not difficult.

We now sketch the first step. Let $H_0(x)$ denote the integrand in (4.17), so that $H_0 = \int H_0(x)\,dx$. The formal generator of Lorentz rotations is

$$M = M_0 + M_I = \int xH_0(x) + \int x{:}P(\phi(x)){:}\,dx.$$

The local Lorentzian with which we are concerned is

$$M(g_1, g_2) = \varepsilon H_0 + H_0(g_1) + H_I(g_2)$$

where $H_0(g_1) = \int H_0(x)g_1(x)\,dx$. We require that $0 < \varepsilon$ and that g_1 and g_2 be nonnegative C_0^{∞} functions. In the second step one requires more, for example that $\varepsilon + g_1(x) = x$ and $g_2(x) = x$ in some local space region. This region is necessarily contained in the interval $[\varepsilon, \infty)$, and general space time regions are considered only in the fourth step.

Although the Lorentzians M, M_0 and M_I are each unbounded from below, $M(g_1, g_2)$ is semibounded, as we shall see, and this fact means that the methods of Section 7 can be used to study it. We begin by decomposing $H_0(g_1)$ as a sum of a diagonal and an off-diagonal term:

$$\begin{aligned} H_0(g_1) &= \int v_D(k, l)a^*(k)a(l)\,dk\,dl \\ &\quad + \int v_{OD}(k, l)[a^*(k)a^*(l) + a(-k)a(-l)]\,dk\,dl \\ &= H_0{}^D(g_1) + H_0{}^{OD}(g_1), \end{aligned}$$

where

$$v_D = 2^{-1}(2\pi)^{-1}(\mu(k)\mu(l) + kl + m^2)\mu(k)^{-\frac{1}{2}}\mu(l)^{-\frac{1}{2}}g_1{}^{\sim}(-k + l)$$

$$v_{OD} = 4^{-1}(2\pi)^{-1}(-\mu(k)\mu(l) - kl + m^2)\mu(k)^{-\frac{1}{2}}\mu(l)^{-\frac{1}{2}}g_1{}^{\sim}(-k - l),$$

as in the proof of Theorem 4.4.

We require the following information about this decomposition.

LEMMA 8.5. (a) $v_{0D} \in L_2(\mathbf{R}^2)$.

(b) v_D *is the kernel of a nonnegative operator and* $\varepsilon\mu(k)\delta(k-l) + \beta v_D$ *is the kernel of a positive self-adjoint operator, for* $\beta \geqslant 0$. *These operators are real in configuration space.*

Proof. (a) is a routine calculation based on the facts that $|g_1^{\sim}(k+l)| \leqslant \text{const}\, \mu(k+l)^{-n}$ for any n and that $\mu(k) - |k| = O(|k|^{-1})$ as $k \to \infty$. We omit further details. (b) is proved using a finite sequence of Kato perturbations. Let $v_\beta = \varepsilon\mu(k)\delta(k-l) + \beta v_D$ and let V_β and V_D denote the operators with kernels v_β and v_D respectively. V_D is a sum of three terms having the form

$$A^* M_{g_1} A$$

in configuration space, where M_{g_1} is multiplication by $g_1 \geqslant 0$. Hence $0 \leqslant V_D$. Moreover for γ sufficiently small, but chosen independently of β,

$$\gamma V_D \leqslant \tfrac{1}{2} V_0 \leqslant \tfrac{1}{2}(V_0 + \beta V_D) = \tfrac{1}{2} V_\beta$$

and so

$$V_{\beta+\gamma} = V_\beta + \gamma V_D$$

is a Kato perturbation, in the sense of bilinear forms. Consequently if V_β is self-adjoint, so is $V_{\beta+\gamma}$, and $\mathscr{D}(V_{\beta+\gamma}^{\frac{1}{2}}) = \mathscr{D}(V_\beta^{\frac{1}{2}})$. A finite induction starting with $V_0 = V_0^*$ now shows that V_β is self-adjoint, for all $\beta \geqslant 0$. ∎

LEMMA 8.6. $H_0{}^D(g_1)$ *is nonnegative and* $\varepsilon H_0 + \beta H_0{}^D(g_1)$ *is self-adjoint, for all* $\beta > 0$.

Proof. The sum is taken as a sum of bilinear forms (in order to avoid the slightly more complicated estimates required to define the sum of operators). One appeals to Lemma 8.5 and general properties of the biquantization map, e.g. $V_D \to H_0{}^D(g_1)$ or one repeats the arguments used in the proof of Lemma 8.5.

Let

$$M(\beta) = \varepsilon H_0 + H_0{}^D(g_1) + \beta H_0{}^{0D}(g_1) + H_I(g_2).$$

Then $M(0)$ is self-adjoint and semibounded by the same construction as was applied to $H(g)$ in Chapter 7. Again we go from $M(0)$ to $M(1) = M(g_1, g_2)$ be a finite sequence of Kato bilinear form perturbations. By Lemma 8.5 and Corollary 4.3,

$$\pm \gamma H_0{}^{0D}(g_1) \leqslant (\varepsilon m_0/2)N + I \leqslant (\varepsilon/2)H_0 + I$$

for γ sufficiently small. We assert that

$$(\varepsilon/2)H_0 \leqslant \tfrac{3}{4}M(\beta) + c(\beta)$$

where $c(\beta)$ is a constant depending on β, $0 \leqslant \beta \leqslant 1$. Using the assertion, we show by induction that $M(\gamma)$, $M(2\gamma)$, ... and $M(1) = M(g_1, g_2)$ are all self-adjoint. Moreover

$$\mathscr{D}(M(g_1, g_2)^{\frac{1}{2}}) = \mathscr{D}(M(0)^{\frac{1}{2}}) = \mathscr{D}(H_D{}^{\frac{1}{2}}).$$

The assertion is equivalent to showing that

$$(\varepsilon/3)H_0 + H_0{}^D(g_1) + \beta H_0{}^{0D}(g_1) + H_I(g_2).$$

is semibounded. Again by the methods of Section 7,

$$(\varepsilon/6)H_0 + (1-\beta)H_0{}^D(g_1) + H_I(g_2)$$

is semibounded. Thus it is sufficient to consider the remainder

$$(\varepsilon/6)H_0 + \beta H_0(g_1).$$

To show that this operator is semibounded, we introduce a momentum cut-off in $H_0(g_1)$. The contribution from the low momentum region is semi-bounded by undoing the Wick order (cf. Section 5). The contribution from the high momentum region is a sum of a diagonal part, which is nonnegative (cf. Lemma 8.5) and an off-diagonal part, which has an L_2 kernel. This kernel has a small norm if the momentum cut-off is sufficiently large, and in this case the off-diagonal term is dominated by $(\varepsilon/6)H_0$, by Corollary 4.3. This completes the construction of the locally correct self-adjoint generator for Lorentz rotations. See Cannon and Jaffe (1970), Rosen (1972a) and Klein (1971).

The second step is the most difficult. One verifies the equation

$$(tD_x + xD_t)\phi = [M, \phi]. \tag{8.6}$$

The field $\phi(x, t) = e^{itH(g)}\phi(x)\,e^{-itH(g)}$ is not taken at time zero, and before the commutator can be evaluated, M must be commuted with the unitary operator $e^{itH(g)}$. The commutator $[M, e^{itH(g)}]$ cannot be evaluated explicitly in terms of Wick monomials. However, the part of $[M, e^{itH(g)}]$ which cannot be evaluated does not contribute to $[M, \phi]$, by virtue of its localization. The calculations to justify (8.6) are extensive, in order to deal with questions of operator domains. See Cannon and Jaffe (1970) and Rosen (1972a).

The key to the third step is to give a covariant definition of the local algebras $\mathfrak{A}(B)$. Let $f \in \mathscr{D}(B)$, the class of real C^∞ functions with support in B, let $\alpha_1, \ldots, \alpha_n$ be real numbers and consider

$$\phi(f) = \int \phi(x, t) f(x, t)\, dx\, dt \tag{8.7}$$

$$\phi(f, t) = \int \phi(x, t) f(x, t)\, dx \tag{8.8}$$

$$\alpha_1 \phi(f, t_1) + \ldots + \alpha_n \phi(f, t_n) \tag{8.9}$$

$$\pi(f, t) = \int \pi(x, t) f(x, t)\, dx. \tag{8.10}$$

For $g \equiv 1$ on a sufficiently large set (the domain of dependence of the region B), the time integration in (8.7) converges strongly, and all four operators above are symmetric and defined on $\mathscr{D}(H(g))$.

THEOREM 8.7. *The operators* (8.7)–(8.10) *are essentially self-adjoint on any core for* $H(g)^{\frac{1}{2}}$.

If a sequence $\{A_n\}$ of self-adjoint operators converges strongly to a self-adjoint limit A on a core for A then the unitary operators e^{itA_n} converge strongly to e^{itA} (Kato, 1966b). Using this fact, one can show that the operators (8.7) and (8.9) generate the same von Neumann algebra, $\mathfrak{A}_1(B)$ and that $\mathfrak{A}_1(B) \supset \mathfrak{A}(B)$. To show that $\mathfrak{A}_1(B) \subset \mathfrak{A}(B)$, recall that a self-adjoint operator A commutes with a bounded operator C provided $C\mathscr{D} \subset \mathscr{D}(A)$ and $CA = AC$ on $\mathscr{D}$, for some core $\mathscr{D}$ of A. Equivalent is the condition that C commute with all bounded functions of A. Also equivalent is the relation $CA = AC$ on $\mathscr{D}(A)$. We choose $\mathscr{D} = \mathscr{D}(H(g))$. If C commutes with all operators of the form (8.8), it also commutes on $\mathscr{D}(H(g))$ with all operators of the form (8.9). Hence $\mathfrak{A}(B)' \subset \mathfrak{A}_1(B)'$ and so

$$\mathfrak{A}_1(B) = \mathfrak{A}_1(B)'' \subset \mathfrak{A}(B)'' = \mathfrak{A}(B).$$

This proves the following result, which gives the covariant definition of $\mathfrak{A}(B)$.

THEOREM 8.8. $\mathfrak{A}(B)$ *is the von Neumann algebra generated by bounded functions of operators of the form* (8.7).

Proof of Theorem 8.7. We consider first the sharp time operators (8.8) and (8.10). Using the unitary operator $U(t) = \exp(itH(g))$, we transform $\phi(f, t)$ and $\pi(f, t)$ into the time zero operators $\phi(f, 0)$ and $\pi(f, 0)$. Since

$U(t)$ maps a core for $H(g)^{\frac{1}{2}}$ into another core for $H(g)^{\frac{1}{2}}$, we may suppose that $t = 0$. For $A = \phi(f, 0)$ or $A = \pi(f, 0)$, we have

$$\begin{aligned}\|A(H(g) - \zeta)^{-\frac{1}{2}}\| \leqslant \|A(N + I)^{-\frac{1}{2}}\| \, \|(N + I)^{\frac{1}{2}}(H_0 + I)^{-\frac{1}{2}}\| \\ \times \|(H_0 + I)^{\frac{1}{2}}(H(g) - \zeta)^{-\frac{1}{2}}\|.\end{aligned} \tag{8.11}$$

The first factor on the right is bounded by Corollary 4.3. The second factor is bounded since $N \leqslant m^{-1}H_0$ and since N and H_0 commute. To bound the third factor, we appeal to the semiboundedness in Theorem 7.4. Let

$$\mathscr{D}_1 = \mathscr{D}(H_0) \cap \mathscr{D}(H_I(g))$$
$$\mathscr{D}_{\frac{1}{2}} = (H(g) - \zeta)^{\frac{1}{2}}\mathscr{D}_1.$$

Since $\mathscr{D}_1$ is a core for $H(g)$ (Theorem 7.4), $\mathscr{D}_1$ is also a core for $H(g)^{\frac{1}{2}}$ and hence $\mathscr{D}_{\frac{1}{2}}$ is dense in $\mathscr{F}$. By Theorem 7.4, with g replaced by $2g$, we have

$$0 \leqslant \tfrac{1}{2}(H_0 + H_I(2g) + \text{const})$$

on $\mathscr{D}_1 \times \mathscr{D}_1$. Hence

$$\tfrac{1}{2}H_0 \leqslant H_0 + H_I(g) + \text{const}$$

and

$$0 \leqslant (H(g) - \zeta)^{-\frac{1}{2}}H_0(H(g) - \zeta)^{-\frac{1}{2}} \leqslant \text{const}$$

for ζ real and sufficiently negative. The latter inequality holds on $\mathscr{D}_{\frac{1}{2}} \times \mathscr{D}_{\frac{1}{2}}$. Since $\mathscr{D}_{\frac{1}{2}}$ is dense, the third factor in (8.11) is a bounded operator.

Now let $\mathscr{D}_0$ be any core for $H(g)^{\frac{1}{2}}$. From (8.11), we see that

$$(A \restriction \mathscr{D}_0)^- = \left(A \restriction \mathscr{D}(H(g)^{\frac{1}{2}})\right)^- \supset (A \restriction \mathscr{D})^-,$$

where $\mathscr{D}$ is the domain introduced in Chapters 3 and 4. By Lemma 5.4, A is essentially self-adjoint on $\mathscr{D}$ and hence on $\mathscr{D}_0$.

Next we consider the time averaged operators (8.7) and (8.9), which we again denote by A. Integrating (8.11) above over t, we obtain the bound

$$\left\|A(H(g) - \zeta)^{-\frac{1}{2}}\right\| + \left\|\dot{A}(H(g) - \zeta)^{-\frac{1}{2}}\right\| \leqslant \text{const} \tag{8.12}$$

where $\dot{A} = i[H(g), A]$ is defined as a bilinear form on $\mathscr{D}(H(g)) \times \mathscr{D}(H(g))$. Theorem 8.7 is now a consequence of the following general result. ∎

THEOREM 8.9. *Let H be a positive self-adjoint operator and let A be a symmetric operator defined on $\mathscr{D}(H^{\frac{1}{2}})$. Suppose that*

$$\|AR^{\frac{1}{2}}\| + \|\dot{A}R^{\frac{1}{2}}\| \leqslant \text{const} \tag{8.13}$$

where $R = (H + I)^{-1}$ and $\dot{A} = i[H, A]$. Then A is essentially self-adjoint on any core for $H^{\frac{1}{2}}$.

Proof. First we construct a self-adjoint extension C for $A_0 = A \restriction \mathscr{D}(H)$, using the remark following Proposition 7.1. Let $H = \int \lambda \, dE_\lambda$ be the spectral resolution for H and let $C_n = E_n A_0 E_n$. C_n is bounded (by (8.13)) and symmetric, and hence self-adjoint. We assert that $\|(C_n - A)R\| \to 0$. Then $C_n \to A_0$ strongly on the dense domain $\mathscr{D}(H)$, verifying condition (2) of Proposition 7.1. To prove the assertion, we use the identity

$$RA = AR + i R\dot{A}R,$$

valid on the domain $\mathscr{D}(H)$. We have

$$\begin{aligned}
\|(C_n - A)R\| &\leqslant \|E_n A(1 - E_n)R\| + \|(I - E_n)AR\| \\
&\leqslant \|AR^{\frac{1}{2}}\| \, \|(I - E_n)R^{\frac{1}{2}}\| + \|RA(I - E_n)AR\|^{\frac{1}{2}} \\
&\leqslant O(n^{-\frac{1}{2}}) + (\|RA(I - E_n)RA\| + \|RA(I - E_n)R\dot{A}R\|)^{\frac{1}{2}} \\
&\leqslant O(n^{-\frac{1}{2}}) + O(n^{-1/4}) \to 0.
\end{aligned}$$

Let $K_n = (C_n \pm iy)^{-1}$. Next we show that $\|(H + I)K_n R\| \leqslant 1$ for y sufficiently large. The operators C_n and K_n leave the domain of H invariant and so for any vector $\theta \in \mathscr{F}$, $K_n R\theta = \psi \in \mathscr{D}(H)$. Thus

$$\begin{aligned}
&\langle \psi, (C_n \mp iy)(H + I)^2 (C_n \pm iy)\psi \rangle \\
&\qquad \geqslant y^2 \langle \psi, (H + I)^2 \psi \rangle - y \, |\langle \psi, (H + I)\dot{C} + \dot{C}(H + I)\psi \rangle| \\
&\qquad \geqslant (y^2 - \text{const}\, y)\langle \psi, (H + I)^2 \psi \rangle.
\end{aligned}$$

For y^2 sufficiently large, $y^2 - \text{const}\, y \geqslant 1$, and

$$\langle \theta, RK_n^*(H + I)^2 K_n R\theta \rangle \leqslant \|\theta\|^2$$

as asserted.

The next step is to prove the convergence of the resolvents K_n. For $\psi \in D(H)$,

$$\begin{aligned}
\|(K_n - K_m)\psi\| &= \|K_n(C_m - C_n)K_n\psi\| \\
&\leqslant \|K_n\| \, \|(C_n - C_m)R\| \, \|(H + I)K_m R\| \, \|(H + I)\psi\|.
\end{aligned}$$

The second factor tends to zero as $n \to \infty$ and the remaining factors are bounded as $n \to \infty$. Thus the resolvents converge strongly. By Proposition 7.1 and the remark following it, $K = \lim_{n \to \infty} K_n$ is the resolvent of a self-adjoint operator C. It is easy to see that $C \supset A_0$ since for $\theta \in \mathscr{D}(H)$ and $\psi \in \mathscr{F}$,

$$\langle K\psi, (C \pm iy)\theta \rangle = \langle \psi, \theta \rangle = \lim_n \langle K_n \psi, (C_n + iy)\theta \rangle = \langle K\psi, (A_0 \pm iy)\theta \rangle.$$

Since the range of K is dense, $(C \pm iy)\theta = (A_0 \pm iy)\theta$.

Let $\theta \in \mathscr{D}(C)$. We show that $\theta \in \mathscr{D}(A_0{}^{-})$ and that $C\theta = A_0{}^{-}\theta$. Thus A is essentially self-adjoint on $\mathscr{D}(H)$, and since $AR^{\frac{1}{2}}$ is a bounded operator, we see that A is also essentially self-adjoint on any core for $H^{\frac{1}{2}}$.

Let $\chi = (C_n - iy)\theta$ and let $P_j = (1 + H/j)^{-1}$. Then $\theta_j = P_j\theta \in \mathscr{D}(H) = \mathscr{D}(A_0)$ and $\theta_j \to \theta$. Furthermore

$$(A_0 - iy)\theta_j = \lim_n (C_n - iy)P_jK_n\chi$$

since $K_n\chi \to K\chi = \theta$, since $\|(C_n - A_0)R\| \to 0$ and since $(H + I)P_j$ is bounded. However

$$\begin{aligned}\|(C_n - iy)P_jK_n\chi - P_j\chi\| &= \|(C_n - iy)[P_j, K_n]\chi\| \\ &= j^{-1}\|(C_n - iy)K_nP_j\dot{C}_nP_jK_n\chi\| \\ &\leqslant j^{-1}\|\dot{C}_nP_j\|\ \|K_n\chi\| \\ &= O(j^{-1})\|\dot{C}_nR^{\frac{1}{2}}\|\ \|(H + I)^{\frac{1}{2}}P_j\| \\ &= O(j^{-1})\,O(j^{\frac{1}{2}}) = O(j^{-\frac{1}{2}}).\end{aligned}$$

Thus

$$\begin{aligned}\|(A_0 - iy)\theta_j - \chi\| &\leqslant \limsup_n \|(C_n - iy)\theta_jK_n\chi - \chi\| \\ &\leqslant \|(P_j - I)\chi\| + O(j^{-\frac{1}{2}}).\end{aligned}$$

Hence $\theta \in \mathscr{D}(A_0{}^{-})$ and

$$A_0{}^{-}\theta = \chi + iy\theta = C\theta,$$

which completes the proof. ∎

IV Boson Quantum Field Models: Part III. Further Developments

PART III. FURTHER DEVELOPMENTS

9. LOCALLY NORMAL REPRESENTATIONS OF THE OBSERVABLES

The local algebras, as constructed in Section 8, are not acting on the Hilbert space of physical particles. On the physical Hilbert space, as in the laboratory, the states have a simple asymptotic description for large values of $|t|$. One observes isolated particles or clusters formed as bound states of several elementary particles. Because they are widely separated, the elementary particles or bound states do not interact, and they behave asymptotically like free particles. We present here the functional analysis preparation for the construction of the physical Hilbert space $\mathscr{F}_{\text{ren}}$ in Section 10. On $\mathscr{F}_{\text{ren}}$, the above asymptotic description of the states should be valid. We begin by listing without proof three general results. A state on a C^*-algebra is by definition a positive linear functional ω which satisfies the normalization condition $\omega(I) = 1$.

THEOREM 9.1. *Let $\mathfrak{A}$ be a C^*-algebra and let ω be a state on $\mathfrak{A}$. There is a Hilbert space $\mathscr{H}_\omega$, a representation π_ω of $\mathfrak{A}$ by operators on $\mathscr{H}_\omega$ and a vector $\Omega \in \mathscr{H}_\omega$ such that*

$$\omega(A) = \langle \Omega, \pi_\omega(A)\Omega \rangle.$$

The vectors $\pi_\omega(\mathfrak{A})\Omega$ are dense in $\mathscr{H}_\omega$.

THEOREM 9.2. *With $\mathfrak{A}$ and ω as above, let $\{\sigma_\lambda : \lambda \in \Lambda\}$ be a group of * automorphisms of $\mathfrak{A}$, and suppose that $\omega \,.\, \sigma_\lambda(A) = \omega(A)$ for all A in $\mathfrak{A}$ and all λ in Λ. Then there is a unique unitary operator U_λ such that $U_\lambda \Omega = \Omega$ and*

$$U_\lambda \pi_\omega(A) U_\lambda^* = \pi_\omega(\sigma_\lambda(A)).$$

Furthermore, $\lambda \to U_\lambda$ is a representation of the group Λ.

THEOREM 9.3. *Let ω be a state on a von Neumann algebra $\mathfrak{A}$. The following four conditions are equivalent.*

(i) *There is a trace class matrix T such that $\omega(A) = \mathrm{Tr}(TA)$ for all $A \in \mathfrak{A}$.*

(ii) *ω is a (possibly infinite) convex combination of the vector states $\omega_\theta(A) = \langle \theta, A\theta \rangle$.*

(iii) *ω is ultraweakly continuous.*

(iv) *If $\{A_\alpha\}$ is a monotone increasing net of operators, bounded from above, then $\sup_\alpha \omega(A_\alpha) = \omega(\sup_\alpha A_\alpha)$.*

Definition. Such an ω is called a normal state. The operator T in (i) is a density matrix for ω.

COROLLARY 9.4. *Let $\{\omega_n\}$ be a norm convergent net of normal states defined on a von Neumann algebra $\mathfrak{A}$. Then the limit state is normal also.*

Proof. Let ω be the limit state, let $\{A_\alpha\}$ be a monotone net of operators as in (iv), and let $A = \sup_\alpha A_\alpha$. Then

$$\omega(A - A_\alpha) = (\omega - \omega_n)(A) + \omega_n(A - A_\alpha) + (\omega_n - \omega)(A_\alpha).$$

We are given a bound on $\|A\|$ and $\sup_\alpha \|A_\alpha\|$, and so if n is chosen large enough, the first and third terms are bounded by $\varepsilon/3$. For fixed n and large α, the second term is also bounded by $\varepsilon/3$, and this completes the proof. ∎

THEOREM 9.5. *If ω is a normal state on $\mathfrak{A}$, then π_ω is normal (preserves the ultraweak topology).*

Proof. Let θ_n be a sequence of vectors in $\mathscr{H}_\omega$. If $\theta_n = \pi_\omega(C_n)\Omega$, $C_n \in \mathfrak{A}$, then the ultraweak continuity of $\mathfrak{A}(B) \ni A \to \langle \theta_n \pi_\omega(A)\theta_n \rangle = \omega(C_n{}^* A C_n)$ follows from that of $\omega \restriction \mathfrak{A}$. Since the vectors $\pi_\omega(C_n)\Omega$ are dense in $\mathscr{H}_\omega$, normality of the above functional for arbitrary θ_n in $\mathscr{H}_\omega$ is a consequence of Corollary 9.4. Thus the functional

$$A \to \sum_{n=1}^{j} \langle \theta_n, \pi_\omega(A)\theta_n \rangle$$

is normal and providing $\Sigma \|\theta_n\|^2 < \infty$, the same is true for the infinite sum by Corollary 9.4. Hence π_ω is normal. ∎

We now specialize to the C^*-algebra $\mathfrak{A}$ of local observables. A state ω of $\mathfrak{A}$ is called locally normal if the restriction $\omega \restriction \mathfrak{A}(B)$ is normal for each of the local subalgebras $\mathfrak{A}(B)$. Similarly a representation π_ω of $\mathfrak{A}$ is locally normal if each of the restrictions $\pi_\omega \restriction \mathfrak{A}(B)$ is normal. A sequence (or net) $\{\omega_n\}$ of states is locally norm convergent if the sequence $\{\omega_n \restriction \mathfrak{A}(B)\}$ is norm convergent for each B. In this case $\{\omega_n\}$ is w^* convergent to a state ω of $\mathfrak{A}$. The w^* convergence follows from the facts that $\|\omega_n\| = 1$ and that $\{\omega_n\}$ is w^* convergent on the norm dense subset $\cup_B \mathfrak{A}(B)$.

Remark. A local norm limit of locally normal states is locally normal. A locally normal state ω yields a locally normal representation. The first statement follows from Corollary 9.4. The second is a corollary to the proof of Theorem 9.5.

We note that the algebra $\mathfrak{A}$ has the following two properties:

(P1) $\mathfrak{A}$ is the norm closure of an increasing sequence of separable factors $\mathfrak{A}(B)$.

(P2) For $B^- \subset \operatorname{Int} C$, $\mathfrak{A}(C) \cap \mathfrak{A}(B)'$ contains a type I_∞ factor.

Haag, *et al.* (1970) have studied similar properties of C^*-algebras. In their terminology, a C^*-algebra is a sequential funnel if it is the norm closure of an increasing sequence of factors.

We work with the time zero algebras $\mathfrak{A}(B)$ of Chapter 6 and the space region $B_n = (-n, n)$. Araki (1963) proved (P1).

THEOREM 9.6. *Let B be an interval. The time zero algebra $\mathfrak{A}(B)$ is a factor.*

Proof. We prove below that

$$(\mathfrak{A}(B) \cup \mathfrak{A}(\sim B))'' = \mathscr{B}(\mathscr{F}). \tag{9.1}$$

Then $\mathfrak{A}(B)' \cap \mathfrak{A}(\sim B)' = \{\lambda I\}$. However $\mathfrak{A}(B) \subset \mathfrak{A}(\sim B)'$, and consequently

$$\mathfrak{A}(B)' \cap \mathfrak{A}(B) = \{\lambda I\},$$

so $\mathfrak{A}(B)$ is a factor. ∎

The proof of (9.1) is contained in the following two propositions.

PROPOSITION 9.7. $(\mathfrak{A}(B) \cup \mathfrak{A}(\sim B))'' = \mathfrak{A}''$.

Proof. Let χ_B be the characteristic function of the interval B. For any test function f,

$$f = \chi_B f + (1 - \chi_B) f$$

and each of the terms $\chi_B f$ and $(1 - \chi_B) f$ can be approximated in L_2 by a sequence of localized test functions:

$$g_n \to \chi_B f; \qquad \operatorname{supp} g_n \subset B$$

$$h_n \to (1 - \chi_B) f; \qquad \operatorname{supp} h_n \subset \operatorname{Int}(\sim B).$$

It follows that

$$\exp(i\phi(g_n)) \exp(i\phi(h_n)) \to \exp(i\phi(f))$$

and so

$$\exp(i\phi(f)) \in (\mathfrak{A}(B) \cup \mathfrak{A}(\sim B))''.$$

Because of the extra singularity of the kernel of π, this argument does not apply to $\exp(i\pi(f))$ in general, but it does apply if f is zero on the boundary points a and b of $B = (a, b)$. Let

$$g_{a,n}(x) = g(n(x - a)) f(a)$$

where $g \in C_0^\infty$ and g is chosen so that $g(0) = 1$, and let $g_{b,n}$ be similarly defined. Then

$$\begin{aligned} \|\mu_x^{\frac{1}{2}} g_{a,n}\|^2 &= |f(a)|^2 \int \mu(k)\, n^{-2}\, |g(k/n)^\sim|^2\, dk \\ &= |f(a)|^2\, n^{-1} \int \mu(nk)\, |g(k)^\sim|^2\, dk \\ &\to |f(a)|^2 \int k\, |g(k)^\sim|^2\, dk. \end{aligned}$$

Let L_a denote the above limit. The sequence $\|\mu_x^{\pm} g_{a,n}\|$ converges weakly to zero. Using these facts, one can show that

$$\exp\bigl(i\pi(g_{a,n})\bigr) \to \exp(-L_a/2) \neq 0 \tag{9.2}$$

in the weak operator topology, using a Taylor series expansion to define the exponential. Consequently

$$\exp\bigl(i\pi(f - g_{a,n} - g_{b,n})\bigr) \in \bigl(\mathfrak{A}(B) \cup \mathfrak{A}(\sim B)\bigr)'',$$

by the argument given for $\phi(f)$, and using the convergence (9.2),

$$\exp\bigl(i\pi(f)\bigr) \in \bigl(\mathfrak{A}(B) \cup \mathfrak{A}(\sim B)\bigr)''.$$

This completes the proof. ∎

PROPOSITION 9.8. $\mathfrak{A}'' = \mathscr{B}(\mathscr{F})$.

Proof. We use the fact that Ω_0 is a cyclic vector for $\mathfrak{A}$ and we show that $\mathfrak{A}' = \{\lambda I\}$.

If $\{e_i\}$ is an orthonormal basis in the space of test functions, one can show that

$$N = \sum_{i=1}^{\infty} a^*(e_i^-)a(e_i).$$

Let

$$N(n) = \sum_{i=1}^{n} a^*(e_i^-)a(e_i).$$

Then $N(n) \to N$ strongly on a core for N, and so $e^{itN(n)} \to e^{itN}$ strongly. One can show that $e^{itN(n)} \in \mathfrak{A}''$, and then $e^{itN} \in \mathfrak{A}''$ also. We use the fact that Ω_0 is a simple eigenvector for N for the eigenvalue zero.

Since $\mathfrak{A}'$ is a von Neumann algebra, $\mathfrak{A}'$ is generated by its orthogonal projections. Let E be a projection in $\mathfrak{A}'$. If $E\Omega_0 = 0$ then $E\mathfrak{A}\Omega_0 = 0$ and since $\mathfrak{A}\Omega_0$ is dense in $\mathscr{F}$, $E = 0$. If $E\Omega_0 \neq 0$ then $E\Omega_0$ is an eigenvector for N with eigenvalue zero, since

$$e^{itN}E\Omega_0 = Ee^{itN}\Omega_0 = E\Omega_0$$

for all $t \in R$. Since zero is a simple eigenvalue, $E\Omega_0$ is proportional to Ω_0. Consequently Ω_0 is in the range of E, and so $E\Omega_0 = \Omega_0$. Thus $EA\Omega_0 = AE\Omega_0 = A\Omega_0$ for all $A \in \mathfrak{A}$, and since Ω_0 is cyclic, $E = I$. Hence $\mathfrak{A}' = \{\lambda I\}$ and $\mathfrak{A}'' = \mathscr{B}(\mathscr{F})$. ∎

To verify (P2) we choose $f \neq 0$ in $\mathscr{S}_{\mathbf{R}}(\mathbf{R})$ with support in $C \sim B$ and let $\mathfrak{M}$ be the von Neumann algebra generated by the operators $\exp(it\phi(f) + is\pi(f))$. By the von Neumann uniqueness theorem in the case of one degree of freedom, $\mathfrak{M}$ is a factor of type I_∞. (See Simon's Chapter.) ∎

THEOREM 9.9. *Locally normal representations of* $\mathfrak{A}$ *are faithful.*

Proof. Let π be a locally normal representation. Then $(\text{kernel}\,\pi) \cap \mathfrak{A}(B)$ is an ultraweakly closed two-sided ideal in $\mathfrak{A}(B)$. Since a factor contains no ultraweakly closed two-sided ideals (Dixmier, 1957, p. 43, Cor. 3), $\pi \restriction \mathfrak{A}(B)$ is one-to-one. A one–one representation is isometric, and this is true for each B. By (P1), π is thus isometric on a dense subalgebra of $\mathfrak{A}$, and hence isometric on $\mathfrak{A}$. This completes the proof. ∎

Remark. It is not difficult to show that $\mathfrak{A}$ is simple, using (P1) and (P2). In fact (P2) has the consequence that any nonzero projection $E \in \mathfrak{A}(B)$ is infinite in $\mathfrak{A}(C)$, and hence not in the kernel of an arbitrary representation of $\mathfrak{A}$.

LEMMA 9.10. *The Hilbert space* $\mathscr{H}_\omega$ *constructed from a locally normal state* ω *of* $\mathfrak{A}$ *is separable.*

Proof. Since $\mathfrak{A}(B)$ is separable in the ultraweak topology and $\pi_\omega \restriction \mathfrak{A}(B)$ is normal, $\pi_\omega(\mathfrak{A}(B))\Omega^-$ is a separable subspace of $\mathscr{H}_\omega$. Since a countable family of such subspaces span $\mathscr{H}_\omega$ by (P1), $\mathscr{H}_\omega$ is separable also. ∎

THEOREM 9.11. *Let* ω *be a locally normal state on* $\mathfrak{A}$. *Then* $\pi_\omega \restriction \mathfrak{A}(B)$ *is unitarily implemented, for each bounded region* B.

Proof. We only use properties (P1) and (P2) of $\mathfrak{A}$. By Theorem 9.9, $\mathfrak{A}(B)$ and $\pi_\omega(\mathfrak{A}(B))$ are isomorphic von Neumann algebras and by (P2) $\mathfrak{A}(B)'$ and $\pi_\omega(\mathfrak{A}(B))'$ contain infinite families of orthogonal equivalent projections, constructed within $\mathfrak{A}(C)$ and $\pi_\omega(\mathfrak{A}(C))$ respectively. Since $\mathscr{H}_\omega$ is separable, $\mathfrak{A}(B)'$ and $\pi_\omega(\mathfrak{A}(B))'$ are of countable type. Consequently the theorem follows from Dixmier (1957). ∎

COROLLARY 9.12. *Let* ω *be a locally normal state of* $\mathfrak{A}$ *and let* Λ *be the subgroup of the Lorentz group leaving* ω *fixed*: $\omega \circ \sigma_\lambda = \omega$, $\lambda \in \Lambda$. *Then the unitary representation* U_λ *of Theorem* 9.2 *is strongly continuous on* Λ.

Proof. σ_λ is locally implemented by continuous unitary groups, by the proof of Theorem 8.1. Thus the same is true for $\pi_\omega \circ \sigma_\lambda$ by Theorem 9.10. Hence

$U_\lambda \pi_\omega(A) U^* \Omega = U_\lambda \pi_\omega(A)\Omega$ depends continuously on λ, for A in some local algebra. Since the vectors $\bigcup_B \pi_\omega(\mathfrak{A}(B))\Omega$ are dense in $\mathscr{H}_\omega$, the proof is complete. ∎

10. THE CONSTRUCTION OF THE PHYSICAL VACUUM

We apply the theory of Section 9 to the physical vacuum state ω. The vacuum state ω which we construct is locally normal and translation invariant. We conjecture that it is invariant under the full Lorentz group. On the renormalized Hilbert space $\mathscr{F}_{\text{ren}} = \mathscr{H}_\omega$, the space time automorphism $\sigma_{t,x}$ is implemented by a unitary operator $U(t, x) = \exp(itH - ixP)$ and $H = \lim_{g \to 1} H(g)$. All but two of the Wightman axioms have been verified for the field operators acting on $\mathscr{F}_{\text{ren}}$ (Glimm and Jaffe, In press). The missing axioms are the uniqueness and the Lorentz invariance of the vacuum. (The spectrum condition, also a Wightman axiom, is known to be equivalent to the Lorentz invariance of the vacuum for $P(\phi)_2$ (Streater, 1972)).

The construction of ω begins with vacuum state $\Omega_g \in \mathscr{F}$ for the Hamiltonian $H(g)$. Let

$$E(g) = \inf \text{spectrum}\,(H_0 + H_I(g)). \tag{10.1}$$

$E(g)$ is the vacuum energy of $H(g)$. We renormalize, by subtracting this vacuum energy from $H(g)$, and it is also convenient to change notation, and call the renormalized operator $H(g)$, so that now

$$H(g) = H_0 + H_I(g) - E(g). \tag{10.2}$$

The spectrum of $H(g)$ now begins at zero.

THEOREM 10.1. *Zero is a simple eigenvalue of* $H(g)$.

The proof of this theorem is not difficult, but it depends on a periodic approximation $H(g, V)$ to $H(g)$, and since we have not set up the notation to handle $H(g, V)$, we refer the reader to Glimm and Jaffe (1972).

Let Ω_g be the eigenvector of $H(g)$ for the eigenvalue zero. We call Ω_g the vacuum state of $H(g)$. As $g \to 1$, the vacuum states Ω_g converge weakly to zero, according to formal perturbation theory. In order to obtain the vacuum state ω, we take limits in the sense of C^*-algebra states, rather than in the sense of Hilbert space vectors. Let

$$\omega_g(A) = \langle \Omega_g, A\Omega_g \rangle. \tag{10.3}$$

We select a sequence $g_n(\cdot) \to 1$ and we average the state ω_{g_n} over space translations, over a distance depending on n and increasing to infinity as $n \to \infty$, see Glimm and Jaffe (1970a). (For the periodic space cut-off approximation, this spatial averaging of the vacuum is unnecessary, see Glimm and Jaffe (1971a). Let ω_n be the resulting averaged state. Then ω_n is normal, and hence locally normal, by construction. The most difficult step in the construction of ω is the next theorem.

THEOREM 10.2. *Let B be a bounded region. The sequence $\{\omega_n \restriction \mathfrak{A}(B)\}$ of states of $\mathfrak{A}(B)$ lies in a norm compact set.*

Given the theorem, we select a subsequence $\omega_{n(j)}$, which converges in norm on each local subalgebra $\mathfrak{A}(B)$. Let ω be the limit. Then ω is locally normal by the remark to Theorem 9.5. Because of the space translation invariance built into the sequence ω_n by the space averaging, it is easy to show that ω is invariant under space translation. It is even easier to show that ω is invariant under time translations, since each state ω_g is invariant:

$$\omega_g(A) = \omega_g(e^{itH(g)} A\, e^{-itH(g)}).$$

Thus Theorems 9.1, 9.2, 9.9, 9.11 and Corollary 9.12 are applicable.

We call $\mathscr{H}_\omega = \mathscr{F}_{\text{ren}}$ the physical Hilbert space, and $\Omega \in \mathscr{F}_{\text{ren}}$ is the physical vacuum. Let H and P be the infinitesimal generators of the time and space translation unitary groups. H is the Hamiltonian operator, and all cut-offs are now removed. P is the momentum operator. Since the unitary operators in the translation group leave Ω fixed (cf. Theorem 9.2),

$$H\Omega = 0 = P\Omega.$$

It is not difficult to relate the spectrum of H to that of $H(g)$:

$$\sigma(H) \subset \limsup_{g \to 1} \sigma(H(g))$$

and consequently

$$0 \leqslant H.$$

Beyond the missing Wightman axioms, the major open problem for the $P(\phi)_2$ model concerns the spectrum of the mass operator $(H^2 - P^2)^{\frac{1}{2}}$ and the construction of scattering states.

We now sketch some of the ideas in the proof of Theorem 10.2. The first step is an estimate on $E(g)$ in (9.1).

THEOREM 10.3. *Let* $0 \leqslant g \leqslant 1$ *and let* g *have compact support. There is a constant* M *independent of* g *such that*

$$-M \times (\text{diam supp}\, g + 1) \leqslant E(g).$$

Proof. Let $b(x)$ be the configuration space annihilation operator, so that $b(f) = \int b(x) f(x)$. Let

$$N_i = \int_{i-\frac{1}{2}}^{i+\frac{1}{2}} b^*(x) b(x)\, dx$$

$$N_{\text{loc}} = \sum_{i=-\infty}^{\infty} N_i e^{-m|i|/2}.$$

With g supported in a fixed interval and $d_{\nu,l}$ defined as in the proof of Theorem 7.4, we assert that

$$\|d_{\nu,l}(N_{\text{loc}} + I)^{-p}\| \leqslant O(\kappa_\nu^{-\frac{1}{4}}). \tag{10.4}$$

Our proof of Theorem 7.4 is valid with $\varepsilon N_{\text{loc}}$ replacing N_0 and (10.4) replacing (7.5). Thus

$$0 \leqslant \varepsilon N^{\text{loc}} + H_I(g) + O(1).$$

Summing this inequality over translates and using the bound $N \leqslant m^{-1} H_0$ yields Theorem 10.3. The proof of (10.4) is not difficult, see Glimm and Jaffe (1971b). ∎

The next step in the proof of Theorem 10.2 is to pass from an estimate on $E(g)$ to an estimate on a local free energy operator $H_{0,\zeta}$. Let ζ be a non-negative C_0^∞ function and let

$$H_{0,\zeta} = \int b^*(x)\zeta(x) k(x-y) \zeta(y) b(y)\, dx\, dy$$

where k is the kernel of the operator μ_x. ∎

THEOREM 10.4. *There is a constant* $M = M(\zeta)$, *independent of* n, *such that*

$$\omega_n(H_{0,\zeta}) \leqslant M(\zeta) \tag{10.5}$$

and $M(\zeta)$ *is invariant under space translations.*

Formal proof. We use the fact that diam supp $g_n = O(n)$. Thus $E(g_n) = O(n)$ and also $E(2g_n) = O(n)$. Hence

$$0 \leqslant H_0 + 2H_I(g_n) + O(n),$$

and so

$$0 \leqslant \tfrac{1}{2}H_0 + H_I(g_n) + O(n),$$

$$\tfrac{1}{2}H_0 \leqslant H(g_n) + O(n),$$

and

$$\omega_{g_n}(H_0) = \langle \Omega_{g_n}, H_0 \Omega_{g_n} \rangle \leqslant 2\langle \Omega_{g_n}, (H(g_n) + O(n))\Omega_{g_n} \rangle \leqslant O(n).$$

Let $\zeta_j(x) = \zeta(x - j)$ be the translate of ζ, and let

$$H_0' = \Sigma' H_{0,\zeta_j}$$

where the sum Σ' runs over a subset of the integers. Then $H_0' \leqslant \text{const}\, H_0$ and so $\omega_{g_n}(H_0') \leqslant O(n)$. Recall that ω_n was obtained from ω_{g_n} by an average over space translates over a distance $O(n)$. We choose Σ' to run over $O(n)$ integers, namely the integers lying in the above set of space translates. Then one sees that

$$\omega_n(H_{0,\zeta}) = O(n^{-1}) \int_0^1 \omega_{g_n}(\sigma_a(H_0'))\, da$$

where σ_a is the space translation automorphism. The above bounds on $\omega_{g_n}(H_0')$ apply equally well to $\omega_{g_n}(\sigma_a(H_0'))$, and so $\omega_n(H_{0,\zeta})$ is bounded, as required. ■

The inequality (10.5) defines the set of states in Theorem 10.2. The most difficult component of Theorem 10.2 is to show that the set of states satisfying inequalities of the form (10.5) is norm compact, on restriction to $\mathfrak{A}(B)$. We now explain one ingredient of the proof, namely Rellich's theorem. Rellich's theorem asserts that the set of functions having support in a bounded region C, and having L_2 derivatives with norm at most one is compact in L_2. Rellich's theorem is also valid for L_2 derivatives of fractional order. Now the inequality in Theorem 10.4 asserts that the wave functions making up the state ω_n have L_2 derivatives of order $\frac{1}{2}$, and locally there is a bound on these L_2 derivatives. Thus a subsequence ω_{n_j} can be selected so that the sequence of wave functions occurring in ω_{n_j} converges in the Fock space norm, on restriction to any local region.

However this analysis does not complete the proof. The localization of states is expressed by the Newton–Wigner localization, that is by the operator $b^*(x)$, while the localization of operators is expressed through the field operator

$$\phi(x) = 2^{-\frac{1}{2}}\mu_x^{-\frac{1}{2}}(b^*(x) + b(x)).$$

If $k^{\pm}(x, y)$ is the kernel of $\mu_x^{\pm\frac{1}{2}}$, then

$$k^{\pm}(x, y) = O \exp\left(-(m - \varepsilon)|x - y|\right)$$

as $|x - y| \to \infty$. Thus the difference between the relativistic localization of the operators $\phi(x)$ and $\pi(x)$ and the Newton Wigner localization of states is small.

Let $\mathfrak{A}_{NW}(B)$ be the von Neumann algebra generated by the operators

$$\exp\left(i\phi(\mu_x^{\frac{1}{2}}f) + i\pi(\mu_x^{-\frac{1}{2}}g)\right)$$

where f and g range over test functions with support in B. The localization of operators $A \in \mathfrak{A}_{NW}(B)$ agrees with the localization of states, and one uses Rellich's theorem to show that the sequence $\{\omega_n \restriction \mathfrak{A}_{NW}(B)\}$ has a norm convergent subsequence, for each bounded region B.

Any bounded operator on Fock space can be expanded as an infinite series in the monomials

$$\int :\phi(x_1) \ldots \phi(x_m)\pi(y_1) \ldots \pi(y_n): w(x, y)\, dx\, dy.$$

Given the region B and given $\delta > 0$, we choose a large region $B_1 = B_1(\delta, B) \supset B$, and we use the $\phi - \pi$ expansion to express certain $A \in \mathfrak{A}(B)$ as a sum,

$$A = A_1 + A_2, \tag{10.7}$$

where

$$A_1 \in \mathfrak{A}_{NW}(B_1), \qquad |\omega_n(A_2)| \leqslant \delta\|A\|. \tag{10.8}$$

It is the small tail (10.6) which leads to the small tail A_2 in (10.7). Theorem 10.2 follows without difficulty from the expansion (10.7–8) and the norm compactness property for the Newton–Wigner localized sequences $\{\omega_n \restriction \mathfrak{A}(B_1)\}$. See Glimm and Jaffe (1970a) for details. ∎

11. FORMAL PERTURBATION THEORY AND MODELS IN THREE SPACE–TIME DIMENSIONS

In three or more space–time dimensions the degree of ultraviolet divergence depends on the degree of the polynomial P. Thus we put aside general polynomial interactions and study the superrenormalizable case ${\phi_3}^4$. The model ${\phi_3}^6$ is more difficult; it is renormalizable but not super renormalizable, and it is equal in difficulty to ${\phi_4}^4$. To say that a model is super renormalizable means that the ultraviolet, or large k, infinities can be written as a polynomial in the coupling constant whose coefficients are divergent integrals. The coupling constant is the coefficient of H_I in the Hamiltonian. The problem of Section 9 and 10, the limit $g \to 1$, involves a divergence proportional to diam supp g in each order of the coupling constant, and hence displays in the infinite volume limit a nonsuperrenormalizable phenomenon.

The $P(\phi)_2$ model is also superrenormalizable. To see this, let $H(g) = H_0 + \lambda H_I(g)$, $0 < \lambda$. The only infinities in $H(g)$ are those removed by the Wick ordering in H_I, and they have the form

$$\lambda c^j = \lambda \left[(4\pi)^{-1} \int \frac{dk}{\mu(k)} \right]^j, \qquad j \leqslant \tfrac{1}{2} \deg P, \tag{11.1}$$

as in (5.11). The $P(\phi)_2$ model is the only model with such simple ultraviolet renormalizations. The next most difficult model is the two-dimensional Yukawa interaction of bosons and fermions. This model involves ultraviolet divergent shifts in the vacuum energy and in the Boson mass. Nonetheless, the Yukawa_2 model is now understood practically at the same level of completeness as $P(\phi)_2$, see Glimm and Jaffe (1972). The Yukawa_2 model is intermediate in difficulty between $P(\phi)_2$ and $\phi_3{}^4$, about which only preliminary technical results are known.

In this section we give a simple picture on a formal level of renormalization in the $\phi_3{}^4$ model. This formal picture is justified by a rigorous mathematical analysis. Glimm (1968) established that $H(g)$ is a densely defined symmetric operator. Hepp (1970) and Fabrey (1971) constructed the time zero algebra $\mathfrak{A}$ and its local subalgebras $\mathfrak{A}(B)$. While these steps may sound easy, and are trivial for the $P(\phi)_2$ model, they are in fact difficult because of the renormalizations involved. Eventually, our program is to construct a $\phi_3{}^4$ model by proving the results known in $P(\phi)_2$. See also Eckmann and Osterwalder (1971) and Eckmann (1972). Thus some main steps yet to be established are:

(a) Positivity of $H(g)$: $0 \leqslant H(g)$

(b) Self-adjointness of $H(g)$: $H(g) = H(g)^*$

(c) To show that $H(g)$ implements locally a * automorphism σ of $\mathfrak{A}$ with finite propagation speed.

In dealing with $\phi_3{}^4$, we are motivated by perturbation theory which comprises the formal calculations mentioned above. It is convenient to use Friedrichs graphs to describe Wick monomials in our calculations. The Friedrichs graph of the Wick monomial W of (4.6) has a single vertex and $m + n$ legs, the m creating legs pointing to the left and the n annihilating legs pointing to the right.

$W =$ 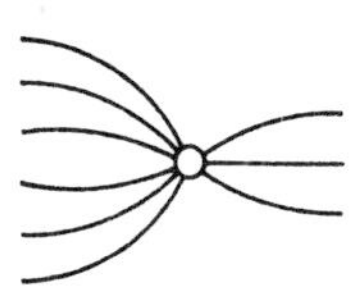 (11.2)

m creating legs $\quad$ n annihilating legs.

Thus in the language of Friedrichs graphs,

$\phi(x) =$ 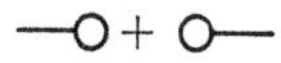 (11.3)

$:\phi^2(x): =$ 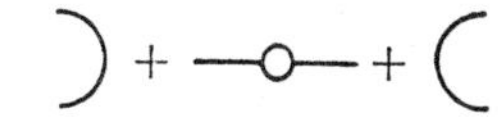 (11.4)

$:\phi^4(x): =$  (11.5)

The commutation relations $[a, a^*] = \delta I$ can also be expressed in Friedrichs graphs:

$a\,a^* =$ 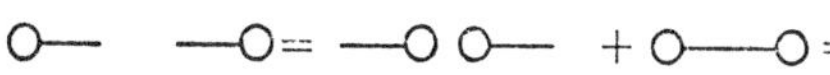 $= a^*a + \delta I.$

Here the contracted term, δI, has two vertices and a line joining them. More generally in the Wick expansion (4.12), we introduce a graph for U_r by contracting r creating legs from the vertex of W_1 to r annihilating legs from the vertex of W_2. We note in passing that the factor $r!\binom{j}{r}\binom{m}{r}$ in (4.14) is just the number of ways of performing these contractions. As a special case of (4.12), we have

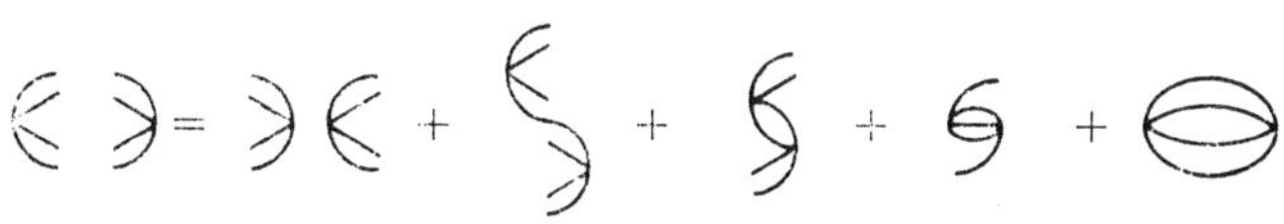 (11.6)

The first term is the Wick ordered term, and it has the least singular kernel. The last term is the fully contracted term. It is a multiple of the identity operator and its kernel is the most singular of the terms above.

The Friedrichs graph, for example (11.2), together with the kernel w uniquely determines the operator. For multivertex graphs, as in (11.6), there is a kernel for each vertex. Generally the choice of the kernel is implied by the context, and then the graph itself uniquely determines the operators. In our discussion of perturbation theory, the kernels come from H_I, and are proportional to (5.2), (5.4) or (5.8), depending on which cut-offs are employed in H_I.

For a kernel w of degree m, n, let

$$\gamma w(k, k') = \frac{w(k, k')}{\sum_{i=1}^{m} \mu(k_i) - \sum_{j=1}^{n} \mu(k_j')} \tag{11.7}$$

and let ΓW be the Wick monomial of degree m, n with kernel γw. The proper interpretation of the singularity in the denominator is important for some aspects of perturbation theory, for example in distinguishing between incoming and outgoing states, but it is unrelated to the ultraviolet divergences we are studying, and so we take the simplest prescription, which is the principal value. The merit of the Γ operation is that

$$[H_0, \Gamma W] = W, \tag{11.8}$$

or in other words, Γ is an inverse to $\operatorname{ad} H_0$,

$$\operatorname{ad} H_0 : X \to [H_0, X] \tag{11.9}$$

$$\operatorname{ad} H_0 \circ \Gamma : X \to X. \tag{11.10}$$

In graph notation, we write a Γ near the vertex to indicate the Γ operation, so that

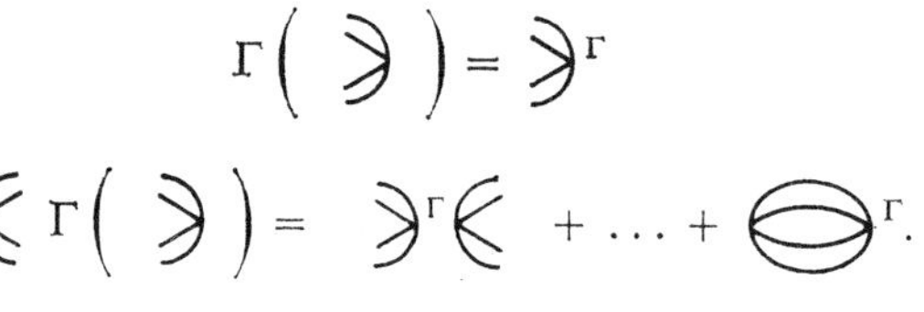

(11.11)

Motivated by the wave operator from potential scattering, we seek an operator T which intertwines H and H_0,

$$HT = TH_0. \tag{11.12}$$

For T to exist, H and H_0 must have the same spectrum. We adjust the spectrum by adding counterterms permitted by the physical interpretation of H. With

$$H = H_0 + \lambda H_I(g) - \tfrac{1}{2}\delta m^2 \int :\phi^2(x): g^2(x)\, dx - E(g), \tag{11.13}$$

we seek to determine the renormalization constants $E(g)$ and δm^2 as well as T. Let us expand formally

$$T = I + \lambda T_1 + O(\lambda^2)$$

$$E(g) = \lambda E_1 + \lambda^2 E_2 + O(\lambda^3)$$

$$\delta m^2 = \lambda \delta m_1{}^2 + \lambda^2 \delta m_2{}^2 + O(\lambda^3)$$

so that

$$E(g) = \inf \operatorname{spectrum} \left(H_0 + \lambda H_I(g) - \tfrac{1}{2}\delta m^2 :\phi^2(g^2):\right).$$

Let

$$\Omega(g) = \Omega_0 + \lambda \Omega_1 + O(\lambda^2).$$

We show that $E_1 = 0$. Since $\langle \Omega(g), \Omega_0 \rangle \neq 0$,

$$E(g) = \frac{\langle \Omega_0, (H_0 + \lambda H_I(g) - \frac{1}{2}\delta m^2 \colon \phi^2(g^2)\colon)\Omega(g)\rangle}{\langle \Omega(g), \Omega_0 \rangle}$$

$$= \frac{\lambda \langle \Omega_0, (H_I(g) - \frac{1}{2}\delta m_1{}^2 \colon \phi^2(g^2)\colon)\Omega_0\rangle + O(\lambda^2)}{1 + O(\lambda)}. \qquad (11.14)$$

Since H_I and $:\phi^2:$ are Wick ordered,

$$E_1 = \langle \Omega_0, (H_I(g) - \frac{1}{2}\delta m_1{}^2 \colon \phi^2(g)\colon)\Omega_0\rangle = 0.$$

A similar analysis shows that $\delta m_1 = 0$, assuming that the coefficient of $:\phi^2:$ in H_I is zero. Substituting these values in (11.12), we have

$$(H_0 + \lambda H_I(g))(I + \lambda T_1) = (1 + \lambda T_1)H_0 + O(\lambda^2),$$

or

$$\lambda H_I(g) = \lambda[T_1, H_0].$$

Thus we set

$$T = I - \lambda \Gamma H_I(g) + O(\lambda^2).$$

Then

$$\Omega_g = T\Omega_0/\|T\Omega_0\| = \frac{\Omega_0 - \lambda \Gamma H_I(g)\Omega_0 + O(\lambda^2)}{\|\Omega_0 - \lambda \Gamma H_I(g)\Omega_0 + O(\lambda^2)\|}$$

$= \dfrac{\Omega_0 - \lambda \Gamma H_I(g)\Omega_0 + O(\lambda^2)}{\left(1 + \lambda^2\left(\Gamma\right.\right.}$ 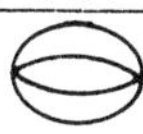 $\dfrac{}{\left.\left.\Gamma\right) + O(\lambda^2)\right)^{\frac{1}{2}}}$

$= \Omega_0 - \lambda$  $\Gamma\, \Omega_0 + O(\lambda^2)$.

Substituting in (11.14), we have

$E(g) = \dfrac{-\frac{1}{2}\delta m^2 \lambda^2 \langle \Omega_0 \colon \phi^2(g^2)\colon \Omega_0\rangle - \lambda^2 \langle \Omega_0, H_I(g)\right.}{1 + O(\lambda^2)}$  $\Gamma\, \Omega_0\rangle + O(\lambda^3)$

$= -\lambda^2 \langle \Omega_0, H_I(g)$  $\Gamma\, \Omega_0 \rangle + O(\lambda^3)$

$= -\lambda^2$ 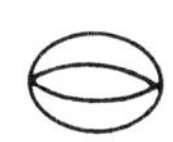 $\Gamma + O(\lambda^3)$

and so

$E_2 = -$ 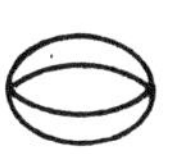 Γ.

Let

$-\frac{1}{2}\delta m^2 \colon \phi^2(g^2)\colon =$  $+$ 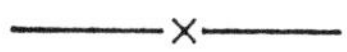 $+$ 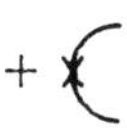

Similarly it can be shown that $\delta m_2{}^2$ is chosen so that the kernel of the second of the following three operators

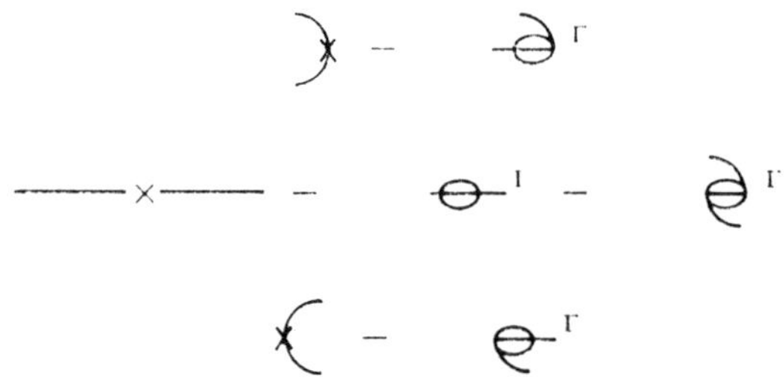

vanish at zero momentum. With this choice of $\delta m_2{}^2$, all three operators above have finite valued kernels, in the limit $\kappa = \infty$. We now determine T to second order:

$$H(g)T - TH_0 = -\lambda^2(H_I\Gamma H_I + E_2 + \tfrac{1}{2}\delta m_2{}^2 :\phi^2(g^2):) + [H_0, T_2].$$

Thus

$$T_2 = \Gamma(H_I\Gamma H_I + E_2 + \tfrac{1}{2}\delta m^2 :\phi^2(g^2):). \tag{11.15}$$

Now $H_I\Gamma H_I$ is not defined. The terms with three and four contractions have kernels which are divergent integrals. However in (11.15), these infinities are cancelled. Thus E_2 cancels the fully contracted term and $\frac{1}{2}\delta m^2 :\phi^2(g^2):$ cancels the infinite part of the terms with three contractions. The calculation of E and δm^2 to higher order reveals one further infinite counterterm,

$E_3 = \Gamma$  Γ .

In addition to the infinities E and δm^2, the $\phi_3{}^4$ model has an infinite wave function renormalization, Λ. From (11.15) and the choice of E_2, we see that $\langle T_2\Omega_0, \Omega_0\rangle = \langle \Omega_2, \Omega_0\rangle = 0$. Thus

$\|T\Omega_0\|^2 = 1 + \lambda^2\,\Gamma$ 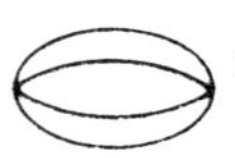 $\Gamma + O(\lambda^3) = 1 + \lambda^2\Lambda_2 + O(\lambda^3)$

and Λ_2 is logarithmically infinite. When computed to higher order, one finds

$$\|T\Omega_0\|^2 = \exp(\lambda^2\Lambda_2 + \dots)$$

and $\Lambda_3, \dots$ are all finite. The fact that Λ_2 is infinite means that T maps out of Fock space. The range of T lies in the renormalized Fock space

$\mathscr{F}_{\text{ren}}$. The change of Hilbert space can be described mathematically as a change to an inequivalent representation of the canonical commutation relations. Eckmann (1972) proved that the representations of a fixed local algebra $\mathfrak{A}(B)$ on $\mathscr{F}_{\text{ren}}$ are independent (in the sense of unitary equivalence) of the space cut-off g, provided $g \equiv 1$ on a neighborhood of B.

TWO NOTES

1. In general, we use real valued functions when working with the position variable x and we use complex valued functions when working with the Fourier transform, or momentum variable k. However the Hilbert spaces on which the canonical commutation relations are represented are always complex.

2. Minlos (1959) proved that a measure is countably additive on the dual of a nuclear space, provided the measure satisfies a weak continuity condition. For Gaussian measures, the proof is simpler. We follow the presentation of Gel'fand and Vilenkin (1961). The existence of a Schrödinger representation, diagonalizing the time zero field operators is possible only in two or three space time dimensions, according to formal considerations, for fields with conventional nonlinear interactions. For free fields or fields with linear interactions there is no such restriction. The Schrödinger representation is implicit in many discussions of free fields, see for example Heisenberg and Pauli (1930) and von Neumann (1932). von Neumann (ca. 1934–1935) constructed a Gaussian measure over an abstract (pre-) Hilbert space $\mathscr{H}$, and showed that the measure was countably additive, when defined on the algebraic dual $\mathscr{H}^* \supset \mathscr{H}$. This result defined the Hilbert space on which the Schrödinger representation exists; the result was not published and was reviewed posthumously in 1961. Friedrichs (1953) outlined the construction of the Schrödinger representation in detail. He used a Gaussian measure and a Hermite expansion to define the inner product in $L_2(d_Bq)$; the commutation relations are given in unbounded form and the unitary equivalence with the Fock representation can be read from Friedrichs' formulas. The connection with Kolmogoroff's theorem and the construction of a measure space with a countably additive measure was promised in a later publication. (Cf. Friedrichs and Shapiro 1957). In 1956, Segal gave a self contained and mathematically complete account of the Schrödinger representation from an abstract point of view. He established the canonical commutation relations in the integrated, or Weyl form. We note finally that fields with nonlinear interaction lead to the consideration of non-Gaussian measures on Q.

V The Particle Structure of the Weakly Coupled $P(\varphi)_2$ Model and Other Applications of High Temperature Expansions: Part I. Physics of Quantum Field Models

THE PARTICLE STRUCTURE OF THE WEAKLY COUPLED $P(\varphi)_2$ MODEL AND OTHER APPLICATIONS OF HIGH TEMPERATURE EXPANSIONS, PART I:

PHYSICS OF QUANTUM FIELD MODELS

James Glimm*
Courant Institute
New York, N.Y.

Arthur Jaffe†
Harvard University
Cambridge, Mass.

Thomas Spencer‡
Courant Institute
New York, N.Y.

1. FIVE YEARS OF MODELS

1.1. An Overview

Constructive quantum field theory has moved rapidly in the past few years. A continuation of this progress clearly requires the introduction of new formal ideas, methods and/or points of view. In these lectures, we will review past progress, present some new results and point out possible directions for future work. In this Part I, we emphasize the basic ideas and concepts, while in Part II we present in full detail some core results from our program. In particular, we present the vacuum cluster expansion and estimates to establish its convergence.

For the two dimensional $\mathcal{P}(\varphi)$ model, a fairly detailed structure is known, and this structure is in qualitative agreement with basic ideas of physics. From the Haag-Ruelle axioms (verified, for instance, for small coupling), we know that fields describe discrete mass particles. An isometric scattering operator expresses the interactions between these particles. We know that the vacuum state is unique for certain coupling constants (e.g., small coupling, or a large - or in some cases, a nonzero - external field). Dobrushin and Minlos have announced that there are multiple phase solutions for even $\mathcal{P}(\varphi) + m_0^2 \varphi^2$ models with $\lambda >> m_0^2$. Symmetry breaking plays a key role in current theories of weak interactions, hence the interest in this phenomenon. There is no direct experimental evidence for or against occurence of broken symmetries in elementary particle physics, since the interparticle coupling constants cannot be varied experimentally (in distinction to the case of statistical mechanics where we can, for example, turn off a magnetic field). Consequently the definitive argument in favor of broken symmetries may come from constructive quantum field theory.

The Yukawa$_2$ (Y_2) and φ_3^4 models are less highly developed. Yet many of the formal ideas developed for $\mathcal{P}(\varphi)_2$ models appear to apply to superrenormalizable models in general. Clearly then, one set of problems is to develop stronger techniques, to make these ideas applicable to Y_2, φ_3^4 and Y_3 . We propose, in fact, four groups of problems.

I. Physical Properties. One important direction for future work is to develop further the physics of existing quantum field models. The particle structure program, bound states, resonances and scattering present interesting problems. Likewise, the long distance and infrared behavior of our models contains much physics. The general particle structure program is: Which interaction polynomials and which coupling constants give rise to which particles, bound states and resonances? How do the masses and half lives depend on the coupling constants? How do cross sections behave asymptotically? We discuss these problems further in Section 1.5 and Chapter 3.

The long distance behavior of our models pertains to the existence of multiple phases, to the existence of a critical point and to the scaling behavior of the models at a critical point. We ask: Does the φ_2^4 model have a critical point? Does it admit scaling properties with anomalous dimensions? What parameters describe the critical point? We discuss these questions further below and in Section 1.5.

II. Four Dimensions (Renormalizable Models). A second important direction is the question of four space-time dimensions, or in other words how to deal with renormalizable interactions, since there are no super-renormalizable interactions in four dimensions. Clearly this is our most challenging goal, to prove the existence of, for example, φ_4^4 . Our present methods have been tied to superrenormalizability (4 - ϵ dimensions) and for $\epsilon = 0$ new ideas are required. We ask: Can an understanding of the renormalization group be an aid to removing the $\epsilon = 0$ ultraviolet cutoff? Do the ideas in the lectures of Symanzik yield insight into charge renormalization? We discuss these questions further in Section 1.5.

III. Simplification. Aside from these two major directions, there is the question of simplifying the present methods. Clearly the major need for simplification concerns problems with ultraviolet divergences, and a major goal of such a program would be to improve the techniques and isolate their essential elements in order to make tractable more complicated superrenormalizable models, such as Y_3, or even Y_2.

IV. Esthetic Questions. Furthermore, there are esthetic or foundational questions. For example, the Schrödinger representation $\mathcal{H} = \mathcal{L}_2(dq)$ exists for $P(\varphi)_2$ models ; what is the fermion representation corresponding to this non-Gaussian boson measure on $\mathcal{S}'$? What are the properties of the path space measures in models with interaction? Related are interesting, but purely mathematical questions motivated by field theory, which we do not pursue here.

In this connection, we remark that the drive toward simplicity and elegance is important and also has been quite successful in the $P(\varphi)_2$ model. However, we emphasize here those methods that admit (or appear to admit) generalization to other more singular interactions. The reason for this emphasis is two-fold. First, we believe that, in the long run, our ability to handle more singular problems will determine the extent to which the model program has succeeded. Second, we believe that a premature emphasis on the simplicity and elegance of the details can divert energy away from central issues, and thereby delay or obstruct progress.

1.2. Survey of Results

To begin, we review the status of the φ_2^4, Y_2 and φ_3^4 models. We give a chronological summary in Figure 1, plotting models of increasing complexity versus results of increasing complexity. In this chart, we enter the years in which these results were proved.

In Figure 2 we give details and references for various φ_2^4 models. The results quoted for $\lambda/m_0^2 \ll 1$, also hold for $\lambda P(\varphi)_2$ models with $\lambda/m_0^2 \ll 1$. We make several comments: The Wightman axioms require a unique ground state (vacuum), namely the existence of a single vector, invariant under inhomogeneous Lorentz transformations. Alternatively, we consider the C^* algebra vacuum state of a finite volume theory, and its infinite volume limits. Each infinite volume state yields a representation, and a Hilbert space vacuum vector. Uniqueness of the vacuum, as required for the Wightman axioms, refers to vectors in this Hilbert space, and is equivalent to irreducibility of the representation. The infinite volume vacuum state is determined as a limit of finite volume states. The latter are determined by parameters in the energy density $\mathcal{H}(x)$ and the boundary conditions. If the parameters in $\mathcal{H}(x)$ alone are sufficient to specify a unique vacuum, independent of the boundary conditions, then there is said to be a unique phase, and otherwise there are multiple phases. Convergent cluster expansions [Gl Ja Sp 1, 2] yield for certain couplings both a unique vacuum and a single phase.

In a $P(\varphi)_2$ theory satisfying the Wightman axioms, except for the uniqueness of the vacuum, the decomposition theorem of Bratteli [Br 1] allows us to decompose the observables and recover a unique vacuum. The local perturbation estimate [Gl Ja IV] and a result of Streater [St] ensure the spectral condition for the decomposed theory. In this manner we arrive at a Wightman theory for an arbitrary $\lambda\varphi^4 + \frac{1}{2}m_0^2\varphi^2 - \mu\varphi$ interaction. In the case $\mu \neq 0$, the Lee-Yang theorem shows that the decomposition is unnecessary [Gr Si, Si II].

In Figure 3, we have details and references for Y_2, and clearly much work needs to be done to bring it to the level of φ_2^4.

3 = 1973

0 = 1970

9 = 1969

4 ≤ 1964

Result	$P(\varphi)_2$	Y_2	φ^4_3	φ^3_4	Y_3
Critical Point					
Asymptotic Completeness					
Resonances					
Bound States	3				
Broken Symmetries	3				
Analyticity in Coupling	3				
Perturbation Theory Asymptotic	3				
Single Particle States	3				
Mass Gap	2				
Wightman Axioms, $V \to$ Convergence	2				
Euclidean Formulation	1	2			
Wightman Functions	1				
Physical Representation	9	1			
Haag-Kastler Axioms	9				
Equations of Motion	9	1			
Space Time Covariance	8	0			
$H_V = H_V^*$	8	9			
$0 \le H_V$	5	8	2		
Local Observables	4	7	1		
H_V	4	7	8	0	

Figure 1. Main Results and Year Established

	$\frac{\lambda}{m_0^2} \ll 1$, $\frac{\lvert\mu\rvert}{m_0^2} \ll 1$. Also $\lambda P(\varphi)_2$.	$\mu = 0$	$\mu \neq 0$	$\lvert\mu\rvert$ large
Renormalized Charge, Anomalous Dimensions, Renormalization Group				
Critical Points				
S-Unitary (Asymptotic Completeness)				
Bound States Resonances		Chapter 3		
Multiple Phases (Broken Symmetries)	No: Gl Ja Sp 1	Yes: $\lambda \gg m_0^2$ (Dob Min)	No: Gr Si, Si II	→
Analyticity of the Schwinger Fns.	Re $\lambda > 0$, $0 < \lvert\lambda\rvert < \lambda_0$, $\lvert\mu\rvert < \mu_0$ Gl Ja Sp 2		Re $\lambda > 0$ Sp 2	→
One Particle States and S-Matrix	Gl Ja Sp 1			
Measure dq	New 1, Fr 2	Fr 2	→	→
Schrödinger Repn.	Fr 2			Fr 2
Perturbation Theory is Asymptotic	Di 3			Di 3
m	$m > 0$ Gl Ja Sp 1	monotonic in m_0 Gu Ro Si 3, Si I	monotonic in $\lvert\mu\rvert$ Gr Si	$m > 0$ Sp 2
Verify Wightman Axioms	Gl Ja Sp 1	Nel 5, Gu Ro Si 3, Gl Ja IV, Os Sch 3, Br 1, St	→ Gr Si, Si II	→ → Sp 2
Formulate Euclidean Axioms	Sy 2, Nel 3, Os Sch 3	→	→	→
Physical Representation	Gl Ja III-IV, 5	→	→	→
Haag-Kastler Axions	Gl Ja I-II, Ca Ja Ro 1-3	→	→	→
Preliminary	Ja 1, 2, Nel 1, Gl 1 Se 1	→	→	→

Figure 2. Details for $\left(\lambda\varphi^4 + \frac{1}{2} m_0^2 \varphi^2 - \mu\varphi\right)_2$

Bound States	
Particle Structure	
Schrödinger representation	
Wightman axioms	
Euclidean Fields and Feynman-Kac formula	Os Sch 4
Properties of Currents	McB 1
Equations of Motion	Di 1
Physical Representation Exists	Sch 2
Space - Time Covariance	Gl Ja 10
$0 \leq H_V = H_V^*$	Gl 2,3, Gl Ja 9, 11

Figure 3. Details for Y_2

We conjecture that the Euclidean fields given by Osterwalder and Schrader [Os Sch 4], and by Ozkaynak [Oz] for higher spin, can be used as a starting point for a cluster expansion, yielding for small coupling: the $V \to \infty$ limit, the Wightman axioms, a unique vacuum, and single particle states. We also hope that Euclidean gauge fields will lead to an understanding of function space integrals for higher spin. In particular, we mention as problems the factorization of the integration over the symmetry (gauge) group, and the renormalization scheme of Faddeev and Popov [Fa Po, Sa St].

The φ_3^4 interaction is at an even more primitive stage. The φ_3^4 renormalized Hamiltonian is bounded from below [Gl Ja 8]. Similar estimates [Fe 2] show that in a finite space time volume, approximate Schwinger functions have $\kappa \to \infty$ limits, see Chapter 4. We hope that the Wightman axioms will soon be verified for φ_3^4 .

There are several recent developments which do not fit into the above description and we mention in particular: Federbush's study of the generalized Yukawa_2 model, $P(\varphi) + \bar{\psi}\psi Q(\varphi)$, see [Fed] the work of Albeverio and Hoegh-Krohn on bounded nonpolynomial interactions [AHK 1 - 4], and recent work on the polaron model by Gross[Gro 1] and by Fröhlich [Fr 1]. Furthermore Fröhlich's recent results on $P(\varphi)_2$ models [Fr 2] yield the Markov property, DLR equations and cyclicity of the vacuum (for time zero fields) when $\lambda \ll m_0^2$ or $|\mu| \gg m_0^2, \lambda$.

There are three main unifying techniques, yielding the results charted above:

1. Euclidean Formulation. The use of imaginary time in constructive quantum field theory dates to Symanzik's Euclidean field formulation and to Nelson's positivity proof for the $\varphi^4_{2,V}$ Hamiltonian. This approach was used in most of the key technical estimates of the $P(\varphi)_2$ and φ^4_3 boson models, and in proving finite propagation speed for the Y_2 model. The covariant formulation in terms of Euclidean fields or Green's functions is important as a conceptual advance yielding Euclidean axioms, the relativistic Feynman-Kac formula and the Markov property [Nel 4, Os Sch 3, Sy 1, Nel 3, Fel 1]. The covariant formulation is important as a technical advance in proving estimates [Gu 1,Gl Ja 8], in simplifying estimates [Gu Ro Si 1, Si], and in reducing the number of properties which must be verified in order to construct a Wightman-Haag-Ruelle theory [Nel 3, Os Sch 2]. Furthermore, the Euclidean framework makes clear the connection of field theory with statistical mechanics, a central tool in the heuristic approach of Fisher and Wilson and in the work of Guerra, Rosen and Simon. The lattice approximation [Ko Wi, Gu Ro Si 3] then becomes natural, and it relates boson interactions to an Ising model with continuous spin and nearest neighbor interaction.

2. Correlation Inequalities [Gu Ro Si 3, Nel 5, Gr Si, SiII]. The convergence of the lattice approximation and the existence of the thermodynamic limit leads to the generalization of correlation inequalities and the Lee-Yang theorem from statistical mechanics to quantum field theory. These methods appear especially suited to boson interactions and are discussed in the lectures of Guerra, Rosen, Simon, Nelson and in Chapter 3 of these lectures.

3. Cluster Expansions [Gl Ja Sp 1, 2]. These high temperature expansions provide (within their region of convergence - see Figure 6) the most detailed information about $P(\varphi)_2$ models. In combination with low temperature expansions (which we believe to exist), the expansions should converge for all interactions

sufficiently far from a critical point. In fact the inductive expansions for $P(\varphi)_2$ yield an analysis of an arbitrary interval $[0, a]$ of the spectrum of the Hamiltonian H. Related are the inductive bounds for $P(\varphi)_2$ [Gl Ja IV] and for φ_3^4 [Gl Ja 8]. The bounds for φ_3^4 have not yet been refined to yield such detailed results as are known for $P(\varphi)_2$. In contrast to the expansions, the bounds hold without restriction on the magnitude of the coupling constants.

1.3. The Goldstone Structure of $P(\varphi)_2$

We now return to the $P(\varphi)_2$ model, and give a more detailed discussion. There is a very simple picture, partly due to Goldstone, and now part of the physics folklore, which accounts for the fact that some interaction polynomials P produce degenerate vacuums, while others do not. See [Go, Go Sa We, Jo L], and for recent discussions see [Co 1, Wi]. Classically, the (Euclidean) ground state of the $P(\varphi)$ model is the function $\Phi \in \mathcal{S}'(R^2)$ which minimizes the classical (Euclidean) energy

$$\int [\nabla \Phi(x)^2 + P(\Phi)]dx \quad .$$

This function Φ is a constant, equal to the value ξ which minimizes $P(\xi)$. Quantum mechanically, the (Euclidean) ground state is a probability distribution on $\mathcal{S}'(R^2)$, centered (in some sense) about the classical minimum. The quantum mechanical picture requires corrections, including Wick ordering and the higher order mass counterterms, see Coleman and Weinberg [Co We]. In $P(\Phi)_2$ models, Hepp obtained the classical limit $\hbar \to 0$ explicitly, see [He 3]. The Goldstone picture also states that (i) the physical mass m is the curvature of P at its minimum and (ii) in the case of a double (or multiple) minimum, multiple phases exist. In our discussion, we include in P the mass in the free Hamiltonian H_0, but because we ignore quantum corrections beyond (bare) Wick order, our discussion is approximate. Let $\frac{1}{2}m_0^2$ be the coefficient of Φ^2 in P. We consider the polynomials

(a) $\quad P(\xi) = \sum_{j=1}^{2n} c_j \xi^j \; ; \; c_{2n} > 0 \, , \; \frac{1}{2}m_0^2 = c_2 >> c_j \quad , \; j \neq 2 \quad .$

(b) $\quad P(\xi) = \lambda \xi^4 + \frac{1}{2}m_0^2 \xi^2 - \mu\xi \; , \; \lambda > 0 \quad , \; \mu \neq 0 \quad .$

(c) $\quad P(\xi) = P_0(\xi) - \mu\xi \; ; \; P_0$ bounded below, $\mu >> 1 \quad .$

(d) $\quad P(\xi) = \lambda P_0(\xi) + \frac{1}{2}m_0^2 \xi^2 \; ; \; \lambda > 0 \, , \; m_0^2 < 0 \, , \; P_0$ even.

(e) $\quad P(\xi) = \lambda \xi^4 \; ; \quad \lambda > 0 \quad .$

See Figure 4.

In case (b) a cubic term may be added through the trivial transformation $\xi \to \xi + \text{const}$. In (c), ξ may be replaced by any odd power ξ^{2j+1}, $2j+1 < \deg \mathcal{P}_0$. In (d), a polynomial $\xi^6 + \cdots$ of higher degree may have more than two minima.

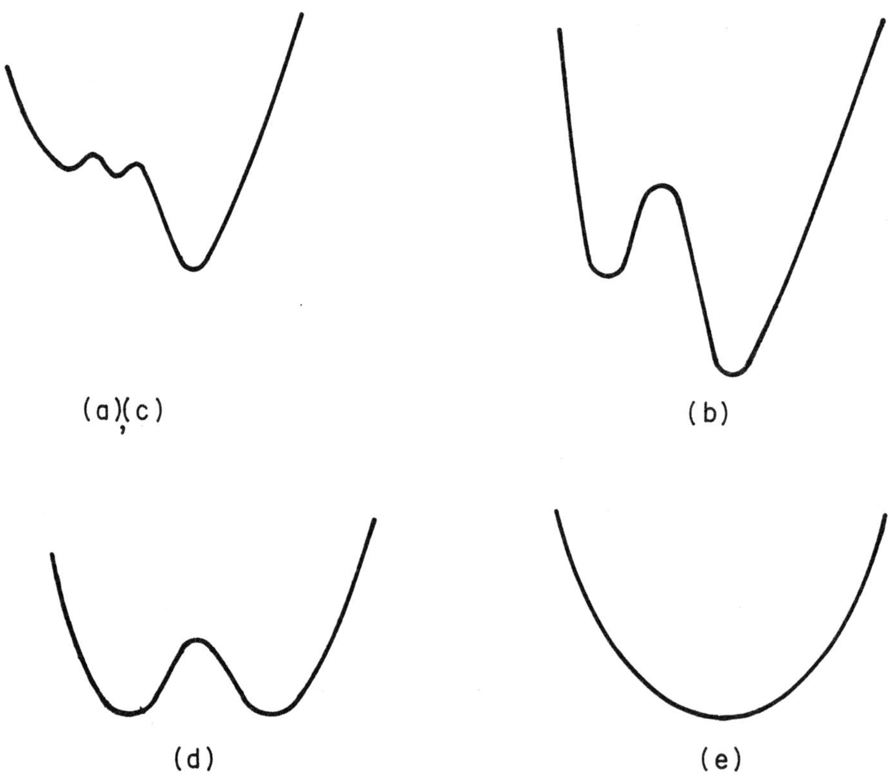

Figure 4. Various interaction polynomials $\mathcal{P}(\xi)$.

In cases (a) - (c) the polynomial $P(\xi)$ has a unique minimum with positive curvature. This suggests the existence of a single phase, and the existence of massive particles. Case (d) has a double minimum, corresponding to the symmetry $\varphi \to -\varphi$, with positive curvature at each minimum. Thus we expect two pure phases, with positive mass particles in each pure phase. The two vacuum states for the two pure phases are localized approximately at minima $\xi = \pm a$. Thus we expect in the vacuum states for these two phases that the expectation value of the field, $\langle\varphi\rangle$, approximately equals $+a$ or $-a$. Case (e) is the limiting case between Case (d) and a particular Case (a). Here $P(\xi)$ has a unique minimum at $\xi = 0$, with curvature zero. The Goldstone picture suggests the presence of zero mass particles, and is the Goldstone picture of a critical point. These figures yield the mean field approximation, which we believe gives the correct qualitative, rather than quantitative, picture. Thus the existence of the limiting case (e) suggests the existence of critical values of the coupling constants for which $m = 0$. In this connection, Figure 5 represents the Goldstone conjecture for even $\lambda P_0(\Phi) + \frac{1}{2}m_0^2\Phi^2$ models. R. Baumel [Ba] remarked that changing the definition of Wick ordering (e.g., Wick ordering in a bare vacuum with a different bare mass) permits the same H to be written with different bare parameters m_0, P. Thus the location of the critical point, when written in these parameters, can be changed, so the sign of m_0^2 at the critical point is not significant.

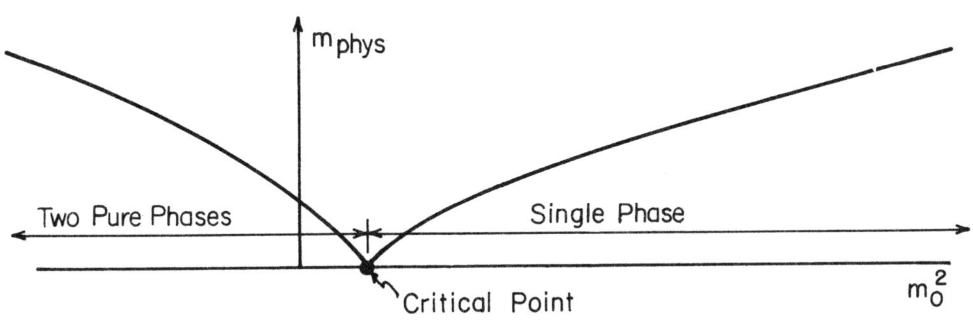

Figure 5. Conjectured symmetry breaking in even $\lambda P_0(\phi) + \frac{1}{2}m_0^2\phi^2$

Results. We give the rigorous results proved in cases (a) - (e). In case (a), convergent cluster expansions show that the mass operator $M = (H^2 - P^2)^{1/2}$ has isolated eigenvalues at 0 and $m > 0$, [Gl Ja Sp 1]. These results have been generalized to case (c) [Sp 2]. A single phase exists, the vacuum is unique and the mass $M = m$ eigenvalue is nondegenerate in the sense that the Lorentz group acts irreducibly on the one particle space. Any additional mass spectrum in the interval $(m, 2m)$ should be discrete and would describe bound states.

In case (b), Griffiths and Simon [Gr Si] extend the Lee-Yang theorem to φ_2^4 and use this result to show [SiII] that the ground state is unique. For λ large and μ small, the structure of the mass spectrum above zero is not known.

In case (d), Dobrushin and Minlos [Dob Min] have announced the existence of at least two phases.

Continuous Symmetry Breaking. Another form of the classical picture [Go Sa We] concerns a continuous symmetry group, rather than the discrete reflection group $\varphi \to -\varphi$ that we discussed above. In the continuous case, the minimum of $P(\xi)$ occurs on a manifold of dimension greater than zero, and translation along this manifold leaves $P(\xi)$ constant. The Goldstone picture now states that in the case of broken symmetry ($\delta \neq 0$ as defined below) particles of mass zero occur. Thus these particles have a mass given by the minimum curvature of $P(\xi)$ at its minimum.

In the physics literature, this broken symmetry is defined in terms of the conserved current, $\partial^\mu j_\mu = 0$, associated with the symmetry group. For classical field theories, a standard variational argument establishes the existence of the conserved current. The generator of the symmetry group is $Q = \int j_0 d\vec{x}$, and the Goldstone picture concerns the vacuum expectation value

$$\delta = \langle [iQ, \varphi] \rangle \quad .$$

If $\delta \neq 0$, discrete zero mass particles (Goldstone bosons) are believed to exist. In case that the symmetry group of automorphisms can not be unitarily implemented,

Kastler, Robinson and Swieca have shown that the mass spectrum extends down to zero [Ka Ro Sw], and Swieca [Sw] has shown that zero mass particles exist.

Two simple examples of such conserved currents are (1) a zero mass free field, with energy density $\pi^2 + (\nabla\varphi)^2$, invariant under translations $\varphi \to \varphi + \text{const.}$, and (2) a multicomponent field invariant under orthogonal transformations in the space of components.

In case (1), $j_\mu = \partial_\mu \varphi$, the parameter $\delta = 1$ and the Goldstone bosons are just the zero mass free particles. (We note this argument is not applicable in two dimensions where zero mass free scalar fields do not exist.)

Case (2) corresponds to Figure 4(d) with a rotational symmetry about the $\xi = 0$ axis. If our field has two components φ_i, then $\delta_i = \pm \langle \varphi_j \rangle$ is the vacuum expectation value of the field components. The Goldstone bosons occur when the vector, formed of the components δ_i, does not vanish i.e., $\vec{\delta} \neq 0$. In terms of $\vec{\varphi}$, the broken symmetry condition requires that the vacuum expectation $\langle \vec{\varphi} \rangle \neq 0$. Let us suppose that $\langle \vec{\varphi} \rangle \neq 0$, and let $\vec{n}$ be the unit vector in the $\langle \vec{\varphi} \rangle$ direction. We decompose $\vec{\varphi}$ into longitudinal and transverse parts, $\vec{\varphi} = \vec{\varphi}_L + \vec{\varphi}_T$, where

$$\vec{\varphi}_L = (\vec{\varphi} \cdot \vec{n})\vec{n} \quad , \quad \vec{\varphi}_T \cdot \vec{n} = 0 \quad .$$

The conventional picture is that along its trough, $\mathcal{P}(\cdot)$ has zero curvature and displays zero mass excitations, while in the orthogonal direction the curvature of $\mathcal{P}(\cdot)$ is positive and one expects massive particles. Thus one expects exponential clustering for $\vec{\varphi}_L$,

$$\langle \vec{\varphi}_L(\vec{x}) \vec{\varphi}_L(\vec{y}) \rangle \leq O(e^{-md}) \quad , \quad m > 0 \ ,$$

but polynomial clustering for $\vec{\varphi}_T$.

As we remarked above, two dimensions displays especially singular infra red behavior. In fact, Coleman [Co 2] has proved that for $d = 2$, a continuous symmetry as above always yields $\delta = 0$. Thus the usual Goldstone picture never ensures the existence of zero mass particles. The basic ingredient of Coleman's proof is the distribution character of the two point function. This precludes zero mass particles appearing in a scalar two point function and forces $\delta = 0$. His argument does not, however, prevent the mass spectrum from going down to zero. Therefore we ask: Does the $\lambda(\vec{\varphi}^{\,2})^2$ model with $\lambda >> m_0^2$ have a mass gap? We conjecture that the mass gap for small λ vanishes as λ is increased to the critical point λ_c , and that the mass gap is zero for $\lambda \geq \lambda_c$. In other words, for $\lambda \geq \lambda_c$ we expect that neither Goldstone bosons nor a mass gap occur.

1.4 Field Theory and Statistical Mechanics

The equivalence of relativistic quantum field theory with statistical mechanics has a long history. Older work includes both the Landau-Ginzberg theory and Symanzik's program to construct Euclidean models. Recent work includes that of Fisher, Wilson, Griffiths and a number of lecturers at this conference. We mention here some selected aspects of this correspondence for boson quantum fields. Ideas of this nature in models with fermions have not yet proved fruitful.

The Partition Function. Let dq denote the Euclidean measure for a boson quantum field model. The partition function

$$Z\{J\} = \int e^{\Phi(J)} dq$$

is the generating function for Schwinger functions, and has been studied in $P(\varphi)_2$ models by Fröhlich [Fr 2]. As mentioned above, the Euclidean field model has a natural approximation by a continuous spin ferromagnetic Ising lattice with nearest neighbor interaction, see for instance [Ko Wi]. The convergence of the lattice approximation [Gr Ro Si 3] and the approximation of φ^4 by spin 1/2 Ising models [Gr Si] sharpens this correspondence, see also [New 2]. The one point Schwinger function $\int \phi(x)\, dq$, which parameterizes symmetry breaking in the Goldstone picture above, corresponds to spontaneous magnetization in the Landau-Ginzberg theory. The coupling constant λ/m_0^2, which measures the deviation from a free theory, corresponds to the inverse temperature $\beta = (kT)^{-1}$. In this way the Goldstone picture of the vacuum corresponds to a picture of phase transitions in many body systems. The existence of a gap in the spectrum of H, and exponential clustering, corresponds to a finite correlation length $\xi = m^{-1}$ in the many body system.

One Particle Structure. We define G{J} by

$$G\{J\} = \ln Z\{J\} - \int \Phi(J)\, dq \quad ,$$

and then $G\{J\}$ is the generating function for the connected (truncated) Euclidean Green's functions. The one particle structure is displayed by an entropy principle (Legendre transformation)

$$\Gamma\{A\} = \inf_J \; [-J\cdot A + G\{J\}] \quad ,$$

or in differential form,

$$\Gamma\{A\} = -J\cdot A + G\{J\} \quad ,$$

where J is determined by $A(x) = \delta G\{J\}/\delta J(x)$. This transformation was introduced in statistical mechanics by De Dominicis and Martin [De Ma], in quantum field theory by Jona-Lasinio [Jo L], and was developed by Symanzik [Sy 4]. The analysis of $\Gamma\{A\}$ in quantum field models [Gl Ja 13] may complement our study of the spectrum of the Hamiltonian by expansions described below. The functional $\Gamma\{A\}$ generates the (amputated, one particle irreducible) vertex functions. These functions are directly related to the magnitude of interparticle forces, i.e., the physical charge.

Bound States. In Chapter 3 we study the presence and absence of bound states in certain quantum field models. Our results in Section 3.3 about the absence of bound states in pure φ_2^4 models depend on methods both from field theory and from statistical mechanics. We use high temperature expansions from field theory (see below). We also use an idea of Lebowitz from statistical mechanics to obtain two-particle clustering for the four point vertex function.

Conversely, in Section 3.4 we sketch a proof that bound states occur in φ_2^4 models in a strong external field. We remark that in statistical mechanics, bound state excitations appear in the transfer matrix for large values of chemical potential μ.

High Temperature Expansions. The high temperature expansions in statistical mechanics yield the existence of the thermodynamic limit and high temperature analyticity, i.e., the absence of phase transitions. These Kirkwood-Salsburg or

Mayer-Montroll expansions converge for T/T_c sufficiently large, where T_c is the critical temperature. Related to these expansions are the virial expansions which converge for large values of the chemical potential μ, and which also yield analyticity (absence of phase transitions), see [Ru]. In field theory, the cluster expansions play an analagous role. They converge for large inverse coupling m_0^2/λ [Gl Ja Sp 1, 2] (large T) and for large external field [Sp 2] (large μ). As a result, the cluster expansions establish the existence of the infinite volume limit in field theory, and the existence of a single phase with a unique vacuum vector. These high temperature expansions do not, in general, arise from Kirkwood-Salsburg (or other) integral equations, but have a wider range of validity. We have, however, obtained Kirkwood-Salsburg equations for the partition function Z, see Chapter 6 of Part II. These integral equations are a useful tool in our proof of analyticity of the Schwinger functions.

In addition to yielding information about the vacuum, the high temperature expansions give us detailed information about the spectrum of the Hamiltonian H, e.g., the particle structure and the presence or absence of bound states, see Chapters 2, 3 of these lectures. We remark that these more detailed field theory techniques may yield insights into statistical mechanics.

Low Temperature Expansions. The Peierls argument [Pe] is the basic proof of the existence of phase transitions at low temperatures. The proof considers the energy associated with boundaries (contours) separating up spins from down spins. For temperatures T sufficiently below T_c, it is energetically favorable to have spins all up or all down. Griffiths, Dobrushin and others have modified and extended these results, see for example [Dob 1-3, Gi, Gr 1, ML, Min Sin 1-2]. In particular, the contour methods yield exponential clustering in pure phases of low temperature spin systems. Some continuous spin systems have been studied [Bo Gr].

We regard these methods as convergent low temperature expansions. We believe that such low temperature contour expansions exist in quantum field models.

They should converge sufficiently far from the critical point. (Such an expansion may have been used in the proof of the announced result [Dob Min].) We believe that low temperature expansions exist independent of whether multiple phases exist. In a pure phase, we believe that they exhibit exponential clustering and thus are useful to investigate particle structure.

In Figure 6 we show our conjectured region of convergence of the high temperature (cluster) expansion and presumed low temperature (contour) expansion in the φ^4 model. For models such as $\lambda\varphi^4 - \mu\varphi$, $\mu \gg \lambda$, in which symmetry breaking does not occur, the regions of convergence of the high and low temperature expansions may overlap.

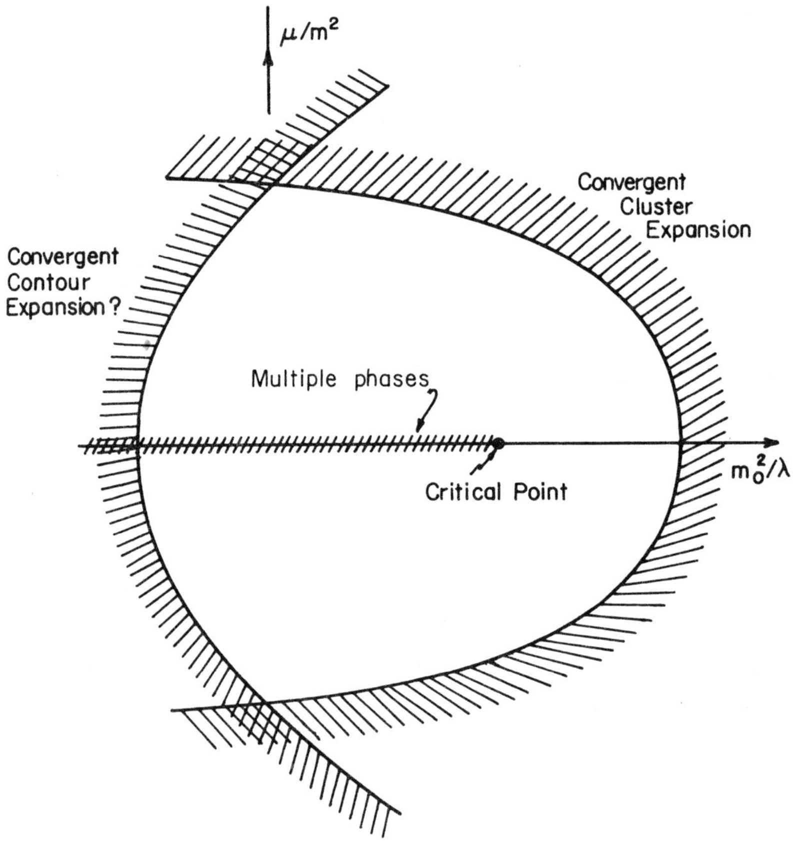

Figure 6. Presumed convergence of cluster and contour expansions.

Correlation Inequalities and the Lee-Yang Theorem. These methods yield the convergence of the Schwinger functions for even $P(\varphi)_2$ models, [GuRoSi 3, Nel 5], and a unique phase for $(\varphi^4 + m_0^2\varphi^2 - \mu\varphi)_2$ models, $\mu \neq 0$, [GrSi, Si II]. These methods and related developments are included in the lectures of Guerra, Nelson, Rosen, Simon, to which we refer the reader for further discussion.

1.5. Some Problems

We discuss several open problems for $P(\varphi)_2$. In addition, problems closely related to other sections are mentioned throughout the lectures.

Asymptotic completeness. In a pure φ^4 model with small coupling, is the S-matrix unitary? Can this be proved (see Chapter 3)? In models with bound states, does the inclusion of the bound state in and out fields for mass spectrum below $2m$ yield asymptotic completeness? Related to these questions is the possibility of performing a cluster expansion in asymptotic fields, as suggested formally by the LSZ expansion of the scattering matrix or the Yang-Feldman equations.

Asymptotic Perturbation Theory. It is known that the Euclidean Green's functions are asymptotic to all orders in the coupling, in the region of convergent cluster expansions, [Di 3]. We conjecture that the S-matrix is asymptotic to its Feynman perturbation series, $S = I + \lambda S_1 + \cdots$. Since $S_1 \neq 0$, this would yield $S \neq I$. We conjecture that the physical mass m is asymptotic in the coupling constant expansion,

$$m = m_0 + \lambda^2 m_2 + \lambda^3 m_3 + \cdots + \lambda^n m_n + O(\lambda^{n+1}) \quad .$$

Cluster Expansions. The high temperature expansions in Part II are based on expanding the Gibbs factor e^{-V_0} in the Gaussian measure $d\Phi$ into $e^{-V_0} = 1 + (e^{-V_0} - 1)$. They yield Kirkwood-Salsburg equations for Z, and related expansions for Z times the Schwinger functions, $ZS(x_1, \cdots, x_n)$. Do these expansions generalize in a natural way to yield particle structure? What is the optimal convergence domain for these expansions?

Contour Expansions. We believe that a low temperature contour expansion exists and converges, independent of whether a $P(\varphi)$ model has an internal symmetry. What is this expansion? Does it yield the existence of the infinite volume limit, of particles and of other properties of the mass spectrum.

Analyticity. In Part II, we show that the $\lambda P(\varphi)_2$ Schwinger functions are analytic in a half circle $0 < |\lambda| < \lambda_0$, $\mathrm{Re}\,\lambda > 0$. What is the complex domain into which the Schwinger functions can be continued? In statistical mechanics, the Lee-Yang theorem is used to extend the analyticity domain of high temperature (small λ/m_0^2 , large m_0^2/λ) expansions and of virial (large μ) expansions. Are the Schwinger functions for $\lambda\varphi^4 - \mu\varphi$ real analytic in λ , μ except for a cut from λ_{crit} to ∞ ? In other words, are the Schwinger functions real analytic in all of Figure 6, except for a cut along the line of multiple phases? What is the domain of complex analyticity? For $\lambda > 0$ and $\mathrm{Re}\,\mu \neq 0$, the pressure is analytic [Sp 2].

Haag-Doplicher-Roberts axioms. Is duality, the missing HDR axiom, valid for $P(\varphi)_2$ models? The HDR analysis of superselection sectors applies only in three and four dimensions, but duality is still presumably true for $P(\varphi)_2$.

Critical Points. If a critical point exists (See Figure 5) how do m , $\langle\varphi\rangle$, etc., behave in a neighborhood of it? Do the mass, spontaneous magnetization, etc., vary with power laws (given by critical exponents)? For $\lambda < \lambda_{crit}$, m is monotone in m_0^2 [Gu Ro Si 3]. Is the mass monotone above the critical point? Since Coleman has shown that $\delta = 0$ for $\lambda(\vec{\varphi}^{\,2})^2_2$ models, do multiple phases exist for this model? Is there more than one phase at the critical point for $\lambda\varphi^4_2$? Do zero mass particles occur in φ^4_2 at the critical point? (We remark that zero mass particles do not occur in the two point function, since it is a tempered distribution [Gl Ja IV].) What is the locus of multiple phases for a φ^6 or φ^8 model, etc? Do critical manifolds exist for these models?

Structure Analysis. With our control over the particle spectrum, we have the ingredients to carry out the particle structure analysis of Green's functions, as proposed by Symanzik [Sy 1]. It is also of interest to perform a structure analysis of models in statistical mechanics. As a first step, one can prove the existence and analyticity of the generating functional for one particle irreducible (IPI) Green's functions, as given in [Gl Ja 13]. These vertex parts are important in the study of symmetry breaking and of the renormalization group. In the former direction, Jona-Lasinio has an effective potential which one believes gives corrections to the mean field Goldstone picture of Section 1.3. Such potentials have by studied heuristically in [Co We]. In what sense is the mean field or the effective potential model a limit of quantum field theory?

<u>Anomalous Dimensions</u>. An extremely interesting circle of problems concerns the more refined aspects of $P(\varphi)_2$ models at the critical point. These ideas also make close contact with ideas of high energy theorists. The short distance behavior of $P(\varphi)_2$ and $\varphi_3^4(g)$ models is canonical, and a rigorous proof should follow from the local perturbation estimates [Gl Ja IV, Fel 2]. Since these estimates hold for all λ, they hold in particular at a critical point for $P(\varphi)_2$, giving a logarithmic singularity. On the other hand, the long distance behavior at the critical point for $P(\varphi)_2$ models is not canonical, since $\langle\varphi(x)\varphi(y)\rangle \to$ const. as $|x - y| \to \infty$. Consequently, we do not expect that any $P(\varphi)_2$ model we have constructed is scale invariant. In fact, a scale invariant vacuum would ensure that scale transformations are unitarily implemented. This would ensure in turn that the long and short distance scaling properties were the same.

Let us assume that a critical point exists. Then we conclude that the theory at the critical point must contain a fundamental length. This length characterizes the distance at which the small distance asymptotic behavior is replaced by the long distance asymptotic behavior. Scale transformations change this length, so if a critical point exists, there are continuously many zero mass theories related to one another by scaling. One can attempt to force scale invariance by performing an infinite scale transformation. Do such limits exist? Some of the problems raised here are unresolved for the three dimensional Ising model, and a serious effort might start with this case.

<u>The Renormalization Group</u>. Above we parameterized zero mass $P(\varphi)_2$ theories by a fundamental length. An alternative description is based on the renormalization group, which itself has intrinsic interest. Can the Callen-Symanzik equations be used to investigate the long distance behavior of $P(\varphi)_2$ models?

2. FROM ESTIMATES TO PHYSICS

How do we obtain physical properties of particles from our expansions and bounds? In this lecture we show how properties of the one particle states follow from known cluster expansions. These basic estimates for quantum field models exhibit the decoupling $\exp(-d/\xi)$ of disjoint regions in Euclidean phase space. In two space-time dimensions $(d = 2)$, cluster expansions yield space-time decoupling, as in Part II. For $d = 3$, related bounds yield phase space decoupling and the positivity of φ_3^4.

We recall that the theory of a single type of particle with mass m has the energy-momentum spectrum

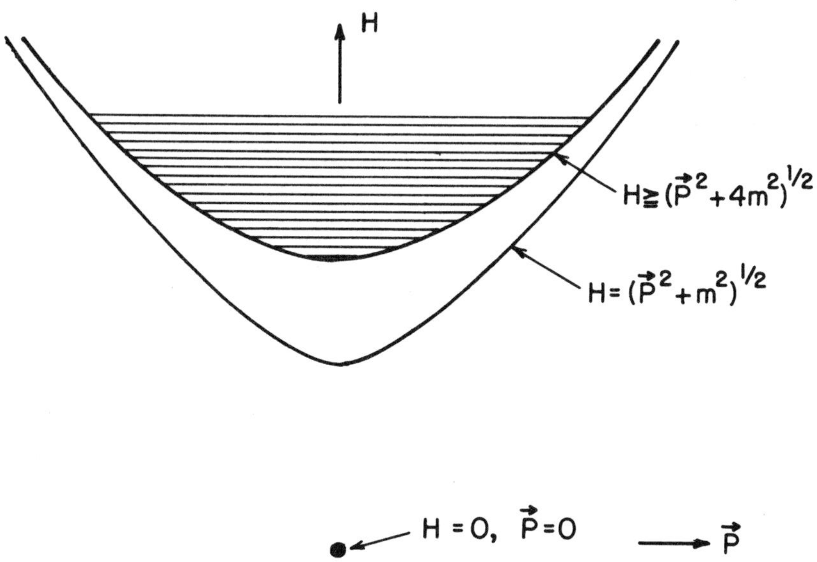

divided into three disjoint parts, the vacuum $\vec{P} = 0$, $H = 0$, the one particle hyperboloid $H^2 - \vec{P}^2 = m^2$ and the continuum $H^2 - \vec{P}^2 \geq (2m)^2$. The two particle states with momentum $\vec{p}_1$, $\vec{p}_2$ are conveniently parameterized by the relative momentum $p_R = \vec{p}_1 - \vec{p}_2$ and the total momentum $\vec{p}_T = \vec{p}_1 + \vec{p}_2$. The invariant mass for the two particle states is $2^{1/2}(\mu_1\mu_2 - \vec{p}_1 \cdot \vec{p}_2 + m^2)^{1/2}$, which for $\vec{p}_R = 0$ equals $2m$.

The mass operator $M = (H^2 - \vec{P}^2)^{1/2}$ has the corresponding spectrum

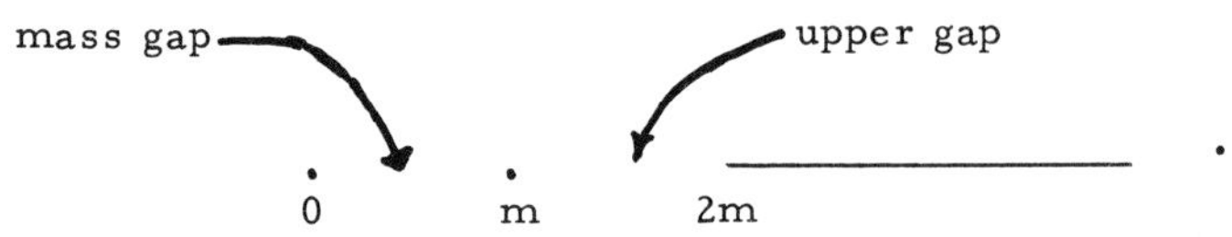

The eigenspace of 0 is the <u>vacuum</u>, and the eigenspace of m is defined to be the <u>one particle space</u>.

In order to establish spectral properties of H and M we use estimates proved by cluster expansions:

(1) Uniform vacuum cluster estimates yield convergence as the volume $\Lambda \to R^2$, and cluster estimates carry over to the infinite volume limit.

(2) The limiting Schwinger functions (for real coupling constants) satisfy the Osterwalder-Schrader axioms, and hence yield a Wightman theory. The cluster property of the vacuum (asymptotic factorization) yields uniqueness of the vacuum vector.

(3) The vacuum cluster expansion bounds the exponential decay to a factorizing vacuum and determines the mass gap. From the one particle cluster expansion, we obtain the upper mass gap and an isolated eigenvalue $M = m$.

In Section 2.1 we give some simple functional analysis. We apply these results in Section 2.2 to establish (1) - (3) above.

2.1. Functional Analysis

Let $0 \leq H = H^*$ and let E_a be the spectral projection for $[0, a]$. Let $\mathcal{D}$ be a dense subset of $\mathcal{H}$, and let $\mathcal{D}_0 \subset \mathcal{D}$ be given.

Proposition 2.1.1. Suppose that for each $\theta \in \mathcal{D}$, there exists $\chi \in \mathcal{D}_0$ and $\epsilon > 0$ such that

$$\langle \theta - \chi , e^{-tH}(\theta - \chi)\rangle \leq M_\theta e^{-(a+\epsilon)t} \quad . \tag{2.1.1}$$

Then $E_a \mathcal{D}_0$ is dense in $E_a \mathcal{H}$ and

$$\langle \theta - \chi , e^{-tH}(\theta - \chi)\rangle \leq \|\theta - \chi\|^2 e^{-(a+\epsilon)t} \quad .$$

Proof. For $\theta \in \mathcal{D}$,

$$\begin{aligned} \|E_a(\theta - \chi)\| &= \|E_a e^{tH} e^{-tH}(\theta - \chi)\| \\ &\leq e^{at}\langle \theta - \chi , e^{-2tH}(\theta - \chi)\rangle^{1/2} \\ &\leq M_\theta^{1/2} e^{-\epsilon t} \longrightarrow 0 \quad . \end{aligned}$$

Thus $E_a \mathcal{D}_0 = E_a \mathcal{D}$, which is dense in $E_a \mathcal{H}$. By applying the Schwarz inequality n times,

$$\begin{aligned} \langle \theta - \chi , e^{-tH}(\theta - \chi)\rangle &\leq \|\theta - \chi\| \, \|e^{-tH}(\theta - \chi)\| \\ &\leq \|\theta - \chi\|^{2-2^{-n}} . e^{-(a+\epsilon)t} M_\theta^{2^{-n}} \\ &\longrightarrow \|\theta - \chi\|^2 e^{-(a+\epsilon)t} \quad . \end{aligned}$$

We now let $\mathcal{H}$ be a Hilbert space carrying a unitary representation $U(a, \Lambda)$ of the inhomogeneous Lorentz group. Let $\mathcal{H}_0 \subset \mathcal{H}$ be a subspace of bounded energy and momentum, $P_0 \leq a$, $|P| \leq b$. Let $U(a)\mathcal{H}_0 \subset \mathcal{H}_0$ and $(\cup U(a, \Lambda)\mathcal{H}_0)^- = \mathcal{H}$.

Proposition 2.1.2. If $\mathcal{H}_0 \neq \{0\}$ contains a cyclic vector for the space translation subgroup $U(\vec{a})$, then the spectrum of $M \restriction \mathcal{H}$ contains exactly one point and $U(a, \Lambda)$ is irreducible on $\mathcal{H}$.

Proof. The family $U(\vec{a})$ is maximal abelian on $\mathcal{H}_0$, so any commuting operator is a function of $\vec{P}$. In particular the energy momentum spectrum is a set of the form $\{H(\vec{P}), \vec{P}\}$, and by Lorentz invariance $H = (\vec{P}^2 + \bar{m}^2)^{1/2}$ for some $\bar{m}$. (Here we assume the nontriviality of $\mathcal{H}_0$.) Thus $M = \bar{m}$ on $\mathcal{H}_0$ and by Lorentz $M = \bar{m}$ on $\mathcal{H}$. Since reducibility would be accompanied by multiplicity in the mass spectrum, the representation $U(a, \Lambda)$ is irreducible.

2.2. Relevance to Physics

The Schwinger functions with a space time cutoff h are given by

(2.2.1)
$$S_h(x_1, \cdots, x_n) = \int \Phi(x_1)\cdots\Phi(x_n)dq_h$$

where dq_h is the measure

(2.2.2)
$$dq_h = \frac{e^{-V(h)}d\Phi}{\int e^{-V(h)}d\Phi} ,$$

$d\Phi$ is the Gaussian measure with mean zero and covariance $(-\Delta + m_0^2)^{-1} = C$, and

$$V(h) = \lambda \int : \mathcal{P}(\Phi(x)) : h(x)dx$$

is the $\mathcal{P}(\varphi)_2$ Euclidean action. If $h(x)$ is the characteristic function for a set $\Lambda \subset R^2$ with area $|\Lambda|$, then $V(h)$ is the action for Λ . We denote the corresponding Schwinger functions S_Λ .

We state the vacuum cluster expansion, which bounds the rate of asymptotic factorization of the vacuum state. Let A be a function of Euclidean fields,

(2.2.3)
$$A = \prod_{i=1}^{n} \int : \Phi(x)^{n_i} : f_i(x)dx$$

where $f_i(x)$ is either an $L_2(R^2)$ function, or else $\delta_s(t)f_i(\vec{x})$, where $f_i(\vec{x})$ is $L_2(R)$. Let $\{\Delta\}$ be a cover of R^2 by unit lattice squares Δ , and define suppt. A as the smallest union of Δ's containing suppt. $f_1 \cup \cdots \cup$ suppt. f_n . In the following, A , B have the form (2.2.3).

Theorem 2.2.1 (Vacuum Cluster Expansion [Gl Ja Sp 1, 2].) Let $\gamma = m_0 - \epsilon$, with $\epsilon > 0$. Consider $\lambda\mathcal{P}(\varphi)_2$ models with $\lambda < \lambda(\epsilon, \mathcal{P}, m_0)$. Let d = dist. {suppt. A , suppt. B} . Then there exists a constant $M_{A,B}$ such that

$$\left|\int AB\,dq_h - \int A\,dq_h \int B\,dq_h\right| \le M_{A,B}e^{-\gamma d} \qquad (2.2.4)$$

uniformly in h.

[The constant $M_{A,B}$ can be bounded explicitly in terms of the f_i. We suppose each f_i is supported in a single Δ_i (an arbitrary A is a sum of such localized A's). We let $N(\Delta)$ be the sum of n_i's, for suppt. $f_i \in \Delta$. Let K_1 be given and let

$$\eta = \prod_{\Delta} (K_1 N(\Delta)!) \quad .$$

We define for a localized A,

$$\|A\| = \eta \prod_{i=1}^{n} \|f_i\| \quad . \qquad (2.2.5)$$

Here $\|f\| = \|f\|_{L_2(R^2)}$ for $f \in L_2(R^2)$, or if $f(x) = \delta_s(t)f(\vec{x})$, $\|f\| = \|f(\vec{x})\|_{L_2(R)}$. Let us assume $n_i \le \bar{n}$. Then for K_1 sufficiently large,

$$M_{A,B} \le \|A\| \; \|B\| \quad . \qquad (2.2.6)$$

If it is not the case that $n_i \le \bar{n}$, we obtain (2.2.6) with η^3 replacing η in (2.2.5). Also

$$\int A\,dq_h \le \|A\| \qquad (2.2.7)$$

uniformly in h.]

Theorem 2.2.2. The Schwinger functions $S_\Lambda(x_1, \cdots, x_n)$ converge in $\mathcal{S}'(R^{2n})$ as $\Lambda \to R^2$, to limits $S(x_1, \cdots, x_n)$ obeying the Osterwalder-Schrader axioms.

Proof. As explained in the lectures of Osterwalder and Nelson, it is sufficient to prove convergence as $h \to 1$, and a simple φ bound that follows from the vacuum cluster expansion. Here we establish convergence. Let $0 \le h_0 \le h_1 \le 1$ be two space-time cutoffs, with $h_1 - h_0$ supported on a bounded set Γ. Let $A = \Phi(f_1) \cdots \Phi(f_n)$, let suppt. $A \subset$ suppt. h_0 and let $d = \mathrm{dist}\,(\Gamma, \text{suppt. } A)$. Let $\bar{\Gamma}$ be the set of lattice squares intersecting Γ.

Define the function

$$F(\alpha) = \int A\, dq_\alpha$$

where $dq_\alpha = dq_{h_\alpha}$ and $h_\alpha = \alpha h_1 + (1-\alpha) h_0$. Then by differentiating (2.2.2) we obtain

$$\left| \int A\, dq_1 - \int A\, dq_0 \right| = \left| \int_0^1 F'(\alpha)\, d\alpha \right|$$

$$= \int_0^1 d\alpha \sum_{\Delta \subset \bar{\Gamma}} \left[\int A V(\Delta)\, dq_\alpha - \int A\, dq_\alpha \int V(\Delta)\, dq_\alpha \right] .$$

By Theorem 2.2.1 and (2.2.6), the sum above is

$$\sup_\Delta \; M_{A, V(\Delta)} e^{-\gamma' d} \le O(1) e^{-\gamma' d} ,$$

for $\gamma' = m_0 - 2\epsilon$. We let h_0, h_1 be characteristic functions of sets $\Lambda_0, \Lambda_1 \subset R^2$. Then $d \to \infty$ as $\Lambda_0, \Lambda_1 \to R^2$ and we obtain the desired convergence.

The λ dependence of the cluster expansion shows immediately that $S(f_1, \cdots, f_n)$ are continuous functions of λ. In fact, in Part II we use the cluster expansion to establish analyticity in λ in the half circle $0 < |\lambda| < \lambda_0$, $\mathrm{Re}\,\lambda > 0$.

Theorem 2.2.3. (Mass Gap). For $\lambda P(\varphi)_2$ models with small coupling λ, the vacuum Ω spans the subspace of energy less than $\gamma = m_0(1-\epsilon)$.

Proof. Let $A = \Phi(f_1) \cdots \Phi(f_n)$, suppt. A compact, and let θ_A be the vector in the relativistic Hilbert space $\mathcal{H}$ associated with the Euclidean function A. The plan is to apply Proposition 2.1.1 with $\mathcal{D}$ the dense subset of $\mathcal{H}$ spanned by such

θ_A , and with $\mathcal{D}_0$ the subspace of $\mathcal{D}$ spanned by Ω .

Since the cluster estimates are uniform in h , they carry over to the infinite volume limit $h = 1$. We choose A_d to be the translate of A in the (Euclidean) time direction. Thus by the vacuum cluster expansion, Theorem 2.2.1, with $\gamma = m_0 - \epsilon$,

$$\langle \theta_A , e^{-dH} \theta_A \rangle = \int A^- A_d dq$$

$$= \int A^- dq \int A dq + O(M_A e^{-\gamma d})$$

$$= |\langle \theta_A , \Omega \rangle|^2 + O(M_A e^{-\gamma d}) \quad .$$

In other words, if $\theta_A^\perp = \theta_A - \langle \Omega , \theta_A \rangle \Omega$ is the component of θ_A orthogonal to Ω ,

$$\langle \theta_A^\perp , e^{-dH} \theta_A^\perp \rangle \leq O(M_A e^{-\gamma d}) \quad .$$

The theorem now follows as planned.

We have established stronger cluster properties, which provide an analysis of arbitrary intervals of the energy spectrum [Gl Ja Sp 1]. These expansions are defined inductively, rather than in closed form or in the form of Kirkwood-Salsburg equations. We now state the $n = 1$ expansion, or one particle cluster expansion. Let $\gamma = 2(m_0 - \epsilon)$ with $\epsilon > 0$.

Theorem 2.2.4. (One Particle Cluster Expansion). Given $\epsilon > 0$ consider $\lambda P(\varphi)_2$ models with $\lambda < \lambda(\epsilon , m_0 , P)$. Then given θ_A as above, there exists an $L_2(R)$ function h such that for $\chi = \langle \Omega , \theta_A \rangle \Omega + \varphi(h)\Omega$, we have

$$\langle \theta_A - \chi , e^{-tH}(\theta_A - \chi) \rangle \leq M_A e^{-\gamma t} \quad .$$

We apply Proposition 2.1.1 once again. We choose $\mathcal{D}$ as in the previous example, and $\mathcal{D}_0$ the span of $\{\Omega, \varphi(h)\Omega\}$, where $h \in L_2(R)$.

Corollary 2.2.5. The vectors $E_{2m_0-\epsilon}\mathcal{D}_0$ span states of energy $< 2m_0-\epsilon$.

Theorem 2.2.6. (Upper Mass Gap). For $\lambda P(\varphi)_2$ models with small coupling, the mass operator M has eigenvalues 0, m and no other spectrum in $[0, 2m_0-\epsilon]$.

Proof. Let $E = E_{2m_0-\epsilon}(I-E_0)$, let $\mathcal{K}_0 = E\mathcal{H}$ and let $\mathcal{K}$ equal the union of the Lorentz translates of $\mathcal{K}_0$. Below we obtain a cyclic vector χ for the space translation subgroup on $\mathcal{K}_0$. By Proposition 2.1.2, the spectrum of M on $\mathcal{K}$ contains exactly one point (unless $\mathcal{K}_0 = \{0\}$). We show $\mathcal{K}_0 \neq 0$: The two point function converges in $\mathcal{S}'$ to the free two point function as $\lambda \to 0$, using the λ dependence of the cluster bounds. Since the free theory has one particle states with mass m_0, the interacting theory must have spectrum in a neighborhood of m_0, for λ sufficiently small. Thus $\mathcal{K}_0 \neq 0$. M has the eigenvalues 0 and $m \in [m_0 - \epsilon, m_0 + \epsilon]$, and no other spectrum in $[0, 2m_0 - \epsilon]$.

To complete the proof we construct χ. Let $h_1 \in \mathcal{S}(R)$, $\tilde{h}_1 > 0$. We show that $E\varphi(h_1)\Omega$ is cyclic on $\mathcal{K}_0$. Let $h_a(x) = h(x-a)$. Then

$$U(\vec{a})\varphi(h)\Omega = \varphi(h_a)\Omega$$

and

$$\begin{aligned}\varphi(h_1 * h_2)\Omega &= \varphi\left(\int h_1(\cdot - a)\, h_2(a)\, da\right)\Omega \\ &= \int da\, h_2(a)\varphi(h_{1a})\Omega \\ &= \int da\, h_2(a)\, U(\vec{a})\varphi(h_1)\Omega \ .\end{aligned}$$

Since E and $U(\vec{a})$ commute,

$$E\varphi(h_1 * h_2)\Omega = \int da\, h_2(a)\, e^{-iPa}\, E\varphi(h_1)\Omega$$

lies in the span of translates of $E\varphi(h_1)\Omega$. Since $(h_1 * h_2)^{\sim} = \tilde{h}_1\tilde{h}_2$ are dense in C_0^∞ as h_2 ranges over C_0^∞, $\chi = E\varphi(h_1)\Omega$ is cyclic for $U(\vec{a})$ on $\mathcal{K}_0$. Here we have also used Corollary 2.2.5 to identigy $\mathcal{K}_0$ with the span of $E\varphi(f)\Omega$.

3. BOUND STATES AND RESONANCES

3.1 Introduction

An important problem in physics is how particles form composites, namely bound states and resonances. In atomic physics, familiar consequences of Coulomb forces and the Schrödinger Hamiltonian are atoms and molecules: their existence and their scattering. The spectrum of atomic and molecular Hamiltonians has been the subject of extensive mathematical analysis.

The realm of nuclear and elementary particle structure includes qualitatively similar ideas, but without detailed justification. Thus a crucial physical question is whether a particular quantum field model does or does not have bound states. For instance: Do mesons bind nucleons to form stable nuclei? Are nucleons bound states of quarks? Are the ρ and the η mesons really π meson resonances?

Little is known about such important questions in quantum field theory. In fact, no quantum field models are known to have bound states, and heuristic calculations based on perturbation theory and the Bethe-Salpeter equation are inconclusive.

In this lecture we give a physical picture of when to expect or not to expect bound states in $P(\varphi)_2$ models with weak coupling or a strong external field. We prove the absence of two particle bound states in weakly coupled, pure φ^4 models. We outline an argument to prove the presence of bound states in the presence of a strong external field, and certain other models.

Bound states are eigenvalues of the mass operator M, introduced in Chapter 2. Two particle bound states lie below the two particle continuum; otherwise no energetic reason would prevent their decay into free particles. (The decay of bound states in the mass continuum may, however, be forbidden by additional selection rules included in the interaction.) On the other hand, there is no physical interpretation of continuous mass spectrum in the spectral interval $[0, 2m)$. Hence none is believed to exist, and two particle bound states may occur in the "bound state interval" $(m, 2m)$ of the mass spectrum, as illustrated in Figure 7.

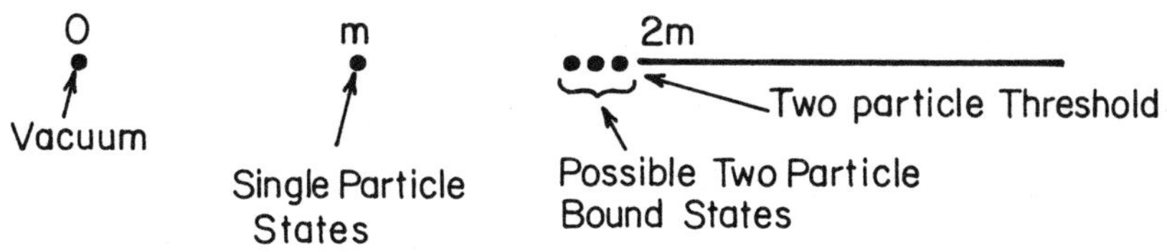

Figure 7. Spectrum of the mass operator M

In an even theory, e.g., φ^4, we can decompose the Hilbert space according to whether states are even or odd under the symmetry $\varphi \to -\varphi$. States with an even number of particles lie in the even subspace. Restricted to the odd subspace, M has the spectrum of Figure 8.

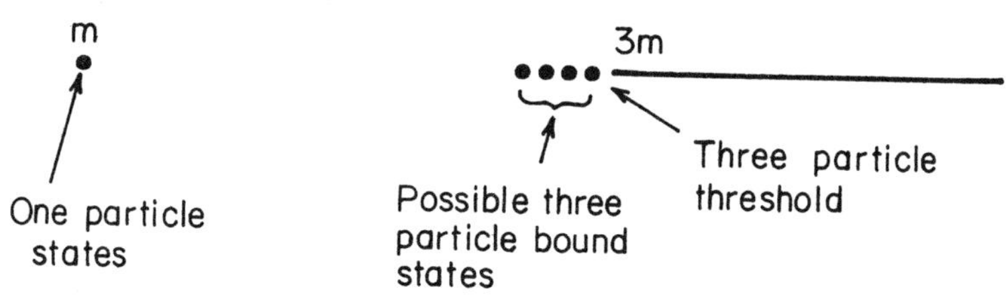

Figure 8. Mass Spectrum on the Odd Subspace of an Even Theory.

The resolvent $(M - z)^{-1} = R(z)$ of the mass operator is an analytic function of z, for $\mathrm{Im}\, z \neq 0$. It has a pole at each eigenvalue of M (particles and bound states) and presumably a cut starts at each n-particle threshold. The question of resonances concerns the analytic properties of $R(z)$ (or suitable matrix elements) after continuation across a threshold cut. A complex pole, close to the cut, is called a resonance. Such a pole appears in the scattering of particles as a peak in the cross section. Another interpretation of a resonance is an unstable particle. The real part of the position of the pole determines the mass of the resonance, while the distance to the real axis determines the lifetime. It is a challenging question to make a detailed investigation of resonances, and to determine: Are there coupling constants for which $\mathcal{P}(\varphi)_2$ models have resonances?

The presence or absence of composite particle states depends on whether the interparticle forces are attractive or repulsive. We pose the related questions: Does the mutual interaction of two particles raise or lower their energy, compared with the state in which they are asymptotically far apart? If the energy is raised, binding does not occur. If the energy is lowered below the continuum, we expect a bound state. In Section 3.2 we motivate our point of view on this question by perturbation theory. In Section 3.3, we use cluster estimates and correlation inequalities to study the same question. In Section 3.4, we show how binding occurs.

Our picture of a two particle bound state is best understood in terms of the relative momentum $\vec{p}_R$. We describe three kinds of forces: attractive, repulsive and dispersive. The attractive and repulsive forces are self explanatory. The dispersive effect arises from the curvature of the mass hyperboloid. A state of two free particles, with $\vec{p}_T = 0$, has a total energy $(4m^2 + \vec{p}_R^{\,2})^{1/2}$, and in general, for small momentum, a two particle state has energy $2m + O(\vec{p}_R^{\,2} + \vec{p}_T^{\,2})$. This raising of the energy away from zero momentum is what we call the dispersive force. For bound states to occur, the attractive force must dominate the repulsive and dispersive forces.

We introduce a parameter δ to measure the spread of the bound state wave packet. For a momentum space distribution concentrated in $|\vec{p}_R| \le \delta$, we have a configuration space spreading of order δ^{-1}. For weak coupling, we expect increased spreading in configuration space, as a bound state grows in size and disappears into the continuum. Thus we expect $\delta \to 0$ as $\lambda \to 0$. The binding forces have characteristic dependences on δ and λ: The dispersive effect is $O(\delta^2)$. In $\mathcal{P}(\varphi)_2$ models, we find in perturbation theory that attractive and repulsive effects are $O(\delta)$, times the appropriate dimensionless coupling constants λ_j/m_0^2. We discuss the balance of these forces in Section 3.4.

3.2 Formal Perturbation Theory

For a $\lambda\varphi^4$ interaction, the first order shift in the two particle energy is given by the Feynman diagram

,

which is positive for $\lambda > 0$. In second order, we find the shift has two sorts of contributions, a second order mass shift with the disconnected Feynman diagrams

and a second order attractive (negative) contribution of the form

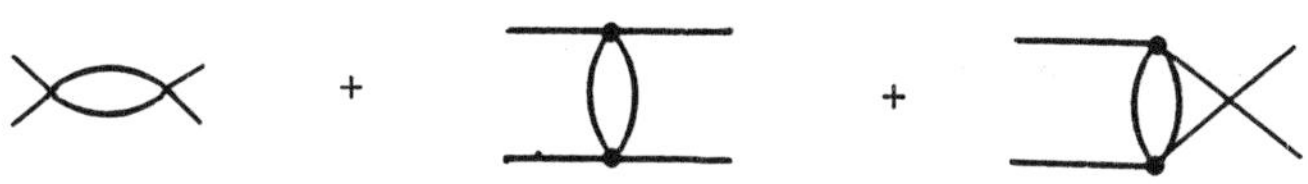

The first order repulsive shift dominates for small λ. Thus we do not expect two particle bound states to occur in weakly coupled φ_2^4 models, and we establish this in Section 3.3.

We remark that the mass shift diagrams above represent the second order mass renormalization of single particle states, i.e., the shift from m_0 to m_2. Of course, to second order, we measure our n-particle forces (energy shifts) with respect to nm_2. We do not include vacuum energy shifts, since they are eliminated by considering perturbations of the exact (coupling λ) ground state.

If we consider three particle interactions, in lowest order, diagrams of the form

give an attractive three body force. However the diagram

gives a repulsive effect in the two particle subsystems. Since the two body force is first order, and the three body force is second order, we expect the repulsive force to dominate at small coupling. A three particle unstable state (resonance) is possible.

With a φ^3 interaction, the lowest order two body force is attractive

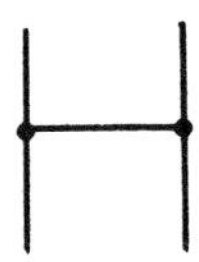

Similarly, n body forces in lowest order are attractive. For instance, in third order we have diagrams of the form

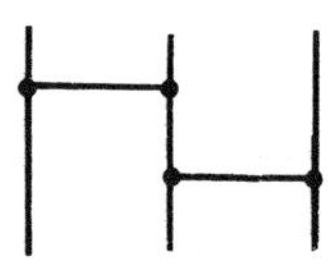

These attractive forces complement the attractive forces in two body subsystems, i. e., in the three body case,

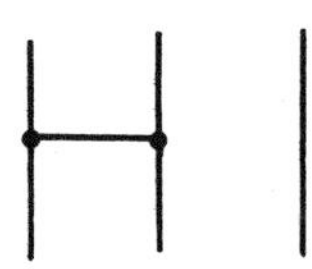

Thus we expect two particle bound states, and bound states of three or more particles if a selection rule prevents their decay. Otherwise, the attractive many body forces should yield many body resonances.

Of course, a pure φ^3 theory does not exist, because the energy is unbounded from below. However, if the φ^3 term in an interaction has a coefficient much larger than the other coupling constants, we expect that the φ^3 effects will dominate. Thus the above qualitative discussion applies to the $\lambda_1\varphi^3 + \lambda_2 P(\varphi)_2$ model, where $\lambda_1 >> \lambda_2$. In this case we expect bound states, and in particular, two particle bound states.

Closely related is the case of a $\mathcal{P}(\varphi)_2$ model in an external field, i.e., the $\mathcal{P}(\varphi)_2 - \mu\varphi$ model. By the transformation $\varphi \to \varphi + \text{const.}$ (implemented locally by $\exp(i\int \pi)$) we can eliminate the external field. For instance, the $\lambda\varphi^4 - \mu\varphi$ model is transformed into a $\lambda\varphi^4 + a\varphi^3 + b\varphi^2$ model, where $4\lambda a^3 + am^2 = \mu$. The mass term b also grows with μ, but by scaling it can be reduced to unity. Thus we <u>conjecture</u>: Bound states exist in the φ^4 model with a strong external field, $\mu >> \lambda$.

A similar analysis applies to an arbitrary $\lambda\mathcal{P}(\varphi)_2 - \mu\varphi$ model. Transforming away the external field, we add to $\mathcal{P}$ a lower degree polynomial. For μ large, the dominant coefficients have degree 2 and 3. The degree 2 term gives a mass shift, while the degree 3 term yields an attractive potential in lowest order. Thus we <u>conjecture</u>: Bound states exist in $\lambda\mathcal{P}(\varphi)_2$ models in external fields with $\mu \gg \lambda$.

<u>Question</u>: Do bound states occur in Y_2 models? We conjecture that this is the case.

3.3 On the Absence of Bound States

We consider the weakly coupled $\lambda\varphi_2^4$ model, and we prove that two particle bound states do not occur.

Theorem 3.3.1. Let λ/m_0^2 be sufficiently small in the $\lambda\varphi_2^4$ model. Then the mass operator $M = (H^2 - P^2)^{1/2}$ has no spectrum in the two particle bound state interval $(m, 2m)$.

From the uniqueness of the vacuum, we infer that the symmetry $\varphi \to -\varphi$ can be unitarily implemented, and that the Hilbert space $\mathcal{H}$ decomposes into even and odd subspaces $\mathcal{H}_e$, $\mathcal{H}_o$ each invariant under $U(a, \Lambda)$ and φ. Our theorem depends on three facts: Cluster expansions [Gl Ja Sp 1] reduce the problem to the consideration of the two point function for $\mathcal{H}_o$, and the four point function for $\mathcal{H}_e$. Second, an inequality that Lebowitz [Leb 2] proved for Ising models excludes the possibility that mass spectrum in the interval $(0, 2m)$ occurs in the four point function. Finally, cluster bounds exclude mass spectrum in the interval $(m, 2m)$ in the two point function.

The condition of weak coupling in Theorem 3.3.1 concerns the rate γ, of exponential decay $e^{-\gamma d}$, in the error term of the two particle cluster expansion. We show in [Gl Ja Sp 1] that $\gamma \to 3m_0$ and $m \to m_0$ as $\lambda/m_0^2 \to 0$. In Theorem 3.3.1 we require that λ/m_0^2 be sufficiently small to ensure $\gamma \geq 2m$.

More generally, we obtain for even $P(\varphi)_2$ models a larger mass gap on the odd subspace, as suggested in Figure 8 above.

Theorem 3.3.2. Consider an even $\lambda P(\varphi)_2$ model. Given $\epsilon > 0$, let λ/m_0^2 be sufficiently small to ensure $\gamma \geq 3m_0 - \epsilon$, for the rate γ of exponential decay for the error in the two particle cluster expansion. Then $M \restriction \mathcal{H}_o$ has no spectrum in the interval $(m, 3m_0 - \epsilon)$.

Let dq be the Euclidean measure for the $\lambda\varphi_2^4$ model, and for a function A of the Euclidean field Φ let

$$\langle A \rangle \equiv \int A \, dq \quad .$$

Proposition 3.3.3. For the $\lambda\varphi_2^4$ model,

$$\text{(3.3.1)} \quad \begin{aligned} &\langle \Phi(x_1)\Phi(x_2)\Phi(x_3)\Phi(x_4)\rangle - \langle \Phi(x_1)\Phi(x_2)\rangle\langle \Phi(x_3)\Phi(x_4)\rangle \\ &\quad \le \langle \Phi(x_1)\Phi(x_3)\rangle\langle \Phi(x_2)\Phi(x_4)\rangle + \langle \Phi(x_1)\Phi(x_4)\rangle\langle \Phi(x_2)\Phi(x_3)\rangle \ . \end{aligned}$$

Remark. Since $\langle \Phi(x)\rangle = 0$, this inequality states that the connected (truncated) four point function is negative. This bound is special to φ^4 models. In fact the philosophy of Section 3.2 suggests the presence of two particle bound states in $\varphi^6 - \varphi^4$ models.

The key inequality due to Lebowitz concerns independent spin variables $\sigma_i = \pm 1$ for a ferromagnetic Ising model. The energy of a spin configuration $\underset{\sim}{\sigma} = \{\sigma_i\}$, $1 \le i \le n$, is

$$H(\underset{\sim}{\sigma}) = -\sum_{i<j} J_{ij}\,\sigma_i\sigma_j \ ,$$

where $J_{ij} \ge 0$. For a function $f(\underset{\sim}{\sigma})$, let

$$\langle f \rangle = Z^{-1} \sum_{\underset{\sim}{\sigma}} f(\underset{\sim}{\sigma})\, e^{-H(\underset{\sim}{\sigma})} \ ,$$

where

$$Z = \sum_{\underset{\sim}{\sigma}} e^{-H(\underset{\sim}{\sigma})} \ .$$

Lebowitz proves [Leb 2]

$$\langle \sigma_i\sigma_j\sigma_k\sigma_\ell\rangle - \langle \sigma_i\sigma_j\rangle\langle \sigma_k\sigma_\ell\rangle \le \langle \sigma_i\sigma_k\rangle\langle \sigma_j\sigma_\ell\rangle + \langle \sigma_i\sigma_\ell\rangle\langle \sigma_j\sigma_k\rangle \ .$$

The inequality (3.3.4) follows immediately, since Griffiths and Simon [Gr Si] have proved that the Euclidean φ_2^4 model is a limit of Ising models of the above form, where $\Phi(x)$ can be expressed as a limit of a sum of spin variable σ_i.

We recall that the relativistic time zero field $\varphi(f)$ equals the time zero Euclidean field $\Phi(f, t=0)$. We let $f_i \in \mathcal{S}(R)$ and define

$$(3.3.2) \quad \theta(f_1, f_2) = \varphi(f_1)\varphi(f_2)\Omega - \langle \Omega, \varphi(f_1)\varphi(f_2)\Omega\rangle \Omega \quad ,$$

where Ω is the vacuum vector.

Corollary 3.3.4. The vectors $\theta(f_1, f_2)$ have energy $\geq 2m$.

Proof. It is no loss of generality to choose f_i real and positive. By the Feynman-Kac formula,

$$\langle \theta, e^{-tH}\theta\rangle = \langle \varphi(f_1)\varphi(f_2)\Omega, e^{-tH}\varphi(f_1)\varphi(f_2)\Omega\rangle - |\langle \Omega, \varphi(f_1)\varphi(f_2)\Omega\rangle|^2$$

$$= \langle \Phi(g_1)\Phi(g_2)\Phi(g_3)\Phi(g_4)\rangle - \langle \Phi(g_1)\Phi(g_2)\rangle \langle \Phi(g_3)\Phi(g_4)\rangle \quad ,$$

where $x = (x_1, x_0)$ and

$$g_1(x) = f_1(x_1)\delta(x_0) \quad , \qquad g_2(x) = f_2(x_1)\delta(x_0) \quad ,$$

$$g_3(x) = f_1(x_1)\delta_t(x_0) \quad , \qquad g_4(x) = f_2(x_1)\delta_t(x_0) \quad .$$

By (3.3.1) and the Feynman-Kac formula,

$$\langle \theta, e^{-tH}\theta \rangle \leq \prod_{i=1}^{2} \langle \varphi(f_i)\Omega, e^{-tH}\varphi(f_i)\Omega\rangle$$

$$+ |\langle \varphi(f_1)\Omega, e^{-tH}\varphi(f_2)\Omega\rangle|^2 \quad .$$

Since $\langle \Omega, \varphi\Omega\rangle = 0$, the spectrum condition yields $|\langle \varphi(f_i)\Omega, e^{-tH}\varphi(f_j)\Omega\rangle| \leq O(1)e^{-tm}$ and therefore

$$\langle \theta, e^{-tH}\theta\rangle \leq O(1)e^{-2tm} \quad ,$$

completing the proof. We remark that only vacuum cluster expansions are necessary to this point.

Next we state a result [Gl Ja Sp 1] which follows from the two particle cluster expansion. We let $\epsilon > 0$, and we let E be the spectral projection for the energy interval $[0, 3m_0 - \epsilon]$ in an even $\lambda P(\varphi)_2$ model. We assume λ/m_0^2 sufficiently small to ensure a decay rate $\gamma = 3m_0 - \epsilon$ in the two particle cluster expansion.

Proposition 3.3.5. With the above assumptions, linear combinations of the vector Ω and $e^{tH}E\theta(f_1, f_2)$ are dense in $E\mathcal{H}_e$. Also the vectors $E\varphi(f)\Omega$ are dense in $E\mathcal{H}_o$.

We remark that in [Gl Ja Sp 1] we prove a weaker result for $E\mathcal{H}_o$, namely that vectors $e^{tH}E\varphi(f)\Omega$ span $E\mathcal{H}_o$. A simple modification of Theorem 4.2, [Gl Ja Sp 1] can be used to bring first degree polynomials in the n-particle cluster expansion to time zero. This yields Proposition 3.3.5, for $n = 2$.

Proof of the Theorems. Suppose that $M \restriction \mathcal{H}_e$ has mass spectrum in the interval $(m, 2m)$. By Lorentz invariance, there is a nonzero vector $\psi \in \mathcal{H}_e$ corresponding to that spectral interval and with energy $< 2m$. By Proposition 3.3.5, ψ is a limit of sums of vectors $e^{tH}E\theta(f_1, f_2)$. By Corollary 3.3.4, $\psi = 0$, proving Theorem 3.3.1 on $\mathcal{H}_e$.

Finally we show $M \restriction E\mathcal{H}_o$ has only one point in its spectrum, namely m. By Proposition 3.3.5, the vectors $\mathcal{K}_0 = \{E\varphi(f)\Omega\}$ span $E\mathcal{H}_o$. We let $\mathcal{K}$ be the closure of the union of Lorentz translates of $\mathcal{K}_0$. Our assertion then follows by Proposition 2.1.2. Theorems 3.3.2 and 3.3.1 then follow by Lorentz invariance.

3.4 On the Presence of Bound States

The ideas of Section 3.2 suggest the presence of bound states in certain $P(\varphi)_2$ models. We give two methods to establish the existence of mass spectrum in the two particle bound state interval $(m, 2m)$. As we mentioned above, there is no physical interpretation of continuous spectrum in this interval, so the existence of spectrum presumably ensures the existence of eigenvalues, i.e., bound states.

Variational Method. The first method is to choose an approximate bound state wave function θ, with the properties: (i) $\|\theta\| \geq 1$; (ii) θ is orthogonal to the vacuum and one particle states; and (iii) $\langle\theta, M\theta\rangle < 2m$. Since $M \leq H$, we may replace the bound on $\langle\theta, M\theta\rangle$ by

$$\langle\theta, H\theta\rangle < 2m \quad . \tag{3.4.1}$$

In a theory with weak coupling, the cluster expansion shows that the low momentum part of the mass interval $(m, 2m)$ is spanned by vectors

$$\theta = \alpha\Omega + a^*(f)\Omega + e^{tH} a^*(f_1) a^*(f_2)\Omega \quad , \tag{3.4.2}$$

see [Gl Ja Sp 1] and Section 3.3. Here a^* is a time zero creation operator. If, in addition, $P(\varphi)$ is even, we may choose $f = 0$ and $\alpha = -\langle\Omega, a^*(f_1)a^*(f_2)\Omega\rangle$. Alternatively, we can replace a^* by the time zero field φ in (3.4.2).

With this variational method, we eliminate H from $\langle\theta, H\theta\rangle$ by using $H\Omega = 0$ and the canonical commutation relations. For instance,

$$[H, a^*(f)] = a^*(\mu f) + [H_I, a^*(f)] \quad , \qquad \text{where} \quad \mu = (\vec{p}^{\,2} + m_0^2)^{1/2} \quad .$$

If $\chi = a^*(f)\Omega$, then

$$\begin{aligned} \langle\chi, H\chi\rangle &= \langle f, \mu f\rangle_{L_2} + \langle\Omega, a^*(\mu f)\, a(\bar{f})\Omega\rangle \\ &\quad + \langle a^*(f)\Omega, [H_I, a^*(f)]\Omega\rangle \end{aligned} \tag{3.4.3}$$

We estimate vacuum expectation values of Wick ordered monomials

$$W = \int a^*(x_1) \cdots a(x_n) w(x)\, dx \quad ,$$

by the cluster expansion [Gl Ja Sp 1]. In fact, before estimation, we expand $\langle \Omega, W\Omega \rangle$ using integration by parts, to isolate low order dependence in the coupling λ, see Chapter 4. For instance, in second order, we obtain a second order mass-shift correction to $\langle f, \mu f \rangle_{L_2}$.

In this manner, we need not calculate the physical mass m exactly, but we can obtain explicitly the relevant low order corrections to m_0. (Here we assume that m is asymptotic to m_0.) Furthermore, let us assume that f is scaled to give momentum localization $O(\delta)$, namely $f(\vec{p}) = \delta^{-1/2} h(\vec{p}/\delta)$. (In Section 3. 1 we explained that $\delta \to 0$ as $\lambda \to 0$.) Then

$$\langle f, \mu f \rangle = m_0 \|f\|^2 + O(\delta^2) \quad ,$$

which exhibits the momentum dispersion about $\vec{p} = 0$ of the single particle state. Similarly, the second order mass correction will equal $m_2 \lambda^2 \|f\|^2 + O(\lambda^2 \delta^2)$.

We sketch our proof for the $\lambda(\varphi^6 - \varphi^4)$ interaction. We take

$$\theta = a^*(f)^2 \Omega - \langle \Omega, a^*(f)^2 \Omega \rangle \Omega \quad , \qquad \text{with } \|f\|_{L_2} = 2^{-1/4}$$

which satisfies (i), (ii) above. We study

$$\langle \theta, H\theta \rangle = \langle \theta, \{a^*(\mu f) a^*(f) + a^*(f) a^*(\mu f)\} \Omega \rangle + \left\langle \theta, \left[\lambda \int \varphi^6 - \lambda \int \varphi^4, a^*(f)^2\right] \Omega \right\rangle \quad ,$$

and integrate by parts. We isolate, in closed form, all terms of degree 0, 1 or 2 in λ. The mass terms have the form

$$2\{m_0 + \lambda^2 m_2 + O(\delta^2 + \lambda^3 + \delta^2 \lambda^2)\} \quad .$$

The attractive contribution from diagrams of the form

lowers the energy by $-O(\delta\lambda)$. Other contributions are $O(\lambda^2\delta)$ or higher order. We choose $\delta = \lambda^{1+\epsilon}$. Then for small λ, the decrease in energy $-O(\delta\lambda) = -O(\lambda^{2+\epsilon})$ dominates the dispersive effect $O(\delta^2) = O(\lambda^{2+2\epsilon})$ and the repulsive effects $O(\lambda^2\delta + \lambda^3) \leq O(\lambda^3)$. The operator parts of these estimates result from a variant of the cluster expansion. This completes our sketch of the proof that bound state spectrum exists in the weakly coupled $\lambda(\varphi^6 - \varphi^4)$ model.

Similar arguments should hold for $\lambda\varphi^6$. In this case, however, the attraction is $O(\lambda^2\delta)$. We must therefore isolate the fourth order mass shift and we set $\delta = \lambda^{2+\epsilon}$. For the interaction $\lambda\varphi^3 + \lambda^6\varphi^4$, we must orthogonalize θ to the one particle states (at least to third order in λ). We would then isolate the fourth order mass renormalization and take $\delta = \lambda^{2+\epsilon}$. We thank B. Simon for observing that an even theory is technically simpler.

Cluster Method. In an even $P(\varphi)_2$ model, for θ of the form (3.3.2),

$$\langle \theta, e^{-tH}\theta\rangle = \langle \Phi(g_1)\Phi(g_2)\Phi(g_3)\Phi(g_4)\rangle_C + \langle \Phi(g_1)\Phi(g_3)\rangle\langle \Phi(g_2)\Phi(g_4)\rangle + \langle \Phi(g_1)\Phi(g_4)\rangle\langle \Phi(g_2)\Phi(g_3)\rangle \, .$$

where $\langle\cdot\rangle_C$ denotes the connected (truncated) part. Thus $\langle\theta, e^{-tH}\theta\rangle$ exhibits the two particle decay $O(e^{-2mt})$, unless

$$\langle \Phi(g_1) \cdots \Phi(g_4)\rangle_C \geq O(e^{-2(m-\epsilon)t}) \, . \tag{3.4.4}$$

Using the Bethe-Salpeter equation, we can isolate in $\lambda(\varphi^6 - \varphi^4)$ models a slowly decaying part of $\langle \Phi(g_1)\cdots\Phi(g_4)\rangle_C$, given by (positive) φ^4 contributions. We propose using cluster expansions to estimate the errors. The inequality (3.4.4) would establish the existence of mass spectrum on $\mathcal{H}_e$ in the interval $(0, 2m-\epsilon]$. This proposed calculation appears interesting. However, unlike the variational proof above, we presently have no error estimates using this method. Conversely, we remark that the existence of two-particle bound state spectrum in a weakly coupled even $P(\varphi)_2$ model (as established by the variational method) ensures (3.4.4).

4. PHASE SPACE LOCALIZATION AND RENORMALIZATION

4.1 Results for φ_3^4

In a series of related papers, we have given convergent expansions [Gl Ja Sp 1, 2] and convergent upper bounds [Gl Ja IV, 8] for quantum field models. These expansions and bounds deal with the problem of removing cutoffs $\varkappa$, Λ, namely in taking infinite volume limits in phase space. Most of this conference has dealt with the $\Lambda \to R^2$ limit. However the $\varkappa \to \infty$ limit in Y_2 and in higher dimensional models presents the most challenging problems, for both physics and for mathematics; we hope these ultraviolet problems will be the focus of increasing attention in constructive field theory. In this section we describe the results for φ_3^4: Let $d\Phi_C$ be the Gaussian measure with covariance C, and let $d\Phi$ denote the choice $C = (-\Delta + m_0^2)^{-1}$.

Let $V(\Lambda, \varkappa)$ denote the Euclidean action, the sum of the φ^4 interaction V_I and the counterterms V_C. Then

$$V_I = \lambda \int_{\Lambda \subset R^3} :\Phi_\varkappa^4: \, dx$$

and V_C are the Green's function counterterms given in second and third order perturbation theory. The partition function for the action $V = V_I + V_C$, namely

$$Z(\Lambda, \varkappa) = \int e^{-V(\Lambda, \varkappa)} \, d\Phi \quad ,$$

contains the ultraviolet divergent counterterms.

Theorem 4.1.1 [Gl Ja 8]. For $0 \le \lambda$

$$Z(\Lambda, \varkappa) \le e^{O(|\Lambda|)} \quad , \tag{4.1.1}$$

uniformly in $\varkappa$. For λ bounded, (4.1.1) is uniform in λ also.

We now let $H(\upsilon)$ denote the renormalized φ_3^4 Hamiltonian, defined formally by

$$H(\mathcal{V}) = H_0 + \lambda \int_{\mathcal{V} \subset R^2} :\varphi^4: \, d\vec{x} - \frac{1}{2}\delta m_2^2 \int :\varphi^2: \, d\vec{x} - E_2 - E_3 \quad .$$

Here δm_2^2, E_2 and E_3 are the Hamiltonian counterterms in second and third order perturbation theory. (These counterterms differ by a constant and a transient from the Green's function counterterms, see [Gl Ja 8].)

Corollary 4.1.2. The Hamiltonian $H(\mathcal{V})$ is bounded from below by a constant proportional to the volume $|\mathcal{V}|$,

$$0 \le H(\mathcal{V}) + O(|\mathcal{V}|) \quad . \tag{4.1.2}$$

The corollary follows from the theorem and the fact that $\langle \Omega(\mathcal{V},\varkappa), \Omega_0 \rangle \neq 0$. In fact

$$\langle \Omega_0, e^{-tH(\mathcal{V},\varkappa)} \Omega_0 \rangle = e^{-tE(\mathcal{V},\varkappa) - A(\mathcal{V},\varkappa) + T(\mathcal{V},\varkappa,t)} \quad ,$$

where $E(\mathcal{V},\varkappa)$ is the partially renormalized vacuum energy, vanishing in second and third order, and convergent as $\varkappa \to \infty$ for fixed volume. The constant A in $e^{-A(\mathcal{V},\varkappa)} = |\langle \Omega_0, \Omega(\mathcal{V},\varkappa) \rangle|^2$ is the logarithmically divergent wave function renormalization constant, independent of t. Also $T(\mathcal{V},\varkappa,t)$ is a transient that is bounded as $\varkappa \to \infty$ and $o(1)$ as $t \to \infty$.

As $|\mathcal{V}| \to \infty$, the constants $E(\mathcal{V},\varkappa)$, $A(\mathcal{V},\varkappa)$ and $T(\mathcal{V},\varkappa,t)$ diverge. The second order, i.e., the ultraviolet divergent, part of $A(\mathcal{V},\varkappa)$ has been cancelled in $Z(\Lambda,\varkappa)$.

These results have been extended by Joel Feldman [Fel 2], who proved

Theorem 4.1.3. The finite volume partition function $Z(\Lambda,\varkappa)$ and

$$Z(\Lambda,\varkappa) S(\Lambda,\varkappa; f_1,\dots,f_n) = \int \Phi(f_1)\dots\Phi(f_n) e^{-V(\Lambda,\varkappa)} \, d\Phi \tag{4.1.3}$$

converge as $\varkappa \to \infty$. The limits are continuous in λ and satisfy

$$|Z(\Lambda) S(\Lambda; f_1, \cdots, f_n)| \le n! \left(\prod_i \|f_i\| \; e^{O(|\Lambda|)} \right) \tag{4.1.4}$$

for a Schwartz space norm $\|\cdot\|$.

From continuity in λ and $Z(\Lambda) = 1$ for $\lambda = 0$, we conclude that for Λ fixed and λ sufficiently small,

$$Z(\Lambda) > 1/2 \quad .$$

Thus for Λ fixed and λ small, the approximate Schwinger functions $S(\Lambda, f_1, \ldots, f_n)$ do not vanish identically and

$$|S(\Lambda; f_1, \ldots, f_n)| \le n! \prod_i \|f_i\| \,. \tag{4.1.5}$$

<u>Corollary</u> 4.1.4 [Fe 2]. For small λ, volume Λ Schwinger functions are the moments of a unique measure on $\mathcal{S}'(R^3)$, namely

$$\begin{aligned} dq &= \lim_{\kappa \to \infty} Z(\Lambda, \kappa)^{-1} e^{-V(\Lambda, \kappa)} \, d\Phi \\ &= \lim_{\kappa \to \infty} dq_{\Lambda, \kappa} \quad . \end{aligned}$$

The corollary is based on a study of the perturbation of Z in an external Euclidean field, namely on the study of the generating functional

$$Z(h) = \int e^{\Phi(h)} \, dq_\Lambda$$

for the (disconnected) Schwinger functions. This functional was studied in $P(\varphi)_2$ by Fröhlich [Fr 2]; see also [Gl Ja 13].

Of course, we are interested in the $\Lambda \to R^3$ limit of these Schwinger functions $S(\Lambda; \cdot)$ and of measures dq_Λ, in order to obtain the full relativistic theory. We conjecture that the Kirkwood-Salsburg equations of Part II can be generalized to φ_3^4 and yield the limit. In fact the local estimates of Theorem 4.1.3 and Corollary 4.1.4 are exactly the type of local estimates which the cluster expansion for $P(\varphi)_2$ uses as input. We conjecture that $Z(\Lambda) \ge \exp(-O|\Lambda|)$ for small λ. We expect that such estimates lead to the Wightman axioms for φ_3^4.

4.2. Elementary Expansion Steps

The proof of the estimates for $P(\varphi)_2$, as well as those for φ_3^4, results from the use of four elementary identities and bounds concerning the non-Gaussian measure

$$e^{-V(\Lambda,\varkappa)}\, d\Phi_C \quad . \tag{4.2.1}$$

The four steps are

I. Change of covariance C.

II. Change of exponent V.

III. Wick ordering bound.

IV. Integration by parts.

The four steps are combined to yield expansions or bounds. The difficult part of the construction is generally the question of how to combine these steps to isolate the desired property of the model, at the same time to ensure convergence. We use three expansion techniques:

a) Explicit expansions. We prescribe definite elementary steps to yield an expansion, as the expansion for ZS in Part II.

b) Neumann series. The Kirkwood-Salsburg equations of Part II, obtained by explicit expansion of Z, yield a Neumann series $(I - \mathcal{K})^{-1} = \Sigma_n \mathcal{K}^n$ for their solution.

c) Inductively defined expansions. We prescribe for each possible term (integral) in our expansion, rules to expand it into a sum of similar terms.

There is considerable freedom in the definition of our expansions and bounds. The inductively defined expansions leave the widest lattitude of choice, since they are not tied to recovering an expansion expressible in closed form, or to obtaining the inverse of an operator. In addition, the inductive expansions and bounds yield the most detailed information about our models, including the positivity of φ_3^4 [Gl Ja 8] and the φ-bounds for all couplings [Gl Ja IV]. These expansions and bounds are not tied to the use of particular boundary conditions on the covariance

C, but have more general validity.

We now give the elementary steps; the first two steps are merely the fundamental theorem of calculus:

I. Change of Covariance. Let $C_\alpha = \alpha C_1 + (1-\alpha)C_0$ be a family of interpolating covariances, and let

$$d\Phi_{C_1} = d\Phi_{C_0} + \int_0^1 d\alpha \frac{d}{d\alpha} d\Phi_{C_\alpha}$$

$$= d\Phi_{C_0} + \frac{1}{2}(C_1 - C_0) \cdot \Delta_\Phi \int_0^1 d\alpha \, d\Phi_{C_\alpha} .$$

This formula has been used to deal with the infinite volume limit, see Part II. It is established by integration by parts on function space [Di Gl]; see also the proof of IV below. We do not use Step I in this chapter.

II. Change of V. Let $V_\alpha = \alpha \in [0,1]$ be a differentiable family of interpolating Euclidean actions. Then

$$e^{-V_1} = e^{-V_0} + \int_0^1 \frac{d}{d\alpha} e^{-V_\alpha} d\alpha$$

$$(4.2.2) \qquad = e^{-V_0} - \int_0^1 \frac{dV_\alpha}{d\alpha} e^{-V_\alpha} d\alpha .$$

We use this identity to lower an upper momentum cutoff in the action V, in the positivity proofs for $P(\varphi)_2$ [See Gl Ja 7, IV] and φ_3^4 [Gl Ja 8], and we call this formula the perturbation or Duhamel identity.

Iterating (4.2.2) leads to the unrenormalized perturbation series. With an ultraviolet cutoff, this series diverges because of the $O(n!^2)$ diagrams arising in n^{th} order. For example, with one degree of freedom,

$$\int e^{-q^2 - \lambda q^4} dq \neq \sum_{n=0}^{\infty} \frac{(-\lambda)^n}{n!} \int e^{-q^2} q^{4n} dq ,$$

since the series on the right side diverges. It is therefore necessary to truncate perturbation theory, for which we use step III below.

III. Wick Bound. For $:\Phi_\kappa^4:$, we have

$$e^{-V(\Lambda,\kappa)} \leq \begin{cases} e^{O(\log^2 \kappa)|\Lambda|} & d=2 \\ e^{O(\kappa^2)|\Lambda|} & d=3 \end{cases} \tag{4.2.3}$$

This bound follows by integrating

$$:\Phi_\kappa^4: \;=\; (\Phi_\kappa - 3c_\kappa)^2 - 6c_\kappa^2 \;\geq\; -6c_\kappa^2$$

over the space time volume Λ. Here

$$c_\kappa = \int \Phi_\kappa(x)^2\, d\Phi = C_\kappa(x,x) \leq \begin{cases} O(\log \kappa) & d=2 \\ O(\kappa) & d=3 \end{cases}$$

This Wick bound is used to raise the lower momentum cutoff ρ in the exponent $V(\Lambda,\kappa,\rho)$. Our expansions terminate if $\kappa = \rho$.

IV. Integration by Parts.

$$\int \Phi(x)\, F(\Phi)\, d\Phi_C = \int dy \int C(x,y) \frac{\partial F}{\partial \Phi(y)}\, d\Phi_C \;. \tag{4.2.4}$$

We use this integration by parts formula to exhibit the cancellation of the divergent renormalization counterterms in $V_1 - V_0$ of (4.2.2). In [Gl Ja IV, 8] we use other forms of (4.2.4), called there the pull through and contraction formulas.

It is easy to establish (4.2.4) by studying finite dimensional approximations to the function space integral. We choose a Gaussian measure $d\Phi_N$ converging to $d\Phi_C$,

$$d\Phi_N = N \exp\left(-\frac{1}{2} \sum_{i,j} q_i C_{ij}^{-1} q_j\right)\left(\prod_k dq_k\right)$$

$$\equiv \nu_N dq_N \;,$$

where C_{ij} is the covariance matrix and N is a normalization constant. Ordinary integration by parts then yields

$$\int \sum_j C_{ij}^{-1} q_j F(q)\, d\Phi_N = -\int F(q) \frac{\partial \nu_N}{\partial q_i} dq_N = \int \frac{\partial F(q)}{\partial q_i} d\Phi_N .$$

Inverting C,

$$\int q_j F(q)\, d\Phi_N = \int \sum_i C_{ji} \frac{\partial F}{\partial q_i} d\Phi_N ,$$

which converges to (4.2.4) as $d\Phi_N \to d\Phi_C$.

For Wick ordered monomials, we obtain similarly

$$\int :\Phi(x_1)\dots\Phi(x_n): F(\Phi)\, d\Phi_C = \int :\Phi(x_2)\dots\Phi(x_n): \int dx_1'\, C(x_1, x_1') \frac{\partial F}{\partial \Phi(x_1')} d\Phi_C .$$

As an example, we integrate by parts one $\Phi(x)$ factor in a simple expression,

$$\int :\Phi^4(x): e^{-\int :\Phi^4: dz} d\Phi_C = -4\int dxdy \int :\Phi^3(x): C(x-y) :\Phi^3(y): e^{-\int :\Phi^4: dz} d\Phi_C .$$

Further integration by parts yields

$$\int :\Phi^4(x): e^{-\int :\Phi^4: dz} d\Phi_C = 4! \int dx\,dy\, C(x-y)^4 \int e^{-\int :\Phi^4: dz} d\Phi_C \tag{4.2.5}$$

$$+ \text{ other terms} .$$

4.3. Synthesis of the Elementary Steps

We have two basic aims in combining the elementary expansion steps. First, we desire convergent expansions in a given space-time or phase space volume. Second, we desire polynomial decoupling of different localization regions. In Part II we present the vacuum cluster expansion for $P(\varphi)_2$ models in full detail. In that case, momentum localization is unnecessary (no ultraviolet divergences occur) and our localization regions are unions of unit lattice squares in space time. With no cutoff, distant regions decouple exponentially, and the decay rate determines the physical mass. In this section we present the basic ideas of phase space localization which we used to deal with the ultraviolet divergent φ_3^4 model, and which yield the results summarized in Section 4.1. For smooth cutoffs in momentum space, we obtain polynomial decoupling.

For simplicity we discuss the partition function Z. We fix the volume Λ and investigate how Z depends on the ultraviolet cutoff κ. In order to truncate the perturbation expansions, we introduce a lower cutoff ρ into the action V. To ensure that $V(\kappa,\rho)$ is bounded from below, we introduce the momentum cutoffs in a symmetric fashion: each momentum component in $V(\kappa,\rho)$ lies in the interval $[\rho,\kappa]$.

We perform our expansions on integrals of the form

$$\int R(\Phi)\, e^{-V(\kappa,\rho)}\, d\Phi \quad , \tag{4.3.1}$$

where R is a polynomial function of Φ. Each expansion step replaces (4.3.1) by a sum of similar terms. We use a high momentum (perturbation) expansion step to lower κ, and we use a low momentum (truncation) expansion step to raise ρ. At the start of the expansion, $\rho = 0$ and $\kappa = \kappa_0$. The expansion terminates when $\rho = \kappa$, i.e., $V(\kappa,\rho) = 0$, and (4.3.1) is reduced to a sum of Gaussian integrals $\int R\, d\Phi$. We estimate this sum uniformly in κ_0.

The rules for alternating the expansion steps are somewhat complicated. The main idea is to obtain a small contribution from each high energy vertex in R, by performing explicit renormalization cancellations. We avoid the $(n!)^r$ number of terms which would arise from iterating (4.2.2), by truncating the perturbation expansion.

The high momentum expansion. We use Step II to lower κ, taking $V_1 = V(\bar{\kappa}, \rho)$, $V_0 = V(\underline{\kappa}, \rho)$. The first term in (4.2.2) has the desired form. The second term has the same upper cutoff and a new vertex $\delta V = dV_\alpha/d\alpha$ in R. Since δV has a lower momentum cutoff at $\underline{\kappa}$, we desire that δV contributes a convergence factor $\underline{\kappa}^{-\epsilon}$ to our final estimates. We obtain the proof of this fact only after performing the renormalization cancellation of the divergent counterterms δV_C in δV. Using Step IV, we integrate by parts δV_I, namely the ϕ^4 part of δV. We also integrate any new V_I produced in R as a result of differentiating the exponent. We thus obtain in closed form (and we cancel) the ultraviolet divergent part of δV. For instance, in (4.2.5), we displayed the second order vacuum energy contribution. The third order vacuum and the mass counterterms occur among the "other terms" in (4.2.5). The vacuum energy contributions cancel exactly with the corresponding counterterm in δV_C. The mass renormalization diagram, after cancellation, leaves a remainder $O(\underline{\kappa}^{-\epsilon})$. The remaining "other terms" from this procedure are convergent and so contribute $\underline{\kappa}^{-\epsilon}$ to the final estimate.

In this manner, Steps II and IV combine to yield one order (i.e., one δV) in a renormalized perturbation expansion. Because of the large number of terms, it is necessary to truncate this expansion after introducing $\bar{\kappa}^{\delta}$ vertices δV.

The low momentum expansion. We truncate the perturbation series by raising the lower cutoff in the exponent. We use Step III to raise ρ from $\underline{\rho}$ to $\bar{\rho}$. The Wick bound is an L_∞ estimate on path space, so we expect to apply it in space-time cubes Δ on which (4.2.3) remains bounded, i.e., cubes for which

$$\bar{\rho}^{2} |\Delta| \leq O(1) \quad .$$

This restriction means that the localization length $L = |\Delta|^{1/3}$ satisfies

(4.3.2) $$L \le O(\bar{\rho}^{-2/3}) \quad ,$$

and defines our phase space localization. On the other hand, the uncertainty principle requires that

(4.3.3) $$O(\underline{\rho}^{-1}) \le L$$

for the localization to be proper, i.e., for the spreading $O(\underline{\rho}^{-1})$ of the wave packet (due to momentum localization) to be less than L. We note that (4.3.2) - (4.3.3) are compatible. This compatibility is actually another aspect of super-renormalizability. For the φ_4^4 model, (4.3.2) would be replaced by $L \le O(\bar{\rho}^{-1})$, for which our estimates are borderline.

Our analysis has shown that we must treat separately cubes Δ belonging to a space-time cover $\mathscr{D}$. (Also some Δ's tend to zero as $\kappa_0 \to \infty$.) Thus we actually deal with upper and lower cutoff functions $\kappa(\Delta)$, $\rho(\Delta)$ which are functions of Δ.

Furthermore, we remark that the Wick bound (4.2.3) deals only with the pure low momentum part of $\delta V = V(\kappa, \underline{\rho}) - V(\kappa, \bar{\rho})$. This low momentum part equals $V(\bar{\rho}, \underline{\rho})$, i.e., all momenta are less than $\bar{\rho}$. The cross terms in δV must also be removed in order to raise the lower cutoff in the exponent. We dominate the cross terms by the low momentum terms and by $V(\kappa, \bar{\rho})$, the new exponent. This procedure has some complications, but poses no essential difficulty, see [Gl Ja 8]. In φ_4^4, however, it is such cross terms which yield the charge renormalization divergences, our biggest challenge.

Independence of Phase Cells. Finally we remark that the relevant distance parameter that we must use with phase space localization is the scaled distance d,

$$d = \text{Euclidean distance} \times \text{lower momentum cutoff}.$$

For smooth momentum cutoffs, scaling standard estimates gives $O(d^{-n})$ decay of correlations between different phase cells with proper localization. Any such decay for $n > 3$ is sufficient to control distance factors in sums over phase cells (whose diameter goes to zero as $\varkappa_0 \to \infty$). We remark that in the limiting theory without ultraviolet cutoff, we expect to recover exponential decoupling and a mass gap.

REFERENCES

[AHK 1] S. Albeverio and R. Hoegh-Krohn, Uniqueness of the physical vacuum and the Wightman functions in the infinite volume limit for some non-polynomial interactions, Commun. Math. Phys. 30, 171-200 (1973).

[AHK 2] ______, The scattering matrix for some non-polynomial interactions, I and II, Helv. Phys. Acta, to appear.

[AHK 3] ______, The Wightman axioms and the mass gap for strong interactions of exponential type in two dimensional space-time, Oslo reprint.

[AHK 4] ______, Asymptotic series for the scattering operator and asymptotic unitarity of the space cut-off interactions, Oslo preprint.

[Ba] R. Baumel, Private communication, see also [GuRoSi 3].

[Bo Gr] A. Bortz and R. Griffiths, Phase transitions in anisotropic classical Heisenberg ferromagnets, Commum. Math. Phys. 26, 102-108 (1972).

[Br 1] O. Bratelli, Conservation of estimates in quantum field theory, Commun. Math. Phys., to appear.

[Br 2] ______, Local norm convergence of states on the zero time boson fields, Courant preprint.

[Ca] J. Cannon, Continuous sample paths in quantum field theory, Rockefeller preprint.

[CaJa] J. Cannon and A. Jaffe, Lorentz covariance of the $(\varphi^4)_2$ quantum field theory, Commun. Math. Phys. 17, 261-321 (1970).

[Co 1] S. Coleman, Dilitations, Lectures given at the 1971 International Summer School of Physics "Ettore Majorana," to appear.

[Co 2] ______, There are no Goldstone bosons in two dimensions, Commun. Math. Phys. 31, 259-264 (1973).

[CoWe] S. Coleman and E. Weinberg, Radiative corrections as the origin of spontaneous symmetry breaking, Phys. Rev. D 7, 1888-1910, (1973).

[ConTh] F. Constantinescu and W. Thalheimer, Euclidean Green's functions for Jaffe fields, Frankfurt preprint.

[De Ma] C. De Dominicis and P. Martin, Stationary entropy principle and renormalization in normal and superfluid systems, I, II. Jour. Math. Phys. 5, 14-30, 31-59 (1964).

[Di 1] J. Dimock, Estimates, renormalized currents and field equations for the Yukawa_2 field theory, Ann. Phys. 72, 177-242 (1972).

[Di 2] ______, Spectrum of local Hamiltonians in the Yukawa_2 field theory, J. Math. Phys. 13, 477-481 (1972).

[Di 3] ______, Perturbation series asymptotic to Schwinger functions in $P(\varphi)_2$, Courant preprint.

[DiGl] J. Dimock and J. Glimm, Measures on the Schwartz distribution space and application to $P(\varphi)_2$ field theories, Courant preprint.

[Dob 1] R. L. Dobrushin, Gibbsian random fields for lattice systems with pairwise interactions, Funct. Anal. and Appl., 2, 292-301 (1968).

[Dob 2] ______, The problem of uniqueness of a Gibbsian random field and the problem of phase transitions, Funct. Anal. and Appl., 2, 302-312 (1968).

[Dob 3] ______, The description of the random field by its conditional distributions and its regularity conditions, Theor. Prob. and Appl., 13, 201-229 (1968).

[DobMin] R. Dobrushin and R. Minlos, Construction of a one dimensional quantum field via a continuous Markov field, Moscow preprint.

[DHR] S. Doplicher, R. Haag and J. Roberts, Fields, observables and gauge transformations II, Commun. Math. Phys., 15, 173-200 (1969).

[Ec] J.-P. Eckmann, Representations of the CCR in the $(\varphi^4)_3$ model: independence of space cutoff, Commun. Math. Phys. 25, 1-61 (1972).

[EcOs] J.-P. Eckmann and K. Osterwalder, On the uniqueness of the Hamiltonian and of the representation of the CCR for the quartic boson interaction in three dimensions, Helv. Phys. Acta 44, 884-909 (1971).

[FaPo] L. Faddeev and V. Popov, Feynman diagrams for the Yang-Mills field, Phys. Letters 25B, 29-30 (1967).

[Fed] P. Federbrush, Positivity for some generalized Yukawa models in one space dimension, Michigan reprint.

[Fe 1] J. Feldman, A relativistic Feynman-Kac formula, Nuclear Physics B 52, 608-614 (1973).

[Fe 2] ______, The $\lambda\varphi_3^4$ field theory in a finite volume, Harvard preprint.

[FiWi] M. Fisher and K. Wilson, Critical exponents in 3.99 dimensions, Phys Rev. Lett. 28, 240-243 (1972).

[Fr 1a] J. Fröhlich, On the infrared problem in a model of scalar electrons and massless, scalar bosons, Annales de l'Institut Henri Poincare, to appear.

[Fr 1b] ______, Existence of dressed one electron states in a class of persistent models, Fortschritte der Physik, to appear.

[Fr 2] ______, Schwinger functions and their generating functionals, Harvard preprint.

[FoKaGi] C. Fortuin, P. Kastelyn and J. Ginibre, Correlation inequalities on some partially ordered sets, Commun. Math. Phys. 22, 89-103 (1971).

[Gi] J. Ginibre, On some recent work of Dobrushin, in: Systemes a un nombre infini de degrés de liberté, ed. by L. Michel and D. Ruelle, CNRS, Paris, 1970.

[Gl 1] J. Glimm, Boson fields with nonlinear self-interaction in two dimensions, Commun. Math. Phys. 8, 12-25 (1968).

[Gl 2] ______, Yukawa coupling of quantum fields in two dimensions, I, Commun. Math. Phys. 5, 343-386 (1967).

[Gl 3] ______, Yukawa coupling of quantum fields in two dimensions, II, Commun. Math. Phys. 6, 120-127 (1967).

[Gl 4] ______, Boson fields with the $:\varphi^4:$ interaction in three dimensions, Commun. Math. Phys. 10, 1-47 (1968).

[GlJa I] J. Glimm and A. Jaffe, A $\lambda(\varphi^4)_2$ quantum field theory without cutoffs I, Phys. Rev., 176, 1945-1951 (1968).

[GlJa II] ______, The $\lambda(\varphi^4)_2$ quantum field theory without cutoffs II, The field operators and the approximate vacuum, Ann. Math. 91, 362-401 (1970).

[GlJa III] ______, The $\lambda(\varphi^4)_2$ quantum field theory without cutoffs III, The physical vacuum, Acta Math. 125, 203-261 (1970).

[GlJa IV] ______, The $\lambda(\varphi^4)_2$ quantum field theory without cutoffs IV, Perturbations of the Hamiltonian, J. Math. Phys. 13, 1558-1584 (1972).

[GlJa 5] ______, The energy momentum spectrum and vacuum expectation values in quantum field theory II, Commun. Math. Phys. 22, 1-22 (1971).

[GlJa 6] ______, What is renormalization? in Symposium on Partial Differential Equations, Berkeley 1971, American Mathematical Society, Providence, 1973.

[GlJa 7] ______, Positivity and self-adjointness of the $P(\varphi)_2$ Hamiltonian, Commun. Math. Phys. 22, 253-258 (1971).

[GlJa 8] ______, Positivity of the φ_3^4 Hamiltonian, Fortschritte der Physik 21, 327-376 (1973).

[GlJa 9] ______, Self-adjointness of the Yukawa$_2$ Hamiltonian, Ann. of Phys. 60, 321-383 (1970).

[GlJa 10] ______, The Yukawa$_2$ quantum field theory without cutoffs, J. Funct. Analysis 7, 323-357 (1971).

[GlJa 11] ______, Quantum field models, in: Statistical mechanics and quantum field theory, ed. by C. de Witt and R. Stora, Gordon and Breach, New York, 1971.

[GlJa 12] ______, Boson quantum field models, in: Mathematics of contemporary physics, ed. by R. Streater, Academic Press, New York, 1972.

[GlJa 13] ______, Entropy principle for vertex functions in quantum field models.

[GlJaSp 1] J. Glimm, A. Jaffe and T. Spencer, The Wightman axioms and particle structure in the $P(\varphi)_2$ quantum field model, Ann. Math., to appear.

[GlJaSp 2] ______, The particle structure of the weakly coupled $P(\varphi)_2$ models and other applications of high temperature expansions, Part II: The cluster expansion, these Erice lectures.

[Go] J. Goldstone, Field theories with "superconductor" solutions, Nuovo Cimento 19, 154-164 (1961).

[GoSaWe] J. Goldstone, A. Salam and S. Weinberg, Broken symmetries, Phys. Rev. 127, 965-970 (1962).

[Gr 1] R. Griffiths, Phase transitions, in: Statistical mechanics and quantum field theory, Les Houches 1970, ed. by C. De Witt and R. Stora, Gordon and Breach, New York, 1971.

[Gr 2] ______, Rigorous results for Ising ferromagnets for arbitrary spin, J. Math. Phys., 10, 1559-1565 (1969).

[GrHuSh] R. Griffiths, C. Hurst and S. Sherman, Concavity of magnetization of an Ising ferromagnet in a positive external field, J. Math. Phys. 11, 790-795 (1970).

[GrSi] R. Griffiths and B. Simon, The $(\varphi^4)_2$ field theory as a classical Ising model, Commun. Math. Phys. to appear.

[Gro 1] L. Gross, The relativistic polar without cutoffs, Commun. Math. Phys. 31, 25-74 (1973).

[Gro 2] ______, Existence and uniqueness of physical ground states, J. Funct. Anal. 10 52-109 (1972).

[Gro 3] ______, Logarithmic Sobolev inequalities, Cornell preprint.

[Gro 4] ______, Analytic vectors for representations of the canonical commutation relations and non-degeneracy of ground states, Cornell preprint.

[Gu 1] F. Guerra, Uniqueness of the vacuum energy density and Van Hove phenomenon in the infinite volume limit for two dimensional self-coupled Bose fields, Phys. Rev. Letts. 28, 1213 (1972).

[GuRu] F. Guerra and P. Ruggiero, A new interpretation of the Euclidean-Markov field in the framework of physical Minkowski space-time, Salerno preprint.

[GuRoSi 1] F. Guerra, L. Rosen and B. Simon, Nelson's symmetry and the infinite volume behaviour of the vacuum in $P(\varphi)_2$, Commun. Math. Phys. 27, 10-22 (1972).

[GuRoSi 2] ______, The vacuum energy for $P(\varphi)_2$: infinite volume limit and coupling constant dependance, Commun. Math. Phys. 29, 233-247 (1973).

[GuRoSi 3] ______, The $P(\varphi)_2$ Euclidean quantum field theory as classical statistical mechanics, Ann. Math., to appear.

[He 1] K. Hepp, Théorie de la renormalisation, Springer-Verlag, Heidelberg 1969 .

[He 2] ______, Renormalization theory, in Statistical mechanics and quantum field theory, ed. by C. De Witt and R. Stora, Gordon and Breach, New York, 1971.

[He 3] ______, Erice lectures, 1973.

[HKSi] R. Hoegh-Krohn and B. Simon, Hypercontractive semi-groups and two dimensional self-coupled Bose fields, J. Funct. Anal. 9, 121-180 (1972).

[Ja 1] A. Jaffe, The dynamics of a cutoff $\lambda\varphi^4$ field theory, Princeton University Thesis.

[Ja 2] ______, Existence theorems for a cutoff $\lambda\varphi^4$ field theory, in: Mathematical theory of elementary particles, ed. by R. Goodman and I.Segal, M.I.T. Press, 1966.

[JaMcB] A. Jaffe and O. McBryan, What constructive quantum field theory has to say about currents, in: Proceedings of a conference on currents, Princeton,1971.

[JoL] G. Jona-Lasinio, Relativistic field theories with symmetry-breaking solutions, Nuovo Cimento Letters 34, 1790-1795 (1964).

[KaRoSw] D. Kastler, D. Robinson and A. Swieca, Conserved currents and associated symmetries; Goldstone's theorem. Commun. Math. Phys. 2, 108-120 (1966).

[KoWi] J. Kogut and K. Wilson, The renormalization group and the ϵ expansion, Phys. Reports, to appear.

[Leb 1] J. Lebowitz, Bounds on the correlations and analyticity properties of ferromagnetic Ising spin systems, Commun. Math. Phys. 28, 313-321 (1972).

[Leb 2] ______, GHS and other inequalities, Yeshiva preprint.

[Leb Pe] J. Lebowitz and O. Penrose, Decay of correlations, Yeshiva preprint.

[ML] A. Martin-Löf, Mixing properties, differentiability of the free energy and the central limit theorem for a pure phase in the Ising model at low temperature, Commun. Math. Phys. 32, 75-92 (1973).

[McB 1] O. McBryan, The vector currents in the Yukawa_2 quantum field theory with SU_3 symmetry, Harvard thesis.

[McB 2] ______, Generators for the Lorentz group in the $P(\varphi)_2$ theory, Toronto preprint.

[MinSin 1] R.A. Minlos and Y.G. Sinai, The phenomenon of "phase separation" at low temperatures in some lattice models of a gas I, Math Sbornik, Tom 73 (115), 335-395 (1967).

[MinSin 2] ______, The phenomenon of "phase separation" at low temperatures in some lattice models of a gas II, Trans. Moscow Math. Soc., 19, 121-196 (1968).

[Nel 1] E. Nelson, A quartic interaction in two dimensions, in: Mathematical theory of elementary particles, ed. by R. Goodman and I. Segal, M.I.T. Press, 1966.

[Nel 2] ______, Quantum fields and Markoff fields, in: Proceedings of Summer Institute of Partial Differential Equations, Berkeley 1971, Amer. Math. Soc., Providence, 1973.

[Nel 3] ______, Construction of quantum fields from Markoff fields, J. Funct. Anal. 12, 97-112 (1973).

[Nel 4] ______, The free Markoff field, J. Funct. Anal. 12, 211-227 (1973).

[Nel 5] ______, These Erice notes.

[New 1] C. Newman, The construction of stationary two-dimensional Markoff fields with applications to quantum field theory, J. Funct. Anal., to appear.

[New 2] ______, Zeroes of the partition function for generalized Ising systems, Courant preprint.

[OsSch 1] K. Osterwalder and R. Schrader, On the uniqueness of the energy density in the infinite volume limit for quantum field models, Helv. Phys. Acta, 45, 746-754 (1972).

[OsSch 2] ______, Feynman-Kac formula for Euclidean Fermi and Bose fields, Phys. Rev. Letts. 29, 1423-1425 (1972).

[OsSch 3] ______, Axioms for Euclidean Green's functions, Commun. Math. Phys. 31, 83-112 (1973).

[OsSch 4] ______, Euclidean Fermi fields and a Feynman-Kac formula for boson-fermion models, Helv. Phys. Acta, to appear.

[Os] K. Osterwalder, Duality for free Bose fields, Commun. Math. Phys. 29, 1-14 (1973).

[Oz] H. Ozkaynak, Euclidean fields for particles of arbitrary spin, Harvard preprint.

[Pa] Y. Park. Local Lorentz transformations of the $P(\varphi)_2$ model, Indiana preprint.

[Pe] R. Peierls, On Ising's model of ferromagnetism, Proc. Camb. Phil. Soc. 32, 477-481 (1936).

[Ro 1] L. Rosen, A $\lambda\varphi^{2n}$ field theory without cutoffs, Commun. Math. Phys. 16, 157-183 (1970).

[Ro 2] ______, The $(\varphi^{2n})_2$ quantum field theory: higher order estimates, Comm. Pure App. Math. 24, 417-457 (1971).

[Ro 3] ______, The $(\varphi^{2n})_2$ quantum field theory: Lorentz covariance, J. Math. Anal. and Appl. 38, 276-311 (1972).

[Ro 4] ______, Renormalization of the Hilbert space in the mass shift model, J. Math. Phys. 13, 918-927 (1972).

[Ru] D. Ruelle, Statistical mechanics, Benjamin, New York, 1969.

[SaSt] A. Salam and J. Strathdee, A renormalizable gauge model of lepton interactions, Nuovo Cimento, 11A, 397-435 (1972).

[Sch 1] R. Schrader, Yukawa quantum field theory in two space-time dimensions without cutoff, Annals of Physics 70, 412-457 (1972).

[Sch 2] ______, A remark on Yukawa plus boson self-interaction in two space-time dimensions, Commun. Math. Phys. 21, 164-170 (1971).

[Schw] S. Schweber, An introduction to relativistic quantum field theory, Row, Peterson and Co., New York, 1961.

[Se 1] I. Segal, Notes toward the construction of nonlinear relativistic quantum fields I; the Hamiltonian in two space-time dimensions as the generator of a C*-automorphism group, P.N.A.S., 57, 1178-1183 (1967).

[Se 2] ______, Construction of nonlinear local quantum processes I, Ann. Math. 92, 462-481 (1970).

[Si I] B. Simon, Correlation inequalities and the mass gap in $P(\varphi)_2$ I, Domination by the two point function, Commun. Math. Phys. 31, 127-136 (1973).

[Si II] ______, Correlation inequalities and the mass gap in $P(\varphi)_2$ II, Uniqueness of the vacuum for a class of strongly coupled theories, Toulon preprint.

[Sl] A. Sloan, The relativistic polaron without cutoffs in two space dimensions, Cornell thesis.

[Sp 1] T. Spencer, Perturbation of the $P(\varphi)_2$ quantum field Hamiltonian, J. Math. Phys. 14, 823-828 (1973).

[Sp 2] ______, The mass gap for the $P(\varphi)_2$ quantum field model with a strong external field.

[St] R. Streater, Connection between the spectrum condition and the Lorentz invariance of $P(\varphi)_2$, Commun. Math. Phys. 26, 109-120 (1972).

[St Wi] R. Streater and A. Wightman, PCT, spin and statistics and all that, Benjamin, New York, 1964.

[Sw] J. Swieca, Range of forces and broken symmetries in many-body systems, Commun. Math. Phys. 4, 1-7 (1967).

[Sy 1] K. Symanzik, On the many-particle structure of Green's functions in quantum field theory I, J. Math. Phys. 1, 249-273 (1960).

[Sy 2] ______, A modified model of Euclidean quantum field theory, N.Y.U. preprint, 1964.

[Sy 3] ______, Euclidean quantum field theory, in: Local quantum theory, proceedings of the International School of Physics "Enrico Fermi" Course 45, ed. by R. Jost, Academic Press, New York, 1969.

[Sy 4] ______, Renormalizable models with simple symmetry breaking I, Symmetry breaking by a source term, Commun. Math. Phys. 16, 48-80 (1970).

[Sy 5] ______, Small distance behaviour in field theory and power counting, Commun. Math. Phys. 18, 227-246 (1970).

[Sy 6] ______, Small distance behaviour, analysis and Wilson expansions, Commun. Math. Phys. 23, 49-86 (1971).

[Sy 7] ______, Small distance behaviour in field theory, in: Strong interaction physics, Springer tracts in modern physics, Vol. 57, Springer-Verlag, New York, 1971.

[Wi] A. Wightman, Lecture at the 1972 Coral Gables Conference, Gordon and Breach Sci. Pub., to appear.

* Supported in part by the National Science Foundation, Grant NSF-GP-24003.

† Supported in part by the National Science Foundation, Grant NSF-GP-40354X.

‡ Supported in part by the Alfred P. Sloan Foundation.

VI The Particle Structure of the Weakly Coupled $P(\varphi)_2$ Model and Other Applications of High Temperature Expansions: Part II. The Cluster Expansion

THE PARTICLE STRUCTURE OF THE WEAKLY COUPLED $P(\phi)_2$ MODEL AND OTHER APPLICATIONS OF HIGH TEMPERATURE EXPANSIONS

Part II: The cluster expansion

James Glimm*
Courant Institute, NYU
New York, N.Y.

Arthur Jaffe†
Harvard University
Cambridge, Mass.

Thomas Spencer‡
Courant Institute, NYU
New York, N.Y.

Contents

*Supported in part by the National Science Foundation, Grant NSF-GP-24003.

†Supported in part by the National Science Foundation, Grant NSF-GP- 40354X

‡Supported in part by the Alfred P. Sloan Foundation.

§1. INTRODUCTION

From the point of view of physics, there are essentially two problems in the study of quantum field models. The first problem is to find the phenomena of interest to physics (e.g. particles, bound states, resonances, broken symmetries, and asymptotic power series expansions) in the two and three dimensional quantum field models.[1] This problem includes obtaining the mathematical structure underlying these phenomena, and determining qualitatively which interactions and/or parameter values give rise to these phenomena and which do not. We believe this problem is feasible at the present time; indeed the initial progress is the subject of the present lectures. The second problem is the construction of nontrivial fields in four space time dimensions. Here it seems that some new ideas are needed to supplement those which have already been used in the construction of quantum fields.

These two problems are also interesting from the point of view of mathematics. In addition we mention a third problem, which is to place the solutions already constructed in a more general conceptual framework. In particular, elliptic equations with an infinite number of independent variables, Markov fields, non Gaussian integrals over functions of several variables, and C*-algebras arise naturally in the construction of quantum fields. The quantum field results may suggest directions for further development of these theories.

In Part II of these lectures, we give in detail a high temperature (cluster) expansion, in a form which yields an exponential cluster property for the Schwinger functions, as well as analyticity in the coupling constant λ, for λ in a bounded sector

$$0 < |\lambda| < \lambda_0 , \qquad -\pi/2 < \arg \lambda < \pi/2 ,$$

for the $P(\phi)_2$ quantum field theory. Other applications of this type of expansion to the study of quantum field models were developed in part I of these lectures; see also [4,5,6,14].

The expansion itself is related to the virial and cluster expansions [13, Chapter 4] of statistical mechanics. The analogy between statistical mechanics and $P(\Phi)$ quantum field theories has been known for some time. For a discussion of the relation of statistical mechanics to quantum field theory, see the lectures of Guerra, Rose

[1] For three space time dimensions, it is still necessary to complete the construction of the quantum fields. However enough progress has been made to eliminate any serious doubt as to their eventual existence.

and Simon, as well as Part I of these lectures.

These lectures develop a cluster expansion in the case of small coupling. We believe expansion techniques will apply for any parameter values which are sufficiently far from the critical point. Our belief is based partly on low temperature expansions developed by Peierls and by Minlos and Sinai [10] in statistical mechanics as well as work of Dobrushin and Minlos mentioned in part I. Furthermore Spencer [14] has extended the expansion presented here to the case of large external field.

Within their region of convergence expansion methods tend to give the most detailed information available. In fact these methods yield the main information known about the energy spectrum, above the energy level $E = 0$ of the vacuum. Correlation inequalities, the Lee-Yang theorem and other analyticity methods complement these techniques and often yield results outside the region of convergence. See [7,10,14] and other lectures of this series.

For the cluster expansion, the relevant formula from statistical mechanics is the following expansion for the density of the Gibbs ensemble:

$$(1.1)\quad \prod_{i<j} e^{-\beta V(x_i-x_j)} = \prod_{i<j}\left[1+\left(e^{-\beta V(x_i-x_j)}-1\right)\right]$$

$$= \sum_{\Gamma}\ \prod_{(i,j)\in\Gamma}\left(e^{-\beta V(x_i-x_j)}-1\right) .$$

Here Γ is a set of distinct unordered pairs (i,j), i.e. Mayer graphs, and the sum extends over all such graphs. This formula expresses the interaction between the distinct i and j $(i\neq j)$ particles, $e^{-\beta V(x_i-x_j)}$, as a sum of a zero interaction term, 1, for which there is no coupling between the particles and a perturbation $e^{-\beta V(x_i-x_j)}-1$ which is small at high temperatures $kT=\beta^{-1}$. Heuristically, the role of the Gibbs density in the $P(\phi)_2$ field theory is replaced by the measure

$$(1.2)\qquad e^{-\int[\mathcal{H}_0(x)+\lambda P(\Phi(x))]dx}\ \prod_{x\in R^2} d\Phi(x) .$$

Here

$$(1.3)\qquad \mathcal{H}_0(x) = \tfrac{1}{2}\left(\nabla\Phi(x)^2 + m_0^2\Phi(x)^2\right)$$

and the formal expression

$$(1.4)\qquad e^{-\int \mathcal{H}_0(x)dx} \prod_{x\in R^2} d\Phi(x)$$

denotes the Gaussian measure on $\mathcal{S}'(R^2)$ with mean zero and covariance

$$(1.5)\qquad C_{\emptyset} = (-\Delta + m_0^2)^{-1} .$$

We also use the notation $d\Phi_C$ for the Gaussian measure with covariance C, so that $(1.4) = d\Phi_{C_{\emptyset}}$. We recognize (1.2) as the canonical Gibbs density for transverse vibrations of an elastic membrane subject to the nonlinear restoring force

$$F = -m_0^2\Phi(x) - \lambda P'(\Phi(x)) ,$$

after integration over the momentum variables $\Pi(x)$.

In (1.2), the coupling between distinct points comes entirely from the $\nabla\Phi$ term in $\mathcal{H}_0(x)$ and so $e^{-\int\nabla\Phi^2}$ in (1.2) plays the role of $e^{-\beta V}$ in (1.1). Our cluster expansion is constructed in the spirit of (1.1). In the completely decoupled theory, the Laplacian in

$$\int \mathcal{H}_0(x)dx = \tfrac{1}{2}\int \Phi(x)\left(-\Delta + m_0^2\right)\Phi(x)dx$$

is replaced by zero. The resulting ultralocal theory [9,12] seems to be very singular relative to the theory defined by (1.2). We reduce and control the singularity of the difference between the coupled and decoupled theories in two separate steps. The first step introduces a lattice structure into R^2 and into the expansion generalizing (1.1). We do not introduce the lattice into the measure (1.2) and so the resulting expansion is an expansion for the continuum infinite volume $P(\Phi)_2$ theory, rather than for a lattice approximation to this theory. Let Γ denote a set of lattice lines, joining nearest neighbor lattice points in Z^2, let Δ_Γ be the Laplace operator with Dirichlet[2] boundary conditions on Γ, and let

$$(1.6)\qquad C_\Gamma = (-\Delta_\Gamma + m_0^2)^{-1} .$$

[2] From the point of view of this discussion, Neumann data might seem more appropriate, but we use Dirichlet data because it is technically easier.

Then $d\Phi_{C_\Gamma}$ plays the role of the decoupled measure, with decoupling along the curve Γ.[3] In summary, the lattice structure gives discrete variables in the sum and product

$$\sum_\Gamma \prod_{(i,j)\in\Gamma}$$

in (1.1), even when this formula is applied to the continuum $P(\Phi)_2$ model.

The second step regularizes the differences corresponding to

$$e^{-\beta V(x_i - x_j)} - 1$$

in (1.1). The difference between two Gaussian measures can be expressed as

$$d\Phi_{C_1} - d\Phi_{C_2} = \int_0^1 \frac{d}{ds}\, d\Phi_{C(s)}$$

where

$$C(s) = sC_1 + (1-s)C_2 \ .$$

This formula applies equally well to the non Gaussian measures

$$e^{-\lambda\int P(\Phi(x))dx}\ d\Phi_C$$

of interest to us. There is a simple formula for the evaluation of $\frac{d}{ds}\, d\Phi_{C(s)}$, namely [1]

$$(1.7) \qquad \frac{d}{ds}\int F d\Phi_{C(s)} = \frac{1}{2}\int \left(C'(s)\cdot\Delta_\Phi\right) F\ d\Phi_{C(s)}$$

where

$$(1.8) \qquad C'(s)\cdot\Delta_\Phi F = \frac{d}{ds}\int C(s,x,y)\ \frac{\delta^2}{\delta\Phi(x)\delta\Phi(y)}\ F\ dxdy$$

$$= \int [C_1(x,y) - C_2(x,y)]\ \frac{\delta^2}{\delta\Phi(x)\delta\Phi(y)}\ F\ dxdy$$

and $C(s,x,y)$ is the kernel of $C(s)$.

The proof of this formula contains an integration by parts in

---

[3]In comparison with (1.1), Γ should be replaced by $(Z^2)^* \sim \Gamma$, where $(Z^2)^*$ is the set of all nearest neighbor bonds in Z^2.

the function space integral. In the case of interest to us, $C_1 = C_{\Gamma_1}$ and $C_2 = C_{\Gamma_2}$, and the integration by parts regularizes[4] $\frac{d}{ds}\, d\Phi_{C(s)}$.

[4] Formally, $\frac{d}{ds}\, d\Phi_{C(s)} = J(\Phi) d\Phi_{C(s)}$, where the Jacobean J is

$$J(\Phi) = \frac{1}{2} \int :\Phi(x)[C(s)^{-1}C'(s)C(s)^{-1}](x,y)\Phi(y):dx\, dy$$

$$= -\frac{1}{2} \int :\Phi(x)[\frac{d}{ds}\, C^{-1}(s)](x,y)\Phi(y):dx\, dy \ .$$

Since C_{Γ_1} and C_{Γ_2} are mutually singular, for $\Gamma_1 \neq \Gamma_2$, some singularity in $J(\Phi)$ should be anticipated.

§2. THE MAIN RESULTS

For small values of λ/m_0^2, we prove an exponential cluster property on the Schwinger functions. We prove the cluster property in a finite volume with bounds independent of the volume. From these results, it follows easily that the Schwinger functions converge in the infinite volume limit, and that the infinite volume Schwinger functions satisfy an exponential cluster property. (They are also independent of the boundary conditions.) See [6]. Using the Osterwalder-Schrader reconstruction theorem, the infinite volume $P(\phi)_2$ field theory is constructed from the Schwinger functions, and in this theory the Wightman axioms are satisfied and the physical mass is strictly positive. We also show that the Schwinger functions are analytic in λ, for λ in the bounded sector

$$(2.1) \qquad 0 < |\lambda| < \varepsilon , \qquad -\pi/2 < \arg \lambda < \pi/2 .$$

Let $\mathcal{C}$ be the set of convex combinations of Dirichlet covariance operators (1.6). The important properties of the measures $d\Phi_C$, $C \in \mathcal{C}$, are summarized in §9 (as preparation for the estimate of §10). In order to make elementary definitions, we mention merely that for Λ a bounded measurable set and for $C \in \mathcal{C}$,

$$(2.2) \qquad V(\Lambda) = \int_\Lambda :P(\Phi(x)):dx \in L_p(\mathcal{S}', d\Phi_C)$$

and

$$(2.3) \qquad e^{-\lambda V(\Lambda)} \in L_p(\mathcal{S}', d\Phi_C)$$

for all $p \in [1,\infty)$ and $\operatorname{Re} \lambda \geq 0$. (See Prop. 9.3, Th. 9.5.) In (2.2) the polynomial P is arbitrary, while in (2.3), P is semibounded. The Wick ordering in (2.2) is taken relative to the fixed measure $d\Phi_{C_\emptyset}$ of (1.4), even when $C \neq C_\emptyset$. To simplify the discussion, we take Λ to be a union of lattice squares, and in the limits $\Lambda \to \infty$, we further restrict Λ to be a square centered at the origin.[5] Let

$$(2.4) \qquad Z(\Lambda) = Z(\Lambda,C) = \int e^{-\lambda V(\Lambda)} d\Phi_C .$$

In Theorem 6.1 below we show that $Z(\Lambda) \neq 0$. Let

$$(2.5) \qquad dq_\Lambda = dq_{\Lambda,C} = Z(\Lambda,C)^{-1} e^{-\lambda V(\Lambda)} d\Phi_C .$$

[5] More general cutoffs, as required for the proof of Euclidean covariance of the limit $\Lambda \to \infty$, can be introduced by inserting a nonnegative L_∞ function h in the integral (2.2). Such an h does not affect the subsequent discussion.

The finite volume Schwinger functions are by definition the moments of the measure dq_Λ:

$$(2.6) \qquad S_\Lambda(x_1,\ldots,x_n) = \int \Phi(x_1)\ldots\Phi(x_n)\, dq_\Lambda \ .$$

S_Λ is defined as a tempered distribution in $\mathcal{S}'(R^{2n})$. To see this, take a test function of the form $w(x) = w_1(x_1)\ldots w_n(x_n)$ and integrate against (2.6). The n+1 factors

$$\Phi(w\ ) = \int \Phi(x_i)w_i(x_i)\, dx_i \qquad \text{and} \qquad e^{-\lambda V(\Lambda)}$$

in the integrand are bounded by Hölder's inequality, (2.3), and the following consequence of Prop. 9.3:

$$\|\Phi(w_i)\|_{L_p(\mathcal{S}')} \leq M_p \|w_i\|_{L_2(R^2)} \ .$$

In addition to the monomials in (2.4), it is useful to integrate products of Wick ordered polynomials, namely

$$(2.7) \qquad A = \int :\Phi(x_1)^{n_1}:\ldots:\Phi(x_j)^{n_j}:w(x_1,\ldots,x_j)dx \ .$$

We assume $w \in \mathcal{S}(R^{2j})$, although weaker bounds would suffice. We define suppt A to be the intersection of all closed subsets $C \subset R^2$ with

$$(2.8) \qquad \text{suppt } w \subset C \times\ldots\times C \qquad \text{(j factors)}$$

Theorem 2.1. Let λ belong to the closure of the half circle (2.1) and let ε/m_0^2 be sufficiently small. Let A and B be functions on $\mathcal{S}'$ of the form (2.7). Let d be the width of a strip in R^2 separating suppt A and suppt B. There is a constant $M = M_{A,B}$ and a positive constant m independent of A and B such that

$$\left| \int AB\, dq_{\Lambda,C_\emptyset} - \int A\, dq_{\Lambda,C_\emptyset} \int B\, dq_{\Lambda,C_\emptyset} \right| \leq M_{A,B}\, e^{-md} ,$$

uniformly in Λ, as $\Lambda \to \infty$. Furthermore, M is independent of translations in either A or B.

Theorem 2.2. Let λ belong to the closure of the half circle (2.1), and let ε/m_0^2 be sufficiently small. For A of the form (2.7), $\left|\int A\, dq_{\Lambda,C}\right|$ is bounded uniformly in λ and in Λ, as $\Lambda \to \infty$.

From analyticity of the finite volume Schwinger functions, from

Vitali's theorem and from the convergence, as $\Lambda \to \infty$, for λ real and small, we have

<u>Corollary 2.3</u>. The Schwinger functions are analytic in λ in (2.1) for ε/m_o^2 small.

§3. THE CLUSTER EXPANSION

The proof of the main theorems are based on a cluster expansion, which we now derive. Let $\mathcal{B}$ be a set of line segments in R^2. We are interested in two examples: Either $\mathcal{B} = (Z^2)^*$, the set of all lattice lines (bonds) joining nearest neighbor lattice sites in Z^2 or $\mathcal{B} = (Z^2)^* \sim \Gamma$ with Γ a finite subset of $(Z^2)^*$. We identify a subset $\Gamma \subset \mathcal{B}$ with the subset

$$\Gamma = \bigsqcup_{b\in\Gamma} b \subset R^2 .$$

The subsets $\Gamma \subset \mathcal{B}$ label terms in our expansion. The term labeled by Γ is decoupled across

$$\Gamma^c = \mathcal{B} \sim \Gamma \tag{3.1}$$

and formed by choosing a Gaussian measure with Dirichlet covariance on Γ^c. Thus the lines b in Γ^c are called Dirichlet lines. For the lines $b \in \Gamma$ there are differences between coupled and decoupled measures as in (1.1). These differences are expressed in terms of derivatives by the fundamental theorem of calculus. Thus the $b \in \Gamma$ are called derivative lines.

The covariance operators we consider are convex combinations of the operators C_Γ. For each $b \in \mathcal{B}$, we introduce a parameter $s_b \in [0,1]$ to measure the strength of the coupling across b. The parameter value $s_b = 0$ corresponds to zero Dirichlet data on b, hence to zero coupling across b, while $s_b = 1$ corresponds to full coupling across b. With

$$s = (s_b)_{b\in\mathcal{B}} , \tag{3.2}$$

we define the multiparameter family of covariance operators

$$C(s) = \sum_{\Gamma\subset\mathcal{B}} \prod_{b\in\Gamma} s_b \prod_{b\in\Gamma^c} (1-s_b) C_{\Gamma^c} . \tag{3.3}$$

Since the coefficients in (3.3) are the terms in the expansion of

$$1 = \prod_{b\in\mathcal{B}} 1 = \prod_{b\in\mathcal{B}} [s_b + (1-s_b)] ,$$

(3.3) is a convex sum as asserted. The free covariance is now denoted

$$C_\emptyset = C(1,1,\ldots) = (-\Delta + m_o^2)^{-1}$$

and the completely decoupled covariance is

$$C_{\mathcal{B}} = C(0,0,\ldots) = (-\Delta_{\mathcal{B}} + m_o^2)^{-1} .$$

The Schwinger functions and the partition function are, in a natural way, functions of s, and we use the notation

$$(3.4)\qquad \begin{cases} Z(s)S_s(x) = Z(\Lambda,s)S_{\Lambda,s}(x) = \int \Pi\Phi(x_i)e^{-\lambda V(\Lambda)}d\Phi_s \\ Z(s) = Z(\Lambda,s) = \int e^{-\lambda V(\Lambda)}d\Phi_s \end{cases}$$

where

$$d\Phi_s = d\Phi_{C(s)}$$

The goal of the cluster expansion is to express coupled quantities ($s_b \equiv 1$) in terms of decoupled (and hence finite volume) quantities. The decoupled quantities have $s_b = 0$ for many $b \in \mathcal{B}$, and in order to formulate them, it is convenient to define

$$s(\Gamma) = \left(s(\Gamma)_b\right)_{b\in\mathcal{B}}$$

by the formula

$$s(\Gamma)_b = \begin{cases} s_b, & b \in \Gamma \\ 0, & b \notin \Gamma . \end{cases}$$

For Γ finite, $s(\Gamma)$ specifies Dirichlet boundary conditions at large distances (on Γ^c), while s can be thought of as giving general boundary conditions on Γ^c, and agreeing with $s(\Gamma)$ on Γ. Thus the following definition is a way of saying that F is independent of the boundary conditions at ∞.

Definition 3.1. A function F(s) is called regular at infinity if for each s,

$$(3.6)\qquad F(s) = \lim_{\{\Gamma\nearrow\mathcal{B}:\ \Gamma\ \text{finite}\}} F\left(s(\Gamma)\right) .$$

Proposition 3.1. The functions (3.4) are regular at infinity. Here S converges in $\mathcal{S}'$, and $\mathcal{B} \subseteq (Z^2)^*$.

Because Λ is fixed and bounded, the limit (3.6) is elementary. We omit the proof, since it is a consequence of techniques to be developed in Chapters 7,9.

The first step in the cluster expansion is to apply the fundamental theorem of calculus to the finite number of nonzero parameters in

$F(s(\Gamma))$. Define

$$\partial^{\Gamma} = \prod_{b\in\Gamma} \frac{d}{ds_b} \tag{3.7}$$

and for two coordinate values s and σ define order coordinatewise, so that

$$\sigma \leq s \Leftrightarrow \sigma_b \leq s_b \ , \qquad \forall b \in \mathcal{B}.$$

<u>Proposition 3.2</u>. Let $F(s)$ be smooth and regular at infinity. Then

$$F(s) = \sum_{\{\Gamma\subset\mathcal{B}:\Gamma \text{ is finite}\}} \int_{0\leq\sigma\leq s(\Gamma)} \partial^{\Gamma} F(\sigma(\Gamma))\, d\sigma \ . \tag{3.8}$$

<u>Proof</u>. Let $G(s)$ denote the right side of (3.8). We assert that $F(s(B)) = G(s(B))$, for B any finite subset of $\mathcal{B}$. Now $G(s(B))$ is just the sum in (3.8), restricted to sets $\Gamma \subset B$. Since $F(s)$ is regular at infinity, convergence of the sum in (3.8) follows from the assertion. Also (3.8) follows by limits in the assertion, as $B \nearrow \mathcal{B}$.

For a function $f(s_b)$ of a single variable, let

$$(\delta^b f)(s_b) = f(s_b) - f(0) = \int_0^{s_b} \partial^b f(\sigma_b)\, d\sigma_b$$

$$(E_0^b f)(s_b) \quad = \quad f(0) \ .$$

Then $I = E_0^b + \delta^b$ (fundamental theorem of calculus) and so

$$I = \prod_{b\in B} (E_0^b + \delta^b) = \sum_{\Gamma\subset B} E_0^{B\sim\Gamma} \delta^{\Gamma} \tag{3.9}$$

where

$$\delta^{\Gamma} = \prod_{b\in\Gamma} \delta^b \ , \qquad E_0^{B\sim\Gamma} = \prod_{b\in B\sim\Gamma} E_0^b \ .$$

It is easy to see that (3.9) yields the desired identity $F(s(B)) = G(s(B))$.

The next step in the cluster expansion is a factorization and partial resummation of (3.8). Write

$$R^2 \sim \Gamma^c = X_1 \cup X_2 \cup \dots \cup X_r \tag{3.10}$$

so that each X_i is a union of connected components such that $X_i \cap X_j = \emptyset$ for $i \neq j$.

Definition 3.2. A function $F(\Lambda,s)$ decouples at $s=0$ provided

$$(3.11)\qquad F(\Lambda,s(\Gamma)) = \prod_{i=1}^{r} F(\Lambda\cap X_i,\ s(\Gamma\cap X_i))$$

whenever (3.10) holds.

Proposition 3.3. Given (3.10), the measures $d\Phi_{s(\Gamma)}$ and $e^{-\lambda V(\Lambda)}d\Phi_{s(\Gamma)}$ each factor into a product of r measures, and the ith factor measure is defined on $\mathcal{S}'(X_i)$. To avoid regularity questions, we assume $\mathcal{B}\subseteq (Z^2)^*$.

Sketch of Proof. Since $C(s(\Gamma))$ has zero Dirichlet data on Γ^c, $C(s(\Gamma))$ is the direct sum of the operators

$$C(s(\Gamma)) \restriction L_2(X_i) \quad .$$

The factorization of $d\Phi_{s(\Gamma)}$ follows from this fact, and can be seen from the standard formula for evaluation of a Gaussian integral of a polynomial (Theorem 9.1). Because $:P(\Phi(x)):$ is a local function, $\exp(-\lambda V(\Lambda))$ also factors, and so does $\exp(-\lambda V)d\Phi_{s(\Gamma)}$.

Corollary 3.4. The functions ZS and Z of (3.4) decouple at s=0.

The cluster expansion represents F as a sum of products of contributions from connected graphs. We start with some set X_o of interest (e.g. $X_o = \{x_1,\ldots,x_n\}$ in (3.4)). The graphs which do not meet X_o are resummed, giving a single factor of F in an exterior region.

We now carry out this resummation in more detail. Substitute (3.11) in the expansion (3.8). This yields

$$(3.12)\qquad F(\Lambda,s) = \sum_{\Gamma}\prod_{i=1}^{r}\int_0^{s(\Gamma_i)} \partial^{\Gamma_i}F(\Lambda\cap X_i,\sigma(\Gamma_i))\,d\sigma$$

where $\Gamma_i = \Gamma\cap X_i$. If the X_i are connected, this is the sum over products of connected graphs. Now we choose X_1 in (3.10) to be the union of all components meeting X_o and let X_2 be the union of the remaining components. The resummation consists of holding X_1 and Γ_1 fixed, while summing over all choices of Γ_2. Explicitly,

$$F(\Lambda,s) = \sum_{X_1,\Gamma_1} \int_0^{s(\Gamma_1)} \partial^{\Gamma_1} F(\Lambda\cap X_1,\sigma(\Gamma_1))\,d\sigma$$

$$\times \sum_{\Gamma_2} \int_0^{s(\Gamma_2)} \partial^{\Gamma_2} F(\Lambda\cap X_2,\sigma(\Gamma_2))\,d\sigma \;.$$

The Γ_2 sum runs over all finite sets Γ_2 of bonds in $\mathcal{B}\sim X_1^-$. For this reason, the Γ_2 sum can be evaluated by (3.8) as $F(\Lambda\cap X_2, s(\mathcal{B}\sim X_1^-))$. Setting $X = X_1^-$ and writing Γ for Γ_1, the expansion has the form

$$(3.13)\quad \begin{cases} F(\Lambda,s) = \sum\limits_X K(X_o,X)F(\Lambda\sim X, s(\mathcal{B}\sim X)) \\ K(X_o,X) = \sum\limits_\Gamma \int_0^{s(\Gamma)} \partial^\Gamma F(\Lambda\cap X,\ \sigma(\Gamma))\ d\sigma \;. \end{cases}$$

In these sums, X ranges over finite unions of closed lattice squares while Γ ranges over finite subsets of $\mathcal{B}$ such that

(i) Each component of $X\sim\Gamma^c$ meets X_o

(ii) $\Gamma\subset \mathrm{Int}\ X$.

If no such Γ exists, for a given X, then $K(X_o,X) \equiv 0$.

<u>Theorem 3.5</u>. Let X_o be bounded and let F be smooth, regular at infinity and decouple at $s=0$. Then the cluster expansion (3.13) holds.

<u>Example 1</u>. Let $\mathcal{B} = (Z^2)^*$ be the set of all lattice lines and let $X_o = \{x_1,\ldots,x_n\}$. We substitute $ZS = F$. See (3.4). Observe that $s=1$

$$(3.14)\qquad F(\Lambda\sim X, s(\mathcal{B}\sim X)) = \int e^{-\lambda V(\Lambda\sim X)} d\Phi_{s(\mathcal{B}\sim X)}$$

$$= Z(\Lambda\sim X, s(\mathcal{B}\sim X))$$

$$= Z(\Lambda\sim X, C_{\partial X}) \equiv Z_{\partial X}(\Lambda\sim X)$$

since a change in the data within X does not affect the integral. Thus if we divide by Z, (3.13) yields

$$(3.15)\quad S_\Lambda(x) = \sum_{X,\Gamma} \int \partial^\Gamma \int \Pi_i\ \Phi(x_i)\ e^{-\lambda V(\Lambda\cap X)}\ d\Phi_{s(\Gamma)}\ ds(\Gamma) \times Z_{\partial X}(\Lambda\sim X)/Z(\Lambda)\ .$$

If we write $X \sim \Gamma^c$ as a union of connected components, the integral in (3.15) can be factored, as in (3.12).

Example 2. Let $\Gamma_1 \subset (Z^2)^*$ and $\Gamma_2 = \Gamma_1 \sim b_1$ be finite sets of lattice lattice bonds, where b_1 is the first element of Γ_1 in some lexicographic order of $(Z^2)^*$. We use the cluster expansion (in all bonds $b \neq b_1$) to study the difference $Z_{\Gamma_1} - Z_{\Gamma_2}$ where

$$Z_\Gamma = \int e^{-\lambda V(\Lambda)}\ d\Phi_{C_\Gamma}$$

and so we take $\mathcal{B} = (Z^2)^* \sim b_1$, $X_0 = b_1$. Z is now a function of the pair (s,s_{b_1}). We define

$$F(\Lambda,s) = \begin{cases} Z(\Lambda,s(\Gamma^c),0) - Z(\Lambda,s(\Gamma^c),1)\ , & b_1 \subset \Lambda \\ Z(\Lambda,s(\Gamma^c),0) & , \quad b_1 \cap \Lambda = \emptyset \end{cases}$$

Ths definition is arranged so that

$$F(s=1) = Z_{\Gamma_1} - Z_{\Gamma_2}\ , \quad \text{for} \quad b \subset \Lambda\ .$$

Also F is smooth, regular at infinity and decouples at s = 0. (The last property depends on the exclusion of b_1 from $\mathcal{B}$.) We note that F is independent of the variables s_b for $b \in \Gamma_2$, and so $\partial^{b_j} F = 0$, for $b \in \Gamma_2$. Thus in all nonzero terms, Γ_2 consists of Dirichlet lines in the cluster expansion for F, and in (3.13), we may impose the further restriction on $\sum_\Gamma$:

(iii) $\Gamma \cap \Gamma_2 = \emptyset$.

From (3.13), we have

$$F(\Lambda,s=1) = Z_{\Gamma_1}(\Lambda) - Z_{\Gamma_2}(\Lambda)$$

$$= \sum_X K(b_1,\Gamma_1,X)F(\Lambda\sim X,\ s(\mathcal{B}\sim X))$$

$$= \sum_X K(b_1,\Gamma_1,X)Z_{\Gamma_1 \cup X^*}(\Lambda\sim X)$$

Here X^* is the set of lattice lines in X and the last line is justified by the fact that the second factor above is independent of s_b, $b \in \text{Int } X$. We multiply and divide by

$$Z_{\partial\Delta}(\Delta)^{|\Lambda \cap X|}$$

where Δ is a single lattice square. Since

$$Z_{\partial\Delta}(\Delta)^{|\Lambda \cap X|}\, Z_{\Gamma_1 \cup X^*}(\Lambda\sim X) = Z_{\Gamma_1 \cup X^*}(\Lambda) ,$$

we obtain (with a new K)

$$(3.16)\quad Z_{\Gamma_1} = Z_{\Gamma_1\sim b_1} + \sum_X K(b_1,\Gamma_1,X)Z_{\Gamma_1 \cup X^*}(\Lambda)$$

and now

$$(3.17)\quad K(b_1,\Gamma_1,X) = Z_{\partial\Delta}(\Delta)^{-|\Lambda \cap X|}\int \sum_\Gamma \partial^{\Gamma \cup b_1} Z(\Lambda \cap X, s(\Gamma \cup b_1))\, ds(\Gamma \cup b_1).$$

These equations have a structure intermediate between the Kirkwood-Salsburg and the Mayer-Montroll equations. They will be studied in §6 in order to obtain bounds on Z_Γ/Z, i.e. the second factor in (3.15).

§4. CLUSTERING AND ANALYTICITY: PROOF OF THE MAIN RESULTS

We prove the cluster property and analyticity -- the main results of these lectures -- in this section, using as hypothesis convergence of the cluster expansion, (3.15). Let $T(x,\Lambda,X,\Gamma)$ denote the X-Γ term in (3.15), so that

$$S_\Lambda(x) = \sum_{X,\Gamma} T(x,\Lambda,X,\Gamma) . \tag{4.1}$$

For each test function $w \in \mathcal{S}(R^{2n})$, the series

$$\int S_\Lambda(x)\, w(x) \quad x = \sum_{X,\Gamma} \langle w,T\rangle \tag{4.2}$$

converges absolutely. The rate of convergence is governed by $|X|$, the area of X. With $K > 0$, we prove that

$$\sum_{\{X,\Gamma:|X|\geq D\}} |\langle w,T\rangle| \leq |w|\, e^{-K(D-n)} . \tag{4.3}$$

where $|w|$ is some (n-dependent) $\mathcal{S}$-norm on w.

<u>Theorem 4.1</u>. Let $K > 0$ be given. Let m_0 be large and ε be small (depending on K) and let λ belong to the closure of (2.1). There is an $\mathcal{S}$-norm w such that (4.3) holds uniformly in λ, m_0, Λ and D, and $|w|$ is invariant under translations in any of its variables.

The proof allows integrands A of the form (2.7) to be substituted in (3.15). Thus Theorem 2.2 and Corollary 2.3 follow from the case $K = 1$, $D = 1$ of (4.3).

Theorem 2.1 also follows from the convergence of the cluster expansion. The proof for the two point function with an even interaction is conceptually easier, but still illustrates the main idea, so we consider that case first.

<u>Proof of Theorem 2.1</u>, for $n = 2$, P even, assuming Theorem 4.1. Because P, the interaction polynomial in (1.2), is even, $\Phi(x) \to -\Phi(x)$ is a symmetry of the theory. For Gaussian integrals defined by the factorizing measure $\exp(-\lambda V)\, d\Phi_{C(s(\Gamma))}$ (cf. Proposition 3.3), more is true:

$$\Phi(x) \to \sigma(x)\, \Phi(x) \tag{4.4}$$

is a symmetry, where

$$\sigma(x) = \pm 1; \quad \sigma(x) = \text{const. on each } X_i .$$

Because of the symmetry $\Phi \to -\Phi$, $S_\Lambda(x_1,\dots,x_n) = 0$ for n odd. Similarly,

$$\int \Pi\, \Phi(x_i)\, e^{-\lambda V(\Lambda\cap X)}\, d\Phi_{C(s(\Gamma))} = 0$$

unless each of the connected components $X_1,\ldots,X_r$ contains an even number of x_i's. Since $\partial^{\Gamma} 0 = 0$, the same restriction $T(x,\Lambda,X,\Gamma) = 0$ applies unless each X_j has an even number of x_i's. Since each X_j has at least one x_i by (i) of §3, each X_j must have at least two x_i's. Thus for n = 2, there is only one X_j. In other words $X\sim\Gamma^c$ is connected. Let d be defined by Theorem 2.1 and let $w = w_1 \otimes w_2$. Because X is connected and suppt $w_i \cap X \neq \emptyset$, we have $d \leq |X| + 1$. Thus by (4.2-3),

$$\left|\int S_\Lambda(x_1,x_2)w_1(x_1)w_2(x_2)dx\right| \leq |w|e^{-K(D-2)} \leq M_w e^{-Kd}$$

Since the one-point functions $S_\Lambda(x_1)$ vanish for P even, this bound completes the proof of Theorem 2.1.

Proof of Theorem 2.1 (general case) assuming Th. 4.1. The idea, as before, is to reduce the cluster expansion to terms involving only one connected component $X_1 = X$. The terms with two or more connected components are to vanish, because of some symmetry. Since the general case (P not even) does not have such a symmetry, we follow Ginibre [2] and introduce a new theory containing an artificial symmetry.

To construct the new theory, let $d\Phi^*_{C^*}$ denote an isomorphic copy of the measure $d\Phi_C$, $(C^* \cong C)$ defined on an isomorphic copy $\mathcal{S}'^*$ of $\mathcal{S}'$. The new theory has free measure $d\Phi_C \times d\Phi^*_{C^*}$, covariance

$$C \otimes I + I \otimes C^* = C^\sim$$

and normalized physical measure

$$Z^{\sim -1} e^{-V(\Lambda)} e^{-V(\Lambda)^*} d\Phi_C \times d\Phi^*_{C^*} = dq^\sim$$

and a field

$$\Phi^\sim = \Phi \otimes I + I \otimes \Phi^* .$$

This theory is invariant (even) under the symmetry $\Phi \leftrightarrow \Phi^*$, which interchanges the two factors.

We now apply the cluster expansion to the expression

$$Z^\sim \int (A-A^*)(B-B^*)dq^\sim$$

The covariance operators which arise in this expansion have the form

$$C(s)^\sim = C(s) \otimes I + I \otimes C(s)^*$$

so that the symmetry $\Phi \leftrightarrow \Phi^*$ is preserved in the Gaussian measure of each term of the expansion. (However the expansion is modified by choosing $\mathcal{B} = (Z\ )^* \sim \Gamma$, where Γ is the set of lattice lines in two connected sets, one containing suppt A = suppt $A-A^*$, and the other containing suppt B. Thus for each component X_i, $X_i \supset$ suppt A or $X_i \cap$ suppt A = $\emptyset$, and the same for suppt B. With this restriction, there are at most two components. Since the n in (4.3) is a bound on the number of components, as one can check, n = 2 in (4.3).) Now consider a term in (3.15) with components $X_1, X_2, \ldots$ satisfying

$$\text{suppt } A \subset X_i \ , \quad \text{suppt } B \subset X_j \ , \quad i \neq j \ . \tag{4.5}$$

For this term, the symmetry $\Phi \leftrightarrow \Phi^*$ can be applied separately in each X_i. However A-A* is odd for the X_i symmetry, so terms satisfying (4.5) must vanish. The nonvanishing terms, which violate (4.5) contain a component $X_1 = X$ with $d \le 0|X|$, and so for $d \ge 4$,

$$\left|\int (A-A^*)(B-B^*)dq^{\sim}\right| \le M_{AB}e^{-md} \ .$$

Expanding the integrand on the left yields four terms, and in each term the dq integral factors. We evaluate the result as

$$2\left|\int ABdq_{\Lambda,C} - \int Adq_{\Lambda,C}\int Bdq_{\Lambda,C}\right| ,$$

and Theorem 2.1 follows.

§5. CONVERGENCE: THE MAIN IDEAS

There are two main ideas in part II of these lectures. The first is the formula (3.17) giving the cluster expansion for the Schwinger functions. The second is the estimates which lead to the convergence of this expansion, uniformly as $\Lambda \to \infty$. In this section we state these estimates as a series of three propositions, and using them, give the proof of Theorem 4.1. The simpler estimates are then proved, while the harder ones are postponed for later sections. The most difficult of these estimates is contained in Prop. 5.3. Since it is in some sense the core of the paper, we discuss the ideas involved in its proof at the end of this section.

The proof of convergence depends on estimates of the following types:

(a) Combinatoric estimates to count the number of terms in the expansion, and especially to count or bound the number of terms of some special type.

(b) Single particle estimates, to bound kernels of covariance operators, and their derivatives, $\partial^{\Gamma} C(s)$.

(c) Estimates on function space integrals. Typically estimates of type (a) and (b) will be components of an estimate of type (c).

Returning to specifics, (4.3) or (3.15) is a sum in which each term is a product of two factors: a ratio of partition functions times a function space integral. The first proposition is purely combinatoric - type (a)- and it counts the number of terms with $|X|$ = area X held fixed. The second and third propositions bound the two factors making up a term in (3.15). These latter two propositions are hybrids, since their proof uses estimates of types (a), (b) and (c).

Proposition 5.1. The number of terms in (3.15-3.16) with a fixed value of $|X|$ is bounded by $e^{K_1|X|} \equiv O(1)e^{19|X|}$.

Proposition 5.2. There is a constant K_2, independent of λ in the closure of (2.1), and of Λ and m_0, such that for ε small and m_0 large,

$$\left|Z_{\partial X}(\Lambda \sim X)/Z(\Lambda)\right| \leq e^{K_2|X|} .$$

Proposition 5.3. There is a constant K_3 and a norm $|w|$ on test functions such that for any $K > 0$, for any Λ and for λ in the closure of (2.1) with m_0 large (the m_0 bound depends on K)

$$\left| < \int \partial^{\Gamma} \int \prod_{i=1}^{n} \Phi(x_i)\, e^{-\lambda V(\Lambda)} d\Phi_{C(s(\Gamma))}\, ds(\Gamma), w > \right| \leq e^{-K|\Gamma| + K_3|\Lambda|}\, |w| .$$

(The m_0 bound depends on K, and $|w|$ is invariant under translation in any of the variables of w.)

Remark. Wick polynomials, as in (2.7), are also allowed in the integrand above.

Proof of Theorem 4.1. We replace Λ by $\Lambda \cap X$ in Prop. 5.3. For the X in (3.15), we have $X = \bigcup_{i=1}^{r} X_i^-$ with $r \leq n$ and X_i connected. Moreover

$$\Gamma \subset \bigcup_{i=1}^{r} \text{Int } X_i^-$$

and so "many" of the lattice lines in X_i^- belong to Γ. In fact since $X_i^- \sim \Gamma^c = X_i$ is connected,

(5.1) $\quad |X_i| - 1 \leq 2|\Gamma \cap \text{Int } X_i^-|$ and $|X| - n \leq 2|\Gamma|$.

Thus we replace the upper bound in Prop. 5.3 by

$$e^{-K(|X|-n)} \; |w| \; ,$$

with a new choice of K and $|w|$. Theorem 4.1 follows directly from this bound combined with Prop. 5.1 and 5.2.

Proof of Proposition 5.1. We consider (3.15). First we bound the number of ways of choosing the component X_i containing a particular x_j, $1 \leq j \leq n$. We identify each lattice square Δ with a lattice point located at its center, and we identify each lattice line $b \in \mathcal{B}$ lying between two squares Δ and Δ' with the line joining the center of Δ to the center of Δ'. Hence we must count the number of ways of drawing a connected graph with nearest neighbor bonds. We assert that each such graph may be constructed by starting at x_j and drawing an oriented path formed of unit line segments which traverses each segment at most twice. In fact if we regard the lattice sites as islands and the line segments as bridges, this simply follows from the solution to the Konigsberg bridge problem[6]. The number of connected paths of length ℓ formed from lattice line segments and starting at x_j is at most 4^{ℓ}. Since $\ell \leq 8|X_i|$, the number of choices for X_j is at most $O(1)2^{16|X_i|}$. Since the number of choices for Γ is at most $4^{|X|}$, the number of choices for pairs X,Γ is at most

$$O(1)2^{|X|}4^{|X|}\prod_i 2^{16|X_i|} = O(1)2^{19|X|} ,$$

since $2^{|X|}$ bounds the number of choices of $|X_i|$ with $|X|$ given.
This completes the proof. The case (3.16) is similar.

Proposition 5.2 is proved in §6 using equations related to the

[6] We thank O. Bratteli for this observation. For a solution of the Konigsberg bridge problem see, D.W. Blackett, Elementary Topology, Academic Press, New York, 1967, page 159.

Kirkwood-Salsburg equations. We only remark here that the change in partition function involves both a change in the interaction region and a change in covariance ($\Lambda \sim X \to \Lambda$ and $C_{\partial X} \to C$).

Discussion of Prop. 5.3. Each derivative d/ds_b in ∂^Γ differentiates either the measure or the integrand. Derivatives of the measure are evaluated by (1.7), and produce s-dependent kernels $C'(s)$ in the integrand. Repeated differentiation yields

(a) A sum of terms coming from the iterated functional derivatives $\partial^2/\partial\Phi^2$ in (1.7).

(b) A sum over terms arising from possible repeated derivatives $\partial^{\Gamma_1}C(s)$ of each $C'(s)$ introduced by (1.7).

Each derivative d/ds_b also produces convergence in one of two ways. Differentiation of the measure introduces a kernel C', and

(c) C' and C are small in the sense that

$$(c_1)\ \|C'(x,y)\|_{L_p} \le O(m_0^{-\varepsilon})$$

$$(c_2)\ 0 \le C'(x,y) = \partial^b C(x,y) \le e^{-m_0(\mathrm{dist}(x,b)+\mathrm{dist}(y,b))}$$

$$(c_3)\ 0 \le C(x,y) \le e^{-m_0|x-y|}, \quad |x-y| \ge 1 .$$

Repeated differentiation of $C(s)$ produces the further convergence

(d) $\partial^\Gamma C$ is small in the sense that

$$(d_1)\ \|\partial^\Gamma C(x,y)\|_{L_p} \le O(m_0^{-\varepsilon|\Gamma|})$$

$$(d_2)\ 0 \le \partial^\Gamma C(x,y) \le e^{-m_0 d}$$

where d is the length of the shortest path connecting x and y and passing through each $b \in \Gamma$. In fact from the Weiner integral representation for $(-\Delta+m_0^2)^{-1}$, $\partial^\Gamma C(s)$ is the integral over Wiener paths from x to y passing through each $b \in \Gamma$, see §7-8.

We use (c_2) to control (a). In fact by (c_2), only x and y near b contribute to the Laplacian $C'(s)\cdot\Delta_\Phi$ of (1.7). By the nature of the expansion, there is at most one Laplacian Δ_Φ per bond $b \in \mathcal{B}$, hence a bounded power, $\Delta_\Phi{}^n$, locally. Evaluation of Δ_Φ^n leads to $K(n) < \infty$ terms, and this evaluation, repeated over disjoint local regions leads to $e^{O(\mathrm{Vol})}$ terms. We use (d_2) to control (b). The basic convergence, $m_0^{-\varepsilon|\Gamma|}$, comes from (d_1).

After performing these steps, there remains a function space integral of the form $\int Re^{-\lambda V}d\Phi$. Then

$$| \int Re^{-\lambda V(\Lambda)} d\Phi | \le \left(\int R^2 \, d\Phi \right)^{1/2} \left(\int e^{-2 Re \lambda V(\Lambda)} d\Phi \right)^{1/2}$$

and each factor on the right is $e^{0(\mathrm{Vol})}$. Now (c_3) produces the decoupling of distant local regions in R^2.

With some practice, estimates such as Proposition 5.3 can be done by inspection. In Sections 7-9 we develop some general machinery, the goal of which is to make these estimates more routine.

§6. AN EQUATION OF KIRKWOOD-SALSBURG TYPE

Equations (3.16-3.17) can be rewritten as a Banach space equation

$$\rho = Z(\Lambda)1 + \mathcal{K}\rho \tag{6.1}$$

with a unique solution

$$\rho = (1-\mathcal{K})^{-1} Z(\Lambda)1$$

satisfying the bound

$$\|\rho\| \leq \|(I-\mathcal{K})^{-1}\| \, |Z(\Lambda)| \leq 4|Z(\Lambda)| \tag{6.2}$$

This bound is essentially Proposition 5.2, as we shall see.

Let $\mathcal{X}$ be the Banach space of functions f defined on finite subsets $\Gamma \subset (Z^2)^*$. For $f \in \mathcal{X}$ let $(f_n)_{n\geq 0}$ denote the restriction of f to n element subsets. The norm on $\mathcal{X}$ is defined to be

$$\|f\| = \sup_{\substack{n,\,\Gamma \\ |\Gamma|=n}} 2^{-n} |f_n(\Gamma)| .$$

We define $\rho_\Lambda \equiv (\rho_{\Lambda,n})_{n\geq 0}$ to be the function

$$\Gamma \to Z_\Gamma(\Lambda) .$$

Theorem 6.1. Let $|\lambda| \leq \varepsilon$, $\mathrm{Re}\,\lambda \geq 0$ where $\varepsilon > 0$ is small and let m_0 be sufficiently large. Then $Z(\Lambda) \neq 0$, and ρ_Λ defined above is the unique solution in $\mathcal{X}$ of equation (6.1). Moreover ρ_Λ satisfies the bounds (6.2).

By the dominated convergence theorem, $Z_{\partial\Delta}(\Delta) \to 1$ as $\varepsilon \to 0$. Thus for small ε,

$$\tfrac{1}{2} \leq |Z_{\partial\Delta}(\Delta)| \leq 2 .$$

This restriction fixes the ε in Theorem 6.1, and in fact is the only restriction on ε in these lectures.

Proof of Proposition 5.2, assuming Theorem 6.1.

$$\begin{aligned} |Z_{\partial X}(\Lambda\sim X)/Z(\Lambda)| &= |Z_{X^*}(\Lambda)/Z(\Lambda)| \; |Z_{\partial\Delta}(\Delta)|^{-|\Lambda\cap X|} \\ &\leq 2^{|X^*|}\|(1-\mathcal{K})^{-1}\| \; |Z_{\partial\Delta}(\Delta)|^{-|\Lambda\cap X|} \\ &\leq e^{K_2|X|} . \end{aligned}$$

Proof of Theorem 6.1. Let $1 = (1,0,0,\ldots) \in \mathcal{X}$, and define $\mathcal{K}$ by the equations

$$(\mathcal{K} f)_n(\Gamma) = f_{n-1}(\Gamma \sim b_1) + \sum_{m=1}^{\infty} \sum_{X, |X|=n} K(b_1,\Gamma,X)\, f_{|X^* \cup \Gamma|}(X^* \cup \Gamma) \tag{6.4}$$

for $n \geq 2$ where the sum over X (as in (3.13) and (3.17)) ranges over connected unions of lattice squares such that $b_1 \in X$. For $n = 1$, we omit the first term on the right of (6.4) and for $n = 0$, $(\mathcal{K} f)_0 \equiv 0$.

We now show that $\mathcal{K}$ is a contraction, $\|\mathcal{K}\| \leq 3/4$, in order to obtain (6.2). It suffices to show that

$$\frac{1}{2} + \sup_{\Gamma \subset \mathcal{B}} 2^{-|\Gamma|} \sum_{m=1}^{\infty} \sum_{X, |X|=m} |K(b_1,\Gamma,x)|\, 2^{|\Gamma \cup X|} \leq \frac{3}{4} \tag{6.5}$$

The proof of (6.5) is similar to the proof of Theorem 4.1 given in §5. In particular it depends on Propositions 5.1 and 5.3. By Proposition 5.1 there are at most $e^{K_1|X|}$ terms for each fixed value of $|X|$. From (3.17) each term has at least one derivative ∂^{b_1}. Using this fact and (5.1) the bound from Propositions 5.1 and 5.3 is

$$|K(b_1,\Gamma,X)| \leq e^{-K(|X|+1)}\, e^{K_3|X|} .$$

For K sufficiently large (6.5) follows. We complete the proof by showing $Z(\Lambda) \neq 0$. By our choice of ε, $Z_{\Lambda^*} = |Z_{\partial\Lambda}(\Delta)|^{|\Lambda|} \neq 0$. Thus $\rho_\Lambda \neq 0$ and by (6.2), $Z(\Lambda) \neq 0$.

§7. COVARIANCE OPERATORS

The basic facts for the kernel $C_\emptyset(x,y)$ of $C_\emptyset = (-\Delta+m_0^2)^{-1}$ are

(a) $C_\emptyset(x,y)$ is a function of the difference $|x-y|$.

(b) $0 < C_\emptyset(x,y)$

(c) for $|x-y| \geq 1$, $C_\emptyset \leq e^{-m_0(1-\delta)|x-y|}$

(d) as $|x-y| \to 0$, $C_\emptyset \sim \log|x-y|$

(e) for $r > 2/(2-\delta)$, $\delta > 0$, $C_\emptyset \in \mathcal{B}^{loc}_{r,\delta}(R^4)$.

By definition, the space $\mathcal{B}^{loc}_{r,\delta}(R^4)$ is the space of distributions $\psi \in \mathcal{S}'$ which satisfy

$$\|\zeta\psi\|_{\mathcal{B}_{r,\delta}} = \|(1+k^2)^{\delta/2}(\zeta\psi)^{\sim}\|_{L_r} < \infty$$

whenever $\zeta \in C_0^\infty$. For r and r' dual indices and for $1 \leq r' \leq 2$, $\mathcal{B}^{loc}_{r,\delta}$ is a space of functions with fractional $L_{r'}$ derivatives, by the Hausdorff-Young inequality. See [8, Chapter 2] for a general theory of such spaces. (e) follows from the fact that as a function of x-y, $C_\emptyset$ has the Fourier transform $(k^2+m^2)^{-1}$.

Property (a), which is based on the translation invariance of the Laplacian, does not extend to the operators $C \in \mathcal{C}$. Recall that $\mathcal{C}$ is the set of all convex combinations of the Dirichlet covariance operators (1.6), or in other words, operators of the form C(s). Properties (b) and (c) are valid for all $C \in \mathcal{C}$, while (d) is valid as an upper bound for all $C \in \mathcal{C}$.

<u>Proposition</u> 7.1. The kernel $C(x,y)$ of $C \in \mathcal{C}$ satisfies

$$0 \leq C(x,y) \leq C_\emptyset(x,y) .$$

<u>Proof</u>. Let $dz^T_{x,y}$ be the conditional Wiener density on paths $z(\tau)$ starting at x at $\tau = 0$ and ending at y at $\tau = T$. Let J^T_b be the function

$$J^T_b(z) = \begin{cases} 0 & \text{if } z(\tau) \in b, \quad 0 \leq \tau \leq T \\ 1 & \text{otherwise} \end{cases} \tag{7.1}$$

defined on Wiener paths. Then C_Γ and $C(s)$ have the representations

$$C_\Gamma(x,y) = \int_0^\infty e^{-m_0^2 T} \int \Pi_{b\in\Gamma} J^T_b(z)\, dz^T_{x,y}\, dT \tag{7.2}$$

$$C(s)(x,y) = \int_0^\infty e^{-m_0^2 T} \int \Pi_{b\in\mathcal{B}}(s_b + (1-s_b)J^T_b)\, dz^T_{x,y}\, dT . \tag{7.3}$$

See [0, §5.3, 6.1 and 6.3].

The inequality of Proposition 7.1 comes from substituting

$$0 \leq (s_b + (1-s_b)J_b^T) \leq 1$$

in the formula above for C(s).

The next estimate combines the exponential decay (c) with a local bound coming from (d). We label each lattice square $\Delta = \Delta_j \subset R^2$ by the lattice point $j \in z^2$ in the lower right corner of Δ. Any lattice square (or any set $X \subset R^2$), will be identified with the multiplication on $L_2(R^2)$ by its characteristic function. Now let $j = (j_1, j_2) \in Z^4$ be a pair of lattice points. We define the localized covariance operator

$$C(j) = \Delta_{j_1} C \Delta_{j_2} \tag{7.4}$$

and the distance

$$d(j) = \text{Dist}\ (\Delta_{j_1}, \Delta_{j_2}) , \tag{7.5}$$

which measures the nonlocality of C(j).

Proposition 7.2. For $1 \leq q < \infty$ and $\delta > 0$, there is a constant $K_4(q,\delta)$ independent of $m_0 \geq 1$ and $C \in \mathcal{C}$ such that

$$\| C(j) \|_{L_q(\Delta_{j_1} \times \Delta_{j_2})} \leq K_4 m_0^{-2/q} \exp(-m_0(1-\delta)d(j)) .$$

Proof. Because of Proposition 7.1, we may assume $C = C_\emptyset$. For $d(j) > 0$, necessarily $d(j) \geq 1$, and the required bound follows from property (c). The factor $m_0^{-2/q}$ in the proposition comes from choosing δ in (c) smaller than the δ in the proposition. Next consider the case $d(j) = 0$ (i.e. equal or adjacent squares). We have

$$\| C(x,y) \|_{L_q(\Delta_{j_1} \times \Delta_{j_2})} \leq K_3 \| C(\cdot,0) \|_{L_q(R^2)} \leq K_4(q) m_0^{-2/q} .$$

The last inequality follows from the fact that $(-\Delta+1)^{-1}(x,0) \in L_q$ for $q \in [1,\infty)$ and from the identity

$$(-\Delta+m_0^2)^{-1}(x,0) = (-\Delta+1)^{-1}(m_0 x, 0) .$$

Remark. The differentiated covariance $\partial^\Gamma C$ also satisfies the bound of Proposition 7.1, as we can see from (7.3) and the inequality

$$0 \le \frac{d}{ds_b}(s_b + (1-s_b)J_b) = 1-J_b \le 1 .$$

Hence $\partial^\Gamma C$ also satisfies the bound of Proposition 7.2.

Local regularity of the covariance $C \in \mathcal{C}$ is used to justify the definition and basic formal properties (e.g. integration by parts) of Gaussian integrals and it is used in the bound $|Z| \le e^{O|\Lambda|}$. We obtain fractional derivatives of $C \in \mathcal{C}$, with bounds slightly weaker than (e).

Proposition 7.3. $C \in \mathcal{B}^{loc}_{2,\delta}$ for $0 \le \delta < 1/2$, with bounds independent of C.

Proof. We start with the observation that

$$0 \le -\Delta \le -\Delta_\Gamma .$$

Consequently $C_\emptyset^{-1/2} C_\Gamma C_\emptyset^{-1/2}$ is a bounded operator, with norm at most one. By convex combinations, $\|B\| \le 1$ also, where

$$B = C_\emptyset^{-1/2}\, C\, C_\emptyset^{-1/2} .$$

Let

$$A = C_\emptyset^{-(1/4)+\varepsilon} \zeta C_\emptyset^{1/2},$$

for $\zeta \in C_0^\infty(R^2)$. We use the fact that $A\, A^*$ is Hilbert Schmidt to calculate the Hilbert Schmidt norm

$$\text{(7.6)} \quad \|(-\Delta+m_0^2)^{-\varepsilon+1/4}\, \zeta\, C\, \zeta(-\Delta+m_0^2)^{-\varepsilon+1/4}\|^2_{HS}$$

$$= \text{Tr } A\, B\, A^*A\, B\, A^* \;=\; \text{Tr } B\, A^*A\, B\, A^*A$$

$$\le \text{Tr } (B\, A^*A)^* B\, A^*A = \text{Tr } A^*A\, B^2 A^*A$$

$$\le \|B\|^2\, \text{Tr } (A^*A)^2 \;\le\; \|A^*A\|^2_{HS} .$$

This inequality proves the proposition (and slightly more).

The reader may skip the next result, since it is not needed in what follows.

Proposition 7.4. For any $q \in (4/3,\infty)$, there is a $\delta = \delta(q) > 0$ with $C \in \mathcal{B}^{loc}_{q,\delta}$ and the bounds are independent of $C \in \mathcal{C}$.

Proof. With $k = (k_1,k_2) \in R^4$, define

$$f(k) = 1+k^2 , \quad g(k) = (1+k_1^2)(1+k_2^2) ,$$

$$h(k) = (\zeta\, C\, \zeta)^\sim(k) .$$

Then

$$g^{-\varepsilon-1} \in L_1 , \quad f/g \in L_\infty .$$

By Proposition 7.1 and the fact that $\zeta C_\emptyset \zeta \in L_1$,

$$h \in L_\infty$$

and by (7.6),

$$g^{-\varepsilon+1/4} h \in L_2 \quad \text{and} \quad h \in L_2 .$$

We apply these facts to the integral

$$(7.7) \quad \int |f^{\delta/2} h|^q = \int |g^{-\varepsilon+1/4} h| \, |\tfrac{f}{g}|^{\delta q/2} |g|^{(\delta q/2)+\varepsilon-1/4} |h|^{q-1} .$$

First take $\delta > 0$ and suppose that $h \in L_{q_1}$, $q_1 > q > 4/3$. The factors in (7.7) belong to L_2, $L_{4+\varepsilon}$, and $L_{q_1/q-1} \cap L_\infty$. For ε small and q near q_1 , this implies $h \in L_q$ also. Since $h \in L_q$ for q in an interval, $h \in L_q$ for $q \in (4/3,\infty]$. Now with $h \in L_q$, $q > 4/3$, (7.7) shows that $f^{\delta/2} h \in L_q$ also, for small δ. This completes the proof.

<u>Proposition</u> 7.5. For $1 \le q < \infty$,

$$c(x) = \lim_{y\to x} C(x,y) - C_\emptyset(x,y) \in L_q^{loc} .$$

There is a constant $K_5(q)$, independent of $m_0 \ge 1$ and $C \in \mathcal{C}$ such that for any lattice square Δ,

$$\| c \|_{L_q(\Delta)} \le K_5 \, m_0^{-1/q} .$$

<u>Proof</u>. By scaling, we may take $m_0 = 1$, as in Proposition 7.2. Let $\Gamma = \mathcal{B}$ be the set of all lattice lines. For $x \notin \Gamma$,

$$\begin{aligned} 0 \le - c(x) &\le C_\emptyset(x,x) - C_\Gamma(x,x) \\ &\le O(1 + |\log \operatorname{dist}(x,\Gamma)|) . \end{aligned}$$

This inequality completes the proof. It is proved in [7] as follows. For $x \notin \Gamma$, $y \notin \Gamma$,

$$(-\Delta_y + m_0^2)[C_\emptyset(x,y) - C_\Gamma(x,y)] = 0 .$$

Hence by the maximum principle, and the fact that $C_\Gamma(x,y) = 0$ for $y \in \Gamma$,

$$0 \le C_\emptyset(x,y) - C_\Gamma(x,y) \le \sup_{y\in\Gamma} C_\emptyset(x,y) \le O(1+|\log \operatorname{dist}(x,\Gamma)|).$$

§8. DERIVATIVES OF COVARIANCE OPERATORS

For the differentiated covariance operator $\partial^\gamma C$, there should be a strong decay, $\sim e^{-m_0 d}$, where d is the length of the shortest path in R^2 joining x to y and passing through each lattice line segment $b \in \gamma$. This can be seen by inspection from the Wiener integral representation

$$(8.1)\quad (\partial^\gamma C(s))(x,y) = \int_0^\infty e^{-m_0^2 T} \int \Pi_{b\in\gamma}(1-J_b^T) \times \Pi_{b\in\mathcal{B}\sim\gamma}[s_b + (1-s_b)J_b^T]\, dz_{x,y}^T\, dT \,:$$

We need the improved bounds on $\partial^\gamma C$ for two reasons. The first is to localize x and y, with γ given. For this purpose

$$(8.2)\qquad d(j,\gamma) = \sup_{b\in\gamma} \{\mathrm{Dist}(\Delta_{j_1},b) + \mathrm{Dist}(\Delta_{j_2},b)\}$$

is sufficient, as a crude lower bound on d.

We now explain the second use of bounds on $\partial^\gamma C$. Let $\mathcal{P}(\Gamma)$ be the set of all partitions π of the set of lattice line segments Γ. In Proposition 5.3, we are called on to bound $\partial^\Gamma \int F\, d\Phi_s$, which by Leibnitz' rule and by (1.7) is just

$$(8.3)\qquad \partial^\Gamma \int F\, d\Phi_s = \sum_{\pi\in\mathcal{P}(\Gamma)} \int \left(\Pi_{\gamma\in\pi} \tfrac{1}{2}\, \partial^\gamma C\cdot\Delta_\Phi\right) F\, d\Phi_s \,.$$

The second use of the bounds on $\partial^\gamma C$ is to control $\sum_{\pi\in\mathcal{P}(\Gamma)}$. As in §7, we also find a factor $m_0^{-0|\gamma|}$, which yields the overall convergence of the expansion.

<u>Proposition</u> 8.1. Let $1 \le q < \infty$ and let m_0 be sufficiently large. There are constants $K_6(q,\gamma)$ and $K_7(q)$, independent of m_0, such that

$$(8.4)\quad \|\partial^\gamma C\|_{L_q(\Delta_{j_1}\times\Delta_{j_2})} \le K_6(q,\gamma)\, m_0^{-|\gamma|/2q} \exp(-m_0\, d(j,\gamma)/2)$$

$$(8.5)\qquad \sum_{\pi\in\mathcal{P}(\Gamma)} \Pi_{\gamma\in\pi} K_6(q,\gamma) \le e^{K_7(q)|\Gamma|} \,.$$

<u>Proof</u>. We use the Wiener integral representation (8.1) for $\partial^\gamma C$. The proof consists of estimates on the Wiener measure of paths $z(\tau)$ which cross the lattice lines $b \in \gamma$ in some definite order together with combinatoric arguments to count the number of ways the lines

$b \in \gamma$ can be so ordered.

Let $L(\gamma)$ be the set of all possible linear orderings of the lines $b \in \gamma$, and for $\ell \in L(\gamma)$, let $\mathcal{W}(\ell)$ be the set of Wiener paths which cross all lines $b \in \gamma$, and whose order of first crossing is ℓ. Then

$$(8.6) \qquad 0 \le \partial^\gamma C(s) \le \int_0^\infty e^{-m_0^2 T} \int \prod_{b \in \gamma} (1-J_b^T(z))\, dz_{x,y}^T \, dT = \partial^\gamma C_\emptyset ,$$

and

$$(8.7) \qquad \partial^\gamma C_\emptyset(x,y) = \sum_{\ell \in L(\gamma)} \int_0^\infty e^{-m_0^2 T} \int_{\mathcal{W}(\ell)} dz_{x,y}^T \, dT .$$

Let $b_1, b_2, \ldots$ be the elements of γ, as ordered by ℓ. Let b_2' be the first of the b's not touching $b_1 = b_1'$, let b_3' be the first of the b's after b_2' and not touching b_2', etc. Set

$$a_j = \mathrm{Dist}(b_{j+1}', b_j') , \qquad 1 \le j \le m$$

and define

$$|\ell| = \textstyle\sum_{i=1}^m a_i .$$

If there is no such b_2', we set $|\ell| = 0$, by convention.

With these definitions, we bound the $\ell \in L(\gamma)$ term in (8.7), for $|\ell| \ge 1$, by

$$\int e^{-m_0^2 T} \int_{\sum t_i = T} \Pi_i (2\pi t_i)^{-1} \exp\left(-\tfrac{1}{2} \textstyle\sum_{i=1}^m \frac{a_i^2}{t_i}\right) dt\, dT$$

$$\le K_8^m \sup_{T, \sum t_i = T} e^{-(m_0^2-2)T} e^{-\frac{1}{3}\sum_{i=1}^m a_i^2/t_i}$$

since

$$\int_{\sum t_i = T} dt \le e^T , \qquad \int e^{-T}\, dT \le 1 ,$$

and for all $a_i \ge 1$,

$$(2\pi t_i)^{-1} e^{-a_i^2/6t_i} \le K_8 < \infty ,$$

thereby defining K_8.

Using the method of Lagrange multipliers to evaluate the maxima, we bound the $\ell \in L(\gamma)$ term by

$$K_8^m \exp\left((m_0^2-2)^{1/2} |\ell|\right) ,$$

for $|\ell| \ge 1$. For $|\ell| = 0$, we use the remark following Prop. 7.2.

There is an entirely similar estimate, based on the distance $d(j,\gamma)$ of (8.2), and taking geometric means of these two bounds yields

$$(8.8) \quad \|\partial^\gamma C\|_{L_q(\Delta_1\times\Delta_2)} \le \sum_{\ell\in L(\gamma)} K_8^{|\gamma|} e^{-m_0|\ell|/(2+\delta)} e^{-m_0 d(j,\gamma)/(2+\delta)}$$

for m_0 large. If $|\ell| \ge 1$ for all $\ell \in L(\gamma)$, then we can include a factor $m_0^{-|\gamma|}$ on the right side of (8.8), by increasing δ. If $|\ell| < 1$ for some ℓ, then $|\ell| = 0$, and in this case $|\gamma| \le 4$. With $|\gamma| \le 4$ and $d(j,\gamma) \ge 1$, we can still include the factor $m_0^{-|\gamma|}$ in (8.8) by increasing δ. Finally for $|\gamma| \le 4$ and $d(j,\gamma) = 0$, the factor $m_0^{-|\gamma|/2q} \le m_0^{-2/q}$ in (8.4) comes from scaling, as in Proposition 7.2.

We define

$$(8.9) \qquad K_6(q,\gamma) = K_4 \sum_{\ell\in L(\gamma)} K_8^{|\gamma|} e^{-m_0|\ell|/(2+\delta)}$$

With this definition, (8.4) follows; in the case $d(j,\gamma) = 0$ and $|\ell| = 0$ for some ℓ, $K_6 \ge K_4$, and we use the bound of Proposition 7.2 to establish (8.4).

We complete the proof by establishing (8.5) as a separate proposition.

<u>Proposition</u> 8.2. For m_0 sufficiently large,

$$(8.10) \qquad \sum_{\pi\in\mathcal{P}(\Gamma)} \prod_{\gamma\in\pi} \sum_{\ell\in L(\gamma)} e^{-m_0|\ell|/3} \le e^{K_9|\Gamma|} .$$

<u>Proof</u>. Let $\mathcal{L}(\Gamma)$ be the set of linear orderings defined on subsets of Γ. Thus $L(\Gamma) \subsetneq \mathcal{L}(\Gamma)$. As before, we define $|\ell|$ for $\ell \in \mathcal{L}(\Gamma)$. We assert that the number of $\ell \in \mathcal{L}(\Gamma)$ with $|\ell| \le r$ is bounded by

$$(8.11) \qquad |\Gamma|\, e^{K_{10}(r+1)} .$$

Using (8.11), we complete the proof. Let $A_\ell = \exp(-m_0|\ell|/3)$. Expanding $\sum \prod \sum A_\ell$ in (8.10), we get a sum of terms of the form

$$A_{\ell_1} A_{\ell_2} \cdots A_{\ell_j}$$

where the ℓ_j are distinct elements of $\mathcal{L}(\Gamma)$. Adding all terms of this form, we bound (8.10) by

$$\sum \prod \sum A_\ell \le \sum A_{\ell_1}\cdots A_{\ell_j} = \prod_{\ell\in\mathcal{L}(\Gamma)} (1+A_\ell)$$

$$\le \prod_{\ell\in\mathcal{L}(\Gamma)} \exp A_\ell = \exp \sum_{\ell\in\mathcal{L}(\Gamma)} A_\ell \le \exp\,(O(1)|\Gamma|) .$$

Here in the last expression, we used the bound (8.11) to estimate $\sum_{\ell\in\mathcal{L}(\Gamma)} A_\ell$ and we choose m_0 sufficiently large.

Next we establish (8.11). Suppose the integer part $[a_i]$ of the distances a_i, are given. We choose $b_1 = b_1'$ in $|\Gamma|$ ways, and we choose the b's between b_1' and b_2' in $O(1)$ ways, since they all must overlap b_1. Next b_2' is chosen in $O(1)$ $[a_1]$ ways, namely from the lattice line segments b with

$$[a_1] \le \mathrm{Dist}(b,b_1') < [a_1]+1 .$$

Continuing in this fashion, we choose all the b's in

$$|\Gamma| \ \Pi_i \ O(1) \ [a_i] \le |\Gamma| \ e^{O(1) \sum [a_i]} \le |\Gamma| e^{O(1)r}$$

ways. Finally we count the number of choices of the $[a_i]$. This is the number of ways of choosing integers $r_i \ge 1$ with $\sum r_i \le r$, namely 2^r. In fact suppose $\sum r_i = r$, and we distribute the r units in $\sum r_i$ as follows: The first 1 goes into a_1 (no choice). The second 1 goes to a_1 or a_2 (one binary choice). If the jth 1 goes to a_i, the j+1st goes to a_i or a_{i+1} (one binary choice). Thus there are r-1 binary choices, or 2^{r-1} ways to choose r_i with $\sum r_i = r \ge 1$. Summing $j = \sum r_i$ gives $\sum_{j=1}^{r} 2^{j-1} = 2^j - 1$. Finally, we get one more choice from $|\ell| = 0$ (no a_i's).

§9. GAUSSIAN INTEGRALS

The integral of a polynomial with respect to a Gaussian measure can be evaluated in closed form. The closed form expression is a sum, and each term in the sum is labelled by a graph. We will encounter complicated polynomials of high degree, and the resulting graphs will also be complicated. However we present some very simple estimates for such polynomials; the structure of these estimates can be seen easily from the associated graphs.

We define a localized monomial to be a polynomial of the form

$$R = \int \prod_{i=1}^{r} :\Phi(x_i)^{n_i}: \; w(x)\, dx \,, \tag{9.1}$$

where $w(x)$ is supported in a product $\Delta_{j_1} \times \ldots \times \Delta_{j_r}$ of lattice squares. We also require $w \in L_{1+\varepsilon}$ and it is convenient but not essential to assume a bound

$$n_i \le \bar{n} \,, \quad 1 \le i \le r \,. \tag{9.2}$$

the bound (9.2) does not restrict r, nor the total degree of R. Polynomials which arise naturally are not usually of this form because the kernels w are not localized. However any polynomial can be written as a sum of localized monomials.

Associated with R of (9.1) is a graph G(R) consisting of r vertices and at the ith vertex, we draw n_i legs. See Figure 1.

Fig. 1. $G\left(\int :\Phi(x_1)^4: \; :\Phi(x_2)^4: \; w(x)\, dx\right)$.

In order to evaluate $\int R \, d\Phi_C$, we integrate by parts:

$$\int \Phi(x) R \, d\Phi_C = \int C(x,y) \frac{\delta R}{\delta \Phi(y)} \, d\Phi_C \, dy \,.$$

This formula can be proved by passing to the Fock space[7] of the measure $d\Phi_C$, expanding Φ as a sum of a creation and an annihilation operator and using the canonical commutation relations. See also Theorem 9.1 below. We integrate by parts to reduce the degree of the monomial R we want to integrate. After $(\sum_{i=1}^{r} n_i)/2$ partial integrations, the monomial is replaced by a sum of constants, and since

[7] See for example Theorem 3.5 of J. Glimm and A. Jaffe, Boson quantum field models, in Mathematics of Contemporary Physics. Ed. by R. Streater, Academic Press, New York, 1972.

$\int d\Phi_C = 1$, the integral is evaluated explicitly.

In applying this procedure, we encounter Φ's in a Wick ordered factor $:\Phi(x_i)^{n_i}:$ in R. For such Φ's, we use the formula

$$(9.3)\quad \int :\Phi(x)^n:\, R\, d\Phi_C = (n-1)\, c(x) \int :\Phi(x)^{n-2}:\, R\, d\Phi_C + \int :\Phi(x)^{n-1}:\, C(x,y)\, \frac{\delta R}{\delta\Phi(y)}\, d\Phi_C\, dy\ ,$$

with

$$c(x) = C(x,x) - C_\emptyset(x,x)$$

defined by Proposition 7.5. The first term arises from the difference between the covariance $C_\emptyset$ in : : and the covariance in $\int \ldots d\Phi_C$. The second term is exactly as before.

The integration by parts formula (9.3) has a simple expression in terms of graphs. In case the $\Phi(x)$ is a factor in $:\Phi(x_i)^{n_i}:$, we label the terms on the right side of (9.3) by drawing a line connecting one leg of the x_i-vertex to a distinct leg at the same or a distinct vertex. The graph with a line from the x_i to the x_j vertex labels each of the n_j terms

$$(9.4)\quad \int C(x_i,x_j)\, :\Phi(x_i)^{n_i-1}:\ :\Phi(x_j)^{n_j-1}:\ \prod_{\ell\neq i,j} :\Phi(x_\ell)^{n_\ell}:\, w(x)\, dx$$

coming from a single integration by parts in (9.1). See Figure 2.

Fig. 2. Integration by Parts.

As an example, we evaluate the integral of Figure 2. After four integrations by parts, we have

$$\iint :\Phi(x_1)^4:\ :\Phi(x_2)^4:\ w\, dx\, d\Phi_C = 4! \int C(x_1,x_2)^4 w\, dx + 2!\binom{4}{2}^2 \int c(x_1)\, c(x_2)\, C(x_1,x_2)^2 w\, dx + \binom{4}{2}^2 \int c(x_1)^2 c(x_2)^2 w\, dx\ .$$

The absolute value of the first term is bounded by

$$4!\ \left|\int C(x_1,x_2)^4\, w(x_1,x_2)\, dx\right| \leq 4!\ \|C\|^4_{L_8(\Delta_{j_1}\times\Delta_{j_2})}\ \|w\|_{L_2} \leq 4!\ K_4 m_0^{-1}\, e^{-m_0(1-\delta)4d(j)}\ \|w\|_{L_2}$$

if suppt $w \subset \Delta_{j_1}\times\Delta_{j_2}$. In the last line, we used Proposition 7.2 to bound $\|C\|_{L_q(\Delta_{j_1}\times\Delta_{j_2})}$. See Figure 3.

Fig. 3. Evaluation of a Gaussian Integral.

To evaluate $\int R\, d\Phi_C$ in the general case, we form the set $\mathcal{V}(R)$ of all vacuum graphs. A vacuum graph is a graph obtained from $G(R)$ by joining pairwise distinct legs until all legs are so joined. If the total number of legs is odd, $\mathcal{V}(R) = \emptyset$ (and $\int R\, d\Phi_C = 0$). For each $G \in \mathcal{V}(R)$, define

$$(9.5)\quad I(G,w,C) = \int w(x)\, \Pi_{\ell'}\, c(x_{i(\ell')})\Pi_\ell\, C(x_{i_1(\ell)},x_{i_2(\ell)})\, dx\ .$$

The first product, $\Pi_{\ell'}$, runs over all lines ℓ' of G which connect two legs of a single vertex, labeled $i(\ell')$. The second product runs over all lines ℓ joining pairs $(i_1(\ell),i_2(\ell))$ of distinct vertices.

We now state the formula for evaluation of Gaussian integrals. The above discussion of integration by parts and its graphical interpretation gives the formal derivation.

Theorem 9.1. Let w be localized and let $w \in L_q$ for some $q > 1$. Then $R \in L_r(\mathcal{S}', d\Phi_C)$, $r \in [1,\infty)$, and

$$(9.6)\qquad \int R\, d\Phi_C = \sum_{G\in\mathcal{V}(R)} I(G,w,C)\ .$$

The proof has a combinatoric aspect (sketched above) and an analytic aspect. The latter is an approximation argument, which uses a modified (momentum cutoff) form of (9.6) to prove that the approximants converge. For this reason, we bound the integrals $I(G,w,C)$ before sketching the proof of Theorem 9.1.

We consider a function of the form

$$F(x_1,\ldots,x_n) = \Pi_\ell\, F_\ell(x_{i_1(\ell)},x_{i_2(\ell)})\ \Pi_i\, \Delta_i(x_i)$$

where $\Delta_i(x)$ is the characteristic function of a set of area 1, and suppose that for each index ℓ,

$$1 \le i_1(\ell) < i_2(\ell) \le n\ .$$

Lemma 9.2. With the above notation,

$$\|F\|_{L_q} \le \Pi_\ell\, \|F_\ell\|_{L_{q_\ell}(\Delta_{i_1(\ell)},\Delta_{i_2(\ell)})}$$

provided that for each i,

$$q^{-1} \geq \sum \left\{ q_\ell^{-1};\ i_1(\ell) = i \text{ or } i_2(\ell) = i \right\} .$$

Proof: With the substitution $F \to F^q$, we are reduced to the case $q = 1$. We apply Hölder's inequality successively in each of the n variables (vertices) of F, and we use the language of graphs, in order to visualize this process. Each factor F_ℓ corresponds to a line joining the $i_1(\ell)$ vertex to the $i_2(\ell)$ vertex. Let

$$\mathcal{L}_i = \{\ell : i_1(\ell) < i = i_2(\ell)\}$$
$$\mathcal{L}_i' = \{\ell : i_1(\ell) = i < i_2(\ell)\} .$$

The hypothesis concerning q_ℓ is

$$\sum_{\ell \in \mathcal{L}_i \cup \mathcal{L}_i'} q_\ell^{-1} \geq 1$$

($\mathcal{L}_i = \emptyset = \mathcal{L}_i'$ is also allowed, trivially.) Thus

$$\int_{\Delta_i} \prod_{\ell \in \mathcal{L}_i} |F_\ell(x_{i_1(\ell)}, x_i)| \prod_{\ell \in \mathcal{L}_i'} \|F_\ell(x_i, \cdot)\|_{L_{q_\ell}(\Delta_{i_2(\ell)})} dx_i$$

$$\leq \prod_{\ell \in \mathcal{L}_i} \|F_\ell(x_{i_1(\ell)}, \cdot)\|_{L_{q_\ell}(\Delta_i)} \prod_{\ell \in \mathcal{L}_i'} \|F_\ell\|_{L_{q_\ell}(\Delta_{i_1(\ell)} \times \Delta_{i_2(\ell)})}$$

by Hölder's inequality. A finite induction on i (decreasing from $i = n$ to $i = 1$) now completes the proof.

Remark. An obvious modification allows some of the factors F_ℓ to depend on a single variable, i.e. $F_\ell = F_\ell(x_{i(\ell)})$.

Sketch of proof of Theorem 9.1. We introduce momentum cutoffs, $\Phi(x_i) \to \int \Phi(y)\, j_k(y-x_i)\, dy$, and omit the x-integration. This replaces R by a polynomial cylinder function $R_k(x)$, based on a finite dimensional subspace of $\mathcal{S}$. $R_k(x)$ is integrable by the definition of $d\Phi_C$, and for $R_k(x)$ the formulas (9.3) and (9.6) are valid by the above combinatorial arguments. Thus (9.6) holds for $R_k = \int R_k(x)\, w(x)\, dx$, and can be used to evaluate the right side of the inequality

$$\int |R_k - R_{k'}|\, d\Phi_C \leq \left(\int (R_k - R_{k'})^2\, d\Phi_C \right)^{1/2} .$$

Removing the momentum cutoff in one linear factor Φ at a time, the estimates of Propositions 7.2 and 7.3 are sufficient. For details see [1].

We remark that the stronger hypothesis of [1], $C \in \mathcal{B}^{loc}_{q,\delta(q)}$, $\forall q > 1$ (used in [1] to control C-Wick ordering) is not required for the present lectures, since only $C_\emptyset = (-\Delta+m_0^2)^{-1}$ Wick ordering is used. See also Proposition 7.5.

Proposition 9.3. Let w be localized. Then

$$(9.7) \quad \left| \int R \, d\Phi_C \right| \le \| w \|_{L_p} \sum_{G \in \mathcal{V}(R)} \Pi_{\ell'} \| c \|_{L_q(\Delta_{j(\ell')})} \times \Pi_\ell \| C \|_{L_q(\Delta_{j_1(\ell)}, \Delta_{j_2(\ell)})}$$

where $q \ge p'\bar{n}$, $\bar{n}$ is defined by (9.2) and $j(\ell) = j_{i(\ell)}$ is the localization of the ith vertex.

Proof. By Theorem 9.1, it is sufficient to bound each integral I(G,w,C). We use the Hölder inequality to separate w from the c and C factors. The integral of the c and C factors is bounded by Lemma 9.2 and the remark following it. This completes the proof.

There are $O(\sum n_i/2)!$ graphs in $\mathcal{V}(R)$. Because of the exponential decay in C as $|x-y| \to \infty$, in general most of these graphs are very small. Efficient estimates must take advantage of this fact. For each lattice square Δ, we define

$$N(\Delta) = \sum \{n_i : \Delta_{j_i} = \Delta\}$$

as the number of legs (linear factors $\Phi(x_i)$) of R localized in Δ.

Theorem 9.4. Let w be localized. There is a constant K_{11} such that

$$\left| \int R \, d\Phi_C \right| \le \| w \|_{L_p} [\Pi_\Delta N(\Delta)! \, (K_{11} m_0^{-1/2} q)^{N(\Delta)}]$$

for $q = p'\bar{n}$, $\bar{n}$ defined by (9.2).

Proof. We bound the L_q norm in (9.7) by Propositions 7.2 and 7.5. Having done this, it is sufficient to show that

$$\sum_{G \in \mathcal{V}(R)} \Pi_\ell \, e^{-m_0(1-\delta)d(j(\ell))} \le \Pi_\Delta (\text{const.}^{N(\Delta)} N(\Delta)!) .$$

Let ν denote a leg of R. Let Δ_ν be the square in which ν is localized and given $G \in \mathcal{V}(R)$, and let Δ'_ν be the square of the leg joined to ν by G. Also let $d(\nu) = \text{dist}(\Delta_\nu, \Delta'_\nu)$. Then the sum over $G \in \mathcal{V}(R)$ can be written as a sum over the choices Δ'_ν for each ν and a sum over the $N(\Delta'_\nu)$ possible contractions in Δ'_ν, for each ν.

The summand is independent of the choice of possible contractions within each square Δ ($\nu \leftrightarrow \Delta'_\nu$ held fixed). To estimate this sum we have only to count the number of terms. Since an arbitrary term can be obtained (nonuniquely, in general) from a single given term by permutation of the legs in each square Δ, there are at most $\Pi_\Delta N(\Delta)!$ terms in this sum. Hence it remains to show that

$$\sum_{\{\Delta'_\nu\}} \Pi_\nu \; e^{-m_0(1-\delta)d(\nu)/2} \leq \Pi_\nu(\text{const.}) ,$$

since each $d(j(\ell))$ occurs as a $d(\nu)$ for exactly two ν's. The summation index $\{\Delta'_\nu\}$ is simply a set of functions from legs to lattice squares. We increase the left side by enlarging the set of summation indices to include all such functions. Then we can interchange the sum and product, obtaining

$$\sum_{\{\Delta'_\nu\}} \Pi_\nu \; e^{-cd(\nu)} \leq \Pi_\nu \sum_{\{\Delta'_\nu :\ \nu \text{ fixed}\}} e^{-cd(\nu)} = \Pi_\nu \;\; \text{const.}$$

We note that K_{11} is independent of s and m_0, for m_0 bounded away from zero.

<u>Theorem</u> 9.5. Let Λ be a union of lattice squares and let $\text{Re}\,\lambda \geq 0$. Then $e^{-V(\Lambda)} \in L_p(\mathcal{S}', d\Phi_C)$ for all $p \in [1,\infty)$. There is a constant K_{12} independent of C such that

$$|Z| \equiv \left| \int e^{-\lambda V} \, d\Phi_C \right| \leq e^{K_{12}|\Lambda|} . \tag{9.8}$$

With $\text{Re}\,\lambda$ bounded and m_0 bounded away from zero, K_{10} can also be chosen independent of λ and m_0.

<u>Remark</u>. The simple proof of [1, §II.3, p. 22-27] is self contained. In fact, using a slight generalization Theorem 9.4 (incorporating Proposition 7.4), the proof of (9.8) is nearly identical to the (standard) proof that $e^{-V(\Lambda)} \in L_1(\mathcal{S}', d\Phi_C)$. This proof is close in structure to that of [3].

<u>Corollary</u> 9.6. For $p > 1$ and $q \geq p'\bar{n}$,

$$\left| \int R e^{-V(\Lambda)} \, d\Phi_C \right| \leq e^{K_{12}|\Lambda|} \, \|w\|_{L_p} \left[\Pi_\Delta \; N(\Delta)! \, (2K_{11} \, m_0^{-1/2q})^{N(\Delta)} \right].$$

<u>Proof</u>. By the Schwartz inequality,

$$\left| \int R \, e^{-V(\Lambda)} \, d\Phi_C \right| \leq \left(\int R^2 \, d\Phi_C \right)^{1/2} \left(\int e^{-2V(\Lambda)} \, d\Phi_C \right)^{1/2} .$$

The factors on the right are estimate by Theoresm 9.4 and 9.5.

Theorem 9.7. Let w be a localized kernel in L_p, $p > 1$, let Λ be a union of lattice squares and let $F = Re^{-V(\Lambda)}$ in (1.7). Then (1.7) is valid.

Sketch of Proof: In order to present the formal ideas, we suppose first that F is a polynomial. Then $\int F \, d\Phi_{C(s)}$ is given explicitly in terms of graphs by (9.5) - (9.6). Differentiating with respect to s_b in these formulas yields

$$(9.9) \quad \frac{d}{ds_b} \int F \, d\Phi_{C(s)} = \sum_{G \in \mathcal{V}(F)} \int \sum_{\ell} \left(\frac{d}{ds_b} C_\ell\right) \Pi_{\ell' \neq \ell} C_{\ell'} \, w \, dx$$

where C_ℓ denotes $C(x_{i_1(\ell)}, x_{i_2(\ell)})$.

The product $\Pi_{\ell' \neq \ell}$ in effect removes one line from the vacuum graph $G \in \mathcal{V}(F)$. Equivalently, one could remove from F the two legs joined by the line ℓ. However removing legs from F is the same as removing linear factors from F, or the same as differentiating F with respect to these linear factors. Thus we see that $\sum_\ell$, the sum over lines removed from $G \in \mathcal{V}(F)$, is equivalent to a sum over mixed second derivatives of F with respect to pairs of linear factors. Such a sum is just $\frac{1}{2} \Delta_\Phi F$, so we identify the right side of (9.9) as

$$\frac{1}{2} \int \left[\left(\frac{d}{ds_b} C\right) \cdot \Delta_\Phi\right] F \, d\Phi_{C(s)} \, .$$

The proof in the general case, $F = R\, e^{-V}$, is based on approximations, starting with F a polynomial. The control over these approximations is given by Corollary 9.6. For details see [1].

§10. CONVERGENCE: THE PROOF COMPLETED

Proof of Proposition 5.3. Without loss of generality, the kernel w is localized, and in this case we take $\|w\| = \|w\|_2$. The expression we want to estimate is

$$(10.1)\qquad < \int \partial^\Gamma \int \Pi_{i=1}^n \Phi(x_i)\, e^{-\lambda V(\Lambda)}\, d\Phi_{s(\Gamma)}\, ds(\Gamma),\, w> \quad .$$

Let $\mathcal{P}(\Gamma)$ be the set of all partitions π of Γ. By Leibnitz' rule and (1.7), (10.1) equals

$$(10.2)\quad <\int \sum_{\pi\in \mathcal{P}(\Gamma)} \int \left(\Pi_{\gamma\in\pi} \tfrac{1}{2}\, \partial^\gamma C\cdot\Delta_\Phi\right) \Pi\, \Phi(x_i) e^{-\lambda V(\Lambda)}\, d\Phi_{s(\Gamma)}\, ds(\Gamma), w>$$

where $C = C(s(\Gamma))$.

As in (7.3), we define

$$\partial^\gamma C(j_\gamma) = \Delta_{j_{1,\gamma}}\, \partial^\gamma C\, \Delta_{j_{2,\gamma}}$$

where $j = (j_{1,\gamma}, j_{2,\gamma}) \in Z^4$, so that the two derivatives in $\partial^\gamma C(j_\gamma)\cdot\Delta_\Phi$ are localized in Δ_{j_1} and Δ_{j_2} respectively, and

$$\partial^\gamma C = \sum_{j_\gamma} \partial^\gamma C(j_\gamma) \ .$$

We substitute this identity into (10.2) and expand.

The resulting sum is now indexed by localizations $\{j_\gamma\}$ and partitions $\pi \in \mathcal{P}(\Gamma)$. For a given term let $M = M(\pi, \{j_\gamma\})$ be the number of terms resulting from the differentiations Δ_Φ in (10.2). By Corollary 9.6 each of the resulting terms can be estimated by

$$\|w'\|_{L_p}\, e^{K_{12}|\Lambda|}\, \Pi_\Delta\, N(\Delta)!\, (2K_{11})^{N(\Delta)}$$

for $m_0 \geq 1$. Here w' is the w of (10.1) multiplied by the kernels $\partial^\gamma C$ arising in (10.2). From Proposition 8.1 and Lemma 9.2 we have (for $p < 2$, and q large),

$$\begin{aligned} \|w'\|_{L_p} &< \|w\|_{L_2}\, \|\Pi_{\gamma\in\pi}\, \partial^\gamma C(j_\gamma)\|_{L_q} \\ &\leq \|w\|_{L_2}\, m_0^{-|\Gamma|/2q}\, \Pi_{\gamma\in\pi}\, K_6(q,\gamma)\, e^{-m_0 d(j_\gamma,\gamma)/2} \ . \end{aligned}$$

Now using (8.5) to control the sum over $\pi \in \mathcal{P}(\Gamma)$ we can bound

(10.2) by

$$\|w\|_{L_2} e^{K_7|\Gamma|} m_0^{-|\Gamma|/2q} \sum_{\{j_\gamma\}} \max_{\pi\in\mathcal{P}(\Gamma)} M \prod_{\gamma\in\pi} e^{-m_0 d(j_\gamma,\gamma)/2} \prod_\Delta N(\Delta)!$$

The proof of Proposition 5.3 follows from two lemmas which control M and the sum over $\{j_\gamma\}$ respectively. Let $M(\Delta)$ be the number of elements in the set $\{j_{i,\gamma}: \Delta_{j_{i,\gamma}} = \Delta,\ i = 1 \text{ or } 2,\ \gamma \in \pi\}$.

Lemma 10.1. There exists a constant K_{13}, independent of m_0, such that

$$M \le e^{K_{13}|\Gamma|} \prod_\Delta (M(\Delta)!)^p$$

and

$$\prod_\Delta N(\Delta)! \le e^{K_{13}|\Gamma|} \prod_\Delta (M(\Delta)!)^p .$$

where p is the degree of the interaction polynomial P.

Lemma 10.2. Given $\pi \in \mathcal{P}(\Gamma)$ and $r > 0$, there exists a constant K_{14}, independent of m_0, such that

$$\sum_{\{j_\gamma\}} \prod_{\gamma\in\pi} e^{-m_0 d(j_\gamma,\gamma)/2} \prod_\Delta M(\Delta)!^r \le e^{K_{14}|\Gamma|} .$$

Proof of Lemma 10.1: Let $N_0(\Delta)$ be the number of x_i, $1\le i\le n$, which are localized in Δ. The number of terms resulting from $M(\Delta)$ differentiations in Δ is bounded by

$$(N_0(\Delta)+1)(N_0(\Delta)+p+1) \ldots (N_0(\Delta)+p(M(\Delta)-1)+1) .$$

Since $N_0(\Delta) \le \sum_\Delta N_0(\Delta) = n$, we have M, the total number of terms resulting from all $\partial/\partial\Phi(y)$ differentiations, bounded by

$$M \le \prod_\Delta p^{pM(\Delta)} (N_0(\Delta)+1+pM(\Delta))^{N_0(\Delta)+1} M(\Delta)!^p ;$$

using the inequalities

$$(a+b)! \le (a+b)^a (b!) \quad \text{and} \quad (ab)! \le a^{ab}(b!)^a .$$

Furthermore with $N(\Delta)$, as defined in §9, the number of legs in Δ, after differentiation, we have

$$N(\Delta) \le N_0(\Delta) + (p-1)M(\Delta)$$

and so $\Pi N(\Delta)!$ is bounded as above.

Proof of Lemma 10.2: The sum $\sum_{\{j_\gamma\}}$ is controlled by the exponentially decreasing distance factor, so it is sufficient to show

$$\Pi_\Delta M(\Delta)!^r \leq \Pi_\gamma e^{\text{const.}|\gamma|} e^{\text{const.}\sum_\gamma d(j,\gamma)}$$

with constants independent of m_0, λ, $\{j_\gamma\}$ and π.

Recall that $d(j,\gamma)$, defined by (8.2), contains the distance from $j_{1,\gamma}$ and $j_{2,\gamma}$ to some $b \in \gamma$. Thus for fixed Δ, there are at most $O(1)r^2$ values of γ within a fixed partition π such that

$$\Delta_{j_\nu,\gamma} = \Delta, \qquad \nu = 1 \text{ or } 2, \tag{10.4}$$

and $d(j,\gamma) \leq r$. By definition there are $M(\Delta)$ γ's which satisfy (10.4). The most distant half $(= M(\Delta)/2)$ of these γ's must also satisfy

$$M(\Delta)^{1/2} \leq \text{const. } d(j,\gamma) + \text{const.}$$

because the γ's are nonoverlapping. Hence

$$M(\Delta)^{3/2} \leq \text{const. } \sum_\gamma \{d(j,\gamma) : \Delta_{j_\nu,\gamma} = \Delta\} + \text{const.}$$

and so the proof is completed by the inequality

$$\begin{aligned}\Pi_\Delta M(\Delta)!^r &\leq \exp\left(r \sum_\Delta M(\Delta) \ln M(\Delta)\right)\\ &\leq \exp\left(O \sum_\Delta \{M(\Delta)^{1+\delta} : M(\Delta) > 0\}\right)\\ &\leq \exp\left(O \sum_\gamma d(j,\gamma)\right) \exp\left(O|\Gamma|\right).\end{aligned}$$

REFERENCES

0. Z. Ciesielski, Lectures on Brownian motion, heat conduction and potential theory, Aarhus Universitet, 1965.
1. J. Dimock and J. Glimm. Measures on the Schwartz distribution Space and Applications to $P(\phi)_2$ field theories.
2. J. Ginibre. General formulation of Griffiths inequalities. Comm. Math. Phys. 16 (1970) 310-328.
3. J. Glimm and A. Jaffe. The $\lambda(\phi)_2^4$ quantum field theory without cutoffs. III The physical vacuum. Acta Math. 125 (1970) 203-261.
4. J. Glimm and A. Jaffe. The $\lambda(\phi)_2^4$ quantum field theory without cutoffs, IV. Perturbations of the Hamiltonian. J. Math. Phys. 13 (1972) 1568-1584.
5. J. Glimm and A. Jaffe. Positivity of the ϕ_3^4 Hamiltonian. Fort. der Physik. To appear.
6. J. Glimm, A. Jaffe and T. Spencer. The Wightman axioms and particle structure in the $P(\phi)_2$ quantum field model. To appear.
7. F. Guerra. L. Rosen and B. Simon. The $P(\phi)_2$ quantum field theory as classical statistical mechanics.
8. L. Hörmander. Linear Partial Differential Operators. Springer-Verlag, Berlin, 1964.
9. J. Klauder. Ultralocal scalar field models. Comm. Math. Phys. 18 (1970) 307-318.
10. J. Lebowitz and O. Penrose. Decay of correlations. Preprint.
11. R. Minlos and Ja. Sinai. The phenomenon of phase separation at low temperatures in some lattice models of a gas II. Trans. Moscow Math. Soc. Vol. 19 (1968), 121-196.
12. C. Newman. Ultralocal quantum field theory in terms of currents. Comm. Math. Phys. 26 (1972) 169-204.
13. D. Ruelle. Statistical Mechanics. Benjamin, New York, 1969.
14. T. Spencer. The mass gap for the $P(\phi)_2$ quantum field model with a strong external field. Preprint.
15. K. Wilson and J. Kogut. The renormalization group and the ε-expansion. Phys. Reports, to appear.

VII Particles and Bound States and Progress Toward Unitarity and Scaling

PARTICLES AND BOUND STATES AND PROGRESS TOWARD UNITARITY AND SCALING

James Glimm[1]
Rockefeller University
New York, N.Y. 10021

Arthur Jaffe[2]
Harvard University
Cambridge, MA 02138

Abstract

We present a survey of recent developments in constructive quantum field theory.

Introduction. The program of constructive quantum field theory starts with an approximate field theory whose existence is known and constructs a Lorentz covariant limit as the approximations are removed [50]. Frequently it has been convenient to work in the path space world of imaginary time [14,15,42]. The Osterwalder-Schrader axioms [34,35] give sufficient conditions on the Euclidean Green's functions (i.e. the path space theory in the case of bosons) to allow analytic continuation back to real (Minkowski) time and a verification of all Wightman axioms.

This program has been carried out in a number of models in space-time dimension $d < 4$. Once a model has been constructed, the interesting questions involve its detailed properties and how these properties depend on the parameters. For space-time dimension $d = 2$, considerable insight has been obtained into several models including the Sine-Gordon equation [11]. For $d = 3$, recent work yields the first nontrivial model [9]. We describe some of these recent results below, but first we propose a program for $d = 4$.

Consider a lattice ϕ_4^4 model, with lattice spacing ε. Using correlation inequalities of Lebowitz [18,30,47] or the Lee Yang theorem [32,33], we can bound the n-point Schwinger functions

[1]Supported in part by the National Science Foundation under Grant MPS 74-13252.

[2]Supported in part by the National Science Foundation under Grant MPS 73-05037.

$$S^{(n)}(x_1,\dots,x_n) = \int \phi(x_1)\dots\phi(x_n)dq ,$$

defined as moments of the measure

$$dq = \frac{e^{-V(\phi)}\prod_x d\phi(x)}{\int e^{-V(\phi)}\prod_x d\phi(x)} ,$$

$$V(\phi) = \sum_x [\tfrac{1}{2} : (\nabla\phi)^2 + m_0^2\phi^2 : + \lambda : \phi^4(x) :] .$$

In fact $S^{(n)}$ is bounded by a sum of products of two point functions. Thus a sufficient condition for the existence of a ϕ^4 field theory (by the method of compactness and subsequences) is a bound, uniform in the lattice spacing ε, on the lattice approximation two point function. In order to prevent the identical vanishing of all n-point functions in the limit $\varepsilon \to 0$, we hold the physical mass fixed, and since $d = 4$, we perform a field strength renormalization.

Field strength renormalization assumes the existence of a one particle pole in the Fourier transform of the two point function, of strength $Z = Z(\varepsilon) \neq 0$. We define

$$S^{(n)}_{ren} = Z^{-n/2}S^{(n)} ,$$

and as above $S^{(n)}_{ren}$ is bounded in terms of $S^{(2)}_{ren} = Z^{-1}S^{(2)}$. Thus the essential missing steps are (i) existence of the one particle pole, so that Z is defined and not zero, for $\varepsilon > 0$, and (ii) control over $Z(\varepsilon)$ as $\varepsilon \to 0$. Since one expects $Z(\varepsilon) \to 0$ (infinite field strength renormalization), it is necessary to show that the rest of the mass spectrum in $S^{(2)}$ has a spectral weight converging to zero (in each bounded mass interval) with ε, so that $S^{(2)}_{ren}$ remains bounded.

The coupling constant renormalization, $\lambda = \lambda(\varepsilon) \to \infty$ as $\varepsilon \to 0$, should not be required for existence of the $\varepsilon \to 0$ limit, but if it is not performed, the limit should be a free field, for $d = 4$. See [24] for further discussion.

In the case $d = 2,3$, we can apply these same ideas to construct the "scaling limit" of (superrenormalizable) ϕ^4 models. In this case we do not require a cutoff for the approximate theories, but vary the bare parameters so $Z \to 0$ and $\lambda \to \infty$, while the physical mass remains fixed and the dimensionless charge approaches its critical value (characterized by the onset of symmetry breaking). We now consider the case $d = 2$ in more detail.

The scaling limit in ϕ_2^4 [24]. Consider an interaction Lagrangian

$$L^{\text{Int}} = -\frac{1}{2}\sigma\int :\phi^2:dx - \lambda\int :\phi^4:dx .$$

Here σ is related to the bare mass m_0, but does not equal m_0^2, since the free Lagrangian will contain a mass parameter also; see the appendix of [20] for a discussion. It is expected that a critical value $\sigma_c = \sigma_c(\lambda)$ of σ exists so that as $\sigma \downarrow \sigma_c$, the physical mass m goes to zero. The scaling limit combines the limit $\sigma \downarrow \sigma_c$ with an infinite scale transformation,

$$\lambda \to s\lambda, \quad \sigma \to s\sigma, \quad m^2 \to sm^2,$$

so that m is held fixed. The correct choice of s is given by the conditions

$$\lambda/\sigma \to \lambda/\sigma_c(\lambda) = (\lambda/\sigma(\lambda))_c ,$$

and in this limit $\lambda \to \infty$, $\sigma_c(\lambda) \to \infty$. Note that $\lambda/\sigma_c(\lambda)$ is dimensionless, hence a pure number independent of λ. In this limit all unrenormalized correlation functions converge (after passage to a subsequence). In taking the scaling limit, we also perform a field strength renormalization, so the one particle pole has residue 1. We expect $Z(\sigma) \to 0$ in the scaling limit.

As before, control over the renormalized two point function is sufficient for the existence of the scaling limit, and again the missing steps are (i), (ii) above. Other sufficient conditions are (α) negativity of the six point vertex function $\Gamma^{(6)} \leq 0$, or (β) better than one particle decay of the inverse propagator $-\Gamma^{(2)}(x)$ (which for $x \neq 0$ equals the proper self energy part $\Pi(x)$) or (γ) absence of level crossings and control over CDD zeros of the momentum space two point function [24], and also [23]. Thus (α)-(γ) are central open issues.

A numerical analysis of the ϕ_1^4 model (anharmonic oscillator) is consistent with the validity of parts of (γ) [28]. An explicit calculation shows that $\Gamma^{(6)} < 0$ in the one-dimensional Ising model [39]. A related correlation inequality $S_T^{(6)} \geq 0$ (positivity of the connected part of $S^{(6)}$) has been announced by Cartier; recently Percus [38] and Sylvester [47] have given proofs. Computer studies of the Ising model by Sylvester indicate furthermore that $(-1)^{n+1} S_T^{(2n)} \geq 0$, [48]. The question of correlation inequalities will be discussed at greater length in the lecture of Simon.

In addition to its relation to the problem of constructing ϕ_4^4, the scaling limit is important in the renormalization group approach to the study of critical exponents; see for example [37]. In this connection we note that a number of critical exponents have been bounded from below by their canonical (mean field) values [17]. Furthermore we prove a priori bounds on renormalized coupling constants [21].

The problems (i) and (ii) above arise also in the scaling limit of the d-dimensional Ising model, $2 \leq d \leq 4$. For I_2, asymptotic calculations using Toeplitz determinants indicate that the required bounds are in fact satisfied [31, 49]. The model I_1 is already scale invariant. A formal interchange of limits identifies the scaling limit of I_d with the infinitely scaled ϕ^4_d model. This statement provides a basis for the idea that spin 1/2 Ising model critical exponents are independent of the details of the lattice structure and are equal to those defined by a ϕ^4 field theory. It also suggests that $\phi^6, \phi^8, \ldots$ tri and multi-critical points are associated with higher spin Ising model tri and multi-critical points.

Particles and Unitarity [25,26]. Let $M = (H^2 - \vec{P}^2)^{\frac{1}{2}}$ be the mass operator. For weakly coupled $P(\phi)_2$ models, the spectrum of M is completely known in the interval $[0, 2m-\varepsilon]$. At zero, M has a simple eigenvalue with eigenvector Ω, the vacuum state. At m, M has an isolated eigenvalue, and the corresponding eigenspace, the space of one particle states, carries an irreducible representation of the Poincaré group. M has no other spectrum below $2m-\varepsilon$; here $\varepsilon \to 0$ and $m \to m_0$ (the bare mass) as the coupling tends to zero. The existence of an isometric S matrix and n-particle in and out states then follows from the Haag-Ruelle scattering theorem.

These results about particles (spectrum of M) are proved using a "cluster expansion," similar to the high temperature expansions in statistical mechanics [25,26]. These expansions, furthermore, give spectral information about the mass (generalized) eigenvectors on any bounded spectral interval, for λ sufficiently small. The main consequence of these expansions is the fact that the Euclidean correlation functions decay (become uncorrelated) as the points separate into clusters, and rate of decay is exponential in the separation distance. Let

$$X = x_1, \ldots, x_n, \qquad Y = y_1, \ldots, y_m$$

$$\Phi(X) = \prod_{i=1}^{n} \phi(x_i), \qquad \Phi(Y) = \prod_{j=1}^{m} \phi(y_j).$$

A typical estimate is

(1) $$\int \Phi(X)\Phi(Y)d\phi - \int \Phi(X)d\phi \int \Phi(Y)d\phi = O(e^{-md})$$

where $m > 0$ is independent of X,Y and

$$d = \mathrm{dist}(X,Y)$$

is the separation distance. The subtraction in (1) can be recognized as a Euclidean vacuum subtraction. Let $\Omega_E \equiv 1$ be the Euclidean vacuum state and let $< , >$ be the Euclidean inner product defined by the measure $d\phi$. Let P_0 be the orthogonal projection onto Ω_E. Then (1) can be written

(2) $$<\Phi(X),(I-P_0)\Phi(Y)> = O(e^{-md}) .$$

In fact, the exponential decay rate in (1),(2) is related to the spectrum of M. Define m_1 as the infimum over the exponential decay rates for different choices of X,Y. Then m_1 is the one particle mass (mass gap) in the spectrum of M.

We generalize (2) by replacing the projection $I-P_0$ with the projection onto the orthogonal complement of the subspace spanned by polynomials of degree $< n$. We then expect decay rates $m_n > m_1$. Let P_n denote the orthogonal projection onto the subspace spanned by Euclidean vectors

$$|X_n> \equiv (I - \sum_{i=0}^{n-1} P_i)\phi(x_1)\ldots\phi(x_n)\Omega .$$

The projection P_n can be written in terms of a kernel $P_n(X,Y)$, and in Dirac notation

(3) $$P_n = \int dXdY |X_n>P_n(X,Y)<Y_n| ,$$

where the integration extends over $R^{nd} \times R^{nd}$ and where the kernel $P_n(X,Y)$ is the inverse (in the sense of integral operators) to the kernel $<X_n|Y_n>$.

The kernel

$$P_1(X,Y) = -K_1(X,Y) = -\Gamma^{(2)}(x-y)$$

is the familiar inverse propagator, while the kernel $P_2(X,Y)$ is related to the Bethe-Salpeter kernel $K_2(X,Y)$, namely

$$P_2(X,Y) = -K_2(X,Y) + \frac{1}{2} K_1 \otimes K_1 .$$

In terms of Feynman graphs, K_2 is the connected part of $-P_2$. More generally, we can define an n-body Bethe-Salpeter kernel $K_n(X,Y)$ as the connected part of $-P_n$. In fact, P_n can be written as a sum of tensor products of Bethe-Salpeter kernels, generalizing the expressions for P_1, P_2 above:

$$(4) \qquad P_n(X,Y) = \sum_{r=1}^{n} \sum_{|\alpha_1|+\ldots+|\alpha_r|=n} (-1)^r \binom{n}{\alpha}^{-2} K_{|\alpha_1|} \otimes \ldots \otimes K_{|\alpha_r|} ,$$

where $\binom{n}{\alpha} \equiv n! (\prod_{i=1}^{r} |\alpha_i|!)^{-1}$ and the sum extends over all cluster decompositions into elastic channels. For instance, the term with $r=1$ is just $-K_n$, while the term with $r=n$ equals $(-1)^n (n!)^{-1} \Gamma \otimes \ldots \otimes \Gamma$.

The existence of the Bethe-Salpeter kernel K_n, for $n=1,2,3$ is proved in [19,45,23]. We will publish an analysis of (4). A much deeper question than existence of K_n is the magnitude of the exponential decay rates associated with these kernels. Let

$$|X_{n,r}> \equiv (I - \sum_{i=0}^{n-1} P_i) \phi(x_1) \ldots \phi(x_r) \Omega .$$

We seek estimates generalizing (2) of the form

$$(5) \qquad < X_{n,r} | Y_{n,r} > = O(e^{-m_n d}),$$

i.e. Euclidean cluster properties, or decay estimates

$$(6) \qquad K_n(X,Y) = O(e^{-\bar{m}_n d}) .$$

In order to analyze the decay of K_n, Spencer has derived a new cluster expansion [45] which generalizes [26] by giving higher particle subtractions. Explicitly in the case $n=2$, for weak coupling even $P(\phi)_2$ models, Spencer proves that $\bar{m}_2 \geq 4m(1-\varepsilon)$, where $\varepsilon \to 0$ as the coupling tends to zero. He and Zirilli expect that the decay rate for the two body Bethe-Salpeter kernel will provide information related to asymptotic completeness, up to the threshold $\bar{m}_2$, i.e. for $M \leq 4m(1-\varepsilon)$ [46]. Similar methods give the decay rate for the part of $S_T^{(4)}$ that is two particle irreducible in each channel [2]; this amplitude is obtained by a second Legendre transformation.

If one imagines an extension of this structure analysis to arbitrary n, it appears that the allowed size of the weak coupling region would be n-dependent, and tend to zero as n tends to infinity. The question of dealing with the particle structure away from weak coupling is also of great importance. In this case we only have results for the ϕ^4 interaction, which is repulsive, as described in the following section.

Bound States. For $d = 2,3$, bound states should occur for weak as well as strong attractive forces; they should be missing for repulsive forces. For single phase ϕ^4 models, it is known that no even bound states can occur [44,7,26]. We expect that odd bound states are also missing, cf [24]. For weak $(\phi^6-\phi^4)_2$ models, it is known that mass spectrum occurs in the bound state interval $(m,2m)$ [26]. The Bethe-Salpeter equation combined with improved decay estimates above should allow a complete analysis of the bound state problem for weak coupling $P(\phi)_2$ models; partial results have been obtained [46]. An interesting question is whether bound states occur in ϕ^4 models with symmetry.

Fermions and Many Body Systems. The original construction of the infinite volume limit for the d=2 Yukawa model was given by Schrader [41]; see also [14]. Some portions of this construction have been derived in a Euclidean formalism [43,1]. Aside from the increased simplicity which may accompany a covariant Euclidean construction, the Euclidean formalism is important as a natural framework for a cluster expansion and a study of particles.

Federbush [5,6] has simplified the Dyson-Lenard proof of the stability of matter. The methods were suggested in part by constructive field theory techniques, including techniques he previously employed for the Yukawa_2 model. [6] contains a cluster expansion, which should have a number of applications.

Phase Transitions. Fröhlich [10] has shown that the Euclidean decomposition of the measure $d\phi$ into time translation invariant components coincides with the direct integral decomposition of the associated quantum fields into pure phases.

Three Dimensions. The original semiboundedness proof for the ϕ_3^4 Hamiltonian [16] and related Schwinger function bounds [8,36] were given in a finite volume. A cluster expansion for ϕ_3^4 has been established by Feldman and Osterwalder [9], yielding the first nontrivial d=3 model of the Osterwalder-Schrader axioms. We can hope that further progress will soon bring ϕ_3^4 to the level of understanding we have for d=2 .

REFERENCES

1. D. Brydges, Boundedness below for fermion model theories. Preprint.

2. C. Burnap, private communication.

3. J. Dimock, The $P(\phi)_2$ Green's functions: smoothness in the coupling constant.

4. J.-P. Eckmann, J. Magnon and R. Seneor, Decay properties and Borel summability for Schwinger functions in $P(\phi)_2$ theories. Commun. Math. Phys. To appear.

5. P. Federbush, A new approach to the stability of matter I,II. J. Math Phys. to appear.

6. ___________, The semi-Euclidean approach in statistical mechanics I. Basic expansion steps and estimates II. The custer expansion, a special example. Preprint.

7. J. Feldman, On the absence of bound states in the $\lambda\phi_2^4$ quantum field model without symmetry breaking. Canadian J. Phys. 52, 1583-1587 (1974).

8. _________, The $\lambda\phi_3^4$ field theory in a finite volume, Commun. Math. Phys. 37, 93-120 (1974).

9. J. Feldman and K. Osterwalder, The Wightman axioms and mass gap for ϕ_3^4, these proceedings.

10. J. Fröhlich, Schwinger functions and their generating functionals, II. Adv. Math, to appear.

11. __________, The quantized "Sine-Gordon" equation with a nonvanishing mass term in two space-time dimensions. Preprint.

12. J. Glimm, The mathematics of quantum field theory. Adv. Math. To appear.

13. ________, Analysis over infinite dimensional spaces and applications to quantum field theory. Proceedings Int. Congress Math., 1974.

14. J. Glimm and A. Jaffe, Quantum field models, in: Statistical mechanics and quantum field theory, ed. by C. de Witt and R. Stora, Gordon and Breach, New York, 1971.

15. ____________________, Boson quantum field models, in: Mathematics of contemporary physics, ed. by R. Streater, Academic Press, New York, 1972.

16. ____________________, Positivity of the ϕ_3^4 Hamiltonian, Fort. d. Physik, 21, 327-376 (1973).

17. ____________________, ϕ^4 quantum field model in the single phase region: Differentiability of the mass and bounds on critical exponents, Phys. Rev. D10, 536-539 (1974).

18. ____________________, A remark on the existence of ϕ_4^4. Phys. Rev. Lett. 33, 440-442 (1974).

19. ____________________, The entropy principle for vertex functions in quantum field models, Ann. l'Inst. H. Poincaré, 21, 1-26 (1974).

20. ____________________, Critical point dominance in quantum field models, Ann. l'Inst. H. Poincaré, 21, 27-41 (1974).

21. J. Glimm and A. Jaffe, Absolute bounds on vertices and couplings, Ann. l'Inst. H. Poincaré, 22, to appear.

22. ____________________, On the approach to the critical point, Ann. l'Inst. H. Poincaré, 22, to appear.

23. ____________________, Two and three body equations in quantum field models, Preprint.

24. ____________________, On three-particle structure of ϕ^4 and the infinite scaling limit. Preprint.

25. J. Glimm, A. Jaffe and T. Spencer, The Wightman axioms and particle structure in the $P(\phi)_2$ quantum field model. Ann. Math. 100, p. 585-632 (1974).

26. ______________________________, The particle structure of the weakly coupled $P(\phi)_2$ model and other applications of high temperature expansions, in: Constructive quantum field theory, Ed. by G. Velo and A. Wightman, Springer-Verlag, Berlin, 1973.

27. F. Guerra, L. Rosen and B. Simon, Correlation inequalities and the mass gap in $P(\phi)_2$ III. Mass gap for a class of strongly coupled theories with nonzero external field. Preprint

28. D. Isaacson, Private communication.

29. A. Jaffe, States of constructive field theory. Proceedings of 17th International Conference on high energy physics, London, 1974. J.R. Smith, editor, pp. I-243 to I-250.

30. J. Lebowitz, GHS and other inequalities. Commun. Math. Phys. 35, 87-92 (1974).

31. B. McCoy and T. Wu, The two dimensional Ising model. Harvard University Press, Cambridge, 1973.

32. C. Newman, Inequalities for Ising models and field theories which obey the Lee-Yang theorem. Commun. Math. Phys. To appear.

33. _________, Moment inequalities for ferromagnetic Gibbs distributions. Preprint.

34. K. Osterwalder and R. Schrader, Axioms for Euclidean Green's functions, I. Commun. Math. Phys. 31, 83-112 (1973).

35. ______________________________, Axioms for Euclidean Green's functions, II. Preprint.

36. Y. Park, Lattice approximation of the $(\lambda\phi^4-\mu\phi)_3$ field theory in a finite volume. Preprint.

37. G. Parisi, Field theory approach to second order phase transitions in three and two dimensional systems. Cargèse Summer School, 1973.

38. J. Percus, Correlation inequalities for Ising spin lattices. Preprint.

39. J. Rosen, Private communication.

40. J. Rosen and B. Simon, Fluctuations in $P(\phi)_1$ processes. Preprint.

41. R. Schrader, Yukawa quantum field theory in two space time dimensions without cutoff. Ann. Phys. 70, 412-457 (1972).

42. B. Simon, The $P(\phi)_2$ Euclidean quantum field theory. Princeton University Press, Princeton, 1974.

43. E. Seiler, Schwinger functions for the Yukawa model in two dimensions with space-time cutoff.

44. T. Spencer, The absence of even bound states in ϕ_2^4. Commun. Math. Phys., 39, 77-79 (1974).

45. __________, The decay of the Bethe Salpeter kernel in $P(\phi)_2$ quantum field models. Preprint.

46. T. Spencer and F. Zirilli, private communication.

47. G. Sylvester, Representations and inequalities for Ising model Ursell functions, Commun. Math. Phys., to appear.

48. ____________, private communication.

49. C. Tracey and B. McCoy, Neutron scattering and the correlation functions of the Ising model near T_c. Phys. Rev. Lett. 31, 1500-1504 (1973).

50. A. Wightman, Introduction to some aspects of the relativistic dynamics of quantized fields, in: 1964 Cargèse Summer School Lectures, Ed. by M. Lévy, Gordon and Breach, New York (1967), p. 171-291.

DISCUSSION

Masuo Suzuki (comment): I hope that by using your inequalities you can obtain such qualitative results as the dependence of critical exponents upon the dimensionality and potential-range parameter in your ϕ^4 model. In the ferromagnetic Ising model, I proved inequalities such as $\gamma(d) \geq \gamma(d+1) \geq \ldots$ and $\nu(d) \geq \nu(d+1) \geq \ldots$, using Griffiths' inequalities. (Physics Letters 38A (1972) 23.)

VIII Critical Problems in Quantum Fields

CRITICAL PROBLEMS IN QUANTUM FIELDS*

James Glimm[1]
Rockefeller University
New York, New York 10021

Arthur Jaffe[2]
Harvard University
Cambridge, Mass. 02138

RESUME Les liaisons entre le problème de la construction des champs quantiques non triviaux à quatre dimensions et le problème du comportement au point critique à quatre dimensions sont expliqués.

ABSTRACT The connections between the problem of constructing non trivial quantum fields in four dimensions and the problem of critical point behaviour in four dimensions are explained.

* Presented at the International Colloquium on Mathematical Methods of Quantum Field Theory, Marseille, June 1975.

1. Supported in part by the National Science Foundation under Grant MPS 74-13252.

2. Supported in part by the National Science Foundation under Grant MPS 73-05037.

The last two years have seen considerable progress in our understanding of the mathematical structure of quantum fields. In two areas, the progress has been close to definitive, and the problems may be largely resolved in the near future. These areas are (a) the construction of more singular superrenormalizable models: Yukawa_2 and ϕ_3^4 and (b) the detailed structure of $P(\phi)_2$ models which are close to free theories, namely particles, bound states, analyticity, unitarity in subspaces of bounded energy, and phase transitions. In two other areas there has been progress, but the progress is far from being definitive. These areas are (c) the structure away from the neighborhood of free theories, and especially near a critical point and (d) results which pertain indirectly to the construction of four dimensional models. For the results in areas (a) and (b), we merely list recent references, and we then turn to the open problems, including (c) and (d).

Yukawa_2-Euclidean methods [Br 1, Sei, McB 1,2, Br 2, Sei-Si]

ϕ_3^4 - weak coupling expansions [Fe-Os, Ma-Sen]

$P(\phi)_2$ - scattering [Sp 1, Sp-Zi]

$P(\phi)_2$ - analyticity [EMS]

$P(\phi)_2$ - phase transitions [GJS 2,3]

The central problem of constructive quantum field theory has not changed over many years (cf. [St-Wi; p. 168]): the construction of nontrivial quantum fields in four dimensions. We explain how this problem is related to critical point theory in four dimensions, and how a number of simpler problems (of independent interest, and involving two or three dimensional quantum fields) are related to this central problem.

The simplest four dimensional interactions, ϕ_4^4 and Yukawa_4 are renormalizable, but not superrenormalizable. This means that the bare and physical coupling constants are dimensionless. In addition to this dimensionless constant, the field theory is parametrized by two or more parameters with dimension of $(\text{length})^{-1}$. Namely, there are one or more masses and an ultraviolet cutoff κ. To make the exposition explicit, we choose the ultraviolet cutoff as a lattice, and then $\kappa^{-1} = \varepsilon$ is the lattice spacing.

The goal of the construction is to take the limit $\kappa \to \infty$, i.e. $\varepsilon \to 0$. Because scaling is a unitary transformation, and because scaling multiplies all lengths by an arbitrary parameter s, the theory with ε small and mass $m = 1$ is equivalent to the theory with $\varepsilon = 1$ and mass small. In this equivalence, the test functions also scale, and so if we choose $\varepsilon = 1$, a typical test function will have support on a set of large diameter $O(m^{-1})$. Thus if we choose $\varepsilon = 1$, we must focus on the long distance behavior, i.e. on the distance scale $O(m^{-1})$ in a theory with small mass. It follows that the limit $\kappa \to \infty$, $\varepsilon \to 0$ which removes the ultraviolet cutoff is equivalent to the limit $m \to 0$ with $\varepsilon = 1$, if in this latter limit we consider the behavior on the distance scale $O(m^{-1})$. This latter limit (correlation length $= m^{-1} \to \infty$) and distance scale is traditionally considered in critical point theories, namely the "scaling limit" in statistical mechanics. Thus we see that the critical point limit, with fixed lattice spacing $\varepsilon = 1$, is equivalent to the removal of the ultraviolet cutoff and to the construction of a (continuum) quantum field ($\varepsilon = 0$). Since the long distance (infrared) singularities are worse in two and three dimensions, we see that critical point theories in two and three dimensions provide a very realistic test for the mathematical difficulties presented by four dimensions. Indeed the two and three dimensional infrared behavior is typical of nonrenormalizable field theories. A simplification of the two and three dimensional problem (and one which we hope will prove to be minor) is that the critical point can be approached by Lorentz covariant fields satisfying Wightman axioms, in place of the lattice theories introduced above, see [GJ5]. For this reason, in two and three dimensions, the spectral representation of the two point function and (presumably) the particle structure and S-matrix theory can be used as tools to study the theories which are approaching the critical point.

To construct the critical point limit, there are four essential steps:

(i) mass renormalization

(ii) field strength renormalization

(iii) uniform estimates up to the critical point

(iv) nontriviality of the limit.

The first three steps concern existence for this question we would be happy to allow a compactness principle and selection of a convergent subsequence, while hoping that the full sequence converged also. This follows principles well accepted in other branches of mathematics (e.g. partial differential equations) where questions of existence and uniqueness are often studied by separate methods. The last step (nontriviality) depends upon the correct choice of charge renormalization. We will see below that for the ϕ^4 interaction each step can be studied independently of the others.

We now examine each of these four steps in turn. We will see which portions have been solved, which portions seem feasible for study at the present time, which steps are highly interesting in their own right, independently of their role in a possible construction of ϕ_4^4, and which portions seem to present essential difficulties and whose resolution will presumably require essentially new ideas.

The first step, mass renormalization, is the step nearest to completion. For a $\lambda\phi^4 + \frac{1}{2} m_0^2\phi^2$ theory (or more generally for an even $P(\phi)$ theory) the physical mass m is a monotonic function of m_0 for a single phase theory [GRS]. This statement also pertains to a lattice theory (as required for the four dimensional program) if the mass is defined as the exponential decay rate of the two point function. For a ϕ_2^4 theory at least, the mass $m(m_0^2)$ is differentiable [GJ 2] for $m > 0$. The analysis of [Ba] suggests that $m(m_0^2)$ is continuous for $m = 0$; in the lattice case this has been rigorously established [J. Ro 2].

Assuming that the ϕ_2^4 mass is continuous for $m \geq 0$ in a single phase $\lambda\phi^4 + \frac{1}{2} m_0^2\phi^2$ theory, then the mass renormalization is defined as the inverse function

$$m_0^2 = m_0^2(m,\lambda).$$

To see that m may take on all values, $0 \leq m < \infty$, we argue by continuity. For $m_0^2 \to \infty$, $m \to \infty$ also [GJS 1], and so we require a critical theory $(m = 0)$ at the end $m_0^2 = m_{0,c}^2$ of the single phase region, with

$$m(m_0^2) \searrow 0 \quad \text{as} \quad m_0^2 \searrow m_{0,c}^2 .$$

To be explicit, we define

$$M(m_0^2) = \lim_{|x-y|\to\infty} \langle\phi(x)\phi(y)\rangle^{1/2}$$

and

$$m_{0,c}^2 = \sup_{m_0^2}\{m_0^2 \mid M(m_0^2) > 0 \quad \text{or} \quad m(m_0^2) = 0\} .$$

The existence of a phase transition for a ϕ^4 lattice theory [Nel] and for ϕ_2^4 [GJS 2,3] shows that $m_{0,c}^2$ is finite. Combining this fact with the method of [GJ 2, Ba], it can be shown that $m \searrow 0$ as $m_0^2 \searrow m_{0,c}^2$ [J.Ro 2], at least in the lattice case. We summarize the problems of this section under the name: existence of the critical point.

For ϕ_3^4, one expects a similar structure for phase transitions. Assuming this conjecture and using the decay at infinity of the zero mass free field, it follows that $M(m_{0,c}^2) = 0$, but the question of whether $m(m_{0,c}^2) = 0$ remains open. For ϕ_2^4 and for a lattice theory the reasoning concerning M does not apply. In two dimensions the zero mass free field two point function does not decay at infinity, and in a lattice theory, the absence of a Lehmann spectral formula means that the free field is not known to bound the lattice two point function.

For the Yukawa interaction, none of the above results have been obtained. Major steps for the ϕ^4 interaction depend on correlation inequalities, which are presumably not valid for the Yukawa interaction. For the pseudoscalar Yukawa interaction, a phase transition associated with a breaking of the $\phi \to -\phi$ symmetry may be expected on formal grounds. For the scalar Yukawa theory, should one expect an absence of phase transitions and $m_{0,c}^2 = -\infty$? What about cases closer to strong interaction physics, such as one charged and one neutral fermion coupled to three mesons (charged $\pm 1, 0$) ? In general the problem here is: to locate the critical point. This problem is important because renormalizable fields (e.g. ϕ_4^4, Y_4) are equivalent to lattice or ultraviolet cutoff fields studied in the critical point limit. From this point of view, one reason for studying phase transitions in field theory is as an aid in locating the critical points.

The second step is to introduce the renormalized field $\phi_{ren} = Z^{-\frac{1}{2}}\phi$, where Z is defined in terms of the spectral representation for the two point function:

$$\langle\phi(x)\phi(y)\rangle^{\sim} = \frac{Z}{p^2+m^2} + \int_{a>m^2} \frac{d\rho(a)}{p^2+a} .$$

The essential problem is to show that $Z \neq 0$ for $m > 0$, or in other words to show that for each field in a noncritical theory, there is a corresponding elementary particle. Furthermore we expect only delta function contributions to $d\rho(a)$ (bound states) below the two particle threshold $a = (2m)^2$, and for an even $P(\phi)$ theory, the same should be true below the three particle threshold, because of the $\phi \to -\phi$ symmetry. In this more general form, the problem could be called Hunziker's theorem for field theory.

We now split the discussion of Hunziker's theorem into two independent paths, which we call the repulsive route and the general route. The repulsive route seeks to make maximum use of the special features of the ϕ^4 interaction, in particular of the presumably repulsive forces in this field theory. Since the ϕ^4 terms should dominate for a $P(\phi)$ critical point which is not near a tricritical point, we expect that the main results obtained in the repulsive route should be valid for general $P(\phi)$ theories near critical points which are not tri, or multicritical.

The repulsive route makes essential use of correlation inequalities. An example is the absence of even bound states for single phase ϕ^4 theories [GJS 1, Fe1, Sp 2]. However correlation inequalities by themselves cannot control the critical point behavior, because the ferromagnetic spin ½ Ising model on the Caley tree lattice has anomalous approach to the critical point [Zi]. Thus correlation inequalities must be used in conjunction with the lattice structure, as reflected in the Euclidean invariance and Hamiltonian structure of ϕ_2^4 and ϕ_3^4. A proposed correlation inequality, $\Gamma^{(6)} \leq 0$, implies that $d\rho(a)$ is supported above the three particle threshold, in the interval $[(3m)^2, \infty)$, and that $Z \gneq 0$ [GJ 5]. Thus $\Gamma^{(6)} \leq 0$ would completely settle step two, following the repulsive route. Lowest order perturbation theory suggests that $\Gamma^{(6)} \leq 0$ for weak coupling, and $\Gamma^{(6)} \leq 0$ has been checked for the one dimensional Ising model [J. Ro 1, Ro-Sy]. In view of its importance here and in step three below, further investigation of this inequality would be very desirable. Numerical calculations in some simple cases support the conjecture $\Gamma^{(6)} \leq 0$ [Is - Mar].

The general route is contained in and is substantially equivalent to the problem of asymptotic completeness. In fact the problem of step two -- the existence of a (discrete mass) particle at the bottom of the energy spectrum -- is equivalent at higher energies to the absence of continuous mass spectrum beyond that associated with multiparticle states. For weak coupling and bounded energies, the problem has been solved using cluster expansions. (In [Sp - Zi], energies up to the three particle threshold are allowed for even $P(\phi)$ interactions). For other regions of convergence of the cluster expansion, large external field [Sp 1] or low temperature [GJS 2,3], the situation is expected to be the same. For weak coupling but arbitrary energies, the present methods do not apply.

Outside of the region of convergence of the cluster expansion, the problem seems to involve all major elements of structure of the field theory, including the structure of the vacuum, bound states and superselection sectors [DHR]. The relation of bound states and superselection sectors to asymptotic completeness is well known, since extra bound states as well as extra elementary particles in some extra charge one superselection sector give rise to extra multiparticle continuous spectrum. The relation of the vacuum structure of phase transitions to solitons and superselection rules in two dimensions is contained in [Go-Ja, DHN, Fr 3]. It is an old question to ask whether the Goldstone picture provides a qualitatively correct picture of phase transitions, and now we ask whether the ideas of [Go-Ja, DHN, Fr 3] are sufficient to describe all superselection sectors for $P(\phi)_2$ fields. Can the reasoning be reversed in the sense that $Z=0$ and suppt $\rho(a) = [m^2, \infty)$ would imply existence of a new superselection sector?

On a less etherial level, we ask whether each pure phase for a $P(\phi)$ interaction can be obtained by an appropriate choice of boundary conditions, as is the case in statistical mechanics. Can the

FKG inequalities be used to prove convergence and Euclidean invariance of the infinite volume limit, for general $P(\phi)$ theories? See the introduction to [GJS 3] for a further discussion of the Goldstone picture of phase transitions.

The third step is a bound, uniform as the critical point is approached, on the renormalized n-point Schwinger functions. We follow the repulsive route in our reliance on correlation inequalities. For the ϕ^4 interaction, a correlation inequality reduces this problem to a bound on the renormalized two point Schwinger function

$$S^{(2)} = \langle\phi_{ren}(x)\phi_{ren}(y)\rangle = Z^{-1}\langle\phi(x)\phi(y)\rangle ,$$

[GJ 1]. The reduction applies to all cases (ϕ_2^4, ϕ_3^4 and lattice ϕ_4^4) considered here. It is convenient to choose the scale parameters so that $m=1$ and $\varepsilon\to 0$, and then a sufficient bound on $S^{(2)}$ is e.g.

$$\int S^{(2)}(x)dx \leq \text{const.}$$

with a constant independent of ε, or more generally, $|S^{(2)}|_{\mathscr{S}'} \leq$ const. for some $\mathscr{S}'$-norm$|\cdot|_{\mathscr{S}'}$ independent of ε. In the absence of level crossings, the required bound on $S^{(2)}$ is equivalent to a bound on CDD zeros [GJ 5]. The conjectured inequality, $\Gamma^{(6)} \leq 0$, would imply an absence of bound states (and thus of bound state level crossings), and of CDD zeros below the three particle threshold [GJ 5]. Thus $\Gamma^{(6)} \leq 0$ would bound $S^{(2)}$ and complete the third step. This application of the $\Gamma^{(6)} \leq 0$ inequality was derived in the context of the ϕ_2^4 interaction. The methods extend without change to the ϕ_3^4 interaction. The adaptations of these methods to the lattice ϕ_4^4 interaction is an open problem.

As a concluding remark on the repulsive route, we mention that considerable progress has been made in deriving new correlation inequalities and in finding interrelations between, and simplified proofs of, old ones. See [Sy, New 1,2, El-New, Du-New, El-Mo].

Unfortunately, we have little to say about a possible general route for the third step. In particular there is no argument for believing (or disbelieving) that the phenomena considered above -- absence of bound states and of CDD zeros below the three particle threshold -- occur in the scaling limit approach to a ϕ^6 tricritical point. An absence of level crossings between bound states and the elementary particle and a bound on CDD zeros away from the elementary particle mass might be a general picture, and would yield a bound on the two point function, but not on the general n point functions. In the approach to a general $P(\phi)$ critical point which is not near a tricritical point, the ϕ^4 (repulsive) critical behavior should dominate. Is there any argument (even heuristic) which can be used to discuss bound states, CDD zeros and/or bounds on the renormalized two point function in this region, other than the proposed $\Gamma^{(6)}$ inequality?

The fourth step is the nontriviality of the limit. We hold m fixed (for example $m=1$) throughout the discussion and consider first $d=2,3$ dimensions, then $d=1$ and finally $d=4$. Starting with a lattice field theory with $\lambda < \infty$ and $\varepsilon > 0$, we have the definition $(d \leq 3)$

$$\phi^4\text{-field theory} = \lim_{\varepsilon\to 0} .$$

We believe that

$$\text{Ising model} = \lim_{\lambda\to\infty} \quad \text{(lattice spacing } \varepsilon)$$

$$\text{scaling limit } \phi^4 \text{ field theory} = \lim_{\lambda\to\infty} \lim_{\varepsilon\to 0}$$

$$\text{scaling limit Ising model} = \lim_{\varepsilon\to 0} \lim_{\lambda\to\infty} .$$

It is reasonable to conjecture that the ε and λ limits above can be interchanged and thus that the scaling limits of the ϕ^4 field theory and the Ising model coincide. This conjecture is a variant of the universality principle for critical exponents in statistical mechanics. Because the Ising critical exponents are known to be nontrivial for $d=2,3$, we can expect the scaling limit for ϕ_2^4, ϕ_3^4 to be nontrivial.

For $d=1$, all steps one-four have been completed [Is], including interchange of the ε-λ limits. The one dimensional Ising model is already scale invariant, and so the $\varepsilon \to 0$ limit has a trivial form. Control over the $\lambda \to \infty$ limit is obtained from an analysis of anharmonic oscillator eigenvalues and eigenfunctions in a neighborhood of the critical point $m_0^2 = -\infty$.

In $d=4$ dimensions the situation is somewhat different from $d < 4$. In terms of the Callan-Symanzik equations, the sign of the crucial function $\beta(\lambda)$ is reversed. This change in the sign of β has its origin in the fact that λ is dimensionless (and thus scale invariant). In terms of the above constructions, the scale invariance means that λ is not taken to infinity by an infinite scale transformation. Rather λ, the bare charge, must be chosen (renormalized) to yield some desired value λ_{phys} of the physical charge. We define the physical charge by

$$\lambda_{phys} = \lambda_{phys}(\lambda,\varepsilon m) = -\varepsilon^3 Z^{-2} \chi^{-4} \sum_{x_1,x_2,x_3} \langle \phi(x_1)\cdots\phi(x_4) \rangle^T ,$$

where $\langle \ \rangle^T$ denotes the connected, Euclidean Green's function (Ursell function). By Lebowitz' inequality,

$$0 \leq \lambda_{phys} .$$

We have taken advantage of the scale invariance of λ_{phys} to write it as a function of the scale invariant parameters λ and εm. Recalling that $\lambda = \infty$ is the Ising model, we define

$$\lambda_I(\varepsilon m) = \lambda_{phys}(\infty, \varepsilon m) .$$

Also note that $\lambda = 0$ is a free lattice field, and

$$0 = \lambda_{phys}(0,\varepsilon m) .$$

To simplify the discussion of renormalization, we suppose that $\lambda_{phys}(\lambda,\varepsilon m)$ is monotone increasing as a function of λ for fixed εm. (However, we have no argument to support such an hypothesis.)

We claim that λ_{phys} should be continuous in λ and εm. We assume that $G^{(4)} \equiv \langle\phi(x_1)\cdots\phi(x_4)\rangle^T Z^{-2}$ is continuous in λ and εm. Then upper bounds on the two point function, suggested by perturbation theory, substituted in the inequalities of [GJ 3] yield an integrable upper bound for $|G^{(4)}| = -G^{(4)}$, independent of λ and εm, for m fixed, $m > 0$. Continuity of λ_{phys} follows from the Lebesgue bounded convergence theorem.

By definition, charge renormalization is the inverse function,

$$\lambda = \lambda(\lambda_{phys}, \varepsilon m),$$

and by continuity, we can choose $\lambda = \lambda(\varepsilon m)$ so that $\lambda_{phys} = \lambda_{phys}(\lambda(\varepsilon m), \varepsilon m))$ approaches any desired value in the interval

$$[0, \lambda_I(0)],$$

as $\varepsilon m \to 0$, $m \neq 0$. Nontriviality of the Ising model (in its critical point limit) is the statement that $\lambda_I(0) \neq 0$. We conclude that the ϕ_4^4 fields constructed here should be nontrivial if and only if the critical behavior of the Ising model is.

According to conventional ideas, $\lambda(\lambda_{phys}, \varepsilon m) \to \infty$ as $\varepsilon m \to 0$ in order to ensure $\lambda_{phys} \neq 0$ (infinite charge renormalization). In order to discuss the long and short distance scaling limits of the ϕ_4^4 field, we also suppose

$$\lambda(\lambda_{phys}, \varepsilon m) \nearrow \infty$$

as $\varepsilon m \searrow 0$.

In the context of the Callan-Symanzik equations, one changes m_0^2, followed by a scale transformation to keep m fixed. The decrease of

m_0^2 is called long distance scaling; the increase of m_0^2 is called short distance scaling. According to conventional ideas, there are two fixed points to this transformation, the points $\lambda_{phys} = 0$, $\lambda_{phys} = \lambda_I(0)$. The zero mass theories associated with these fixed points are scale invariant.

At the endpoint $\lambda_{phys} = 0$ (assuming χ is finite), the field is Gaussian [Newman]. Presumably it is the free field, invariant under the above transformation group (the renormalization group). At the endpoint $\lambda = \lambda_I(0)$, we expect the field theory to coincide with the long distance scaling limit of the Ising model.

We now consider λ_{phys} lying in the interval $(0, \lambda_I(0))$. For such a theory, according to conventional ideas, the short distance behavior is governed by the fixed point $\lambda_{phys} = \lambda_I(0)$, while the long distance behavior is governed by the fixed point $\lambda_{phys} = 0$. We show that λ_{phys} is monotone increasing in its dependence on m_0^2. Since λ_{phys} is dimensionless, and hence unchanged under scale transformations, this also shows that λ_{phys} decreases under long distance renormalization group transformations and increases under short distance transformations, i.e. $\beta \geq 0$.

Consider two values of the bare mass, m_0, m_0^* satisfying $(m_0^*)^2 < m_0^2$. Let $m^* < m$ be the corresponding masses. By definition

$$\lambda_{phys} = \lim_{\varepsilon \to 0} \lambda_{phys}(\lambda(\lambda_{phys}, \varepsilon m), \varepsilon m)$$

$$\lambda_{phys}^* = \lim_{\varepsilon \to 0} \lambda_{phys}(\lambda(\lambda_{phys}, \varepsilon m), \varepsilon m^*) .$$

Since ε is a dummy variable, we replace it in λ_{phys}^* by $\varepsilon m/m^*$

$$\lambda_{phys}^* = \lim_{\varepsilon \to 0} \lambda_{phys}(\lambda(\lambda_{phys}, \varepsilon m^2/m^*), \varepsilon m) .$$

Since $m/m^* > 1$, we have by monotonicity of λ_{phys} in λ and monotonicity of λ in εm that

$$\lambda_{phys}^{*} \leq \lambda_{phys} .$$

This completes the proof.

The statement that the charge renormalization is infinite is equivalent to the statement that the lattice ϕ_4^4 field is free in its critical point behavior (e.g. $Z \to 1$ as $\varepsilon \to 0$, with $\lambda = \text{const.} < \infty$). This is, of course, an open problem.

The existence of ϕ_3^4 and $P(\phi)_2$ fields suggests that the critical point scaling limit exists for the corresponding lattice fields; in the $P(\phi)_2$ case, tri- and multi-critical point limits should also exist. More generally, we summarize the discussion up to this point by asserting that a Euclidean quantum field is the critical point scaling limit of a corresponding lattice field. In the limit of strong physical coupling, the lattice field is replaced by an Ising model.

An alternate approach to nontriviality of the ϕ_4^4 field theory could be based on existence of the classical limit $\hbar \to 0$. Scattering for $\hbar = 0$ is known to be nontrivial. We thank Raczka for this comment.

Turning away from the construction of ϕ_4^4 via critical point theory, we note that recent work [Co, Fr 1] solves the (renormalizable but not superrenormalizable) massive Thirring model in two dimensions. Does this solution provide insight into the problems of charge and wave-function renormalization? Can other solvable two dimensional models be used as a starting point to prove existence of fields, for interactions of the form (explicitly solvable) + (superrenormalizable)?

Most thinking in contemporary particle physics uses nonabelian gauge fields and a Higgs mechanism as an ingredient. There are many problems here, including a proof of the existence of a Higgs mechanism, even in the lattice case.

We conclude by mentioning two other problems in mathematical physics which may be related to the critical and nonrenormalizable infrared problems considered above. First, the approach of Kolmogoroff to a statistical theory of turbulence uses scaling arguments to deduce exponents governing (short distance) asymptotic behavior. The second problem is the divergence of the virial expansion for the transport coefficients. Here the problem is infrared (slow decay of large time correlations) and infrared. In some cases, the leading divergences can be resummed, and the leading nonanalytic dependence on the density explicitly determined, on a formal level, cf. [Ha - Co].

Bibliography

[Ba] G. Baker, Self-interacting boson quantum field theory and the thermodynamic limit in d dimensions, J. Math. Phys. 16 (1975) 1324-1346.

[Bi-Wi] J. Bisognano and E. Wichmann, On the duality condition for a Hermitian scalar field, J. Math. Phys. 16 (1975) 985-1007.

[Br 1] D. Brydges, Boundedness below for fermion model theories, I and II.

[Br 2] __________, Cluster expansions for fermion fields by the time dependent Hamiltonian approach.

[Co] J. Coleman, The quantum sine-Gordon equation as the massive Thirring model.

[DHN] R. Dashen, B. Hasslacher and A. Niveau, Nonperturbative methods and extended-hadron models in field theory I, II, III, Phys. Rev. D 10 (1974) 4114-4142.

[DHR] S. Doplicher, R. Haag and J. Roberts, Fields, observables and gauge transformations, Commun. Math. Phys. 13 (1969) 1-23.

[Du-New] F. Dunlop and C. Newman, Multicomponent field theories and classical rotators.

[EMS] J.-P. Eckmann, J. Magnen and R. Seneor, Decay properties and Borel summability for Schwinger functions in $P(\phi)_2$ theories, Commun. Math. Phys. 39 (1975) 251-272.

[El-Mo] R. Ellis and J. Monroe, A simple proof of GHS and further inequalities, Commun. Math. Phys. 41 (1975) 33-38.

[El-New] R. Ellis and C. Newman, A tale of two inequalities: Concave force implies concave magnetization.

[Fel] J. Feldman, On the absence of bound states in the $\lambda\phi_2^4$ quantum field model without symmetry breaking, Canad. J. Phys. 52 (1974) 1583-1587.

[Fel-Os] J. Feldman and K. Osterwalder, The Wightman axioms and the mass gap for weakly coupled $(\phi^4)_3$ quantum field theories.

[Fr 1] J. Fröhlich, The quantized Sine-Gordon equation with a nonvanishing mass term in two space-time dimensions.

[Fr 2] J. Fröhlich, The pure phases, the irreducible quantum fields and dynamical symmetry breaking in the Symanzik-Nelson positive quantum field theories.

[Fr 3] ________, New super-selection sectors ('solition states') in two dimensional Bose quantum field models.

[GJ 1] J. Glimm and A. Jaffe, A remark on the existence of ϕ_4^4. Phys. Rev. Lett. 33 (1974) 440-441.

[GJ 2] ________, ϕ_2^4 quantum field model in the single phase region: Differentiability of the mass and bounds on critical exponents. Phys. Rev. D 10 (1974) 536-539.

[GJ 3] ________, Absolute bounds on vertices and couplings, Ann. Inst. H. Poincaré, 22 (1975).

[GJ 4] ________, On the approach to the critical point, Ann. Inst. H. Poincaré, 22 (1975).

[GJ 5] ________, Three particle structure of ϕ^4 interactions and the scaling limit. Phys. Rev. D 11 (1975).

[GJS 1] J. Glimm, A. Jaffe, and T. Spencer, The particle structure of the weakly coupled $P(\phi)_2$ models and other applications of high temperature expansions. In: Constructive Quantum Field Theory, G. Velo and A. Wightman (eds.), Springer-Verlag, Berlin, 1973.

[GJS 2] ________, Phase transitions for ϕ_2^4 quantum fields. To appear.

[GJS 3] ________, A cluster expansion for the ϕ_2^4 quantum field theory in the two phase region. To appear.

[Go-Ja] J. Goldstone and R. Jackiw, Quantization of nonlinear waves. Phys. Rev. D 11 (1975) 1486-1498.

[GRS] F. Guerra, L. Rosen and B. Simon, The $P(\phi)_2$ Euclidean quantum field theory as classical statistical mechanics, Ann. Math. 101 (1975) 111-259.

[Ha-Co] E. Hauge and E. Cohen, Divergences in non-equilibrium statistical mechanics and Ehrenfest's wind tree model. Arkiv for det Fysiske Seminar i Trondheim, No. 7 - 1968.

[Is] D. Isaacson, Private communication.

[Is-Mar] D. Isaacson and D. Marchesin, Private communication.

[Kl-La] A. Klein and L. Landau, An upper bound on the energy gap in the $(\lambda\phi^4+\sigma\phi^2)_2$ model.

[McB 1] O. McBryan, Finite mass renormalizations in the Euclidean Yukawa_2 field theory, Commun. Math. Phys. To appear.

[McB 2] _________, Volume dependence of Schwinger functions in the Yukawa_2 quantum field theory.

[Ma-Sen] J. Magnen and R. Seneor, The infinite volume limit of the ϕ_3^4 model.

[Nel] E. Nelson, Probability and Euclidean field theory. In: Constructive quantum field theory, G. Velo and A. Wightman (eds.) Springer-Verlag, Berlin, 1973.

[New 1] C. Newman, Inequalities for Ising models and field theories which obey the Lee-Yang theorem, Commun. Math. Phys. 41 (1975) 1-10.

[New 2] _________, Gaussian correlation inequalities for ferromagnets.

[J. Ro 1] J. Rosen, The number of product-weighted lead codes for ballots and its relation to the Ursell functions of the Ising model, J. Comb. Anal. To appear.

[J. Ro 2] _________, Private communication.

[Ro-Sy] J. Rosen and G. Sylvester, Private communication.

[Sei] E. Seiler, Schwinger functions for the Yukawa model in two dimensions with space-time cutoff, Commun. Math. Phys.

[Sei-Si] E. Seiler and B. Simon, On finite mass renormalization in the two dimensional Yukawa model.

[Sp 1] T. Spencer, The mass gap for the $P(\phi)_2$ quantum field model with a strong external field, Commun. Math. Phys. 39 (1974) 63-76.

[Sp 2] T. Spencer, The absence of even bound states in ϕ_2^4, Commun. Math. Phys. 39 (1974) 77-79.

[Sp 3] ________, The decay of the Bethe-Salpeter kernel in $P(\phi)_2$ quantum field models, Commun. Math. Phys. To appear.

[Sp-Zi] T. Spencer and F. Zirilli, Private communication.

[St-Wi] R. Streater and A. Wightman, PCT, Spin and Statistics and all that, Benjamin, New York, 1964.

[Sy] G. Sylvester, Continuous-spin inequalities for Ising ferromagnets.

[Zi] J. Zittartz, New type of phase transition, International Symposium on Mathematical Problems in Theoretical Physics, Kyoto, January 1975.

IX Existence of Phase Transitions for φ_2^4 Quantum Fields

Existence of Phase Transitions for φ_2^4 Quantum Fields

James Glimm [1]
Rockefeller University
New York, N.Y. 10021

Arthur Jaffe [2]
Harvard University
Cambridge, Mass. 02138

Thomas Spencer [2,3]
Harvard University
Cambridge, Mass. 02138

RESUME L'existence des transitions de phase pour les champs quantiques $\lambda\varphi_2^4$ dans la région $\lambda \gg 1$ de couplage nu est établie.
La brisure de symétrie pour l'interaction $\lim \mu \to \pm 0$ $(\varphi^4 - \mu\varphi)$ est aussi démontrée. On fait la distinction entre les transitions de phase et la brisure de symétrie.

ABSTRACT The existence of phase transitions for $\lambda\varphi_2^4$ quamtum fields in the region $\lambda \gg 1$ of bare coupling is established. Symmetry breaking for the interaction $\mu \to \pm 0$ $(\varphi^4 - \mu\varphi)$ is alsoproved and the distinction between phase transitions and symmetry breaking is emphasized.

1 - New Results

We prove the existence of phase transitions for $\lambda\varphi_2^4$ quantum fields in the region $\lambda \gg 1$ of bare coupling. The same methods apply in principle to even $\lambda P(\varphi)_2$ models. We demonstrate the existence of long range order in the (even) $P(\varphi)_2$ theory defined with zero Dirichlet boundary data. (However, we restrict attention in this talk to φ_2^4.) We also prove the existence of symmetry breaking for the interaction

$$\lim_{\mu \to \pm 0} (\varphi^4 - \mu\varphi) .$$

As in statistical mechanics, where phase transitions may occur without symmetry breaking [4], we expect phase transitions in certain quantum field models which do not possess a symmetry group, such as the interaction

$$(\varphi^2 - \sigma^2)^4 + \epsilon\varphi^3 - \mu\varphi ,$$

with $\sigma >> 1$, $\epsilon << 1$, $\mu = \mu(\epsilon, \sigma)$. Thus we emphasize this distinction between phase transitions and symmetry breaking.

In a separate article [5], we give a cluster expansion for strong (bare) coupling of even φ_2^4 models. This expansion allows us to construct two pure phases, each satisfying the Wightman and Osterwalder-Schrader axioms, with a unique vacuum and with a mass gap.

In contrast to our detailed study based on the cluster expansion [5], we present at this conference a simple, direct proof that phase transitions occur. The details of this talk will be published separately [6]. An alternative approach to the problem of phase transitions has been announced in [7], but the proof has not appeared.

Theorem 1. Consider the $\lambda :\varphi_2^4:_{m_0^2} + \frac{1}{2} m_0^2 :\varphi^2:_{m_0^2}$ theory with Wick ordering mass m_0, bare mass m_0, and zero Dirichlet boundary conditions. For λ/m_0^2 sufficiently large, there is long range order (lack of clustering).

Theorem 2. Consider the model

$$\lim_{\mu \to 0+} (\lambda :\varphi_2^4:_{m_0^2} + \tfrac{1}{2} m_0^2 :\varphi^2:_{m_0^2} - \mu\varphi)$$

with Wick ordering mass m_0 and bare mass m_0. For λ/m_0^2 sufficiently large, there is symmetry breaking, i.e.

$$\lim_{\mu \searrow 0} \langle \varphi \rangle > 0 ,$$

where $\langle \cdot \rangle$ denotes the vacuum expectation value. Likewise the model defined by $\mu \to 0-$ has $\langle\varphi\rangle < 0$.

Our proof of these theorems is based on a Peierls argument, similar to the proof of phase transitions in statistical mechanics. The basic idea is to study the average field

$$\varphi(\Delta) = \int_{\Delta} \varphi(x)dx$$

where the average is taken over a unit square Δ in Euclidean space-time. The average (low momentum) field dominates the description of phase transitions, while the error

$$\delta\varphi(x) = \varphi(x) - \varphi(\Delta), \qquad x \in \Delta ,$$

the "fluctuating field" is estimated in terms of the kinetic part of the action, $\frac{1}{2}(\nabla\varphi)^2$. Technically, we use φ^j bounds to establish the estimates which give the convergent Peierls expansion, and show the probability of "flipping" values of $\varphi(\Delta)$ is small.

In place of repeating the material in [6], we explain the classical (mean field) approximation to the φ^4 theory. This classical picture is the basis for our convergent expansions about the mean field.

2. Classical Approximation

Consider a quantum field defined by the Euclidean action density

$$\tfrac{1}{2}:(\nabla\varphi)^2:_{a^2} + :\mathcal{U}(\varphi):_{a^2} = :\tfrac{1}{2}(\nabla\varphi)^2 + \tfrac{1}{2}a^2\varphi^2:_{a^2} + :P(\varphi):_{a^2}$$

Here $: \;:_{a^2}$ denotes Wick ordering with respect to mass a, and by convention we include a bare mass a in the free part of the action, $\frac{1}{2}:(\nabla\varphi)^2 + a^2\varphi^2:$.
The classical approximation for the ground state of the field φ is obtained by regarding $\frac{1}{2}(\nabla\varphi)^2$ as a kinetic term and $\mathcal{V} \equiv \frac{1}{2}a^2\varphi^2 + P(\varphi)$ as a potential term.
Then in the classical approximation the vacuum expectation (mean) $\langle\varphi\rangle$ of φ equals φ_c, a value of φ which minimizes $\mathcal{V}$. The classical mass m_c is given by

$$m_c^2 = \mathcal{V}''(\varphi_c) = a^2 + P''(\varphi_c) .$$

In other words the classical low mass states of φ are those of a free field with action density

$$\mathcal{V}_c = :\tfrac{1}{2}(\nabla\varphi)^2 + \tfrac{1}{2}m_c^2(\varphi - \varphi_c)^2:_{m_c^2} .$$

For convenience, we choose the constant in P so that $P(0) = 0$. (The same then holds for $\mathcal{V}$.)

We expect the classical approximation to be accurate (up to higher order quantum corrections) for those interaction polynomials $\mathcal{V}$ such that

(i) $\mathcal{V} - \mathcal{V}_c$ is small for $\varphi - \varphi_c$ small,

and

(ii) $a^2 = m_c^2$.

We say that an interaction $:P:_{a^2}$ satisfying (i) and (ii) is classical.

To understand the conditions (i) and (ii) concretely, we write $\mathcal{V}(\varphi)$ in terms of its Taylor series about $\varphi = \varphi_c$, namely

$$\mathcal{V}(\varphi) = \mathcal{V}(\varphi_c) + \tfrac{1}{2}m_c^2(\varphi - \varphi_c)^2 + \sum_{i\geq 3} \lambda_i(\varphi - \varphi_c)^i ,$$

where

$$\lambda_i = (i!)^{-1}\mathcal{V}^{(i)}(\varphi_c)$$

$$= (i!)^{-1}P^{(i)}(\varphi_c) , \qquad i \geq 3 .$$

In particular, condition (i) is satisfied if

$$(1) \qquad |\lambda_i/m_c^2| \ll 1 ,$$

where $i \geq 3$.

To achieve (ii) will normally require Wick reordering, and in preparation, we calculate the a dependence of the Wick constant

$$c(a^2) = \frac{1}{(2\pi)^2} \int \frac{d^2p}{p^2 + a^2} .$$

Then

$$-\frac{d}{da^2} c(a^2) = \frac{1}{(2\pi)^2} \int \frac{d^2p}{(p^2 + a^2)^2} = (4\pi a^2)^{-1} ,$$

where we interpret this formula as a $\kappa \to \infty$ limit of cutoff equations in which $p^2 \leq \kappa^2$. We expect that (ii) will be satisfied after Wick reordering if

(2) $$(\lambda_i/m_c^2)\ell n^{i/2}(m_c^2/a^2) \ll 1\,, \qquad i \geq 3\,.$$

In the following section we carry out this choice for the φ_2^4 model. The classical approximation is also referred to as the Goldstone approximation or the mean field approximation.

3. The φ^4 Interaction

The conventional definition of the φ^4 interaction is

(3) $$:\mathcal{V}:_{m_0^2} = \lambda:\varphi^4:_{m_0^2} + \tfrac{1}{2}m_0^2:\varphi^2:_{m_0^2}$$

$$= :P(\varphi):_{m_0^2} + \tfrac{1}{2}m_0^2:\varphi^2:_{m_0^2}\,.$$

The weak coupling region $\lambda/m_0^2 \ll 1$ satisfies (1) and (2), and hence is also a classical region. In this region $\varphi_c = 0$, $m_c = m_0$. Thus the classical picture of weakly coupled φ^4 is a field with mean zero, and particles of mass $m_c = m_0$. The $\varphi \to -\varphi$ symmetry preserves $\langle\varphi\rangle = 0$ as an exact identity, but we expect quantum corrections to give a physical mass

(4) $$m = m_c(1 + O(\lambda/m_c^2)), \qquad \lambda/m_c^2 \to 0\,.$$

In fact the weak coupling region is well understood from the cluster expansion [8], which yields a Wightman-Osterwalder-Schrader theory for $\lambda/m_0^2 \ll 1$, and $m - m_c = o(\lambda/m_c^2)$.

We now turn our attention to the region $\lambda/m_0^2 \gg 1$. In this region $:\mathcal{V}:_{m_0^2}$ given by (3) is clearly not classical, since both (1) and (2) fail. In order to obtain a classical interpretation, we rewrite (3) in terms of a new Wick ordering mass a satisfying

$$a^2 >> \lambda >> m_0^2. \tag{5}$$

Then we write (3) as a new polynomial $:\mathcal{V}_1(\varphi):_{a^2}$ satisfying $\mathcal{V}_1(0) = 0$. Thus

$$:\mathcal{V}(\varphi):_{m_0^2} + \text{const.} = :\mathcal{V}_1(\varphi):_{a^2} \tag{6}$$

$$= :\lambda\varphi^4 - \left(\frac{6\lambda}{4\pi} \ln \frac{a^2}{m_0^2} - \tfrac{1}{2}m_0^2\right)\varphi^2:_{a^2}$$

Here

$$\frac{1}{4\pi} \ln \frac{a^2}{m_0^2} = c(m_0^2) - c(a^2) .$$

Likewise

$$:\mathcal{P}_1(\varphi):_{a^2} = :\lambda\varphi^4 - \left(\frac{6\lambda}{4\pi} \ln \frac{a^2}{m_0^2} + \tfrac{1}{2}a^2 - \tfrac{1}{2}m_0^2\right)\varphi^2:_{a^2} \tag{7}$$

and

$$\mathcal{V}_1 = \mathcal{P}_1 + \tfrac{1}{2}a^2\varphi^2 .$$

By (5), the coefficient of φ^2 in $\mathcal{V}_1$ is negative, so $\mathcal{V}_1$ has a double minimum at $\varphi = \varphi_c = \pm\sigma$. Here m_c and σ are related by

$$m_c^2 = 8\lambda\sigma^2 = \frac{6\lambda}{\pi}\,\ell n\frac{a^2}{m_0^2} - 2m_0^2 . \tag{8}$$

We now choose a so that $m = a$, as can be achieved by letting x solve the equation

$$x = \frac{\lambda}{m_0^2}\,\frac{6}{\pi}\,\ell n\, x - 2 .$$

For λ/m_0^2 sufficiently large, this equation has exactly two solutions. The larger solution determines a by the relation $x = (a/m_0)^2$. The smaller solution is spurious in the sense that is gives an interaction satisfying (ii) but not (i).

Next we perform a scale transformation so the classical mass becomes one. Since the Wick ordering mass transforms similarly, it also becomes 1. Thus after the scale transformation. we obtain an interaction polynomial $:\mathcal{V}_2:_1$ given by

$$:\mathcal{V}_2:_1 = :\frac{1}{8\sigma^2}\varphi^4 - \frac{3}{4}\varphi^2:_1 + \tfrac{1}{2}:\varphi^2:_1 \tag{9}$$

$$= :P_2:_1 + \tfrac{1}{2}:\varphi^2:_1 .$$

By (8), we see that $\sigma \gg 1$. Thus the interaction (3), in the strong coupling region $\lambda/m_0^2 \gg 1$, is equivalent to the weakly coupled φ^4 interaction (9), with a negative quadratic term, with bare mass 1 and with Wick mass 1. For the interaction (9), we find that

$$(10) \qquad \varphi_c = \pm\sigma, \qquad m_c = 1,$$

$$\lambda_3 = \pm(2\sigma)^{-1}, \quad \lambda_4 = (8\sigma^2)^{-1}.$$

Thus for σ large, both (1) and (2) are satisfied and (9) is classical. It exhibits the two phase classical approximation to strongly coupled φ_2^4, since φ_c has two possible mean field values. In our second paper [5], we present a systematic expansion about the classical field φ_c, combined with a Peierls argument to select a given phase. We find that in each of two pure phases, the physical mass is positive.

References and Footnotes

1. Supported in part by the National Science Foundation under Grant MPS 74-13252.

2. Supported in part by the National Science Foundation under Grant MPS 73-05037.

3. On leave from Rockefeller University, New York, NY 10021.

4. S. A. Priogov and Ya. G. Sinai, Phase transitions of the first kind for small perturbations of the Ising model, Funct. Anal. and its Appl. 8, 21-25 (1974). (English Trans.)

5. J. Glimm, A. Jaffe and T. Spencer, A cluster expansion in the wo phase region, in preparation.

6. J. Glimm, A. Jaffe and T. Spencer, Phase transitions for φ_2^4 quantum fields, Commun. Math. Phys., to appear.

7. R. Dobrushyn and R. Minlos, Construction of a one-dimensional quantum field via a continuous Markov field, Funct. Anal. and its Appl. 7, 324-325 (1973). (English Trans.)

8. J. Glimm, A. Jaffe and T. Spencer, The Wightman Axioms and particle structure in the $P(\varphi)_2$ quantum field model, Ann. Math. 100, 585-632 (1974).

X Critical Exponents and Renormalization in the φ^4 Scaling Limit

Critical Exponents and Renormalization in the ϕ^4 Scaling Limit

J. Glimm[+]
Rockefeller University
New York, New York 10021

A. Jaffe[++]
Harvard University
Cambridge, Mass. 02138

Abstract

For dimensions $d \leq 3$, the ϕ^4 scaling limit defines a nonrenormalizable field theory. The standard relations between critical exponents and renormalization are presented. Arguments supporting the existence of the scaling limit are based on correlation inequalities and the numerical values of Ising model exponents, $2\eta \lneq \eta_E$ for d=2,3.

Contents

[+] Supported in part by the National Science Foundation under Grant 74 - 13252.

[++] Supported in part by the National Science Foundation under Grant 75 - 21212.

1. Formulation of the Problem

The Euclidean action density

$$:(\nabla\phi^2 + \lambda\phi^4 + \sigma\phi^2 + \mu\phi): \tag{1.1}$$

determines a quantum field theory for $d=1,2,3$ [7,21,22,4,20,26, 6,23]. The Wick order is defined with a bare mass 1. Dirichlet boundary conditions and (for d=3) additional mass counterterms are implied in (1.1). For large positive values of σ, the field has a unique state Ω and a positive mass

$$m = m(\lambda,\sigma,\mu) > 0, \tag{1.2}$$

[13,4,20]. For d=2 (and presumably for d=3), for $\mu=0$, and for large negative value of σ, the vacuum is degenerate [14]. We define the critical value of σ, $\sigma_c=\sigma_c(\lambda)$ as the supremum of the values of σ for which either the vacuum is degenerate, or for which (1.2) fails.

On the interval $(\sigma_c(\lambda),\infty)$, $m(\lambda,\sigma,0)$ is monotone increasing and Lipschitz continuous [16,9] in σ, for d=2 at least, and guided by theorems concerning lattice fields [1,24] (and arbitrary $d\geq 2$), we expect that $m\searrow 0$ as $\sigma\searrow\sigma_c$. This is the only mathematically rigorous statement which can be made about the critical point: for ϕ_2^4 fields (and Ising models), $\sigma_c < \infty$ while for lattice ϕ^4 fields (and Ising models) $m(\sigma)\searrow 0$ as $\sigma\searrow\sigma_c$. In addition to the corresponding continuum problem, as mentioned above, the uniqueness of the vacuum at $\sigma=\sigma_c$ and the conjecture

$$m(\lambda,\sigma_c,0) = 0$$

are open problems, both for Ising models ($d\geq 3$), and lattice and continuum fields ($d\geq 2$). The d=2 Ising model, I_2, is of course a special case, because the existence of a closed form solution makes the detailed critical structure accessible [2].

Because of the absence of a mathematical theory of critical behavior, the remainder of our discussion will be mainly on a heuristic level. Let $\langle\ldots\rangle = \int \ldots d\phi$ be the Euclidean vacuum expectation associated with the quantum field defined by (1.1). The two point Schwinger function

$$S^{(2)}(x,y) = \int \phi(x)\phi(y)d\phi$$

can be represented, according to the Lehmann spectral theorem, in the form

$$S^{(2)}(x,y) = \int C_a(x-y)d\rho_a, \qquad (x \neq y)$$

where C_a is the convolution inverse to $(-\Delta+a)$. Necessarily,

$$m^2 = \inf \text{suppt}\, d\rho_a$$

and we make the further assumption (absence of bound states in $S^{(2)}$) that

$$S^{(2)}(x,y) = Z_3 C_{m^2}(x-y) + \int_{a \geq (3m)^2} C_a(x-y)d\rho_a \tag{1.3}$$

The evidence in favor of this assumption will be presented below. In particular, we take $\mu=0$, $\sigma>\sigma_c$, since otherwise bound states may be expected. Here the wave function renormalization constant Z_3 is defined by (1.3).

The problem of critical exponents is to understand the leading singularity of the long distance behavior of the field ϕ, as $\sigma \to \sigma_c$, in particular as expressed in such formulae as

$$m \sim \text{const} \left(\frac{\sigma-\sigma_c}{\sigma_c}\right)^{\nu} \tag{1.4}$$

$$\chi \equiv S^{(2)}(x-y)dx \sim \text{const} \left(\frac{\sigma-\sigma_c}{\sigma_c}\right)^{-\gamma} \tag{1.5}$$

$$Z_3 \sim \text{const} \left(\frac{\sigma-\sigma_c}{\sigma_c}\right)^{\zeta_3} \tag{1.6}$$

$$S^{(2)}(x-y)\Big|_{\sigma=\sigma_c} \sim \text{const}\,|x-y|^{2-d-\eta} \quad \text{as } |x-y|\to\infty \tag{1.7}$$

2. The Scaling and Critical Point Limits

The ϕ^4 field theory, $d\leq 3$, has two intrinsic length scales. The longer length scale is the correlation length,

$$\xi = m^{-1}$$

and this length governs the long distance decay of the correlation functions

$$S^{(n)} = \int \phi(x_1)\ldots\phi(x_n)d\phi$$

In fact for n=2,

$$S^{(2)}(x-y) \sim Z_3 m^{d-2}(mr)^{(1-d)/2}e^{-mr}, \quad mr\to\infty$$

assuming (1.3). The short distance behavior of a ϕ^4 field is canonical (free) for $d\leq 3$, cf. [8,10]. The short length scale is the distance scale on which this canonical behavior becomes dominant. For a lattice field or Ising model, the short length scale could be the lattice spacing. The lattice spacing, of course, functions as an ultraviolet cutoff, but we will see below that the short distance scale is <u>always</u> an ultraviolet cutoff. In the case of a Lorentz covariant ϕ^4 continuum field theory, the short distance scale cuts off the nonrenormalizable infrared singularities, and substitutes a Lorentz covariant short distance canonical behavior, for d=2,3. It could happen (for example in the continuum limit of lattice theories) that the lattice spacing is very short relative to the above Gaussian short distance scale. This case, in which there are two short distance scales and one long distance scale, will not be discussed further.

In the case of two length scales, one can always be eliminated by a scale transformation, and only the dimensionless ratio

$$\varepsilon = \frac{\text{short distance scale}}{\text{long distance scale}} \tag{2.1}$$

has an intrinsic signifiance. The scale transformation U_s is a unitary operator connecting the Hilbert spaces of the fields (1.1) having the same value of dimensionless bare charge

$$g = \lambda\sigma^{(d-4)/2} \tag{2.2}$$

and differing values of σ. U_s is defined to multiply all lengths by the factor s, so that

$$U_s\phi(x)U_s^{-1} = s^{(2-d)/2}\phi(s^{-1}x) \equiv \psi_s(x) \tag{2.3}$$

defines a field with the new parameter values

$$\lambda_s = s^{d-4}\lambda, \qquad \sigma_s = s^{-2}\sigma$$

and Wick ordering mass s^{-1}. (A reWick ordering then gives the action of the field ψ_s the form (1.1).)

There are two distinguished choices for the parameter s, namely

$$s = 1 \qquad \text{or} \qquad s = \varepsilon$$

In the first choice, s=1, the short distance scale is fixed at one; this choice defines the <u>unscaled</u> theory. In the second choice, s=m, the long distance scale ξ = 1/mass, is fixed at one; this choice defines the <u>scaled</u> theory. We note ε is the mass of the unscaled theory and the lattice spacing or short distance cutoff of the scaled theory.

It follows that there are three distinguished distance intervals:

$$r \le \text{short distance scale} \tag{2.4a}$$

$$\text{short distance scale} \le r \le \text{long distance scale} \tag{2.4b}$$

$$\text{long distance scale} \le r \tag{2.4c}$$

We call these intervals (a) the canonical (or free, or Gaussian) interval, (b) the scaling (or critical) interval, and (c) the one particle interval. In the two extreme intervals, i.e. in the canonical and one particle intervals, the behavior of ϕ is well understood. In cases where these two intervals dominate, (high or low temperature) there is a convergent cluster expansion, and ϕ is well understood on all distance scales [13,14,15].

Near the critical point, ε becomes small, and the critical, or scaling interval (2.4b) dominates. As $\varepsilon \to 0$, we may construct three limiting theories.

i) The critical point is defined by $\varepsilon \to 0$, $\sigma \searrow \sigma_c$ in the unscaled theories. This limit is characterized by

$0 = m$

$0 \neq$ short distance scale.

ii) The scaling limit is defined by $\varepsilon \to 0$ in the scaled theories, and is characterized by

$1 = m$

$0 =$ short distance scale

In this limit, we use the renormalized field $\phi_{ren} = Z_3^{-1/2}\phi$ in place of ϕ.

iii) The scaled critical point is defined by a long distance scaling of i) or a short distance scaling, $m \to 0$, of ii), and is characterized by

$m = 0 =$ short distance scale

These two definitions of iii) are related by an interchange of order of limits.

Alternately, we may say that i) eliminates (2.4c), ii) eliminates (2.4a) and iii) eliminates both, in either order. To check this interchange in the order of limits, we analyze its influence on η, using Ising model exponents. In the limit iii),

$$S^{(2)} = r^{2-d-\eta} \tag{2.5}$$

and η as defined in (1.7), refers to the long distance scaling of i).

We now assert, as a scaling hypothesis, that the decay rate (2,5) holds for $S^{(2)}$ for the entire interval (2.4b). As a check on this assertion, we use it to compute (in the unscaled theory)

$$\chi = \int S^{(2)}(r)dr \sim \int_{|r|\leq\varepsilon^{-1}} r^{2-d-\eta}dr \sim \varepsilon^{-2+\eta}$$

In (1.4-6), the unscaled theory is understood, and so

$$\varepsilon \sim \left(\frac{\sigma-\sigma_c}{\sigma_c}\right)^{\nu}; \quad Z_3 \sim \varepsilon^{\zeta_3/\nu}; \quad \chi \sim \varepsilon^{-\gamma/\nu}$$

Thus the consistency check follows from the identity

$$\gamma = (2-\eta)\nu \tag{2.6}$$

valid for Ising model exponents, d=1,2,3.

We next consider the scaled theory. At the short distance end of (2.4b),

$$S^{(2)}_{ren}(\varepsilon) = Z_3^{-1}S^{(2)}_{free}(\varepsilon) = Z_3^{-1}\varepsilon^{2-d} = \varepsilon^{2-d-\zeta_3/\nu} \tag{2.7}$$

neglecting logarithms in d=2 dimensions. Thus

$$\begin{aligned} S^{(2)}_{ren}(r) &= \varepsilon^{2-d-\zeta_3/\nu}(r/\varepsilon)^{2-d-\eta} \\ &= \varepsilon^{\eta-\zeta_3/\nu}r^{2-d-\eta} \end{aligned} \tag{2.8}$$

on (2.4b) by the scaling hypothesis. We note that the (rigorous) inequalities

$$2\nu - \zeta_3 \leq \gamma \leq 2\nu - \eta\nu$$

[5,9] imply that

$$\eta - \zeta_3/\nu \leq 0 \tag{2.9}$$

so that (2.8) does not vanish as $\varepsilon\to 0$.

We further require equality in (2.9), as is known for I_2, and as follows from the existence of the limit ii), in the special case (2.8). Then in the scaling limit ii), we have

$$S^{(2)}_{ren}(r) \cong r^{2-d-\eta}, \quad r << 1 \tag{2.10}$$

and scaling from short distances gives $S^{(2)}_{ren}(r) = r^{2-d-\eta}$ for all r. Thus the two definitions of the limit iii) agree, at least in their two point function. In particular the two definitions of η agree. In summary we can say that interchange of limits in iii) follows from a scale relation, such as (2.5), valid over the entire scaling interval (2.4b), and that this hypothesis can be checked against known values of the exponents.

A comprehensive exposition of the scaling behavior and the theory of the renormalization group is given in [28].

3. Renormalization of the $\phi^2(x)$ Field

The field $\phi^2(x)$ requires both an additive and a multiplicative renormalization. The additive renormalization is Wick ordering in the physical vacuum, defined by

$$:\phi^2(x): \; = \; \phi^2(x) \; - \; <\phi^2(x)> \tag{3.1}$$

(These Wick dots do not coincide with those of (1.1), but in the term $\sigma\int:\phi^2:$ the difference is a constant, and of no consequence). The multiplicative renormalization

$$\begin{aligned} (:\phi^2(x):)_{ren} &= Z_E^{-1/2}:\phi^2(x): \\ &\neq \; : \; (\phi_{ren})^2(x) \; : \end{aligned} \tag{3.2}$$

is <u>not</u> obtained by renormalizing each factor ϕ in $\phi^2(x)$. Rather we define the exponents

$$<:\phi^2(x)::\phi^2(y):>\Big|_{\sigma=\sigma_c} \sim \text{const}\,|x-y|^{4-2d-\eta_E} \quad \text{as} \quad |x-y|\to\infty \tag{3.3}$$

$$C_H = \int<:\phi^2(x)::\phi^2(y):>d(x-y) \sim \left(\frac{\sigma-\sigma_c}{\sigma_c}\right)^{-\alpha} = \varepsilon^{-\alpha/\nu} \tag{3.4}$$

Here C_H is the specific heat, and if $Z(\sigma,\mu)$ is the partition function for (1.1), then

$$C_H = \frac{\partial^2}{\partial\sigma^2}\frac{\ln Z}{V}$$

$$\chi = \frac{\partial^2}{\partial\mu^2}\frac{\ln Z}{V}$$

As in §2, we make a scaling hypothesis, that (3.3) is valid over the scaling interval (2.4b), and then the identity in the unscaled theory

$$\varepsilon^{-\alpha/\nu} = \int_{1\leq|r|\leq\varepsilon^{-1}} r^{4-2d-\eta_E}dr + \int_{|r|\leq 1} r^{4-2d}dr$$

$$= \varepsilon^{-(4-d-\eta_E)}$$

implies

$$\alpha/\nu = 4-d-\eta_E = \begin{matrix} 0, & d=2 & \text{Ising} \\ 2 & d=3 & \text{Ising} \end{matrix} \tag{3.5}$$

This identity is analogous to the identity (2.6) for the ϕ-two point function. As in §2, a matching of the canonical form of the ϕ^2 two point function on (2.4a) with its scaling form on (2.4b) implies $\zeta_E/\nu = \eta_E$. In fact we assume in the unscaled theory that

$$<:\phi^2(x)::\phi^2(y):> \sim c_a r^{4-2d} \qquad \text{on (2.4a)}$$

$$\sim c_b(\varepsilon)r^{4-2d-\eta_E} \qquad \text{on (2.5b)}$$

Because $\sigma=\sigma_c$ is a regular point for the canonical part of ϕ, we may take $c_a=c_a(\varepsilon)\big|_{\varepsilon=0}$ to be independent of ε. Equality of the two asymptotic forms at r=1 implies $c_b=c_a$ is also regular in ε. It follows that in the scaled theory

$$<:\phi^2(x)::\phi^2(y):> \sim c_b\varepsilon^{4-2d}(r/\varepsilon)^{4-2d-\eta_E} \tag{3.6a}$$

and

$$<:\phi^2:_{ren}:\phi^2(y):_{ren}> \sim c_b\varepsilon^{\eta_E-\zeta_E/\nu}r^{4-2d-\eta_E} \tag{3.6b}$$

on (2.4b). We now define $Z_E = \varepsilon^{\zeta_E/\nu}$ and

$$\zeta_E/\nu = \eta_E \tag{3.7}$$

The fact that

$$\eta_E = \begin{cases} 2, & d=2 \\ .8, & d=3 \end{cases} > 2\eta = \begin{cases} 1/2, & d=2 \\ .082, & d=3 \end{cases} \tag{3.8}$$

implies the following important result:

$$:(\phi_{ren})(x)^2: = 0 \tag{3.9}$$

in the scale limit. Here we obtain the values for η_E from (3.5).

Theorem 3.1. For the scaled theories,

$$||\int :(\phi_{ren})(x)^2:dx||^2_{L_2} \sim \varepsilon^{\eta_E - 2\eta}$$

Proof. We use (3.6)-(3.7) as follows:

$$||\int_\Lambda :(\phi_{ren})(x)^2:dx||^2_{L_2} = \varepsilon^{\eta_E - 2\eta} ||\int_\Lambda :\phi(x)^2:_{ren} dx||^2_{L_2}$$

$$= O(1)\varepsilon^{\eta_E - 2\eta} \int_{\varepsilon \le |r| \le 1} r^{4-2d-\eta_E} dr$$

$$= O(1)\varepsilon^{\eta_E - 2\eta}$$

In more intuitive language, $:(\phi_{ren})^2:$ and $:\phi^2:_{ren}$ differ by an infinite multiple. Since the constant Z_E is defined so that the larger, $:\phi^2:_{ren}$, is finite, the smaller, $:(\phi_{ren})^2:$, must vanish.

4. Existence of the Scaling Limit

Existence of the scaling limit, in the weakest sense, means bounds uniform in ε on the renormalized Schwinger functions

$$S^{(n)}_{\text{ren}} = Z_3^{-n/2} S^{(n)}$$

so that by compactness, a convergent subsequence may be selected. By the explicit introduction of an invariant mean, the limit may be taken to be covariant under the translation subgroup. The Osterwalder-Schrader reconstruction theorem [21,22] then guarantees the existence of a scaling limit field theory which is at least translation covariant.

The Lebowitz correlation inequalities bound $S^{(n)}_{\text{ren}}$ by a sum of products of two point functions [11]. Thus existence follows from a uniform bound on $S^{(2)}_{\text{ren}}$. The required bounds on $S^{(2)}_{\text{ren}}$ are implied [12] by a conjectured correlation inequality

$$\Gamma^{(6)}(xxx\ yyy) \leq 0 \tag{4.1}$$

Here $\Gamma^{(6)}(x_1,\ldots,x_6)$ is the six point vertex, or direct correlation function. In graphical language, $\Gamma^{(6)}$ is one particle irreducible, and in (4.1) we may take $\Gamma^{(6)}$ in either its unamputated or its amputated form. Let $G^{(n)}$ denote the n-point truncated (Ursell) function. Choosing the unamputated form for the Γ's, and defining

$$\Gamma^{(2)} = -(G^{(2)})^{-1}$$

(convolution inverse), we have the explicit formulae

$$\Gamma^{(4)} = G^{(4)}$$

$$\Gamma^{(6)} = G^{(6)} + \frac{1}{2} \sum_{\text{Permutations}} (3!)^{-2} G^{(4)}(x_{i_1},\ldots x_{i_3},z) \times$$

$$\Gamma^{(2)}(z,z')G^{(4)}(a',x_{i_4},\dots x_{i_6})$$

The $G^{(n)}$ are connected parts of the $S^{(n)}$.

The identity $\phi^2 \equiv$ const. in the scaling limit, allows an explicit calculation of $\Gamma^{(6)}(xxx\ yyy)$. We first consider the more elementary calculation

$$G^{(4)}(xxyy) = S^{(4)}(xxyy) - S^{(2)}(xx)S^{(2)}(yy) - 2S^{(2)}(xy)^2$$

$$= \int(\phi^2(x) - \langle\phi^2(x)\rangle)(\phi^2(y) - \langle\phi^2(y)\rangle)d\phi - 2S^{(2)}(xy)^2$$

$$= \int :\phi^2(x)::\phi^2(y):d\phi - 2S^{(2)}(xy)^2 = -2S^{(2)}(xy)^2$$

For $\Gamma^{(6)}$ we have

$$\Gamma^{(6)}(xxxyyy) = -24\ G^{(2)}(xy)^3 \tag{4.2}$$

(4.2) holds only for $\varepsilon=0$, and thus does not establish (4.1) as $\varepsilon\to 0$, but it certainly makes the inequality (4.1) highly plausible, at least for small ε. Other tests of (4.1) include the one-dimensional Ising model, in which

$$\Gamma^6(x_1,\dots,x_6) = -24\ G^{(2)}(x_1x_6)G^{(2)}(x_2x_5)G^{(2)}(x_3x_4)$$

for

$$x_1 \le x_2 \le \dots \le x_6$$

[25], and numerical studies of the d=1 ϕ^4 field theory -- i.e., the anharmonic oscillator [19].

We remark that (4.1) has one other consequence: an absence of bound states in the two point function [12]. In particular if (4.1) holds, then $S^{(2)}$ has the form (1.3), with $0<Z_3\le 1$. Moreover in case (4.1) holds, there are no CDD zeros, which means that $\Gamma^{(2)} = -(G^{(2)})^{-1}$ has a 3m decay rate [12]. In particular the scaling limit has neither three particle bound states nor CDD zeros. Even bound states are also excluded by a general argument [3,27], which also applies for $\sigma>\sigma_c$.

The absence of bound states may be special to $\sigma>\sigma_c$, $\mu=0$. In fact for the two dimensional Ising model with $T<T_c$, it is suggested [15] that as $\mu \searrow 0$, the number of bound states becomes infinite, and fills in to form a continuum threshold. Here we appeal to the explicit solution of the Ising model for $T<T_c$, $\mu=0$. If the Ising and ϕ^4 theories have identical scaling limits, then a similar phenomena should occur in ϕ^4 field theory for $\sigma \lesssim \sigma_c$, $\mu \searrow 0$.

For d=1, existence of the scaling limit was established [17] on the basis of asymptotic estimates on the eigenvalues and eigenvectors of the anharmonic oscillator. This scale limit coincides with the one dimensional Ising model, proving universality in this case.

5. The Josephson Inequality

The Josephson inequality states

$$2-\alpha \leq d\nu \tag{5.1}$$

[18]. For I_2, (5.1) is an equality, and for I_3, (5.1) is correct as an inequality, and nearly (or possibly) correct as an equality. For the Gaussian (free field) case, the exponents are

$$\nu = 1/2, \quad \alpha = 2-d/2, \quad 0 = \eta = \eta_E = \zeta_3 = \zeta_E, \quad \gamma = 1$$

and (5.1) is again an equality.

Following a similar calculation in [9] based on Lebowitz's inequality, we have

$$\begin{aligned} \varepsilon^{-\alpha/\nu} \sim C_H &\leq \int \langle\phi(x)\phi(y)\rangle^2 dx \\ &\leq O(1) \int_{|r|\leq\varepsilon^{-1}} r^{4-2d-2\eta} dr \\ &\leq O(\varepsilon^{-(r-d-2\eta)}) \qquad \text{for } d = 2,3,4. \end{aligned}$$

Thus

$$\alpha \leq (4-d-2\eta)\nu \qquad d = 2,3,4. \tag{5.2}$$

Here we have used the scaling hypothesis for the two point function. Without this hypothesis, the same reasoning implies $\alpha \leq (4-d)\nu$ as a rigorous inequality, d=2,3.

Combined with (5.1), (5.2) reproduces the inequality $\nu \geq (2-\eta)^{-1} \geq \frac{1}{2}$ Furthermore of α/ν is strictly less than its Gaussian value, 4-d, then ν is strictly greater than its canonical value, 1/2. From Lebowitz's inequality, we have the (rigorous) inequality

$$\eta_E \geq 2\eta \geq 0 = \eta_{E\ \text{Gaussian}} = \eta_{\text{Gaussian}} \tag{5.3}$$

for d=1,2,3.

We complete this section with a formal derivation of (5.1), based on field theory ideas. The Hamiltonian and particle structure occur as hypotheses, and use is made of Lebowitz inequality. From (1.4), we have

$$\frac{dm}{d\sigma} \sim \left(\frac{\sigma-\sigma_c}{\sigma_c}\right)^{\nu-1} = \varepsilon^{1-1/\nu}$$

Let $<1|$ be the one particle zero momentum state. Then

$$m = <1|H|1>$$

and

$$\frac{dm}{d\sigma} = <1|\frac{dH}{d\sigma}|1>$$

Equating these two expressions gives

$$\varepsilon^{1-1/\nu} \sim <1|\int :\phi^2(x):d\vec{x}|1>$$

and because $m\int_0^{1/m} e^{-mt}dt = 1-e^{-1}$ and $m = \varepsilon$,

$$\varepsilon^{-1/\nu} \sim <1|\int_{|x_o|\leq m^{-1}} :\phi^2(x):dx|1> \tag{5.4}$$

The right side of (5.4) will be bounded by a Schwarz inequality. To justify this step, we use two approximations

$$<1| \to \text{const} \int_{|x|\leq m^{-1}} \phi(x)d\vec{x}\Omega \tag{5.5}$$

$$\text{const} = \Big(\int_{\substack{|\vec{x}|\leq m^{-1}\\ |\vec{y}|\leq m^{-1}}} S^{(2)}(x-y)d\vec{x}\ d\vec{y}\Big)^{-1/2}$$

and

$$\int_{|x_o|\leq m^{-1}} :{}^2(x):dx \to \int_{|x|\leq m^{-1}} :\phi^2(x):dx \tag{5.6}$$

With these approximations, $\langle 1|\varepsilon L_4$, with an L_4 norm bounded independently of ε. In fact $||\langle 1|||^4_{L_4}$ is a four point function divided by a product of two two point functions. The four point function can be expanded as

$$S^{(4)} = G^{(4)} - \sum G^{(2)} G^{(2)}$$

and by Lebowitz inequality

$$0 \le -G^{(4)} \le G^{(2)} G^{(2)}$$

Here the two point functions $G^{(2)}$ are bounded by the two point functions occurring in the denominator, and so $\langle 1|\varepsilon L_4$ as asserted.

Applying the Schwarz inequality to (5.4) yields

$$\varepsilon^{-1/\nu} \le O\Big(\int_{|x|,|y|\le m} \langle :\phi^2(x)::\phi^2(y):\rangle dx\, dy\Big)^{1/2}$$

$$\sim (\varepsilon^{-d-\alpha/\nu})^{1/2}$$

Thus

$$1/\nu \le \tfrac{1}{2}(d + \alpha/\nu)$$

and (5.1) follows.

Because of the wave function renormalization in the scaling limit, we include a discussion of the substitution (5.5). Let $\langle 1_{appx}|$ denote the right side of (5.5). From the scaling hypothesis, we find that

$$\text{const} = O(\varepsilon^{(d+2-\eta)/2}) = Z_3^{-1/2} O(\varepsilon^{d/2})$$

in (5.5). The factor $Z_3^{-1/2}$ replaces ϕ by ϕ_{ren}, and ensures that the one particle contribution to ϕ_{ren} has unit strength. The factor is proportional to

$$\Big|\Big|\int_{|\vec{x}|\le m^{-1}} \phi_{free}(x)\, d\vec{x}\,\Omega_{free}\Big|\Big|^{-1}$$

and guarantees that the one particle portion of $\langle 1_{appx}|$ has strength $O(1)$, uniformly as $\varepsilon \to 0$.

References

1. G. BAKER, Self-interacting boson quantum field theory and the thermodynamic limit in d dimensions, J. Math. Phys. 16, 1324 - 1346 (1975).
2. E. BAROUCH, B. MCCOY, C. TRAY and T. T. WU, The spin-spin correlation functions for the two dimensional Ising-model Exact theory in the scaling limit, to appear.
3. J. FELDMAN, On the absence of bound states in the ϕ^4 quantum field model without symmetry breaking, Cand. J. Phys. 52, 1583 - 1587 (1974).
4. J. FELDMAN and K. OSTERWALDER, The Wightman axioms and the mass gap for weakly coupled $(\phi)^4_3$ quantum field theories, to appear.
5. M. FISHER, Rigorous inequalities for critical point correlation exponents, Phys. Rev. 180, 594 - 600 (1969).
6. J. FRÖHLICH, Existence and analyticity in the bare parameters of the $|\lambda(\vec{\phi}\,\vec{\phi})^2 - \sigma\phi_1^2 - \mu\phi_1|$ quantum field models, I. Manuscript.
7. J. GLIMM and A. JAFFE, The $(\lambda\phi^4)_2$ quantum field theory without cutoffs III. The physical vacuum, Acta Math. 125, 203 - 261 (1970).
8. ——————, The $(\lambda\phi^4)_2$ quantum field theory without cutoffs IV. Perturbation of the Hamiltonian, J. Math. Phys. 13, 1558 - 1584 (1972).
9. ——————, ϕ^4_2 quantum field theory in the single phase region: Differentiability of the mass and bounds on critical exponents, Phys. Rev. D10, 536 - 539 (1974).
10. ——————, Two and three body equations in quantum field models, Commun. Math. Phys. 44, 293 - 320 (1974).
11. ——————, A remark on the existence of ϕ^4_4. Phys. Rev. Lett. 33, 440 - 442 (1974).
12. ——————, Three particle structure of ϕ^4 interactions and the scaling limit, Phys. Rev. D11, 2816 - 2827 (1975).
13. J. GLIMM, A. JAFFE, and T. SPENCER, The particle structure of the weakly coupled $P(\phi)_2$ models and other applications of high temperature expansions. In: Constructive quantum field theory, G. Velo and A. Wightman (eds.) Springer Verlag, Berlin, 1973.

14. J. GLIMM, A. JAFFE, and T. SPENCER, Existence of phase transitions for ϕ_2^4 quantum fields, Commun. Math. Phys. To appear.

15. ————, A cluster expansion for the ϕ_2^4 quantum field theory in the two phase region. In preparation.

16. F. GUERRA, L. ROSEN, and B. SIMON, In: Constructive quantum field theory, G. Velo and A. Wightman (eds.) Springer Verlag, Berlin, 1973.

17. D. ISAACSON, The critical behavior of the autoharmonic oscillator, NYU Thesis.

18. B. JOSEPHSON, Inequality for the specific heat, I Derivation, II Applications, Proc. Phil. Soc. 92, 269 - 284 (1967).

19. D. MARCHESIN, Work in progress.

20. J. MAGNEN and R. SENEOR, The infinite volume limit of the ϕ_3^4 model, Ann. Inst. H. Poincaré, to appear.

21. K. OSTERWALDER and R. SCHRADER, Axioms for Euclidean Green's functions, Commun. Math. Phys. 31, 83 - 112 (1974).

22. ————, Axioms for Euclidean Green's functions II, Commun. Math. Phys. 42, 281 - 305 (1975).

23. Y. PARK, Uniform bounds of the pressure of the $\lambda\phi_3^4$ field model. Preprint.

24. J. ROSEN, Mass renormalization for $\lambda\phi_2^4$ Euclidean lattice field theory.

25. J. ROSEN, Private communication.

26. E. SEILER and B. SIMON, Nelson's symmetry and all that in the Yukawa_2 and ϕ_3^4 field theories. Preprint.

27. T. SPENCER, The absence of even bound states in ϕ_2^4. Commun. Math. Phys. 39, 77 - 79 (1974).

28. E. BREZIN, J. C. LEGUILLORE and J. ZINN-JUSTIN, Field theoretical approach to critical phenomena. In: Phase transitions and critical phenomena, Vol. VI., Ed. by Domb and Green, Academic Press, New York, to appear.

XI A Tutorial Course in Constructive Field Theory

A TUTORIAL COURSE IN CONSTRUCTIVE FIELD THEORY

James Glimm[1]
The Rockefeller University
New York, N.Y. 10021

Arthur Jaffe[2]
Harvard University
Cambridge, MA 02138

---

1. Supported in part by the National Science Foundation under Grant PHY76-17191.
2. Supported in part by the National Science Foundation under Grant PHY75-21212.

1.1 INTRODUCTION

Prior to 1970, the major focus of constructive field theory was the mathematical framework required to establish the existence of quantum fields [1, 2]. Since 1970, the emphasis has gradually shifted, first toward verifying physical properties of the known models, and more recently toward bringing constructive field theory closer to the mainstream of physics [3]. In fact, by 1973 it had become more or less clear that the mathematical framework developed to give the first examples in $d = 2, 3$ space-time dimensions would be adequate to study $d = 4$. However, it was also clear that to solve the ultraviolet problem in $d = 4$, it would be necessary to incorporate into mathematical physics a deeper physical understanding of the questions being studied. In particular ideas of scaling and of critical behavior may be useful to select and analyze a suitable nontrivial critical point as the first step in dealing with the $d = 4$ ultraviolet problem. An infinite scaling transformation connects the problem of removing the ultraviolet cutoff with the problem of existence of scaling behavior at the critical point.

In these lectures we describe some results for boson (generally φ^4) fields. For $d = 4$, scaling arguments (e.g. the renormalization group) indicate that the φ^4 theory is not free at high energy. Thus the study of the φ_4^4 theory is a strong coupling problem which must be approached independent of perturbation theory. See, e.g. [4, 5] for a discussion of the mathematical aspects of this program. On the other hand, asymptotic freedom suggests that $d = 4$, nonabelian gauge field theories may be free at high energy and thus amenable to perturbation theory. For this reason gauge fields should play an important role in the future of the mathematical study of field theory, and the lecture of Osterwalder provides an introduction to that topic.

1.2 e^{-tH} AS A FUNCTIONAL INTEGRAL

It is now easy to formulate the connection between Euclidean field theory (functional integrals) and quantum mechanics (Hilbert space and Hamiltonians). We give a complete mathematical presentation of this connection for boson fields in the accompanying article, "Functional Integral Methods in Quantum Field Theory" [6]. Here we give a brief summary.

Assume we are given a functional integral $d\mu(\varphi)$, i.e. a measure $d\mu(\varphi)$ on generalized functions $\varphi \in \mathscr{S}'(R^d)$. Here φ is a classical field, and given a functional $F(\varphi)$, we define the expectation $\langle F \rangle$ by

$$(1.1) \qquad \langle F \rangle = \int F(\varphi) d\mu(\varphi) .$$

We require the normalization condition

$$(1.2) \qquad \langle 1 \rangle = \int d\mu(\varphi) = 1 ,$$

i.e. $d\mu(\varphi)$ is a probability measure. Of particular interest is the choice $F(\varphi) = \exp(i\varphi(f))$, which leads to the Fourier transform of $d\mu(\varphi)$, which we denote

$$(1.3) \qquad S\{f\} = \int e^{i\varphi(f)} d\mu(\varphi) .$$

$S\{f\}$ is also known as the generating functional.

We now wish to define a Hilbert space $\mathcal{H}$ of quantum mechanics in $d - 1$ space dimensions, associated with $d\mu(\varphi)$. In order to define $\mathcal{H}$, the measure $d\mu$ must satisfy three simple requirements, which we expect from every such functional integral which arises in physics. The vectors in $\mathcal{H}$ (at least a dense set of vectors) will be functionals $A = A(\varphi)$, for example $A = \varphi(f)$ or $A = \exp(i\varphi(f))$, and for certain $f \in \mathscr{S}$. The fundamental formula which relates the Hamiltonian H to functional integrals is

$$(1.4) \qquad \boxed{\langle A, e^{-tH} A \rangle_{\mathcal{H}} = \int \overline{\vartheta A}\, A_t d\mu(\varphi)} ,$$

where ϑ denotes time inversion, $\overline{A}$ denotes complex conjugation and A_t denotes time translation. In (1.4) A plays a dual role. On the left side, A is a vector in the Hilbert space of quantum mechanics $\mathcal{H}$, while on the right side $A(\varphi)$ is a functional of the classical field. The importance of (1.4) is that any question about H can be reduced to a question about a functional integral. For example: Does H have a unique vacuum? Does H have a mass gap? Does H have bound states and what are their masses? Do the asymptotic states span $\mathcal{H}$ (asymptotic completeness)? etc. We start with a brief explanation of (1.4), see [6] for details.

Let us start by mentioning the three requirements on $S\{f\}$ or $d\mu$ needed to obtain $\mathcal{H}$. (The generalization to include spinor or tensor fields poses no fundamental difficulty.) The conditions are

(1) Invariance
(2) Reflection Positivity
(3) Regularity

(1) We require invariance of $S\{f\}$ under a one parameter group of time translations $f(\vec{x}, s) \to f_\tau(\vec{x}, t-\tau)$ where $\tau \in R$ (or $\tau = n\delta$, $n = 0, \pm 1, \pm 2, \cdots$, i, e, lattice translations) and also invariance under time inversion $f(\vec{x}, t) \to (\vartheta f)(\vec{x}, t) = f(\vec{x}, -t)$. Then $S\{f_\tau\} = S\{\vartheta f\} = S\{f\}$.

(2) Let $f^{(j)}$, $j = 1, 2, \cdots, r$ be any sequence of real functions which vanish unless $t > 0$. Then reflection positivity is the requirement $S\{f_i - \vartheta f_j\} = M_{ij}$ is a positive matrix. In other words, if $A = \sum_{j=1}^{r} c_j \exp(i\varphi(f^{(j)}))$, then

$$\int \overline{\vartheta A}\, A d\mu \geq 0 . \tag{1.5}$$

This condition was discovered by Osterwalder and Schrader [7].

(3) The regularity assumption is important for a continuous time translation group, in which case $S\{f_\tau\} \to S\{f\}$ as $\tau \to 0$.

From (1-3) we obtain $\mathcal{H}$ and e^{-tH} (in the case of a continuous time translation group). In the case of discrete time translations we obtain $\mathcal{H}$ and a transfer matrix K, with K^t playing the role of e^{-tH}. In [6] we assume invariance under the full Euclidean group on R^d (rotations, translations and reflections in hyperplanes) and we obtain a stronger result, namely the construction of a full relativistic field theory.

In fact, assuming (2), we may define a scalar product on functionals $A(\varphi)$ at positive time, e.g. functionals of the form

$A(\varphi)$ in (1.5). We let

$$\langle A, B\rangle_{\mathcal{H}} \equiv \int \overline{\vartheta A}\, B d\mu(\varphi) \ . \tag{1.6}$$

The Hilbert space $\mathcal{H}$ is the completion of these positive time $A(\varphi)$ in the scalar product (1.6), and it is these positive time functionals $A(\varphi)$ which are allowed in (1.4). Time translation is defined by

$$A_t = \sum c_j \exp(i\varphi(f^{(j)}))_t = \sum c_j \exp(i\varphi(f_t^{(j)})) \ , \tag{1.7}$$

so for $t \geq 0$, A_t is also a positive time functional. Thus the formula

$$\langle A, R(t)A\rangle_{\mathcal{H}} = \int \overline{\vartheta A}\, A_t d\mu \tag{1.8}$$

defines an operator $R(t)$ on $\mathcal{H}$. As a consequence of (1) - (2), it follows that

$$0 \leq R(t) = R(t)^* \leq I, \quad R(t+s) = R(t)R(s) \ , \tag{1.9}$$

see for example [6]. For $t = 1$, $R(1) = K$ is the self adjoint transfer matrix, and $R(t) = K^t$. In case that time is continuous, then (3) ensures that $R(t) = e^{-tH}$, where H is a positive, self adjoint Hamiltonian, $0 \leq H = H^*$. Furthermore, since the functional $A(\varphi) = 1$ satisfies $A_t = A = 1$, it follows that $R(t)1 = 1$ and 1 is a ground state for H, $H1 = 0$. The vector 1 in $\mathcal{H}$ is generally denoted by Ω, so $H\Omega = 0$.

1.3 EXAMPLES

A. Gaussian Examples

Consider the case of full Euclidean symmetry. The simplest example is the Gaussian measure $d\mu_0$ corresponding to the free field of mass $\sigma^{\frac{1}{2}}$

$$S_0\{f\} = \int e^{i\varphi(f)} d\mu_0(\varphi) = e^{-\frac{1}{2}\langle f, (-\Delta+\sigma)^{-1} f\rangle} . \tag{1.10}$$

In other words, $d\mu_0(\varphi)$ is the unique Gaussian measure with mean zero and covariance $(-\Delta + \sigma)^{-1}$. The only question to verify is whether (1.10) satisfies reflection positivity. This fact follows from

$$(1.11) \qquad \langle \vartheta f, (-\Delta + \sigma)^{-1} f \rangle = \left(\int_0^\infty \| e^{-t\mu} (2\mu)^{-\frac{1}{2}} f(\cdot, t) \|_{L_2} dt \right)^2$$

$$\geq 0 ,$$

where f vanishes unless $t \geq 0$ and where $\mu^2 = -(\vec{\nabla})^2 + \sigma$, where $\vec{\nabla}$ is the gradient in the $\vec{x}$ directions.

We can approximate R^d by a lattice Z^d, in which case we obtain

$$(1.12) \qquad d\mu_{0,\delta} = \lim_{\Lambda \nearrow Z^d_\delta} \frac{1}{N(\delta, \Lambda)} e^{-\frac{1}{2} \sum_{(n,n)} \delta^{d-2} (\varphi(i) - \varphi(i'))^2 - \frac{1}{2} \sigma \sum_i \delta^d \varphi(i)^2} \prod_i d\varphi(i).$$

In (1.12), Λ denotes a finite subset of the lattice Z^d_δ with lattice spacing δ. The sum and product over i extend over lattice sites in Λ, while the sum $\sum_{(n,n)}$ extends over nearest neighbor lattice sites in Λ. The measure $d\varphi(i)$ is Lebesgue measure, and $N(\delta, \Lambda)$ is chosen so $\int d\mu_{0,\delta} = 1$. The limit $\Lambda \nearrow Z^d$ exists in the sense of convergence of $S_{0,\delta,\Lambda}\{f\}$.

We can rewrite (1.12) as

$$(1.13) \qquad d\mu_{0,\delta} = \lim_{\Lambda \nearrow Z^d_\delta} e^{\beta \sum_{(n,n)} \varphi(i)\varphi(i')} \prod_i d\nu_0(\varphi(i)) ,$$

where $\beta = \delta^{d-2}$ and $d\nu_0(\varphi)$ is a Gaussian. Thus $d\mu_{0,\delta}$ is a ferromagnetic, nearest neighbor spin system with inverse temperature β and a single spin distribution function $d\nu_0(\varphi)$. The ferromagnetic coupling arises from the off-diagonal part of the gradient term in (1.12).

The limit $\delta \to 0$ for (1.12) - (1.13) exists and is (1.10). (This limit, however, must be taken in the form $S_{0,\delta}\{f\} \to S\{f\}$, since the measures $d\mu_{0,\delta}$ do not converge in the usual sense of convergence of measures on a finite dimensional space.) Thus formally

(1.14) $$d\mu_0 = N^{-1}\int e^{-\frac{1}{2}\int[(\nabla\varphi(x))^2+\sigma\varphi(x)^2]dx} \prod_{x\in R^d} d\varphi(x) ,$$

but the mathematically meaningful form of (1.14) is (1.10).

B. Non-Gaussian Examples

In the case of invariance of $d\mu$ under continuous time translations, we obtain non-Gaussian examples from non-relativistic quantum mechanics ($d = 1$) and nonlinear quantum fields ($d = 2, 3$). In the case of nonrelativistic quantum mechanics with one degree of freedom, the measure

(1.15) $$d\mu(q) = \frac{1}{N} e^{-\int_{-\infty}^{\infty} V(q(t))dt} d\mu_0(q)$$

formally satisfies (1-3). The quantum mechanics Hamiltonian is just

(1.16) $$H = -\frac{1}{2}\frac{d^2}{dq^2} + \frac{1}{2}q^2 + V(q) - E$$

$$= H_0 + V - E ,$$

i.e., the perturbation of the harmonic oscillator Hamiltonian H_0 by the potential V. Here E is a constant chosen so $H \geq 0$, $H\Omega = 0$. The formula (1.15) is exactly the Feynman-Kac functional integral representation of the ground state of H. In fact integration of $A(q(0))$ over $d\mu(q)$ is expectation of A in the state Ω:

(1.17) $$\int A(q_0)d\mu(q) = \lim_{T\to\infty} \frac{1}{N_T}\int e^{-\int_{-T}^{T} V(q(t))dt} A(q(0))d\mu_0(q)$$

$$= \lim_{T\to\infty} \frac{\langle e^{-TH}\Omega_0, Ae^{-TH}\Omega_0\rangle}{\|e^{-TH}\Omega_0\|^2}$$

$$= \langle \Omega, A\Omega\rangle .$$

Here Ω_0 is the ground state of H_0 and Ω is the ground state of H.

In quantum field theory, the known examples for $d = 2, 3$ arise from interaction energy densities $P(\varphi(x)) = V(x)$ for polynomials P. For instance the $\lambda\varphi_2^4$ model has the measure

$$(1.18) \qquad d\mu = \lim_{\Lambda \nearrow R^2} \frac{1}{N_\Lambda} e^{-\int_\Lambda V(x)dx} d\mu_0$$

$$= \lim_{\Lambda \nearrow R^2} \frac{1}{N_\Lambda} e^{-\int_\Lambda :\varphi^4(x):dx} d\mu_0 .$$

The Wick ordering of V is necessary since for $d > 1$, the measure $d\mu_0$ is concentrated on certain distributions, rather than on continuous functions as in the case of $d = 1$ (e.g. Wiener measure). The main work in this approach to proving existence of nonlinear quantum fields lies in establishing the existence of measures such as (1.18).

On a lattice, the local interaction V(x) in (1.18) does not effect the nearest neighbor term of the measure (1.12)-(1.13). Rather, V(x) contributes to the distribution of a single spin $d\nu(\varphi)$. Thus a lattice $P(\varphi)_d$ model has a measure

$$(1.19) \qquad d\mu_\delta = \lim_{\Lambda \nearrow Z_\delta^d} e^{\delta^{d-2} \sum_{(n,n)} \varphi(i)\varphi(i')} \prod_i d\nu(\varphi(i)) ,$$

with

$$d\nu(\varphi) = \frac{1}{N} e^{-\delta^d V(\varphi)} d\nu_0(\varphi) .$$

Here N is an appropriate normalization constant. For example, the lattice $\lambda\varphi_d^4$ model has

$$(1.20) \qquad d\nu(\varphi) = \frac{1}{N} e^{-\delta^d \lambda (\varphi^2 - c)^2} d\nu_0(\varphi) ,$$

where

$$c = 3\int \varphi^2 d\mu_0(\varphi) = \begin{cases} O(\ln \delta^{-1}) & d = 2 \\ O(\delta^{-(d-2)}) & d > 2 \end{cases}$$

is the Wick ordering constant.

1.4 APPLICATIONS OF THE FUNCTIONAL INTEGRAL REPRESENTATION

The usefulness of formula (1.4) is that it provides a computational framework to answer questions concerning the spectrum of H. For instance consider the question: Is the ground state Ω unique? In other words, we ask whether $e^{-tH}\theta - P_\Omega\theta$ converges to zero as $t \to \infty$. Here P_Ω is the projection onto Ω,

$$(1.21) \qquad P_\Omega\theta = |\Omega\rangle\langle\Omega|\theta\rangle = \int \theta(\varphi)d\mu(\varphi) .$$

Equivalently, is there a dense set of positive time functionals $\theta(\varphi)$ such that

$$(1.22) \quad F_\theta(t) \equiv \langle\theta, e^{-tH}(1 - P_\Omega)\theta\rangle = \int \overline{\vartheta\theta}\, \theta_t d\mu - \left|\int \theta d\mu\right|^2$$

converges to zero as $t \to +\infty$? If so, then Ω is the unique ground state of H. On the other hand if there is a positive time functional θ for which $F_\theta(t) \not\to 0$, then Ω is degenerate. In terms of the measure $d\mu(\varphi)$, the uniqueness of the vacuum Ω is equivalent to ergodicity of $d\mu(\varphi)$ under the group of time translations.

If, more generally, there are exactly r normalized, orthogonal vacuum states Ω_i, $i = 1, 2, \cdots, r$, then the function

$$F(t) = \int \overline{\vartheta\theta}\, \theta_t d\mu - \sum_{i=1}^{r} \left|\int \Omega_i \theta d\mu\right|^2$$

would converge to zero as $t \to +\infty$, rather than (1.22).

We see later that in the case of $\lambda\varphi_2^4$ or $\lambda\varphi_3^4$ quantum field models, the vacuum Ω is unique for $\lambda \ll 1$. However, Ω is degenerate for $\lambda \gg 1$. The construction of $d\mu(\varphi)$ outlined above, yields for $\lambda \gg 1$ an even mixture of vacuum states and $\langle\varphi\rangle = 0$. By introducing boundary conditions to select a particular vacuum, we may obtain solutions breaking the $\varphi \to -\varphi$ symmetry, $\langle\varphi\rangle \neq 0$, and with a unique vacuum. See the lectures of Fröhlich and Spencer for further discussion of phase transitions, and in particular a discussion of continuous symmetry breaking.

A second question concerning H is the existence of a mass gap, i.e. a gap in the spectrum corresponding to massive particles. The occurrence of a gap (0, m) in the spectrum is equivalent to

$$|F_\theta(t)| \le O(1)e^{-mt}, \tag{1.23}$$

where the constant O(1) depends on θ. Thus again the spectral properties of H are reduced to asymptotic decay rates of certain functional integrals. The proof of such decay rates in models has been established by expansion methods or by using correlation inequalities as described below.

In a theory which is even (e.g. a φ^4 model in which the symmetry $\varphi \to -\varphi$ is not broken) we can decompose the Hilbert space $\mathcal{H} = \mathcal{H}_e + \mathcal{H}_o$ into subspaces even or odd under the transformation $\varphi \to -\varphi$. The vacuum lies in $\mathcal{H}_e$, while the one particle states lie in $\mathcal{H}_o$. Let m denote the bottom of the spectrum on $\mathcal{H}_o$. On $\mathcal{H}_e$, we thus expect a mass gap of magnitude m', where $m < m' \le 2m$. (Since two particle scattering states occur in $\mathcal{H}_e$, the Hamiltonian will always have spectrum throughout the interval $[2m, \infty)$.) The statement $m' = 2m$ is the statement that two particle bound states do not occur in $\mathcal{H}_e$. This is equivalent to

$$|F_\theta(t)| \le O(1)e^{-2mt} \tag{1.24}$$

as θ ranges over a dense set of $\mathcal{H}_e$. We discuss this further in the next section.

Finally, in order to analyze the bound states or scattering of several particles, it is useful to study kernels (e.g. exact Bethe-Salpeter kernels) which characterize the Hamiltonian for n-body processes. Such kernels have a functional integral representation, and a detailed study has been made by Spencer and Zirilli [7, 8] in the case n = 2. (See also [2, 3, 9]

1.5 ISING, GAUSSIAN AND SCALING LIMITS

We briefly mention the qualitative structure of the φ_d^4 lattice quantum field model of §1.3, in its dependence on the parameters δ, λ, σ. In particular, we discuss the measures $d\mu_\delta$ defined in (1.19)-(1.20); an analogous discussion could be given for $P(\varphi)$ models. See [4, 5, 10].

To begin with, consider the (λ, δ) parameter space with fixed σ. For δ fixed, we study $\lambda \to 0$ and $\lambda \to \infty$, the minimum and maximum coupling. It is clear that the $\lambda \to 0$ limit of (1.19)-(1.20) is Gaussian, in fact $d\mu_\delta \to d\mu_{0,\delta}$. (In every case we define convergence as convergence

$$(1.25) \quad S_\delta\{f\} = \int e^{i\varphi(f)} d\mu_\delta(\varphi) \to S_{0,\delta}\{f\} = \int e^{i\varphi(f)} d\mu_{0,\delta}(\varphi)$$

of generating functionals.) On the other hand, for $\lambda \to \infty$, with δ fixed, the measure $d\nu(\varphi)$ becomes concentrated at the points where $|\varphi| = c^{\frac{1}{2}}$, i.e. $\varphi = \pm c^{\frac{1}{2}}$. Since the integral of $d\mu_\delta$ is normalized to one, in this limit

$$(1.26) \qquad d\mu_\delta = \lim_{\Lambda \nearrow Z_\delta^d} e^{\beta \sum\limits_{(n,n)} \varphi(i)\varphi(i')} \prod_i d\nu(\varphi_i)$$

where $d\nu(\varphi) = \frac{1}{N}(\delta(\varphi - c^{\frac{1}{2}}) + \delta(\varphi + c^{\frac{1}{2}}))$. In other words, $d\mu_\delta$ is an Ising model with lattice spacing δ and spin φ normalized to take the values $\pm c^{\frac{1}{2}}$. The mathematical existence of this Ising limit was established [11]. Furthermore, for $d = 2$ (or with $d = 3$ and the proper choice $\sigma = \sigma(\delta) = O(\ell n\, \delta^{-1})$ to ensure massrrenormalization, the $\delta \to 0$ limit can be taken with λ fixed. This continuum limit yields the Euclidean φ^4 model.

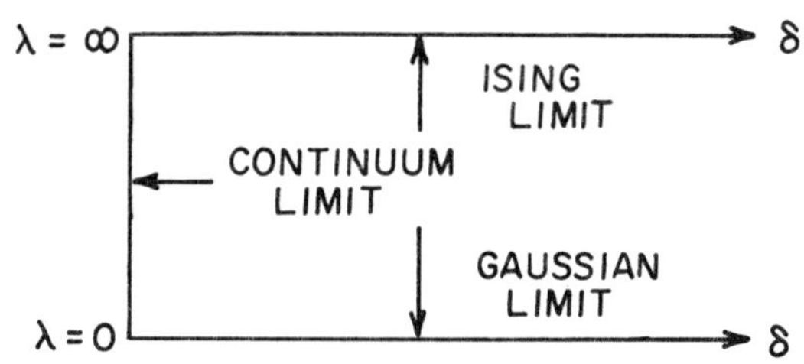

We next modify this picture slightly by fixing the mass gap m. On the lattice, m is defined as the gap in the spectrum of $-\ell n K$ (K is the transfer matrix); if $\delta = 0$, m is defined as the gap in the spectrum of H. Since $m = m(\lambda, \delta, \sigma)$, we achieve this by choosing σ in such a way that we remain in the single phase region and such that m = const. The required continuity of m follows by [12, 13]. We now plot the projection of such a $\sigma = \sigma(\lambda, \delta)$ surface in the (λ, δ) plane.

One can now ask whether the $\delta \to 0$ limit of Ising models exist, and whether the $\lambda \to \infty$ limit of continuum models exists. The first is a scaling limit of the Ising model, the second is a scaling limit of the $\lambda\varphi^4$ model. In the continuum theory ($d = 2, 3$), increasing λ with σ fixed would result in a phase transition and $m \to 0$. Thus in the scaling limit, with m fixed, it follows that $\sigma(\lambda, \delta = 0) \to \infty$ as $\lambda \to \infty$, i.e. there is an infinite change of scale. Also the scaling limit is formally an infinite scaling of a $m = 0$ (critical) theory. We conjecture that both the $\lambda \to \infty$ and the $\delta \to 0$ scaling limits exist, and that they agree. (See §1.5.)

In studying this limit, it is also useful to consider curves in the (λ, δ) plane with constant unrenormalized, dimensionless charge $g_0 \doteq \lambda\delta^{4-d}$. For $d < 4$ (the superrenormalizable case) these curves lead to the scaling limit ($\lambda = \infty$, $\delta = 0$) discussed above, i.e. strong coupling. For $d > 4$ (the nonrenormalizable case) these curves lead to $\lambda = \delta = 0$, i.e. weak coupling.

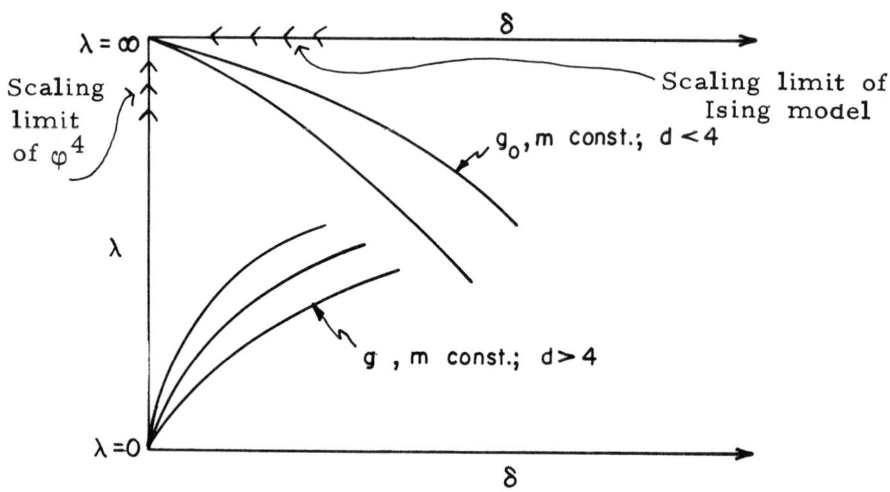

This picture leads us to conjecture that for $d > 4$, the $\delta \to 0$ limit for g_0 bounded is Gaussian (trivial), but a nontrivial theory could result with $g_0 \to \infty$ as $\delta \to 0$ (charge renormalization). See §2.5 for further discussion, and also [20, 10].

1.6 MAIN RESULTS

We sketch the main results for the φ^4 model, φ^4 lattice fields and Ising models, some of which we discuss in detail in the next two chapters. There are two main methods to derive these results: correlation inequalities (discussed in Chapter 2) and expansion techniques (discussed in Chapter 3). The correlation inequalities for the φ^4 model express in part the repulsive character of the forces in these models (in the single phase region). We obtain mathematical proofs for portions of the critical behavior in these models, as well as an initial analysis of the elementary particle and bound state problems.

A. The critical point has a conventional φ^4 structure. For $\sigma \gg 0$, there is a unique phase, independent of boundary conditions, and for $\sigma \ll 0$ and $d \geq 2$, there are at least two pure phases, depending on the boundary conditions [14, 15, 16] The critical value $\sigma_c = \sigma_c(\lambda, \delta)$ is defined as the largest (and presumably the only) value of σ for which $m = m(\sigma) \to 0$ as $\sigma \searrow \sigma_c$. This critical σ_c exists [13] and $m(\sigma)$ is monotonic increasing for $\sigma \geq \sigma_c$ [17]. For $d \geq 3$ (and presumably for $d = 2$ also) $m(\sigma_c) = 0$, and for $\sigma = \sigma_c$ there is a unique phase and zero magnetization [13]. For $\sigma \geq \sigma_c$, the physical charge g (defined as the amputated connected four point function evaluated at zero momentum) is finite and bounded, uniformly as $\sigma \searrow \sigma_c$, and $\lambda \to \infty$. (See §2.3.) Furthermore the n-point Schwinger functions $S^{(n)}$ satisfy a Gaussian upper bound

$$0 \leq S^{(n)}(x_1 \cdots x_n) \leq \sum_{\text{pairing}} \prod_{\text{pairs}} S^{(2)}(x_i x_{i'}) ,$$

see [18, 19]. Closely related are critical exponent bounds of the form

$$\text{Gaussian exponent} \leq \varphi\text{-exponent} \leq \varphi^2\text{-exponent} .$$

For example with η the anomalous dimension of the field φ and η_E the anomalous dimension of the field $:\varphi^2(x):$,

$$0 \leq 2\eta \leq \eta_E .$$

See [20, 21, 22] for other recent exponent inequalities.

B. Particles do not form even bound states, for $\sigma \geq \sigma_c$. In the single phase region, particles exist for a.e. $m > 0$ [21]. (Presumably they exist for all $m > 0$, but at least for $d = 2, 3$, the particles should not exist for $m = 0$, i.e. $\sigma = \sigma_c$.) Here particles are poles in the two point function at Minkowsky momenta, or δ-functions in a Källen-Lehmann representation for the two point function or (at least in the Euclidean invariant case of a continuum field) an Ornstein-Zernike decay rate

$$\langle\varphi(x)\varphi(y)\rangle \sim Zr^{(1-d)/2}e^{-mr}$$

as $r = |x - y| \to \infty$. These particles do not form even bound states (with energies below the two particle continuum) [23, 24] and there are indications that they do not form odd bound states (with energies below the three particle continuum) [23, 24].

C. A heuristic interchange of the $\delta \to 0$ (continuum) limit and the $\lambda \to \infty$ (Ising) limit "shows" the identical critical point behavior [25, 19]. Combining this idea with the known spectral properties of the $d = 2$ Ising suggests that for φ_2^4 continuum or lattice fields, with $\sigma < \sigma_c$, the elementary particle is actually a two soliton bound state. Furthermore in this picture the binding energy goes to zero (relative to the soliton mass) as $\sigma \nearrow \sigma_c$ and the field strength renormalization $Z \to 0$, and it probably vanishes faster than the strength of the two soliton continuum, thus suggesting that because of the solitons, the intermediate renormalization is correct, but the mass shell renormalization is incorrect, for $\sigma \nearrow \sigma_c$, $d = 2$, unless the mass shell renormalization includes all sectors, with the soliton as elementary particle. Introducing a small external field $-\mu\varphi(x)$ in the action, we have argued [26] that for $d = 2$ and $\sigma < \sigma_c$, but $\sigma \approx \sigma_c$, the limit $\mu \to 0$ introduces many bound states which coalesce to form a two soliton continuum at $\mu = 0$. For $\sigma \ll \sigma_c$ the same reasoning suggests many resonances coalescing to form continuous spectrum.

D. Within its region of convergence, the cluster expansion allows a nearly complete analysis of the field theory. To begin with, one can prove convergence of the infinite volume limit and uniqueness of the vacuum. (In a two phase region, suitable boundary conditions are required to select a pure phase.) The convergence is valid for λ complex in a sector about $\lambda = 0$ and it follows that the correlation functions are also analytic in λ and other parameters, for $\lambda \neq 0$. For the case of a $\lambda\varphi_2^4$ interaction in the region $\lambda << 1$, the perturbation series about $\lambda = 0$ is Borel summable to the exact solution [27]. For general $P(\varphi)_2$ interactions in the region (5.1), the perturbation series about $\lambda = 0$ for the Euclidean and Minkowsky correlation functions and for the S-matrix is also asymptotic [28, 29, 30]. The particles (whose scattering is described by the S-matrix) are also constructed from the cluster expansion [15]. Criteria (in terms of P) for existence or nonexistence of weak coupling $P(\varphi)_2$ bound states are given in [31] following earlier work of [8]. Here the cluster expansion permits the study of the exact Bethe-Salpeter equation, and shows that the low order terms (ladder approximation + $\cdots$) give the dominant effects. Up to some energy level, this analysis of the Bethe-Salpeter equation also shows asymptotic completeness [8]. Unfortunately, the energies allowed by present techniques apparently do not reach up to the two soliton threshold, in the two phase region.

2.1 CORRELATION INEQUALITIES

In this section we derive some basic correlation inequalities and state some others. In the following section we derive some consequences of these inequalities, and finally we consider the conjectured inequality $\Gamma^{(6)} \leq 0$.

For positive integers $a_1, \cdots, a_n$ define

(2.1) $$\varphi_A = \varphi(x_1)^{a_1} \cdots\cdot \varphi(x_n)^{a_n} .$$

Theorem 2.1: For a lattice $(\lambda\varphi^4 - \mu\varphi)_d$ quantum field with $\lambda, \mu \geq 0$,

(GKS 1) $$0 \leq \langle\varphi_A\rangle$$

(GKS 2) $$0 \leq \langle\varphi_A\varphi_B\rangle - \langle\varphi_A\rangle\langle\varphi_B\rangle .$$

These are the first and second Griffiths (Griffiths, Kelly, Sherman) inequalities. Since they say that certain quantities are positive, they are preserved under limits, e.g. $\delta \to 0$ or

$\lambda \to \infty$, whenever such limits exist. The solution of the $P(\varphi)_2$ ultraviolet problem [1] was extended by Guerra, Rosen and Simon to lattice cutoffs [17] in order to prove GKS and related inequalities for these models. See [32] for φ_3^4.

Proof of GKS 1: We write on a finite lattice

$$(2.4) \quad \langle \varphi_A \rangle = \int \varphi_A d\mu_\delta = \int \varphi_A e^{\beta \Sigma \varphi(i)\varphi(i') + \mu \Sigma \varphi(i)} \prod_i d\nu(\varphi(i)) \, .$$

Note that $d\nu(\varphi) = d\nu(-\varphi)$, so

$$(2.5) \qquad \int \varphi(i)^{a_i} d\nu(\varphi(i)) = \begin{cases} 0 \text{ if } a_i \text{ odd} \\ \text{positive if } a_i \text{ even} \end{cases}$$

The basic idea of the proof is to expand the exponentials in (2.4) in power series and then factor the resulting integrals over lattice sites. Using (2.5), we obtain a sum of products of positive terms, and hence (2.4) is positive. We then take the limit as the finite lattice increases to Z^d.

Proof of GKS 2: The basic idea is to use the technique of duplicate variables. Let φ, ψ be independent, lattice fields. Define an expectation of functionals $A = A(\varphi, \psi)$ by

$$(2.6) \qquad \langle A \rangle = \int A(\varphi, \psi) d\mu(\varphi)_\delta d\mu(\psi)_\delta \, .$$

Here

$$d\mu(\varphi)_\delta = e^{\beta \Sigma \varphi(i)\varphi(i')} \prod_i d\nu(\varphi(i))$$

$$d\mu(\psi)_\delta = e^{\beta \Sigma \psi(i)\psi(i')} \prod_i d\nu(\psi(i)) \, ,$$

and for simplicity we let $\mu = 0$. Define

$$(2.7) \qquad t = \varphi + \psi, \quad q = \varphi - \psi$$

as the even and odd combinations of φ, ψ, under interchange of φ and ψ. Note

(2.8) $$\langle t\rangle = \langle\varphi\rangle + \langle\psi\rangle = 2\langle\varphi\rangle\,,$$

$$\langle q\rangle = 0\,,$$

$$d\mu(\varphi)_\delta d\mu(\psi)_\delta = e^{\frac{1}{2}\beta\Sigma(t(i)t(i')+q(i)q(i'))}\prod_i d\nu\left(\frac{t(i)+q(i)}{2}\right)d\nu\left(\frac{t(i)-q(i)}{2}\right)\,.$$

First we remark that $d\nu(t+q)d\nu(t-q)$ is even under the transformation $t \to -t$ and also under $q \to -q$. Thus

(2.9) $$\int q^\alpha t^\beta d\nu(t+q)d\nu(t-q) = \begin{cases} 0 \text{ if } \alpha \text{ or } \beta \text{ odd} \\ \text{positive if } \alpha, \beta \text{ both even}\,. \end{cases}$$

Following the proof of GKS 1, and using (2.9), we find that for all A, B,

(2.10) $$0 \le \langle q_A t_B\rangle\,.$$

To complete the proof of GKS 2, we write

(2.11) $$\langle\varphi_A\varphi_B\rangle - \langle\varphi_A\rangle\langle\varphi_B\rangle = \langle\varphi_A(\varphi_B - \psi_B)\rangle$$

$$= 2^{-|A|-|B|}\langle(t+q)_A[(t+q)_B - (t-q)_B]\rangle.$$

But $(t+q)_B(t-q)_B$ is a polynomial in t, q with positive coefficients. Hence (2.10) shows that (2.11) is positive.

The proof above follows the presentation of Sylvester [33], which we recommend for proofs of other correlation inequalities. We now state three inequalities:

<u>Theorem 2.2</u>: For a lattice $(\lambda\varphi^4 - \mu\varphi)_d$ quantum field theories with $\lambda, \mu \ge 0$,

(2.12) $$\langle t_A t_B\rangle - \langle t_A\rangle\langle t_B\rangle \ge 0\,,$$

(2.13) $$\langle q_A q_B\rangle - \langle q_A\rangle\langle q_B\rangle \ge 0\,,$$

(2.14) $$\langle q_A t_B \rangle - \langle q_A \rangle \langle t_B \rangle \le 0 .$$

The inequalities (2.14) were first proved for the Ising model by Lebowitz [34], and have a number of interesting consequences. We remark that two special cases of (2.14) are

(2.15) $$\langle \varphi(x_1)\varphi(x_2)\varphi(x_3)\rangle_T \le 0 , \quad \mu \ge 0 ,$$

(2.16) $$\langle \varphi(x_1)\varphi(x_2)\varphi(x_3)\varphi(x_4)\rangle_T \le 0 , \quad \mu = 0 .$$

Here $\langle \quad \rangle_T$ denotes the truncated (connected) expectation values, defined by

(2.17) $$\langle \varphi(f)^n \rangle_T = \frac{d^n}{d\alpha^n} \ell n \langle e^{\alpha\varphi(f)} \rangle \Big|_{\alpha=0} ,$$

and extended to $\langle \varphi(f_1) \cdots \varphi(f_n) \rangle_T$ by multilinearity. We obtain (2.15), the Griffiths-Hearst-Sherman inequality, by expanding $\langle t_1 q_2 q_3 \rangle - \langle t_1 \rangle \langle q_2 q_3 \rangle \le 0$. The inequality (2.16) follows from evaluating $\langle t_1 t_2 q_3 q_4 \rangle - \langle t_1 t_2 \rangle \langle q_3 q_4 \rangle \le 0$ in case $\mu = 0$.

2.2 ABSENCE OF EVEN BOUND STATES

In a single phase, even φ^4 model, i.e. for $\sigma > \sigma_c$, we now show that the Hamiltonian, restricted to $\mathcal{H}_{even}$, has no spectrum in the interval $(0, 2m)$, i.e. two particle bound states do not exist. We remark that $\mathcal{H}_{even}$ is spanned by vectors Ω, $\varphi(f_1) \cdots \varphi(f_n)\Omega$, $n = 2, 4, \cdots$, where suppt f_j is contained in $t > 0$.

Theorem 2.3: Consider a φ^4 field or Ising model with zero external field and $\sigma > \sigma_c$, and let A and B have an even number of elements. Then

$$\langle \varphi_A \varphi_B \rangle - \langle \varphi_A \rangle \langle \varphi_B \rangle \le \sum_{\substack{A_1 \subset A,\ A_1 \text{ odd} \\ B_1 \subset B,\ B_1 \text{ odd}}} \langle \varphi_{A_1} \varphi_{B_1} \rangle \langle \varphi_{A-A_1} \varphi_{B-B_1} \rangle$$

Corollary 2.4: Under the hypothesis of Theorem 3.3, there are no even bound states with energy below the two particle threshold.

Proof: Let Ω be the vacuum state, unique since it is assumed that $\sigma > \sigma_c$. We write $x = x_1, \cdots, x_d$ as

$$x = (t, \vec{x})$$

with $\vec{x} \in R^{d-1}$. In particular if

$$A + s = \{(t + s, \vec{x}) : t, \vec{x} \in A\},$$

then the equation (1.4), namely

$$\langle \varphi_A e^{-sH} \varphi_B \rangle = \langle \vartheta \varphi_A \varphi_{B+s} \rangle,$$

is valid when the times in A precede the times in B. In particular we choose A to have only negative times $t \le 0$, and B, chosen as

$$B = \{(-t, \vec{x}) : (t, \vec{x}) \in A\}$$

then has only positive times. With this choice of A and B, and with P_Ω the projection onto the vacuum state, we recognize

$$\langle \varphi_{A+s} \varphi_{B+s} \rangle - \langle \varphi_{A+s} \rangle \langle \varphi_{B+s} \rangle = \| e^{-sH} (I - P_\Omega) \varphi_A \Omega \|^2$$

so that the Theorem 2.3 gives a bound on the decay rates which occur in

$$e^{-sH}(I - P_\Omega).$$

For A_1 odd, $\varphi_{A_1}\Omega$ is perpendicular to the vacuum $((\Omega, \varphi_{A_1} \Omega) = \langle \varphi_{A_1} \rangle = 0)$, and so $\langle \varphi_{A_1 - s} \varphi_{B_1 + s} \rangle$ has as its slowest exponential decay rate, m, by definition the mass of the theory. Thus by definition of m,

$$\langle \varphi_{A_1 - s} \varphi_{B_1 + s} \rangle \le C_{A_1, B_1} e^{-ms}$$

for some constant C_{A_1, B_1} depending on A_1 and B_1. The same bound holds for $\langle \varphi_{(A - A_1) - s} \varphi_{(B - B_1) + s} \rangle$, and so by Theorem 2.3,

$$\| e^{-sH}(I - P_\Omega)\varphi_A \Omega \|^2 \le \text{const.}\, e^{-2ms} .$$

Thus there are no even states, except Ω, with energy below 2m, hence in particular no even bound states in this energy range.

Proof of Theorem 2.3:

$$\langle t_A q_B \rangle = \sum_{\substack{A=A_1 \cup A_2 \\ B=B_1 \cup B_2}} (-1)^{|B_2|} \langle \varphi_{A_1} \psi_{A_2} \varphi_{B_1} \psi_{B_2} \rangle$$

$$= \sum (-1)^{|B_2|} \langle \varphi_{A_1} \varphi_{B_1} \rangle \langle \varphi_{A_2} \varphi_{B_2} \rangle$$

$$\le \langle t_A \rangle \langle t_B \rangle = \sum (-1)^{|B_2|} \langle \varphi_{A_1} \rangle \langle \varphi_{A_1} \rangle \langle \varphi_{B_1} \rangle \langle \varphi_{B_2} \rangle .$$

We drop zero terms from the right hand side (A_2 or B_2 odd), and from the left hand side (only one of A_2, B_2 odd). For the terms with B_2 even and the partition nontrivial, we combine right and left sides and eliminate from the inequality, using Theorem 2.2. The terms remaining yield Theorem 2.3.

2.3 BOUND ON g

Define the dimensionless φ^4 coupling constant by

$$g = -m^{d-4} \chi^{-4} \int \langle \varphi(x_1) \cdots \varphi(x_4) \rangle_T \, dx_1 dx_2 dx_3$$

where

$$\chi = \int \langle \varphi(x_1) \varphi(x_2) \rangle dx_1 = \int_0^\infty \frac{d\rho(a)}{a} .$$

By GKS 1, $\chi \ge 0$. For a massive, single phase, even φ^4 interaction $g \ge 0$ by (2.16). We now assume in addition that the proper field strength renormalization has been performed;

in the case of an isolated particle of mass m, this means $d\rho(a) = \delta(a - m^2)da + d\sigma(a)$, where inf suppt $d\sigma > m^2$. We then prove an upper bound on g.

<u>Theorem 2.5</u> [35]: Under the above assumptions,

$$0 \leq g \leq \text{const.} ,$$

where the dimensionless constant is independent of all parameters (e.g. λ, σ).

We outline the proof. For details, see the original paper. We use the basic inequality GKS 2 to derive (writing 1 for $\varphi(x_1)$, etc.)

$$(2.18)\; 0 \leq \langle 1234 \rangle - \langle 12 \rangle \langle 34 \rangle = \langle 1234 \rangle_T + \langle 13 \rangle \langle 24 \rangle + \langle 14 \rangle \langle 23 \rangle .$$

By (2.16), $\langle 1234 \rangle_T \leq 0$ and

$$0 \leq -\langle 1234 \rangle_T \leq \langle 13 \rangle \langle 24 \rangle + \langle 14 \rangle \langle 23 \rangle .$$

After symmetrization over the choices of variables,

$$(2.19)\; -\langle 1234 \rangle_T \leq (\langle 13 \rangle \langle 24 \rangle + \langle 14 \rangle \langle 23 \rangle)^{\frac{1}{3}} (\langle 12 \rangle \langle 34 \rangle + \langle 13 \rangle \langle 24 \rangle)^{\frac{1}{3}}$$

$$\times (\langle 14 \rangle \langle 23 \rangle + \langle 12 \rangle \langle 34 \rangle)^{\frac{1}{3}} .$$

From elementary properties of the Green's function for the Poisson operator (i.e. $\ker(-\Delta + a)^{-1}(x, y)$) we find

$$\langle xy \rangle = \int_0^\infty \ker(-\Delta + a)^{-1}(x, y) d\rho(a)$$

$$\leq \text{const.}\, \chi \, |x - y|^{-d} \exp(-m|x - y|/2) .$$

Inserting this in our bound (2.19) for $-\langle 1234 \rangle_T$ gives

$$g \leq \text{const.}\, m^{-4} \chi^2 .$$

Since

$$\chi = \int_{m^2}^{\infty} \frac{d\rho(a)}{a} \geq m^{-2},$$

we obtain $g \leq$ const. as claimed.

Observe that the final bound does not depend on m, and hence also holds in the limit $m \to 0$. Hence the critical point (which for $d < 4$ should be an infrared stable fixed point of the renormalization group) occurs for finite g.

2.4 BOUND ON $dm^2/d\sigma$ AND PARTICLES

Here we consider a canonical, single phase φ^4 model (i.e. without field strength renormalization). We establish

$$\frac{dm^2(\sigma)}{d\sigma} \leq Z(\sigma) \tag{2.20}$$

from which our next result follows by approximation methods:

Theorem 2.6: (See [21].) For almost every value of m, particles exist, i.e. $Z \neq 0$.

Proof of (2.20): Consider $\Gamma(p) = -S(p)^{-1}$, where S(p) is the Fourier transform of $\langle\varphi(x)\varphi(0)\rangle$. Note

$$S(p) = \frac{Z}{p^2 + m^2} + \int \frac{d\sigma(a)}{p^2 + a},$$

and

$$Z^{-1} = -(d\Gamma/dp^2)_{p^2=-m^2}. \tag{2.21}$$

Since $\Gamma = 0$ on the one particle curve $p^2 = -m^2(\sigma)$, $\nabla\Gamma$ must be orthogonal to the vector $(dm^2/d\sigma, 1)$ in the $-p^2, \sigma$ space. Thus for $p^2 = -m^2$,

$$0 = -\frac{\partial\Gamma}{\partial p^2}\frac{dm^2}{d\sigma} + \frac{\partial\Gamma}{\partial\sigma} = Z^{-1}\frac{dm^2}{d\sigma} + \frac{\partial\Gamma}{\partial\sigma}.$$

The desired inequality follows from

<u>Theorem 2.7</u>: Under the above assumptions,

(2.22) $$\left(\frac{\partial\Gamma}{\partial\sigma}\right)_{p^2=-m^2} \geq -1 .$$

<u>Proof</u>: Let $\chi(p) = \int \langle\varphi(x)\varphi(0)\rangle e^{-px}dx.$ Then

$$-\frac{d\chi(p)}{d\sigma} = \tfrac{1}{2}\int\int[\langle xozz\rangle - \langle xo\rangle\langle zz\rangle]dze^{-px}dx$$

by (2.16), (2.18),

$$\leq \int\int\langle xz\rangle\langle yz\rangle e^{-p(x-z)}e^{-pz}dxdz$$

$$= \chi(p)^2 .$$

Thus

$$0 \leq \frac{d\chi(p)^{-1}}{d\sigma} \leq 1 .$$

However $\chi(p) = -\Gamma(p)|_{p^2=-m^2}$, so (2.22) is proved.

2.5 THE CONJECTURE $\Gamma^{(6)} \leq 0$

The unamputated six point vertex function is defined by

(2.23) $$\Gamma^{(6)}(xxxyyy) = \langle xxxyyy\rangle_T + \int\langle xxxz\rangle_T\Gamma(zz')\langle z'yyy\rangle_T dzdz'$$

$$+ 9\int\langle xxyz\rangle_T\Gamma(zz')\langle z'xyy\rangle_T dzdz' .$$

The conjecture

(2.24) $$\Gamma^{(6)}(xxxyyy) \leq 0$$

has a number of interesting consequences: e.g., the absence of three particle bound states in the propagator, the existence of the scaling limit, and certain bounds on critical exponents, see [25, 5, 20].

There is some evidence for (2.24) in single phase, even φ^4 models. For example it is true in perturbation theory (i.e. for $\sigma >> 0$ or high temperature). It holds in the one dimensional Ising model [36] and numerical studies indicate that it holds for the anharmonic oscillator [37]. There is a heuristic argument that it holds near σ_c. However, some good new idea is needed to <u>prove</u> (2.24).

In this section we illustrate some uses of (2.24). For example, we have

<u>Theorem 2.8</u>: If (2.24) holds, then

$$0 \le \Gamma(x) \le e^{-3m|x|}, \quad |x| \to \infty . \tag{2.25}$$

<u>Remark</u>: The bound (2.25) excludes spectrum in $\Gamma(x)$ in the interval (0, 3m), and hence spectrum in $d\sigma(a)$ in the interval (m, 3m). Thus no three particle bound states occur in the propagator, i.e. in the states spanned by $\varphi(x)\Omega$.

<u>Outline of Proof</u>: We use the integration by parts formula [25]

$$\int \varphi(x)A(\varphi)d\mu(\varphi) = \langle \varphi(x)A \rangle \tag{2.26}$$

$$= \int dyS(x-y)[\langle \frac{\delta A}{\delta\varphi(y)} \rangle - \langle \mathcal{V}'(y)(I-P_1)A \rangle] .$$

Here $\mathcal{V} = \lambda :\varphi^4:$ is the interaction, and $P_1 A = \int \varphi(z)\Gamma(z-z')\langle \varphi(z')A \rangle dzdz'$. From (2.26), it follows that for $x \neq 0$,

$$\Gamma(x-y) = \langle \mathcal{V}'(x)(I-P_1)\mathcal{V}'(y) = \lambda^2 \langle \varphi^3(x)(I-P_1)\varphi^3(y) \rangle , \tag{2.27}$$

see [25]. Expanding (2.27), and using (2.23),

$$\lambda^{-2}\Gamma(x-y) = 6\langle xy \rangle^3 + 9\langle xxyy \rangle_T \langle xy \rangle \tag{2.28}$$

$$- 9\int \langle xxyz \rangle_T \Gamma(zz')\langle z'xyy \rangle_T dzdz' + \Gamma^{(6)}(xxxyyy).$$

The first term in (2.28) is $o(e^{-m|x-y|})^3$, for $|x - y| \to \infty$. The second term is negative. The third term, also has a three particle decay, which can be established using the absence of two particle bound states in $\langle xxyz\rangle_T$, see [25]. Thus (2.24) results in

$$\Gamma(x - y) \le e^{-3m|x-y|}, \quad |x - y| \to \infty .$$

The positivity of $\Gamma(x - y)$ follows from the fact that it is the Fourier transform of a Herglotz function. This completes the outline of the proof.

We finish this section with the statement of another consequence of (2.24), and an Ornstein-Zernicke upper bound

(2.29) $$\langle \varphi(x)\varphi(0)\rangle \le K \min(|x|, m^{-1})^{-(d-2+\eta)} e^{-m|x|} .$$

Theorem 2.8 [20]: Assume (2.24), (2.29) and $\lambda < \infty$. Then

$$\eta \le \begin{cases} .4 & d = 4 \\ .8 & d = 3 \\ 1.2 & d = 2 \end{cases} .$$

Also $\eta = 0$, $Z^{-1} < \infty$ for $d \ge 5$.

Corollary 2.9 [20]: Assume (2.24). If the $\delta \to 0$ limit of the $\lambda\varphi^4_d$ lattice field theory is Euclidean invariant for $g_0(\delta) = \lambda\delta^{4-d} \le$ const. (finite charge renormalization), then the limit is a free field for $d \ge 6$.

3. CLUSTER EXPANSIONS

3.1 THE REGION OF CONVERGENCE

The cluster expansion, in field theory as in statistical mechanics, provides almost complete information for parameter values away from critical, and it provides only limited information for parameter values near critical. In statistical mechanics, this expansion is a variant of the virial, high temperature and low temperature (Peierls' contour) expansions. These names distinguish various regions of the coupling constants, and expansion parameters. In field theory, convergence of the cluster expansion is known for the corresponding parameter values. In particular, for a two dimensional $P(\varphi)$ field theory, the cluster expansion is convergent in the following asymptotic regions [14, 15, 38, 26, 39].

$$P(\varphi) = \lambda :P_0(\varphi): + \sigma\varphi^2, \quad \lambda \to 0 \tag{3.1}$$

$$P(\varphi) = \lambda :P_0(\varphi): - \mu\varphi, \quad \mu \to \infty \tag{3.2}$$

$$P(\varphi) = \lambda :\varphi^4: + \sigma\varphi^2, \quad \sigma \to -\infty \text{ or } \lambda \to +\infty, \tag{3.3}$$

or more generally, whenever P, expanded about a suitable global minimum φ_c of $P(\varphi)$ has a dominant quadratic term. For the Yukawa_2 and φ_3^4 interactions, convergence of the expansion is known in the high temperature region ($\lambda \to 0$ as in (3.1)) [40-43].

3.2 THE ZEROTH ORDER EXPANSION

The expansion is adapted from the virial expansion of statistical mechanics. In the zeroth order, all couplings are removed. We divide Euclidean space time into cells (lattice cubes) and then remove the coupling between distinct cells. In the zeroth approximation, all correlations factor,

$$\langle \varphi(x_1)\cdots\varphi(x_n)\rangle_0 = \prod_{\Delta=\text{cell}} \langle \prod_{x_j\in\Delta} \varphi(x_j)\rangle_0 .$$

Consequently the long distance behavior is trivial and all states have infinite energy in the zeroth approximation.

To define the zeroth approximation, let $\partial\Delta$ be the boundary of the cell Δ.. Then formally

$$\mathcal{A}_0(\varphi) = \infty \sum_\Delta \int_{\partial\Delta} \varphi^2(x)dx + \mathcal{A} \tag{3.4}$$

is the action defining the zeroth approximation, if $\mathcal{A}$ is the action of the full theory. To rewrite this expression in mathematical language, we combine the $\partial\Delta$ term in $\mathcal{A}_0$ with the gradient term in $\mathcal{G}$,

$$(3.5) \qquad \infty \sum_{\Delta} \int_{\partial\Delta} \varphi^2 dx + \int \nabla\varphi^2(x)dx = \langle \varphi - \Delta_D \varphi \rangle ,$$

where $-\Delta_D$ is the Laplace operator with zero Dirichlet boundary conditions on all cube faces $\partial\Delta$.

3.3 THE PRIMITIVE EXPANSION

Graphical expansions in statistical mechanics are generated by the identity

$$(3.6) \qquad \prod_{i<j} e^{-V(r_i - r_j)} = \prod_{i<j} [1 - e^{-V(r_i - r_j)} - 1)]$$

$$= \sum_{\text{sets of pairs}} \prod_{\text{pairs in set}} (e^{-V(r_i - r_j)} - 1) .$$

Now a pair $\{r_i, r_j\}$ is represented by the line segment connecting r_i to r_j, so that a set of pairs is a graph. Thus (3.6) is a graph expansion. Before adapting (3.6) to field theory, we modify it slightly. Let $V^{(s)}$ be a one parameter family of potentials, $0 \le s \le 1$, with

$$V^{(0)} = 0 , \quad V^{(1)} = V .$$

Then

$$e^{-V} - 1 = \int_0^1 \frac{d}{ds} e^{-V^{(s)}} ds ,$$

and introducing a parameter s_{ij} for each pair $\{r_i, r_j\}$, we have

$$(3.7) \qquad \prod_{i<j} e^{-V(r_i - r_j)} = \sum_{\Gamma} \int \prod_{\{ij\}\in\Gamma} \frac{d}{ds_{ij}} e^{-V^{(s_{ij})}} ds_{ij}$$

where Γ is a graph constructed from pairs $\{i, j\}$. In (3.7), the s_{ij} for $\{i, j\}$ not in Γ are evaluated at $s = 0$. Writing

$$s^{\Gamma} = \{ s_{ij} : \{i, j\} \in \Gamma \}$$

$$s^{\Gamma^c} = \{ s_{ij} : \{i, j\} \notin \Gamma \} ,$$

this formula can be expressed in the compact notation

$$\prod_{i<j} e^{-V(r_i - r_j)} = \sum_{\Gamma} \int \partial s^{\Gamma} \prod_{\{i,j\} \in \Gamma} e^{-V(s_{ij})} ds^{\Gamma} \Big|_{s^{\Gamma^c}=0} . \tag{3.8}$$

To adapt this formula to field theory, we only need to choose a multiparameter family $s^{\Gamma} = \{ s^{\gamma} : \gamma \in \Gamma \}$ interpolating between the totally decoupled Laplacian Δ_D and the ordinary Laplacian Δ. This can easily be done, and we only mention that it is convenient to choose the s-dependence to be linear in the inverse operators $(-\Delta + m_0^2)^{-1}$. In field theory, (3.8) becomes

$$\langle F(\varphi) \rangle = \sum_{\Gamma} \int \partial s^{\Gamma} \langle F(\varphi) \rangle_s ds^{\Gamma} \Big|_{s^{\Gamma^c}=0} . \tag{3.9}$$

Here $F(\varphi)$ is some function of φ and $\langle \quad \rangle_s$ is the expectation defined by the interpolating "Laplacian" $\Delta(s)$. Observe that $s^{\gamma} = 0$ corresponds to Dirichlet data on γ, and no coupling, while $s^{\gamma} > 0$ allows coupling across Γ. Thus the hypercube faces $\gamma \in \Gamma$ transmit coupling from one cube to its neighbors, while the faces $\gamma \in \Gamma^c$ do not.

3.4 FACTORIZATION AND PARTIAL RESUMMATION

In graphical language, we are concerned here with the cancellation of disconnected vacuum contributions. In (3.9), the expectation $\langle \cdot \rangle$ is normalized so that $\langle 1 \rangle = 1$, while $\langle \cdot \rangle_s$ is normalized by the same denominator, so that

$$\langle 1 \rangle_s = \int e^{-\int P(\varphi) dx} d\varphi_{0,s} \Big/ \int e^{-\int P(\varphi) dx} d\varphi_0$$

$$= Z(s)/Z$$

where $d\varphi_{0,s}$ is the interpolating Gaussian measure defined by the parameter s. It follows that the expansion (3.9) defines disconnected graphs.

Suppose $F(\varphi)$ depends on $\varphi(x)$ only for x in some set $S \subset R^d$. In (3.9), we consider the term with Γ fixed. Because of the Dirichlet data on Γ^c, this term decouples across Γ^c. More precisely suppose Γ^c divides R^d into a certain number of connected components. Let Y be the closure of the union of the components which meet S; graphically, Y is the region connected to S by Γ. Then the expectation $\langle F(\varphi)\rangle_s \big|_{s_{\Gamma^c}=0}$ factors across ∂Y, allowing us to write it as a product of an "inside" expectation associated with Y times an "outside" expectation associated with $R^d \sim Y$.

Now let us fix Y in place of Γ and sum over all Γ which yield this fixed Y. This sum factorizes into independent sums over $\Gamma \cap Y$ and $\Gamma \sim Y$. The sum over $\Gamma \sim Y$ of the outside expectations has no constraints, and sums to a partition function $Z(\sim Y)$ for the region $\sim Y$, defined with Dirichlet data on ∂Y. (This resummation is just the reverse of the expansion (3.9), but for the region $\sim Y$ in place of R^d.) In a similar fashion we define $Z(Y)$, the partition function for the region Y, with Dirichlet data on ∂Y. Let

$$\langle F(\varphi)\rangle_Y \, Z(Y)/Z$$

denote the sum over $\Gamma \cap Y$ of the inside expectations (including the overall normalization denominator Z). Then (3.9) becomes

$$\langle F(\varphi)\rangle = \sum_Y \langle F(\varphi)\rangle_Y \frac{Z(Y)Z(\sim Y)}{Z} . \tag{3.10}$$

In this expression, all disconnected vacuum components have been resummed into the single factor $Z(Y)Z(\sim Y)/Z$. Somewhat lengthy estimates, including a Kirkwood-Salzburg equation, prove the convergence of (3.10) for the parameters and interactions indicated in §3.1.

3.5 TYPICAL APPLICATIONS

To prove exponential decay of correlations, i.e. a positive mass, we use the method of duplicate variables. Thus we write

$$|\langle F(\varphi)G(\varphi)\rangle - \langle F(\varphi)G(\varphi)\rangle| = 2^{-1}|\langle [F(\varphi) - F(\psi)][G(\varphi) - G(\psi)]\rangle|$$

$$\leq \sum_{Y} |\langle [F(\varphi) - F(\psi)][G(\varphi) - G(\psi)]\rangle_{Y}|$$

$$\times \frac{Z(Y)Z(\sim Y)}{Z} .$$

Here we choose Y to be the closure of the union of the connected components containing the support of F alone. Howevery Y's which do not also meet the support of G then vanish, by the $\varphi \to \psi$ symmetry. The convergence proof shows that the remaining terms have the desired exponential decay rate.

To establish the upper mass gap and the isolated one particle hyperboloid, the cluster expansion can be applied to the vertex functions $\Gamma^{(n)}$ in place of the Schwinger functions considered above. Similarly a cluster expansion in the Bethe-Salpeter kernel K defined graphically by

(3.11) 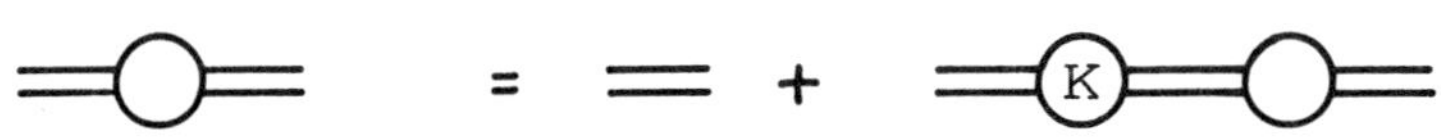

shows that K has three particle (and for even theories four particle) decay. In Minkowski momentum space, this means that K has an analytic continuation up to a neighborhood of the three or four particle threshold. Returning to (3.11), the only singularities of the four point function up to these energies come from the two particle cut. Hence asymptotic completeness up to these energies follows.

To extend these methods to the two phase region (3.3), we choose a length $L >> 1$ and write

$$\overline{\varphi} = L^{-d} \int_{L\text{-cube}} \varphi(x)dx$$

$$\sigma(\overline{\varphi}) = \operatorname{sgn} \overline{\varphi} .$$

Then σ is an Ising variable, defined on the lattice $(LZ)^d$. An expansion in Peierls contours isolates large regions of R^d in which σ is a constant. In these regions, the field φ is concentrated within a single well of the W-shaped potential, and a Gaussian approximation to this well is possible. This Gaussian approximation is controlled by a cluster expansion. Thus the final expansion is a double expansion, first in Peierls contours and then Dirichlet contours, as in (3.10).

REFERENCES

[1] J. Glimm and A. Jaffe, Quantum Field Models, in Statistical Mechanics, C. DeWitt and R. Stora (eds.) 1970 Les Houches Lectures, Gordon and Breach Science Publishers, New York, 1971.

[2] __________, Boson Quantum Field Theory, in Mathematics of Contemporary Physics, R. Streater (ed.), London Mathematical Society, Academic Press, 1972.

[3] J. Glimm, A. Jaffe and T. Spencer, The Particle Structure of the Weakly Coupled $P(\varphi)_2$ Model and Other Applications of High Temperature Expansions, Part I. Physics of Quantum Field Models, in Constructive Quantum Field Theory, G. Velo and A. S. Wightman (eds.), 1973 Erice Lectures, Springer Lecture Notes in Physics, Vol. 25, Springer-Verlag, 1973.

[4] J. Glimm and A. Jaffe, Critical Problems in Quantum Fields, presented at the International Colloquium on Mathematical Methods of Quantum Field Theory, Marseille, June 1975.

[5] __________, Critical Exponents and Renormalization in the φ^4 Scaling Limit, to appear in Quantum Dynamics: Models and Mathematics, L. Streit (ed.), Springer.

[6] __________, Functional Integral Methods in Quantum Field Theory, proceedings of the 1976 Cargèse Summer Institute.

[7] T. Spencer, The Decay of the Bethe-Salpeter Kernel in $P(\varphi)_2$ Quantum Field Models, Commun. Math. Phys. 44, 143-164 (1975).

[8] T. Spencer and F. Zirilli, Scattering States and Bound States in $\lambda P(\varphi)_2$, Commun. Math. Phys. 49, 1-16 (1976).

[9] J. Glimm and A. Jaffe, Particles and Bound States and Progress Toward Unitarity and Scaling, presented at the International Symposium on Mathematical Problems in Theoretical Physics, Kyoto, January 23-29, 1975.

[10] R. Schrader, A Constructive Approach to φ_4^4. I. Commun. Math. Phys. 49, 131-153 (1976); II. preprint; III. Commun. Math. Phys. 50, 97-102 (1976).

[11] J. Rosen, The Ising Model Limit of φ^4 Lattice Fields, preprint, 1976.

[12] J. Glimm and A. Jaffe, The φ_2^4 Quantum Field Model in the Single Phase Region: Differentiability of the Mass and Bounds on Critical Exponents, Phys. Rev. D10, 536-539 (1974).

[13] O. McBryan and J. Rosen, Existence of the Critical Point in φ^4 Field Theory, preprint 1976.

[14] J. Glimm, A. Jaffe and T. Spencer, The Particle Structure of the Weakly Coupled $P(\varphi)_2$ Model and Other Applications of High T mperature Expansions, Part II. The Cluster Expansion, in Constructive Quantum Field Theory, G. Velo and A. S. Wightman (eds.), 1973 Erice Lectures, Springer Lecture Notes in Physics, Vol. 25, Springer-Verlag, 1973.

[15] __________, The Wightman Axioms and Particle Structure in the $P(\varphi)_2$ Quantum Field Model, Ann. Math. 100, 585-632 (1974).

[16] J. Fröhlich, B. Simon and T. Spencer, Infrared Bounds, Phase Transitions and Continuous Symmetry Breaking, Commun. Math. Phys. 50, 79-95 (1976).

[17] F. Guerra, L. Rosen, B. Simon, The $P(\varphi)_2$ Euclidean Quantum Field Theory as Classical Statistical Mechanics, Ann. Math. 101, 111-259 (1975).

[18] J. Glimm and A. Jaffe, A Remark on the Existence of φ_4^4, Phys. Rev. Lett. 33, 440-442 (1974).

[19] C. Newman, preprint.

[20] J. Glimm and A. Jaffe, Particles and Scaling for Lattice Fields and Ising Models, Commun. Math. Phys. 51, 1-14 (1976).

[21] __________, Critical Exponents and Elementary Particles, Commun. Math. Phys., to appear.

[22] R. Schrader, preprint.

[23] J. Feldman, On the Absence of Bound States in the $\lambda\varphi_2^4$ Quantum Field Model without Symmetry Breaking, Canad. J. Phys. 52, 1583-1587 (1974).

[24] T. Spencer, The Absence of Even Bound States for $\lambda(\varphi^4)_2$, Commun. Math. Phys. 39, 77-79 (1974).

[25] J. Glimm and A. Jaffe, Three Particle Structure of φ^4 Interactions and the Scaling Limit, Phys. Rev. D11, 2816-2827 (1975).

[26] J. Glimm, A. Jaffe and T. Spencer, A Convergent Expansion about Mean Field Theory, Part I. The Expansion, Ann. Phys. 101, 610-630 (1976).

[27] J.-P. Eckmann, J. Magnen, and R. Sénéor, Decay Properties and Borel Summability for the Schwinger Functions in $P(\varphi)_2$ Theories, Commun. Math. Phys. 39, 251-271 (1975).

[28] J. Dimock, The $P(\varphi)_2$ Green's Functions: Asymptotic Perturbation Expansion, Helv. Phys. Acta 49, 199-216 (1976).

[29] K. Osterwalder and R. Sénéor, The Scattering Matrix is Non-Trivial for Weakly Coupled $P(\varphi)_2$ Models, Helv. Phys. Acta 49, 525-534 (1976).

[30] J.-P. Eckmann, H. Epstein and J. Fröhlich, Asymptotic Perturbation Expansion for the S-Matrix and the Definition of Time-Ordered Functions in Relativistic Quantum Field Models, Ann. de l'Inst. H. Poincaré 25, 1-34 (1976).

[31] J. Dimock and J.-P. Eckmann, Spectral Properties and Bound State Scattering for Weakly Coupled $P(\varphi)_2$ Models, to appear in Annals of Physics.

[32] Y. Park, Lattice Approximation of the $(\lambda\varphi^4 - \mu\varphi)_3$ Field Theory in a Finite Volume, J. Math. Phys. 16, 1065-1075 (1975).

[33] G. Sylvester, Continuous-Spin Ising Ferromagnets, Ph.D. thesis, M.I.T., 1976; J. Stat. Phys., to appear.

[34] J. Lebowitz, GHS and Other Inequalities, Commun. Math. Phys. 35, 87-92 (1974).

[35] J. Glimm and A. Jaffe, Absolute Bounds on Vertices and Couplings, Ann. de l'Inst. H. Poincaré 22, 1-11 (1975).

[36] J. Rosen, Mass Renormalization for Lattice $\lambda\varphi_2^4$ Fields, preprint, 1976.

[37] D. Marchesin, private communication.

[38] T. Spencer, The Mass Gap for the $P(\varphi)_2$ Quantum Field Model with a Strong External Field, Commun. Math. Phys. 39, 63-76 (1974).

[39] J. Glimm, A. Jaffe and T. Spencer, A Convergent Expansion about Mean Field Theory, Part II. Convergence of the Expansion, Annals of Phys. 101, 631-669 (1976).

[40] J. Magnen and R. Sénéor, The Infinite Volume Limit of the φ_3^4 Model, Ann. de l'Inst. H. Poincaré 24, 95-159 (1976).

[41] J. Feldman and K. Osterwalder, The Wightman Axioms and the Mass Gap for Weakly Coupled $(\varphi^4)_3$ Quantum Field Theories, Annals of Phys. 97, 80-135 (1976).

[42] J. Magnen and R. Sénéor, Wightman Axioms for the Weakly Coupled Yukawa Model in Two Dimensions, to appear in Commun. Math. Phys.

[43] A. Cooper and L. Rosen, The Weakly Coupled Yukawa_2 Field Theory: Cluster Expansion and Wightman Axioms, preprint.

Contemporary Physicists

Gian-Carlo Rota
David Sharp

Editors

Birkhäuser
Boston · Basel · Stuttgart

James Glimm
Arthur Jaffe

Collected Papers
Volume 2

Constructive Quantum Field Theory

Selected Papers

Birkhäuser
Boston · Basel · Stuttgart
1985

James Glimm
Courant Institute
New York, N.Y.
USA

Arthur Jaffe
Harvard University
Cambridge, Mass.
USA

Library of Congress Cataloging in Publication Data

Glimm, James.
Constructive quantum field theory.

(Contemporary physicists)
Reprint of articles originally published 1968–1980.
Includes bibliographies.
1. Quantum field theory—Addresses, essays, lectures.
2. Statistical physics—Addresses, essays, lectures.
I. Jaffe, Arthur, 1937– . II. Title. III. Series.
QC174.46.G584 1985 530.1'43 84–21525
ISBN 3-7643-3272-7 (Basel)
ISBN 0-8176-3272-7 (Boston)

CIP-Kurztitelaufnahme der Deutschen Bibliothek

Glimm, James:
Collected papers / James Glimm ; Arthur Jaffe. –
Boston ; Basel ; Stuttgart : Birkhäuser
ISBN 3-7643-3273-5
(Contemporary physicists)

NE: Jaffe, Arthur:; Glimm, James: [Sammlung];
Jaffe, Arthur: [Sammlung]

Vol. 2. Constructive quantum field theory :
selected papers. – 1985.
ISBN 3-7643-3272-7

All rights reserved.
No part of this publication may be reproduced, stored in a retrieval system, or transmitted in any form or by any means, electronics, mechanical, photocopying or otherwise, without the prior permission of the copyright owner.

© 1985 Birkhäuser Boston Inc.

Printed in Switzerland
ISBN 0-8176-3273-5 (complete set) 3-7643-3273-5 (complete set)
ISBN 0-8176-3271-9 (Vol. 1) 3-7643-3271-9 (Vol. 1)
ISBN 0-8176-3272-7 (Vol. 2) 3-7643-3272-7 (Vol. 2)

James Glimm/Arthur Jaffe

Collected Papers – Volume 2

I Construction of $P(\varphi)_2$

II Phase Cell Localization and $\varphi_3{}^4$ Stability

III Phase Transitions Exist

IV Phase Transitions and Critical Behavior

V Particle Structure

VI Bounds on Coupling Constants

VII Confinement and Instantons

VIII Reflection Positivity

Collected Papers – Volume 1

I Construction of $P(\varphi)_2$

Introduction

The series of papers constructing the $\lambda\phi_2^4$ quantum field model without cutoffs gave the first example of a quantum field model with a nontrivial scattering matrix. In the first paper, a self adjoint and locally correct Hamiltonian H(B) is constructed for each bounded region B of space-time. The quantum field $\phi(x,t) = \exp(itH(B))\phi(x)\exp(-itH(B))$ can be defined in terms of the time zero value $\phi(x)$. This field is studied in the second paper and is shown to be a self adjoint operator which is independent of B as long as B contains the causal shadow of the space-time point (x,t). The intuitive idea is to define H(B) on the time-zero surface as the correct Hamiltonian for the intersection of this surface with the interior of the light cones which eminate from points in B. Since in a relativistic theory influence propagates at a finite speed (at most the speed of light), local field expressions such as $\phi(x,t)$ can be constructed from a locally correct generator H(B) of time translations. In addition the ground state $\Omega(B)$ for H(B) is shown to exist. The eigenstates of H(B), however, depend on B. Thus expectations of ϕ in $\Omega(B)$ do depend on B.

The Wightman axioms require the existence of a global Hamiltonian H, not just the locally correct H(B)'s, as well as a ground state Ω for H. The construction of these objects is carried out in paper III, using C*-algebra methods. The basic estimate shows the finiteness of the vacuum energy shift per unit volume due to the interaction (after renormalization). As a consequence it follows that local functions $A(\phi)$ of the space-time field $\phi(x,t)$ have convergent power series expansions in the time zero fields $\phi(x,0)$ and their canonical conjugate fields $\pi(x,0)$. Using this expansion, the vacuum states $\{\Omega(B)\}$ are shown to have a subsequence which converges when B increases to R^2. For this subsequence, expectations $< \Omega(B), A\Omega(B) >$ converge to limits $< \Omega, A\Omega >$ which define the vaccum Ω and Hamiltonian H. The question of uniqueness is quite subtle and depends on the value of λ. While first developed for the ϕ^4 interaction, the methods also were applied to the $P(\phi)_2$ and Yukawa_2 models. Here P denotes a polynomial, bounded from below to ensure classical stability of the energy.

The paper on the Wightman axioms uses Euclidean functional integrals to establish all the Wightman axioms for a subset of the $\lambda P(\phi) + \phi^2$models constructed by the Hamiltonian approach above, namely those with weak coupling $0 < \lambda \ll 1$. The method relies on a convergent expansion to decouple different space-time regions in expectations of products of fields. For this reason the expansion has become known as the cluster expansion. Uniqueness of the ground state Ω follows for λ in its domain of convergence.

The cluster expansion is sufficiently powerful to investigate the spectrum of $\exp(-tH)$, and hence the spectrum of the Hamiltonian H. To study the spectrum of H on the interval $[0,\varepsilon]$ one proves $< A,(I - P_\varepsilon)\exp(-tH)A > \ \leq O(1)\exp(-\varepsilon t)$ as $t \to \infty$. Here P_ε are the spectral projections of H, which are constructed by an inductive procedure using the cluster expansion.

Boson quantum field Hamiltonians are equivalent to Schrödinger operators acting on an infinite dimensional space. They have the form $H = -\Delta + V$, where Δ is an infinite dimensional Laplacian, where V is a multiplication operator, and where H is bounded from below. Let $E_0 \leq E_1 \leq E_2 \leq \cdots$ denote the eigenvalues of a finite dimensional approximation to H. The cluster expansion method gives a tool to study the gaps $E_{i+1} - E_i$ between the smallest eigenvalues of such operators, uniformly in the dimension of the space on which they act. This is a difficult mathematical problem. Independent of the significance for physics, mathematicians may well wish to study spectral properties of such Laplacians further by these methods. In fact the heat kernel $\exp(-tH)$ can be considered as generating a random process labelled by time. Such random processes arise both in pure mathematics (geometry and topology as well as analysis) and in many fields of application, such as electrical engineering, stochastic control theory and presumably econometrics and population biology. The random field (labeled by space as well as time) has the infinite dimensional Laplacian H as its generator. It would be reasonable to guess that its abstract theory and its applications would be as rich as those of the random process indexed by time alone.

Bibliography

A $\lambda(\phi^4)_2$ quantum field theory without cutoffs I, Phys. Rev. 176, 1945–1951 (1968).

The $\lambda(\phi^4)_2$ quantum field theory without cutoffs II. The field operators and the approximate vacuum, Ann. Math. 91, 362–401 (1970).

The $\lambda(\phi^4)_2$ quantum field theory without cutoffs III. The physical vacuum, Acta Math. 125, 204–267 (1970).

The Wightman axioms and particle structure in the $P(\phi)_2$ quantum field model, with T. Spencer, Ann. Math. 100, 585–632 (1974).

A $\lambda\varphi^4$ Quantum Field Theory without Cutoffs. I

James Glimm*

Department of Mathematics, Massachusetts Institute of Technology, Cambridge, Massachusetts 02139

AND

Arthur Jaffe†

Lyman Laboratory of Physics, Harvard University, Cambridge, Massachusetts 02138

(Received 20 May 1968)

We construct the dynamics for a local quantum field theory in two-dimensional space-time. Our model describes a scalar field φ with a $\lambda\varphi^4$ self-interaction. We rely upon the fact that, in the Heisenberg picture, influence propagates at the speed of light. The resulting dynamics is independent of any cutoff, and hence the theory is formally Lorentz-covariant.

I. INTRODUCTION

In this paper we obtain the dynamics for a local quantum field theory in two-dimensional space-time. We study a scalar $\lambda\varphi^4$ self-interaction, and no cutoffs appear in the solution. Formally, our model is Lorentz-covariant, and a covariant local quantum field theory should result.

This $\lambda\varphi^4$ model is simple because the only momentum-space divergences are removed by Wick-ordering the Hamiltonian density. The vacuum energy has a divergence linear in the volume in each order of perturbation theory, as is expected in any theory with vacuum polarization. In perturbation theory the scattering matrix for this interaction is nontrivial. However, the perturbation series is known to diverge when summed to all orders,[1] so we cannot solve the theory in this manner.

We construct the dynamics in the Heisenberg picture which means that we have a map σ_t which takes the field operators from time zero to time t. The main advantage of the Heisenberg picture, is that the volume divergence does not enter in this formulation.

In order to prove that the theory exists, we introduce a spatial cutoff in the interaction Hamiltonian. Thus the formal Hamiltonian expression

$$H(g)=H_0+\lambda\int_{t=0} :\varphi^4(x): dx$$

is replaced by the operator on Fock space,

$$H(g)=H_0+\int_{t=0} :\varphi^4(x): g(x)dx$$

$$=H_0+H_I(g), \tag{1.1}$$

where the space-dependent coupling constant $g(x)$ is a smooth, non-negative function which vanishes outside a large interval. We derive an estimate for fixed $g(x)$,

$$-[H_0^{1/2},[H_0^{1/2},H_I(g)]]\leqq \epsilon H_0^2+b, \tag{1.2}$$

where ϵ is any positive number and $b(\epsilon)$ a suitably large number. This estimate, along with the previously known estimate[2,3]

$$-H_I(g)\leqq \epsilon H_0+b, \tag{1.3}$$

and a knowledge of the self adjointness of H_0 and $H_I(g)$, allows us to prove that the total Hamiltonian $H(g)$ is self adjoint. The connection between estimates (1.2) and (1.3) and self adjointness leads to a general theory of singular perturbations developed elsewhere.[4]

We then discuss the Heisenberg picture dynamics

$$\sigma_t(A)=e^{iH(g)t}Ae^{-iH(g)t},$$

of an observable A associated with a bounded region of space at $t=0$. The locality of the Hamiltonian density is essential to our method. As physicists have long known, it ensures that influence propagates in the Heisenberg picture at the speed of light. A local Hamiltonian density, therefore, assures a local theory on the formal level. Guenin pointed out that in perturbation theory the finite propagation speed is not destroyed by a space cutoff.[5] Segal states this result as a rigorous theorem, under the hypothesis that the total and the interaction Hamiltonians both are self adjoint.[6] In other words $\sigma_t(A)$ does not depend on $g(x)$, provided that $g(x)$ equals a constant, i.e., the desired coupling constant λ, for x in some suitably large region. By choosing an appropriate $g(x)$ for each bounded region of space-time, we can patch together a time translation for any operator. A similar argument shows that the theory is local. Since we have proved that the Hamiltonian $H(g)$ is self adjoint, we have constructed a local quantum field theory without a cutoff.

* Supported in part by the National Science Foundation under Grant No. GP-7477.

† Supported in part by the Air Force Office of Scientific Research.

[1] A. M. Jaffe, Commun. Math. Phys. **1**, 127 (1965).

[2] E. Nelson, in *Mathematical Theory of Elementary Particles*, edited by R. Goodman and I. Segal (The M.I.T. Press, Cambridge, Mass., 1966).

[3] J. Glimm, Commun. Math. Phys. **8**, 12 (1968); P. Federbush (to be published).

[4] J. Glimm and A. M. Jaffe, Commun. Pure Appl. Math. (to be published).

[5] M. Guenin, Commun. Math. Phys. **3**, 120 (1966).

[6] I. Segal, Proc. Natl. Acad. Sci. U.S. **57**, 1178 (1967).

II. ESTIMATES ON THE HAMILTONIAN

Let $\mathfrak{F}$ be the Fock space for a massive, neutral scalar field in two-dimensional space-time. The elements of $\mathfrak{F}$ are sequences of functions on momentum space. Let the annihilation and creation operators be normalized by the relation

$$[a(k),a^*(k')]=\delta(k-k'), \tag{2.1}$$

so that the free-field Hamiltonian is

$$H_0=\int a^*(k)a(k)\omega(k)dk,$$

where

$$\omega(k)=(k^2+m^2)^{1/2}.$$

The $t=0$ field

$$\varphi(x)=\frac{1}{(2\pi)^{1/2}\sqrt{2}}\int e^{-ikx}[a^*(k)+a(-k)]\omega(k)^{-1/2}dk \tag{2.3}$$

is used to construct the spatially cut-off interaction Hamiltonian

$$H_I(g)=\int :\varphi^4(x):g(x)dx$$

$$=\sum_{j=0}^{4}\binom{4}{j}\int a^*(k_1)\cdots a^*(k_j)a(-k_{j+1})\cdots$$

$$\times a(-k_4)\tilde{g}(\sum_{i=1}^{4}k_i)\prod_{i=1}^{4}\omega(k_i)^{-1/2}dk_i. \tag{2.4}$$

The total Hamiltonian for the spatially cut-off interaction is

$$H(g)=H_0+H_I(g). \tag{2.5}$$

We also need the number operator

$$N=\int a^*(k)a(k)dk, \tag{2.6}$$

and the domain

$$D_0=\bigcap_{n=0}^{\infty}D(H_0^n). \tag{2.7}$$

We let $\mathfrak{S}(R^1)$ denote the Schwartz space of infinitely differentiable functions, which, along with all their derivatives, vanish faster than any inverse polynomial at infinity.

Theorem 2.1: For any $\epsilon>0$ and for fixed $g\in\mathfrak{S}(R^1)$ there is a constant b such that as bilinear forms on $D_0\times D_0$

$$-[H_0^{1/2},[H_0^{1/2},H_I(g)]]\leq\epsilon H_0^2+b \tag{2.8}$$

and

$$-[N,[N,H_I(g)]]\leq\epsilon N^3+b. \tag{2.9}$$

Let us state two lemmas, which then yield the theorem.

Lemma 2.2: Let W be an operator of the form

$$W=\int w(k)a^*(k_1)\cdots a(-k_m)dk. \tag{2.10}$$

Then

$$\|(N+I)^{-j/2}W(N+I)^{-(m-j)/2}\| \leq \text{const}\ \|w\|_{L^2},\quad |j|\leq m \tag{2.11}$$

$$\|(H_0+I)^{-1}[H_0^{1/2},[H_0^{1/2},W]](H_0+I)^{-1}(N+I)^{-(m-4)/2}\| \leq\text{const}\ \|\omega^{1/2}(\sum_{i=1}^{m}k_i)w\|_{L^2}, \tag{2.12}$$

and

$$\|[H_0^{1/2},[H_0^{1/2},W]](N+I)^{-m/2}\| \leq\text{const}\ \|\sum_{i=1}^{m}\omega(k_i)w\|_{L^2}. \tag{2.13}$$

Lemma 2.3: Let W be as above, and suppose that $w(k)=0$ whenever $|k_i|>\kappa$ for any i. Then

$$\|[H_0^{1/2},[H_0^{1/2},W]](N+I)^{-(m-2)/2}\| \leq\text{const}\ \kappa^2\|w\|_{L^2}. \tag{2.14}$$

Proof of Theorem 2.1: Introduce the $t=0$ field φ_κ with an ultraviolet cutoff:

$$\varphi_\kappa(x)=\frac{1}{(2\pi)^{1/2}\sqrt{2}}\int_{-\kappa}^{\kappa}e^{-ikx}[a^*(k)+a(-k)][\omega(k)]^{-1/2}dk.$$

The interaction $H_{I,\kappa}(g)$ is defined by

$$H_{I,\kappa}=\int :\varphi_\kappa^4(x):g(x)dx, \tag{2.15}$$

and formally

$$H_I=\lim_{\kappa\to\infty}H_{I,\kappa}. \tag{2.16}$$

If we write H_I as a sum of five operators of the form W in (2.10), then by Lemma 2.2 taken for the case $m=4$

$$\|(H_0+1)^{-1}[H_0^{1/2},[H_0^{1/2},W]](H_0+1)^{-1}\| \leq\text{const}\ \|\omega^{1/2}(\sum_{i=1}^{4}k_i)w\|_{L^2}. \tag{2.17}$$

Since the kernel $w(k)$ has an over-all factor $\tilde{g}(\sum_{i=1}^4 k_i)$, where $\tilde{g}$ is the Fourier transform of the spatial cutoff $g(x)$, the fast decrease of $\tilde{g}(k)$ ensures that $\omega^{1/2}(\sum_{i=1}^4 k_i)w$ is in L^2. Thus the kernel for the corresponding cutoff interaction term w_κ approximates w in the sense that

$$\|\omega^{1/2}(\sum_{i=1}^{4}k_i)(w-w_\kappa)\|_{L^2}\to 0, \tag{2.18}$$

as $\kappa\to\infty$. This is true for each W making up $H_I(g)$, so we infer that there exists a κ_0 such that for $\kappa>\kappa_0$,

$$\|(H_0+I)^{-1}[H_0^{1/2},[H_0^{1/2},H_I-H_{I,\kappa}]]\times(H_0+I)^{-1}\|\leqq\tfrac{1}{2}\epsilon. \quad (2.19)$$

By Lemma 2.3, we have the relations

$$-[H_0^{1/2},[H_0^{1/2},H_{I,\kappa}]]\leqq\text{const}\,\kappa_0^2(N+I)\leqq\tfrac{1}{2}\epsilon H_0^2+\text{const}, \quad (2.20)$$

on the domain $D_0\times D_0$. Thus combining (2.20) with the estimate on the tail (2.19),

$$-[H_0^{1/2},[H_0^{1/2},H_I]]=-[H_0^{1/2},[H_0^{1/2},H_I-H_{I,\kappa}]]-[H_0^{1/2},[H_0^{1/2},H_{I,\kappa}]]\leqq\epsilon H_0^2+b,$$

which is (2.8). In order to prove (2.9), we expand $H_I(g)$ once more as a sum of terms W of the form (2.10). Since $[N,W]=\text{const}\,W$, and W is a fourth-order operator with an L^2 kernel, we infer from (2.11) that

$$|(\psi[N,[N,W]]\psi)|\leqq\text{const}\,(\psi,(N+I)^2\psi)\leqq\tfrac{1}{5}\epsilon(\psi,N^3\psi)+\tfrac{1}{5}b(\psi,\psi). \quad (2.21)$$

As $H_I(g)$ has 5 such terms,

$$[N,[N,H_I]]\leqq\epsilon N^3+b,$$

which completes the proof of the theorem.

Proof of Lemmas 2.2 and 2.3: Let θ be an n-particle wave function. Then $(a(k)\theta)(k_1,\cdots,k_{n-1})$ is an $(n-1)$-particle state defined by

$$(a(k)\theta)(k_1,\cdots,k_{n-1})=(\sqrt{n})\theta(k,k_1,\cdots,k_{n-1}). \quad (2.22)$$

Using (2.22) and the Schwarz inequality, it is easy to establish the bound (2.11).

Note that

$$([H_0^{1/2},a(k)]\theta)(k_1,\cdots,k_{n-1})=\lambda_1(k;k_1,\cdots,k_{n-1})(a(k)\theta)(k_1,\cdots,k_{n-1})\equiv(a_1(k)\theta)(k_1,\cdots,k_{n-1}), \quad (2.23)$$

where

$$\lambda_1(k;k_1,\cdots,k_{n-1})=[\omega(k)+\sum_{i=1}^{n-1}\omega(k_i)]^{1/2}-[\sum_{i=1}^{n-1}\omega(k_i)]^{1/2}. \quad (2.24)$$

For convenience, we have defined a modified annihilation operator $a_1(k)$, which decreases the particle number by one.

For $x\geq0$, we see that

$$(1+x)^{1/2}-1\leq x^{1/2},$$

so

$$|\lambda_1(k;k_1,\cdots,k_{n-1})|\leq[\omega(k)]^{1/2}. \quad (2.25)$$

Furthermore, for $x\geq0$,

$$(1+x)^{1/2}-1\leq\tfrac{1}{2}x,$$

so

$$|\lambda_1(k;k_1,\cdots,k_{n-1})|\leq\tfrac{1}{2}\omega(k)[\sum_{i=1}^{n-1}\omega(k)]^{-1/2}\leq\tfrac{1}{2}\omega(k)\frac{1}{m^{1/2}(n-1)^{1/2}}. \quad (2.26)$$

In the same way

$$([H_0^{1/2},[H_0^{1/2},a(k)]]\theta)(k_1,\cdots,k_{n-1})=\lambda_2(k;k_1,\cdots,k_{n-1})(a(k)\theta)(k_1,\cdots,k_{n-1})\equiv(a_2(k)\theta)(k_1,\cdots,k_{n-1}). \quad (2.27)$$

Here

$$|\lambda_2(k;k_1,\cdots,k_{n-1})|\leq\begin{cases}\omega(k)\\ \text{const}\;\omega^2(k)/n\end{cases}. \quad (2.28)$$

Taking adjoints,

$$[H_0^{1/2},a^*(k)]=-a_1^*(k), \quad (2.29)$$

and

$$[H_0^{1/2},[H_0^{1/2},a^*(k)]]=a_2^*(k). \quad (2.30)$$

Note that the double commutation $[H_0^{1/2},[H_0^{1/2},W]]$ can be expressed as a sum of two kinds of terms. There will be m contributions arising from double commutators of $H_0^{1/2}$ with one of the m creation or annihilation operators, and these will be of the form of the original W in (2.10), but they will have one $a(k)$ or one $a^*(k)$, replaced by an $a_2(k)$ or $a_2^*(k)$. There will also be $\frac{1}{2}m(m-1)$ terms arising from the commutator $H_0^{1/2}$ with two distinct creators or annihilators of W. These terms will be of the form of the right-hand side of (2.10) but will involve two of the modified operators $a_1(k)$ or $-a_1^*(k)$.

In order to prove (2.12), we use the definitions (2.22)–(2.24) and (2.27). In order to estimate the matrix element $|(\psi,[H_0^{1/2},[H_0^{1/2},W]]\theta)|$, we use the bound (2.25) and the first of the inequalities (2.28). As in the proof of (2.11), the Schwarz inequality then yields

$$\|[H_0^{1/2},[H_0^{1/2},W]](N+I)^{-m/2}\|\leq\text{const}\,\|\sum_{i=1}^{m}\omega(k_i)w\|_{L^2}, \quad (2.31)$$

which is (2.13). The proof of (2.14) proceeds in a similar fashion, using the second set of bounds for the $|\lambda_j|$.

In order to prove (2.12), we investigate the operator

$$X=(H_0+I)^{-1}W(H_0+I)^{-1}(N+I)^{-(m-4)/2}. \quad (2.32)$$

By following the proof of Lemma 2.41 and Theorem 2.43 of Ref. 7, we find that

$$\|X\|\leq\text{const}\,\|E^{-1/2}w\|_{L^2},$$

where

$$E=\begin{cases}\sup_{i\neq j}\omega(k_i)\omega(k_j) & \text{if } m>1\\ \omega(k_i) & \text{if } m=1.\end{cases}$$

[7] J. Glimm, Commun. Math. Phys. 5, 343 (1967).

Replacing the operator W by $\hat{W}=[H_0^{1/2},[H_0^{1/2},W]]$, we find that for the corresponding $\hat{X}$,

$$\|\hat{X}\| \leq \text{const } \|E^{-1/2} \sum_{i=1}^{m} \omega(k_i) w\|_{L^2}.$$

Each $\omega(k_i)$ is dominated by a constant multiple of $E^{1/2}$, unless k_i is very much larger than the magnitude of all the other momenta, or unless $m=1$. The former can occur only when $\sum_{i=1}^m k_i$ is large in magnitude, and in either case

$$\omega(k_i) \leqq \text{const } E^{1/2}\omega^{1/2}(\sum_{i=1}^{m} k_i).$$

Thus

$$\|\hat{X}\| \leqq \text{const } \|\omega^{1/2}(\sum_{i=1}^{m} k_i) w\|_{L^2},$$

which was to be proved.

III. SELF ADJOINTNESS OF THE INTERACTION HAMILTONIAN

For a real spatial cutoff $g(x)$ in the Schwartz space $\mathfrak{S}(R^1)$, the interaction part of the Hamiltonian $H_I(g)$ is self adjoint. This result was proved by Lanford, and also by Doplicher and Jaffe. Statement of these theorems are given in Wightman.[8] With a slightly different formulation, the result has been announced by Segal.[6]

We initially define $H_I(g)$ on the domain

$$D_0 = \bigcap_{n=0}^{\infty} D(H_0^n),$$

and we will prove that its closure is self adjoint. It follows from (2.11) that H_I is well defined on D_0.

Theorem 3.1: If $g \in \mathfrak{S}(R^1)$ is real, then

$$H_I(g) = \int_{t=0} :\varphi^4(x): g(x) dx \tag{3.1}$$

is essentially self adjoint on D_0.

Let us introduce a domain D_1 obtained by applying any polynomial of the $t=0$ fields $\varphi(f_i)$, $f_i \in \mathfrak{S}(R^1)$, to the no-particle state Ω_0. Clearly $D_1 \subset D_0$, and any vector Ω in D_1 is an entire vector for $\varphi(f)$, which means that the power series

$$\sum_{n=0}^{\infty} \frac{\|\varphi(f)^n \Omega\|}{n!} z^n \tag{3.2}$$

defines an entire function of z. Since D_1 is dense in Fock space, a result of Nelson,[9] shows that for real f, $\varphi(f)$ is essentially self adjoint on D_1. A similar argument can be made for the canonically conjugate $t=0$ fields $\pi(f)$.

Let $\mathfrak{M}$ denote the von Neumann algebra of operators generated by the spectral projections of all the $t=0$ fields $\varphi(f)$, $f \in \mathfrak{S}(R^1)$. The algebra $\mathfrak{M}$ is maximal Abelian. In other words, a bounded operator which commutes with all operators in $\mathfrak{M}$ is itself in $\mathfrak{M}$.

Let us consider $\varphi(f)$ for supp $f \subset O \subset R^1$, where O is an open region of space. (The support of a function is the smallest closed set outside of which the function vanishes identically.) Define $\mathfrak{A}(O)$ as the von Neumann algebra of operators generated by the spectral projections of all the fields $\varphi(f)$ and $\pi(f)$, supp $f \subset O$.

Since

$$\varphi(x,t) = \int \left(\Delta(x-y,\, t)\pi(y) - \frac{\partial \Delta}{\partial t}(x-y,\, t)\varphi(y) \right) dy, \tag{3.3}$$

and $\Delta(x,t)$ vanishes outside the light cone, we infer that

$$e^{iH_0t}\mathfrak{A}(O)e^{-iH_0t} \subset \mathfrak{A}(O_t), \tag{3.4}$$

where O_t is the region O expanded by t.

Theorem 3.2: If $g(x) \in \mathfrak{S}(R^1)$ is real and has its support in an open interval O, then for the H_I of (3.1)

$$\exp(iH_I(g)t) \in \mathfrak{A}(O) \cap \mathfrak{M}.$$

We now prove two lemmas.

Lemma 3.3: Let T be any operator with domain D_1 such that

$$TD_1 \subset D(\varphi(f)^n), \tag{3.5}$$

$$TD_1 \subset D(T|_{D_1}{}^*), \tag{3.6}$$

and

$$[T, \varphi(f)^n] D_1 = 0. \tag{3.7}$$

Then

$$\mathfrak{M} D_1 \subset D(T|_{D_1}{}^-) \tag{3.8}$$

and

$$[T^-, \mathfrak{M}] D_1 = 0. \tag{3.9}$$

Proof: For $\Omega \in D_1$, from (3.5) and (3.7) we have

$$T\varphi(f)^n \Omega = \varphi(f)^n T\Omega.$$

But by (3.6), for real f

$$\|T\varphi(f)^n\Omega\|^2 = (T\Omega, \varphi(f)^{2n} T\Omega) = (T^* T\Omega, \varphi(f)^{2n}\Omega) \leqq \|T^*T\Omega\| \, \|\varphi(f)^{2n}\Omega\|.$$

Thus the convergent power series (3.2) shows that for Ω in D_1,

$$T^- e^{i\varphi(f)}\Omega = e^{i\varphi(f)} T\Omega. \tag{3.10}$$

It is clear that (3.10) is still valid with $\exp(i\varphi(f))$ replaced by strong limits of sums of such exponentials, and hence (3.8) and (3.9).

Lemma 3.4: Let $\mathfrak{M}$ be a maximal Abelian algebra of bounded operators on a Hilbert space $\mathfrak{H}$ with a cyclic vector Ω_0. Let T be a symmetric operator with domain

[8] A. S. Wightman, in *1964 Cargèse Summer School Lectures*, edited by M. Lévy (Gordon and Breach, Science Publishers, Inc., New York, 1967).

[9] E. Nelson, Ann. Math. **70**, 572 (1959); H. J. Borchers and W. Zimmermann, Nuovo Cimento **31**, 1047 (1963).

$\mathfrak{M}\Omega_0$, and let T commute with $\mathfrak{M}$. Then T is essentially self adjoint.

Proof: Without loss of generality, $\mathfrak{M}=L_\infty(X)$ and $\mathfrak{H}=L_2(X)$ for some measure space X, and Ω_0 is the function 1. Let $t=T\Omega_0$. Then $t\in L_2$ and T is multiplication by t, with domain L_∞. Let $f\in L_2$ and suppose $tf\in L_2$ also and let

$$f_n(x)=\begin{cases} f(x) & \text{if } |f(x)|\leqq n \\ 0 & \text{otherwise.}\end{cases}$$

Then $f_n\in L_\infty=D(T)$ and $f_n\to f$, $tf_n\to tf$ in L_2 norm by the bounded convergence theorem. Thus $\{f,tf\}$ is in the graph of the closure of T. Thus the closure of T is self adjoint, and T is essentially self adjoint.

Remark: Let T_n be a sequence of operators with the property of T in the lemma. Then $T_n\to T$ strongly on the domain $\mathfrak{M}\Omega_0$ if and only if $T_n\Omega_0\to T\Omega_0$.

Proof of Theorems 3.1 and 3.2: We apply the Lemmas 3.3 and 3.4 with the case $T=H_I(g)$, $\mathfrak{M}$ in Lemma 3.4 as in Lemma 3.3, the Hilbert space Fock space, and Ω_0 the Fock no-particle state. The hypotheses (3.5) and (3.6) can be verified by a direct computation.[10] Thus $H_I(g)$ is essentially self adjoint on $D_1\subset D_0$, and hence $H_I(g)$ is essentially self adjoint on D_0.

If we assume that supp $g\subset O$, then as O is an open interval, supp $g\subset O_1$ where O_1 is O contracted by some small amount $\epsilon>0$. Since $H_I(g)$ commutes with $\mathfrak{M}$, and $\mathfrak{M}$ is maximal Abelian, $\exp(iH_I(g)t)\in\mathfrak{M}$. Furthermore the argument in the proof of Lemma 3.3, can be repeated to show that $H_I(g)$ commutes with $\mathfrak{A}(O_1')$, where O_1' is the complement of the closure of O_1.[11] Since $\mathfrak{A}(R^1)$ is irreducible and $H_I(g)$ commutes with $\mathfrak{A}(O_1')$, $\exp(iH_I(g)t)\in\mathfrak{A}(O_2)$ where O_2 is O_1 expanded by any amount $\epsilon'>0$.[12] Taking $\epsilon'<\epsilon$, we have $\exp(iH_I(g)t)\in\mathfrak{A}(O)$, which completes the proof.

IV. SELF ADJOINTNESS OF THE TOTAL HAMILTONIAN

Theorem 4.1: (a) For real $g(x)\in\mathfrak{S}(R^1)$, the total Hamiltonian $H(g)=H_0+H_I(g)$ is self adjoint with the domain $D(H(g))=D(H_0)\cap D(H_I(g))$.

(b) The total Hamiltonian $H(g)$ is essentially self adjoint on the domain

$$D_0=\bigcap_{n=0}^{\infty} D(H_0{}^n).$$

In order to prove the self adjointness of H, we combine the estimates of Sec. II, the self adjointness of $H_I(g)$ proved in Sec. III, and a singular perturbation theory developed elsewhere.[4] We need the following result which is a special case of Theorem 8 of Ref. 4.

Theorem 4.2: Under the hypotheses (i)–(iii) below, the operator $H=H_0+H_I$ is self adjoint.

(i) Both H_0 and H_I are self adjoint. The domain

$$D_0=\bigcap_{n=0}^{\infty} D(H_0{}^n)$$

is contained in the domain of H_I, and H_I is essentially self adjoint on D_0.

(ii) Let $N\geqq 0$ be a positive self adjoint operator, commuting with H_0, and such that $N\leqq$ const H_0. Suppose that the operators $(I+N)^{-1}H_I(I+N)^{-1}$ and $(I+N)^{+1}H_I(I+N)^{-3}$ are bounded.

(iii) Suppose that for any $\epsilon>0$, there exists a number b such that as bilinear forms on $D_0\times D_0$,

$$-H_I\leqq\epsilon N+bI, \tag{4.1}$$

$$-[H_0{}^{1/2},[H_0{}^{1/2},H_I]]\leqq\epsilon H_0{}^2+bI, \tag{4.2}$$

and

$$-[N,[N,H_I]]\leqq\epsilon N^3+bI. \tag{4.3}$$

Proof of Theorem 4.1: In order to prove that H is self adjoint, we apply Theorem 4.2 in the case that H_0 is the free Hamiltonian, N is the number operator, and H_I is the interaction Hamiltonian $H_I(g)$. Thus we need to verify (i–iii). Condition (i) was dealt with in Theorem 3.1, while condition (ii) is a consequence of (2.11). In Refs. 2, and 3, it is shown that for any $\epsilon>0$, there is a number b such that

$$-H_I(g)\leqq\epsilon H_0+bI.$$

By following that proof, but using the smoothing operator $\exp(-tN)$, in place of $\exp(-tH_0)$, one arrives at the estimate (4.1) required in (iii). The remaining estimates (4.2) and (4.3) were established in Theorem 2.1. Thus we conclude from Theorem 4.2 that $H(g)$ is self adjoint on the domain $D(H_0)\cap D(H_I(g))$.

We now show that $H(g)$ is essentially self adjoint on D_0. We first show that $H(g)$ is essentially self adjoint on $D_2=D(H_0)\cap D(N^2)$. By (2.11) it is clear that the domain of $H(g)$ contains D_2. For $\psi\in D(H(g))=D(H_0)\cap D(H_I(g))$, consider $\psi_n\in D_2$ defined by

$$\psi_n=n(nI+N)^{-1}\psi. \tag{4.4}$$

Thus $\|\psi_n-\psi\|+\|H_0\psi_n-H_0\psi\|\to 0$ as $n\to\infty$. We study

$$H_I\psi_n-H_I\psi=-N(nI+N)^{-1}H_I\psi+n[H_I,\ (nI+N)^{-1}]\psi.$$

Since $N(nI+N)^{-1}$ is a uniformly bounded sequence converging to zero on the dense set $D(N)$, it converges to zero and $\|N(nI+N)^{-1}H_I\psi\|\to 0$ as $n\to\infty$. But for the other term

$$\begin{aligned} n[H_I,\ (nI+N)^{-1}]\psi &= [H_I,\ (nI+N)^{-1}](nI+N)n(nI+N)^{-1}\psi \\ &= (nI+N)^{-1}[N,H_I]n(nI+N)^{-1}\psi \\ &= (nI+N)^{-1}(I+N)(I+N)^{-1}[N,H_I] \\ &\quad\times(I+N)^{-1}n(nI+N)^{-1}(I+N)\psi. \end{aligned}$$

[10] A. M. Jaffe, J. Math. Phys. 7, 1250 (1966).
[11] A. M. Jaffe, Ann. Phys. (N. Y.) 32, 127 (1965).
[12] H. Araki, J. Math. Phys. 5, 1 (1964).

Note that as $n\to\infty$, $n(nI+N)^{-1}(I+N)$ converges strongly to $(I+N)\psi$, that by (2.11), $(I+N)^{-1}[N,H_I]\times(I+N)^{-1}$ is bounded, and that $(nI+N)^{-1}(I+N)$ converges strongly to zero. Thus $\|[H_I,(nI+N)^{-1}]\psi\|\to 0$ as $n\to\infty$, and so $\|H_I\psi_n-H_I\psi\|\psi\to 0$. Thus we see that $H(g)$ is the closure of $H(g)$ restricted to D_2, so $H(g)$ is essentially self adjoint on D_2.

Let D_2 be a Hilbert space with the norm

$$(\|\psi\|_{D_2})^2=\|\psi\|^2+\|H_0\psi\|^2+\|N^2\psi\|^2.$$

From (2.11) we infer that

$$\|H\psi\|\leq\text{const}\|\psi\|_{D_2},$$

so that $H(g)$ is essentially self adjoint on any subset of D_2 which is dense in the Hilbert space D_2. For $\psi\in D_2$, $\psi_\lambda=\exp(-\lambda H_0)\psi\in D_0=\bigcap_{n=0}^{\infty}D(H_0^n)$, and $\|\psi_\lambda-\psi\|_{D_1}\to 0$ as $\lambda\to 0$. Thus $H(g)$ is essentially self adjoint on D_0.

V. REMOVING THE SPATIAL CUTOFF AND LOCALITY

For the reader's convenience, we sketch a proof of Segal's theorem[6] that the self adjointness of $H(g)$ allows the removal of the spatial cutoff. In fact, if A is a bounded function of the free fields localized in a bounded region of space at $t=0$, then

$$\sigma_t(A)=e^{itH(g)}Ae^{-itH(g)}$$

is independent of $g(x)$ provided that $g(x)=\lambda$, the desired coupling constant, on a sufficiently large interval, depending on t. Furthermore, if A is localized in the region of space O, then $\sigma_t(A)$ is localized in the region O_t, where O_t is the region O expanded by t. (We have taken the velocity of light to be one.) In other words, the time translation σ_t gives rise to a local theory. If one chooses for the operator A a spectral projection of the $t=0$ field $\varphi(f)$, one can piece together the time translation operator for the fields themselves.

In section IV, we showed that $H=H_0+H_I$, which is sum of two self adjoint operators, is itself self adjoint. As a consequence of this fact, the Trotter product formula[13] says that for all ψ

$$e^{itH}\psi=\lim_{n\to\infty}(e^{itH_0/n}e^{itH_I/n})^n\psi.$$

Thus

$$\sigma_t(A)\psi=\lim_{n\to\infty}(e^{iH_0t/n}e^{iH_It/n})^nA\,(e^{-iH_It/n}e^{-iH_0t/n})^n\psi.$$

Let O be the region defined by $|x|<M$, $t=0$, and let $A\in\mathfrak{A}(O)$, where $\mathfrak{A}(O)$ is defined in Sec. III.

Given an arbitrary, positive ϵ, split $g(x)$ into two infinitely differentiable parts

$$g(x)=g_1(x)+g_2(x),$$

where supp $g_1(x)\subset O_\epsilon$, and supp $g_2\cap O_{\epsilon/2}$ is empty. Write

$$H_I(g)=H_I(g_1)+H_I(g_2),$$

so that as a consequence of theorems 3.1 and 3.2, $H_I(g_1)$ and $H_I(g_2)$ commute, and

$$\exp[iH_I(g)t/n]=\exp[iH_I(g_1)t/n]\exp[iH_I(g_2)t/n].$$

Furthermore,

$$\exp[iH_I(g_1)t/n]\in\mathfrak{A}(O_\epsilon),$$

and $\exp[iH_I(g_2)t/n]$ commutes with $\mathfrak{A}(O_{\epsilon/4})$. Therefore,

$$A_1(t)=\exp(iH_0t/n)\exp[iH_I(g)t/n]A\times\exp[-iH_I(g)t/n]\exp(-iH_0t/n)$$

depends on $g(x)$ only in the region O_ϵ, and by the free propagation property (3.4),

$$A_1\in\mathfrak{A}(O_{(t/n)+\epsilon}).$$

We continue step by step, and after n steps we conclude that

$$A_n(t)=[\exp(iH_0t/n)\exp(iH_I(g)t/n)]^nA\times[\exp(-iH_I(g)t/n)\exp(-H_0t/n)]^n$$

depends on $g(x)$ only in the region $O_{t+n\epsilon}$, and

$$A_n(t)\in\mathfrak{A}(O_{t+n\epsilon}).$$

Since ϵ can be chosen arbitrarily, $A_n(t)$ depends on $g(x)$ only in the region $\bar{O}_t$, the closure of O_t, and

$$A_n(t)\in\bigcap_{\epsilon>0}\mathfrak{A}(O_{t+\epsilon}).$$

Thus $A_n(t)$ commutes with any local observable B localized in open region of space O' such that O' and O_t are disjoint. As this is true for each n, it is true for

$$\sigma_t(A)=\text{strong}\lim_{n\to\infty}A_n(t).$$

Hence $\sigma_t(A)$ is *local*, and it depends on $g(x)$ only in the region $\bar{O}_t$, where we choose $g(x)=\lambda$. We therefore conclude that the spatial cutoff has been removed and the resulting theory is local.

VI. THEORY IN A BOX

We can consider a somewhat different cutoff theory, namely the $g\varphi^4$ theory in a periodic box. This gives a cut-off interaction which is translation invariant, and therefore it is useful for the study of the vacuum state.[14] In a finite interval, but with no ultraviolet cutoff, we prove that the total Hamiltonian is self adjoint and has a complete set of normalizable eigenstates.

The theory in volume V is constructed by taking a Fock space $\mathfrak{F}_V$ of functions defined on the momentum

[13] H. F. Trotter, Proc. Am. Math. Soc. **10**, 545 (1959); E. Nelson, J. Math. Phys. **5**, 332 (1964).

[14] A. M. Jaffe and R. T. Powers, Commun. Math. Phys. **7**, 218 (1968); J. Glimm and A. M. Jaffe (to be published).

space lattices Γ_V, $\Gamma_V \times \Gamma_V$, etc., where

$$\Gamma_V = \{k:\ k = 2\pi n/V,\ n = 0, +1, \pm 2, \cdots\}. \quad (6.1)$$

Thus the free Hamiltonian is

$$H_{0,V} = \sum_{k \in \Gamma V} a_V^*(k) a_V(k) \omega(k), \quad (6.2)$$

and the $t=0$ field is

$$\varphi_V(x) = \frac{1}{(2V)^{1/2}} \times \sum_{k \in \Gamma_V} e^{-ikx}[a_V^*(k) + a_V(-k)]\omega(k)^{-1/2}. \quad (6.3)$$

The interaction Hamiltonian $H_{I,V}$ is defined by

$$H_{I,V} = g \int_V :\varphi_V(x)^4: dx, \quad g \geqq 0 \quad (6.4)$$

and $H_{I,V}$ is self adjoint as was shown in Sec. III for $H_I(g)$. The total Hamiltonian in a box is

$$H_V = H_{0,V} + H_{I,V}.$$

It is possible to regard $\mathfrak{F}_V$ as a subspace of $\mathfrak{F}$ defined by functions on R, $R \times R$, etc., which are piecewise constant between lattice sites. The correspondence between a_V and a is

$$a_V(k) = \left(\frac{V}{2\pi}\right)^{1/2} \int_0^{2\pi/V} a(k+l) dl, \quad k \in \Gamma_V. \quad (6.5)$$

Therefore the estimates of Sec. II are valid for H_V, and so H_V is self adjoint on the domain $D(H_{0,V}) \cap D(H_{I,V})$, and H_V is essentially self adjoint on the domain

$$D_0 = \bigcap_{n=0}^{\infty} D(H_{0,V}^n).$$

Theorem 6.1: The spectrum of H_V is discrete with finite multiplicity, so H_V has a complete set of normalizable eigenstates.

Proof: Let $\Gamma_{\kappa,V} = \Gamma_V \cap \{k: |k| \leqq \kappa\}$, so that an ultraviolet cut-off field $\varphi_{\kappa,V}(x)$ is obtained from φ_V by summing (6.4) over $\Gamma_{\kappa,V}$. Then a sharply cut-off Hamiltonian $H_{\kappa,V} = H_{0,V} + H_{I,\kappa,V}$ comes from

$$H_{I,\kappa,V} = g \int_V :\varphi_{\kappa,V}(x)^4: dx.$$

The operator $H_{\kappa,V}$ has pure discrete spectrum with finite multiplicity.[15] Furthermore, $H_{I,\kappa,V}$ is the sum of five expressions of the form $W_{\kappa,V}$ of (2.10), such that $\|w_{\kappa,V} - w_{\kappa',V}\|_{L^2} \to 0$ as $\kappa, \kappa' \to \infty$. Therefore the bounded operators $(I+N_V)^{-1} H_{I,\kappa,V} (I+N_V)^{-1}$ converge in norm as $\kappa \to \infty$ to $(I+N_V)^{-1} H_{I,V} (I+N_V)^{-1}$. We now appeal to Corollary 10 of Ref. 4 to infer that the resolvents of $H_{\kappa,V}$ converge in norm as $\kappa \to \infty$ to the resolvent of H_V. Since each $R_{\kappa,V} = (H_{\kappa,V} + c)^{-1}$ is compact, so is their uniform limit R_V, and the theorem is proved.

We further see that the projections onto each eigenvalue $\lambda_n(\kappa, V)$ of $H_{\kappa,V}$ converge as $\kappa \to \infty$. This is a consequence of Theorem IV. 3.16 of Kato.[16] Thus, in particular the vacuum vectors for $H_{\kappa,V}$ converge in Fock space to a vacuum vector for H_V.

VII. CONCLUSIONS

We have shown the existence of a local time translation for a two-dimensional $\lambda\varphi^4$ theory without cutoffs. The method relies on basic inequalities satisfied by the first and second power of the Hamiltonian. Similar estimates of all the higher powers of the Hamiltonian proved useful to investigate the existence of vacuum-expectation values in a $\lambda\varphi^4$ theory with a sharp momentum-space cutoff.[15] It does not seem likely, in our case, that H_0^3 is dominated by higher powers of H, since the ground state of H must be in the domain of H^n, for all n, and the first-order perturbation correction to the Fock vacuum is in $D(H_0)$, but not in $D(H_0^{3/2})$.

[15] A. M. Jaffe, Ph.D. thesis, Princeton University (to be published).

[16] T. Kato, *Perturbation Theory for Linear Operators* (Springer-Verlag, Berlin, 1966).

The $\lambda(\varphi^4)_2$ quantum field theory without cutoffs: II. The field operators and the approximate vacuum

JAMES GLIMM[1] and ARTHUR JAFFE[2]

TABLE OF CONTENTS

1. Introduction

1.1. *Discussion of the results.* We study a simple model for quantum field theory and we solve the equations of motion. Our model describes bosons with a non-linear φ^4 self-interaction in two dimensional space time (a $(\varphi^4)_2$ interaction for short). The essential simplification in this model is the reduction in the number of dimensions, which renders finite many (but not all) of the standard divergent integrals of quantum field theory. We verify most of the Haag-Kastler axioms for quantum field theory for this model. The remaining axiom is valid on the level of formal perturbation theory[3], as are the Wightman axioms.

The equation of motion in the Schrödinger picture is

$$i\frac{\partial\theta}{\partial t} = H\theta$$

[1] Supported by the New York State Science and Technology Foundation, Grant SSF-(8)-8.

[2] Alfred P. Sloan Foundation Fellow. Supported in part by the Air Force Office of Scientific Research and by the Schweizerische Nationalfonds.

[3] *Added in proof.* The missing axiom, Lorentz covariance, is estabilished in **[30]**.

where the state θ of the quantum field is a vector of unit length in a space $\mathcal{F}$, and H, the hamiltonian operator, is a self-adjoint operator on $\mathcal{F}$. The solution of this Schrödinger equation is

$$\theta_S(t) = e^{-itH}\theta_S(0)$$

and this solution describes the dynamics in the Schrödinger picture. The observables are represented by certain self-adjoint operators A_S on $\mathcal{F}$ and the inner product $\big(\theta_S(t), A_S\theta_S(t)\big)$ is the expected value for the measurement of the quantity A at the time t if the quantum field is in the state $\theta_S(0)$ at the time zero. The Schrödinger picture observable A is independent of time.

In the Heisenberg picture, the state θ_H in $\mathcal{F}$ is independent of time (and equals $\theta_S(0)$) while the observable $A_H(t) = e^{itH}A_H(0)e^{-itH}$ depends on time. We take $A_H(0) = A_S$ and then

$$\big(\theta_S(t), A_S\theta_S(t)\big) = \big(\theta_H, A_H(t)\theta_H\big) ,$$

so that the expected values of measurements in the Heisenberg and Schrödinger pictures agree.

As is typical for quantum field theory problems, the hamiltonian H for the $(\varphi^4)_2$ interaction is not given as a self-adjoint operator, but rather as a singular formal expression (some of the terms having infinite coefficients) with no vectors obviously in its domain, other than $\theta = 0$. For this model it is easier to construct the dynamics in the Heisenberg picture, and this was accomplished in [8, 9]. The bounded observables $A_H(0)$ generate a C^* algebra $\mathfrak{A}$. (See § 4.) The observable $A_H(t)$ belongs to $\mathfrak{A}$ for all t and so $\mathfrak{A}$ is invariant under t. The map σ_t: $A_H(0) \rightarrow A_H(t)$ defines an automorphism of $\mathfrak{A}$ and σ is a one parameter group $\sigma_{t+s} = \sigma_t\sigma_s$.

In § 3 of this paper we establish the existence of self adjoint field operatos $\varphi(f)$,

$$\varphi(f) = \int\varphi(x, t)f(x, t)dxdt$$

acting on the Fock-Hilbert space $\mathcal{F}$. They are local fields, so that $\varphi(f)$ and $\varphi(h)$ commute when the test functions f and h have supports which are space like separated. In other words, the field in one space time region can affect measurements in a second space time region only if a light signal can be sent from one region to the other. We construct a domain in $\mathcal{F}$ which is invariant under the field operator $\varphi(f)$ and prove that the field has a unique self adjoint extension when defined on this domain. The Heisenberg field equation

$$(\Box + m_0^2)\varphi(x, t) = -4\lambda : \varphi(x, t)^3 :$$

is satisfied, see § 3.2.

The hamiltonian operator $H(g)$ is self adjoint [8, 9] and semi-bounded [21, 7.]. We show that the bottom of the spectrum of $H(g)$ is an isolated eigenvalue (§ 2.2) and that it has multiplicity one (§ 2.3). Thus if

$$E_g = \inf \{\text{spectrum}\,(H(g))\} \tag{1.1.1}$$

there is a vector $\Omega_g \in \mathcal{F}$ such that

$$H(g)\Omega_g = E_g\Omega_g\,, \qquad \|\Omega_g\| = 1\,, \tag{1.1.2}$$

and Ω_g is unique up to a scalar multiple.

In the next paper in this series [29] we establish the existence of a physical vacuum, of the renormalized hamiltonian operator H and of the Schrödinger picture dynamics. None of these three will exist on the Hilbert space $\mathcal{F}$; in fact there is a general theorem due to Haag [10, 12] which says roughly that in order for these objects to exist, the Hilbert space $\mathcal{F}$ must be replaced by a new Hilbert space $\mathcal{F}_{\text{ren}}$.

1.2. *Notation.* The Fock space $\mathcal{F}$ is the Hilbert space completion of the symmetric tensor algebra over $L_2(R)$,

$$\mathcal{F} = \mathfrak{S}(L_2(R)) = \textstyle\bigoplus_{n=0}^{\infty} \mathcal{F}_n\,, \tag{1.2.1}$$

where $\mathcal{F}_n$ is the space of n non-interacting particles.

$$\mathcal{F}_n = L_2(R) \otimes_s L_2(R) \otimes_s \cdots \otimes_s L_2(R) \qquad (n \text{ factors}) \tag{1.2.2}$$

The variable $k \in R$ denotes momentum. For $\psi = \{\psi_0, \psi_2, \cdots\} \in \mathcal{F} = \mathcal{F}_0 \oplus \mathcal{F}_1 \oplus \cdots$, we have $\|\psi\|^2 = \sum_{n=0}^{\infty} \|\psi_n\|^2$. The no particle space $\mathcal{F}_0$ is the complex numbers, and

$$\Omega_0 = \{1, 0, 0, \cdots\} \in \mathcal{F}\,, \tag{1.2.3}$$

is the (bare) vacuum or (bare) no-particle state vector. We define operators N and H_0 by

$$(N\psi)_n = n\psi_n\,, \tag{1.2.4}$$

and

$$(H_0\psi)_n(k_1, \cdots, k_n) = \textstyle\sum_{j=1}^{n} \mu(k_j)\psi_n(k_1, \cdots, k_n)\,, \tag{1.2.5}$$

where

$$\mu(k) = (k^2 + m_0^2)^{1/2}\,. \tag{1.2.6}$$

Here N is the number of particles operator, and H_0 is the free energy operator (the free hamiltonian). The rest mass of the non-interacting particles is m_0, and $\mu(k)$ is the energy of a free particle with momentum k.

We use the standard annihilation and creation operators $a(k)$ and $a^*(k)$,

$$(a(k)\psi)_{n-1}(k_1, \cdots, k_{n-1}) = n^{1/2}\psi_n(k, k_1, \cdots, k_{n-1})\,.$$

As a convenient minimal domain for $a(k)$, we use the set $\mathfrak{S}$ of vectors ψ in $\mathcal{F}$ with $\psi_n = 0$ for large n and $\psi_n \in \mathcal{S}(R^n)$ for all n.

$$\begin{aligned}(a^*(k)\psi)_{n+1}(k_1, \cdots, k_{n+1}) \\ = (n+1)^{-1/2} \textstyle\sum_{j=1}^{n+1} \delta(k - k_j)\psi_n(k_1, \cdots, \hat{k}_j, \cdots, k_{n+1})\end{aligned} \tag{1.2.7}$$

Here the variable k_j is omitted. While $a^*(k)$ is not an operator, it is a densely defined bilinear form on $\mathfrak{S}\times\mathfrak{S}$. For a function or distribution b we can define a bilinear form

$$\begin{aligned}B = \int b(k_1, \cdots, k_m; k_1', \cdots, k_n')a^*(k_1) \cdots a^*(k_m) \\ \times a(-k_1') \cdots a(-k_n')dk_1 \cdots dk_n' .\end{aligned} \tag{1.2.8}$$

The integration helps in (1.2.8) and B is not only a bilinear form, but often an operator. This is the case if, for example, b is the kernel of a bounded operator B_0 from $\mathcal{F}_n$ to $\mathcal{F}_m$. In this case

$$\| (N + I)^{-\alpha/2}B(N + I)^{-\beta/2} \| \leqq \text{const.} \| B_0 \| , \tag{1.2.9}$$

provided that $m + n \leqq \alpha + \beta$. The constant depends only on α, β, m and n. Intuitively we think of B as being dominated by $N^{(n+m)/2}$; in particular B is an operator on $D(N^{(m+n)/2})$, the domain of $N^{(n+m)/2}$. The inequality (1.2.9) is one of our basic estimates and in using it we will often dominate $\| B_0 \|$ by the Hilbert Schmidt norm $\| B_0 \| \leqq \| b \|_2$.

By definition the field $\varphi(x)$ is

$$\varphi(x) = (4\pi)^{-1/2} \int e^{-ikx}\mu(k)^{-1/2}\{a^*(k) + a(-k)\}dk . \tag{1.2.10}$$

Since the kernels $b(k) = e^{\pm ikx}\mu(k)^{-1/2}$ are not in L_2, the B_0 are not bounded operators. In fact $\varphi(x)$ is a bilinear form and not a densely defined operator. We also define the bilinear form

$$\pi(x) = i(4\pi)^{-1/2}\int e^{-ikx}\mu(k)^{1/2}\{a^*(k) - a(-k)\}dk , \tag{1.2.11}$$

the conjugate momentum to $\varphi(x)$. For real $f(x)$, $g(x)$ such that $\mu(k)^{-1/2}\tilde{f}(k) \in L_2$ and $\mu(k)^{1/2}\tilde{g}(k) \in L_2$, the bilinear forms $\varphi(f)$ and $\pi(g)$ define operators whose closures on $\mathfrak{D}(N^{1/2})$ are self-adjoint. They satisfy the canonical commutation relations

$$e^{i\pi(g)}e^{i\varphi(f)} = e^{-i(f,g)}e^{i\varphi(f)}e^{i\pi(g)} . \tag{1.2.12}$$

It is furthermore possible to define polynomial functions of the field $\varphi(x)$, the Wick polynomials $:\varphi(x)^n:$ (see [28] and also [5, 3] for a definition of the Wick dots : :). Explicitly, as a bilinear form on $D(N^{n/2}) \times D(N^{n/2})$,

(1.2.13) $$:\varphi(x)^n: = \sum_{j=0}^{n} \binom{n}{j} b_x(k_1, \cdots, k_n) a^*(k_1) \cdots a^*(k_j) a(-k_{j+1}) \cdots a(-k_n)$$

where

$$b_x(k_1, \cdots, k_n) = (4\pi)^{-n/2} \{\mu(k_1) \cdots \mu(k_n)\}^{-1/2} e^{-i(k_1 + \cdots + k_n)x} .$$

Thus for real $f(x) \in \mathcal{S}$, the bilinear form $:\varphi^n:(f) = \int :\varphi^n(x): f(x)dx$ has a kernel proportional to $\{\mu(k_1) \cdots \mu(k_n)\}^{-1/2} \tilde{f}(k_1 + \cdots + k_n) \in L_2$. Thus from (1.2.9) we conclude that $:\varphi^n:(f)$ defines a symmetric operator on the domain $\mathfrak{D}(N^{n/2})$. It was shown in [8] that $:\varphi^n:(f)$ is essentially self-adjoint on this domain. This result was proved independently by Lanford and by Segal. (See for example [25, p. 82, lines 4–15].)

2. The cutoff hamiltonian $H(g)$

2.1. *Introduction.* The hamiltonian $H(g)$ acts on $\mathcal{F}$ and can be written in terms of the field operator $\varphi(x)$ as

(2.1.1) $$\begin{aligned} H(g) &= H_0 + \int :\varphi(x)^4: g(x)dx \\ &= H_0 + H_{I,g} , \end{aligned}$$

where $H_0 = H(0)$ is the free hamiltonian, and $0 \leqq g$. Let

$$C^\infty(H_0) = \{\psi : \psi \in D(H_0^n),\ n = 1, 2, \cdots\}$$

be the set of C^∞ vectors for H_0. It was shown in [8–9] that $H(g)$ and $H_{I,g}$ are essentially self-adjoint on $C^\infty(H_0)$, that

(2.1.2) $$\mathfrak{D}(H(g)) = \mathfrak{D}(H_0) \cap \mathfrak{D}(H_{I,g}) ,$$

and that there are finite a, $b = b(g)$ such that

(2.1.3) $$\| H_0\psi \| + \| H_{I,g}\psi \| \leqq a \| (H(g) + b)\psi \|$$

for all $\psi \in \mathfrak{D}(H(g))$.

We find that it is convenient to introduce a periodic approximation in configuration space. Under this approximation, the Fourier transform, or momentum space variable $k \in R$ is replaced by a discrete variable $k \in \Gamma_V$,

$$\Gamma_V = \left\{k : k = \frac{2\pi n}{V},\ n = 0, \pm 1, \pm 2, \cdots\right\} .$$

Thus we define $\mathcal{F}_V$, the Fock space for volume V, as

$$\begin{aligned} \mathcal{F}_V &= \mathfrak{S}(l_2(\Gamma_V)) \\ &= C \oplus l_2(\Gamma_V) \oplus [l_2(\Gamma_V) \otimes_s l_2(\Gamma_V)] \cdots . \end{aligned}$$

We identify $\mathcal{F}_V$ with the subspace of $\mathcal{F}$ consisting of piecewise constant functions which are constant on each cube of volume $(2\pi/V)^j$ centered about a lattice point

$$\{k_1, \cdots, k_j\} \in \Gamma_V \times \Gamma_V \cdots \times \Gamma_V = \Gamma_V^j .$$

The periodic annihilation and creation operators $a_V(k)$ and $a_V^*(k)$ can be extended from $\mathcal{F}_V$ to $\mathcal{F}$ through the formulas

$$a_V(k) = \left(\frac{V}{2\pi}\right)^{1/2}\int_{-\pi/V}^{\pi/V} a(k + l)dl \tag{2.1.4}$$

and

$$a_V^*(k) = \left(\frac{V}{2\pi}\right)^{1/2}\int_{-\pi/V}^{\pi/V} a^*(k + l)dl . \tag{2.1.5}$$

Similarly the periodic field $\varphi_V(x)$ and the periodic hamiltonian H_V can be extended to act on $\mathcal{F}$ through the formulas

$$\varphi_V(x) = (2V)^{-1/2} \sum_{k \in \Gamma_V} e^{-ikx}\big(a_V^*(k) + a_V(-k)\big)\mu(k)^{-1/2} , \tag{2.1.6}$$

and

$$H_V = H_{0,V} + H_{I,V} . \tag{2.1.7}$$

Here

$$H_{I,V} = \lambda\int_{-V/2}^{V/2} :\varphi_V(x)^4: dx , \tag{2.1.8}$$

and in (2.1.7) we take

$$H_{0,V} = \int a^*(k)a(k)\mu(k_V)dk , \tag{2.1.9}$$

with k_V a lattice point close to k,

$$k_V \in \Gamma_V, \ | k - k_V | \leqq \pi/V . \tag{2.1.10}$$

Note the absence of a V in the a^* and a in (2.1.9). On $\mathcal{F}_V$, this definition of $H_{0,V}$ agrees with the standard definition $\sum_{k \in \Gamma_V} a_V^*(k)a_V(k)\mu(k)$.

The operators $H_{I,V}$ and H_V are essentially self adjoint on $C^\infty(H_{0,V})$, and

$$\mathcal{D}(H_V) = \mathcal{D}(H_{0,V}) \cap \mathcal{D}(H_{I,V}) . \tag{2.1.11}$$

For all $\psi \in D(H_V)$,

$$\| H_{0,V}\psi \| + \| H_{I,V}\psi \| \leqq a \, \| (H_V + b)\psi \| \tag{2.1.12}$$

where b depends on V [8–9]. On $\mathcal{F}_V$, the operator H_V has a compact resolvent.

We want to approximate $H(g)$ by operators with compact resolvents on $\mathcal{F}_V$, so we define

$$\begin{aligned} H(g)_V &= H_{0,V} + \int :\varphi_V(x)^4: g(x)dx \\ &= H_{0,V} + H_{I,g,V} . \end{aligned} \tag{2.1.13}$$

As in [8, 9] we can show that $H(g)_V$ and $H_{I,g,V}$ are essentially self-adjoint on $C^\infty(H_{0,V})$, and that

$$\mathfrak{D}(H(g)_V) = \mathfrak{D}(H_{0,V}) \cap \mathfrak{D}(H_{I,g,V}) . \tag{2.1.14}$$

Furthermore, for all $\psi \in \mathfrak{D}(H(g)_V)$,

$$\| H_{0,V}\psi \| + \| H_{I,g,V}\psi \| \leqq a \| (H(g)_V + b)\psi \| . \tag{2.1.15}$$

In this case both g and V serve as volume cutoffs, and the constant $b = b(g, V)$ can be chosen independently of V for fixed g. On the space $\mathcal{F}_V$, the operator $H(g)_V$ has a compact resolvent.

Our hamiltonians are semi-bounded and for each $\varepsilon > 0$, there is a constant b such that

$$0 \leqq \varepsilon H_0 + H_{I,g} + b , \tag{2.1.16}$$

$$0 \leqq \varepsilon H_{0,V} + H_{I,V} + b , \tag{2.1.17}$$

and

$$0 \leqq \varepsilon H_{0,V} + H_{I,g,V} + b , \tag{2.1.18}$$

see [21, 7, 2]. In (2.1.18), the b can be chosen to be independent of V. Taking $\varepsilon = \frac{1}{2}$, we have $\frac{1}{2}H_0 \leq H(g) + b$, which implies that for all $\psi \in \mathfrak{D}(H(g)^{1/2})$,

$$\| H_0^{1/2}\psi \| \leqq \sqrt{2} \| (H(g) + b)^{1/2}\psi \| . \tag{2.1.19}$$

Here we must choose b at least $\frac{1}{2} | E_{2g} |$, where E_{2g} is the vacuum energy for the cutoff $2g$.

2.2. *The existence of a vacuum.* We prove the existence of a vacuum vector Ω_g for $H(g)$, and in the following section we prove that the vacuum is unique. Since the hamiltonian $H(g)$ is bounded from below, we can define the vacuum energy E_g to be the infimum of the spectrum of $H(g)$. We also refer to E_g as the lower bound of $H(g)$. We show that E_g is an isolated point in the spectrum. In a relativistic theory, the gap between the ground state and the first excited state is the mass of the interacting particle. For this reason we say that $H(g)$ has a mass gap. A vacuum vector $\Omega_g \in \mathcal{F}$ is defined as a normalized eigenvector of $H(g)$ corresponding to the eigenvalue E_g.

$$H(g)\Omega_g = E_g\Omega_g, \quad \| \Omega_g \| = 1 . \tag{2.2.1}$$

Our result is

THEOREM 2.2.1. *There exists a vacuum vector Ω_g for $H(g)$. For any $\varepsilon > 0$, the operator $H(g)$, restricted to the spectral interval $[E_g, E_g + m_0 - \varepsilon]$ is compact.*

In this section we also prove

THEOREM 2.2.2. *The approximate hamiltonian $H(g)_V$ has a vacuum vector $\Omega_{g,V}$. Any sequence of volumes V_j tending to infinity has a subsequence V_l such that*

$$\Omega_g = \lim_l \Omega_{g,V_l} \tag{2.2.2}$$

exists and satisfies (2.2.1).

Let $E_{g,V}$ be the lower bound of $H(g)_V$ on $\mathcal{F}_V$. Since $H(g)_V$ has a compact resolvent on $\mathcal{F}_V$, there is a vacuum vector $\Omega_{g,V} \in \mathcal{F}_V$ for $H(g)_V \mid \mathcal{F}_V$. We now see that $E_{g,V}$ is the lower bound for $H(g)_V$ on $\mathcal{F}$, so that $\Omega_{g,V}$ is a vacuum vector for $H(g)_V$.

Let $\mathcal{F}_V^\perp$ be the orthogonal complement of $\mathcal{F}_V$. Since $H(g)_V$ leaves $\mathcal{F}_V$ invariant and is self-adjoint, $H(g)_V$ also leaves $\mathcal{F}_V^\perp$ invariant.

LEMMA 2.2.3. *The lower bound of $H(g)_V$ on $\mathcal{F}_V^\perp$ is $E_{g,V} + m_0$, where m_0 is the rest mass of the Fock space bosons.*

This lemma shows that $\Omega_{g,V}$ is a vacuum for $H(g)_V$.

PROOF. We have an orthogonal decomposition in the single particle space

$$\mathcal{F}_1 = L_2(R) = \mathcal{F}_{1V} \oplus \mathcal{F}_{1V}^\perp \,. \tag{2.2.3}$$

Here $\mathcal{F}_{1V} = \mathcal{F}_1 \cap \mathcal{F}_V$ consists of functions piecewise constant on intervals centered at lattice points. Thus we may write

$$\mathcal{F} = \bigoplus_{j=0}^\infty \mathcal{F}^{(j)} \,, \qquad \mathcal{F}_V^\perp = \bigoplus_{j=1}^\infty \mathcal{F}^{(j)} \,, \tag{2.2.4}$$

where $\mathcal{F}^{(j)}$ consists of vectors with exactly j particles from $\mathcal{F}_{1V}^\perp$ and

$$\mathcal{F}^{(j)} = \mathcal{F}_{1V}^\perp \otimes_s \cdots \otimes_s \mathcal{F}_{1V}^\perp \otimes_s \mathcal{F}_V \,. \tag{2.2.5}$$

In this tensor product decomposition there are j factors $\mathcal{F}_{1V}^\perp$. The hamiltonian $H(g)_V$ leaves each subspace $\mathcal{F}^{(j)}$ invariant, and on $\mathcal{F}^{(j)}$ we have $H(g)_V = I \otimes A + B \otimes I$, where $A = H(g) \mid \mathcal{F}_V$ and B is a sum of j copies of $H_{0,V}$, each acting on a single factor $\mathcal{F}_{1V}^\perp$. Since

$$jm_0 \leqq B \,, \tag{2.2.6}$$

the lemma follows from this decomposition.

LEMMA 2.2.4. *For $V \leqq \infty$, and for b sufficiently large we have*

$$\mathfrak{D}(H_0) \subset \mathfrak{D}(H_0^{1/2}) \cap \mathfrak{D}(N) \subset \mathfrak{D}\big(H(g)_V + b)^{1/2}\big) \,, \tag{2.2.7}$$

and

$$\mathfrak{D}(H_0) \subset \mathfrak{D}\big([(N + I)^{-1}(H(g)_V + b)]^-\big) \,. \tag{2.2.8}$$

PROOF. We take b large enough so that $H(g)_V + b$ is positive, see (2.1.18). By (1.2.9) and (2.1.14)

$$\mathfrak{D}(H_0) \cap \mathfrak{D}(N^2) \subset \mathfrak{D}(H_0) \cap \mathfrak{D}(H_{I,g,V}) = \mathfrak{D}\big(H(g)_V\big) \subset \mathfrak{D}\big((H(g)_V + b)^{1/2}\big) \,.$$

Thus for $\psi \in \mathfrak{D}(H_0) \cap \mathfrak{D}(N^2)$,

$$\begin{aligned} \| (H(g)_V + b)^{1/2}\psi \|^2 &= (\psi, (H(g)_V + b)\psi) \\ &\leqq (\psi, (H_{0,V} + b)\psi) \\ &\quad + \| (N + I)^{-1} H_{I,g,V} (N + I)^{-1} \| \, \| (N + I)\psi \|^2 . \end{aligned}$$

Since $(H(g)_V + b)^{1/2}$ is a closed operator, we can extend this inequality by continuity. As N and $H_{0,V}$ commute, the inequality extends by continuity to all $\psi \in \mathfrak{D}(H_0^{1/2}) \cap \mathfrak{D}(N) \supset \mathfrak{D}(H_0)$. The proof of (2.2.8) is similar.

LEMMA 2.2.5. *Let z be non-real or real and sufficiently negative. Then as V tends to infinity,*

$$\| (H(g)_V - zI)^{-1} - (H(g) - zI)^{-1} \| = O(V^{-1}) , \tag{2.2.9}$$

and $E_{g,V} \to E_g$.

PROOF. (See also [9, Cor. 10(b)].) Let us fix g and z and suppress g when possible. In [8] we have shown that $H(g)$ is essentially self-adjoint on $C^\infty(H_0)$. Thus vectors of the form $\chi = (H - z)\psi$, $\psi \in C^\infty(H_0)$, are dense in $\mathcal{F}$. On such vectors

$$\begin{aligned} \{(H_V - z)^{-1} - (H - z)^{-1}\}\chi &= (H_V - z)^{-1}\{(H - z)\psi - (H_V - z)\psi\} \\ &= (H_V - z)^{-1}(H - H_V)(H - z)^{-1}\chi \\ &= (H_V - z)^{-1}(N + I)(N + I)^{-1}(H - H_V)(N + I)^{-1} \\ &\quad \times (N + I)(H - z)^{-1}\chi . \end{aligned}$$

For $\theta \in \mathcal{F}$,

$$\begin{aligned} &| (\theta, \{(H_V - z)^{-1} - (H - z)^{-1}\}\chi) | \\ &\leqq \| (N + I)(H_V - \bar{z})^{-1} \| \, \| \theta \| \, \| (N + I)^{-1}(H - H_V)(N + I)^{-1} \| \\ &\quad \| (N + I)(H - z)^{-1} \| \, \| \chi \| . \end{aligned} \tag{2.2.10}$$

Using (2.1.15), we find that $\| (N + I)(H_V - \bar{z})^{-1} \|$ is bounded uniformly in V, since

$$\begin{aligned} \| (N + I)(H_V - \bar{z})^{-1}\psi \| &\leqq \text{const.} \, \| (H_{0,V} + I)(H_V - \bar{z})^{-1}\psi \| \\ &\leqq \text{const.} \, \| H_V(H_V - \bar{z})^{-1}\psi \| \\ &\quad + \text{const.} \, \| (H_V - \bar{z})^{-1}\psi \| , \end{aligned}$$

where the constants can be chosen independently of V. In a similar fashion, (2.2.3) shows that $(N + I)(H - z)^{-1}$ is a bounded operator. Thus from (2.2.10), and the fact that the χ are dense, we infer

$$\| (H_V - z)^{-1} - (H - z)^{-1} \| \leqq \text{const.} \, \| (N + I)^{-1}(H - H_V)(N + I)^{-1} \| , \tag{2.2.11}$$

with a constant independent of V.

The difference $H - H_V = (H_0 - H_{0,V}) + (H_{I,g} - H_{I,g,V})$, and for large V,

$$\| (N + I)^{-1/2}(H_0 - H_{0,V})(N + I)^{-1/2} \| = O(V^{-1}) . \tag{2.2.12}$$

This is a simple direct computation, using $|\mu(k_V) - \mu(k)| = O(V^{-1})$. For the interaction terms, we use (1.2.10) to estimate

$$\|(N+I)^{-1}(H_{I,g} - H_{I,g,V})(N+I)^{-1}\| . \tag{2.2.13}$$

The kernel $b(k_1, \cdots, k_4)$ corresponding to a monomial in $H_{I,g}$ is

$$b(k_1, \cdots, k_4) = \binom{4}{j} \prod_{i=1}^{4} \mu(k_i)^{-1/2} \tilde{g}(k_1 + k_2 + \cdots + k_4) , \qquad 0 \leqq j \leqq 4 .$$

The kernel $b_V(k_1, \cdots, k_4)$ for the corresponding monomial in $H_{I,g,V}$ is obtained by replacing the factor $\prod_{i=1}^{4} \mu(k_i)^{-1/2}$ with $\prod_{i=1}^{4} \mu(k_{iV})^{-1/2}$. Inspection of $b - b_V$ shows that as V tends to infinity, $\|b - b_V\|_{L^2} = O(V^{-1})$, from which we conclude that (2.2.13) is $O(V^{-1})$. The convergence of the resolvents follows from (2.2.11–13). The limit $E_{g,V} \rightarrow E_g$ follows from the convergence of the resolvents, since for large positive b,

$$(E_{g,V} + b)^{-1} = \|(H(g)_V + b)^{-1}\| .$$

PROOF OF THEOREMS 2.2.1 AND 2.2.2. Let $f(x)$ be a smooth positive function with support in the interval $[-\varepsilon, m_0 - \varepsilon]$. Then $f(H(g)_V - E_{g,V}) \mid \mathcal{F}_V$ is compact, since the resolvent of $H(g)_V \mid \mathcal{F}_V$ is compact on $\mathcal{F}_V$. By Lemma 2.2.3, $f(H(g)_V - E_{g,V}) \mid \mathcal{F}_V^{\perp} = 0$. Thus $f(H(g) - E_{g,V})$ is compact on the full Fock space $\mathcal{F}$.

By Lemma 2.2.5, the resolvents $(H(g)_V - E_{g,V} - z)^{-1}$ converge in norm as $V \rightarrow \infty$. Therefore

$$\|f(H(g)_V - E_{g,V}) - f(H(g) - E_g)\| \rightarrow 0 ,$$

since $f(H(g) - E_g)$ is a bounded function of $(H(g) - E_g - z)^{-1}$ which vanishes at infinity. Since the uniform limit of compact operators is compact, $H(g)$ restricted to the spectral interval $[E_g, E_g + m_0 - \varepsilon]$ is compact. This means furthermore that only a finite number of eigenvalues of $H(g)_V$ converge to E_g. Theorem 3.16 of [18, p. 212] shows that the projection onto the corresponding set of eigenvectors of $H(g)_V$ converge as $V \rightarrow \infty$. Since $\Omega_{g,V}$ is an eigenvector of $f(H(g)_V - E_{g,V})$, a subsequence of the $\Omega_{g,V}$ converge to a limit as V tends to ∞. For this limit

$$\begin{aligned}(E_g + b)^{-1}\Omega_g &= \lim_{l\rightarrow\infty} (E_{g,V_l} + b)^{-1}\Omega_{g,V_l} \\ &= \lim_{l\rightarrow\infty} (H(g)_{V_l} + b)^{-1}\Omega_{g,V_l} \\ &= (H(g) + b)^{-1}\Omega_g ,\end{aligned}$$

by Lemma 2.2.5. Hence $\Omega_g \in \mathfrak{D}(H(g))$, $H(g)\Omega_g = E_g\Omega_g$, and Ω_g is a vacuum vector for $H(g)$.

In the following section we will see that Ω_g and $\Omega_{g,V}$ are unique except for an arbitrary phase multiple $e^{i\theta}$, and that there is a natural choice for this arbitrary phase. With this choice, we then will prove that the $\Omega_{g,V}$ converge to Ω_g as $V \rightarrow \infty$.

2.3. *Uniqueness of the vacuum.*

THEOREM 2.3.1. *The vacuum vector Ω_g for $H(g)$ is unique.*

In other words E_g, the lower bound of $H(g)$ is a simple eigenvalue. Using the Feynman-Kac formula we show that the bounded operator $e^{-H(g)}$ has a positive, ergodic kernel. We also show that a positive ergodic kernel ensures that the eigenvalue $e^{-E_g} = \| e^{-H(g)} \|$ is simple.

Let $\mathcal{H} = L^2(Q, df)$ be a Hilbert space. We say that a bounded operator A has a *strictly positive kernel* provided that

$$0 < (\psi, A\chi) \tag{2.3.1}$$

whenever ψ and χ are non-negative L_2 functions with non-zero norms. Such an operator transforms a function χ, $\chi \geqq 0$, $\| \chi \| \neq 0$, into a function $A\chi$ which is strictly positive almost everywhere.

A weaker requirement is that A have a *positive, ergodic kernel.* We say that A has a positive, ergodic kernel if for each ψ, χ as above $0 \leqq (\psi, A\chi)$ and

$$0 < (\psi, A^j\chi) \tag{2.3.2}$$

for some j, depending on ψ and χ. Clearly every A with a strictly positive kernel has a positive, ergodic kernel.

THEOREM 2.3.2. *Let A have a positive ergodic kernel, and suppose that $\| A \|$ is an eigenvalue of A. Then $\| A \|$ is a simple eigenvalue and the corresponding eigenvector can be chosen to be a strictly positive function.*

Remark. Classical theorems of this sort make assumptions about A which are strong enough to prove both the existence of an eigenvalue and its uniqueness. In 1912, Jentzsch [16] showed that on a compact interval, any symmetric integral operator with a strictly positive, continuous kernel has a maximum eigenvalue, and it is simple. For positive matrices see Perron [22] and Frobenius [4].

PROOF. Since A maps positive functions into positive functions, it also maps real functions into real functions. If $\psi \in \mathcal{H}$ satisfies $A\psi = \| A \| \psi$, then so do Re ψ and Im ψ. Therefore without loss of generality we may assume that ψ is real. Since $\| A^j \| = \| A \|^j$, and $A^j\psi = \| A \|^j \psi$, we infer that

$$\| A^j \| \, \| \psi \|^2 = (\psi, A^j\psi) \leqq (| \psi |, A^j | \psi |) \leqq \| A^j \| \, \| \psi \|^2 ,$$

and

$$(\psi, A^j\psi) = (| \psi |, A^j | \psi |) .$$

Writing $\psi = \psi^+ - \psi^-$, where ψ^+ and ψ^- are the positive and negative parts of ψ,

$$(\psi^+, A^j\psi^+) - (\psi^+, A^j\psi^-) - (\psi^-, A^j\psi^+) + (\psi^-, A^j\psi^-)$$
$$= (\psi^+, A^j\psi^+) + (\psi^+, A^j\psi^-) + (\psi^-, A^j\psi^+) + (\psi^-, A^j\psi^-) ,$$

or

$$(\psi^+, A^j\psi^-) + (\psi^-, A^j\psi^+) = 0 . \tag{2.3.3}$$

Unless $\psi^+ = 0$ or $\psi^- = 0$, each term of (2.3.3) could be made strictly positive by choosing an appropriate j. Thus either ψ^+ or ψ^- must vanish, and we may choose the eigenvector ψ to be non-negative.

If $\chi \geqq 0$, $\| \chi \| \neq 0$, then for some j, $0 < (\chi, A^j\psi) = \| A \|^j (\chi, \psi)$. This proves that $\chi\psi$ is not zero almost everywhere, and that ψ is strictly positive almost everywhere.

Finally, if ψ and χ were linearly independent eigenvectors of A with the eigenvalue $\| A \|$, then we could repeat the above argument with the component of χ orthogonal to ψ. This would yield two positive, orthogonal eigenvectors, which would be impossible, and the proof is complete.

Let $\varphi(h) = \int \varphi(x)h(x)dx$ denote the smeared, time zero free field operators. The spectral projections of the $\varphi(h)$, or the functions $\exp\big(i\varphi(h)\big)$ generate a maximal abelian algebra $\mathfrak{M}$ of bounded operators on $\mathcal{F}$. Let Q be the spectrum of $\mathfrak{M}$. The no particle vector $\Omega_0 \in \mathcal{F}$ is a cyclic vector for $\mathfrak{M}$, namely $\mathcal{F} = (\mathfrak{M}\Omega_0)^-$. Therefore we may introduce a measure df on Q so that $\mathcal{F}$ is unitarily equivalent to $L^2(Q, df)$ and so that the equivalence carries $\mathfrak{M}$ into $L_\infty(Q)$ and takes Ω_0 into the function 1.

LEMMA 2.3.3. *With $\mathcal{F}$ represented as $L^2(Q, df)$, e^{-H_0} has a positive, ergodic kernel.*

PROOF. Let ψ and χ be non-negative. Write $\psi = \psi_1 + \psi_2$, where ψ_1 is the component of ψ along Ω_0. Thus the L_1 norm of ψ is given by $\| \psi \|_1 = (\psi, \Omega_0) = (\psi_1, \Omega_0)$. Note $\| \psi \|_1 \neq 0$ whenever ψ is non-zero, and $\| e^{-tH_0}\psi_2 \| \leqq e^{-tm_0} \| \psi_2 \|$, where m_0 is the boson mass. Thus

$$(\psi, e^{-tH_0}\chi) \geqq \| \psi \|_1 \| \chi \|_1 - \| \psi_2 \| \| \chi_2 \| e^{-tm_0} . \tag{2.3.4}$$

By choosing t sufficiently large, (2.3.4) is positive, which proves (2.3.2). If

$$e^{-tm_0} < \frac{1}{2} \frac{\| \psi \|_1 \| \chi \|_1}{\| \psi_2 \| \| \chi_2 \|} = \frac{1}{2} \frac{\| \psi \|_1 \| \chi \|_1}{(\| \psi \|^2 - \| \psi \|_1^2)^{1/2}(\| \chi \|^2 - \| \chi \|_1^2)^{1/2}} \tag{2.3.5}$$

then

$$(\psi, e^{-tH_0}\chi) \geqq \frac{1}{2} \| \psi \|_1 \| \chi \|_1 . \tag{2.3.6}$$

We need to show that $0 \leqq (\psi, e^{-tH_0}\chi)$ for all t. In fact, it is sufficient to prove this for a dense set of non-negative ψ and χ.

Let us consider an approximate free energy operator

$$H_{0,K,V} = \int_{-K}^{K} a^*(k)a(k)\mu(k_V)dk \ .$$

For vectors $\psi \in C^\infty(H_0)$, as $K, V \to \infty$, $\| H_{0,K,V}\psi - H_0\psi \| \to 0$. Since H_0 is essentially self-adjoint on $C^\infty(H_0)$, the resolvents of $H_{0,K,V}$ converge strongly [18, p. 429]. Thus the semigroup convergence theorem [18, p. 502] ensures that for all $\psi \in \mathcal{F}$, $\| e^{-tH_{0,K,V}}\psi - e^{-tH_0}\psi \| \to 0$ as $K, V \to \infty$, and the convergence is uniform on compact sets of t. Therefore we need only show that for a dense set of non-negative ψ and χ, $0 \leqq (\psi, e^{-tH_{0,K,V}}\chi)$. Let $F(x_1, \cdots, x_n)$ be a non-negative, infinitely differentiable function with compact support, and let

$$\psi = F(\varphi(f_1), \cdots, \varphi(f_n))\Omega_0 \ , \tag{2.3.8}$$

where $f_1, \cdots, f_n$ are real. The set of all such vectors are dense in $\mathcal{F}^+$, the non-negative vectors in $\mathcal{F}$. Furthermore, we define

$$\psi_{K,V} = F(\varphi_{K,V}(f_1) \cdots \varphi_{K,V}(f_n))\Omega_0 \ , \tag{2.3.9}$$

where $\varphi_{K,V}(x)$ is defined by restricting the sum in (2.1.6) to those $k \in \Gamma_{K,V} = \Gamma_V \cap \{k: |k| \leqq K\}$. Then $\psi_{K,V} \in \mathcal{F}^+_{K,V} \subset \mathcal{F}^+$ where $\mathcal{F}_{K,V}$ is the Fock space corresponding to the modes $k \in \Gamma_{K,V}$. For any vector $\chi \in C^\infty(H_0)$,

$$\| \varphi_{K,V}(f)\chi - \varphi(f)\chi \| \to 0 \ , \qquad \text{as } K, V \to \infty \ ,$$

and as $C^\infty(H_0)$ is a core for $\varphi(f)$, the resolvents of $\varphi_{K,V}(f)$ converge strongly to the resolvent of $\varphi(f)$[18, p. 429]. Thus the semigroup convergence theorem [18, p. 502] ensures that for each $\chi \in \mathcal{F}$, s real,

$$\| e^{is\varphi_{K,V}(f)}\chi - e^{is\varphi(f)}\chi \| \to 0 \ , \qquad \text{as } K, V \to \infty \ ,$$

and the convergence is uniform for compact sets of s. By (2.3.9)

$$\psi_{K,V} = \int \tilde{F}(s_1, \cdots, s_n) \exp\left(i \textstyle\sum_{j=1}^{n} s_j\varphi_{K,V}(f_j)\right)ds_1 \cdots ds_n \ ,$$

and $\tilde{F}(s_1, \cdots, s_n)$ vanishes rapidly at infinity, so we conclude that

$$\| \psi_{K,V} - \psi \| \to 0 \ , \qquad \text{as } K, V \to \infty \ .$$

Thus for such vectors ψ, χ,

$$(\psi, e^{-tH_0}\chi) = \lim_{K,V\to\infty} (\psi_{K,V}, e^{-tH_{0,K,V}}\chi_{K,V})$$

and we need only show that

$$0 \leqq (\psi_{K,V}, e^{-tH_{0,K,V}}\chi_{K,V}) \ . \tag{2.3.10}$$

However on $\mathcal{F}_{K,V}$, $H_{0,K,V} = \sum_{k\in\Gamma_{K,V}} a_V^*(k)a_V(k)\mu(k) = \sum_{k\in\Gamma_{K,V}} H_{0,K,V}$, so $e^{-tH_{0,K,V}} =$

$\prod_{k \in \Gamma_{K,V}} e^{-tH_{0,K,V}}$. Each operator $e^{-tH_{0,K,V}}$ is known by explicit computation to have a strictly positive kernel [21], so (2.3.10) holds and the proof is complete.

LEMMA 2.3.4. *With $\mathcal{F}$ represented as $L^2(Q,df)$, the operator $e^{-H(g)}$ has a positive, ergodic kernel.*

Remark. We expect that e^{-H_0} and $e^{-H(g)}$ have strictly positive kernels.

PROOF. As in Lemma 2.3.3, formula (2.3.7), we consider $H(g)_{K,V} = H_{0,K,V} + H_{I,g,K,V}$. The approximate interaction $H_{I,g,K,V}$ is constructed with $\varphi_{K,V}$ in place of φ_V. Since $C^\infty(H_0)$ is a core for $H(g)$, we can argue as in the previous lemma that for all $\psi \in \mathcal{F}$,

$$\exp\left(-tH(g)_{K,V}\right)\psi \longrightarrow \exp\left(-tH(g)\right)\psi , \qquad \text{as } K, V \longrightarrow \infty .$$

Thus we need only prove that for ψ, χ as in Lemma 2.3.2,

$$0 < \varepsilon < (\psi_{K,V}, e^{-tH(g)_{K,V}}\chi_{K,V}) , \tag{2.3.11}$$

and that for sufficiently large t, the constant $\varepsilon = \varepsilon(\psi, \chi, t, K, V)$ can be chosen independently of K and V.

On $\mathcal{F}_{K,V}$ we have an explicit representation of $\exp\left(-tH(g)_{K,V}\right)$ given by the Feynman-Kac integral formula [17].

$$\begin{aligned}&\left(\psi_{K,V}, \exp\left(-tH(g)_{K,V}\right)\chi_{K,V}\right)\\ &\quad= \int_{C_{K,V}} \exp\left(-\int_0^t H_{I,g,K,V}(q(s))ds\right)\psi_{K,V}(q(0))\chi_{K,V}(q(t))dq(\cdot) .\end{aligned} \tag{2.3.12}$$

Here q denotes a point in the spectrum of the modes

$$q_V(k) = a_V(k) + a_V(-k) + a_V^*(k) + a_V^*(-k) ,$$

and

$$q'_V(k) = a_V(k) - a_V(-k) + a_V^*(k) - a_V^*(-k) ,$$

for $k \in \Gamma_{K,V}$, and $C_{K,V}$ is the path space for these modes. Since $\exp(-tH_{0,K,V})$ has a strictly positive kernel, (2.3.12) exhibits $\exp\left(-tH(g)_{K,V}\right)$ explicitly as an operator with a strictly positive kernel. Thus (2.3.11) is valid, and taking the limit K, $V \longrightarrow \infty$ shows that

$$0 \leqq (\psi, e^{-tH(g)}\chi) . \tag{2.3.13}$$

We now establish a uniform lower bound on ε in (2.3.11) to prove that for t sufficiently large (2.3.13) is strictly positive. Given any positive M we can split the integral (2.3.13) into two parts. Let $C_{K,V}^{(1)}$ be those paths such that the exponent in the Feynman-Kac formula satisfies $-\int_0^t H_{I,g,K,V}(q(s))ds \geqq -M$, and let $C_{K,V}^{(2)}$ be the complementary set of paths. Thus

$$
\begin{aligned}
(\psi_{\mathrm{K},V}, e^{-tH(g)_{\mathrm{K},V}}\chi_{\mathrm{K},V}) &\geqq e^{-M}\int_{C^{(1)}_{\mathrm{K},V}} \psi_{\mathrm{K},V}(q(0))\chi_{\mathrm{K},V}(q(t))dq(\cdot) \\
&= e^{-M}\Big\{(\psi_{\mathrm{K},V}, e^{-tH_{0,\mathrm{K},V}}\chi_{\mathrm{K},V}) \\
&\quad - \int_{C^{(2)}_{\mathrm{K},V}} \psi_{\mathrm{K},V}(q(0))\chi_{\mathrm{K},V}(q(t))dq(\cdot)\Big\}.
\end{aligned} \tag{2.3.14}
$$

First we choose t by (2.3.5) so that (2.3.6) holds. Then for sufficiently large K and V (depending on t),

$$
\begin{aligned}
(\psi_{\mathrm{K},V} e^{-tH_{0,\mathrm{K},V}}\psi_{\mathrm{K},V}) &\geqq \frac{1}{2}(\psi, e^{-tH_0}\chi) \\
&\geqq \frac{1}{4}\|\psi\|_1\|\chi\|_1 .
\end{aligned}
$$

Thus (2.3.14) becomes

$$
\begin{aligned}
&(\psi_{\mathrm{K},V}, e^{-tH(g)_{\mathrm{K},V}}\chi_{\mathrm{K},V}) \\
&\quad \geqq e^{-M}\Big\{\frac{1}{4}\|\psi\|_1\|\chi\|_1 - \int_{C^{(2)}_{\mathrm{K},V}} \psi_{\mathrm{K},V}(q(0))\chi_{\mathrm{K},V}(q(t))dq(\cdot)\Big\}.
\end{aligned}
$$

Let Pr denote the measure on path space, so that by the Hölder inequality

$$
\begin{aligned}
&\Big|\int_{C^{(2)}_{\mathrm{K},V}} \psi_{\mathrm{K},V}(q(0))\chi_{\mathrm{K},V}(q(t))dq(\cdot)\Big| \\
&\quad \leqq (\Pr\{C^{(2)}_{\mathrm{K},V}\})^{(r-1)/r}\|\psi_{\mathrm{K},V}(q(0))\chi_{\mathrm{K},V}(q(t))\|_r ,
\end{aligned}
$$

where $1 < r < 2$. By the smoothing property of $\exp(-tH_{0,\mathrm{K},V})$ for sufficiently large t [7],

$$
\|\psi_{\mathrm{K},V}(q(0))\chi_{\mathrm{K},V}(q(t))\|_r \leqq \|\psi_{\mathrm{K},V}\|_2\|\chi_{\mathrm{K},V}\|_2 ,
$$

and for K and V sufficiently large, this is dominated by $2\|\psi\|_2\|\chi\|_2$. Thus with the choices so far made for K, V, t, M,

$$
\begin{aligned}
(\psi_{\mathrm{K},V}, e^{-tH(g)_{\mathrm{K},V}}\chi_{\mathrm{K},V}) &\geqq e^{-M}\Big\{\frac{1}{4}\|\psi\|_1\|\chi\|_1 \\
&\quad -2\|\psi\|_2\|\chi\|_2(\Pr\{C^{(2)}_{\mathrm{K},V}\})^{(r-1)/r}\Big\}, \\
&\geqq \frac{1}{8}e^{-M}\|\psi\|_1\|\chi\|_1 > \varepsilon(\psi, \chi, t) > 0 ,
\end{aligned}
$$

provided in addition that

$$
\Pr\{C^{(2)}_{\mathrm{K},V}\} \leq \left(\frac{\|\psi\|_1\|\chi\|_1}{16\|\psi\|_2\|\chi\|_2}\right)^{r/(r-1)} . \tag{2.3.15}
$$

We now show that for M sufficiently large, (2.3.15) is satisfied and therefore the lemma is proved.

$$\begin{aligned}\Pr\{C^{(2)}_{\mathrm{K},V}\} &= \Pr\left\{M \leqq \int_0^t H_{I,g,\mathrm{K},V}(q(s))ds\right\} \\ &= \Pr\left\{1 \leqq M^{-1}\int_0^t H_{I,g,\mathrm{K},V}(q(s))ds\right\} \\ &\leqq M^{-2}\int_{C_{\mathrm{K},V}}\left|\int_0^t H_{I,g,\mathrm{K},V}(q(s))ds\right|^2 dq(\cdot)\ .\end{aligned}$$

Replacing the integral over s by a limit of Riemann sums, we obtain a bound in terms of Wiener integrals depending on a finite number of times.

$$\begin{aligned}\Pr\{C^{(2)}_{\mathrm{K},V}\} \leqq \lim_{n\to\infty}(t/nM)^2\sum_{i,j=1}^n \int_{C_{\mathrm{K},V}} & H_{I,g,\mathrm{K},V}(q(ti/n)) \\ & \times H_{I,g,\mathrm{K},V}(q(tj/n))dq(\cdot)\ .\end{aligned}$$

By the definition of the Wiener integral, this expression can be evaluated in terms of no-particle expectation values, and it equals

$$\lim_{n\to\infty}(t/nM)^2\sum_{i,j=1}^n(\Omega_0,\ H_{I,g,\mathrm{K},V}\exp(-\mid i-j\mid tH_0/n)\cdot H_{I,g,\mathrm{K},V}\Omega_0)\ .$$

By the Schwarz inequality

$$\begin{aligned}\Pr\{C^{(2)}_{\mathrm{K},V}\} &\leqq (t/M)^2\,\|\,H_{I,g,\mathrm{K},V}\Omega_0\,\|^2 \\ &\leqq (tD/M)^2\ ,\end{aligned}$$

for some constant D independent of K and V. Thus we choose

$$M \geqq Dt\left(\frac{16\,\|\,\psi\,\|_2\,\|\,\chi\,\|_2}{\|\,\psi\,\|_1\,\|\,\chi\,\|_1}\right)^{r/2(r-1)}\ .$$

Combining Lemma 2.3.4 with Theorem 2.3.2 yields a proof of Theorem 2.3.1. Clearly the same proof applies to $H(g)_V$, to show that its vacuum is unique.

COROLLARY 2.3.5. *Let $\Omega_{g,V}$ be the vacuum for $H(g)_V$ with its phase determined by the requirement*

$$(\Omega_0,\ \Omega_{g,V}) > 0\ . \tag{2.3.16}$$

Then $\lim_{V\to\infty}\Omega_{g,V} = \Omega_g$ exists, Ω_g is the vacuum for $H(g)$, and

$$(\Omega_0,\ \Omega_g) > 0\ . \tag{2.3.17}$$

PROOF. A sequence Ω_{g,V_j} with $V_j \to \infty$ has a convergent subsequence by Theorem 2.2.2, converging to a vacuum for $H(g)$. The phase (2.3.16) fixes the phase (2.3.17) so every convergent subsequence has the same limit Ω_g. Thus the $\Omega_{g,V}$ converge to Ω_g, as required.

COROLLARY 2.3.6. *The vacuum Ω_g is a cyclic vector for $\mathfrak{M}$.*

PROOF. The function Ω_g is positive for almost all Q, and $\mathfrak{M} = L_\infty(Q)$ in the $L_2(Q,\ df)$ representation of $\mathcal{F}$.

3. The Heisenberg picture field operators

3.1. *Introduction.* In the Heisenberg picture operators have the time dependence

$$A(t) = e^{itH(g)}A(0)e^{-itH(g)} . \tag{3.1.1}$$

This definition of the dynamics contains the cutoff function $g(x)$ explicitly. For an important class of operators $A(0)$, however, $A(t)$ is independent of $g(x)$ provided that $g(x) = \lambda$, the coupling constant, on a suitably large set. For example, we take $A(0)$ to be an observable representing a measurement performed in some region B of space (at time $t = 0$). Then $A(t)$ represents the same measurement performed at time t. A hamiltonian with no ultraviolet cutoff, such as $H(g)$, propagates information with at most the speed of light. Therefore if $g(x) = \lambda$ on an interval containing B, and t is sufficiently small, the fact that $g(x)$ does not equal λ everywhere will never be recorded by a measurement $A(t)$. See Figures 1–2.

For each localized observable $A(0)$ and each t, we make an appropriate choice for $g(x)$. Therefore (3.1.1) provides the correct dynamics for the $(\varphi^4)_2$

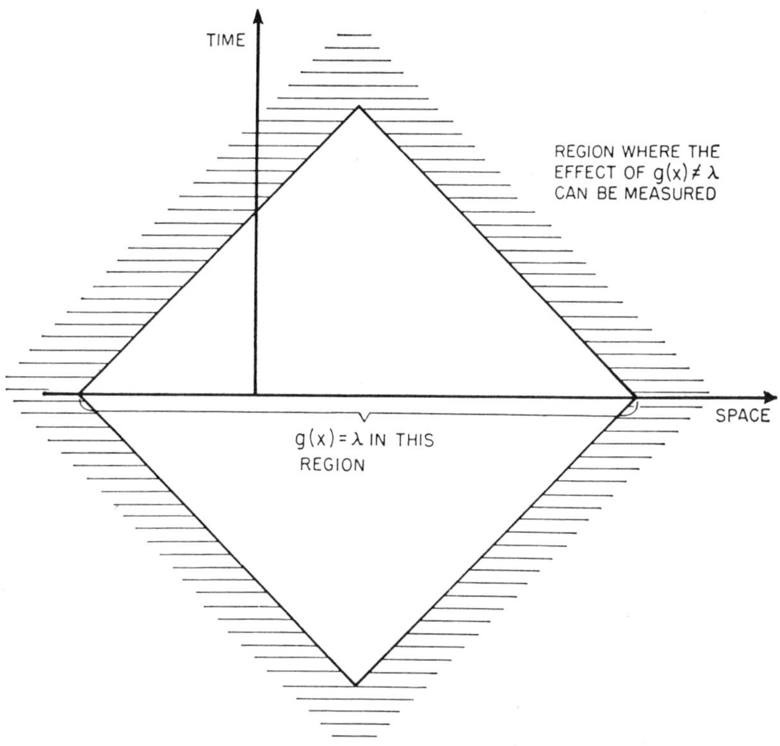

FIG. 1 Information propagates at the speed of light for a local hamiltonian

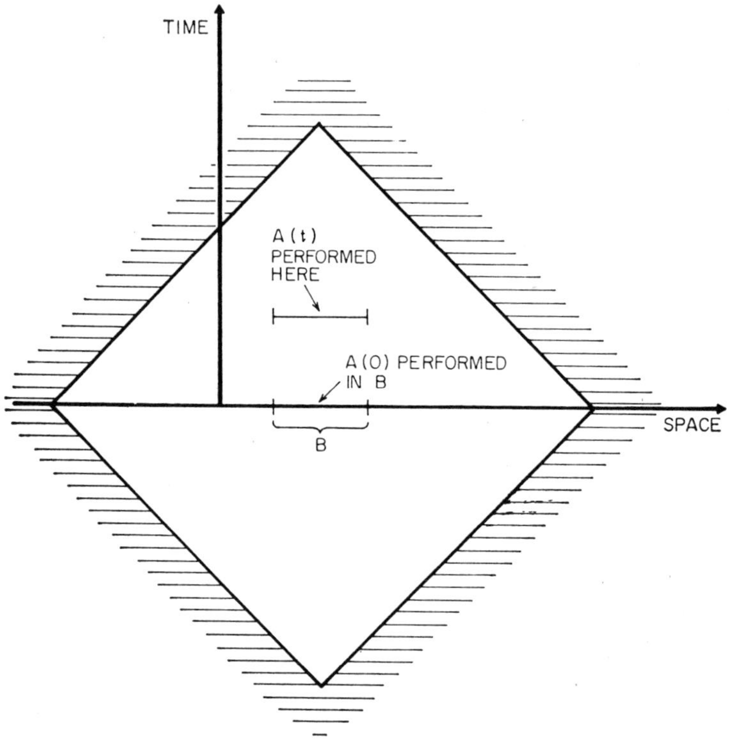

FIG. 2 Measurements performed inside the square obey laws of physics independent of any cutoff

quantum field theory with all the cutoffs removed. The mathematical proof of these statements is the main result of [8–9]. This method of removing the spatial cutoff was first proposed by Guenin. In [8] we used a theorem of Segal [26] to the effect that the above finite propagation speed follows from the Trotter product formula. In order to apply the Trotter product formula, we proved in [8] that $H(g)$ is self-adjoint. See [8] for further discussion.

In this section we discuss the field operators $\varphi(x, t)$ or

$$\varphi(f) = \int \varphi(x, t) f(x, t) dx dt \, . \tag{3.1.2}$$

The operator $\varphi(f)$ is unbounded but for real f it is self-adjoint. The field $\varphi(x, t)$ is not an operator, but it is a bilinear form. It is possible to regard the field at a given time t as an amplitude for a collection of oscillators, one at each point of space x. Then the field strength $\varphi(x, t)$(field deformation) measures the amplitude of the oscillator at x at time t, and $\pi(x, t) = \dot{\varphi}(x, t)$ represents the momentum of the corresponding oscillator. The kinetic energy density for the oscillators is $\frac{1}{2}\!:\pi(x, t)^2:$, while the total energy density is

$$\frac{1}{2}(:\pi(x,t)^2: + :(\nabla\varphi(x,t))^2: + m_0^2 :\varphi(x,t)^2:) + \lambda :\varphi(x,t)^4: ,$$

where $\lambda \geqq 0$. For both the classical and the quantum field theory, the positivity of the energy ensures that in states of finite energy, the field strength $\varphi(x, t)$ has one L_2 derivative as a function of x. For the quantum case, the precise mathematical meaning is that for $\psi \in \mathfrak{D}((H(g) + b)^{1/2})$,

$$\left|\int(\psi, \varphi(x,t)\psi)(D_x f)(x,t)dxdt\right| \leq \text{const.}\ \|f(\cdot, t)\|_2 (\psi, (H(g)+b)\psi) .$$

In particular, in one dimensional space a Sobolev inequality ensures that a function with one L_2 derivative is continuous. Hence the expectation value of the field strength $(\psi, \varphi(x, t)\psi)$ is a continuous function of x, for vectors ψ with finite energy $H(g)$. See Lemma 3.2.1.

We see that integration helps in (3.1.2) because $\varphi(f)$ is an operator while $\varphi(x, t)$ is a bilinear form. Actually the time integration is not required and for real f,

$$A(t) = \int\varphi(x,t)f(x)dx \tag{3.1.3}$$

is also a self-adjoint operator depending continuously on t. We expect that this is a special feature of the two dimensional model we are considering and that sharp time fields will not be operators in four dimensions. For this reason, basic physical concepts have been formulated in terms of the time averaged fields (3.1.2) rather than the sharp time fields (3.1.3). For example, Wightman's axioms for a quantum field theory are expressed in terms of the operators (3.1.2), and we will show that many of his axioms are satisfied for our model.

The time integration in (3.1.2) presents some new difficulties (for example in the proof of self-adjointness or of locality) which would not occur if we considered only the sharp time operators (3.1.3). An advantage of the time averaged field is that products $\varphi(f_1) \cdots \varphi(f_n)$ can be defined on vectors with finite energy (Corollary 3.2.5). In fact we will construct a dense domain which the localized field operator $\varphi(f)$ leaves invariant, and on which $\varphi(f)$ is essentially self-adjoint for real f.

3.2. *An invariant domain for localized fields.* In this section we study the Heisenberg picture field localized in a region of space time $\mathcal{B}$. We find that $\varphi(x, t)$ is a bilinear form and that for real f, $\varphi(f)$ is a densely defined symmetric operator. We start with the region $\mathcal{B}$, a bounded open subset of space time. We require that $H(g)$ be a hamiltonian for $\mathcal{B}$. This means that the spatial cutoff $g(x)$ equals the coupling constant λ on a sufficiently large interval to contain the domain of dependence of $\mathcal{B}$. In other words, assuming that the velocity of light is one, for every point (y, t) in $\mathcal{B}$

$$g(x) = \lambda , \qquad \text{if } |x - y| \leqq |t| . \tag{3.2.1}$$

It is convenient to deal with the field $\varphi_g(x, t) = e^{itH(g)}\varphi(x)e^{-itH(g)}$, and its time derivative

$$\pi_g(x, t) = e^{itH(g)}\pi(x)e^{-itH(g)} = \dot{\varphi}_g(x, t) .$$

The time zero field $\varphi(x)$ and its conjugate momentum $\pi(x)$ were defined in (1.2.10–11). We shall see that for $(x, t) \in \mathcal{B}$, $\varphi_g(x, t)$ is independent of g, and equals the field $\varphi(x, t)$. Thus all the cutoffs have been removed in the definition of $\varphi(x, t)$. For each C^∞ function $f(x, t)$ with support in $\mathcal{B}$, we show that

$$\varphi(f) = \int \varphi(x, t)f(x, t)dxdt \tag{3.2.2}$$

is an operator whose domain contains

$$\mathcal{D}_g = C^\infty(H(g)) . \tag{3.2.3}$$

In fact $\mathcal{D}_g$ is an invariant domain,

$$\varphi(f)\mathcal{D}_g \subset \mathcal{D}_g , \tag{3.2.4}$$

so that $\mathcal{D}_g \subset C^\infty(\varphi(f))$. We note that this invariant domain may depend on the region $\mathcal{B}$ in which the field is localized. For $\psi \in \mathcal{D}_g$ the expectation value

$$(\psi, \varphi(x_1, t_1) \cdots \varphi(x_n, t_n)\psi) \tag{3.2.5}$$

is a Schwartz distribution in $\mathcal{D}'(\mathcal{B} \times \cdots \times \mathcal{B})$.

If $f(x, t)$ is a function in $\mathcal{S}$, then $\varphi_g(f)$ still is defined on $\mathcal{D}_g$ and leaves it invariant. The expectation values (3.2.5) of $\varphi_g(x, t)$ are tempered distributions. However, the fields $\varphi_g(f)$ may depend on g.

LEMMA 3.2.1. *The field $\varphi_g(x, t)$ is a bilinear form on $\mathcal{D}((H(g) + b)^{1/2}) \times \mathcal{D}((H(g) + b)^{1/2})$ continuous in x and t. Namely for $\psi \in \mathcal{D}((H(g) + b)^{1/2})$, $(\psi, \varphi(x, t)\psi)$ is a continuous function. Furthermore*

$$\left|\int(\psi, \varphi_g(x, t)\psi)D_x f(x)dx\right| \leqq \text{const.}\, \|f\|_2 (\psi, (H(g) + b)\psi) . \tag{3.2.6}$$

PROOF. The free field $\varphi(x, 0)$ is the sum of two expressions of the form (1.2.8). The kernels $b(k)$ are not in L_2 and so (1.2.9) does not help. We have however, $b(k)\mu(k)^{-1/2} = e^{\pm ikx}\mu(k)^{-1} \in L_2$. The estimate (1.2.9) has been generalized [6] to cover such kernels, giving us

$$\| (H_0 + I)^{-1/2}\varphi(x, 0)(H_0 + I)^{-1/2} \| \leqq \text{const.}\, \| b\mu^{-1/2} \|_2 < \infty . \tag{3.2.7}$$

Thus for $\psi \in \mathfrak{D}((H(g)+b)^{1/2})$, $e^{-itH(g)}\psi \in \mathfrak{D}((H(g)+b)^{1/2}) \subset \mathfrak{D}(H_0^{1/2})$, by (2.1.19) and so $(\psi, \varphi_g(x,t)\psi) = (e^{-itH(g)}\psi, \varphi(x,0)e^{-itH(g)}\psi)$ is defined and

$$|(\psi, \varphi_g(x,t)\psi)| \leqq \text{const.}\ \| b\mu^{-1/2} \|_2 (\psi, (H(g)+b)\psi)\ .$$

Since $\| b(k)\mu(k)^{-1/2}k\tilde{f}(k) \|_2 \leqq \| b\mu^{-1/2}k \|_\infty \| f \|_2 \leqq \text{const.}\ \| f \|_2$, the inequality (3.2.6) holds.

Let us write b_x for b to denote the dependence of b on x. Then $\| (b_x - b_y)\mu^{-1/2} \|_2$ is a function of $(x-y)$ only and it tends to zero as $|x-y| \to 0$. Since

$$\begin{aligned} |(\psi, (\varphi_g(x,t) - \varphi_g(y,t))\psi| &\leqq \text{const.}\ \| (1+H_0)^{1/2}\psi \|^2 \| (b_x - b_y)\mu^{-1/2} \|_2 \\ &\leqq \text{const.}\ \| (H(g)+b)^{1/2}\psi \|^2 \| (b_x - b_y)\mu^{-1/2} \|_2 \end{aligned} \tag{3.2.8}$$

we have continuity with respect to x. Also $\| (H(g)+b)^{1/2}(e^{itH(g)} - e^{isH(g)})\psi \| \to 0$, as $|t-s| \to 0$. Thus

$$\begin{aligned} &|(\psi, (\varphi_g(x,t) - \varphi_g(x,s))\psi)| \\ &\quad \leqq \text{const.}\ \| (H_0+I)^{1/2}(e^{itH(g)} - e^{isH(g)})\psi \|\ \| b_x\mu^{-1/2} \|_2 \\ &\qquad \times \{\| (H_0+I)^{1/2}e^{itH(g)}\psi \| + \| (H_0+I)^{1/2}e^{isH(g)}\psi \|\} \\ &\quad \leqq \text{const.}\ \| (H(g)+b)^{1/2}(e^{itH(g)} - e^{isH(g)})\psi \|\ \| b_x\mu^{-1/2} \|_2 \\ &\qquad \times \| (H(g)+b)^{1/2}\psi \| \to 0 \qquad \text{as } |t-s| \to 0. \end{aligned} \tag{3.2.9}$$

From (3.2.8–9) we see that $\varphi(x,t)$ is jointly continuous in x and t.

Probably $\pi(x,t)$ is a bilinear form on $\mathfrak{D}((H(g)+b)^{3/2}) \times \mathfrak{D}((H(g)+b)^{3/2})$ continuous in x and t, but our estimates are not strong enough to prove this.

The functions $f(x,t)$ in $\mathcal{S}(R^2)$ determine bounded measures $d\nu = f(x,t)dxdt$, so $\varphi_g(f) = \int \varphi_g(x,t)d\nu$ is a bilinear form. If $d\nu_n \to d\nu$ in the w^* topology for measures, then $\int \varphi_g d\nu_n \to \int \varphi_g d\nu$ in the weak sense that for $\psi \in \mathfrak{D}((H(g)+b)^{1/2})$,

$$\left(\psi, \int \varphi_g d\nu_n \psi\right) \to \left(\int \psi, \int \varphi_g d\nu \psi\right)\ . \tag{3.2.10}$$

We define the sharp time field

$$A_g(t) = \int \varphi_g(x,t)f(x,t)dx \tag{3.2.11}$$

and

$$B_g(t) = \int \pi_g(x,t)f(x,t)dx \tag{3.2.12}$$

LEMMA 3.2.2. *Let $f \in \mathcal{S}(R^2)$ be real. Then $A_g(t)$ and $B_g(t)$ define self-adjoint operators, and their domains include $\mathfrak{D}((H(g)+b)^{1/2})$. With a constant c independent of g,*

(3.2.13) $\|A_g(t)\psi\| + \|B_g(t)\psi\| \leq c\{\|f(\cdot, t)\|_2 + \|D_x f(\cdot, t)\|_2\}\|(H(g) + b)^{1/2}\psi\|$, *for all* $\psi \in \mathfrak{D}((H(g) + b)^{1/2})$.

PROOF. It is sufficient to consider $\varphi(f_t) = \int \varphi(x)f(x, t)dx$ in place of $A_g(t)$, as they are unitarily equivalent *via* the unitary operator $e^{-itH(g)}$, and this unitary leaves $\mathfrak{D}((H(g) + b)^{1/2})$ invariant. The same is true for $\pi(f_t)$. By (2.1.19)

$$\|(N + I)^{1/2}\psi\|^2 \leq m_0^{-1}\|(H_0 + I)^{1/2}\psi\|^2 \leq 2m_0^{-1}\|(H(g) + b)^{1/2}\psi\|^2 ,$$

so we need only prove that

(3.2.14) $\|\varphi(f_t)\psi\| + \|\pi(f_t)\psi\| \leq c\{\|f(\cdot, t)\|_2 + \|D_x f(\cdot, t)\|_2\}\|(N + I)^{1/2}\psi\|$.

The lemma now follows from (1.2.9). For example, $\pi(f_t)$ is the sum of two operators of the form (1.2.8) with kernels $b(k) = \mu(k)^{1/2}\int f(x, t)e^{\pm ikx}dx$, satisfying

$$\begin{aligned}\|b\|_2 &= \|(-D_x^2 + m_0^2)^{1/4}f(\cdot, t)\|_2 \\ &\leq \text{const.}\left(\|f(\cdot)\|_2 + \|D_x f(\cdot, t)\|_2\right) .\end{aligned}$$

The kernel for $\varphi(f_t)$ can be bounded by the $\|f(\cdot, t)\|_2$ norm alone. The estimate (3.2.13) now follows from (3.2.14). The self-adjointness of $A_g(t)$ and $B_g(t)$ can be proven by showing that $\varphi(f_t)$ and $\pi(f_t)$ are self-adjoint. But (3.2.14) ensures that every vector in $\mathcal{F}$ with a finite number of particles is analytic for $\varphi(f_t)$ and for $\pi(f_t)$, so these operators are essentially self-adjoint on the domain of vectors with a finite number of particles. Hence they are uniquely determined by their definition on that domain.

We now explain the sense in which the integral (3.2.12) converges, since we did not show that $\pi_g(x, t)$ was a bilinear form. If $\psi \in \mathfrak{D}((H(g) + b)^{1/2})$, then $\psi \in \mathfrak{D}((N + I)^{1/2})$ and

(3.2.15) $$(\psi, \tilde{\pi}(k)\psi) = \mu(k)^{1/2}\{(a(k)\psi, \psi) + (\psi, a(-k)\psi)\}$$

is a slowly increasing, locally summable function, and hence a tempered distribution. Thus $(\psi, \pi(x)\psi)$ is by definition the distribution Fourier transform of (3.2.15), and hence a tempered distribution. Finally (3.2.12) is the weak integral

$$(\psi, B_g(t)\psi) = \int f(x, t)(e^{-itH(g)}\psi, \pi(x)e^{-itH(g)}\psi)dx .$$

THEOREM 3.2.3. *Let* $\psi \in \mathfrak{D}((H(g) + b)^{1/2})$, *and let* $f(x, t)$ *be a real function in* $\mathcal{S}(R^2)$. *Then the vectors* $A(t)\psi$ *and* $B(t)\psi$ *are strongly continuous and are rapidly decreasing functions of* t. *The integrals* $\int A_g(t)\psi dt = \varphi_g(f)\psi$, *and* $\int B_g(t)\psi dt = \pi_g(f)\psi$ *exist and define* $\varphi_g(f)$ *and* $\pi_g(f)$ *as closed symmetric opera-*

tors with domains containing $\mathfrak{D}((H(g)+b)^{1/2})$. *We have*

$$\|\varphi_g(f)\psi\| + \|\pi_g(f)\psi\| \leqq c\int\{\|f(\cdot,t)\|_2 + \|D_x f(\cdot,t)\|_2\}dt \times \|(H(g)+b)^{1/2}\psi\|, \tag{3.2.16}$$

with a constant c independent of f and g.

PROOF. We write

$$|f(\cdot,t)|_1 = c\|f(\cdot,t)\|_2 + c\|D_x f(\cdot,t)\|_2 \tag{3.2.17}$$

and

$$A_g(s)\psi - A_g(t)\psi = (I - e^{i(t-s)H(g)})A_g(s)\psi + e^{itH(g)}\int\varphi(x)(f(x,s)-f(x,t))dx e^{-isH(g)}\psi + A_g(t)(e^{i(t-s)H(g)} - I)\psi .$$

Thus by (3.2.13),

$$\|A_g(s)\psi - A_g(t)\psi\| \leqq \|(I - e^{i(t-s)H(g)})A_g(s)\psi\| + |f(\cdot,s) - f(\cdot,t)|_1 \|(H(g)+b)^{1/2}\psi\| + |f(\cdot,t)|_1 \|(e^{i(t-s)H(g)} - I)(H(g)+b)^{1/2}\psi\| \to 0$$

as $t \to s$. This proves the continuity. The rapid decrease is ensured by (3.2.13) and the fact that $f \in \mathcal{S}$. A similar argument works for $B_g(t)\psi$. The integrals defining $\varphi_g(f)\psi$ and $\pi_g(f)\psi$ now exist, (3.2.16) follows from integrating (3.2.13). Since $A_g(t)$ and $B_g(t)$ are self adjoint, for $\psi \in \mathfrak{D}((H(g)+b)^{1/2}) \subset \mathfrak{D}(A_g(t)) \cap \mathfrak{D}(B_g(t))$,

$$(\psi, \varphi_g(f)\psi) = \int(\psi, A_g(t)\psi)dt = \int(A_g(t)\psi, \psi)dt$$

is real, and similarly for π. Symmetric operators are closable and we now define $\varphi_g(f)$ and $\pi_g(f)$ as the closure of the above operators on the domain

$$\mathfrak{D}((H(g)+b)^{1/2}) . \tag{3.2.18}$$

Remarks. (a). The integrals defining $\varphi_g(f)\psi$ and $\pi_g(f)\psi$ are strong Riemann integrals, $\varphi_g(f)$ is a strong limit of operators of the form

$$A_g(t_1)\Delta t_1 + \cdots + A_g(t_n)\Delta t_n . \tag{3.2.19}$$

Conversely using the continuity of $A_g(t)\psi$, we see that an operator of the form (3.2.19) is a strong limit of a sequence of $\varphi_g(f_j)$, and the f_j can be chosen with the norm

$$|f|_1 = c\int\{\|f(\cdot,t)\|_2 + \|D_x f(\cdot,t)\|_2\}dt \tag{3.2.20}$$

uniformly bounded. For both limits the convergence occurs on the domain

(3.2.18) and similar considerations apply to π_g. Furthermore $\varphi_g(f)$ and $\pi_g(f)$ can be defined whenever $|f|_1 < \infty$.

(b). Using (2.1.19) to estimate H_0, we have from (3.2.16),

$$\|\varphi_g(f)\Omega_g\| + \|\pi_g(f)\Omega_g\| \leqq |f|_1\left(\left|\frac{1}{2}E_{2g} - E_g\right| + 1\right)^{1/2}, \tag{3.2.21}$$

but the bound on the right grows in the diameter of the support of g. We derive a better estimate in [29, § 3.3].

THEOREM 3.2.4. *Let $|f|_1 + |D_t f|_1$ defined in* (3.2.20) *be finite. On the domain $\mathfrak{D}((H(g)+b)^{3/2})$, the field φ_g satisfies the differential equation*

$$\begin{aligned}(D_t\varphi_g)(f) &= -\varphi_g(D_t f)\\ &= \pi_g(f) = [iH(g), \varphi_g(f)]\,.\end{aligned} \tag{3.2.22}$$

PROOF. The first equality in (3.2.22) is the definition of a distribution derivative, and is not to be proved. Part of the theorem is to show that the commutator is defined on $\mathfrak{D}((H(g)+b)^{3/2})$. We write out the difference quotient for $D_t f$, $(\Delta_\varepsilon f)(x, t) = \{f(x, t+\varepsilon) - f(x, t)\}/\varepsilon$ which converges to $D_t f$ in the norm $|f|_1$. Thus by (3.2.16), for $\psi \in \mathfrak{D}((H(g)+b)^{1/2})$,

$$\|\varphi_g(\Delta_\varepsilon f)\psi - \varphi_g(D_t f)\psi\| \to 0\,.$$

We have for $\psi \in \mathfrak{D}((H(g)+b)^{3/2})$,

$$\begin{aligned}\varphi_g(\Delta_\varepsilon f)\psi &= \varepsilon^{-1}(I - e^{i\varepsilon H(g)})\int\varphi_g(x, t-\varepsilon)f(x, t)dx\psi dt\\ &\quad + \varepsilon^{-1}\int A(t)(e^{i\varepsilon H(g)} - I)\psi dt\,.\end{aligned} \tag{3.2.23}$$

By Lemma (3.2.2) the last term in (3.2.23) converges as $\varepsilon \to 0$ and its limit is

$$i\int A(t)H(g)\psi dt = i\varphi(f)H(g)\psi\,.$$

Since $\varphi_g(\Delta_\varepsilon f)\psi$ also converges, the remaining term in (3.2.23) necessarily converges also to a limit ψ_1.

For $\chi \in \mathfrak{D}(H(g))$,

$$\begin{aligned}(\chi, \psi_1) &= \lim_{\varepsilon\to 0}\left(\chi, \varepsilon^{-1}(I - e^{i\varepsilon H(g)})\int\varphi_g(x, t-\varepsilon)f(x, t)dx\psi\right)dt\\ &= \left(iH(g)\chi, \varphi_g(f)\psi\right)\,.\end{aligned}$$

Since $H(g) = H(g)^*$, we conclude that $\varphi_g(f)\psi \in \mathfrak{D}(H(g))$ and $\psi_1 = -iH(g)\varphi_g(f)\psi$. Therefore $-\varphi_g(D_t f)\psi = i[H(g), \varphi_g(f)]\psi$.

We still need to prove that $-\varphi_g(D_t f) = \pi_g(f)$, and it is sufficient to prove this as a bilinear form on $\mathfrak{D}_g \times \mathfrak{D}_g$. From the above equation we have

$$i(\psi, \varphi_g(D_t f)\psi) = \int\Big(H(g)\psi(t), \int\varphi(x, 0)f(x, t)dx\psi(t)\Big)dt \tag{3.2.24}$$
$$- \int dt\Big(\int\varphi(x, 0)f(x, t)dx\psi(t), H(g)\psi(t)\Big),$$

where $\psi(t) = e^{-itH(g)}\psi$. Since $\psi(t) \in \mathfrak{D}(H(g)) = \mathfrak{D}(H_0) \cap \mathfrak{D}(H_{I,g})$, and

$$\| H_{I,g}(\psi(t) - \psi(s)) \| \leqq a \| (H(g) + b)(\psi(t) - \psi(s)) \| \to 0 ,$$

as $|t - s| \to 0$, we may substitute $H_0 + H_{I,g}$ for $H(g)$ and consider each term separately. It is known that $H_{I,g}$ and $\int\varphi(x, 0)f(x, t)dx$ commute; see [8, Ths. 3.1–3.2] for example. Thus $H_{I,g}$ contributes zero to (3.2.24). Furthermore the identity

$$\Big(H_0\psi, \int\varphi(x, 0)f(x, t)dx\psi\Big) - \Big(\int\varphi(x, 0)f(x, t)dx\psi, H_0\psi\Big)$$
$$= \Big(\psi, - i\int\pi(x, 0)f(x, t)dx\psi\Big)$$

is a standard computation valid for any $\psi \in \mathfrak{D}(H_0)$, for example $\psi = \psi(t)$. Thus (3.2.24) yields

$$i(\psi, \varphi_g(D_t f)\psi) = \int\Big(\psi(t), -i\int\pi(x, 0)f(x, t)dx\psi(t)\Big)dt$$
$$= (\psi, - i\pi_g(f)\psi) ,$$

and the theorem is proved.

Remarks. (a). In exactly the same fashion one proves that

$$(D_{tt}\varphi_g)(f) = (D_t\pi_g)(f) = -\pi_g(D_t f)$$
$$= [iH(g), \pi_g(f)]$$

if $| D_{tt}f |_1$ is also finite. The commutator is a bilinear form on $\mathfrak{D}_g \times \mathfrak{D}_g$, namely

$$(D_{tt}\varphi_g)(f) = (D_{xx}\varphi_g)(f) - m_0^2\varphi(f) - 4\int :\varphi_g^3(x, t:f(x, t)g(x)dxdt . \tag{3.2.25}$$

Here we define the $:\ :$ product by

$$:\varphi_g^3(x, t): = e^{iH(g)t}:\varphi^3(x):e^{-iH(g)t} , \tag{3.2.26}$$

which we now prove is an operator valued distribution. First we note that $\int:\varphi^3(x):f(x, t)dx$ is a sum of monomials in creation and annihilation operators with kernels in L_2, and their L_2 norms are continuous in t. Thus by (1.2.9) $\int:\varphi^3(x):f(x, t)g(x)dx$ is a bilinear form on $\mathfrak{D}(N) \times \mathfrak{D}(N)$. By (2.1.3), $(H(g) + b)^{-1}\int:\varphi^3(x):f(x, t)g(x)dx(H(g) + b)^{-1}$ is a bounded operator, norm continuous in t. Thus on $\mathfrak{D}(H(g))\times\mathfrak{D}(H(g))$, $:\varphi_g^3:(f)=\int:\varphi_g^3(x,t):f(x,t)dxdt$ is defined as a bilinear form. Hence (3.2.25) holds as an equation for bilinear forms on $\mathfrak{D}(H(g)) \times \mathfrak{D}(H(g))$. But each term except the last is an operator defined on $\mathfrak{D}((H(g) + b)^{1/2})$. Thus $:\varphi_g^3:(f)$ is actually an operator on

$\mathfrak{D}((H(g)+b)^{1/2})$, and in fact for real $f(x, t)$ it is essentially self adjoint, as we shall see in § 3.3.

Furthermore, on $\mathfrak{D}((H(g)+b)^{1/2}) \times \mathfrak{D}((H(g)+b)^{1/2})$ each term in (3.2.25) except the last is a bilinear form which is a distribution of order two. Thus the same is true for $(\psi, :\varphi^3(x, t): \psi)$.

We have used (3.2.26) to define the cube of the interacting field. It would be interesting to determine whether this definition agrees with conventional notions involving the separation of points.

We shall see in this section and the following one that the $g(x)$ in equation (3.2.25) can be removed if $f(x, t)$ has compact support and $g(x) = \lambda$ on a sufficiently large set. Then (3.2.25) becomes $(\Box + m_0^2)\varphi(f) = -4\lambda :\varphi^3:(f)$, which is a non-linear equation for a self-adjoint operator valued distribution.

(b). The identity $\pi_g(f) = [iH(g), \varphi_g(f)]$ implies

$$B_g(t) = i[H(g), A_g(t)] \tag{3.2.27}$$

provided that the right and left sides of (3.2.27) make sense and are continuous in t. They are certainly defined and are continuous as bilinear forms on $\mathfrak{D}((H(g)) \times \mathfrak{D}(H(g))$. To see that (3.2.27) makes sense as operators on $\mathfrak{D}((H(g)+b)^{3/2})$ we need only show that $A_g(t)$ maps $\mathfrak{D}((H(g)+b)^{3/2})$ into $\mathfrak{D}(H(g))$. We choose a sequence $f_j(x, t) = f(x, t_0)\delta_j(t-t_0)$ where δ_j is a sequence of smooth functions converging to δ in the w^* topology on measures and with the norms $|f_j|_1$ uniformly bounded. Then the bilinear forms converge, which means that the inner products

$$(\theta, iH(g)\varphi_g(f_j)\psi) = (\theta, \varphi_g(f_j)iH(g)\psi) + (\theta, \pi_g(f_j)\psi)$$

converge for $\theta \in \mathfrak{D}(H(g))$. However the norms

$$\begin{aligned} \| H(g)\varphi_g(f_j)\psi \| &\leqq \| \varphi(f_j)H(g)\psi \| + \| \pi(f_j)\psi \| \\ &\leqq |f|_1 \, \| (H(g)+b)^{3/2}\psi \| \end{aligned}$$

are uniformly bounded, and so the inner products converge for all $\theta \in \mathcal{F}$. Thus the limit $A(t)\psi = \text{weak} \lim \varphi(f_j)\psi$ is in $\mathfrak{D}(H(g)^*) = \mathfrak{D}(H(g))$ which proves (3.2.27) on the domain $\mathfrak{D}((H(g)+b)^{3/2})$.

COROLLARY 3.2.5. *Let* $f \in \mathcal{S}$. *Then* $\mathfrak{D}_g = C^\infty(H(g)) \subset C^\infty(\varphi_g(f))$, *and* $\varphi_g(f)\mathfrak{D}_g \subset \mathfrak{D}_g$.

PROOF. Using Theorem 3.2.4, we prove by induction on m that $\varphi_g(f)\mathfrak{D}_g \subset \mathfrak{D}(H(g)^m)$ and that for $\psi \in \mathfrak{D}_g$,

$$H(g)^m\varphi_g(f)\psi = \varphi_g(f)H(g)^m\psi + \sum_{j=1}^m \binom{m}{j} i^j\varphi_g(D_t^j f)H(g)^{m-j}\psi \,.$$

This formula is a special case of the standard identity

$A^m B \supset \sum_{j=1}^m \binom{m}{j} [(adA)^j B] A^{m-j}$. Thus

$$\| H(g)^m \varphi_g(f)\psi \| \leqq \sum_{j=1}^m \binom{m}{j} | D_t^j f |_1 \, \| (H(g) + b)^{m-j+(1/2)} \psi \| , \tag{3.2.28}$$

where $|f|_1$ is the norm (3.2.20).

THEOREM 3.2.6. *Let $f \in C_0^\infty(\mathcal{B})$, that is f is C^∞ with support in the open region of space time $\mathcal{B}$. Let $H(g)$ be a hamiltonian for $\mathcal{B}$, so that $g = \lambda$ on a large set. Then*

$$\varphi_g(f) = \varphi(f) \qquad \textit{and} \qquad \pi_g(f) = \pi(f)$$

are independent of g/λ.

PROOF. The spectral projections $E^t(\lambda)$ of the sharp time field

$$A_0(t) = \int \varphi(x, t) f(x, t_0) dx = \int \lambda dE^t(\lambda)$$

are given by the formula $E^t(\lambda) = e^{itH(g)} E^0(\lambda) e^{-itH(g)}$ and are independent of g by [8]. Thus $A_0(t)$ is independent of g and so is $A(t)$. By (2.2.7), for all g,

$$\mathcal{D}_0 = \mathcal{D}(H_0^{1/2}) \cap \mathcal{D}(N) \subset \mathcal{D}((H(g) + b)^{1/2}) \subset \mathcal{D}(\varphi_g(f)) ,$$

so that $\varphi_g(f) \mid \mathcal{D}_0 = \varphi(f) \mid \mathcal{D}_0$ is independent of g. Thus to complete the proof, we only need to show that the domain of $\varphi_g(f) = (\varphi_g(f) \mid \mathcal{D}((H(g) + b)^{1/2}))^-$ is independent of g. Since $H(g)$ is essentially self-adjoint on $C^\infty(H_0) \subset \mathcal{D}_0$, [8], so is $(H(g) + b)^{1/2}$. Thus by (3.2.16) $(\varphi(f) \mid \mathcal{D}_0)^- \supset \varphi_g(f) \mid \mathcal{D}((H(g) + b)^{1/2})$. Therefore $(\varphi(f) \mid \mathcal{D}_0)^- = \varphi_g(f)$, so $\varphi_g(f) = \varphi(f)$ is independent of g. Similarly $\pi(f)$ is independent of g.

THEOREM 3.2.7. *Let $\psi \in \mathcal{D}_g$, with $H(g)$ a hamiltonian for $\mathcal{B}$. Then*

$$(\psi, \varphi(x_1, t_1) \cdots \varphi(x_n, t_n)\psi)$$

is a distribution in $\mathcal{D}'(\mathcal{B} \times \cdots \times \mathcal{B})$.

PROOF. This follows from our previous estimates (3.2.16) and (3.2.28).

3.3. *Essential self-adjointness.* The main result of this section is the proof that for real test functions $f = f(x, t)$ with compact support, the field $\varphi(f)$ is self-adjoint, and essentially self-adjoint on $\mathcal{D}_0 = \mathcal{D}(H_0^{1/2}) \cap \mathcal{D}(N)$, or on any $\mathcal{D}_g$ where $H(g)$ is a hamiltonian for the support of f. We see furthermore that if f is real and $|f|_1$ defined by (3.2.20) is finite, then $\varphi_g(f)$ is self-adjoint and essentially self-adjoint on $\mathcal{D}_g$.

The proof has three main steps. First, we assume that f is a regular function of t; in that case we use an analytic vector argument to show that

$\varphi_g(f)$ is essentially self-adjoint on $\mathfrak{D}_g$. As a second step, we take limits in the resolvents $(\varphi_g(f) - z)^{-1}$ as f tends to a more general function. In this way, we obtain a self-adjoint operator $\varphi_g(f)$. As a third step, we show that $\varphi_g(f)$ is essentially self-adjoint on $\mathfrak{D}(H(g))$.

The regularity we impose on f is the requirement that its Fourier transform be a smooth function with compact support, or more generally that for the norm $|f|_1$ of (3.2.20) there exist constants $\alpha = \alpha(f)$, $\beta = \beta(f)$ such that

$$|D_t^r f| \leqq \alpha\beta^r , \qquad r = 0, 1, 2 \cdots . \tag{3.3.1}$$

For a vector ψ, we consider the condition

$$\| (H(g) + b)^r \psi \| \leqq ad^r , \qquad r = 0, 1, \cdots . \tag{3.3.2}$$

LEMMA 3.3.1. *Assume* (3.3.1) *and* (3.3.2). *Then*

$$\| (H(g) + b)^r \varphi_g(f)\psi \| \leqq \alpha adc(b + \beta)^r \tag{3.3.3}$$

for some constant c independent of r, ψ and f.

PROOF. By (3.2.28) and (3.3.1–2)

$$\begin{aligned} \| (H(g) + b)^r \varphi_g(f)\psi \| &\leqq \textstyle\sum_{j=0}^r \binom{r}{j} c\alpha\beta^j \| (H(g) + b)^{r-j+1}\psi \| \\ &\leqq \textstyle\sum_{j=0}^r \binom{r}{j} c\alpha\beta^j ad^{r-j+1} \\ &= \alpha adc(d + \beta)^r . \end{aligned}$$

LEMMA 3.3.2. *Assume* (3.3.1) *and* (3.3.2). *Then ψ is an analytic vector for $\varphi_g(f)$. In particular for real $f(x, t)$, $\varphi_g(f)$ is essentially self-adjoint on $\mathfrak{D}_g$.*

PROOF. We apply the preceding lemma successively. We see that multiplication by $\varphi_g(f)$ changes the constants a and d of (3.3.2) as follows: $a \to \alpha adc$, $d \to d + \beta$. Thus

$$\begin{aligned} &\| (H(g) + b)^r \varphi_g(f)^k \psi \| \\ &\qquad \leqq a(\alpha c)^k d(d + \beta)(d + 2\beta) \cdots (d + (k - 1)\beta)(d + k\beta)^r , \end{aligned}$$

and

$$\| \varphi_g(f)^k \psi \| \leqq aB^k k!$$

for some constant B, which proves that ψ is analytic for $\varphi_g(f)$. The essential self-adjointness of $\varphi_g(f)$ follows from a theorem of Nelson [20].

We can draw more information from (3.3.3). If we write $H(g) + b = \int \lambda dE(\lambda)$, then (3.3.2) is equivalent to $\psi \in \text{Range } E(d)$ and (3.3.3) ensures that

$$\varphi_g(f) \text{ Range } E(d) \subset \text{Range } E(d + \beta) . \tag{3.3.4}$$

Because $\varphi_g(f)$ is self-adjoint, we also have

$$\varphi_g(f) \operatorname{Range}\big(I - E(d)\big) \subset \operatorname{Range}\big(I - E(d - \beta)\big) . \tag{3.3.5}$$

These two inclusions have simple physical interpretations. We imagine that $\varphi_g(f)$ is written as a sum of two operators, one creating physical wave packets associated with $H(g)$, and the other annihilating them. Because of (3.3.1) the wave packets have energy at most β, and so $\varphi_g(f)$ can increase or decrease the total energy, $H(g)$, by at most β.

We note that $\varphi_g(f)$ is essentially self-adjoint on the domain $\bigcup_l \operatorname{Range} E(l)$, by the proof of Lemma 3.3.2 and the remarks above.

Our next step is to take limits with respect to f in the resolvents $R = R(f, z) = \big(\varphi_g(f) - z\big)^{-1}$. As preparation, we now prove that R preserves regularity, which means

$$\| \big(H(g) + b\big)^{1/2} R(f, z)\psi \| \leqq M \| \big(H(g) + b\big)^{1/2} \psi \| . \tag{3.3.6}$$

LEMMA 3.3.3. *Let f be real and satisfy* (3.3.1). *Then the estimate* (3.3.6) *holds for* $\operatorname{Im} z \neq 0$. *The constants M and b depend only on z, g, and $|f|_1$.*

PROOF. To prove this lemma, we obtain uniform estimates on approximating operators R_n. If (3.3.6) holds for R_n, with M independent of n, and

$$R = \text{strong} \lim R_n , \tag{3.3.7}$$

then (3.3.6) also holds for R. In fact $\| \big(H(g) + b\big)^{1/2}\psi \|$ defines a norm on the domain $\mathfrak{D}\big(\big(H(g) + b\big)^{1/2}\big) = \mathcal{H}_1$ which makes it into a Hilbert space. The inequality (3.3.6) in equivalent to R_n being a bounded operator on $\mathcal{H}_1$, and the norm $\| R_n \|_{1,1}$ of R_n as an operator from $\mathcal{H}_1$ to itself is defined by

$$\| R_n \|_{1,1} = \| \big(H(g) + b\big)^{1/2} R_n \big(H(g) + b\big)^{-1/2} \| \leqq M . \tag{3.3.8}$$

From the strong convergence (3.3.7) on $\mathcal{F}$, we conclude that on a dense set of vectors in $\mathcal{H}_1$, R_n converges weakly to R. Since the R_n are uniformly bounded on $\mathcal{H}_1$, $R_n \to R$ in weak operator convergence on $\mathcal{H}_1$. Thus the norm $\| R \|_{1,1}$ is bounded by the lim sup of the $\| R_n \|_{1,1}$ and (3.3.6) holds for R.

Let

$$H(g) + b = \int \lambda dE(\lambda) . \tag{3.3.9}$$

We approximate $\varphi_g(f)$ by the bounded self-adjoint operator $C_n = E(n)\varphi_g(f)E(n)$. From (3.3.4) it is clear that $C_n \to \varphi_g(f)$ on vectors with compact support in the energy. Since $\varphi_g(f)$ is essentially self-adjoint on this domain, the resolvents also converge strongly [18, p. 429] $\lim_{n\to\infty} R_n(z) = \lim_{n\to\infty} (C_n - z)^{-1} \to R(z)$, proving (3.3.7).

We now show that (3.3.8) holds, and it is sufficient to prove

$$\| (H(g) + b)^{1/2} R_n (H(g) + b)^{-1/2} \psi \| \leqq M \| \psi \| \tag{3.3.10}$$

for ψ in the dense set $\mathfrak{D}_g$. Since $(H(g) + b)^{-1/2}$ and R_n both map $\mathfrak{D}_g$ onto $\mathfrak{D}_g$, we need only prove that on $\mathfrak{D}_g \times \mathfrak{D}_g$,

$$\begin{aligned}(H(g) + b) &\leqq M^2(C_n - \bar{z})(H(g) + b)(C_n - z) \\ &= M^2(C_n - x)(H(g) + b)(C_n - x) + (My)^2(H(g) + b) \\ &\quad + iM^2 y[H(g), C_n] ,\end{aligned} \tag{3.3.11}$$

where $z = x + iy$. As the first term is positive, it is sufficient to show that

$$0 \leqq ((My)^2 - 1)(H(g) + b) + M^2 y[iH(g), C_n] . \tag{3.3.12}$$

But

$$\begin{aligned}[iH(g), C_n] &= E(n)[iH(g), \varphi_g(f)]E(n) \\ &= E(n)\pi_g(f)E(n) .\end{aligned}$$

By Theorem 3.2.3,

$$\begin{aligned}|(\psi, [iH(g), C_n]\psi)| &= |(E(n)\psi, \pi_g(f)E(n)\psi)| \leqq \|\psi\| \, \|\pi_g(f)E(n)\psi\| \\ &\leqq \|\psi\| \, |f|_1 \, \|(H(g) + b)^{1/2}E(n)\psi\| \\ &\leqq \frac{1}{2} |f|_1 \Big\{\varepsilon(\psi, (H(g) + b)\psi) + \frac{1}{\varepsilon}(\psi, \psi)\Big\} ,\end{aligned}$$

for any $\varepsilon > 0$. Furthermore, the norm $|f|_1$ of (3.2.20) can be chosen independent of b for large b, since $H(g) + b_1 \leqq H(b) + b_2$ for $b_1 \leqq b_2$. Therefore (3.3.12) is valid as long as

$$0 \leqq \Big\{(My)^2 - 1 - \frac{1}{2} M^2 y \varepsilon |f|_1\Big\}(H(g) + b) - \frac{M^2 y}{2\varepsilon} |f|_1 .$$

For each $|f|_1, |y| \neq 0$, we can pick M large so that $(My)^2 > 3$, ε small so that $M^2 y \varepsilon |f|_1/2 < 1$, and b large so that the inequality is valid. This completes the proof.

We now show that the resolvents of approximate field operators converge. We use the spectral projections $E(n)$ defined by (3.3.9) to cut off the field. If $\varphi_g(f)$ is a closed symmetric field operator, then $E(n)\varphi_g(f)E(n)$ is a bounded, self-adjoint approximation to $\varphi_g(f)$.

LEMMA 3.3.4. *Let $\{f_n\}$ be a sequence of real functions satisfying* (3.3.1) *with β depending on n. If the graphs $\mathfrak{G}(\varphi_g(f_n))$ converge to the graph of a densely defined operator, if*

$$\| (H(g) + b)^{-1/2} \{\varphi_g(f_n) - \varphi_g(f_m)\} (H(g) + b)^{-1/2} \| \to 0 \tag{3.3.13}$$

and if the norms $|f_n|_1$ are uniformly bounded, then the resolvents

$$R_n(z) = (C_n - z)^{-1} , \tag{3.3.14}$$

of

$$C_n = E(n)\varphi_g(f_n)E(n) \tag{3.3.15}$$

converge strongly to the resolvent of a self-adjoint operator C.

PROOF. This result is a special case of [9, Th. 5 and Cor. 6]. See that paper for notation. Recall that $E(n)$ is defined by (3.3.9). Note that

$$\begin{aligned} &\| (H(g) + b)^{-1/2}(C_n - \varphi_g(f_n))(H(g) + b)^{-1/2} \| \\ &\quad \leqq \| (E(n) - I)(H(g) + b)^{-1/2} \| \{ \| \varphi_g(f_n)(H(g) + b)^{-1/2} \| \\ &\qquad + \| (H(g) + b)^{-1/2}\varphi_g(f_n) \| \} \\ &\quad \leqq n^{-1/2}\{ \| \varphi_g(f_n)(H(g) + b)^{-1/2} \| + \| (H(g) + b)^{-1/2}\varphi_g(f_n) \| \} . \end{aligned} \tag{3.3.16}$$

By Theorem 3.2.3, $\varphi_g(f_n)(H(g) + b)^{-1/2}$ and its adjoint are bounded with $\| \varphi_g(f_n)(H(g) + b)^{-1/2} \| \leqq c \, |f_n|_1$, which is bounded uniformly in n. Then

$$\| (H(g) + b)^{-1/2}(C_n - \varphi_g(f_n))(H(g) + b)^{-1/2} \| = O(n^{-1/2}) \tag{3.3.17}$$

and so by (3.3.13), $\| (H(g) + b)^{-1/2}(C_n - C_m)(H(g) + b)^{-1/2} \| \to 0$. The required uniform boundedness of the resolvents $\| (C_n - z)^{-1} \|_{1,1} \leqq$ const. follows from Lemma 3.3.3.

We now discuss when the hypotheses of Lemma 3.3.4 are satisfied. If the $\varphi_g(f_n)$ converge strongly on a dense domain, then the graphs converge. The $\varphi_g(f_n)$ will converge on $\mathfrak{D}_g$ if $f_n \to f$ in the norm $|\cdot|_1$; they will also converge for some sequence f_n

$$f_n \to f(\cdot, t_1)\delta(t - t_1) + \cdots + f(\cdot, t_j)\delta(t - t_j) \tag{3.3.18}$$

with $f(\cdot, t_i) \in \mathcal{S}(R)$. We can choose $f_n(x, t)$ to have the form $f_n(x, t) = \sum_{r=1}^{j} f(x, t_r)\delta_n(t - t_r)$ where $\delta_n(t) \geqq 0$ has support in $|t| \leqq n^{-1}$, and $\int \delta_n(t)dt = 1$. For such a sequence $|f_n|_1$ is uniformly bounded in n. From (3.2.9) we see that w^* convergence of the δ_n as bounded measures implies (3.3.13). Thus the hypotheses are satisfied for the sequence (3.3.18). They are also satisfied if the f_n converge in the norm $|\cdot|_1$, and every f with finite $|f|_1$ is the limit of such a sequence.

THEOREM 3.3.5. *Let f be real and $|f|_1$ finite. Then $\varphi_g(f)$ is self-adjoint and essentially self-adjoint on $\mathfrak{D}_g$. A real linear combination of sharp time fields with real test functions in $\mathcal{S}$, $A_g(t_1) + \cdots + A_g(t_r)$, is also essentially self-adjoint on $\mathfrak{D}_g$.*

PROOF. The two cases are similar and we only consider $\varphi_g(f)$. We first prove that the C of Lemma 3.3.4 extends $\varphi_g(f)$

$$C \supset \varphi_g(f) . \tag{3.3.19}$$

This is a consequence of [9], but we give the following explicit proof. As in the proof of Theorem 3.2.6, we have

$$\varphi_g(f) = (\varphi_g(f) \mid \mathfrak{D}_g)^- = (\varphi_g(f) \mid \mathfrak{D}_0)^- , \tag{3.2.20}$$

where $\mathfrak{D}_0 = \mathfrak{D}(H_0^{1/2}) \cap \mathfrak{D}(N)$. Let $R_n(z)$ be defined by (3.3.14), where f_n approximates f and satisfies the hypotheses of Lemma 3.3.4. Thus $R(z) = \lim_{n\to\infty} R_n(z)$ exists and is the resolvent of a self-adjoint operator C. For $\psi \in \mathfrak{D}_g$, $C_n\psi = E(n)\varphi_g(f_n)E(n)\psi \to \varphi_g(f)\psi$, and convergence can be shown on $\mathfrak{D}((H(g) + b)^{3/2})$. For $\chi = (\varphi_g(f) - z)\psi$, $R(z)\chi = \lim_{n\to\infty} R_n(z)\chi = \lim_{n\to\infty} R_n(z)(C_n - z)\psi = \psi$. Thus

$$(C - z)^{-1}(\varphi_g(f) - z) \mid \mathfrak{D}_g = I \mid \mathfrak{D}_g ,$$

and by (3.3.20), (3.3.19) is valid.

We now show that $\varphi_g(f)$ is equal to C, which completes the proof. We need only show that if $\psi \in \mathfrak{D}(C)$, then $\psi \in \mathfrak{D}(\varphi_g(f))$. We first notice that

$$R(z)\mathfrak{D}((H(g) + b)^{1/2}) \subset \mathfrak{D}((H(g) + b)^{1/2}) \tag{3.3.21}$$

and that (3.3.8) is valid for $R(z)$. The argument for this is the same as the proof of Lemma 3.3.3, but the approximate operator $C_n = E(n)\varphi_g(f_n)E(n)$ replaces the C_n of the former proof. The remaining calculation is the same since the $|f_n|_1$ are assumed uniformly bounded.

We now introduce the smoothing operator

$$P_j = \left(1 + \frac{1}{j}(H(g) + b)^2\right)^{-1} \tag{3.3.22}$$

with the properties $\| P_j \| \leqq 1$,

$$\text{strong } \lim_{j\to\infty} P_j = I , \tag{3.3.23}$$

and for $r \leqq 2$,

$$\| P_j(H(g) + b)^r \| = \| (H(g) + b)^r P_j \| \leqq j^{r/2} . \tag{3.3.24}$$

Let $\psi \in \mathfrak{D}(C)$ and $\psi = R(z)\chi$. Then $P_j\psi = \psi_j \to \psi$ and $\psi_j \in \mathfrak{D}(H(g)) \subset \mathfrak{D}(\varphi_g(f))$. If $\varphi_g(f)\psi_j$ converges, then ψ is in the domain of the closed operator $\varphi_g(f)$, so we prove this.

$$\begin{aligned}(\varphi_g(f) - z)\psi_j &= (\varphi_g(f) - z)P_j\psi \\ &= (\varphi_g(f) - z)P_jR(z)\chi \\ &= (\varphi_g(f) - z)R(z)P_j\chi + (\varphi_g(f) - z)[P_j, R(z)]\chi .\end{aligned} \tag{3.3.25}$$

The last equality is valid since $P_j\chi \in \mathfrak{D}(H(g)^2) \subset \mathfrak{D}((H(g) + b)^{1/2})$, and by (3.3.21) $R(z)P_j\chi \in \mathfrak{D}((H(g) + b)^{1/2}) \subset \mathfrak{D}(\varphi_g(f))$. Since C extends $\varphi_g(f)$, $(\varphi_g(f) - z)\psi_j = P_j\chi + (\varphi_g - z)[P_j, R(z)]\chi$. As $P_j\chi \to \chi = (C - z)\psi$, to conclude that $\psi \in \mathfrak{D}(\varphi_g(f))$, we need to show that

(3.3.26) $$\Lambda_j = (\varphi_g - z)[P_j, R(z)]\chi \to 0 .$$

We now argue that

(3.3.27) $$\Lambda_j = \lim_{n\to\infty} (C_n - z)[P_j, R_n(z)]\chi ,$$

where C_n and R_n are defined in (3.3.14–15). Since $(C_n - z)R_n(z) = (C - z)R(z) = I$, we need only prove the existence of the limit (3.3.27) with the commutator removed. As observed in the first paragraph of the proof, for $\psi \in \mathfrak{D}((H(g) + b)^{3/2})$, $\| C_n\psi - C\psi \| = \| C_n\psi - \varphi_g(f)\psi \| \to 0$. Since $P_jR(z)\chi \in \mathfrak{D}(H(g)^2) \subset \mathfrak{D}((H(g) + b)^{3/2})$, as $n \to \infty$, $(C_n - z)P_jR(z)\chi \to (\varphi_g(f) - z)P_jR(z)\chi$. Also $R_n(z)\chi \to R(z)\chi$, and by Theorem 3.2.3 and (3.3.24), $\| (C_n - z)P_j \| \leqq \text{const.}\, | f_n |_1 j^{1/4}$, which is bounded uniformly in n. Therefore

$$(C_n - z)P_jR_n(z)\chi \to (\varphi_g(f) - z)P_jR(z)\chi ,$$

and (3.3.27) is established. Thus $\Lambda_j = \lim_{n\to\infty} \Lambda_{j,n}$, where

$$\begin{aligned}\Lambda_{j,n} &= (C_n - z)[P_j, R_n(z)]\chi \\ &= (C_n - z)R_n(z)P_j[P_j^{-1}, (C_n - z)]P_jR_n(z)\chi \\ &= \frac{1}{j}P_j[(H(g) + b)^2, C_n - z]P_jR_n(z)\chi \\ &= \frac{1}{j}P_j\{(H(g) + b)[H(g), C_n] + [H(g), C_n](H(g) + b)\}P_jR_n(z)\chi \\ &= \frac{-i}{j}P_j\{(H(g) + b)E(n)\pi_g(f_n)E(n) \\ &\qquad + E(n)\pi_g(f_n)E(n)(H(g) + b)\}P_jR_n(z)\chi .\end{aligned}$$

Now by (3.3.24) $\| P_j(H(g) + b) \| = \| (H(g) + b)P_j \| \leqq j^{1/2}$ and

$$\begin{aligned}\| E(n)\pi_g(f_n)E(n)P_j \| + \| P_jE(n)\pi_g(f_n)E(n) \| &\leqq \text{const.}\, | f_n |_1 j^{1/4} \\ &\leqq \text{const.}\, j^{1/4} ,\end{aligned}$$

as the $| f_n |_1$ are assumed uniformly bounded. The constant is independent of j and n. Therefore $\| \Lambda_{jn} \| \leqq \text{const.}\, j^{-1/4}$, and

$$\lim_{j\to\infty} \| \Lambda_j \| \leqq \lim_{j\to\infty} \lim_{n\to\infty} \| \Lambda_{jn} \| = 0 .$$

Thus (3.3.26) is established and the proof is complete.

3.4. *The field as a tempered distribution.* In the previous sections we studied the field $\varphi_g(f)$ corresponding to the hamiltonian $H(g)$. We found that if $\varphi_g(f)$ is localized, namely if f has compact support in $\mathcal{B}$ and $H(g)$ is a hamiltonian for $\mathcal{B}$, then $\varphi_g(f) = \varphi(f)$ is independent of the spatial cutoff g. In this section we show that there is a cutoff independent field $\varphi(f)$ defined for all

$f \in \mathcal{S}$, and $\varphi(f)$ agrees with the previous one when f has compact support. The domain of $\varphi(f)$ includes $\mathfrak{D}_0 = \mathfrak{D}(H_0^{1/2}) \cap \mathfrak{D}(N)$, and on this domain $\varphi(f)$ is a tempered distribution.

LEMMA 3.4.1. *Let* $\mathfrak{D}_0 = \mathfrak{D}(H_0^{1/2}) \cap \mathfrak{D}(N)$. *For* $\psi \in \mathfrak{D}_0$, $\big(\psi, \varphi(x, t)\psi\big)$ *is a continuous, polynomially bounded function and*

$$\left|\int(\psi, \varphi(x, t)\psi)D_x f(x, t)dx\right| \leqq O(|t|)\, \|f(\cdot, t)\|_2 \big(\psi, (H_0 + N^2 + I)\psi\big) . \tag{3.4.1}$$

PROOF. We divide space time into a number of similar regions with a partition of unity. Let $\zeta(x, t)$ be a C^∞ function satisfying

$$0 \leqq \zeta(x, t) \leqq 1 , \tag{3.4.2}$$

$$\operatorname{supp} \zeta \subset \{|x| \leqq 1, |t| \leqq 1\} . \tag{3.4.3}$$

and such that

$$\sum_{i,j} \zeta_{ij}(x, t) = \sum_{i,j} \zeta(x - i, t - j) = 1 . \tag{3.4.4}$$

Thus if $f(x, t) \in \mathcal{S}$, $f = \sum_{i,j} f_{ij} = \sum_{i,j} f\zeta_{ij}$ with $f_{ij}(x, t)$ a C^∞ function with support in the cube

$$\mathcal{B}_{ij} = \{(x, t): |x - i| \leqq 1, |t - j| \leqq 1\} . \tag{3.4.6}$$

We also pick a C^∞ function $g_0(x)$ such that

$$g_0(x) = \lambda, \text{ if } |x| \leqq 2 , \tag{3.4.7}$$

and

$$g_0(x) = 0, \text{ if } |x| \geqq 3 . \tag{3.4.8}$$

Thus $H(g_{ij})$ is a hamiltonian for $\mathcal{B}_{ij}$ when

$$g_{ij}(x) = g_0\big((x - i)/(|j| + 1)\big) . \tag{3.4.9}$$

Furthermore

$$\|(I + N)^{-1}H_{I,g_{ij}}(I + N)^{-1}\| = O(j) , \tag{3.4.10}$$

as the kernels of operators contributing to $H_{I,g_{ij}}$ have L_2 norms with are $O(j)$. For $(x, t) \in \mathcal{B}_{ij}$ and $\psi \in \mathfrak{D}_0$, we have by Lemma 3.2.1, that $\big(\psi, \varphi(x, t)\psi\big)$ is continuous and

$$\big|\big(\psi, \varphi(x, t)\psi\big)\big| \leqq \text{const.} \big(\psi, \big(H(g_{ij}) + b\big)\psi\big) \tag{3.4.11}$$

where the constant is independent of x, t, i, and j. Here b is proportional to the lower bound of $H(g_{ij})$, see (2.1.19). In [29, § 5] we prove that the lower bound of $H(g_{ij})$ is of order the diameter of the support of g_{ij}, namely $O(j)$.

Thus (3.4.11) gives the bound for $(x, t) \in \mathcal{B}_{ij}$, $\psi \in \mathfrak{D}_0$

$$
\begin{aligned}
&|(\psi, \varphi(x, t)\psi)| \\
(3.4.12)\quad &\leq \text{const.}\{\| H_0^{1/2}\psi \|^2 + \| (N+I)\psi \|^2 \| (I+N)^{-1} H_{I,g_{ij}} (I+N)^{-1} \| \\
&\qquad + b \| \psi \|^2\} \\
&\leq \text{const.}\,(\psi, (H_0 + N^2 + I)\psi) O(j) ,
\end{aligned}
$$

by (3.4.10) and the above discussion of b. Since $O(j) = O(|t|)$, we have proved polynomial boundedness. Thus, as in Lemma 3.2.1,

$$\left| \int (\psi, \varphi(x, t)\psi) D_x f(x, t) dx \right| \leq O(|t|) \| f(\cdot, t) \|_2 (\psi, (H_0 + N^2 + I)\psi) ,$$

which yields (3.4.1).

We now define the sharp time fields

$$A(t) = \int \varphi(x, t) f(x, t) dx \tag{3.4.13}$$

and

$$B(t) = \int \pi(x, t) f(x, t) dx . \tag{3.4.14}$$

LEMMA 3.4.2. *Let $f \in \mathcal{S}(R^2)$ be real. Then $A(t)$ and $B(t)$ define self-adjoint operators, and their domain includes $\mathfrak{D}_0$. For $\psi \in \mathfrak{D}_0$, $\| A(t)\psi \| + \| B(t)\psi \| \leq | f(\cdot, t)|_2 \| F\psi \|$ where $F = (H_0 + N^2 + I)^{1/2}$, and $| f(\cdot, t) |_2 = c(1 + |t|)\{\| f(\cdot, t) \|_2 + \| D_x f(\cdot, t) \|_2\}$.*

PROOF. The proof is similar to that of Lemmas 3.2.2 aud 3.4.1.

THEOREM 3.4.3. *Let $f(x, t) \in \mathcal{S}$ be a real function in $\mathcal{S}$. The vectors $A(t)\psi$ and $B(t)\psi$, $\psi \in \mathfrak{D}_0$, are continuous and rapidly decreasing in t. Their integrals over t exist and define closed symmetric operators $\varphi(f)$ and $\pi(f)$ with domains containing $\mathfrak{D}_0$. The fields $\varphi(f)$, $\pi(f)$, $A(t)$ and $B(t)$ are all independent of $g(x)$. For $\psi \in \mathfrak{D}_0$ we have $\| \varphi(f)\psi \| + \| \pi(f)\psi \| \leq | f |_2 \| F\psi \|$, where $| f |_2 = \int | f(\cdot, t) |_2 dt$ and $F = (H_0 + N^2 + I)^{1/2}$.*

PROOF. This proof is based on the proofs of Lemma 3.4.1, Theorem 3.2.3, and Theorem 3.2.6. The fields $\varphi(f)$ and $\pi(f)$ are defined as their closures on $\mathfrak{D}_0$.

3.5. *Locality.* In this section we derive locality of the field operators. Locality means that two field operators $\varphi(f)$ and $\varphi(h)$ commute provided the supports of f and h are spacelike separated. In other words, whenever

$$(x, t) \in \operatorname{supp} f ,$$
$$(y, s) \in \operatorname{supp} h ,$$

we have

$$|x - y| > |t - s| .$$

Under this hypothesis a signal originating in supp f (caused, for example, by the process of performing the measurement of $\varphi(f)$) cannot be recorded by the measurement of $\varphi(h)$. Thus one expects that the measurement of $\varphi(f)$ does not interfere with the measurement of $\varphi(h)$, and that the joint measurement of $\varphi(f)$ and $\varphi(h)$ can be performed in either order. The precise mathematical statement that the measurements can be performed in either order is that $\varphi(f)$ and $\varphi(h)$ commute.

For any closed operator A, a core $\mathfrak{D}$ of A is defined to be a dense domain contained in $\mathfrak{D}(A)$ such that $A = (A \mid \mathfrak{D})^-$. Self-adjoint operators A and B commute if and only if for any spectral projection E of B, and core $\mathfrak{D}$ of A, $E\mathfrak{D} \subset \mathfrak{D}(A)$ and for $\psi \in \mathfrak{D}$, $EA\psi = AE\psi$.

THEOREM 3.5.1. *If* supp f *and* supp h *are spacelike separated,* $\varphi_g(f)$ *and* $\varphi_g(h)$ *commute.*

PROOF. Let $A_g(t)$ and $B_g(t)$ be the sharp time fields obtained from the test functions f and h. First we prove that $A_g(s)$ and $B_g(t)$ commute. For any open set $\mathcal{I}$ in space, we define the algebra $\mathfrak{A}(\mathcal{I})$ as the weak closure of the bounded functions of the $t = 0$ fields

$$\int \varphi(x, 0)f(x)dx \qquad \text{and} \qquad \int \pi(x, 0)f(x)dx$$

as f runs over the C^∞ functions with support in $\mathcal{I}$. If $\mathcal{I}$ and $\mathcal{J}$ are disjoint open sets, then elements of $\mathfrak{A}(\mathcal{I})$ and $\mathfrak{A}(\mathcal{J})$ commute, and it was shown in [8] that

$$e^{i\sigma H(g)}\mathfrak{A}(\mathcal{I})e^{-i\sigma H(g)} \subset \mathfrak{A}(\mathcal{I}_\sigma) , \tag{3.5.1}$$

where $\mathcal{I}_\sigma$ is the set of all points in space with distance less than $|\sigma|$ from $\mathcal{I}$. The proof in [8] is valid whether or not $g(x) = \text{const.}$ on the set $\mathcal{I}$.

If $\mathcal{I}$ is a small neighborhood of supp $f \cap \{\text{time} = s\}$ and $\mathcal{J}$ is similarly defined with respect to h at time t, then $\mathcal{I}$ and $(\mathcal{J})_{t-s}$ are disjoint. Since the bounded functions of $A_g(s)$ belong to $e^{isH(g)}\mathfrak{A}(\mathcal{I})e^{-isH(g)}$ and the bounded functions of $B_g(t)$ belong to

$$e^{itH(g)}\mathfrak{A}(\mathcal{J})e^{-itH(g)} \subset e^{isH(g)}\mathfrak{A}\big((\mathcal{J})_{s-t}\big)e^{-isH(g)} ,$$

$A_g(s)$ and $B_g(t)$ commute.

Let E be a spectral projection of $A_g(s)$ and let $\psi \in \mathfrak{D}_g$, a core for $\varphi_g(h)$. Then $E\psi \in \mathfrak{D}\big(B_g(t)\big)$ for all t and

$$(\varphi_g(h)\theta, E\psi) = \int (\theta, B_g(t)E\psi)dt = \Big(\theta, E\int B_g(t)\psi dt\Big) = (\theta, E\varphi_g(h)\psi)$$

for all $\theta \in \mathfrak{D}_g$. Thus

$$E\psi \in D(\varphi_g(h)^*) = \mathfrak{D}(\varphi_g(h)) ,$$
$$\varphi_g(h)E\psi = E\varphi_g(h)\psi ,$$

and $A_g(s)$ commutes with $\varphi_g(h)$. Now let F be a spectral projection for $\varphi(h)$. Then $F\psi \in \mathfrak{D}(A(s))$ for all s and

$$(\varphi_g(f)\theta, F\psi) = \left(\theta, \int A_g(s)F\psi ds\right) = \left(\theta, F\int A_g(s)\psi ds\right)$$

as before, so that $F\psi \in \mathfrak{D}(\varphi_g(f)^*) = \mathfrak{D}(\varphi(f))$ and $\varphi_g(f)F\psi = F\varphi_g(f)\psi$. Therefore, $\varphi_g(f)$ and $\varphi_g(h)$ commute.

3.6. *Space time covariance.* Space time covariance means that the field transforms in the expected fashion under the space time translation α, τ,

(3.6.1) $$\varphi(x, t) \rightarrow \varphi(x + \alpha, t + \tau) .$$

By its very definition (3.1.1) and § 3.4, the field transforms correctly under time translation. Let $U(\alpha)$ be the unitary operator on $\mathfrak{F}$ which implements the free field space translation $x \rightarrow x + \alpha$. By definition, $U(\alpha)$ acts on each vector θ_j in the j particle subspace $\mathfrak{F}_j$ by

(3.6.2) $$U(\alpha)\theta(k_1, \cdots, k_j) = e^{+i\alpha(k_1+\cdots+k_j)}\theta(k_1, \cdots, k_j) .$$

We use the convention that $U(\alpha)$ is the Schrödinger picture operator. The active transformation $V(\alpha) = U(-\alpha)$ commonly occurs in the literature. On a suitable domain,

(3.6.3) $$U(-\alpha)a^*(k)U(+\alpha) = e^{-ik\alpha}a^*(k) ,$$

(3.6.4) $$U(-\alpha)a(k)U(+\alpha) = e^{ik\alpha}a(k) ,$$

and from the definition (1.2.11) we have $U(-\alpha)\varphi(x)U(+\alpha) = \varphi(x + \alpha)$. Now $U(\alpha)$ does not commute with $H(g)$, and in fact $U(-\alpha)H(g)U(+\alpha) = H(g')$, where $g'(x) = g(x - \alpha)$. But $\varphi(x, t)$ is independent of the space cutoff function $g(y)$ provided that $g(y) = \lambda$ for $|x-y| > |t|$. Thus if $g(y) = \lambda$ for $|x+\alpha-y| > |t|$,

$$\begin{aligned} U(-\alpha)\varphi(x, t)U(+\alpha) &= U(-\alpha)e^{itH(g)}\varphi(x)e^{-itH(g)}U(\alpha) \\ &= e^{itH(g')}\varphi(x + \alpha)e^{-itH(g')} \\ &= \varphi(x + \alpha, t) \end{aligned}$$

on a suitable domain, for example on $\mathfrak{D}_0 \times \mathfrak{D}_0$. Thus $U(\alpha)$ implements the space translation for fields $\varphi(f)$.

4. The algebra of local observables

To each open region $\mathfrak{B}$ of space time, we associate a C^* algebra $\mathfrak{A}(\mathfrak{B})$ in such a way that the self-adjoint elements of $\mathfrak{A}(\mathfrak{B})$ are exactly the operators corresponding to experiments which may be performed in $\mathfrak{B}$. A number of authors have formulated postulates for quantum field theory in terms of such algebras, see Segal [23], Haag [10], Araki [1], or Haag and Kastler [11]. The requirements of [11] for a local quantum theory are

(a) To each bounded open region $\mathcal{B}$ of space time, there is an associated C^* algebra $\mathfrak{A}(\mathcal{B})$ containing the identity.

(b) Isotony: If $\mathcal{B}_1 \supset \mathcal{B}_2$, then $\mathfrak{A}(\mathcal{B}_1) \supset \mathfrak{A}(\mathcal{B}_2)$.

(c) Locality: If $\mathcal{B}_1$ and $\mathcal{B}_2$ are space like separated, then $\mathfrak{A}(\mathcal{B}_1)$ commutes with $\mathfrak{A}(\mathcal{B}_2)$.

(d) The algebra of local observables $\mathfrak{A}$ is defined as the norm closure of the union of the $\mathfrak{A}(\mathcal{B})$. The algebra is *primitive*; in other words, it has a faithful, irreducible representation.

(e) Lorentz covariance: Let $\{a, \Lambda\}$ be an element of the inhomogeneous Lorentz group $\mathcal{G}$. Then there is a representation $\sigma_{\{a,V\}}$ of $\mathcal{G}$ by a group of $*$ automorphisms of $\mathfrak{A}$, such that for a bounded region $\mathcal{B}$

$$\sigma_{\{a,\Lambda\}}(\mathfrak{A}(\mathcal{B})) = \mathfrak{A}(\{a, \Lambda\}\mathcal{B}) . \tag{4.1}$$

In this section we consider several possible definitions for the algebra $\mathfrak{A}(\mathcal{B})$. The different definitions undoubtedly lead to different C^* algebras. In order to arrive at a natural and aesthetic definition, we prove that all reasonable candidates for $\mathfrak{A}(\mathcal{B})$ have the same weak closure; we take this weakly closed algebra as the definition of $\mathfrak{A}(\mathcal{B})$. A further justification of the use of the weak closure in defining $\mathfrak{A}(\mathcal{B})$ is the fact proved in [29] that the physical representation π of $\mathfrak{A}$ is weakly continuous on each local subalgebra $\mathfrak{A}(\mathcal{B})$.

In general terms, the $\mathfrak{A}(\mathcal{B})$ is formed from bounded functions of the $\varphi(f)$, where $\operatorname{supp} f \subset \mathcal{B}$. We make this notation precise as follows.

Definition. $\mathfrak{A}(\mathcal{B})$ is the weakly closed operator algebra generated by the operators

$$\{F(\varphi(f)) \colon F \in L_\infty, \operatorname{supp} f \subset \mathcal{B}, |f|_1 < \infty\} .$$

The definition is unchanged if we replace L_∞ by some $*$ algebra which is dense in the weak operator topology. It is also unchanged if we replace the class of test functions by another (for example $\mathfrak{D}(\mathcal{B})$) having the same closure in the $|\cdot|_1$ norm. In fact, if $|f_n - f|_1 \to 0$, then $(\varphi(f_n) - z)^{-1} \to (\varphi(f) - z)^{-1}$ in the strong operator topology by Lemma 3.3.4 and by the semigroup convergence theorem [18, p. 502] $e^{i\varphi(f_n)} \to e^{i\varphi(f)}$. Thus $e^{i\varphi(f)}$ and $F(\varphi(f))$ belong to the weak closure if each f_n is admitted as a test function in defining $\mathfrak{A}(\mathcal{B})$.

The same algebra $\mathfrak{A}(\mathcal{B})$ is generated by the bounded functions of sharp time fields $A(t) = \int \varphi(x, t) f(x, t) dx$, $f \in \mathfrak{D}(\mathcal{B})$. In fact, using a sequence f_n such as (3.3.18), we have the resolvents converging $(\varphi(f_n) - z)^{-1} \to (A(t) - z)^{-1}$, and so $F(A(t)) \in \mathfrak{A}(\mathcal{B})$. Thus the sharp time fields generate a smaller algebra.

However, if $f \in \mathfrak{D}(\mathfrak{B})$, we can approximate $\varphi(f)$ by $A(t_1)\Delta t_1 + \cdots + A(t_n)\Delta t_n$ with strong convergence on $\mathfrak{D}_g$. By [18, pp. 429, 502] or by Lemma 3.3.4 the resolvents converge, so $F(\varphi(f))$ belongs to the weakly closed algebra generated by the bounded functions of finite linear combinations of the sharp time fields. We now see that all such $F(\varphi(f))$ belong to the algebra generated by the bounded functions of the sharp time fields themselves. Let F be a bounded operator commuting with $A(t_1), A(t_2), \cdots$, and $A(t_n)$. Then by the spectral theorem, F commutes with $A(t_1)\Delta t_1 + \cdots + A(t_n)\Delta t_n$ on the domain $\mathfrak{D}_g$, which by Theorem 3.3.5 is a core for $A(t_1)\Delta t_1 + \cdots + A(t_n)\Delta t_n$. Thus F commutes with $A(t_1)\Delta t_1 + \cdots + A(t_n)\Delta t_n$. Thus the commutant of $A(t_1)\Delta t_1 + \cdots + A(t_n)\Delta t_n$ is larger than that of $A(t_1), \cdots, A(t_n)$, and the double commutant smaller. Therefore, the sharp time fields generate $\mathfrak{A}(\mathfrak{B})$ as asserted.

THEOREM 4.1. *With our definition of* $\mathfrak{A}(\mathfrak{B})$, *the axioms* (a)–(e) *are satisfied with the possible exception*[4] *of covariance under Lorentz rotations.*

PROOF. The properties (a) and (b) are obvious, and (c) is a consequence of Theorem 3.5.1. To prove (d) we observe that $F(A(0)) \in \mathfrak{A}$ if $F \in L_\infty$ and $A(0)$ is a time zero field operator. Furthermore $(A(t) - A(0))/t$ converges strongly on $\mathfrak{D}_g$ to $B(0) = \int \pi(x, 0) f(x, 0) dx$.

Since $B(0)$ is essentially self-adjoint on $\mathfrak{D}_g$, the resolvent of the difference quotients converge. Thus $F(B(0)) \in \mathfrak{A}$ also. It is known that any bounded operator commuting with all the $F(A(0))$ and the $F(B(0))$ is a multiple of the identity, which proves (d).

Let $(\alpha, t) = a$ be a space time translation, and let $A \in \mathfrak{A}(\mathfrak{B})$. We define

$$\sigma_a(A) = U(-\alpha) e^{itH(g)} A e^{-itH(g)} U(+\alpha) ,$$

where $g(x) = \lambda$ on a sufficiently large interval. It is clear that this σ_a preserves operator norm, and thus extends to $\mathfrak{A}$ by continuity. The verification of (4.1) for σ_a follows from the space time covariance of the field φ.

COURANT INSTITUTE, NEW YORK UNIVERSITY
HARVARD UNIVERSITY

REFERENCES

[1] H. ARAKI, *A lattice of von Neumann algebras associated with the quantum theory of a free Bose field*, J. Math. Phys. **4** (1963), 1343–1362.

[2] P. FEDERBUSH, *A partially alternative derivation of a result of Nelson*, J. Math. Phys. **10** (1969), 50–52.

[3] K. FRIEDRICHS, Perturbation of Spectra in Hilbert Space, American Mathematical Society, Providence, 1965.

[4] *Added in proof.* This possible exception is removed in [30].

[4] G. FROBENIUS, *Über Matrizen mit positiven Elementen*, Sitzungsber. d.k. Preuss. Akad. d. Wissen., Berlin, (1908), 471-476.

[5] L. GÅRDING and A. WIGHTMAN, *Fields as operator-valued distributions in relativistic quantum theory*, Arkiv för Fysik, **28** (1964), 129-184.

[6] J. GLIMM, *Yukawa coupling of quantum fields in two dimensions*, Comm. Math. Phys. **5** (1967), 343-386.

[7] ———, *Boson fields with non-linear self-interaction in two dimensions*, Comm. Math. Phys. **8** (1968), 12-25.

[8] ——— and A. JAFFE, *A $\lambda\varphi^4$ quantum field theory without cutoffs*: I, Phys. Rev. **176** (1968), 1945-1951.

[9] ———, *Singular perturbations of self adjoint operators*, Comm. Pure Appl. Math. **22** (1969), 401-414.

[10] R. HAAG, *On quantum field theories*, Mat.-Fys. Medd. Kong. Danske Videns. Selskab, **29** (1955), No. 12.

[11] ——— and D. KASTLER, *An algebraic approach to quantum field theory*, J. Math. Phys. **5** (1964), 848-861.

[12] D. HALL and A. WIGHTMAN, *A theorem on invariant functions with applications to relativistic quantum field theory*, Mat.-Fys. Medd. Kong. Danske Videns. Selskab, **31** (1957), No. 5.

[13] A. JAFFE, *Divergence of perturbation theory for Bosons*, Comm. Math. Phys. **1** (1965), 127-149.

[14] ———, Constructing a $\lambda(\varphi^4)_2$ quantum field theory, Varenna lectures, 1968.

[15] ——— and R. POWERS, *Infinite volume limit of a $\lambda\varphi^4$ field theory*, Comm. Math. Phys. **7** (1968), 218-221.

[16] R. JENTZSCH, *Über Integralgleichungen mit positiven Kern*, J. für d. reine u. angew. Math. **141** (1912), 235-244.

[17] M. KAC, Probability and Related Topics in Physical Sciences, Interscience Publishers, New York, 1959.

[18] T. KATO, Perturbation Theory for Linear Operators, Springer Verlag, New York, 1966.

[19] M. NAIMARK, Normed Rings, Noordhoff, Groningen, 1964.

[20] E. NELSON, *Analytic vectors*, Ann. of Math. **70** (1959), 572-615.

[21] ———, "A quartic interaction in two dimensions", in Mathemtical Theory of Elementary Particles, R. Goodman and I. Segal, Editors, M.I.T. Press, Cambridge, 1966.

[22] O. PERRON, *Zür Theorie der Matrices*, Math. Ann. **64** (1907), 248-263.

[23] I. SEGAL, *Postulates for general quantum mechanics*, Ann. of Math. **48** (1947), 930-948.

[24] ———, *Foundations of the theory of dynamical systems of infinitely many degrees of freedom*, Mat.-Fys. Medd. Kong. Danske Videns. Selskab, **31** (1959), No. 12.

[25] ———, "Nonlinear functions of random distributions and generalized normal products", in Proceedings of the Conference on Functional Integration and Constructive Quantum Field Theory, edited by I. Segal, April 1966.

[26] ———, *Notes towards the construction of non-linear relativistic quantum fields*: I. *The hamiltonian in two dimensions as the generator of a C^* automorphism group*, Proc. Nat. Acad. Sci. U.S.A. **57** (1967), 1178-1183.

[27] S. WEINBERG, *High energy behavior in quantum field theory*, Phys. Rev. **118** (1960), 838-849.

[28] G. C. WICK, *The evaluation of the collision matrix*, Phys. Rev. **80** (1950), 268-272.

[29] J. GLIMM and A. JAFFE, *The $\lambda(\varphi^4)_2$ quantum field theory without cutoffs*: III. *The physical vacuum*, to appear.

[30] J. CANNON and A. JAFFE, *Lorentz covariance of the $\lambda(\varphi^4)_2$ quantum field theory*, to appear.

(Received March 3, 1969)
(Revised June 30, 1969)

THE $\lambda(\phi^4)_2$ QUANTUM FIELD THEORY WITHOUT CUTOFFS

III. *The physical vacuum*

BY

JAMES GLIMM and ARTHUR JAFFE

Courant Institute of Mathematical Sciences, New York University, New York, New York (1) *and Lyman Laboratory of Physics, Harvard University, Cambridge, Massachusetts* (2)

Contents

1. Introduction

In this series of papers we construct a quantum field theory model. This model describes a spin-zero boson field ϕ with a nonlinear ϕ^4 selfinteraction in two dimensional space time. The corresponding classical field equation is

$$\frac{\partial^2\phi}{\partial t^2}-\frac{\partial^2\phi}{\partial x^2}+m_0^2\phi+4\lambda\phi^3=0. \tag{1.1}$$

The classical field ϕ is by definition a real valued function of x and t which is a solution to (1.1). The quantum field ϕ is also a function of x and t, but its values $\phi(x, t)$ are densely defined bilinear forms on some Hilbert space. The quantum field ϕ is a solution to (1.1),

(1) Supported in part by the New York State Science and Technology Foundation, Grant SSF-(8)-8.

(2) Alfred P. Sloan Foundation Fellow. Supported in part by the Air Force Office of Scientific Research, AF 49(638)-1380.

14 – 702903 *Acta mathematica*. 125. Imprimé le 22 Octobre 1970.

provided that ϕ^3 is suitably interpreted. The quantum field $\phi(x, t)$ depends continuously on x and t and the derivatives in (1.1) are taken in the sense of distributions.

Certain averages of the quantum field

$$\phi(f) = \int \phi(x, t)\, f(x, t)\, dx\, dt \tag{1.2}$$

and

$$\phi(f, t) = \int \phi(x, t)\, f(x)\, dx \tag{1.3}$$

are self adjoint operators on the Hilbert space. In addition, the quantum field ϕ satisfies the canonical equal time commutation relations

$$\left[\phi(f, t), \frac{\partial}{\partial t}\phi(f, t)\right] = i \int f(x)^2 dx. \tag{1.4}$$

The construction of a quantum field ϕ satisfying (1.1)–(1.4) was one of the main results of the first papers in this series [10, 12] (denoted hereafter as I, II). General background material can be found in [9, 17, 19].

The field ϕ constructed in I and II is a bilinear form on the Fock space $\mathcal{F}$, the Hilbert space for noninteracting bosons. The Cauchy data for the quantum field

$$\phi(x, 0) \quad \text{and} \quad \left(\frac{\partial}{\partial t}\phi\right)(x,0) \tag{1.5}$$

are the same Cauchy data that give a solution to the free field equation

$$\frac{\partial^2}{\partial t^2}\phi_{\text{free}} - \frac{\partial^2}{\partial x^2}\phi_{\text{free}} + m_0^2\phi_{\text{free}} = 0, \tag{1.6}$$

and that satisfy (1.4). In other words,

$$\phi(x, 0) = \phi_{\text{free}}(x, 0) \tag{1.7}$$

and

$$\left(\frac{\partial \phi}{\partial t}\right)(x, 0) = \left(\frac{\partial}{\partial t}\phi_{\text{free}}\right)(x, 0). \tag{1.8}$$

It is expected that this quantum field theory of Iand II, with ϕ realized on the Fock space $\mathcal{F}$, will not possess all the properties conventionally assumed in quantum field theory. For example, it seems that there will not exist a Hamiltonian operator H to generate time translations

$$\phi(x, t+\tau) = e^{iH\tau}\phi(x, t)\, e^{-iH\tau}.$$

Also there will not exist a vacuum vector (a translation invariant ground state of H). The cause of these difficulties is made clear by the circle of ideas and results known as Haag's theorem [16, 24]. Haag's theorem states that if the field $\phi \neq \phi_{\text{free}}$ satisfies (1.7)–(1.8) and if there is a Hamiltonian H for ϕ, then H will not possess a unique Euclidean invariant vacuum vector.

In order to understand and deal with these difficulties—the lack of a Hamiltonian operator H on $\mathfrak{F}$ and of a vacuum vector Ω in $\mathfrak{F}$—we first consider the cutoff field equation

$$\frac{\partial^2}{\partial t^2}\phi - \frac{\partial^2}{\partial x^2}\phi + m_0^2\phi + 4\lambda g\phi^3 = 0. \tag{1.9}$$

Here the spatial cutoff $g(x)$ is a smooth, positive function that equals one on some bounded set and that vanishes off some larger bounded set. Corresponding to the cutoff field equation (1.9), there is a Hamiltonian operator

$$H(g) = H_0 + \lambda\int :\phi^4(x):g(x)\,dx - E_g = H_0 + H_{I,g} - E_g. \tag{1.10}$$

The operator H_0 is the free field Hamiltonian and it corresponds to the free field equation (1.6). The interaction energy operator $H_{I,g}$ contains the spatial cutoff g. The operator $H_0 + H_{I,g}$ is a self adjoint operator [10, 11] and is bounded from below [22, 8]. The constant E_g is chosen so that

$$0 = \inf\,\{\text{spectrum } H(g)\}.$$

The constant E_g is one of the standard renormalization counterterms of quantum field theory. E_g is called the self energy of the vacuum. E_g is finite because of the spatial cutoff and because of the limitation to only one space dimension [13]. Both g and the restriction to a single space variable serve to reduce the singularity of the perturbing operator $H_{I,g}$.

In I, II we constructed the Heisenberg picture dynamics for (1.1). We started from the locally correct equation (1.9) for which the solution is

$$\phi_g(x,t) = e^{iH(g)t}\,\phi(x,0)\,e^{-iH(g)t}. \tag{1.11}$$

The solution ϕ_g is independent of $g(\cdot)$ provided $g(y)=1$ for $|y| \leq |x| + |t|$. Thus by patching together different local solutions (1.11), corresponding to $g=1$ on different sets, we constructed a solution ϕ to (1.1). This solution is associated with a time translation automorphism σ_τ implemented locally by $H(g)$ as in (1.11).

$$\sigma_\tau(\phi(x,t)) = \phi(x, t+\tau). \tag{1.12}$$

While a vacuum vector Ω does not exist in $\mathfrak{F}$, we showed in II that the approximate

Hamiltonian $H(g)$ does have a vacuum. In fact, zero is an isolated eigenvalue of $H(g)$ with multiplicity one. Thus there is a vector Ω_g in $\mathcal{F}$ such that

$$H(g)\Omega_g = 0, \quad \|\Omega_g\| = 1, \tag{1.13}$$

and Ω_g is unique up to a scalar multiple.

We wish to obtain the Hamiltonian H and the vacuum Ω corresponding to (1.1) by taking the limit $g(\cdot)\to 1$. We cannot use the most conventional types of limits because as $g(\cdot)\to 1$,

$$E_g \to -\infty \tag{1.14}$$

[13]. Furthermore, perturbation theory predicts that

$$\Omega_g \to 0, \tag{1.15}$$

in the sense of weak convergence on $\mathcal{F}$. In fact (1.15) can be compared to the simpler weak limit in $L_2(R)$,

$$g(\cdot)/\|g\|_2 \to 0.$$

Instead of studying these quantities, we define the expectation values

$$\omega_g(A) = (\Omega_g, A\Omega_g), \tag{1.16}$$

where A is a bounded function of the field operators (1.2) or (1.3). The set of all such A form a C^* algebra $\mathfrak{A}$ of bounded operators on $\mathcal{F}$, and $\omega_g \in \mathfrak{A}^*$, the dual space of $\mathfrak{A}$. Furthermore, ω_g is positive and has norm one, and thus ω_g is a *state* in the sense of C^* algebras. We use a limiting process as $g(\cdot)\to 1$ to obtain a limiting expectation value $\omega(A)$, and then $\omega \in \mathfrak{A}^*$ is necessarily a state also.

According to the Gelfand–Segal construction, ω comes from a vector Ω in some new Hilbert space $\mathcal{F}_{\text{ren}}$ in the sense that

$$\omega(A) = (\Omega, \pi(A)\Omega), \quad \|\Omega\| = 1, \tag{1.17}$$

where

$$\pi: A \to \pi(A) = A_{\text{ren}}$$

is a representation of $\mathfrak{A}$ as operators on $\mathcal{F}_{\text{ren}}$. The Schrödinger picture dynamics exists on $\mathcal{F}_{\text{ren}}$. In the Schrödinger picture, the time translation is given by a one parameter family of unitary operators

$$U(t) = e^{-iHt}. \tag{1.18}$$

If

$$A \to A(t) = \sigma_t(A)$$

is the Heisenberg picture dynamics constructed in I, II, then

$$\pi(\sigma_t(A)) = e^{iHt}\,\pi(A)\,e^{-iHt}. \tag{1.19}$$

Here H is positive and Ω is a vacuum for H,

$$H\Omega = 0. \tag{1.20}$$

The operator H is the renormalized Hamiltonian, and it is the limit of the $H(g)$ in the sense that

$$(\pi(B)\,\Omega,\, e^{iHt}\,\pi(A)\,\Omega)$$

is obtained through a limit as $g(\cdot)\to 1$ of

$$(B\Omega_g,\, e^{iH(g)t}\,A\Omega_g).$$

We call Ω the physical vacuum and π the physical representation. The vectors in $\mathcal{F}_{\text{ren}}$ are called physical vectors.

The space translations are also given by a one parameter group, commuting with the $U(t)$ and leaving Ω fixed.

$$\pi(\sigma_{x,t}(A)) = e^{iHt - iPx}\,\pi(A)\,e^{-iHt+iPx}, \qquad e^{iPx}\,\Omega = \Omega. \tag{1.21}$$

It is well known that certain representations of $\mathfrak{A}$ (coming from states of $\mathfrak{A}$ via the Gelfand–Segal construction) cannot be extended to the unbounded field operators $\phi(f)$ of (1.2), (1.3). Such states and representations seem to be totally unsuited for use in physics. The deepest result of this paper is the fact that the physical representation π is not one of these pathological representations. We now formulate this result in a stronger and more precise form.

The C^* algebra $\mathfrak{A}$ is an inductive limit. Let $\mathcal{B}$ be a bounded region of space time (or a bounded region of space at time zero). Let $\mathfrak{A}(\mathcal{B})$ be the weakly closed (von Neumann) algebra generated by bounded functions of the operators (1.2) or (1.3), but with f restricted to have support in $\mathcal{B}$. Then $\mathfrak{A}$ is defined as the norm closure of

$$\bigcup_B \mathfrak{A}(\mathcal{B}).$$

We prove that π is locally equivalent to the representation of $\mathfrak{A}$ as operators on the Fock space $\mathcal{F}$; in brief π is *locally Fock*. For each bounded region $\mathcal{B}$, there is a unitary operator U_B,

$$U_B\colon \mathcal{F}\to \mathcal{F}_{\text{ren}} \tag{1.22}$$

such that for all A in $\mathfrak{A}(\mathcal{B})$,

$$\pi(A) = A_{\text{ren}} = U_B A U_B^*. \tag{1.23}$$

We have remarked that there is no vacuum vector Ω in $\mathcal{F}$. It immediately follows from the locally Fock property that for each bounded region $\mathcal{B}$ there is a vector Ω_B in $\mathcal{F}$ which serves as a vacuum vector for the algebra $\mathfrak{A}(\mathcal{B})$. In other words, for all $A \in \mathfrak{A}(\mathcal{B})$,

$$\omega(A) = (\Omega, A_{\text{ren}}\Omega) = (\Omega_B, A\Omega_B). \tag{1.24}$$

The vector Ω_B could be called a local vacuum. The vector Ω_B is not unique, but one choice would be $U_B^*\Omega$. The unitary U_B is also not unique.

An important consequence of the locally Fock property of π is the fact that the field ϕ exists on the physical Hilbert space $\mathcal{F}_{\text{ren}}$. Let $\phi(f)$ be self adjoint on $\mathcal{F}$. This is the case if, for example, f is a real, twice differentiable function with compact support [II]. Then for supp. $f \subset \mathcal{B}$ and for real s,

$$s \to \pi(e^{is\phi(f)}) = U_B\, e^{is\phi(f)}\, U_B^*$$

is a weakly continuous one parameter group of unitary operators on $\mathcal{F}_{\text{ren}}$. We let

$$\pi(\phi(f)) = \phi_{\text{ren}}(f) = U_B\phi(f)\, U_B^*$$

be the self adjoint operator that is the infinitesimal generator of the group.(1) Furthermore,

$$\phi_{\text{ren}}(x,t) = U_B\phi(x,t)\, U_B^*, \quad (x,t) \in \mathcal{B},$$

is a densely defined bilinear form depending continuously on x and t, and ϕ_{ren} is a solution to (1.1)–(1.4). All the local properties of ϕ established in II go over immediately to ϕ_{ren} For example, ϕ_{ren} is local, so that $\phi_{\text{ren}}(f_1)$ and $\phi_{\text{ren}}(f_2)$ commute if the supports of f_1 and f_2 are spacelike separated. Also, $\phi_{\text{ren}}(x, t)$ transforms correctly under the space-time translation group.

$$\phi_{\text{ren}}(x+\alpha, t+\tau) = e^{iH\tau - iP\alpha}\phi_{\text{ren}}(x,t)\, e^{-iH\tau+iP\alpha}.$$

The Haag–Kastler axioms [15] are valid for $\mathfrak{A}$ [12, 27] and carry over to $\pi(\mathfrak{A})$ on $\mathcal{F}_{\text{ren}}$.

The main technical step in proving the locally Fock property of π is an estimate on E_g,

$$-MV \leqslant E_g.$$

Here M is a positive constant and V is essentially the length of the support of the spatial cutoff $g(\cdot)$. From this estimate it easily follows that

$$\omega_g\left(\frac{H_0}{V}\right) \leqslant \text{const.}$$

(1) Our definition of ϕ_{ren} is $Z_3^{\frac{1}{2}}$ times the usual renormalized field. The field strength renormalization constant Z_3 is chosen to normalize the one particle states produced by the field ϕ. Perturbation theory predicts that Z_3 is strictly positive in our model. Hence $Z_3^{-\frac{1}{2}}$ should be finite and neglecting Z_3 is merely a matter of convenience.

We use space translation averaging in the limiting process which defines ω, and therefore we are able to conclude that

$$\omega(H_0^{\text{loc}}) \leqslant \text{const.},$$

where H_0^{loc} resembles a local energy operator for free particles. An estimate of this nature, valid for a sequence ω_n of Fock states implies that

$$\omega_n \restriction \mathfrak{A}(\mathcal{B})$$

lies in a *norm* compact subset of the dual $\mathfrak{A}(\mathcal{B})^*$. We note that the set of all states is w^* compact, and that any state $\varrho \in \mathfrak{A}^*$ can be reached as a w^* limit point of Fock states [5, 7]. In particular, the pathological states referred to above are w^* limits of Fock states. However, a norm convergent limit of Fock states is normal on $\mathfrak{A}(\mathcal{B})$, and this excludes the unwanted states. In order to complete the construction of U_B, we use the result of Araki [1] that $\mathfrak{A}(\mathcal{B})$ is a factor of type III, for any bounded open set $\mathcal{B}$ at time zero. We also use a result of Griffin [14] that isomorphisms between separable factors of type III are unitarily implementable.

There are several properties of $\Omega \in \mathcal{F}_{\text{ren}}$ which we have not established. For example, the uniqueness of Ω is one of the Wightman axioms. Also Ω should be in the domain of any polynomial in the field operators $\phi_{\text{ren}}(f)$.

The main phenomenon that we encounter in this paper is the necessity of changing Hilbert spaces. This is a common feature of quantum field theory. In all cases this phenomenon can be traced to the fact that the fundamental objects, such as the Hamiltonian or energy operator, are given in the form

$$(\text{Self adjoint operator}) + (\text{Perturbation})$$

with the perturbation exceedingly singular. (In the $(\phi^4)_2$ theory, the perturbation is a bilinear form, but not an operator on $\mathcal{F}$. In fact the perturbation is so singular that the total Hamiltonian, as a bilinear form on Fock space, is unbounded below. It does not yield an operator on Fock space, but on $\mathcal{F}_{\text{ren}}$ the renormalized Hamiltonian is a positive self adjoint operator.) Two important reasons for this phenomenon are the translation invariance of the Hamiltonian and the fact that the interaction involves an infinite number of degrees of freedom.

The results of this paper extend to the interaction $P(\phi)_2$, where P is a semibounded polynomial. L. Rosen [28] has proved for this model that the cutoff Hamiltonian $H(g) = H_0 + H_{I,g}$ is essentially self adjoint and has a vacuum vector. We expect that the conclusions of section 2 of this work remain valid for certain models, even when the estimates on the vacuum self energy derived in section 5 fail to hold. In particular, we expect that if the vacuum energy in a finite volume has a logarithmic (but not a linear) ultraviolet divergence, the corresponding physical representation is still locally Fock. In two dimensional space-time, the interaction $P(\phi)+Q(\phi)\bar{\psi}\psi$ has this property. Here $P(\phi)$ is a polynomial bounded from below and $Q(\phi)$ is a polynomial of lower degree than P. The two dimensional Yukawa interaction $\bar{\psi}\psi\phi$ also has a logarithmic vacuum energy divergence so we expect that the theory is locally Fock.

The logical order of the following sections is 5, 3, 4, 2. We have rearranged this order to present results, first, followed by details of increasing complexity.

2. The physical representation and renormalization

In this section we construct the physical vacuum vector Ω and the corresponding Hilbert space $\mathcal{F}_{\text{ren}}$. On $\mathcal{F}_{\text{ren}}$ we construct the renormalized self adjoint Hamiltonian H, and the self adjoint momentum operator P. The space time translations are implemented by the unitary group

$$U(\alpha, \tau) = e^{iH\tau - iP\alpha} \tag{2.1}$$

and the vacuum satisfies

$$H\Omega = P\Omega = 0. \tag{2.2}$$

Furthermore, H and P commute and $H \geqslant 0$.

We construct Ω as a limit of the vacuum vectors Ω_g of the cutoff Hamiltonians $H(g)$. For concreteness, we take $g(x)$ to be a nonnegative C_0^∞ function equal to one on the interval $[-3, 3]$. Let

$$g_n(x) = g(x/n) \tag{2.3}$$

and for $A \in \mathfrak{A}$, let

$$\omega_n(A) = \frac{1}{n}\int (\Omega_{g_n}, \sigma_\alpha(A)\Omega_{g_n}) h(\alpha/n)\, d\alpha, \tag{2.4}$$

where σ_α is the space translation automorphism (see II, sections 3.6 and 4). The function $h(\cdot) \geqslant 0$ is C^∞, has support in $[-1, 1]$ and

$$\int h(x)\, dx = 1 = \frac{1}{n}\int h(\alpha/n)\, d\alpha. \tag{2.5}$$

By a general compactness principle for states on a C^* algebra, there is a w^* convergent subnet ω_{n_j} of the states $\{\omega_n\}$.

Thus for each $A \in \mathfrak{A}$,

$$\omega_{n_j}(A) \to \omega(A).$$

The limiting state ω can be used to construct the inner product of a new Hilbert space $\mathcal{F}_{\text{ren}}$, and the operators $A \in \mathfrak{A}$ can be realized as operators on $\mathcal{F}_{\text{ren}}$. The advantage of working on the Hilbert space $\mathcal{F}_{\text{ren}}$ is that the space–time automorphisms σ_α are given by unitary operators (2.1). We state some results.

THEOREM 2.1. *Let ω be a w^* limit point of the sequence $\omega_n \in \mathfrak{A}^*$. There is a subsequence of the ω_n which converges w^* to ω. There is a separable Hilbert space $\mathcal{F}_{\text{ren}}$, a * isomorphism π of $\mathfrak{A}$, a continuous unitary representation U of R^2, and a vector $\Omega \in \mathcal{F}_{\text{ren}}$ such that*

$$\omega(A) = (\Omega, \pi(A)\Omega) = (\Omega, A_{\text{ren}}\Omega) \tag{2.6}$$

$$U(a)\pi(A)U(-a) = \pi(\sigma_a(A)) \tag{2.7}$$

and

$$U(a)\Omega = \Omega. \tag{2.8}$$

The spectrum of the generator H of time translations is contained in the interval $[0, \infty)$.

The existence of $\mathcal{F}_{\text{ren}}$, π and Ω satisfying (2.6) is the Gelfand–Segal theorem [21]. We will prove Theorem 2.1 using the locally Fock property of π, which also leads to the properties of ϕ_{ren} explained in section 1.

THEOREM 2.2. *Let ω be a w^* limit point of the sequence $\{\omega_n\}$ and let π be the corresponding representation of $\mathfrak{A}$. Let $\mathcal{B}$ be a bounded region of space time or of space at $t=0$. Then there is a unitary operator $U_{\mathcal{B}}$: $\mathcal{F} \to \mathcal{F}_{\text{ren}}$ such that for $A \in \mathfrak{A}(\mathcal{B})$*

$$\pi(A) = U_{\mathcal{B}} A U_{\mathcal{B}}^*. \tag{2.9}$$

In short, we say that ω and π are *locally Fock*.

The locally Fock property of ω rests on the following theorem.

THEOREM 2.3. *Let $\mathcal{B}$ be a bounded region of space time or of space at $t=0$. Then the sequence*

$$\omega_n \restriction \mathfrak{A}(\mathcal{B}) \in \mathfrak{A}(\mathcal{B})^* \tag{2.10}$$

lies in a norm compact subset of $\mathfrak{A}(\mathcal{B})^*$. Any limit point $\omega \restriction \mathfrak{A}(\mathcal{B})$ is normal.

The proof of Theorem 2.3 will be given in the following sections. We introduce some von Neumann algebra terminology [2]. A state ϱ of a von Neumann algebra $\mathfrak{M}$ is called *normal* if

$$\sup \varrho(A_\alpha) = \varrho\,(\sup A_\alpha) \tag{2.11}$$

for each monotone increasing generalized sequence $\{A_\alpha\}$, $A_\alpha \in \mathfrak{M}$, provided the sequence A_α is bounded from above, so that $\sup A_\alpha$ exists.

Let us consider linear functionals on $\mathfrak{M}$ of the form $A \to l(A)$,

$$l(A) = \sum_{i=1}^{\infty} (A\theta_i, \psi_i), \tag{2.12}$$

where
$$\sum_{i=1}^{\infty} \|\theta_i\|^2 < \infty, \quad \sum_{i=1}^{\infty} \|\psi_i\|^2 < \infty. \tag{2.13}$$

Ultraweak convergence in $\mathfrak{M}$ is defined by convergence of all linear functionals of the form (2.12)–(2.13), and conversely each ultraweakly continuous linear functional on $\mathfrak{M}$ can be represented in the form (2.12)–(2.13) [2, page 40]. A state ϱ is ultraweakly continuous if and only if it is normal, and every ultraweakly continuous linear functional can be written as a linear combination of four normal states [2, page 54]. The norm of $l \in \mathfrak{M}^*$ defined by
$$\|l\| = \sup \{|l(A)|: A \in \mathfrak{M}, \|A\| \leqslant 1\}. \tag{2.14}$$

The A in (2.14) can be restricted to a subalgebra $\mathfrak{M}_0$ of $\mathfrak{M}$ which is weakly or ultraweakly dense in $\mathfrak{M}$, by Kaplansky's density theorem [2, page 46, page 43].

We apply these concepts to the von Neumann algebra $\mathfrak{A}(\mathcal{B})$ and the weakly dense subalgebra $\mathfrak{A}_0(\mathcal{B})$ generated (algebraically) by the operators
$$\{e^{i\phi(f)}, e^{i\dot{\phi}(f)}: \text{supp } f \subset \mathcal{B}, f \in C^{\infty}\}, \tag{2.15}$$

where $\mathcal{B}$ is a region of space time or of space at $t=0$.

Proof of Theorem 2.2. We first prove that $\pi \restriction \mathfrak{A}(\mathcal{B})$ is ultraweakly continuous. The right and left multiplications
$$A \to AC \quad \text{and} \quad A \to CA$$

are ultraweakly continuous on $\mathcal{B}(\mathcal{J})$. For C_1 and C_2 in some $\mathfrak{A}(\mathcal{B}_1)$, with $\mathcal{B}_1 \supset \mathcal{B}$,
$$A \to C_1^* A C_2 \in \mathfrak{A}(\mathcal{B}_1)$$

is ultraweakly continuous. By Theorem 2.3, $\omega \restriction \mathfrak{A}(\mathcal{B}_1)$ is normal, and hence ultraweakly continuous. Thus
$$A \to \omega(C_1^* A C_2) = (\pi(C_1)\,\Omega, \pi(A)\,\pi(C_2)\,\Omega) \tag{2.16}$$

is ultraweakly continuous. Since the vectors $\pi(C_i)\Omega$ run over a dense subset of $\mathcal{J}_{\text{ren}}$, and since $\pi(A)$ is bounded, the map
$$A \to \pi(A), \quad A \in \mathfrak{A}(\mathcal{B})$$

is ultraweakly–weakly continuous. (This map is also ultraweakly continuous.)

Since $\mathfrak{A}(\mathcal{B})$ is ultraweakly separable, $\pi(\mathfrak{A}(\mathcal{B}))$ has a countable, weakly dense subset.

Since the weak and strong closures of a linear subspace of $\mathfrak{F}_{\text{ren}}$ coincide, $\pi(\mathfrak{A}(\mathcal{B}))$ has a countable, strongly dense subset. But

$$\bigcup_{\mathcal{B}} \pi(\mathfrak{A}(\mathcal{B}))\,\Omega$$

is dense in $\mathfrak{F}_{\text{ren}}$, so $\mathfrak{F}_{\text{ren}}$ is separable also.

In proving the theorem, we replace $\mathcal{B}$ by a bounded open set at time zero (the domain of dependence of $\mathcal{B}$), since this change merely enlarges $\mathfrak{A}(\mathcal{B})$. For such a $\mathcal{B}$ at time zero, Araki has shown that $\mathfrak{A}(\mathcal{B})$ is a factor of type III [1]. His proof requires a computation that a certain operator is not of trace class, and he shows that the operator in question is closed but not everywhere defined due to the conditions at the boundary of $\mathcal{B}$. Thus his proof which is given in three spatial dimensions is valid in one spatial dimension, and is somewhat easier in the latter case.

The kernel of $\pi \restriction \mathfrak{A}(\mathcal{B})$ is a weakly closed two sided ideal. Since there are no proper weakly closed two sided ideals in any factor, we see that $\ker(\pi \restriction \mathfrak{A}(\mathcal{B}))$ equals zero and $\pi \restriction \mathfrak{A}(\mathcal{B})$ is an isomorphism. According to a theorem of E. Griffin [14], $\pi \restriction \mathfrak{A}(\mathcal{B})$, as an isomorphism between separable type III factors, is unitarily implemented. This completes the proof.

We remark that the argument of the above paragraph combined with the fact that there are no norm closed two sided ideals in a factor of type III shows that $\mathfrak{A}$ is simple, and thus π is an isomorphism of $\mathfrak{A}$.

Proof of Theorem 2.1. On the metric space $\mathfrak{A}^*(\mathcal{B})$, compactness implies sequential compactness. Thus using Theorem 2.3, we can find a subsequence ω_{n_j} with

$$\|(\omega - \omega_{n_j}) \restriction \mathfrak{A}(\mathcal{B})\| \to 0.$$

Using the diagonal process, the subsequence can be chosen independently of $\mathcal{B}$.

The existence of the unitary $U(a)$ satisfying (2.7)–(2.8) follows from the Gelfand–Segal construction if for each space–time translation σ_a,

$$\omega(\sigma_a(A)) = \omega(A). \tag{2.17}$$

Thus we must show that ω is a fixed point of σ_a^*. Since ω is norm continuous, it is sufficient to take A in the dense subalgebra

$$\bigcup_{\mathcal{B}} \mathfrak{A}(\mathcal{B}).$$

Thus we let A belong to $\mathfrak{A}(\mathcal{B})$ for some region $\mathcal{B}$, and we choose n sufficiently large so that for all $(x, t) \in \mathcal{B}$,

$$n > |x| + |t|.$$

We define $\mathcal{B}_{|\tau|}$ as the set of points within the distance $|\tau|$ of $\mathcal{B}$. With the above restrictions, $H(g_n)$ is a Hamiltonian for $\mathcal{B}_{|\tau|+|\alpha|}$, provided that $|\alpha| < n$ and $|\tau| < n$. The restriction on α is satisfied if

$$\alpha \in \operatorname{supp} h(\cdot/n).$$

Under these restrictions $$\sigma_\alpha(\sigma_\tau(A)) \in \mathfrak{A}(\mathcal{B}_{|\tau|+|\alpha|})$$

and $$\sigma_\alpha(\sigma_\tau(A)) = \sigma_\tau(\sigma_\alpha(A)) = e^{iH_n t}\sigma_\alpha(A)\, e^{-iH_n\tau},$$

see II. We have set $$H_n = H(g_n) = H_0 + H_{I,g_n} - E_{g_n}. \tag{2.18}$$

Since $$e^{-iH_n\tau}\Omega_{g_n} = \Omega_{g_n},$$

we see that for α in supp $h(\cdot/n)$ and $|\tau| < n$,

$$(\Omega_{g_n}, \sigma_\alpha(\sigma_\tau(A))\,\Omega_{g_n}) = (\Omega_{g_n}, \sigma_\alpha(A)\,\Omega_{g_n})$$

and $$\omega_n(\sigma_\tau(A)) = \omega_n(A).$$

The formula (2.17) for time translations follows immediately.

In order to prove (2.17) for space translations, we substitute $\sigma_\beta(A)$ for A in (2.4) and perform a change of variables $\alpha \to \alpha - \beta$ in the integration. This leads to the estimate

$$|\omega_n(\sigma_\beta(A) - A)|$$
$$= \frac{1}{n}\left|\int (\Omega_{g_n}, \sigma_\alpha(A)\,\Omega_{g_n})\,\{h((\alpha-\beta)/n) - h(\alpha/n)\}\,d\alpha\right| \leqslant \|A\| \int \left| h\left(\alpha - \frac{\beta}{n}\right) - h(\alpha)\right| d\alpha.$$

This vanishes as $n \to \infty$, so that for $A \in \mathfrak{A}(\mathcal{B})$,

$$\omega(\sigma_\beta(A)) = \omega(A),$$

(2.17) is valid, and $U(a)$ does exist.

We prove now the strong continuity of $U(a)$. The vectors $\pi(A)\Omega$ are dense in $\mathcal{F}_{\text{ren}}$ for A in $\bigcup_{\mathcal{B}}\mathfrak{A}(\mathcal{B})$ for bounded space–time regions $\mathcal{B}$. If $\sigma_a(A)$ is in some $\mathfrak{A}(\mathcal{B})$, we have

$$U(a)\,\pi(A)\,\Omega = \pi(\sigma_a(A))\,\Omega$$

by (2.7) and (2.8). By (2.9),

$$U(a)\,\pi(A)\,\Omega = U_{\mathcal{B}}\,\sigma_a(A)\,U_{\mathcal{B}}^*\,\Omega. \tag{2.19}$$

Since σ_a is implemented locally on $\mathcal{F}$ by unitary operators of the form

$$\exp(iH(g)\,\tau)\,\exp(-iP\alpha),$$

the right side of (2.19) is strongly continuous in a. This proves the strong continuity of $U(a)$ on a dense set of vectors, which is sufficient.

The remaining assertion of the theorem is that $H \geqslant 0$. Let

$$H = \int \mu dE(\mu)$$

be the spectral resolution of H. Then we define

$$F(\tau) = \omega(A^* \sigma_\tau(A)) = (\pi(A)\,\Omega,\, e^{iH\tau}\pi(A)\,\Omega) = \int e^{i\mu\tau} d\,\|E(\mu)\,\pi(A)\Omega\|^2, \tag{2.20}$$

In order to prove that H has nonnegative spectrum, we must show that only the half line $[0, \infty)$ contributes to the integral (2.20). For A in some local algebra $\mathfrak{A}(\mathcal{B})$, we infer from Theorem 2.3 that $F(\tau)$ is the limit through a subsequence of the similar functions

$$F_n(\tau) = \omega_n(A^* \sigma_\tau(A)) = \frac{1}{n} \int (\sigma_\alpha(A^*)\,\Omega_{g_n},\, e^{iH_n\tau} \sigma_\alpha(A)\,\Omega_{g_n}) \times h(\alpha/n)\, d\alpha,$$

where H_n is defined in (2.18). Since $H_n \geqslant 0$, $F_n(\tau)$ has a Fourier transform $\tilde{F}_n(\mu)$ with support in the half line $[0, \infty)$. The subsequence $F_{n_j}(\tau)$ converges pointwise to F, and is uniformly bounded,

$$|F_{n_j}(\tau)| \leqslant \|A\|^2.$$

Hence the F_{n_j} converge to F as tempered distributions, and consequently the Fourier transforms $\tilde{F}_{n_j}$ converge as tempered distributions also. Their limit $\tilde{F}$ must therefore have support in $[0, \infty)$, and so only the interval $[0, \infty)$ contributes to (2.20). As the vectors $\pi(A)\Omega$, $A \in \mathfrak{A}(\mathcal{B})$, are dense in $\mathcal{F}_{\text{ren}}$, this shows that $H \geqslant 0$.

3. Local number operators

In this section we define local number operators N_B, $N_{\tau,B}$ and $N_{\tau,\zeta}$ and we investigate their properties. Each of these operators is the biquantization of an operator on the one particle space. We reduce the study of such quantized operators to the study of the corresponding one particle operators. We analyze the one particle operators in detail, and thereby arrive at results for the quantized operators. In Section 3.3 we assume the estimates derived in Section 5, and we show that expectation values of the above local number operators in the approximate vacuum states ω_n are bounded uniformly in n. We shall use such results in Section 4.

3.1. One particle operators. The one particle space consists of Lebesque square integrable functions $L_2(R^1)$. We study operators c on $L_2(R^1)$ that are defined on the dense domain

$S(R^1)$, the Schwartz space of rapidly decreasing C^∞ functions with rapidly decreasing derivatives. We require that c be a bounded transformation from $S(R^1)$ to $L_2(R^1)$,

$$\|cf\|_2 \leq \|f\|_s, \tag{3.1.1}$$

where $\|\cdot\|_s$ denotes some Schwartz space norm. Such an operator c can be represented by a tempered distribution kernel. The Fourier transform is an operator of this type.

$$f(x) = (2\pi)^{-\frac{1}{2}} \int \tilde{f}(p)\, e^{-ipx}\, dp. \tag{3.1.2}$$

Corresponding to the physical interpretation of position and momentum, we say that Fourier transformation connects the momentum representation with the configuration space (position) representation. This physical identification corresponds to the introduction of Newton–Wigner position coordinates.

To each bounded operator $c_p: S(R^1) \to L_2(R^1)$ in the momentum representation, there is a bounded operator c_x: $S(R^1) \to L_2(R^1)$ in the configuration space representation, where

$$(c_x f)(x) = (2\pi)^{-\frac{1}{2}} \int (c_p \tilde{f})(p)\, e^{-ipx}\, dp. \tag{3.1.3}$$

Conversely given c_x there is a corresponding c_p.

We now mention a few other examples. Let $\mu(p)$ be the energy of a particle of mass m_0. The operator $c_p = \mu(p)^\tau$ of multiplication by

$$\mu(p)^\tau = (p^2 + m_0^2)^{\tau/2} \tag{3.1.4}$$

maps $S(R^1)$ onto $S(R^1)$.

The configuration space operator μ_x^τ corresponding to $\mu(p)^\tau$ is convolution by a kernel k_τ. Here $k_\tau(x)$ is C^∞ except for $x=0$. If $\tau/2$ is a nonnegative integer, k_τ has support at $x=0$. Otherwise $k_\tau(x)$ decreases exponentially at infinity. Explicitly [25, page 185],

$$k_\tau(x) = \frac{2^{\tau/2+1}}{\Gamma\left(-\frac{\tau}{2}\right)} \left(\frac{m_0}{|x|}\right)^{(\tau+1)/2} \int_0^\infty e^{-m_0|x|\cosh t} \cosh\left(\frac{t}{2} + \tau t\right) dt, \tag{3.1.5}$$

from which we see that for $\tau \geq -1$,

$$\left|\frac{d^n}{dx^n} k_\tau(x)\right| \leq O(e^{-m_0|x|}), \quad \text{as } |x| \to \infty, \tag{3.1.6}$$

for $n = 0, 1, 2, \ldots$ For $\tau < -1$, estimates of the form (3.1.6) hold if m_0 is replaced by $m_0 - \varepsilon$, for any $\varepsilon > 0$.

A third example of an operator C_x which we will need is multiplication by the characteristic function $E(x)$ of an interval $\mathcal{B}=[a, b]$. The Fourier transformed operator c_p is convolution by

$$\left(\frac{2}{\pi}\right)^{\frac{1}{2}}(e^{i(a+b)/2})\frac{\sin((a-b)p/2)}{p}=(2\pi)^{-\frac{1}{2}}\tilde{E}(p). \tag{3.1.7}$$

A fourth example of an operator c_x is multiplication by a C_0^∞ function $\zeta(x)$. The corresponding c_p is convolution by $(2\pi)^{-\frac{1}{2}}\tilde{\zeta}(p)\in S(R^1)$. Both c_x and c_p map $S(R^1)$ into itself.

Thus the operator

$$c_x=\zeta(x)\,\mu_x^\tau\zeta(x) \tag{3.1.8}$$

is a bounded transformation of $S(R^1)$ into itself, and it is a localization of the operator μ_x^τ. We also want a localized version of μ_x^τ with a sharp boundary. Thus we wish to replace $\zeta(x)$ in (3.1.8) by $E(x)$.

THEOREM 3.1.1. *Let $\tau<\frac{1}{4}$. Then $E(x)\mu_x^\tau E(x)$ is a bounded operator from $S(R^1)$ to $L_2(R^1)$. Let $\tau<\frac{1}{2}$. If $\zeta(x)$ is a positive C_0^∞ function equal to one in a neighborhood of the support of $E(x)$, and if $\varepsilon>0$, then as bilinear forms on $S(R^1)\times S(R^1)$,*

$$E(x)\,\mu_x^\tau E(x)\leqslant \text{const. } \zeta(x)\,\mu_x^{2\tau+\varepsilon}\zeta(x), \tag{3.1.9}$$

where the constant depends only on τ, ε and supp E.

Proof. It is sufficient to prove that for $\tau<\frac{1}{4}$, $\varepsilon>0$, and $f\in S(R^1)$,

$$(f, E\mu_x^{2\tau}Ef)\leqslant \text{const. } (f, \mu_x^{4\tau+\varepsilon}f). \tag{3.1.10}$$

In that case
$$\|E\mu_x^\tau Ef\|^2\leqslant\|\mu_x^\tau Ef\|^2=(f, E\mu_x^{2\tau}Ef)\leqslant \text{const. } \|\mu_x^{2\tau+\varepsilon/2}f\|^2, \tag{3.1.11}$$

which shows that $E(x)\,\mu_x^\tau E(x)$ is an operator and a bounded transformation from $S(R^1)$ to $L_2(R^1)$. Furthermore, $f=\zeta f+(1-\zeta)f$ and

$$Ef=E\zeta f.$$

Hence
$$(f, E\mu_x^\tau Ef)=(\zeta f, E\mu_x^\tau E\zeta f)\leqslant \text{const. } (f, \zeta\mu_x^{2\tau+\varepsilon}\zeta f),$$

which is (3.1.9).

In order to prove (3.1.10), we write

$$(f, E\mu_x^\tau Ef)=(f, E\mu_x^\tau f)+(f, E[\mu_x^\tau, E]f).$$

Since
$$|(f, E\mu_x^\tau f)|\leqslant\|f\|\,\|\mu_x^\tau f\|\leqslant \text{const. } \|\mu_x^\tau f\|^2,$$

we infer that
$$(f, E\mu_x^\tau Ef)\leqslant \text{const. } (f, \mu_x^{2\tau}f)+|(f, E[\mu_x^\tau, E]f)|. \tag{3.1.12}$$

We now estimate
$$(f, E[\mu_x^\tau, E]f) = (\tilde{f}, g), \tag{3.1.13}$$

where
$$g(p) = \frac{1}{(2\pi)} \int \tilde{E}(p-q)\,(\mu(q)^\tau - \mu(s)^\tau)\,\tilde{E}(q-s)\,\tilde{f}(s)\,dq\,ds.$$

Since for $\tau \leqslant 1$,
$$|\mu(q)^\tau - \mu(s)^\tau| \leqslant \text{const. } \mu(q-s)^\tau,$$

and since by (3.1.4),
$$|\tilde{E}(p)| \leqslant \text{const. } \mu(p)^{-1},$$

we have
$$|g(p)| \leqslant (h * |\tilde{f}|)(p), \tag{3.1.14}$$

where
$$h(p) = \text{const.} \int \mu(p-q)^{-1} \mu(q)^{-1+\tau} dq. \tag{3.1.15}$$

We now show that for any $\varepsilon > 0$, there is a constant such that $h(p) \leqslant \text{const. } \mu(p)^{-1+\tau+\varepsilon}$. We break up the integral in (3.1.15) into two regions:

$$\text{I}: |q| \leqslant \tfrac{1}{2}|p|$$

and
$$\text{II}: |q| > \tfrac{1}{2}|p|.$$

In region I, $\mu(p-q) \geqslant \text{const. } \mu(p)$, so that

$$\int_{\text{I}} \mu(p-q)^{-1} \mu(q)^{-1+\tau} dq \leqslant \text{const. } \mu(p)^{-1} \int_{\text{I}} \mu(q)^{-1+\tau} dq \leqslant \text{const. } \mu(p)^{-1+\tau},$$

since
$$\int_{\text{I}} \mu(q)^{-1+\tau} dq \leqslant \text{const.} \int_1^{|p|} |q|^{-1+\tau} dq \leqslant \text{const. } \mu(p)^\tau.$$

On the other hand, for $\varepsilon > 0$,

$$\mu(q)^{-\varepsilon/2} \leqslant \text{const. } \mu(p-q)^{-\varepsilon/2} \mu(p)^{\varepsilon/2},$$

so that
$$\int_{\text{II}} \mu(p-q)^{-1} \mu(q)^{-1+\tau} dq \leqslant \text{const. } \mu(p)^{\varepsilon/2} \int_{\text{II}} \mu(p-q)^{-1-\varepsilon/2} \mu(q)^{-1+\tau+\varepsilon/2}\, dq.$$

But in the region II, $\mu(q)^{-1} \leqslant 2\mu(p)^{-1}$, so

$$\int_{\text{II}} \mu(p-q)^{-1} \mu(q)^{-1+\tau} dq \leqslant \text{const. } \mu(p)^{(\varepsilon/2)-1+\tau+\varepsilon/2},$$

from which we conclude that
$$h(p) \leqslant \text{const. } \mu(p)^{-1+\tau+\varepsilon}. \tag{3.1.16}$$

We now bound (3.1.13) by writing for $\nu \geqslant 0$,

$$|(\tilde{f}, g)| \leqslant \|\tilde{f}\|_{2-\nu/(1+\nu)} \|g\|_{2+\nu} \leqslant \|\tilde{f}\|_{2-\nu/(1+\nu)} \|h * |\tilde{f}|\,\|_{2+\nu}, \tag{3.1.17}$$

where subscripts denote L_p norms. By the Hausdorff–Young inequality

$$\|h * |\tilde{f}|\,\|_{2+\nu} \leqslant \|\tilde{h}(x)\,|\tilde{f}|^{\sim}(x)\|_{2-\nu/(1+\nu)}$$

and by the Hölder inequality

$$\leqslant \|\tilde{h}(x)\|_{(4+2\nu)/\nu} \|f\|_2,$$

since

$$\||\tilde{f}|^{\sim}\|_2 = \||\tilde{f}|\|_2 = \|f\|_2.$$

Again applying the Hausdorff–Young inequality

$$\|h * |\tilde{f}|\|_{2+\nu} \leqslant \|h\|_{1+\nu/(4+\nu)} \|f\|_2, \tag{3.1.18}$$

and by (3.1.16), $h \in L_{1+\nu/(4+\nu)}$ if

$$(1-\tau)\left(1 + \frac{\nu}{4+\nu}\right) > 1.$$

This can be satisfied if

$$\tau < \tfrac{1}{2} \tag{3.1.19}$$

by choosing ν sufficiently large so that $\nu > (1-2\tau)/4\tau$, or in other words

$$\tau < \frac{\nu}{4+2\nu}. \tag{3.1.20}$$

For such a value of ν, we have by (3.1.17)

$$|(\tilde{f}, g)| \leqslant \text{const. } \|\tilde{f}\|_{2-\nu/(1+\nu)} \|f\|_2.$$

On the other hand, we shall prove that for certain ν satisfying (3.1.20) and $\varepsilon > 0$,

$$\|\tilde{f}\|_{2-\nu/(1+\nu)} \leqslant \text{const. } \|\mu_x^{\tau+\varepsilon/2} f\|_2, \tag{3.1.21}$$

so that

$$|(f, E[\mu_x^\tau, E]\,f)| = |(\tilde{f}, g)| \leqslant \text{const. } \|\mu_x^{\tau+\varepsilon/2} f\|_2 \|f\|_2 \leqslant \text{const. } (f, \mu_x^{2\tau+\varepsilon} f). \tag{3.1.22}$$

Since this τ is restricted to $\tau < \frac{1}{2}$, the inequalities (3.1.22) and (3.1.12) complete the proof of (3.1.10). In order to establish (3.1.21), we write

$$\|\tilde{f}\|_{2-\nu/(1+\nu)}^{2-\nu/(1+\nu)} = \int |\tilde{f}(p)|^{2-\nu/(1+\nu)} dp = \int |\mu(p)^\sigma \tilde{f}(p)|^{2-\nu/(1+\nu)} \mu(p)^{-\sigma(2-\nu/(1+\nu))} dp$$

$$\leqslant \|\mu(p)^\sigma \tilde{f}\|_2^{2-\nu/(1+\nu)} \|\mu(p)^{-2\sigma+\sigma\nu/(1+\nu)}\|_{2(1+\nu)/\nu}.$$

15 – 702903 *Acta mathematica*. 125. Imprimé le 23 Octobre 1970.

The function $\mu(p)^{-2\sigma+\sigma\nu/(1+\nu)}$ is in $L_{2(1+\nu)/\nu}$ if

$$\sigma\left(2-\frac{\nu}{1+\nu}\right)\left(2\cdot\frac{1+\nu}{\nu}\right)>1.$$

This is true for any σ satisfying

$$\sigma>\frac{\nu}{4+2\nu}.$$

Since ν may be any positive number satisfying (3.1.20), we can choose ν so that

$$\tau+\varepsilon/2>\frac{\nu}{4+2\nu}>\tau,$$

and then choose $\sigma=\tau+\varepsilon/2$. Hence

$$\|\tilde{f}\|_{2-\nu/(1+\nu)}\leqslant \text{const. } \|\mu(p)^{\tau+\varepsilon/2}\tilde{f}\|_2=\text{const. } \|\mu_x^{\tau+\varepsilon/2}f\|_2,$$

which is (3.1.21). This completes the proof of the theorem.

For E, ζ, τ as above, the operators

$$E\mu_x^\tau E \text{ and } \zeta\mu_x^\tau\zeta$$

are positive operators with domain $S(R^1)$. Let c_τ and s_τ denote their respective Friedrichs extensions. We note that for

$$f\in \mathcal{D}(s_{2\tau+\varepsilon}^{\frac{1}{2}}),$$

we have

$$f\in \mathcal{D}(c_\tau^{\frac{1}{2}})$$

and for a constant depending only on supp E, τ and ε,

$$\|c_\tau^{\frac{1}{2}}f\|^2\leqslant \text{const. } \|s_{2\tau+\varepsilon}^{\frac{1}{2}}f\|. \tag{3.1.23}$$

This is a consequence of the fact that inequality (3.1.9) extends by continuity to the closure of $S(R^1)$ in the norm

$$|||f|||_1=\|\mu_x^{\tau+\varepsilon/2}\zeta f\|,$$

which is the domain of $s_{2\tau+\varepsilon}^{\frac{1}{2}}$. Furthermore,

$$|||f|||_2=\|\mu_x^{\tau/2}Ef\|\leqslant \text{const. } |||f|||_1$$

so that the domain of $c_\tau^{\frac{1}{2}}$ includes the domain of $s_{2\tau+\varepsilon}^{\frac{1}{2}}$.

THEOREM 3.1.2. *Let $0<\tau<\frac{1}{2}$ and let c_τ be the Friedrichs extension of*

$$E(x)\,\mu_x^\tau E(x),$$

where $E(x)$ is the characteristic function for a bounded interval $\mathcal{B}$. Then c_τ commutes with the orthogonal projection of $L_2(R^1)$ onto $L_2(\mathcal{B})$, and

$$[c_\tau \restriction L_2(\mathcal{B})]^{-1}$$

is a compact operator.

Proof. The projection of $L_2(R^1)$ onto $L_2(\mathcal{B})$ is given by multiplication by $E(x)$. Clearly $E(x)\mu_x^\tau E(x)$ commutes with multiplication by $E(x)$ on $\mathcal{S}(R^1)$, since $E(x)^2=E(x)$. By continuity, this extends as a bilinear form to the domain $\mathcal{D}(c_x^{\frac{1}{2}})\times\mathcal{D}(c_x^{\frac{1}{2}})$, and so holds on $\mathcal{D}(c_\tau)$ as an operator equality.

In order to show that c_τ has a compact resolvent on $L_2(\mathcal{B})$, it is sufficient to show that the resolvent of $c_\tau^{\frac{1}{2}}$ is compact on $L_2(\mathcal{B})$. The latter is true if whenever $\mathcal{D}$ is a set of vectors such that

$$\mathcal{D}\subset\mathcal{D}(c_\tau^{\frac{1}{2}}\restriction L_2(\mathcal{B})) \tag{3.1.24}$$

and

$$\sup_{f\in\mathcal{D}}\|c_\tau^{\frac{1}{2}}f\|<\infty, \tag{3.1.25}$$

then $\mathcal{D}$ has a compact closure. It is sufficient to replace $\mathcal{D}(c_\tau^{\frac{1}{2}}\restriction L_2(\mathcal{B}))$ in (3.1.24) by any core of $c_\tau^{\frac{1}{2}}\restriction L_2(\mathcal{B})$. By definition of the Friedrichs extension, $\mathcal{S}(R^1)$ is a core for $c_\tau^{\frac{1}{2}}$ and $E(x)\,\mathcal{S}(R^1)$ is a core for $c_\tau^{\frac{1}{2}}\restriction L_2(\mathcal{B})$.

It is convenient to imbed $L_2(\mathcal{B})$ in $L_2(R^1)$ and to write a nonlocal expression for $c_\tau^{\frac{1}{2}}$. For $f\in E(x)\,\mathcal{S}(R^1)$, we have

$$\|c_\tau^{\frac{1}{2}}f\|^2=\|\mu_x^{\tau/2}E(x)f\|^2=\|\mu_x^{\tau/2}f\|^2, \tag{3.1.26}$$

where we call this nonlocal since $\mu_x^{\tau/2}f\in L_2(R^1)$. We remark that (3.1.26) implies a continuity condition on f. Let $f_a(x)=f(x+a)$, so that

$$\|f_a-f\|^2=\int|\tilde f(p)\,(e^{-ipa}-1)|^2\,dp.$$

Since for $0<\tau\leqslant 1$,
$$|(e^{-ipa}-1)|\leqslant 2\,|pa|^\tau\leqslant 2\,|a|^\tau\mu(p)^\tau,$$

we infer that
$$\|f_a-f\|\leqslant 2\,|a|^\tau\left(\int|\tilde f(p)|^2\mu(p)^{2\tau}dp\right)^{\frac{1}{2}}=2\,|a|^\tau\,\|\mu_x^\tau f\|.$$

Thus a uniform bound (3.1.25) for $f\in\mathcal{D}$, gives

$$\sup_{f\in\mathcal{D}}\|f_a-f\|\leqslant\text{const.}\,|a|^\tau,$$

or equicontinuity of the functions in $\mathcal{D}$. It is a classical result that an equicontinuous set $\mathcal{D}$ of L_2 functions with support in a bounded interval has a compact closure $\mathcal{D}^-$. This completes the proof of the theorem.

3.2. Quantized operators on Fock space. We first review (1) some standard notation. The Fock space $\mathcal{F}$ for noninteracting bosons is the Hilbert space completion of the direct sum of n-particle Fock spaces,

$$\mathcal{F} = \bigoplus_{n=0}^{\infty} \mathcal{F}_n. \tag{3.2.1}$$

Here $\mathcal{F}_0$ is the complex numbers and $\mathcal{F}_n$ is the completion in $L_2(R^n)$ of the n-fold symmetric tensor product of $\mathcal{F}_1 = L_2(R_1)$. Thus a vector $f = \{f_n\} \in \mathcal{F}$ is a sequence of n particle vectors in $\mathcal{F}_n$, $\|f\|^2 = \sum_{n=0}^{\infty} \|f_n\|^2$. As for the one particle space, we represent n particle vectors in the momentum representation $f_n(k_1, ..., k_n)$ or in the configuration representation $f_n(x_1, ..., x_n)$.

$$f_n(x_1, \dots, x_n) = (2\pi)^{-n/2} \int e^{-ikx} \tilde{f}(k_1, \dots, k_n)\, dk, \tag{3.2.2}$$

where $kx = \sum_{i=1}^{n} k_i x_i$ and $dk = \prod_{i=1}^{n} dk_i$. Fourier transformation is an isomorphism of $\mathcal{F}_n$ and of $\mathcal{F}$. We use a domain $\mathcal{D}_0 \subset \mathcal{F}$ of well behaved vectors, The vectors in $\mathcal{D}_0$ have a finite number of particles with wave functions in the Schwartz space $\mathcal{S}(R^n)$,

$$\mathcal{D}_0 = \{f : f \in \mathcal{F},\ f_n = 0 \text{ for large } n, \text{ and } f_n \in \mathcal{S}(R_n)\} \tag{3.2.3}$$

Fourier transformation maps $\mathcal{D}_0$ onto itself.

We define the annihilation operator $a(k)$ on $\mathcal{D}_0$, which maps $\mathcal{F}_n$ onto $\mathcal{F}_{n-1}$. For $f \in \mathcal{D}_0$,

$$(a(k) f)_{n-1}(k_1, \dots, k_{n-1}) = n^{\frac{1}{2}} f_n(k, k_1, \dots, k_{n-1}), \tag{3.2.4}$$

so that $a(k)\mathcal{D}_0 \subset \mathcal{D}_0$. Clearly $\quad [a(k), a(k')]\,\mathcal{D}_0 = 0$

and any product $a(k_1) \dots a(k_\beta)$ is defined on the domain $\mathcal{D}_0$.

The adjoint $a(k)^*$ of $a(k)$ has domain zero as an operator, but it is a densely defined bilinear form on $\mathcal{D}_0 \times \mathcal{D}_0$. Even though $a(k)^*$ is not a densely defined operator, we follow convention and call it the creation operator for a particle with momentum k.

Any monomial of creation and annihilation operators

(1) Formulations of Fock space and operators on it may be found in many places. Convenient for our point of view is the systematic treatment of Kristensen, Mejlbo and Poulsen [20]. Creation and annihilation operators as bilinear forms were also considered by Galindo [6]. Our use in [10] of weak integrals of bilinear forms has been questioned [23], and for this reason we present the elementary details leading to formula (3.2.7). We note that every Wightman field $\phi(x)$ is a densely defined bilinear form with C^∞ dependence on x [3].

$$a^*(k_1) \dots a^*(k_\alpha) a(k_1) \dots a(k_\beta), \tag{3.2.5}$$

with the creation operators to the left of the annihilators, is a bilinear form on $\mathcal{D}_0 \times \mathcal{D}_0$ with values

$$(f, a^*(k_1) \dots a^*(k_\alpha)\, a(p_1) \dots a(p_\beta)\, g) \in \mathcal{S}(R^{\alpha+\beta}). \tag{3.2.6}$$

Thus if $c_{\alpha\beta}(k;p) \in \mathcal{S}'(R^{\alpha+\beta})$ is a tempered distribution,

$$C_{\alpha\beta} = \int c_{\alpha\beta}(k;p)\, a^*(k_1) \dots a^*(k_\alpha)\, a(p_1) \dots a(p_\beta)\, dk\, dp \tag{3.2.7}$$

is a bilinear form on $\mathcal{D}_0 \times \mathcal{D}_0$ that exists as a weak integral of bilinear forms. The $C_{\alpha\beta}$ of (3.2.7) is a *Wick ordered monomial* and is the quantization of the tempered distribution $c_{\alpha\beta}(k;p)$.

If $c_{\alpha\beta}(k;p)$ is the kernel of a bounded operator $c_{\alpha\beta}$ from $\mathcal{S}(R^\beta)$ to $L_2(R^\alpha)$, then the bilinear form $C_{\alpha\beta}$ on $\mathcal{D}_0 \times \mathcal{D}_0$ determines an unbounded operator on $\mathcal{F}$ with domain $\mathcal{D}_0$. In order to establish this, we note that if $i_{\gamma\gamma}$ is the identity operator on $L_2(R^\gamma)$, then $c_{\alpha\beta} \otimes i_{\gamma\gamma}$ is a bounded operator from $\mathcal{S}(R^\beta) \otimes \mathcal{S}(R^\gamma)$ into $L_2(R^{\alpha+\gamma})$, and it extends uniquely to a bounded operator from $\mathcal{S}(R^{\beta+\gamma})$ into $L_2(R^{\alpha+\gamma})$.

Symmetrization is a projection on L_2 and on $\mathcal{S}$. Thus the symmetrization of $c_{\alpha\beta} \otimes i_{\gamma\gamma}$ in the $\beta+\gamma$ initial variables and in the $\alpha+\gamma$ final variables yields a bounded operator from $\mathcal{S}(R^{\beta+\gamma})$ to $L_2(R^{\alpha+\gamma})$, and an unbounded operator from $\mathcal{F}_{\beta+\gamma}$ to $\mathcal{F}_{\alpha+\gamma}$ with domain $\mathcal{S}(R^{\beta+\gamma}) \cap \mathcal{F}_{\beta+\gamma}$. Finally, $C_{\alpha\beta}$ is the sum over γ of the symmetrizations of $((\alpha+\gamma)!/\gamma!)\,((\beta+\gamma)!/\gamma!)^{\frac{1}{2}}$ $c_{\alpha\beta} \otimes i_{\gamma\gamma}$ with domain $\mathcal{D}_0$.

If c_{11} is a bounded operator from $\mathcal{S}(R^1)$ to $L_2(R^1)$, then the operator C_{11} has a closure on $\mathcal{F}$ if the operator c_{11} has a closure as an unbounded operator from $L_2(R^1)$ to $L_2(R^1)$. If $c_{\alpha\beta}$ is bounded from $L_2(R^\beta)$ to $L_2(R^\alpha)$, then $C_{\alpha\beta}$ has a closure with domain containing $\mathcal{D}(N^{(\alpha+\beta)/2})$, where N is the total number of particles operator defined by

$$(Nf)_n = n f_n. \tag{3.2.8}$$

If $\|c_{\alpha\beta}\|$ is the norm of $c_{\alpha\beta}$ as a transformation from $L_2(R^\beta)$ to $L_2(R^\alpha)$ then

$$\|(N+I)^{-\alpha/2} C_{\alpha\beta} (N+I)^{-\beta/2}\| \leqslant \|c_{\alpha\beta}\|. \tag{3.2.9}$$

This we derive from the bound

$$\begin{aligned} |(f_{\alpha+\gamma}, C_{\alpha\beta}\, g_{\beta+\gamma})| &\leqslant \left(\frac{(\alpha+\gamma)!}{\gamma!}\,\frac{(\beta+\gamma)!}{\gamma!}\right)^{\frac{1}{2}} \|c_{\alpha\beta}\|\,\|f_{\alpha+\gamma}\|\,\|g_{\beta+\gamma}\| \\ &\leqslant \|c_{\alpha\beta}\|\,(\alpha+\gamma)^{\alpha/2}\,\|f_{\alpha+\gamma}\|\,(\beta+\gamma)^{\beta/2}\,\|g_{\beta+\gamma}\|. \end{aligned}$$

If $c_{\alpha\beta}$ maps $\mathcal{S}(R^\beta)$ continuously into $\mathcal{S}(R^\alpha)$, then $C_{\alpha\beta}$ maps $\mathcal{D}_0$ into $\mathcal{D}_0$.

By Fourier transformation we define the annihilation operator $A(x)$ for a particle at the space point x,

$$A(x) = (2\pi)^{-\frac{1}{2}} \int a(p)\, e^{-ipx}\, dp, \tag{3.2.10}$$

corresponding to the quantization of $c_{01}(p) = (2\pi)^{-1/2} \exp(-ipx)$. In the configuration space representation, for $f \in \mathcal{D}_0$,

$$(A(x)f)_{n-1}(x_1, \ldots, x_{n-1}) = n^{\frac{1}{2}} f_n(x, x_1, \ldots, x_{n-1}). \tag{3.2.11}$$

As above, we can quantize a tempered distribution $c_{\alpha\beta}(x; y)$ yielding

$$C_{\alpha\beta} = \int c_{\alpha\beta}(x; y)\, A^*(x_1) \ldots A^*(x_\alpha)\, A(y_1) \ldots A(y_\beta)\, dx\, dy, \tag{3.2.12}$$

a bilinear form on $\mathcal{D}_0 \times \mathcal{D}_0$. These $C_{\alpha\beta}$ have the same properties as the $C_{\alpha\beta}$ in the momentum representation discussed above.

We also use the Fock space $\mathcal{F}(\mathcal{B})$ defined over the one particle space $L_2(\mathcal{B})$ where $\mathcal{B}$ is an interval in R^1.

$$\mathcal{F}(\mathcal{B}) = \bigoplus_{n=0}^{\infty} \mathcal{F}_n(\mathcal{B}), \tag{3.2.13}$$

where $\mathcal{F}_n(\mathcal{B})$ is the Hilbert space completion of the n-fold symmetric tensor product of $L_2(\mathcal{B}) = \mathcal{F}_1(\mathcal{B})$. If $\mathcal{B} \subset \mathcal{B}_1$, then $\mathcal{F}(\mathcal{B}) \subset \mathcal{F}(\mathcal{B}_1)$. If $\mathcal{B} = \mathcal{B}_1 \cup \mathcal{B}_2$ and $\mathcal{B}_1 \cap \mathcal{B}_2 = \varnothing$, then

$$\mathcal{F}(\mathcal{B}) = \mathcal{F}(\mathcal{B}_1) \otimes_s \mathcal{F}(\mathcal{B}_2), \tag{3.2.14}$$

where $\otimes_s$ denotes the completion of the symmetric tensor product.

We now specialize to study the biquantization C of operators c on the one particle space $\mathcal{F}_1$. For simplicity, we do not write c_{11} or C_{11}. We consider operators c studied in Section 3.1; they are defined on the domain $\mathcal{S}(R^1)$ and are continuous transformations of $\mathcal{S}(R^1)$ into $L_2(R^1)$. Then on the domain $\mathcal{D}_0$,

$$C = \int c(x, y)\, A^*(x)\, A(y)\, dx\, dy, \tag{3.2.15}$$

where $c(x, y)$ is the kernel of c.

We give some examples which we call local number operators:

$$N_{\mathcal{B}} \text{ is the quantization of multiplication by } E(x). \tag{3.2.16}$$

$$N_\tau \text{ is the quantization of } \mu_x^\tau. \tag{3.2.17}$$

$$N_{\tau,\zeta} \text{ is the quantization of } \zeta(x)\mu_x^\tau \zeta(x). \tag{3.2.18}$$

$$N_{\tau,\mathcal{B}} \text{ is the quantization of } E(x)\mu_x^\tau E(x), \text{ for } \tau < \tfrac{1}{4}. \tag{3.2.19}$$

In the case $\tau = 1$, the operator $N_{1,\zeta}$ provides a local energy operator H_0^{loc}, discussed in Section 1.

THEOREM 3.2.1. *The operator* $C \restriction \mathcal{D}_0$ *of* (3.2.15) *has any of the following properties if and only if the operator* $c \restriction \mathcal{S}(R^1)$ *has the same property.*

(a) *The operator is essentially self adjoint.*

(b) *The operator is positive.*

(c) *The operator is positive and the Friedrichs extension* (*or closure*) *has a compact resolvent.*

Proof. (a) If $C \restriction \mathcal{D}_0$ has a self adjoint closure, the domain of the self adjoint closure contains a dense set of analytic vectors. Since C commutes with the projection of $\mathcal{F}$ onto $\mathcal{F}_1$, there is a dense set of one particle analytic vectors, obtained as limits of one particle states in $\mathcal{S}(R^1)$. Thus they are in the domain of the closure of c, and c is essentially self adjoint. Conversely, if $c \restriction \mathcal{S}(R^1)$ has a self adjoint closure, the domain of the closure c^- contains a total set of analytic unit vectors $e_i(x)$, $i=1,2,\ldots$ The vectors $e_{i_1}\ldots_{i_n}=e_{i_1}\otimes_s\ldots\otimes_s e_{i_n}$ are n-particle vectors in the domain of the closure of $C \restriction \mathcal{D}_0$. Clearly they are C^∞ vectors for C^-, since

$$C^- e_{i_1}\ldots_{i_n} = c^- e_{i_1}\otimes_s\ldots\otimes_s e_{i_n} + \ldots + e_{i_1}\otimes_s\ldots\otimes_s c^- e_{i_n}.$$

Furthermore if, $$\|(c^-)^m e_i\| \leqslant \beta_i^m m!,$$

then $$\|(C^-)^m e_{i_1}\ldots_{i_n}\| \leqslant n^m \beta^m m!, \qquad \beta=\max\{\beta_{i_1},\ldots,\beta_{i_n}\}.$$

Thus the $e_{i_1\ldots i_n}$ are analytic vectors for C^- and $C \restriction \mathcal{D}_0$ is essentially self adjoint.

(b) If C is positive, c is clearly positive. Conversely, suppose that c is positive and $f_n(x_1,\ldots,x_n)\in\mathcal{D}_0\cap\mathcal{F}_n$. Then for each $x_2,\ldots,x_n$, $f_n(x_1,x_2,\ldots,x_n)\in\mathcal{S}(R^1)$ and

$$\int \overline{f_n}(x_1,x_2,\ldots,x_n)\, c(x,y)\, f_n(y,x_2,\ldots,x_n)\,dx\,dy \geqslant 0$$

is a function in $\mathcal{S}(R^{n-1})$. Thus integrating over $x_2,\ldots,x_n$ we have shown that c is a positive operator on each n particle coordinate and hence C is positive on $\mathcal{D}_0\cap\mathcal{F}_n$, and so on $\mathcal{D}_0$.

(c) Let C_F and c_F be the Friedrichs extensions of $C \restriction \mathcal{D}_0$ and $c \restriction \mathcal{S}(R^1)$ respectively. It is sufficient to show that $c_F^{\frac12}$ has a compact resolvent if and only if $C_F^{\frac12}$ does. Note that $\mathcal{D}_0$ is a core for $C_F^{\frac12}$ and $\mathcal{S}(R^1)$ is a core for $c_F^{\frac12}$. The Friedrichs extension C_F commutes with the projection P_1 onto $\mathcal{F}_1$ as a bilinear form on $\mathcal{D}(C_F^{\frac12})\times\mathcal{D}(C_F^{\frac12})$, and hence as an operator on $\mathcal{D}(C_F)$. Thus $C_F^{\frac12}$ commutes with P_1, and if $C_F^{\frac12}$ has a compact resolvent, so does $c_F^{\frac12}=P_1C_F^{\frac12}P_1$. Conversely, if $c_F^{\frac12}$ has a compact resolvent, we choose orthonormal eigenvectors e_i of $c_F^{\frac12}$ corresponding to eigenvalues $0\leqslant\lambda_i$, $\lambda_i\to\infty$. Since $e_i=\lim_{j\to\infty}e_{ij}$ where $e_{ij}(x)\in\mathcal{S}(R^1)$ and

$$\lim_{j,k\to\infty} (e_{ij} - e_{ik}, ce_{ij} - e_{ik}) = 0, \tag{3.2.20}$$

it follows that $e_{i_1 \dots i_n} = e_{i_1} \otimes_s \dots \otimes_s e_{i_n}$ is in the domain of $C_F^{\frac{1}{2}}$, and hence is an eigenvector with eigenvalue $\lambda_{i_1} + \dots + \lambda_{i_n}$. These eigenvectors are total in $\mathcal{J}$, so that $C_F^{\frac{1}{2}}$ has been diagonalized. It has discrete spectrum and eigenvalues

$$\sum_{i=0}^{\infty} n_i \lambda_i \tag{3.2.21}$$

for nonnegative integers n_i, only a finite number being non-zero. Since $\lambda_i \to \infty$, the eigenvalues have no point of accumulation except ∞ and hence $C_F^{\frac{1}{2}}$ has a compact resolvent. This completes the proof.

COROLLARY 3.2.2. (a) *The operators N_B and N are essentially self adjoint on $\mathcal{D}_0$. The operators $N_{\tau,\zeta}$ for $\zeta(x) \geqslant 0$ and $N_{\tau,B}$ for $\tau < \frac{1}{4}$ are positive.*

(b) *If $0 < \tau < \frac{1}{4}$, the Friedrichs extension of $N_{\tau,B}$ has a compact resolvent on $\mathcal{J}(\mathcal{B})$.*

(c) *Let $\mathcal{B}$ be a bounded interval and*

$$\mathcal{B} = \bigcup_{j=1}^{J} \mathcal{B}_j.$$

Then for $0 < \tau < \frac{1}{4}$ the Friedrichs extension of

$$\sum_{j=1}^{J} N_{\tau,B_j}$$

has a compact resolvent on $\mathcal{J}(\mathcal{B})$.

Proof. We need only prove (b) and (c). Since the Friedrichs extension of $E\mu_x^\tau E$ commutes with the projection of $L_2(R^1)$ onto $L_2(\mathcal{B})$ (see the proof of Theorem 3.1.2) a similar argument shows that the Friedrichs extension of $N_{\tau,B}$ commutes with the orthogonal projection P_B of $\mathcal{J}$ onto $\mathcal{J}(\mathcal{B})$. The corollary now follows from part (c) of the theorem, which can be proved for operators $C \restriction P_B \mathcal{D}_0$ and $c \restriction E\mathcal{S}(R^1)$ on $\mathcal{J}(\mathcal{B})$ and $L_2(\mathcal{B})$. The compactness of the resolvent of $(E\mu_x^\tau E)_F$ is proved in Theorem 3.1.2. This completes the proof of (b). The proof of (c) follows from the compactness on $L_2(\mathcal{B})$ of the resolvent of

$$\Big(\sum_{j=1}^{J} E_j \mu_x^\tau E_j\Big)_F, \tag{3.2.22}$$

where E_j is the characteristic function of $\mathcal{B}_j$. Noting that $\|f\|^2 \leqslant \sum_{j=1}^{J} \|E_j f\|^2$, we follow the proof of Theorem 3.1.2 in order to establish the compactness of the resolvent of the operator (3.2.22).

3.3. Vacuum expectation values of local number operators. In this section we assume the results of Section 5, and give estimates on the vacuum expectation values of local number operators. We first recall some definitions. In terms of the $A(x)$ and $A(x)^*$, the local fields $\phi(x)$ and $\pi(x)$ are defined by

$$\phi(x)=\frac{1}{\sqrt{2}}\mu_x^{-\frac{1}{2}}(A^*(x)+A(x)) \tag{3.3.1}$$

$$\pi(x)=\frac{i}{\sqrt{2}}\mu_x^{\frac{1}{2}}(A^*(x)-A(x)), \tag{3.3.2}$$

and are $\mathcal{S}(R^1)$ valued bilinear forms on $\mathcal{D}_0\times\mathcal{D}_0$.

In Section 2 we defined the approximate vacuum ω_n by

$$\omega_n(A)=\frac{1}{n}\int(\Omega_{g_n},\sigma_\alpha(A)\,\Omega_{g_n})\,h(\alpha/n)\,d\alpha. \tag{3.3.3}$$

THEOREM 3.3.1. (a) *Let $\mathcal{B}$ be a bounded interval in R_1. Then*

$$\omega_n(N_{\mathcal{B}})\leqslant M, \tag{3.3.4}$$

where M is a translation invariant constant, independent of n.

(b) *If $0\leqslant\zeta(x)\in C_0^\infty$, and $\tau\leqslant 1$, then*

$$\omega_n(N_{\tau,\zeta})\leqslant M, \tag{3.3.5}$$

where M is a translation invariant constant, independent of n.

(c) *If $f=\bar{f}\in S(R^1)$, then*

$$\omega_n(\phi(f)^2+\pi(f)^2)\leqslant|f|^2,$$

where $|f|$ is a Schwartz space norm of f, independent of n.

(d) *If $\tau<\frac{1}{4}$*

$$\omega_n(N_{\tau,\mathcal{B}})\leqslant M, \tag{3.3.6}$$

where the constant M is translation invariant and independent of n.

Proof. (a) Let $\mathcal{B}=[a,b]$ and

$$c_n(x,y)=\frac{1}{n}\int E(x-\alpha)\,E(y-\alpha)\,h(\alpha/n)\,d\alpha\delta(x-y),$$

so that

$$\omega_n(N_{\mathcal{B}})=(\Omega_{g_n},C_n\Omega_{g_n}). \tag{3.3.7}$$

Since $$\left|\int E(x-\alpha)\,E(y-\alpha)\,h(\alpha/n)\,d\alpha\right| \leqslant \|h\|_\infty\,\|E\|_1 = \|h\|_\infty\,|b-a|,$$

$c_n(x, y)$ is the kernel of an operator on $\mathcal{J}_1$ with norm bounded by

$$\text{const.}\,\frac{1}{n}\,|b-a|.$$

Thus by (3.2.9), $$\|(N+I)^{-\frac{1}{2}}C_n(N+I)^{-\frac{1}{2}}\| \leqslant \text{const.}\,\frac{1}{n}\,|b-a|$$

and by (3.3.7) $$\omega_n(N_B) \leqslant \text{const.}\,\frac{1}{n}\,|b-a|\,\omega_n(N) \leqslant \text{const.}\,|b-a|$$

by Theorem 5.1.

(b) We follow the above proof with

$$c_n(x, y) = \frac{1}{n}\int \zeta(x-\alpha)\,k_\tau(x-y)\,\zeta(y-\alpha)\,h(\alpha/n)\,d\alpha.$$

Let $\zeta_\alpha = \zeta(\cdot - \alpha)$. We note that for $f \in S(R^1)$,

$$\begin{aligned}(f, c_n f) &= \frac{1}{n}\int (f, \zeta_\alpha \mu_x^\tau \zeta_\alpha f)\,h(\alpha/n)\,d\alpha \leqslant \frac{1}{n}\,\|h\|_\infty \left\{\int (f, \mu_x^{\tau/2}\zeta_\alpha^2 \mu_x^{\tau/2} f)\,d\alpha\right. \\ &\quad \left. + \int |(f, [\zeta_\alpha, \mu_x^{\tau/2}]\,[\mu_x^{\tau/2}, \zeta_\alpha]\,f)|\,d\alpha + 2\int |(f, \mu_x^{\tau/2}\zeta_\alpha[\zeta_\alpha, \mu_x^{\tau/2}]\,f)|\,d\alpha\right\} \\ &\leqslant \frac{1}{n}\,\|h\|_\infty \left\{\int \|\zeta_\alpha \mu_x^{\tau/2} f\|^2\,d\alpha + \int \|[\mu_x^{\tau/2}, \zeta_\alpha]\,f\|^2\,d\alpha + 2\int \|[\mu_x^{\tau/2}, \zeta_\alpha]\,f\|\,\|\zeta_\alpha \mu_x^{\tau/2} f\|\,d\alpha\right\} \\ &\leqslant \frac{2}{n}\,\|h\|_\infty \int (\|\zeta_\alpha \mu_x^{\tau/2} f\|^2 + \|[\mu_x^{\tau/2}, \zeta_\alpha]\,f\|^2)\,d\alpha \\ &= \frac{2}{n}\,\|h\|_\infty \left(\|\zeta\|_2^2\,\|\mu_x^{\tau/2} f\|^2 + \int \|[\mu_x^{\tau/2}, \zeta_\alpha]\,f\|^2\,d\alpha\right).\end{aligned}$$

The commutator is a bounded operator, since

$$\begin{aligned}|([\mu_x^{\tau/2}, \zeta]\,f)^\sim\,(p)| &\leqslant (2\pi)^{-\frac{1}{2}}\int |(\mu(p)^{\tau/2} - \mu(q)^{\tau/2})\,\tilde{\zeta}(p-q)\,\tilde{f}(q)|\,dq \\ &\leqslant \text{const.}\int (p-q)^{\tau/2}\,|\tilde{\zeta}(p-q)\,\tilde{f}(q)|\,dq\end{aligned}$$

and therefore $$\|[\mu_x^{\tau/2}, \zeta]\,f\|_2 \leqslant \text{const.}\,\|(\mu^{\tau/2}\,|\tilde{\zeta}|)^\sim\|_\infty\,\|f\|_2 \leqslant M\|f\|_2,$$

as $\tilde{\zeta}$ is rapidly decreasing.

Let $B=[a,b]$ be a neighborhood of the support of ζ, let E be the characteristic function of B, and let $E_\alpha = E(\cdot - \alpha)$. Then

$$\int \|[\mu_x^{\tau/2}, \zeta_\alpha]\, E_\alpha f\|^2\, d\alpha \leqslant \text{const.} \int \|E_\alpha f\|^2\, d\alpha \leqslant \text{const.}\ (b-a)\,\|f\|^2.$$

Furthermore, the operator

$$[\mu_x^{\tau/2}, \zeta_\alpha]\,(I-E_\alpha) = \zeta_\alpha\, \mu_x^{\tau/2}\,(I-E_\alpha)$$

is an integral operator with a kernel dominated by const. $e^{-m_0(|x-\alpha|+|y-\alpha|)}$. (See (3.1.6).) Thus the operator

$$K_\alpha = [\mu_x^{\tau/2}, \zeta_\alpha]\,(I-E_\alpha)\,(1+(x-\alpha)^2)$$

has an L_2 kernel and is consequently bounded with a bound M_1 independent of α. It follows that

$$\int \|[\mu_x^{\tau/2}, \zeta_\alpha]\,(I-E_\alpha)\, f\|^2\, d\alpha = \int \|K_\alpha\,(1+(x-\alpha)^2)^{-1} f\|^2\, d\alpha$$

$$\leqslant M_1^2 \int \|(1+(x-\alpha)^2)^{-1} f\|^2\, d\alpha = M_1^2 \int (1+t^2)^{-2}\, dt\, \|f\|^2.$$

Combining these results

$$(f, c_n f) \leqslant \frac{1}{n}\ \text{const.}\ (\|\mu_x^{\tau/2} f\|^2 + \|f\|^2) \leqslant \frac{1}{n}\ \text{const.}\ \|\mu_x^{\tau/2} f\|^2.$$

Therefore there is a constant M such that

$$\frac{M}{n}\,\mu_x^{\tau} - c_n$$

is a positive operator, and hence by Theorem 3.2.1 b we have after biquantization

$$\frac{M}{n}\, N_\tau - C_n \geqslant 0.$$

Thus for $\tau \leqslant 1$,
$$\omega_n(C_n) \leqslant \frac{M}{n}\,\omega_n(N_\tau) \leqslant \frac{M}{n}\ \text{const.}\ \omega_n(H_0).$$

By Theorem 5.1 and Remark 2, equation (5.10),

$$\omega_n(C_n) \leqslant \text{const.}$$

(c) Since

$$\phi(f)^2 + \pi(f)^2 = :\phi(f)^2: + :\pi(f)^2: + \|\mu_x^{-\frac{1}{2}} f\|^2 + \|\mu_x^{\frac{1}{2}} f\|^2, \tag{3.3.8}$$

we need only consider the Wick ordered terms in (3.3.8). Each such term has j annihilators, $j=0, 1, 2,$ and $2-j$ creators. We now show that it is sufficient to consider the terms with $j=1$. Consider, for example, the pure creation term $(A^*(\mu_x^{\frac{1}{2}} f))^2$. We have

$$|\omega_n((A^*(\mu_x^{\frac{1}{2}} f))^2)| \leqslant \omega_n(A^*(\mu_x^{\frac{1}{2}} f)\, A(\mu_x^{\frac{1}{2}} f)) + \tfrac{1}{2}\, \|\mu_x^{\frac{1}{2}} f\|^2.$$

Similarly $$\omega_n(\phi(f)^2 + \pi(f)^2) \leqslant \text{const. } \omega_n(A^*(\mu_x^{\frac{1}{2}} f)\, A(\mu_x^{\frac{1}{2}} f)) + |f|^2 \qquad (3.3.9)$$

for a suitable norm $|f|$. We estimate (3.3.9) as in the proof of parts (a) and (b).

(d) From Theorem 3.1.1, and Theorem 3.2.1b, we conclude that on $\mathcal{D}_0 \times \mathcal{D}_0$,

$$N_{\tau, B} \leqslant \text{const. } N_{2\tau+\varepsilon, \zeta}, \qquad (3.3.10)$$

where $\varepsilon > 0$. Hence (3.3.10) extends to the domain $\mathcal{D} \times \mathcal{D}$, where $\mathcal{D}$ is the domain of the square root of the Friedrichs extension of $N_{2\tau+\varepsilon, \zeta}$. The desired inequality now follows from part (b) of this theorem, by choosing ε sufficiently small so that $2\tau + \varepsilon \leqslant 1$.

4. Norm compactness of the approximate vacuums

In this section we use the bounds on the local number operators derived in Section 3. We give a proof of Theorem 2.3, and hence complete the proof of Theorems 2.1 and 2.2. We restate our result as Theorem 4.1, in a somewhat more general form.

THEOREM 4.1. *Let ω_n be a sequence of normal states defined on the algebra $\mathfrak{A}$ and suppose that for some $\tau > 0$ and for each $\zeta(x) \geqslant 0$ in C_0^∞,*

$$\omega_n(N_{\tau, \zeta}) \leqslant M, \qquad (4.1)$$

where $M = M(\zeta)$ a translation invariant constant independent of n. Then the sequence $\omega_n \restriction \mathfrak{A}(\mathcal{B})$ lies in a norm compact subset of the dual $\mathfrak{A}(\mathcal{B})^*$, and any limit point $\omega \restriction \mathfrak{A}(\mathcal{B})$ is normal.

Remarks. 1. The state ω_n defined in (2.4) is a normal state on $\mathfrak{A}$, since the vector state $(\Omega_{q_n}, \cdot\, \Omega_{q_n})$ is normal and the space translation automorphism σ_α in (2.4) is implemented by a one parameter continuous unitary group of operators on $\mathcal{F}$. Hence σ_α extends to a strongly continuous automorphism of all bounded operators on $\mathcal{F}$.

2. In Theorem 3.3.1 we established the bound (4.1) for all $\tau \leqslant 1$ in the case that the ω_n are the approximate vacuum states defined in (2.4) for the $(\phi^4)_2$ interaction. Thus a proof of Theorem 4.1 provides a proof of Theorem 2.3.

3. If the inequality (4.1) holds for some $\tau > 0$, then for $\sigma \leqslant \tau$, a similar inequality holds.

$$\omega_n(N_{\sigma,\zeta}) \leqslant m_0^{\tau-\sigma}\,\omega_n(N_{\tau,\zeta}).$$

If $\sigma < \frac{1}{2}\tau$, and $\sigma < \frac{1}{4}$, then for a bounded interval $\mathcal{B}$ we conclude from (4.1) and the proof of Theorem 3.3.1 (d) that

$$\omega_n(N_{\sigma,\mathcal{B}}) \leqslant M, \tag{3.3.6a}$$

for a translation invariant constant $M = M(\mathcal{B})$ that is independent of n. The estimate (3.3.6(a)), for some $\sigma > 0$, is the crucial estimate in the proof of Theorem 4.1.

The method of proof is to use (3.3.6(a)) to replace the sequence ω_n by a regularized sequence ω_n^ε that is close to ω_n in norm.

$$\|\omega_n - \omega_n^\varepsilon\| \leqslant \varepsilon. \tag{4.2}$$

In the regularized state ω_n^ε, there will be only a finite number of particles in each bounded region of space, and that number will grow at most as a power of the diameter of the region. These bounds on ω_n^ε will be uniform in n.

The algebra $\mathfrak{A}_0(\mathcal{B})$ defined in (2.15) is weakly dense in $\mathfrak{A}(\mathcal{B})$. Since

$$\|(\omega_n^\varepsilon - \omega_m^\varepsilon) \restriction \mathfrak{A}(\mathcal{B})\| = \|(\omega_n^\varepsilon - \omega_m^\varepsilon) \restriction \mathfrak{A}_0(\mathcal{B})\|,$$

it is sufficient to prove that any sequence $\omega_n^\varepsilon \restriction \mathfrak{A}_0(\beta)$ has a norm convergent subsequence.

The next step in the proof is to approximate $C \in \mathfrak{A}_0(\mathcal{B})$ by an operator C^ε such that

$$|\omega_n^\varepsilon(C - C^\varepsilon)| \leqslant \varepsilon\|C\|, \tag{4.3}$$

where C^ε is independent of n. The approximation C^ε of C is obtained by expanding the exponentials in C in power series, each term of which is a product of unbounded field operators. Because of the regularity of ω_n^ε, the expectation value $\omega_n^\varepsilon(C)$ has a corresponding infinite series expansion which converges uniformly in n. To obtain C^ε, we neglect the tail of this expansion and we also localize each term in a finite interval of space. The localization also leads to a small error, uniform in n, and satisfying (4.3).

Finally it will be possible to apply the compactness of the resolvent of the localized number operator $N_{\tau,\mathcal{B}}$ proved in Corollary 3.2.2 (c), in order to prove that the twice regularized expectation values $\omega_n^\varepsilon(C^\varepsilon)$ have a subsequence such that

$$|(\omega_{n_i}^\varepsilon - \omega_{n_j}^\varepsilon)(C^\varepsilon)| \leqslant o(1)\,\|C\|. \tag{4.4}$$

From (4.2)–(4.4) we will conclude that Theorem 4.1 is valid.

We prove Theorem 4.1 in the case where $\mathcal{B}$ is an open interval of space whose closure is contained in (0, 1). This restriction on $\mathcal{B}$ is for convenience only. Let

$$X_j = [j, j+1).$$

The local number operators N_{X_j}, $j=0, \pm 1, \ldots$, are commuting self adjoint operators. Let $P_{l,j}$ denote the spectral projection of the self adjoint operator N_{X_j} for the spectral interval $[0, l]$.

If $0 \leqslant \zeta(x)$ is a C_0^∞ function equal to one on X_0, then $\zeta_j(x) = \zeta(x+j)$ equals one on X_j and

$$N_{X_j} \leqslant N_{0,\zeta_j}.$$

By (3.3.6 (a)) in the case $\tau = 0$,

$$\omega_n(N_{X_j}) \leqslant M_1. \tag{4.5}$$

The constant M_1 is independent of n and of j. In the following, we introduce constants $M_2, \ldots$, independent of n and of C in $\mathfrak{A}$.

From (4.5) we have

$$\omega_n(I - P_{l,j}) \leqslant M_1 l^{-1}. \tag{4.6}$$

We use this fact to construct the approximate vacuum ω_n^ε, by keeping a finite number of particles in each interval X_j. For some constant M_2 to be chosen later, we define

$$l = l(j) = M_2 \varepsilon^{-2}(1+j^2). \tag{4.7}$$

We define the projection

$$P = \prod_{j=-\infty}^{\infty} P_{l(j),j}, \tag{4.8}$$

and we define the linear functional $\omega_n^\varepsilon(\cdot)$ on $\mathfrak{A}$ by

$$\omega_n^\varepsilon(C) = \omega_n(PCP). \tag{4.9}$$

By (4.8) and the fact that the $P_{l,j}$ are commuting projections, we have

$$(I-P) \leqslant \sum_{j=-\infty}^{\infty} (I - P_{l(j),j}). \tag{4.10}$$

Thus

$$\omega_n(I-P) \leqslant \sum_{j=-\infty}^{\infty} \omega_n(I - P_{l(j),j})$$

and by (4.6) – (4.7)

$$\leqslant \varepsilon^2 \left(\frac{M_1}{M_2}\right) \sum_{j=-\infty}^{\infty} (1+j^2)^{-1}.$$

We now choose M_2 large, so that

$$\omega_n(I-P) \leqslant (\varepsilon/3)^2. \tag{4.11}$$

In order to verify that ω_n and ω_n^ε are close in norm, we write

$$\omega_n(C) - \omega_n^\varepsilon(C) = \omega_n((I-P)\,C(I-P)) + \omega_n((I-P)\,CP) + \omega_n(PC(I-P)). \tag{4.12}$$

Note that
$$|\omega_n(A^* B)|^2 \leqslant \omega_n(A^* A)\,\omega_n(B^* B). \tag{4.13}$$

Since
$$\|P\| = 1 = \|I-P\|,$$

and
$$(I-P)^2 = (I-P),$$

we have
$$|\omega_n(C) - \omega_n^\varepsilon(C)| \leqslant 3\,\|C\|\,[\omega_n(I-P)]^{\frac{1}{2}}.$$

Thus from (4.11) we conclude that

$$|\omega_n(C) - \omega_n^\varepsilon(C)| \leqslant \varepsilon\,\|C\|.$$

In terms of the states,
$$\|\omega_n - \omega_n^\varepsilon\| \leqslant \varepsilon. \tag{4.14}$$

We now study an operator $C \in \mathfrak{A}_0(\mathcal{B})$. Thus

$$C = \sum_{j=1}^{J} \alpha_j e^{i\phi(f_j) + i\pi(h_j)}. \tag{4.15}$$

The test functions f_j and h_j are smooth and their support is contained in $\mathcal{B}$. For such a C, it is obvious what an expansion in terms of field operators means. First we expand C in terms of annihilation and creation operators A and A^*. We write

$$C = \sum_{\alpha,\beta=0}^{\infty} C_{\alpha\beta}, \tag{4.16}$$

where $C_{\alpha\beta}$ is a Wick ordered monomial of degree α in creation operators A^* and of degree β in annihilation operators A. Thus

$$C_{\alpha\beta} = \int c_{\alpha\beta}(x_1, \dots, x_\alpha, x_{\alpha+1}, \dots, x_{\alpha+\beta})\,A^*(x_1) \dots A^*(x_\alpha)\,A(x_{\alpha+1}) \dots A(x_{\alpha+\beta})\,dx. \tag{4.17}$$

The kernel $c_{\alpha\beta}(x)$ is symmetric in the α creation variables $x_1, \dots, x_\alpha$ and in the β annihilation variables $x_{\alpha+1}, \dots, x_{\alpha+\beta}$. We note several properties of the operators $C_{\alpha\beta}$, the kernels $c_{\alpha\beta}(x)$, and the corresponding operator $c_{\alpha\beta}$ from $\mathfrak{F}_\beta$ to $\mathfrak{F}_\alpha$.

Suppose that the kernel $c_{\alpha\beta}(x)$ has support in the region

$$x_i \in \mathcal{B}_i, \qquad i = 1, \dots, \alpha+\beta,$$

for intervals $\mathcal{B}_i \subset R^1$. Then if

$$N_+ = \prod_{i=1}^{\alpha} N_{B_i}, \qquad N_- = \prod_{i=\alpha+1}^{\alpha+\beta} N_{B_i}, \tag{4.18}$$

we have

$$\|(N_+ + I)^{-\frac{1}{2}} C_{\alpha\beta} (N_- + I)^{-\frac{1}{2}}\| \leqslant \|c_{\alpha\beta}\|. \tag{4.19}$$

The bound (4.19) is a localization of (3.2.9), and is a consequence of the local action of $A^*(x)$ and $A(x)$ on the Newton–Wigner wave functions $f_n(x_1, \ldots, x_n) \in \mathfrak{J}_n$. (See Section 3.2.) We also note that for $\psi_\alpha \in \mathfrak{J}_\alpha$, $\psi_\beta \in \mathfrak{J}_\beta$,

$$\|c_{\alpha\beta}\| = \sup_{\substack{\|\psi_\alpha\|=1 \\ \|\psi_\beta\|=1}} |(\alpha!\,\beta!)^{-\frac{1}{2}} (\psi_\alpha, C_{\alpha\beta}\psi_\beta)| \tag{4.20}$$

where $C_{\alpha\beta}$ is the quantization (4.17) of $c_{\alpha\beta}$ on Fock space. Thus for

$$0 \leqslant \gamma \leqslant \min(\alpha, \beta),$$

we infer from (4.20) that

$$|(\psi_\alpha, C_{\alpha-\gamma,\beta-\gamma}\psi_\beta)| \leqslant \|\psi_\alpha\|\,\|\psi_\beta\| \left\{\frac{\alpha!}{\gamma!}\frac{\beta!}{\gamma!}\right\}^{\frac{1}{2}} \|c_{\alpha-\gamma,\beta-\gamma}\| \tag{4.21}$$

LEMMA 4.2. *For a bounded operator C of the form* (4.16) – (4.17),

$$\|c_{\alpha\beta}\| \leqslant 2^{\alpha+\beta}\|C\|. \tag{4.22}$$

Proof. Note that

$$(\psi_\alpha, C_{\alpha\beta}\psi_\beta) = (\psi_\alpha, \{C - \sum_{\gamma=1}^{\min(\alpha,\beta)} C_{\alpha-\gamma,\beta-\gamma}\}\psi_\beta)$$

By (4.20) – (4.21)

$$\|c_{\alpha\beta}\| \leqslant (\alpha!\,\beta!)^{-\frac{1}{2}}\left\{\|C\| + \sum_{\gamma=1}^{\min(\alpha,\beta)} \frac{(\alpha!\,\beta!)^{\frac{1}{2}}}{\gamma!}\|c_{\alpha-\gamma,\beta-\gamma}\|\right\}$$

$$= \|C\|\left\{(\alpha!\,\beta!)^{-\frac{1}{2}} + \sum_{\gamma=1}^{\min(\alpha,\beta)} \frac{1}{\gamma!}\frac{\|c_{\alpha-\gamma,\beta-\gamma}\|}{\|C\|}\right\}. \tag{4.23}$$

We reason by induction on $\min(\alpha, \beta)$. If $\min(\alpha, \beta) = 0$, then the sum in (4.23) does not occur and

$$\|c_{\alpha\beta}\| \leqslant \|C\|.$$

If $\min(\alpha, \beta) \neq 0$, then by the induction hypothesis (4.22) is valid for each $\|c_{\alpha-\gamma,\beta-\gamma}\|$ in the sum on the right side of (4.23). Thus

$$\|c_{\alpha\beta}\| \leqslant \|C\|\left\{(\alpha!\,\beta!)^{-\frac{1}{2}} + \sum_{\gamma=1}^{n}\frac{1}{\gamma!}2^{\alpha+\beta-2\gamma}\right\} \leqslant \|C\|\left\{(\alpha!\,\beta!)^{-\frac{1}{2}} + \sum_{\gamma=1}^{\infty} 2^{\alpha+\beta}\frac{4^{-\gamma}}{\gamma!}\right\}$$

$$= \|C\|\{(\alpha!\,\beta!)^{-\frac{1}{2}} + 2^{\alpha+\beta}(e^{\frac{1}{4}} - 1)\} \leqslant 2^{\alpha+\beta}\|C\|,$$

and this completes the proof of Lemma 4.2.

We want to show that $c_{\alpha\beta}(x)$ is exponentially small at infinity in each of the variables x_i. For this purpose we define a localization multi-index $L=\{L_1, L_2, ..., L_{\alpha+\beta}\}$. We let L_k be an integer corresponding to the localization of the coordinate x_k in the interval X_{L_k}. We use the localization indices L to localize the operator $C_{\alpha\beta}$. We define

$$c^L_{\alpha\beta}(x) = \text{Sym.}\, c_{\alpha\beta}(x) \prod_{k=1}^{\alpha+\beta} E(X_{L_k}), \quad L=\{L_k\}, \tag{4.24}$$

where $E(X_j)$ is the characteristic function for the interval X_j. The symmetrization in (4.24) symmetrizes the kernel in the α creation variables $(x_1, ..., x_\alpha)$ and in the β annihilation variables $(x_{\alpha+1}, ..., x_{\alpha+\beta})$. The corresponding localized operator $c^L_{\alpha\beta}$ from $\mathfrak{F}_\beta$ to $\mathfrak{F}_\alpha$ has the kernel $c^L_{\alpha\beta}(x)$. We denote by $C^L_{\alpha\beta}$ the localized, quantized operator arising from $c^L_{\alpha\beta}$.

$$C^L_{\alpha\beta} = \int c^L_{\alpha\beta}(x)\, A^*(x_1) \dots A^*(x_\alpha)\, A(x_{\alpha+1}) \dots A(x_{\alpha+\beta})\, dx. \tag{4.25}$$

We define
$$N^L_+ = \prod_{k=1}^{\alpha} N_{X_{L_k}} \tag{4.26}$$

and
$$N^L_- = \prod_{k=\alpha+1}^{\alpha+\beta} N_{X_{L_k}}. \tag{4.27}$$

Then by (4.19)
$$\|(N^L_+ + I)^{-\frac{1}{2}} C^L_{\alpha\beta} (N^L_- + I)^{-\frac{1}{2}}\| \leqslant \|c^L_{\alpha\beta}\|. \tag{4.28}$$

We note that the kernels $c_{\alpha\beta}(x)$ from operators $C\in\mathfrak{A}_0(\mathcal{B})$ are functions in $\mathcal{S}(R^{\alpha+\beta})$. Thus each term in (4.24) before symmetrization is a bounded operator from $L_2(R^\beta)$ to $L_2(R^\alpha)$ with a norm bounded by $\|c_{\alpha\beta}\|$. Hence the same holds for $c^L_{\alpha\beta}(x)$ and

$$\|c^L_{\alpha\beta}\| \leqslant \|c_{\alpha\beta}\|.$$

Thus by Lemma 4.2,
$$\|c^L_{\alpha\beta}\| \leqslant 2^{\alpha+\beta}\|C\|. \tag{4.29}$$

We now define a measure $D=D(L)$ of the total distance from the origin of the localization L in $C^L_{\alpha\beta}$.

$$D = D(L) = \sum_{k=1}^{\alpha+\beta} |L_k|. \tag{4.30}$$

LEMMA 4.3. *There is a constant M_4 independent of C such that for $C\in\mathfrak{A}_0(\mathcal{B})$,*

$$\|c^L_{\alpha\beta}\| \leqslant M_4^{\alpha+\beta} e^{-m_0 D(L)} \|C\|. \tag{4.31}$$

This lemma improves the bound (4.29). We postpone the proof to the end of this section. We remark, however, that the basis for the localization (4.31) is the fact that when

16 – 702903 *Acta mathematica*. 125. Imprimé le 23 Octobre 1970.

C is expanded in terms of the fields $\phi(y)$ and $\pi(y)$, only points $y\in\mathcal{B}$ enter the expansion. The expansion (4.17) approximately preserves this localization, and (4.31) is a quantitative estimate of the loss of locality that results from using the A^*, A expansion.

We now study the operator $PC^L_{\alpha\beta}P$ where P is the projection defined in (4.8). By definition P projects onto a subspace of $\mathfrak{F}$ with a finite number $l(j)$ of particles in each interval X_j. Since the localization L restricts the kernel $c^L_{\alpha\beta}(x)$ to have compact support, the operator $C^L_{\alpha\beta}$ creates or annihilates particles in a bounded region. Hence by (4.25),

$$PC^L_{\alpha\beta}P \tag{4.32}$$

is a bounded operator.

LEMMA 4.4. *Given $\varepsilon_1>0$, there exists a constant M_5 such that for all $C\in\mathfrak{A}_0(\mathcal{B})$,*

$$\|PC^L_{\alpha\beta}P\|\leqslant M_5^{\alpha+\beta}e^{-(m_0-\varepsilon_1)D(L)}\|C\|.$$

Proof. We first prove that given $\varepsilon_1>0$, there is a constant M_6 such that

$$\|P(N^L_+ +I)^{\frac{1}{2}}P\|\,\|P(N^L_- +I)^{\frac{1}{2}}P\|\leqslant M_6^{\alpha+\beta}e^{\varepsilon_1 D(L)}.$$

By the definition of P and the fact that each N_{X_j} commutes with P, we have

$$\|P(N^L_+ +I)^{\frac{1}{2}}P\|\,\|P(N^L_- +I)^{\frac{1}{2}}P\|\leqslant\prod_{k=1}^{\alpha+\beta}(l(L_k)+1)^{\frac{1}{2}}. \tag{4.33}$$

For any $\varepsilon_1>0$, there is a constant M_6 such that

$$(l(L_k)+1)^{\frac{1}{2}}=\{M_2\varepsilon^{-2}(1+L_k^2)+1\}^{\frac{1}{2}}\leqslant M_6e^{\varepsilon_1|L_k|}.$$

Thus
$$\prod_{k=1}^{\alpha+\beta}(l(L_k)+1)^{\frac{1}{2}}\leqslant M_6^{\alpha+\beta}e^{\varepsilon_1 D(L)}.$$

Using (4.33),
$$\|P(N^L_+ +I)^{\frac{1}{2}}P\|\,\|P(N^L_- +I)^{\frac{1}{2}}P\|\leqslant M_6^{\alpha+\beta}e^{\varepsilon_1 D(L)}.$$

Using estimate (4.28), we find that

$$\begin{aligned}\|PC^L_{\alpha\beta}P\|&=\|P(N^L_+ +I)^{\frac{1}{2}}(N^L_+ +I)^{-\frac{1}{2}}C^L_{\alpha\beta}(N^L_- +I)^{-\frac{1}{2}}(N^L_- +I)^{\frac{1}{2}}P\|\\&\leqslant\|P(N^L_+ +I)P\|\,\|P(N^L_- +I)^{\frac{1}{2}}P\|\,\|(N^L_+ +I)^{-\frac{1}{2}}C^L_{\alpha\beta}(N^L_- +I)^{-\frac{1}{2}}\|\\&\leqslant M_6^{\alpha+\beta}e^{\varepsilon_1 D(L)}\|c^L_{\alpha\beta}\|\end{aligned}$$

and by Lemma 4.3,
$$\leqslant(M_4M_6)^{\alpha+\beta}e^{-(m_0-\varepsilon_1)D(L)}\|C\|,$$

to complete the proof.

Our next estimate bounds $D(L)$ from below.

LEMMA 4.5. *If P and $C^L_{\alpha\beta}$ are defined as in* (4.8) *and* (4.25), *and if*

$$PC^L_{\alpha\beta}P \neq 0, \tag{4.34}$$

then there is a constant M_7 such that

$$D(L)+1 \geqslant M_7(\alpha^{\frac{4}{3}}+\beta^{\frac{4}{3}}). \tag{4.35}$$

Proof. For fixed values of α and β, we can minimize (4.30) by choosing the $|L_k|$ as small as possible. However, (4.34) will hold only if $C^L_{\alpha\beta}$ both creates and annihilates no more than $l(j)$ particles in each region X_j. For this reason it is convenient to count the number of particles created and annihilated in the region X_j for the localization L. We define $L_+(j)$ and $L_-(j)$ to be these numbers:

$$L_+(j) = \text{the number of } L_k = j, \quad \text{for } k=1, 2, \ldots, \alpha.$$

$$L_-(j) = \text{the number of } L_k = j, \quad \text{for } k=\alpha+1, \ldots, \alpha+\beta.$$

These numbers must satisfy

$$\sum_{j=-\infty}^{\infty} L_+(j) = \alpha, \quad \text{and} \quad \sum_{j=-\infty}^{\infty} L_-(j) = \beta. \tag{4.36}$$

In addition, (4.34) will be valid only if

$$L_+(j) \leqslant l(j) = M_2\varepsilon^{-2}(1+j^2), \tag{4.37}$$

and

$$L_-(j) \leqslant l(j) = M_2\varepsilon^{-2}(1+j^2). \tag{4.38}$$

In order to minimize $D(L)$ consistent with (4.34), we choose

$$L_\pm(j) = \begin{cases} l(j) = M_2\varepsilon^{-2}(1+j^2), & \text{if } |j| < J_\pm \\ 0, & \text{if } |j| > J_\pm. \end{cases} \tag{4.39}$$

The maximum value J_+ or J_- of any localization index $|L_k|$ in L, and the number of indices with this maximum value, are fixed by the requirements (4.36). Using (4.39) we conclude that there is a constant M_8 (approximately equal to $M_2\varepsilon^{-2}/3$) such that (excluding the trivial case $J_+=0$)

$$\alpha = \sum_{|j|<J_+} M_2\varepsilon^{-2}(1+j^2) + L_+(J_+) \leqslant M_8 J_+^3, \tag{4.40}$$

and similarly (excluding the trivial case $J_-=0$),

$$\beta \leqslant M_8 J_-^3. \tag{4.41}$$

For our choice of L, there is a constant M_9, independent of α and β, such that

$$D(L) = \sum_{k=1}^{\alpha+\beta} |L_k| = \sum_{|j| \leqslant J_+} |j| L_+(j) + \sum_{|j| \leqslant J_-} |j| L_-(j) \geqslant M_9 (J_+^4 + J_-^4),$$

where we have used (4.39). Thus by (4.40) – (4.41) which bound J_+ and J_-,

$$D(L) \geqslant M_9 M_8^{-\frac{4}{3}} (\alpha^{\frac{4}{3}} + \beta^{\frac{4}{3}}).$$

If $J_+ = 0$, then $\alpha \leqslant M_2 \varepsilon^{-2}$ and if $J_- = 0$, then $\beta \leqslant M_2 \varepsilon^{-2}$. The proof is thus complete.

These estimates are useful for constructing the operator C^ε and for analyzing the error $C - C^\varepsilon$. Let us define the set of localization indices

$$\mathfrak{L}_M = \{L\colon |L_k| \leqslant M,\ k = 1, 2, \ldots, \alpha+\beta\}. \tag{4.42}$$

If $L \in \mathfrak{L}_M$, then all the particles that are created or annihilated by $C_{\alpha\beta}^L$ are in the region of space

$$Z = Z_M = \bigcup_{|j| \leqslant M} X_j. \tag{4.43}$$

We define for $C \in \mathfrak{A}_0(\mathcal{B})$,

$$C^\varepsilon = \sum_{\alpha, \beta = 0}^{M} \sum_{L \in \mathfrak{L}_M} C_{\alpha\beta}^L, \tag{4.44}$$

where $M = M(\varepsilon)$ will be chosen later, independently of C.

Lemma 4.6 (a). *There exists a constant M_{10} independent of M such that for $C \in \mathfrak{A}_0(\mathcal{B})$ and for C^ε defined in* (4.44),

$$\|PC^\varepsilon P\| \leqslant M_{10} \|C\|. \tag{4.45}$$

(b) *Given $\varepsilon > 0$, there exists $M = M(\varepsilon)$ sufficiently large such that for all $C \in \mathfrak{A}_0(\mathcal{B})$*

$$\|P(C - C^\varepsilon) P\| \leqslant \varepsilon \|C\|, \tag{4.46}$$

and

$$|\omega_n^\varepsilon (C - C^\varepsilon)| \leqslant \varepsilon \|C\|. \tag{4.47}$$

Proof. By Lemma 4.4 with $\varepsilon_1 = \varepsilon$,

$$\|PC^\varepsilon P\| \leqslant \sum_{\alpha, \beta = 0}^{M} \sum_{L \in \mathfrak{L}_M}' M_5^{\alpha+\beta} e^{-(m_0 - \varepsilon) D(L)} \|C\|,$$

where $\sum_L'$ denotes the sum over those L for which $PC_{\alpha\beta}^L P \neq 0$, namely the L which satisfy the hypotheses of Lemma 4.5. Thus

$$\|PC^\varepsilon P\| \leqslant M_3 \|C\| \sum_{\alpha,\beta=0}^{M} \sum_{L\in\mathcal{L}_M} M_5^{\alpha+\beta} \exp\{-\tfrac{1}{2}m_0 D(L) - M_{11}(\alpha^{\frac{4}{3}}+\beta^{\frac{4}{3}})\}, \tag{4.48}$$

where $M_{11} = M_7(\frac{1}{2}m_0 - \varepsilon)$ and $M_3 = e^{m_0-\varepsilon}$. The sum over all L converges exponentially fast, for

$$\sum_L \exp\{-\tfrac{1}{2}m_0 D(L)\} = \left[\sum_{i=-\infty}^{\infty} e^{-m_0|i|/2}\right]^{\alpha+\beta} = (M_{12})^{\alpha+\beta}. \tag{4.49}$$

The sum over the tail, $L \notin \mathcal{L}_M$, is characterized by the fact that for some k, $|L_k| > M$. Thus

$$\sum_{L\notin\mathcal{L}_M} \exp\{-\tfrac{1}{2}m_0 D(L)\} \leqslant (\alpha+\beta)\, e^{-Mm_0/2}(M_{12})^{\alpha+\beta}. \tag{4.50}$$

Using the estimate (4.49) in (4.48) shows that

$$\|PC^\varepsilon P\| \leqslant M_3 \|C\| \sum_{\alpha,\beta=0}^{M} (M_5 M_{12})^{\alpha+\beta} \exp\{-M_{11}(\alpha^{\frac{4}{3}}+\beta^{\frac{4}{3}})\}, \tag{4.51}$$

which converges faster than exponentially as $M\to\infty$. Hence (4.48) is bounded independently of $M = M(\varepsilon)$.

To establish (4.46), we write

$$\|P(C-C^\varepsilon)P\| \leqslant \sum_{\alpha,\beta=0}^{M} \sum_{L\notin\mathcal{L}_M} + \sum_{\substack{\alpha>M\\ \beta\geqslant 0}} \sum_L + \sum_{\substack{0\leqslant\alpha\leqslant M\\ \beta>M}} \sum_L \|PC^L_{\alpha\beta}P\|. \tag{4.52}$$

From (4.50) we conclude that the first sum in (4.52) is bounded by const. $\exp(-m_0 M/2)$, which converges to zero exponentially fast as $M\to\infty$. The remaining two terms in (4.52) converge to zero faster than exponentially as $M\to\infty$, as a consequence of the estimate on the tail of (4.51). Hence by choosing $M = M(\varepsilon)$ sufficiently large, we can assure that

$$\|P(C-C^\varepsilon)P\| \leqslant \varepsilon \|C\|.$$

Lastly, we remark that

$$|\omega_n^\varepsilon(C-C^\varepsilon)| = |\omega_n(P(C-C^\varepsilon)P)| \leqslant \|P(C-C^\varepsilon)P\| \leqslant \varepsilon\|C\|, \tag{4.53}$$

to complete the proof.

We next study the convergence of the sequence $\{\omega_n^\varepsilon\}$ of approximate states applied to the approximate operators C^ε. In other words, we study the convergence of the states ω_n on the bounded operators $PC^\varepsilon P$.

As a first reduction, we restrict our attention to the Fock space of a bounded region. Let Z be the interval defined in (4.43). Then if Z_1 is the complement of Z in R^1,

$$R^1 = Z \cup Z_1 \tag{4.54}$$

and by (3.2.14)
$$\mathcal{F}=\mathcal{F}(Z)\otimes_s \mathcal{F}(Z_1) \tag{4.55}$$

With this decomposition we write

$$P=(P_0\otimes I)\,(I\otimes P_1). \tag{4.56}$$

We now study the local number operator N_{τ,X_j} defined in (3.2.19).

Here we choose $\mathcal{B}=X_j$ and we restrict τ to be less than $\frac{1}{4}$. Since the Friedrichs extension $(N_{\tau,X_j})_F$ of $N_{\tau,X_j}\restriction\mathcal{D}_0$ commutes with the projection onto $\mathcal{F}(X_i)$ for all i, it follows that the various $(N_{\tau,X_j})_F$ commute. Also $(N_{\tau,X_j})_F$ commutes with each N_{X_i} and with the spectral projections $P_{l,i}$. Thus $(N_{\tau,X_j})_F$ commutes with $P_0\otimes I$, with $I\otimes P_1$, and with P. We now let N_{τ,X_j} denote the Friedrichs extension of $N_{\tau,X_j}\restriction\mathcal{D}_0$.

As a consequence of the commutativity with P,

$$\omega_n^\varepsilon(N_{\tau,X_j})=\omega_n(PN_{\tau,X_j}P)=\omega_n(N^{\frac{1}{2}}_{\tau,X_j}PN^{\frac{1}{2}}_{\tau,X_j})\leqslant\omega_n(N_{\tau,X_j}).$$

We choose $\tau>0$ sufficiently small so that (3.3.6 a) is valid. (See Remark 3 following Theorem 4.1.) Hence

$$\omega_n^\varepsilon(N_{\tau,X_j})\leqslant \text{const.}$$

We now choose M as in (4.42) – (4.44) and define

$$N_{\tau,M}=\Big(\sum_{|j|\leqslant M}N_{\tau,X_j}\Big)_F. \tag{4.57}$$

Thus
$$\omega_n^\varepsilon(N_{\tau,M})\leqslant\text{const.} \tag{4.58}$$

We note that $N_{\tau,M}$ commutes with each $P_0\otimes I$ and with $I\otimes P_1$. Also $N_{\tau,M}$ leaves $\mathcal{F}(Z)$ invariant, and by Corollary 3.2.2 c,

$$N_{\tau,M}\restriction\mathcal{F}(Z)$$

has a compact resolvent.

We now study the sequence $\{\omega_n^\varepsilon\}$ of functionals restricted to $\mathfrak{B}(Z)$, the algebra of all bounded operators on the Hilbert space $\mathcal{F}(Z)$. Each functional ω_n^ε is normal, since ω_n is normal by assumption in Theorem 4.1, and P is a projection. The normal functional

$$\omega_n^\varepsilon\restriction\mathfrak{B}(Z)$$

has the form
$$\omega_n^\varepsilon(A)=\mathrm{Tr}\,(\Lambda_nA),\quad A\in\mathfrak{B}(Z), \tag{4.59}$$

where Λ_n is a positive trace class operator in $\mathfrak{B}(Z)$ with

$$\|\omega_n^\varepsilon\restriction\mathfrak{B}(Z)\|=\mathrm{Tr}\,(\Lambda_n)\leqslant 1. \tag{4.60}$$

(See [2].) The operator $N_{\tau,M}$ in (4.57) can be restricted to $\mathcal{F}(Z)$, and thus by (4.58)

$$\omega_n^\varepsilon (N_{T,M} \restriction \mathcal{J}(Z)) \leqslant M. \tag{4.61}$$

We wish to study the convergence of

$$\omega_n^\varepsilon (C^\varepsilon) = \omega_n (PC^\varepsilon P).$$

We note that $PC^\varepsilon P = P_0 C^\varepsilon P_0 \otimes P_1$ where we regard C^ε on the right side as an operator on $\mathcal{J}(Z)$. Thus

$$\omega_n^\varepsilon (C^\varepsilon) = \operatorname{Tr} (\Lambda_n P_0 C^\varepsilon P_0).$$

The desired convergence of the $\omega_n^\varepsilon (C^\varepsilon)$ now follows from

LEMMA 4.7. *Let $\mathfrak{B}$ be the algebra of all bounded operators on some Hilbert space $\mathcal{H}$, and let N be a positive (unbounded) operator with N^{-1} a compact element of $\mathfrak{B}$. Then the set*

$$\{\Lambda : \Lambda \in \mathfrak{B},\ 0 \leqslant \Lambda \leqslant I,\ \operatorname{Tr} (\Lambda N) \leqslant 1\} = \mathfrak{J}_N$$

is compact in the trace norm.

Proof. Choose an orthonormal basis $\{e_i\}$ for $\mathcal{H}$ consisting of eigenvectors of N and let b_i be the corresponding eigenvalues. For Λ in $\mathfrak{J}_N$, we have

$$\operatorname{Tr} (\Lambda N) = \operatorname{Tr} (N^{\frac{1}{2}} \Lambda N^{\frac{1}{2}}) = \sum_i \lambda_{ii} b_i$$

if (λ_{ij}) is the matrix representing the operator Λ. If $P = \Lambda^{\frac{1}{2}}$ and if ϱ_{ij} is the corresponding matrix, then

$$\operatorname{Tr} (\Lambda N) = \sum_{i<j} (b_i + b_j) |\varrho_{ij}|^2 + \sum_i b_i |\varrho_{ii}|^2.$$

Let Λ_n be a sequence in $\mathfrak{J}_N$ and let $P_n = \Lambda_n^{\frac{1}{2}}$. Since $b_i \to \infty$ as $i \to \infty$, we see that a subsequence P_n of the P's converge to a limit P in the Hilbert–Schmidt norm. For a trace class operator A, the trace norm $\|A\|_1$ is given by

$$\|A\|_1 = \sup_U \operatorname{Tr} (UA),$$

where the supremum runs over all unitary operators. Thus if $\Lambda = P^2$,

$$\|\Lambda_n - \Lambda\|_1 = \sup_U \operatorname{Tr} (U(\Lambda_n - \Lambda)) = \sup_U \{\operatorname{Tr} (U(P_n - P) P_n) + \operatorname{Tr} (UP(P_n - P))\}$$

$$\leqslant \sup_U \|U(P_n - P)\|_2 \|P_n\|_2 + \sup_U \|UP\|_2 \|P_n - P\|_2 = (\|P_n\|_2 + \|P\|_2) \|P_n - P\|_2 \to 0,$$

with
$$\|P\|_2^2 = \sum_{i,j} |\varrho_{ij}|^2$$

denoting the Hilbert–Schmidt norm, and n belonging to the subsequence.

We apply this lemma to the case $\mathcal{J} = \mathcal{J}(Z)$ and $\mathfrak{B} = \mathfrak{B}(Z)$. We use the operator $N_{\tau,M}$ of (4.57) to give an

$$N = (N_{\tau,M} + I)$$

for Lemma 4.7. The states $\omega_n^\varepsilon \restriction \mathfrak{B}(Z)$ satisfy the hypotheses of the lemma by (4.59) – (4.61). By (4.58) and Corollary 3.3.2 (c) the operator N has the desired properties. Thus there is a norm convergent subsequence ω_{n_j} satisfying

$$|(\omega_{n_i} - \omega_{n_j})\,(PC^\varepsilon P)| \leqslant o(1)\,\|PC^\varepsilon P\| \leqslant o(1)\,\|C\|$$

by (4.45). For this subsequence

$$|(\omega_{n_i}^\varepsilon - \omega_{n_j}^\varepsilon)\,(C^\varepsilon)| \leqslant o(1)\,\|C\|, \tag{4.62}$$

which is the announced bound (4.4).

In order to complete the proof of Theorem 4.1, we write for $C \in \mathfrak{A}_0(\mathcal{B})$.

$$|(\omega_{n_i} - \omega_{n_j})\,(C)| \leqslant |(\omega_{n_i} - \omega_{n_i}^\varepsilon)\,(C)| + |\omega_{n_i}^\varepsilon(C - C^\varepsilon)| + |(\omega_{n_i}^\varepsilon - \omega_{n_j}^\varepsilon)\,(C^\varepsilon)| + |\omega_{n_j}^\varepsilon(C^\varepsilon - C)|$$
$$+ |(\omega_{n_j}^\varepsilon - \omega_{n_j})\,(C)| \leqslant 4\,\varepsilon\,\|C\| + o(1)\,\|C\|. \tag{4.63}$$

Here we have chosen n_i, n_j to belong to the convergent subsequence in (4.62), and we used (4.14), (4.46) and (4.62) to dominate (4.63). Since

$$\|(\omega_{n_i} - \omega_{n_j}) \restriction \mathfrak{A}_0(\mathcal{B})\| = \|(\omega_{n_i} - \omega_{n_j}) \restriction \mathfrak{A}(\mathcal{B})\|,$$

we conclude that the subsequence in (4.63) is norm convergent on $\mathfrak{A}(\mathcal{B})$.

It is known and easy to prove that the norm limit of a sequence of normal states is normal. Thus any norm limit ω of the ω_n is normal on $\mathfrak{A}(\mathcal{B})$, and Theorem 4.1 is established.

We now return to prove Lemma 4.3. We analyze the expansion of $C \in \mathfrak{A}_0(\mathcal{B})$ in terms of the local, time zero fields. We start by proving a useful lemma.

Let α, β, δ, ε, μ, ν be nonnegative integers satisfying $\alpha + \beta = \delta + \varepsilon$, $\mu \leqslant \delta$, $\nu \leqslant \varepsilon$ and $\alpha - \delta = \nu - \mu$.

LEMMA 4.8. *Let $c_{\delta\varepsilon}$ be a bounded operator from $L_2(R^\varepsilon)$ to $L_2(R^\delta)$ with bound $\|c_{\delta\varepsilon}\|$. Let k_C and k_A be Hilbert–Schmidt operators on $L_2(R^\mu)$ and on $L_2(R^\nu)$ respectively, with Hilbert–Schmidt norms*

$$\|k_C\|_2 \text{ and } \|k_A\|_2.$$

Then

$$\int k_A(x_1, \ldots, x_\nu; w_1, \ldots, w_\nu)\, c_{\delta\varepsilon}(x_{\nu+1}, \ldots, x_\alpha, z_1, \ldots, z_\mu; w_1, \ldots, w_\nu, y_{\mu+1}, \ldots, y_\beta)$$
$$\times\, k_C(z_1, \ldots, z_\mu; y_1, \ldots, y_\mu)\, dw\, dz$$

is the kernel of a bounded operator $c_{\alpha\beta}$ from $L_2(R^\beta)$ to $L_2(R^\alpha)$ with norm $\|c_{\alpha\beta}\|$ satisfying

$$\|c_{\alpha\beta}\| \leq \|k_A\|_2\, \|c_{\delta\varepsilon}\|\, \|k_C\|_2.$$

Proof. We introduce the variables

$$x_a = (x_1, \ldots, x_\nu), \qquad x_b = (x_{\nu+1}, \ldots, x_\alpha), \qquad x = (x_a, x_b),$$
$$y_a = (y_1, \ldots, y_\mu), \qquad y_b = (y_{\mu+1}, \ldots, y_\beta), \qquad y = (y_a, y_b)$$
$$w = (w_1, \ldots, w_\nu), \qquad z = (z_1, \ldots, z_\mu).$$

Then for $f_\alpha \in \mathfrak{J}_\alpha$, $g_\beta \in \mathfrak{J}_\beta$,

$$|(f_\alpha, c_{\alpha\beta} g_\beta)| = \left| \int f_\alpha(x)\, k_A(x_a; w)\, c_{\delta\varepsilon}(x_b, z; w, y_b)\, k_C(z; y_a)\, g_\beta(y)\, dx\, dy\, dw\, dz \right|.$$

By the definition of $\|c_{\delta\varepsilon}\|$,

$$|(f_\alpha, c_{\alpha\beta} g_\beta)| \leq \|c_{\delta\varepsilon}\| \int \|f_\alpha(x_a, \cdot)\|_2\, \|k_C(\cdot\, ; y_a)\|_2\, \|k_A(x_a; \cdot)\|_2\, \|g_\beta(y_a, \cdot)\|_2\, dx_a\, dy_a,$$

and by the Schwartz inequality

$$|(f_\alpha, c_{\alpha\beta} g_\beta)| \leq \|c_{\alpha\beta}\|\, \|k_C\|_2\, \|k_A\|_2\, \|f_\alpha\|\, \|g_\beta\|,$$

to complete the proof.

We now expand $C \in \mathfrak{A}_0(\mathcal{B})$ in terms of ϕ and π fields. Let

$$C_\gamma = \sum_{\alpha+\beta=\gamma} C_{\alpha\beta}.$$

Then
$$C_\gamma = \sum_{\mu+\nu=\gamma} B_{\mu\nu},$$

where
$$B_{\mu\nu} = \int b_{\mu\nu}(y) \colon \phi(y_1) \ldots \phi(y_\mu)\, \pi(y_{\mu+1}) \ldots \pi(y_{\mu+\nu}) \colon dy.$$

The locality of $C \in \mathfrak{A}_0(\mathcal{B})$ means that

$$\operatorname{supp}\, b_{\mu\nu}(y_1, \ldots, y_{\mu+\nu}) \in \mathcal{B} \times \mathcal{B} \times \ldots \times \mathcal{B}.$$

The kernels $b_{\mu\nu}(y)$ are symmetric in the μ variables $y_1, \ldots, y_\mu$ corresponding to the ϕ's, and also in the ν variables $y_{\mu+1}, \ldots, y_{\mu+\nu}$ corresponding to the π's. For fixed $\mu+\nu=\gamma$, there are exactly $\gamma+1$ different $B_{\mu\nu}$'s and for fixed $\alpha+\beta=\gamma$, there are exactly $\gamma+1$ different $C_{\alpha\beta}$'s. We now specify the relation between them. First we inspect the simple case $\alpha+\beta=\gamma=\mu+\nu=1$. From (3.9) we conclude that

$$c_{10}(x)=\frac{1}{\sqrt{2}}\mu_x^{-\frac{1}{2}}b_{10}(x)+\frac{i}{\sqrt{2}}\mu_x^{\frac{1}{2}}b_{01}(x) \tag{4.64}$$

and

$$c_{01}(x)=\frac{1}{\sqrt{2}}\mu_x^{-\frac{1}{2}}b_{10}(x)-\frac{i}{\sqrt{2}}\mu_x^{\frac{1}{2}}b_{01}(x). \tag{4.65}$$

We also have the inverse relation

$$b_{10}(x)=\frac{1}{\sqrt{2}}\mu_x^{\frac{1}{2}}c_{10}(x)+\frac{1}{\sqrt{2}}\mu_x^{\frac{1}{2}}c_{01}(x) \tag{4.66}$$

and

$$b_{01}(x)=\frac{-i}{\sqrt{2}}\mu_x^{-\frac{1}{2}}c_{10}(x)+\frac{i}{\sqrt{2}}\mu_x^{-\frac{1}{2}}c_{01}(x). \tag{4.67}$$

The transformations (4.64)–(4.67) by themselves are valid for a wide variety of $b_{01}(x)$, $b_{10}(x)$; they are defined, for instance on tempered distributions. These relations do not require that the $b_{01}(x)$ and $b_{10}(x)$ be local. We now make the locality explicit by means of a C_0^∞ function $\zeta(x)$ which has its support in X_0 and which equals one on a neighborhood of $\mathcal{B}$. We rewrite (4.66) as

$$b_{10}(x)=\frac{1}{\sqrt{2}}\zeta(x)\mu_x^{\frac{1}{2}}c_{10}(x)+\frac{1}{\sqrt{2}}\zeta(x)\mu_x^{\frac{1}{2}}c_{01}(x), \tag{4.68}$$

and similarly for $b_{01}(x)$. We note that the operators μ_x^τ, which were introduced in Section 3.1 have integral kernels $k_\tau(x)$ satisfying (3.1.5)–(3.1.6).

$$(\mu_x^\tau f)(x)=\int k_\tau(x-z)\,f(z)\,dz.$$

By substituting (4.68) into (4.64) – (4.65), we obtain the equations

$$c_{10}(x)=(K_+c_{10})(x)+(K_-c_{01})(x) \tag{4.69}$$

and

$$c_{01}(x)=(K_-c_{10})(x)+(K_+c_{01})(x), \tag{4.70}$$

where we define

$$K_\pm=\tfrac{1}{2}\mu_x^{-\frac{1}{2}}\zeta\mu_x^{\frac{1}{2}}\pm\tfrac{1}{2}\mu_x^{\frac{1}{2}}\zeta\mu_x^{-\frac{1}{2}}. \tag{4.71}$$

Let $k_\pm(x, z)$ be the (tempered distribution) kernel of $K_\pm$. For $x\notin X_0=[0, 1)$, the property (3.1.6) of k_τ implies that $k_\pm(x, z)$ is a C^∞ function of x and z and that for a constant M_{14},

$$|k_{\pm}(x,z)| \leqslant M_{14} \exp(-m_0|x| - m_0|z|). \tag{4.72}$$

We note that the locality requirement

$$\operatorname{supp} b_{10}(x) \subset \mathcal{B}, \qquad \operatorname{supp} b_{01}(x) \subset \mathcal{B},$$

imposes implicit restrictions on—and relations between—$c_{01}(x)$ and $c_{10}(x)$; for instance, if either one is nonzero, they must both be nonzero. Thus we cannot expect, and we do not find, that the identity transformation (4.69)–(4.70) yields the identity on all functions as in (4.64)–(4.67). In fact for the kernels of (4.71), we have

$$k_{+}(x,z) \neq \delta(x-z), \quad \text{and} \quad k_{-}(x,z) \neq 0.$$

We now define localized kernels and estimate their norms. Let

$$k^i_{\pm}(x,z) = \begin{cases} k_{\pm}(x,z), & x \in X_i \\ 0 \quad , & \text{otherwise,} \end{cases} \tag{4.73}$$

and let $K^i_{\pm}$ be the operator with kernel $k^i_{\pm}$.

LEMMA. 4.9. *For some constant M_{15}, the operator norm of K^i_{+} is bounded by*

$$\|K^i_{+}\| \leqslant M_{15} e^{-m_0|i|}, \tag{4.74}$$

and the Hilbert–Schmidt norm of k^i_{-} is bounded by

$$\|k^i_{-}\|_2 \leqslant M_{15} e^{-m_0|i|}. \tag{4.75}$$

Proof. If $i \neq 0$, the bound (4.72) immediately yields

$$\|K^i_{\pm}\| \leqslant \|k^i_{\pm}\|_2 \leqslant M_{15} \exp(-m_0|i|),$$

so we need only consider the case $i = 0$. Since

$$\mu_x^{-\frac{1}{2}} \zeta \mu_x^{\frac{1}{2}} = \zeta + \mu_x^{-\frac{1}{2}} [\zeta, \mu_x^{\frac{1}{2}}],$$

the lemma follows from (4.71) and an estimate on the Hilbert–Schmidt norm of the operator

$$A = \mu_x^{-\frac{1}{2}} [\zeta, \mu_x^{\frac{1}{2}}].$$

Let $a(x,z)$ be the kernel of A. The Fourier transform of $a(x,z)$ is

$$\tilde{a}(p,q) = \mu(p)^{-\frac{1}{2}} \tilde{\zeta}(p-q) \{\mu(q)^{\frac{1}{2}} - \mu(p)^{\frac{1}{2}}\},$$

where $\tilde{\zeta}$ is the Fourier transform of ζ. Since

$$|\mu(p)^{\frac{1}{2}}-\mu(q)^{\frac{1}{2}}|=\frac{|\mu(p)-\mu(q)|}{\mu(p)^{\frac{1}{2}}+\mu(q)^{\frac{1}{2}}}\leqslant\frac{\mu(p-q)}{\mu(p)^{\frac{1}{2}}+\mu(q)^{\frac{1}{2}}}$$

we have
$$|\tilde{a}(p,q)|\leqslant\frac{\mu(p-q)\,|\tilde{\zeta}(p-q)|}{\mu(p)^{\frac{1}{2}}\{\mu(p)^{\frac{1}{2}}+\mu(q)^{\frac{1}{2}}\}}\in L_2.$$

Thus
$$\|a\|_2=\|\tilde{a}\|_2<\infty,$$

and the proof is complete.

Proof of Lemma 4.3. Let $L=\{L_1\}$ be a localization index in the case $\gamma=1$, which localizes the single particle in X_{L_1}. Then

$$c^L_{10}(x)=(K^{L_1}_{+}c_{10})(x)+(K^{L_1}_{-}c_{01})(x) \tag{4.76}$$

and
$$c^L_{01}(x)=(K^{L_1}_{-}c_{10})(x)+(K^{L_1}_{+}c_{01})(x). \tag{4.77}$$

The bound of Lemma 4.3 for c^L_{10} follows from

$$\|c^L_{10}\|\leqslant\|K^{L_1}_{+}c_{10}\|+\|K^{L_1}_{-}c_{01}\|\leqslant\|K^{L_1}_{+}\|\,\|c_{10}\|+\|K^{L_1}_{-}c_{01}\| \tag{4.78}$$

$$\leqslant\|K^{L_1}_{+}\|\,\|c_{10}\|+\|k^{L_1}_{-}\|_2\,\|c_{01}\|$$

by Lemma 4.8

$$\leqslant M_{15}(\|c_{10}\|+\|c_{01}\|)\exp(-m_0|L_1|)$$

by Lemma 4.9

$$\leqslant 2M_{15}\exp(-m_0|L_1|)\,\|C\|=2M_{15}e^{-m_0D(L)}\|C\|$$

by Lemma 4.2. A similar result holds for c^L_{01}.

We now generalize this method to deal with any $c^L_{\alpha\beta}$, and we need an expansion similar to (4.76)–(4.77). We first obtain such an expansion for operators C_γ which have the form

$$C_\gamma=\sum_{\alpha+\beta=\gamma}C_{\alpha\beta}=:C_1^\gamma:, \tag{4.79}$$

where C_1 is a degree one expression

$$C_1=A^*(c_{10})+A(c_{01})=\phi(b_{10})+\pi(b_{01}), \tag{4.80}$$

and supp $b_{10}\subset\mathcal{B}$, supp $b_{01}\subset\mathcal{B}$. For our special case

$$C_\gamma=\sum_{\alpha+\beta=\gamma}\int A^*(x_1)\dots A^*(x_\alpha)A(x_{\alpha+1})\dots A(x_{\alpha+\beta})\binom{\alpha+\beta}{\alpha}$$
$$\times c_{10}(x_1)\dots c_{10}(x_\alpha)c_{01}(x_{\alpha+1})\dots c_{01}(x_{\alpha+\beta})\,dx \tag{4.81}$$

or
$$C_\gamma = \sum_{\mu+\nu=\gamma} \int :\phi(y_1)\ldots\phi(y_\mu)\,\pi(y_{\mu+1})\ldots\pi(y_{\mu+\nu}): \binom{\mu+\nu}{\nu}$$
$$\times\, b_{10}(y_1)\ldots b_{10}(y_\mu)\, b_{01}(y_{\mu+1})\ldots b_{01}(y_{\mu+\nu})\,dy. \tag{4.82}$$

We now apply (4.69) to the creation variables, expanding $c_{10}(x_1), \ldots, c_{10}(x_\alpha)$. We apply (4.70) to the annihilation variables, expanding $c_{01}(x_{\alpha+1}) \ldots c_{01}(x_{\alpha+\beta})$. The kernels $c_{\alpha\beta}(x)$ for (4.79)
$$c_{\alpha\beta}(x) = \binom{\alpha+\beta}{\alpha} \prod_{j=1}^{\alpha} c_{10}(x_j) \prod_{j=\alpha+1}^{\beta} c_{01}(x_j) \tag{4.83}$$
can be written
$$c_{\alpha\beta}(x) = \binom{\alpha+\beta}{\alpha} \prod_{j=1}^{\alpha} \{(K_+ c_{10})(x_j) + (K_- c_{01})(x_j)\} \prod_{j=\alpha+1}^{\alpha+\beta} \{(K_+ c_{01})(x_j) + (K_- c_{10})(x_j)\}. \tag{4.84}$$

In a more compact form
$$c_{\alpha\beta}(x) = \sum_{\delta+\varepsilon=\alpha+\beta} \int l_{\alpha\beta,\,\delta\varepsilon}(x,z)\, c_{\delta\varepsilon}(z)\,dz, \tag{4.85}$$

where the kernel $l_{\alpha\beta,\,\delta\varepsilon}(x,z)$ is a sum of tensor products of the kernels $k_\pm(x,z)$ times some numerical factors. We can write
$$l_{\alpha\beta,\,\delta\varepsilon}(x,z) = \text{Sym.} \sum_{\mu=0}^{\delta} \binom{\delta}{\mu} \binom{\varepsilon}{\alpha-\mu} \prod_{j=1}^{\mu} k_+(x_j, z_j)$$
$$\times \prod_{j=\mu+1}^{\alpha} k_-(x_j, z_{j+\delta-\mu}) \prod_{j=\alpha+1}^{\delta+\alpha-\mu} k_-(x_j, z_{j+\mu-\alpha}) \prod_{j=\delta+\alpha-\mu+1}^{\delta+\varepsilon} k_+(x_j, z_j) \tag{4.86}$$

where the symmetrization operation symmetrizes the kernel in the α creation variables $x_1, \ldots, x_\alpha$ and in the β annihilation variables $x_{\alpha+1}, \ldots, x_{\alpha+\beta}$.

We note some properties of the monomials contributing to the sum (4.86). In each kernel $k_+(x_j, z_j)$ that occurs, either z_j is a creation variable for $c_{\delta\varepsilon}$ and x_j is a creation variable for $c_{\alpha\beta}$, or else z_j is an annihilation variable for $c_{\delta\varepsilon}$ and x_j is an annihilation variable for $c_{\alpha\beta}$. In other words, the K_+ operators in (4.86) do not connect creation variables to annihilation variables. On the other hand, the kernels $k_-(x_j, z_{j'})$ connect an annihilation (creation) variable in $c_{\delta\varepsilon}$ to a creation (annihilation) variable in $c_{\alpha\beta}$. In majorizing (4.86) we will later use the operator norm to dominate k_+ kernels, and Lemma 4.8 to deal with the k_- kernels.

The next step in our argument is to prove that the transformation (4.85)–(4.86) which we derived for a class of $c_{\alpha\beta}(x)$ defined in (4.83), is correct for any $c_{\alpha\beta}(x)$ that might occur in the expansion of an operator $C \in \mathfrak{A}_0(\mathcal{B})$. We note that by a generalized polarization identity, (4.85)–(4.86) is valid for kernels arising from operators

$$C_\gamma = :C_1^{(1)} C_1^{(2)} \dots C_1^{(\gamma)}:, \tag{4.87}$$

where
$$C_1^{(j)} = A^*(c_{10}^{(j)}) + A(c_{01}^{(j)}) = \phi(b_{10}^{(j)}) + \pi(b_{01}^{(j)}), \quad j = 1, 2, \dots, \gamma, \tag{4.88}$$

and supp $b_{10}^{(j)} \subset \mathcal{B}$, supp $b_{01}^{(j)} \subset \mathcal{B}$.

Here we use the identity (for real $a_1, a_2 \dots, a_n$),

$$2^{n-1} n!\, a_1 a_2 \dots a_n = \sum_{\varepsilon_j = \pm 1} \varepsilon_2 \dots \varepsilon_n (a_1 + \varepsilon_2 a_2 + \dots + \varepsilon_n a_n)^n. \tag{4.89}$$

Kernels that are finite sums of kernels of the operators (4.87) are the only kernels $c_{\alpha\beta}(x)$ that occur in the expansion of operators $C \in \mathfrak{A}_0(\mathcal{B})$. Hence the identity transformation (4.85)–(4.86) holds for all our kernels.

We now return to the localized kernels $c_{\alpha\beta}(x)$ and the proof of Lemma 4.3. Using (4.85) we can write

$$c_{\alpha\beta}^L(x) = \sum_{\delta+\varepsilon=\alpha+\beta} \int l_{\alpha\beta,\,\delta\varepsilon}^L(x, z)\, c_{\delta\varepsilon}(z)\, dz, \tag{4.90}$$

where the localized kernel $l_{\alpha\beta,\,\delta\varepsilon}^L(x, z)$ is defined by

$$l_{\alpha\beta,\,\delta\varepsilon}^L(x, z) = \mathrm{Sym} \sum_{\mu=0}^{\delta} \binom{\delta}{\mu} \binom{\varepsilon}{\alpha-\mu} \prod_{j=1}^{\mu} k_+^{L_j}(x_j, z_j) \prod_{j=\mu+1}^{\delta+\alpha-\mu} k_-^{L_j}(x_j, z_{j'}) \prod_{j=\delta+\alpha-\mu+1}^{\delta+\varepsilon} k_+^{L_j}(x_j, z_j) \tag{4.91}$$

and $L = \{L_1, \dots, L_{\alpha+\beta}\}$ is the localization multi-index. Each kernel (4.91), before symmetrization, is the sum of $\delta + 1$ terms with a numerical coefficient dominated by $2^{\delta+\varepsilon} = 2^{\alpha+\beta}$. The kernels from one term, obtained by fixing μ, are a tensor product of k_+ kernels and of k_- kernels. The k_- factor

$$\prod_{j=\mu+1}^{\delta+\alpha-\mu} k_-^{L_j}(x_j, z_{j'}) = \prod_{j=\mu+1}^{\alpha} k_-^{L_j}(x_j, z_{j+\delta-\mu}) \prod_{j=\alpha+1}^{\delta+\alpha-\mu} k_-^{L_j}(x_j, z_{j+\mu-\alpha})$$

is the tensor product of Hilbert–Schmidt kernels and has the form of a tensor product of kernels $k_C k_A$ of Lemma 4.8. The first product in (4.91) of k_+ kernels is a bounded operator O_α on $\mathcal{J}_\alpha$ and the second product in (4.91) of k_+ kernels is a bounded operator O_β on $\mathcal{J}_\beta$. Thus applying Lemma 4.8 to the $k_C k_A$ kernels and using the operator norm on the k_+ kernels, we find that each term $l_{\alpha\beta,\,\delta\varepsilon}^L(x, z)_\mu$ in (4.91) with fixed μ is a kernel such that

$$\int l_{\alpha\beta,\,\delta\varepsilon}^L(x, z)_\mu\, c_{\delta\varepsilon}(z)\, dz$$

is the kernel of an operator from $L_2(R^\beta)$ to $L_2(R^\alpha)$ with a norm dominated by

$$2^{\alpha+\beta} \|O_\alpha\|\, \|k_C\|_2\, \|k_A\|_2\, \|O_\beta\|\, \|c_{\delta\varepsilon}\| = 2^{\alpha+\beta} \prod_{j=1}^{\alpha_1} \|K_+^{L_j}\| \prod_{j=\alpha_1+1}^{\delta+\alpha+\alpha_1} \|k_-^{L_j}\|_2 \prod_{j=\delta-\alpha-\alpha_1+1}^{\delta+\varepsilon} \|K_+^{L_j}\|\, \|c_{\delta\varepsilon}\|.$$

By Lemma 4.9, this is less than

$$2^{\alpha+\beta} M_{15}^{\alpha+\beta} \prod_{j=1}^{\alpha+\beta} e^{-m_0 |L_j|} \|c_{\delta\varepsilon}\| = (2 M_{15})^{\alpha+\beta} e^{-m_0 D(L)} \|c_{\delta\varepsilon}\|,$$

and by Lemma 4.2, $\qquad \leqslant (4 M_{15})^{\alpha+\beta} e^{-m_0 D(L)} \|C\|.$

Summing over μ in (4.91) and over $\delta+\varepsilon=\alpha+\beta$ in (4.90) yields fewer than

$$(\delta+1)(\alpha+\beta+1) \leqslant (\alpha+\beta+1)^2 \leqslant e^{\alpha+\beta}$$

such terms. Since symmetrization does not increase our bound,

$$\|c_{\alpha\beta}^L\| \leqslant (4 e M_{15})^{\alpha+\beta} e^{-m_0 D(L)} \|C\|,$$

which completes the proof of Lemma 4.3.

5. The vacuum self energy per unit volume is finite

The lower bound E_g on the operator $\hat{H}(g) = H_0 + H_{I,g}$ is called the vacuum energy, because

$$\hat{H}(g)\,\Omega_g = E_g \Omega_g, \tag{5.1}$$

and

$$E_g = (\Omega_g, \hat{H}(g)\,\Omega_g)$$

is the value of the energy operator $\hat{H}(g)$ in its vacuum state. This actually represents the *shift* in vacuum energy between the problem with and without interaction,

$$E_g = E_g - E_0,$$

where $E_0 = 0$ is the lower bound of H_0. The scale of $\hat{H}(g)$ is arbitrarily specified so that

$$(\Omega_0, \hat{H}(g)\,\Omega_0) = 0. \tag{5.2}$$

It is customary and convenient to add a constant to the Hamiltonian $\hat{H}(g)$ so that its vacuum energy is zero

$$H(g) = \hat{H}(g) - E_g, \tag{5.3}$$

and

$$H(g)\,\Omega_g = 0. \tag{5.4}$$

This shift is one of the standard renormalizations of quantum field theory, and we shall call $H(g)$ the renormalized Hamiltonian. We omit the renormalization to the mass m_0 and to the coupling constant λ since perturbation theory indicates that they are finite for our

$(\varphi^4)_2$ model, even in the limit $g(x)\to 1$. We note that it is the renormalized Hamiltonians $H(g)$, and not the unrenormalized Hamiltonians $\hat{H}(g)$ that have a limit as $g(x)\to 1$. In Section 2 we proved that as $g(x)\to 1$,

$$e^{-iH(g)t}\to e^{-iHt}$$

in a certain weak sense, and $0\leqslant H$.

In this section, we bound the rate at which E_g can diverge. According to perturbation theory, E_g is proportional to the volume of space in which the particles interact. However, the perturbation expansion for E_g can be shown by methods such as [18] not to converge and at best to be asymptotic. Thus we cannot get a rigorous bound on E_g from perturbation theory. Instead, we use a modification of Nelson's method [22]. It is this method which produced the estimates (cf. II, (2.1.16–18)) stating that the vacuum self energy for a fixed volume cutoff is finite [22, 8, 4].

THEOREM 5.1. *Let* $0\leqslant g(x)\leqslant 1$ *and for some constant* M_0,

$$\left|\frac{dg(x)}{dx}\right|\leqslant M_0.$$

Then

$$-M\leqslant E_g\leqslant 0, \tag{5.5}$$

where M is a positive constant proportional to the volume of the set $(\operatorname{supp} g)_1$, *that is the set of points within distance one of* $(\operatorname{supp} g)$.

Remarks 1. This theorem states that the average vacuum energy density, namely the vacuum self energy per unit volume, is finite for the theory with $g(x)=1$. Thus it shows that perturbation theory, though inapplicable, actually predicts the correct answer.

2. Furthermore, the theorem assures us that the expected number of (bare) particles per unit volume and the expected free energy per unit volume are both finite in the vacuum Ω. We define these expectation values as limits as $g(x)\to 1$ of the expectation value in Ω_g. Since

$$(\Omega_g, N\Omega_g)\leqslant\frac{1}{m_0}(\Omega_g, H_0\Omega_g), \tag{5.6}$$

it is sufficient to bound the free energy per unit volume. Since $H_0=2\hat{H}(g)-\hat{H}(2g)$,

$$(\Omega_g, H_0\Omega_g)\leqslant(\Omega_g, (2\hat{H}(g)-E_{2g})\,\Omega_g)=2E_g-E_{2g}. \tag{5.7}$$

In particular, if we choose g to be $g_n(x)$ defined in (2.3), then (5.5) says that for M independent of n,

$$-nM\leqslant E_{2g_n}. \tag{5.8}$$

Therefore, by (5.6–7)
$$(\Omega_{g_n}, N\Omega_{g_n}) \leqslant \text{const. } n, \tag{5.9}$$

$$(\Omega_{g_n}, H_0\Omega_{g_n}) \leqslant \text{const. } n, \tag{5.10}$$

$$\omega_n(N) \leqslant \text{const. } n, \tag{5.11}$$

and
$$\omega_n(H_0) \leqslant \text{const. } n. \tag{5.12}$$

3. A similar result holds for the Hamiltonians H_V defined in II, (2.1.7). Namely, there is a positive constant M independent of V, such that as $V \to \infty$

$$-MV \leqslant H_V. \tag{5.13}$$

Our proof yields (5.13) as well as (5.5), although a somewhat easier proof may be based on a modification of Federbush's proof for V fixed [4, 19].

4. The theorem also holds for any interaction Hamiltonian in two dimensions of the form

$$H_{I,g} = \int :P(\varphi(x)):\, g(x)\, dx,$$

where P is a polynomial of even degree whose leading coefficient is positive.

5. The same proof also shows that for $g(x)$ satisfying the hypotheses of the theorem, and for $\varepsilon > 0$, there exists a constant M proportional to $(\operatorname{supp} g)_1$ such that

$$0 \leqslant \varepsilon N + H_{I,g} + M.$$

Proof of Theorem 5.1. We use a partition of unity $\zeta_i(x)$ constructed from a positive C^∞ function $\zeta(x)$ with support in the interval $|x| < 1$, such that

$$\zeta_i(x) = \zeta(x-i), \tag{5.14}$$

and
$$\sum_i \zeta_i(x) = 1. \tag{5.15}$$

Thus we decompose g and $H_{I,g}$ into a sum of local parts

$$g_i = \zeta_i g \tag{5.16}$$

$$H_{I,g} = \sum_i H_{I,g_i} \tag{5.17}$$

Since we use the Feynman Kac formula to bound $\hat{H}(g)$, we wish to approximate $\hat{H}(g)$ by a Hamiltonian with a finite number of modes. Thus we study the Hamiltonian

17 – 702903 *Acta mathematica*. 125. Imprimé le 26 Octobre 1970.

$$H(g)_{\mathrm{K},V} = H_{0,\mathrm{K},V} + H_{I,g,\mathrm{K},V} \tag{5.18}$$

$$H_{0,\mathrm{K},V} = \int_{-\mathrm{K}}^{\mathrm{K}} a^*(k)\, a(k)\, \mu(k_V)\, dk \tag{5.19}$$

$$H_{I,g,\mathrm{K},V} = \lambda \int :\varphi_{\mathrm{K},V}(x)^4:\, g(x)\, dx \tag{5.20}$$

and

$$\varphi_{\mathrm{K},V}(x) = \left(\frac{1}{2V}\right)^{\frac{1}{2}} \sum_{k\in\Gamma_V} e^{-ikx}\,(a_V^*(k) + a_V(-k))\, \mu(k)^{-\frac{1}{2}}\, \xi(k/\mathrm{K}). \tag{5.21}$$

See II, section 2.1 for the notation used above. Here $\xi(x) \geqslant 0$ is a fixed C_0^∞ function, equal to one on a neighborhood of zero, and vanishing for $|x| > 1$. As in II, section 2.2, we find that

$$\hat{H}(g)_{\mathrm{K},V} \,|\, \mathfrak{F}_{\mathrm{K},V}^{\perp} \geqslant E_{\mathrm{K},V}, \tag{5.22}$$

where $E_{\mathrm{K},V}$ is the lower bound of $\hat{H}(g)_{\mathrm{K},V}$ on the full Fock space $\mathfrak{F}$. Therefore, we need only find the lower bound of $\hat{H}(g)_{\mathrm{K},V} \,|\, \mathfrak{F}_{\mathrm{K},V}$, which we now study. For simplicity of notation, we sometimes suppress the V cutoffs and define

$$H_I(\mathrm{K}, j) = H_{I,g_j,\mathrm{K},V}. \tag{5.23}$$

Thus it is sufficient to obtain a lower bound for

$$\{H_{0,\mathrm{K},V} + \sum_j H_I(\mathrm{K}, j)\} \,|\, \mathfrak{F}_{\mathrm{K},V} \tag{5.24}$$

which is independent of K and V, and which is at most proportional to the number of nonzero terms $H_I(K, j)$. On $\mathfrak{F}_{\mathrm{K},V}$

$$H_{0,\mathrm{K},V} = \sum_{\substack{k\in\Gamma_V \\ |k|\leqslant \mathrm{K}}} a_V^*(k)\, a_V(k)\, \mu(k) \tag{5.25}$$

We note that $H_I(\mathrm{K}, j)$ is the sum of five monomials in creation and annihilation operators, each of which has the kernel proportional to

$$b_{\mathrm{K},j}(k_1, \ldots, k_4) = \lambda (2V)^{-2} (g\zeta_j)^{\sim} \left(\sum_{l=1}^{4} k_l\right) \prod_{l=1}^{4} [\mu(k_l)^{-\frac{1}{2}}\, \xi(k_l/\mathrm{K})], \tag{5.26}$$

corresponding to an expansion in creation and annihilation operators for volume V,

$$H_I(\mathrm{K}, j) = \sum_{k_1,\ldots,k_4\in\Gamma_V} \sum_{\alpha=0}^{4} \binom{4}{\alpha} b_{\mathrm{K},j}(k_1, \ldots, k_4)\, a_V^*(k_1) \ldots a_V^*(k_\alpha)\, a_V(-k_{\alpha+1}) \ldots a_V(-k_4). \tag{5.27}$$

We decompose the single particle Fock space $\mathfrak{F}_{V,1}$ of $\mathfrak{F}_V$ into an orthogonal direct sum

$$\mathfrak{F}_{V,1} = \bigoplus_i \mathfrak{F}_{V,1,i}, \tag{5.28}$$

where $\mathfrak{F}_{V,1,i}$ consists of the Fourier series of functions vanishing off the interval

$$[i-\tfrac{1}{2}, i+\tfrac{1}{2}] \cap [-V/2, V/2]. \tag{5.29}$$

Furthermore, we have the Parseval equality

$$\int_{-V/2}^{V/2} f(x)\, h(x)\, dx = \sum_{k\in\Gamma_V} \tilde{f}(k)\, \tilde{h}(-k) = \sum_{k,\, k'\in\Gamma_V} \tilde{f}(k)\, \tilde{h}(k')\, \delta(k; -k'),$$

where

$$\delta(k; k') = \begin{cases} 1, & k = k' \\ 0, & k \neq k'. \end{cases}$$

From this, we see that

$$\sum_{k\in\Gamma_V} \tilde{f}(k)\, \tilde{h}(-k) = \sum_{i,j} \int_{-V/2}^{V/2} f_i(x)\, h_j(x)\, dx = \sum_i \int_{-V/2}^{V/2} f_i(x)\, h_i(x)\, dx$$

$$= \sum_i \sum_{k\in\Gamma_V} (f_i)^{\sim}(k)\, (h_i)^{\sim}(-k). \tag{5.30}$$

Thus (5.30) provides a decomposition into localized parts,

$$f_i \in \mathfrak{F}_{V,1,i}.$$

Using this decomposition, we can decompose any r-fold tensor product

$$\mathfrak{F}_{V,1} \otimes_s \mathfrak{F}_{V,1} \otimes_s \cdots \otimes_s \mathfrak{F}_{V,1},$$

into an orthogonal direct sum, and corresponding to (5.30) we have

$$\sum_{k_j\in\Gamma_V} \tilde{f}(k_1, \ldots, k_r)\, \tilde{h}(-k_1, \ldots, -k_r) = \sum_i \sum_{k_j\in\Gamma_V} (f_i)^{\sim}(k_1, \ldots, k_r)\, (h_i)^{\sim}(-k_1, \ldots, -k_r), \tag{5.31}$$

where $i = \{i_1, \ldots, i_r\}$ is a localization index and

$$f_i \in \mathfrak{F}_{V,1,i_1} \otimes_s \cdots \otimes_s \mathfrak{F}_{V,1,i_r}.$$

We therefore can write

$$b_{K,j}(k_1, \ldots, k_4) = \sum_{i_1, \ldots, i_4} b_{K,j,i_1,\ldots,i_4}(k_1, \ldots, k_4) \tag{5.32}$$

corresponding to the direct sum (5.28), with

$$b_{K,j,i_1,\ldots,i_4} \in \mathfrak{J}_{V,1,i_1} \otimes_s \cdots \otimes_s \mathfrak{J}_{V,1,i_4}.$$

Let H_I (K, j, i_1, ..., i_4) be the part of H_I(K, j) with the kernels (5.32). We notice that the kernel (5.32) has five localization indices. If the indices are incompatible, that is, if any of the differences $|j-i_l|$ is large, then the kernel must be small. We need the same result for a somewhat more general kernel. For $k_l \in \Gamma_V$, let

$$b_{K,j,s}(k_1, \ldots, k_4) = b_{K,j}(k_1, \ldots, k_4) \exp\{-\sum_{l=1}^{4} s_l \mu(k_l)\}, \tag{5.33}$$

where $s = \{s_1, \ldots, s_4\}$.

LEMMA 5.2. *For each integer r, each $\varepsilon > 0$, and each positive number T, there is a constant $c = c(r, T)$ such that for $0 \leqslant s_l \leqslant T$, and $1 \leqslant \Lambda \leqslant K$,*

$$\|b_{\Lambda,j,s,i} - b_{K,j,s,i}\|_{2,V} \leqslant c\Lambda^{-\frac{1}{2}+\varepsilon} \prod_{l=1}^{4} (1 + |j - i_l|)^{-r}. \tag{5.34}$$

Note $i = \{i_1, \ldots, i_4\}$. Note that $b_{\Lambda,j,s,i}$ is the ith component of $b_{\Lambda,j,s}$ defined in (5.33). Here the norm $\|\cdot\|_{2,V}$ refers to the Fourier series L_2 norm, which equals $(V/2\pi)^2$ times the usual L_2 norm of (5.26) with each k_l replaced by k_{lV}.

Before proving this lemma, we make some general remarks. Let $f(x)$ be a function with a smooth Fourier transform. That is, assume that for all α,

$$\|D^\alpha \tilde{f}\|_2 < \infty. \tag{5.35}$$

This corresponds to a rapidly decreasing $f(x)$. Now let

$$f_V(x) \in L_2([-V/2, V/2]) \tag{5.36}$$

be a periodic function with Fourier coefficients

$$V^{-\frac{1}{2}} \tilde{f}(k), \quad k \in \Gamma_V.$$

Then

$$f_V(x) = \sum_{n=-\infty}^{\infty} f(x + nV), \tag{5.37}$$

as can be checked by computing the Fourier coefficient of the right side of (5.37). In fact (5.37) is valid for $f \in L_1$. Using the rapid decrease of $f(x)$, we have for any integer r, a bound on the norms,

$$\|(1 + |x|)^r f_V\|_{2,V} \tag{5.38}$$

in $L_2([V/2, V/2])$, which is independent of V, as long as V is bounded away from zero. Furthermore, the norms (5.38) can be estimated in terms of the norms (5.35).

Proof of Lemma 5.2. We have $b_{K,j,s}$ given by (5.33). Multiplication of the $b_{K,j,s}$ by $\exp(ij\sum_{l=1}^4 k_l)$ maps the $(i_1, \ldots, i_4)$ component of $b_{K,j,s}$ into the $(i_1-j, \ldots, i_4-j)$ component of

$$\lambda(2V)^{-2}(g(\cdot+j)\zeta)^{\sim}\Big(\sum_{l=1}^{4} k_l\Big)\prod_{l=1}^{4}\mu(k_l)^{-\frac{1}{2}}\xi(k_l/K)\exp\Big(-\sum_{l=1}^{4} s_l\mu(k_l)\Big).$$

These functions are of the form f_V, and we now see that the corresponding functions f satisfy estimates of the form (5.35), which are independent of K and j. Let $P=k_1+\ldots+k_4$, and

$$F(K,\Lambda,s)=(1+P^2)^{-\frac{1}{2}}\prod_{l=1}^{4}[\mu(k_l)^{-\frac{1}{2}}e^{-s_l\mu(k_l)}]\Big\{\prod_{j=1}^{4}\xi(k_j/K)-\prod_{j=1}^{4}\xi(k_j/\Lambda)\Big\}.$$

Then $D^rF(K,\Lambda,s)\in L_2$ and

$$\Big\|D^r\Big\{(b_{K,j,s}-b_{\Lambda,j,s})\exp\Big(ij\sum_{l=1}^{4}k_l\Big)\Big\}\Big\|_2$$

$$\leqslant\sum_{\alpha}\binom{r}{\alpha}\|D^{\alpha}[(1+P^2)^{\frac{1}{2}}(g(\cdot+j)\zeta)^{\sim}(P)]\|_{\infty}\times\|D^{r-\alpha}F(K,\Lambda,s)\|_2. \tag{5.39}$$

We observe that $\xi(k/K)=1=\xi(k/\Lambda)$

unless $|k|>\text{const.}\,\Lambda.$

Thus $D^{p-\alpha}F(K,\Lambda,s)=0$ unless at least one of the k's has magnitude greater than const. Λ. By checking the order of convergence of the integrals, we see that for each r, the L_2 norm of $D^{r-\alpha}F(K,\Lambda,s)$ is $O(\Lambda^{-\frac{1}{2}+\varepsilon})$, independent of K and s as long as they lie in their allowed ranges. (See [26].)[1] In order to see that the L_∞ norms in (5.39) are bounded independently of j, we note that

$$|D^{\beta}(1+P^2)^{\frac{1}{2}}|\leqslant\text{const.}\,(1+|P|)$$

for a constant independent of β if $|\beta|\leqslant|r|$, and $|r|$ is fixed. Thus for $|\alpha|\leqslant r$.

$$\begin{aligned}\|D^{\alpha}\{(1+P^2)^{\frac{1}{2}}(g(\cdot+j)\zeta)(P)\}\|_{\infty}&\leqslant\sup_{\beta\leqslant|r|}\text{const.}\,\|(1+|P|)D^{\beta}(g(\cdot+j)\zeta)^{\sim}(P)\|_{\infty}\\&\leqslant\sup_{\beta\leqslant|r|}\text{const.}\,(\|x^{\beta}g(\cdot+j)\zeta\|_1+\|D(x^{\beta}g(\cdot+j)\zeta)\|_1)\\&\leqslant\text{const.}\sup_{\beta\leqslant r}((1+M_0)\|x^{\beta}\zeta\|_1+\|D(x^{\beta}\zeta)\|_1)\end{aligned}\tag{5.40}$$

by the assumed bounds on $g(x)$ and $|Dg(x)|$. Since $\zeta\in C^{\infty}$ and has compact support, (5.40) is dominated by $c=c(|r|, M_0, T)$. Thus (5.39) satisfies a bound

$$\Big\|D^r\Big\{(b_{K,j,s}-b_{\Lambda,j,s})\exp\Big(i\sum_{l=1}^{4}k_lj\Big)\Big\}\Big\|_2\leqslant c(r,T)\Lambda^{-\frac{1}{2}+\varepsilon},$$

---

[1] A direct calculation is also possible.

which is independent of j and K. This is an estimate of the form (5.35), so the corresponding norms (5.38) satisfy similar bounds. The rapid decrease of these functions could be written as the estimate (5.34), which completes the proof.

We now bound $|E_g|$ through the formula

$$|E_g| = t^{-1} \log \|e^{-tH(g)}\| = t^{-1} \log \{ \sup_{\|\theta\|=1=\|\chi\|} |(\theta, e^{-tH(g)}\chi)| \}. \tag{5.41}$$

Since

$$\lim_{\mathrm{K}, V \to \infty} E_{g, \mathrm{K}, V} = E_g,$$

it is sufficient to bound $|E_{g,\mathrm{K},V}|$ uniformly in K and V. We use the Feynman Kac formula to represent the operator $\exp(-tH(g)_{\mathrm{K},V})$ on $\mathfrak{F}_{\mathrm{K},V}$. For $\theta, \chi \in \mathfrak{F}_{\mathrm{K},V}$

$$(\theta, e^{-tH(g)_{\mathrm{K},V}}\chi) = \int_{C_{\mathrm{K},V}} \exp\left[-\int_0^t H_{I,g,\mathrm{K},V}(q(s))\,ds\right] \bar{\theta}(q(0))\,\chi(q(t))\,dq(\cdot). \tag{5.42}$$

Here $C_{\mathrm{K},V}$ is the path space corresponding to $\mathfrak{F}_{\mathrm{K},V}$. It consists of continuous paths $q(\cdot)$ taking values $q(s)$ in a Euclidean space of high dimension—one dimension for each mode permitted by the cutoffs K and V. The integral is a Wiener type integral on $C_{K,V}$ coming from $H_{0,\mathrm{K},V}$. The transition probabilities of the Wiener process are the kernel of $\exp(-tH_{0,\mathrm{K},V})$. We bound the inner product (5.42) with Hölder's inequality, as in [22, 8]. For some sufficiently large numbers p and t chosen independently of g, K and V, we have

$$|E_{g,\mathrm{K},V}| \leq (pt)^{-1} \log \left\{ \int_{C_{\mathrm{K},V}} \exp - \left[p \int_0^t H_{I,g,\mathrm{K},V}(q(s))\,ds \right] dq(\cdot) \right\}. \tag{5.43}$$

Thus we must show that

$$\int_{C_{\mathrm{K},V}} \exp\left[-p\int_0^t H_{I,g,\mathrm{K},V}(q(s))\,ds\right] dq(\cdot) \tag{5.44}$$

grows at most exponentially in the number of j's, with the exponential constant independent of K and V. We have

$$:\varphi_{\Lambda,V}(x)^4: = (\varphi_{\Lambda,V}^2(x) - 3\,C_0)^2 - 6\,C_0^2,$$

where C_0 is the no particle expectation value

$$C_0 = (\Omega_0, \varphi_{\Lambda,V}(x)^2\,\Omega_0) = \frac{1}{2V} \sum_{k \in \Gamma_V} \frac{1}{\mu(k)}\,\xi^2(k/\Lambda) \leq C(\log \Lambda)$$

for $\Lambda \geq 2$. Thus

$$-6\,C^2 \lambda \|\zeta\|_1 (\log \Lambda)^2 \leq H_I(\Lambda, j),$$

and for a new constant c_1 depending on t

$$1 - c_1(\log \Lambda)^2 \leqslant \int_0^t H_I(\Lambda, j)\,(q(s))\,ds. \tag{5.45}$$

Let

$$\delta H(\Lambda, j) = H_I(\mathrm{K}, j) - H_I(\Lambda, j) \tag{5.46}$$

and

$$I(\Lambda, j) = \int_0^t \delta H(\Lambda, j)\,(q(s))\,ds. \tag{5.47}$$

Also let Pr denote the measure defined by the integral

$$\int_{C_{\mathrm{K},V}} dq(\cdot).$$

We define $\mathfrak{J}$ as the set of integers j for which $g\zeta_j \neq 0$. Let $\mathfrak{J}_0$ denote any subset of $\mathfrak{J}$. We consider a finite sequence $\{\Lambda_j : j \in \mathfrak{J}_0\}$ such that

$$2 \leqslant \Lambda_j \leqslant \mathrm{K}.$$

We define for such a sequence $\{\Lambda_j\}$

$$P(\{\Lambda_j\}) = Pr\{|I(\Lambda_j, j)| \geqslant 1, \text{ for all } j \in \mathfrak{J}_0\}. \tag{5.48}$$

The main step remaining is to prove

LEMMA 5.3. *For constants c, d independent of* K *and* V,

$$P(\{\Lambda_j\} \leqslant \prod_{j \in \mathfrak{J}_0} [c \exp(-d\,\Lambda_j^{\frac{1}{2}-\varepsilon})]. \tag{5.49}$$

Let us for the moment assume Lemma 5.3 and derive the bound on (5.37). We represent $C_{\mathrm{K},V}$ as a disjoint union of measurable subsets X_ν, and we estimate the integral by the maximum of its integrand on each X_ν.

$$\left|\int_{C_{\mathrm{K},V}} F(q(\cdot))\,dq(\cdot)\right| \leqslant \sum_\nu Pr\{X_\nu\} \sup_{q(\cdot)\in X_\nu} |F(q(\cdot))|. \tag{5.50}$$

Let us assign to each path $q(\cdot) \in C_{\mathrm{K},V}$ a sequence of integers ν_j with $j \in \mathfrak{J}$. We define

$$\nu_j(q(\cdot)) = \nu_j \tag{5.51}$$

as the smallest integer in the interval $[d_1, \mathrm{K}]$ such that

$$|I(\nu_j, j)\,(q(\cdot))| \leqslant 1. \tag{5.52}$$

Here $d_1 > 2$ is a constant which we choose depending only on the c and d in Lemma 5.3.

We take d_1 and K to be integers. Since $I(\mathrm{K}, j)=0$, there is a smallest integer ν_j with the required property.

For each sequence of integers $\{\nu_j\}$, namely

$$\nu_j \in [d_1, \mathrm{K}], \quad j \in \mathfrak{J},$$

let $X(\{\nu_j\})$ be the set of all paths in $C_{\mathrm{K},V}$ which, by (5.51), correspond to the sequence $\{\nu_j\}$. Then $C_{\mathrm{K},V}$ is a disjoint union of measurable sets

$$C_{\mathrm{K},V} = \bigcup_{\{\nu_j\}} X(\{\nu_j\}), \tag{5.53}$$

where the various ν_j range over $[d_1, \mathrm{K}]$.

Furthermore, we note that

$$\int_0^t H_{I,g,\mathrm{K},V}(q(s))\,ds = \sum_{j\in\mathfrak{J}} \left\{ \int_0^t H_I(\nu_j, j)\,(q(s))\,ds + I(\nu_j, j) \right\}, \tag{5.54}$$

and by (5.45)

$$1 - c_1 (\log \nu_j)^2 \leqslant \int_0^t H_I(\nu_j, j)\,(q(s))\,ds.$$

Therefore for $q(\cdot) \in X(\{\nu_j\})$, (5.54) yields

$$-c_1 \sum_{j\in\mathfrak{J}} (\log \nu_j)^2 \leqslant \int_0^t H_{I,g,\mathrm{K},V}(q(s))\,ds. \tag{5.55}$$

To each path $q(\cdot) \in C_{\mathrm{K},V}$ we now assign a second sequence of integers $\{\Lambda_j\}$, where j ranges over a subset $\mathfrak{J}_0$ of $\mathfrak{J}$. Let $\mathfrak{J}_0$ consist of those k for which

$$\nu_k > d_1 + 1, \tag{5.56}$$

and for $k \in \mathfrak{J}_0$ define

$$\Lambda_k = \nu_k - 1.$$

To each subset $X(\{\nu_j\})$ of $C_{\mathrm{K},V}$, there corresponds one sequence $\{\Lambda_k\}$. For $k \in \mathfrak{J}_0 \subset \mathfrak{J}$,

$$|I(\Lambda_k, k)\,(q(\cdot))| \geqslant 1,$$

and so

$$Pr\{X(\{\nu_j\})\} \leqslant P(\{\Lambda_k\}). \tag{5.57}$$

Thus using (5.50), (5.55), (5.57) and Lemma 5.3, the desired integral (5.44) is bounded by

$$\sum_{\{\nu_j\}} Pr(X\{\nu_j\}) \exp\Big[c_1 p \sum_{j\in\mathfrak{J}} \log^2 \nu_j\Big] \leqslant \sum_{\{\nu_j\}} \prod_{j\in\mathfrak{J}_0} [c \exp(-d\Lambda_j^{\frac{1}{4}-\varepsilon})] \prod_{j\in\mathfrak{J}} [\exp(c_1 p \log^2 \nu_j)]. \tag{5.58}$$

Each ν_j in the above sum ranges over $d_1 \leqslant \nu_j \leqslant \mathrm{K}$, and we now treat the various cases in which some of the ν_j equal d_1. Let M be the number of elements of $\mathfrak{J}$ and M_0 be the number of elements of $\mathfrak{J}_0$, leaving $(M-M_0)$ of the ν_j equal to d_1. Since there are 2^M subsets of $\mathfrak{J}$, $\mathfrak{J}_0$ can be chosen in 2^M different ways. Restricting the sum (5.58) to $\{\nu_j\}$ which give one fixed subset $\mathfrak{J}_0$, this restricted sum is bounded by

$$\exp\,[(M-M_0)\,c_1 p \log^2 d_1] \sum_{\{\Lambda_j\}} \prod_{j\in\mathfrak{J}_0} [\exp\,(\log c + c_1 p \log^2 \nu_j - d\Lambda_j^{\frac{1}{4}-\varepsilon})]$$
$$\leqslant \exp\,[(M-M_0)\,c_1 p \log^2 d_1] \exp\,[M_0 d_2]. \tag{5.59}$$

Here we have chosen the arbitrary constant d_1 sufficiently large so that for $\nu > d_1$,

$$c_1 p \log^2 \nu + \log c < \tfrac{1}{2} d(\nu-1)^{\frac{1}{4}-\varepsilon} = \tfrac{1}{2} d\Lambda^{\frac{1}{4}-\varepsilon},$$

in which case

$$\sum_{\Lambda=d_1}^{K-1} \exp\,(-\tfrac{1}{2} d\Lambda^{\frac{1}{4}-\varepsilon}) \leqslant \sum_{\Lambda=0}^{\infty} \exp\,(-\tfrac{1}{2} d\Lambda^{\frac{1}{4}-\varepsilon}) = \exp\,[d_2].$$

We are considering the case of M_0 different Λ_j's so the bound (5.59) follows. Thus if

$$a = d_2 + c_1 p \log^2 d_1,$$

(5.59) is dominated by $\exp\,[aM]$ and (5.44) is less than

$$2^M \exp\,[aM],$$

which proves the theorem.

Proof of Lemma 5.3. For any set of positive even integers l_j,

$$P(\{\Lambda_j\}) \leqslant \int_{C_{K,V}} \prod_{j\in\mathfrak{J}_0} I(\Lambda_j, j)^{l_j} dq(\cdot). \tag{5.60}$$

We approximate the Riemann integral defining $I(\Lambda_j, j)$ by Riemann sums. In order to bound (5.60) it is sufficient to do so with $I(\Lambda_j, j)$ replaced by a Riemann sum. We replace all $L=\sum_j l_j$ factors by Riemann sums and then expand. The result will be a sum of M^L terms, if each Riemann sum has M terms. Each term is a product of L factors, and each factor occurs at a sharp time. To write down a typical term, we choose mesh times $s_1, \ldots, s_L$, and for the αth factor we choose an index $j(\alpha)$. The index j is chosen exactly l_j times. Then (5.60) is replaced by

$$P(\{\Lambda_j\}) \leqslant t^L \sup \int_{C_{K,V}} \delta H(\Lambda_{j(1)}, j(1))\,(q(s_1)) \ldots \delta H(\Lambda_{j(L)}, j(L))\,(q(s_L))\,dq(\cdot), \tag{5.61}$$

where the supremum is taken over all choices of mesh times, mesh lengths and all choices of $j(\alpha)$ consistent with the l_j. We can rearrange the factors in the integral, so that is no loss of generality to assume that

$$s_1 < s_2 < \ldots < s_L.$$

By the definition of the Wiener integral, (5.61) equals the no particle expectation value

$$t^L \sup (\Omega_0, \exp(-s_1 H_{0,\mathrm{K},V})\, \delta H(\Lambda_{j(1)}, j(1)) \exp(-(s_2 - s_1) H_{0,\mathrm{K},V})$$
$$\ldots \exp(-(s_L - s_{L-1}) H_{0,\mathrm{K},V})\, \delta H(\Lambda_{j(L)}, j(L))\, \Omega_0). \tag{5.62}$$

The next step is to evaluate (5.62) as a sum of elementary terms. Each elementary term will itself be a sum of explicit functions. This reduction of (5.62) is accomplished by repeated application of the commutation relation

$$[a_V(-k), a_V^*(k')] = \delta(-k; k') = \begin{cases} 1 & \text{if } -k = k' \\ 0 & \text{otherwise} \end{cases}$$

and

$$[a_V(-k), a_V(k')] = 0,$$

for $k, k' \in \Gamma_V$. We use this relation to bring all the annihilation operators $a_V(k)$ to the right and the creation operators $a_V(k)$ to the left. Furthermore, these commutation relations yield the relations for $k \in \Gamma_{\mathrm{K},V}$,

$$a_V(k) \exp(-s H_{0,\mathrm{K},V}) = \exp(-s H_{0,\mathrm{K},V})\, a_V(k) \exp[-s\mu(k)],$$

which are used to bring the annihilation operators to the right past the factors $\exp(-sH_{0,\mathrm{K},V})$. Since

$$a_V(k)\,\Omega_0 = 0,$$

moving the annihilation operators to the right does reduce (5.62) to the desired sum of elementary terms. Furthermore, each variable k must occur in a delta function $\delta(-k; k')$ in any non zero term.

In order to keep track of the distinct terms and the commutation rules used, we introduce a graph G for each term. The value of the term will be a number $N(G)$ assigned to G. Thus we shall write

$$(\Omega_0, \prod_{\alpha=1}^{L} \{\exp[-(s_\alpha - s_{\alpha-1}) H_{0,\mathrm{K},V}]\, \delta H(\Lambda_{j(\alpha)}, j(\alpha))\}\, \Omega_0) = \sum_G N(G), \tag{5.63}$$

where $s_0 = 0$, and where the product of noncommuting operators is defined by

$$\prod_{\alpha=1}^{L} \{A_\alpha\} = A_1 A_2 \ldots A_L.$$

We construct G as follows. We start with L vertices, labeled by the index α, $1 \leqslant \alpha \leqslant L$. From each vertex we draw four lines (called legs), labeled by (α, ϱ), $1 \leqslant \varrho \leqslant 4$. Each factor

$\delta H(\Lambda_{j(\alpha)}, j(\alpha))$ in (5.63) is the difference between two interaction Hamiltonians, and as such is the sum of five monomials as in (5.27). For each factor we choose one such monomial with ν_α creators and $4-\nu_\alpha$ annihilators. This monomial will correspond to a vertex in G, and we draw ν_α legs from the αth vertex to the left, and $4-\nu_\alpha$ legs pointing to the right. Each leg (α, ϱ) corresponds to a momentum variable $k_{\alpha\varrho}$ in the kernel $B(\Lambda_{i(\alpha)}, j(\alpha))$ at vertex α. Since our factors are noncommuting, we order the vertices linearly from left to right, as in (5.63), with vertex 1 on the left and vertex L on the right. Each time that the use of the commutation relations introduces a delta function, we say that the variables $-k_{\alpha\varrho}$ and $k_{\alpha'\varrho'}$ occurring in it are *contracted*. Graphically we join together the annihilation leg (α, ϱ) and the creation leg (α', ϱ') which correspond to these variables. We have already noticed that for a nonzero term, each variable occurs in exactly one delta function, and so for the corresponding graph each creating leg (pointing left) is connected to a unique annihilating leg (pointing right). Thus only terms with $\nu_1=0$, $\nu_L=4$ will contribute. Since annihilating legs all moved to the right, the annihilating legs at vertex α are always contracted with creating legs on their right.

The possible graphs are all graphs which can be constructed in this manner. Each graph consists of L vertices and $2L$ lines (each line being two joined legs). Each vertex is the endpoint of four lines and each line joins two distinct vertices. This last fact results from Wick ordering the interaction. Evidently there are fewer than

$$(4L)!! = (4L-1)(4L-3)\dots 3\cdot 1$$

possible graphs.

We let $N(G)$ denote the contribution to (5.63) from a given graph G and this is obtained as follows: To the vertex α of G we assign the kernel

$$B(\Lambda_{j(\alpha)}, j(\alpha); k_{\alpha 1}, \dots, k_{\alpha 4}) = b_{\mathrm{K},j(\alpha)}(k_{\alpha 1}, \dots, k_{\alpha 4}) - b_{\Lambda_{j(\alpha)},j(\alpha)}(k_{\alpha 1}, \dots, k_{\alpha 4}), \tag{5.64}$$

where $b_{\mathrm{K},j}$ is defined in (5.26). We write down the product of the L kernels (5.64) for G and multiply this by a product of $2L$ delta functions

$$\delta(-k_{\alpha\varrho}; k_{\alpha'\varrho'}),$$

one delta function for each line, obtained by pairing the annihilator (α, ϱ) with the creator (α', ϱ'), $\alpha'>\alpha$. We then multiply by a product of $2L$ energy factors

$$\exp[-(s_{\alpha'}-s_\alpha)\mu(k_{\alpha\varrho})],$$

one for each line, which arise from the commutations past $\exp(-sH_{0,\mathrm{K},V})$ factors. These factors could actually be included in the kernels $B(\Lambda_{j(\alpha)}, j(\alpha))$ by defining for each graph G

$$\hat{B}(\Lambda_{j(\alpha)}, j(\alpha)) = B(\Lambda_{j(\alpha)}, j(\alpha)) \exp\left[-\sum_{\varrho=1}^{4} (s_{\alpha'} - s_\alpha)\, \mu(k_{\alpha\varrho})/2\right]. \tag{5.65}$$

The numerical value of G is the sum of this expression over the $4L$ variables $k_{\alpha\varrho} \in \Gamma_{K,V}$ times some numerical factors.

$$N(G) = \sum_{\substack{k_{\beta\sigma} \in \Gamma_{K,V} \\ 1 \leqslant \beta \leqslant L \\ 1 \leqslant \sigma \leqslant 4}} \left[\sum_{\alpha=1}^{L} \binom{4}{\nu_\alpha} \hat{B}(\Lambda_{j(\alpha)}, j(\alpha); k_{\alpha\varrho})\right] \left[\prod_{\substack{1 \leqslant \alpha \leqslant L \\ 1 \leqslant \varrho \leqslant 4}} \delta(-k_{\alpha\varrho}; k_{\alpha'\varrho'})\right]. \tag{5.66}$$

Let us now motivate the type of bound we expect (in the simple case $s_1 = \ldots = s_L = 0$). We expect that different regions of localization will be independent, by the localization property of free fields (cluster property). Since each localization index j must be chosen exactly l_j times, exact localization would mean for $s_1 = \ldots = s_L = 0$ that (5.63) equals

$$\prod_{j \in \mathfrak{J}_0} (\Omega_0, \delta H(\Lambda_j, j)^{l_j} \Omega_0), \tag{5.67}$$

which can be shown to be dominated by

$$\prod_{j \in \mathfrak{J}_0} [c^{l_j} (2\,l_j)!\, \Lambda_j^{-l_j/2}], \tag{5.68}$$

for some constant c independent of K, V, $\mathfrak{J}_0$, Λ_j and l_j. In fact, we will show that although the localization is not exact, it is approximate in the sense that (5.68) actually bounds (5.63), and this holds even in the case that the s_j are nonzero. In other words, the action of $\exp(-sH_0)$ does not destroy, in any essential way, the localization of the particles.

LEMMA 5.4. *The expectation value* (5.63) *is dominated by*

$$\prod_{j \in \mathfrak{J}_0} [c^{l_j} (2\,l_j)!\, \Lambda_j^{-l_j(1-\varepsilon)/2}],$$

where the constant c depends only on T providing the bounds

$$0 \leqslant g(x) \leqslant 1,$$

$$|Dg(x)| \leqslant M_0,$$

and

$$0 \leqslant s_\alpha \leqslant T$$

are satisfied.

Let us assume this lemma and complete the proof of Lemma 5.3. We apply Lemma 5.4 to (5.63), and get a bound on (5.61) independent of the variables in the supremum. Thus

$$Pr(\{\Lambda_j\}) \leqslant (ct)^L \prod_{j \in \mathfrak{J}_0} (2l_j)!\Lambda_j^{-l_j(1-\varepsilon)/2}.$$

We now choose l_j to be the positive even integer nearest to $(4ct/\Lambda_j^{\frac{1}{2}-\varepsilon})^{-\frac{1}{2}} = O(\Lambda_j^{(1-\varepsilon)/4})$. Then using Sirling's formula to estimate the factorials, with $d \leqslant (4ct)^{-\frac{1}{2}}$, with a new constant c_1, and with a new ε,

$$Pr(\{\Lambda_j\}) \leqslant \prod_{j \in \mathfrak{J}_0} [c_1 \exp(-d\Lambda_j^{\frac{1}{4}-\varepsilon})].$$

This completes the proof.

Proof of Lemma 5.4. By repeated application of the Schwarz inequality in the variables $k_{\alpha\varrho}$, $N(G)$ in (5.66) can be dominated by

$$|N(G)| \leqslant 2^{4L} \prod_{\alpha=1}^{L} \|\hat{B}(\Lambda_{j(\alpha)}, j(\alpha))\|_2 = 2^{4L} \prod_{j \in \mathfrak{J}_0} \|\hat{B}(\Lambda_j, j)\|_2^{l_j} \leqslant \prod_{j \in \mathfrak{J}_0} [c^{l_j}\Lambda_j^{-l_j(1-\varepsilon)/2}],$$

by Lemma 5.2. However, we need a sharper bound (5.82) which reflects the localization of particles. Thus we classify the graphs in a way which reflects this property.

For a particular graph G, each vertex α, $1 \leqslant \alpha \leqslant L$, has an associated localization number $j(\alpha) \in \mathfrak{J}_0$. For $\varrho = 1, 2, 3, 4$, let us define $j(\alpha, \varrho)$ by

$$j(\alpha, \varrho) = j(G; \alpha, \varrho) = j(\alpha'), \tag{5.69}$$

where the ϱth leg at vertex α is connected in G to the ϱ'th leg at vertex α'. We can also define

$$\delta(\alpha, \varrho) = j(\alpha) - j(\alpha, \varrho), \tag{5.70}$$

which measures the difference in configuration space localization between the two vertices connected by the pair of legs (α, ϱ), (α', ϱ') corresponding to a line in G.

The functions (5.69) uniquely specify the functions (5.70) and vice versa, for any graph G contributing to (5.63). However, (5.69) does not uniquely specify the graph. We now show that there are at most

$$\prod_{j \in \mathfrak{J}_0} [4^{l_j}(2l_j)!] \tag{5.71}$$

out of the (fewer than $(4L)!!$) different graphs G with the same function $j(G, \alpha, \varrho)$.

Let us suppose that in G there are

$$l_{ij} = l_{ij}(G)$$

lines connecting a j-localized vertex to an i-localized vertex, and to simplify the following formulae, we count each line with $i = j$ twice. Thus

$$\sum_{j \in \mathfrak{J}_0} l_{ij} = 4l_i, \tag{5.72}$$

and
$$\sum_{i,j\in\mathcal{J}_0} l_{ij} = 4L. \tag{5.73}$$

Note that the G's which contribute to (5.65) are exactly those for which the off diagonal l_{ij}'s vanish, and $l_{ij}=4l_i\delta_{ij}$.

With the functions $j(\alpha, \varrho)$ given and with i and j held fixed, there are two sets of l_{ij} legs specified, and for the graphs we are considering each leg in the first set must be contracted to a leg in the second set. For a vertex α of a leg in the first set, $j(\alpha)=j$ and for a vertex α' of a leg in the second set, $j(\alpha')=i$. If $i\neq j$ then these two sets of legs are disjoint and the contractions can be made in at most $(l_{ij})!$ ways. If $i=j$ then the two sets of legs coincide and the contractions can be made in at most $(l_{ii})!!$ ways. Once the contractions have been made for all $i\leqslant j$, the graph G is uniquely determined, and so there are at most

$$\prod_{\substack{i<j\\ i,j\in\mathcal{J}_0}} [l_{ij}!] \prod_{i\in\mathcal{J}_0} [l_{ii}!!] \tag{5.74}$$

graphs with given functions $j(\alpha, \varrho)$.

If l is even
$$l! \leqslant 2^l[(\tfrac{1}{2}l)!]^2$$
$$l!! \leqslant 2^{l/2}(\tfrac{1}{2}l)!$$

and there are similar estimates for l_{ij} odd.

Using (5.72) – (5.73), we thus have that (5.74) is dominated by

$$\prod_{i,j\in\mathcal{J}_0} [2^{l_{ij}/2}(\tfrac{1}{2}l_{ij})!] = 4^L \prod_{j\in\mathcal{J}_0}\{\prod_{i\in\mathcal{J}_0}[(\tfrac{1}{2}l_{ij})!]\} \leqslant 4^L \prod_{j\in\mathcal{J}_0}[(2l_j)!],$$

which is the desired bound (5.71).

Thus
$$\Big|(\Omega_g, \prod_{\alpha=1}^{L} \delta H(\Lambda_{j(\alpha)}, j(\alpha))\,\Omega_0)\Big| \leqslant \{\prod_{j\in\mathcal{J}_0}[(4)^{l_j}(2l_j)!]\} \sum_{\{\delta(\alpha,\varrho)\}} \sup |N(G)|. \tag{5.75}$$

Here the supremum is over all graphs G with fixed $j(G, \alpha, \varrho)=j(\alpha, \varrho)$, or fixed $\delta(\alpha, \varrho)$, and the sum is over all possible integers $\delta(\alpha, \varrho)$ of graphs contributing to (5.63). We now bound $N(G)$ in (5.66).

Each of the $2L$ independent sums over momentum variables $k_{\alpha\varrho}\in\Gamma_V$ can be regarded as an inner product in the one particle space $\mathfrak{F}_{V,1}$, or the equivalent inner product in the space of Fourier series on $[-V/2, V/2]$. We now make an expansion by writing the inner product on $\mathfrak{F}_{V,1}$ as a sum of inner products on the $\mathfrak{F}_{V,1,i}$, using (5.28). Because of the support properties, many terms vanish giving as in (5.31)

$$N(G) = \sum_{\substack{i_{\beta\sigma}, k_{\beta\sigma}\\ 1\leqslant\beta\leqslant L\\ 1\leqslant\sigma\leqslant 4}} \prod_{\alpha=1}^{L} \hat{B}(\Lambda_{j(\alpha)}, j(\alpha), i_{\alpha\varrho}; k_{\alpha\varrho}) \binom{4}{\nu_\alpha} \prod_{\substack{1\leqslant\alpha\leqslant L\\ 1\leqslant\varrho\leqslant 4}} \delta(-k_{\alpha\varrho}; k_{\alpha'\varrho'})\,\delta(i_{\alpha\varrho}; i_{\alpha'\varrho'}). \tag{5.76}$$

The sum runs over all possible choices of the localization indices $i_{\beta\sigma}$ and the momenta $k_{\beta\sigma}$. We now apply the Schwarz inequality repeatedly in the $2L$ independent $k_{\alpha\varrho}$ variables. This gives the bound

$$|N(G)| \leqslant \sum_{\substack{i_{\beta\sigma} \\ 1\leqslant\beta\leqslant L \\ 1\leqslant\sigma\leqslant 4}} \prod_{\alpha=1}^{L} 16\, \|\hat{B}(\Lambda_{j(\alpha)}, j(\alpha), i_{\alpha\varrho})\|_2 \prod_{\substack{1\leqslant\alpha\leqslant L \\ 1\leqslant\varrho\leqslant 4}} \delta(i_{\alpha\varrho}; i_{\alpha'\varrho'}). \tag{5.77}$$

We choose V sufficiently large to enclose the support of the spatial cutoff $g(x)$. Thus by Lemma 5.2 independently of V,

$$\|\hat{B}(\Lambda_{j(\alpha)}, j(\alpha), i_{\alpha 1}, \ldots, i_{\alpha 4})\|_2 \leqslant c\,\Lambda_{j(\alpha)}^{-(1-\varepsilon)/2} \prod_{1\leqslant\varrho\leqslant 4} (1+|j(\alpha)-i_{\alpha\varrho}|)^{-r}, \tag{5.78}$$

where c depends on λ, M_0, r and T. Using the fact that each index $j=j(\alpha)$ occurs l_j times in G, we have by (5.77)–(5.78),

$$|N(G)| \leqslant \prod_{j\in\mathcal{J}_0} [16\, c^{l_j} \Lambda_j^{-l_j(1-\varepsilon)/2}] \sum_{\substack{i_{\beta\sigma} \\ 1\leqslant\beta\leqslant L \\ 1\leqslant\sigma\leqslant 4}} \prod_{\substack{1\leqslant\alpha\leqslant L \\ 1\leqslant\varrho\leqslant 4}} (1+|j(\alpha)-i_{\alpha\varrho}|)^{-r}\, \delta(i_{\alpha\varrho}; i_{\alpha'\varrho'}). \tag{5.79}$$

Since each leg (α, ϱ) is paired with another leg (α', ϱ'), the factors

$$(1+|j(\alpha)-i_{\alpha\varrho}|)^{-r}\, \delta(i_{\alpha\varrho}; i_{\alpha'\varrho'})$$

occur in pairs,

$$\begin{aligned}(1+|j(\alpha)-i_{\alpha\varrho}|)^{-r} &(1+|j(\alpha')-i_{\alpha'\varrho'}|)^{-r}\, \delta(i_{\alpha\varrho}; i_{\alpha'\varrho'})^2 \\ &= (1+|j(\alpha)-i_{\alpha\varrho}|)^{-r} (1+|j(\alpha')-i_{\alpha\varrho}|)^{-r}\, \delta(i_{\alpha\varrho}; i_{\alpha'\varrho'}),\end{aligned} \tag{5.80}$$

where

$$\delta(\alpha, \varrho) = j(\alpha) - j(\alpha') = j(\alpha) - j(\alpha, \varrho). \tag{5.81}$$

As

$$(1+|x|)^{-2}(1+|y|)^{-2} \leqslant (1+|x-y|)^{-1}(1+|x|)^{-1},$$

(5.80) is dominated by

$$(1+|\delta(\alpha, \varrho)|)^{-r/2}(1+|j(\alpha)-i_{\alpha\varrho}|)^{-r/2}.$$

Hence by (5.79)

$$|N(G)| \leqslant \prod_{j\in\mathcal{J}_0} (16\, cc_1)^{l_j} \Lambda_j^{-l_j(1-\varepsilon)/2} \prod_{\substack{1\leqslant\alpha\leqslant L \\ 1\leqslant\varrho\leqslant 4}} (1+|\delta(\alpha, \varrho)|)^{-r/2}, \tag{5.82}$$

where we have chosen $r>2$ and used the fact that

$$\sum_{\substack{i_{\beta\sigma} \\ 1\leqslant\beta\leqslant L \\ 1\leqslant\sigma\leqslant 4}} \prod_{\substack{1\leqslant\alpha\leqslant L \\ 1\leqslant\varrho\leqslant 4}} (1+|j(\alpha)-i_{\alpha\varrho}|)^{-r/2} = c_1^L,$$

with

$$c_1 = [\sum_{n=-\infty}^{\infty} (1+|n|)^{-r/2}]^4.$$

The bound (5.82) gives the dependence of G on the localization of its vertices. Thus by (5.75) and (5.82),

$$\left|\left(\Omega_0, \prod_{\alpha=1}^{L} \delta H(\Lambda_{j(\alpha)}, j(\alpha))\, \Omega_0\right)\right| \leqslant \prod_{j\in \mathcal{J}_0} [(64\, cc_1^2)^{l_j} (2l_j)!\, \Lambda_j^{-l_j(1-\varepsilon)/2}],$$

which establishes the lemma.

References

[1]. ARAKI, H., The type of von Neumann algebra associated with the free field. *Progr. of Theor. Phys.*, 32 (1964), 956–965.

[2]. DIXMIER, J., *Les algèbres d'opérateurs dans l'espace Hilbertien (algèbres de von Neumann).* Gauthiers-Villars, Paris, 1957.

[3]. EPSTEIN, H., GLASER, V. & JAFFE, A., Nonpositivity of the energy density in quantized field theories. *Nuovo Cimento,* 36 (1965), 1016–1022.

[4]. FEDERBUSH, P., A partially alternate derivation of a result of Nelson. *J. Math. Phys.*, 10 (1969), 50–52.

[5]. FELL, J., The dual spaces of C^*-algebras. *Trans. Amer. Math. Soc.*, 94 (1960), 365–403.

[6]. GALINDO, A., On a class of perturbations in quantum field theory. *Proc. Nat. Acad. Sci. (U.S.A.)*, 48 (1962), 1128–1134.

[7]. GLIMM, J., A Stone–Weierstrass theorem for C^*-algebras. *Ann. of Math.*, 72 (1960), 216–244.

[8]. —— Boson fields with nonlinear selfinteraction in two dimensions. *Commun. Math. Phys.*, 8 (1968), 12–25.

[9]. —— *Varenna summer school-lectures, 1968.* Academic Press.

[10]. GLIMM, J. & JAFFE, A., A $\lambda(\phi^4)_2$ quantum field theory without cutoffs I. *Phys. Rev.*, 176 (1968), 1945–1951.

[11]. —— Singular perturbations of self adjoint operators. *Comm. Pure Appl. Math.*, 22 (1969), 401–414.

[12]. —— The $\lambda(\phi^4)_2$ quantum field theory without cutoffs II. The field operators and the approximate vacuum. *Ann. of Math.*, 91 (1970). 362–401.

[13]. —— An infinite renormalization of the Hamiltonian is necessary. *J. Math. Phys.*, 10 (1969), 2213–2214.

[14]. GRIFFIN, E., Some contributions to the theory of rings of operators II. *Trans. Amer. Math. Soc.*, 79 (1955), 389–400.

[15]. HAAG, R. & KASTLER, D., An algebraic approach to quantum field theory. *J. Math. Phys.*, 5 (1964), 848–861.

[16]. HALL, D. & WIGHTMAN, A., A theorem on invariant functions with applications to relativistic quantum field theory. *Mat.-Fys. Medd. Kongl. Danske Vidensk. Selskab,* 31 (1957) No. 5.

[17]. HEPP, K., Théorie de la renormalisation, Springer Verlag, New York, 1969,

[18]. JAFFE, A., Divergence of perturbation theory for bosons. *Commun. Math. Phys.*, 1 (1965), 127–149.

[19]. —— *Varenna summer school lectures, 1968.* Academic Press.

[20]. KRISTENSEN, P., MEJLBO, L. & POULSEN, E., Tempered distributions in infinitely many dimensions I. *Commun. Math. Phys.*, 1 (1965), 175–214.

[21]. NAIMARK, M., *Normed rings.* Noordhoff, Groningen, 1959.

[22]. NELSON, E., A quartic interaction in two dimensions. In *Mathematical theory of elementary particles,* edited by R. Goodman and I. Segal, M.I.T. Press 1966.

[23]. SEGAL, I., Nonlinear functions of weak processes I. *J. Functional Anal.*, 4 (1969), 404–456.

[24]. STREIT, L., A generalization of Haag's theorem. *Nuovo Cimento*, to appear.
[25]. WATSON, G., *A treatise on the theory of Bessel functions*. Cambridge 1944.
[26]. WEINBERG, S., High energy behavior in quantum field theory. *Phys. Rev.*, 118 (1960), 838–849.
[27]. CANNON, J. & JAFFE, A., Lorentz covariance of the $\lambda\phi_2^4$ quantum field theory. *Commun. Math. Phys.*, 17 (1970), 261–321.
[28]. ROSEN, L., A $\lambda\phi^{2n}$ field theory without cutoffs. *Commun. Math. Phys.*, 16 (1970), 157–183.

Received November 7, 1969

The Wightman axioms and particle structure in the $\mathcal{P}(\phi)_2$ quantum field model[1]

By JAMES GLIMM[2], ARTHUR JAFFE[3] and THOMAS SPENCER[4]

Abstract: The $\mathcal{P}(\phi)_2$ quantum field model with small coupling constant satisfies the Wightman axioms. The spectrum of the mass operator $M = (H^2 - P^2)^{1/2}$ contains two isolated eigenvalues: 0 and $m > 0$. Hence the model describes the scattering of particles of mass $m > 0$. The proof contains an n-particle cluster expansion and an analysis of the subspace of states with energy less than $(n + 1)m(1 - \varepsilon)$.

CONTENTS

1. The particle structure

Nontrivial quantum fields exist in two space time dimensions, [3]–[6], [8]–[9]. Physicists use similar fields in four dimensions to describe elementary particles. For this reason it is important to establish qualitative properties of these fields, in order to demonstrate that they are suitable for the description of physical processes, such as the scattering of particles.

We establish two such results. For sufficiently weak $\mathcal{P}(\phi)_2$ interactions we verify the Wightman axioms. These axioms guarantee the principles of

[1] This paper incorporates the results presented previously in two unpublished manuscripts, references [10], [11].

[2] Supported in part by NSF Grant GP-24003.

[3] Supported in part by NSF Grant GP-31239X and by the Air Force Office of Scientific Research, Contract AF 44620-70-C 0030.

[4] Supported in part by the Alfred P. Sloan Foundation.

quantum mechanics and special relativity. Secondly, we establish the existence of particles and the existence of a scattering matrix S for these models. Single particles appear as an isolated eigenvalue $m > 0$ of the mass operator M, as described below.

Our analysis of the spectrum of the energy and momentum operators involves an n particle cluster expansion. We establish estimates for this expansion that yield the existence of an infinite volume limit for the theory, the proof of the remaining Wightman axioms, and the analysis of the particle structure. These estimates are an outgrowth of a study of local Hamiltonians [6], [7]. However the results and estimates of this paper are independent of [3]-[7].

A general requirement of quantum mechanics, and one of the Wightman axioms, is that the states are given by unit vectors θ in a Hilbert space $\mathcal{H}$. We believe that the Hilbert space $\mathcal{H}$ for a weak $\mathcal{P}(\phi)_2$ interaction has a Schrödinger representation

$$\mathcal{H} = L_2(\mathcal{S}'(R^1), dq) . \tag{1.1}$$

Here the Schwartz distribution space $\mathcal{S}'(R^1)$ is the configuration space of a classical field, and dq is a countably additive non-Gaussian Borel measure on $\mathcal{S}'(R^1)$. In this representation the time zero field $\phi_0(f), f \in \mathcal{S}(R^1)$, is simply multiplication by the linear coordinate function $f \in \mathcal{S}(R^1)$. Thus for $F \in L_2(\mathcal{S}'(R^1), dq)$ and $q \in \mathcal{S}'(R^1)$

$$\phi_0(f)F(q) = \left(\int f(\vec{x})q(\vec{x})d\vec{x}\right)F(q) = \langle q, f\rangle F(q) . \tag{1.2}$$

Hence $\phi_0(f)$ measures the strength or position of the field in the coordinate direction f, just as the position operator q_j in ordinary quantum mechanics measures the position of the jth particle. The moments of dq, namely

$$W_0(\vec{x}_1, \cdots, \vec{x}_n) = \int \phi_0(\vec{x}_1) \cdots \phi_0(\vec{x}_n)dq \in \mathcal{S}'(R^n) , \tag{1.3}$$

are the time-zero vacuum expectation values.

Relativistic covariance is expressed in terms of a unitary representation U of the inhomogeneous Lorentz group on $\mathcal{H}$. The energy operator $P_0 \geqq 0$ and the momentum operator P_1 are defined to be self-adjoint infinitesimal generators of the time and space translation subgroups respectively. The space-time field operator $\phi(f)$, $f \in \mathcal{S}(R^2)$, is defined by a time evolution of ϕ_0 in the Heisenberg picture, namely,

$$\phi(f) = \int e^{itP_0}\phi_0(f(\cdot, t))e^{-itP_0}\, dt . \tag{1.4}$$

A particle may be thought of as a stable isolated disturbance of the quantum field. Let Ω denote the vacuum state vector in $\mathcal{H}$. Then $P_0\Omega = 0$

$= P_1\Omega$ and $e^{-itP_0}\Omega = \Omega$, so Ω is an equilibrium state under the time evolution subgroup. In the representation (1.1) above, Ω is the function

$$\Omega = 1 \in L_2(\mathcal{S}'(R^1), dq) , \tag{1.5}$$

so the expectation defined by Ω is the probability measure dq on the space $\mathcal{S}'(R^1)$ of classical field configurations. We think of Ω, or dq, as the undisturbed state of the field. Thus any state $\theta \neq \Omega$ is a disturbed state, and the distinguishing feature of a particle is its stability: In isolation from other particles, the structure of a particle is preserved under time evolution. According to this point of view, two or more particles can be identified in a state of a quantum field only if the particles are sufficiently separated so that they are not interacting. To find a particle in an arbitrary state θ, we consider the time dependent state

$$\theta(t) = U(t)\theta = e^{-itP_0}\,\theta$$

into which (or from which) θ evolves. As $|t| \to \infty$, we expect the particles to separate, and so they can be identified in $\theta(t)$. These are the particles before ($t \to -\infty$) or after ($t \to +\infty$) the interaction, and they are called the "in" particles and the "out" particles of the state θ. It follows that particles are properties of the $|t| \to \infty$ asymptotes of $\theta(t)$. In particular they provide a set of labels for these asymptotes. The Haag-Ruelle scattering theory [20] establishes the validity of this picture for certain states θ, under further hypotheses to be verified below.

From general spectral theory, however, the $|t| \to \infty$ asymptotes are determined by the (generalized) eigenfunctions of P_0. Since P_0 is the energy operator, particles provide a natural set of labels for the energy eigenfunctions. Since P_0 and P_1 commute we can choose the eigenfunctions to be joint eigenfunctions for P_0 and P_1. In this case the associated particles have definite momentum. The in and out particles are associated with two distinct sets of eigenfunctions and provide two distinct (in and out) representations of the spectral resolution of P_0. The in and out particles, eigenfunctions and spectral resolutions are interchanged by the scattering operator, i. e., S matrix.

Let $\sigma(P_0, P_1) \subset R^2$ denote the joint spectrum of the operators P_0, P_1. Positivity of the energy and Lorentz covariance ensure that $\sigma(P_0, P_1)$ is a Lorentz invariant subset of the forward cone,

$$0 \leqq P_0 , \qquad 0 \leqq P_0^2 - P_1^2 .$$

According to the Wightman axioms, the origin of R^2 is a simple eigenvalue of P_0 and P_1, and the corresponding eigenfunction is the vacuum state Ω.

By special relativity, single particles have a mass $(P_0^2 - P_1^2)^{1/2}$. Thus the mass operator, defined to be

$$M = (P_0^2 - P_1^2)^{1/2} ,$$

has an isolated eigenvalue $m > 0$ in the case the field describes a particle with positive mass. States with two or more asymptotic particles would have $M \geqq 2m$. If the in and out asymptotic particle states span $\mathcal{H}$, the theory is said to be asymptotically complete, and S is a unitary mapping of $\mathcal{H}$.

In this paper we establish this qualitative picture,* with the exception of asymptotic completeness, for weak $\mathcal{P}(\phi)_2$ interactions. We state our results in detail in the following section, and reduce the problem to estimates. In Chapters 2–4, we prove these estimates by an inductively defined expansion. We establish convergence of the expansion by introducing a family of norms on the terms in our expansion, and proving that each expansion step is norm decreasing. These techniques are related to expansions used in statistical mechanics.

The spatially cutoff $\mathcal{P}(\phi)_2$ model is defined by the Fock space Hamiltonian

$$H(g, \lambda, m_0) = H_0(m_0) + \lambda \int_{R_1} : \mathcal{P}(\phi_0(\vec{x}, m_0)) : g(\vec{x}) d\vec{x} . \tag{1.6}$$

Fock space $\mathcal{F} = L_2(\mathcal{S}'(R^1), d\phi)$ is defined by the Gaussian measure $d\phi$ with mean zero and covariance $\mu^{-1} = (-d^2/dx_1^2 + m_0^2)^{-1/2}$, see [1]. Also : : denotes Wick ordering with respect to $d\phi$. We choose $\lambda\mathcal{P}(\cdot)$ bounded from below and $0 \leqq g(\cdot) \leqq 1$ with g measurable and of compact support. Then $H(g, \lambda, m_0)$ is essentially self-adjoint and inf spectrum H is a simple eigenvalue of H, with eigenstate Ω_g. We define

$$W(f_1, \cdots, f_n) = \langle \Omega_g, \phi(f_1) \cdots \phi(f_n) \Omega_g \rangle , \tag{1.7}$$

where $f_i \in \mathcal{S}(R^2)$ and where the time translation (1.4) is given by $H(g, \lambda, m_0)$.

We take advantage of the fact that positivity of the energy, $P_0 \geqq 0$, permits an analytic continuation to the region $\operatorname{Im} t \leqq 0$, and consequently a substitution $t \longrightarrow -it$. This substitution converts the original hyperbolic field equation, e.g. $\Box\phi + m^2\phi + \phi^3 = 0$, into an elliptic equation, namely $-\Delta\phi + m^2\phi + \phi^3 = \delta/\delta\phi$, and it converts the Schrödinger equation $i\partial/\partial t = P_0$ into a heat equation $\partial/\partial t = -P_0$. The covariance group also undergoes an analytic continuation, and for the elliptic, or imaginary, time fields, the covariance group is the Euclidean group; hence these fields are called Euclidean fields [22], [15], [18], [13].

* Note added in proof: The presence or absence of bound states (depending on $\mathcal{P}$) is established in [24], [25], [26]. The Bethe Salpeter kernel, which relates to asymptotic completeness, will be studied in forthcoming papers.

The Feynman-Kac formula [22], [15], [2] provides an explicit construction of Euclidean field Φ in terms of the $t=0$ field ϕ_0. In the Feynman-Kac formula, we consider paths which are functions of the (imaginary) time parameter t, and which take values in the space of classical field configurations, namely $\mathcal{S}'(R^1)$. A convenient (and covariant) choice for the space of all such paths is $\mathcal{S}'(R^2)$. The Euclidean field, denoted $\Phi(f)$, $f \in \mathcal{S}(R^2)$, is simply the linear coordinate function

$$\Phi(f)(q) = \langle q, f \rangle \,, \qquad q \in \mathcal{S}'(R^2)$$

defined on the path space $\mathcal{S}'(R^2)$. Thus $\Phi(f)$ measures the position, or strength, of the field in the coordinate direction f. We will consider measures dq and $d\Phi$, defined on the path space $\mathcal{S}'(R^2)$. These measures have sufficient regularity to allow the definition of the sharp time field

$$\Phi(\tau, f) \equiv \Phi(\delta_\tau \otimes f) = \int \Phi(\tau, \vec{x}) f(\vec{x}) d\vec{x}$$

for $f \in \mathcal{S}(R^1)$. The connection between Φ and ϕ_0 is given by the formula

$$\Phi(0, \vec{x}) = \phi_0(\vec{x}) \,.$$

The Euclidean-Fock space is $L_2(\mathcal{S}'(R^2), d\Phi)$, where $d\Phi$ is the Gaussian measure with mean zero and covariance $(-\Delta + m_0^2)^{-1}$. Let h be a function with compact support in R^2, and let $0 \leqq h \leqq 1$. Define the action

$$V(h) = \lambda \int_{R^2} : \mathcal{P}(\Phi(x)) : h(x) dx \tag{1.8}$$

as a function on $\mathcal{S}'(R^2)$, where : : denotes Wick order with respect to the measure $d\Phi$, as in [9], [1]. Let dq_h be the cutoff measure defined by

$$dq_h = \frac{e^{-V(h)} d\Phi}{\int e^{-V(h)} d\Phi} \,.$$

If $h \to 1$ through a sequence of rectangular cutoffs, the limit

$$dq = \lim_{h \to 1} dq_h \,, \tag{1.9}$$

exists at least for λ/m_0^2 small, and defines dq [17].[5] The cutoff Schwinger function

$$S_h(x_1, \cdots, x_n) = \int \Phi(x_1) \cdots \Phi(x_n) dq_h \,, \tag{1.10}$$

is by definition the nth moment of the cutoff measure dq_h. The approximate Schwinger function S_h is well defined as a tempered distribution and, in

[5] The construction [17] of the measure dq assumes Theorem 1.1.7, and consequently the $n=0$ case of Theorem 1.1.11. We do not use dq in the proof of Theorem 1.1.7, but we assume its existence for our general exposition.

fact, integration over the space variable alone is sufficient to define S_h. Let $h_j = g(\vec{x})\chi_j(t)$, where χ_j is the characteristic function of the interval $[-j, j]$, and let $\hat{x} = (ix_0, \vec{x})$. Then the Feynman-Kac formula yields

$$W(\hat{x}_1, \cdots, \hat{x}_n) = \lim_{j\to\infty} S_{h_j}(x_1, \cdots, x_n) . \tag{1.11}$$

Finally, we note that a unitary scaling transformation on $L_2(\mathcal{S}'(R^2), d\Phi)$ carries $\Phi(x, m_0) \to \Phi(sx, s^{-1}m_0)$ and $V(\lambda, m_0) \to V(s^{-2}\lambda, s^{-1}m_0)$. Thus λ/m_0^2 is a dimensionless coupling constant, and a basic hypothesis of the entire paper is that λ/m_0^2 is chosen to be small.

1.1. The main results

The $\mathcal{P}(\phi)_2$ quantum field model is constructed as a limit of space cutoff theories, as the limit in $\mathcal{S}'(R^{2n})$ of the $W(x_1, \cdots, x_n)$ defined in (1.7) as $g\to 1$, or the limit $S_h(x_1, \cdots, x_n)$ defined in (1.10) as $h \to 1$.

THEOREM 1.1.1. *Let λ/m_0^2 be sufficiently small. The $\mathcal{P}(\phi)_2$ quantum field model exists as a limit of space cutoff theories and satisfies the Wightman axioms. Moreover the Wightman functions are continuous in λ $(\lambda \geq 0)$ in $\mathcal{S}'$ for λ sufficiently small.*

Remark. For the fourth order interaction $\mathcal{P}(\xi) = \xi^4 + m^2\xi^2 + \mu\xi$, with $\mu \neq 0$, Griffiths and Simon have eliminated the weak coupling hypothesis of this theorem [12], [21]. Our proof does not depend in an essential way on the boundary conditions while the Griffiths-Simon proof uses Dirichlet conditions.

Let

$$m = \inf \text{spec}\cdot M \restriction \Omega^{\perp} .$$

We call m the physical mass.

THEOREM 1.1.2. *Let λ/m_0^2 be sufficiently small. The physical mass m is positive and m is an isolated discrete eigenvalue for M. Thus M has no spectrum in an interval (m, m_1), $m_1 > m$. As $\lambda \to 0$ with m_0 fixed, $m \to m_0$ and $m_1 \to 2m_0$.*

The eigenspace corresponding to the eigenvalue m of M is the space of one particle states. If $f \in \mathcal{S}(R^2)$ has fourier transform $f^{\sim}$ which satisfies

$$\text{suppt } f^{\sim}(p_0, p_1) \subset \{p: 0 < p_0^2 - p_1^2 < m_1^2\} , \tag{1.1.1}$$

then $\phi(f)\Omega$ is a single particle state, i.e. an eigenstate of M with eigenvalue m.

THEOREM 1.1.3. *Let λ/m_0^2 be sufficiently small. The single particle states are spanned by the states $\phi(f)\Omega$, where f satisfies* (1.1.1).

According to conventional ideas, we expect only discrete spectrum for M in the interval $(m, 2m)$, and only continuous spectrum in the interval

$[2m, \infty]$.[6] The discrete spectrum (if any) in $(m, 2m)$ corresponds to bound states. Each discrete (nonzero) eigenvalue of M has infinite multiplicity, because its eigenspace carries a unitary representation of the Lorentz group. If the unitary representation is irreducible, then the eigenvalue refers to a single elementary particle or bound state.

THEOREM 1.1.4. *Let λ/m_0^2 be sufficiently small. The mass m representation of the Lorentz group on the single particle space is irreducible.*

Again let f satisfy (1.1.1) and define f_t by the equation

$$f_t^{\sim}(p) = \exp\left(it\mu(\vec{p})\right)f^{\sim}(p)$$

where $\mu(\vec{p}) = (\vec{p}^2 + m^2)^{1/2} = (p_1^2 + m^2)^{1/2}$. According to the Haag-Ruelle scattering theory [20] the multiparticle asymptotic in and out states are spanned by the vectors

$$\lim_{t\to\pm\infty} e^{-itP_0}\phi(f_t^{(1)}) \cdots \phi(f_t^{(n)})\Omega \,, \tag{1.1.2}$$

where the $f^{(j)}$ have nonoverlapping supports in momentum space. This restriction on the supports is necessary in two or three space-time dimensions for the proof of convergence of (1.1.2) as $t \to \pm\infty$, see [14].

Let $\mathcal{H}_{\rm in}$ and $\mathcal{H}_{\rm out}$ be the set of in and out states. By the Haag-Ruelle theory, Theorems 1.1.2 and 1.1.3 ensure the existence of an isometric scattering operator S from $\mathcal{H}_{\rm in}$ and $\mathcal{H}_{\rm out}$. A major open question is asymptotic completeness, that is, whether $\mathcal{H}_{\rm in} = \mathcal{H} = \mathcal{H}_{\rm out}$.

We derive Theorems 1.1.2–1.1.4 from an analysis of the bounded portions of the energy spectrum. We now consider test functions $h \in L_2(R^1)$ at the time zero, and let $\phi(h)$ denote the corresponding time-zero fields. Let

$$\phi_r(h_r) = \int :\phi(\vec{x}_1) \cdots \phi(\vec{x}_r):\, h_r(\vec{x}_1, \cdots, \vec{x}_r)d\vec{x} \,,$$

$h_r \in L_2(R^r)$, be a time-zero Wick monomial of degree r. Let E_γ denote spectral projection for the energy region[7] $P_0 \leqq \gamma = (n+1)m_0(1-\varepsilon)$.

THEOREM 1.1.5. *Let n and $\varepsilon > 0$ be given, and let λ/m_0^2 be sufficiently small. Then $E_\gamma\mathcal{H}$ is spanned by vectors of the form*

$$\Omega,\ e^{tP_0}E_\gamma\phi(h_1)\Omega,\ \cdots,\ e^{tP_0}E_\gamma\phi(h_n)\Omega \,. \tag{1.1.3}$$

In the case $n = 1$, we prove a stronger result, by setting $t = 0$.

THEOREM 1.1.6. *Let $\delta > 0$ be given, and let λ/m_0^2 be sufficiently small.*

---

[6] For an even theory (e.g., ϕ^4), the symmetry $\phi \to -\phi$ generates a selection rule (the even and odd subspaces). On the odd subspace, discrete spectrum may be expected in the interval $(m, 3m)$.

[7] The energy bound and decay rate can be expressed intrinsically as $\gamma = (n+1)m(1-\varepsilon')$, since Theorem 1.1.2 ensures $m \to m_0$ as $\lambda/m_0^2 \to 0$.

Then $\mathcal{H}$ is spanned by vectors of the form

$$\Omega, \ \phi(h)\Omega \tag{1.1.4}$$

together with vectors of energy $P_0 \geqq 2m_0(1-\delta)$.

Remark. In Chapter 4 we establish a weaker result than Theorem 1.1.6, where we also include in the spanning states all vectors of total momentum $|P_1| \geqq \sqrt{3}\, m_0(1-\delta)$. This weaker version (Theorem 1.1.6′) is sufficient for the proof of Theorems 1.1.2, 1.1.3, and 1.1.4. Then using Theorem 1.1.2, Theorem 1.1.6 follows as stated.[8]

Proof of Theorems 1.1.2, 1.1.3, and 1.1.4, assuming Theorems 1.1.1, 1.1.5, and 1.1.6′. Let E be the projection onto states with energy $P_0 \leqq 2m_0(1-\varepsilon)$ and momentum $|P_1| \leqq \sqrt{3}\, m_0(1-\varepsilon)$. Let F be the projection onto states spanned by (1.1.4). By Theorem 1.1.6′

$$\mathcal{H} = \text{Range}\,(1-E) + \text{Range}\, F$$

and so Range E is spanned by the vectors

$$\Omega, \ E\phi(h)\Omega\,, \qquad h \in C_0^\infty(R^1)\,.$$

Let $h_a(x) = h(x-a)$. Then

$$e^{-iaP_1}\phi(h)\Omega = \phi(h_a)\Omega$$

and so

$$\phi(h_1 * h_2) = \int h_1(a) e^{-iaP_1}\phi(h_2)\Omega\, da$$

is in the span of the space translates of $\phi(h_2)\Omega$. Since E and P_1 commute, $E\phi(h_1 * h_2)\Omega$ is in the span of the space translates of $E\phi(h_2)\Omega$.

By Theorem 1.1.1 the two point function approaches the free field two point function as $\lambda \to 0$. If we express the truncated two point function in momentum space we see that for small λ the spectrum of P_0 is nonempty in the interval $(0, m_0 + \varepsilon)$ and $E(\Omega^\perp) \neq 0$. We now choose h_2 so that $\tilde{h}_2(\vec{p}) > 0$. With h_2 fixed, the functions $h_1 * h_2$ are dense in $C_0^\infty(R^1)$ so for any $h \in \mathcal{S}(R^1)$, $E\phi(h)\Omega$ lies in the span of translates of $E\phi(h_2)\Omega$. Thus

$$E\phi(h_2)\Omega - \langle\Omega, \phi(h_2)\Omega\rangle\Omega$$

is a cyclic vector for the space translation group acting on $E(\Omega^\perp)$, and so the space translation group acting on this subspace has multiplicity one.

[8] The geometry of the one particle hyperboloid ensures that states with energy $P_0 \leqq 2m_0(1-\varepsilon)$ have momentum $|P_1| \leqq \sqrt{3}\, m_0(1-\varepsilon)$, as follows: By Theorem 1.1.2, $m \geqq m_0(1-\varepsilon)$. For states orthogonal to the vacuum Ω,

$$P_1^2 \leqq P_0^2 - m^2 \leqq 3m_0^2(1-\varepsilon)^2\,.$$

Since $P_1^2 = 0$ for the vacuum, $P_1^2 \leqq 3m_0^2(1-\varepsilon)^2$ in both cases. Thus the extra hypothesis in Theorem 1.1.6′ is not needed, and Theorem 1.1.6 follows.

Therefore the energy P_0 on this subspace is a function of the momentum. By Lorentz invariance, $P_0^2 = P_1^2 + (m')^2$ for some m', and we easily see that $m = m'$. The mass operator acting on $E(\Omega^\perp)$ therefore has a single eigenvalue m in $(0, 2m_0(1 - \varepsilon))$. Since M is Lorentz invariant, M acting on $\mathcal{H}$ has a single eigenvalue in $(0, 2m_0(1 - \varepsilon))$.

By Theorem 1.1.5 with $n = 0$, we see that $m \geqq m_0 - \varepsilon$ hence Theorem 1.1.2 is complete. Theorem 1.1.4 follows from the fact that reducibility produces multiplicity in the representation of the space translation subgroup. To prove Theorem 1.1.3, we observe that $E\phi(h)\Omega = \phi(f)\Omega$ for some $f \in L_2(R^2)$ satisfying (1.1.1). Thus the one particle states in the range of E are spanned by the states $\phi(f)\Omega$, where f satisfies (1.1.1). The theorem follows by Lorentz invariance.

The proof of Theorem 1.1.1 depends on a cluster property (uniform in h) for the measures dq_h or its moments S_h. The cluster property states that at widely separated points x and y, the strengths $\Phi(x)$ and $\Phi(y)$ of the field are independent up to some error $O(\exp[-m|x - y|])$. Two random variables $X(\omega)$ and $Y(\omega)$ are independent when the factorization property

$$\int F(X)G(Y)d\omega = \int F(X)d\omega \int G(X)d\omega$$

holds for functions F and G. The cluster property asserts that this identity holds up to an exponentially small error. The n-particle cluster expansion isolates explicitly n-particle corrections to the factorization; the remaining error $o(\exp[-nm|x - y|])$ is smaller than an n-particle decay. We use such cluster properties to establish Theorems 1.1.5–1.1.6′.

For functions of the Euclidean field in this identity, we consider monomials of the form

$$Q(x_0) = \int d\bar{x}\, w(x) \textstyle\prod_{i=1}^{r} :\Phi(x_i)^{n_i}: \tag{1.1.5}$$

and

$$Q = \int dx_0 Q(x_0)\,. \tag{1.1.6}$$

Q is said to be localized if $w(x)$ is supported in a product $\Delta_{j_1} \times \cdots \times \Delta_{j_n} \subset R^{2r}$ where Δ_{j_i} is the unit lattice square in R^2 centered at the lattice side $j_i \in Z^2$. We assume a bound on the n_i in (1.1.5), $n_i < \bar{n}$.[9] Any choice $\bar{n} \geqq \deg P$ is sufficient. Since there is no restriction on r, Q still may have an arbitrarily large degree.

For a localized monomial Q or $Q(x_0)$, let $N(\Delta)$ be the local degree, i.e. the sum of the n_i for which $\Delta_{j_i} = \Delta$. The estimates in this paper are expres-

[9] We may eliminate this hypothesis, i.e., take $\bar{n} = \infty$, by replacing $N(\Delta)!$ by $(N(\Delta)!)^3$ in (1.1.7), (1.1.8), and (2.4.5).

sed in terms of norms. Define

$$\|Q\| = K_1[\prod_\Delta K_2^{2N(\Delta)} N(\Delta)!]\|w\|_2 , \tag{1.1.7}$$

$$\|Q(x_0)\| = K_1[\prod_\Delta K_2^{2N(\Delta)} N(\Delta)!]\|w(x_0, \cdot)\|_2 . \tag{1.1.8}$$

If Q is not localized, or is a general polynomial, it is written as a sum of localized monomials Q_i, and then we set

$$\|Q\| = \sum_i \|Q_i\| . \tag{1.1.9}$$

For two localized monomials Q_1 and Q_2, with disjoint localization squares $\{\Delta_{ji}^{(1)}\}$ and $\{\Delta_{jl}^{(2)}\}$ respectively, let d be the width of the widest strip bounded by two parallel lines, and separating the $\{\Delta_{ji}^{(1)}\}$ from the $\{\Delta_{jl}^{(2)}\}$

THEOREM 1.1.7. *Let m_0, K_2 and K_1 be sufficiently large and let λ be sufficiently small. There are constants $0 < m_1$ and $O(1)$ independent of h, Q_1 and Q_2 such that*

$$\left|\int Q_1Q_2dq_h - \int Q_1dq_h\int Q_2dq_h\right| \leqq O(1)e^{-m_1d}\|Q_1\|\,\|Q_2\| . \tag{1.1.10}$$

Proof of Theorem 1.1.1 assuming Theorem 1.1.7. In order to establish the Wightman axioms we verify the equivalent Osterwalder-Schrader axioms for the Schwinger functions [18], [23]. By the estimates of this paper, S_h is bounded in $\mathcal{S}'(R^{2n})$ uniformly as $h \to 1$. According to the discussion of [19, Section 6] it thus is sufficient to prove convergence of the cutoff Schwinger functions S_h. Convergence as $h \to 1$ of the real time vacuum expectation values (1.1.7) follows: The uniform bound on $W(x_1, \cdots, x_n)$ in $\mathcal{S}'(R^{2n})$ yields a uniform bound on their analytic continuations to Euclidean points. Thus convergence of the Schwinger functions and Vitali's theorem yield real-time convergence. The convergence (uniform in λ) of the Schwinger functions in λ ensures that the Schwinger functions are continuous in λ. Continuity of the Wightman functions follows from the uniform bound [6] on their analytic continuation and Vitali's theorem.

By the uniform boundedness of S_h in $\mathcal{S}'(R^{2n})$, we may prove convergence on a dense set of test functions. For test functions $f_i \in C_0^\infty(\mathcal{O})$ and for cutoff functions h_1 and h_2, we assert that

$$\left|\int \prod_i \Phi(f_i)dq_{h_1} - \int \prod_i \Phi(f_i)dq_{h_2}\right| \leqq 0(1)e^{-m_1d/2} \tag{1.1.11}$$

where $d = \text{dist}\,(\mathcal{O}, \text{suppt}\, h_1 - h_2)$, and then Theorem 1.1.1 follows. To prove (1.1.11) let $g = h_2 - h_1$, and define $g_i = g\chi_{\Delta_i}$, where χ_{Δ_i} is the characteristic function of Δ_i. Also we use the notation

$$\langle F\rangle_h = \int F dq_h ,$$

and set $Q = \prod_i \Phi(f_i)$. Then

$$\langle Q\rangle_{h_2} - \langle Q\rangle_{h_1} = \int_0^1 d\alpha \frac{d}{d\alpha} \langle Q\rangle_{h_1+\alpha g}$$

$$(1.1.12) \qquad = -\int_0^1 d\alpha [\langle Q V(g)\rangle_{h_1+\alpha g} - \langle Q\rangle_{h_1+\alpha g}\langle V(g)\rangle_{h_1+\alpha g}]$$

$$= -\int_0^1 d\alpha \sum_{i\in Z^2} [\langle Q V(g_i)\rangle_{h_1+\alpha g} - \langle Q\rangle_{h_1+\alpha g}\langle V(g_i)\rangle_{h_1+\alpha g}] \,.$$

We apply Theorem 1.1.7 to each term in the sum to obtain

$$|\langle Q\rangle_{h_2} - \langle Q\rangle_{h_1}| \leqq \sum_{i\in Z^2\cap \operatorname{supp} g} O(1)e^{-m_1 \operatorname{dist}(\mathcal{O},\Delta_i)}$$
$$\leqq O(1)e^{-m_1 d/2} \,.$$

As a special case, the cutoff Schwinger functions S_h converge as $h \to 1$.

Remark. The $h = 1$ fields can also be constructed from the Wightman reconstruction theorem starting from the Wightman functions as well as from the GNS construction using C^* algebra states. These latter constructions coincide with each other [6], [5] and with the Osterwalder-Schrader construction employed above.

We now reformulate Theorems 1.1.5 and 1.1.6 in terms of path space. In the monomial $Q(x_0)$ of (1.1.5), suppose $x_0 = (x_{1,0}, \cdots, x_{r,0})$ where $x_{1,0} \leqq \cdots \leqq x_{r,0}$. Let

$$\eta_l = x_{l+1,0} - x_{l,0} \geqq 0$$

and let

$$(1.1.13) \qquad \theta = \int d\vec{x}w(x){:}\,\phi_0(\vec{x}_{1,1})^{n_1}{:}\, e^{-\eta_1 P_0} \cdots e^{-\eta_{r-1}P_0}{:}\,\phi_0(\vec{x}_{r,1})^{n_r}{:}\,\Omega \,.$$

One version of the Feyman-Kac formula is the identity

$$\int Q(x_0)dq = \langle \Omega, \theta\rangle$$

where the inner product on the right side is in $\mathcal{H}$.

THEOREM 1.1.8. *For $Q(x_0)$ of the form* (1.1.5),

$$\left|\int Q(x_0)\,dq_h\right| \leqq O(1)\,||Q(x_0)||\,,$$

uniformly as $h \to 1$.

Remark. A special case of the theorem is the bound on the n-point Schwinger functions

$$|S_h(f_1 \otimes \cdots \otimes f_n)| \leqq n! \prod |f_i|_{\mathcal{S}}$$

where $|\cdot|_{\mathcal{S}}$ is a Schwartz space norm (in this case an L_2 norm) independent of h. This bound verifies the Osterwalder-Schrader axiom EO'' of [23] for weakly coupled $\mathcal{P}(\phi)_2$ models.

COROLLARY 1.1.9. *Let Q and θ be related as in* (1.1.13). *Then*

$$\|\theta\| \leqq O(1)\|Q(x_0)\| = O(1)\left[\prod_\Delta K_2^{2N(\Delta)} N(\Delta)!\right]\|w(x_0, \cdot)\| .$$

Proof. We write $\|\cdot\|_{K_2}$ to indicate the dependence of the norm on K_2. Apply the theorem to

$$Q_1 = Q(x_{i0} - x_{10})Q(x_{10} - x_{i0})^- ,$$

where Q has the form (1.1.5), to obtain

$$\|Q(x_0)\|^2 = \int Q_1 dq \leqq O(1)\|Q_1\|_{K_2} \leqq O(1)\|Q(x_0)\|_{2K_2}^2 .$$

We let A and B denote monomials of type (1.1.5) of degree $|A|$ and $|B|$ and with kernels w_A and w_B respectively. We assume that A is localized at times $x_{i0} \leqq 0$, and that B is localized at times $x_{i0} \geqq 0$. Let B_T denote the translate of B at time $T > 0$, so that B_T is localized at times $\geqq T$.

In the next two results, we take all times $x_{1,0} = t = \cdots = x_{r,0}$ equal in $Q(x_0)$. To indicate this restriction, we write $Q(t)$ in place of $Q(x_0)$.

THEOREM 1.1.10. *Let $\varepsilon > 0$, $T > 0$ be given, and let B have support in momentum space in the region $|P_1| \leqq \sqrt{3}\, m_0(1 - \varepsilon)$. Then there is a*

$$Q(T) = \alpha(T) + \int \Phi(0, \vec{x}) h_T(\vec{x}) d\vec{x} ,$$

chosen independently of B, such that

$$\alpha(T) \in R^1 , \quad h_T \in L_2(R^1) , \quad \|Q(T)\| \leqq O(1)\|A\| ,$$

and

$$\left|\int (A - Q(T))B_T dq\right| \leqq O(1)\|A\|\,\|B\| e^{-2Tm_0(1-\varepsilon)} .$$

THEOREM 1.1.11. *Let n and $\varepsilon > 0$ be given and let $\gamma = (n + 1)m_0(1 - \varepsilon)$. For any A, we may choose n^{th} degree polynomials $Q_n(t_j)$ of the form* (1.1.5) *with $r = 1$, so that*

$$\sum_{t_j > 0} e^{\gamma t_j}\|Q_n(t_j)\| \leqq O(1)\|A\| \tag{1.1.14}$$

and so that

$$Q = \sum_{0 < t_j < T} Q_n(t_j) \tag{1.1.15}$$

satisfies

$$\left|\int (A - Q)B_T dq_h\right| \leqq O(1)\|A\|\,\|B\| e^{-\gamma T} , \tag{1.1.16}$$

uniformly in h. We shall choose $t_j = j$, $j \in Z$.

Proof of Theorem 1.1.7, assuming Theorems 1.1.8 and 1.1.11. Take

$n = 0$ in Theorem 1.1.11 Then Q is a polynomial of degree zero, i.e., a constant. From the choice $B = I$, it follows that

$$Q = \int A dq_h + O(1) \| A \| e^{-m_0(1-\varepsilon)T} .$$

After rotating and translating the strip separating Q_1 and Q_2 in Theorem 1.1.7, and after choosing $A = Q_1$, $B_T = Q_2$ in Theorem 1.1.11, the proof follows from Theorem 1.1.8, (1.1.16) and the above determination of Q.

Proof of Theorem 1.1.5, assuming Theorem 1.1.11, Corollary 1.1.9. We choose

$$\beta \equiv (n+1)m_0(1-\delta) < \gamma \equiv (n+1)m_0(1-\varepsilon) .$$

Let $\theta(t_j)$ be the vector (1.1.13) corresponding to $Q_n(t_j)^-$ and let

$$\psi(t_j) = e^{t_j P_0} E_\gamma \theta(t_j)$$
$$\psi = \sum_j \psi(t_j) .$$

Since

$$\begin{aligned} \sum_j \| \psi(t_j) \| &\leqq \sum_j e^{t_j \gamma} \| \theta(t_j) \| \\ &\leqq \sum_j e^{\gamma t_j} \| Q_n(t_j) \| \leqq O(1) \| A \| , \end{aligned}$$

the series defining ψ converges, and $\| \psi \| \leqq O(1) \| A \|$. To prove Theorem 1.1.5 it suffices to show that

$$E_\beta(\theta_A - \psi) = 0$$

where θ_A is the vector (1.1.13) corresponding to A. Let

$$\psi_T = \sum_{0 < t_j < T} \psi(t_j) ,$$

and then $\psi_T \to \psi$. Moreover, if $T_1 \leqq T_2 \leqq \infty$,

$$\begin{aligned} \| e^{-T_1 P_0}(\psi_{T_1} - \psi_{T_2}) \| &\leqq \sum_{T_1 \leqq t_j < T_2} \| e^{(t_j - T_1)P_0} E_\gamma \theta(t_j) \| \\ &\leqq \sum_{T_1 \leqq t_j < T_2} e^{(t_j - T_1)\gamma} \| \theta(t_j) \| \\ &\leqq O(1) e^{-\gamma T_1} \| A \| . \end{aligned} \tag{1.1.17}$$

Here we used Corollary 1.1.9 and (1.1.14). Consequently,

$$\begin{aligned} \| E_\beta(\theta_A - \psi) \|^2 &\leqq e^{\beta T} \langle \theta_A - \psi, E_\beta e^{-TP_0}(\theta_A - \psi) \rangle \\ &\leqq e^{\beta T} \langle \theta_A - \psi, e^{-TP_0}(\theta_A - \psi) \rangle \\ &\leqq O(1) \| A \|^2 e^{-(\gamma - \beta)T} + e^{\beta T} | \langle \theta_A - \psi_T, e^{-TP_0} \theta_A \rangle | \\ &\quad + e^{\beta T} | \langle \theta_A - \psi_T, e^{-TP_0} \psi_T \rangle | . \end{aligned} \tag{1.1.18}$$

We bound the second term above as follows:

$$\begin{aligned} \langle \theta_A - \psi_T, e^{-TP_0} \theta_A \rangle &= \int (A - Q)^- A_T dq \\ &\quad + \sum_{0 < t_j < T} \langle \theta(t_j), (E_\gamma - I) e^{-(T - t_j)P_0} \theta_A \rangle . \end{aligned} \tag{1.1.19}$$

Hence by (1.1.14)-(1.1.16) and $\|(E_\gamma - I)e^{-(T-t_j)P_0}\| \leqq e^{-(T-t_j)\gamma}$,

$$
\begin{aligned}
|\langle\theta_A - \psi_T, e^{-TP_0}\theta_A\rangle| &\leqq O(1)\|A\|^2 e^{-\gamma T} + \|A\| \textstyle\sum_{0<t_j<T} e^{-(T-t_j)\gamma}\|\theta(t_j)\| \\
&\leqq O(1)\|A\|^2 e^{-\gamma T} .
\end{aligned} \tag{1.1.20}
$$

Similarly, we bound the third term in (1.1.18):

$$
\begin{aligned}
\langle\theta_A - \psi_T, e^{-TP_0}\psi_T\rangle = &\textstyle\sum_{0<t_j<T} \langle\theta_A - \psi_{T-t_j}, e^{-(T-t_j)P_0}\theta(t_j)\rangle \\
&- \textstyle\sum_{0<t_j<T} \langle\psi_T - \psi_{T-t_j}, e^{-(T-t_j)P_0}E_\gamma\theta(t_j)\rangle \\
&+ \textstyle\sum_{0<t_j<T} \langle\theta_A - \psi_{T-t_j}, e^{-(T-t_j)P_0}(E_\gamma - I)\theta(t_j)\rangle.
\end{aligned} \tag{1.1.21}
$$

We bound the first term in (1.1.21) using (1.1.19) with $T - t_j$ in place of T and $\theta(t_j)$ in place of θ_A. In place of (1.1.20), we obtain

$$
\begin{aligned}
&|\textstyle\sum_{0<t_j<T} \langle\theta_A - \psi_{T-t_j}, e^{-(T-t_j)P_0}\theta(t_j)\rangle| \\
&\leqq \textstyle\sum_{0<t_j<T} O(1)\|A\|\,\|Q(t_j)\|e^{-(T-t_j)\gamma} \leqq O(1)\|A\|^2 e^{-\gamma T} .
\end{aligned} \tag{1.1.22}
$$

For the second term in (1.1.21), we use (1.1.17) with $T_1 = T - t_j$, $T_2 = T$, obtaining the bound $O(1)\|A\|^2 e^{-\gamma T}$. For the third term in (1.1.21), we use

$$
\begin{aligned}
&|\textstyle\sum_{0<t_j<T} \langle\theta_A - \psi_{T_j}, e^{-(T-t_j)P_0}(E_\gamma - I)\theta(t_j)\rangle| \\
&\leqq \textstyle\sum_{0<t_j<T} \{\|\theta_A\| + \|\psi_{T-t_j}\|\}e^{-(T-t_j)\gamma}\|\theta(t_j)\| \\
&\leqq O(1)\|A\| \textstyle\sum_{0<t_j<T} e^{-(T-t_j)\gamma}\|Q(t_j)\| \\
&\leqq O(1)\|A\|^2 e^{-T\gamma} .
\end{aligned}
$$

Combining these bounds,

$$
|\langle\theta_A - \psi_T, e^{-TP_0}\psi_T\rangle| \leqq O(1)\|A\|^2 e^{-\gamma T} .
$$

Thus using (1.1.20), we obtain for (1.1.18) the bound

$$
\|E_\beta(\theta_A - \psi)\|^2 \leqq O(1)\|A\|^2 e^{-(\gamma-\beta)T} .
$$

Thus $E_\beta(\theta_A - \psi) = 0$. Since the states θ_A are dense in $\mathcal{H}$, the proof is complete.

Proof of Theorem 1.1.6′, assuming theorem 1.1.10. We use Theorem 1.1.10 in place of Theorem 1.1.11. The only new complication is the restriction on the momentum support of B. Let A be supported in the region

$$
|P_1| \leqq \sqrt{3}\, m_0(1 - \delta) . \tag{1.1.23}
$$

The proof consists of approximating $E_\beta\theta_A$ by $E_\beta\theta_Q$ where Q is a linear function of the time-zero field. We choose a function $\zeta \in \mathcal{S}(R^1)$ with Fourier transform $\zeta^\sim(P_1)$ supported in the region

$$
|P_1| \leqq \sqrt{3}\, m_0(1 - \varepsilon) , \tag{1.1.24}
$$

and equal to one on (1.1.23). Let $Q(T)$, α_T, h_T be chosen by Theorem 1.1.10 and let

$$Q(T, \zeta) = \alpha_T + \int \Phi(0, \vec{x})(h_T * \zeta)(\vec{x}) d\vec{x} .$$

By Theorem 1.1.10

$$\left| \int (A - Q(T) Q(T, \zeta)_{\bar{T}}) dq \right| \leqq O(1) \| A \|^2 e^{-T\gamma} , \tag{1.1.25}$$

where $\gamma = 2m_0(1 - \varepsilon)$. We introduce the vectors $\theta_T = \theta_{Q(T)}$ and $\theta_{T,\zeta} = \theta_{Q(T,\zeta)}$. Then (1.1.25) can be written

$$| \langle \theta_A - \theta_T, e^{-TP_0} \theta_{T,\zeta} \rangle | \leqq O(1) \| A \|^2 e^{-T\gamma} . \tag{1.1.26}$$

Furthermore

$$Q(T) - Q(T, \zeta) = \Phi(0, h_T * (\delta - \zeta)) ,$$

is supported in momentum space on the complement of (1.1.23). For this reason, and because $P_1^2 \leqq P_0^2$, we see that $\theta_T - \theta_{T,\zeta}$ is a state whose energy spectrum satisfies the inequality

$$\nu \equiv \sqrt{3}\, m_0(1 - \delta) \leqq P_0 . \tag{1.1.27}$$

Then

$$\begin{aligned} &\langle \theta_A - \theta_T, e^{-TP_0}(\theta_A - \theta_T) \rangle \\ &\leqq O(1) \| A \|^2 e^{-\gamma T} + | \langle \theta_A - \theta_T, e^{-TP_0}(\theta_T - \theta_{T,\zeta}) \rangle | \\ &\quad + | \langle \theta_A - \theta_T, e^{-TP_0} \theta_{T,\zeta} \rangle | . \end{aligned} \tag{1.1.28}$$

Here the first term results from Theorem 1.1.10 with $B = A$. The third term is bounded by (1.1.26). For the second term, we use (1.1.27) and Theorem 1.1.8 to obtain the bound $O(1) \| A \|^2 e^{-\nu T}$. Thus for $\nu' < \nu$,

$$\begin{aligned} \| E_{\nu'}(\theta_A - \theta_T) \|^2 &\leqq e^{\nu' T} \| E_{\nu'} e^{-TP_0/2}(\theta_A - \theta_T) \|^2 \\ &\leqq e^{\nu' T} \| e^{-TP_0/2}(\theta_A - \theta_T) \|^2 \\ &\leqq O(1) \| A \|^2 e^{-(\nu - \nu')T} \longrightarrow 0 \end{aligned} \tag{1.1.29}$$

as $T \to \infty$. This proves Theorem 1.1.6′ with $\beta = 2m_0(1 - \delta)$ replaced by ν, and thus it proves Theorem 1.1.2 with $m_1 \to 2m_0$ replaced by $m_1 \geqq \nu$. However, from Theorem 1.1.2 we still conclude that $m \to m_0$ as $\lambda \to 0$, and so we may suppose that $m > m_0(1 - \varepsilon)$. With this extra information, we can improve on (1.1.27) as follows:

$$\| e^{-TP_0}(\theta_T - \theta_{T,\zeta}) \| \leqq O(1) \| A \|^2 e^{-T\beta'} ,$$

where

$$\beta' = m_0((1 - \varepsilon)^2 + 3(1 - \delta)^2)^{1/2} > \beta .$$

Now we choose $\nu' = \beta$ in (1.1.29) and obtain

$$\begin{aligned} \| E_\beta(\theta_A - \theta_T) \|^2 &\leqq e^{\beta T} \langle \theta_A - \theta_T, E_\beta e^{-TP_0}(\theta_A - \theta_T) \rangle \\ &\leqq O(1) \| A \|^2 e^{-(\gamma - \beta)T} \longrightarrow 0 . \end{aligned}$$

This completes the proof.

Proof of Theorem 1.1.8, assuming Theorem 1.1.10. Choose $T = 0$ and $B = I$ in Theorem 1.1.11. Then $Q = 0$, so Theorem 1.1.8 is a special case of Theorem 1.1.10.

We are now reduced to proving Theorems 1.1.10 and 1.1.11. These results are obtained from the cluster expansion, to which the remainder of the paper is devoted.

2. Inserting contours

2.0 Introduction

The proofs of Theorems 1.1.10 and 1.1.11 are based on an inductively defined expansion for $\int AB_T dq_h$, described in this and subsequent chapters. We establish the convergence of this expansion by introducing a norm on the set of terms in each expansion step and by showing that each expansion step is norm decreasing. In our expansion, we use a space-time cover by squares of unit area. Thus our expansions involve a characteristic length, and convergence of our expansions requires both sufficiently small λ and sufficiently large m_0. Our Theorems 1.1.1–1.1.6, the consequences of our expansions, involve no such length. Hence, by scaling, they apply whenever the dimensionless coupling λ/m_0^2 is sufficiently small.

In order to define the expansion we introduce measures on $\mathcal{S}'(R^2)$ which factor over disjoint regions of space-time. If C is a continuous positive definite bilinear form on $\mathcal{S}(R^2)$, let $d\Phi_C$ denote the Gaussian measure on $\mathcal{S}'(R^2)$ of mean zero and covariance C. The family of covariances $\mathcal{C}$ that we consider are constructed from the free covariance $C_0 = (-\Delta + m_0^2)^{-1}$ by the application of two elementary operations: Diagonalization and convex sums. Let $Y \subset R^2$ be a union of unit lattice squares Δ, centered at integer coordinates, and let Y also be the operator of multiplication by the characteristic function of the set Y. If C is a bounded operator on $L_2(R^2)$ we define its diagonalization with respect to Y to be

$$C_Y = YCY + (1 - Y)C(1 - Y) .$$

We say ∂Y is a complete contour in C_Y. The convex sum of two covariance operators C' and C'' is

$$C(s) = sC' + (1 - s)C'' , \qquad s \in [0, 1] . \tag{2.0.1}$$

If $s \in (0, 1)$ and C' or C'' has complete (or partial) contour ∂Y we say $C(s)$ has ∂Y as a partial contour.

If ∂Y is a complete contour in Y, the measure $d\Phi_C$ factors across ∂Y. This means that if polynomials A and B are localized in Y and $\sim Y$ respectively (i. e., their localizations are separated by ∂Y), then

$$\int ABe^{-V(h)}d\Phi_C = \int Ae^{-V(Yh)}d\Phi_C \cdot \int Be^{-V((1-Y)h)}d\Phi_C$$

$$= \frac{\int Ae^{-V(h)}d\Phi_C \cdot \int Be^{-V(h)}d\Phi_C}{\int e^{-V(h)}d\Phi_C} . \tag{2.0.2}$$

In other words, the Gaussian measure $d\Phi_C$ factors across each of its complete contours, for $C \in \mathcal{C}$. Also the measure

$$dq = dq_C = \frac{e^{-V}d\Phi_C}{\int e^{-V}d\Phi_C}$$

factors across each complete contour, and

$$\int ABdq_C = \int Adq_C \int Bdq_C .$$

We say that (2.0.2) "factors out" the polynomial (or vertices) A.

In this chapter we express $\int AB_T e^{-V}d\Phi$ as a sum of terms labeled by indices α or α',

$$\sum_\alpha b_\alpha \int \left(\prod_{i=1}^{n_\alpha} a_i\right) B_T e^{-V} d\Phi_{C_\alpha} + \sum_{\alpha'} b_{\alpha'} \int e^{-V} d\Phi_{C_{\alpha'}} . \tag{2.0.3}$$

Here C_α belongs to $\mathcal{C}$, b_α and $b_{\alpha'}$ are constants, the a_i are annihilation operators, to be defined later, and $n_\alpha \leqq n$. We call this part of our expansion *the increasing contour expansion*. We generate the expansion with the elementary identity

$$\int Fd\Phi_{C'} = \int Fd\Phi_{C''} + \int_0^1 ds \frac{d}{ds} \int Fd\Phi_{C(s)} , \tag{2.0.4}$$

where $C(s) = sC' + (1-s)C''$, as in (2.0.1). If C'' has ∂Y as a complete contour, we say that (2.0.4) inserts the complete contour ∂Y. If $C'' = C_0$, we say that (2.0.4) removes all contours. For suitably regular functions F on $\mathcal{S}'(R^2)$, integration by parts yields (see [1])

$$\frac{d}{ds} \int Fd\Phi_{C(s)} = \frac{1}{2} \int_0^1 ds \int (\delta C \cdot \Delta_\Phi) Fd\Phi_{C(s)} . \tag{2.0.5}$$

Here $\delta C = dC(s)/ds = C' - C''$,

$$(\delta C \cdot \Delta_\Phi)F = \int \delta C(x, y) \frac{\partial^2 F}{\partial\Phi(x)\partial\Phi(y)} dxdy , \tag{2.0.6}$$

and $\delta C(x, y)$ is the integral kernel of the operator δC.

Three general types of integrands F occur in our expansion. For the first class of terms,

$$F = R(\Phi)B_T e^{-V(h)}, \tag{2.0.7}$$

where $R(\Phi)$ is a polynomial in Φ localized at times earlier than T. Such polynomials $R(\Phi)$ arise, for example, from derivatives $(\delta C \cdot \Delta_\Phi)Ae^{-V}$. In these terms we factor out $R(\Phi)$ by inserting a contour that separates the localization of $R(\Phi)$ from the localization of B_T. The term in (2.0.4) with the complete contour contributes to the first sum in (2.0.3), while the second term in (2.0.4) is expanded further. The new contour is chosen to enclose previous contours, so the contours are "increasing".

The second type of integrand is

$$F = R(\Phi)e^{-V(h)},$$

where $R(\Phi)$ is a polynomial whose localizations include times greater than T. In this case, factoring out $R(\Phi)$, yields a contribution to the second sum in (2.0.3), plus a term to be expanded further. The third class of F has annihilators; we deal with these terms in a slightly more complicated way. We obtain (2.0.3) by an iteration of these steps, described in detail in Section 2.3.

Each application of (2.0.6) will yield a small factor in our estimates, either $O(m_0^{-1})$ or $O(\lambda)$, and this is the key mechanism yielding convergence of our expansion.

In Chapter 3 we give an expansion of the form (2.0.3) where now $C_\alpha = C_{\alpha'} = C_0$. We obtain this expansion by alternating the increasing contour expansion and the formula (2.0.4) for removing contours.

Finally, in Chapter 4, we define the full n-particle cluster expansion. For the full expansion, we apply the Chapter 3 expansion and we use an identity for the Euclidean field, $\Phi = a + a^*$. Integration by parts yields

$$a = \Phi + K\frac{\partial}{\partial\Phi} \tag{2.0.8}$$

where K is an integral operator defined in Chapter 4. Hence we obtain in (2.0.3),

$$\int\left(\textstyle\prod_{i=1}^{n_\alpha} a_i\right)B_T e^{-V}\,d\Phi = \int\left[\textstyle\prod_{i=1}^{n_\alpha}(\Phi + K\partial/\partial\Phi)\right]B_T e^{-V}\,d\Phi\,.$$

The pure Φ term has the desired form and contributes to the polynomial Q in Theorem 1.1.11. Terms involving $K\partial/\partial\Phi$ yield new vertices, and we apply the Chapter 3 expansion to these terms. Iterating this procedure, we obtain the desired cluster expansion. We summarize the structure of the expansion in the following block diagram:

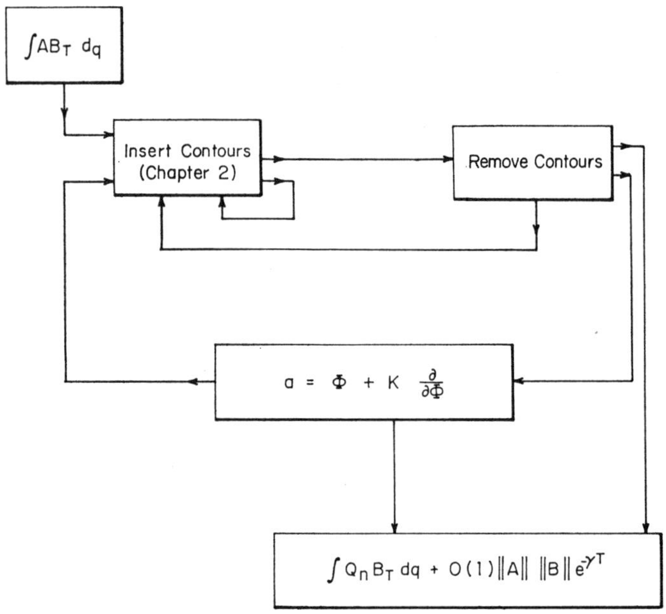

FIGURE 1. *The n-particle cluster expansion*, an example of an "inductive construction". The Chapter 3 expansion includes the two blocks for inserting contours and for removing contours. In Chapter 4, all blocks are combined.

2.1 The covariance operators

In this section we introduce notation for the covariance operators of class $\mathcal{C}$, and we prove estimates on the behavior of the kernels δC which appear in (2.0.5). We consider two subsets $\mathcal{C}_1$, $\mathcal{C}_2$ of $\mathcal{C}$. The first subset, $\mathcal{C}_1$, occurs in the increasing contour expansion of this chapter. Given an element $C_* \in \mathcal{C}_2$, and finite sequences

$$\mathcal{Y}_j = \{Y_1, \cdots, Y_j\}\,, \qquad s = \{s_1, \cdots, s_j\}\,, \tag{2.1.1}$$

of expanding contours ∂Y_i, $Y_i \subseteqq Y_{i+1}$, and parameters $s_i \in (0, 1)$, we define $C_*(\mathcal{Y}_j, s) = C_j$ by recursion:

$$C_1 = s_1 C_* + (1 - s_1)(C_*)_{Y_1}\,,$$

and

$$C_i = s_i C_{i-1} + (1 - s_i)(C_{i-1})_{Y_i}\,, \qquad i \geqq 2\,. \tag{2.1.2}$$

Let $Y_0 = \emptyset$, and let $\delta Y_k = Y_k \sim Y_{k-1}$ be the increment added by Y_k to Y_{k-1}. Clearly $C_i(x, y) = C_{i-1}(x, y)$, unless x and y are separated by ∂Y_i. Let $y \notin Y_j$

and $x \in Y_j$. By induction we see for $x \in \delta Y_k$,

$$C_j(x, y) = S_{k,j} C_*(x, y) = \left(\prod_{i=k}^{j} s_i\right) C_*(x, y) . \tag{2.1.3}$$

Here

$$S_{k\,j} = \prod_{i=k}^{j} s_i . \tag{2.1.4}$$

A factor s_i enters the product $S_{k,j}$ for each contour ∂Y_i among $\partial Y_1, \cdots, \partial Y_j$ which separate x and y.

The increment $\delta C_j = (d/ds_j)C_j$ vanishes unless x and y are separated by δY_j. Let

$$\delta S_1 \equiv 1; \quad \delta S_{k,j} \equiv \frac{d}{ds_j} S_{k.j} = \prod_{i=k}^{j-1} s_i = S_{k.j-1} , \tag{2.1.5}$$

for $j > 1$. If $x \in Y_j \cap \delta Y_k$, $y \notin Y_j$, then

$$\delta C_j(x, y) = \delta S_{k.j} C_*(x, y) .$$

When possible, we suppress j in $S_{k,j} = S_k$. We say that C_* is a background covariance with contours Y_j superimposed.

PROPOSITION 2.1.1. *Let Δ and Δ' be squares of unit area or intervals of unit length in R^2. Let $d =$ dist.(Δ, Δ'), $\varepsilon > 0$, $p < \infty$ and let $C_j \in \mathcal{C}_1$ have the form above, with j contours ∂Y_i. Then*

$$\sum_{\Delta,\Delta'} \|\delta C_j\|_{L_p(\Delta\times\Delta')} e^{m_0(1-\varepsilon)d} \leqq O(1) \sum_{k=1}^{j} \delta S_{k,j} |\delta Y_k| , \tag{2.1.6}$$

and for $d \geqq 1$, $p = \infty$ is allowed in (2.1.6).

Proof. We note that $|C_*(x, y)| \leqq C_0(x, y)$ so we may bound C_* by C_0 in (2.1.2). Also note C_0 is locally L_p, and for $|x - y|$ bounded away from zero, $C_0(x - y) \leqq O(1) \exp(-m_0(1 - \varepsilon/2)|x - y|)$. Since δC_j vanishes unless δY_j separates x and y, we can choose $\Delta \in \delta Y_k$, $\Delta' \notin Y_j$. We fix Δ and sum over Δ'. By (2.1.6) and the exponential bound on C_0,

$$\sum_{\Delta'} \|\delta C_j\|_p e^{m_0(1-\varepsilon)d} \leqq O(1) \delta S_{k,j} .$$

Summing Δ over δY_k gives the area $|\delta Y_k|$, and summing k yields (2.1.6).

To complete our characterization of $\mathcal{C}_1$, we describe the subset $\mathcal{C}_2$ of $\mathcal{C}$ which contains C_*. The class $\mathcal{C}_2$ is generated by alternately diagonalizing the covariance C, and forming a convex linear combination of C with C_0. The first step is to diagonalize C_0. Thus given

$$\mathcal{X}_l = \{X_1, \cdots, X_l\}; \qquad u = \{u_1, \cdots, u_l\}$$

with $u_i \in (0, 1)$, $u_l \in (0, 1]$, we define

$$C_j = (1 - u_j)C_0 + u_j(C_{j-1})_{X_j} , \tag{2.1.7}$$

and $C(\mathcal{X}_l, u) = C_l$. If $u_l = 1$, we say that C_l has a complete contour X_l.

The family of sets $X_1, \cdots, X_l$ is not monotone in general, in contrast to the Y_i's in (2.1.6). Also $C_*(\mathcal{X}_j, s) \equiv C(\mathcal{X}_l, \mathcal{Y}_j, u, s) \equiv C(u, s)$.

The increment

$$\delta C_j = \delta C_j(x, y) = \frac{d}{du_j} C_j = (C_{j-1})_{X_j} - C_0$$

equals $- C_0$ if ∂X_j separates x and y. We define

$$U_{j,l} = U_j = \prod_{i=j}^{l} u_i; \quad \delta U_{j,l} = \prod_{i=j}^{l-1} u_i = U_{j,l-1}; \quad \delta U_{l,l} = 1 . \tag{2.1.8}$$

Then we determine the form for δC_l by induction: For $C_l = C(\mathcal{X}_l, u) \in \mathcal{C}_2$,

$$\delta C_l(x, y) = - \delta U_{j,l} C_0(x, y) \tag{2.1.9}$$

when x and y are separated by ∂X_j, but are not separated by ∂X_r for any $r > j$. Also $\delta C_l = 0$ if no contour separates x and y.

PROPOSITION 2.1.2. *Let $p < \infty$, $\varepsilon > 0$, $d = \mathrm{dist}(\Delta, \Delta')$ and let $C_l \in \mathcal{C}_2$ have the form above with l contours ∂X_i. Then*

$$\sum_{\Delta,\Delta'} \|\delta C_l\|_{L_p(\Delta\times\Delta')} e^{m_0(1-\varepsilon)d} \leqq O(1) \sum_{j=1}^{l} \delta U_{j,l} |X_j| . \tag{2.1.10}$$

Proof. We modify the proof of Proposition 2.1.1 as follows: Fix Δ inside X_j and Δ' outside X_j. We sum over all Δ, Δ' that are not separated by ∂X_r, for any $r > j$. Denoting this restricted sum $\sum^{(j)}$, we obtain as before,

$$\sum_{\Delta,\Delta'}^{(j)} \|\delta C_l\|_p e^{m_0(1-\varepsilon)d} \leqq O(1) \delta U_{j,l} |X_j| .$$

Since δC_l vanishes unless some contour separates Δ and Δ', the sums $\sum^{(l)}$, $\sum^{(l-1)}, \cdots, \sum^{(1)}$ exhaust all $R^2 \times R^2$ for which δC_l does not vanish.

2.2 Change of covariance

In this section we formulate the basic identity

$$\int F d\Phi_{C'} - \int F d\Phi_{C''} = \frac{1}{2}\int_0^1 ds \int (\delta C \cdot \Delta_\Phi) F d\Phi_{C(s)} , \tag{2.2.1}$$

where $\delta C = (d/ds)C(s) = C' - C''$, in terms of annihilation operators. We also specify the general form of the functions F that arise in the expansion.

For each $C \in \mathcal{C}_1$, let t_C be the largest time of any contour in C, i.e.,

$$t_C = \sup\{x_0 \,|\, x \in \partial Y_j \text{ or } x \in \partial X_j\} ,$$

where ∂Y_i and ∂X_j are contours of $C = C(\mathcal{Y}, \mathcal{X})$. Let

$$\mu = \mu_{\vec{x}} = \left(- \frac{d^2}{d\vec{x}^2} + m_0^2 \right)^{1/2} , \tag{2.2.2}$$

and let $\mu(\vec{p}) = (\vec{p}^2 + m_0^2)^{1/2}$. For an operator A on L_2, let $A(x, y)$ denote its integral kernel.

LEMMA 2.2.1. *For each* z_0 *such that* $t_C \leqq z_0 \leqq y_0$ *and* $x_0 \leqq z_0$,

$$\delta C(x, y) = \int \delta C(x, z)[e^{-(y_0 - z_0)\mu}](\vec{z}, \vec{y}) d\vec{z} . \tag{2.2.3}$$

Proof. By the definition of $\mathcal{C}_1$ in Section 2.1, if $t_C \leqq z_0 \leqq y_0$, then $\delta C(x, z)$ equals $C_0(x, z)$ times a function that is independent of z. Thus it suffices to establish (2.2.3) with δC replaced by C_0. For real p,

$$(2\mu(p_1))^{-1}\int_{-\infty}^{\infty} e^{ip_0 t - \mu(p_1)|t|} dt = \mu(p_1)^{-1}\mathrm{Re}(ip_0 + \mu(p_1))^{-1} = (p_0^2 + p_1^2 + m_0^2)^{-1} ,$$

so

$$C_0(x, z) = \left[\frac{e^{-|z_0 - x_0|\mu}}{2\mu}\right](\vec{x}, \vec{z})$$

and (2.2.3) follows.

It is convenient to introduce the time z_0 annihilation operator defined on $L_2(\mathcal{S}'(R^2), d\Phi)$ by

$$\begin{aligned} a(z) &= \int_{y_0 \geqq z_0} dy[(2\mu)^{-1} \exp(- y_0 - z_0)\mu](\vec{z}, \vec{y}) \partial/\partial\Phi(y) \\ &= \int_{y_0 \geqq z_0} dy C_0(z, y) \partial/\partial\Phi(y) . \end{aligned} \tag{2.2.4}$$

The annihilation operator is simply a derivative, $\partial/\partial\Phi$, forward in time. We distinguish it from $\partial/\partial\Phi$ for the purpose of defining our expansion.

We note that $a(z)$ has a relativistic localization, so that $[a(z_0, \vec{z}), a(\vec{z}_0, \vec{y})^*]$ is proportional to $\mu^{-1}(\vec{y} - \vec{z})$. By Lemma 2.2.1 for each $z_0 \geqq t_{C(s)}$,

$$\begin{aligned} \frac{1}{2}\delta C \cdot \Delta_\Phi &= \int_{x_0 \leqq z_0 \leqq y_0} dxdy\, \delta C(x, y)\partial^2/\partial\Phi(x)\partial\Phi(y) \\ &\quad + \frac{1}{2}\int_{x_0, y_0 \leqq z_0} dxdy\, \delta C(x, y)\partial^2/\partial\Phi(x)\partial\Phi(y) \\ &= \int_{x_0 \leqq t_{C(s)}} dxd\vec{z}\, \delta\hat{C}(x, z)\partial/\partial\Phi(x) a(z) \\ &\quad + \frac{1}{2}\int_{x_0, y_0 \leqq z_0} dxdy\, \delta C(x, y)\partial^2/\partial\Phi(x)\partial\Phi(y) . \end{aligned} \tag{2.2.5}$$

Here $\delta\hat{C}(x, z) = (2\mu_{\vec{z}}\delta C)(x, z)$. We also note that $\delta C(x, z)$ and $\delta\hat{C}(x, z)$ vanish unless x_0 or z_0 is less than $t_{C(s)}$.

Suppose that δC_j (or δC_l) are as in Proposition 2.1.1 (or 2.1.2). Furthermore assume that if $x \in \Delta$, $y \in \Delta'$, then

$$x_0 \leqq t_{C(s)} \leqq z_0 - 1 \leqq y_0 - 1 .$$

PROPOSITION 2.1.1′ (2.1.2′). *Under the above hypotheses,* $\delta\hat{C}_j$ *(or* $\delta\hat{C}_l$*) satisfies* (2.1.6) (*or*(2.1.10)).

Proof. For Δ, Δ' as above, $\delta C(x, y)$ is a function of x, s, u times $C_0(x, y)$. Thus $\delta\hat{C}(x, y)$ is the same function of x, s, u times

$$[e^{-|y_0-x_0|\mu}](\vec{x}, \vec{y}) ,$$

which is bounded by $O(1) \exp\left(-m_0(1-\varepsilon/2)|x-y|\right)$. The propositions now follow as previously.

The integrands in (2.2.1) will all have the form

$$F_\alpha = \int dx d\vec{y} \left(\prod_{j=1}^{n_a} a(y_j)\right)\left(\prod_i :\Phi(x_i)^{n_i}:\right) e^{-V} w(\alpha, x, y) \tag{2.2.6}$$

and the kernels w may also depend on parameters s, u. In addition, certain of the vertices: $\Phi(x_i)^{n_i}$: may occur at sharp times (e. g., A or B_T vertices) and for such vertices the x_i integration extends only over the space variable $\vec{x}_i$. For convenience, we place all annihilation operators at equal times y_{0j} in $Z + 1/2$. (i. e., at the time boundary of some lattice square) with $y_{0_\alpha} = y_{01} = y_{02} = \cdots = y_{0n_a}$. Moreover, the annihilators occur at a later time than any contour in C, that is, $y_0 > t_{C(u,s)}$.

In addition to the formulas (2.2.1), (2.2.5) for a change in the covariance, we have a second basic formula to advance the time of an annihilation operator. For $y_0 \leqq z_0$,

$$\begin{aligned} a(y) &= \int d\vec{z}\,[e^{-(z_0-y_0)\mu}](\vec{y}, \vec{z})a(z_0, \vec{z}) \\ &\quad + \int_{y_0}^{z_0} dx_0 \int d\vec{x}[(2\mu)^{-1}e^{-(x_0-y_0)\mu}](\vec{y}, \vec{x})\partial/\partial\Phi(x) . \end{aligned} \tag{2.2.7}$$

For a product of annihilation operators, as in (2.2.6), we apply (2.2.7) iteratively to each factor $a(y_i)$. Note that by advancing an annihilator a time $d = z_0 - y_0$, the first term in (2.2.7) contributes a factor $O(e^{-m_0 d})$ to the expansion, and the second term, after differentiating $e^{-V(h)}$, contributes a factor $O(\lambda)$. These factors yield convergence of the expansion.

It simplifies the expansion to allow (2.2.7) in the case of $n_a = 0$ annihilators. In this case, (2.2.7) yields the identity $1 = 1$, i. e., no change is produced by such an expansion step. We note that the inductive construction of Section 2.3 specifies a time y_0 for the annihilation operators even in the case $n_a = 0$. For the initial term, at the start of the expansion, $y_0 = 0$, by convention.

In order to take advantage of the fact that $C_0(x, y)$ vanishes exponentially at large values of $|x - y|$, we localize all variables x and $\vec{y}$. Let Δ_i denote the lattice square of unit length, centered at $i \in Z^2$, and let I_i denote the interval of unit length centered at $i \in Z$. We require that $w(x, y)$ in (2.2.6) be supported in a product of squares $x_\nu \in \Delta_{i_\nu}$ and intervals $\vec{y}_\nu \in I_{i_\nu}$.

Such an F is called localized, and a general F is a sum of localized integrands. The application of a derivative $\partial/\partial\Phi$ localized in $\Delta \in R^2$, to a localized integrand F, can be written as a sum of derivatives with respect to each of the variables in F localized in Δ. In the language of graphs, this is the sum over all possible contractions in Δ. In the use of formulas (2.2.1), (2.2.5), and (2.2.7), we imply a sum over all localizations and contractions for factors $\Phi(x)$, $\partial/\partial\Phi(x)$, and $a(x)$. Thus each term resulting from these formulas will have uniquely defined localizations and contractions.

2.3. The increasing contour expansion

The increasing contour expansion is based on an iterated application of (2.2.1), (2.2.5), and (2.2.7). The purpose of this expansion is to factor out vertices (polynomials in Φ) arising from A and from derivatives $\partial/\partial\Phi$ applied to e^{-V}. We achieve this factorization by inserting a contour enclosing the lattice squares in which the vertices are localized. We distinguish between the derivative $\partial/\partial\Phi$ and the annihilation operator a, in that we do not factor out vertices arising from

$$\left(\prod_{i=1}^{n_a} a(y_i)\right) B_T e^{-V} .$$

The expansion is defined so that a contour never encloses an annihilator. We achieve this by using (2.2.7) to advance the time of each annihilator beyond the localization times of all vertices that we wish to factor out. We now explain the expansion in detail.

Throughout the increasing contour construction, the terms have the form

$$m_\alpha = \int ds\, du \int F_\alpha(u, s)\, d\Phi_{C(u,s)} , \tag{2.3.1}$$

where F has the form (2.2.6) and where $C(u, s)$ is an element of $\mathcal{C}_1$, defined in Section 2.1. The integration over u, s ranges over the partial contours in $C(u, s)$. Each expansion step replaces a term of the form (2.3.1) by a sum of similar terms. This sum ranges over the localizations of F and the contractions arising from (2.2.5) and (2.2.7) described in the previous section.

We divide our terms into two categories: Leading terms (l-terms) and remainder terms (r-terms). We define an r-term by the condition $y_0 \geqq T$, where y_0 is the time of the annihilators in F (see (2.2.6) and Section 2.4). The expansion of an r-term yields only r-terms. A term is called a leading term if it is not an r-term. The expansion of an l-term will yield both l- and r-terms.

Since annihilators advanced beyond the time T do not contribute to convergence, we define

$$t = \min\{y_0, T\},$$

to aid the discussion of convergence in Section 2.4.

Our expansion is defined so that the kernel w of F factors in all l-terms, $w = w_1 \otimes w_B$, where w_1 is independent of B. At the start of the increasing contour expansion, $C(u, s)$ in (2.3.1) equals C_*, some element of $\mathcal{C}_2$, say $C_* = C(\mathcal{X}_l, u)$. We let $t_* \geqq t_{C_*}$ be the time y_0 at the start of the construction. When the increasing contour expansion terminates for a particular term, the covariance for that term has the form $(C_*)_{Y_j} = C_*(Y_j)$ of an element of $\mathcal{C}_2$ with a complete contour. Thus $C_*(Y_j) = C(\mathcal{X}_{l+1}, u)$ where $X_{l+1} = Y_j$ and $u_{l+1} = 1$.

The increasing contour construction then has the general form

$$\int du \int F(u)\, d\Phi_{C_*} = \sum\nolimits_\alpha \int du \int F_\alpha(u)\, d\Phi_{C_*(Y_j(\alpha))},$$

which we now specify further.

Given a term m_α of the form (2.3.1), we give the next step of the expanding contour construction. The construction terminates for the term in question under one of the following two conditions:

T1. The construction terminates for an l-term m_α which contains no vertices at times earlier than T, contains at most n annihilators ($n_\alpha \leqq n$) and for which C_α contains a complete contour.

T2. The construction terminates for an r-term m_α which contains no vertices, contains no annihilators and for which C_α contains a complete contour.

If m_α does not satisfy T1–T2, we either:

I. Insert a contour using (2.2.1) and (2.2.5), or

II. Advance the time of annihilation operators using (2.2.7).

To apply (2.2.1) to a term m_α, let $C' = C_\alpha = C(\mathcal{X}_l, \mathcal{Y}_j)$ and set $C'' = C'_{Y_{j+1}} = C(\mathcal{X}_l, Y_{j+1})$. The new contour ∂Y_{j+1} is defined as follows: Let $N(\Delta)$ be the degree of the monomial in F_α localized in the lattice square Δ, as in Section 1.1. Let $S_j(\Delta)$ be the square centered at Δ of length $2[N_j(\Delta)^{1/4}]+1$, where $[x]$ denotes the integral part of x. Then

(2.3.2) $$Y_{j+1} = Y_j \cup \bigcup\nolimits_{\Delta \in \mathcal{D}} S_j(\Delta),$$

where

$$\mathcal{D} = \{\Delta : N_j(\Delta) \neq 0\},$$

if m_α is an r-term, and

$$\mathcal{D} = \{\Delta\colon N_j(\Delta) \neq 0, \text{ and time } \Delta \leqq T\},$$

if m_α is an l-term. Let t_{j+1} be the maximum time of Y_{j+1}.

We apply (2.2.1) and (2.2.7) in detail as follows:

I. Insert the contour ∂Y_{j+1} using (2.2.1) with $C'' = (C_\alpha)_{Y_{j+1}}$ as above if

a) $n_a \leqq n$; and

b) $t_{j+1} < y_0$; and

c) $y_0 < T - 1$ or $y_0 = +\infty$.

For l-terms we express $\delta C \cdot \Delta_\Phi$ as in (2.2.5). For r-terms we use (2.0.5) so that no annihilators occur. Note (b) guarantees that ∂Y_{j+1} does not enclose any annihilators.

II. Use (2.2.7) to advance the time of $0 \leqq n_\alpha \leqq n+1$ annihilators from the time y_0 to the time z_0, if

a) $n_a = n + 1$ and $y_0 < T - 1$; or

b) $y_0 < t_{j+1}$ and $y_0 < T - 1$; or

c) $T - 1 \leqq y_0 < +\infty$.

In cases (a) and (b), choose $z_0 = y_0 + 1$, and in case (c) choose $z_0 = +\infty$.

Remark. The contour ∂Y_{j+1} is defined not only to contain the localization of extraneous monomials, permitting them to be factored out, but also ∂Y_{j+1} is defined to avoid divergences resulting from monomials of high degree localized in a single square Δ. Although we can not avoid high values of $N(\Delta)$ in all terms, Y_{j+1} is constructed so that if $N(\Delta)$ is large, each additional contraction or derivative occurs over a long distance, giving rise to a compensating factor of $\exp(-m_0 d(\Delta, \partial Y_{j+1}))$.

Throughout the increasing contour expansion, note that $n_a \leqq n + 1$. For a given elementary expansion step II note that new vertices are the only vertices localized in the time strip $R^1 \times [y_0, T]$. New vertices are defined to be those vertices arising from contraction to the exponent within a given expansion step. Hence there are at most n_a possible contractions to new vertices arising from the application of (2.2.7) in a single expansion step.

The expanding contour construction is completely defined by iterating I–II until termination. In the terms for which the expansion terminates (and thus have a complete contour) we factor out the vertices in the interior of the complete contour, using (2.0.2). In l-terms, this factorization leaves unchanged the vertices of B_T and the annihilators present. In r-terms, no vertices or annihilators remain after the factorization. Our expansion now takes the form

$$\int du \int F(u)\, d\Phi_{C_*}$$

(2.3.3)
$$= \sum_\alpha \int du ds \int d\bar{x}\, d\bar{y}\, w_l'(\alpha, u, s, x, y) \int \left(\prod_{i=1}^{n_\alpha} a(y_i) \right) B_T(x) e^{-V} d\Phi_{C_\alpha}$$
$$+ \sum_\alpha \int du\, ds\, w_r'(\alpha, u, s) \int e^{-V} d\Phi_{C_\alpha} .$$

Here $C_\alpha = C_*(\mathcal{Y}_j, s)$ has a complete contour ∂Y_j, depending on α. Also $B_T(x)$ is the density for B_T.

$$B_T = \int d\bar{x}\, w_B(\bar{x})\, B_T(x) .$$

The first sum in (2.3.3) is the sum of l-terms, while the second sum in (2.3.3) is the sum of r-terms. The summation over α indexes the various terms. For l-terms, it includes summation over n_α, $0 \leqq n_\alpha \leqq n$. For both l-terms and r-terms it labels summation over the various complete contours ∂Y_j and over the various contractions. The numerical kernels w_l', w_r' are functionals of the factored vertices, as given by (2.0.2). Namely,

(2.3.4)
$$w_l'(\alpha, u, s, x, y) = \frac{\int P_l(\alpha, u, s, x, y) e^{-V(Y_j)} d\Phi_{C_\alpha}}{\int e^{-V(Y_j)} d\Phi_{C_\alpha}} ,$$
$$w_r'(\alpha, u, s) = \frac{\int P_r(\alpha, u, s) e^{-V(Y_j)} d\Phi_{C_\alpha}}{\int e^{-V(Y_j)} d\Phi_{C_\alpha}} ,$$

and $P_l(\alpha, \cdot)$ and $P_r(\alpha, \cdot)$ are polynomials localized in $Y_j = Y_j(\alpha)$. We have used (2.0.1) to factor $\int \exp(-V(\sim Y_j)) d\Phi_{C_\alpha}$ from both numerator and denominator.

We obtain the desired expansion, by making the change

$$d\Phi_{C_\alpha} \longrightarrow d\Phi_{C_*(Y_j(\alpha))} ,$$

in (2.3.4), where $\partial Y_j(\alpha)$ is the complete contour in the term labeled by α. We write for the remainder terms

(2.3.5)
$$\int e^{-V} d\Phi_{C_\alpha} = \int e^{-V} d\Phi_{C_*(Y_j(\alpha))} \left(\frac{\int e^{-V(Y_j)} d\Phi_{C_\alpha}}{\int e^{-V(Y_j)} d\Phi_{C_*(Y_j(\alpha))}} \right) ,$$

and for the l-terms,

$$\int \left(\prod_{i=1}^{n_\alpha} a(y_i)\right) B_T(x) e^{-V} d\Phi_{C_\alpha}$$

$$(2.3.6) \qquad = \int \left(\prod_{i=1}^{n_\alpha} a(y_i)\right) B_T(x) e^{-V} d\Phi_{C_*(Y_j(\alpha))} \left(\frac{\int e^{-V(Y_j)} d\Phi_{C_\alpha}}{\int e^{-V(Y_j)} d\Phi_{C_*(Y_j(\alpha))}} \right).$$

We absorb the ratios in (2.3.5), (2.3.6) into the coefficients w' in (2.3.4), and we include this rearrangement as part of the elementary step producing the term with the complete contour ∂Y_j. Thus the increasing contour expansion is

$$\int du \int F(u) d\Phi_{C_*}$$

$$(2.3.7) \qquad = \sum_\alpha \int du \int d\bar{x} d\bar{y} w_l(\alpha, u, x, y) \int \left(\prod_{i=1}^{n_\alpha} a(y_i)\right) B_T(x) e^{-V} d\Phi_{C_*(X_{l+1}(\alpha))}$$

$$+ \sum_\alpha \int du\, w_r(\alpha, u) \int e^{-V} d\Phi_{C_*(X_{l+1}(\alpha))} ,$$

where $\partial Y_j = \partial Y_j(\alpha) = \partial X_{l+1}(\alpha)$ is the inserted contour, and where

$$(2.3.8) \qquad \begin{aligned} w_l(\alpha, u, x, y) &= \frac{\int ds \int P_l(\alpha, u, s, x, y) e^{-V(Y_j)} d\Phi_{C_\alpha}}{\int e^{-V(Y_j)} d\Phi_{C_*(Y_j)}} , \\ w_r(\alpha, u) &= \frac{\int ds \int P_r(\alpha, u, s) e^{-V(Y_j)} d\Phi_{C_\alpha}}{\int e^{-V(Y_j)} d\Phi_{C_*(Y_j)}} . \end{aligned}$$

2.4. Norms for convergence

In this section we establish the convergence of the increasing contour expansion. Each elementary step in the expansion expresses a term

$$(2.4.1) \qquad m_\alpha = \int du\, ds \int dx\, d\bar{y} \prod_{i=1}^{n_\alpha} a(y_i) \prod_{i=1}^{r} : \Phi(x_i)^{n_i} : e^{-V} w_\alpha d\Phi_{C_\alpha}$$

as a sum of terms of the same form. Here $w_\alpha = w(\alpha, u, s, x, y)$ is defined by (2.3.4), and $C_\alpha = C(\mathcal{X}_l, \mathcal{Y}_j, u, s)$.

Let I_k be the set of indices α which label the terms m_α at the kth stage of the increasing contour expansion. Thus for each k we have

$$\int AB_T e^{-V} d\Phi = \sum_{\alpha \in I_k} m_\alpha .$$

Also let $M_k = \{m_\alpha : \alpha \in I_k\}$ be the set of terms in the sum. Each step of the expansion can be regarded as a mapping

$$M_k \longrightarrow M_{k+1}$$

of sets of expansion terms.

For each term m_α, we introduce a norm $\|m_\alpha\|$, and then we define a norm on M_k by

$$\|M_k\| = \textstyle\sum_{\alpha \in I_k} \|m_\alpha\| . \tag{2.4.2}$$

The proof of convergence consists in showing that for each k,

$$\|M_{k+1}\| \leqq \|M_k\| .$$

This decrease in norm is ensured by the small factors, $O(\lambda)$ or $O(m_0^{-1})$, introduced by each step in our expansion, as well as by an appropriate choice of norms $\|\cdot\|$. Unfortunately this simple structure in the convergence proof is made possible only by the choice of a complicated norm.

To define our norms, we introduce the free module W_k generated by the symbols $\alpha \in I_k$. In other words, W_k is the vector space of functions defined on I_k. For a function of the form $\alpha \to c_\alpha m_\alpha$ in W_k, we define the norm $\sum |c_\alpha| \, \|m_\alpha\|$, where $\|m_\alpha\|$ is given by the formula

$$\begin{aligned} \|m_\alpha\| &= \mathcal{K} K_1^\nu e^{\gamma t} \mathfrak{N} \\ &\times \int du\, ds \exp\big[\textstyle\sum_{k=1}^{l} U_{k,l} |X_k| + \sum_{k=1}^{j} S_{k,j} |\delta Y_k|\big] \|w(\alpha, u, s, y_0, \cdot)\|_2 . \end{aligned} \tag{2.4.3}$$

The constants $\mathcal{K}$, ν, etc., are explained below. Then

$$\|\alpha \longrightarrow c_\alpha m_\alpha\| = \textstyle\sum_{\alpha \in I_k} |c_\alpha| \, \|m_\alpha\| .$$

For initial and final terms we shall see that there is a simple relation between this norm and the norm defined by (1.1.7), (1.1.8). The purpose of this section is to show that the increasing contour expansion is norm decreasing; that is, $\sum_{\alpha \in I_k} \|m_\alpha\|$ is decreasing in k. In Chapters 3 and 4 we show that the full n-particle expansion is norm decreasing so that the norm of the initial term

$$\int A B_T e^{-V} d\Phi$$

dominates the sum of the norms of the final terms. This bound implies Theorem 1.1.11, as we shall see in Chapter 4.

Now we explain (2.4.3) in detail. The indices α (or the corresponding terms m_α) are divided into five categories and for each category we specify values of $\mathcal{K}$, $\mathfrak{N}$, ν, t above. Two of these categories do not occur until later chapters. In the cases where w does not depend on u or s, the sum $\sum_{k=1}^{l}$ or $\sum_{k=1}^{j}$ runs over an empty set of indices, and equals zero, so the integral $\int du$ or $\int ds$ can be omitted from the norm.

The first category is the initial term $m_\alpha = \int AB_T dq$. There are three categories of partially expanded terms, denoted P_1, P_2, and P_3, and there are final terms. The P_1 terms arise in Chapter 2. They have a covariance $C = C_0$ or $C \neq C_0$, $C \in \mathcal{C}_1$, but no complete contour. The P_2 terms occur at the end of the Chapter 2 expansion. They have a covariance $C \neq C_0$, $C \in \mathcal{C}_2$, and C has one complete contour. The leading, or $l - P_2$ terms have no vertices before time T, and the remainder, or $r - P_2$ terms have no vertices at all. The P_3 terms occur at the end of the Chapter 3 expansion. They have covariance $C = C_0$, $n_a > 0$ annihilators, and vertices as in P_2 terms. The final terms occur at the end of the Chapter 4 expansion, and also at the end of the Chapter 3 expansion, in the case of $n_a = 0$ annihilators.

The time y_0 is chosen to increase as the expansion progresses. (The expansion step (4.5) is exceptional, and allows a decrease in y_0.) For initial terms, $y_0 = 0$, and for remainder terms, $y_0 = +\infty$. For leading terms, $y_0 \geqq 0$ is the largest time of any annihilator. Recall that y_0 is defined by step II of the inductive expansion which advances y_0 even when $n_a = 0$. Recall also $t = \min\{y_0, T\}$. We choose $\gamma = (n+1)m_0(1-\varepsilon)$.

We select ν to decrease as the expansion progresses. For each change in category, ν decreases by one. Thus for leading terms, $\nu = 6$ (initial), $\nu = 5$ (P_1), etc. For remainder terms, $\nu = 4$ (P_1), $\nu = 3$ (P_2), etc. There are no initial or P_3 remainder terms. The definition of the categories is summarized in Table 1 below.

TABLE 1. Categories of expansion terms.

	ν	Covariance	Vertices and Annihilators
Initial	6	$C=C_0$	AB_T not in form required for final terms
P_1	5 if l 4 if r	$C=C_0$; or $C\neq C_0$, $C\in\mathcal{C}_1$, no complete contour	
P_2	4 if l 3 if r	$C\neq C_0$, $C\in\mathcal{C}_2$ one complete contour	
P_3	3	$C=C_0$	$N_a>0$ annihilators
Final-l	2	$C=C_0$	$B_T\times$\{polynomial of degree $\leq n$ at sharp time $t<T$\}
Final-r	1	$C=C_0$	No vertices, no annihilators

We take $\mathcal{K} = 1$ except for P_1 terms. For a P_1 term, $\mathcal{K} = \mathcal{K}_1\mathcal{K}_2\mathcal{K}_3\mathcal{K}_4$ is a product of four factors, and each $\mathcal{K}_i$ has the role of transferring convergence from strongly convergent steps (i. e., norm decreasing steps) to steps which would otherwise have some limited increase in norm.

The factor $\mathcal{K}_1$ contains the convergence which arises from the contractions over long distances. Explicitly,

$$\mathcal{K}_1 = e^{\varepsilon m_0 D/2} ,$$

where D is the sum of the contraction distances arising since the last insertion of a complete contour. Thus if we consider the Chapter 2 expansion alone, D is the sum of all contraction distances in the expansion. Hence $\mathcal{K}_1$ increases throughout the increasing contour construction and returns to one at the end of it.

$\mathcal{K}_2$ and $\mathcal{K}_3$ decrease throughout the increasing contour construction, and return to one at the end of it. $\mathcal{K}_2$ compensates for the insufficient convergence which results from increasing the time t in the presence of $0 \leqq n_a < n+1$ annihilators in the expansion step II. Explicitly, if $\delta t = t' - t$ is the increase in t, then

$$\mathcal{K}_2 = e^{-(1-\varepsilon) m_0 \Sigma \delta n} , \tag{2.4.4}$$

where $\delta n = (n + 1 - n_a)\delta t$, and the sum runs over all expansion steps since the last insertion of a complete contour. We also define

$$\mathcal{K}_3 = e^{-|Y_j|} ,$$

$$\mathcal{K}_4 = K_2^j ,$$

where j is the number of $\mathcal{Y}_j$ contours in

$$C = C(\mathcal{X}_l, \mathcal{Y}_j, u, s) ,$$

and we define $\mathfrak{N} = \prod_\Delta \mathfrak{N}(\Delta)$ and

$$\mathfrak{N}(\Delta) = \begin{cases} K_2^{2N(\Delta)} N(\Delta)! \\ K_2^{N(\Delta)} N(\Delta)! . \end{cases} \tag{2.4.5}$$

Here the first choice is made for all initial and final terms and also for leading terms with Δ beyond the time T. Otherwise the second choice is made.

Remark. Let $||\cdot||_0$ denote the norm defined in (1.1.7), (1.1.8). For initial and final terms m_α we have

$$||m_\alpha|| = K_1^\nu e^{\gamma t} ||m_\alpha||_0 ,$$

where ν, t depend on the term m_α.

THEOREM 2.4.1. *Let ε and n be given. Let m_0, K_2 and K_1 be sufficiently large and let λ be sufficiently small. Then the increasing contour expansion (2.3.7) is norm decreasing.*

Proof. It is sufficient to show that if $m_\alpha = \sum m'_{\alpha'}$ represents an elementary expansion of the term m_α as defined by I or II, we have

$$\sum \| m'_{\alpha'} \| \leqq \| m_\alpha \| .$$

We choose the constants of our theorem in the following order: m_0, K_2, K_1, and then λ.

LEMMA 2.4.2. *The elementary step* II *of advancing* n_α *annihilators from time* $y_0 < T - 1$ *to* $y'_0 = y_0 + 1$ *is norm decreasing.*

Proof. Since $y'_0 < T$, m_α, as well as each $m'_{\alpha'}$, is a leading term. A term $m'_{\alpha'}$ has $n'_{\alpha'} = k$ annihilators at the time $y'_0 = y_0 + 1$ and $n_\alpha - k$ contractions in the strip $[y_0, y_0 + 1]$. The sum extends over the various possible contractions and their localizations of the k annihilators and over the possible values of k, $k = 0, \cdots, n_\alpha$.

To obtain the kernel $w'_{\alpha'}$ of $m'_{\alpha'}$, we define the function

$$v = (\otimes_{i=1}^{k} e^{-\mu})\left(\prod_{i=1}^{n_\alpha - k} e^{-\sigma_i \mu}\right) w_\alpha$$

with n_α factors $e^{-\sigma\mu}$ acting on the $\bar{y}$ variables of w_α. Of these factors, there is one factor $e^{-\mu}$ for each of the k annihilators advanced by one unit of time, and one factor $e^{\sigma_i \mu}$ for each of the $n_\alpha - k$ contractions, $0 \leqq \sigma_i \leqq 1$. The kernel $w'_{\alpha'}$ is obtained as follows: Multiply v by the characteristic function of a product L' of squares and intervals to provide the localization of each variable. If $k < n_\alpha$, at least one contraction to the exponent occurs in each term, so we multiply v by a numerical factor $\leqq O(\lambda)$ given by the coefficients of the interaction polynomial. If two or more contractions to the same vertex occur, identify the corresponding variables of v.

Let $\sum d'$ be the sum of the distances for this step, as determined by L' and the localization of w_α. Then

$$D' = D + \sum d' , \qquad \sum d' \geqq k. \tag{2.4.6}$$

Furthermore, $w'_{\alpha'}$ satisfies the estimate

$$\| w'_{\alpha'} \|_{L_2} \leqq O(1) e^{-m_0 (\Sigma d')(1-\varepsilon/4)} \| w_\alpha \|_{L_2} \tag{2.4.7}$$

where $O(1) \to 0$ as $m_0 \to \infty$, for $k = n_\alpha > 0$ and $O(1) \to 0$ as $\lambda \to 0$ for $k < n_\alpha$. This estimate follows from the exponential decay of $\exp(-\sigma\mu)$ (cf. [6, Lemma 3.2]).

From the definition (2.4.4),

$$\mathcal{K}'_2 e^{\gamma y'_0} = e^{n_\alpha m_0 (1-\varepsilon)} \mathcal{K}_2 e^{\gamma y_0} . \tag{2.4.8}$$

Moreover

$$\mathcal{K}'_1 = \mathcal{K}_1 e^{\varepsilon m_0 \Sigma d'/2} , \qquad \mathcal{K}'_3 = \mathcal{K}_3,\ \mathcal{K}'_4 = \mathcal{K}_4 .$$

It follows that

$$\mathcal{K}' e^{\gamma y'_0} \| w'_{\alpha'} \|_{L_2} \leqq O(1) e^{-m_0 \varepsilon \Sigma d'/4} \mathcal{K} e^{\gamma y_0} \| w_\alpha \|_{L_2} . \tag{2.4.9}$$

For $k = n_a$, we use $\sum d' \geqq k$ in proving (2.4.9). For $k < n_a$, $O(1) \to 0$ as $\lambda \to 0$ and since λ is chosen after m_0, a factor $e^{n_a m_0(1-\varepsilon)}$ can be absorbed into $O(1)$. For $k = n_a$, the factors $N(\Delta)$ remain unchanged so

$$\| m'_{\alpha'} \| \leqq O(1) e^{-\varepsilon m_0 \Sigma d'/4} \| m_\alpha \| .$$

The sum over localizations is controlled using the exponential decay in $\sum d'$, and so for m_0 sufficiently large

$$\sum_{\alpha';\, k=n_a} \| m'_{\alpha'} \| \leqq \frac{1}{2} \| m_\alpha \| . \tag{2.4.10}$$

For $k < n_a$, all $n_a - k$ contractions occur in squares in which $N(\Delta) = 0$. Thus

$$\mathfrak{N}' \leqq K_2^{p(n_a-k)} (p(n_a - k))!\, \mathfrak{N} , \tag{2.4.11}$$

where $p = \deg\ (\mathcal{P} = V(h))$ in the action (1.8). The new factor $K_2^{p(n_a-k)} (p(n_a - k))!$ can be absorbed by $O(1)$ in (2.4.9), since λ is chosen after K_2. Thus for λ sufficiently small,

$$\sum_{\alpha';\, k<n} \| m'_{\alpha'} \| \leqq \frac{1}{2} \| m_\alpha \| , \tag{2.4.12}$$

which combined with (2.4.10), completes the proof of the lemma, for $n_a > 0$. For $n_a = 0$ there is only one term in the expansion of m_α. The norm is unchanged because the decrease in $\mathcal{K}_2$ compensates for the increase in $e^{\gamma t}$.

LEMMA 2.4.3. *The elementary step* II *of advancing annihilators from a time* $y_0 \in [T - 1, T]$ *is norm decreasing.*

Proof. We describe the modifications of the proof above. The principal modification concerns contractions to B_T. Each term $m'_{\alpha'}$ is now a remainder term and so $k = 0$, i.e., the annihilators are advanced to time $+\infty$, and necessarily contract. The sum over contraction localizations includes all lattice squares with times greater than y_0. The kernel $w'_{\alpha'}$ of $m'_{\alpha'}$ satisfies the bound (2.4.7). For contractions to new vertices, the proof is as before. If an annihilator contracts to B_T, the kernel is obtained by identifying the contracted variables in

$$v = w_B \otimes \left(\bigotimes_{i=1}^{n_a} e^{-\sigma_i \mu} w_1 \right) .$$

Here we use the factorization property for the kernel of leading terms (e.g., m_α) to write w_α as

$$w_\alpha = w_B \otimes w_1 ,$$

where w_B is the kernel for the observable B_T. For contractions to B_T, $\sigma_i \geqq 1$ and for σ_i bounded away from zero, $e^{-\sigma_i \mu}$ is a bounded operator from

L_2 to L_∞, with norm $O(\exp(-\sigma_i m_0(1-\varepsilon))) = o(1)$ as $m_0 \to \infty$, and (2.4.7) follows from this fact.

For all remainder terms, $t' = T$, so $t' \leqq t + 1$ and (2.4.8) is replaced by

$$\mathcal{K}_2' e^{\gamma t'} \leqq e^{n_a m_0(1-\varepsilon)} \mathcal{K}_2 e^{\gamma t} , \tag{2.4.13}$$

while (2.4.6) and (2.4.9) are unchanged.

Let $j_1(\Delta)$ be the number of annihilators which contract in Δ, and let $j_0(\Delta)$ be the number which contract to B_T in Δ, so that

$$0 \leqq j_0(\Delta) \leqq j_1(\Delta)$$

and

$$\textstyle\sum_\Delta j_1(\Delta) = n_a \leqq n + 1 . \tag{2.4.14}$$

Thus

$$\begin{aligned} N'(\Delta) &\leqq N(\Delta) + (p-1)(j_1(\Delta) - j_0(\Delta)) - j_0(\Delta) \\ &\leqq N(\Delta) + p j_1(\Delta) - j_0(\Delta) , \end{aligned}$$

and using (2.4.14), and $(a+b)! \leqq 2^{a+b} a!\, b!$,

$$\textstyle\prod_\Delta N'(\Delta)! \leqq O(1) 2^{|B|} \prod_\Delta (N(\Delta) - j_0(\Delta))! . \tag{2.4.15}$$

Here $O(1)$ depends on p and n, but not on the degree of B,

$$|B| \equiv \textstyle\sum_{\Delta \subset [y_0, \infty] \times R} N(\Delta) .$$

Recall that definition (2.4.5) on $\mathfrak{N}$ changes from a leading term m_α to remainder term $m'_{\alpha'}$, and that the change is a factor K_2 per monomial degree, for squares with times greater that T. We take $K_2 \geqq 2$ so that the product of the K_2 factors dominates the factor $2^{|B|}$. Thus

$$\mathfrak{N}' \leqq O(1) \mathfrak{N} K_2^{pn_a} \prod_\Delta \frac{(N(\Delta) - j_0(\Delta))!}{N(\Delta)!} . \tag{2.4.16}$$

Here $O(1)$ is independent of $|B|$. This inequality is used in place of (2.4.11). Furthermore if $j_0(\Delta) = j_1(\Delta)$ for all Δ, i.e., if all contractions are to B_T, then the factor $K_2^{pn_a}$ can be omitted from (2.4.16).

For fixed values of $j_0(\Delta)$ and $j_1(\Delta)$, the $j_0(\Delta)$ annihilators contracting to B_T in Δ can be chosen in at most

$$\binom{n_a}{j_0} \leqq (n+1)^{n+1} = O(1)$$

ways, and the corresponding B_T vertices and contractions can be chosen in

$$\binom{N(\Delta)}{j_0} j_0! \leqq O(1) \frac{N(\Delta)!}{(N(\Delta) - j_0(\Delta))!}$$

ways. For fixed $j_1(\Delta)$, the sum over choices of $j_0(\Delta)$ yields at most

$$\prod_\Delta (j_1(\Delta)+1) \leqq \prod_\Delta e^{j_1(\Delta)} \leqq e^{n+1} = O(1)$$

terms. The value of $j_1(\Delta)$ is determined by the contraction localizations and the sum over localizations is bounded as before. The estimate now proceeds as before, with

$$O(1)\mathfrak{N}'\prod_\Delta \frac{N(\Delta)!}{(N(\Delta)-j_0(\Delta))!} \leqq O(1)\mathfrak{N} .$$

In case $j_0(\Delta) < j_1(\Delta)$ for some Δ, the constants $O(1)$ depend on K_2, but since λ is chosen after K_2, $O(1)$ may be absorbed by the $O(1)$ in (2.4.7). Thus for λ small,

$$\sum_{\alpha'} \|m'_{\alpha'}\| \leqq \|m_\alpha\| .$$

In case $j_0(\Delta) = j_1(\Delta)$ for all Δ, the constant $O(1)$ is independent of m_0, K_2, K_1 and λ, and so for m_0 large

$$\sum_{\alpha'} \|m'_{\alpha'}\| \leqq \|m_\alpha\| ,$$

which completes the proof of the lemma.

PROPOSITION 2.4.4. *The elementary step* I *of inserting a contour in* m_α *decreases norms by a factor of at least* 3/4.

Proof. The expansion takes the form

$$m_\alpha = m''_\alpha + \sum_{\alpha'} m'_{\alpha'}$$

where m''_α is the P_2 term (with one complete contour) and each m'_α is a P_1 term with only partial contours. The sum over α' extends over the two terms of (2.2.5), over the various contractions, and over their localizations.

Let the covariance m_α be $C = C(\mathcal{X}_l, \mathcal{Y}_j, u, s)$. Then the covariance of each term $m'_{\alpha'}$ is

$$C' = C(\mathcal{X}_l, \mathcal{Y}_{j+1}, u, s') \in \mathcal{C}_1 , \tag{2.4.17}$$

where

$$\mathcal{Y}_{j+1} = \{\mathcal{Y}_j, Y_{j+1}\} , \qquad s' = \{s, s_{j+1}\} .$$

The term m''_α has $\partial Y_{j+1} = \partial X_{l+1}$ as a complete contour, and it has the covariance

$$C_2 = C(\mathcal{X}_{l+1}, u) \in \mathcal{C}_2 \tag{2.4.18}$$

where

$$\mathcal{X}_{l+1} = \{\mathcal{X}_l, X_{l+1} = Y_{j+1}\} , \qquad u_{l+1} = 1 .$$

In this term, (2.3.5) or (2.3.6) is used to replace $C(\mathcal{X}_l, \mathcal{Y}_{j+1})$ by $C(\mathcal{X}_{l+1})$. We now divide the proof of the proposition into two lemmas.

LEMMA 2.4.5. *Let $m'_{\alpha'}$ be as above. Then*

$$\sum_{\alpha'} \|m'_{\alpha'}\| \leqq \frac{1}{4}\|m_\alpha\| .$$

Consider first the P_1 terms $m'_{\alpha'}$, and divide these terms into four classes $m'_{\alpha',i}$, $i = 1, 2, 3, 4$, according to whether outside Y_{j+1} there is a new annihilator ($i = 1, 2$) or a new contraction ($i = 3, 4$) and whether the new contraction inside Y_{j+1} occurs to the exponent ($i = 1, 3$) or to an existing vertex ($i = 2, 4$). (Recall that $\delta C'(x, y)$ vanishes unless ∂Y_{j+1} separates x and y.) Consider first the case $i = 1$ (outside, annihilator; inside, contraction to the exponent (that is a new vertex).) The kernel of $m'_{\alpha',1}$ is

$$w'_{\alpha',1} = w(\alpha, u, s, x, y_0, \vec{y})\delta\hat{C}_{j+1}(x_{j+1}, y_{j+1})v(x_{j+1}, y_{j+1}) .$$

Here $v(x_{j+1}, y_{j+1})$ is a product of a numerical coefficient, $O(\lambda)$, from the new vertex, times the characteristic function of $\Delta_1 \times \Delta_2$ and Δ_1 and Δ_2 provide the localization for the variables x_{j+1} and y_{j+1}. Thus

$$\|w'_{\alpha',1}(u, s', y'_0, \cdot)\|_2 = \|v\|_\infty \|\delta\hat{C}\|_{L_2(\Delta_1\times\Delta_2)}\|w\|_2 .$$

Let $d = \text{dist.}(\Delta_1, \Delta_2)$ be the contraction distance. By construction, $d \geqq 1$. By Proposition 2.1.1′ and the bound $\|v\|_\infty \leqq O(\lambda)$,

$$\sum_{\Delta_1,\Delta_2} e^{m_0(1-\varepsilon)d}\|w'_{\alpha',1}\|_2 \leqq O(\lambda)\sum_{k=1}^{j+1}\delta S_{k,j+1}|\delta Y_k|\,\|w_\alpha\|_2 . \tag{2.4.19}$$

For all cases $i = 1, \cdots, 4$, either $y'_0 = y_0$ and $\delta n = 0$ or else $n_a = 0$ and $\delta n = (n + 1)(t' - t)$. Thus $\mathcal{K}'_2 e^{\gamma t'} = \mathcal{K}_2 e^{\gamma t}$ and one can check that

$$\mathcal{K}' e^{\gamma t'} = K_2 e^{\varepsilon m_0 d/2 - |\delta Y_{j+1}|}\mathcal{K} e^{\gamma t} . \tag{2.4.20}$$

Returning to the case $i = 1$, we have

$$N'(\Delta) = N(\Delta) , \qquad \Delta \neq \Delta_1$$
$$N'(\Delta_1) \leqq N(\Delta_1) + p - 1 .$$

If $N(\Delta_1) = 0$ then $N'(\Delta_1) \leqq (p - 1) = O(1)$ and if $N(\Delta_1) \neq 0$ then $N(\Delta_1) \leqq d^4$ and

$$\begin{aligned} N'(\Delta_1)! &\leqq (N(\Delta_1) + p)^p N(\Delta_1)! \\ &\leqq O(1)e^{\varepsilon m_0 d/2}N(\Delta_1)! . \end{aligned} \tag{2.4.21}$$

Combining this bound with (2.4.20) gives

$$\mathcal{K}' e^{\gamma t'}\mathfrak{N}' \leqq O(1)e^{-|\delta Y_{j+1}| + \varepsilon m_0 d}\mathcal{K} e^{\gamma t}\mathfrak{N} \tag{2.4.22}$$

where $O(1)$ depends on K_2 and m_0. Take $\varepsilon < 2/5$, and substitute (2.4.19) and (2.4.22) in the sum over $\|m'_{\alpha',1}\|$. With

$$A = \sum_{k=1}^{j} S_{k,j}|\delta Y_k| + |\delta Y_{j+1}| ,$$
$$B = \sum_{k=1}^{l} U_{k,l}|X_k| ,$$

$$\sum_{\alpha'} \| m'_{\alpha',1} \| \leqq O(\lambda) e^{-|\delta Y_{j+1}|} \mathcal{K} e^{\gamma t} \mathfrak{N} \int du\, ds' A e^{A s_{j+1} + B} \| w_\alpha \|_2 \, .$$

Because w does not depend on s_{j+1}, and because

$$ds' = ds\, ds_{j+1} \, ,$$

the ds_{j+1} integration occurs in the form

$$\int_0^1 ds_{j+1} A e^{A s_{j+1}} \leqq 2e^A \, ,$$

and it follows that

$$\sum_{\alpha'} \| m'_{\alpha',1} \| \leqq O(\lambda) e^{-|\delta Y_{j+1}|} \mathcal{K} e^{\gamma t} \mathfrak{N} \int du\, ds\, e^{A+B} \| w_\alpha \|_2 = O(\lambda) \| m_\alpha \| \, . \tag{2.4.23}$$

Consider next the terms $m'_{\alpha',2}$ with an annihilator outside Y_{j+1} and a contraction to an existing vertex inside. The kernel $w'_{\alpha',2}$ is a localization of

$$w(\alpha, u, s, y_0, \bar{y}, x) \delta \hat{C}_{j+1}(x_i, y_{j+1})$$

where x_i is one of the variables from x. With the localizations $x_i \in \Delta_1$ and $y_{j+1} \in \Delta_2$, we use the bound

$$\| w'_{\alpha',2} \|_2 \leqq \| \delta \hat{C}_{j+1} \|_{L_\infty(\Delta_1 \times \Delta_2)} \| w_\alpha \|_2 \, .$$

By construction, the existing vertices have a distance at least 1 to the contour ∂Y_{j+1}, and so the contraction distance d is at least 1.

By Proposition 2.1.1, with ε replaced by $3\varepsilon/8$ and $p = \infty$,

$$\sum_{\Delta_1, \Delta_2} e^{m_0(1-\varepsilon/2)d} \| w'_{\alpha',2} \|_2 \leqq O(e^{-m_0\varepsilon/8}) \Big(\sum_{k=1}^{j+1} \delta S_{k,j+1} | \delta Y_k | \Big) \| w_\alpha \|_2 \, . \tag{2.4.24}$$

For $i = 2$, (2.4.21) is replaced by

$$N'(\Delta) = N(\Delta) \, , \qquad \Delta \neq \Delta_1$$
$$N'(\Delta_1) = N(\Delta_1) - 1 \, ,$$

which yields, in place of (2.4.22),

$$\mathcal{K}' e^{\gamma t'} \mathfrak{N}' \leqq O(1) N(\Delta_1)^{-1} e^{-|\delta Y_{j+1}| + \varepsilon m_0 d/2} \mathcal{K} e^{\gamma t} \mathfrak{N} \, . \tag{2.4.25}$$

The sum over α' includes a sum over the $N(\Delta_1)$ possible contractions in Δ_1, as well as the sum over Δ_1 and Δ_2. Using (2.4.24) and (2.4.25), we follow the derivation of (2.4.23) to obtain

$$\sum_{\alpha'} \| m'_{\alpha',2} \| \leqq O(e^{-m_0\varepsilon/8}) \| m_\alpha \| \, .$$

The remaining cases $i = 3, 4$ are bounded similarly, and we obtain

$$\sum_{\alpha'} \| m'_{\alpha'} \| \leqq O(1) \| m_\alpha \| \tag{2.4.26}$$

where $O(1) \leqq 1/4$ for m_0 large and $\lambda = \lambda(m_0, K_2)$ small.

We now consider m''_α in the expansion. The following lemma completes the proof of the proposition.

LEMMA 2.4.6.

$$\|m''_\alpha\| \leqq \frac{1}{2}\|m_\alpha\| .$$

Proof. The term m''_α is a P_2 term, and so for this term $\mathcal{K}'' = 1$. Furthermore, $t'' = t$, and

$$\mathfrak{N}''(\Delta) = \begin{cases} 0, \ \Delta \subset Y_{j+1} \\ \mathfrak{N}(\Delta), \ \Delta \not\subset Y_{j+1} \end{cases}$$

since m''_α contains only B_T vertices. We write

$$\mathfrak{N}_1 = \prod_{\Delta \subset Y_{j+1}} K_2^{N(\Delta)} , \qquad \mathfrak{N}_2 = \prod_{\Delta \subset Y_{j+1}} N(\Delta)! .$$

Thus $\mathfrak{N} = \mathfrak{N}''\mathfrak{N}_1\mathfrak{N}_2$ and

(2.4.27) $$\mathcal{K}'' K_1^{\nu''} e^{\gamma t} \mathfrak{N}'' \leqq K_1^{\nu-1} e^{\gamma t} \mathfrak{N}'' .$$

We claim (i) there is a constant $K_3 > 0$ such that if $m_0 \geqq 1$, if $K_2 = K_2(m_0, K_3)$ is sufficiently large and if $\lambda \leqq 1$, then

(2.4.28) $$\|w''_\alpha(u, y_0, \cdot)\| \leqq \mathfrak{N}_1^{1/2}\mathfrak{N}_2 e^{K_3|Y_{j+1}|}\int ds \|w_\alpha(u, s, y_0, \cdot)\|_2 ,$$

(ii) for $K_2 = K_2(m_0, K_3)$ sufficiently large,

(2.4.29) $$e^{(K_3+1)|Y_{j+1}|} \leqq \frac{1}{2}\mathfrak{N}_1^{1/2}\mathcal{K} .$$

By (2.4.27)-(2.4.29), with $K_1 \geqq 1$, $\|m''_\alpha\| \leqq (1/2)\|m_\alpha\|$. Note that the factor $e^{|X_{l+1}|}$ which occurs in $\|m''_\alpha\|$ but not in $\|m_\alpha\|$ is controlled by (2.4.29), since $X_{l+1} = Y_{j+1}$.

We first establish (2.4.29). Recall that

(2.4.30) $$\mathcal{K} = K_2^j \exp\left[\varepsilon m_0 D/2 - (1-\varepsilon)m_0 \sum \delta n - |Y_{j+1}|\right] .$$

Since at most $n + 1$ annihilators are advanced in time (by step II) to allow the insertion of some contour ∂Y_r with $|\delta Y_r| \geqq 1$, one can check that

(2.4.31) $$\sum \delta n \leqq (n+1)\sum_{r=1}^{j+1}|\delta Y_r| \leqq (n+1)|Y_{j+1}| .$$

Let us consider a nonempty increment

$$\delta Y_r = Y_{r+1} \sim Y_r \neq \emptyset$$

in Y_r. Except when Y_r is enlarged to contain A or B_T vertices,

$$\begin{aligned}\delta Y_r &= \textstyle\bigcup_i (S_{r+1}(\Delta_i) \sim S_r(\Delta_i)) \\ &= \textstyle\bigcup_i \delta S_r(\Delta_i) .\end{aligned}$$

Hence such an increase occurs because in certain squares Δ_i, the length $2[N_r(\Delta_i)^{1/4}] + 1$ of $S_r(\Delta_i)$ has increased; see Section 2.3. We define v_k to be the smallest integer such that $[v_k^{1/4}] = k$. Then the regions $\delta S_r(\Delta_i)$ contributing to δY_r are characterized by an increase in $N_r(\Delta_i)$, namely,

$$N_r(\Delta_i) < v_{k_i} \leqq v_{k'_i} \leqq N_{r+1}(\Delta_i) .$$

With this notation,

$$|S_r(\Delta_i)| = (2k_i + 1)^2 \text{ and } S_{r+1}(\Delta_i) = (2k'_i + 1)^2 ,$$

so

$$|\delta Y_r| \leqq O(1) \textstyle\sum_i ((k'_i)^2 - k_i^2) .$$

However, $N(\Delta_i)$ increases only for steps in which there is a contraction to the exponent in Δ, and for each contraction the increase in $N(\Delta_i)$ is at most $p - 1$. Such contractions arise either from advancing the time of $n + 1$ (or fewer) annihilators, or from the (at most two) contractions from $\delta C \cdot \Delta_\Phi$. Thus we see that there are at most $\max\{n + 1, 2\} \leqq n + 2$ squares Δ_i with $\delta S_r(\Delta_i) \neq \varnothing$. It follows that

$$\begin{aligned} \delta v_{k_i} &= v_{k'_i} - v_{k_i} \leqq \delta N(\Delta_i) \\ &= N_{r+1}(\Delta_i) - N_r(\Delta_i) \leqq (p - 1)(n + 2) , \end{aligned}$$

and so for $k \geqq k_{\min}(p, n)$, we have $k'_i = k_i + 1$, and $(k'_i)^2 - k_i^2 = 2k_i + 1$. Hence

$$|\delta Y_r| \leqq O(1) \textstyle\sum_i k_i \leqq O(1) \sum_i v_{k_i}^{1/4} .$$

For each Δ_i there must have been $(v_{k'_i} - v_{k_i})/p$ contractions to the exponent in Δ_i, whose contraction distances were at least k_i. Thus the corresponding contribution δD_r to D satisfies

$$\delta D_r \geqq \sum_i k_i \frac{|v_{k'_i} - v_{k_i}|}{p} \geqq \varepsilon_1 \sum_i v_{k_i} ,$$

where ε_1 depends only on p. Thus for any constant $\alpha > 0$,

$$|\delta Y_r| \leqq \alpha \delta D_r + O(\alpha^{-1}) . \tag{2.4.32}$$

The exceptions to the derivation of (2.4.32) occur in the steps in which Y_r is enlarged to contain either the A or the B_T vertices. We consider the two cases:

$$N_{j+1}(\Delta_i) \leqq \frac{1}{2} N_{r+1}(\Delta_i) , \text{ and } N_{j+1}(\Delta_i) > \frac{1}{2} N_{r+1}(\Delta_i) .$$

In the first case, the reduction in the degree of the A or B_T monomial in Δ, namely

$$N_{r+1}(\Delta_i) - N_{j+1}(\Delta_i) \geqq \frac{1}{2} N_{r+1}(\Delta_i) ,$$

implies there were at least $(1/2)N_{r+1}(\Delta_i)$ contractions to vertices in Δ_i. Since each corresponding contraction distance is at least $\varepsilon_1 N_{r+1}(\Delta_i)^{1/4}$, we recover (2.4.32) in this case.

In the second case, $N_{j+1}(\Delta_i) \geqq (1/2)N_{r+1}(\Delta_i)$, recall that $N(\Delta_i) = N_{j+1}(\Delta_i)$ by definition. Thus

$$|S_{r+1}(\Delta_i) - S_r(\Delta_i)| \leqq O(N(\Delta_i)^{1/2})$$

and for $K_2 = K_2(K_3)$ sufficiently large,

$$\exp\left[(K_3 + 2 + (n+1)m_0)|\delta Y_r|\right] \leqq \mathfrak{N}_1^{1/2} .$$

Hence in all cases, summing (2.4.32) over r yields

$$\exp\left[(K_3 + 2 + (n+1)m_0)|Y_{j+1}|\right] \leqq \exp\left[\alpha D + O(j+1)\alpha^{-1}\right]\mathfrak{N}_1^{1/2} . \tag{2.4.33}$$

Substituting (2.4.31) and (2.4.33) into (2.4.30) we obtain (2.4.29) if $K_2 = K_2(\varepsilon, m_0, K_3)$ is large enough.

We now establish (2.4.28). The kernel w_α'' is given by (2.3.8), and for an l-term,

$$P_l''(u, s, x_B, y) = w_B(x_B)\int w_1(\alpha, u, s, y_0, \vec{y}, x)\prod_i : \Phi(x_i)^{n_i}: dx$$

is the path space monomial containing all vertices at times earlier than T, and factored out of the path space integral in (2.3.4) and (2.3.8). For $m_0 \geqq 1$, $\lambda \leqq 1$, there is a constant K_3, independent of C, m_0 and λ such that

$$1 \leqq \|e^{-V(Y)}\|_{L_2(d\Phi_C)} \leqq e^{K_3|Y|} \tag{2.4.34}$$

and

$$\|P_l''(u, s, y, x_B)\|_{L_2(d\Phi_C)} \leqq \left[\prod_\Delta K_3^{N(\Delta)} N(\Delta)!\right]\|w(\alpha, u, s, y, x_B, \cdot)\|_{L_2} . \tag{2.4.35}$$

These are standard estimates for such integrals (see [1]). Thus

$$\begin{aligned} |w_\alpha''(u, x_B, y)| &= \left|\int ds \int P_l'' e^{-V(Y_{j+1})} d\Phi_C\right| \\ &\leqq e^{K_3|Y_{j+1}|}\left[\prod_\Delta K_3^{N(\Delta)} N(\Delta)!\right]\int ds \|w(\alpha, u, s, y, x_B, \cdot)\|_{L_2} , \end{aligned}$$

which yields (2.4.28), for $K_2 \geq K_3^2$.

Since our expansion does not terminate after a finite number of steps, we must show that the sum (2.3.7) is convergent. We show that the sum of all terms not in final form tends to zero. This follows from two facts. First, for each term m_α,

$$|m_\alpha| \leqq \|m_\alpha\|_{2K_2} e^{O(\mathrm{Vol})} ,$$

where the volume of interaction, denoted Vol, is the area of the set of points within unit distance of suppt h. Here we use (2.4.34), (2.4.35) and a Schwarz inequality to factor out vertices. Also $O(\text{Vol}) \to 0$ as $\lambda \to 0$. Secondly, a fraction $\beta > 0$ of the elementary expansion steps is the step I of inserting a contour. By Propositions 2.4.4, after $O(L)$ expansion steps,

$$\sum' \|m_\alpha\| \leqq \left(\frac{3}{4}\right)^{O(L)} \|AB_T\| ,$$

where $\sum'$ denotes the sum over terms not in final form. The desired bound then follows.

3. Contours removed

In this chapter we remove all contours from the Chapter 2 expansion. In Chapter 2, we factored out the vertices at times earlier than T from the integrand F. Starting with the same integral, $\int F d\Phi_{C_*}$, the goal of the present expansion is still to factor out vertices at times earlier than T, but also to return to the Gaussian measure $d\Phi_{C_*}$. The new expansion is still defined inductively, and the terms still have the form (2.2.6). There are two elementary steps. The first is the Chapter 2 expansion, (2.3.7), the increasing contour construction. The second elementary step is the use of (2.2.1) and (2.2.5) to remove contours. These two steps are applied in alternation until the expansion terminates.

To describe the process of removing contours in more detail, we consider a P_2 term m_α, arising at the end of an increasing contour expansion, i.e., in (2.3.7). If $C_* = C(\mathfrak{X}_l, u)$, and if Y_j is the complete contour in m_α, then $X_{l+1} = Y_j$ and in (2.2.1) $s = u_{l+1}$, $C' = C_{*X_{l+1}}$ and $C'' = C_0$. The first term in (2.2.1) is a P_3 term if $n_a > 0$, and a final term if $n_a = 0$. The expansion of this chapter terminates for this term. The second term in (2.2.1) is written as a sum (over possible localizations and contractions), and for the purpose of defining norms the terms in this sum are regarded as P_1 terms. In these terms, we make the new choice $C_* = C(\mathfrak{X}_{l+1}, u)$, $\mathfrak{X}_{l+1} = \{\mathfrak{X}_l, X_{l+1}\}$, and we take $j = 0$ and $Y_0 = \varnothing$. Then $\mathcal{K}$, defined by m_α, equals one and

$$\mathcal{K}' = e^{\varepsilon m_0 d'/2} ,$$

where d' is the contraction distance in (2.2.1).

The expansion now has the form

$$\int F d\Phi_{C_0} = \sum_\alpha \int d\bar{x}\, d\bar{y}\, w_L(\alpha, x, y) \int \prod_{i=1}^{n_a} a(y_i) B_T(x) e^{-V} d\Phi_{C_0} \tag{3.1}$$
$$+ \sum_\alpha w_R(\alpha) \int e^{-V} d\Phi_{C_0} .$$

The summation in (3.1) extends over the various terms generated by our expansion. This includes summation over $0 \leqq n_a \leqq n$ annihilators and over various space time localizations.

THEOREM 3.1. *Let* ε, n *be given. Let* m_0, K_2, K_1 *be sufficiently large and let* λ *be sufficiently small. Then the expansion* (3.1) *is norm decreasing.*

Remark. The expansion (3.1) is also norm decreasing if we start with a P_1 or P_2 term in place of an initial term.

Proof. It is sufficient to prove that each elementary step of the expansion decreases norms, and using Theorem 2.4.1 it is sufficient to show that removing the contours decreases norms. It is sufficient to consider this step applied to a single P_2 term m_α.

Consider the first (P_3 or final) term, m'_α, in (2.2.1). We have

$$||m'_\alpha|| = K_1^{\nu'} e^{rt'} \mathfrak{N}' ||w'||_2 \leqq K_1^{-1} ||m_\alpha|| \leqq \frac{1}{2} ||m_\alpha|| , \tag{3.2}$$

since $\nu' \leqq \nu - 1$, $t' = t$, $\mathfrak{N}'(\Delta) = \mathfrak{N}(\Delta)$ and

$$\begin{aligned} |w'(\alpha, x, y)| &= \left| \int du\, w(\alpha, u, x, y) \right| \\ &\leqq \int du\, e^A |w(\alpha, u, x, y)| , \end{aligned} \tag{3.3}$$

with

$$A = \sum_{k=1}^{l} U_{k,l} |X_k| + |X_{l+1}| = \sum_{k=1}^{l=1} \delta U_{k,l+1} |X_k| . \tag{3.4}$$

Consider next the P_1 terms $m'_{\alpha'}$ which result from applying (2.2.1) and (2.2.5) to m_α as above. Consider the term which introduces an annihilator $a(y_{n_a+1})$ localized in a lattice interval Δ_2. The kernel $w'_{\alpha'}$ of $m'_{\alpha'}$ is

$$w'_{\alpha'}(u', y_0, \vec{y}', x) = w(\alpha, u, y_0, \vec{y}) \delta \hat{C}_{l+1}(x, y_{n_a+1}) v(x)$$

where $y' = \{y, y_{n_a+1}\}$, $\delta\hat{C}_{l+1}$ is given by the discussion following (2.1.7) and (2.2.5), and $v(x)$ is $O(\lambda)$ times the characteristic function of a lattice square Δ_1. As in (2.4.19), we obtain

$$\sum_{\Delta_1 \Delta_2} e^{m_0(1-\varepsilon/2)d'} ||w'_{\alpha'}||_2 \leqq O(\lambda A) ||w_\alpha||_2 . \tag{3.5}$$

Here Proposition 2.1.2 replaces 2.1.1 and A is defined in (3.4). Since

$$N(\Delta'_1) \leqq p - 1 = O(1) ,$$

$$||m'_{\alpha'}|| = O(1) K_1^{\nu'} e^{\varepsilon m_0 d'/2} e^{rt} \mathfrak{N} \int du' e^{Au_{l+1}} ||w'_{\alpha'}||_2 ,$$

where $O(1)$ depends on K_2 and $du' = du\, du_{l+1}$. Taking $\varepsilon \leqq 1$ and substituting (3.5) in the above equation yields

$$\begin{aligned}\sum_{\alpha'}\|m'_{\alpha'}\| &\leqq O(\lambda)K_1^{\nu'}e^{rt}\mathfrak{N}\int du'\,Ae^{Au_{l+1}}\|w_\alpha\|_2 \\ &\leqq O(\lambda)K_1^{\nu'}e^{rt}\mathfrak{N}\int du\,e^{A}\|w_\alpha\|_2 \\ &= O(\lambda)\|m_\alpha\|\,.\end{aligned} \tag{3.6}$$

Here $O(\lambda)$ depends on K_1 and K_2, but since λ is chosen after K_1 and K_2, $O(\lambda)\leqq 1/4$ for λ small.

The remaining terms $m'_{\alpha'}$ have an outside contraction $\partial/\partial\Phi(y_{n_\alpha+1})$ to a vertex, and are bounded similarly. Combined with (3.2) and (3.6), this completes the proof. We note that the expansion (3.1) is convergent, as follows similarly to the proof of convergence of (2.3.7).

4. The n-particle cluster expansion

The n-particle cluster expansion leads to Theorems 1.1.8–1.1.11. In this expansion, the annihilators from the Chapter 3 expansion are replaced by Euclidean fields. In the case $n=1$, the Euclidean fields occur at time zero only.

The n-particle expansion is defined inductively, by alternating two elementary expansion steps. The first expansion step is the Chapter 3 expansion. The second elementary step uses a modified form of (2.2.7), namely,

$$a(y)=\Phi(y)-\int_{z_0\leqq y_0}dz[(2\mu)^{-1}e^{-(y_0-z_0)\mu}](\vec{y},z)\frac{\partial}{\partial\Phi(z)}\,, \tag{4.1}$$

to convert an annihilator at the time y_0 into a field at the same time. The proof of (4.1) involves an integration by parts in the function space integral $\int F d\Phi_{C_0}$, and it depends on the explicit form of the covariance C_0. Hence (4.1) is valid when applied to an integrand in such an integral, but it is not valid as an operator on F. We write (4.1) as

$$a(y)=\Phi(y)+\left(K\frac{\partial}{\partial\Phi}\right)(y)$$

where K is the integral operator defined in (4.1).

In order to obtain a generalization of (4.1) valid for more than one annihilator, let

$$:\Phi(y_1,\cdots,y_r):\;=\;:\Phi(y_1)\cdots\Phi(y_r):\,,$$

$$(K\Phi)(y_1,\cdots,y_r)=\prod_{i=1}^r\left(K\frac{\partial}{\partial\Phi}\right)(y_i)$$

for $r>1$ and let $\Phi(\varnothing)=1$,

$$\left(K\frac{\partial}{\partial\Phi}\right)(\varnothing)=1\,.$$

For J a subset of $\{1, \cdots, n_a\}$ with complement $\sim J$, let y_J and $y_{\sim J}$ denote the corresponding subsets of $(y_1, \cdots, y_{n_a})$. Then

$$\sum_{i=1}^{n_a} a(y_i) = \sum_{J\subset\{1,\cdots,n_a\}} :\Phi(y_J):\left(K\frac{\partial}{\partial\Phi}\right)(y_{\sim J}) . \tag{4.2}$$

The derivatives in each monomial act on F, but not on the Wick ordered coefficient $:\Phi(y_J):$. Again integration by parts occurs in (4.2), and (4.2) is valid when applied to an integrand. The term with $J = \{1, 2, \cdots, n_a\}$ in (4.2) is a final term, and the other terms are P_1 terms. The full expansion, obtained by alternating these two steps, has the form

$$\begin{aligned}\int AB_T e^{-V} d\Phi_{C_0} = \sum_{y_0 \leq T} \sum_{i=0}^{n} \int d\bar{y}\, w_i(y) \int :\Phi(y_1) \cdots \Phi(y_i): B_T e^{-V} d\Phi_{C_0} \\ + w_r \int e^{-V} d\Phi_{C_0} ,\end{aligned} \tag{4.3}$$

once we group together terms with common values of i and y_0.

THEOREM 4.1. *Let ε and n be given. Let m_0, K_2 and K_1 be sufficiently large, and let λ be sufficiently small. Then the n-particle cluster expansion* (4.3) *is norm decreasing.*

Proof. We show that each expansion step, applied to a single term m_α, is norm decreasing. Since the Chapter 3 expansion is norm decreasing, it is sufficient to consider the step (4.2). This step is similar to, but easier than the analogous step in the proof of Theorem 4.2 below, so we do not give the details.

Proof of Theorems 1.1.8 *and* 1.1.11. Divide (4.3) by $\int e^{-V} d\Phi_{C_0}$ and set

$$Q_n(y_0) = \sum_{i=0}^{n} \int d\bar{y}\, w_i(y) \int :\Phi(y_1) \cdots \Phi(y_i): .$$

(By removal of the Wick ordering in $Q_n(y_0)$, we may apply the norms of (1.1.8).)
Then as $y_0 = t$ for $y_0 \leqq T$,

$$\int AB_T dq = \sum_{t \leqq T} \int Q_n(t) B_T dq + w_r . \tag{4.4}$$

Let $||\cdot||_0$ denote the Chapter 1 norm (1.1.7), (1.1.8), and let $||\cdot||$ denote the norm (2.4.3). Then for $t \leq T$,

$$\begin{aligned} ||Q_n(t)B_T|| &= e^{\gamma t} ||Q_n(t)||_0 ||B||_0 , \\ ||w_r|| &= K_1 e^{\gamma T} |w_r| , \\ ||AB_T|| &= K_1^4 ||A||_0 ||B||_0 . \end{aligned}$$

Hence by Theorem 4.1,

$$\sum_{t \leqq T} e^{\gamma t} \| Q_n(t) \|_0 \| B \|_0 + e^{\gamma T} | w_r | \leqq \sum_{t \leqq T} \| Q_n(t) B_T \| + \| w_r \|$$
$$\leqq \| AB_T \| = K_1^4 \| A \|_0 \| B \|_0 .$$

Theorem 1.1.11 follows from this inequality, while Theorem 1.1.8 is the special case $B = I$, $n = 0$.

In the case $n = 1$, the identity

$$a(y) = \int d\vec{z} \, [e^{y_0 \mu}](\vec{y}, \vec{z}) \Phi(\vec{z}, 0) - \int_{z_0 \leqq y_0} dz [(2\mu)^{-1} e^{(y_0 - |z_0|)\mu}](\vec{y}, \vec{z}) \frac{\partial}{\partial \Phi}(z) , \tag{4.5}$$

for $y_0 \geqq 0$, replaces (4.1). This identity also involves integration by parts, and is valid when applied to the integrand F in $\int F d\Phi_{C_0}$. The one-particle expansion is

$$\int AB_T dq = w_0 \int B_T dq + \int d\vec{y} \, w_1(\vec{y}) \int \Phi(\vec{y}, 0) B_T dq + w_r . \tag{4.6}$$

Theorem 1.1.10 is implied by the following result.

THEOREM 4.2. *Let B have support in the momentum range $|P_1| \leqq \sqrt{3}\, m_0(1 - \varepsilon)$. Let m_0, K_2 and K_1 be sufficiently large and let λ be sufficiently small. Then the one-particle cluster expansion* (4.6) *is norm decreasing.*

The restriction on the momentum support of B applies to the Fourier transform

$$B^{\sim}(\vec{p}) = \int d\vec{x} \, e^{-i\vec{p}\vec{x}} \, U(\vec{x}) B U(\vec{x})^* .$$

It is important to work with a translation invariant Hamiltonian, in order that time translation preserve subspaces of bounded momentum. The estimates of Theorems 1.1.8, 1.1.11, 3.1 and 4.1 are uniform in the space cutoff h, and the path space integrals converge as $h \to 1$. Thus these theorems hold in the infinite volume limit, and we proceed to work in the limit $h = 1$. An alternative would be to use a periodic space cutoff; see [8] for example. In either case, the Hamiltonian is translation invariant and conserves momentum. By momentum conservation, a nonzero contribution to $\int AB_T dq$ from $A^{\sim}(\vec{p})$ arises only for $\vec{p}$ in the momentum range $|\vec{p}| \leqq \sqrt{3}\, m_0(1 - \varepsilon)$. Thus, in (4.6), $\Phi^{\sim}(\vec{p}, 0)$ can be multiplied by a function $\zeta^{\sim}(\vec{p})$ which equals one on the set $|\vec{p}| \leqq \sqrt{3}\, m_0(1 - \varepsilon)$. We take $\zeta^{\sim}$ to be smooth and supported on the set $|\vec{p}| \leqq \sqrt{3}\, m_0(1 - \varepsilon/2)$.

The main difference between (4.5) and (4.1) is the occurrence of the unbounded operator kernel $\exp(y_0 \mu)$, $y_0 \geqq 0$, in (4.5). We use the fact that there is a gap between the energy bounding $\mu(\vec{p})$, for $|\vec{p}| \leqq \sqrt{3}\, m_0(1 - \varepsilon/2)$, and the energy $2m_0(1 - \delta)$ given by the decay rate of Theorem 1.1.11 with

$n = 1$ and ε replaced by $\delta < \varepsilon$. In fact for such $\vec{p}$,

$$\mu(\vec{p}) \leqq m_0(1 + 3(1 - \varepsilon/2)^2)^{1/2} = 2m_0(1 - \varepsilon') , \tag{4.7}$$

defining $\varepsilon' = \varepsilon'(\varepsilon)$. Then

$$\varepsilon' = \frac{3}{8}\varepsilon - \frac{3}{32}\varepsilon^2 + \frac{1}{2}(\varepsilon')^2 > \frac{9}{32}\varepsilon ,$$

so we choose $\delta = \varepsilon/4$, ensuring $\delta < \varepsilon'$.

Proof of Theorem 4.2. Each expansion step is norm decreasing. To show this, we again need consider only the step (4.5) applied to a single term m_α. This m_α is a P_3 term, of the form

$$m_\alpha = \int d\vec{y}\, w_\alpha(\vec{y}) \int a(y)\, B_T\, dq .$$

Let I_i denote the interval $|z - i| \leqq 1/2$, and also its characteristic function. The expansion step (4.5) is the identity

$$m_\alpha = \sum\nolimits_{i \in Z} m_i + \sum\nolimits_{\substack{j \in Z^2 \\ j_0 \leqq y_0}} m_j$$

where

$$m_i = \int_{I_i} w'_\alpha(\vec{z}, 0) \int \Phi(\vec{z}, 0)\, B_T\, dq$$

$$m_j = \int_{\Delta_j} dz\, w''_\alpha(z) \int \frac{\partial V}{\partial \Phi(z)} B_T\, dq$$

$$w'_\alpha(z) = [e^{(y_0 - |z_0|)\mu}](\zeta^* w_\alpha)(\vec{z})$$

$$w''_\alpha(z) = [(2\mu)^{-1} e^{(y_0 - |z_0|)\mu}](\zeta^* w_\alpha)(\vec{z})$$

and $\zeta^\sim$ is the C^∞ function chosen above. Here m_i is a final term and m_j is a P_1 term.

LEMMA 4.3. *Let ε'' satisfy $\delta < \varepsilon'' < \varepsilon'$ and let $w \in L_2(I_0)$. Let $\sigma = 0, 1$. Then there is a constant $K_4 = K_4(m_0)$ such that*

$$\| I_i \mu^{-\sigma} e^{t\mu}(\zeta^* w) \|_2 \leqq K_4(i^2 + 1)^{-1} e^{2tm_0(1 - \varepsilon'')} \| w \|_2 .$$

Proof. By Fourier analysis, and (4.7),

$$\begin{aligned}
\| I_i \mu^{-\sigma} e^{t\mu}(\zeta^* w) \| &\leqq O(1)(i^2 + 1)^{-1} \left\| \left(-\frac{\partial^2}{\partial \vec{p}^2} + 1 \right) \mu^{-\sigma} e^{t\mu(\vec{p})} \zeta^\sim w^\sim \right\| \\
&\leqq O(t^2)(i^2 + 1)^{-1} e^{2tm_0(1-\varepsilon')} \left\| \left(-\frac{\partial^2}{\partial \vec{p}^2} + 1 \right) w^\sim \right\|_2 \\
&= O(1)(i^2 + 1)^{-1} e^{2tm_0(1-\varepsilon'')} \| (\vec{z}^2 + 1) w \|_2 \\
&\leqq O(1)(i^2 + 1)^{-1} e^{2tm_0(1-\varepsilon'')} \| w \|_2 ,
\end{aligned}$$

because suppt $w \subset I_0$.

To compare the norms of m_i and m_j with m_α, we note that $\mathcal{K} = 1 = \mathcal{K}'$, and $\mathfrak{N}' = O(1)\mathfrak{N}$ where $O(1)$ is a constant depending on K_2. The value of t decreases in this expansion step, so that $t' = 0$ for the m_i term and $t' = j_0 + (1/2)$ for m_j. Thus using Lemma 4.3,

$$\begin{aligned} \|m_i\| &= O(1) K_1^2 (i^2 + 1)^{-1} e^{\gamma t} \mathfrak{N} \|w_B\|_2 \|w_\alpha\|_2 \\ &= O(1) K_1^{-1} (i^2 + 1)^{-1} \|m_\alpha\| \end{aligned}$$

if w_α is supported in I_0, and a translate of this estimate in the general case. Here $O(1)$ depends on m_0 and K_2. Since K_1 is chosen after m_0 and K_2,

$$\sum_i \|m_i\| \leqq \frac{1}{2} \|m_\alpha\|$$

for K_1 sufficiently large.

The estimates on m_j are similar, with the observation that the new vertex $\partial V/\partial\Phi$ has a kernel containing a factor $O(\lambda)$. We use small λ, chosen after K_1, to ensure

$$\sum_j \|m_j\| \leqq \frac{1}{4} \|m_\alpha\| .$$

This completes the proof of Theorem 4.2.

Finally, we note that the expansions (4.3) and (4.6) converge, as follows from two facts. First, the $n = 0$ expansion converges, since the Chapter 3 expansion converges. Hence Theorem 1.1.8 applies and $|m_\alpha| \leqq O(1)\|m_\alpha\|$. Second, each application of the expansion (4.2) or (4.5) decreases by a factor 3/4, as does the Chapter 3 expansion. Thus the sum $\sum |m_\alpha|$ over terms not in final form is decreasing.

References

[1] J. Dimock and J. Glimm, Measures on Schwarz distribution space and applications to $\mathcal{P}(\phi)_2$ field theories, Adv. Math. **12** (1974), 58-83.

[2] J. Feldman, A relativistic Feynman-Kac formula, Nuclear Physics **B 52** (1973), 608-614.

[3] J. Glimm and A. Jaffe, The $\lambda\phi_2^4$ quantum field theory without cutoffs I, Phys. Rev. **176** (1968), 1945-1951.

[4] ———, The $\lambda\phi_2^4$ quantum field theory without cutoffs, II. The field operators and the approximate vacuum, Ann. of Math. **91** (1970), 362-401.

[5] ———, The $\lambda\phi_2^4$ quantum field theory without cutoffs, III. The physical vacuum, Acta Math. **125** (1970), 203-267.

[6] ———, The $\lambda\phi_2^4$ quantum field theory without cutoffs, IV. Perturbations of the Hamiltonian, J. Math. Phys. **13** (1972), 1558-1584.

[7] ———, Positivity of the ϕ_3^4 Hamiltonian, Fortschritte der Physik, **21** (1973), 327-375.

[8] ———, Quantum field models, in *Statistical Mechanics and Quantum Field Theory*, Ed. by C. de Witt and R. Stora, Gordon and Breach, New York, 1971.

[9] ———, Boson quantum field models, in *Mathematics of Contemporary Physics*, Ed. by R. Streater; Academic Press, New York, 1972.

[10] ———, The n-particle cluster expansion for the $\mathcal{P}(\phi)_2$ quantum field model, unpublished.

[11] J. GLIMM and T. SPENCER, The Wightman axioms and the mass gap for the $\mathcal{P}(\phi)_2$ quantum field theory, unpublished.

[12] R. GRIFFITHS and B. SIMON, The ϕ_2^4 field theory as acla ssical Ising model, Commun. Math. Phys. **33** (1973), 145-164.

[13] F. GUERRA, L. ROSEN and B. SIMON, The $\mathcal{P}(\phi)_2$ Euclidean quantum field theory as classical statistical mechanics, to appear in Ann. of Math.

[14] K. HEPP, On the connection between Wightman and LSZ quantum field theory, in *Axiomatic Field Theory*, Ed. by M. Chretien and S. Deser, Gordon and Breach, New York, 1966.

[15] E. NELSON, Quantum fields and Markoff fields, A.M.S. Symp. on Partial Differential Equations, Berkeley, 1971.

[16] ———, Construction of quantum fields from Markoff fields, J. Funct. Anal. **12** (1973), 97-112.

[17] C. NEWMAN, The construction of stationary two dimensional Markoff fields with an application to quantum field theory, J. Funct. Anal. **14** (1973), 44-61.

[18] K. OSTERWALDER and R. SCHRADER, Axioms for Euclidean Green's functions, Commun. Math. Phys. **31** (1973), 83-112.

[19] ———, Euclidean Fermi fields and a Feynman-Kac formula for boson-fermion models, Helv. Phys. Acta, **46** (1973), 277-302.

[20] D. RUELLE, On the axiomatic condition in quantum field theory, Helv. Phys. Acta **35** (1962), 147-163.

[21] B. SIMON, Correlation inequalities and the mass gap in $\mathcal{P}(\phi)_2$, II. Uniqueness of the vacuum for a class of strongly coupled theories, to appear in Ann. of Math.

[22] K. SYMANZIK, Euclidean quantum field theory, in *Local Quantum Theory*, 1968 Rendiconti della Scuola internazionale di Fisica, "E. Fermi", XLV Corso, R. Jost, Editor, Academic Press, New York, 1969.

[23] K. OSTERWALDER and R. SCHRADER, Axioms for Euclidean Green's function II, preprint.

[24] J. GLIMM, A. JAFFE and T. SPENCER, In: Lecture Notes in Physics Vol. 25, Ed. by G. Velo and A. Wightman, Springer Verlag, Berlin, 1973.

[25] J. FELDMAN, On the absence of bound states in the $\lambda(\varphi^4)_2$ quantum field model without symmetry breaking. To appear in Canadian Journal of Physics.

[26] T. SPENCER. The absence of even bound states for $\lambda(\varphi^4)_2$, to appear in Commun. Math. Phys.

(Received October 16, 1972)
(Revised June 28, 1973)

II Phase Cell Localization and φ_3^4 Stability

Introduction

The fundamental idea in all the work above was to understand in detail the consequences of certain Lagrangians. The methods always led to consideration of degrees of freedom in the interaction that could be localized approximately in phase space cells of unit size. These degrees of freedom could be analyzed in detail, and it was possible to prove that different regions of phase space were approximately independent. On the basis of this independence, renormalization could be studied independently in each local space-time volume, as well as the exponential decay rates in the cluster expansions of Section 1.

This philosophy emerged slowly as a systematic technique. One simple illustration of the idea was a new proof of the finite vacuum energy per unit volume in $P(\phi)_2$ Hamiltonians. The positivity bound involves localization in momentum space, and O(1) localization in space-time. The independence of different space-time volumes yields a bound on the energy density. These methods were developed further to analyze perturbations of the $P(\phi)_2$ Hamiltonians by local field operators. In the fourth paper in the ϕ_2^4 series, the authors bound the vacuum enery of $H \pm \phi$. As a consequence, they prove self adjointness of the field operators in the representation defined by the physical vacuum. A general method to study perturbing operators of H, based on estimates of commutators between ϕ and H, is also developed in that paper.

These methods were practice to develop techniques sufficiently powerful to analyze the three-dimensional problem. The fundamental estimate on the positivity of the ϕ_3^4 Hamiltonian is proved in the last paper in this section, certainly the most complicated paper by the authors. In this construction the authors estimate those degrees of freedom which are localized simultaneously on specific length scales and on specific momentum scales. The estimates for $\exp(-tH)$ are defined inductively on successive length or momentum scales, and they apply for all values of the ϕ^4 coupling constant. This philosophy also appears at the basis of the renormalization group method. Using block spin transformations and scaling, the ideas in this paper have been implemented from such a point of view.

Bibliography

Positivity and self adjointness of the $P(\phi)_2$ Hamiltonian, Commun. Math. Phys. 22, 253–258 (1971).

The $\lambda\phi_2^4$ quantum field theory without cutoffs IV. Perturbations of the Hamiltonian, J. Math. Phys. 13, 1568–1584 (1972).

Positivity of the ϕ_3^4 Hamiltonian, Fort. Physik 21, 327–376 (1973).

Positivity and Self Adjointness of the $P(\phi)_2$ Hamiltonian

JAMES GLIMM*

Courant Institute of Mathematical Sciences, New York University, New York, New York

ARTHUR JAFFE**

Lyman Laboratory of Physics, Harvard University, Cambridge, Massachusetts

Received May 15, 1971

Abstract. We give a new proof that the locally correct Hamiltonian $H(g)$ is self adjoint, and that the vacuum energy $E(g) = \inf \text{spectrum} H(g)$ satisfies $-O(D) \leqq E(g)$, where $0 \leqq g \leqq 1$ and $D = \text{diam. supp.} g$.

An existence theorem has been proved for boson quantum fields with polynomial self interactions in two space time dimensions, and many basic properties of these quantum field models have been established. A self contained account of this theory is presented in [1]. A principal step in the construction of the field theory is to show that the Hamiltonian (energy operator) for a bounded space time region is bounded from below and self adjoint. The original proof of semiboundedness was given by Nelson [2] for the ϕ^4 theory. It was extended by Glimm [3] to a different type of space cutoff and to a positive polynomial $P(\phi)$ interaction. The authors [4] obtained a volume independent bound on the vacuum energy per unit volume. The original proof of self adjointness was given by the authors [5] for the ϕ^4 theory and by Rosen [6] for the $P(\phi)$ theory. Subsequent simplifications have been given [7–12]. In this note we present an easy proof of self adjointness and of the volume independent lower bound. The previous simplifications did not yield the volume independent lower bound. See [1] for notation.

Let

$$H(\kappa) = H_{o\,V} + H_I(\kappa) = H_{0,V} + \int :P(\phi_{\kappa,V})(x): g(x)\, dx\,,$$

where g is a function with compact support and $0 \leqq g \leqq 1$. We use the fact that the operators $H(\kappa)$, $H_{o,V}$, $H_I(\kappa)$ are each self adjoint and bounded from below.

---

* Supported in part by Air Force Office of Scientific Research, Contract AF49(638) –1719.

** On leave at Princeton University. Alfred P. Sloan Foundation Fellow. Supported in part by Air Force Office of Scientific Research, Contract F44620–70–C–0030.

Theorem 1. *$H(\kappa)$ is bounded from below, uniformly in κ, and V for $V \geqq 1$. Moreover* $\exp(-tH(\kappa))$ *converges in norm as $\kappa \to \infty$ and $V \to \infty$, uniformly in t, for t bounded away from zero and infinity.*

The proof of this result depends on the fact that certain perturbation expansions for $H(\kappa)$ converge in the limit of high energy. See also [11]. We use the Duhamel formula to generate such an expansion:

$$e^{-tH(\kappa)} = e^{-tH(\kappa')} - \int_0^t e^{-sH(\kappa')}(H(\kappa) - H(\kappa'))\, e^{-(t-s)H(\kappa)} ds\,. \tag{1}$$

To simplify the formulas below, we define

$$H(\sigma, s) = \begin{cases} H(\kappa') & \text{for} \quad \sigma < s \\ H(\kappa) & \text{for} \quad \sigma > s \end{cases}$$

and write the integrand in (1) as

$$\left(e^{-\int_0^t H(\sigma,s)d\sigma} \delta H(s)\right)_+ \tag{2}$$

where the subscript $+$ denotes a time ordering and $\delta H(s)$ inserts $\delta H = H(\kappa) - H(\kappa')$ at the time s. The formula (1) follows from differentiating $F(t, s) = \exp(-sH(\kappa'))\exp(-(t-s)(H(\kappa)))$ with respect to s and integrating the result from 0 to t. Since $F(t, s)$ is strongly continuous in s for $0 \leqq s \leqq t$, and it is norm differentiable with the bounded and norm continuous derivative (2) in $0 < s < t$, the equality (1) is valid whenever the integral in (1) exists. To prove Theorem 1, we use an iterated Duhamel formula, and we bound the resulting integrands. The bounds on the integrands are independent of the Duhamel formula and justify its use.

Let $p = \deg P$ and $\kappa_j = \exp(j^{2/p})$. By undoing the Wick ordering in H_I, we have the cutoff dependent bound

$$-O(j) = -O(\log \kappa_j)^{p/2} \leqq H_I(\kappa_j)\,. \tag{3}$$

Let

$$h_j = \begin{cases} H(\kappa_j) & \text{if} \quad \kappa_j \leqq \kappa \\ H(\kappa) & \quad\quad \kappa_j \geqq \kappa \end{cases}$$

and let $\delta h_j = H(\kappa) - H(\kappa_j)$. Iterating (1) we obtain

$$\exp(-tH(\kappa)) = \sum_{n=0}^{\infty} (-1)^n \int ds \left(e^{-\int_0^t H(\sigma,s)d\sigma} \prod_{i=1}^{n} \delta h_i(s_i)\right)_+ \tag{4}$$

where the time integration extends over the domain

$$0 \leqq s_1 \leqq s_2 \leqq \cdots \leqq s_n \leqq t \tag{5}$$

and $H(\sigma, s) = h_j$ if $s_{j-1} < \sigma < s_j$. The series terminates for $\kappa_n \geqq \kappa$.

In the representation of Fock space for which the operators $H_I(\kappa)$ are all diagonal, $\exp(-tH_{o,V})$ is an integral operator with a positive kernel. We take absolute values of operator kernels in this representation. We use the Trotter product formula for each factor $\exp(-(s_{j+1}-s_j)h_j)$ in (4) together with the lower bound (3) and the upper bound

$$|\delta h_i| \leqq \kappa_i^{-1/4} + \kappa_i^{1/4}(\delta h_i)^2 = f_i$$

to obtain

$$\left\langle \theta, \left(e^{-\int_0^t H(\sigma,s)\,d\sigma} \prod_{i=1}^n \delta h_i(s_i) \right)_+ \theta \right\rangle \leqq \|\theta\|^2 e^{O(n)t} \left\| \left(e^{-\int_0^t H_{o,V}\,d\sigma} \prod_{i=1}^n f_i(s_i) \right)_+ \right\| . \tag{6}$$

The time ordered product above is easily seen to be a bounded operator, and it is norm continuous on the open domain $0<s_1<\cdots<s_n<t$. To obtain a bound on the closed domain (5), we use the estimate

$$\| f_j(N+I)^{-p} \| \leqq O(\kappa_j^{-1/4}) = 0(\exp(-\tfrac{1}{4}j^{2/p})) . \tag{7}$$

Here N is the number of particles operator, see [1]. The maximum length $s_j - s_{j-1}$ of the time intervals in the time ordered product is at least $t/(n+1)$. We commute np powers of N onto the maximum time interval, and use (7) to show

$$\left\| \left(e^{-\int_0^t H_{o,V}\,d\sigma} \prod_{i=1}^n f_i(s_i) \right)_+ \right\| \leqq a^n (n!)^p \, \|N^{pn} e^{-tH_o/(n+1)}\| \prod_{j=1}^n \kappa_j^{-1/4}$$
$$\leqq \left(\frac{c}{t}\right)^{pn} (n!)^{2p} \exp(-bn^{1+2/p})$$

for certain positive constants a, b, c. Hence (6) is bounded uniformly in s and so the integral in (4) exists. This justifies (4) and yields the bound

$$\exp(-tH(\kappa)) \leqq \sum_{n=1}^{\infty} \exp(an \log n - bn^{1+2/p}) < \infty , \tag{8}$$

with new constants a and b. These constants are independent of κ and are bounded uniformly if $V>1$ and if t is bounded away from zero and ∞. This completes the proof that $H(\kappa)$ is bounded from below. The same estimates yield norm convergence of the semigroup $\exp(-tH(\kappa))$ as $\kappa \to \infty$, and the limit $V \to \infty$ is also easily controlled. It follows from Theorem 1 and the convergence of $H(\kappa)$ on the dense set $C^\infty(H_o)$, that the limit of $\exp(-tH(\kappa))$ is a semigroup $\exp(-tH)$ with self adjoint generator H, see [1].

We now establish a volume independent lower bound for $H(\kappa)$. Let $A(x)$ be the configuration space annihilation operator, so $A(x)$ annihilates a particle localized (in the Newton-Wigner sense) at x. We let

$$N_i = \int_{i-1/2}^{i+1/2} A(x)^* A(x)\,dx$$

and

$$N_{\text{loc}} = \sum_{i=-\infty}^{\infty} N_i e^{-m|i|/2}$$

be localized number operators. We assert that for g supported in a fixed interval B,

$$\|f_j(N_{\text{loc}}+I)^{-p}\| \leqq O(\kappa_j^{-1/4}). \tag{9}$$

Our proof of Theorem 1 is valid with $\varepsilon N_{\text{loc}}$ replacing H_o and (9) replacing (7). Thus

$$0 \leqq \varepsilon N_{\text{loc}} + H_I(\kappa) + O(1)\,\mathrm{I} \tag{10}$$

Summing (10) over translates and using the bound $N \leqq \text{const}\, H_o$ yields

Theorem 2. *Let $0 \leqq g \leqq 1$ and let $D = \text{diam.supp.}g$. Then*

$$0 \leqq H(\kappa) + 0(D),$$

as $D \to \infty$, and $O(D)$ is uniform in κ and $V \geqq 1$.

Proof. We only need to establish (9), and without loss of generality we may assume that $\text{supp}\, g \subset (-\frac{1}{8}, \frac{1}{8})$. Let $w_j(x_1, \ldots, x_r)$ be the configuration space kernel of a Wick monomial W_j contributing to f_j, $0 < r \leqq 2p$. Let

$$w_{j,i_1,\ldots,i_r}(x) = w_j(x) \prod_{\nu=1}^{r} E_{i_\nu}(x_\nu)$$

where $E_i(x)$ is the characteristic function of the interval $[i-1/2, i+1/2]$. We assert that

$$\begin{aligned} \|W_{j,i_1,\ldots,i_r}\theta\| &\leqq \text{const}\, \|w_{j,i_1,\ldots,i_r}\|_2 \left\| \prod_{\nu=1}^{r} (N_{i_\nu}+I)^{1/2}\theta \right\| \\ &\leqq O(\kappa_j^{-1/4}) \exp(-m|i_1| - \cdots - m|i_r|) \left\| \prod_{\nu=1}^{r} (N_{i_\nu}+I)^{1/2}\theta \right\|. \end{aligned} \tag{11}$$

Since $\|(N_i+I)^{1/2}(N_{\text{loc}}+I)^{-1/2}\| \leqq \exp(m|i|/2)$, summing (11) over the i_ν proves (9). The first inequality in (11) is elementary, since N_i measures the number of particles in $[i-\frac{1}{2}, i+\frac{1}{2}]$. To bound the L_2 norm of

$w_{j,i_1,\ldots,i_r}$, we represent the kernel as

$$w_{j,i_1,\ldots,i_r} = \left(\prod_{\nu=1}^{r} K_{\nu,i_\nu}\right) w_j \tag{12}$$

where K_{ν,i_ν} is an integral operator on the variable x_ν of w_j. Then

$$\|w_{j,i_1,\ldots,i_r}\|_2 \leqq \|w_j\|_2 \prod_{\nu=1}^{r} \|K_{\nu,i_\nu}\|$$

$$\leqq O(\kappa_j^{-1/4}) \prod_{\nu=1}^{r} \|K_{\nu,i_\nu}\| .$$

Let $\mu_x^2 = \left(-\dfrac{d^2}{dx^2} + m^2\right)$ and let ζ be multiplication by a C^∞ function with support in $(-1/4, 1/4)$, such that for all x and y, $\zeta(x)\,\chi_\kappa(x-y)\,g(y) = \chi_\kappa(x-y)\,g(y)$. Here the ultraviolet cutoff in $H_I(\kappa)$ is introduced by the cutoff field $\phi_\kappa(x) = (\chi_\kappa * \phi)(x)$ and χ_κ is supported in $[-\kappa^{-1}, \kappa^{-1}]$. Then we take

$$K_{\nu,i_\nu} = E_{i_\nu}(x_\nu)\,\mu_{x_\nu}^{-1/2}\,\zeta(x_\nu)\,\mu_{x_\nu}^{1/2},$$

and (12) is valid. The operators $\mu_x^{\pm 1/2}$ are given by convolution with a distribution $k_\pm(x)$ that is C^∞ except at $x=0$, with derivatives $0(1)\exp(-m|x|)$ as $|x| \to \infty$. An easy computation then shows that $\|K_{\nu,i}\| \leqq O(1)e^{-m|i|}$. This completes the proof of Theorem 2.

References

1. Glimm, J., Jaffe, A.: Field theory models. In: Dewitt, C., Stora, R. (Eds.) 1970 Les Houches Lectures. New York: Gordon and Breach Science Publ. 1971.
2. Nelson, E.: A quartic interaction in two dimensions. In: Goodman, R., Segal, I. (Eds.): Mathematical theory of elementary particles. Cambridge: MIT Press 1966.
3. Glimm, J.: Boson fields with nonlinear self interaction in two dimensions. Commun. math. Phys. **8**, 12—25 (1968).
4. — Jaffe, A.: The $\lambda(\phi^4)_2$ quantum field theory without cutoffs III: The physical vacuum. Acta Math. **125**, 203—267 (1970).
5. — — A $\lambda\phi^4$ quantum field theory without cutoffs. I. Phys. Rev. **176**, 1945—51 (1968).
6. Rosen, L.: A $\lambda\phi^{2n}$ field theory without cutoffs. Commun. math. Phys. **16**, 157—183 (1970).
7. Segal, I.: Construction of nonlinear local quantum processes: I. Ann. Math. **92**, 462—481 (1970).
8. Høegh-Krohn, R., Simon, B.: Hypercontractive semigroups and two dimensional self-coupled bose fields. J. Functional Analysis, To appear.
9. Masson, D., McClary, W.: On the self adjointness of the $(g(x)\,\phi^4)^2$ Hamiltonian. Commun. math. Phys. **21**, 71—74 (1971).

10. Konrady, J.: Almost positive perturbations of positive self adjoint operators. To appear.
11. Federbush, P.: A convergent expansion for the resolvent of $:\phi^4:_{1+1}$. To appear.
12. Masson, D.: Essential self adjointness of semi bounded operators: An extension of the Kato-Rellich theorem. To appear.

James Glimm
Courant Institute of Mathematical Sciences
New York University
251 Mercer Street
New York, N.Y. 10012/USA

Arthur Jaffe
Lyman Laboratory of Physics
Harvard University
Cambridge, Mass. 02138, USA

The $\lambda\varphi_2^4$ Quantum Field Theory without Cutoffs. IV. Perturbations of the Hamiltonian

James Glimm*
Courant Institute of Mathematical Sciences, New York University, New York, New York

and

Arthur Jaffe†
Lyman Laboratory of Physics, Harvard University, Cambridge, Massachusetts

(Received 3 March 1972)

We introduce an inductive method to estimate the shift δE in the vacuum energy, caused by a perturbation δH of the $\mathcal{P}(\varphi)_2$ Hamiltonian H. We prove that if δH equals the field bilinear form $\varphi(x, t)$, then δE is finite. We show that the vacuum expectation values of products of fields (Wightman functions) exist and are tempered distributions. They determine, via the reconstruction theorem, essentially self-adjoint field operators $\varphi(f)$, for real test functions $f \in \mathcal{S}(R^2)$. We also bound the perturbation of the $\mathcal{P}(\varphi)_2$ Hamiltonian by a polynomial $(\mathcal{P}_1(\varphi))(h) = \delta H$, so long as $\mathcal{P} + \mathcal{P}_1$ is formally positive. In that case, and with $\|h\|_\infty \le 1$, δE is bounded by const$(1 + \text{diam supp } h)$.

1. THE MAIN RESULTS

We derive new estimates on the vacuum energy in the $\mathcal{P}(\varphi)_2$ quantum field model. These estimates permit us to establish one of the missing Wightman axioms for $\mathcal{P}(\varphi)_2$.

Theorem A: The Wightman functions or vacuum expectation values

$$\langle \Omega, \varphi(x_1, t_1)\varphi(x_2, t_2)\cdots\varphi(x_n, t_n)\Omega\rangle, \tag{1.1}$$

for the $\mathcal{P}(\varphi)_2$ model exist as tempered distributions in $\mathcal{S}'(R^{2n})$.

Our estimates bound the shift in the vacuum energy due to a local perturbation of the Hamiltonian H. The simplest such estimate is

Theorem B: The field $\varphi(x, t)$ is a bilinear form on the domain $\mathfrak{D}(H^{1/2}) \times \mathfrak{D}(H^{1/2})$, and $H + \varphi(x, t)$ is bounded from below.

Other new results for the $\mathcal{P}(\varphi)_2$ quantum field model are established in Theorem 1.1–Corollary 1.3. The methods and estimates of this paper appear to have other applications. For example, it should be possible to show that the Wightman functions (1.1) have exactly the singularities predicted by perturbation theory. Furthermore, the methods used here appear useful in the study of the more singular φ_3^4 model.

Solutions to the $\mathcal{P}(\varphi)_2$ model were constructed in earlier work of the authors and of Rosen. They are known to satisfy the Haag–Kastler axioms for quantum fields, and most of the Wightman axioms, as was shown by the authors, Cannon and Rosen. See Refs. 1 and 2 for references, notation, and proofs.

The Hamiltonian operator in the $\mathcal{P}(\varphi)_2$ theory is defined as the limit of approximate, or cutoff, Hamiltonians $H(g)$,

$$H(g) = H_0 + H_1(g) - E(g)I. \tag{1.2}$$

Here $H_0 = H(0)$ is the free Hamiltonian for bosons of mass $m > 0$, and

$$H_1(g) = \int :\mathcal{P}(\varphi(x)): g(x)dx. \tag{1.3}$$

The space cutoff g has compact support, and

$$0 \le g(x) \le 1. \tag{1.4}$$

The vacuum energy $E(g)$ is a finite constant (divergent as $g \to 1$) and chosen so that

$$0 = \inf \text{spectrum } H(g). \tag{1.5}$$

Certain uniform estimates on $H(g)$ allow us to pass to a limit $g = 1$ and so obtain a Hamiltonian without cutoffs,

$$H = \lim H(g), \tag{1.6}$$

see Refs. 1 and 3.

We now describe our new uniform estimates for cutoff Hamiltonians. We use these estimates to establish properties of the theory without cutoffs ($g \equiv 1$), such as Theorems A and B.

Case 1. Perturbations of arbitrary degree: Let h be a real function $\|h\|_\infty \le 1$, $D = 1 +$ diam. supp. $h < \infty$. Let $p = \max\{\deg\mathcal{P}, \deg\mathcal{P}_1\}$, and let

$$0 \le \mathcal{P}(\zeta)g(x), \qquad 0 \le \mathcal{P}(\zeta)g(x) + \mathcal{P}_1(\zeta)h(x) \tag{1.7}$$

for all real ζ, x. If $\deg\mathcal{P} = \deg\mathcal{P}_1$ and $\mathcal{P} \ne \mathcal{P}_1$, we assume that the coefficient of the degree p term in $\mathcal{P}g + \mathcal{P}_1 h$ is bounded away from zero on a neighborhood of supp h. Let

$$\delta H = \int :\mathcal{P}_1(\varphi(x)): h(x)dx = \delta H(h), \tag{1.8}$$

$$\delta E = \inf \text{spectrum}\{H(g) + \delta H\} = \delta E(g, h). \tag{1.9}$$

Case 2. Linear perturbations: Let h be a real C^∞ function with

$$\| \mu^{-(1/2)-\epsilon} h \|_2 \le 1 \tag{1.10}$$

for $\epsilon = (8p)^{-1}$, $p = \deg \mathcal{P}$, and let $D = 1 + \text{diam supp } h < \infty$. Let

$$\delta H = \varphi(h) = \int \varphi(x) h(x) dx, \tag{1.11}$$

and let

$$\delta E = \delta E(g, h) = \inf \text{spectrum} \{H + \varphi(h)\}. \tag{1.12}$$

Theorem 1.1: There exists a constant M, independent of g and h, such that in Cases 1 and 2

$$|\delta E| \le MD. \tag{1.13}$$

Remarks: In Case 2, if $1 \le \|\mu^{-1/2-\epsilon}h\|_2 < \infty$, we prove (1.13) with $M = M(\|\mu^{-1/2-\epsilon}h\|_2)$. Formal reasoning indicates that $\epsilon = \frac{1}{2}$ is possible; but a proof requires improvements in Sec. 3B.

Theorems A and B are corollaries to Theorem 1.1.

We consider space-time averaged fields

$$\varphi(f) = \int \varphi(x,t) f(x,t) dx dt$$

in the theory with the cutoff Hamiltonian $H(g)$ and vacuum Ω_g, $H(g)\Omega_g = 0$. The g-cutoff vacuum expectation values are tempered distributions.[4] The analog to Theorem A in the cutoff model is:

Theorem A_g: There exist translation invariant Schwartz space norms $|\cdot|_r$, independent of g, such that

$$|\langle \Omega_g, \varphi(f_1) \cdots \varphi(f_n) \Omega_g \rangle| \le |f_1|_1 \cdots |f_n|_n. \tag{1.14}$$

The new aspect of (1.14) is a bound that is independent of g. In order to ensure space translation invariance of a limiting vacuum, we average the vacuum expectation values $\langle \Omega_g, \varphi(f_1) \cdots \varphi(f_n)\Omega_g\rangle$ over space translations in an interval of length 0 (diam supp g). Since the norms $|\cdot|_r$ are translation invariant, the space averaged expectation values are also bounded by $|f_1|_1 \cdots |f_n|_n$. By the methods of Refs. 3 and 5 we pass to a subsequence as $g \to 1$. As in Refs. 3 and 5, the limiting vacuum expectation values (1.1) satisfy the bound (1.14) with $g = 1$. Furthermore, the limiting vacuum expectation values (1.1) are obtained as limits as $g \to 1$ of a subsequence of the space averaged $\langle \Omega_g, \varphi(f_1) \cdots \varphi(f_n)\Omega_g\rangle$. Applying the Schwartz nuclear theorem to (1.14) proves that the vacuum expectation values are tempered distributions. Thus Theorem A follows from Theorem A_g, which we establish below.

We remark that the estimate of Theorem 1.1 is also valid with a space cutoff given by a periodic box of volume V. (See Refs. 1 and 5 for notation.) In Cases 1_V and 2_V we define δH and δE as in (1.8)-(1.12), with an additional dependence on V. We assume supp $h \subset [-\frac{1}{2}V, \frac{1}{2}V]$.

Theorem 1.1_V: There is a constant M independent of g, h, and V such that in Cases 1_V and 2_V

$$|\delta E| \le MD.$$

Our proof in this paper of Theorem 1.1, combined with the methods of Ref. 5, Sec. 2 that deal with estimates in a periodic box, prove Theorem 1.1_V. Likewise, we can establish a relevant Theorem A_V, and we pass to the $V = \infty$ limit in the vacuum expectation values. With this method, it is not necessary to average the vacuum expectation values

$$\langle \Omega_V, \varphi_V(f_1) \cdots \varphi_V(f_n)\Omega_V\rangle$$

over space translations in order to ensure translation invariance of the limiting vacuum Ω.

We now assume Theorem 1.1 and turn to the proof of Theorems A_g and B. We use

Lemma 1.1: Let $0 \le H = H^*$. Let A and $\dot{A} = [iH, A]$ be real bilinear forms with domain $\mathcal{C}^\infty(H) \times \mathcal{C}^\infty(H)$. Let $R = (H + I)^{-1}$, and let

$$\| R^{1/2} A R^{1/2} \| + \| R^{1/2} \dot{A} R^{1/2} \| \le M. \tag{1.15}$$

Then A uniquely determines a symmetric operator (also called A) with domain $\mathfrak{D}(H)$ and for a universal constant α,

$$\|AR\| \le \alpha M. \tag{1.16}$$

Proof: Let $R(y) = [H + (y+1)I]^{-1}$. Then

$$R^{1/2} = \frac{1}{\pi} \int_0^\infty \lambda^{-1/2} R(\lambda) d\lambda.$$

On $\mathcal{C}^\infty(H) \times \mathcal{C}^\infty(H)$,

$$\begin{aligned} AR^{1/2} &= R^{1/2}A + \frac{1}{\pi}\int_0^\infty \lambda^{-1/2}[A, R(\lambda)] d\lambda \\ &= R^{1/2}A - \frac{i}{\pi}\int_0^\infty \lambda^{-1/2} R(\lambda) \dot{A} R(\lambda) d\lambda. \end{aligned} \tag{1.17}$$

By (1.15), $R(\lambda)\dot{A}R(\lambda)$ is bounded, norm continuous in λ, with norm bounded by $(\lambda+1)^{-1}M$. Hence the integral over λ in (1.17) is norm convergent, and by (1.17)

$$\begin{aligned} |\langle \theta, AR\theta\rangle| &\le \|R^{1/2}AR^{1/2}\| \|\theta\|^2 + O(M)\|\theta\|^2 \\ &\le \alpha M \|\theta\|^2. \end{aligned}$$

The constant α depends only on the integral of $\lambda^{-1/2}(\lambda+1)^{-1}$. Lemma 1.1 then follows by the Riesz representation theorem.

Proof of Theorem A_g: Let g be fixed. We apply Theorem 1.1, Case 2, for an arbitrary real element $h/\|\mu^{-(1/2)-\epsilon}h\|_2$ of $\mathcal{C}_0^\infty(R^1)$. This yields

$$\begin{aligned} \pm\varphi(h) &\le \text{const} \|\mu^{-(1/2)-\epsilon}h\|_2 [H(g) + I] \\ &\le \text{const} \|h\|_1 [H(g) + I] \end{aligned} \tag{1.18}$$

with constants independent of g, but possibly depending on D. By taking a partition of unity, $1 = \sum \zeta_i$, where ζ_i is the translate of a fixed $\mathcal{C}_0^\infty$ function, we obtain

$$\pm\varphi(h\zeta_i) \le \text{const} \|h\zeta_i\|_1 [H(g) + I],$$

and summing

$$\pm\varphi(h) \le \text{const} \|h\|_1 [H(g) + I],$$

with a constant independent of g and h. For real $f \in \mathcal{S}(R^2)$, we integrate over time to obtain

$$\pm\varphi(f) \le \text{const} \|f\|_1 [H(g) + I] \tag{1.18$'$}$$

with a constant independent of f, g. Likewise

$$\pm \dot{\varphi}(f) = \mp \varphi(D_t f) \le \text{const} \|D_t f\|_1 [H(g) + I],$$

so by Lemma 1.1

$$\|\varphi(f) R_g\| \le \alpha(\|f\|_1 + \|D_t f\|_1), \tag{1.19}$$

where $R_g = [H(g) + I]^{-1}$. On the domain $\mathfrak{C}^\infty(H(g)) \times \mathfrak{C}^\infty(H(g))$,

$$\langle \chi, [H(g) + I]^r \varphi(f)\theta \rangle$$
$$= \sum_{j=0}^{r} \binom{r}{j} i^j \langle \chi, \varphi(D_t^j f)[H(g) + I]^{r-j}\theta \rangle. \tag{1.20}$$

Thus by (1.19)

$$\|(H(g) + I)^r \varphi(f)\theta\| \le |f|_{r+1} \|(H(g) + I)^{r+1}\theta\|, \tag{1.21}$$

where

$$|f|_{r+1} = \alpha 2^{r+1} \sum_{j=0}^{r+1} \|D_t^j f\|_1 \tag{1.22}$$

is a translation invariant Schwartz space norm. The bound (1.14), and hence Theorem A_g follows from (1.21) and (1.22).

The Hilbert space $\mathcal{H}_{\text{GNS}}$ associated with the infinite volume limit is defined by means of the GNS construction, see Refs. 3 and 5. The limiting vacuum expectation values (1.1) can also be used to construct an infinite volume Hilbert space $\mathcal{H}_W$, by the Wightman reconstruction theorem. The vacuum expectation values in these two theories coincide. Hence by the uniqueness portion of the reconstruction theorem, $\mathcal{H}_W$ can be identified as a subspace of $\mathcal{H}_{\text{GNS}}$.

Proposition 1.1: $\mathcal{H}_W = \mathcal{H}_{\text{GNS}}$: That is, the Wightman and GNS constructions yield the same infinite volume Hilbert space. In particular the cyclic subspace $\mathfrak{D}$ generated from Ω by polynomials in $\varphi(f)$, $f \in \mathcal{S}(R^2)$, is dense in $\mathcal{H}_{\text{GNS}}$.

Proof: Let $\mathcal{K}_{\alpha\beta}$ be the subset of the real elements $\mathcal{S}_R(R^2)$ of $\mathcal{S}(R^2)$ satisfying

$$\|D_t^n f\|_1 \le \alpha\beta^n.$$

Clearly $\cup_{\alpha\beta}\mathcal{K}_{\alpha\beta}$ is dense in $\mathcal{S}_R(R^2)$. Let $\mathfrak{F}_{\alpha\beta}$ be the linear span of vectors

$$e^{i\varphi(f_1)} \cdots e^{i\varphi(f_n)}\Omega$$

for $f_j \in \mathcal{K}_{\alpha\beta}$, $n = 0, 1, \cdots$. By definition $\cup_{\alpha\beta}\mathfrak{F}_{\alpha\beta}$ is dense in $\mathcal{H}_{\text{GNS}}$. Using (1.19) and (1.20), we establish for $f_j \in \mathcal{K}_{\alpha\beta}$,

$$\|\varphi(f_1) \cdots \varphi(f_k)\Omega\| \le B^k k! \tag{1.23}$$

with a constant $B = B(\alpha, \beta)$. (See Ref. 4, p. 389). We now prove that $\mathfrak{D}$ is dense. For any χ we define

$$f(\lambda) = \langle \chi, e^{i\lambda_1\varphi(f_1)} \cdots e^{i\lambda_n\varphi(f_n)}\Omega \rangle$$

as an iterated power series, summing first λ_1, then $\lambda_2, \cdots$. These series converge on account of (1.23). Thus if $\chi \perp \mathfrak{D}$, $f(\lambda) = 0$. In that case $\chi \perp \cup_{\alpha\beta}\mathfrak{F}_{\alpha\beta}$, so $\chi = 0$ and $\mathfrak{D}$ is dense.

We give three short corollaries to Theorem 1.1 which establish further properties of the space-time averaged fields $\varphi(f)$.

Corollary 1.1: Let $f, D_t f \in \mathcal{L}_1$. Then $\varphi(f)$ is an operator on $\mathfrak{D}(H)$ and

$$\|\varphi(f)(H + I)^{-1}\| \le \alpha(\|f\|_1 + \|D_t f\|_1). \tag{1.24}$$

Corollary 1.2: Let f be real, and let $f, D_t f, D_{tt} f \in \mathcal{L}_1$. Then $\varphi(f)$ is essentially self-adjoint on any core for H.

Corollary 1.3: Let $f \in \mathcal{S}(R^2)$ be real, and let $\varphi(f)$ be the operator obtained from the vacuum expectation values (1.1) by the reconstruction theorem. [The domain of $\varphi(f)$ is $\mathfrak{D}$.] Then $\varphi(f)$ is essentially self-adjoint.

Proofs: By (1.18′), for real f,

$$\pm \langle \theta_g, \varphi(f)\theta_g \rangle \le O(1)\|f\|_1 \langle \theta_g, [H(g) + I]\theta_g \rangle \tag{1.25}$$

for $\theta_g = \varphi(f_1) \cdots \varphi(f_n)\Omega_g \in \mathfrak{F}$. Let

$$\theta = \varphi(f_1) \cdots \varphi(f_n)\Omega \in \mathfrak{D} \subset \mathcal{H}_{\text{GNS}} = \mathcal{H}_W = \mathfrak{F}_{\text{ren}}.$$

We recall that the vacuum expectation value $\langle \theta, \varphi(f)\theta \rangle$ is obtained as a limit of a subsequence from space translation averages of $\langle \theta_g, \varphi(f)\theta_g \rangle$, and similarly $\langle \theta, H\theta \rangle$ is obtained as a limit from the cutoff expectation values $\langle \theta_g, H(g)\theta_g \rangle$. Thus we obtain in the limit $g = 1$,

$$\pm \langle \theta, \varphi(f)\theta \rangle \le O(1)\|f\|_1 \langle \theta, (H + I)\theta \rangle.$$

Furthermore, we have $\dot{\varphi}(f) = -\varphi(D_t f)$ on $\mathfrak{D} \times \mathfrak{D}$, as a consequence of the corresponding identity in the cutoff theory. Hence Corollary 1.1 follows by Lemma 1.1.

By Proposition 1.1, the domain $\mathfrak{D}$ is dense. Since $\mathfrak{D}$ is invariant under $\exp(itH)$, $\mathfrak{D}$ is a core for H. Hence Corollary 1.3 is a consequence of Corollary 1.2. We infer Corollary 1.2 from Corollary 1.1 and a general result:

Theorem 1.2: Let $0 \le H = H^*$ and $R = (H + I)^{-1}$. Let A be a symmetric operator with domain $\mathfrak{D}(H)$. Suppose that

$$\|AR\| + \|\dot{A}R\| < \infty, \tag{1.26}$$

where $\dot{A} = [iH, A]$ is defined in $\mathfrak{D}(H^2) \times \mathfrak{D}(H^2)$. Then A is essentially self-adjoint on any core for H.

Proof: Let $H = \int \lambda dE_\lambda$ be the spectral resolution of H. Let $C_n = E_n A E_n$. Then C_n is a bounded, self-adjoint operator with resolvent $K_n = (C_n - iy)^{-1}$. We first prove that K_n converges to the resolvent of a self-adjoint operator C. We then show that A^- extends C, so A is essentially self-adjoint on $\mathfrak{D}(H)$. The estimate $\|AR\| \le O(1)$ then yields essential self-adjointness on any core for H.

(a) Graph convergence: We note that given (1.26), it follows by (1.17) that $R^{1/2}AR^{1/2}$ is bounded. Since $\|(E_n - I)R^{1/2}\| \le O(n^{-1/2})$ as $n \to \infty$,

$$\|R(C_n - A)R\| \leq O(n^{-1/2}). \quad (1.27)$$

Furthermore, using (1.27) and the identity $C_n R^2 = RC_n R - iR\dot{C}_n R$, we have

$$\|(C_n - A)R^2\| \leq O(n^{-1/2}), \quad n \to \infty. \quad (1.28)$$

Hence the operators C_n have a strong graph limit, a closed operator C. (See Ref. 1, Sec. 6.2). By (1.28) C agrees with A on $\mathfrak{D}(H^2)$. Since AR is bounded, and C is closed, C agrees with A on $\mathfrak{D}(H)$.

(b) *Smoothness preserving:* For real y, $|y|$ sufficiently large, we prove

$$\|(H + I)K_n R\| \leq 1. \quad (1.29)$$

In fact, for $\theta \in \mathfrak{D}(H)$ and $|y| - \|\dot{A}R\| \geq 1$,

$$\begin{aligned}\|(H + I)(C_n - iy)\theta\| &\geq \|(C_n - iy)(H + I)\theta\| - \|\dot{C}_n\theta\| \\ &\geq (|y| - \|\dot{C}_n R\|)\|(H + I)\theta\| \\ &\geq \|(H + I)\theta\|,\end{aligned}$$

proving (1.29).

(c) *Resolvent convergence:* For $\theta \in \mathfrak{D}(H)$, $\|(K_n - K_m)\theta\|^2 = \langle K_n\theta, (K_n - K_m)\theta\rangle - \langle K_m\theta, (K_n - K_m)\theta\rangle$ By (1.29), for $|y|$ sufficiently large,

$$\begin{aligned}\|(K_n - K_m)\theta\|^2 &\leq 2\|R(C_n - C_m)R\|\,\|(H + I)\theta\|^2 \\ &\leq o(1)\|(H + I)\theta\|^2\end{aligned}$$

as $n, m \to \infty$ by (1.27). Since $C_n = C_n^*$, we have $\|K_n\| \leq |y|^{-1}$, for y real. Hence for $|y|$ sufficiently large,

$$\text{s-lim}\, K_n = K. \quad (1.30)$$

By (a) and (c), $C_n = C_n^*$ has a graph limit C and a strong resolvent limit. We infer from Ref. 1, Theorem 6.2.3 that $C = C^*$ and $\lim K_n = K$ is the resolvent of C. Furthermore, since H is self-adjoint, the uniform bound (1.29) ensures $K: \mathfrak{D}(H) \to \mathfrak{D}(H)$ and

$$\|(H + I)KR\| \leq 1.$$

(d) *Essential self-adjointness:* We prove that $C \subset A^-$. Hence $C = A^-$. Let $\theta \in \mathfrak{D}(C)$ or $\theta = K\chi$. Choose $\chi_n \in \mathfrak{D}(H)$, $\chi_n \to \chi$. Since K is bounded, $\theta_n = K\chi_n \to \theta$. By (c), $\theta_n \in K\mathfrak{D}(H) \subset \mathfrak{D}(H) = \mathfrak{D}(A)$. By (a), C and A agree on $\mathfrak{D}(H)$. Thus

$$(A - iy)\theta_n = (C - iy)\theta_n = \chi_n \to \chi.$$

Hence $\theta \in \mathfrak{D}(A^-)$ and $A^-\theta = C\theta$.

Proof of Theorem B: In the case of the space–time averaged field operator $\varphi(f)$, we have identified $\varphi(f)$, acting on $\mathfrak{D} \subset \mathcal{H}_{\text{GNS}} = \mathcal{H}_W$, with a restriction of the self-adjoint field operator $\varphi(f)$ introduced in Ref. 3. In the case of the bilinear form $\varphi(x, t)$, we do not attempt such an identification. Bilinear forms, in contrast with self-adjoint operators, do not have natural maximal domains of definition. The domains for $\varphi(x, t)$ introduced in Ref. 3 could very possibly meet $\mathfrak{D}$ in the zero vector alone. Thus we proceed as follows.

By Theorem A and the space–time translation invariance of the vacuum, there exists a bilinear form $\varphi(x, t)$ on $\mathfrak{D} \times \mathfrak{D}$ such that

$$\langle\theta, \varphi(x, t)\chi\rangle \in \mathcal{C}^\infty(R^2), \quad (1.31)$$

and

$$\int \langle\theta, \varphi(x, t)\chi\rangle f(x, t)\,dx\,dt = \langle\theta, \varphi(f)\chi\rangle, \quad (1.32)$$

for all θ, $\chi \in \mathfrak{D}$ and all $f \in \mathcal{S}(R^2)$. It is with this definition of $\varphi(x, t)$ that we prove Theorem B.

Let χ be a real, positive $\mathcal{C}_0^\infty$ function of total integral one and $\chi_\kappa(x) = \kappa\chi(\kappa x)$. In one space dimension, $\|\mu^{-1/2-\epsilon}\delta_x\|_2 \leq$ const, where δ_x is the Dirac measure concentrated at x. Then

$$\|\mu^{-1/2-\epsilon}(\chi_\kappa * \delta_x)\|_2 \leq \|\mu^{-(1/2)-\epsilon}\delta_x\|_2 \leq M, \quad \|\chi_\kappa\|_1 \leq 1, \quad (1.33)$$

$$\|\mu^{-(1/2)-\epsilon}(\chi_\kappa * \delta_x - \delta_x)\| \leq O(\kappa^{-\epsilon/2}), \quad (1.34)$$

$$\|\mu^{-(1/2)-\epsilon}(\delta_x - \delta_{x+a})\| \leq O(|a|^{\epsilon/2}). \quad (1.35)$$

We let $f_\kappa(x, t) = h_\kappa(x)\alpha_\kappa(t)$. By the remark below Theorem 1.1, we have for vectors θ_g as in (1.25)

$$\begin{aligned}\pm\langle\theta_g, \varphi(f_\kappa)\theta_g\rangle \leq{}& (2M)^{-1}\|\mu^{-(1/2)-\epsilon}h_\kappa\|_2\|\alpha_\kappa\|_1 \\ &\times\langle\theta_g, [H(g) + \text{const}]\theta_g\rangle. \quad (1.36)\end{aligned}$$

As in the proof of Corollary 1.1, we space average and pass to a limit $g \to 1$ to obtain for $\theta \in \mathfrak{D}$

$$\begin{aligned}\pm\langle\theta, \varphi(f_\kappa)\theta\rangle \leq{}& (2M)^{-1}\|\mu^{-(1/2)-\epsilon}h_\kappa\|_2\|\alpha_\kappa\|_1 \\ &\times\langle\theta, (H + \text{const})\theta\rangle. \quad (1.37)\end{aligned}$$

We now choose $h_\kappa = \chi_\kappa * \delta_x$ and $\alpha_\kappa = \chi_\kappa * \delta_t$. Passing to the limit $\kappa \to \infty$, we have (1.31)–(1.37)

$$\pm\langle\theta, \varphi(f_\kappa)\theta\rangle \to \pm\langle\theta, \varphi(x, t)\theta\rangle \leq \tfrac{1}{2}\langle\theta, (H + \text{const})\theta\rangle. \quad (1.38)$$

By taking limits of θ, (1.38) extends to $\mathfrak{D}(H^{1/2}) \times \mathfrak{D}(H^{1/2})$, and defines $\varphi(x, t)$ on that domain. We note by (1.35) that $\varphi(x, t)$ so defined is continuous in (x, t), and hence satisfies (1.32). This completes the proof of Theorem B.

The remainder of this paper is devoted to proving Theorem 1.1. For convenience, we suppress the κ cutoff or g cutoff and write

$$H(h) = H(g) + \delta H(h), \quad \delta E = \text{inf spectrum } H(h).$$

Both H_0 and $H(h)$ are self-adjoint operators on Fock space, and each has a unique ground state vector:

$$H_0\Omega_0 = 0, \quad H(h)\Omega(h) = \delta E\Omega(h).$$

The vectors Ω_0 and $\Omega(h)$ are not orthogonal. See Refs. 1 and 4 for proofs. It follows that $\delta E(h)$ is given by the exact formula

$$\delta E(h) = -\lim_{T\to\infty}(1/T)\log\langle\Omega_0, e^{-TH(h)}\Omega_0\rangle.$$

Our proof of Theorem 1.1. is based on estimating

$$\langle\Omega_0, e^{-TH(h)}\Omega_0\rangle \quad (1.39)$$

and bounding it by exp (MDT). The best bound previously established was $|\delta E| \leq O(1 + \text{diam supp } g)$, which diverges as $g \to 1$. The stronger bound proved here estimates the finite correction δE to the divergent vacuum energy $E(g)$ of (1.2).

We base our estimate of (1.39) on function space integration and the Feynman–Kac formula. Our new estimates can also be proved on Fock space (Q space) without appealing to probability theory methods. With some simple modifications to the present proof, we can eliminate the path space methods and substitute methods of Refs. 6 and 2.

In our proof of a uniform bound on δE, a major idea is to take advantage of localization in configuration space, as well as in momentum space. We isolate the contributions to $\delta E(h)$ from parts of the interaction localized near the support of h. We bound these contributions by the inductive construction of Sec. 2. We bound (1.39) by a sum of terms; each term represented by a graph. Each step in the induction replaces one graph by a sum of graphs with new vertices and new lines. At the end of each inductive step we decide for each graph whether to terminate the induction or whether to continue the expansion for that graph. If the inductive construction continues, the new vertices and lines introduce convergence factors into the expansion. One possibility is that the new vertices have a large lower momentum cutoff κ in some new vertex and thus contribute the small tail $O(\kappa^{-\delta})$ of a convergent integral to our estimates. Such vertices are said to be localized in high momentum regions. The remaining new vertices are localized in space–time regions far from the vertices of the original graph, so that at least one line connects vertices with space–time localizations separated by a large Euclidean distance d. By construction, d grows rapidly for successive inductive steps. In Sec. 3, we show that each line so localized in space–time contributes a convergence factor exp $(-md/2)$ to our estimates of the corresponding graph. Here m is the mass of the particles in H_0 (bare mass). The small factors $\kappa^{-\delta}$ or $\exp(-md/2)$ lead to the proof of convergence of our inductive estimate. In particular they dominate the large number of graphs that occur and yield a convergent sum of small terms to bound δE.

2. THE INDUCTIVE CONSTRUCTION

A. Introduction

The proof of Theorem 1.1 is based upon an inductively constructed upper bound to (1.39). The inductive construction consists of repeating three basic substeps:

(a) the graph expansion,
(b) the squaring operation,
(c) the path space construction.

The estimates which prove convergence of the inductive construction are established in Secs. 3 and 4.

It is convenient to define

$$H_{\mathrm{II}} = H_{\mathrm{I}} + \delta H. \tag{2.1}$$

We also introduce a momentum cutoff indexed by κ into the interaction Hamiltonian H_{II}. Let

$$H_{\mathrm{I},\kappa} = \int :\mathcal{P}(\varphi_\kappa(x)):g(x)dx,$$
$$H_{\mathrm{II},\kappa} = H_{\mathrm{I},\kappa} + \int \mathcal{P}_1(\varphi_\kappa(x)):h(x)dx.$$

We now describe the localization in space and time of functions on path space. We index the space–time localization by $i = \{i_0, i_1\} \in \mathbb{Z}^+ \times \mathbb{Z}$. We define the space–time square $\Delta_i = [i_0, i_0 + 1) \times [i_1, i_1 + 1)$. Here the first factor $[i_0, i_0 + 1)$ is a time interval, and the second factor is a space interval. We define the space–time localized interaction Hamiltonian $H_{I,i}$ as a time-dependent operator on Fock space

$$H_{\mathrm{I},i}(\mathrm{t}) = H_{\mathrm{I}}[\zeta_i(\cdot, t)g(\cdot)]$$

or as a function on path space,

$$H_{\mathrm{I},i}(t, q(t)) = H_{\mathrm{I}}[\zeta_i(\cdot, t)g(\cdot)](q(t)).$$

Here ζ_i is the characteristic function of Δ_i. Thus

$$H_{\mathrm{I}} = \sum_i H_{\mathrm{I},i}. \tag{2.2}$$

Similarly we define δH_i, $H_{\mathrm{II},i}$. Also we may localize the interaction both in space–time and in momentum space. We define

$$H_{\mathrm{I},\kappa,i} = \int :\mathcal{P}(\varphi_\kappa(x)): \zeta_i g(x)dx,$$

and similarly we define $H_{\mathrm{II},\kappa,i}$.

Let X be a union of space intervals $[j, j + 1)$, $j \in \mathbb{Z}$. We define

$$H_{\mathrm{I}}(X) = \sum_{i_1 \subset X} H_{\mathrm{I}}(\zeta_i g), \quad H(X) = H_0 + H_I(X),$$

and

$$E(X) = \inf \text{spectrum}\{H_0 + H_{\mathrm{I}}(X)\}.$$

(Note that we now omit the vacuum energy from H.) Furthermore, we introduce sets $Y \subset R^2$ which are unions of squares $\Delta_i \subset R^2$. Let $Y(t)$ be the time slice of Y at time t. Let $|X|$ be the length of X and let $|Y|$ be the area of Y.

With the hypothesis of Theorem 1.1, we establish the inequality

$$-MD \leq \delta E. \tag{2.3}$$

The constant M is independent of g and h. For Theorem 1.1, Case 2, the bound (2.3) ensures

$$0 \leq H - E(g) \pm \varphi(h) + MD.$$

Thus if Ω is the vacuum for H,

$$\delta E \leq \langle \Omega, \{H - E(g) + \varphi(h)\}\Omega\rangle$$
$$\leq \langle \Omega, \varphi(h)\Omega\rangle \leq \langle \Omega, \{H - E(g) + MD\}\Omega\rangle \leq MD,$$

proving $|\delta E| \leq MD$. For Case 1 of the theorem, the desired inequality $|\delta E| \leq MD$ follows by an interchange in the roles of H and $H + \delta H$ in the proof of (2.3). The bound (2.3) is a consequence of two inequalities.

Proposition 2.1: There is a constant a_0 such that for all intervals X with $|X| \geq a_0$,

$$E(g) \leq E(\sim X) + |X|. \tag{2.4}$$

There is a constant M such that for all intervals X_0 containing supp h,

$$\inf\{E(\sim X) + |X| : X_0 \subset X\} \leq E(g) + \delta E(g,h) + M|X_0|. \quad (2.5)$$

The constants M and a_0 are independent of g and h, subject to the conditions of Theorem 1.1.

We choose $\max\{D, a_0\} \leq |X_0| \leq \max\{D, a_0\} + 1$. By Proposition 2.1,

$$E(g) \leq E(g) + \delta E(g,h) + M|X_0|. \quad (2.6)$$

Thus with a new constant M, we have proved (2.3) and hence Theorem 1.1.

In Secs. 2 and 3 we prove (2.5). In Sec. 4 we modify this proof to establish (2.4). We now describe the inductive construction used to bound (1.39). At each step in the inductive construction, we dominate (1.39) by a sum of terms of the form

$$\int ds \int_Q R(s) \exp[-\int H_{\text{II}}(\sim Y(\sigma),\sigma)(q(\sigma))d\sigma]\,dq(\cdot), \quad (2.7)$$

where R is a polynomial on path space and where $Y(\sigma)$ defines a localization of the exponent, as above. The connection between the Fock space inner product (1.39) and the path space integral (2.7) is given by the Feynman–Kac formula. Each inductive step replaces (2.7) by a sum of similar terms, and our final bound dominates (1.39) by the sum of a convergent series.

The first part of each inductive step, the graph expansion, is an expansion of (2.7) using the pull through formula. Each term in the expansion is represented by a graph. For some terms, the polynomial R is a constant. The associated graphs have no external legs and are called vacuum graphs. For these graphs, the inductive construction terminates. In terms for which the polynomial R is not a constant, new vertices have been introduced during the graph expansion, as will be explained below.

The second part of each inductive step is the squaring operation, in which we replace R by a positive polynomial R', $|R| \leq R'$. The fact that R' is positive makes the next path space construction possible, and the fact that R' is a polynomial allows us to perform the next graph expansion step.

The third portion of each inductive step is the path space construction, which we use to bound (and remove) those parts of the exponent that are localized near the localization of R. We use the near positivity of $\int H_{\text{I},i}(q(s))ds$ to give an upper bound for localized portions of the exponent. This upper bound has the form of a convergent sum of polynomials on path space. Hence we remove part of the exponent and replace R by a sum of new polynomials R'.

In the first inductive step, we omit the graph expansion and the squaring operation and begin with the path space construction, see Sec. 2D. We then repeat the three substeps and arrive at a bound for (2.7) as a sum of vacuum graphs. Each vacuum graph is multiplied by a factor

$$\int_Q \exp(-\int_0^T H_{\text{I}}(\sim Y(\sigma))(q(\sigma))d\sigma)$$

$$= \langle \Omega_0, e^{-H(\sim Y(0))}e^{-H(\sim Y(1))}\cdots e^{-H(\sim Y(T))}\Omega_0\rangle$$

$$\leq \exp(-\int_0^T E(\sim Y(\sigma))d\sigma).$$

The vacuum graphs are small, for reasons given in Sec. 1.

B. The Graph Expansion

The graph expansion is generated by the pull through formula

$$a(k)e^{-tH} = e^{-t(H+\mu)}a(k) - \int_0^t ds\, e^{-s(H+\mu)}[a(k), H_{\text{I}}]e^{-(t-s)H}. \quad (2.8)$$

Here $H = H(X) = H_0 + H_{\text{I}}(X)$ and $\mu = (k^2 + m^2)^{1/2}$. To derive (2.8) we set

$$F(s) = \int e^{-s(H+\mu)}a(k)e^{-(t-s)H}f(k)\,dk,$$

where $f \in L_2$, and we evaluate

$$F(0) = F(t) - \int_0^t F'(s)ds.$$

By Rosen's higher-order estimates, $F(s)$ is continuously norm differentiable on $[0,t]$. From the more elementary first order estimates, one can show that $F(s)$ is strongly differentiable and norm bounded on $[0,t]$ and continuously norm differentiable on the open interval $(0,t)$.

We call our expansion the graph expansion because each term in it can be labeled by a graph; the graphs provide a convenient way to visualize the expansion. In order to accommodate our time-dependent Hamiltonians, we rewrite (2.8) as an anti-time-ordered integral

$$\overline{T}a(k,0)\exp\left(-\int_0^t H(\sigma)d\sigma\right) = \overline{T}a(k,t)e^{-t\mu}\exp\left(-\int_0^t H(\sigma)d\sigma\right) - \overline{T}\int_0^t ds[a(k,s)e^{-s\mu}, H_{\text{I}}(X,s)]\exp\left(-\int_0^t H(\sigma)d\sigma\right). \quad (2.9)$$

Here $\overline{T}$ denotes the anti-time ordering. Later times, denoted by larger values of σ, are written to the right and precede in the operator product. The time parameter s in $a(k,s)$ and $H_{\text{I}}(X,s)$ indicate that the operators $a(k)$ or $H_{\text{I}}(X)$ occur at time s. We replace X by the space–time region Y, so that $H(\sigma) = H(Y(\sigma))$, and then (2.9) remains valid.

We say that (2.9) is the sum of a term with $a(k)$ "pulled through the exponent" and a term in which $a(k)$ "contracts to the exponent." When we use (2.9) to generate the graph expansion, (2.9) occurs multiplied by a polynomial in the fields R. New terms occur, arising from commutators between a and R, and we call these terms "contractions to existing vertices." For instance, if $R(s)$ equals a product of monomials $R_1(s_1)\cdots R_n(s_n)$, then for $t_1 \leq t, s_1, \ldots, s_n \leq t_2$,

$$\overline{T}a(k,t)R(s)\exp\left(-\int_{t_1}^{t_2} H(\sigma)d\sigma\right) = \overline{T}a(k,t_2)e^{-(t_2-t)\mu}R(s)\exp\left(-\int_{t_1}^{t_2} H(\sigma)d\sigma\right) - \overline{T}R(s)\int_{t_1}^{t_2} ds'[a(k,s')e^{-(s'-t)\mu}, H_{\text{I}}(x,s')] \times \exp\left(-\int_{t_1}^{t_2} H(\sigma)d\sigma\right)$$

$$+ \overline{T} \sum_{j=1}^{n} [a(k, s_j) e^{-(s_j - t)\mu}, R_j(s_j)] \times \prod_{i \neq j} R_i(s_i) \exp\left(- \int_{t_1}^{t_2} H(\sigma)\, d\sigma\right). \quad (2.10)$$

We apply (2.10) to the basic term in the inductive construction, which we can write either as an integral on path space or as an anti-time ordered integral

$$\int ds \int_Q R(s) \exp\left[- \int H_{\mathrm{I}}(Y(\sigma)) d\sigma\right] dq(\cdot) = \int ds \left\langle \Omega_0, \overline{T} R(s) \exp\left(- \int_0^T H(Y(\sigma))\, d\sigma\right) \Omega_0 \right\rangle. \quad (2.11)$$

The integration over $s_1 \le s_2 \le \cdots \le s_n$ in (2.11) arises from previous use of (2.10). In this section we consider the integrands of the ds integration in (2.11). We choose a linear factor φ in R and write it as the sum of an annihilation operator a and a creation operator a^*. Then we use (2.10) to move a to the right and the adjoint of (2.10) to move a^* to the left. The terms in which a or a^* are pulled through equal zero, since $a\Omega_0 = 0$. The remaining terms from (2.10) are functions of φ, but not π, and hence they can be written as integrals on path space.

We now explain the graph terminology. Vertices in our graphs arise from monomials φ^n which occur in the interaction Hamiltonian H_{II}. These vertices may be produced in R during the graph expansion, the squaring operation, or the path space construction. In applying our expansions, we always use the representation (2.2) to express H_{I} as a sum of parts $H_{I,i}$ localized in space-time cubes Δ_i. Hence each vertex in our graphs has a space-time localization index i. Each vertex also occurs at a definite time, corresponding to the time s' in the integrand of (2.10). Later vertices are placed to the right, in accordance with the anti-time ordering. A vertex arising from φ^n has n legs connected to it. Of these legs, j are contracted (connected) to legs of other vertices and $n - j$ are not contracted. The uncontracted legs have no specified directions and label linear fields φ in φ^n. The contracted legs necessarily point to the left or to the right, according to whether the vertex to which they contract lies at an earlier or at a later time. The legs pointing to the left are creation legs, and those pointing to the right are annihilation legs. These legs label creation operators $a(k)^*$ and annihilation operators $a(-k)$, respectively. Each use of the commutation relations $[a(-k), a(l)^*] = \delta(k + l)$, replaces R by a new monomial. The graph of the new monomial R' is formed by contracting the legs of $a(-k)$ and $a(l)^*$ in the graph of R. Two such contracted *legs* are called a *line*. Each use of the pull through formula (2.10) also replaces R by a sum of new monomials. The second term on the right of (2.10) has a new graph formed by introducing a new vertex at time s', and one leg of this vertex is contracted to the leg of $a(k, s')$. We also say that this new vertex results from the contraction of the leg of $a(k, s')$ to the exponent.

Associated with each vertex is a kernel. If the vertex comes from φ^n and if it has j contracted legs, then the kernel is

$$v(k_1, \ldots, k_j) = [a^{\#}(k_1), [a^{\#}(k_2), \cdots, [a^{\#}(k_j), \varphi^n] \cdots]],$$

where $a^{\#}(k) = a(-k)$ or $a^*(k)$. If $j = n$, then v is fully contracted.

We label the vertices of the graph of R by a variable ν, $\nu = 1, 2, \ldots, N$, and the momenta $k_{r,\nu}$ at vertex ν. For a fully contracted φ^n vertex we introduce the kernel

$$v_\nu(k_{1,\nu}, \ldots, k_{n,\nu}) = (\zeta_i g)^{\sim}\left(\sum_{r=1}^{n} k_{r,\nu}\right) \times \prod_{r=1}^{n} [\mu(k_{r,\nu})^{-1/2} \tilde{\chi}(k_{r,\nu}\kappa^{-1})], \quad (2.12)$$

where ζ_i gives the space-time localization of the vertex and $\tilde{\chi}(\cdot\kappa^{-1})$ is the upper momentum cutoff function for the $H_{\mathrm{I},\kappa,i}$ vertex. In the path space construction of Sec. 2D, we will encounter lower cutoff vertices of the form $H_{\mathrm{I},i} - H_{\mathrm{I},\kappa,i}$. For such a vertex, we replace

$$\prod_{r=1}^{n} \tilde{\chi}(k_{r,\nu}\kappa^{-1}) \quad \text{by} \quad 1 - \prod_{r=1}^{n} \tilde{\chi}(k_{r,\nu}\kappa^{-1}) \quad (2.13)$$

in (2.12). For such vertices we define the lower momentum cutoff $\kappa(v_\nu) = \max\{1, \kappa\}$.

A φ^n vertex which is not fully contracted has a kernel obtained as follows: Multiply (2.12) by a Wick ordered product of fields $\tilde{\varphi}(k)$, one field for each uncontracted momentum variable. Then integrate this product over the uncontracted momenta.

We also introduce energy factors $\mu(k_{r,\nu})^{\epsilon_{r,\nu}} = \mu_{r,\nu}^{\epsilon_{r,\nu}}$ and define the kernel

$$\bar{v}_\nu = v_\nu \prod_r \mu_{r,\nu}^{\epsilon_{r,\nu}}. \quad (2.14)$$

We use these factors μ^ϵ in order to transfer powers of the energy from one kernel to another. Hence they satisfy

$$\prod_{\nu, r} \mu_{r,\nu}^{\epsilon_{r,\nu}} = 1, \quad (2.15)$$

and then $(\prod_\nu v_\nu) = (\prod_\nu \bar{v}_\nu)$.

We choose $-\frac{1}{2} \le \epsilon_{r,\nu}$ and

$$\epsilon_\nu = \sum_r \epsilon^+_{r,\nu} < \tfrac{1}{2},$$

where $\epsilon^+_{r,\nu} = \max\{0, \epsilon_{r,\nu}\}$. We note that for $\delta > 0$,

$$\|\bar{v}_\nu\|_2 \le O[\kappa(v_\nu)^{-(1/2)r\epsilon_\nu + \delta}].$$

In addition to the vertices, each line (two contracted legs) has associated a function of the momenta and an integration over the momenta of the two contracted legs. Suppose the annihilation leg with momentum $k_{r,\nu}$ of vertex ν (localized at time $s_\nu = s_{r,\nu}$) contracts with the creation leg with momentum $k_{r',\nu'}$ of vertex ν' (localized at time $s_{r',\nu'}$). Associated with this line is the factor

$$\delta(k_{r,\nu} + k_{r',\nu'}) \exp[-(s_{r',\nu'} - s_{r,\nu})\mu_{r,\nu}] = \delta(k_{r,\nu} + k_{r',\nu'}) \exp(-\delta s_{r,\nu}\mu_{r,\nu}), \quad (2.16)$$

where $0 \le \delta s_{r,\nu}$ is the time difference of the contraction. The factor $\exp(-\delta s\mu)$ introduces a time localization. We insert this factor in the kernel by defining

$$\hat{v}_\nu = \bar{v}_\nu \exp(-\tfrac{1}{2}\sum_r \delta s_{r,\nu}\mu_{r,\nu}). \quad (2.17)$$

Hence we assign half the time localization factor for a given line to each of the two vertices which the line connects.

A polynomial $\hat{R}(s)$ is then constructed from its graph by the formula

$$\hat{R}(s) = \overline{T}\int[\prod_\nu \hat{v}_\nu(k_{r,\nu}; q(\cdot))][\prod_{\text{lines}} \delta(k_{r,\nu} + k_{r',\nu'})]dk. \tag{2.18}$$

We define the polynomial

$$R(s) = BL\hat{R}(s), \tag{2.19}$$

where B and L are constants, depending on the inductive construction. B is a product of constants e^j, e^{J_0}, and η from (2.32) or (2.34), while L is a product of distance localization factors l_ν, coming from (2.22) or (2.24).

The polynomial R and the kernels $\hat{v}_\nu$ contributing to it are unbounded operators in the anti-time ordered formula (2.11). When this formula is interpreted in path space, R and the kernels $\hat{v}_\nu$ become multiplication operators on path space. If $\hat{v}_\nu$ is fully contracted, then it is a constant as a function on path space. Similarly, if the graph of R has no uncontracted legs, then R is a constant as a function on path space. Such a graph is called a vacuum graph.

To complete the graph expansion, we must specify which legs of the polynomial are to be pulled through and which are not, since the infinite series expansion obtained by pulling through all legs appears to diverge. We treat on a different footing the vertices which were present at the start of the graph expansion part of the given inductive step. We pull through all these legs until they contract to yield lines. We do not pull through legs of vertices added in the course of the graph expansion of the given inductive step. Thus the graph expansion in a given inductive step yields a finite number of terms, and at the end of a given graph expansion step all vertices in R with uncontracted legs are new (pull through) vertices introduced during that graph expansion step.

C. The Squaring Operation

At the conclusion of the graph expansion for (2.11), we have replaced (2.11) by a sum of similar terms. We now replace R in each term of the form (2.11) by a positive polynomial on path space. We make this replacement for two reasons: During the PS construction we take absolute values, which could replace the polynomial R on path space by the nonpolynomial function $|R|$. In that case we could not perform the next inductive step, since the graph expansion is only defined for polynomial R. The second reason to modify R is to improve our estimates in Sec. 3 for the pull through vertices in R (vertices introduced by contractions to the exponent). For this reason we sometimes choose to replace R by a polynomial of high degree.

A simple estimate for R is obtained by repeated application of the Schwarz inequality to (2.18). We obtain

$$|R| \le \prod_\nu \|\hat{v}_\nu\|_2 \le \prod_\nu \|\bar{v}_\nu\|_2, \tag{2.20}$$

where the $\mathcal{L}_2$ norms $\|\hat{v}\|_2$ or $\|\bar{v}\|_2$ are $\mathcal{L}_2$ norms in the contracted momentum variables. Unless v_ν is fully contracted, $\|\bar{v}_\nu\|_2$ depends on the uncontracted legs as a function on path space. In Theorem 3.1 we improve (2.20) to take into account the space–time localization of the vertices. We prove

$$|R| \le \prod_\nu (\gamma e^{-md_\nu/3}\|\bar{v}_\nu\|_2), \tag{2.21}$$

where d_ν is the contraction distance for vertex v_ν, as defined below in Sec. 3B.

We use the bound (2.21) as the first step in our squaring operation. We give the graphical interpretation of $\|\bar{v}_\nu\|_2$. Suppose the set $\mathcal{C}$ of contracted legs $\bar{v}_\nu$ has $\bar{n}$ elements. We start with two identical $\bar{v}_\nu$ vertices and contract the $\bar{n}$ legs of the first set $\mathcal{C}$ to the $\bar{n}$ legs of the second set $\mathcal{C}$. We then take the square root, yielding the graph of $\|\bar{v}_\nu\|_2$. In case that $\bar{v}_\nu$ is fully contracted, $\|\bar{v}_\nu\|_2$ is the square root of a vacuum diagram. It is a constant on path space, so we do not need to square: Otherwise, $\|\bar{v}_\nu\|_2$ is the square root of a polynomial function on path space.

In the latter case, we use the bound $ab \le a(1 + b^n)$ with $a = \exp(-md_\nu/6) \le 1$ and $b = \gamma\exp(-md_\nu/6)\|\bar{v}\|_2$ to obtain

$$\gamma e^{-md_\nu/3}\|\bar{v}_\nu\|_2 \le l_\nu + (l_\nu\|\bar{v}_\nu\|_2)^{n_\nu}. \tag{2.22}$$

Here $l_\nu = \gamma e^{-md_\nu/6}$, and the squaring exponent n_ν is a positive even integer, chosen below. Since n_ν is even, $\|\bar{v}_\nu\|_2^{n_\nu}$ is a polynomial on path space.

We define the graph of the constant term l_ν in (2.22) to be a single vertex with no legs. We define the kernel corresponding to this vertex to equal one. The constant l_ν becomes a factor in L of (2.19), reflecting the history of previous contractions to $\bar{v}_\nu$. It assigns a space–time localization factor for each such contraction.

The second term in (2.22) yields a graph in which n_ν pull through vertices have uncontracted legs. The contracted legs of this graph are all localized in a single square Δ, the localization of the original vertex $\hat{v}_\nu$. Therefore these lines contribute no small localization factors. Each vertex, however, is accompanied by l_ν, which becomes a factor in L for the polynomial R.

Let $\hat{v}_\nu$ be a pull through vertex with uncontracted legs. We define $d_c(v_\nu)$ as the (integer part of the) Euclidean distance of the contraction giving rise to the vertex $\hat{v}_\nu$. We choose n_ν as follows: Either

$$n_\nu = 2 \quad \text{or} \quad n_\nu = [4d_c(v_\nu)]^6. \tag{2.23}$$

To choose between the two possibilities, consider a fixed cube Δ and apply (2.22) successively, in order of decreasing value of $d_c(v_\nu)$, to the kernels $\hat{v}_\nu$ localized in Δ. Each application of (2.22) yields two terms. Thus the j-fold application to j vertices in Δ yields 2^j terms. In these terms n_ν enters only when the second term in (2.22) is selected at least once. For the $(j+1)$th application of (2.22) we choose $n_\nu = 2$ in the above case, that is when the second term in (2.22) has been selected at least once in the first j applications of (2.22). Otherwise, we choose $n_\nu = [4d_c(v_\nu)]^6$. With N vertices in Δ we obtain 2^N terms. In each of these terms we define $d_c(\Delta) = d_c(v_\nu)$, where $\hat{v}_\nu$ is the vertex for which $n_\nu = [4d_c(v_\nu)]^6$.

For uniformity of notation, we use the bound

$$\gamma e^{-md_\nu/3}\|\bar{v}_\nu\|_2 \le l_\nu\|\bar{v}_\nu\|_2, \tag{2.24}$$

for fully contracted vertices in (2.21).

D. The Path Space Construction

The path space integrals

$$\int R(s) \exp\Big[-\int H_{\mathrm{II}}(q(\sigma))\, d\sigma\Big] dq(\cdot) \tag{2.25}$$

cannot be evaluated or estimated directly. On the other hand, integrals of polynomials on path space can be estimated easily. The purpose of the path space construction is to bound (2.25) by a sum of integrals of polynomials on path space. The term by term integration of the exponential series in (2.25) converges only if $p \le 2$. In the case of linear perturbations we make a power series expansion of $\exp[-\int \delta H(q(\sigma))\, d\sigma]$. Otherwise, we use an asymptotic expansion of $\exp[-\int H_{\mathrm{II}}(q(\sigma)) d\sigma]$ to obtain our bound. We use the formal positivity of H_{II} and undo the Wick ordering in low momentum regions. This bounds the low momentum part of the exponent from above by a constant. The high momentum tail is bounded by a convergent sum of polynomials on path space.

More precisely, for Theorem 1.1, Case 1, we write

$$H_{\mathrm{II},i} = H_{\mathrm{II},i,\kappa} + \delta H_{\mathrm{II},i,\kappa}. \tag{2.26}$$

By undoing the Wick ordering in the low momentum part of $H_{\mathrm{II},i,\kappa}$, we obtain

$$-M_1(\log\kappa)^{p/2} + 2 \le H_{\mathrm{II},i,\kappa}. \tag{2.27}$$

The constant M_1 is independent of g, h, i and κ (see, e.g., Refs. 1 and 2.) It is convenient to choose a sequence of cutoffs $\kappa_j = \exp[(j/M_1)^{2/p}]$, so that by (2.27) and the localization of $H_{\mathrm{II},i}$ in a time interval of unit length,

$$-j + 2 \le \int H_{\mathrm{II},i,\kappa_j}\,(q(\sigma))\, d\sigma. \tag{2.28}$$

We bound $\exp(-\int H_{\mathrm{II},i} d\sigma)$ by writing path space Q as a countable disjoint union $Q = \cup_{j \ge j_0} Q_j$, where for $j > j_0$,

$$Q_j = \{Q : -j \le \int H_{\mathrm{II},i} d\sigma \le -j+1\}. \tag{2.29}$$

Then

$$Q_{j_0} = \{Q : -j_0 \le \int H_{\mathrm{II},i} d\sigma\}. \tag{2.30}$$

By (2.28), paths in Q_j, for $j > j_0$, yield the bound

$$1 \le |\int \delta H_{\mathrm{II},i,\kappa_j}\, d\sigma| = R_{i,j}. \tag{2.31}$$

Hence for $n \ge 1$, we have $1 \le (R_{i,j})^n$ on Q_j. For $j > j_0$, we choose $n = n(j)$ to be the largest even integer less than $\kappa_j^{1/8p}$. We define $n(j_0) = 0$. Then

$$\exp(-\int H_{\mathrm{II},i} d\sigma) \le \sum_{j \ge j_0} e^j R_{i,j}^{n(j)}. \tag{2.32}$$

For Theorem 1.1, Case 2, we write

$$\int \delta H_i\, d\sigma = \int \varphi(h)_i\, d\sigma = \int \varphi(h)_i (q(\sigma))\, d\sigma$$

and expand

$$\exp(-\int \delta H_i\, d\sigma) \le e^{J_0} + \eta \sum_{k \ge J_0} \frac{1}{(2k)!} [\int \varphi(h)_i\, d\sigma]^{2k}. \tag{2.33}$$

Here η is a universal constant, independent of J_0. We apply (2.32) to $H_{\mathrm{I},i}$ rather than $H_{\mathrm{II},i}$. Since $H_{\mathrm{II},i} = H_{\mathrm{I},i} + \delta H_i$, we have

$$\exp\Big(-\int H_{\mathrm{II},i}\, d\sigma\Big) \le \Big(e^{J_0} + \eta \sum_{k \ge J_0} \frac{1}{(2k)!} (\int \varphi(h)_i\, d\sigma)^{2k}\Big) \times \sum_{j \ge J_0} e^j R_{i,j}^{n(j)}. \tag{2.34}$$

The "path space construction" is the use of (2.32) or (2.34) in squares Δ_i. We also say that we "remove portions of the exponent" from (2.25). Our choice of $j_0 = j_0(\Delta_i)$ will be made in this and later sections.

We begin the inductive construction (the first inductive step) by applying (2.32) or (2.34) in each square $\Delta_i \in \mathfrak{D}_0$, where $\mathfrak{D}_0$ is a given union of squares Δ. For our proof of (2.5), we choose

$$\mathfrak{D}_0 = X_0 \times [0, T],$$

so the perturbation δH is localized in $\mathfrak{D}_0$. In these squares we choose $j_0(\Delta_i) = J_0$. After this initial use of the PS construction, only H_{I} remains in the exponent, δH has been completely removed. Further application of the PS construction introduces new H_{I} vertices into graphs, but no new δH vertices.

For each term in the PS construction we decide when to terminate the PS portion of an inductive step. An individual term has the form (2.25) and is labeled by a single graph. For each term we make some definitions: We let $\mathfrak{D}_e$ be the set of squares Δ_i with a nonzero exponent. For a square Δ, we let

$$d_0(\Delta) = \mathrm{dist}(\Delta, \mathfrak{D}_0), \qquad d_e(\Delta) = \mathrm{dist}(\Delta, \mathfrak{D}_e),$$

where $\mathrm{dist}(X, Y)$ is the Euclidean distance between X and Y. We let $n(\Delta)$ equal the number of vertices with uncontracted legs that are localized in Δ. If $n(\Delta) \ne 0$, let $\Gamma(\Delta)$ be the square of side length $8n(\Delta)^{1/3}$, centered at Δ, and with sides parallel to the space and time axes. Note that $n(\Delta)$, $\mathfrak{D}_e$, and $d_e(\Delta)$ change during the inductive construction. We let $n(\Delta)_r$ denote the value of $n(\Delta)$ at the end of the rth inductive step, etc.

Our condition for terminating the PS construction of a given inductive step (and hence for terminating the inductive step) is the following:

$$\text{If } n(\Delta) \ne 0, \qquad \text{then } \Gamma(\Delta) \cap \mathfrak{D}_e = \emptyset, \tag{2.35}$$

for every square Δ in our space-time cover. If (2.35) is not met in some square Δ', we apply (2.32) to remove the exponent in each $\Delta \subset \Gamma(\Delta') \cap \mathfrak{D}_e$. We say that Δ' has forced the PS construction in Δ, and we choose $j_0(\Delta)$ in this case as

$$j_0(\Delta) = [J_1 n(\Delta')^{1/6}] + 1. \tag{2.36}$$

Here $[\cdot]$ denotes "integral part." Also we choose J_1 sufficiently large so that for $x \ge 1$ we have

$$(15)^3 x^2 \le \tfrac{1}{2} \exp(x^{1/3p} M_2),$$

where $M_2 = (8p)^{-1}(J_1/M_1)^{2/p}$. By (2.36), if $n(\Delta) \ne 0$,

$$\begin{aligned}(15)^3 n(\Delta')^2 &\le \tfrac{1}{2} \exp[n(\Delta')^{1/3p} M_2] \\ &\le \kappa_{j_0(\Delta)}^{\;1/8p} \le n_{ps}(\Delta) \le n(\Delta).\end{aligned} \tag{2.37}$$

Remark: Note $n(\Delta) = n_{pt}(\Delta) + n_{ps}(\Delta) + n_h(\Delta)$, where $n_{pt}(\Delta)$ is the number of PT vertices in Δ with uncontracted legs, $n_{ps}(\Delta)$ is the number of PS vertices in Δ

with uncontracted legs, and $n_h(\Delta)$ is the number of uncontracted $\varphi(h)$ vertices in Δ. Here we call the vertices from $R_{i,j}^n$ in (2.32) PS vertices in Δ_i, and we call the vertices from $\int \varphi(h)_i \, d\sigma$ in (2.34), the $\varphi(h)$ vertices in Δ_i.

Proposition 2.2: If $\mathfrak{D}_0 \cap \mathfrak{D}_e = \emptyset$ and $n(\Delta) \neq 0$ during the PS contruction, then

$$d_0(\Delta) \leq n(\Delta)^{1/3}. \tag{2.38}$$

Proof: Step 1: Assume that the proposition holds at some point during the PS construction of the rth inductive step. We show that the proposition remains valid after the application of PS construction in any square Δ and, hence, at the end of the inductive step. If $n_{pt}(\Delta) \neq 0$, the PT vertices in Δ were present throughout the rth PS construction. Hence, by assumption, $d_0(\Delta) \leq n_{pt}(\Delta)^{1/3} \leq n(\Delta)^{1/3}$ and (2.38) is valid. On the other hand, suppose $n_{ps}(\Delta) \neq 0$ and Δ' forced the PS construction in Δ. Since $\Gamma(\Delta')$ has diagonal length $\sqrt{2}\, 8n(\Delta')^{1/3} \geq \text{dist}(\Delta', \Delta)$ and since $d_0(\Delta') \leq n(\Delta')^{1/3}$, the triangle inequality yields

$$\begin{aligned} d_0(\Delta) &\leq n(\Delta') + \text{dist}(\Delta, \Delta') \\ &\leq (1 + \sqrt{2}\, 8) n(\Delta')^{1/3} \leq 15 n(\Delta')^{1/3}. \end{aligned} \tag{2.39}$$

In this case (2.38) follows by (2.37). Since $\Delta \notin \mathfrak{D}_0$, $n_h(\Delta) = 0$ and step 1 is proved.

Step 2: The proposition holds for the first PS construction. In the configuration immediately after removing the exponent from $\mathfrak{D}_0$, if $n(\Delta) \neq 0$, then $\Delta \in \mathfrak{D}_0$ and $d_0(\Delta) = 0$. Hence the assertion follows by step 1.

Step 3: Assuming the proposition holds for the rth PS construction, we establish the proposition for the $(r+1)$th PS construction. By the definition of $\Gamma(\Delta)$, whenever $n(\Delta)_r \neq 0$,

$$d_0(\Delta) \leq n(\Delta)_r^{1/3} \leq \tfrac{1}{2} d_e(\Delta)_r. \tag{2.40}$$

We consider the start of the $(r+1)$th PS construction. Only PT vertices v formed in the graph expansion of the $(r+1)$th inductive step can have uncontracted legs. These vertices have been squared during the squaring operation of the $(r+1)$th step. Hence if such a vertex v is localized in Δ, we have $n(\Delta) \geq [4d_c(\Delta)]^6$ by (2.23). Suppose v was formed by the contraction to the exponent of v' localized in Δ'. By the definition of the graph expansion, the vertex v' must be present at the end of the rth PS construction. Thus $n(\Delta')_r \neq 0$, and by (2.40) we have $d_0(\Delta') \leq d_e(\Delta')_r$. Note also $d_c(v) = \text{dist}(\Delta, \Delta') \leq d_c(\Delta)$ and $d_e(\Delta')_r \leq d_c(\Delta)$. Hence by the triangle inequality,

$$\begin{aligned} d_0(\Delta) &\leq d_0(\Delta') + \text{dist}(\Delta, \Delta') \leq d_e(\Delta')_r + d_c(\Delta) \\ &\leq 2d_c(\Delta) \leq n(\Delta)^{1/6} \leq n(\Delta)^{1/3}, \end{aligned} \tag{2.41}$$

establishing (2.38) for Δ. By step 1 the proof of the proposition is complete. We have also proved

Corollary 2.1: At the end of the PS construction of each inductive step, (2.40) holds in every Δ with $n(\Delta) \neq 0$.

Let $n(\Delta) \neq 0$ during the PS portion of some inductive step. We assign a square

$$\Delta\# \in \mathfrak{D}_0, \quad n(\Delta\#)_1 \neq 0, \tag{2.42}$$

and an integer $k(\Delta) \in \mathfrak{Z}^+$ to Δ in the following manner: If $\Delta \in \mathfrak{D}_0$ let $\Delta\# = \Delta$, $k(\Delta) = 0$. If $\Delta \notin \mathfrak{D}_0$, suppose we are considering the rth PS construction. In case $n_{ps}(\Delta) \neq 0$, let $\Delta^{(1)}$ be the square that forced the PS construction in Δ, and we consider $\Delta^{(1)}$ at the time (during the rth PS construction) that the exponent is removed in Δ. In case $n_{ps}(\Delta) = 0$, then $n_{pt}(\Delta) \neq 0$, and we choose any PT vertex v in Δ with the maximum contraction distance, i.e., $d_c(v) = d_c(\Delta)$. We define $\Delta^{(1)}$ to be the square with the vertex $v^{(1)}$ that contracted to the exponent forming v, and we consider $\Delta^{(1)}$ at the end of the $(r-1)$th inductive step. Continuing in this fashion, we assign $\Delta^{(k)}$ to $\Delta^{(k-1)}$ until we arrive at a square $\Delta^{(k)} \in \mathfrak{D}_0$ (and the first inductive step). We let $\Delta\# = \Delta^{(k)}$, $k(\Delta) = k \geq r$.

Let

$$N_0 = \min\{n(\Delta)_1 : \Delta \in \mathfrak{D}_0, n(\Delta)_1 \neq 0\}. \tag{2.43}$$

We note that in the proof of (2.5), either $n_h(\Delta) \geq 2J_0$ or $n_{ps}(\Delta) \geq \frac{1}{2}\kappa_{J_0}^{1/8p}$ for $\Delta \in \mathfrak{D}_0$, $n(\Delta)_1 \neq 0$. Thus $N_0 \to \infty$ as $J_0 \to \infty$.

Corollary 2.2: If $n(\Delta) \neq 0$ during the PS construction of the rth inductive step,

$$n(\Delta) \geq N_0^{2^r}. \tag{2.44}$$

Proof: If $n_{ps}(\Delta) \neq 0$, we have by (2.37) that $0 \neq n(\Delta^{(1)})^2 \leq n(\Delta)$. If $n_{ps}(\Delta) = 0$, we have by Corollary 2.1 and Eq. (2.23),

$$0 \neq n(\Delta^{(1)})_{r-1}^2 \leq d_e(\Delta^{(1)})_{r-1}^6 \leq d_c(\Delta)^6 \leq n(\Delta).$$

By induction we obtain

$$n(\Delta) \geq n(\Delta\#)_1^{2^k} \geq N_0^{2^r}, \tag{2.45}$$

since $n(\Delta\#)_1 \neq 0$.

Corollary 2.3: For any PT vertex v,

$$d_c(v) \geq N_0. \tag{2.46}$$

Proof: Let v be a PT vertex formed by the contraction of a vertex v' in Δ' during the rth inductive step. By Corollary 2.1, $d_c(v) \geq d_e(\Delta')_{r-1} \geq n(\Delta')_{r-1}^{1/3}$ and, by Corollary 2.2, $d_c(v) \geq N_0$.

E. Special Procedures for $\varphi(h)$ Vertices

We modify the inductive construction slightly in order to improve our estimates on $\varphi(h)$ vertices. These vertices are introduced with the first path space construction by applying (2.34). They are contracted during the first graph expansion. All the $\varphi(h)$ vertices are localized in $\mathfrak{D}_0$; but they may contract to vertices localized in any square Δ.

The Schwarz inequality of Sec. 2C introduces into our estimates the factors $\|\mu^{-1/2}h\|_2$. We modify Sec. 2C to arrive at constants $\|\mu^{-1/2-\epsilon}h\|_2$ required for Theorem 1.1. The extra factor $\mu^{-\epsilon}$ is obtained from the convergence of the $\mathfrak{L}_2$ norm of the kernels of H_I or from the integration over some times s. Formally the method yields $\epsilon = \frac{1}{2}$, while the bound should fail for $\epsilon > \frac{1}{2}$. We give the proof for $\epsilon = (8p)^{-1}$.

In the graph expansion, a $\varphi(h)$ vertex contracts either to an H_I vertex or to another $\varphi(h)$. If $\varphi(h)$ contracts to an H_I, we multiply the kernel $v_\nu(k)$ of the contracted $\varphi(h)$ vertex by $\mu(k)^{-\epsilon}$, and we multiply the kernel of H_I vertex by μ^ϵ. Hence our kernels, with the $\mu^{\pm\epsilon}$ factors, have the general form (2.17) that we estimate in Sec. 3. We choose $\epsilon = (8p)^{-1}$. [To deal with $\epsilon \geq (2p)^{-1}$, it would be necessary to integrate over times s in such graphs.]

The remaining $\varphi(h)$ vertices contract to each other. The polynomial of a corresponding connected component of the graph of $\hat{R}(s)$ is

$$r_{\nu\nu'} = \int v_\nu(k) v_{\nu'}(-k) e^{-\delta s\mu}\, dk ds_\nu ds_{\nu'}, \tag{2.47}$$

where the $s_\nu, s_{\nu'}$ integrals are restricted to the time localization of the vertices and also $0 \leqq \delta s$. We perform the s_ν integration. We obtain a denominator μ^{-1} and two terms coming from the two endpoints of integration. Thus we obtain two graphs, of the form (2.47), each with only one time integration and with the kernels $\mu^{-1/2} v_\nu$ replacing v_ν. The application of (2.21) to such graphs yields for the integrand of the $s_{\nu'}$ integration

$$|r_{\nu\nu'}(s_{\nu'})| \leq 2 l_\nu^2 \|\mu^{-1/2} v_\nu\|_2 \, \|\mu^{-1/2} v_{\nu'}\|_2. \tag{2.48}$$

3. ESTIMATES

A. Estimates on Kernels

In this section we derive estimates on kernels in the inductive construction. These estimates deal quantitatively with localization in configuration and in momentum space.

Lemma 3.1: Let $0 \leq \alpha, \beta$, $\alpha^2 + \beta^2 < m^2$, $0 < t$, $p \in R^s$, and

$$f(x,t) = D_t^j D_x^k \int \mu^\nu e^{-\mu t + ipx} dp.$$

Then for $\lambda = [\nu + j + |k| + s]$,

$$\left| \left(e^{\beta|x|} - \sum_{n=0}^{\lambda} \frac{1}{n!} (\beta|x|)^n \right) f(x,t) \right| \leq O(e^{-\alpha t}). \tag{3.1}$$

Remarks: The function $f(x,t)$ is smooth except possibly at $x = 0, t = 0$. Our estimates are uniform for $t > 0$. For $|x| > \epsilon > 0$ we have

$$|f(x,t)| \leq O(e^{-\alpha t - \beta|x|}) \tag{3.2}$$

by the lemma. For $t > \epsilon > 0$, our proof of (3.1) proves (3.2). We take the limit $t \to 0$ and define $f(x,0)$ which satisfies (3.1) and (3.2) with $\alpha = 0$, $\beta < m$. Furthermore, if $\lambda < 0$, then $f(x,t)$ is continuous and (3.2) holds for all (x,t).

Proof: We establish the case $s = 1$, $\nu = j = k = 0$. Then $\lambda = 1$. The proof in general is similar. We use the Cauchy formula to estimate the derivatives of $\exp[-t\mu(p)]$

$$D_p^n e^{-t\mu(p)} = \frac{n!}{2\pi i} \oint \frac{e^{-t\mu(\zeta)}}{(\zeta - p)^{n+1}} d\zeta, \tag{3.3}$$

and we integrate over a circle centered at p. For $|p| \leq m$, we choose a circle of radius $\beta + \epsilon$, where $\alpha^2 + (\beta+\epsilon)^2 < m^2$. We note that if $\mathrm{Re} z \geq 0$, then

$$(\mathrm{Re} z)^{1/2} \leq \mathrm{Re}(z^{1/2}). \tag{3.4}$$

Thus on our circle of integration $(\mathrm{Re}\mu^2)^{1/2} \leq \mathrm{Re}\mu$ and

$$\alpha \leq [p^2 + m^2 - (\beta+\epsilon)^2]^{1/2} \leq \mathrm{Re}\mu. \tag{3.5}$$

Thus (3.3) yields for $|p| \leq m$,

$$|D_p^n e^{-t\mu(p)}| \leq n!(\beta+\epsilon)^{-n} e^{-\alpha t}.$$

For $|p| > m$, we use a circle of radius $|p|$. On this circle $m \leq (\mathrm{Re}\mu^2)^{1/2} \leq \mathrm{Re}\mu$, and

$$|D_p^n e^{-t\mu(p)}| \leq n! |p|^{-n} e^{-mt}.$$

Expanding $e^{\beta|x|}$ in (3.1) in a power series and using $\|x^r f\|_\infty \leq \|D^r \tilde{f}\|_1$,

$$\begin{aligned} &|\{e^{\beta|x|} - (1 + \beta|x|)\} f(x,t)| \\ &\leq \sum_{n=2}^{\infty} \frac{1}{n!} (\beta|x|)^n |f(x,t)| \\ &\leq \sum_{n=2}^{\infty} \frac{\beta^n}{n!} \|D_p^n e^{-t\mu(p)}\|_1 \\ &\leq \sum_{n=2}^{\infty} 2m \left(\frac{\beta}{\beta+\epsilon}\right)^n e^{-\alpha t} + 2\int_m^\infty dp \left(\frac{\beta}{p}\right)^2 \\ &\quad \times e^{-mt} \sum_{n=0}^{\infty} \left(\frac{\beta}{|p|}\right)^n \\ &\leq O(e^{-\alpha t}) + O(e^{-mt}) \int_m^\infty p^{-2} dp \leq O(e^{-\alpha t}). \end{aligned}$$

Lemma 3.2: Let e_i be the characteristic function of $[i, i+1]$. Then for $0 \leq t$, $\alpha^2 + \beta^2 < m^2$, $\mu_x = \mathfrak{F}\mu(p)\mathfrak{F}^{-1}$, we have

$$\|e_i \exp(-t\mu_x) e_j\| \leq O(e^{-\alpha t - \beta|i-j|}).$$

Proof: If $|i - j| \leq 2$, the bound follows from $|e_i| \leq 1$, $\|e^{-t\mu_x}\| \leq e^{-tm}$. If $|i - j| > 2$, the kernel $k_t(x,y)$ of $e_i \exp(-t\mu_x) e_j$ is $k_t(x,y) = e_i(x) f_t(x-y) e_j(y)$ where f_t is the Fourier transform of $\exp[-t\mu(p)]$. Note that k_t has support in the region $|x - y| > 1$. We apply Lemma 3.1 with $\alpha^2 + (\beta+\epsilon)^2 < m^2$, $j = k = \nu = 0$,

$$\begin{aligned} |k_t(x,y)| &\leq O(1) e^{-\alpha t - \beta|x-y| - \epsilon|x-y|} e_i(x) e_j(y) \\ &\leq O(e^{-\alpha t - \beta|i-j|}) e^{-\epsilon|x-y|}. \end{aligned}$$

Since $\exp(-\epsilon|x-y|)$ is the kernel of a bounded operator, Lemma 3.2 follows.

Lemma 3.3: Let $\zeta \in \mathfrak{C}_0^\infty$, $\beta < m$, $-1 \leq \tau \leq 1$. Then

$$\|e_i \mu_x^{-\tau} \zeta \mu^\tau\| \leq O(e^{-\beta|i|}).$$

Proof: We let $e = \sum_{|j| \leq N} e_j$, where $[-N, N]$ is sufficiently large to include the support of ζ. We write $w_i = e_i \mu^{-\tau} \zeta \mu^\tau$ as

$$w_i = w_i e + w_i(I - e)$$

and bound each term. Let $r(x,y)$ be the kernel of $w_i(I-e)$. By Lemma 3.1, for $|i|$ sufficiently large,

$$|r(x,y)| \leq O(e^{-\beta|i| - \epsilon|x-y|}).$$

Thus for $|i|$ large,

$$\|w_i(I-e)\| \le O(e^{-\beta|i|}).$$

We now assume $\tau \ge 0$, and note that $\|w_i\| \le$ const, since

$$w_i = e_i\zeta + e_i\mu^{-\tau}[\zeta, \mu^\tau],$$

and $[\zeta, \mu^\tau]$ is a bounded operator. In the momentum representation, the kernel $h(p,p')$ of $[\zeta, \mu^\tau]$ satisfies

$$|h(p,p')| = |\tilde{\zeta}(p-p')\{\mu(p)^\tau - \mu(p')^\tau\}| \le O(1)\mu(p-p')^\tau|\tilde{\zeta}(p-p')|,$$

which is a rapidly decreasing function of $p-p'$ and, hence, the kernel of a bounded operator. We use this to bound $\|w_i\|$ and $\|w_i e\|$ for $|i| \le O(1)$. If $\tau < 0$, we use the representation

$$w_i = e_i\zeta + [\mu^{-\tau}, \zeta]\mu^\tau.$$

For $|i|$ large, we bound the kernel $r_i(x,y)$ of $w_i e$ as follows:

$$r_i(x,y) = e_i(x)\int f_-(x-z)\zeta(z)f_+(z-y)dz e(y),$$

where $f_\pm$ is the Fourier transform of $\mu(p)^{\pm\tau}$. For $|x|$ sufficiently large, $f_-(x-z)\zeta(z) = h_x(z)$ is a $\mathcal{C}_0^\infty$ function of z, and, by Lemma 3.1, $D_z^n h_x(z)$ is bounded by $O(e^{-\beta|x|})$, $|x| > R$. Furthermore,

$$\int f_-(x-z)\zeta(z)f_+(z-y)dz = \langle f_{+,y}, h_x\rangle$$

is the evaluation of a tempered distribution f_+, translated by y, with the $\mathcal{C}_0^\infty$ function $h_x(\cdot)$. Since translation is continuous in $\mathcal{S}'$, $f_{+,y}e(y)$ varies over a bounded set in $\mathcal{S}'$ as y varies over the support of e. Hence for some Schwartz space norm $\|\cdot\|_s$ independent of x, y,

$$|e(y)\langle f_{+,y}, h_x\rangle| \le \|h_x\|_s \le O(e^{-\beta|x|}),$$

for $|x|$ sufficiently large. Thus for $|i|$ sufficiently large

$$|r_i(x,y)| \le O(e^{-\beta|i|})e_i(x)e(y),$$

and, since $e_i(x)e(y)$ is the kernel of a bounded operator,

$$\|w_i e\| \le O(e^{-\beta|i|}),$$

to complete the proof.

We now apply these bounds to a φ^n vertex $\hat{v}_\nu$ of the form (2.17). Suppose $\hat{v}_\nu$ has $\bar{n}$ contracted legs and is localized in Δ_{i_0}. (Thus the time $s_{r,\nu}$ of the vertex ν lies in $[i_0, i_0+1)$.) Let E_j be the localization projection

$$E_j = \prod_{r=1}^{\bar{n}} e_{j_r}(x_r),$$

and $\sum E_j = I$.

Proposition 3.1: Let $\alpha_r^2 + \beta_r^2 < m^2$. Then

$$\|E_j\hat{v}_\nu\|_2 \le \|\bar{v}_\nu\|_2\gamma \prod_{r=1}^{\bar{n}} (e^{-(1/2)\alpha_r\delta s_{r,\nu} - (1/2)\beta_r|j_r - i_1|}). \quad (3.6)$$

Proof: Let $\delta S = \frac{1}{2}\sum_r \mu_{r,\nu}\delta s_{r,\nu}$. For $0 < \delta$, δ sufficiently small, we have by Lemma 3.2,

$$\|E_j\hat{v}_\nu\|_2 \le \sum_{j'} \|E_j e^{-\delta S}E_{j'}\bar{v}_\nu\|_2$$
$$\le O(1)\sum_{j'}\prod_{r=1}^{\bar{n}} (e^{-(1/2)\alpha_r\delta s_{r,\nu} - (1/2)(\beta_r+\delta)|j_r - j_r'|})\|E_{j'}\bar{v}_\nu\|_2. \quad (3.7)$$

In order to display the localization of $\bar{v}_\nu$, we write

$$\bar{v}_\nu = \left(\prod_{r=1}^{\bar{n}} K_{r,i_1}\right)\bar{v}_\nu = \bar{K}\bar{v}_\nu. \quad (3.8)$$

Here

$$K_{r,i_1} = \mu_{x_r}^{-(1/2)+\epsilon_{r,\nu}}\zeta_{i_1}(x_r)\mu_{x_r}^{(1/2)-\epsilon_{r,\nu}},$$

$\zeta_{i_1} \in \mathcal{C}_0^\infty$, and ζ_{i_1} equals one on a neighborhood of $[i_1, i_1+1]$, sufficiently large so (3.8) holds. In fact, we take ζ_{i_1} to be the translate of a fixed $\mathcal{C}_0^\infty$ function, so the $\mathcal{L}_p$ norms of its derivatives are independent of i_1. By Lemma 3.3,

$$\|E_{j'}\bar{K}\| \le O(1)\prod_{r=1}^{\bar{n}} e^{-(1/2)(\beta_r+\delta)|j_r' - i_1|}.$$

Thus performing the sum in (3.7) yields (3.6).

We remark that for an optimal choice of (α, β) satisfying $\alpha^2 + \beta^2 = (m-\epsilon)^2$,

$$e^{-\alpha t - \beta|x|} \le e^{-(m-\epsilon)d},$$

where $d = (t^2 + x^2)^{1/2}$ is the Euclidean distance.

B. Estimates on Graphs

We use the estimates of Sec. 3A to establish estimates on a single polynomial $\hat{R}_\alpha$. We consider kernels $\hat{v}_\nu$ of the form (2.17), where $\nu = 1, 2, \ldots, N$ labels the vertices of the graph of $\hat{R}_\alpha$. Each kernel $\hat{v}_\nu(k_{1,\nu}, \ldots, k_{s,\nu}, q(\cdot))$ is a function of contracted lines with momenta $k_{r,\nu}$ and of the path space variables $q(\cdot)$ for uncontracted legs.

Each vertex has a space–time localization and a lower momentum cutoff $\kappa(v_\nu)$. The norm $\|\hat{v}_\nu\|_2 = \|\hat{v}(\cdot\,; q(\cdot))\|_2$ is the $\mathcal{L}_2$ norm of $\hat{v}$ in the contracted momentum variables, and it is a function on path space. (It is a constant on path space for fully contracted graphs.)

Let the line (i.e., the pair of contracted legs) with momentum $k_{r,\nu}$ connect vertices v_ν and $v_{\nu'}$ with space–time localizations Δ and Δ', respectively. We define the contraction distance $d_{r,\nu}$ of this line as the Euclidean distance between Δ and Δ'. We define the total contraction distance $d(v_\nu) = d_\nu$ of the vertex v_ν as

$$d_\nu = d(v_\nu) = \sum_r d_{r,\nu}, \quad (3.9)$$

where the sum extends over the contracted legs at vertex ν. Given $\epsilon > 0$, we introduce the localization factor l_ν of the vertex v_ν:

$$\gamma e^{-md_\nu/3}. \quad (3.10)$$

Here γ is a sufficiently large constant, chosen below.

Let $\hat{R}_\alpha(s)$ be a polynomial of the form (2.18).

Theorem 3.1: Given $0 < \epsilon$, there exists a constant γ such that

$$|\hat{R}_\alpha(s)| \le \left(\prod_{\text{vertices}} \gamma\|\bar{v}_\nu\|_2\right)\left(\prod_{\text{lines}} e^{-(m-\epsilon)d_{r,\nu}}\right) \le \prod_\nu(\gamma e^{-md_\nu/3}\|\bar{v}_\nu\|_2). \quad (3.11)$$

Remarks: For a fully contracted graph, $\|\tilde{v}_\nu\|_2 \leqq O(\kappa(v_\nu)^{-(1/2)+\epsilon_\nu+\delta})$, $\delta > 0$. Thus for a new constant γ,

$$|\hat{R}_\alpha(s)| \leq \prod_\nu [\gamma\kappa(v_\nu)^{-(1/2)+\epsilon_\nu+\delta} e^{-md_\nu/3}]. \tag{3.12}$$

With our choice $\epsilon_\nu \leq \frac{1}{8}$ in Sec. 2E, fully contracted graphs arising in the inductive construction satisfy

$$|\hat{R}_\alpha(s)| \leq \prod_\nu [\gamma\kappa(v_\nu)^{-1/4} e^{-md_\nu/3}]. \tag{3.13}$$

The proof of Theorem 3.1 follows from Proposition 3.1 in an elementary fashion. We localize each leg in space, and apply the Schwarz inequality to obtain

$$|\hat{R}| \leq \sum_{i_\nu} \prod_\nu \|E_{i_\nu} \hat{v}_\nu\|_2.$$

Using the fact that each pair of contracted legs has the same space–time localization, we bound the summation over i_ν with Proposition 3.1. We choose $(\alpha_{r,\nu}, \beta_{r,\nu})$ to optimize our bound and so obtain (3.11).

C. Combinatoric Estimates

At the conclusion of the inductive construction, we obtain an upper bound for (1.39) as a sum of constants on path space, with each constant represented by a vacuum graph. We now estimate this sum, which involves among other things, counting the number of terms in the expansion. We assign a positive constant $c(v_\nu)$ to each vertex v_ν in the inductive construction; $c(v_\nu)$ is called a combinatoric factor.

One use of combinatoric factors is to estimate sums by supremums. For instance,

$$|\sum_\alpha a_\alpha| \leq \sup_\alpha |c_\alpha a_\alpha|,$$

provided $\sum c_\alpha^{-1} \leq 1$. We also use the combinatoric factors to absorb the constants e^j or η in (2.34), except those coming from squares $\Delta \in \mathfrak{D}_0$. Since each vertex is localized in some square Δ, the time integration in (2.11) extends over a region of volume at most one. Thus we replace the integral over s by the supremum norm of the integrand. We note that for each $\varphi(h)$–$\varphi(h)$ contraction, the integral over one $\varphi(h)$ time has already been estimated and is not a part of the ds integration in (2.11), see Sec. 2E. We have

$$\begin{aligned}\langle\Omega_0, e^{-T(H_0+H_I+\delta H)}\Omega_0\rangle &= \int_Q \exp\left(-\int_0^T H_{II}(q(\sigma))d\sigma\right)dq(\cdot)\\ &\leq \sum_\alpha \int ds R_\alpha(s) \exp\left(-\int_0^T H_I(\sim Y_\alpha(\sigma))(q(\sigma))d\sigma\right)\\ &\leq \sum_\alpha \int ds |R_\alpha(s)| \exp\left(-\int_0^T E(\sim Y_\alpha(\sigma))d\sigma\right).\end{aligned} \tag{3.14}$$

Each polynomial $R_\alpha(s)$ is defined in (2.19) and is estimated in Sec. B:

$$|R_\alpha(s)| \leq B_\alpha L_\alpha \prod_\nu (l_\nu \|\tilde{v}_\nu\|_2).$$

The constants in B_α, with the exception of factors e^{J_0} from applying (2.32) or (2.34) in squares $\Delta \in \mathfrak{D}_0$, are parts of combinatoric factors $c(v_\nu)$. With $c(v_\nu)$ as chosen below we will obtain

$$\langle\Omega_0, e^{-T(H_0+H_I+\delta H)}\Omega_0\rangle \leq e^{2J_0|X_0|T} \sup_\alpha \prod_\nu (c(v_\nu)l_\nu\|\tilde{v}_\nu\|_2) \times \exp\left(-\int_0^T E(\sim Y_\alpha(\sigma))d\sigma\right). \tag{3.15}$$

The factor $\exp(2J_0|X_0|T)$ bounds the product of the constants e^{J_0} from squares $\Delta \in \mathfrak{D}_0$. We also prove

Proposition 3.2: For N_0 of (2.43) sufficiently large, we have

$$\prod_\nu (c(v_\nu)l_\nu\|\tilde{v}_\nu\|_2) \leq e^{-N(\alpha)} \leq \exp\left(-\int_0^T |Y_\alpha(\sigma)|d\sigma - |X_0|T\right), \tag{3.16}$$

for every graph R_α in (3.14), where $N(\alpha)$ is the number of vertices in R_α.

The inequality (2.5) follows immediately. By (3.15) and (3.16), we have

$$\begin{aligned}&\langle\Omega_0, e^{-T(H_0+H_I+\delta H)}\Omega_0\rangle\\ &\leq e^{(2J_0+1)|X_0|T} \sup_\alpha \exp\left(-\int_0^T \{E(\sim Y_\alpha(\sigma)) + |Y_\alpha(\sigma)|\}d\sigma\right)\\ &\leq e^{(2J_0+1)|X_0|T} \sup_{X\supset X_0} e^{-T\{E(\sim X)+|X|\}}.\end{aligned} \tag{3.17}$$

Thus, by the formula above (1.39),

$$-E(g) - \delta E(g,h) \leq M|X_0| - \inf_{X\supset X_0} \{E(\sim X) + |X|\},$$

where $M = 2J_0 + 1$, which proves (2.5).

We now assign the combinatoric factors $c(v_\nu)$ to vertices v_ν. We denote these factors c_{pt}, c_{ps}, or c_h for PT, PS, or $\varphi(h)$ vertices, respectively.

Proposition 3.3: Let $\beta_0 > 0$ be given, and let N_0 of (2.43) be sufficiently large. There are constants β, N, independent of β_0 and N_0 such that

$$c_{pt}(v) \leq \beta d(v)^N, \tag{3.18}$$

$$c_{ps}(v) \leq \beta d(v)^N \kappa(v)^{1/8}, \tag{3.19}$$

$$c_h(v) \leq \beta_0 d(v)^N \tag{3.20}$$

and such that (3.5) is valid.

Proof: We assign a finite product of combinatoric factors to each vertex. Each factor in the product has the form $O(1)d(v)^n\kappa(v)^\epsilon$ and, hence, so does the product. Therefore, we only need to keep track of which powers $\kappa(v)^\epsilon$ occur, and we need to show that some factor for $\varphi(h)$ vertices can be made small.

1. *The Graph Expansion*

We start by assigning combinatoric factors to a given graph expansion step. This is our major assignment of combinatoric factors, since we count the number of graphs R_α that are formed during a graph expansion step. At the start of a graph expansion of a particular term, $n(\Delta)$ vertices in Δ have uncontracted legs. These "old" vertices contract to each other during the graph expansion, and they also contract to the exponent to form "new" PT vertices. The old vertices may contract further with the new vertices; but the new vertices do not contract to the exponent. A given PS or $\varphi(h)$ vertex can contract during at most two inductive steps. We assign combinatoric factors to vertices from the time they are introduced until the graph expansion when they fully contract.

If the rth leg of the νth vertex contracts to a vertex localized in Δ_j, we set $j = j(r,\nu)$. Consider the set of all graphs in which the contraction localizations $j = j(r, \nu)$ of the old vertices take on fixed values. We count the number of such graphs $N(\{j(r,\nu)\})$, and then we sum over the allowed $\{j(r,\nu)\}$. Suppose that vertices v and v', localized in Δ and Δ', respectively, are connected by a line with contraction distance dist $(\Delta, \Delta') \le \min\{d(v), d(v')\}$. Since

$$\sum_{\Delta'} \mathrm{dist}(\Delta, \Delta')^{-3} = \beta_1 < \infty,$$

a combinatoric factor $\beta_1 d(v)^3$ assigned to each leg is sufficient to deal with each sum over contraction localizations.

Thus we may consider a fixed set of $\{j(r,\nu)\}$. Suppose that in a given graph expansion step l_{ij} old legs localized in Δ_i contract to legs localized in Δ_j, and that l'_{ij} new legs localized in Δ_i contract to legs localized in Δ_j. (Either l_{ij} or l'_{ij} is zero.) The contraction localizations $\{j(r,\nu)\}$ determine the $\{l_{ij}\}$ and the $\{l'_{ij}\}$. However the $\{j(r,\nu)\}$ do not determine the number of new vertices, their degrees, nor the number of contracted legs at each new vertex. Since new vertices need not be fully contracted, the number of new vertices in Δ_i with legs contracting to Δ_j lies between l'_{ij}/p and l'_{ij}. Each new vertex is a monomial φ^n, $1 \le n \le p$, and l'_{ij} values of n may be chosen in $p^{l'_{ij}}$ ways for a total of at most $l'_{ij}\, p^{l'_{ij}}$ possible choices of the number and degree of new vertices in Δ_i. We assign a combinatoric factor $\beta_2 = p^{2p}$ per new PT vertex, or at least $\beta_2^{l'_{ij}/p} = p^{2l'_{ij}}$ to count new vertices in Δ_i, and we assign a factor 2^p per vertex to count whether each of the (at most) p legs at a new vertex contracts. For new vertices, the localization factor $\beta_1 d(v)^3$ per leg deals with the sum over localizations, as above. Thus we now may fix the contraction localizations $\{j(r,\nu)'\}$ of the new vertices as well as $\{j(r,\nu)\}$. Let $l^{\#}_{ij}$ denote l_{ij} or l'_{ij}.

For new or old vertices, the $l^{\#}_{ij}$ legs from Δ_i to Δ_j can be connected in $l^{\#}_{ij}!$ ways if $i \ne j$ or in $l_{ii}!! \le (l_{ii}!)^{1/2}$ ways if $i = j$. We need to count the contractions from Δ_i to Δ_j or from Δ_j to Δ_i but not both, so

$$N\{j, j'\} \le \prod_{i,j} (l^{\#}_{ij}!)^{1/2}. \tag{3.21}$$

If $l'_{ij} \ne 0$, we divide $(l'_{ij}!)^{1/2} \le (l'_{ij})^{l'_{ij}/2}$ equally among the l'_{ij} legs and, thus, assign a factor $(l'_{ij})^{p/2}$ per new PT vertex. In the $(k+1)$th inductive step, by Corollary 2.1, any PT vertex v in Δ_i contributing to l'_{ij} satisfies

$$l'_{ij} \le pn(\Delta_j)_k \le pd_e(\Delta_j)_k^3 \le pd(v)^3.$$

Hence a factor $[pd(v)^3]^{p/2}$ per new PT vertex in Δ_i dominates the contribution of Δ_i to (3.21).

We now consider the case of old vertices $l_{ij} \ne 0$. Let m_{il}, $l = 1, 2, 3$ denote the number of uncontracted legs in Δ_i at the end of the kth inductive step and belonging to PS vertices ($l = 1$), $\varphi(h)$ vertices ($l = 2$), or PT vertices ($l = 3$). Then $m_{i1} \le pn_{ps}(\Delta_i)_k$, $m_{i2} \le n_h(\Delta_i)_k$, and $m_{i3} \le (p-1)n_{pt}(\Delta_i)_k$. Also

$$l_i = \sum_j l_{ij} = m_{i1} + m_{i2} + m_{i3},$$

$$\prod_j (l_{ij}!)^{1/2} \le (m_i!)^{1/2} \le \prod_{l=1}^{3} (9m_{il})^{m_{il}/2}. \tag{3.22}$$

We assign the factor $(9m_{il})^{m_{il}/2}$ among the vertices of type l localized in Δ_i. For $l = 1$, we have $m_{i1} \le pn_{ps}(\Delta_i) \le \kappa(v)^{1/8p}$. We divide the factor $m_{i1}^{1/2}$ equally among the PS vertices in Δ_i, yielding a factor

$$m_{i1}^{p/2} \le O(1)\kappa(v)^{1/16} \tag{3.23}$$

per vertex.

For $\varphi(h)$ vertices, we assign the factor

$$3m_{i2}^{1/2} \le 3n_h(\Delta_i)^{1/2} \tag{3.24}$$

per vertex, as well as a factor 2 per vertex for the two terms formed by the time integration for each φ–φ contraction (see Sec. 2E).

Finally, we consider PT vertices m_{i3}. We assert that there exists a constant β_3, depending only on p, such that

$$m_{i3} \le \beta_3 d_c(\Delta_i)^6/3. \tag{3.25}$$

Our combinatoric factor (3.22) for $l = 3$ is bounded by

$$[O(1)d_c(\Delta_i)]^{\beta_3 d_c(\Delta_i)^6}. \tag{3.26}$$

The squaring exponent (2.23) of Sec. 2C ensures that if $m_{i3} \ne 0$, then at least $d_c(\Delta_i)^6$ old PT vertices v, each with $d_c(\Delta_i) = d_c(v)$, are localized in Δ_i. We assign a combinatoric factor $O(1)d_c(v)^{\beta_3}$ to each old PT vertex, which dominates (3.26).

We now prove (3.25). Let N_k be the number of new PT vertices produced in the graph expansion of the kth inductive step, and which retain uncontracted legs after squaring. Hence by (2.23),

$$m_{i3} \le (p-1)n_{pt}(\Delta_i)_k \le (p-1)\{2N_k + [4d_c(\Delta_i)]^6\}.$$

But the vertices N_k are produced by contractions to the exponent, so

$$N_k \le p \sum_{\Delta_j} n(\Delta_j)_{k-1}, \tag{3.27}$$

where the sum extends over squares Δ_j satisfying $n(\Delta_j)_{k-1} \le d_e(\Delta_j)_{k-1}^3$, by Corollary 2.1. Since the vertices remain after squaring, $d_e(\Delta_j)_{k-1}^3 \le d_c(\Delta_i)^3$, and at most $4d_c(\Delta_i)^2$ squares Δ_j enter the sum (3.27). Hence $N_k \le 4pd_c(\Delta_i)^5$, proving (3.25).

2. *The Squaring Operation*

The squaring operation (2.22) introduces two terms; so a combinatoric factor 2 is sufficient for each application of (2.22). We assign this factor to the vertex v' which contracts to the exponent forming the new PT vertex v requiring the use of (2.22). Each vertex v' can form at most p new PT vertices during the graph expansion, and we assign a combinatoric factor 2^p per vertex to deal with the squaring operation.

3. *The Path Space Construction*

Let us first consider (2.32). If $j > j_0$, we assign a combinatoric factor $\exp(\beta_4)$ per PS vertex. This factor $\exp[\beta_4 n_{ps}(\Delta)]$ dominates both e^j and the factor $O(1)j^2$ for summation over j. In fact, $j = O(1)$

$(\log\kappa_j)^{p/2} \le O(1)\kappa_j^{1/8p}$, and $n_{ps}(\Delta) \ge \kappa_j^{1/8p}$ if the lower cutoff in Δ is κ_j. Thus for β_4 sufficiently large,

$$O(1)j^2e^j \le \exp(\beta_4\kappa_j^{1/8p}) \le \exp[\beta_4 n_{ps}(\Delta)].$$

In case $j = j_0$, there are no PS vertices in Δ_i. For $\Delta_i \subset \mathfrak{D}_0$, we include the factor $\exp(J_0)$ explicitly in our bound (3.15). For $\Delta_i \not\subset \mathfrak{D}_0$, we assign $\exp[j_0(\Delta_i)]$ as a combinatoric factor for the vertices localized in Δ_j, where Δ_j forced the PS construction in Δ_i. By (2.35), the $n(\Delta_j)$ vertices in Δ_j with uncontracted legs force the PS construction in at most $64n(\Delta_j)^{2/3}$ squares Δ_i. [This is the area of $\Gamma(\Delta_j)$.] By (2.36), we assign at most $\exp(64n(\Delta_j)^{2/3}\{[J_1 n(\Delta_j)^{1/6}] + 1\})$ among the vertices in Δ_j. Hence a factor $O(1)$ per vertex is sufficient.

Finally we consider (2.33), the part of (2.34) not analyzed above. The factor $\exp(J_0)$ is kept explicitly in (3.15). For $2k$ $\varphi(h)$ vertices we have a factor $\eta/(2k)! \le \eta k^{-2k}$ or a factor

$$O(1)n_h(\Delta_i)^{-1} \tag{3.28}$$

per vertex. A combinatoric factor $O(1)$ per vertex dominates the sum over k. We choose $N_0 \le n_h(\Delta_i)$ sufficiently large so that (3.28) dominates (3.24), and their product is sufficiently small to ensure (3.20). This completes the proof of the proposition.

Proof of Proposition 3.2: We use Theorem 3.1 and Proposition 3.3. For a PT, PS, or $\varphi(h)$ vertex v_ν, we have

$$c(v_\nu)l_\nu\|\bar{v}_\nu\|_2 \le \begin{cases} \beta\gamma d(v_\nu)^{N}e^{-md(v_\nu)/6}, & \text{PT} \\ \beta\gamma d(v_\nu)^{N}e^{-md(v_\nu)/6}\kappa(v_\nu)^{-1/8}, & \text{PS} \\ \beta_0\gamma d(v_\nu)^{N}e^{-md(v_\nu)/6}\|\mu^{-(1/2)-1/8p}h\|_2, & \varphi(h)\end{cases}$$

We use Theorem 3.1 to give the estimate $O[\kappa(v_\nu)^{-(1/2)+1/8+\delta}]$ for PS vertices, the $\kappa^{1/8}$ coming from Sec. 2E, and this dominates the $O[\kappa(v_\nu)^{1/8}]$ growth of (3.19). By Corollary 2.2, $\kappa(v_\nu)$ for PS vertices tends to infinity with N_0 of (2.43). Similarly, for PT vertices, $d(v_\nu)\to\infty$ as $N_0\to\infty$ by Corollary 2.3. We choose β_0 sufficiently small and N_0 sufficiently large so

$$c(v_\nu)l_\nu\|\bar{v}_\nu\|_2 \le e^{-1}.$$

To complete the proof of (3.16), we show that each polynomial R_α has a graph with $N(\alpha)$ vertices, where

$$N(\alpha) \ge \int_0^T |Y_\alpha(\sigma)|d\sigma - |X_0|T.$$

The right side is the area in $\sim\mathfrak{D}_0$ in which the exponent has been removed. Let $\Delta \not\subset \mathfrak{D}_0$ and let Δ' force the PS construction in Δ. We count the vertices in Δ'. The square $\Gamma(\Delta')$ has area $64n(\Delta')^{2/3}$ and if $N_0 > (64)^3$, then by Corollary 2.2, $64n(\Delta')^{2/3} < n(\Delta')$. Hence R_α has more vertices than squares in $\sim\mathfrak{D}_0$ with the exponent removed.

4. COMPLETION OF THE BOUND ON δE

Theorem 4.1: There exists a_0 such that for all intervals X with $|X| \ge a_0$,

$$E(g) \le E(\sim X) + |X|. \tag{4.1}$$

We use the inductive construction to bound $E(\sim X)$ and prove (4.1). In this section we describe the modifications to Secs. 2 and 3 required to establish (4.1), i.e., (2.4) and, hence, to complete the proof of Proposition 2.1 and Theorem 1.1.

A. The Inductive Construction

Let $X_a = X_a(g)$ minimize $E(\sim X) + |X|$ under the restriction that X is an interval of length $|X| \ge a$. Then, for $|X| \ge a$,

$$-E(\sim X) \le -E(\sim X_a) + |X| - |X_a|. \tag{4.2}$$

We assume that (4.1) is false, namely for $a \ge a_0$,

$$-E(g) \le -E(\sim X_a) - |X_a|. \tag{4.3}$$

We use the inductive construction and (4.2) and (4.3) to bound $E(\sim X_a)$; we thereby obtain a contradiction, establishing (4.1). We bound

$$\int \exp\left(-\int_0^T H_I(\sim X_a)(q(\sigma))d\sigma\right)dq(\cdot). \tag{4.4}$$

The only change in the inductive construction occurs at the start of the first inductive step, i.e., the start of the first PS construction. Roughly, the idea is to begin the inductive construction by inserting and removing the factor $\exp(-H_I(X_a))$ in the integrand of (4.4). For this purpose we use the inequality

$$1 \le e^{-r+s} + (r/s)^{2n}, \tag{4.5}$$

valid for real r, for $s > 0$ and for $n \in \mathfrak{Z}^+$. We choose $n = [|X_a|^{1/8p}]$, $s = \frac{1}{2}|X_a|$, and

$$r = \int_{i_0}^{i_0+1} H_I(X_a)(q(\sigma))d\sigma. \tag{4.6}$$

We begin the first PS construction by inserting (4.5) and (4.6) in the integrand of (4.4) at each time i_0 such that $0 \le i_0 \le T$, $i_0 \in \tau\mathfrak{Z}^+$, where τ is a positive integer to be determined later. Thus we apply (4.5) and (4.6) in the union of squares

$$\mathfrak{D}_1 = \{\Delta_i : i_1 \in X_a, i_0 \in \tau\mathfrak{Z}^+, i_0 \le T\},$$

producing $2^{[T/\tau]}$ terms. For each such term, let $\mathfrak{D}_0$ be the subset of $\mathfrak{D}_1$ in which the exponent is removed, i.e., the squares in which the second term in (4.5) is selected. Let $\mathfrak{D}_0'$ be the complementary subset of $\mathfrak{D}_1$ in which the exponent remains. Then $\mathfrak{D}_1 = \mathfrak{D}_0 \cup \mathfrak{D}_0'$, and each square $\Delta_i \in \mathfrak{D}_0$ contains $2[|X_a|^{1/8p}]$ vertices, each of the form $2|X_a|^{-1}H_{I,i}$. We call these vertices PS vertices, we include the factor $2|X_a|^{-1}$ with the kernel of these PS vertices, and we assign the function $\kappa(v) = |X_a|$ to such PS vertices. Then $\kappa(v)$ plays the role of a lower cutoff $n = [\kappa(v)^{1/8p}]$ and $\|v\|_2 \le O(\kappa(v)^{-1/2})$.

After starting the PS construction in this manner and obtaining PS vertices in $\mathfrak{D}_0$, we continue the inductive construction as in Sec. 2. With our definition of the PS vertices in $\mathfrak{D}_0$, the estimates of Sec. 2 and 3 concerning the inductive construction apply. In the case of (2.43), $N_0 = 2n \ge 2a_0^{1/8p} \to \infty$ as $a_0 \to \infty$.

At the end of the inductive construction, we will obtain $\exp[-E(\sim X)]$, which we bound by (4.2) or (4.3). Here X is the space region (at some time) in which the exponent is removed. The inequalities (4.2) or (4.3) apply if X is an interval and either $|X| \ge a$ or $X = \emptyset$, respectively. We now verify these conditions for any X appearing at the end of the inductive construction. If $\Delta_i \not\subset X_a \times [0, T]$, then $d_0(\Delta_i)$ exceeds the horizontal, or equal time distance from Δ_i to $X_a \times [0, T]$. By Corollary 2.1, if $n(\Delta_i)_r \ne 0$, then at the end of the rth PS construction $d_e(\Delta_i)_r > d_0(\Delta_i)$. Thus we ensure that X has the desired form if whenever the exponent is removed from $\Delta \in \mathfrak{D}_0'$, it is also removed from all other squares in $\mathfrak{D}_0'$ with the same time localization as Δ.

Suppose that Δ' forces the PS construction in $\Delta \in \mathfrak{D}_0'$. If $d_0(\Delta') \ge 8^{-1}\,\mathrm{dist}(\Delta, \Delta') + 8^{-1}|X_a|$, then by (2.38)

$$n(\Delta')^{1/3} \ge d_0(\Delta') \ge 8^{-1}\,\mathrm{dist}(\Delta, \Delta') + 8^{-1}|X_a|.$$

Thus Δ' forces the PS construction in a square $\Gamma(\Delta')$ with side length $8n(\Delta')^{1/3} \ge \mathrm{dist}(\Delta, \Delta') + |X_a|$ and centered at Δ'. This $\Gamma(\Delta')$ includes all squares in $\mathfrak{D}_0'$ that are localized at the same time as Δ.

On the other hand, if $d_0(\Delta') \le 8^{-1}\,\mathrm{dist}(\Delta, \Delta') + 8^{-1}|X_a|$, let $d_0(\Delta') = \mathrm{dist}(\Delta', \Delta'')$, $\Delta'' \in \mathfrak{D}_0$. Then

$$\begin{aligned}\mathrm{dist}(\Delta, \Delta') &\ge \mathrm{dist}(\Delta, \Delta'') - \mathrm{dist}(\Delta'', \Delta') \\ &\ge \tau - d_0 \ge \tau - 8^{-1}\,\mathrm{dist}(\Delta, \Delta') - 8^{-1}|X_a|.\end{aligned}$$

Suppose $\tau = 10|X_a|$. Thus

$$|X_a| \le 8^{-1}\,\mathrm{dist}(\Delta, \Delta'). \tag{4.7}$$

Let $d_1(\Delta')$ be the distance between the space localization of Δ' and the center of X_a. {If $\Delta' = \Delta_i$ and $X_a = [\alpha, \beta]$, then $d_1(\Delta') = |i_1 - \frac{1}{2}(\alpha + \beta)|$.} Half the side length of $\Gamma(\Delta')$ is greater than $2^{-1/2}\,\mathrm{dist}(\Delta, \Delta')$. We show that

$$d_1(\Delta') + \tfrac{1}{2}|X_a| \le 2^{-1/2}\,\mathrm{dist}(\Delta, \Delta'), \tag{4.8}$$

so Δ' forces the PS construction in all squares in $\mathfrak{D}_0'$ localized at the same time as Δ. By the triangle inequality, $d_1(\Delta') \le d_0(\Delta') + \frac{1}{2}|X_a|$; so by (4.7),

$$\begin{aligned}d_1(\Delta') + \tfrac{1}{2}|X_a| &\le d_0(\Delta') + |X_a| \\ &\le \tfrac{1}{8}\,\mathrm{dist}(\Delta, \Delta') + \tfrac{9}{8}|X_a| \\ &\le \tfrac{3}{8}\,\mathrm{dist}(\Delta, \Delta') \le 2^{-1/2}\,\mathrm{dist}(\Delta, \Delta').\end{aligned}$$

Thus we choose $\tau = \tau(a) = 10|X_a|$ to ensure the desired geometry.

B. The Combinatorial Estimates

The combinatoric factors for the graph expansion and the squaring operation are assigned as in chapter 3. In the path space construction, we now consider the combinatoric factors from the application of (4.5) and (4.6) in $\mathfrak{D}_1$. Let $\mathcal{I}_1 = \{i_0 : 0 \le i_0 \le T, i_0 \in \tau\mathbb{Z}_+\}$. For a vacuum graph R_α at the end of the inductive construction, let $\mathcal{I}_0 = \mathcal{I}_0(\alpha) \subset \mathcal{I}_1$ be the subset of times for which the second term of (4.5) was chosen. Let $\mathcal{I}_0'(\alpha) \subset \mathcal{I}_1$ be the complementary subset of times for which the first term of (4.5) was chosen. Let $\mathcal{I}_0'' \subset \mathcal{I}_0'$ be the subset of times in $\mathcal{I}_0'$ when the PS construction was not used in X_a. We note that $Y_\alpha(\sigma)$ in (3.14) has the form

$$Y_\alpha(\sigma) = \begin{cases}\emptyset, & [\sigma] \in \mathcal{I}_0''(\alpha) \\ X, X \supset X_a, & [\sigma] \notin \mathcal{I}_0''(\alpha).\end{cases} \tag{4.9}$$

Thus $H_I(\sim Y_\alpha(\sigma)) = H_I(g), H_I(\sim X)$ in these respective cases.

Each application of (4.5) yields two terms, so we assign a combinatoric factor 2 to each time $i_0 \in \mathcal{I}_1$. If $i_0 \in \mathcal{I}_0$, there are vertices in X_a at time i_0 and we assign the factor 2 to any such vertex. If $i_0 \in \mathcal{I}_0' \sim \mathcal{I}_0''$, we assign the factor 2 to a vertex in the square forcing the PS construction, as with the factor e^s below. The remaining factor is $\exp(|\mathcal{I}_0''|\log 2)$, where $|\mathcal{I}_0''|$ is the number of elements of $\mathcal{I}_0''$.

The other relevant factor is the constant $\exp(\frac{1}{2}|X_a|) = e^s$ from the first term of (4.5). If $i_0 \in \mathcal{I}_0' \sim \mathcal{I}_0''$, then $\exp(\frac{1}{2}|X_a|)$ is a combinatoric factor $\exp(\frac{1}{2})$ per square $\Delta = \Delta_{i_0, i_1}$ for $i_1 \in X_a$. We distribute this factor among the vertices in Δ' that forced the PS construction in Δ. Since $n(\Delta')$ vertices in Δ' force the PS construction in at most $64n(\Delta')^{2/3}$ squares, we assign at most a factor $O(1)$ per vertex in Δ'. For times $i_0 \in \mathcal{I}_0''$, we obtain a factor $\exp(\frac{1}{2}|X_a||\mathcal{I}_0''|)$ in our expansion.

Thus the inductive construction yields

$$\begin{aligned}\langle\Omega_0, e^{-T[H_0+H_I(\sim X_a)]}\Omega_0\rangle &= \int e^{-H_I[\sim(X_a)(q(\sigma))]d\sigma}\,dq(\cdot) \\ &\le \sum_\alpha \int ds\,|R_\alpha(s)|\exp\Big(-\int_0^T H_I(\sim Y_\alpha(\sigma))(q(\sigma))d\sigma\Big) \\ &\le \sup_\alpha e^{|\mathcal{I}_0''(\alpha)|\{\log 2 + \frac{1}{2}|X_a|\}} \\ &\quad\times \prod_\nu (c(v_\nu) l_\nu \gamma \|\bar{v}_\nu\|_2)\exp\Big(-\int_0^T E(\sim Y_\alpha(\sigma))d\sigma\Big).\end{aligned} \tag{4.10}$$

For $|\mathcal{I}_0''(\alpha)|$ time intervals $\sigma \in [i_0, i_0 + 1)$, we bound $-E(\sim Y_\alpha(\sigma)) = -E(g)$ by (4.3). For the remaining time intervals we use (4.2). Thus

$$\begin{aligned}&\exp\Big(-\int E(\sim Y_\alpha(\sigma))d\sigma\Big) \\ &\quad\le e^{-|\mathcal{I}_0''(\alpha)||X_a|}e^{-TE(\sim X_a)}\exp\Big(\int_0^T |Y_\alpha(\sigma) \sim X_a|d\sigma\Big).\end{aligned} \tag{4.11}$$

By Proposition 3.2, for N_0 sufficiently large,

$$\prod_\nu (c(v_\nu) l_\nu \gamma \|\bar{v}_\nu\|_2) \le e^{-N(\alpha)}.$$

In our case

$$N(\alpha) \ge \int_0^T \{|Y_\alpha(\sigma) \sim X_a|\}d\sigma + \{T/\tau - |\mathcal{I}_0''(\alpha)|\}. \tag{4.12}$$

The bound $N(\alpha) \ge \int_0^T \{|Y_\alpha(\sigma) \sim X_a|\}d\sigma$ follows as in the proof of Proposition 3.2, neglecting squares in $\mathfrak{D}_1$. In a strip $X_a \times [i_0, i_0 + 1]$, $i_0 \in \mathcal{I}_0$, there is at least one vertex. For a square $\Delta \in X_a \times \mathcal{I}_0' \sim \mathcal{I}_0''$, we count one of the $n(\Delta')$ vertices in Δ', where Δ' forced the PS construction in Δ. Thus we count at least one vertex for each time interval in $\mathcal{I}_1 \sim \mathcal{I}_0''$. There are $T/\tau - |\mathcal{I}_0''|$ such time intervals, establishing (4.12).

From (4.10)–(4.12), we find

$$\begin{aligned}&\langle\Omega_0, e^{-T[H_0+H_I(\sim X_a)]}\Omega_0\rangle \\ &\quad\le \sup_\alpha e^{-TE(\sim X_a)}e^{-T/\tau}e^{-|\mathcal{I}_0''(\alpha)|\{-\log 2 - 1 + \frac{1}{2}|X_a|\}} \\ &\quad\le e^{-T[E(\sim X_a) + \tau^{-1}]},\end{aligned}$$

if a_0 (and hence $|X_a| \ge a \ge a_0$ and N_0) are sufficiently large.

Taking the logarithm, dividing by T and letting $T \to \infty$, the left side converges to $-E(\sim X_a)$. This gives the contradiction $0 \leq -\tau^{-1}$, and proves (4.1).

5. A CORRECTION

Our proof[5] of the spectrum condition $P^2 \leq H^2$ contains a gap, as was pointed out by Fröhlich and Faris. Namely, we required the Lorentz rotated Hamiltonian in a periodic box to have a simple ground state, which does not follow, as claimed, from standard methods. The remaining results in Ref. 5 are either independent of this gap or are proved (and improved) by the present paper, with the exception of the estimate

$$\pm \nabla\varphi(h) \leq \text{const}\|h\|_2(H + I).$$

This bound is Theorem 1.1, Case 2, $\epsilon = \frac{1}{2}$, and it should follow from the methods of the present paper.

---

* Supported in part by the National Science Foundation, Grant GP-24003.

† Supported in part by the Air Force Office of Scientific Research, Contract AF 44620-70-C0030 and the National Science Foundation, Grant GP-31239X.

1 J. Glimm and A. Jaffe, *Quantum Field Theory Models*, in the 1970 Les Houches lectures, edited by C. Dewitt and R. Stora (Gordon and Breach, New York, 1971).

2 J. Glimm and A. Jaffe, "Boson Quantum Field Models," in the 1971 London lectures, edited by R. Streater (to appear).

3 J. Glimm and A. Jaffe, Acta Math **125**, 203 (1970).

4 J. Glimm and A. Jaffe, Ann. Math. **91**, 362 (1970).

5 J. Glimm and A. Jaffe, J. Math. Phys. **11**, 3335 (1970); Commun. Math. Phys. **22**, (1971).

6 J. Glimm and A. Jaffe, Commun. Math. Phys. **22**, 253 (1971).

Positivity of the φ_3^4 Hamiltonian

James Glimm[1])

Courant Institute of Mathematical Science, New York University, New York, N. Y.

Arthur Jaffe[2])

Lyman Laboratory of Physics, Harvard University, Cambridge, Mass.

Abstract

The renormalized φ_3^4 Hamiltonian is bounded from below by a constant proportional to the volume.

Contents

---

[1]) Supported in part by the National Science Foundation, under Grant NSF GP 24002.
[2]) Supported in part by the National Science Foundation, under Grant NSF GP 31239 X and by the AFOSR under Contract AF 44620—70—C0030.

1. Introduction

1.1. Discussion

We prove that the renormalized φ_3^4 Hamiltonian is bounded from below by a constant proportional to the volume. The renormalization constants in our Hamiltonian $H(g)$, with a spatial cutoff g, are given by perturbation theory; they are the ultraviolet divergent contributions to the ground state energy and to the rest mass of a single particle. After subtraction of a finite vacuum energy constant $E(g)$, we obtain a positive, densely defined Hamiltonian. The analogous positivity result is well known for the two dimensional $\mathscr{P}(\varphi)_2$ and Yukawa$_2$ Hamiltonians. A subsequent development of the φ_3^4 model might proceed along the lines of the $\mathscr{P}(\varphi)_2$ and Yukawa$_2$ models. In this case the lower bound on $H(g)$ would be a major step toward constructing a self adjoint operator $H(g)$, which defines dynamics with finite propagation speed.

The main difficulty in the positivity proof lies with the divergences. Since the ultraviolet divergences become more severe with increasing numbers of space dimensions, our proof of positivity for the φ_3^4 Hamiltonian is considerably more difficult than for the two dimensional models. The only "superrenormalizable" boson self interactions are $\mathscr{P}(\varphi)_2$ and φ_3^4. The next most singular self interaction in three dimensions is φ_3^6 which is comparable in difficulty to φ_4^4. Our present methods seem to be marginal for such "renormalizable, but not superrenormalizable" interactions, and new methods are presumably needed for them.

The Hamiltonian $H(g)$ is defined as a limit (as $\varkappa \to \infty$) of approximate, momentum cutoff Hamiltonians $H(g, \varkappa)$. The cutoff operators are self adjoint and bounded from below on Fock space. The first result about the φ_3^4 Hamiltonian was the proof [*4*] that the sequence $\{H(g, \varkappa)\}$ yields a densely defined limiting operator $H(g)$. The domain of $H(g)$ lies in a Hilbert space $\mathscr{F}(g)$ disjoint from Fock space. The non Fock representation of the fields on $\mathscr{F}(g)$ was studied by HEPP [*9*], FABREY [*3*], ECKMANN and OSTERWALDER [*2*], and ECKMANN [*1*].

In the present paper we show that $\{H(g, \varkappa)\}$ is bounded from below, uniformly in $\varkappa$, yielding a lower bound for $H(g)$. Let g be a $C_0^\infty(R^2)$ function, $0 \leqq g \leqq 1$, and let $A(g)$ denote the area of the set of points within distance 1 of the support of g. Let $E(g, \varkappa)$ be the vacuum energy of $H(g, \varkappa)$, namely

$$E(g, \varkappa) = \inf \text{spectrum } H(g, \varkappa). \tag{1.1.1}$$

Theorem 1.1.1. *There exists a positive constant, independent of g and $\varkappa$, such that*

$$-\text{const } A(g) \leqq E(g, \varkappa). \tag{1.1.2}$$

With g held fixed, it follows from [*4*] that $E(g, \varkappa)$ is bounded from above. We expect, but have not proved, that as $g \to 1$,

$$E(g, \varkappa) \leqq \text{const } A(g). \tag{1.1.3}$$

The bound (1.1.3) is used in the $\mathscr{P}(\varphi)_2$ and Yukawa$_2$ models, in passing to limit $g = 1$.

The cutoff operator $H(g, \varkappa)$ has a unique vacuum vector $\Omega(g, \varkappa)$, i.e. $E(g, \varkappa)$ is a simple eigenvalue of $H(g, \varkappa)$. The uniqueness proof [*5*] shows that $\Omega(g, \varkappa)$ is not orthogonal to the vacuum vector Ω_0 of H_0. Hence,

$$-E(g, \varkappa) = \lim_{t\to\infty} \frac{\log \langle \Omega_0, \exp(-t\,H(g,\varkappa))\,\Omega_0\rangle}{t}\,. \tag{1.1.4}$$

We prove Theorem 1.1.1, by showing that there exists a constant 0(1), independent of g, $\varkappa$ and t, such that for $t \geqq 1$,

$$\langle \Omega_0, \exp(-tH(g,\varkappa))\,\Omega_0\rangle \leqq \exp(0(1)\,tA(g))\,. \tag{1.1.5}$$

We use two basic expansions to estimate (1.1.5). Our high momentum expansion, the "perturbation-contraction" expansion, is given in Chapter 2. This expansion is a renormalized perturbation expansion for $\langle \exp(-tH(g,\varkappa))\rangle$, which exhibits the renormalization cancellations in an explicit fashion. The perturbation-contraction expansion is insufficient by itself, since the perturbation series for $\langle \exp(-t\,H(g,\varkappa))\rangle$ diverges. The second expansion, the Wick expansion, gives a bound on the low momentum part of the interaction. It is based on the Wick ordering bound

$$-0(\varkappa^2) \leqq \, : \varphi_x{}^4(\boldsymbol{x}):, \tag{1.1.6}$$

obtained from: $\varphi_\varkappa{}^4: = \varphi_\varkappa{}^4 - 6\varphi_\varkappa{}^2 \langle\varphi_\varkappa{}^2\rangle + 3\,\langle\varphi_x{}^2\rangle^2$, where $\langle\varphi_\varkappa{}^2\rangle = \langle\Omega_0, \varphi_x(\boldsymbol{x})^2\,\Omega_0\rangle = 0(\varkappa)$. Combining these two methods, we obtain (1.1.5).

The basic idea of our method is to perform the perturbation-contraction expansion for portions of the Hamiltonian that are localized in sufficiently small space regions and time intervals. We use (1.1.6) to obtain a convergent expansion for $\exp(-tH)$ in each region, and we show that different regions are essentially independent. We localize the fields $\varphi(\boldsymbol{x})$ in H_I, or in other words, we localize the particles that enter the interaction. For our present problem, we find it necessary to localize the particles simultaneously in configuration space and in momentum space, namely in phase space. Of course, such localization can not be exact, but it can be given by some function f so that both f and its Fourier transform $f^\sim$ are rapidly decreasing. We call such localization, polynomial localization in phase space. The exact form of our localization function $\zeta(x, k)$ is given in Section 1.2.

The diameter of our localization region is limited by the uncertainty principle

$$O(1) \leqq \delta x\,\delta p\,, \tag{1.1.7}$$

for each position and momentum coordinate. In particular, at high values of the momentum, $|p| > M$, we can impose a small spreading, $O(M^{-1})$, of the configuration space localization due to the momentum, space localization. Thus if $M = O(\varkappa)$, the minimum volume of space localization volume compatible with the uncertainty principle is $O(\varkappa^{-s})$, where s is the number of space dimensions. Since our actual estimates deal with the localized action, we allow for time intervals of length δt as well, where $0(1) \leqq \delta E\,\delta t$. The minimum space-time volume $|\Delta|$ for a space time localization cube Δ is

$$|\Delta| = O\big((\delta x)^s\,\delta t\big) = O(\varkappa^{-s-1})\,. \tag{1.1.8}$$

In (1.1.5), the exponent $S = O(1)\,tA(g)$ has the dimensions of action. In perturbation theory, for a superrenormalizable interaction, the vacuum energy satisfies

$$|E_\varkappa| \leqq A(g)\,O(\varkappa^s)\,, \tag{1.1.9}$$

28*

as $\varkappa \to \infty$. Since the vacuum energy is the most divergent renormalization constant, the bound on S, neglecting renormalization cancellations, is

$$|S_\varkappa| \leqq tA(g)\, O(\varkappa^s). \tag{1.1.10}$$

We show that $S_\varkappa$ is approximately additive,

$$S_\varkappa \approx \sum S_\varkappa(\Delta),$$
$$S_\varkappa(\Delta) \approx \big(|\Delta|/tA(g)\big)\, S_\varkappa \leqq 0\,(|\Delta|\,\varkappa^s), \tag{1.1.11}$$

by (1.1.10). We conclude from (1.1.8), that the cubes Δ of minimum size satisfy $S_\varkappa(\Delta) \leqq O(\varkappa^{-1})$, as $\varkappa \to \infty$. Associated with the approximate additivity of S_x is the approximate factorization of the physical vacuum state Ω with a single factor $\Omega(\Delta)$ for each region Δ. Corresponding to the small contribution, $S(\Delta)$, of Δ to S, is the fact that at most $O(1)$ bare particles, with momenta greater than $O(|\Delta|^{-1/(s+1)})$, contribute to the factor $\Omega(\Delta)$ and to Ω. We use the bound (1.1.11), along with convergence from the perturbation-contraction expansion in Δ. These ideas yield (1.1.5), and in principle they apply to superrenormalizable theories in general.
For interactions that are renormalizable, but not superrenormalizable, these ideas become marginal, and hence their applicability is not clear. For instance, the vacuum energy in this case satisfies $|E| \leqq O\big(A(g)\,\varkappa^{s+1}\big)$ in perturbation theory. Hence $S_x(\Delta) \approx O(1)$ for cubes Δ of minimum size, and we lose control over the approximations in the above analysis. As in the present paper, it seems that the portions of H_I which couple high and low momentum modes will be the most difficult to control. (Various such problems arise in Chapters 2 and 3 here.) Indeed, this part of H_I gives rise to the mass, field strength and charge renormalizations. However, we note that this portion of H_I is superrenormalizable, although H_I itself is not.
Our bound for $\langle \exp(-tH(g, \varkappa))\rangle$ is defined by an inductive construction [6]. At each stage of the proof we dominate $\langle \exp(-tH)\rangle$ by a sum of terms of the form

$$\int ds \langle TR(s) \exp\big(-\int_0^t d\sigma\, H(\sigma, s)\big)\rangle, \tag{1.1.12}$$

where $R(s)$ is a polynomial in the fields, $H(\sigma, s)\big)$ is a time dependent Hamiltonian, and T is a time ordering operator for time parameters s and σ. Each inductive step or substep replaces (1.1.12) by a sum of similar terms bounding (1.1.12) from above. The induction continues for a particular term until $H(\sigma, s) \equiv 0$ for that term. If $H(\sigma, s) = 0$, then (1.1.12) has the form of a sum of vacuum Feynman graphs. Our expansion is defined in Chapters 2 and 3, and we estimate the sum of terms (1.1.12) in Chapters 4, 5 and 6.

1.2. Localization of Wick polynomials

For mathematical properties of operators and forms on Fock space, see [7—8]. We define a Euclidean field $\varphi(x)$, for $x \in R^3$, by

$$\varphi(x) = (2\pi)^{-3/2} \int e^{-ikx}\, \varphi(k)^\sim\, dk,$$
$$\varphi(k)^\sim = (k^2 + m^2)^{-1/2}\, \{A(k)^* + A(-k)\}.$$

Here the Euclidean annihilation and creation operators $A(k)$ satisfy $[A(k), A(k')] = 0$, $[A(k), A(k')^*] = \delta^3(k - k')$. This Euclidean field reduces at time zero to the usual time zero free field,

$$\varphi(\boldsymbol{x}) \equiv \varphi(\boldsymbol{x}, x_0 = 0) = (2\pi)^{-1} 2^{-1/2} \int e^{-i\boldsymbol{k}\boldsymbol{x}} \mu(\boldsymbol{k})^{-1/2} \{a(\boldsymbol{k})^* + a(-\boldsymbol{k})\} \, d\boldsymbol{k},$$

defining $a(\boldsymbol{k})$ in terms of $A(k)$. Then $[a(\boldsymbol{k}), a(\boldsymbol{k}')^*] = \delta^2(\boldsymbol{k} - \boldsymbol{k}')$.
For suitable functions ζ, defined on the relativistic phase space $R^3 \times R^3$, we define a localized Euclidean field $\varphi(\zeta)$ by

$$\varphi(\zeta) = (2\pi)^{-3/2} \int e^{-ikx} \zeta(x, k) \varphi(k)^{\sim} \, dk \, dx. \tag{1.2.1}$$

For a Wick polynomial $:\varphi^j:$ we permit j distinct mo mentum localizations, but only one space-time localization. We define

$$:\varphi^j(\zeta): = (2\pi)^{-3j/2} \int e^{-i\left(\sum_{l=1}^{j} k_l\right)x} \zeta(x, k) :\prod_{l=1}^{j} \varphi^{\sim}(k_l): \, dk \, dx, \tag{1.2.2}$$

where $\zeta(x, k)$ is a function of $x \in R^3$ and $k = (k_1, \ldots, k_j) \in R^{3j}$. If (1.2.2) is hermitian, $\zeta(x, k) = \zeta(x, -k)$. In later sections of the pa per, we set $m = 1$ for convenience.
We choose our momentum function η to be a C^∞ function on R^1 satisfying

$$\begin{aligned} \eta(\varrho) &= 1, & |\varrho| &\leqq \frac{1}{2} \\ 0 \leqq \eta(\varrho) &\leqq 1, & \frac{1}{2} &\leqq \varrho \leqq 2 \\ \eta(\varrho) &= 0, & 2 &\leqq \varrho. \end{aligned}$$

For $k \in R^3$ and a vector cutoff $\alpha \in (R^+)^3$, we set

$$\eta_\alpha(k) = \prod_{l=0}^{2} \eta(k^{(l)}/\alpha^{(l)}), \tag{1.2.3}$$

where l indexes the components of the momentum. With $\alpha^{(l)} = 0$, we define $\eta(k^{(l)}/\alpha^{(l)}) = 0$. With $\alpha^{(l)} \leqq \beta^{(l)}$, the momentum cutoff

$$\eta_{\alpha,\beta}(k) = \eta_\beta(k) - \eta_\alpha(k) \tag{1.2.4}$$

provides an upper and lower cutoff in momentum space. In space time, we localize in cubes $\Delta \subset R^3$, with characteristic function χ_Δ. We allow $\zeta(x, k)$ in (1.2.1) – (1.2.2) to be a linear combination of functions of the form

$$h(x) \chi_\Delta(x) \prod_{r=1}^{j} \eta_{\alpha_r,\beta_r}(k_r) \tag{1.2.5}$$

where $h(x)$ is a space time cutoff. We define for a polynomial $:\varphi^j:$,

$$\lambda = \max_{l,r} \{\alpha_r^{(l)}, 2\}$$

as the maximum lower momentum cutoff.

It is convenient to restrict the allowed classes of cubes Δ and intervals $[\alpha^{(l)}, \beta^{(l)}]$. The allowed cubes have the form $2^{-j}\Delta_0 + 2^{-j}z$, where $z \in Z^3$, and Δ_0 is the cube with volume $|\Delta_0| = 1$, centered at the origin. For a space time cutoff h, we define vol (h) as the number of translates $\Delta_0 + z$, $z \in Z^3$, that intersect support h.
The allowed momentum cutoffs $\alpha^{(l)}$, $\beta^{(l)}$ are chosen from the sequence $\{M_0, M_1, \ldots, \infty\}$ where $M_0 = 0$, M_1 is chosen sufficiently large, and

$$M_{j+1} = M_j^{l+\nu} = M_1^{(l+\nu)^j}. \tag{1.2.6}$$

In later chapters we choose $\nu > 0$, sufficiently small.
According to the uncertainty principle, particles created by (1.2.1) or (1.2.2) are localized in a cube with length

$$|\Delta|^{1/3} + M_j^{-1} \tag{1.2.7}$$

where M_j is the lower momentum cutoff for the particle in question and hence determines spreading. If

$$M_j^{-1} \leqq |\Delta|^{1/3}, \tag{1.2.8}$$

we call the localization proper, and otherwise we call it improper. We introduce the localization ratio

$$b = \max\{1, |\Delta|^{-1/3} M_j^{-1}\}, \tag{1.2.9}$$

and in both cases the particles are localized in cubes of length

$$0(1)\, b\, |\Delta|^{1/3}. \tag{1.2.10}$$

We are also interested in measuring the effective strength of the force between two particles localized in different cubes Δ, Δ', and in momentum ranges with lower cutoffs M_j, $M_{j'}$. This force is expressed in terms of the scaled distance d, which we define to be

$$d = \max\{1, \text{dist.}(\Delta, \Delta'), M_j \,\text{dist.}(\Delta, \Delta'), M_{j'}\,\text{dist.}(\Delta, \Delta')\}. \tag{1.2.11}$$

In case both lower momentum cutoffs are zero, the scaled distance d reduces to max $\{1, \text{dist.}(\Delta, \Delta')\}$, where dist. (Δ, Δ') is the Euclidean distance. We have chosen a dimensionally inhomogeneous definition, to emphasize the ultraviolet aspects of d.

1.3. The renormalized Hamiltonian

We study Hamiltonians of the form

$$H(g, \varkappa) = H_0 + \int :\mathscr{P}\big(\varphi_\varkappa(\boldsymbol{x}), g(\boldsymbol{x})\big): d\boldsymbol{x},$$

where $\mathscr{P}$ is a quartic interaction plus counterterms. In particular we bound $\langle \Omega_0, \exp(-t\, H(g, \varkappa))\, \Omega_0\rangle$. It is convenient to relate this expression to the Euclidean field $\varphi(x)$ of Section 1.2., for we then obtain a perturbation series directly in terms of Euclidean Feynman propagators, $(k^2 + m^2)^{-1}$. We use the "relativistic Feynman-Kac formula" investigated by SYMANZIK *[11]* and NELSON *[10]*,

$$\begin{aligned}
&\langle \Omega_0, \exp(-t\, H(g, \varkappa))\, \Omega_0\rangle \\
&\quad = \langle \Omega_0, \exp\Big(-\int_0^t dx_0 \int d\boldsymbol{x} :\mathscr{P}(\varphi_x(x)\, g(\boldsymbol{x})):\Big)\, \Omega_0\rangle \\
&\quad = \langle \Omega_0, \exp(-V)\, \Omega_0\rangle .
\end{aligned}$$

Thus we define the Hamiltonian by giving the renormalized, Euclidean action, $V = V_I + V_C$. Here V_I and V_C are Wick polynomials in the Euclidean field $\varphi(x)$. If the renormalization constants are time independent, if the space time cutoff of Section 1.2. has the form

$$h(x) = g(\boldsymbol{x})\, \chi_{[0,t]}(x_0)\,, \tag{1.3.1}$$

and if the momentum cutoff η of Section 1.2. is independent of time and energy, then $\exp(-V)$ has an interpretation in terms of a time independent Hamiltonian. We call such a cutoff in V a "Hamiltonian cutoff", and we assume $g(\boldsymbol{x}) \in C_0{}^\infty(R^2)$, or else that $h(\boldsymbol{x})$ is the characteristic function of a union of translates of Δ_0. Our estimates for $\exp(-V)$ also apply for $h(x) \in C_0{}^\infty(R^3)$. We begin our proof of Theorem 1.1.1 with a Hamiltonian cutoff, but in the inductive construction we replace ζ by the more general cutoffs of Section 1.2.

It is convenient to define the counterterm V_C in a slightly different manner than suggested by the Hamiltonian. Rather, we use the vacuum energy renormalization usually associated with the Euclidean Green's functions

$$\langle \Omega_0, \varphi(x_1) \dots \varphi(x_n) \exp(-V) \Omega_0 \rangle .$$

We show below that our Green's function renormalization differs from the Hamiltonian renormalization by a wave function renormalization constant, and a transient that vanishes as $t \to \infty$. Hence we obtain estimates on the Hamiltonian. The advantage of using a Green's function counterterm, rather than a local Hamiltonian counterterm, is that we obtain exact cancellation of the second and third order vacuum graphs in our expansion of $\exp(-V)$.

For a cutoff ζ, we define

$$V_I = :\varphi^4(\zeta):. \tag{1.3.2}$$

Then the vacuum energy counterterms are

$$E_2(\zeta) = -\frac{1}{2} \langle V_I{}^2 \rangle_{\Omega_0}, \tag{1.3.3}$$

$$E_3(\zeta) = (1/6) \langle V_I{}^3 \rangle_{\Omega_0},$$

where the expectation is in the (Euclidean) no particle state Ω_0. If

$$\zeta(x, k) = \zeta_0(\boldsymbol{x}, \boldsymbol{k})\, \chi_{[0,t]}(x_0) = h(x)\eta(\boldsymbol{k})\,, \tag{1.3.4}$$

then the energy integrals in (1.3.3) are elementary and can be evaluated by residues. This gives

$$E_2(\zeta) = t\bar{E}_2(\zeta_0) + \Lambda_2(\zeta_0) + T_2(\zeta_0, t) \tag{1.3.5}$$

where we now identify $\bar{E}_2$ as the Hamiltonian counterterm in second order, Λ_2 as the wave function renormalization in second order, and $T_2(\zeta_0, t)$ as a transient. In fact, for the cutoff (1.3.4),

$$V_I = \int_0^t dx_0\, :\varphi^4(\zeta_0, x_0):, \tag{1.3.6}$$

and if $H_I = :\varphi^4(\zeta_0, x_0 = 0):$, we obtain

$$\begin{aligned} \bar{E}_2(\zeta_0) &= -\langle H_I H_0{}^{-1} H_I \rangle_{\Omega_0} \\ &= 4!\,2^{-4} (2\pi)^{-13/2} \int (\mu_1\mu_2\mu_3\mu_4)^{-1} (\mu_1 + \cdots \mu_4)^{-1} |\zeta_0(\textstyle\sum \boldsymbol{k}_i, \boldsymbol{k})|^2\, d\boldsymbol{k}\,, \end{aligned} \tag{1.3.7}$$

where $\mu(k) = (k^2 + m^2)^{1/2}$, and where $\tilde{\zeta}_0$ denotes the partial Fourier transform of $\zeta_0(\boldsymbol{x}, \boldsymbol{k})$ in $\boldsymbol{x}$. Similarly, we find that

$$\Lambda_2(\zeta_0) = \langle H_I H_0^{-2} H_I \rangle_{\Omega_0}, \tag{1.3.8}$$

$$T_2(\zeta_0, t) = -\langle H_I H_0^{-2} e^{-tH_0} H_I \rangle_{\Omega_0}. \tag{1.3.9}$$

Thus we see that $\bar{E}_2$ is the exact second order Hamiltonian counterterm for the vacuum energy. Since Λ_2 enters (1.3.5) independent of t, it does not provide a Hamiltonian counterterm. Rather, it is the second order contribution to the wave function renormalization. Let $\Omega(\zeta_0)$ be the ground state of

$$\begin{aligned} H(\zeta_0) &= H_0 + H_I(\zeta_0) + C(\zeta_0) \\ &= H_0 + H_I(\zeta_0) - \bar{E}_2(\zeta_0) - \bar{E}_3(\zeta_0) - \bar{M}_2(\zeta_0). \end{aligned} \tag{1.3.10}$$

Then we define the wave function renormalization as

$$Z(\zeta_0) = |\langle \Omega_0, \Omega(\zeta_0)\rangle|^2 = \exp\big(-\Lambda(\zeta_0)\big),$$

and in second order perturbation theory, $\Lambda(\zeta_0) = \Lambda_2(\zeta_0)$. Finally, we note that $T_2(\zeta_0, t)$ of (1.3.9) is bounded uniformly in $\varkappa$ for t bounded away from zero, and $T_2(\zeta_0, t) \to 0$ exponentially as $t \to \infty$.
Similarly, we find that

$$E_3(\zeta) = t\bar{E}_3(\zeta_0) + \Lambda_3(\zeta_0) + T_3(\zeta_0, t), \tag{1.3.11}$$

where $\bar{E}_3(\zeta_0)$ is the third order Hamiltonian counterterm for the vacuum energy $\langle H_I H_0^{-1} H_I H_0^{-1} H_I \rangle$, where $\Lambda_3(\zeta_0)$ is the third order wave function renormalization constant, and where $T_3(\zeta_0, t) \to 0$ exponentially as $t \to \infty$. The respective ultraviolet behavior is $0(\log \varkappa)$ for $\bar{E}_3$ and $0(1)$ for Λ_3 and T_3.
Next we consider the mass renormalization counterterm, which is logarithmically divergent in second order. In a cube Δ, the momentum cutoff $\eta = \eta(k)$ has the form

$$\eta(k) = \prod_{i=1}^{4} \eta_i(k_i),$$

cf. (1.2.5). Furthermore $\eta_1 = \eta_2 = \eta_3 = \eta_4$, as we will see in (2.7). We define

$$\delta m^2(\eta) = -4^2 \times 6 \times (2\pi)^{-9} \int \delta(k_2 + k_3 + k_4) \left(\prod_{i=2}^{4} \mu_i^{-2}\right) \eta_i(-k_i)\, \eta_i(k_i)\, dk_2\, dk_3\, dk_4 \tag{1.3.12}$$

and

$$M_2(\zeta) = \frac{1}{2} \Sigma_\Delta\, \delta m^2(\eta) :\varphi^2_{\eta_1}(h^2\chi_\Delta):. \tag{1.3.13}$$

In the case of a Hamiltonian cutoff, $M_2(\zeta)$ is an integral (from 0 to t), and the time zero integrand, denoted $\bar{M}_2(\zeta_0)$ is the Hamiltonian counterterm.
With this definition of $\bar{M}_2(\zeta_0)$, the renormalized Hamiltonian is given by (1.3.10).
Similarly we define the renormalized action V by

$$\begin{aligned} V(\zeta) &= V_I(\zeta) + V_C(\zeta) \\ V_C(\zeta) &= -E_2(\zeta) - E_3(\zeta) - M_2(\zeta). \end{aligned} \tag{1.3.14}$$

Our bound for $\exp\big(-V(\zeta)\big)$ thus differs from the corresponding bound for

$\exp\left(-tH(\zeta_0)\right)$ by a scalar factor. With the definitions above,

$$\langle\Omega_0, e^{-tH(g,\varkappa)}\,\Omega_0\rangle \tag{1.3.15}$$
$$= \langle\Omega_0, e^{-V}\,\Omega_0\rangle \exp\left(-\Lambda_2(\zeta_0) - \Lambda_3(\zeta_0) - T_2(\zeta_0, t) - T_3(\zeta_0, t)\right) \leqq \langle\Omega_0, e^{-V}\,\Omega_0\rangle,$$

for $\varkappa$ sufficiently large.

Finally, we note that the Hamiltonian interpretation of $\exp(-V)$ requires that the momentum cutoff η in (1.3.4) be energy independent. For Theorem 1.1.1, we desire a uniform bound on $\exp\left(-V(\zeta)_\varkappa\right)$ where $\varkappa$ is the upper cutoff in the momentum only. In the remainder of the paper, however, it is convenient to have an overall upper cutoff $\varkappa$ in both the energy and in the momentum. The inductive proof of Chapters 2—6 is equally valid if we fix the cutoff in the momentum at $\varkappa$, and we let $\varkappa_0$ denote the energy cutoff. (Note our cutoffs (1.2.3) can be chosen separately in each component.) The resulting bound is uniform in $\varkappa_0$ and $\varkappa$, and

$$\exp\left(-V(\zeta)_{\varkappa_0,\varkappa}\right) \leqq \exp O\left(tA(g)\right). \tag{1.3.16}$$

Furthermore, with $\varkappa$ fixed, $V(\zeta_{\varkappa_0,\varkappa})$ is bounded from below, uniformly in $\varkappa_0$, by a standard Wick ordering bound. This yields $\exp\left(-V(\zeta)^{\varkappa_0,\varkappa} \to \exp\left(-V(\zeta_0)_\varkappa\right)\right.$ as $\varkappa_0 \to \infty$, and hence the bound (1.3.16) applies to the Hamiltonian as desired: $\exp\left(-V(\zeta_0)_\varkappa\right) \leqq \exp O\left(tA(g)\right)$.

2. A renormalized perturbation expansion

We use two simple identities, which we call the *perturbation formula* and the *contraction formula* to generate an asymptotic expansion for $\langle\exp(-tH)\rangle$. We call this the perturbation contraction (P—C) expansion. The P formula generates a series in the renormalized interaction $V = V_I + V_C$. The C formula exhibits the cancellation of divergences between the unrenormalized interaction V_I and the counterterms V_C. The P—C expansion is a partially renormalized expansion. In later stages of the expansion we allow certain uncancelled logarithmic divergences. They occur only in terms that have strongly convergent factors, introduced earlier in the expansion, which dominate these divergences.

We define our expansion for $\langle\exp(-V)\rangle$ inductively, and hence the name *inductive construction*. Each step consists of three main parts: A high momentum (P—C) expansion given in this chapter, a low momentum (Wick) expansion in Chapter 3 and combinatoric estimates in Chapter 4 which bound the number of terms generated during these expansions. After these three steps, we decide for each term whether to terminate the expansion, or whether to return to the high momentum expansion and another inductive step. When the induction terminates for a particular term, we use the estimates of Chapters 5 and 6 to bound the term. Each such term is labeled by a Feynman graph G, and the term is an integral $I(G)$. These integrals are the familiar momentum space Feynman integrals that arise in perturbation calculations. These integrals occur, however, with a restricted momentum range and in particular we utilize the convergence from a lower cutoff. In this manner we convert an asymptotic series in the coupling constant into a convergent bound. Our final bound for $\langle\exp(-tH)\rangle$ is a convergent sum of upper bounds on Feynman graphs.

The purpose of the high momentum (P—C) expansion is to lower the upper momentum cutoff in the exponent V. For this purpose, let ζ_s be a family of cutoffs ζ, where $0 \leqq s \leqq 1$, and ζ_0 has a lower value of the upper momentum cutoff than ζ_1.

The perturbation formula (P formula) is

$$e^{-V(\zeta_1)} = e^{-V(\zeta_0)} - \int_0^1 ds \frac{d}{ds} V(\zeta_s)\, e^{-V(\zeta_s)}, \tag{2.1}$$

valid since $V(\zeta_{s_0})$ and $V(\zeta_{s_1})$ are both functions of the Euclidean field, and hence commute. Thus the P formula expresses $\exp(-V)$ as the sum of a term with a lowered cutoff and a term involving dV/ds. We call dV/ds a perturbation vertex (P vertex). If we wish to assert that we applied (2.1) during the rth inductive step, we call dV/ds a P_r vertex, and we say that the vertex is introduced in the $P_r - C_r$ expansion.

We exhibit the renormalization cancellations in $V = V_I + V_C$ only when V occurs as a P vertex. For this purpose we use the contraction formula (C formula),

$$A(k)\, e^{-V} = e^{-V} A(k) + [V, A(k)]\, e^{-V}, \tag{2.2}$$

or its adjoint. The C formula is valid because both V and $[V, A]$ are functions of the Euclidean field, and hence commute. We call $[V, A(k)]$ or $[A(k)^*, V]$ a contraction vertex (C vertex).

The perturbation formula (2.1) remains valid upon multiplication by a polynomial R in the Euclidean fields,

$$Re^{-V(\zeta_1)} = Re^{-V(\zeta_0)} - \int ds\, R \frac{dV(\zeta_s)}{ds} e^{-V(\zeta_s)}. \tag{2.3}$$

In the contraction formula, multiplication by R yields extra contracted terms, but no new vertices,

$$A(k)\, Re^{-V} = Re^{-V} A(k) + \{[A(k), R] + [V, A(k)]\, R\}\, e^{-V}. \tag{2.4}$$

We use (2.4) to move certain annihilation operators to the right, until they give $A(k)\,\Omega_0 = 0$ plus a sum of contracted terms. We use its adjoint to move certain $A(k)^*$'s similarly to the left in our expansion.

At each point in the inductive construction, we have an upper bound for $\langle \exp(-V) \rangle$ as a sum of terms. Each inductive step or substep replaces a single term by a sum of terms, each of which has the form

$$\sup_s |\langle R(s)\, e^{-V(\zeta_s)} \rangle|. \tag{2.5}$$

The $P-C$ expansion is a finite sequence of substeps. In each substep we add a single P vertex, localized in a space time cube Δ, and we exhibit certain renormalization cancellations for this vertex. The expansion is specified once we give (a) the prescription for adding one P vertex and (b) the prescription for terminating the expansion.

Associated with each term (2.5) is a cover $\mathscr{D}$ of space time by nonoverlapping cubes Δ. For each cube $\Delta \in \mathscr{D}$ there is an upper cutoff $M_{u(\Delta)}$, a lower cutoff $M_{l(\Delta)}$ and an intensity $s(\Delta)$. The intensity lies in $[0, 1]$, and interpolates between the upper cutoffs $M_{u(\Delta)}$ and $M_{u(\Delta)-1}$. The s-dependent momentum cutoff for one leg is

$$\eta_{M_l, M_u}(s, k) = s\eta_{M_l, M_u}(k) + (1 - s)\, \eta_{M_l, M_{u-1}}(k), \tag{2.6}$$

while the cutoff ζ_s in $\exp\big(-V(\zeta_s)\big)$ is

$$\zeta(x, k) = \Big(\prod_i \eta_{M_{l(\Delta)}, M_{u(\Delta)}}\big(s(\Delta), k_i\big)\Big) (h\chi_\Delta)^\sim \big(\sum_i k_i\big) = \eta_s(k)\,(h\chi_\Delta)^\sim (\Sigma k_i) \tag{2.7}$$

in the cube Δ. At the beginning of the first inductive step we choose $s(\Delta) = 1$, $M_{l(\Delta)} = 0$, $M_{u(\Delta)-1} \leqq \varkappa \leqq M_{u(\Delta)}$.
The space time cover at the start of the first inductive step is

$$\{\Delta_0 + \alpha : \alpha \in Z^3\}. \tag{2.8}$$

Next we explain the substep of adding a single P vertex in Δ. With l, u, s and ζ given. we apply (2.1) with s replaced by

$$s'(\Delta) \in [0, s(\Delta)],$$

provided $s(\Delta) > 0$. If $s(\Delta) = 0$, the first term in (2.1) was selected in the previous application of (2.1) in Δ. We then replace $s(\Delta)$ by 1, $u(\Delta)$ by $u(\Delta) - 1$ and proceed as before.
Next we prescribe the renormalization cancellations. (A) First we use (2.4) to contract all legs of the new P vertex. (B) In most cases, the cancellation is completed by contracting all legs of all new C vertices formed by (A). We make three exceptions to (B):
(B 1) If the P vertex in (A) is a mass counterterm in V_C, then the legs of the new C vertices from A are not contracted, namely (B) is not performed.
(B 2) If the P vertex in (A) contracts three times to one new C vertex and once to a second new C vertex, a mass subdiagram is formed. In order to permit cancellation with the counterterm in (B 1), we contract the remaining leg of the mass subdiagram, i.e. the fourth leg of the first new C vertex, but do not contract the three remaining legs of the second new C vertex.
(B 3) If the P vertex contracts three times to old vertices and once to a new C vertex in (A), then we apply (2.4) "in reverse". We write this term as a term in which a single P leg is uncontracted, plus terms in which all four legs contract to old vertices. This completes the C expansion associated with a single P vertex.
Now we describe which terms are combined to cancel, yielding a convergent remainder. The second and third order vacuum energy dagrams formed from a P vertex and one or two C vertices cancel exactly with the δE_2 and δE_3 counterterm contributions to the P vertex, see (1.3.3). Graphically, these terms are

P ≣ C and 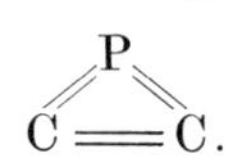

Turning to the mass renormalization, we combine the counterterm in the P vertex with the mass term containing a P vertex contracted three times to a new C vertex, graphically

—P≡C— + —M—.

Let the P vertex be localized in Δ and the C vertex in Δ'. Since Δ and Δ' belong to the same cover, they do not overlap. If $\Delta \neq \Delta'$, the counterterm is zero and the combined mass subgraph has a kernel

$$w_{\Delta,\Delta'}(k_1, k_5) = \text{const.} \int \alpha\beta\gamma\, dk_2\, dk_3\, dk_4\,.$$

If $\Delta = \Delta'$, it has a kernel

$$w_{\Delta,\Delta}(k_1, k_5) = \text{const.} \int (\alpha\beta\gamma - \bar{\alpha}\bar{\beta}\bar{\gamma})\, dk_2\, dk_3\, dk_4. \tag{2.10}$$

In both cases,

$$\alpha = \left(\frac{d}{ds}\eta_s\right)(k_1, k_2, k_3, k_4)\,\eta_s(k_2, k_3, k_4, k_5),$$

$$\beta = \bar{\beta} = (h\chi_\Delta)^\sim(-P - k_1)\,(h\chi_{\Delta'})^\sim(P + k_5)$$

$$\gamma = (\mu_1\mu_2{}^2\mu_3{}^2\mu_4{}^2\mu_5)^{-1}$$

$$\bar{\alpha} = \frac{1}{2}\frac{d}{ds}\eta_s(0, p_2, p_3, p_4)\,\eta_s(p_2, p_3, p_4, 0)\,\eta_1(k_1)\,\eta_1(k_5)$$

$$\bar{\gamma} = \left(\mu_1\,\mu(p_2)^2\,\mu(p_3)^2\,\mu(p_4)^2\,\mu_5\right)^{-1},$$

and $\mu_i = \mu(k_i)$, $\quad p_i = k_i - (1/3)P$, $\quad P = k_2 + k_3 + k_4$.

Finally, we specify when to terminate the $P-C$ expansion in a particular inductive step. Let $n_r(\Delta)$ be the number of P vertices that are introduced during the the rth inductive step and are localized in Δ. Let $\mathscr{D}_e$ be the set of cubes for which the exponent is nonzero. We permit a new P_r vertex in Δ provided Conditions (a)—(c) below are satisfied. If there are no cubes Δ satisfying Conditions (a)—(c), the $P-C$ expansion terminates for this inductive step, and the inductive construction continues according to Chapter 3.

(a) $\Delta \in \mathscr{D}_e$, $\quad u(\Delta) \geqq 2$

(b) $n_1(\Delta) \leqq M^\varepsilon_{u(\Delta)}$, if $r = 1$,

$\quad n_r(\Delta) \leqq |\Delta|^{-\varepsilon/2}$, if $r > 1$,

(c) $\sup\limits_{\Delta' \in \mathscr{D}_e} M^\varepsilon_{u(\Delta')}\, d(\Delta, \Delta')^{-4} \leqq c M^\varepsilon_{u(\Delta)}$, if $r = 1$.

Here $d(\Delta, \Delta') = 1 + r(\Delta, \Delta')$, and $r(\Delta, \Delta')$ is the distance from the center of Δ to the center of Δ'. The constant c is

$$c = \sum_{\alpha \in Z^3} d(\Delta_0, \Delta_0 + \alpha)^{-4}.$$

The Condition (c) ensures that the upper cutoff does not vary too rapidly from cube to cube.

Let $\mathscr{D}_a$, $\mathscr{D}_c$ be the sets of cubes violating (a), (c). We claim that the sup over $\Delta' \in \mathscr{D}_e$ in Condition (c) equals the sup over $\Delta' \in \mathscr{D}_e \sim \mathscr{D}_c$. (We note that $\mathscr{D}_e \neq \mathscr{D}_c$, by nature of the expansion.) In fact, suppose $r = 1$, suppose a cube Δ' stops the expansion in Δ through Condition (c), and suppose Δ'' stops the expansion in Δ'. Then

$$d(\Delta, \Delta'') \leqq 1 + r(\Delta, \Delta') + r(\Delta', \Delta'') \leqq d(\Delta, \Delta')\, d(\Delta', \Delta''). \tag{2.11}$$

Thus

$$\begin{aligned} M^\varepsilon_{u(\Delta)} &\leqq \left(d(\Delta, \Delta')^4\, c\right)^{-1} M^\varepsilon_{u(\Delta')} \\ &\leqq \left(d(\Delta, \Delta')^4\, d(\Delta', \Delta'')^4\, c^2\right)^{-1} M^\varepsilon_{u(\Delta'')} \\ &\leqq \left(d(\Delta, \Delta'')^4\, c\right)^{-1} M^\varepsilon_{u(\Delta'')}, \end{aligned}$$

and so Δ'' also stops the expansion in Δ. This transitivity property proves our assertion.

At the end of the $P_1 - C_1$ expansion, each cube $\Delta' \in \mathscr{D}_e \sim \mathscr{D}_c$ has $M^\varepsilon_{u(\Delta')}$ vertices, or else $\Delta' \in \mathscr{D}_a$. If $\mathscr{D}_a \neq \mathscr{D}_e$, there exist $\Delta' \in \mathscr{D}_e \sim \mathscr{D}_c$ with $M^\varepsilon_{u(\Delta')}$ P_1 vertices. We "assign"

$$\frac{1}{2} M^\varepsilon_{u(\Delta')} \left(d(\Delta, \Delta')^4 c\right)^{-1} \geqq \frac{1}{2} M^\varepsilon_{u(\Delta)} \tag{2.12}$$

of the Δ' vertices to Δ, and we "assign" $1/2\ M^\varepsilon_{u(\Delta')}$ of the Δ' vertices to Δ'. In this way, each cube $\Delta \in \mathscr{D}_e$ has at least $1/2\ M^\varepsilon_{u(\Delta)}$ vertices assigned to it, each with a maximum lower cutoff of at least $M_{u(\Delta)-1}$. We count the vertices assigned to Δ in $n_1(\Delta)$ in Chapters 3 and 4, and we treat these vertices as if they were in Δ. The case $\mathscr{D}_a = \mathscr{D}_e$ occurs only in the first inductive step, and it is handled separately in Chapter 3.

3. The low momentum expansion

3.1. Introduction

The lower bound for H rests on two expansions. Each expansions bounds a term

$$\sup_s \langle R(s)\, e^{-V(s)} \rangle \tag{3.1.1}$$

by a sum of similar terms, where $R(s)$ and $V(s)$ are polynomials in the fields, and the expectation is taken in the no particle state.

The high momentum expansion (P—C expansion) was given in Chapter 2. After cancellation of the renormalization counterterm, we obtain a convergence factor from each integration over a high momentum P vertex in the polynomial R.

The low momentum expansion of this chapter results in raising the lower momentum cutoff in the exponent V, and the expansion has two parts. The first, the Wick construction, is based on the fact that $:\varphi^4:$ is formally positive. By removing the Wick ordering in $:\varphi^4:$, we bound the low momentum part of exp $(-:\varphi^4:)$ from above.

We obtain convergence of the Wick construction by introducing Wick vertices (polynomials) into R in order to estimate the subsets of path space on which $:\varphi^4:$ is very negative. The Wick construction gives an upper bound for (3.1.1) and it depends on positivity. The polynomial $R(s)$ in (3.1.1) may be assumed positive, for we can use the "squaring" bound

$$|R(s)| \leqq \frac{1}{2} \{\delta^{-1} + \delta R(s)^2\} \tag{3.1.2}$$

where the constant δ depends on $R(s)$. We postpone the choice of δ to Chapter 4. (See Case S.) It is sometimes convenient to regard the constant term $(2\delta)^{-1}$ in (3.1.2) as a graph: we define this term as a *squaring graph*. This graph has vertices, but no legs. It has one squaring vertex for each P vertex in $R(s)$ introduced since the previous squaring. These squaring vertices merely serve a bookkeeping purpose. By convention, if G is a squaring graph, we let $I(G) = 1$ and multiply it by the constant $(2\delta)^{-1}$.

The second part of the low momentum expansion, the low-momentum contraction, is performed after completion of the Wick expansion. This expansion reduces the degree of the polynomial $R(s)$. The purpose of this expansion is to avoid squaring the low momentum vertices too many times. The convergence associated with the low momentum regions in early inductive steps is limited, and each use of the squaring formula (3.1.2) increases the divergent combinatoric factors. These factors, introduced in Chapter 4, are used to estimate the number of terms in our expansion.

3.2. The Wick construction

For any space time cube Δ', we define $l(\Delta')$ to be the largest integer for which

$$|\Delta'|\, M_{l(\Delta')}^2 \leqq 1 . \tag{3.2.1}$$

Note that for $|\Delta'| = 1$, $l(\Delta') = 0$ and $M_{l(\Delta')} = 0$, as at the start of the induction. In the Wick construction, we are given a cube Δ', and a cutoff ζ in the exponent V in (3.1.1). Furthermore in Δ', ζ has a lower momentum cutoff at $M_{l(\Delta)}$. Here Δ is some cube containing Δ', and so $M_{l(\Delta)} \leqq M_{l(\Delta')}$. The purpose of the Wick construction is to increase the lower momentum cutoff in Δ' from $M_{l(\Delta)}$ to $M_{l(\Delta')}$. Let ζ' be the new cutoff function with the desired lower momentum cutoff in Δ', and define $\delta\zeta$ and δV by

$$\zeta = \zeta' + \delta\zeta$$

$$V(\zeta) = V(\zeta') + \delta V .$$

Our basic inequality is

$$e^{-V(\zeta)} \leqq K_1 e^{-V(\zeta')} + (1-\chi)\, e^{-V(\zeta)}. \tag{3.2.2}$$

Here K_1 is a constant, independent of V, ε and M_1 and chosen below (3.2.22), and χ is the characteristic function of the subset of path space on which

$$e^{-\delta V} \leqq K_1 .$$

For cubes below the minimum size

$$|\Delta'|\, M_{u(\Delta')}^2 \leqq 1 , \tag{3.2.3}$$

$M_{l(\Delta')} = M_{u(\Delta')}$ and $\zeta' = O$ on Δ'. In this case

$$-O(1) = -O(|\Delta'|)\, M_{u(\Delta')}^2 \leqq \delta V$$

and so $e^{-\delta V} \leqq K_1$ on all of path space. Thus for cubes satisfying (3.2.3), $\chi \equiv 1$ and the exponent is completely removed in Δ' by (3.2.2).
In the remaining case, with (3.2.3) violated, we bound $1-\chi$ from above by a polynomial. To do this, we analyse δV, using the decomposition

$$\varphi(x, t) = \varphi_1(x, t) + \varphi_2(x, t),$$

where φ_1 has lower and upper momentum cutoffs at $M_{l(\Delta)}$ and $M_{l(\Delta')}$ respectively, while φ_2 has lower and upper momentum cutoffs at $M_{l(\Delta')}$ and $M_{u(\Delta)}$ respectively. The estimate

$$-O(1) = -O(|\Delta'|)\, M_{l(\Delta')}^2 \leqq \int_{\Delta'} :\varphi_1^4:\, dx\, dt , \tag{3.2.4}$$

which will govern the analysis of δV, provides the reason for choosing $M_{l(\Delta')} \sim |\Delta'|^{-1/2}$ as the crossover momentum between φ_1 and φ_2, and hence for the definition (3.2.1). On the other hand the requirement that φ_2 (i.e. the field in $V(\zeta')$) should be properly localized in Δ' is satisfied if $M_{l(\Delta')} \geqq |\Delta'|^{-1/3}$. The inequality

$$|\Delta'|^{-1/3} \leqq M_{l(\Delta')} \sim |\Delta'|^{-1/2}$$

is the formal reason that the Wick construction is useful. For the $\varphi_4{}^4$ interaction the corresponding inequality is marginal.

In the analysis of δV, we replace φ_1 by its average value $\bar{\varphi}_1$, plus an error. Explicitly,

$$\bar{\varphi}_1 = |\Delta'|^{-1} \int_{\Delta'} \varphi_1(x, t)\, dx\, dt \tag{3.2.5}$$

$$\delta\varphi_1(x, t) = \varphi_1(x, t) - \bar{\varphi}_1, \qquad (x, t) \in \Delta'. \tag{3.2.6}$$

(This method is similar to the treatment of some infrared problems in quantum field theory). We estimate the average field $\bar{\varphi}_1$ by classical methods, i.e. Hölder's inequality, and we find that the error term $\delta\varphi$ has an effective low momentum cutoff. In Chapter 6, we obtain a convergence factor $b^{-3(1+\nu)/2}$ associated with each $\delta\varphi_1$ leg in R.

Let

$$A = \Big(\int_{\Delta'} \varphi_1{}^4\, dx\, dt\Big)^{1/4} = ||\varphi_1||_4. \tag{3.2.7}$$

By Hölder's inequality,

$$|\bar{\varphi}_1| = |\Delta'|^{-1} \Big|\int_{\Delta'} \varphi_1\, dx\, dt\Big| \leqq |\Delta'|^{-1/4} A. \tag{3.2.8}$$

We expand δV in terms of φ_1 and φ_2. For simplicity in our formulas below, we omit the space-time cutoff h and the measure $dx\, dt$ for integrations over the space-time domain Δ'.

$$\delta V = \delta V_I(\Delta') + \delta V_C(\Delta') \tag{3.2.9}$$

$$\begin{aligned} \delta V_I &= \int :\varphi_1{}^4: + 4\int :\varphi_1{}^3\varphi_2: + 6\int :\varphi_1{}^2\varphi_2{}^2: + 4\int :\varphi_1\varphi_2{}^3: \\ \delta V_C &= O(\log M_{u(\Delta')}) \left(\int :\varphi_1{}^2: + 2\int :\varphi_1\varphi_2:\right) \\ &\quad + O(\log M_{l(\Delta')}) \left(\int :\varphi_1{}^2: + 2\int :\varphi_1\varphi_2: + \int :\varphi_2{}^2:\right) \\ &\quad + O(|\Delta'|\, M_{l(\Delta')} \log M_{u(\Delta')}). \end{aligned}$$

The (positive) constant $O(\log M_{u(\Delta')})$ arises from the term $\delta m^2 \int :\varphi^2: - \delta m^2 \int :\varphi_2{}^2:$, while the (positive) constant $O(\log M_{l(\Delta')})$ arises from raising the lower cutoff in δm^2. The change in the second order vacuum energy counterterm is bounded by $O(|\Delta'|\, M_l \log M_l)$ see Proposition 6.3.1. This constant also dominates the change in the third order vacuum energy, $O(|\Delta'| \log M_l)$, as we will see in Chapter 6.

We require the relation

$$\log M_{u(\Delta')} \leqq \varepsilon^{-1} O(\log |\Delta'|^{-1}), \tag{3.2.10}$$

to be justified in Section 3.3. Also,

$$-O(M_{l(\Delta')}) \leqq :\varphi_1{}^2:,$$

so by (3.2.1)

$$-O(|\Delta'|^{1/2}) \leqq \int :\varphi_1{}^2:. \tag{3.2.11}$$

Using (3.2.10)—(3.2.11), we have

$$\delta V_C \geqq \varepsilon^{-1} O(\log |\Delta'|^{-1}) \{\int :\varphi_1\varphi_2: + O(1) \int :\varphi_2^2: + O(|\Delta'|^{1/2})\} \tag{3.2.12}$$
$$\geqq |\Delta'|^{-1/2} \{\int \varphi_1\varphi_2 + O(1) \int :\varphi_2^2: + O(|\Delta'|^{1/2})\}$$

where we use

$$\int :\varphi_1\varphi_2: = \int \varphi_1\varphi_2 + O(|\Delta'|^{1/2}),$$

and we choose M_1 sufficiently large, depending on ε. Choosing M_1 large here forces $|\Delta'|$ to be small, see Section 3.3. Similarly, removing the Wick ordering in the φ_1 parts of δV_1 yields

$$\begin{aligned}\delta V_I \geqq \int \varphi_1^4 &+ 4 \int \varphi_1^3\varphi_2 + 6 \int \varphi_1 :\varphi_2^2: + 4 \int \varphi_1 :\varphi_2^3: \\ &+ O(|\Delta'|^{-1/2}) \int \varphi_1^2 + O(|\Delta'|^{-1/2}) \int \varphi_1\varphi_2 \\ &+ O(|\Delta'|^{-1/2}) \int :\varphi_2^2: + O(1).\end{aligned} \tag{3.2.13}$$

We substitute the bound

$$\frac{1}{2} \int \varphi_1^4 + O(|\Delta'|^{-1/2}) \int :\varphi_1^2 \geqq O(1) \tag{3.2.14}$$

to remove the pure φ_1^2 term. Thus

$$\begin{aligned}\delta V \geqq \frac{1}{2} \int \varphi_1^4 &+ 4 \int \varphi_1^3\varphi_2 + 6 \int \varphi_1^2 :\varphi_2^2: \\ &+ 4 \int \varphi_1 :\varphi_2^3: + O(|\Delta'|^{-1/2}) \int \varphi_1\varphi_2 \\ &+ O(|\Delta'|^{-1/2}) \int :\varphi_2^2: + O(1),\end{aligned} \tag{3.2.15}$$

so we bound δV from below by $(1/2)A^4 + O(1)$, plus cross terms between Wick ordered high momentum fields and unordered low momentum fields. In order to isolate the average low momentum field we insert $\varphi_1 = \bar{\varphi}_1 + \delta\varphi_1$ in the cross terms to obtain,

$$\delta V \geqq \frac{1}{2} \{A^4 + b_3(\bar{\varphi}_1)^3 + b_2(\bar{\varphi}_1)^2 + b_1\bar{\varphi}_1 + b_0 + O(1)\}. \tag{3.2.16}$$

Here

$$b_3 = 8 \int \varphi_2 \tag{3.2.17}$$
$$b_2 = 24 \int \delta\varphi_1\varphi_2 + 12 \int :\varphi_2^2:$$
$$b_1 = 24 \int (\delta\varphi_1)^2 \varphi_2 + 24 \int \delta\varphi_1 :\varphi_2^2: + 8 \int :\varphi_2^3: + O(|\Delta'|^{-1/2}) \int \varphi_2$$
$$\begin{aligned}b_0 = 8 \int (\delta\varphi_1)^3 \varphi_2 &+ 12 \int (\delta\varphi_1)^2 :\varphi_2^2: + 8 \int \delta\varphi_1 :\varphi_2^3: \\ &+ O(|\Delta'|^{-1/2}) \int \delta\varphi_1\varphi_2 + O(|\Delta'|^{-1/2}) \int :\varphi_2^2:.\end{aligned}$$

Each term in each b_i contains φ_2. Using (3.2.8), we bound (3.2.16) from below by

$$\delta V \geqq \frac{1}{2} \{A^4 - a_3A^3 - a_2A^2 - a_1A - a_0 - O(1)\}, \tag{3.2.18}$$

where

$$a_i = a_i(\Delta') = |\Delta'|^{-i/4} |b_i|, \tag{3.2.19}$$

for $i = 0, 1, 2, 3$. We note that terms in a_i of total degree l in $\delta\varphi_1$ and φ_2 have a coefficient c satisfying

$$|c| \leqq O(|\Delta'|^{-1+l/4}). \tag{3.2.20}$$

Let $\sum_{i=0}^{3} a_i = \sum_{j=1}^{12} \alpha_j$, where each α_j is a monomial, and let

$$P(\Delta') = \sum_{i=1}^{12} \alpha_j(\Delta')^l,$$

where $l = 2(\log_2 |\Delta'|^{-1})^3$.

Lemma 3.2.1. *Let* $\chi(\Delta')$ *be as in* (3.2.1). *Then*

$$1 - \chi(\Delta') \leqq P(\Delta'). \tag{3.2.21}$$

Proof. By (3.2.18), δV is bounded from below on the set of path space on which

$$\sup_{1 \leqq j \leqq 12} |\alpha_j(\Delta')| \leqq 1, \tag{3.2.22}$$

with a lower bound K_2 independent of the cutoffs or the particular cube Δ'. Thus we define $K_1 = \exp(-K_2)$. Furthermore, on the complement of the set (3.2.22), some $|\alpha_j| > 1$, so

$$\{1 - \chi(\Delta')\} \leqq \{1 - \chi(\Delta')\} P(\Delta') \leqq P(\Delta').$$

3.3. The localization cubes

We apply (3.2.2) to (3.1.1) as follows: First we choose a refinement $\mathscr{D}_1$ of $\mathscr{D}$, and we apply (3.2.2) in the cubes Δ' in a subset $\mathscr{D}_1{}^0$ of $\mathscr{D}_1$. We obtain thereby a sum of terms, and we now consider a fixed term in this sum. If the first term is selected in Δ_1, the lower cutoff in the exponent in Δ_1 is $M_{l(\Delta_1)}$. For cubes Δ_1 not of this type, that is for which the second term in (3.2.2) is selected, the lower cutoff in the exponential remains at $M_{l(\Delta)}$, where

$$\Delta_1 \subset \Delta \in \mathscr{D}.$$

Let $\mathscr{D}_1{}'$ be the set of these cubes.

We subdivide the cubes in $\mathscr{D}_1{}'$, obtaining a new refinement $\mathscr{D}_2$ of $\mathscr{D}_1$, and we apply (3.2.2) to the cubes $\mathscr{D}_2 \sim \mathscr{D}_1$. We continue in this fashion with refinements $\mathscr{D}_3$, $\mathscr{D}_4, \ldots$. This process terminates for a given term with a cover $\mathscr{D}_j$ if for each cube $\Delta_j \in \mathscr{D}_j$ the lower momentum cutoff in Δ_j occurs at $M_{l(\Delta_j)}$. For cubes Δ' below the minimum size (3.2.3), the exponent is completely removed in Δ'. Thus the Wick expansion terminates after a finite number of steps.

In each cover, we require adjacent cubes to have nearly uniform size, namely for adjacent cubes Δ_α, Δ_β of the same cover,

$$|\Delta_\alpha| \,/\, |\Delta_\beta| = \frac{1}{8}, 1, \text{ or } 8, \tag{3.3.1}$$

provided the exponent is not zero in Δ_α and Δ_β.

Lemma 3.3.1. *Let Δ be a cube in a cover $\mathscr{D}$ satisfying* (3.3.1). *Let Δ be subdivided into* 8 *cubes, and let other cubes in $\mathscr{D}$ be divided in eighths as necessary in order to get a new cover $\mathscr{D}'$ satisfying* (3.3.1). *Then the number of cubes in $\mathscr{D}' \sim \mathscr{D}$ satisfies*

$$|\mathscr{D}' \sim \mathscr{D}| \leqq O(\log |\Delta|^{-1}).$$

Proof. Let $C(\Delta) = \mathscr{D} \sim \mathscr{D}'$ be the set of cubes that must be divided to obtain $\mathscr{D}'$. Then $|\mathscr{D}' \sim \mathscr{D}| = 8\ |C(\Delta)|$. Let $|\Delta| = 8^{-l}$, and for $0 \leqq i \leqq l$, let $C_i(\Delta)$ be the set of 27 cubes of volume 8^{-i} forming a $3 \times 3 \times 3$ cubic array on our lattice, with Δ in the center cube. (The position of Δ in the center cube Δ_c is determined by obtaining Δ_c by refinement of our original space time cover, and by further refining Δ_c to obtain Δ.) We see that

$$C(\Delta) \subset \bigcup_i C_i(\Delta),$$

as follows: A cube Δ' of volume $8^{-(l-1)}$ need be divided only if it borders Δ, and hence belongs to $C_{l-1}(\Delta)$. If dividing Δ' necessitates dividing Δ'' of volume $8^{-(l-2)}$, then Δ'' must border $C_{l-1}(\Delta)$, and hence belongs to $C_{l-2}(\Delta)$, etc. Hence

$$\begin{aligned} |\mathscr{D}' \sim \mathscr{D}| &\leqq \sum_{i=0}^{l} 8\ |C_i(\Delta)| \\ &\leqq 8 \times 27(l+1) \\ &= O(l) \leqq O(-\log |\Delta|). \end{aligned}$$

We construct $\mathscr{D}_1$ in steps, dividing Δ or its subcubes into eighths, and adding extra cubes to satisfy (3.3.1) at each step. In choosing which cube $\Delta' \subset \Delta \in \mathscr{D}$ to subdivide into eighths, we pick a Δ' of maximum size in Δ. We call this a *primary* subdivision, while the extra cubes added to satisfy (3.3.1) arise from a *secondary* subdivision. We continue this subdivision until each subcube $\Delta' \subset \Delta \in \mathscr{D}$ satisfies $|\Delta'| \leqq n(\Delta)^{-1/2}\ |\Delta|$, for $r > 1$. In the first inductive step we use the extra condition

$$|\Delta_1| \leqq \min\{M_3^{-2}, M_{u(\Delta)}^{-\varepsilon/2}\}, \tag{3.3.2}$$

so that $l(\Delta_1) \geqq 3$. This removes the exponent in cubes $\Delta \in \mathscr{D}_a$.

We define $\mathscr{D}_1{}^0$, the set of cubes in $\mathscr{D}_1$ in which (3.2.2) is applied, as the set of cubes for which $\delta V \neq 0$ and $V(\zeta) \neq V(\zeta')$, cf. (3.2.2).

We now turn to the construction of $\mathscr{D}_2, \mathscr{D}_3, \ldots$ By induction, suppose we are given a sequence of covers $\mathscr{D} = \mathscr{D}_0, \mathscr{D}_1, \ldots, \mathscr{D}_i$ with $\mathscr{D}_{\alpha+1}$ refining $\mathscr{D}_\alpha$, and suppose that we are given a term resulting from the application of (3.2.2) to the appropriate cubes in $\bigcup_{\alpha=1}^{i} \mathscr{D}_\alpha$. To carry out the next step in the Wick expansion, we define $\mathscr{D}_i{}^0$ to be the set of cubes, $\Delta_i \in \mathscr{D}_i$, to which (3.2.2) has been applied, and $\mathscr{D}_i' \subset \mathscr{D}_i{}^0$ the subset of cubes on which the second term was selected. Thus $\mathscr{D}_i'$ is exactly the set of cubes Δ_i in which the lower momentum cutoff lies below $M_{l(\Delta_i)}$. If $\mathscr{D}_i' = \varnothing$, the Wick expansion for the term we are considering is complete. If $\mathscr{D}_i' \neq \varnothing$, we divide each cube $\Delta_i \in \mathscr{D}_i'$ into eighths, and we subdivide other cubes of $\mathscr{D}_i$ as necessary, in order to achieve (3.3.1). The resulting new cover is $\mathscr{D}_{i+1}$. Next we apply (3.2.2) to each cube $\Delta_{i+1} \in \mathscr{D}_{i+1}$ for which the lower momentum cutoff in the exponent lies below $M_{l(\Delta_{i+1})}$. This condition defines $\mathscr{D}_{i+1}^0$. With

$\Delta_{i+1} \subset \Delta_i \in \mathscr{D}_i$, we have

$$M_{l(\Delta_{i+1})} = M_{l(\Delta_i)}$$

before the application of (3.2.2) to cubes in $\mathscr{D}_{i+1}$, and so (3.2.2) is certainly applied in the case $\Delta_{i+1} \subset \Delta \in \mathscr{D}_i'$. This completes the induction and defines a single Wick expansion step.

Next we bound the factor $1 - \chi$ in (3.2.2), in a term for which the Wick expansion is complete. Let $\mathscr{D}_1, \ldots, \mathscr{D}_j$ be the finite sequence of covers associated with this term. Note that $\mathscr{D}_j$ is the cover $\mathscr{D}$ for the $P-C$ expansion of the following inductive step, and that $\mathscr{D}_j' = \varnothing$. We introduce the definitions

$$\mathscr{D}^0 = \bigcup_{i=1}^{j} \mathscr{D}_i{}^0$$

$$\mathscr{D}' = \bigcup_{i=1}^{j-1} \mathscr{D}_i{}' .$$

Then $\mathscr{D}^0$ is the set of cubes in which (3.2.2) has been applied and $\mathscr{D}'$ is subset consisting of cubes in which the second term of (3.2.2) was selected. Let $\mathscr{D}^*$ be the set of cubes in $\mathscr{D}'$ which do not contain smaller cubes in $\mathscr{D}'$. It follows that the cubes in $\mathscr{D}^*$ are disjoint. For $\Delta \in \mathscr{D}' \sim \mathscr{D}^*$, we substitute the upper bound

$$1 - \chi \leqq 1 .$$

For $\Delta \in \mathscr{D}^*$, we use (3.2.21).

For combinatoric purposes, we must count the number of times we apply the expansion (3.2.2). For this reason, we construct a mapping of cubes in $\mathscr{D}^0$ (of the rth inductive step) into P_r and W_r vertices. In the case $r = 1$, we also map onto cubes in $\mathscr{D}_e$. For similar reasons, we map cubes in $\mathscr{D}_j \sim \mathscr{D}$ into P_r and W_r vertices.

Theorem 3.3.2. *The above assignment of cubes in $\mathscr{D}^0 \cup (\mathscr{D}_j \sim \mathscr{D})$ is one to one when mapping into vertices, and maps at most $O(1)$ cubes onto each cube of $\mathscr{D}_e$ for $r = 1$.*

Proof. We first bound the number of cubes in $\mathscr{D}_1 \sim \mathscr{D}$, for $r > 1$. Recall that the definition of $\mathscr{D}_1$ allows no primary subdivision of a cube Δ' with volume less than $|\Delta|\, n(\Delta)^{-1/2}$, where $\Delta \in \mathscr{D}$. Hence if Δ' has a primary subdivision and $|\Delta'| = 8^{-j}$,

$$j \leqq \log |\Delta|^{-1} + \frac{1}{2} \log n(\Delta),$$

and by Lemma 3.3.1

$$|\mathscr{D}_1 \sim \mathscr{D}| \leqq \sum_{\Delta \in \mathscr{D}_e} O(1)\, n(\Delta)^{1/2} \left(\log |\Delta|^{-1} + \frac{1}{2} \log n(\Delta)\right)^2 .$$

If $\Delta \in \mathscr{D}_e$, then $\Delta \notin \mathscr{D}_a$ by (3.3.2). Condition (b) of Chapter 2 ensures $|\Delta|^{-\varepsilon} \leqq n(\Delta)$, and by (3.3.2), $|\Delta| \leqq M_1{}^{-2}$ for $\Delta \in \mathscr{D}_e$. Thus we obtain the bound

$$|\mathscr{D}_1 \sim \mathscr{D}| \leqq \sum_{\Delta \in \mathscr{D}_e} n(\Delta)^{2/3}$$

for $M_1 = M_1(\varepsilon)$ sufficiently large. For $r = 1$, $|\Delta| = 1$, and $\log |\Delta|^{-1} = 0$. Thus the analysis above yields

$$|\mathscr{D}_1 \sim \mathscr{D}| \leqq \sum_{\Delta \in \mathscr{D}_e} \left(\frac{1}{2} M^{\varepsilon}_{u(\Delta)} + O(1)\right),$$

where $O(1)$ depends on M_1 and arises from the sum over cubes in $\mathscr{D}_a$. Thus we can assign each cube in $\mathscr{D}_1 \sim \mathscr{D}$ to a distinct P_r vertex, with the exception for $r = 1$, noted above.

We now assign each cube Δ in $\mathscr{D}_i \sim \mathscr{D}_{i-1}$ to a cube in $\mathscr{D}^*$. We note that $\mathscr{D}^0$ is contained in $\bigcup_{i=1}^{j} (\mathscr{D}_i \sim \mathscr{D}_{i-1})$. We first assign Δ to Δ', where Δ' is the cube in $\mathscr{D}_i'$ whose primary subdivision led to the square Δ. By Lemma 3.3.1, at most $O(1) \log |\Delta'|^{-1}$ cubes Δ are assigned to Δ' in this fashion. We assign Δ' to the smallest cube Δ^* in $\mathscr{D}^*$ contained in Δ'. Each cube Δ^* in $\mathscr{D}^*$ is contained in at most $O(1) \log |\Delta^*|^{-1}$ larger cubes Δ', and any such cube Δ' satisfies $\log |\Delta'|^{-1} \leqq \log |\Delta^*|^{-1}$. Hence at most $O(1)(\log |\Delta^*|^{-1})^2$ cubes Δ are assigned to each $\Delta^* \in \mathscr{D}^*$, and every cube in $\mathscr{D}^0 \sim \mathscr{D}_1$ and in $\mathscr{D}_j \sim \mathscr{D}$ has been assigned to a cube in $\mathscr{D}^*$. For M_1 sufficiently large, $O(1)(\log |\Delta^*|^{-1})^2 \leqq (\log |\Delta^*|^{-1})^3$, and the Theorem follows.

Proposition 3.3.3. *There is an integer $n_1 = n_1(\varepsilon)$ with the following property: Let Δ_1 and Δ_2 be cubes associated with the term we are considering. If $|\Delta_1| < 1$ and if a vertex or exponent is localized in Δ_2, then*

$$|\Delta_1|^{n_1} \leqq |\Delta_2| \{2c + 2c \operatorname{dist}(\Delta_1, \Delta_2)\}^{n_1}$$

Proof. The cubes Δ_1 and Δ_2 may be introduced during different inductive steps. Let Δ_i' be the cube of volume 1 containing Δ_i. By the definition of the first refinement $\mathscr{D}_1$ of $\mathscr{D}$,

$$|\Delta_1| \leqq \frac{1}{2} M_{u(\Delta_1')}^{-\varepsilon},$$

and by condition (c) of Chapter 2,

$$M_{u(\Delta_1')}^{-\varepsilon} \leqq (M_{u(\Delta_2')}^{-\varepsilon})\, c\, d(\Delta_1', \Delta_2')^4 .$$

By (3.2.2), and the fact that $u(\Delta_2) \leqq u(\Delta_2')$,

$$M_{u(\Delta_2')} \geqq M_{u(\Delta_2)} \geqq |\Delta_2|^{-1/2} .$$

As

$$d(\Delta_1', \Delta_2') \leqq 1 + \text{dist.}(\Delta_1, \Delta_2),$$

$$|\Delta_1|^{2/\varepsilon} \leqq |\Delta_2| \{2c + 2c \text{ dist.}(\Delta_1, \Delta_2)\}^{8/\varepsilon},$$

and the proposition follows.

3.4. The low momentum contractions

Lemma 3.4.1. *Let ν be sufficiently small, depending on ε. Let $\mathscr{D}$ be a cover associated with a term at the end of the Wick construction of the rth inductive step. For $\Delta \in \mathscr{D}$, $l(\Delta) \geqq 3r$.*

Proof. In the first inductive step, $l(\Delta) \geqq 3$ by (3.2.1) and (3.3.2). We assume that the lemma is valid at the end of the rth inductive step. In the $(r+1)^{\text{st}}$ $P{-}C$ expansion, we have $n(\Delta) \geqq |\Delta|^{-\varepsilon} P_{r+1}$ vertices in Δ, by virtue of Condition (b) of Chapter 2 and (2.12). During the first refinement $\mathscr{D}_1$ of the $(r+1)^{\text{st}}$ inductive step, Δ is divided into at least $n_{r+1}(\Delta)^{1/2}$ subcubes. If Δ_1 is any such subcube,

$$|\Delta_1| \leqq n_{r+1}(\Delta)^{-1/2} |\Delta| \leqq |\Delta|^{1+\varepsilon/2} \leqq M_{l(\Delta)}^{-2(1+\varepsilon/2)} .$$

Choosing ν so small that $(1 + \nu)^3 \leqq 1 + \varepsilon/2$, we have

$$M_{l(\Delta)}^{1+\varepsilon/2} \geqq M_{3r}^{1+\varepsilon/2} \geqq M_{3r}^{(1+\nu)^3} \geqq M_{3(r+1)} .$$

Thus $|\Delta_1| \leqq M_{r+1}^{-2}$ and the lemma follows by (3.2.1).

To specify the low momentum contractions, suppose that we are in the rth inductive step, and consider an uncontracted field in $R(s)$, created in the ith inductive step. In this field we specify a new cutoff

$$M_{r+2i-1} \tag{3.4.1}$$

for Wick vertices and at

$$M_{r+2i-2} \tag{3.4.2}$$

for P or C vertices. We then contract the part of the field below this cutoff, by moving the creation operator to the left and the annihilation operator to the right. The legs are considered in reverse inductive oder, i.e., in order of decreasing i, and W vertices before P or C vertices. The cutoffs (3.4.1)—(3.4.2) grow faster than 2^{r-i+1}, the number of vertices produced by squaring, but less rapidly than the lower cutoff in the exponent, by Lemma 3.4.1. For this reason, no contractions to the exponent occur.

If $V \neq 0$, the rth inductive step is now complete. If $V = 0$, we contract all legs in the polynomial R in (3.1.1), and then the entire expansion is complete for the term in question.

4. Combinatoric factors

The inductive construction of Chapters 2 and 3 is now complete, and each inductive step yields an upper bound for $\langle\exp(-tH)\rangle$ as a sum of terms of the form $\sup_s \langle R(s)\, e^{-V(s)}\rangle$. Those terms for which the inductive construction has terminated satisfy $V(s) = 0$. For these terms, $R(s)$ (as a function on path space) is a constant, expressed as a sum of elementary integrals $I(G)$ labeled by Feynman graphs G. The integrals $I(G)$ are estimated in Chapters 5 and 6. In this chapter we are concerned with the combinatoric problem of estimating the number of terms in the sum over graphs G, especially the sum for which $I(G)$ satisfies some given upper bound. These estimates, combined with the estimates of Chapters 5 and 6, complete the proof that the $\varphi_3{}^4$ Hamiltonian is bounded from below, Theorem 1.1.1.

We pose the combinatoric estimates of this chapter in terms of combinatoric factors $c(G)$, as explained in [6]. For example,

$$|\sum_G I(G)| \leqq \big(\sum_G c(G)^{-1}\big) \sup_G c(G)\, |I(G)|, \leqq \sup_G c(G)\, |I(G)|,$$

where the combinatoric factors $c(G)$ are positive numbers satisfying $\sum c(G)^{-1} \leqq 1$. We assign combinatoric factors $c(G)$ to certain vertices, lines and legs as the final part of each inductive step. The combinatoric factors are assigned to lines and vertices in each elementary substep of the inductive construction, and the total combinatoric factor for a line or vertex is the product of these elementary factors. In addition to using the combinatoric factors to count terms, we absorb into our upper bounds on the factors $c(G)$, the constant $\delta^{\pm 1}$ in formula (3.2) and the constant K_1 in (3.4).

Theorem 4.1. *The following are upper bounds on combinatoric factors associated with the inductive expansion of Chapters* 2 *and* 3: *Given* ε, ν *there exist* n, M, L, β *such that if* $M_1 \geqq M$, *we have the bounds*

d^n	per line,
$L\lambda^{\beta\varepsilon}$	per P_1 vertex,
$L\|\Delta\|^{-\beta\varepsilon}$	per C, W or P_r vertex, $r > 1$,
1	per squaring vertex,
$b^{3(1+\nu)/2}$	per W leg.

Here β *is independent of* ε *and* ν. *In addition we obtain an overall factor exp* $(0(\mathrm{vol}\, h))$ *for the expansion.*

Proof. We assign the combinatoric factors separately for each of the various substeps that occur in the inductive construction. Here $\lambda \geqq M_{u(\Delta)-1}$ is the maximum lower cutoff for a P_1 vertex, Δ is the localization cube for a vertex, d is the scaled length for a line, and b is the localization ratio. See Section 1.2. for definitions. Since the bounds of the theorem are invariant in form when raised to a bounded power (except the factor per W leg) we need only establish the first four bounds for each of the substeps considered.

Case P. We first consider the substep of using the P expansion formula (2.1) to add a single P vertex localized in Δ. Formula (2.1) gives two terms, one with a decrease in the upper momentum cutoff in Δ, and one with a new P vertex in Δ. We assign a combinatoric factor $c_1 = O(1) \log \lambda$ to the new vertex, and $c_2 = (1 - c_1^{-1})^{-1}$ to the term with a decreased upper momentum cutoff. With this choice, $c_1^{-1} + c_2^{-1} = 1$. We now consider c_1 and c_2 separately. We bound c_1 as follows. For a P_n vertex, $n > 1$, let $\Delta' \supset \Delta$ be the cube in the first inductive step containing Δ. By Condition (b) of Chapter 2, and (3.3.2),

$$M^{\varepsilon}_{u(\Delta)} \leqq M^{\varepsilon}_{u(\Delta')} \leqq |\Delta|^{-2}. \tag{4.1}$$

Thus

$$\log M_{u(\Delta)} \leqq \frac{2}{\varepsilon} \log (|\Delta|^{-1}) + \frac{1}{\varepsilon} \log 2. \tag{4.2}$$

Given $\alpha > 0$, we have after the first Wick expansion,

$$(\log M_{u(\Delta)})^{\alpha} \leqq |\Delta|^{-\varepsilon}, \tag{4.3}$$

for M_1 sufficiently large (depending on ε and α). Our bound for c_1 becomes

$$c_1 \leqq O(1) \log \lambda, \qquad P_1 \text{ vertex}$$

$$c_1 \leqq O(1)\, |\Delta|^{-\varepsilon}, \qquad P_r \text{ vertex}, \quad r > 1.$$

We note that for a sequence λ_j of decreasing cutoffs equal to M_j,

$$\begin{aligned}\sum c_1^{-1} &= \sum_j O(1) (\log M_j)^{-1} \\ &= O(1) (\log M_1)^{-1} \sum_j (1 + \nu)^{-j} \leqq O(1),\end{aligned} \tag{4.5}$$

where the bound $O(1)$ is uniform for $2 \leqq M_1$, but diverges as $\nu \to 0$. (We assume $0 < \nu \leqq 1$.)

We now return to the factor c_2 that we associate with the term in (2.1) with a decreased upper cutoff. In a given inductive step, we may obtain a sequence of successively lower cutoffs from repeated application of (2.1) in a given cube Δ. The product of the c_2 factors for such a sequence is

$$\begin{aligned} \Pi c_2 &= \Pi(1 - c_1{}^{-1})^{-1} \leqq \Pi(1 + 2c_1{}^{-1}) \\ &\leqq \exp\left(2 \sum c_1{}^{-1}\right) \leqq O(1) \end{aligned} \tag{4.6}$$

by virtue of (4.5). Here $O(1)$ diverges as $\nu \to 0$. The product (4.6) of c_2's is a combinatoric factor, and we assign it to the preceding P vertex introduced in Δ during the present inductive step. If there are no such P vertices in Δ, we assign the combinatoric factor as follows: Assume that we are in the rth inductive step. If $r = 1$, we assign the factor $O(1)$ to the cube Δ. Since $|\Delta| = 1$, such factors contribute at most the overall factor $\exp\big(O(\text{vol}\, h)\big)$.

If $r > 1$, then Δ belongs to $\mathscr{D}_j$ for the $(r - 1)^{\text{st}}$ inductive step. Recall that $\mathscr{D}_j$ is the final cover for the Wick construction (3.4). By Theorem 3.3.2, we assign Δ to a P_{r-1} or W_{r-1} vertex.

Case C. Next we consider the case of a single contraction. This may be either the expansion of Chapter 2 that generates new C vertices and cancels the divergent terms, or else it may be the low momentum contraction of Section 3.4 which reduces the number of uncontracted legs.

In dealing with the P—C expansion, it is convenient to associate each C vertex with generating vertices. A P vertex may contract to the exponent to form a C vertex, or it may contract to the exponent to form C', which contracts to the exponent to form the C vertex. We call the vertices P and C' (if it exists) *generating* vertices for C. By the P—C expansion of Chapter 2, in a particular term, at most 16 C vertices have a given P vertex as a generating vertex, and at most 3 C vertices have a given C vertex as a generating vertex. We now divide the possible contraction steps into seven cases.

Case C1. The low momentum contraction expansion of Chapter 3 defines a cutoff in each inductive step, below which legs of existing vertices contract at the end of the inductive step. The first subcase is the restriction of a leg to have momentum above or below this cutoff. Each such decomposition introduces a combinatoric factor 2 per leg or at most 2^4 per vertex.

By Lemma 3.4.1 and (3.4.1)—(3.4.2), every vertex with upper cutoff at M_j is fully contracted at the end of the jth inductive step. Thus the maximum combinatoric factor for a vertex with upper cutoff M_j is 2^{4j}.

For $0 < \nu < 1$,

$$j = \frac{\log\log M_j - \log\log M_1}{\log(1 + \nu)} \leqq \frac{2}{\nu} \log\log M_j. \tag{4.7}$$

Thus

$$2^{4j} \leqq (\log M_j)^{8\nu^{-1}\log 2}. \tag{4.8}$$

If the vertex in question is a P_1 vertex,

$$2^{4j} \leqq M_{j-1}^{\varepsilon} = \lambda^{\varepsilon} \tag{4.9}$$

valid for M_1 sufficiently large (depending on ε and ν). We assign the combinatoric factor λ^{ε} to each P_1 vertex.

If the vertex in question is a C_1 vertex localized in Δ' and with upper cutoff $M_{j'}$, we obtain as above a bound $M^{\varepsilon}_{j'-1}$ on the combinatoric factor, which we assign to the generating P_1 vertex. Let the generating P_1 vertex be localized in Δ and have upper cutoff M_j. By Condition (c) of Chapter 2, applied when the cube Δ was selected for the P_1 vertex,

$$M^{\varepsilon}_{j'} \leqq O(1)\,\{1 + \text{dist.}\,(\Delta, \Delta')\}^4\, M^{\varepsilon}_j. \tag{4.10}$$

Hence

$$2^{4j'} \leqq O(1)\,\lambda^{\varepsilon}\,\{1 + \text{dist.}\,(\Delta, \Delta')\}^4, \tag{4.11}$$

where λ refers to the generating P_1 vertex. Thus (4.11) is dominated by a factor

$$O(1)\,\lambda^{\varepsilon} \tag{4.12}$$

assigned to the generating P_1 vertex and d^4 assigned to each line at the generating P_1 and C_1 vertices.

Finally, we consider the case that the vertex being restricted is a W, P_r or C_r vertex, $r > 1$. Let us assume the vertex has upper cutoff M_j, is localized in Δ, and $M_{u(\Delta)}$ is the upper cutoff in Δ when the vertex is introduced. As in (4.1)—(4.3), and using (4.7),

$$2^{4j} \leqq |\Delta|^{-\varepsilon}, \tag{4.13}$$

for M_1 sufficiently large (depending on ε and ν). In this case we assign the combinatoric factor $|\Delta|^{-\varepsilon}$ to the W, P_r or C_r vertex.

The following C subcases refer to the contraction of two legs to form a line. In every case, we assume that a leg of a vertex localized in Δ, and introduced in the rth inductive step, contracts to a leg of a vertex localized in $\Delta(\alpha)$ and introduced in the $r(\alpha)^{\text{th}}$ inductive step. We assume that the contraction occurs during the ith inductive step. We may refer to the leg in Δ as a "Δ leg", or a "$P_r(\Delta)$ leg", etc., to be more specific. In the following lemma, we consider the scaled distance d of (1.2.11) for the line produced by such a contraction.

Lemma 4.2 (a) *Let $n \geqq 4$. For contraction to a $P_{r(\alpha)}$ or $C_{r(\alpha)}$ leg*

$$\sup_{\substack{r(\alpha)\leqq r\\ \Delta,i}} \sum_{\Delta(\alpha)} d^{-n} \leqq O(1). \tag{4.14}$$

(b) *Let n be sufficiently large (depending on ε). For contraction to a $W_{r(\alpha)}$ leg, $(r > r(\alpha)$, (4.14) is valid.*

(c) *For contraction of a W_r leg to a W_r leg* i.e. $r(\alpha) = r$,

$$\sup_{\Delta,i,r} \sum_{\Delta(\alpha)} |\Delta(\alpha)|^{\varepsilon}\,(b_{\alpha}\,d)^{-3(1+\nu)} \leqq O(1). \tag{4.15}$$

Proof. (a) In the proof, let $\Delta(\alpha) = \Delta'$, $r(\alpha) = r'$, etc. Since $r' \leqq r$, we have by (3.3.1), with the exception of at most $O(1)$ cubes Δ' in each cover $\big(O(1) \leqq 57\big)$,

$$\text{dist.}\,(\Delta,\Delta') > \frac{1}{2}\,|\Delta'|^{1/3}. \tag{4.16}$$

We fix r'. According to (3.2.1.), the lower cutoff in a $P_{r'}$ or $C_{r'}$ leg is greater than $1/2|\Delta'|^{-1/2}$. By (1.2.11),

$$d \geqq \frac{1}{2}\,|\Delta'|^{-1/6}\,|\Delta'|^{-1/3}\,\text{dist.}\,(\Delta, \Delta'). \tag{4.17}$$

Let $\sum^{(j)}$ denote the sum over Δ' satisfying (4.16), and with size $|\Delta'| = 8^{-j}$. Then

$$\begin{aligned}\sum_{\Delta'} d^{-4} &\leqq O(1) + \sum_{j=0}^{\infty} \sum{}^{(j)} d^{-4} \\ &\leqq O(1) + \sum_{j=0}^{\infty} \sum{}^{(j)} (4^6 |\Delta'|)^{2/3} \left(|\Delta'|^{-1/3} \operatorname{dist.} (\Delta, \Delta')\right)^{-4} \\ &\leqq O(1) + O(1) \sum_{j=0}^{\infty} 8^{-2j/3} \leqq O(1). \end{aligned} \tag{4.18}$$

Here we use the fact that for cubes Δ' of a fixed size, and with $\sum'$ the sum over Δ' satisfying (4.16),

$$\sum{}' |\Delta'|^{4/3} \left(\operatorname{dist.} (\Delta, \Delta')\right)^{-4} \leqq O(1),$$

with $O(1)$ independent of Δ, $|\Delta'|$. Thus the constants in (4.18) are independent of Δ, i, r, $r(\alpha)$ and the proof of (a) is complete.

(b) For a $P_r(\Delta)$ or $C_r(\Delta)$ vertex, we have with $O(1)$ exceptions Δ',

$$\operatorname{dist.} (\Delta, \Delta') > \frac{1}{2} |\Delta|^{1/3}. \tag{4.19}$$

The lower cutoff in the $P_r(\Delta)$ or $C_r(\Delta)$ leg is greater than $1/2\ |\Delta|^{-1/2}$, and when (4.19) holds,

$$d > \frac{1}{2} |\Delta|^{-1/6} > \frac{1}{2} |\Delta'|^{-1/6n_1} (2cd)^{-1/6}$$

where the last inequality follows by Proposition 3.3.3. Here $n_1 = n_1(\varepsilon)$ is given by the lemma. Thus if $\sum'$ denotes the sum over cubes Δ' satisfying (4.19),

$$\begin{aligned}\sum_{\Delta'} d^{-n} &\leqq O(1) + O(1) \sum{}' |\Delta'|\, d^{-n+7n_1} \\ &\leqq O(1) \end{aligned} \tag{4.20}$$

for $n \geqq 4 + 7n_1(\varepsilon)$. Here we use the fact that the sum over Δ' ranges over $\mathscr{D}^*$, and $\mathscr{D}^*$ is a set of disjoint cubes (not necessarily a cover).

For a $W_r(\Delta)$ leg, $r > r(\alpha)$, we consider Δ'', the cube from the $P_r - C_r$ expansion that contains Δ. We remark that Δ'' belongs to the same cover as Δ', or to an earlier cover than Δ'. Hence with the exception of $O(1)$ cubes Δ', we have dist. (Δ, Δ') $> 1/2\ |\Delta''|^{1/3}$. The lower cutoff for a $W_r(\Delta)$ vertex is greater than $1/2\ |\Delta''|^{-1/2}$. Hence with these $O(1)$ exceptions, by Proposition 3.3.3,

$$d \geqq \frac{1}{4} |\Delta''|^{-1/6} \geqq \frac{1}{4} |\Delta'|^{-1/6n_1} (2cd)^{-1/6}$$

and the above proof of (4.20) is valid.

(c) Let $M_{j'}$, b' be the lower momentum cutoff and localization ratio for the $W_r(\Delta')$ leg. Then

$$\begin{aligned} b'd &\geqq b' \frac{1}{2} \left(1 + M_{j'} \operatorname{dist.} (\Delta, \Delta')\right) \\ &\geqq \frac{1}{2} \left(1 + |\Delta'|^{-1/3} \operatorname{dist.} (\Delta, \Delta')\right). \end{aligned}$$

Thus

$$\sum_{\Delta'} |\Delta'|^{\varepsilon} (b'd)^{-3(1+\nu)} \leqq O(1) \sum_{\Delta'} |\Delta'|^{\varepsilon} \big(1 + |\Delta'|^{-1/3} \text{ dist. } (\Delta, \Delta')\big)^{-3(1+\nu)}$$

$$\leqq O(1) \sum_{j=0}^{\infty} 8^{-\varepsilon j} \leqq O(1) .$$

Here we use the fact that for cubes Δ' of a fixed size $|\Delta'| = 8^{-j}$,

$$\sum_{\Delta'} \big(1 + |\Delta'|^{-1/3} \text{ dist. } (\Delta, \Delta')\big)^{-3(1+\nu)} \leqq O(1),$$

with $O(1)$ independent of Δ, $|\Delta'|$. The sum over j accounts for cubes of various sizes.
Case C 2. The second subcase is the contraction to the exponent producing a C_i vertex. This has two terms (one from V_I and one from the mass counterterm) for each cube $\Delta(\alpha) \in \mathscr{D}_e$. By Lemma 4.2 (a), a combinatoric factor $O(1)\, d^4$, for the contracted line, is sufficient to fix the cube $\Delta(\alpha)$. We assign d^4 to the line and $O(1)$ to the new C vertex.
Case C 3. The third subcase is the contraction to a $C_{r(\alpha)}$ vertex with upper cutoff M_j. By Lemma 4.2 (a), a combinatoric factor d^4 assigned to the contracted line and $O(1)$ to the $C_{r(\alpha)}$ vertex is sufficient to fix $\Delta(\alpha)$.
We now bound the number of existing C legs in $\Delta(\alpha)$. Let γ be the generating P vertex for α, and β the generating C vertex (if one exists). We may fix the localization cubes $\Delta(\beta)$ and $\Delta(\gamma)$ of β and γ with a combinatoric factor $O(1)\, d^4_{\alpha\beta} d^4_{\alpha\gamma}$ or $O(1)\, d^4_{\alpha\gamma}$. We next bound the number $n_{r(\alpha)}\big(\Delta(\gamma)\big)$ of P vertices introduced in $\Delta(\gamma)$ during the $r(\alpha)^{\text{th}}$ inductive step. In fact

$$n_1\big(\Delta(\gamma)\big) \leqq \lambda(\gamma)^{2\varepsilon} \tag{4.21}$$

$$n_{r(\alpha)}\big(\Delta(\gamma)\big) \leqq |\Delta(\gamma)|^{-\varepsilon},\ r(\alpha) > 1,$$

and each such P vertex generates at most $128 = O(1)$ C legs. The increase in C vertices due to squaring is bounded by a factor $2^{i-r(\alpha)+1}$, and a factor $\big(1 + r(\alpha)^2\big)$ covers the summation over $r(\alpha)$. As in Case C 1, $i \leqq j$. Thus

$$\big(1 + r(\alpha)^2\big)\, 2^{i-r(\alpha)+1} \leqq 2^{i+2} \leqq 2^{j+2}. \tag{4.22}$$

We assign the factor $O(1)\, n_{r(\alpha)}\big(\Delta(\gamma)\big)$ of (4.21) to the generating P vertex γ for α. We bound (4.22) as we bound the factor (4.11) in Case C 1, and as there, we assign it to the generating P vertex for α, and associated lines.
Case C 4. The fourth subcase is the contraction to a $P_{r(\alpha)}$ vertex. By Lemma 4.2. (a), a combinatoric factor d^4 for the line and $O(1)$ for the $P_{r(\alpha)}$ vertex fixes $\Delta(\alpha)$. As in Case C 3, we arrive at a combinatoric factor for the $P_{r(\alpha)}$ vertex of $\lambda(\alpha)^{3\varepsilon}$ if $r = 1$, or $|\Delta(\alpha)|^{-2\varepsilon}$ if $r > 1$.
Case C 5. The fifth subcase is contraction to a $W_{r(\alpha)}$ vertex, $r(\alpha) < r$. By Lemma 4.2 (b), a combinatoric factor $O(1)\, d^n$ assigned to the line is sufficient to fix $\Delta(\alpha)$. We bound the numbers of $W_{r(\alpha)}$ vertices in $\Delta(\alpha)$ by $O(1)\big(\log |\Delta(\alpha)|^{-1}\big)^3$ from applying (3.2.21), times a factor $2^{i-r(\alpha)+1}$ for the increase in vertices due to squaring. As

in Case C 1,

$$O(1)\left(\log|\Delta(\alpha)|^{-1}\right)^3 2^{i-r(\alpha)+1} \leqq |\Delta(\alpha)|^{-\varepsilon}, \tag{4.23}$$

for M_1 sufficiently large, depending on ν and ε, and we assign (4.23) to the W vertex α.

Case C 6. The sixth subcase is contraction to a $W_r(\Delta(\alpha))$ leg. By the low momentum contraction expansion, the Δ leg must be a $W_r(\Delta)$ leg, i.e. if $r = r(\alpha)$ only a W leg contracts to a W. We wish to reorder the sum over cubes so that legs contract from smaller to larger cubes. For each W_r leg, we introduce a factor 2^{i-r+1} which fixes the choice of whether that leg contracts to another W_r leg in the ith inductive step. For the legs that do contract, we have another factor 2 to determine whether they contract to a smaller cube or to a larger cube. The sum over all contractions can now be written as a sum over contractions from small cubes to large cubes, $|\Delta| \leqq |\Delta(\alpha)|$. It is an iterated sum, with a single sum for the possible contractions at each smaller cube. We bound the combinatoric factor 2^{r-i+2} by $|\Delta|^{-\varepsilon}$ per $W_r(\Delta)$ vertex, as in Case C 5 above.

By Lemma 4.2 (c), a combinatoric factor $O(1)\,|\Delta(\alpha)|^{-\varepsilon}\,(db_\alpha)^{3(1+\nu)}$ is sufficient to fix the sum over the cubes $\Delta(\alpha)$. We assign $O(1)\,|\Delta(\alpha)|^{-\varepsilon}$ to the $W_r(\Delta(\alpha))$ vertex, and $d^{3(1+\nu)}$ to the line. Since $|\Delta| \leqq |\Delta(\alpha)|$, we have $b_\alpha \leqq b$. We assign $b_\alpha{}^{3(1+\nu)/2}$ to the $W_r(\Delta(\alpha))$ leg and $b_\alpha{}^{3(1+\nu)/2} \leqq b^{3(1+\nu)/2}$ to the $W_r(\Delta)$ leg.

As in Case C 5, a combinatoric factor $|\Delta(\alpha)|^{-\varepsilon}$ bounds the number of $W_r(\Delta(\alpha))$ legs during the ith inductive step.

Case S. The squaring operation (3.1.2) introduces two terms for each application. We assign a combinatoric factor 2 to each term, which cancels the 1/2 in (3.1.2). If we are in the ith inductive step, we let

$$\delta = O(1)^N \prod_{R(s)} c, \tag{4.24}$$

where N is the number of P_i vertices in $R(s)$, and where the product ranges over the various combinatoric factors c assigned to the vertices, lines and legs of $R(s)$, prior, to the application of (3.1.2). With this choice, the coefficient of each squaring vertex as it occurs in the inductive expansion is $O(1)^{-1}$. We choose this factor $O(1)^{-1}$ sufficiently small, by choosing $O(1)$ in (4.24) sufficiently large, to dominate the combinatoric factors $O(1)$ that are assigned in Cases P and W 1 to squaring vertices. Since a bounded number of such assigments occurs, we may make the total coefficient of a squaring vertex, namely $O(1)^{-1}$ times these combinatoric factors, less than 1. With this choice of δ, the second term in (3.1.2), multiplied by its combinatoric coefficient, is

$$\delta\Big(\prod_{R(s)} c\Big)\, R(s)^2 = O(1)^N \Big(\prod_{R(s)} c\Big)^2 R(s)^2. \tag{4.25}$$

Thus for the purpose of our estimates, we regard the combinatoric factors for vertices, lines and legs of $R(s)$ unchanged in this term, except for an increase by a factor $O(1)$ for each P_i vertex.

Case W 1. This subcase is the application of the Wick bound in (3.2.2) in Δ. Each application of (3.2.2) produces two terms. In order to absorb the constant K_1 in (3.2.2), we bound the combinatoric factor for each such term by $2K_1 = O(1)$. By Theorem 3.2.2, we assign this factor $O(1)$ for Δ to a P vertex or W vertex localized in Δ' in $\mathscr{D} \cup \mathscr{D}^*$.

Case W 2. We finally consider the subcase of applying the bound (3.2.21) in cubes $\Delta^* \in \mathscr{D}^*$. To facilitate the estimates of Chapters 5—6, we express each W vertex introduced by (3.2.21) as a sum of vertices with momenta of each leg restricted to an interval $[M_j, M_{j+1}]$. To control the sum over j, we assign a combinatoric factor $O(1)\,(1 + j^2)^4$ to the W vertex. This factor is bounded by $|\Delta|^{-\varepsilon}$, by (4.13). A factor $O(1) = 12$ controls the sum over the twelve terms α_j^l in P in (3.2.21).

5. Estimating large graphs

5.1. Basic estimates and proof of Theorem 1.1.1.

Our bound for $\langle \exp(-tH)\rangle_0$ now has the form

$$\langle \Omega_0, e^{-tH}\,\Omega_0\rangle \leqq \sup_G c(G)\,|I(G)| \tag{5.1.1}$$

where $I(G)$ is the integral corresponding to a graph G, where G is a vacuum graph occurring in the inductive construction of Chapters 2—3, and where $c(G)$ is the combinatorial coefficient given in Theorem 4.1. In this chapter we establish estimates on $I(G)$, by reducing such bounds to estimates on certain small subgraphs with one to four vertices. We establish the required bounds on the small subgraphs in Chapter 6.

Our main result is an estimate which, combined with (5.1.1) and Theorem 4.1 proves the postivity of the $\varphi_3{}^4$ Hamiltonian, namely Theorem 1.1.1. Let λ, b, d be as in Section 1.2. For each line, we let $s = \max\{1, M\}$, where $M = \min\{\alpha^{(l)}\}$ is the lower momentum cutoff in the line, c.f. (1.2.4). Note that if $d > 1$, then $d < |r|\,s$ where r is the vector displacement between the centers of the localization cubes Δ, Δ'_j connected by the line. Let ε, $\nu = \nu(\varepsilon)$ and $M_1 = M_1(\nu, \varepsilon)$ be as in previous chapters.

Theorem 5.1., *Given $\varepsilon_1 > 0$, sufficiently small, given n_1 and given ε, ν positive but sufficiently small depending on ε_1, there exists constants $\alpha_1 = \alpha_1(\varepsilon_1, n_1, \varepsilon, \nu)$ and $M_1(\varepsilon_1, \varepsilon)$ such that*

$$|I(G)| \leqq \Big(\prod_{\text{lines}} d^{-n_1}\Big)\Big(\prod_{P_1\text{vertices}} \lambda^{-\varepsilon_1}\Big)\Big(\prod_{\text{vertices}} \alpha_1\,|\Delta|^{\varepsilon_1}\Big)\Big(\prod_{w\text{legs}} b^{-3(1+\nu)/2}\Big), \tag{5.1.2}$$

whenever M_1 of (1.2.6) *is larger than $M_1(\varepsilon_1, \varepsilon)$.*

Proof of Theorem 1.1.1. The inductive construction yields a bound

$$\langle \Omega_0, \exp\big(-tH(g,\varkappa)\big)\,\Omega_0\rangle \leqq \sup_G c(G)\,|I(G)|$$

and by Theorem 4.1 and Theorem 5.1,

$$\begin{aligned}&\langle \Omega_0, \exp\,(-tH(g,\varkappa))\,\Omega_0\rangle\\ &\leqq \exp\Big(O\big(\mathrm{vol}(h)\big)\Big) \sup_G \Big(\prod_{\text{lines}} d^{n-n_1}\Big)\Big(\prod_{P_1\text{vertices}} \lambda^{\beta\varepsilon-\varepsilon_1}\Big)\Big(\prod_{\text{vertices}} L\alpha_1|\Delta|^{-\beta\varepsilon+\varepsilon_1}\Big).\end{aligned}$$

(1) Choose ε_1 sufficiently small, then choose $\varepsilon = \varepsilon(\varepsilon_1)$, $\nu = \nu(\varepsilon)$ sufficiently small and $M_1(\varepsilon_1, \varepsilon)$ sufficiently large so that Theorem 5.1 holds.

(2) Choose ε sufficiently small so that $\beta\varepsilon < \varepsilon_1$ in Theorem 4.1. This fixes n, L, M of Theorem 4.1, and $M_1(\varepsilon_1, \varepsilon)$ of Theorem 5.1.

(3) Choose n_1 sufficiently large, so $n < n_1$. This fixes α_1 of Theorem 5.1.
(4) Pick M_1 sufficiently large so that $M_1 \geqq M_1(\varepsilon_1, \varepsilon)$ and $M_1 \geqq M$ as required above. Furthermore large M_1 forces λ to be large and (for $r > 1$) $|\Delta|$ to be small. Thus we can also require that

$$(L\,\alpha_1)^{17}\,\lambda^{-(\varepsilon_1-\beta\varepsilon)} \leqq 1\,, \qquad P_1 \text{ vertices}$$

$$L\,\alpha_1\,|\Delta|^{\varepsilon_1-\beta\varepsilon} \leqq 1\,, \qquad W, P_r, C_r, r > 1\,, \text{ vertices.}$$

For C_1 vertices, $|\Delta| = 1$ and we assign $L\alpha_1$ for each C_1 vertex to the generating P_1 vertex. In Chapter 4, we remarked that each P vertex is the generating P vertex for at most 16 C vertices. Thus our choice of M_1 above is sufficient.
Making these choices yields the bound $\langle\Omega_0, \exp\,(-tH(g,\varkappa))\,\Omega_0\rangle \leqq \exp\,\big(O(t)\,A(g)\big)$, where for a Hamiltonian cutoff vol $(h) = O(t)\,A(g)$, and hence by (1.1.4),

$$-O(1)\,A(g) \leqq E(g,\varkappa)\,,$$

as desired.

5.2. Localization factors

We devote the rest of Chapters 5 and 6 to the proof of Theorem 5.1. First we deal with the localization factor d^{n_1} per line, providing polynomial space time localization in the scaled distance. We note that our momentum cutoffs have compact support, so we cannot obtain exponential space-time localization. For convergent graphs with no momentum cutoff, our proof yields the localization factor $\exp\big((1-\varepsilon)\,|r|\big)$.
We now obtain a majorizing function $\bar{I}(G)$, and constants a_1', n_1' such that

$$|I(G)| \leqq \Big(\prod_{\text{lines}} \alpha_1'\,d^{-n_1}\Big)\,\bar{I}(G)\,, \tag{5.2.1}$$

and

$$\bar{I}(G) \leqq \Big(\prod_{\text{lines}} d^{n_1'}\Big)\Big(\prod_{P_1\text{vertices}} \lambda^{-\varepsilon_1}\Big)\Big(\prod_{\text{vertices}} \alpha_1'\,|\Delta|^{\varepsilon_1}\Big)\Big(\prod_{W\text{legs}} b^{-3(1-\nu)\,2}\Big). \tag{5.2.2}$$

The theorem follows from (5.2.1)—(5.2.2), since n_1' is independent of the choice of n_1.
Let us assume that the graph G contains only P or C vertices. For each vertex in G we define $d_{\max}$ as the maximum of the scaled distances d_j for lines contracting to the vertex. If $d_{\max} > 1$, then $d_{\max} \leqq |r_{\max}|\,s_{\max}$ where $r_{\max}$ and $s_{\max}$ are defined as r and s for the line yielding $d_{\max}$. We let $\mathscr{L}$ be the set of vertices of G with $d_{\max} > 1$. We now prove that,

$$|I(G)| \leqq \Big(\prod_{\mathscr{L}} \alpha_2 d_{\max}^{-4n}\Big)\,\bar{I}(G)\,, \tag{5.2.3}$$

from which (5.2.1) follows. (We here set $n_1 = n$ for simplicity.)
Suppose a vertex in $\mathscr{L}$ is localized in Δ. If $s_{\max} > 2\,|\Delta|^{-1/3}$, we refine Δ by cubes $\{\Delta'\}$, where

$$|\Delta'|^{-1/3} \leqq s_{\max} \leqq 2\,|\Delta'|^{-1/3}. \tag{5.2.4}$$

We write the vertex localized in Δ as a sum of vertices localized in each Δ'. The localization cube enters the kernel for the vertex through the linear dependence of the kernel on the momentum conservation factor $(h\chi_\Delta)^\sim$. Thus the sum over

$O(|\Delta|\,|\Delta'|^{-1})$ translates of $h\chi_{\Delta'}$ yields the required refinement. In this manner we obtain $\prod_{\mathscr{L}} O(|\Delta|\,|\Delta'|^{-1})$ topologically identical graphs G_α whose integrals $I(G_\alpha)$ have the same cutoffs, the same s's as $I(G)$, but where each G_α has a different set of r's and d's. Then

$$I(G) = \sum_\alpha I(G_\alpha). \tag{5.2.5}$$

We label the lines of G by j, and the corresponding line of G_α by (j, α). Likewise we let $r_{j,\alpha}$ be the vector displacement of the jth line in G_α, etc. Since each Δ' is contained in its Δ, for each line $d_j \leqq d_{j,\alpha}$, and for each vertex, $d_{\max} \leqq d_{\max,\alpha}$. Thus for each vertex, if $d_{\max} > 1$, then $d_{\max} < |r_{\max,\alpha}|\, s_{\max}$, for all α.
The kernel of the various $I(G_\alpha)$, for a given $I(G)$, has the form

$$m_\alpha(k) = w(k) \prod_{\text{lines}} \exp\,(ik_j r_{j\alpha}), \tag{5.2.6}$$

where $w(k)$ is independent of α and of the $r_{j\alpha}$. Then

$$I(G_\alpha) = \int m_\alpha(k)\, dk.$$

Multiplication of $I(G_\alpha)$ by a component $r_{j\alpha}^{(l)}$ of $r_{j\alpha}$ is equivalent to replacing w by $i\partial w/\partial k_{j\alpha}^{(l)}$. Thus

$$\left|\left(\prod_{\mathscr{L}} d_{\max}\, r_{\max,\alpha}\, s_{\max}\right)^{4n} I(G_\alpha)\right| \leqq \left|\int \left(\prod_{\text{lines}} e^{ik_j r_{j\alpha}}\right)\left(\prod_{\mathscr{L}} s_{\max}^2\, \Delta_{\max}^2\right)^{4n} w(k)\, dk\right| \tag{5.2.7}$$

where $\Delta_{\max}^2$ is the Laplacian $\sum \partial^2/\partial k_{\max}^{(l)2}$ for the momentum $k_{\max}$ of the line yielding $d_{\max}$.
We now estimate the derivatives of the factor μ^{-2}, η and $(h\chi_{\Delta'})_0^\sim$ in $w(k)$ that appear in (5.2.7). The propagator $\mu(k)^{-2} = (k^2+1)^{-1}$ satisfies for each n_1,

$$|D_j^{n_1}\mu(k_j)^{-2}| \leqq O(1)\,\mu(k_j)^{-2-|n_1|},$$

where

$$D_j^{n_1} = \prod_{l=0}^{2}\left(\frac{\partial}{\partial k_j^{\,l}}\right)^{n_{1,l}}, \sum_l n_{1,l} = |n_1|.$$

This estimate follows from the Cauchy integral representation of μ^{-2}, and the constant is bounded by $O(n_1!)$. On the support of the kernel w,

$$|D_j^{n_1}\mu(k_j)^{-2}| \leqq O(s_j^{-|n_1|})\,\mu(k_j)^{-2} \tag{5.2.8}$$

Likewise, on the support of w, the $\mathscr{S}$ valued momentum cutoff in the jth line $\eta_{\alpha_j,\beta_j}(k_j)$ satisfies

$$|D_j^{n_2}\,\eta_{\alpha_j,\beta_j}(k_j)| \leqq O(1)\, s_j^{-|n_2|} \tag{5.2.9}$$

by (1.2.4) and the fact that $M_1 \geqq 1$. Here $O(1)$ in (5.2.9) depends on $|n_2|$, but is uniform in (α_j, β_j).
Finally, we note that the conservation of momentum factor $(h\chi_{\Delta'})_0^\sim$ that enters w

is $(h\chi_{\Delta'})^\sim$ translated to the origin. With the comparison function

$$F_\Delta(k) = \prod_{l=0}^{2} (1 + |\Delta|^{1/3} |k^{(l)}|)^{-1}, \tag{5.2.10}$$

we have

$$|D^{n_3}(h\chi_{\Delta'})^\sim (k) \leqq O(1)|\Delta'|^{1+|n_3|/3} F_{\Delta'}(k) \leqq O(1)\, (s_{\max}{}^{-|n_3|})\, |\Delta|\, F_\Delta(k), \tag{5.2.11}$$

since $|\Delta'| \leqq |\Delta|$, in which case $|\Delta'|\, F_\Delta \leqq O(1)\, |\Delta|\, F_\Delta$.
Combining (5.2.7)–(5.2.11), we obtain an estimate for $I(G_\alpha)$, namely

$$\left|\left(\prod_{\mathscr{L}} d^{4n}_{\max}\right) I(G_\alpha)\right| \leqq \left[\prod_{\mathscr{L}} (r_{\max,\alpha}\, s_{\max})^{-4n}\right] \bar{I}(G), \tag{5.2.12}$$

$$\bar{I}(G) = \int \bar{w}(k)\, dk,$$

where $\bar{w}$ majorizes w and where $\bar{I}(G)$ will be shown to satisfy (5.2.2). Here $\bar{w}(k)$ is a product of the following factors: (i) a factor $\mu_i^{-2} = \mu(k_i)^{-2}$ for each factor μ_i^{-2} in $w(k)$; (ii) a factor $O(1)\, |\Delta|\, F_\Delta(q)$ bounding each factor $(h\chi_\Delta)^\sim(q)$ in w; (iii) the characteristic function of the support of w; (iv) any constants, such as uncancelled renormalization constants, not absorbed in the combinatoric factors of Chapter 4.
In order to bound $I(G)$, we use (5.2.5) and sum the bounds (5.2.12) over α. Each line with $d = d_{\max,\alpha} > 1$, contracting to a vertex in the set $\mathscr{L}$ and localized in Δ', satisfies

$$|r_{j,\alpha}| > \frac{1}{2} |\Delta'|^{1/3} = O(s^{-1}_{\max}), \tag{5.2.13}$$

where $s_{\max}$ is determined by Δ'. The sum over α is a sum over translates of individual vertices in $\mathscr{L}$. For each such vertex, by (5.2.13),

$$\sum_\alpha r^{-4n}_{\max,\alpha} \leqq O(s^{4n}_{\max}),$$

and for all vertices,

$$\sum_\alpha \prod_{\mathscr{L}} (r^{-4n}_{\max,\alpha}) \leqq \prod_{\mathscr{L}} O(s^{4n}_{\max}). \tag{5.2.14}$$

From (5.2.12) and (5.2.14) we infer (5.2.3), and thus (5.2.1), for a diagram with P and C vertices only.
For a diagram containing W vertices, we follow a similar procedure. The only additional complication comes from the more complicated majorizing function $\bar{w}$ for w, due to the more complicated kernels for W vertices. A single W vertex with $\mathscr{L}$ legs has a kernel $(\mu_1 \cdots \mu_l)^{-1}\, v(k)$, where $v(k)$ is given by (6.2.6–7) and (6.2.12). With $s_{\max}$ defined as above, we define Δ' as in (5.2.4), when $s_{\max} > 2\, |\Delta|^{-1/3}$. Note that if $s_{\max}$ at the W vertex is associated with a $\delta\varphi_1$ leg, $s_{\max} \leqq |\Delta|^{-1/3}$, and we define $\Delta' = \Delta$. We expand the "overall" conservation of momentum factor $(h\chi_\Delta)^\sim(\sum_{I_1}\lambda_j k_j + \sum_{I_2} k_j)$ as a sum over $(h\chi_{\Delta'})^\sim$ of the same momentum. We then proceed as above in the proof of (5.2.3) and (5.2.1), using the bound

$$|D^{n_3}(h\chi_\Delta)^\sim| \leqq O(1)\, s^{-|n_3|}_{\max}\, |\Delta| F_\Delta,$$

if $s_{\max} \leqq |\Delta|^{-1/3}$, in place of (5.2.11). Furthermore, in the kernel $\overline{w}$, we use the bound (6.2.16) for $|v(k)|$ from the kernel of a W vertex. Thus (5.2.1) follows in all cases. In applying our estimates, we include any counterterm used to cancel a divergence (in Chapter 2) as part of the graph being estimated. Thus the exactly cancelled vacuum energy terms

P≣C and 

do not appear in the expansion. The mass term

—P≡C—

may occur by itself in uncancelled form, or in the cancelled form

—P≡C— + —M—,

where the second graph denotes the mass counterterm in V_C, a M_2 type P vertex. The cancelled form of the mass term is convergent, and occurs only if the localization cubes of the P and C vertices coincide, see Proposition 5.3.4. If the cubes of the P and C vertices do not coincide, the uncancelled mass diagram is convergent.

5.3. Decomposing large graphs

The major part of our estimates for large graphs involves decomposing such a graph G into small subgraphs $\{g_j\}$, $j = 1, 2, \ldots$, and estimating $I(G)$ or $\bar{I}(G)$ in terms of the $I(g_j)$ or the $\bar{I}(g_j)$. Each subgraph g_j consists of a set of vertices, a set of "internal lines" (lines that connect these vertices) and a set of "external legs" (legs that either connect g_j to other subgraphs or which are external to G). Each line or leg has a monentum variable. We say $\{g_j\}$ is a decomposition of G into disjoint subgraphs, if $g_j \subset G$, and if each vertex of G lies in exactly one g_j. The integral $I(g_j)$ is a function of the external momenta. We divide these external momenta into two sets, initial (i) and final (f) momenta, according to whether they contract to g_{j+n} with $n < 0$ (initial), or with $n > 0$ (final). Legs external to G may be arbitrarily assigned. (The sets (i) and (f) should not be confused with a and a^* legs of Friedrichs diagrams.) We let $\|I(g)\|_{i,f}$ denote the operator norm of $I(g)$ from L_2 of the initial variables to L_2 of the final variables (Lebesque measure). Diagramatically we denote $I(g_j)$ by a vertex with initial legs i_j and final legs f_j leaving opposite sides,

$I(g_j) =$ 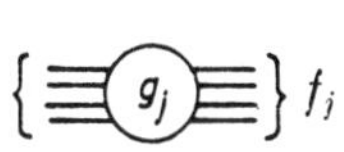

Our basic estimate for decomposing $I(G)$ is

$$\|I(G)\|_{i,f} \leqq \prod_j \|I(g_j)\|_{i_j,f_j}, \tag{5.3.1}$$

valid since $\|A \otimes I\| = \|A\|$ and $\|AB\| \leqq \|A\| \, \|B\|$. Furthermore, each $\|\cdot\|_{i,f}$ can be dominated by the Hilbert Schmidt norm $\|\cdot\|_{H.S.}$ and equals the Hilbert Schmidt norm if $i = \varnothing$ or $f = \varnothing$. Sometimes it is convenient to label the operator norms

$||\cdot||_{r,s}$ where r and s are the number of momenta in i and f respectively. Diagramatically, for example,

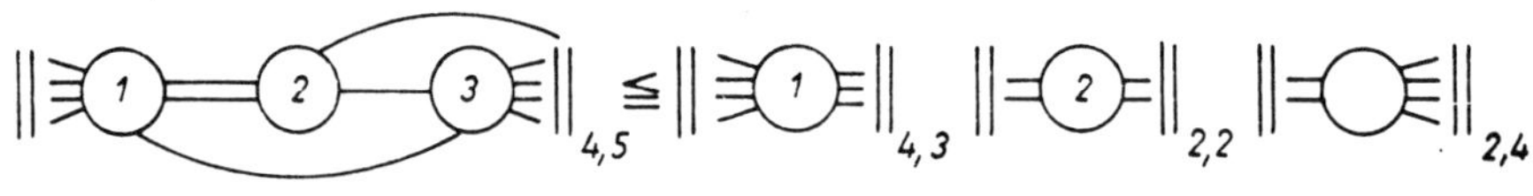

(5.3.2)

We now partition a vacuum diagram G into a union of disjoint subgraphs $\{g_j'\}$. We number these subgraphs in the order that they were formed in the inductive construction, i.e. from earlier to later subgraphs. Each subgraph g_j' consists of either
1) a single W vertex,
or
2) a P vertex, and the new C vertices it generates. If a g_j' consists of a single P vertex and it contracts four times to earlier subgraphs, we call this g_j' *exceptional.* We define a new partition $\{g_j\}$ by combining each exceptional subgraph g_j' with the latest subgraph g_j' to which it contracts.
We call g_j a P_r subgraph if its nonexceptional g_l' contains a P_r vertex (Case 2 above), and this P is called the *leading* P vertex. Otherwise, we call g_j a W subgraph. We further classify the P_l subgraphs as convergent P_C, or logarithmically divergent P_D, according to the number l_j of initial legs on the leading P vertex. Diagrams are P_C for $l_j \leqq 2$ and P_D for $l_j = 3$. According to the $P-C$ expansion of Chapter 2, the only two possible P_D subgraphs are

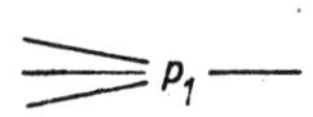 and 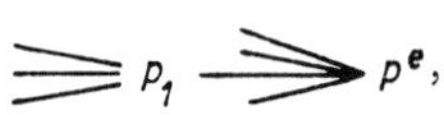 (5.3.3 a, b)

where P^e denotes an exceptional P vertex.
Lemma 5.3.1. Let ε_2 be sufficiently small, and let M_1 of (1.2.6) be sufficiently large, depending on ε, ε_2. There exist constants α_2 and n_2 such that for a P_C subgraph

$$\|\bar{I}(g_j)\|_{1_j,f_j} \leqq \lambda^{-\varepsilon_2} \Big(\prod_{\text{lines}} d^{n_2}\Big) \Big(\prod_{\text{vertices}} \alpha_2 \, |\Delta|^{\varepsilon_2}\Big), \tag{5.3.4}$$

for a P_D subgraph

$$\|\bar{I}(g_j)\|_{i_j,f_j} \leqq (\log \lambda)^4 \Big(\prod_{\text{lines}} d^{n_2}\Big) \Big(\prod_{\text{vertices}} \alpha_2 \, |\Delta|^{\varepsilon_2}\Big), \tag{5.3.5}$$

and for a W or P_r, $r > 1$, subgraph

$$\|\bar{I}(g_j)\|_{i_j,f_i} \leqq \Big(\prod_{\text{vertices}} \alpha_2 \, |\Delta|^{\varepsilon_2}\Big) \Big(\prod_{w\text{legs}} b^{-3(1+\nu)/2}\Big). \tag{5.3.6}$$

The factor λ refers to the leading vertex of the subgraph.

Proof of Theorem 5.1.1. We establish (5.2.2) to complete the proof of the theorem. We use (5.3.1) to bound $\bar{I}(G)$ by a product of factors of the form (5.3.4)—(5.3.6). The bound (5.3.6) on P_r, $r > 1$, or on W subgraphs already has the correct form required by (5.2.2), provided ε_1 is chosen sufficiently small. We note that P_C subgraphs contain at most 53 vertices: 1 leading P_1 vertex, 16 C vertices and 36 exceptional P vertices. Let λ_1 be the maximum lower cutoff for an exceptional $P_1(\Delta_1)$ vertex in a P_C subgraph. By conditon (c) of Chapter 2, applied at the time the leading P_1 vertex is introduced (and thus before the exceptional $P_1(\Delta_1)$. vertex is introduced),

$$\lambda^{-\delta} \leqq (cd^4)^{\delta/(1+\nu)\varepsilon} M_{(u\Delta_1)}^{-\delta/(1+\nu)} \leqq (cd^4)^{\delta/(1+\nu)\varepsilon} \lambda_1^{-\delta}. \tag{5.3.7}$$

Thus we may bound the convergence factor $\lambda^{-\varepsilon_2}$ on the leading vertex in (5.3.4) by a product of factors $\lambda_1^{-\varepsilon_2/37}$, with one factor for each P_1 vertex, times a factor d^{n_2} per line and $O(1)$ per vertex.

For P_D subgraphs, we transfer factors $\lambda^{-\delta}$ from earlier P_C subgraphs in order to dominate the $(\log \lambda)^4$ in (5.3.5) and to obtain convergence for its P_1 vertices. Basically, this method works because a P_D subgraph has at least 3 initial legs and at most 1 final leg, and consequently the number of P_D subgraphs is bounded in terms of the number of P_C subgraphs. We note that initial legs of P_D subgraph must contract to P_C or P_D subgraph.

We define a *chain* to be a finite, connected sequence of lines and vertices in G, chosen so that each line joins two consecutive vertices in the sequence, and so that the order of the vertices in the sequence coincides with the order in which these vertices were created in the inductive construction. We consider chains that start with the final leg of a P_C subgraph, that terminate with an initial leg of a P_D subgraph, and otherwise pass through P_D subgraphs (necessarily of type (5.3.3a). Let l be the number of lines in the chain, and let $\varrho = 3^{-l+1}$ be the *weight* of a chain of length l. We let $\sum^i$ denote the sum over chains in G with a fixed initial leg, and $\sum^f$ denote a sum over chains in G with a fixed final leg.

Lemma 5.3.2. *For any fixed initial leg, or any fixed final leg,*

$$\sum{}^i \varrho \leqq 2, \qquad \sum{}^f \varrho = 1. \tag{5.3.8i, f}$$

Proof. The first inequality is clear, since a chain with a fixed initial leg cannot branch. Thus there is exactly one such chain for each l, and

$$\sum{}^i \varrho \leqq \sum_{l=1}^{\infty} 3^{-l+1} = \frac{3}{2}.$$

Let T be the set of P_D vertices that occur in the set of chains with a fixed final leg. We prove (5.3.8f) by induction on the length l of the chain. If $l = 1$, there is exactly one chain of length one, and (5.3.8f) holds. Let $n(l)$ be the number of chains of length l with a fixed final leg. Assume (5.3.8f) holds for chains of length at most k and add a P_D vertex to chains of length k. Then

$$\sum_{l=1}^{k} \varrho(l)\, n(l) = 1,$$

and after adding a vertex to j chains of length k,

$$\begin{aligned}\sum_{l=1}^{k+1} \varrho(l)\, n(l) &= \sum_{l=1}^{k-1} \varrho(l)\, n(l) + \varrho(k)\big(n(k) - j\big) + 3j\varrho(k+1) \\ &= \sum_{l=1}^{k} \varrho(l)\, n(l) + j\big(3\varrho(k+1) - \varrho(k)\big).\end{aligned}$$

Since $3\varrho(k+1) = \varrho(k)$, the lemma follows.

We now return to the proof of Theorem 5.1. above. For each chain, we take a factor $\lambda^{-\varrho\delta/3}$ from the leading P vertex of the initial P_D subgraph in the product

(5.3.1). By the bound (5.3.8i), each initial P_C subgraph has at most a factor $\lambda^{2\delta/3}$ assigned in this manner among the various P_D subgraphs on chains initiating at the P_C subgraph. We choose δ sufficiently small so that $2\delta/3 < \varepsilon_2/74$; thus the total factor from the initial P_C subgraph that must be assigned is $\lambda^{-\varepsilon_2/74}$, leaving a similar factor for the leading vertex in the P_C subgraph. As in the case of P_C above, we use (5.3.7) to convert $\lambda^{-\varrho\delta/3}$ into a factor $d^{n_2\varrho}$ per line in the chain, $O(1)^\varrho$ per vertex in the chain, and $\lambda_D{}^{-\varrho\delta/3}$ for the final P_D subdiagram. Again, by (5.3.8i), this results in at most a total factor d^{n_2} per line and $O(1)$ per vertex. The identity (5.3.8f) assures that a total factor $\lambda_D{}^{-\delta}$ is assigned to (5.3.3a) and $\lambda_D{}^{-2\delta}$ to (5.3.3b).
In the latter case, $P_D = P_1 - P_1$ and we divide $\lambda_D{}^{-2\delta}$ equally between the two P_D vertices, as above. This convergence dominates the logarithmic divergence in (5.3.5), and completes the proof of (5.2.2).
Next we state the basic lemmas on small subdiagrams to be used in proving lemma 5.3.1. These estimates are proved in Chapter 6. We let $M(1), \ldots, M(4)$ denote the upper cutoffs in the legs of a P or C vertex, numbered so that $M(1) \leqq M(2) \leqq \cdots$.
Proposition 5.3.3. *Let w be the kernel of a $\varphi^4 P$ or C vertex in Δ.*

(a) 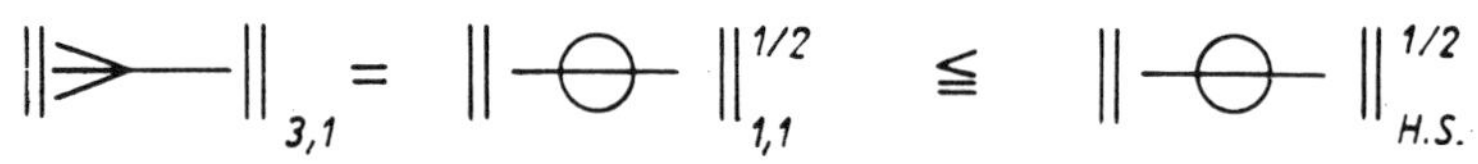

$$\leqq (|\Delta|^{\varepsilon_3}) \left(\log M(2)\right)^{1/2},$$

for $\varepsilon_3 \leqq 1/4$.

(b) 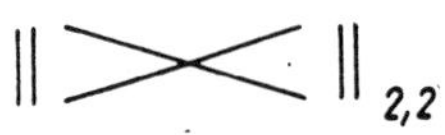 $\leqq O(|\Delta|^{\varepsilon_3})\, \lambda^{-\varepsilon_3}$,

for $\varepsilon_3 < 1/16$.

(c) 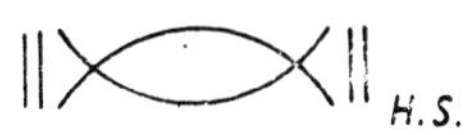 $\leqq O(|\Delta|^{\varepsilon_3})\, \lambda^{-\varepsilon_3}$,

for $\varepsilon_3 < 1/16$.

Proposition 5.3.4. *The uncancelled mass renormalization diagram*

$$M = M(\Delta, \Delta') = -\mathrm{P}\equiv\mathrm{C}-$$

and the mass counterterm

$$\delta M = \delta M(\Delta, \Delta')\, \delta_{\Delta,\Delta'}$$

($\delta_{\Delta,\Delta'} = 0$ for $\Delta \neq \Delta'$ and 1 otherwise) have Hilbert Schmidt norms that are $O(|\Delta|^{1/4}\, |\Delta'|^{1/4} \log M(2)^{1/2})$. Their sum with kernel w, satisfies

$$\|w\|_{H.S.} \leqq O(|\Delta'|^{\varepsilon_3}\, |\Delta|^{\varepsilon_3}\, \lambda^{-\varepsilon_3}\, d^{-n})$$

for some $\varepsilon_3 > 0$. Here d is the scaled distance from Δ to Δ'.
Proposition 5.3.5. *Let w be the kernel of a W vertex, as given by* (3.2.17), (3.2.19), *with each momentum leg restricted to an interval $[M_j, M_{j+1}]$. Let $\varepsilon_3 < 1/4$ and tet ν of* (1.2.6) *be sufficiently small. Then*

$$\|w\| \leqq O(|\Delta|^{\varepsilon_3}) \prod_{\text{legs}} b^{-3(1+\nu)/2}.$$

30*

Proposition 5.3.6. *The two vertex diagram*

$$P \equiv W$$

for contraction between a P vertex in Δ and a W vertex ix Δ′ is bounded by

$$O\Big(|\Delta|^{\varepsilon_3}\, |\Delta'|^{\varepsilon_3}\, \lambda^{-\varepsilon_3} \prod_{\text{legs}} b^{-3(1+\nu)/2}\Big)$$

for some $\varepsilon_3 > 0$, *provided* ε *and* ν *are sufficiently small.*

Proof of Lemma 5.3.1. We list the various possible P_C, P_D, P_r or W subgraphs, and estimate each of them. We first consider a W subgraph. This consists of a single W vertex, and possibly some exceptional P vertices contracted to it. In the case of a single exceptional P vertex contracted four times to the W vertex, we use Proposition 5.3.6. In the remaining cases, we bound the operator norm of the W subdiagram by a Hilbert Schmidt norm.

If a graph G is the union of two graphs G_1 and G_2, then a special case of (5.3.1) is

$$\|I(G)\|_{\text{H.S.}} \leqq \|I(G_1)\|_{\text{H.S.}}\, \|I(G_2)\| \tag{5.3.9}$$

where the norm on $I(G_2)$ is an operator norm, acting from variables internal in G but external in G_2 to variables external in both G and G_2.

As a special case of (5.3.9), we let G_1 be the W vertex in the W subgraph, and G_2 the complementary subgraph of the exceptional P vertices. By construction each P vertex in G_2 contracts to the W subgraph G_1 and has legs external to G. Thus the operator norm of $I(G_2)$ is bounded by (a) and (b) of Proposition 5.3.3. We bound $\|I(G_1)\|_{H.S.}$ by Proposition 5.3.5, yielding

$$\|\bar{I}(g_j)\|_{\text{H.S.}} \leqq \Big(\prod_{W\text{legs}} \log \lambda\Big) \Big(\prod_{\text{vertices}} \alpha_3\, |\Delta|^{\varepsilon_3}\Big) \Big(\prod_{W\text{legs}} b^{-3(1+\nu)/2}\Big). \tag{5.3.10}$$

We recall that after the first Wick construction (and hence for P^e vertices in a W subgraph) we have by (4.2),

$$\begin{aligned} \log \lambda \leqq \log M_{u(\Delta)} &\leqq O(\varepsilon^{-1}) \log (2|\Delta|^{-1}) \\ &\leqq |\Delta|^{\varepsilon_3/2}. \end{aligned} \tag{5.3.11}$$

Here $u(\Delta)$ refers to the point in the inductive construction when the P^e vertex is introduced, and the last inequality follows for M_1 sufficiently large, depending on ε, ε_3. We thus obtain (5.3.6) with $\varepsilon_2 < \varepsilon_3/2$, and our proof for W subgraphs is complete.

We next consider the two P_D subgraphs. The desired bound on (5.3.3a) follows by Proposition 5.3.3 (a). In the case of (5.3.3b), the operator norm of the P_D subgraph is the Hilbert Schmidt norm, and this norm squared is represented by the graph.

(5.3.12)

Splitting (5.3.12) across the horizontal axis into two subgraphs, we bound (5.3.12) by

$$\| -P_1 \equiv P_2 - \|_{\text{H.S.}} \| -P^e \equiv P^e - \|_{\text{H.S.}} \leqq \| -P_1 \equiv \|_{\text{H.S.}} \| \equiv P^e - \|_{\text{H.S.}} \tag{5.3.13}$$

using (5.3.1) two times. We now use Proposition 5.3.3 (a). If P^e is a P_r vertex with $r > 1$, we bound $\log \lambda_e$ by (5.3.11). If P^e is a P_1 vertex, we use Condition (c) of Chapter 2, as in the proof of (5.3.7), to establish $\log \lambda_e \leqq O(1)\, d \log \lambda$, where d is the distance for the line in the P_D subgraph connecting P_1 and P^e. In either case, we obtain (5.3.5).

We next consider P_C subgraphs; these have fewer than 53 vertices. In our estimates below, we obtain a convergence factor $\lambda^{-\varepsilon_3}$ from the leading P vertex (denoted by p) and $O(1)\,|\Delta|^{\varepsilon_3}$ from every vertex localized in Δ. In addition, certain nonleading vertices yield an ultraviolet divergent bound $O(\log \bar{\lambda}_j)$, where $\bar{\lambda}_j$ is the upper cutoff in the P or C vertex labelled by j. For a P_r or C_r vertex, $r > 1$, we have as in (5.3.11),

$$\log \bar{\lambda}_j \leqq \log M_{u(\Delta)} \leqq |\Delta|^{\varepsilon_3/2} \tag{5.3.14}$$

for M_1 sufficiently large. For a $P_1(\Delta)$ or $C_1(\Delta)$ vertex, by Condition (c) of Chapter 2,

$$\log \bar{\lambda}_j \leqq O(1) \log \big(1 + \text{dist.}\,(\Delta, \Delta_l)\big) \log \lambda,$$

where Δ_l is the localization cube for p. Thus by (2.11), and the definition (1.2.11),

$$\log \bar{\lambda}_j \leqq \big(\prod O(1)\, d\big) \log \lambda, \tag{5.3.15}$$

where the product ranges over the scaled distances d for a sequence of lines connecting p to the vertex labelled by j. Since there are at most 52 such vertices j in the P_C subgraph, our logarithmically divergent bounds can be absorbed into a factor d^{n_2} per line, $O(1)$ per vertex, and $\lambda^{\varepsilon_3/2}$ for the vertex p, leaving a convergence factor $\lambda^{-\varepsilon_3/2}$ for p.

Aside from the leading vertex p, our P_C subgraph consists of a set of C vertices to which p contracts (inner C's), a set of C vertices to which the inner C's contract (outer C's) and exceptional P vertices contracting to the inner and outer C vertices. We choose a subgraph $C(g_j)$, called a *core*, contained in g_j. If p contracts two or three times to a single C, the core consists of p and this C vertex; otherwise it consists of p and two inner C vertices. (Since g_j is a P_C subgraph, p has at most two external legs and at least two contractions.) Each C vertex is either a φ^4 vertex or a M vertex (mass renormalization vertex from V_C). We choose C vertices in the core to be φ^4 vertices, as far as possible. Among possible cores with an equal number of vertices and φ^4 vertices, we choose a core with the maximum number of internal lines. The purpose of these restrictions is to reduce the number of cases to be considered when bounding $I(g_j)$. We note that an M type p vertex does not occur alone, but is part of a cancelled mass diagram.

We dominate $\|I(g_j)\|_{\text{H.S.}}$ for a P_C subgraph. We first use (3.5.9) to remove "outer layers" of g_j; while leaving its core. We take $C(g_j) \subset G_1$, and we remove G_2, allowing logarithmically divergent estimates on the norm $\|I(G_2)\|$. We first remove, one by one, P^e or noncore C vertices with external legs, bounding these by Proposition 5.3.3 (a), (b), or Proposition 5.3.4. We then remove an P^e—C or C—C pair, contracting to each other, but not part of the core. If a single φ^4 vertex contracts four times to the pair, we include that vertex with the pair in G_2. We bound the

Hilbert Schmidt norm of $I(G_2)$ using Proposition 5.3.3 (a), (c) or Proposition 5.3.4, and possibly (5.3.1). This yields a convergent or a logarithmically divergent bound. We continue removing noncore singlet P^e or C vertices, and noncore P^e-C or $C-C$ pairs, as far as possible. By the construction of Chapter 2 and the definition of the core, no P^e vertex contracts exclusively to the core. Thus all P^e vertices are removed at this stage, and noncore C vertices contract completely to the core. We finally remove any noncore, M type C vertex with a Hilbert Schmidt norm, using Proposition 5.3.4.

For a two vertex core, there are only two possible cases

$$=\mathrm{p}=\mathrm{C}, \qquad -\mathrm{p}\equiv\mathrm{C}-, \tag{5.3.16}$$

because the third order vacuum diagram

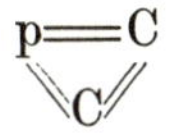

is cancelled exactly and does not occur. The Hilbert Schmidt norm in these cases is bounded by Proposition 5.3.3 (c) or Proposition 5.3.4.

For a three vertex core, $C(g_j)$ must be

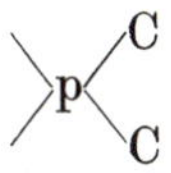

(5.3.17)

with possibly additional contractions between the two C vertices. For two φ^4 type C vertices, an additional φ^4 type C vertex can contract four times to the core, yielding

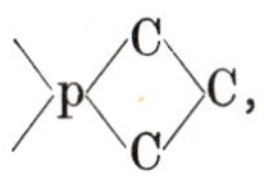

suppressing additional internal lines. We remove the p vertex with a convergent operator norm, using Proposition 5.3.3 (b), with the exception of the case

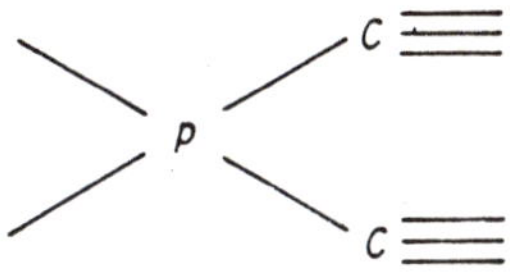

(5.3.18)

considered below. The remaining graph consists of 2 or 3 C vertices only and is connected. By removing one singlet C with an external leg from the 3 C graph, we are reduced to a $C-C$ pair and we obtain a logarithmically divergent bound by Proposition 5.3.3 (a), (b).

The Hilbert Schmidt norm of (5.3.18) squared is

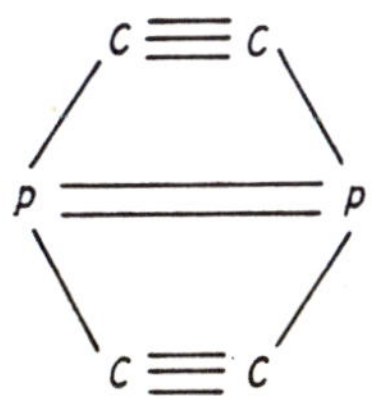

$$\leqslant \| = p = p = \|_{H.S.} \; \| - c \equiv c - \|^2_{H.S.}$$

The desired bound follows by Proposition 5.3.3 (a), (c) and (5.3.1). This concludes our analysis of P_C subgraphs and proves (5.3.4).

Remaining are P_r subgraphs, $r > 1$. We treat these as with P_C or P_D subgraphs, but never use (5.3.15) to bound logarithmic divergences. Since P_r subgraphs contain $P_{r'}$ or $C_{r'}$ vertices, $r' \geqq r$, we use (5.3.14) in every case. This yields the bound (5.3.6). and concludes the proof of Lemma 5.3.1.

6. Estimates of small graphs

6.1. Convolution estimates

We prove the basic estimates on small graphs in this chapter. We start with elementary estimates on convolutions, useful in more general circumstances. In the second section, we bound the Hilbert Schmidt norm of a single W vertex, while in the final section we deal with our remaining estimates.

Proposition 6.1.1. Let $p \in R^n$, $\delta + \lambda > n$, $\nu = \min\{\delta, \lambda, \delta + \lambda - n\}$, $\mu(p) = (p^2 + 1)^{1/2}$. Then

$$\mu^{-\delta} * \mu^{-\lambda} \leqq O(1) \begin{cases} \mu^{-\nu}, \text{ if } \max\{\delta, \lambda\} \neq n \\ \mu^{-\nu} \log \mu, \text{ if } \max\{\delta, \lambda\} = n. \end{cases}$$

Proposition 6.1.2. *Let* $p \in R^1$, $0 \leqq \delta \neq 1$, $0 \leqq \lambda \neq 1$, $\delta + \lambda > 1$, $\nu = \min\{\delta, \lambda, \delta + \lambda - 1\}$, $0 < a \leqq 1$ *and* $\mu_a(p) = \mu(ap)$. *Then*

$$\mu_a{}^{-\delta} * \mu^{-\lambda} \leqq \begin{cases} O(a^{-\delta})\, \mu^{-\nu}, \text{ if } \delta < 1 \\ O(a^{-1})\, \mu^{-\nu}, \text{ if } \lambda < 1 < \delta \\ O(a^{-\nu})\, \mu^{-\nu}, \text{ if } 1 < \lambda,\ 1 < \delta. \end{cases}$$

By scaling, we obtain from Proposition 6.1.2 the

Corollary 6.1.3: *Let* $p \in R^1$,

$$0 < a \leqq b,\ 0 \leqq \delta \neq 1, \qquad 0 \leqq \lambda \neq 1, \delta + \lambda > 1, \nu = \min\{\delta, \lambda, \delta + \lambda - 1\}.$$

(a) *If* $\delta < 1$, *then*

$$a^\delta b^{1-\delta} \mu_a{}^{-\delta} * \mu_b{}^{-\lambda} \leqq O(1)\, \mu_b{}^{-\nu}.$$

(b) *If* $\lambda < 1 < \delta$, *then*

$$a \mu_a{}^{-\delta} * \mu_b{}^{-\lambda} \leqq O(1)\, \mu_b{}^{-\lambda}.$$

(c) *If* $1 < \lambda$, $1 < \delta$, $\nu = \min\{\delta, \lambda\}$, *then*

$$\left(\frac{a}{b}\right)^\nu b \mu_a{}^{-\delta} * \mu_b{}^{-\lambda} \leqq O(1)\, \mu_b{}^{-\nu}.$$

Proofs. (6.1.1) It is sufficient to consider $0 < \delta \leqq \lambda$, for if $\delta < 0$ the proposition follows from the fact that $\nu = \delta$ and

$$\mu(p - q)^{-\delta} \leqq \mu(p)^{-\delta} \mu(q)^{-\delta}.$$

We consider three regions of integration,

$$\text{I:} \quad |q| \leqq \frac{1}{2}|p|; \qquad \text{II:} \quad \frac{1}{2}|p| \leqq |q| \leqq 2|p|;$$

$$\text{III:} \quad 2|p| \leqq |q|.$$

In region I, $\mu(p-q) \geqq O(1)\,\mu(p)$, so

$$\int_I \mu(p-q)^{-\delta}\,\mu(q)^{-\lambda}\,dq \leqq O\big(\mu(p)^{-\delta}\big) \int_I \mu(q)^{-\lambda}\,dq$$

$$\leqq \begin{cases} O(\mu^{-\delta}), & \lambda > n \\ O(\mu^{-\delta}\log\mu), & \lambda = n \\ O(\mu^{-\delta+n-\lambda}), & \lambda < n, \end{cases}$$

the desired bounds for region I.

In region II, $|q-p| \leqq |q| + |p| \leqq 3|p|$ and $|p| \leqq 2|q|$, so

$$\int_{II} \mu(p-q)^{-\delta}\,\mu(q)^{-\lambda}\,dq \leqq O\big(\mu(p)^{-\lambda}\big) \int_{II} \mu(p-q)^{-\delta}\,dq$$

$$\leqq \begin{cases} O\big(\mu(p)^{-\lambda}\big) & \delta > n \\ O(\mu^{-\lambda}\log\mu) & \delta = n \\ O(\mu^{-\lambda+n-\delta}) & \delta < n. \end{cases}$$

If $\delta = n$ and $\lambda > n$, $\mu^{-\nu}\log\mu \leqq O(\mu^{-\nu})$. If $\delta = n = \lambda = \nu$, the bound $O(\mu^{-\lambda}\log\mu)$ also has the desired form.

In region III, $|p-q| \geqq 1/2\,|q| \geqq |p|$. Thus

$$\int_{III} \mu(p-q)^{-\delta}\,\mu(q)^{-\lambda}\,dq \leqq \int_{III} \mu(q)^{-\lambda-\delta}\,dq \leqq O(1)\,\mu(p)^{-\lambda-\delta+n},$$

to complete the proof of (6.1.1).

We note that

$$\mu_a(p)^{-1} \leqq O(a^{-1})\,\mu(p)^{-1} \tag{6.1.2}$$

and so the case $\delta < 1$ follows from Prop. 6.1.1. For $1 < \lambda, \delta$ we have

$$\mu_a^{-\delta} * \mu^{-\lambda} \leqq \mu_a^{-\nu} * \mu^{-\lambda} \leqq O(a^{-\nu})\,\mu_a^{-\nu} * \mu^{-\lambda},$$

since $\mu^{-\delta} \leqq \mu^{-\lambda}$. Thus this case also follows from Prop. 6.1.1. We consider the remaining case, $\lambda < 1 < \delta$. In the region of integration I above,

$$\int_I \mu_a(p-q)^{-\delta}\,\mu(q)^{-\lambda}\,dq \leqq O(1)\,\mu_a(p)^{-\delta} \int_I \mu(q)^{-\lambda}\,dq$$

$$\leqq O\big(\mu_a(p)^{-\delta}\,\mu(p)^{1-\lambda}\big)$$

$$\leqq O(a^{-\delta})\,\mu(p)^{-(\delta+\lambda-1)}$$

which is the desired bound in region I.

In region II, the integral is bounded by

$$O\big(\mu(p)^{-\lambda}\big) \int_{II} \mu_a(p-q)^{-\delta}\,dq \leqq O(a^{-1})\,\mu(p)^{\lambda}$$

if $\delta > 1$, which is the desired bound.

In region III, the integral is bounded by

$$\int_{III} \mu_a(q)^{-\delta} \mu(q)^{-\lambda} dq \leqq O(a^{-1}) \mu(p)^{-\lambda} \int \mu(q)^{-\delta} dq \leqq O(a^{-1}) \mu(p)^{-\lambda},$$

to complete the proof of (6.1.2)

Proposition 6.1.4. *Let* $0 < \delta$, $0 < \lambda$, $\delta + \lambda = n$. *Then for* $p \in R^n$,

$$\int_{|q| \leqq \varkappa} \mu(p - q)^{-\delta} \mu(q)^{-\lambda} dq \leqq O(\log \varkappa).$$

Proof. In regions I, II of the proof (a) above, the above estimates yield a bound $O(1)$, while in region III the bound is $O(\log \varkappa)$.

We introduce comparison functions

$$F(k) = \prod_{l=0}^{2} \mu(k^{(l)})^{-1},$$

$$F_\Delta(k) = \prod_{l=0}^{2} \mu(|\Delta|^{1/3} k^{(l)})^{-1},$$

where the product extends over the three components of the energy and momentum in three space time dimensions. We note that

$$|\chi_\Delta{}^\sim| \leqq O(|\Delta|) F_\Delta .$$

In addition, F_Δ is practically L_1, and Proposition 6.1.1 and Corollary 6.1.3 yield

Proposition 6.1.5 (a) *For* $\delta + \lambda > 1$, $\nu = \min\{\delta, \lambda, \delta + \lambda - 1\}$, $\bar{\nu} = \max\{\delta, \lambda\}$, $\varepsilon > 0$,

$$|\Delta| F_\Delta{}^\delta * F_\Delta{}^\lambda \leqq \begin{cases} O(1) F_\Delta{}^\nu, \text{ if } \bar{\nu} \neq 1 \\ O(1) F_\Delta{}^{\nu - \varepsilon}, \text{ if } \bar{\nu} = 1 \end{cases}$$

(b) *For* $0 \leqq \delta \neq 1$, $0 \leqq \lambda \neq 1$, $\delta + \lambda > 1$, $\nu = \min\{\delta, \lambda, \delta + \lambda - 1\}$, $|\Delta| \leqq |\Delta'|$,

$$|\Delta|^\delta |\Delta'|^{1-\delta} F_\Delta{}^\delta * F_{\Delta'}{}^\lambda \leqq O(1) F_{\Delta'}{}^\nu, \text{ if } \delta < 1,$$

$$|\Delta| F_\Delta{}^\delta * F_{\Delta'}{}^\lambda \leqq O(1) F_{\Delta'}{}^\lambda, \text{ if } \lambda < 1 < \delta,$$

$$|\Delta|^\nu |\Delta'|^{1-\nu} F_\Delta{}^\delta * F_{\Delta'}{}^\lambda \leqq O(1) F_{\Delta'}{}^\nu, \text{ if } 1 < \lambda, 1 < \delta.$$

The last inequality also yields

$$|\Delta|^\nu |\Delta'| F_\Delta{}^\delta * F_{\Delta'}{}^\lambda \leqq O(1) F^\nu,$$

if $1 < \lambda$, $1 < \delta$. We next give bounds for the integral of a product of three F's. Other such inequalities follow using Proposition 6.1.5 and Hölder inequalitites.

Corollary 6.1.6. *Let* $0 < \varepsilon < 1/2$, $0 \leqq \alpha_i < 1/4$,

$$\alpha = \alpha_1 + \alpha_2, \qquad |\Delta| \leqq |\Delta'| \leqq |\Delta''| \leqq 1.$$

Then

$$\int F_\Delta(k + p) F_{\Delta''}(p)^\alpha F_{\Delta'}(k' - p) dp$$

is bounded by $O(1)$ times any of

$$|\Delta|^{-3/4-\varepsilon/2}\,|\Delta'|^{-1/4+\varepsilon/2}\,F_{\Delta'}(k+k')^{(1+\varepsilon)/2}\,F_{\Delta''}(k)^{\alpha_1}\,F_{\Delta''}(k')^{\alpha_2},$$

$$|\Delta|^{-3/4-\varepsilon/2}\,|\Delta'|^{-3/4}\,F(k+k')^{(1+\varepsilon)/2}\,F_{\Delta''}(k)^{\alpha_1}\,F_{\Delta''}(k')^{\alpha_2},$$

or

$$|\Delta|^{-3/4-\varepsilon/2}\,|\Delta'|^{-3/4}\,|\Delta''|^{-\alpha}\,F(k+k')^{(1+\varepsilon)/2}\,F(k)^{\alpha_1}\,F(k')^{\alpha_2}.$$

Proof. By the Schwarz inequality and the second and third inequalities of Proposition 6.1.5 (b)

$$\int F_\Delta(k+p)\,F_{\Delta''}(p)^\alpha\,F_{\Delta'}(k'-p)\,dp$$

$$\leqq \left(\int F_\Delta(k+p)^{1+\varepsilon}\,F_{\Delta'}(k'-p)^{1+\varepsilon}\,dp\right)^{1/2}\left(\int F_\Delta(k+p)^{1-\varepsilon}\,F_{\Delta''}(p)^{2\alpha}F_{\Delta'}(k'-p)^{1-\varepsilon}\,dp\right)^{1/2}$$

$$\leqq O(|\Delta|^{-(1+\varepsilon)/2}\,|\Delta'|^{\varepsilon/2})\,F_{\Delta'}(k+k')^{(1+\varepsilon)/2}\times$$

$$\times\left(\int F_\Delta(k+p)^{2(1-\varepsilon)}\,F_{\Delta''}(p)^{4\alpha_1}\,dp^1\right)^{1/4}\left(\int F_{\Delta'}(k'-p)^{2(1-\varepsilon)}\,F_{\Delta''}(p)^{4\alpha_2}\,dp\right)^{1/4}$$

$$\leqq O(\Delta|^{-3/4-\varepsilon/2}\,|\Delta'|^{-1/4+\varepsilon/2})\,F_{\Delta'}(k+k')^{(1+\varepsilon)/2}\,F_{\Delta''}(k)^{\alpha_1}\,F_{\Delta''}(k')^{\alpha_2}.$$

The second and third inequalities follow from the first, and from the bound $|\Delta|\,F_\Delta \leqq F$, valid for $|\Delta| \leqq 1$.
In case $|\Delta'| = |\Delta''|$, we obtain
Corollary 6.1.7. *As in Corollary* 6.1.6, *if we assume $|\Delta'| = |\Delta''|$ but allow $-1/4 + \varepsilon/2 < \alpha_2 < 1/4$, the bounds of Corollary* 6.1.6 *follow.*

6.2. Bounds on a W vertex

The kernel of an individual vertex has the form

$$w(k_1, \ldots, k_l) = (\mu_1 \ldots \mu_l)^{-1}\,v(k_1, \ldots, k_l) \tag{6.2.1}$$

where v denotes the various cutoff and localization factors. For contraction vertices arising from φ^4, $l = 4$ and

$$v = (h\chi_\Delta)^\sim\left(\sum_i k_i\right)\eta(k_1, \ldots, k_l), \tag{6.2.2}$$

where χ_Δ is the characteristic function of the space time localization region Δ, and η is the momentum cutoff, a product of factors (1.2.4). In the case of a mass renormalization C vertex, $l = 2$ and v includes the mass shift,

$$v = -\frac{1}{2}\,\delta m^2\,(h\chi_\Delta)^\sim\,(k_1+k_2)\,\eta(k_1, k_2). \tag{6.2.3}$$

Perturbation vertices from V_I are characterized by $l = 4$ and $v = (h\chi_\Delta)^\sim\,\delta\eta$, where $\delta\eta$ denotes the change in ultraviolet cutoffs. Mass renormalization P vertices may also occur, and for them the increment $\delta(\delta m^2\eta)$ replaces the factor $\delta m^2\eta$ in the C vertex kernel.

The Wick vertices have several possible kernels, namely those from (3.2.17) and (3.2.19); these may have $l = 1, 2, 3$ or 4. For a particular l, let I_1 denote the set of $\delta\varphi_1$ legs and I_2 the set of φ_2 legs, $|I_1| + |I_2| = l$. Expanding each $\delta\varphi_1$ leg as $\delta\varphi_1 = \varphi_1 - \bar{\varphi}_1$ contributes two terms to the kernel for v. Thus we obtain $2^{|I_1|}$ terms in all, with each term labeled by a subset J_1 of I_1 denoting the legs for which $\bar{\varphi}_1$ is chosen. The momenta of the $\bar{\varphi}_1$ legs do not occur in the momentum conservation factor $(h\chi_\Delta)^\sim$ for the vertex. Rather, the average of the low momentum field φ_1 over the space time cube Δ yields a factor $\chi_\Delta{}^\sim(k)/\chi_\Delta{}^\sim(O)$ which occurs in the kernel for each φ_1 leg. Thus the function $v(k)$ for W vertices equals

$$v(k) = c \sum_{J_1 \subset I_1} \left\{ \prod_{i \in J_1} \frac{\chi_\Delta{}^\sim(k_i)}{\chi_\Delta{}^\sim(0)} \right\} (h\chi_\Delta)^\sim \left(\sum_{i \notin J_1} k_i \right) \eta(k), \tag{6.2.4}$$

where $c = c(l, |I_1|)$ are the constants that arise in (3.2.19), and by (3.2.20)

$$|c| \leqq O(|\Delta|^{-1+l/4}). \tag{6.2.5}$$

We define for $\lambda_i \in [0, 1]$,

$$v_\lambda(k) = c(h\chi_\Delta)^\sim \left(\sum_{i \in I_1} \lambda_i k_i + \sum_{i \in I_2} k_i \right) \eta(k) \prod_{i \in I_1} \frac{\chi_\Delta{}^\sim\big((1 - \lambda_i)\, k_i\big)}{\chi_\Delta{}^\sim(0)}. \tag{6.2.6}$$

Then (6.4) can be written as a sum of $2^{|I_1|}$ evaluations at $\lambda_i = 0$ or $\lambda_i = 1$,

$$v(k) = v_\lambda(k) \,\big|_0^1\big|_0^1 \cdots \big|_0^1. \tag{6.2.7}$$

Proof of Proposition 5.3.5. We estimate the Hilbert Schmidt norm of the kernel $w = w(l, |I_1|)$ for a W vertex, and obtain for $\varepsilon > 0$,

$$\|w\|_{\text{H.S.}}^2 \leqq 0(|\Delta|^{1/2-\varepsilon}) \prod_j b_j^{-3(1+\nu)}. \tag{6.2.8}$$

We note that localization ratio b_i equals 1 for φ_2 legs.
Each momentum variable is restricted to an interval $|k^{(l)}| \in [M_i, M_{i+1}]$, where $M_i = M_{i+1}^{1/1+\nu}$, or possibly $|k^{(l)}| \in [0, M_1]$. If $b > 1$, then $b = M_i^{-1} |\Delta|^{-1/3}$ and

$$M_{i+1} \leqq (b^{-(1+\nu)} |\Delta|^{-1/3}) \, |\Delta|^{-\nu/3}. \tag{6.2.9}$$

We use the estimate (6.2.9) a bounded number of times in establishing (6.2.8). Thus it is convenient to suppress the factor $|\Delta|^{-\nu/3}$, which we absorb into $|\Delta|^{-\varepsilon}$ of (6.2.8), for each use of (6.2.9). We must then choose ν sufficiently small. We let $M(j)$ denote the upper endpoint of the momentum interval for the j*th* momentum variable (leg) k_j.
We divide the I_1 legs into two disjoint sets

$$I_1 = I_1{}' \cup I_1{}''. \tag{6.2.10}$$

Here $I_1{}'$ consists of those legs for which $b_j > 1$ (low momentum legs satisfying $M(j)^{1/1+\nu} < |\Delta|^{-1/3}$), while $I_1{}''$ consists of those $\delta\varphi_1$ legs for which $b_j = 1$ (high momentum legs satisfying $|\Delta|^{-1/3} \leqq \mu_j \leqq |\Delta|^{-1/2}$). Here and later we suppress factors $|\Delta|^{-\nu/3}$, so we replace (6.2.9) by

$$M(j) \leqq b_j^{-(1+\nu)} \, |\Delta|^{-1/3}, \; j \in I_1{}'; \; b_j = 1, \; j \in I_1{}''. \tag{6.2.11}$$

We exhibit the low momentum cancellation implicit in (6.2.7) by writing $v(k)$ as an integral over derivatives of $v_\lambda(k)$.

$$v(k) = \int_0^1 d\lambda_1 \cdots \int_0^1 d\lambda_\alpha \frac{\partial}{\partial\lambda} \cdots \frac{\partial}{\partial\lambda_\alpha} v_\lambda(k) \Big|_0^1 \cdots \Big|_0^1, \tag{6.2.12}$$

where the integral extends over $\{\lambda_j : j \in I_1'\}$, and the evaluation extends over $\{\lambda_j : j \in I_1''\}$. We note that

$$|D^n(h\chi_\Delta)^\sim| \leqq O(|\Delta|^{1+n/3})\, F_\Delta(k), \tag{6.2.13}$$

and

$$|D^n(h\chi_\Delta)^\sim(k)/\chi_\Delta^\sim(0)| \leqq O(|\Delta|^{n/3})\, F_\Delta(k). \tag{6.2.14}$$

We define

$$F_\lambda(k) = \Big[\prod_{i \in I_1'' \sim J_1''} F_\Delta(k_i)\Big]\, F_\Delta(\xi + P) \tag{6.2.15}$$

where $J_1'' \subset I_1''$ is the subset of indices from (6.2.12) for which λ_i, $i \in I_1''$ is evaluated at $\lambda_i = 1$, and where

$$\xi = \sum_{i \in I_1'} \lambda_i k_i, \qquad P = \sum_{i \in J_1'' \cup I_2} k_i.$$

On the support of w, $|\xi| \leqq O(|\Delta|^{-1/3})$. The momentum P is approximately conserved. By (6.2.5)—(6.2.6) and (6.2.12)—(6.2.14) we obtain,

$$|v(k)| \leqq \sum_{J_1''} O\left(|\Delta|^{(l/4)+(|I_1'|/3)}\right) \Big(\prod_{i \in I_1'} \mu_i\Big) \int_0^1 d\lambda_1 \cdots \int_0^1 d\lambda_\alpha\, F_\lambda(k). \tag{6.2.16}$$

We let $w(l, |I|; I_1'', J_1'')$ denote the kernel for a particular choice of J_1'', Suppressing l, $|I_1|$, and I_1'',

$$w = \sum_{J_1''} w(J_1''),\ \|w\|_{\mathrm{H.S.}} \leqq \sum_{I_1'', J_1''} \|w(J_1'')\|_{\mathrm{H.S.}}$$

Suppressing the cutoffs $\eta_{\alpha,\beta}(k)$, we have

$$\|w(J_1'')\|^2_{\mathrm{H.S.}} \leqq O(|\Delta|^{(l/2)+(2|I_1'|/3)}) \sup_\lambda \int dk\, \frac{F_\lambda(k)^2}{D},$$

where $dk = dk_1 \cdots dk_l$ and $D = \prod_{i \in I_1'' \cup I_2} \mu_i^2$. Integrating the I_1' momenta yields a factor per I_1' leg of

$$\int_{|k_i| \leqq M(i)} d^3k_i \leqq O\big(M(i)^3\big) \leqq O(b_i^{-3(1+\nu)}\, |\Delta|^{-1}),$$

where the last inequality follows by (6.2.11). Integrating the momenta in $I_1'' \sim J_1''$ yields a factor per leg

$$\int_{|\Delta|^{-1/3} \leqq |k_i|} F_\Delta(k_i)^2 \mu_i^{-2}\, dk_i \leqq O(|\Delta|^{2/3}) \int F_\Delta(k_i)^2\, dk_i \leqq O(|\Delta|^{-1/3}\, b_i^{-3/(1+\nu)})$$

using Proposition 6.1.5 (b), the lower cutoff $|\Delta|^{-1/3}$ for I_1'' legs, and the fact that $b_i = 1$ for $i \in I_1''$. We show below that for $\epsilon > 0$, integrating the remaining momenta in D gives,

$$\int_{J_1'' \cup I_2} dk \sup_\xi \left\{ \frac{F_\Delta(\xi + P)^2}{\prod_{i \in J_1'' \cup I_2} \mu_i^2} \right\} \leqq O(|\Delta|^{(1/2)\{1-|I_2|-|J_1''|\}-\varepsilon}). \tag{6.2.17}$$

so that using $l = |I_1'| + |I_1''| + |I_2|$, we obtain

$$\|w(J_1'')\|^2_{\mathrm{H.S.}} \leqq O(|\Delta|^{(1/2)+(1/6)(|I_1''|-|J_1''|)-\varepsilon} \left(\prod_{i \in I_1} b_i^{-3(1+\nu)}\right) \leqq O(|\Delta|^{(1/2)-\varepsilon}) \left(\prod_i b_i^{-3(1+\nu)}\right). \tag{6.2.18}$$

To establish (6.2.17), we note that $|I_2| = 1, 2$ or 3 for all W vertices, see (3.2.17). We check (6.2.17) for the various values of $|J_1''| + |I_2|$. If $|J_1''| + |I_2| = 1$, we use the lower cutoff $|\Delta|^{-1/2}$ on the I_2 leg to give

$$\int_{|P| > |\Delta|^{-1/2}} dP \sup_\xi \frac{F_\Delta(\xi + P)^2}{\mu(P)^2} \leqq O(|\Delta|) \int dP \sup F_\Delta(\xi + P)^2 \leqq O(1),$$

since $|\xi| \leqq O(|\Delta|^{-1/3}) \leqq O(|P|^{2/3})$ on the region of integration. If $|J_1''| + |I_2| = 2$, Proposition 6.1.5 (b) similarly yields $O(|\Delta|^{-1/2})$. If $|J_1''| + |I_2| = 3$, Proposition 6.1.4 yields

$$\int dP \sup_\xi F_\Delta(\xi + P)^2 \int \frac{dk_2\, dk_3}{\mu(P - k_2 - k_3)^2\, \mu(k_2)^2\, \mu(k_3)^2} \leqq O(|\Delta|^{-1} \log M_u(\Delta)).$$

As in Chapter 5 estimates, we use $\log M_{u(\Delta)} \leqq O(\log |\Delta|^{-1})$, since W vertices occur in Chapter 3 in small cubes. This yields the desired estimate for $|J_1''| + |I_2| = 3$. If $|J_1''| + |I_2| = 4$, we proceed similarly with variables P, k_1, k_2, k_3. Here $M(1)$, the upper cutoff in k_1 is the smallest upper cutoff, and since $|I_2| = 4$, this momentum is a J_1'' leg with an upper cutoff at $|\Delta|^{-1/2}$. Integrating k_2 and k_3 yields $O(\log M_{u(\Delta)})$ as above. The k_1 integral is

$$\int_{|k_1| < |\Delta|^{-1/2}} \mu(k_1)^{-2}\, dk_1 \leqq O(|\Delta|^{-1/2})$$

so the case $|J_1''| + |I_2| = 4$ is dominated by $O(|\Delta|^{-3/2} \log |\Delta|^{-1})$. This completes the proof of (6.2.19) and Proposition 5.3.5.

6.3. Estimates on other small subgraphs

The following proofs of Proposition 5.3.3 and Proposition 5.3.4 establish stronger bounds than claimed in Chapter 5.

Proof of Proposition 5.3.3. (a) We set

$A =$  $\mu^{1/4-\varepsilon}$, and bound

$$\|A\|_{3,1} = \|A^*A\|_{1,1}^{1/2} \leqq \|A^*A\|_{\text{H.S.}}^{1/2},$$

where

$$A^*A = $$  , $$\|A^*A\|_{\text{H.S.}}^2 = $$ 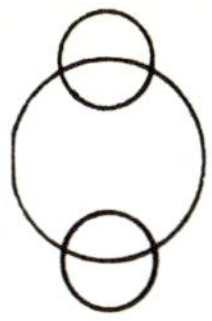 .

The kernel of A^*A is bounded by

$$v(k_1, k_5) = |\Delta|^2 \int dk_2\, dk_3\, dk_4 \, \frac{F_\Delta(k_1 + k_2 + k_3 + k_4)\, F_\Delta(-k_2 - k_3 - k_4 + k_5)}{(\mu_1\mu_5)^{(3/4)+\varepsilon}\, \mu_2{}^2\, \mu_3{}^2\, \mu_4{}^2}.$$

We let $P = k_2 + k_3 + k_4$, k_2 and k_3 be the variables of integration, where the upper momentum cutoff $M(2)$ for k_2 is the smallest upper cutoff among k_2, k_3, k_4. Thus by Propositions 6.1.1 and 6.1.4,

$$\int dk_2\, dk_3 \, \frac{1}{\mu_2{}^2\, \mu_3{}^2\, \mu(P - k_2 - k_5)^2} \leqq O\big(\log M(2)\big) \tag{6.3.1}$$

and by Proposition 6.1.5, for $\delta > 0$,

$$v(k_1, k_5) \leqq O\big(\log M(2)\big)\, |\Delta|\, F_\Delta(k_1 + k_2)^{1-\delta}\, (\mu_1\mu_5)^{-(3/4+\varepsilon)}.$$

Taking L_2 norms gives

$$\|v\|_2{}^2 \leqq O\big(\log^2 M(2)\big)\, |\Delta|^2 \int \frac{F_\Delta(k_1 + k_5)^{2-2\delta}}{(\mu_1\mu_5)^{(3+\varepsilon)/2}}\, dk_1\, dk_5 \leqq O\big(\log^2 M(2)\, |\Delta|\big),$$

proving

$$\|w\mu^{1/4-\varepsilon}\|_{3,1} \leqq O\big(|\Delta|^{1/4} \log^{1/2} M(2)\big).$$

(b—c) We first estimate the kernel $v(k_1, k_2, k_3, k_4)$ of the diagram

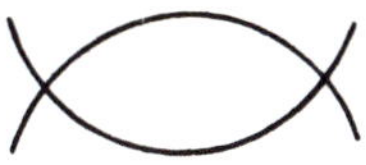

with a factor μ^ν multiplying each external line, and a factor $\mu^{2\lambda}$ multiplying each internal line. We let $K_1 = k_1 + k_2$, $K_2 = k_3 + k_4$, $P = k_5 + k_6$, and assume $\nu < 1$.

$$|v(k_1, k_2, k_3, k_4)| \leqq O(|\Delta|^2)\, (\mu_1\mu_2\mu_3\mu_4)^{-1+\nu} \int dk_5\, dk_6 \, \frac{F_\Delta(K_1 + P)\, F_\Delta(-P + K_2)}{(\mu_5\mu_6)^{2-2\lambda}}$$

$$\leqq O(|\Delta|^2)\, (\mu_1\mu_2\mu_3\mu_4)^{-1+\nu} \int dP \, \frac{F_\Delta(K_1 + P)\, F_\Delta(-P + K_2)}{\mu(P)^{1-4\lambda}}$$

and by Corollary 6.6, for $\lambda < 1/4$,

$$\leqq O(|\Delta|)\, (\mu_1\mu_2\mu_3\mu_4)^{-1+\nu}\, F_\Delta(K_1 + K_2)^{(1+\delta)/2}\, F(K_1)^{(1-4\lambda)/6}\, F(K_2)^{(1-4\lambda)/6}.$$

The L_2 norm of v is dominated by

$$\|v\|_2{}^2 \leqq O(|\Delta|^2) \int dK_1\, dK_2\, dk_1\, dk_4 \frac{F_\Delta(K_1 + K_2)^{1+\delta} F(K_1)^{(1-4\lambda)/3} F(K_2)^{(1-4\lambda)/3}}{[\mu_1 \mu_4 \mu(K_1 - k_1)\mu(K_2 - k_4)]^{2(1-\nu)}}$$

$$\leqq O(|\Delta|^2) \int dK_1\, dK_2 \frac{F_\Delta(K_1 + K_2)^{1-\delta} F(K_1)^{(1-4\lambda)/3} F(K_2)^{(1-4\lambda)/3}}{[\mu(K_1)\, \mu(K_2)]^{1-4\nu}}$$

$$\leqq O(|\Delta|^2) \int dK_1\, dK_2\, F_\Delta(K_1 + K_2)^{1-\delta} [F(K_1)\, F(K_2)]^{(2/3)-(4/3)(\lambda+\nu)}$$

$$\leqq O(|\Delta|^2) \int dK_1\, dK_2\, F_\Delta(K_1 + K_2)^{1+\delta} F(K_1)^{(4/3)-(8/3)(\nu+\lambda)} \leqq O(|\Delta|),$$

provided that $(\lambda + \nu)$ is sufficiently small, namely $4/3 - (8/3)\,(\lambda + \nu) > 1$, or

$$(\lambda + \nu) < \frac{1}{8}.$$

In particular, if $\lambda = \nu$, and if w is the kernel of a P or C vertex,

$$\|w \prod_{i=1}^{4} \mu_i^{1/16-\varepsilon}\|_{2,2} \leqq O(|\Delta|^{1/4}),$$

using

$$\|A\|_{2,2} = \|A^*A\|_{2,2}^{1/2} \leqq \|A^*A\|_{\text{H.S.}}^{1/2}.$$

If $\lambda = \nu$, and w is the kernel of the diagram

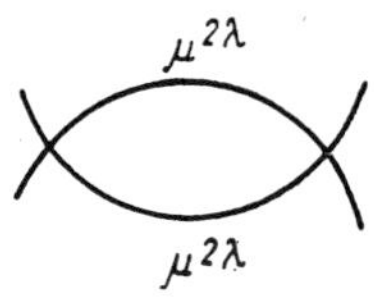

with either P or C vertices, then

$$\|w\Big(\prod_{i=1}^{4} \mu_i^{1/16-\varepsilon}\Big)\|_{\text{H.S.}} \leqq O(|\Delta|^{1/2}).$$

Proof of Proposition 5.3.4. We note that in the notation of Chapter 2, the kernel of the mass renormalization diagram is

$$w_{\Delta,\Delta'}(k_1, k_5) = \text{const.} \int (\alpha\beta\gamma - \bar{\alpha}\bar{\beta}\bar{\gamma})\, dk_2\, dk_3\, dk_4. \tag{6.3.2}$$

For $\Delta = \Delta'$, this is just (2.10), while for $\Delta \neq \Delta'$ we note that the contribution from $\int \bar{\alpha}\bar{\beta}\bar{\gamma}\, dk$ is zero, reducing (6.3.2) to (2.9). This is a consequence of $\chi_\Delta \chi_{\Delta'} = 0$, yielding

$$\int \bar{\beta}_{\Delta,\Delta'}\, dP = \text{const.}\, (h^2 \chi_\Delta \chi_{\Delta'})^\sim\, (k_5 - k_1) = 0.$$

We note that $P = k_2 + k_3 + k_4$, $k_i = p_i + P$,

$$|\mu(k_i)^{-2} - \mu(p_i)^{-2}| \leqq O(1)\, \mu(P)^{3\varepsilon} \{\mu(k_i)^{-2-\varepsilon} + \mu(p_i)^{-2-\varepsilon}\}$$

yielding for $0 \leqq \varepsilon \leqq 1/3$,

$$|\gamma-\bar{\gamma}| \leqq O(1)\,(\mu_1\mu_5)^{-1}\,\mu(P)^{3\varepsilon}\,\{\mu(k_2)^{-2-\varepsilon}+\mu(p_2)^{-2-\varepsilon})\,\mu(k_3)^{-2}\,\mu(k_4)^{-2}+$$
$$+\mu(p_2)^{-2}\left(\mu(k_3)^{-2-\varepsilon}+\mu(p_3)^{-2-\varepsilon}\right)\mu(k_4)^{-2}+$$
$$+\mu(p_2)^{-2}\,\mu(p_3)^{-2}\left(\mu(k_3)^{-2-\varepsilon}+\mu(p_3)^{-2-\varepsilon}\right)\}.$$

In addition we need a bound on $(\alpha-\bar{\alpha})$. The P vertex in a mass diagram arises from lowering the upper cutoff in Δ. Let λ be the corresponding maximum lower cutoff in the P vertex, due to this change of cutoffs $\delta\eta = d/ds\,\eta_s$. For $\delta \geqq 0$, on the support of the cutoffs,

$$|\delta\eta - \delta\bar{\eta}| \leqq O(|P|^\delta\,\lambda^{-\delta}).$$

Since

$$(\alpha-\bar{\alpha}) = \frac{d}{ds}\prod_{\text{legs}}(\eta_s\eta_{s'} - \bar{\eta}_s\bar{\eta}_{s'})|_{s=s'},$$

we have

$$|\alpha - \bar{\alpha}| \leqq O(P^\delta\lambda^{-\delta}).$$

Since $\lambda \geqq M(2)^{1/1+\nu} \geqq M(2)^{1/2}$,

$$|\alpha-\bar{\alpha}| \leqq O\big(|P|^\delta\,\lambda^{-\delta/2}\,M(2)^{-\delta/4}\big)$$
$$\leqq O\,|\Delta'|^{-\delta/3}\,\mu(|\Delta|^{1/3}\,P)^\delta\,\lambda^{-\delta/2}\,M(2)^{-\delta/4}. \tag{6.3.4}$$

We now estimate the Hilbert Schmidt norm of (6.3.2). We assume $|\Delta| \leqq |\Delta'|$. Though our kernel w is asymmetric in Δ and Δ', our estimates are symmetric. Thus

$$|w_{\Delta,\Delta'}(k_1,k_5)| \leqq O(1)\int|\alpha-\bar{\alpha}|\,\bar{\beta}\bar{\gamma}\,dk_2\,dk_3\,dk_4 + \tag{6.3.5}$$
$$+\,O(1)\int\bar{\beta}\,|\gamma-\bar{\gamma}|\,dk_2\,dk_3\,dk_4,$$

using $|\delta\eta| \leqq O(1)$.
We estimate the first term, using (6.3.4), by

$$O(|\Delta|\,|\Delta'|)\,\lambda^{-\delta/2}\,M(2)^{-\delta/4}\int dP\,dp_2\,dp_3\,\frac{F_\Delta(k_1+P)\,F_{\Delta'}(P)^{-\delta}\,|\Delta'|^{-\delta/3}\,F_{\Delta'}(-P+k_5)}{\mu_1\mu_5\mu(p_2)^2\,\mu(p_3)^2\,\mu(p_4)^2}.$$

Using $p_4 = -(p_2+p_3+2P)$ and (6.3.1), we bound this by

$$O(|\Delta|\,|\Delta'|^{1-\delta/3})\,\lambda^{-\delta/2}\,M(2)^{-\delta/4}\log M(2)\,(\mu_1\mu_5)^{-1}\times \tag{6.3.6}$$
$$\times\int dP\,F_\Delta(k_1+P)\,F_{\Delta'}(P)^{-\delta}\,F_{\Delta'}(-P+k_5)$$
$$\leqq O\,(|\Delta|^{1/4-\varepsilon/2}\,|\Delta'|^{1/4-\delta/3})\,\lambda^{-\delta/2}\,F(k_1+k_5)^{(1+\varepsilon)/2}\,F(k_5)^{-\delta}\,(\mu_1\mu_5)^{-1}.$$

The last inequality follows from Corollary 6.1.7 with $\alpha_1 = 0$, $\alpha_2 = -\delta$. In (6.3.6) we dominate the logarithmic divergence in M_i by $M_i^{-\delta/4}$, leaving the convergence factor $\lambda^{-\delta/2}$.

The L_2 norm of (6.3.5), multiplied by $(\mu_1\mu_5)^{1/4-\varepsilon}$, is dominated by

$$O(|\Delta|^{1/4-\varepsilon/2}|\Delta'|^{1/4-\delta/3})\,\lambda^{-\delta/2}\,[\textstyle\int F(k_1+k_5)^{1+\varepsilon}\,(\mu_1\mu_5)^{3/2+6\delta-2\nu}\,dk_1\,dk_5]^{1/2}$$

$$\leq O(\Delta^{1/4-\varepsilon/2}\,|\Delta'|^{1/4-\delta/3})\,\lambda^{-\delta/2}$$

for $\delta(\nu)$ sufficiently small.
Using (6.3.2)—(6.3.3), the second term in (6.3.5) yields a similar bound. Thus given $\varepsilon > 0$, there exists $\delta = \delta(\varepsilon) > 0$ such that

$$\|(\mu_1\mu_5)^{1/4-\varepsilon}\,w_{\Delta,\Delta'}\|_2 \leq O(|\Delta|^{1/8}\,|\Delta'|^{1/8}\,\lambda^{-\delta}). \tag{6.3.7}$$

Proof of Proposition 5.3.6. First assume that $|\Delta| \leq |\Delta'|$. By the Schwartz inequality

$$|P \equiv\equiv\equiv\equiv W|^2 \leq P \equiv\equiv\equiv\equiv P \quad W \equiv\equiv\equiv\equiv W$$

$$\leq \left(O|\Delta|\;M(1)\;M(2)^{3\varepsilon_1}\,\lambda^{-2\varepsilon_1}\,|\Delta'|^{1/2-\varepsilon}\right)\prod_i b_i^{-3(1+\nu)},$$

where we use Proposition 5.3.3 (d) and 5.3.5. Since the W vertex has at least one $\delta\varphi_1$ leg, and that leg has an upper cutoff $O(|\Delta'|^{-1/2})$, we have $M(1) \leq O(|\Delta'|^{-1/2})$. As in the proof of Proposition 5.3.5, there exists ε_2 such that

$$M(2)^{3\varepsilon_1} \leq O(|\Delta'|^{-\varepsilon_2}).$$

Hence for ε_1 sufficiently small, and using $|\Delta| \leq |\Delta'|$, there exists $\varepsilon_3 > 0$ such that

$$|P \equiv\equiv\equiv\equiv W|^2 \leq O(|\Delta|^{2\varepsilon_3}\,|\Delta'|^{2\varepsilon_3})\,\lambda^{-2\varepsilon_1}\prod_i b_i^{-3(1+\nu)}.$$

In case $|\Delta'| \leq |\Delta|$, we modify the above proof as follows: The kernel of the P vertex contains a factor $(h\chi_\Delta)^\sim(P)$, where P is the total momentum. In the estimate on the Hilbert Schmidt norm of the P vertex, we use only $[(h\chi_\Delta)^\sim]^{(1+\varepsilon_3)/2}$. Thus we can, without change of that estimate, include a factor $[(h\chi_\Delta)^\sim(P)]^{(1-\varepsilon_3)/2}$ in the kernel of the W vertex. Following our proof of Proposition 5.3.5, with our modified kernel, this yields an estimate

$$\|(h\chi_\Delta)^\sim(P)^{(1-2\varepsilon_3)}\,w_W\|^2_{\text{H.S.}} \leq O(|\Delta|^{-1+2\varepsilon_3}\,|\Delta'|^{3/2-\varepsilon-2\varepsilon_3})\prod_i b_i^{-3(1+\nu)}.$$

Thus

$$|P \equiv\equiv\equiv\equiv W|^2 \leq O\left(|\Delta|^{2\varepsilon_3}\,|\Delta'|^{3/2-\varepsilon-2\varepsilon_3}\,M(1)\,M(2)^{3\varepsilon_1}\,\lambda^{-2\varepsilon_1}\right)\prod_i b_i^{-3(1+\nu)}$$

and for ε, ε_1, and δ sufficiently small,

$$|P \equiv\equiv\equiv\equiv W|^2 \leq O(|\Delta|^{2\varepsilon_3}\,|\Delta'|^{2\varepsilon_3}\,\lambda^{-2\varepsilon_1})\prod_i b_i^{-3(1+\nu)}$$

as above.

Proposition 6.3.1. *The changes* $\delta E_2(\Delta)$, *and* $\delta E_3(\Delta)$ *in the second and third order counterterms in* δV *of* (3.2.9) *are bounded by* $O(|\Delta|^{1-\delta}\,M(1)\log M(2) \leq O(1)$.

Proof. The second order vacuum energy $E_2(\Delta)$ is a summation of $E_2(\Delta, \Delta')$ over Δ', where $E_2(\Delta, \Delta') = -1/2\,\langle V_I(\Delta)\;V_I(\Delta')\rangle$, and where Δ' ranges over a cover of $\mathscr{D}_\varepsilon$.

As in (5.2.1), we obtain

$$|E_2(\Delta, \Delta')| \leq O(1) \left(\prod_{i=1}^{4} d_i^{-n} \right) I ,$$

where

$$\bar{I} = |\Delta| \, |\Delta'| \int \frac{F_\Delta(P) \, F_{\Delta'}(P) \, dP \, dk_1 \, dk_2 \, dk_3}{\mu(P - k_1 - k_2 - k_3)^2 \, \mu_1^2 \, \mu_2^2 \, \mu_3^2} .$$

By (6.3.1) and Proposition 6.1.5 (b),

$$\bar{I} \leq |\Delta| \, |\Delta'| \, O\big(\log M(2)\big) \int F_\Delta(P) \, F_{\Delta'}(P) \, \mu_1^{-2} \, dk_1 \, dP$$

$$\leq O\big(M(1) \log M(2)\big) \, |\Delta|^{1-\delta} .$$

During the Wick expansion (3.2.9) of δV, $M(1) \leq O(|\Delta|^{-1/2})$, see (3.2.1), and log $M(2) \leq |\Delta|^{-\varepsilon}$, see (5.3.14). Thus $\bar{I} \leq O(1)$, and summing over d^{-4} yields the desired bound on $\delta E_2(\Delta)$, see Lemma 4.2 (a).
The bound on $\delta E_3(\Delta)$ is proved similarly.

References

[1] J.-P. ECKMANN, Commun. Math. Phys., **25**, 1 (1972).
[2] J.-P. ECKMANN and K. OSTERWALDER, Helv. Phys. Acta, **44**, 864 (1971).
[3] J. FABREY, Commun. Math. Phys., **19**, 1 (1970).
[4] J. GLIMM, Commun. Math. Phys., **10**, 1 (1968).
[5] J. GLIMM and A. JAFFE, Ann. Math., **91**, 362 (1970).
[6] J. GLIMM and A. JAFFE, J. Math. Phys., **13**, 1568 (1972).
[7] J. GLIMM and A. JAFFE, Field Theory Models, in *Statistical Mechanics and Quantum Field Theory*, C. Dewitt and R. Stora, editors, 1970 Les Houches Lectures, Gordon and Breach, New York 1971.
[8] J. GLIMM and A. JAFFE, Boson Field Models, in *Mathematics of Contemporary Physics*, R. Streater, editor, Academic Press, New York, 1972.
[9] K. HEPP, Théorie de la Renormalization, Springer Verlag, Berlin 1969.
[10] E. NELSON, Quantum Fields and Markoff Fields, American Mathematical Society Symposium on Partial Differential Equations, 1971, to appear.
[11] K. SYMANZIK, Euclidean Quantum Field Theory, in *Local Quantum Theory*, R. Jost, editor 1968 Varenna Lectures, Academic Press, 1969.

Herausgeber: Prof. Dr. Frank Kaschluhn, Prof. Dr. Artur Lösche, Prof. Dr. Rudolf Ritschl und Prof. Dr. Robert Rompe. Manuskripte sind zu richten an die Schriftleitung: Dr. Lutz Rothkirch, Sektion Physik der Humboldt-Universität zu Berlin, 104 Berlin, Hessische Str. 2. Verlag: Akademie-Verlag GmbH, 108 Berlin, Leipziger Str. 3/4, Fernruf: 220441, Telex-Nr. 0112020, Postscheckkonto: Berlin 35021. Die Zeitschrift „Fortschritte der Physik" erscheint monatlich; Bezugspreis dieses Heftes 10,—. Bestellnummer dieses Heftes: 1027/21/7. — Satz und Druck: VEB Druckhaus „Maxim Gorki", 74 Altenburg, Bez. Leipzig, Carl-von-Ossietzky-Str. 30/31. — Veröffentlicht unter der Lizenznummer 1324 des Presseamtes beim Vorsitzenden des Ministerrates der Deutschen Demokratischen Republik.

III Phase Transitions Exist

Introduction

While phase transitions were known to occur Ising-type lattice spin systems since Peierls' famous 1936 paper, they were believed to exist but had not been established in the corresponding statistical mechanics of classical fields or in continuum quantum field theory. The proof that phase transitions in such systems do exist in these systems is presented in the paper Phase transitions for ϕ_2^4 quantum fields with T. Spencer. The specific example worked out is the $\lambda\phi^4 - \phi^2$ model, with $0 < \lambda \ll 1$. The idea is to treat the low momentum degrees of freedom in each unit space-time volume as an Ising spin system, while the high momentum degrees of freedom are shown to have small effect for $\lambda \ll 1$. The technical tool which enabled the implementation of this phase cell localization idea is the "multiple reflection bounds" developed in that article. (These bounds are also known as "chessboard" estimates.)

The end result of the study is the nonuniqueness of the model for small λ. This nonuniqueness has an intuitive interpretation in terms of the two minima of the W-shaped potential function $\lambda\phi^4 - \phi^2$. Each minimum corresponds to an approximate quantum mechanics ground state. Hence quantum field theory admits two solutions, in distinction to quantum mechanics of a finite number of degrees of freedom which has a nodeless eigenfunction for a unique ground state. The choice of one solution of a mixture is dictated by some boundary condition or other parameter.

The other articles in this group show that the cluster expansion method can be extended to apply to the multiple phase ϕ^4 model. Hence it is possible to obtain information on the mass gap and other spectral issues in the two phase region. This requires an expansion in local phases, as well as the cluster expansion which decouples different volumes. Both these methods have been extended to more general classes of models. Eventually generalizations of these points of view have been used to settle important questions in the theory of phase transitions in statistical physics.

Bibliography

Phase transitions for ϕ_2^4 quantum fields, with T. Spencer, Commun. Math. Phys. 45, 203–216 (1975).

A convergent expansion about mean field theory I. The expansion, with T. Spencer, Ann. Phys. 101, 610–630 (1976).

A convergent expansion about mean field theory II. Convergence of the expansion, with T. Spencer, Ann. Phys. 101, 631–669 (1976).

Phase Transitions for φ_2^4 Quantum Fields*

James Glimm**
Rockefeller University, New York, N.Y. 10021, USA

Arthur Jaffe*** and Thomas Spencer***,†
Harvard University, Cambridge, Mass. 02138, USA

Abstract. Phase transitions for the quantum field interaction $\lambda\varphi^4 + m_0^2\varphi^2$, $m_0^2/\lambda \ll 1$ are established in two dimensional space time.

1. Introduction

We present a direct proof of the existence of phase transitions in quantum field theory. We consider here the simplest interaction for which a phase transition is expected, namely the

$$\lambda\varphi^4 + \tfrac{1}{2} m_0^2 \varphi^2, \quad m_0^2/\lambda \ll 1, \tag{1.1}$$

perturbation of the free field of mass m_0. We give a complete proof in space time dimension $d=2$. Our same methods apply in principle to arbitrary even $P(\varphi)_2$ models without cutoff.

To define the interaction (1.1) for $d=2$ we require Wick ordering. We denote Wick ordering of P with respect to the covariance $(-\Delta + m_0^2)^{-1}$ by $:P:_{m_0}$. Then scaling and re Wick ordering leads to an equivalent theory with the bare mass $O(\sigma)^{-1}$ and the interaction which we study, see [13], is

$$:P(\varphi):_{\sigma^{-1}} = :(\varphi^2 - \sigma^2)^2/\sigma^2:_{\sigma^{-1}}, \quad \sigma \gg 1. \tag{1.2}$$

It is the occurrence of two distinct minima, separated by a large barrier, which suggests the occurrence of phase transitions for the interaction (1.2). The two pure phases are ground states localized (in φ space) near the two minima $\varphi = \pm\sigma$.

In the case we consider, the polynominal $P(\varphi)$ is invariant under the symmetry transformation $\varphi \to -\varphi$, while the pure phases are interchanged by the symmetry. We note, however, that symmetry breaking is a distinct issue from the existence of phase transitions. Just as in statistical mechanics, where phase transitions may occur without symmetry breaking [11], we expect phase transitions in field theory for certain $P(\varphi)$ models which do not possess a symmetry group, such as

---

* Presented at the International Colloquium on Mathematical Methods of Quantum Field Theory, Marseille, June 23–27, 1975.
** Supported in part by the National Science Foundation under Grant MPS 74-13252.
*** Supported in part by the National Science Foundation under Grant MPS 73-05037.
† On leave from Rockefeller University, New York, NY 10021, USA.

for the interaction

$$\sigma^2 P(\varphi) = (\varphi^2 - \sigma^2)^4 + \varepsilon\varphi^3 - \mu\varphi ,$$

for $\sigma \gg 1$, $\varepsilon \ll 1$, and $\mu = \mu(\varepsilon, \sigma)$.

The method we use here to study the interaction (1.2) is to generalize the φ^j bounds [3, 8, 2, 6, 5] in order to estimate perturbations of the measures for the interacting ground state. These estimates follow from the Osterwalder-Schrader positivity and the Euclidean invariance of the theory, combined with bounds we prove in Section 2 on the vacuum energy per unit volume. The existence of long range order (phase transitions and symmetry breaking) follow by a variant of a statistical mechanics argument due to Peierls. As a consequence, we find that the Dirichlet φ^4 theory (1.2) has a non-unique vacuum for $\sigma \gg 1$. The two phases are separated by an external field $-\mu\varphi$ added to $P(\varphi)$ of (1.2).

In a separate article [7], we generalize our convergent weak coupling expansion to yield a cluster expansion for the interaction (1.2). This expansion allows us to introduce boundary conditions (other than zero Dirichlet) which yield two distinct pure phases. In each pure phase we obtain from the cluster expansion a model with a unique vacuum, which satisfies the Wightman axioms with a mass gap. Of course, it would be of interest to pursue the particle structure of the model (1.2) via the cluster expansion.

In contrast to our detailed study based on the cluster expansion, our present paper gives a simple, direct proof that phase transitions occur. An alternative approach to the problem of phase transitions has been announced in [1], but the proof has not appeared.

We now introduce some notation and explain the Peierls argument. Let $\Lambda \subset R^2$ be a square, centered at the origin, and let Δ_Λ be the Laplace operator with zero Dirichlet boundary conditions on the boundary $\partial\Lambda$ of Λ. Let $d\varphi_{\varepsilon,\Lambda}$ denote the Gaussian measure over $\mathscr{S}'(R^2)$ wit mean zero and covariance $(-\Delta_\Lambda + \varepsilon^2)^{-1}$. Let

$$V(\Lambda) = \int_\Lambda :P(\varphi(x)):dx , \tag{1.3}$$

where $:P:$ denotes Wick ordering with respect to the covariance $(-\Delta_\Lambda + \varepsilon^2)^{-1}$. We let

$$\langle A \rangle = \lim_{\Lambda \nearrow R^2} \frac{\int A(\varphi) e^{-V(\Lambda)} d\varphi_{\varepsilon,\Lambda}}{\int e^{-V(\Lambda)} d\varphi_{\varepsilon,\Lambda}} , \tag{1.4}$$

and in order to ensure that P defined by (1.2) is equivalent to (1.1), we also choose $\varepsilon = \sigma^{-1}$.

For P given by (1.2), plus a linear term, the limit $\Lambda \nearrow R^2$ in (1.4) exists for

$$A = \int \varphi(x_1) \dots \varphi(x_r) v(x_1, \dots, x_r) dx , \qquad v \in \mathscr{S} ,$$

and also for $A = \exp Q$ with Q defined by (1.9), as well as for

$$A = \prod \chi_\pm(\Delta_j)$$

defined below.

For each unit lattice square $\Delta = \Delta_j$, centered at $j \in Z^2$, we define the averaged field

$$\varphi(\Delta) = \int_\Delta \varphi(x) dx .$$

Essentially $\varphi(\Delta)$ is the low momentum field in Δ, so we expect that $\varphi(\Delta)$ characterizes phase transitions in Δ. We define a localizing function of the averaged field

$$\chi_{(a,b)}(\Delta) = \begin{cases} 1 & \text{if} \quad \varphi(\Delta) \in (a, b) \\ 0 & \text{otherwise.} \end{cases}$$

We use the particular functions

$$\chi_+ = \chi_{(0,\infty)} \quad \text{and} \quad \chi_- = \chi_{(-\infty,0)}$$

to separate phases and to establish the existence of long range order.

Lattice squares Δ_i, Δ_j are nearest neighbors if $|i-j|=1$. Let $\mathcal{N}$ be a collection of nearest neighbor pairs and let $|\mathcal{N}|$ denote the cardinality of $\mathcal{N}$. Assume there exists a constant $K > 0$ such that

$$\langle \textstyle\prod_{(\Delta,\Delta') \in \mathcal{N}} \chi_-(\Delta)\chi_+(\Delta') \rangle \leqq e^{-K|\mathcal{N}|} . \tag{1.5}$$

Theorem 1.1. *Consider a $P(\varphi)_2$ quantum field theory defined by* (1.3)–(1.4). *If* (1.5) *holds for K sufficiently large and for all $\mathcal{N}$, then for any two lattice squares Δ_i and Δ_j*

$$\langle \chi_+(\Delta_i)\chi_-(\Delta_j) \rangle \leqq 1/8 .$$

Corollary 1. *If the field φ has a $\varphi \to -\varphi$ symmetry (for example if P and the boundary conditions are both even) then $\langle \chi_\pm(\Delta) \rangle = \frac{1}{2}$. In this case,*

$$|\langle \chi_+(\Delta_i)\chi_-(\Delta_j) \rangle - \langle \chi_+(\Delta_i) \rangle \langle \chi_-(\Delta_j) \rangle| \geqq 1/8 ,$$

which establishes the existence of long range order.

In order to establish symmetry breaking, we consider the model

$$\begin{aligned} :P(\varphi, \mu): &= :(\varphi^2 - \sigma^2)^2/\sigma^2: - \mu\varphi \\ &= :P(\varphi): - \mu\varphi , \end{aligned} \tag{1.6}$$

with $:P:$ given by (1.2). The estimates in this paper are uniform for $0 \leqq \mu < \sigma^{-2}$. Then we find

Corollary 2. *For the interaction defined by* (1.3), (1.4), *and* (1.6),

$$\lim_{\mu \searrow 0} \langle \chi_+(\Delta) \rangle_\mu > 0.8 .$$

Proof. For $\mu \neq 0$, the $:P(\varphi, \mu):$ model has a unique vacuum, as a corollary of the Lee-Yang theorem [12]. Thus as $\text{dist}(\Delta, \Delta') \to \infty$,

$$\begin{aligned} \langle \chi_+(\Delta)\chi_-(\Delta') \rangle_\mu &\to \langle \chi_+(\Delta) \rangle_\mu \langle \chi_-(\Delta') \rangle_\mu \\ &= \langle \chi_+(\Delta) \rangle_\mu (1 - \langle \chi_+(\Delta) \rangle_\mu) , \end{aligned}$$

where we use clustering and translation invariance. Thus by the theorem,

$$\langle\chi_+(\Delta)\rangle_\mu(1-\langle\chi_+(\Delta)\rangle_\mu)\leqq 1/8\,,$$

which has the solution $\langle\chi_+(\Delta)\rangle_\mu>0.8$. By FKG inequalities, $\langle\chi_+(\Delta)\rangle_\mu$ is monotone in μ, as $\mu\searrow 0$. Thus $\lim_{\mu\searrow 0}\langle\chi_+(\Delta)\rangle_\mu$ exists and is greater than $0.8>\frac{1}{2}$.

Theorem 1.2. *Let K be given, and let a φ^4 quantum field be defined by* (1.3), (1.4), *and* (1.6). *For σ sufficiently large,* (1.5) *holds for all collections $\mathcal{N}$ of nearest neighbor pairs.*

Proof of Theorem 1.1. Let Λ_0 be a square, centered at the origin, containing Δ_i, Δ_j and large enough (depending on i and j) so that

$$2+\operatorname{dist}(\Delta_i,\Delta_j)\leqq\frac{1}{9}|\partial\Lambda_0|\leqq\operatorname{dist}(\partial\Lambda_0,\{\Delta_i,\Delta_j\})\,. \tag{1.7}$$

A configuration is defined as a map from lattice squares in Λ_0 to ± 1. Given a configuration c, a subset $Y\subset c^{-1}(+1)$ is said to be + connected if any two lattice squares of Y can be connected by a path of nearest neighbor squares belonging to Y. The definition of – connected is analogous. Each configuration introduces a decomposition of Λ_0 into + and — connected components $\{X_k(c)\}$. Let X_i and X_j be components containing Δ_i and Δ_j respectively. Let γ_i be the outermost contour in ∂X_i. Technically, γ_i is defined to be the boundary of the component of $R^2\backslash X_i$ which contains the point at ∞. At least one of the contours γ_i or γ_j must separate Δ_i from Δ_j. Supposing this contour is γ_i, we define

$$\gamma=\gamma_i\backslash\partial\Lambda_0\,,$$

i.e. the portion of γ_i not contained in $\partial\Lambda_0$.

Letting $|\gamma|$ denote the length of γ, we assert that

$$|\gamma_i|+\operatorname{dist}(\gamma,\{\Delta_i,\Delta_j\})\leqq 11|\gamma|\,. \tag{1.8}$$

If $\gamma_i\cap\partial\Lambda_0=\emptyset$, then there is nothing to prove, since the shortest contour with a given radius ($\geqq\operatorname{dist}(\gamma,\{\Delta_i,\Delta_j\})$) is a square, and for this case, the circumference is eight times the radius. Now suppose $\gamma_i\cap\partial\Lambda_0\neq\emptyset$. Then

$$\begin{aligned}\operatorname{dist}(\gamma,\{\Delta_i,\Delta_j\})&\leqq\tfrac{1}{2}\operatorname{dist}(\Delta_i,\Delta_j)\leqq\frac{1}{18}|\partial\Lambda_0|\\&\leqq\tfrac{1}{2}\operatorname{dist}(\partial\Lambda_0,\{\Delta_i,\Delta_j\})\\&\leqq\tfrac{1}{2}|\gamma|\end{aligned}$$

by (1.7). Thus

$$\begin{aligned}|\gamma_i|+\operatorname{dist}(\gamma,\{\Delta_i,\Delta_j\})&\leqq|\gamma|+|\partial\Lambda_0|+\tfrac{1}{2}|\gamma|\\&\leqq 11|\gamma|\,.\end{aligned}$$

Since

$$\prod\nolimits_{\Delta\subset\Lambda_0}(\chi_+(\Delta)+\chi_-(\Delta))=1\,,$$

we have

$$\langle \chi_+(\Delta_i)\chi_-(\Delta_j)\rangle = \sum_c' \langle \chi_+(\Delta_i)\chi_-(\Delta_j) \prod_\Delta \chi_{c(\Delta)}(\Delta)\rangle ,$$

where $\sum_c'$ is the sum over configurations c with $c(\Delta_i)=1$, $c(\Delta_j)=-1$ and $\prod_\Delta$ omits Δ_i and Δ_j. Let $\gamma=\gamma(c)$ be defined as above, and let $\mathcal{N}=\mathcal{N}(\gamma)$ be the set of nearest neighbor pairs bordering γ. Then

$$\begin{aligned}\langle \chi_+(\Delta_i)\chi_-(\Delta_j)\rangle &= \sum_\gamma \sum_{\{c:\gamma(c)=\gamma\}} \langle \chi_+(\Delta_i)\chi_-(\Delta_j) \prod_\Delta \chi_{c(\Delta)}(\Delta)\rangle \\ &\leqq \sum_\gamma \langle \chi_+(\Delta_i)\chi_-(\Delta_j) \prod_{\Delta,\Delta'\in\mathcal{N}(\gamma)} \chi_+(\Delta)\chi_-(\Delta')\rangle \\ &\leqq \sum_\gamma e^{-K|\gamma|} .\end{aligned}$$

The sum ranges over γ's resulting from contours γ_i enclosing Δ_i and separating Δ_i from Δ_j. There are at most $3^{|\gamma_i|}$ such contours of length $|\gamma_i|$, with a fixed starting point. We choose as starting point the point of γ_i lying on the line (Δ_i, Δ_j) and closest to Δ_i. With $|\gamma_i|$ fixed, this can be chosen in at most $11|\gamma|$ ways, by (1.8). Thus $\langle \chi_+(\Delta_i)\chi_-(\Delta_j)\rangle < 1/8$ for K sufficiently large.

In order to prove Theorem 1.2, we study perturbations of the interacting Euclidean measure defined by polynomials of the form

$$Q(\xi, X) \equiv \sum_{\nu=1}^4 Q_\nu(\xi^{(\nu)}, X) , \tag{1.9}$$

where

$$Q_1(\xi^{(1)}, X) = \sigma^{-1} \sum_{\Delta_j\subset X} \xi_j^{(1)} \int_{\Delta_j} (:\varphi^2(x): - \sigma^2) dx , \tag{1.10}$$

$$Q_2(\xi^{(2)}, X) = (\ln\sigma)^{-1} \sum_{\Delta_j\subset X} \xi_j^{(2)} \int_{\Delta_j} :\varphi^2(x) - \varphi(\Delta_j)^2 : dx , \tag{1.11}$$

$$Q_3(\xi^{(3)}, X) = \sum_{\substack{\Delta_i,\Delta_j\subset X\\ |i-j|=1}} \xi_{ij}^{(3)} (\varphi(\Delta_i) - \varphi(\Delta_j)) , \tag{1.12}$$

$$Q_4(\xi_j^{(4)}, \kappa, \kappa', X) = \sigma^{-4}(\kappa+1)^\delta \sum_{\Delta_j\subset X} \xi_j^{(4)} \sum_{j=1}^4 \int_{\Delta_i} :\varphi_\kappa^j(x) - \varphi_{\kappa'}^j(x): dx . \tag{1.13}$$

In (1.13), κ and κ' denote ultraviolet cutoffs, and

$$0 \leqq \kappa \leqq \kappa' \leqq \infty , \qquad \kappa < \infty .$$

For $\kappa=0$, $:\varphi_\kappa^j: \equiv 0$. Let $|X|$ denote the area of the subset $X\subset R^2$. Our main technical estimate is the following theorem, which will be proved in Section 3.

Theorem 1.3. *There is a constant* K_1 *and a* $\delta>0$ *such that for all* Q *defined by* (1.9)–(1.13), *with* $|\xi_\alpha^{(\beta)}| \leqq 1$ *and for* $\langle\cdot\rangle$ *defined by* (1.2)–(1.4), *and* (1.6) *with* σ *sufficiently large*

$$|\langle e^{Q(\xi, X)}\rangle| \leqq e^{K_1(\log\sigma)^2|X|} .$$

Proof of Theorem 1.2, assuming Theorem 1.3. For each pair $(\Delta_i, \Delta_j)\in\mathcal{N}$ we write

$$\chi_+(\Delta_i)\chi_-(\Delta_j) = (\chi_{(0,\sigma/2)} + \chi_{(\sigma/2,\infty)})(\chi_{(-\infty,-\sigma/2)} + \chi_{(-\sigma/2,0)})$$

as a sum of four terms. We shall bound each term for any even M by $[\sigma^{-1}Q_\nu(\varphi)]^M$, $\nu=1,2,3$. By using the Cauchy formula for derivatives in the variable ξ as in [2] we have

$$\sigma^{-M}\langle Q_\nu^M\rangle = \sigma^{-M}\left[\frac{d}{d\xi^{(\nu)}}\right]^M \langle e^Q\rangle \leqq \frac{M!\,\sigma^{-M}}{2\pi}\left|\oint \frac{\langle e^Q\rangle}{[\xi^\nu]^{M+1}} d\xi\right| . \tag{1.15}$$

In each case we choose $M=2[\sigma/8]$ to optimize the above bound. Here $[x]$ denotes the integer part of x.

For instance, we bound the term $\chi_{(\sigma/2,\infty)}(\Delta_i)\chi_{(-\infty,-\sigma/2)}(\Delta_j)$ by

$$[\sigma^{-1}(\varphi(\Delta_i)-\varphi(\Delta_j))]^M=\sigma^{-M}\left[\frac{d}{d\xi_{ij}^{(3)}}\right]^M e^{Q_3}\Bigg|_{\xi=0}.$$

Each of the remaining three terms is bounded by $\chi_{(0,\sigma/2)}(\Delta_i)$ or $\chi_{(-\sigma/2,0)}(\Delta_j)$. Note that on the support of $\chi_{(0,\sigma/2)}(\Delta)$,

$$\begin{aligned}\chi_{(0,\sigma/2)}(\Delta)&\leqq 1\leqq 4/3\left[1-\left(\frac{\varphi(\Delta)}{\sigma}\right)^2\right]\\&\leqq\frac{4}{3}\sigma^{-2}[\sigma^2-\textstyle\int_\Delta:\varphi(x)^2:dx]+\frac{4}{3}\sigma^{-2}[\int_\Delta:\varphi(x)^2:dx-:\varphi(\Delta)^2:+O(\ln\sigma)].\end{aligned}$$

Here $O(\ln\sigma)$ is a Wick ordering constant. Since $\sigma^{-2}O(\ln\sigma)\to 0$ one has for sufficiently large σ either

$$\chi_{(0,\sigma/2)}(\Delta_i)\leqq\sigma^{-M}\left[\frac{4}{\sigma}(\textstyle\int_{\Delta_i}:\varphi(x)^2:dx-\sigma^2)\right]^M$$

or

$$\chi_{(0,\sigma/2)}(\Delta_i)\leqq\sigma^{-2M}[4\textstyle\int_{\Delta_i}:\varphi(x)^2:dx-:\varphi(\Delta)^2:]^M.$$

The factor $\chi_{(-\sigma/2,0)}(\Delta_j)$ is bounded by the same expressions with Δ_j replacing Δ_i.

Now we can apply (1.15) to each pair of squares. Since a single square Δ_i occurs in Q_1 or Q_2 in at most four pairs $(\Delta_i,\Delta_j)\in\mathcal{N}$, the factorials are bounded by

$$(4M!)^{|\mathcal{N}|/4}\leqq([\sigma]!)^{|\mathcal{N}|/4}.$$

The theorem now follows from

$$\begin{aligned}\langle\textstyle\prod\chi_+(\Delta_i)\chi_-(\Delta_j)\rangle&\leqq e^{O(\log\sigma)^2|\mathcal{N}|}[[\sigma]!\sigma^{-\sigma}]^{|\mathcal{N}|/4}\\&\leqq e^{O(\log\sigma)^2|\mathcal{N}|}e^{-\sigma|\mathcal{N}|/4}.\end{aligned}$$

In the last line we have used Stirling's formula (with $n=[\sigma]$),

$$n!\leqq(2\pi n)^{\frac{1}{2}}(n/e)^n.$$

Note that the weakly divergent bound $\exp(O(\ln^2\sigma))$ per square from Theorem 1.3 is bounded by the strongly convergent factor $\exp(-O(\sigma))$.

2. Bounds on Vacuum Energy per Unit Volume

In this section we give upper and lower bounds on the vacuum energy per unit volume. The upper bound is more elementary, and we consider it first. Let $|\partial\Lambda|$ denote the length of the boundary of Λ.

Proposition 2.1. *There is a constant $O(1)$ independent of σ and Λ such that*

$$e^{-O(1)|\Lambda|}\leqq\int e^{-V(\Lambda)}d\varphi_{\sigma^{-1},\Lambda}\tag{2.1}$$

for all Λ satisfying $\sigma|\partial\Lambda|\leqq|\Lambda|$.

Proof. We translate φ by the function $\sigma f(x)$, where

$$0 \leqq f(x) = \sigma^{-2} \int_\Lambda (-\Delta_\Lambda + \sigma^{-2})^{-1}(x, y) dy \leqq 1 .$$

(Note that the Dirichlet Green's function is less than the free Green's function.) By convexity of the exponential,

$$\int e^W d\varphi \geqq \exp\left(\int W d\varphi\right) .$$

Hence for $W(\varphi)$ a Wick ordered polynomial without a constant term,

$$\int W d\varphi = 0 \quad \text{and} \quad \int e^W d\varphi \geqq 1 .$$

Thus we need only compute the constant in the exponent after translation. The purpose of the translation is to eliminate the leading term $O(\sigma^2|\Lambda|)$. There is a constant term associated with $V(\Lambda)$ and a constant associated with the change of measure $d\varphi \to d(\varphi + \sigma f)$. The constant associated with $V(\Lambda)$ of (1.3) is

$$\begin{aligned} \int_\Lambda P(\sigma f(x)) dx &= \sigma^2 \int_\Lambda (f(x)^2 - 1)^2 dx \\ &\leqq 2\sigma^2 \int_\Lambda (1 - f(x)) dx \\ &= 2 \int_{\Lambda \times \Lambda} [(-\Delta + \sigma^2)^{-1} - (-\Delta_\Lambda + \sigma^{-2})^{-1}](x, y) dx dy \\ &\quad + 2 \int_{R^2 \setminus \Lambda} \int_\Lambda (-\Delta + \sigma^{-2})^{-1}(x, y) dx dy . \end{aligned}$$

In the first term, Δ and Δ_Λ agree except on $\partial\Lambda$, so that the difference kernel contributes only along the boundary $\partial\Lambda$. Scaling the length scale σ back to one shows that this term is $O(\sigma|\partial\Lambda|)$. The second term is integrated with one variable in and one variable outside of Λ. Again scaling the length back to one we obtain a contribution $O(\sigma|\partial\Lambda|)$. Thus

$$\int_\Lambda P(\sigma f(x)) dx \leqq O(\sigma|\partial\Lambda|) \leqq O(|\Lambda|) ,$$

where the last inequality follows by our restriction on the volumes.

The constant associated with the translation of the Gaussian measure $d\varphi_{\sigma^{-1}, \Lambda}$ is

$$\tfrac{1}{2}\sigma^2 \langle f, (-\Delta_\Lambda + \sigma^{-2}) f \rangle = \tfrac{1}{2} \int_\Lambda f(x) dx = O(|\Lambda|) ,$$

and the proof for $\mu = 0$ is complete. For $\mu > 0$, we use the Griffiths inequality

$$\frac{d}{d\mu} \int e^{-[V(\Lambda) - \mu\varphi(\Lambda)]} d\varphi \geqq 0$$

to complete the proof.

The lower bound on the vacuum energy concerns the perturbed Euclidean measure

$$\int e^{-[V(\Lambda) + Q(\xi, \Lambda)]} d\varphi_{\sigma^{-1}, \Lambda} . \tag{2.2}$$

This bound, which is slowly divergent as $\sigma \to \infty$ (see Theorem 1.3), is sufficient for the present paper. Bounds uniform in σ may be found in [7]. By Schwarz' inequality, it is sufficient to establish separately the bounds

$$\int e^{4Q_\nu} d\varphi_{\sigma^{-1}, \Lambda} \leqq e^{O(1)|\Lambda|} , \qquad \nu = 2, 3 , \tag{2.3}$$

$$\int e^{-2V(\Lambda)} e^{4Q_\nu} d\varphi_{\sigma^{-1}, \Lambda} \leqq e^{O(1) \ln^2 \sigma |\Lambda|} , \qquad \nu = 1, 4 . \tag{2.4}$$

Proposition 2.2. *There is a constant independent of* σ *and* Λ *such that* (2.3) *and* (2.4) *hold.*

Proof. The main point is that the low momentum modes of Q_2 and Q_3 are suppressed, thereby introducing an effective mass into the covariance. In technical terms, this idea emerges as the following fact: The Laplace operator of a bounded region, with Neumann boundary data, has $\lambda_1=0$ as the lowest eigenvalue, and the next eigenvalue λ_2 is strictly positive. The eigenspace corresponding to $\lambda_1=0$ is the space of functions which are constant on each connected component of the bounded region.

We write Q_3 (and $\xi^{(3)}$) as a sum of four terms so that in a single term, each lattice square Δ occurs only once in the sum over nearest neighbor pairs. By Hölders inequality, it is sufficient to consider a single such term. Let $\mathcal{N}=\{(\Delta,\Delta')\}$ be the corresponding set of nearest neighbor pairs of lattice squares, and let $\Delta_{\mathcal{N}}$ be the Laplace operator with Neumann data on

$$\bigcup_{(\Delta,\Delta')\in\mathcal{N}} \partial(\Delta\cup\Delta').$$

Let $d\varphi_{\mathcal{N}}$ be the Gaussian measure with covariance $(-\Delta_{\mathcal{N}}+\sigma^{-2})^{-1}$. By conditioning [10] and factorization of $d\varphi_{\mathcal{N}}$ across boundary lines, it is sufficient to consider a single pair of lattice squares. The required bound reduces to

$$\int e^{16\,\mathrm{Re}\,\xi(\varphi(\Delta)-\varphi(\Delta'))}d\varphi_{\mathcal{N}}\leqq\exp\left[\tfrac{1}{2}(16\,\mathrm{Re}\,\xi)^2\langle\chi,(-\Delta_{\mathcal{N}}+\sigma^{-2})^{-1}\chi\rangle\right]=O(1),$$

where $\chi=\chi_\Delta-\chi_{\Delta'}$ is the difference of the characteristic functions of Δ and Δ'. To establish the bound $O(1)$, independent of σ, we use the fact that χ is perpendicular to the functions which are constant on $\Delta\cup\Delta'$, and so the lowest eigenvalue $\lambda_1=0$ of $\Delta_{\mathcal{N}}$ does not contribute.

For the case $\nu=2$, it suffices to estimate

$$\int e^{4\xi:\varphi^2(\Delta)-\varphi(\Delta)^2:/\log\sigma}d\varphi_{\mathcal{N}},$$

where the Gaussian measure $d\varphi_{\mathcal{N}}$ has a covariance corresponding to Neumann boundary conditions on the boundary $\partial\Delta$ of the lattice square Δ. Let P_Δ be the orthogonal projection onto the subspace of L_2 spanned by χ_Δ and let

$$A=(\log\sigma)^{-1}(-\Delta_{\mathcal{N}}+\sigma^{-2})^{-1}(\chi_\Delta-P_\Delta).$$

Then by a standard computation (see also [7])

$$\int e^{4\xi:(\varphi^2(\Delta)-\varphi(\Delta)^2):/\log\sigma)}d\varphi_{\mathcal{N}}=|e^{-\frac{1}{2}\,\mathrm{tr}\ln(I-A)}e^{-\frac{1}{2}\,\mathrm{tr}\,A}|\leqq e^{\frac{1}{2}\|A\|_2^2},$$

where $\|\cdot\|_2^2$ is the Hilbert-Schmidt norm and we use $0\leqq A<I$, true for $\sigma\gg 1$. In fact, since $\chi_\Delta-P_\Delta$ annihilates the eigenspace of $\Delta_{\mathcal{N}}$ corresponding to the eigenvalue $\lambda_1=0$, we see that for large σ

$$\|A\|_2^2\leqq\left(\sum_{i=2}^{\infty}\lambda_i^{-2}\right)/\ln^2\sigma<\tfrac{1}{2}.$$

This completes the proof.

For the case $\nu=1$, we note that when $|\xi|\leqq 1$,

$$-O(1)\sigma^2(1+\ln^2\kappa)\leqq:(\varphi_\kappa^2(x)-\sigma^2)^2:+\xi\sigma:\varphi_\kappa^2(x)-\sigma^2:, \qquad (2.5)$$

with Wick ordering with respect to a unit mass. By standard methods the vacuum energy per unit volume corresponding to the polynomial on the right is $O(\sigma^2 \ln^2 \sigma)$, see [9]. The proof of (2.4) for $\mu=0$ then reduces to scaling by σ^{-2}. We introduce the external field $\mu\varphi$ directly into (2.5). Again, for μ sufficiently small, we obtain the lower bound (2.5) by removing Wick ordering, and this completes the proof of (2.4). We omit the discussion of $v=4$.

3. Local Perturbations of the Interaction

In this section we extend the φ^j bounds to localized perturbations of the type $Q(\xi, X)$, and thereby prove Theorem 1.3. We make two modifications to the basic φ^j bound argument [3, 8, 2, 5]. The first is that we consider perturbations of the measure such as

$$\varphi(\Delta)^2 - \int_\Delta :\varphi(x)^2: dx\,,$$

which, while localized in a bounded region Δ, cannot be expressed as a local perturbation of the Hamiltonian. Secondly, by using Euclidean invariance of the physical measure ($\Lambda = R^2$), we show that it suffices to consider the case where X is a large rectangle. In the case X = rectangle, we work in a finite volume and obtain bounds uniform in Λ. This allows the passage to the limit $\Lambda \nearrow R^2$. The φ bounds have the form

$$\pm\varphi(f) \leqq H + O(1)|supp\, f|$$

rather than the cruder estimate

$$\pm\varphi(f) \leqq H + O(1)|\text{diam}\; supp\, f|\,.$$

We do not require estimates on β_∞ [cf. 8, 2] which measures surface effects on the vacuum energy due to removing or adding parts of the interaction associated with a bounded region. (We remark that estimates on β_∞ are poor because of the small mass σ^{-1} in the covariance.) Dirichlet data provide a further complication to such estimates.

Let

$$H_l(\sigma) = H_{0,l}(\sigma) + \sigma^{-2} \int_{-l}^{l} :(\varphi(x)^2 - \sigma^2)^2: dx \tag{3.1}$$

be the Hamiltonian with Dirichlet boundary conditions at $x = \pm l$. Here $H_{0,l}(\sigma)$ is the free Hamiltonian with the same boundary conditions and mass σ^{-1}. Then $H_{0,l}(\sigma)$ and $H_l(\sigma)$ act on the time zero Fock space

$$\mathscr{F} = L^2(\mathscr{S}'(R^1), d\mu_{0,l})\,. \tag{3.2}$$

Here $d\mu_{0,l}$ is the Gaussian measure defined by the condition that the vacuum $\Omega_{0,l}$ for $H_{0,l}$ is represented by the function 1. Let Ω_l be the vacuum for $H_l(\sigma)$ and let $E_l(\sigma)$ be the vacuum energy, so that

$$(H_l(\sigma) - E_l(\sigma))\Omega_l = 0\,.$$

Recall that $H_l(\sigma)$ is a direct sum of a free Hamiltonian associated with the region $|x| \geqq l$ and an operator with compact resolvent associated with the region $|x| \leqq l$, so that the existence of Ω_l follows.

Lemma 3.1. *There is a constant independent of σ and a sequence $l_i \to \infty$ (possibility dependent on σ) such that*

$$-E_{l_i-a}(\sigma) \leqq -E_{l_i}(\sigma) + \text{const}\, a\,, \tag{3.3}$$

for any $a \geqq 0$.

Proof. By Proposition 2.1, there is a constant independent of σ and l such that

$$l \leqq e(l) \equiv -E_l + \text{const}\, l \tag{3.4}$$

for l sufficiently large. Choose l_i to be a one-sided maximum of e, so that for positive $l \leqq l_i$,

$$e(l) \leqq e(l_i)\,.$$

Since $e(l) \to \infty$ as $l \to \infty$, there is a sequence of such l_i, $l_i \to \infty$. This completes the proof.

The Euclidean Fock space is

$$\mathscr{K} = \mathscr{K}(-l, l) = L^2(\mathscr{S}'(R^2), d\varphi_{\sigma^{-1}, \Lambda_l})\,,$$

where $\Lambda_l = R \times [-l, l]$. For each time t there is a natural imbedding $J_t : \mathscr{F} \to \mathscr{K}$ from $\mathscr{F}$ to the time t subspace of $\mathscr{K}$. We also introduce spacetime rectangles

$$Y_{T,l} = [-T, 0] \times [-l, l]$$
$$X_{T,l} = [-T, T] \times [-l, l]\,.$$

For each $a \in (-l, l)$, we also introduce the Fock spaces

$$\mathscr{F}(a, l) = L_2(\mathscr{S}'(R^1), d\mu_{a,l})$$

and

$$\mathscr{K}(-l+2a, l),\ a > 0$$
$$\mathscr{K}(-l, l-2a),\ a > 0\,.$$

The space $\mathscr{K}(-l+2a, l)$ is determined by placing zero Dirichlet data on the line $x = l$ and also on the symmetric line $x = -l + 2a$ obtained by reflecting the line $x = l$ about the line $x = a$, for $a > 0$. The definition of $\mathscr{K}(-l, l-2a)$, $a > 0$, is similar. The measure $d\mu_{a,l}$ is Gaussian, and its covariance is given explicitly in [5]. It is characterized by the following property: Let I_a be the natural injection from polynominals on $\mathscr{S}'(R^1)$ to generalized polynomials on $\mathscr{S}'(R^2)$ defined by the formula

$$I_a \phi(f) = \phi(f \otimes \delta_a)\,.$$

Here $f \in \mathscr{S}(R^1)$ and ϕ on the left above is a field (linear function) defined on $\mathscr{S}'(R^1)$, while ϕ on the right is a field on $\mathscr{S}'(R^2)$. We extend I_a to polynomials,

$$I_a(Q(\phi(f_1), \ldots, \phi(f_n)) = Q(I_a(\phi(f_1)), \ldots I_a(\phi(f_n)))\,,$$

so that I_a can be regarded as an identification map. Then $d\mu_{a,l}$ is characterized by the fact that I_a is isometric from $\mathscr{F}(a, l)$ into $\mathscr{K}(-l+2a, l)$ $(a > 0)$ or $\mathscr{K}(-l, l+2a)$ $(a < 0)$. We use the fact that conditional expectation onto the line $x = a$, $a > 0$, of a function supported in the region $a \leqq x \leqq l$, is independent of the location of Dirichlet boundary data in the region $x<a$. In particular these conditional expectations, defined in $\mathscr{K}(-l+2a, l)$ and in $\mathscr{K}(-l, l)$ coincide, see [5]. In order to give this statement a meaning, we identify the $x=a$ functions of the fields in these two spaces. However, this identification is not a unitary map between the Hilbert spaces.

Proposition 3.2. *Let $l=l_i$ be as in Lemma 3.1. There is a constant independent of ξ, a, σ, and i such that for $a \geqq O(1)\sigma$ and for l sufficiently large,*

$$\|J_{-2}^* e^{-V(Y_{2,l})} e^{+Q(\xi, Y_{2,a})} J_0\|_{\mathscr{F}}^2 e^{E_l} \leqq e^{\text{const}\, a(\ln\sigma)^2}.$$

Proof. Let

$$A = J_{-2}^* e^{-V(Y_{2,l})} e^{-Q(\xi, Y_{2,a})} J_0,$$

and let $\tilde{Q}(\xi, X_{2,a})$ be the sum of $Q(\xi, Y_{2,a})$ and its time reflexion about the x-axis. Then

$$AA^* = J_{-2}^* e^{-V(X_{2,l})} e^{-\tilde{Q}(\xi, X_{2,a})} J_2.$$

Since AA^* is self adjoint and positivity preserving, there is a unit vector $v \in L_\infty(\mathscr{S}'(R^1), d\mu_{0,l})$ such that

$$\begin{aligned} \|A\|^2 = \|AA^*\| &= \lim_{N\to\infty} \langle v, (AA^*)^N v\rangle_{\mathscr{F}}^{1/N} \\ &\leqq \lim_{N\to\infty} \|v\|_{L_\infty}^{2/N} \langle \Omega_{0,l}, (AA^*)^N \Omega_{0,l}\rangle_{\mathscr{F}}^{1/N} \\ &= \lim_{N\to\infty} \langle \Omega_{0,l}, (AA^*)^N \Omega_{0,l}\rangle_{\mathscr{F}}^{1/N}. \end{aligned}$$

The inner product can be written in Euclidean form, yielding

$$\begin{aligned} \|A\|^2 &\leqq \lim_{N\to\infty} \left(\int e^{-V(X_{2N,l})} e^{Q^\sim(\xi, X_{2N,a})} d\varphi_{\sigma^{-1}, \Lambda_l}\right)^{1/N} \\ &\leqq \lim_{N\to\infty} \left(\int e^{-V(X_{2N,l-a})} d\varphi_{\sigma^{-1}, \Lambda_{l-a}}\right)^{1/N} \\ &\quad \cdot \|I_{-a}^* e^{-V(X_{2N,a})} e^{\tilde{Q}(\xi, X_{2N,a})} I_a\|^{1/N} \\ &\leqq e^{-E_{l-a}} \lim_{N\to\infty} \|\ldots\|^{1/N} \\ &\leqq e^{O(a)-E_l} \lim_{N\to\infty} \|\ldots\|^{1/N}, \end{aligned} \tag{3.5}$$

where we use Lemma 3.1. Here the norm $\|\ldots\|$ is the operator norm from $\mathscr{F}(a, l)$ to $\mathscr{F}(-a, l)$. The norm is estimated by an integral. For vectors $v_\pm \in \mathscr{F}(\pm a, l)$, we have by the Schwarz inequality

$$\begin{aligned} |\langle I_{-a} v_-, e^{-V(X_{2N,a})} e^{\tilde{Q}(\xi, X_{2N,a})} I_a v_+\rangle| &\leqq \|(I_{-a} v_-)(I_a v_+)\|_{L_2(d\varphi_{\sigma^{-1}, \Lambda_l})} \\ &\quad \cdot \|e^{-V(X_{2N,a})} e^{\tilde{Q}(\xi, X_{2N,a})}\|_{L_2(d\varphi_{\sigma^{-1}, \Lambda_l})}. \end{aligned}$$

By Proposition 2.2, the second factor is bounded by $\exp[\text{const}(\ln\sigma)^2 Na]$. We bound the first factor as in [5]. For a $\geqq O(1)\sigma$ and l large, we have the hypercontractive bound

$$\int |(I_{-a} v_-)(I_a v_+)|^2 d\varphi_{\sigma^{-1}, \Lambda_l} \leqq \prod_{+,-} \|I_{\pm a} v_\pm\|_{L_2(d\varphi_{\sigma^{-1}, \Lambda_l})}^2$$

Thus the norm $\|\dots\|$ in (3.5) satisfies

$$\|\dots\|^{1/N} \leqq \exp[O(a)(\ln\sigma)^2]$$

and using (3.5), the proof is complete.

Theorem 3.3. *The limit $\Lambda \nearrow R^2$ in* (1.4) *exists for*

$$A = e^{zQ(\xi, X)} \quad \text{and} \quad A = \prod_{\Delta_j} \chi_\pm(\Delta_j).$$

Proof. We prove convergence for $\exp(zQ)$ in two steps. First in the ultraviolet cutoff case $Q = Q_\kappa$, the existence of the limit $\Lambda \nearrow R^2$ for $A = Q_\kappa^n$ exists by Griffiths' inequality. (Because of the use of full Dirichlet data, this argument is special to φ^4.) We expand $\exp(zQ_\kappa)$ in a power series in z. Using term by term convergence, and a φ^j bound to give a uniform bound on the sum, we obtain convergence for $\exp(zQ_\kappa)$. We then remove the momentum cutoff by considering the difference $\exp(zQ_\kappa) - \exp(zQ_{\kappa'})$ as a power series. By Proposition 2.2 (and 3.2) we have a uniform bound which is $o(1)$ as $\kappa, \kappa' \to \infty$. Here we use the bound on Q_4, cf. [6, Chapters 2 and 5].

To prove the convergence for $A = \Pi\chi_\pm(\Delta)$ note that since $\langle \prod_j \exp(i\xi_j \varphi(\Delta))\rangle_\Lambda$ converges so does $\langle \psi(\varphi(\Delta))\rangle$ for $\tilde{\psi} \in L_1$. Using the fact that $\chi_\pm \leqq 1$ we need only establish

$$\langle |(\chi_+ - \psi_r)(\varphi(\Delta))|\rangle_\Lambda \leqq \langle (\chi_+ - \psi_r)^2(\varphi(\Delta))\rangle_\Lambda^{\frac{1}{2}} \to 0$$

uniformly in Λ for a suitable sequence $\tilde{\psi}_r \in L_1$. To deal with the singularities of $\tilde{\chi}_\pm$ it suffices to show

$$\text{(a)} \quad \frac{d^n}{d\xi^n} \langle e^{i\xi\varphi(\Delta)}\rangle_\Lambda$$

$$\text{(b)} \quad \xi^n \langle e^{i\xi\varphi(\Delta)}\rangle_\Lambda$$

are bounded uniformly in ξ and Λ. (a) is an immediate consequence of the φ bound. (b) follows from integration by parts, e.g.

$$\begin{aligned}\langle \xi e^{i\varphi(\Delta)}\rangle_\Lambda &= -i \int_\Delta dx \left\langle \frac{\delta}{\delta\varphi(x)} e^{i\xi\varphi(\Delta)} \right\rangle_\Lambda \\ &= -i \int_\Delta dx \langle e^{i\xi\varphi(\Delta)} \int dy (-\Delta + \sigma^{-2})^{-1}(x, y) : P'(\varphi(y)) : \rangle_\Lambda\end{aligned}$$

and Proposition 3.2.

For the remainder of this paper, we take $\Lambda = R^2$. The bounds of Proposition 3.2 extend to this case, and prove Theorem 1.3 for (sufficiently long) rectangles (cf. [4]). The problem is to allow X to be an arbitrary union of lattice squares. Let T_a denote time translation, $(t, x) \to (t + a, x)$ and let $\mathscr{R}$ denote time inversion, $(t, x) \to (-t, x)$, and let T_a and $\mathscr{R}$ act on functions of the Euclidean field $\varphi(t, x)$.

Proposition 3.4. *Let A, B, and C be functions of φ localized in the time intervals* $(-\infty, -1)$, $(-1, 0)$, *and* $(0, \infty)$ *respectively, and suppose $B \geqq 0$. Then*

$$|\langle ABC\rangle| \leqq \langle C\mathscr{R}C^-\rangle^{\frac{1}{2}} \langle (T_1 A)\mathscr{R}(T_1 A)^-\rangle^{\frac{1}{2}} \lim_{N\to\infty} \langle \prod_{i=1}^N T_{2i}(B\mathscr{R}B^-)\rangle^{1/2N}$$

Proof. We use the positivity condition of Osterwalder-Schrader (positivity of the inner product in the physical Hilbert space). The result then follows from the Schwarz inequality and the density of L_∞ in L_2.

Proof of Theorem 1.3. Let $i=(i_0, i_1)\in Z^2$. Given ξ, let $\xi(i_0, N)$ and $\xi(i, N)$ be multiindices defined by

$$\xi(i_0, N)_j = \begin{cases} \xi_i & \text{if} \quad i_1 = j_1 \quad \text{and} \quad |i_0 - j_0| \leqq N \\ 0 & \text{otherwise} \end{cases}$$

$$\xi(i, N)_j = \begin{cases} \xi_i & \text{if} \quad |i_0 - j_0| \leqq N \quad \text{and} \quad |i_1 - j_1| \leqq N \\ 0 & \text{otherwise}. \end{cases}$$

Consider the case $Q_3=0$. By repeated use of Proposition 3.4 in the x_0 direction and again after Euclidean rotation in the x_1-direction we conclude that when X_N is an $N \times N$ rectangle

$$\langle e^{Q(\xi, X_N)}\rangle \leqq \prod_{i_0} \lim_{N\to\infty} \langle e^{Q(\xi(i_0, N))}\rangle^{1/N}$$

$$\leqq \prod_{i\in X} \lim_{N\to\infty} \langle e^{Q(\xi(i,N))}\rangle^{1/N^2}.$$

By Theorem 3.3 with $\Lambda = \Lambda_l = [-l, l] \times R$ we have

$$\langle e^{Q(\xi(i,N))}\rangle = \lim_{l\to\infty} \langle e^{Q(\xi(i,N))}\rangle_{\Lambda_l}.$$

To express the last term in Fock space let

$$A = J_{-2}^* e^{-V(Y_{2,l})} e^{-Q(\xi(i,N), Y_{2,N})} J_0.$$

Then

$$\langle e^{Q(\xi(i,N)}\rangle_{\Lambda_l} = \langle \Omega_l A^{N/2} \Omega_l \rangle_{\mathscr{F}}$$

$$\leqq \|A\|^{N/2}$$

$$\leqq e^{\text{const } N^2 (\log\sigma)^2}.$$

Here we have applied Proposition 3.2 with $a=N$. Hence combining these estimates we have

$$\langle e^{Q(\xi, X)}\rangle \leqq \prod_{i\in Z^2\cap X} e^{K_1(\log\sigma)^2} = e^{K_1(\log\sigma)^2|X|}.$$

In the case of Q_3 we apply the same argument with appropriate reflections, as in Proposition 3.2.

References

1. Dobrushyn, R., Minlos, R.: Construction of a one-dimensional quantum field via a continuous Markov field. Funct. Anal. Appl. **7**, 324—325 (1973)
2. Fröhlich, J.: Schwinger functions and their generating functionals. II. Adv. Math. (to appear)
3. Glimm, J., Jaffe, A.: The $\lambda(\varphi^4)_2$ quantum field theory without cutoffs. IV. J. Math. Phys. **13**, 1558—1584 (1972)
4. Glimm, J., Jaffe, A.: On the approach to the critical point. Ann. Inst. Henri Poincaré **22**, 13—26 (1975)
5. Glimm, J., Jaffe, A.: φ^j bounds in $P(\varphi)_2$ quantum field models. Proc. of the Colloq. on Math. Methods of Quantum Field Theory, Marseille, June 1975

6. Glimm, J., Jaffe, A.: Two and three body equations in quantum field models. Commun. math. Phys. **44**, 293—320 (1975)
7. Glimm, J., Jaffe, A., Spencer, T.: A cluster expansion for the φ_2^4 quantum field theory in the two phase region (in preparation)
8. Guerra, F., Rosen, L., Simon, B.: Nelson's symmetry and the infinite volume behavior of the vacuum in $P(\varphi)_2$. Commun. math. Phys. **27**, 10—22 (1972)
9. Guerra, F., Rosen, L., Simon, B.: The vacuum energy for $P(\varphi)_2$ infinite volume limit and coupling constant dependence. Commun. math. Phys. **29**, 233—247 (1973)
10. Guerra, F., Rosen, L., Simon, B.: The $P(\varphi)_2$ Euclidean quantum field theory as classical statistical mechanics. Ann. Math. **101**, 111—259 (1975)
11. Pirogov, S. A., Sinai, Ya. G.: Phase transitions of the first kind for small perturbations of the Ising model. Funct. Anal. Appl. **8**, 21—25 (1974) (Engl. trans.)
12. Simon, B., Griffiths, R.: The φ_2^4 field theory as a classical Ising model. Commun. math. Phys. **33**, 145—164 (1973)
13. Glimm, J., Jaffe, A., Spencer, T.: Existence of phase transitions for φ_2^4 quantum fields. Proc. of the Colloq. on Math. Methods of Quantum Field Theory, Marseille, June 1975

Communicated by K. Hepp

(Received June 29, 1975)

A Convergent Expansion about Mean Field Theory
I. The Expansion

JAMES GLIMM*

Rockefeller University, New York, New York 10021

ARTHUR JAFFE†

Harvard University, Cambridge, Massachusetts 02138

AND

THOMAS SPENCER*

Rockefeller University, New York, New York 10021

Received June 2, 1976

We give a convergent expansion for nearly Gaussian quantum field theory in the multiphase region. The expansion combines (1) an expansion in phase boundaries, (2) a cluster expansion, and (3) a perturbation expansion to isolate dominant behavior. We study in detail the ground state of the $\mathscr{P}(\phi)_2 = (\lambda\phi^4 - \phi^2 - \mu\phi)_2$ model, with $|\mu| \leqslant \lambda^2 \ll 1$. The ground state is close to the classical free field, obtained by replacing $\mathscr{P}(\phi)$ by the quadratic mean field polynomial $\mathscr{P}_c(\phi)$, tangent to $\mathscr{P}$ at a global minimum. Selecting one minimum gives a pure phase (ergodic ground state) satisfying the Wightman–Osterwalder–Schrader axioms with a positive mass. We also establish analyticity in λ for $\mu = 0$ in the sector $|\operatorname{Im} \lambda| < \epsilon \operatorname{Re} \lambda \ll 1$, for $\epsilon \ll 1$.

1. FORMAL IDEAS AND THE MEAN FIELD PICTURE

In [18], the authors established a nonuniqueness theorem for quantum fields: Consider an even polynomial $:\lambda\mathscr{P}(\phi) - \phi^2:$ with bare mass $m = 1$, Wick ordering mass 1, and space-time dimension $d = 2$. For λ sufficiently small, there is a phase transition. In other words, there are at least two vacuum states corresponding to distinct phases [18]. In this paper we develop a convergent expansion in two distinct pure phases. We establish the Wightman and the Osterwalder–Schrader

* Supported in part by the National Science Foundation under Grant MPS74-13252.
† Supported in part by the National Science Foundation under Grant MPS75-21212.

axioms with a mass gap in each of two pure phases. We show that the pure phase theory is approximately Gaussian (free), and we give bounds on the deviation of the vacuum state from a Gaussian.

The leading term in our expansion is given by the mean field approximation, in this case a direct sum of two Gaussian (free) theories with mean $\pm\phi_c = \xi_\pm$ and mass m_c. Let us start with the interaction polynomial

$$:\mathscr{P}_1(\phi): = :\lambda\phi^4 - \tfrac{1}{4}\phi^2 + (64\lambda)^{-1}:, \qquad \lambda \ll 1. \tag{1.1}$$

We note that by a scale transformation, a re-Wick ordering, and the addition of a constant, $:\mathscr{P}_1(\phi):$ is reparameterized as the polynomial

$$:\mathscr{P}_2(\phi): = :\lambda\phi^4 + \tfrac{1}{2}\phi^2:, \qquad \lambda \gg 1, \tag{1.2}$$

again with Wick mass 1 (see, e.g., [17]). Thus the strong bare coupling ϕ^4 model (1.2) is equivalent to weak coupling ϕ^4 with a negative mass term. The advantage of the parameterization (1.1) over (1.2) is the small coefficient of ϕ^4, relative to the mass term. Thus we can expect higher-order effects will be small for (1.1), but large if we study (1.2). The λ's in (1.1) and (1.2) do not coincide, and $\lambda_{(1.2)} \to \infty$ as $\lambda_{(1.1)} \to 0$. We study $\mathscr{P}_1$ of (1.1).

The classical values $\phi_c = \pm(8\lambda)^{-1/2}$ minimize $\mathscr{P}_1(\xi)$, and the classical mass m_c is defined by

$$m_c^2 = \mathscr{P}_1''(\phi_c) = 1. \tag{1.3}$$

Expanding $:\mathscr{P}_1(\phi):$ about $\phi_c = \pm(8\lambda)^{-1/2}$ gives

$$:\mathscr{P}_1(\phi): = \lambda:(\phi - \phi_c)^4: \pm(2\lambda)^{1/2} :\phi - \phi_c)^3: + \tfrac{1}{2}:(\phi - \phi_c)^2:.$$

Between the minima $:\mathscr{P}_1(\phi): = 0$ at $\phi = \phi_c = \pm(8\lambda)^{-1/2}$, the polynomial $:\mathscr{P}_1(\phi):$ forms a high potential barrier, since

$$:\mathscr{P}_1(0): = (64\lambda)^{-1} \gg 0.$$

As $\lambda \to \infty$, the width of the barrier increases because $|\phi_c(\lambda)| \to \infty$.

We show here that each classical minimum $\phi_c = \pm(8\lambda)^{-1/2}$ gives rise to a vacuum state. We select one minimum, $\phi_c = (8\lambda)^{-1/2}$, by imposing a positive external field outside a box Λ, and then we take the infinite volume limit. The transition probability between the two minima is small because of the large barrier at $\mathscr{P}_1(0)$ and because the kinetic energy makes large local fluctuations unlikely. For λ sufficiently small, our external field boundary condition chooses the vacuum state localized with $\langle\phi\rangle \sim (8\lambda)^{-1/2}$. The details of our proof are patterned in part after Peierls' proof of phase transition in the Ising model. By symmetry, we also have a solution localized with $\langle\phi\rangle \sim -(8\lambda)^{-1/2}$, and the two solutions are exchanged by the automorphism $\phi \to -\phi$.

Our expansion combines a Peierls-type method with the cluster expansion, developed previously to analyze regions of weak bare coupling, i.e., $\lambda \ll 1$ in (1.2), or large external field [15, 16, 27]. In both previous cases, the cluster expansion was used far from the region of phase transitions. In this single phase region, the cluster expansion has been an effective tool for the detailed investigation of superrenormalizable quantum fields [1, 2, 9, 10, 22, 23] including a description of the mass spectrum and scattering states (up to some threshold) [15, 28, 29], analyticity in the coupling constants [16], the Borel summability of the perturbation series [7], and nontriviality of the S-matrix [4, 6, 25]. Using the expansion presented here, similar questions can be studied in the multiphase weak coupling region ($\lambda \ll 1$ in (1.1)) or the (equivalent) strong bare coupling region ($\lambda \gg 1$ in (1.2)).

In the multiphase region we expect numerous bound states and resonances. As a special feature of two dimensions, soliton states accompany phase transitions [3, 8, 11, 19]. Correspondingly the soliton–antisoliton state should introduce a continuum threshold into the vacuum sector and possible bound state spectrum. Introduction of a small external field breaks the phase transition and should make the soliton sector unstable. It should transform these continuum (two soliton) states into bound states and/or resonances.

In particular, we note that in the two-dimensional Ising model with $T < T_c$, the truncated two-point function has a decay rate $r^{-2}e^{-mr}$ [24, 299], indicative of a continuum threshold. An isolated particle would yield an Ornstein-Zernike $r^{-1/2}\exp(-mr)$ decay in two dimensions. The continuum behavior arises from the fact that the solitons do not form a bound state. (The associated Fermi field is a free field.) Thus the Ising model particle spectrum is missing its "elementary particle," i.e., the isolated hyperboloid-like curve at the bottom of the nonzero energy–momentum spectrum for $T < T_c$.

The single soliton appears in a disjoint state space. Its presence can be seen directly from the fermion solution of I_2 [21] by replacing the periodic boundary conditions of [21] with antiperiodic boundary conditions. Since it is generally believed that the ϕ^4-soliton–antisoliton pairs do form bound states, one could still expect an elementary (boson) particle and an upper mass gap for the two-phase ϕ_2^4 theory, as well as a conventional decay rate, for its two-point function. Returning to the Ising model: As the external field $\mu \searrow 0+$, one could expect only bound state spectrum between m and $2m$ (for $T < T_c$), filling in densely to become a continuum at $\mu = 0$.

Let us now consider the mean field picture for a general interaction

$$\sum_{i=3}^{2n} \lambda_i(\phi - \phi_c)^i + \tfrac{1}{2}m_c^2(\phi - \phi_c)^2 - \mu(\phi - \phi_c),$$

with Wick ordering mass m_c. In the region of parameter space $(|\mu| + \Sigma|\lambda_i|)m_c^{-2} \ll 1$, we expect that the mean field approximation is valid both qualitatively and quantitatively. Qualitatively, the mean field picture should determine the topology of the phase transition hypersurface. Branches (multiple points) of the hypersurface should occur on surfaces of lower dimension, where more than two phases coexist in equilibrium. In general, branches with coexistence of up to n phases should occur for $\mathscr{P}(\phi)_2$ theories with deg $\mathscr{P} = 2n$. These singularities of the phase transition hypersurface should occur for $\mathscr{P}(\phi)_2$ inside (as well as outside) the domain of convergence of the cluster expansion. Aside from these branches, given topologically by the mean field approximation, the phase transition should be regular (real analytic) within the domain of convergence of the cluster expansion. The boundaries (edges) of the phase transition hypersurface have a special significance; they are critical points. They should lie outside the domain of convergence of the cluster expansion. Quantitatively, the validity of the mean field picture means that a perturbation theory, in powers of λ_i/m_c^2, should be asymptotic as $\lambda_i/m_c^2 \to 0$.

As a special case of the above ideas, we mention two problems: (1) To demonstrate the existence of a triple point for the interaction $\phi^6 + \lambda\phi^4 - \sigma\phi^2$ with $\sigma \gg 1$ and $\lambda = \lambda(\sigma)$. (2) To prove the existence of phase transitions without symmetry breaking, for the interaction $\phi^6 + \epsilon\phi^3 - \sigma\phi^2 - \mu\phi$ for $\sigma \gg 1$, $\epsilon \ll 1$ and $\mu = \mu(\epsilon, \sigma) \neq 0$.

2. The Main Results

Our expansion applies to $\lambda\mathscr{P}_{\text{even}}(\phi) - \phi^2 - \mu\phi$ models with $\lambda > 0$ sufficiently small. For simplicity of notation, we restrict attention to the ϕ^4 model in a weak external field. Define the real interaction polynomial

$$\mathscr{P}(\xi) = \lambda\xi^4 - \tfrac{1}{4}\xi^2 - \mu\xi - E_c, \tag{2.1a}$$

where

$$0 < \lambda \ll 1, \qquad |\mu| \leqslant \lambda^2, \qquad \text{and} \qquad \inf_\xi \mathscr{P}(\xi) = 0. \tag{2.1b}$$

We define the interaction in a region Λ by

$$:\mathscr{P}_\Lambda: = \int_\Lambda :\mathscr{P}(\phi(x)): dx. \tag{2.2}$$

Here : : denotes Wick ordering mass equal to one. Let $d(\Phi - \mu_b)$ be the Gaussian measure with mean μ_b and covariance

$$C = (-\Delta + 1)^{-1}. \tag{2.3}$$

We define the finite volume measure $d\phi_\Lambda$ by

$$d\phi_\Lambda \equiv \exp(-:\mathscr{P}_\Lambda: + \tfrac{1}{2}:(\phi - \mu_b)^2_\Lambda:)\, d(\Phi - \mu_b). \tag{2.4}$$

Note that the second term in the exponent of (2.4) cancels the mean and the mass of the measure $d(\Phi - \mu_b)$ in the region Λ. However, a positive external field remains in the region exterior to Λ, which defines our + boundary conditions (for $\mu_b > 0$) as $\Lambda \nearrow R^2$.

For $\mu = 0$, the polynomial (2.1) has two absolute minima $\xi_\pm$ satisfying

$$\phi_c = \xi_+ = -\xi_- = (8\lambda)^{-1/2}. \tag{2.5}$$

For $\mu > 0$, $\mathscr{P}$ has a unique absolute minimum at

$$\phi_c = \xi_+ = \xi_+(\lambda, \mu) = (8\lambda)^{-1/2} + \mu + 0(\mu^2). \tag{2.6a}$$

and a relative minimum at

$$\xi_- = \xi_-(\lambda, \mu) = -(8\lambda)^{-1/2} + \mu + 0(\mu^2), \tag{2.6b}$$

uniformly in λ. At the absolute minimum ξ_+, the classical mass m_c is given by

$$m_c = 1 + (72\lambda)^{1/2}\mu + 0(\mu^2). \tag{2.7}$$

Since $m_c \sim 1$, we have used Wick ordering and bare mass 1 in (2.2)–(2.4). We remark that the classical vacuum energy per unit volume is

$$E_c = -\{(64\lambda)^{-1} + \mu(8\lambda)^{-1/2}\} + 0(\mu^2). \tag{2.8}$$

We now take Λ to be a square centered near the origin, such that $\text{dist}(0, \partial\Lambda) \geqslant \frac{1}{4} |\Lambda|^{1/2}$. We state the main results proved by our new expansion.

THEOREM 2.1. *Let $d\phi_\Lambda$ be as above with $0 \leqslant \mu$ and $\mu_b = \xi_+$. For λ sufficiently small, the infinite volume limit of Schwinger functions*

$$S_n(x_1, ..., x_n) = \left\langle \prod_{i=1}^n \phi(x_i) \right\rangle = \lim_{\Lambda \nearrow R^2} \frac{\int (\prod_{i=1}^n \phi(x_i))\, d\phi_\Lambda}{\int d\phi_\Lambda} \tag{2.9}$$

exist in $\mathscr{S}'(R^{2n})$ and satisfy the Osterwalder–Schrader axioms with exponental clustering. The corresponding quantum field model obeys all the Wightman axioms. It has a unique vacuum, a positive field expectation $\langle\phi\rangle$ and a positive mass $m = m(\lambda, \mu) > 0$.

For $\mu \leqslant 0$ and $\mu_b = \xi_-$, the same results hold except that the field expectation $\langle\phi\rangle$ is negative. For $\mu = 0$, the two choices $\mu_b = \xi_\pm$ give two solutions.

THEOREM 2.2. *Let $0 \leqslant \mu$, $\mu_b = \xi_+$. Then for λ sufficiently small,*

$$\langle \phi(x) \rangle = \xi_+ + 0(\lambda^{3/2}), \tag{2.10a}$$

$$\langle \phi(x)\phi(y) \rangle - \langle \phi(x) \rangle \langle \phi(y) \rangle = C(x-y) + \lambda M_2(x, y), \tag{2.10b}$$

$$\langle \phi(x_1),..., \phi(x_n) \rangle_T = \lambda^{1/2} M_n(x_1 ,..., x_n), \qquad n \geqslant 3, \tag{2.10c}$$

$$\left\langle \left(\int (\phi(x) - \xi_+) f(x)\, dx \right)^n \right\rangle$$

$$= (n-1)!!\, C(f, f)^{n/2} + O(\lambda^{1/2})(\|f\|_{L_1} + C(f, f)^{1/2})^n. \tag{2.10d}$$

Here $C = (-\Delta + 1)^{-1}$, and the functions $M_n(x_1 ,..., x_n)$ are bounded and continuous in λ, x_1, $x_2 ,..., x_n$, including $\lambda = 0$. Also, $\langle \cdot \rangle_T$ denotes the truncated (connected part of the) expectation value. By Theorem 2.1, the M_n are exponentially decreasing functions of the difference variables $x_j - x_{j+1}$. The $0(\lambda^{1/2})$ in (2.10d) depends on n.

For $\mu \leqslant 0$ and $\mu_b = \xi_-$, the above results hold, if we replace ξ_+ by ξ_- in (2.10). The generalized Schwinger functions satisfy corresponding local regularity estimates:

THEOREM 2.3. *Let $0 \leqslant \mu$, $\mu_b = \xi_+$, and λ be sufficiently small. Then*

$$\left\langle \prod_{i=1}^{n} :(\phi(x_i) - \xi_+)^{m_i}: \right\rangle_T = \left\langle \prod_{i=1}^{n} :\psi(x_i)^{m_i}: \right\rangle_{0,T} + \lambda^{1/2} N_n(x_1 ,..., x_n), \tag{2.11}$$

where $\langle \cdot \rangle_0$ denotes a Gaussian expectation of ψ with mean zero and covariance $C = (-\Delta + 1)^{-1}$. Also $N_n(x_1 ,..., x_n)$ is bounded and continuous in λ, $x_1 ,..., x_n$, including $\lambda = 0$, and the N_n are exponentially decreasing functions in the difference variables $x_j - x_{j+1}$.

THEOREM 2.4. *Under the hypothesis of Theorem 2.1, there is a measure $d\phi$ on $\mathscr{S}'(R^2)$ such that*

$$\langle \phi(x_1) \cdots \phi(x_n) \rangle = \int \phi(x_1) \cdots \phi(x_n)\, d\phi, \tag{2.12}$$

and $d\phi$ is ergodic under space or time translations.

Remark 1. The estimates (2.10)–(2.11) give a precise statement bounding the deviation (as $\lambda \to 0$) of $d\phi$ from a Gaussian measure with mean ξ_+ of ξ_- and covariance $C = (-\Delta + 1)^{-1}$.

Remark 2. The bound (2.10b) shows that $d\langle \phi \rangle / d\mu = 1 + 0(\lambda)$. Integration yields

$$\langle \phi \rangle_\mu = \langle \phi \rangle_{\mu=0} + \mu + 0(\lambda\mu), \tag{2.13}$$

where $\langle\phi\rangle_{\mu=0}$ is positive or negative according to whether μ, μ_b are positive or negative.

THEOREM 2.5. *The Schwinger functions* (2.9) *for* $\mu = 0$ *are analytic functions of* λ *in the interior of a sector:*

$$|\operatorname{Im}\lambda| \leqslant \epsilon \operatorname{Re}\lambda \ll 1, \tag{2.14}$$

for ϵ *sufficiently small.*

3. THE EXPANSION IN PHASE BOUNDARIES

For the remainder of Part I we give the expansion which yields our results. In Part II we prove the convergence of our expansion and the theorems stated here. The expansion has three main ingredients: the expansion in phase boundaries given here, the cluster expansion of Section 5, and the asymptotic expansion of Section 6. The expansion in phase boundaries is motivated by low-temperature techniques in classical statistical mechanics, first introduced by Peierls in 1936 [26]. To obtain a mapping of the field variable $\phi(x)$ onto an Ising spin variable σ_i, we cover space-time R^2 by a lattice of unit squares Δ_i, centered at the lattice sites $i \in Z^2$. The boundaries of the Δ_i are (unit) lattice lines. Let $\sigma(\cdot)$ be a function on R^2 which is constant on lattice squares and assumes only the values $\pm$. In other words,

$$\sigma_i \equiv \sigma(\Delta_i) \equiv \sigma(x), \qquad i \in Z^2,\ x \in \Delta_i. \tag{3.1}$$

We impose "plus" boundary conditions on σ_i, namely,

$$\sigma_i = + \qquad \text{for } \Delta_i \not\subset \Lambda. \tag{3.2}$$

A configuration

$$\Sigma = \{\sigma_i\} \tag{3.3}$$

of the spin variable belongs to the $2^{|\Lambda|}$-dimensional state space. Each configuration Σ is in 1–1 correspondence with the set of lattice lines on which $\sigma(x)$ is discontinuous. We call this set of lattice lines the Peierls contour, or phase boundary, and also denote it by Σ. The + boundary conditions ensure that the phase boundary Σ is a union of closed curves, and we let $|\Sigma|$ denote the total length of these curves.

We now consider the averaged field in Δ (the low momentum field in Δ) which we define by

$$\phi(\Delta) \equiv \int_\Delta \phi(x)\, dx \equiv \bar{\phi}(x), \qquad x \in \Delta. \tag{3.4a}$$

We define the fluctuation field by

$$\delta\phi(x) = \phi(x) - \bar{\phi}(x), \tag{3.4b}$$

which we use in later sections to bound the deviations of ϕ from its mean value $\bar{\phi}$.

The basic idea behind the expansion in phase boundaries is to choose $\sigma_i = \operatorname{sgn} \phi(\Delta_i)$ and to write $d\phi_\Lambda$ as a sum of $2^{|\Lambda|}$ measures $d\phi_{\Lambda,\Sigma}$. Here $d\phi_{\Lambda,\Sigma}$ is the restriction of $d\phi_\Lambda$ to the subset of paths with a fixed $\Sigma = \{\sigma_i\}$.

For technical reasons, we smooth out this definition of $d\phi_{\Lambda,\Sigma}$ as follows. Let $\chi_\pm$ be a smooth partition of unity on R^1 defined by

$$\begin{gathered} \chi_+(\xi) = \pi^{-1/2} \int_0^\infty e^{-(\xi-z)^2}\, dz = \chi_-(-\xi), \\ \chi_+(\xi) + \chi_-(\xi) = 1. \end{gathered} \tag{3.5}$$

We set

$$\chi_\Sigma = \prod_{\Delta_i \subset \Lambda} \chi_{\sigma_i}(\phi(\Delta_i)) \tag{3.6}$$

and note that

$$\sum_{\{\sigma_i\}} \chi_\Sigma = 1.$$

The expansion in phase boundaries is merely the identity

$$d\phi_\Lambda = \sum_{\{\sigma_i\}} d\phi_{\Lambda,\Sigma} \tag{3.7}$$

where $d\phi_{\Lambda,\Sigma} = \chi_\Sigma\, d\phi_\Lambda$.

We remark that for a more general $\mathscr{P}(\phi)$ model, $\mathscr{P}$ may have up to $\frac{1}{2} \deg \mathscr{P}$ minima, and there may be up to $\frac{1}{2} \deg \mathscr{P}$ equilibrium phases. We require a spin with $\frac{1}{2} \deg \mathscr{P} = 2s + 1$ components, i.e., at each lattice site we have the possibility

$$\sigma = -s, -s+1, \ldots, (s-1), s. \tag{3.8}$$

Thus, we have an analogy with a "spin s" Ising model, the case $s = \frac{1}{2}$ considered above.

4. Translation of ϕ

For each spin configuration Σ, we translate $\phi(x)$ by a function $g(x) = g(x, \Sigma)$. We choose g so that the translated field

$$\psi(x) = \phi(x) - g(x) \tag{4.1}$$

has a small mean in the measure $d\phi_{\Lambda,\Sigma}$. Ignoring the shift in kinetic energy, we would translate in each square Δ_i to one of the relative mimima, $\xi_\pm$ of $\mathscr{P}(\xi)$. Thus we define an external field

$$h(x) = \sigma(x)\,\xi_+ .$$

The classical magnetism (the response to the external field $h(x)$) is given by

$$g_c(x) = (\eta(-\Delta + \eta)^{-1}h)(x) = \xi_+ + \eta(-\Delta - \eta)^{-1}(h - \xi_+). \tag{4.3}$$

For convenience, we choose $\eta \neq 1$ to be a small constant defined later, independent of λ. Translation by the function $g_c(x)$ causes some technical inconvenience, because $g_c(x)$ has an exponential tail. We modify (4.3) by choosing a C_0^∞ function $\zeta(x)$ satisfying $0 \leqslant \zeta(x) \leqslant 1$ with

$$\begin{aligned} \zeta(x) &= 0, \qquad |x| > \tfrac{1}{2} \\ &= 1, \qquad |x| \leqslant \tfrac{1}{4}. \end{aligned}$$

For a given length scale L, define an approximate classical magnetism by

$$g(x) = \eta_\zeta \int (-\Delta + \eta)^{-1}\,(x - y)\,\zeta((x - y)/L)\,h(y)\,dy, \tag{4.4}$$

where

$$\eta_\zeta^{-1} = \int (-\Delta + \eta)^{-1}\,(y)\,\zeta(y/L)\,dy.$$

LEMMA 4.1. *Let* dist$(x, y) \geqslant \frac{1}{2}L$. *Then* $g(x)$ *and* $(-\Delta + \eta)(g - g_c)(x)$ *are independent of* $h(y)$. *In particular, if* dist$(x,\Sigma) \geqslant \frac{1}{2}L$,

$$g(x) = h(x) \qquad \text{and} \qquad (-\Delta + \eta)(g - g_c)(x) = 0. \tag{4.5}$$

Proof. The locality of $g(x)$ and $(-\Delta + \eta)g(x)$ follow from the localization of the kernel $(-\Delta + \eta)^{-1}\zeta$ in (4.4). Furthermore, $(-\Delta + \eta)\,g_c(x) = \eta h(x)$ is local, and (4.5) then follows.

Since $g(x) = h(x)$ for dist$(x,\Sigma) \geqslant \frac{1}{2}L$, the length scale L provides an upper bound on the width of the phase boundaries. We use L as a characteristic length for the effects of phase boundaries.

Let $d\psi_a$ denote the Gaussian measure with mean zero and covariance $(-\Delta + a)^{-1}$.

LEMMA 4.2. *With* ψ *defined by* (4.1), (4.4),

$$d\phi_{\Lambda,\Sigma} = \chi_\Sigma e^{-F} \exp\left(-{:}\mathscr{P}_\Lambda(\phi){:} + \frac{\eta}{2}\int_\Lambda {:}(\phi - h)^2{:}\,dx + \left(\frac{1-\eta}{2}\right)\int_\Lambda {:}\psi^2{:}\,dx\right) d\psi_1, \tag{4.6}$$

where

$$F = F(\Sigma) = F_1 + F_2 + F_3 + F_4,$$

and

$$F_1 = \tfrac{1}{2}\eta \int (h - g)^2\, dx + \tfrac{1}{2} \int (\nabla g)^2\, dx \tag{4.7a}$$

$$F_2 = \left(\frac{1-\eta}{2}\right) \int_{\sim\Lambda} (g - \xi_+)^2\, dx \tag{4.7b}$$

$$F_3 = \int \psi(x)(-\Delta + \eta)(g - g_c)\, dx \tag{4.7c}$$

$$F_4 = (1 - \eta) \int_{\sim\Lambda} \psi(x)(g(x) - \xi_+)\, dx \tag{4.7d}$$

To simplify the notation we define

$$V(W, \Sigma) = :\mathscr{P}_{\Lambda\cap W}(\phi): - \frac{1}{2}\eta \int_{\Lambda\cap W} :(\phi - h)^2:\, dx - \left(\frac{1-\eta}{2}\right) \int_{\Lambda\cap W} :\psi^2:\, dx \tag{4.8a}$$

and

$$Q(W,\Sigma) = V(W,\Sigma) + F(W,\Sigma), \tag{4.8b}$$

where $F(W,\Sigma)$ *is obtained by restricting the range of integration in* (4.7a), (4.7c) *to* W *and in* (4.7b), (4.7d) *to* $W \sim \Lambda$. *Thus in* (4.8), *and throughout the rest of the paper, the* Λ *dependence is not explicit in the notation. We also define* $Q(\Sigma) = Q(\mathbb{R}^2,\Sigma)$. *By Lemma* 4.1, *if*

$$\operatorname{dist}(\partial W,\Sigma) \geqslant \tfrac{1}{2}L, \tag{4.8c}$$

then

$$Q(W,\Sigma) = Q(W,\Sigma \cap W). \tag{4.8d}$$

Since we are only concerned with $Q(W,\Sigma)$ *in case* (4.8c), *we always have* (4.8d), *which expresses the locality of* Q.

In this notation, (4.6) *becomes*

$$d\phi_{\Lambda,\Sigma} = \chi_\Sigma e^{-Q(\Sigma)} d\psi_1. \tag{4.9}$$

Proof. The basic identity for translation of the Gaussian measure $d\psi_1$ is

$$d(\psi + f)_1 = \exp(-\tfrac{1}{2}\langle f, (-\Delta + 1)f\rangle - \langle \psi, (-\Delta + 1)f\rangle)\, d\psi_1. \tag{4.10}$$

Hence

$$\begin{aligned} d\psi_1 &= d(\phi - g)_1 = d(\phi - \xi_+ + \xi_+ - g)_1 \\ &= \exp(-\tfrac{1}{2}\langle(\xi_+ - g), (-\Delta + 1)(\xi_+ - g)\rangle - \langle(\phi - \xi_+), (-\Delta + 1)(\xi_+ - g)\rangle) d(\phi - \xi_+)_1. \end{aligned}$$

Comparison with (2.4) and (4.6) leads to the identity

$$F = \frac{\eta}{2}\int_{\Lambda} :(\phi - h)^2: dx + \left(\frac{1-\eta}{2}\right)\int_{\Lambda} :(\phi - g)^2: dx - \frac{1}{2}\int_{\Lambda} :(\phi - \xi_+)^2: dx$$

$$- \frac{1}{2}\langle(\xi_+ - g), (-\Delta + 1)(\xi_+ - g)\rangle - \langle \phi - \xi_+, (-\Delta + 1)(\xi_+ - g)\rangle.$$

The term in F linear in ψ is

$$\int_{\Lambda} \psi[-\eta h - (1-\eta) g + \xi_+]\, dx + \int \psi[-\xi_+ + (-\Delta + 1) g]\, dx,$$

which by (4.3) and the identity $h \equiv \xi_+$ on $\sim\Lambda$ equals

$$(1-\eta)\int_{\sim\Lambda} \psi(g - \xi_+)\, dx + \int \psi(-\Delta + \eta)(g - g_c)\, dx.$$

The term in F independent of ψ is

$$\int_{\Lambda} \left\langle \left(\frac{1}{2}\eta h^2 + \frac{1-\eta}{2}\right) g^2 - \frac{1}{2}\xi_+{}^2 \right\rangle dx$$

$$- \frac{1}{2}\langle(\xi_+ - g), (-\Delta + 1)(\xi_+ - g)\rangle + \langle \xi_+, (-\Delta + 1)(\xi_+ - g)\rangle$$

$$+ (1-\eta)\int_{\sim\Lambda} g(g - \xi_+)\, dx + \int g(-\Delta + \eta)(g - g_0)\, dx$$

$$= \left(\frac{1-\eta}{2}\right)\int_{\sim\Lambda} (\xi_+ - g)^2\, dx + \frac{1}{2}\eta\langle h - g, h - g\rangle + \frac{1}{2}\langle \nabla g, \nabla g\rangle.$$

5. The Cluster Expansion

The cluster expansion allows us to take $\Lambda \nearrow R^2$. In order to establish convergence of the cluster expansion, we introduce a localization length scale $l \gg m_0^{-1} \approx m^{-1} \approx 1$. Since the length scale L of the previous section bounds the width of phase boundaries, we also choose $l \ll L$ to display the structure of the pure phase. Thus we choose

$$1 \ll l \ll L. \tag{5.1}$$

We will apply the expansion on l-lattice segments b which bound l-lattice squares joining nearest neighbor points of lZ^2. Let $\mathscr{B}$ denote the set of all l lattice segments (bonds) b, and let Γ denote a subset of $\mathscr{B}$, with $|\Gamma|$ elements. Let

$$C_\Gamma = (-\Delta_\Gamma + 1)^{-1}, \tag{5.2}$$

where Δ_Γ is the Laplace operator with zero Dirichlet boundary conditions on Γ.

For each spin configuration Σ, we define a subset $\mathscr{B}(\Sigma)$ of $\mathscr{B}$ on which we perform the cluster expansion [16]. We interpolate between free and Dirichlet boundary conditions by introducing a parameter $s_b \in [0, 1]$ for each $b \in \mathscr{B}(\Sigma)$. Let $s = \{s_b\}$ and

$$C(s) = \sum_{\Gamma \subset \mathscr{B}(\Sigma)} \left(\prod_{b \in \Gamma^c} s_b\right)\left(\prod_{b \in \Gamma} (1 - s_b)\right) C_\Gamma \tag{5.3}$$

where

$$\Gamma^c = \mathscr{B}(\Sigma) \backslash \Gamma.$$

Also let $d\psi(s)$ denote the Gaussian measure with mean zero and covariance $C(s)$. Then $s_b = 0$ corresponds to zero Dirichlet data on the l-lattice segment b, while $s = 1$ denotes the absence of Dirichlet data on b. If $s_b = 0$ on Γ, we say that Γ is a Dirichlet contour. Let

$$s_\Gamma = \{s_b \colon b \in \Gamma\},$$

$$ds_{\Gamma^c} = \prod_{b \in \Gamma^c} ds_b \Big|_{s_\Gamma = 0},$$

$$\partial_s^{\,\Gamma} = \prod_{b \in \Gamma} (\partial / \partial s_b).$$

The idea of the cluster expansion is to apply the fundamental theorem of calculus to each s_b,

$$F(s_b = 1) = F(s_b = 0) + \int_0^1 ds_b\, \partial_{s_b} F(s),$$

for $b \in \mathscr{B}(\Sigma)$. Thus we obtain the fundamental expansion formula

$$d\phi_{\Lambda,\Sigma} = \sum_{\Gamma \subset \mathscr{B}(\Sigma)} d\phi_{\Lambda,\Sigma,\Gamma} \tag{5.5}$$

where

$$d\phi_{\Lambda,\Sigma,\Gamma} = \int ds_{\Gamma^c}\, \partial_s^{\Gamma^c} \int \chi_\Sigma e^{-Q(R^2,\Sigma)}\, d\psi(s). \tag{5.6}$$

Here χ and Q are defined by (3.6) and (4.8). Note that $s_b = 0$ in (5.6), for $b \in \Gamma$, by the definition of ds_{Γ_c}. Recall that the Λ dependence of $Q(R^2, \Sigma)$ is suppressed. See (4.8).

It remains to define $\mathscr{B}(\Sigma)$, the set of l-lattice lines for the cluster expansion. We choose $\mathscr{B}(\Sigma)$ to be those islands of $+$ or $-$ spins, separated from a phase boundary by at least the width L. Thus on $\mathscr{B}(\Sigma)$ we expect the measure $d\phi_{\Lambda,\Sigma}$ to be nearly Gaussian, with mean $\sim\xi_\sigma$. More precisely, let

$$X = X(\Sigma) = \{\Delta_i : \operatorname{dist}(i,\Sigma) \geqslant L\}, \tag{5.7}$$

and

$$\mathscr{B}(\Sigma) = \mathscr{B} \cap X. \tag{5.8}$$

We apply the expansion (5.5)–(5.6) to a partition function

$$Z(\Lambda) = \int d\phi_\Lambda = \sum_{\Sigma} \sum_{\Gamma \subset \mathscr{B}(\Sigma)} \int d\phi_{\Lambda,\Sigma,\Gamma} \tag{5.9}$$

or to an expectation

$$\begin{aligned} \langle A \rangle &= Z(\Lambda)^{-1} \int A(\phi)\, d\phi_\Lambda, \\ &= Z(\Lambda)^{-1} \sum_{\Sigma} \sum_{\Gamma \subset \mathscr{B}(\Sigma)} \int A\, d\phi_{\Lambda,\Sigma,\Gamma}. \end{aligned} \tag{5.10}$$

In the remainder of the paper we study (5.9)–(5.10).

It is convenient to generalize (5.7)–(5.8) and to define

$$X_r^{\pm} = \{\Delta_i : \sigma_i = \pm, \operatorname{dist}(i,\Sigma) \geqslant rL\}, \tag{5.11}$$

and

$$X_r = X_r^+ \cup X_r^-, \qquad X_r^0 = \mathscr{R}^2 \backslash X_r. \tag{5.12}$$

In general we omit the subscript for $r = 1$. Also let

$$\mathscr{B}(\Sigma, Y) = \mathscr{B}(\Sigma) \cap Y = \mathscr{B} \cap X \cap Y. \tag{5.13}$$

6. The Asymptotic Expansion

We give the asymptotic expansion for $\langle A \rangle_\Lambda = \int A d\phi_\Lambda$ as a function of λ. We use the asymptotic expansion to isolate the dominant contribution as $\lambda \to 0$. The asymptotic expansion is performed before the expansion in phase boundaries and before the cluster expansion.

Let

$$A(\phi - \xi_+) \equiv \prod_{i=1}^{n} \int :(\phi(x) - \xi_+)^{m_i}: f_i(x)\, dx, \tag{6.1}$$

where : : denotes $(-\Delta + 1)^{-1}$ Wick ordering.

In order to state precisely the f dependence of the estimates, we introduce a norm $|||\, A \,|||$ on the functions of the form (6.1). In case suppt. $f_i \subset \Delta_i$, where Δ_i is a unit lattice square, let

$$N(\Delta, A) = \left\{\sum m_i : \Delta_i = \Delta\right\},$$

$$N(A) = \sum_{\Delta} N(\Delta, A) = \sum_i m_i .$$

For some constant K we define

$$|||\, A \,||| = e^{KlN(A)} \left(\prod_{\Delta} N(\Delta, A)!\right)\left(\prod_i \|f_i\|_{L_2(\Delta_i)}\right). \tag{6.2}$$

For general f_i with no restrictions on the support, $|||\, A \,|||$ is defined by linearity, as a sum over $\Delta_1, \ldots, \Delta_n$ of localized A's.

THEOREM 6.1. *Let* $0 \leqslant \mu$, $\mu_b = \xi_+$, $r \in Z_+$, *and* A *as above. Then there are constants* $\alpha_j(A, \Lambda, \lambda, \mu)$ *and a finite sum of functions* A^k *of the form* (6.1) *such that*

$$\int A \, d\phi_\Lambda = \sum_{j=0}^{r} \alpha_j \lambda^{j/2} \int d\phi_\Lambda + \lambda^{(r+1)/2} \int \sum_k A_k \, d\phi_\Lambda .$$

Here α_j and A_k are continuous in $\lambda \in [0, \lambda_0]$, uniformly in Λ, and convergent as $\Lambda \nearrow R^2$, for λ_0 sufficiently small. In (6.1) we may also allow $f_i = \delta(x - a_i)$ in case the a_i are all unequal.

We show later that for functions A of the form (6.1), $\lim \int A d\phi_\Lambda / \int d\phi_\Lambda$ exists as $\Lambda \nearrow R^2$. Thus the theorem shows that

$$\int A \, d\phi = \lim_{\Lambda \nearrow R^2} \frac{\int A \, d\phi_\Lambda}{\int d\phi_\Lambda} = \sum \alpha_j \lambda^{j/2} + O(\lambda^{(r+1)/2}). \tag{6.3}$$

We generate the asymptotic expansion by integration by parts on path space. Before integration by parts, we choose a new variable $\psi = \phi - \xi_+$. We thus translate by the constant function $g = g_0 = h = \xi_+$, and (6.1) can be written

$$A = A(\psi) = \prod_{i=1}^{n} \int {:}\psi(x)^{m_i}{:}\, f_i(x) \, dx. \tag{6.4}$$

The function $F = 0$ in (4.6), so

$$\int A \, d\phi_\Lambda = \int A(\psi) \, e^{-V} \, d\psi_1 \tag{6.5}$$

where

$$V = \int_{\Lambda} V(x)\, dx \tag{6.6}$$

and where

$$V(x) = \lambda{:}\psi(x)^4{:} + \alpha(2\lambda)^{1/2}{:}\,\psi(x)^3{:} + \beta\lambda^{5/2}{:}\psi(x)^2{:}. \tag{6.7}$$

Computation shows that

$$\begin{aligned} \alpha &= (8\lambda)^{1/2}\,\xi_+ = 1 + O(\lambda^{1/2}\mu) \\ \beta &= (18)^{1/2}\,\mu\lambda^{-2}\alpha^{-1} = (18)^{1/2}\,\mu\lambda^{-2} + O(\mu^2\lambda^{-3/2}) = O(1), \end{aligned} \tag{6.8}$$

so the coefficients of V are small for small λ. Note if $\mu = 0$, then $\alpha = 1$, $\beta = 0$.

The formula for integration by parts is

$$\int {:}\psi(x)^m{:}\; R(\psi)\, d\phi_\Lambda = \iint C(x-y)\left[{:}\psi(x)^{m-1}{:}\,\frac{\delta R}{\delta\psi(y)} - R\,\frac{\delta V}{\delta\psi(y)}\right] dy\, d\phi_\Lambda\,. \tag{6.9}$$

Here

$$\delta V/\delta\psi(y) = 4\lambda{:}\psi(y)^3{:} + \alpha(18\lambda)^{1/2}{:}\psi(y)^2{:} + 2\beta\lambda^{5/2}\psi(y). \tag{6.10}$$

To generate the asymptotic series (6.3), we apply (6.9) in order to integrate by parts each factor ψ in the integrand of $\int A d\phi_\Lambda$. We also integrate by parts each factor of ψ produced by such integration, and we continue inductively until either (i) the integrand is constant on path space, or (ii) the integrand has a coefficient $O(\lambda^{(r+1)/2})$ and each Wick power has degree $\leqslant \deg \mathscr{P}$. Finally, we invert the transformation from ψ to ϕ. Thus we replace A by $\sum A_k$ plus constant terms.

The terms from the expansion which are constant on path space (i.e., the α_j) have several interesting properties. First we remark that

$$\alpha_0 = \int A(\psi)\, d\psi \tag{6.11}$$

where $d\psi$ is the Gaussian measure with mean zero and covariance $(-\Delta + 1)^{-1}$. Thus the terms independent of λ are the free field expectations of $A(\psi)$.

The terms α_j for $j > 0$ have an interpretation in terms of Feynman diagrams. In particular, the asymptotic series $\sum \alpha_j\lambda^{j/2}$ is exactly equal to the usual perturbation series for $\langle A(\psi)\rangle$, where the interaction density equals (6.7). (For $-$ boundary conditions, we reverse the sign of the cubic term in (6.7) and take $\psi = \phi - \xi_-$). We note in the special case

$$A = \phi(x) - \xi_+ = \psi(x), \tag{6.12a}$$

that

$$\alpha_0 = \alpha_1 = \alpha_2 = 0. \tag{6.12b}$$

An explicit calculation of α_3 (in the case $\Lambda = R^2$) shows that

$$\alpha_3 = 144\sqrt{2} \int C(x)^3\, dx - 864\sqrt{2} \int C(x)\, C(y)\, C(x-y)^2\, dx\, dy. \tag{6.13}$$

Here we have used the fact that $\int C(x)\, dx = 1$. In the conventional notation, α_3 is represented by the perturbation diagrams

The numerical value of α_3 can be calculated from (6.13). As a corollary of Theorem 6.1, we find

$$\begin{aligned} \langle \phi(x) \rangle &= \xi_+ + \alpha_3 \lambda^{3/2} + O(\lambda^2) \\ &= \phi_c + \alpha_3 \lambda^{3/2} + O(\lambda^2) \end{aligned} \tag{6.14}$$

Thus $\alpha_3 \lambda^{3/2}$ is the first "quantum correction" to the classical field expectation ϕ_c.

We apply our estimates to the functions A obtained from the asymptotic expansion above. Thus it is no loss of generality to assume $m_i \leqslant 4$.

THEOREM 6.2. *Consider* A *of the form* (6.1) *with* $m_i \leqslant 4$. *Then for* λ *sufficiently small,*

$$\left| \int A\, d\phi_\Lambda \right| \leqslant |||\, A \,|||\, (1 + \lambda^{-N(A)/2} e^{-O(\lambda^{-1 \cdot 2})}) \int d\phi_\Lambda . \tag{6.15}$$

Theorems 2.2, 2.3, and 6.1 follow from Theorem 6.2 and the above discussion. In addition the bounds on (2.9) and (2.12) uniform in Λ follow from Theorem 6.2. To obtain the λ independent bounds required for Theorems 2.2 and 2.3 from the l (and hence λ) dependent norm $|||\cdot|||$ in (6.15), it is necessary first to use the asymptotic expansion of this section, and extract enough factors $\lambda^{1/2}$. The resulting bounds are uniform in λ for fixed $N(A)$ and uniform in $N(A)$ for fixed (small) λ.

7. FACTORIZATION AND PARTIAL RESUMMATION

We resume parts of the expansion (5.10) to factor out and cancel vacuum energy contributions. Let $\mathscr{L}(A)$ denote the localization of A, i.e., the set of lattice squares Δ_i contributing to (6.1) and let $\mathscr{L}(A) \subset W \cap \Lambda$. Define

$$R''(A, W, \Sigma, \Gamma) = \int_0^1 ds_{\Gamma^c}\, \partial^{\Gamma^c \subset \mathscr{B}(\Sigma, W)} \int \chi_\Sigma A e^{-Q(W,\Sigma)}\, d\psi(s) \tag{7.1}$$

and

$$Z''(W, \Sigma, \Gamma) = R''(A = I, W, \Sigma, \Gamma). \tag{7.2}$$

Then (5.10) can be written as

$$\langle A \rangle = Z(\Lambda)^{-1} \sum_{\Sigma} \sum_{\Gamma \subset \mathscr{B}(\Sigma)} R''(A, R^2, \Sigma, \Gamma). \tag{7.3}$$

We now exhibit the factorization in (7.3). Let $\{Y_1, ..., Y_\nu\}$ denote the set of connected components of $R^2 \sim \Gamma$. For $s_\Gamma = 0$, the covariance $C(s)$ defined in (5.3) is a direct sum over the Y_i, i.e., it is reduced by the direct sum decomposition.

$$L_2(R^2) = \sum_{i=1}^{\nu} L_2(Y_i).$$

Consequently, the measure $d\psi(s)$ of (7.1) factors. We now see that the integrand factors also.

For convenience, we consider A in (6.1) to be localized, so that suppt $f_i \subset \Delta_i$. The general case is then handled by summation over localizations. With this restriction A factors.

Because the Dirichlet contour Γ is by construction bounded away from the phase boundary Σ, we apply Lemma 4.1 and the discussion following it to express $Q(R^2, \Sigma)$ as a local sum

$$Q(R^2, \Sigma) = \sum_{i=1}^{\nu} Q(Y_i, \Sigma \cap Y_i). \tag{7.4}$$

Finally, defining

$$\chi_{\Sigma \cap Y_i} = \prod_{\Delta_j \subset Y_i} \chi_{\sigma_j}(\phi(\Delta_j)),$$

we have

$$\chi_\Sigma = \prod_{i=1}^{\nu} \chi_{\Sigma \cap Y_i}. \tag{7.5}$$

Thus the integrand in (7.1) factors as asserted, and

$$R''(A, R^2, \Sigma, \Gamma) = \prod_{i=1}^{\nu} R''(A_i, Y_i, \Sigma \cap Y_i, \Gamma \cap Y_i). \tag{7.6}$$

Now we turn to the partial resummation. Fix Σ and Γ in (5.6), (7.3), and define $Y = Y(\Gamma)$ as the closure of the union of all components γ_i which meet the localization of A from (6.1):

$$Y = U_i\{Y_i : \text{suppt}\, f_j \subset Y_i, \text{ some } j\}.$$

We also define

$$\partial Y^{\pm} = \partial Y \cap X^{\pm}$$

and then ∂Y, ∂Y^{+}, ∂Y^{-} each consist of a finite number of nonoverlapping closed contours.

By applying (7.3)–(7.6), we isolate A from the partition function for $\sim Y$. We perform the first partial resummation by holding $Y = Y(\Gamma)$ fixed and summing over all Dirichlet contours Γ compatible with $Y = Y(\Gamma)$. We call this condition a *constraint* on the sum over Γ. Thus we consider

$$\sum_{\{\Gamma \subset \mathscr{B}(\Sigma): Y(\Gamma)=Y\}} R''(A, R^2, \Sigma, \Gamma)$$

$$= \sum_{\{\Gamma \subset \mathscr{B}(\Sigma): Y(\Gamma)=Y\}} R''(A, Y, \Sigma \cap Y, \Gamma \cap Y)\, Z''(\sim Y, \Sigma \sim Y, \Gamma \sim Y). \quad (7.7)$$

The constrained sum (7.7) factors into an inside sum over $\mathscr{B}(\Sigma, Y)$ and an outside sum over $\mathscr{B}(\Sigma, \sim Y)$. The inside sum is

$$R'(A, Y, \Sigma \cap Y) = \sum_{\{\Gamma \subset \mathscr{B}(\Sigma, Y): Y(\Gamma)=Y\}} R''(A, Y, \Sigma \cap Y, \Gamma \cap Y) \quad (7.8)$$

and the outside sum is

$$Z'(\sim Y, \Sigma \sim Y) = \sum_{\{\Gamma \subset \mathscr{B}(\Sigma, \sim Y): \Gamma \supset \partial Y\}} Z''(\sim Y, \Sigma \sim Y, \Gamma \sim Y) \quad (7.9)$$

The outside sum can be evaluated exactly, using (5.5). In fact (7.9) is the partition function

$$Z'(\sim Y, \Sigma \sim Y, \partial Y) \equiv \int e^{-Q(\sim Y, \Sigma \sim Y)} \chi_{\Sigma \sim Y}\, d\psi_{\partial Y}, \quad (7.10)$$

where the subscript ∂Y in the measure $d\psi$ denotes Dirichlet data. Thus we write (7.5) as

$$\langle A \rangle = \sum_{\Sigma} \sum_{\partial Y} R'(A, Y, \Sigma \cap Y) \frac{Z'(\sim Y, \Sigma \sim Y, \partial Y)}{Z(R^2, \varnothing, \varnothing)}. \quad (7.11)$$

Let $N(W)$ denote the set of lattice squares within distance L of a set W. The second partial resummation is over Σ. We hold ∂Y^{+} and ∂Y^{-} fixed ($\partial Y^{\pm} = \partial Y \cap X^{\pm}$), and consequently

$$\Sigma \upharpoonright N(\partial Y^{\pm}) = \pm \quad (7.12)$$

is the constraint on the sum over Σ. With this constraint, the Σ sum also factors across $N(\partial Y)$ into an inside sum and an outside sum. We define

$$R(A, Y, \partial Y^{\pm}) = \sum_{\{\Sigma \cap Y: \sigma_i \upharpoonright N(\partial Y^{\pm})=\pm\}} R'(A, Y, \Sigma \cap Y) \quad (7.13)$$

and

$$Z(Y, N^+, N^-, \Gamma) = \sum_{\{\Sigma\cap Y:\sigma_i\restriction N^\pm=\pm\}} Z'(Y, \Sigma \cap Y, \Gamma \cap Y). \tag{7.14}$$

Here we assume $N^+ \cap N^- = \varnothing$. In particular,

$$Z(Y, N(\partial Y^+), N(\partial Y^-), \partial Y) = \sum_{\{\Sigma\cap Y: N(\delta Y^\pm)=\pm\}} \int e^{-Q(Y,\Sigma\cap Y)} \chi_{\Sigma\cap Y}\, d\psi_{\delta Y}, \tag{7.15}$$

and (7.11) becomes

$$\langle A\rangle = \sum_{\partial Y^\pm} R(A, Y, \partial Y^\pm) \frac{Z(\sim Y, N(\partial Y^+), N(\partial Y^-), \partial Y)}{Z(\Lambda, \varnothing, \varnothing, \varnothing)}. \tag{7.16}$$

Again we can evaluate the outside sum, (7.15). We say that Σ is compatible with a pair $N^\pm$ if $\sigma \restriction N^\pm = \pm$.

LEMMA 7.1. *Let A be a function of ϕ. For any two configurations Σ_1, Σ_2 compatible with $N(\Gamma^\pm)$ we have*

$$\int Ae^{-Q(Y,\Sigma_1)}\, d\psi_\Gamma = \int Ae^{-Q(Y,\Sigma_2)}\, d\psi_\Gamma. \tag{7.17}$$

Proof. We express Q and $d\psi_\Gamma$ in terms of ϕ. If we approximate Dirichlet boundary conditions by a large local mass $\delta m^2(x)$ concentrated in a small neighborhood of Γ, we see that the right and left sides of the above equations become

$$\frac{\int A \exp(-\int \delta m^2(x) :(\phi(x) - g_i(x))^2: dx)\, d\phi_{\Lambda\cap Y}}{\int \exp(-\int \delta m^2(x) :\psi(x)^2: dx)\, d\psi_1}, \tag{7.18}$$

$i = 1, 2$. Near Γ, the compatibility condition and Lemma 4.1 yield $g_1 = g_2$. Since $\delta m^2(x)$ is supported near Γ, the two expressions in (7.18) are equal.

Let $d\phi_{Y\cap\Lambda,\Gamma}$ denote the measure $\exp[-Q(Y,\Sigma)]\, d\psi_\Gamma$ in (7.17). This measure can be represented as in (2.4) as a Gaussian measure with nonzero Dirichlet data $\phi = \pm\xi$ on $\Gamma^\pm$ times a $\mathscr{P}(\phi)$ action factor. Then (7.14) becomes

$$Z(Y, N^+, N^-, \Gamma) = \int \left[\prod_{\Delta\subset N^+} \chi_+(\phi(\Delta))\right]\left[\prod_{\Delta\subset N^-} \chi_-(\phi(\Delta))\right] d\phi_{Y\cap\Lambda,\Gamma}. \tag{7.19}$$

REFERENCES

1. D. BRYDGES, Cluster expansions for fermion fields by the time-dependent Hamiltonian approach, *J. Math. Phys.*, to appear.

2. A. Cooper and L. Rosen, "The Weakly Coupled Yukawa$_2$ Field Theory: Cluster Expansion and Wightman Axioms," preprint.
3. R. Dashen, B. Hasslacher, and A. Neveu, Nonperturbative methods and extended Hadron models in field theory, I, II, III. *Phys. Rev. D.* **10** (1974), 4114–4141.
4. J. Dimock, "The $\mathscr{P}(\phi)_2$ Green's Functions: Asymptotic Perturbation Expansion," preprint.
5. J. Dimock and J. Glimm, Measures on the Schwartz distribution space and applications to quantum field theory, *Adv. Math.* **12** (1974), 58–83.
6. J.-P. Eckmann, H. Epstein, and J. Fröhlich, Asymptotic perturbation expansion for the S-matrix and the definition of time ordered functions in relativistic quantum field models.
7. J.-P. Eckmann, J. Magnen, and R. Sénéor, Decay properties and Borel summability for the Schwinger functions in $\mathscr{P}(\phi)_2$-theories, *Commun. Math. Phys.* **39** (1975), 251–271.
8. L. Faddeev, "Quantization of Solitons," IAS preprint.
9. J. Feldman and K. Osterwalder, The Wightman axioms and the mass gap for weakly coupled $\phi_3{}^4$. *Ann. Phys.*, to appear.
10. J. Feldman and K. Osterwalder, The construction of $\lambda\phi_3{}^4$ quantum field models, *in* "Proceedings of Marseille Conference," June 1975.
11a. J. Fröhlich, New super selection sectors (Soliton-states) in two dimensional Bose quantum field models, *Commun. Math. Phys.* **47** (1976), 269–310.
11b. J. Fröhlich and B. Simon, "Pure States for General $P(\phi)_2$ Theories: Construction, Regularity, and Variational Equality," preprint, 1976.
12. J. Glimm and A. Jaffe, A $\lambda(\phi^4)_2$ Quantum field theory without cutoffs. I. *Phys. Rev.* **176** (1968), 1945–1951.
13. J. Glimm and A. Jaffe, The $\lambda(\phi^4)_2$ quantum field theory without cutoffs. III. The physical vacuum, *Act. Math.* **125** (1970), 203–267.
14. J. Glimm and A. Jaffe, The Yukawa$_2$ quantum field theory without cutoffs, *J. Functional Anal.* **7** (1971), 323–357.
15. J. Glimm, A. Jaffe, and T. Spencer, The Wightman axioms and particle structure in the $\mathscr{P}(\phi)_2$ quantum field model. *Ann. Math.* **100** (1974), 585–632.
16. J. Glimm, A. Jaffe, and T. Spencer, The particle structure of the weakly coupled $\mathscr{P}(\phi)_2$ model and other applications of high temperature expansions. Part II, The cluster expansion, *in* "Constructive quantum field theory" (G. Velo and A. Wightman, Eds.), Lecture notes in Physics, Vol. 25, Springer–Verlag, Berlin, 1973.
17. J. Glimm, A. Jaffe, and T. Spencer, Existence of phase transitions for quantum fields, 1975 Marseille conference.
18. J. Glimm, A. Jaffe, and T. Spencer, Phase transition for $\phi_2{}^4$ quantum fields, *Commun. Math. Phys.* **45** (1975), 203–216.
19. J. Goldstone and R. Jackiw, Quantization of nonlinear waves, *Phys. Rev. D* **11** (1975), 1486–1498.
20. F. Guerra, L. Rosen, and B. Simon, The $\mathscr{P}(\phi)_2$ Euclidean quantum field theory as classical statistical mechanics, *Ann. Math.* **101** (1975), 111–259.
21. E. Lieb, D. Mattis, and T. Schultz, Two-dimensional Ising model as a soluble problem of many fermions. *Mod. Phys.* **36** (1964), 856–871.
22. J. Magnen and R. Sénéor, The infinite volume limit of the $\phi_3{}^4$ model, *Ann. Inst. Henri Poincaré*, to appear.
23. J. Magnen and R. Sénéor, The Wightman axioms for the weakly coupled yukawa model in two dimensions, *Commun. Math. Phys.*, to appear.
24. B. McCoy and T. T. Wu, "The Two-Dimensional Ising Model," Harvard Univ. Press, Cambridge, Mass., 1973.

25. K. OSTERWALDER AND R. SÉNÉOR, The scattering matrix is nontrivial for weakly coupled $\mathscr{P}(\phi)_2$ models, *Helv. Phys. Acta*, to appear.
26. R. PEIERLS, Ising's model of ferromagnetism, *Proc. Cambridge Philos. Soc.* **32** (1936), 477–481.
27. T. SPENCER, A mass gap for the $\mathscr{P}(\phi)_2$ quantum field model with a strong external field, *Commun. Math. Phys.* **39** (1974), 63–76.
28. T. SPENCER, The decay of the Bethe-Salpeter kernel in $\mathscr{P}(\phi)_2$ quantum field models, *Commun. Math. Phys.* **44** (1975), 143–164.
29. T. SPENCER AND F. ZIRILLI, Scattering states and bound states in $\lambda\mathscr{P}(\phi)_2$," *Commun. Math. Phys.* **49** (1976), 1–16.

A Convergent Expansion about Mean Field Theory
II. Convergence of the Expansion

JAMES GLIMM*

Rockefeller University, New York, New York 10021

ARTHUR JAFFE†

Harvard University, Cambridge, Massachusetts 02138

AND

THOMAS SPENCER*

Rockefeller University, New York, New York 10021

Received June 2, 1976

We prove convergence of the mean field cluster expansion defined in the preceding paper and establish exponential clustering in a pure phase selected by suitable boundary conditions.

1. INTRODUCTION

1.1. *Outline of the Estimates*

Convergence of the mean field cluster expansion is established in this paper. The expansion defined in Part I (referred to here as I) has the form

$$\langle A \rangle = \lim_{\Lambda \nearrow R^2} Z(\Lambda)^{-1} \sum_{\Sigma,\Gamma} \int A \, d\phi_{\Lambda,\Sigma,\Gamma}, \tag{1.1.1}$$

cf. (I.5.10). The summation ranges over phase boundaries Σ and Dirichlet contours Γ, and A is a function of the field.

For each term in the expansion, there are two sources of convergence. The

* Supported in part by the National Science Foundation under Grant MPS74-13252.
† Supported in part by the National Science Foundation under Grant MPS 75-21212.

Gaussian approximation provides the first source of convergence. Far from phase boundaries, i.e., in the region X of (I.5.7.), the interaction V has the form

$$V(x) = :\lambda\psi(x)^4: \pm (2\lambda)^{1/2}\alpha :\psi(x): + \lambda^{5/2}\beta :\psi(x)^2: \tag{1.1.2}$$

with coefficients $O(\lambda^{1/2})$. Thus $e^{-V}d\psi_1$ is a small perturbation of the Gaussian ground-state measure $d\psi_1$ with mean zero and mass 1. The convergence of the approximation $e^{-V}d\psi_1$ by $d\psi_1$ is introduced in (1.1.1) by factors λ, $\lambda^{1/2}$, or $\lambda^{5/2}$ in (1.1.2).

The second source of convergence arises from the presence of a phase boundary Σ. For λ small, the existence of a phase boundary means the existence of a large fluctuation in the value of the field ϕ. In fact, only with small probability can $|\phi| \leqslant (32\lambda)^{-1/2}$, due to the potential barrier $\mathscr{P}(0) = O(\lambda^{-1})$ in the energy density $\mathscr{P}(\xi)$, near $\xi = 0$. The resulting small probability is $\exp(-O(\lambda^{-1}))$. This means that when a phase boundary Σ occurs, the field (with large probability) changes from order $\lambda^{-1/2}$ on one side of Σ to order $-\lambda^{-1/2}$ on the other side. This fluctuation yields a large derivative $\nabla\phi$, which results in a small factor $\exp(-O(\nabla\phi)^2) = \exp(-O(\lambda^{-1}))$ arising from the gradient term in the Euclidean action, cf. Section 2.4.

Associated with each convergence factor there is a length scale. The unit squares are used to analyze the fluctuations of the field, whereas the squares of length l are introduced to exhibit the decoupling of the field in the Gaussian approximation. Since $m \sim m_c \sim 1$, we exhibit the exponential clustering of the ground state in X by choosing

$$l = |\log \lambda|^{1/4} \gg 1. \tag{1.1.3}$$

This choice of l gives coupling at most

$$e^{-m_c l} \ll 1$$

between nonadjacent l-lattice squares. (Compare with [15, 16][1] where $m_c = m_0 \gg 1$ and $l = 1$, but again $m_c l \gg 1$.)

The parameter L, chosen as

$$L = |\log \lambda|^2 \gg l, \tag{1.1.4}$$

is defined so that the functional derivatives $\delta/\delta\psi$ which contract to the exponent in $X^0 = R^2\backslash X$ occur over a long distance $L/2$. We choose L large to compensate for the fact that the coefficients of V in X^0 are not small. Thus L provides an upper bound for the distance scale on which phase boundaries influence the Gaussian approximation.

[1] See Part I for reference list.

The key estimate required to establish convergence is

$$\left| \int A \, d\phi_{\Lambda,\Sigma,\Gamma} \right| \leqslant \exp(a\lambda^{1/2} \mid \Lambda \mid - \delta l \mid \Gamma^c \mid - \delta\lambda^{-1/2} \mid \Sigma \mid), \tag{1.1.5}$$

where a, δ are constants independent of λ, Λ, Σ. We establish this bound in Section 3. The main idea is as follows. Evaluate the s derivatives in (I.5.6) (using integration by parts) to obtain a sum of terms of the form

$$\int ARe^{-Q}\chi_\Sigma \, d\psi(s).$$

Here R is a polynomial in ψ obtained from derivatives $\delta/\delta\psi$ applied to e^{-Q} and χ_Σ' denotes either χ_Σ or a possible derivative of χ_Σ. The polynomial R has coefficients of order $\mid \lambda^{1/2} \mid$ since the coefficients of V are of this size within X, where the functional derivatives are localized. By Hölder's inequality such a term is bounded by

$$\left| \int (AR)^q \, d\psi(s) \right|^{1/q} \left[\int e^{-pQ}(\chi_\Sigma')^p \, d\psi(s) \right]^{1/p}. \tag{1.1.6}$$

For illustrative purposes, let us consider the case where there are no phase boundaries, i.e., $\Sigma = \{\sigma_i = 1\}$, and where no derivatives occur, i.e., $\chi_\Sigma' = \chi_\Sigma$. We choose p near 1, independent of λ, and so that q is a large even integer. The first factor in (1.1.6) is a straightforward Gaussian integral which is bounded by $O(e^{-\delta l \mid \Gamma^c \mid})$ as in [16]. For the second factor the key estimate is a bound on the vacuum energy per unit volume, uniform as $\lambda \searrow 0$. By shifting a portion of the (unit) mass of $d\psi_1$ to the exponent the desired vacuum energy bound takes the form

$$\int \exp\left(-p\left[V(\Lambda) + \left(\frac{1-\eta}{2}\right)\int_\Lambda :\psi:^2(x)\,dx\right]\right) d\psi_{(\eta p+1-p)} \leqslant \exp(\text{const} \mid \Lambda \mid).$$

For fixed momentum cutoff κ and fixed $\eta > 0$, the above exponent (as a function of $\phi_\kappa(x)$),

$$:P(\phi_\kappa(x)): - \eta :(\phi_\kappa(x) - \xi_+)^2: \tag{1.1.7}$$

is not uniformly bounded from below as $\lambda \searrow 0$. In fact, when $\phi_\kappa(x) \approx -\xi_+$, (1.1.7) approximately equals $-\eta(2\xi_+)^2 = -(\eta/2)\lambda^{-1}$. In order to obtain a uniform bound on (1.1.7) we must consider the contribution of the factor χ_Σ which constrains the averaged field to be positive. In addition, it is necessary to include two error terms: The first error term occurs because χ_+ is the smoothed out $\bar{\phi}$ localization of (I.2.5). The second error term is $\zeta :\delta\phi^2:$, with ζ small, and measures the deviation of $\phi(x)$ from the averaged field $\bar{\phi}$ constrained by χ_+. In Theorem 3.1.1 we show that

$$:P(\phi_\kappa(x)): - \eta :(\phi_\kappa(x) - \xi_+)^2: - \log \chi_+(\bar{\phi}) \geqslant b \log^2 \kappa - \zeta :(\delta\phi_\kappa):^2 (x). \quad (1.1.8)$$

Here $\zeta > 0$ is given, $\eta \leqslant \eta(\zeta)$ is sufficiently small and $b = b(\zeta)$ is large. The constants η, b, ζ do not, however, depend on λ, $0 \leqslant \lambda \leqslant 1$ and ζ is chosen so that

$$\int \exp(4\zeta:\delta\phi_\kappa{}^2(\Lambda):)\, d\phi_\epsilon \leqslant e^{O(1)|\Lambda|} \quad (1.1.9)$$

uniformly in the momentum cutoff κ and the mass $\epsilon > 0$. This bound is obtained in the following section and relies on the fact that since $\delta\phi(x)$ is orthogonal to the zero momentum mode,

$$\int_\Delta \delta\phi^2(x)\, dx \leqslant \pi^{-2} \int_\Delta (\nabla\phi)^2 (x)\, dx.$$

1.2. *An Estimate on Schwinger Functions*

In this section we state two estimates, which yield the proof of Theorem I.6.2.

PROPOSITION 1.2.1. *Let A have the form* (I.6.1). *There is a $\delta > 0$ such that for λ sufficiently small,*

$$| R(A, Y, \partial Y^\pm)| \leqslant \lambda^{-N(A)/2} \,|||\, A \,|||\, e^{-\delta|Y\cap A|/l}. \quad (1.2.1)$$

If $\partial Y^- = \varnothing$, the factor $\lambda^{-N/2}$ can be replaced by

$$(1 + \lambda^{-N/2} e^{-O(\lambda^{-1/2})}),$$

provided $N(A) > 0$.

Remark. The divergent factor $\lambda^{-N/2}$ arises from the possibility of localizing $\phi(x)$ near ξ_-, i.e., in the minus well of the potential.

PROPOSITION 1.2.2. *Let λ be sufficiently small. Let $N^\pm = N(\partial Y^\pm)$. There is a constant $c < \infty$ such that*

$$\left| \frac{Z(\sim Y, N^+, N^-, \partial Y)}{Z(R^2, \varnothing, \varnothing, \varnothing)} \right| \leqslant \exp(c\lambda^{1/2}(|\, Y \cap \Lambda \,| + |\, N \,| + l^2 \,|\, \partial Y \,|)). \quad (1.2.2)$$

If $N^- \neq \varnothing$, we can multiply the right side by $\exp(-O(\lambda^{-1/2}))$.

Remark. The convergence factor $\exp(-O(\lambda^{-1/2}))$ arises from the small probability for occurrence of a phase boundary.

Proof of Theorem I.6.2. By a simple combinatoric argument (see [16]) there are at most $\exp(O(r/l^2))$ allowed choices of Y such that $|Y| = r$. With ∂Y fixed, there are at most $2^{|Y|/l^2}$ choices of ∂Y^+, ∂Y^-. Thus the series (I.7.15) is majorized by

$$||| A ||| (1 + \lambda^{-N/2} e^{-O(\lambda^{-1/2})}) \sum_{r=1}^{\infty} e^{-\delta r/l} \leqslant ||| A ||| O(l^2)(1 + \lambda^{-N/2} e^{-O(\lambda^{-1/2})}).$$

The factor $O(l^2)$ is absorbed into $||| A |||$ by an increase in the constant K in (I.6.2), and the proof is complete.

2. Estimates on Fluctuations

2.1. *The Momentum Cutoff*

We introduce a momentum cutoff (smoothing) operator $\rho_\kappa : f \to f_\kappa$. In order to preserve the notion of localization in space-time squares Δ, before and after smoothing, we require several properties of the cutoff:

(i) $\rho_\kappa \to I$ as $\kappa \to \infty$.

(ii) ρ_κ is self-adjoint.

(iii) If $x \in \Delta$, then $(\rho_\kappa f)(x)$ depends only on $f(y)$ for $y \in \Delta$.

(iv) $\rho_\kappa e_\Delta = e_\Delta$, where e_Δ is the characteristic function for Δ.

We achieve these properties with ρ_κ defined by the symmetrization of a nonnegative integral kernel $\rho_\kappa'(x, y)$, i.e.,

$$f_\kappa(x) = \int \rho_\kappa'(x, y) f(y)\, dy, \tag{2.1.1}$$

such that for all Δ,

$$\int_\Delta \rho_\kappa'(x, y)\, dx = e_\Delta(y) = \int_\Delta \rho_\kappa'(y, x)\, dx. \tag{2.1.2}$$

Then (ii)–(iv) follow. To obtain ρ_κ', we first choose an even, positive, C^∞ function $w(t)$, supported in the interval $[-\frac{1}{2}, \frac{1}{2}]$, and with total integral 1. Define $\rho_\kappa(x) = \kappa^2 w(\kappa x_0)\, w(\kappa x_1)$ as an approximate Dirac measure. For $x \in \Delta$, let $x_\Delta^{(i)}$, $i = 1, 2, 3, 4$ be the four points (including $x \equiv x_\Delta^{(1)}$) obtained by reflecting about the two sides of Δ nearest to x. Then define

$$\rho_\kappa'(x, y) = \sum_\Delta \sum_{i=1}^{4} \rho_\kappa(x_\Delta^{(i)} - y)\, e_\Delta(x)\, e_\Delta(y). \tag{2.1.3}$$

We check that

$$\int_{\Delta} \rho_\kappa'(x, y)\, dx = \sum_{i=1}^{4} \int \rho_\kappa(x_\Delta^{(i)} - y)\, e_\Delta(y)\, dx$$

$$= \int \rho_\kappa(x - y)\, e_\Delta(y)\, dx = e_\Delta(y).$$

Similarly

$$\int_{\Delta} \rho_\kappa'(y, x)\, dx = e_\Delta(y),$$

so (2.1.2) holds. The inequality

$$\|(f_\kappa - f)\, e_\Delta\|_{L_p(\Delta)} \leqslant O(\kappa^{-\epsilon})\, \|f\|_S \tag{2.1.4}$$

expresses the precise form in which (i) holds. Here $2 \leqslant p < \infty$ and $\|f\|_S$ denotes a Sobolev norm with fractional derivatives in $L_{p'}(\Delta)$. If $\operatorname{dist}(\operatorname{suppt} f, \partial\Delta) \geqslant \kappa^{-1}$ then the operator ρ_κ is a convolution operator and (2.1.4) follows, e.g., using a Fourier transform. Let $\partial = \{x \mid \operatorname{dist}(x, \partial\Delta) \leqslant \kappa^{-1}\}$. By Hölder's inequality

$$\|f\|_{L_p(\partial)} \leqslant \kappa^{-\epsilon}\, \|f\|_{L_{p'}(\partial)}$$

for $p' > p$, which establishes (2.1.4).

The Wick ordering constant for the cutoff field ϕ_κ is

$$\begin{aligned} c_\kappa(x) &= \int \phi_\kappa(x)^2\, d\Phi \\ &= \int \rho_\kappa(x, y)\, \rho_\kappa(x, z)\, C(y - z)\, dy\, dz = (\rho_\kappa C \rho_\kappa)(x, x). \end{aligned} \tag{2.1.5}$$

Furthermore,

$$c_\kappa(x) \leqslant O(\ln \kappa), \tag{2.1.6}$$

since

$$c_\kappa(x) \leqslant \sup_y \int \rho_\kappa(x, z)\, C(y - z)\, dz$$

$$\leqslant 4 \sup_y \int \rho_\kappa(y - z)\, C(z)\, dz = O(\ln \kappa).$$

Finally, we remark that as an operator on $L_2(\Delta)$,

$$\|\, e_\Delta \rho_\kappa e_\Delta \,\| \leqslant 4. \tag{2.1.7}$$

Here we use (2.1.3) and the fact that convolution by $\rho_\kappa(x)$ has operator norm 1. Thus

$$\int_\Delta |f_\kappa|^2 \, dx \leqslant 16 \int_\Delta |f|^2 \, dx. \tag{2.1.8}$$

2.2. *Space Dependent Mass*

In this section we consider covariance operators for a Gaussian measure with a space-dependent mass. Let $\omega(x)$ satisfy

$$0 < \omega(x) \leqslant 1, \qquad \inf_x \omega(x) = \boldsymbol{\omega} > 0, \tag{2.2.1}$$

and let $1 - \omega(x)$ have compact support. In (2.2.2) below, $\omega(x)^{1/2}$ plays the role of a mass. Consider

$$d\psi_\omega(s) = Z^{-1} \exp\left(\tfrac{1}{2} \int (1 - \omega(x)) :\psi(x)^2: dx\right) d\psi_1(s) \tag{2.2.2}$$

where

$$Z = \int \exp\left(\tfrac{1}{2} \int (1 - \omega(x)) :\psi(x)^2: dx\right) d\psi_1(s). \tag{2.2.3}$$

By Jensen's inequality, $Z \neq 0$. In fact,

$$\int e^V \, d\psi \geqslant e^{\int V d\psi}$$

and so

$$Z^{-1} \leqslant \exp\left[O(1) \int (1 - \omega) \, dx\right]. \tag{2.2.4}$$

Thus $d\psi_\omega(s)$ is a Gaussian measure with mean zero and covariance

$$\begin{aligned} C_\omega(s) &= \sum_{j=0}^{\infty} C(s)^{1/2} \, [C(s)^{1/2} \, (1 - \omega) \, C(s)^{1/2}]^j \, C(s)^{1/2} \\ &= [C(s)^{-1} - (1 - \omega)]^{-1}. \end{aligned} \tag{2.2.5}$$

The convergence follows from the fact that $0 \leqslant 1 - \omega(x) \leqslant 1 - \boldsymbol{\omega} < 1$ and $\| C(s) \| \leqslant 1$. For $s = 1$,

$$(-\Delta + 1 - \omega(x))^{-1}$$

$$= \sum_{j=0}^{\infty} (-\Delta + 1)^{-1/2} \, [(-\Delta + 1)^{-1/2} \, (1 - \omega)(-\Delta + 1)^{-1/2}]^j \, (-\Delta + 1)^{-1/2} \tag{2.2.6}$$

Let Δ_N be the Laplace operator with zero Neumann data on some set of (lattice) lines in R^2, and let

$$C_\omega(s, t) = (1 - t)\, C_\omega(s) + t(-\Delta_N + \omega)^{-1}. \tag{2.2.7}$$

We note that

$$-\Delta_N \leqslant -\Delta_\Gamma \tag{2.2.8}$$

so

$$C(s) \leqslant (-\Delta_N + 1)^{-1} \tag{2.2.9}$$

and by (2.2.5)

$$C_\omega(s) \leqslant C_\infty(s) \leqslant (-\Delta_N + \omega)^{-1} \tag{2.2.10}$$

where

$$C_\infty(s) = [C(s)^{-1} - (1 - \omega)]^{-1}.$$

It follows that

$$\frac{d}{dt}\, C_\omega(s, t) = (-\Delta^N + \omega)^{-1} - C_\omega(s) \geqslant 0. \tag{2.2.11}$$

2.3. *The Fluctuation Field*

In this section we bound the fluctuation field $\delta\phi(x)$ and its translate $\delta\psi(x) = \delta\phi(x) - \delta g(x)$. Here $g(x)$ is the approximate classical magnetization (1.4.4). We use the method of conditioning [20], and we start with a direct proof of the conditioning inequality.

PROPOSITION 2.3.1. *Let V be a fixed (s, t independent) polynomial in ψ, and let $:V:$ denote $C_\omega(s, t)$ Wick ordering. Then*

$$\int e^{-:V:}\, d\psi_\omega(s, t) \tag{2.3.1}$$

is increasing in s and t.

Proof. We give the proof for t only. Let $Z(s, t)$ denote (2.3.1). Then

$$\begin{aligned} \frac{dZ(s, t)}{dt} &= -\int e^{-:V:} \frac{d{:}V{:}}{dt}\, d\psi_\omega(s, t) + \int e^{-:V:} \frac{d}{dt}\, d\psi_\omega(s, t) \\ &= \int \left(-\frac{d{:}V{:}}{dt} + \frac{1}{2} \Delta_\psi \cdot \frac{dC_\omega(s, t)}{dt} \right) e^{-:V:}\, d\psi_\omega(s, t) \end{aligned}$$

where

$$\Delta_\psi \cdot K \equiv \int K(x, y) \frac{\delta^2}{\delta\psi(x)\, \delta\psi(y)}\, dx\, dy$$

as in [16]. The term $-d{:}V{:}/dt$ cancels with $2^{-1}(\Delta_\psi \cdot dC_\omega/dt){:}V{:}$. Thus

$$\frac{dZ(s,t)}{dt} = \frac{1}{2}\int \left[\iint \frac{\delta{:}V{:}}{\delta\psi(x)}\frac{d}{dt}c(x,y)\frac{\delta{:}V{:}}{\delta\psi(y)}\,dx\,dy\right] e^{-{:}V{:}}\,d\psi_\omega(s,t).$$

Here $c(x, y)$ is the kernel of $C_\omega(s, t)$. By (2.2.11), $dC_\omega/dt \geqslant 0$, so dc/dt is a positive bilinear form and $[\cdots] \geqslant 0$. Thus Z is increasing in t.

PROPOSITION 2.3.2. *Let* $\zeta \leqslant \pi^2/32$. There is a constant K (independent of ω, $\boldsymbol{\omega}$, s, t, ζ, κ) such that

$$\int \exp\left(\int_Y \zeta{:}(\delta\psi_\kappa)^2{:}\,dx\right) d\psi_\omega(s,t) \leqslant \exp(K\zeta \mid Y \mid).$$

Proof. We may suppose that the exponent is Wick ordered in the covariance $C_\omega(s, t)$. In fact, the modification $s = 1 \to s \leqslant 1$ decreases Wick subtractions and increases the integral. The further changes $t = 0 \to t \geqslant 0$ and $\omega = 1 \to \omega \leqslant 1$ are monotonic in the other direction, and so it is sufficient to examine the effect on the Wick constant when $\omega = 0$, $t = 1$. In this case we use the fact that $\delta\psi_\kappa$ is orthogonal to the constant function e_Δ to conclude that the change in the Wick constant is still $O(\zeta \mid Y \mid)$. Thus Proposition 2.3.1 is applicable. Choose Δ_N to have Neumann data on all unit lattice lines. The value $t = 1$ then provides an upper bound. However, for $t = 1$, the integral factors into a product over unit lattice squares, and so it is sufficient to consider a single lattice square.

Let P_Δ denote the projection in L_2 onto e_Δ. The integral for a single lattice square is given by

$$\exp(-\tfrac{1}{2}\operatorname{tr}\ln(1 - A) - \tfrac{1}{2}\operatorname{tr} A) \leqslant \exp(\text{const} \| A \|_{H.S.}^2)$$

provided $\| A \| \leqslant \frac{1}{2}$ where A is defined by

$$A = \zeta[-\Delta_N + \boldsymbol{\omega}]^{-1}[\rho_\kappa{}^2 - P_\Delta].$$

Since $\rho_\kappa{}^2 - P_\Delta$ annihilates the ground state e_Δ of Δ_N and $-\Delta_N \upharpoonright \{e_\Delta\}^\perp \geqslant \pi^2$, the inverse is uniformly bounded for $0 \leqslant \boldsymbol{\omega} \leqslant 1$. Moreover, because $\| \rho_\kappa{}^2 \| \leqslant 16$ the condition $\| A \| \leqslant \frac{1}{2}$ is satisfied for $\zeta \leqslant \pi^2/32$. The bound on the Hilbert–Schmidt norm of A uniform in $\boldsymbol{\omega}$ follows similarly. (The restriction $\| A \| \leqslant 1 - \epsilon$ was unfortunately omitted in [18].)

Recall that $\psi = \phi - g$ and let $\delta g = g(x) - \int_\Delta g(x)\,dx$, $x \in \Delta$.

PROPOSITION 2.3.3. *Let* $\zeta \leqslant \pi^2/64$. *There is a constant* K (*independent of* ω, $\boldsymbol{\omega}$, s, t, ζ, κ) *such that*

$$\int \exp\left(\int_Y \zeta{:}(\delta\phi_\kappa)^2{:}\,dx\right) d\psi_\omega(s,t) \leqslant \exp\left(\zeta K\left(\int_Y (1 + |\nabla g|^2)\,dx\right)\right).$$

Proof. We expand

$$:\delta\phi_\kappa(x)^2: = :\delta\psi_\kappa(x)^2: + 2\delta\psi_\kappa(x)\,\delta g_\kappa(x) + \delta g_\kappa(x)^2. \tag{2.3.2}$$

We claim

$$(16)^{-1} \int_Y (\delta g_\kappa)^2\,dx \leqslant \int_Y (\delta g)^2\,dx \leqslant \pi^{-2} \int_Y (\nabla g)^2\,dx. \tag{2.3.3}$$

The first inequality follows by (2.1.8) while the second inequality follows from the bound

$$\pi \leqslant -\Delta_N \upharpoonright (\{e_\Delta\}^\perp)$$

as in the proof of Proposition 2.3.2.

A Schwarz inequality allows us to consider separately the $:\delta\psi^2:$ and $\delta\psi$ terms from expansion (2.3.2) if we replace ζ by 2ζ. The $:\delta\psi^2:$ integral is bounded by Proposition 2.3.2. The $\delta\psi$ integral can be explicitly evaluated as

$$\exp\left[8\zeta^2 \int_{Y\times Y} \delta g_\kappa(x)(\rho_\kappa{}^*C\rho_\kappa)(x, y)\,\delta g_\kappa(y)\,dx\,dy\right] \leqslant \exp\left[O(1)\,\zeta^2 \int_Y (\nabla g)^2\,dx\right]. \tag{2.3.4}$$

Here ρ_κ is the momentum cutoff operator and $C = C_\omega(s, t)$. We have used in (2. 3. 4), the bound

$$0 \leqslant E\rho_\kappa C\rho_\kappa E \leqslant E\rho_\kappa(-\Delta_N)^{-1}\rho_\kappa E \leqslant (4/\pi)^2 E$$

valid on $\{e_\Delta\}^\perp$. Here E is the operator of multiplication by the characteristic function of Y.

2.4. *Phase Boundaries: The Leading Term*

The occurrence of phase boundaries, i.e., $\Sigma \neq \varnothing$, is reflected in the factor $\exp(-F)$ in the ground state measure $d\phi_{\Lambda,\Sigma}$ of (4.6). This factor is the effect of the change of variables $\phi \to \psi = \phi - g$ on the kinetic energy term $|\nabla g|^2$ in the action density.

The functions F_i of (I.4.7) are local, when $\operatorname{dist}(\partial Y, \Sigma) \geqslant L/2$. Locality means that in the notation of (I.4.8d),

$$F_i(Y, \Sigma) = F_i(Y, \Sigma \cap Y)$$

is independent of $\Sigma \sim Y$. See also Lemma I.4.1. The dominant contribution to F is F_1, which we now relate to the length $|\Sigma(Y)|$ of the phase boundaries in Y.

PROPOSITION 2.4.1. *Let $\partial Y \subset X$. Then*

$$\lambda^{-1}\,|\,4\ln\lambda\,|^{-4}\,|\,X^0 \cap Y\,| \leqslant (\xi_+ - \xi_-)^2\,|\,\Sigma \cap Y\,| \leqslant 12\eta^{-1}F_1(Y, \Sigma). \tag{2.4.1}$$

Proof. The first inequality follows from the definition (I.5.12) of X^0. For each segment $\gamma \subset \Sigma \cap Y$, let $X^0(\gamma)$ be the set of points within distance L of γ. Then

$$X^0 \cap Y \subseteq \bigcup_{\gamma \in \Sigma \cap Y} X^0(\gamma).$$

In fact, for $\Delta \subset X^0$, we have $\Delta \subset X^0(\gamma)$ for some $\gamma \in \Sigma$ by definition of X^0. Moreover if $\Delta \subset Y$ also, then $\gamma \subset Y$, because $\partial Y \subset X$ implies $\partial Y \cap X^0 = \phi$. Thus

$$(2L)^{-2} \mid X^0 \cap Y \mid \leqslant \mid \Sigma \cap Y \mid. \tag{2.4.2}$$

Since $L = (\ln \lambda)^2$, and since each jump has size $\xi_+ - \xi_- = (2\lambda)^{-1/2} + O(\mu)$, the first inequality holds.

To prove the second inequality, we note

$$g(x) - g(x') = \int_{x'}^{x} \nabla g \, ds$$

where $x \in \Delta$, $x' \in \Delta'$, and Δ, Δ' are neighboring squares with a common side γ. We integrate along a path parallel to the sides of Δ, Δ'. By the triangle and Schwarz inequalities,

$$\begin{aligned} \mid h(x) - h(x')\mid^2 &\leqslant (\mid h(x) - g(x)\mid + \mid g(x) - g(x')\mid + \mid g(x') - h(x')\mid)^2 \\ &\leqslant 3 \left(\mid h(x) - g(x)\mid^2 + \mid h(x') - g(x')\mid^2 + \int_{x}^{x} \mid \nabla g \mid^2 ds\right). \end{aligned}$$

We now integrate over Δ, Δ' and sum over $\mathcal{N}(Y)$, the nearest neighbor squares in Y. Since

$$\sum_{\mathcal{N}(Y)} (h(\Delta) - h(\Delta'))^2 = (\xi_+ - \xi_-)^2 \mid \Sigma \cap Y \mid, \tag{2.4.3}$$

the second inequality follows by (1.4.7a).

2.5. *Phase Boundaries: Error Estimates*

In the above section we found

$$e^{-F_1(Y)} \leqslant e^{-a\eta\lambda^{-1}|\Sigma \cap Y|} \tag{2.5.1}$$

for a constant a. Now we consider the error terms F_2, F_3, and F_4. The main results are contained in two estimates.

PROPOSITION 2.5.1. *Let $\partial Y \subset X$, let ∂Y be a Dirichlet contour, let $d\psi_\omega(s)$ be given by* (2.2.2), *let $0 < \eta_0 \leqslant 1$, and let $\lambda \leqslant \lambda(\eta_0, \omega)$ be sufficiently small. Then*

$$\int e^{-O(\eta_0^{-1}) F_3(Y)} \, d\psi_\omega(s) \leqslant e^{\lambda F_1(Y)}. \tag{2.5.2}$$

PROPOSITION 2.5.2. *Assume the hypotheses of Proposition 2.5.1 and in addition let* $0 \leqslant \zeta < \pi^2/2^7$, *let* $\omega \restriction \sim\Lambda = 1$, $1 \leqslant p$, *and let* $Y_t \subset Y \cap \Lambda$. *Then there exists a constant* K (*independent of* ω, $\boldsymbol{\omega}$, s, t, p, ζ, κ, η, η_0) *such that*

$$\int \exp\left[\zeta \int_{Y_t} :\delta\phi_\kappa{}^2(x): dx - pF_4(Y)\right] d\psi_\omega(s)$$

$$\leqslant \exp(K(|\, Y_t\,| + \zeta F_1(Y)) \exp(p^2(1-\eta)\, F_2(Y)). \tag{2.5.3}$$

LEMMA 2.5.3. *Let* $0 < \eta_0 \leqslant \eta \leqslant 1$, *and let* $\lambda \leqslant \lambda(\eta_0)$ *be sufficiently small. Then*

$$0 \leqslant \eta_\zeta - \eta \leqslant \lambda^2 \tag{2.5.4}$$

Proof. Recall that $\zeta = \zeta(x)$ is introduced below (I.4.3). Since $0 \leqslant \zeta \leqslant 1$,

$$\eta_\zeta - \eta = \eta\eta_\zeta(\eta^{-1} - \eta_\zeta^{-1}) = \eta\eta_\zeta \int (-\Delta + \eta)^{-1}(x)(1 - \zeta(x/L))\, dx \geqslant 0.$$

The integration above is supported in $|\, x\,| \geqslant L/4$, so

$$\eta_\zeta - \eta \leqslant \eta_\zeta\eta \int_{|x|\geqslant L/4} (-\Delta + \eta)^{-1}(x)\, dx$$

$$\leqslant O(e^{-\eta L/4}) \leqslant O(\lambda^2)$$

for small λ, since $L = |\ln \lambda\,|^2$.

LEMMA 2.5.4. *Let* $0 < \eta_0 \leqslant \eta \leqslant 1$ *and let* $\lambda \leqslant \lambda(\eta_0)$ *be sufficiently small. Then*

$$\| g - g_c \|_\infty + \| \nabla(g - g_c)\|_\infty \leqslant \lambda. \tag{2.5.5}$$

Proof. We estimate $g - g_c$.

$$(g - g_c)(x) = (\eta_\zeta - \eta)((-\Delta + \eta)^{-1} h)(x)$$

$$+ \eta_\zeta \int (-\Delta + \eta)^{-1}(x - y)\left[\zeta\left(\frac{x-y}{L}\right) - 1\right] h(y)\, dy.$$

Thus

$$\| g - g_c\|_\infty \leqslant (\eta_\zeta - \eta)\, \|(-\Delta + \eta)^{-1} h\|_\infty$$

$$+ \text{const}\, \| h\|_\infty\, \eta_\zeta\eta^{-1}(1 + \eta L^2)\, e^{-(\eta^{1/2}L/4)}. \tag{2.5.6}$$

By Lemma 2.5.3 and

$$\|(-\Delta + \eta)^{-1} h\|_\infty \leqslant \eta^{-1}\, \| h\|_\infty \leqslant \text{const}\, \eta_0^{-1}\lambda^{-1/2},$$

the first term in (2.5.6) is bounded by $\lambda/4$ for λ small. For the second term in (2.5.6), we follow the proof of Lemma 2.5.3 to bound it by

$$\text{const } \lambda^{-1/2}\eta_0^{-1}\lambda^2 < \lambda/4$$

for small λ, so $\| g - g_c \| \leqslant \lambda/2$. The bound on $\nabla(g - g_c)$ is similar, using the fact $\nabla(-\Delta + \eta)^{-1}$ is a bounded operator from L_∞ to L_∞.

Proof of Proposition 2.5.1. Write $Y = \bigcup_{i=1}^{\nu} Y_i$ as a union of its connected components. Both sides of (2.5.2) factor, so we can assume without loss of generality that Y is connected. Then ∂Y consists of a finite number of closed curves, and each component of $R^2 \sim Y$ has exactly one of these curves for its boundary. In this case, $\Sigma \cap Y$ is a set of phase boundaries in R^2, which determines a new choice of spins, and this new choice of spins agrees with $\{\sigma_i\} = \Sigma$ in Y. Define h_Y, $g_{Y,c}$, and g_Y as in Section I.4, using $\Sigma \cap Y$ in place of Σ. By Lemma I.4.1,

$$\begin{aligned}(-\Delta + \eta)(g_Y - g_{Y,c}) &= (-\Delta + \eta)(g - g_c) \quad &&\text{on } Y \\ &= 0 &&\text{on } \sim Y.\end{aligned} \tag{2.5.7}$$

Thus

$$F_3(Y) = \langle \psi, (-\Delta + \eta)(g_Y - g_{Y,c})\rangle$$

and

$$\int e^{aF_3(Y)}\, d\psi_\omega(s) = \exp[2^{-1}a^2 \| C_\omega(s)^{1/2} (-\Delta + \eta)(g_Y - g_{Y,c})\|^2] \tag{2.5.8}$$

Since

$$C_\omega(s) \leqslant (-\Delta + \omega)^{-1},$$

(2.5.8) is majorized by

$$\exp[2^{-1}a^2\omega^{-1}\langle g_Y - g_{Y,c}, (-\Delta + \eta)(g_Y - g_{Y,c})\rangle].$$

By (2.5.1) and Lemma 2.5.4, the inner product above is bounded by $2\lambda^2 \mid X^0 \cap Y \mid$, which by Proposition 2.4.1, completes the proof.

Proof of Proposition 2.5.2. By conditioning we can take $t = 1$, so that the integral factors. The factors associated with squares $\Delta \subset Y_t$ are bounded by Proposition 2.3.3 because F_4 does not contribute. On squares $\Delta \not\subset \Lambda$, $\delta\phi$ does not contribute since $Y_t \subset \Lambda$, and ω does not contribute since $\omega = 1$ on $\sim\Lambda$. Thus

$$\begin{aligned}&\int e^{-pF_4(Y)}\, d\psi_\omega(s, t = 1) \\ &= \exp\left(\tfrac{1}{2}p^2(1 - \eta)^2 \int_{\sim\Lambda} (g - \xi_+)(-\Delta^N + 1)^{-1} (g - \xi_+)\, dx\right) \\ &\leqslant \exp(p^2(1 - \eta)\, F_2(Y)).\end{aligned}$$

2.6. *Spin Localization*

We give some simple estimates on the spin localization functions $\chi_\pm$ defined in (1.3.5). In fact, we bound the functions,

$$\chi_{a,b}(\xi) = \pi^{-1/2} \int_a^b e^{-(\xi-z)^2}\, dz,$$

which can be used for the spin localization of a more general interaction polynomial.

LEMMA 2.6.1. *Let $\theta < 1$. There is a constant $K = K(\theta)$ such that*

$$\chi_{a,b}(\xi) \leqslant K e^{-\theta(\xi-a)^2}, \qquad \xi \leqslant a \tag{2.6.1}$$

$$\chi_{a,b}(\xi) \geqslant 1 - K[e^{-\theta(\xi-a)^2} + e^{-\theta(\xi-b)^2}], \qquad a \leqslant \xi \leqslant b \tag{2.6.2}$$

$$\chi_{a,b}(\xi) \leqslant K e^{-\theta(\xi-b)^2}, \qquad b \leqslant \xi. \tag{2.6.3}$$

Proof. For $\xi \leqslant a$ (or similarly for $\xi > b$),

$$\begin{aligned}\chi_{a,b}(\xi) &\leqslant e^{-\theta(\xi-a)^2}\pi^{-1/2} \int_a^b e^{-(1-\theta)(\xi-z)^2}\, dz \\ &\leqslant (1-\theta)^{-1/2}\, e^{-\theta(\xi-a)^2}.\end{aligned}$$

For $a \leqslant \xi \leqslant b$, we use $\chi_{a,b} = 1 - \chi_{-\infty,a} - \chi_{b,\infty}$ and the above.

LEMMA 2.6.2. *For $\theta < 1$, there is a constant $K = K_\theta < \infty$ such that for all integers $n \geqslant 1$,*

$$\left| \frac{d^n}{d\xi^n} \chi_{a,b}(\xi) \right| \leqslant Kn!\, (e^{-\theta(\xi-a)^2} + e^{-\theta(\xi-b)^2}).$$

Proof. For $\theta < \eta < 1$, we obtain from Cauchy's integral formula the bound

$$\left| \frac{d^n}{d\xi^n} e^{-(\xi-z)^2} \right| \leqslant Kn!\, e^{-\eta(\xi-z)^2}$$

for real ξ, z. For $\xi \leqslant a$, or $\xi \geqslant b$, the bound of the lemma follows, as in Lemma 2.6.1. For $a \leqslant \xi \leqslant b$ we use the identity

$$\frac{d^n}{d\xi^n} \chi_{a,b} = -\frac{d^n}{d\xi^n} (\chi_{-\infty,a} + \chi_{b,\infty})$$

and proceed as before.

3. The Vacuum Energy Bound

3.1. *The Wick Ordering Lower Bound*

In this section we consider the interaction polynomial

$$:\mathscr{P}(\phi_\kappa(x)): = \lambda :\phi_\kappa(x)^4: - \tfrac{1}{4} :\phi_\kappa(x)^2: - \mu\phi_\kappa(x) - E_c \tag{3.1.1}$$

under the restrictions $0 < \lambda \leqslant \frac{1}{2}$, $0 \leqslant \mu \leqslant \lambda^2$, $E_c = -(64\lambda)^{-1} + O(\lambda^2)$, and with Wick ordering given by the covariance $(-\Delta + 1)^{-1}$. See I.2 for further details.

THEOREM 3.1.1. *Let $\eta \leqslant 10^{-2}$ and $81\eta \leqslant \zeta$. Then there is a constant $b = b(\zeta)$ such that for any $:\mathscr{P}(\phi_\kappa(x)):$ as above, any $x \in \Delta$ and any Σ,*

$$:\mathscr{P}(\phi_\kappa(x)): - \eta :(\phi_\kappa(x) - h(x))^2: - \ln \chi_{\sigma(\Delta)}(\bar{\phi}(\Delta)) \geqslant -b(\zeta) \ln^2\kappa - \zeta :\delta\phi_\kappa(x)^2:. \tag{3.1.2}$$

Proof. We define τ by

$$\begin{aligned}\tau(x) &= :\mathscr{P}(\phi_\kappa(x)): - \eta :(\phi_\kappa(x) - h)^2: + \zeta :\delta\phi_\kappa(x)^2: \\ &= \mathscr{P}(\phi_\kappa(x)) - \eta(\phi_\kappa(x) - h)^2 + \zeta\delta\phi_\kappa(x)^2 - 6\lambda c_\kappa \phi_\kappa^2 \\ &\quad + 3\lambda c_\kappa^2 + \eta c_\kappa - \zeta\delta c_\kappa ,\end{aligned}$$

where c_κ, $\delta c_\kappa = O(\ln \kappa)$ are the Wick constants for ϕ_κ and $\delta\phi_\kappa$. We establish

$$\tau(x) - \ln \chi_{\sigma(\Delta)}(\Delta) \geqslant -O(1) \ln^2\kappa \tag{3.1.3}$$

and the theorem then follows.

Note for any $\epsilon > 0$ (and using $\mu\lambda \leqslant \lambda^3 \leqslant \lambda^2$),

$$\begin{aligned}\lambda(6c_\kappa + \epsilon/4)\, \phi_\kappa^2 + \epsilon\mu\lambda\phi_\kappa &\leqslant \epsilon\lambda^2\phi_\kappa^4 + (2\epsilon)^{-1}(6c_\kappa + \epsilon/4)^2 + 2\epsilon\lambda^3 \\ &\leqslant \epsilon\lambda^2\phi_\kappa^4 + O(\epsilon^{-1}) \ln^2\kappa.\end{aligned}$$

Also $|\epsilon\lambda E_c| \leqslant O(1)$, so

$$\tau(x) \geqslant (1 - \epsilon\lambda)\, \mathscr{P}(\phi_\kappa(x)) - \eta(\phi_\kappa - h)^2 + \zeta(\delta\phi_\kappa)^2 - O(\epsilon^{-1}) \ln^2\kappa. \tag{3.1.4}$$

We now fix $x \in \Delta$ and specify $h(\Delta)$ and $\phi(\Delta)$. This results in a number of cases to establish (3.1.3). We consider in detail $h(\Delta) = -\xi_+$. The estimates for $h(\Delta) = \xi_+$ are identical if we exchange ξ_+ and $-\xi_+$ throughout.

Case 1. $h(\Delta) = -\xi_+$ and $|\phi_\kappa(x) - \xi_+| > \xi_+/3$. In this case $h(\Delta)$ is localized in one well of the W-shaped potential, while $\phi_\kappa(x)$ is localized outside the other well. Write $\mathscr{P}(\xi)$ in the form

$$\mathscr{P}(\xi) = \lambda(\xi - (8\lambda)^{-1/2})^2 (\xi + (8\lambda)^{-1/2})^2 - \mu\xi - \delta E_c ,$$

where $\delta E_c = E_c(\mu) - E_c(\mu = 0)$ is the shift in classical energy density for $\mu > 0$. By (I.2.8),

$$| \delta E_c | \leqslant O(\lambda^{3/2}).$$

Using (I.2.6), we see that if $| \xi - \xi_+ | > \xi_+/3$, then for some $\delta > 0$

$$\begin{aligned} \mathscr{P}(\xi) &\geqslant \delta(\xi + \xi_+)^2 - O(\lambda)(\xi - \xi_-) - O(\lambda), \\ &\geqslant \delta(1 - \epsilon\lambda)(\xi + \xi_+)^2 - O(\epsilon^{-1}). \end{aligned}$$

Substituting this bound into (3.1.4) yields

$$\begin{aligned} \tau &\geqslant [\delta(1 - \epsilon\lambda)^2 - \eta](\phi_\kappa(x) + \xi_+)^2 + \zeta(\delta\phi_\kappa)^2 - O(\epsilon^{-1}) \ln^2 \kappa \\ &\geqslant -O(\epsilon^{-1}) \ln^2 \kappa. \end{aligned} \tag{3.1.5}$$

In the last bound we choose $\epsilon = 10^{-3}$ and $\eta < 10^{-2}$ so the coefficient of $(\phi + \xi_+)^2$ is positive. Since $\chi_\sigma \leqslant 1$ and $-\ln \chi_\sigma \geqslant 0$, (3.1.3) follows.

Case 2. $h(\Delta) = -\xi_+$, $| \phi_\kappa(x) - \xi_+ | \leqslant \xi_+/3$. In this case h and $\phi_\kappa(x)$ are localized in different wells. Recall $\phi_\kappa(x) = \bar{\phi}(x) + \delta\phi_\kappa(x)$; thus in the region with $| \phi_\kappa - \xi_+ | \leqslant \xi_+/3$, either

$$\text{(a) } \delta\phi_\kappa \geqslant \xi_+/3 \qquad \text{or} \qquad \text{(b) } \bar{\phi} \geqslant \xi_+/3.$$

Subcase (a). We bound (3.1.4) from below, omitting the positive term $\mathscr{P}(\phi_\kappa)$. Then

$$\tau \geqslant -\eta(\phi_\kappa + \xi_+)^2 + \zeta(\delta\phi_\kappa)^2 - O(\epsilon^{-1}) \ln^2\kappa.$$

Using $(\phi_\kappa - \xi_+)^2 \leqslant \xi_+^2/9$,

$$(\phi_\kappa + \xi_+)^2 \leqslant 2(\phi_\kappa - \xi_+)^2 + 8\xi_+^2 \leqslant 9\xi_+^2.$$

Thus for $\eta \leqslant \zeta/81$,

$$\begin{aligned} \tau &\geqslant -9\eta\xi_+^2 + \zeta(\delta\phi_\kappa)^2 - O(\epsilon^{-1}) \ln^2\kappa \\ &\geqslant (9^{-1}\zeta - 9\eta)\, \xi_+^2 - O(\epsilon^{-1}) \ln^2\kappa \\ &\geqslant -O(\epsilon^{-1}) \ln^2\kappa. \end{aligned} \tag{3.1.6}$$

Subcase (b). We bound (3.1.4) from below (using $\mathscr{P}(\phi_\kappa) \geqslant 0$) by

$$\begin{aligned} \tau &\geqslant -\eta(\phi_\kappa + \xi_+)^2 + \zeta(\delta\phi_\kappa)^2 - O(\epsilon^{-1}) \ln^2 \kappa \\ &\geqslant -2\eta(\bar{\phi} + \xi_+)^2 + (\zeta - 2\eta)(\delta\phi_\kappa)^2 - O(\epsilon^{-1}) \ln^2 \kappa. \end{aligned} \tag{3.1.7}$$

Note by hypothesis (b) that

$$(\bar{\phi} + \xi_+)^2 \leqslant 20\bar{\phi}^2.$$

Thus for $\eta \leqslant \zeta/2$

$$\tau \geqslant -40\eta\bar{\phi}^2 - O(\epsilon^{-1}) \ln^2\kappa. \tag{3.1.8}$$

We now use Lemma 2.6.1, and the fact that χ_σ and ϕ_κ are localized in opposite wells. Thus for $\theta < 1$,

$$-\ln \chi_{\sigma(\Delta)} \geqslant \theta\bar{\phi}^2 - O(1).$$

Choose $\eta < 1/80$ and $\theta \geqslant \frac{1}{2}$. Then

$$\tau - \ln \chi_{\sigma(\Delta)} \geqslant -O(\epsilon^{-1}) \ln^2\kappa. \tag{3.1.9}$$

The bounds (3.1.5), (3.1.6), and (3.1.9) establish (3.1.3) and the theorem.

We obtain a stronger Wick bound in case χ_σ is replaced by a derivative χ_σ'. In this case $\chi_\sigma'(\bar{\phi})$ has its maximum near the boundary of the localization region, i.e., near near $\bar{\phi} = 0$. However, this region is near the relative maximum $\mathscr{P}(0) = O(\lambda^{-1})$ of the W-shaped potential $\mathscr{P}$. Let $n(\Delta)$ denote the order of the derivative of χ_σ in Δ.

THEOREM 3.1.2. *Let $\zeta > 0$. There exist constants $a = a(\zeta) > 0$ and $b = b(\zeta)$ such that for $\mathscr{P}$, η, σ, and χ as in Theorem* 3.1.1 *and* $n(\Delta) > 0$,

$$\begin{aligned} &:\mathscr{P}(\phi_\kappa(x)): - \eta:(\phi_\kappa(x) - h(x))^2: - \ln |\chi'_{\sigma(\Delta)}(\bar{\phi}(\Delta))| \\ &\qquad \geqslant a\lambda^{-1} - b \ln^2 \kappa - \zeta:\delta\phi_\kappa(x)^2: - \ln(n(\Delta)!) \end{aligned} \tag{3.1.10}$$

Proof. We modify the proof of Theorem 3.1.1. As before, it is sufficient to consider in detail $h(\Delta) = -\xi_+$.

Case 1. $h(\Delta) = -\xi_+$ and $| \phi_\kappa(x) - \xi_+ | \geqslant \xi_+/3$. As in (3.1.5) there is a constant $a_1 > 0$, independent of λ, μ, η, κ such that

$$\tau \geqslant a_1(\phi_\kappa + \xi_+)^2 + \zeta(\delta\phi_\kappa)^2 - O(1) \ln^2\kappa. \tag{3.1.11}$$

We consider two subcases

$$\text{(i) } | \phi_\kappa(x) + \xi_+ | \geqslant \xi_+/3, \qquad \text{(ii) } | \phi_\kappa(x) + \xi_+ | \leqslant \xi_+/3.$$

In subcase (i),

$$\tau \geqslant a_1\xi_+^2/9 - O(1) \ln^2\kappa \geqslant a\lambda^{-1} - O(1) \ln^2\kappa. \tag{3.1.12}$$

In subcase (ii), either

$$\text{(a) } | \delta\phi_\kappa | \geqslant \xi_+/3 \qquad \text{or} \qquad \text{(b) } |\bar{\phi} | \geqslant \xi_+/3.$$

If (a) holds, then

$$\tau \geqslant \zeta\xi_+{}^2/9 - O(1)\ln^2\kappa \geqslant a\lambda^{-1} - O(1)\ln^2\kappa. \tag{3.1.13}$$

If (b) holds, then by Lemma 2.6.2, for $\theta < 1$,

$$\begin{aligned} -\ln|\chi'_{\sigma(\Delta)}| &\geqslant \theta\bar{\phi}^2 - O(1) - \ln(n!) \\ &\geqslant \theta\xi_+{}^2/9 - O(1) - \ln(n!) \end{aligned}$$

where $n = n(\Delta)$ is the order of the derivative of χ in Δ. Thus

$$\tau - \ln|\chi'_{\sigma(\Delta)}| \geqslant a\lambda^{-1} - O(1) - \ln(n!) \tag{3.1.14}$$

Case 2. $h(\Delta) = -\xi_+$, $|\phi_\kappa(x) - \xi_+| \leqslant \xi_+/3$. Either

$$\text{(a) } \delta\phi_\kappa \geqslant \xi_+/3 \qquad \text{or} \qquad \text{(b)} \bar{\phi} \geqslant \xi_+/3.$$

Subcase (a) follows as in Theorem 3.1.1 and subcase (b) follows as in subcase (b) above. This completes the proof in all cases.

3.2. *The Vacuum Energy*

The main technical estimate of this chapter is the vacuum energy bound. With Σ and $n(\Delta)$ given let X' be the union of the lattice squares Δ in which χ_Δ is differentiated, and thus for which $n(\Delta) \neq 0$. For any set Y let

$$Y_c = Y \cap (X^0 \cup X') \cap \Lambda, \qquad Y_d = (Y \sim Y_c) \cap \Lambda$$

denote convergent and weakly divergent regions in the vacuum energy estimates.

THEOREM 3.2.1. *Let $\eta > 0$ be sufficiently small. There are strictly positive constants $a = a(\eta)$, $b = b(\eta)$, and $c = c(\eta)$ with the following property. Let $Q(Y)$ be defined by* (I.4.8). *Let $1 \leqslant p \leqslant 1 + \eta/30$. Then*

$$\begin{aligned} &\int e^{-pQ(Y)} |\chi'_{\Sigma\cap Y}|^p \, d\psi_1(s) \\ &\qquad \leqslant \prod_{\Delta\subset Y} (n(\Delta)!) \exp[-aF_1 - b\lambda^{-1/2}|Y_c| + c\lambda^{1/2}|Y_d|]. \end{aligned}$$

The idea of the proof is to express the high momentum tail $\delta V_\kappa \equiv V - V_\kappa$ in terms of ψ, and then to apply to $V_\kappa \equiv V(\phi_\kappa)$ the Wick bounds of Theorems 3.1.1–3.1.2. The resulting bound is useful, but too crude, and is refined by including effects of first-order perturbation theory in Y_d. In this part of the proof, the perturbation V is expressed in terms of ψ. First we obtain detailed properties of the coefficients of ψ in V, δV_κ. Writing

$$V(\Delta) = \sum_{j=0}^{4} \int_\Delta k_j(x) :\psi^j(x):_1 \, dx, \tag{3.2.1}$$

where: $:_1$ denotes a Wick ordering mass squared of 1, we compute

$$\begin{aligned} k_4 &= \lambda, \qquad k_3 = 4\lambda g, \qquad k_2 = \tfrac{3}{4}(8\lambda g^2 - 1), \\ k_1 &= \tfrac{1}{2} g(8\lambda g^2 - 1) - \mu - \eta(g - h), \\ k_0 &= g^2(\lambda g^2 - \tfrac{1}{4}) - E_c - \mu g - \tfrac{1}{2}\eta(g - h)^2. \end{aligned}$$

LEMMA 3.2.2. *The following bounds are satisfied:*

$$\begin{aligned} | k_j(x)| &\leqslant O(\lambda^{1/2}) + O(1) \,| g(x) - h(x)|, \qquad j > 0 \\ | k_0(x)| &\leqslant O(\lambda) + O(1) \,| g(x) - h(x)|^2 \\ | k_j(x)| &\leqslant O(\lambda^{1/2}) \qquad \text{on } X. \end{aligned}$$

Proof. Note by (I.2.6) and (I.4.2),

$$\| g \|_\infty \leqslant \| h \|_\infty \leqslant O(\lambda^{-1/2}). \tag{3.2.2}$$

Thus $| k_4(x)|$ and $| k_3(x)|$ are $O(\lambda^{1/2})$. We next show

$$| 8\lambda g(x)^2 - 1 | \leqslant O(\lambda) + O(\lambda^{1/2}) \,| g(x) - h(x)|. \tag{3.2.3}$$

In fact, by (I.2.6) and $0 \leqslant \mu \leqslant \lambda^2$,

$$8\lambda h(x)^2 = 1 + O(\lambda).$$

So by (3.2.2)

$$8\lambda g^2 - 1 = 8\lambda(g - h)(g + h) + O(\lambda) \leqslant O(\lambda^{1/2}) \,| g - h | + O(\lambda),$$

proving (3.2.3). Using (3.2.2)–(3.2.3) we bound $| k_2(x)|$, $| k_1(x)|$ as desired. By (I.2.6),

$$E_c + (64\lambda)^{-1} = O(\mu\lambda^{-1/2}) \leqslant O(\lambda).$$

Thus

$$g^2(\lambda g^2 - \tfrac{1}{4}) - E_c = (64\lambda)^{-1}(8\lambda g^2 - 1)^2 + O(\lambda),$$

and using (3.2.3) we bound $| k_0(x)|$, to complete the proof.

For a lattice square $\Delta \subset \Lambda$, the basic polynomial for the Wick bound estimates is,

$$\begin{aligned} W(\Delta) &= V(\Delta) + \tfrac{1}{2}(1-\eta)\int_\Delta :\psi^2:_1 dx \\ &= \int_\Delta :\mathscr{P}(\phi):_1 dx - \tfrac{1}{2}\eta \int_\Delta :(\phi - h)^2:_1 dx \end{aligned} \tag{3.2.4}$$

and we define $W_\kappa(\Delta)$ by the substitution $\phi \to \phi_\kappa$. Let $\delta W_\kappa(\Delta) = W(\Delta) - W_\kappa(\Delta)$. Then

$$\delta W_\kappa(\Delta) = \sum_{j=1}^{4} \int_\Delta \hat{k}_j(x)(:\psi^j(x): - :\psi_\kappa{}^j(x):)\, dx + \sum_{j=1}^{3} \int_\Delta \delta\hat{k}_j(x, \kappa) :\psi_\kappa{}^j(x): d, \tag{3.2.5}$$

where

$$\begin{aligned} \delta\hat{k}_3 &= \lambda(g - g_\kappa) \qquad \delta\hat{k}_2 = 6\lambda(g^2 - g_\kappa{}^2) \\ \delta\hat{k}_1 &= \tfrac{1}{2}g(8\lambda g^2 - 1) - \tfrac{1}{2}g_\kappa(8\lambda g_\kappa{}^2 - 1) - \eta(g - g_\kappa) \\ \delta\hat{k}_0 &= g^2(\lambda g^2 - \tfrac{1}{4}) - g_\kappa{}^2(\lambda g_\kappa{}^2 - \tfrac{1}{4}) - \mu(g - g_\kappa) \\ &\quad - \tfrac{1}{2}\eta[(g-h)^2 - (g_\kappa - h)^2], \\ \hat{k}_j &= k_j, \qquad j \neq 2, \qquad \hat{k}_2 = 6\lambda g^2 - \tfrac{1}{4} - (\eta/2). \end{aligned}$$

LEMMA 3.2.3. *For each $p < \infty$ there is a constant $\epsilon = \epsilon(p) > 0$ such that the following bounds are satisfied*

$$\begin{aligned} |\hat{k}_j(x)| &\leqslant O(1)(1 + |g - h|) \leqslant O(\lambda^{-1/2}) \\ \|\delta\hat{k}_j\|_{L_p(\Delta)} &\leqslant O(\lambda^{-1}\kappa^\epsilon) \\ \delta\hat{k}_j = 0 \text{ on } X; &\qquad |\hat{k}_j(x)| \leqslant O(1) \text{ on } X. \end{aligned}$$

Proof. The bounds on $\hat{k}_j$ are a consequence of the preceding lemma while the L_p bound on $\delta\hat{k}_j$ follows from (2.1.4). Since $g = h$ on X by (I.4.5), $k_j = $ const on lattice squares $\Delta \subset X$ and so by (iv) of Section 2.1, $\delta\hat{k}_j = 0$ on X.

Recall that ω, $d\psi_\omega(s)$, and $C_\omega(s)$ are defined in (2.2.1)–(2.2.5). We require an L_q bound on some fractional derivative of $C_\omega(s)$ as follows from a convergent Neumann series expansion combined with [16, Propositions 7.3–7.4]. In fact,

$$C_\omega(s) = C + C(1-\omega)\, C_\omega(s), \tag{3.2.6}$$

where C belongs to the class $\mathscr{C}$ of [16] of covariance operators with fractional derivatives in L_q (see also [2]). After differentiating the left variable of $C_\omega(s)$ (i.e., C in (3.2.6)), we use the majorization $0 \leqslant 1 - \omega \leqslant 1$ and

$$0 \leqslant C_\omega(s) \leqslant (-\Delta + \omega^2)^{-1}$$

(pointwise as an inequality between kernels of operators) and the fact that $(-\Delta + \omega^2)^{-1}$, as an L_1 convolution operator, preserves L_q norms.

LEMMA 3.2.4. *Let* ω, $d\psi_\omega$, *and* δW_κ *be as above. There are positive constants* $K(\omega)$ *and* δ *with the following property. Let* $\{m(\Delta):\Delta \subset Y \cap \Lambda\}$ *be a set of nonnegative integers and let* $\{\kappa(\Delta):\Delta \subset Y \cap \Lambda\}$ *be a set of positive numbers with* $\kappa(\Delta)^\delta \geqslant \lambda^{-1}$ *for* $\Delta \subset X^0$. *Then*

$$\left| \int \left[\prod_{\Delta \subset Y \cap \Lambda} \delta W_{\kappa(\Delta)}(\Delta)^{m(\Delta)} \right] d\psi_\omega(s) \right| \leqslant \prod_{\Delta \subset Y \cap \Lambda} [m(\Delta)!^2 \, (K(\omega)\, \kappa(\Delta)^{-\delta})^{m(\Delta)}].$$

Proof. As in [13, 5, 16].

LEMMA 3.2.5. *There is a constant* K *such that for* q *a positive even integer and for* Y_t *a subset of* X,

$$\left| \int \prod_{\Delta \subset Y_t} V(\Delta)^q \, d\psi_1(s) \right| \leqslant ((q!)^2 (K\lambda)^{q/2})^{|Y_t|}. \tag{3.2.7}$$

Proof. For $x \in X$, $g = h$ by Lemma I.4.1. Thus by Lemma 3.2.2, the kernels $k_j(x)$ satisfy $\| k_j \|_\infty \leqslant O(\lambda^{1/2})$. Hence, the lemma follows from the standard bound

$$\prod_{\Delta \subset Y_t} \left(K_1{}^q (q!)^2 \sum_{j=1}^{4} \| k_j \|_\infty \right)$$

on (3.2.7).

Next we establish Theorem 3.2.1 with the factor $\lambda^{1/2} \mid Y_d \mid$ replaced by the weaker bound $\mid Y_d \mid$. To take advantage of the fact that the coefficients of V are $O(\lambda^{1/2})$ in X and, ultimately, to recover the improved bound $O(\lambda^{1/2}) \mid Y_d \mid$ of Theorem 3.2.1, we perturb V in each lattice square $\Delta \subset Y_d \subset X$. For this purpose we define

$$V(t, Y) = \sum_{\Delta \subset Y_d} t(\Delta)\, V(\Delta) + V(Y_c).$$

Let $Y_t = \{\Delta : t(\Delta) \neq 0\}$. Since $T_t \subset Y_d \subset X$, the previous lemma applies to this choice of Y_t.

LEMMA 3.2.6. *Under the hypothesis of Theorem* 3.2.1,

$$\int e^{-p(F(Y)+V(t,Y))} \mid \chi'_{\Sigma \cap Y} \mid^p d\psi_1(s)$$

$$\leqslant \left(\prod_{\Delta \subset Y \cap \Lambda} n(\Delta)! \right) \exp[-aF_1(Y) - b\lambda^{-1/2} \mid Y_c \mid + O(1) \mid Y_t \mid).$$

Proof. Let

$$W(t, Y) = V(t, Y) + \tfrac{1}{2}(1-\eta)\int_{Y_t} :\psi^2:_1 \, dx.$$

Then with $\omega = 1 - p(1-\eta)\chi_{Y_t}$,

$$\begin{aligned}
&\int e^{-p(F(Y)+V(t,Y))} \,|\, \chi'_{\Sigma\cap Y} |^p \, d\psi_1(s) \\
&\quad = Z^{-1} \int e^{-p(F(Y)+W(t,Y))} \,|\, \chi'_{\Sigma\cap Y} |^p \, d\psi_\omega(s) \qquad (3.2.8) \\
&\quad \leqslant e^{-p(F_1+F_2)} Z^{-1} \left(\int e^{-q'pF_3}\, d\psi_\omega(s)\right)^{1/q'} \left(\int e^{-pq\left(F_4+W\right)} \,|\, \chi'_{\Sigma\cap Y} |^{pq} \, d\psi_\omega(s)\right)^{1/q}.
\end{aligned}$$

Here q and q' are dual Hölder exponents. We choose

$$q = 1 + \frac{\eta_0}{10}, \qquad p < 1 + \frac{\eta}{10}$$

so that qp, $q^2p^2 < 1 + \eta/2$, and $pq' = 0(\eta_0^{-1})$. The F_3 integral is bounded by Proposition 2.5.1 and Z^{-1} is bounded by (2.2.4), giving

$$\begin{aligned}
&\left|\int e^{-p(F+V(t,Y))} \,|\, \chi'_{\Sigma\cap Y} |^p \, d\psi_1(s)\right| \\
&\qquad \leqslant e^{-(1-\lambda)F_1 - pF_2} e^{K|Y_t|} \left(\int e^{-pq(F_4+W-\ln|\chi'_{\Sigma\cup Y}|)}\, d\psi_\omega(s)\right)^{1/q}
\end{aligned}$$

We bound the W-integral using the standard proof of the linear lower bound on the vacuum energy, see [13, 5]. The basic idea is: Consider the subset of paths for which

$$\begin{aligned}
&W(t, \Delta) + t(\Delta)\, \zeta{:}\delta\phi_\kappa\,(\Delta){:} - \ln |\, \chi'_{\sigma(\Delta)})| \\
&\qquad \leqslant \inf_\phi\, [W_\kappa(t, \Delta) + t(\Delta)\, \zeta{:}\delta\phi_\kappa{}^2(\Delta){:} - \ln |\, \chi'_{\sigma(\Delta)} | - 1].
\end{aligned}$$

On this set it is also true that

$$1 \leqslant |\, \delta W_\kappa(t, \Delta)| \leqslant \delta W_\kappa(t, \Delta)^m$$

for any positive even integer m. Thus

$$\begin{aligned}
I(\kappa, \Delta) &\equiv \inf_\phi\, [W_\kappa(t, \Delta) + t(\Delta)\, \zeta{:}\delta\phi_\kappa{}^2(\Delta){:} - \ln |\, \chi'_{\sigma(\Delta)} |] \\
&\geqslant \inf_\phi \Big[t(\Delta)\left({:}\mathscr{P}_\Delta(\phi_\kappa){:}_1 - \tfrac{1}{2}\eta \int_\Delta {:}(\phi_\kappa - h)^2{:}\, dx + \zeta \int_\Delta {:}\delta\phi_\kappa{}^2{:}\, dx\right) \\
&\qquad - \ln |\, \chi'_{\sigma(\Delta)} |\Big] \\
&\geqslant \begin{cases} -O(\ln^2 \kappa), & n(\Delta) = 0 \\ a\lambda^{-1} - O(\ln^2 \kappa) - \ln(n(\Delta)!), & n(\Delta) > 0 \end{cases}
\end{aligned}$$

by Theorems 3.1.1 and 3.1.2.

We choose $\kappa_i = 2^i$, $i = i(\Delta)$ a positive integer depending on Δ. Let $\sum_i'$ denote the sum over all such $i(\Delta)$, subject to the restriction $\kappa_\Delta^{i(\Delta)} \geqslant \lambda^{-1}$ if $\Delta \subset X^0$. This restriction is made to offset the $O(\lambda^{-1})$ divergence of the coefficients in X^0. Then

$$\begin{aligned}
\int & e^{-pq(F_4+W-\ln|x_\sigma'|)}\, d\psi_\omega(s) \\
&\leqslant \sum_i{}' \int d\psi_\omega(s) \Bigg(\prod_{\Delta\subset Y_t} (\delta W_{\kappa_{i(\Delta)}}(t, \Delta))^{m_i(\Delta)} \\
&\quad \times \exp[-pq(I(2\kappa_{i(\Delta)}, \Delta) - \zeta{:}\delta\phi_\kappa{}^2(\Delta){:})]\Bigg) e^{-pqF_4} \\
&\leqslant \left(\prod_\Delta n(\Delta)!\right) e^{-a\lambda^{-1}|X'\cap Y|} \sum_i{}' \int d\psi_\omega(s) \\
&\quad \times \prod_{\Delta\subset Y_t} (\delta W_{\kappa_{i(\Delta)}}^{m_i(\Delta)} \exp[(pq(O\ln^2\kappa_{i(\Delta)} + \zeta{:}\delta\phi^2_{\kappa_{i(\Delta)}}(\Delta)]) e^{-pqF_4}.
\end{aligned}$$

We separate $e^{pq(\zeta:\delta\phi^2:-F_4)}$ from the rest of the integrand with a Hölder inequality and bound it by Proposition 2.5.2. This gives as an upper bound for (3.2.8),

$$\begin{aligned}
& e^{(-(1-\lambda)+K\zeta)F_1(Y)} e^{K|Y_t|} \left(\prod_\Delta n(\Delta)!\right) e^{-b\lambda^{-1}|X'\cap Y|} \\
&\quad \times \sum_i{}' \left(\int d\psi_\omega(s) \prod_\Delta \delta W_{\kappa_{i(\Delta)}}^{q'm_i(\Delta)}\right)^{1/q'} e^{O\ln^2\kappa_{i(\Delta)}}.
\end{aligned}$$

Here if $i(\Delta)$ assumes its minimum value in Δ, then $m_{i(\Delta)}$ must be zero. Otherwise we choose

$$m_{i(\Delta)} = \kappa_{i(\Delta)}^{\delta/2}/q'.$$

We choose $\zeta < \pi^2/4$ and small enough so that the coefficient of F_1 above is bounded by $-\frac{3}{4}$. The δW integral is estimated by Lemma 3.2.4, and the factorials are bounded by Stirling's formula, leading to the upper bound for (3.2.8),

$$\begin{aligned}
& \exp(-\tfrac{3}{4}F_1(Y)) \exp(K\,|\,Y_t\,| - b\lambda^{-1}\,|\,X'\cap Y\,|) \\
&\quad \times \left(\prod_\Delta n(\Delta)!\right) \sum_i{}' \prod_{\Delta\subset Y_t} \exp(-\kappa_{i(\Delta)}^{\delta/2}/q' + O\ln^2\kappa_{i(\Delta)}),
\end{aligned}$$

where it is understood that $\kappa_{i(\Delta)}^{\delta/2}/q'$ is omitted from the exponent when $i(\Delta)$ assumes its minimum value. Recall that for $\Delta \subset X^0$, $\kappa_{i(\Delta)}^{\delta} \geqslant \lambda^{1/2}$, so that

$$\begin{aligned}
\sum_i{}' \cdots &\leqslant O(1)^{|Y_t|} \exp(O\ln^2(\lambda^{-1})\,|\,Y\cap X^0\,|) \\
&\leqslant O(1)^{|Y_t|} \exp(\lambda^{1/2}F_1(Y))
\end{aligned}$$

using Proposition 2.4.1. Substitution into the above upper bound, another use of Proposition 2.4.1, and the definitions of Y_c and Y_d complete the proof.

Proof of Theorem 3.2.1. We use the identity

$$e^{-pV(\Delta)} = 1 - pV(\Delta) \int_0^1 e^{-pt(\Delta)V(\Delta)}\, dt.$$

Thus with $t \equiv \{t(\Delta): \Delta \subset Y_t\}$ we have

$$\int e^{-pQ(Y)} \,|\, \chi'_{\Sigma\cap Y} \,|^p \, d\psi_1(s)$$
$$\leqslant \sum_{Y_t \subset Y_d} \int_0^1 \int \Big| \prod_{\Delta \subset Y_d} pV(\Delta)\Big| \, e^{-p[F(Y)+V(t,Y)]} \,|\, \chi'_{\Sigma\cap Y} \,|^p \, d\psi_1(s)\, ds\, dt$$
$$\leqslant \sum_{Y_t \subset Y_d} \int \Big| \prod_{\Delta \subset Y_t} pV(\Delta)\Big|^{q'} d\psi_1(q)^{1/q'} \sup_t \Big| \int e^{-pq[F(Y)+V(t,Y)]} \,|\, \chi'_{\Sigma\cap Y} \,|^{pq} \, d\psi_1(s)\Big|^{1/q}. \tag{3.2.9}$$

Here $q \simeq 1 + \eta_0/30$ so that $pq \leqslant 1 + \eta/10$ and $q' = O(\eta_0^{-1})$ is an even integer. Since $Y_d \subset X$, using Lemmas 3.2.5 and 3.2.6, we bound (3.2.9) by

$$\sum_{Y_t \subset Y_d} (K(\eta_0)\, \lambda^{1/2})^{|Y_t|} \Big(\prod_\Delta n(\Delta)!\Big) \exp(-aF_1 - b\lambda^{-1/2} \,|\, Y_c \,|)$$
$$= \Big(\prod_\Delta n(\Delta)!\Big)\Big(\prod_{\Delta \subset Y_d} (1 + K(\eta_0)\, \lambda^{1/2})\Big) \exp(-aF_1 - b\lambda^{-1/2} \,|\, Y_c \,|)$$
$$\leqslant \Big(\prod_\Delta n(\Delta)!\Big) \exp(-aF_1 - b\lambda^{-1/2} \,|\, Y_c \,| + c\lambda^{1/2} \,|\, Y_d \,|).$$

This completes the proof.

3.3. *Deviation from the Gaussian*

To prove convergence of the cluster expansion we need a bound of the form (1.1.5). Consider a monomial A of the form (I.6.1), localized in unit lattice squares Δ_i:

$$A = A(\psi) = \int \prod_i :\psi(x)^{m_i}: f_i(x)\, dx. \tag{3.3.1}$$

Recall that $|||\, A \,|||$ is defined by (I.6.2).

THEOREM 3.3.1. *There are positive constants b, c, d, and a constant K in* (I.6.2) *such that*

$$\left| \partial_s^{\Gamma} \int e^{-Q(Y)} \chi_\Sigma A \, d\psi_1(s) \right| \leqslant ||| A ||| \, e^{-dl|\Gamma|} e^{-b\lambda^{-1/2}|X^0 \cap Y|} e^{c\lambda^{1/2}|Y \cap \Lambda|} \tag{3.3.2}$$

for all λ *sufficiently small and for any consistent choice of* Γ, Σ, *and* Y.

Proof. Let $\mathscr{P}(\Gamma)$ be the set of partitions of the set Γ of l-lattice lines. As in [16], the left side of (3.3.2) equals

$$\sum_{\pi \in \mathscr{P}(\Gamma)} \int \left[\prod_{\gamma \in \pi} \tfrac{1}{2}(\partial_s^{\gamma} C(s)) \cdot \Delta_\psi \right] e^{-Q(Y)} \chi_\Sigma A \, d\psi_1(s). \tag{3.3.3}$$

We write $\partial_s^{\gamma} C(s)$ as a sum of its l-lattice localizations $j = (j_1, j_2) \in lZ^4$,

$$\partial_s^{\gamma} C(s) = \sum_{j_\gamma \in lZ^4} \partial_s^{\gamma} C(s, j_\gamma) \equiv \sum_{j_\gamma \in lZ^4} \chi_{\Delta_{j_{\gamma,1}}} \partial_s^{\gamma} C(s) \, \chi_{\Delta_{j_{\gamma,2}}}.$$

Here $\Delta_{j_{\gamma,1}}$ and $\Delta_{j_{\gamma,2}}$ are l-lattice squares. Consider a term, indexed by π and $J \equiv \{j_\gamma\}_{\gamma \in \pi}$ in the resulting sum. Each such term has the form

$$\sum{}' \int A' W(\pi, J) \, e^{-Q(Y)} \chi_\Sigma' \, d\psi_1(s) \tag{3.3.4}$$

where $\sum'$ ranges over possible derivatives of the various linear factors of A, of the factors $\chi_{\sigma(\Delta)}$ of χ_Σ and of $e^{-Q(Y)}$. Hölder's inequality yields as a bound on (3.3.4)

$$\sum{}' \| A' W(\pi, J) \|_{p'} \, \| \chi_\Sigma' e^{-Q(Y)} \|_p \tag{3.3.5}$$

where p and p' are dual indices. Using Theorem 3.2.1 for $p \simeq 1 + \eta/30$ and for p' an even integer, the second factor is bounded by

$$\left(\prod_{\Delta} n(\Delta)! \right) \exp[-aF_1 - b\lambda^{-1/2} \, | Y_c | + c\lambda^{1/2} \, | Y_d |]. \tag{3.3.6}$$

To bound the first factor in (3.3.5), we need to estimate the derivatives of $C(s)$ as in Proposition 8.1 of [16]. Let

$$d(j, \gamma) = \max\{\operatorname{dist}(j_1, b) + \operatorname{dist}(j_2, b) : b \in \gamma\}$$

and let

$$\| f(x, y) \|_{L_{q_1, q_2}} = \left(\int \left| \int | f(x, y) |^{q_2} \, dy \right|^{q_1/q_2} dx \right)^{1/q_1}.$$

Let γ^* be a shortest path passing through $b \in \gamma$ and let $| \gamma^* |$ be its length.

LEMMA 3.3.2. *Let* $1 \leqslant q_1, \ q_2 \leqslant q_0 < \infty$. There are constants $K_6(q_0, \gamma)$, $K_7(q_0)$, and $\delta > 0$ such that for l sufficiently large (λ sufficiently small)

$$\| \partial_s{}^\gamma C(s) \|_{L_{q_1 q_2}(\Delta_{j_1} \times \Delta_{j_2})} \leqslant l^2 K_6(q_0, \gamma) \exp(-\delta(d(j, \gamma) + | \gamma^* |)) \tag{3.3.7}$$

and

$$\sum_{\pi \in \mathscr{P}(\Gamma)} \prod_{\gamma \in \pi} K_6(q_0, \gamma) \leqslant e^{K_7(q_0)|\Gamma|}. \tag{3.3.8}$$

Proof. Follow [16, 27], except for modifications arising from the use of a large-scale distance and a fixed bare mass equal to one. We can extract a factor $e^{-\delta|\gamma^*|}$ from [16, (8.8)] since $| \gamma^* | \leqslant | l |$ for all $l \in \mathscr{L}(\gamma)$, in the notation of [16].

LEMMA 3.3.3. *Let p, p' be given. There is a* $K = K(p')$ in (I.6.2), *a* q_0 *in Lemma* 3.3.2, *and a* $\delta' > 0$ *such that for all* λ *sufficiently small,*

$$\| A' W(\pi, J) \|_{p'} \leqslant ||| A ||| \, e^{-\delta' l |\Gamma|} e^{l |Y \cap X'|} \prod_{\gamma \in \pi} K_6(q_0, \gamma) \, e^{-\delta' d(j_\gamma, \gamma)}. \tag{3.3.9}$$

Proof of Theorem 3.3.1. Since $Y \cap X' \subset Y_c$, the factor $\exp(l \mid Y \cap X' \mid)$ is controlled by the factor $\exp(-b\lambda^{-1/2} \mid Y_c \mid)$ in (3.3.6). We substitute (3.3.6) and (3.3.9) in (3.3.5). The sum $\sum'$ over derivatives in (3.3.4) and the sum over localizations J is bounded by [16, Lemmas 10.1–10.2]. The sum over $\pi \in \mathscr{P}(\Gamma)$ is dominated by (3.3.8).

Proof of Lemma 3.3.3. We modify [16, p. 239]. The L_{q_1, q_2} norms of (3.3.7), $1 \leqslant q_1, q_2 \leqslant q_0$ allow the use of [16, Lemma 9.2] even for l-lattice squares Δ_{j_ν} of area $l^2 \neq 1$. The other new issues are (a) the factor $l^2 = | \log \lambda |^{1/2}$ in (3.3.7), arising from the use of large lattice squares, (b) the large coefficients of $Q(Y)$ in case $\Delta_{j_\nu} \subset X^0$, and (c) another factor l^4, which is equal to the number of unit lattice square localizations for $\partial^\gamma C$, with the l-lattice localization j_γ held fixed.

The minimum nonzero value taken on by $| \gamma^* |$ and $d(j, \gamma)$ is l. Since $l^6 e^{-\delta l/2} \to 0$ as $\lambda \to 0$, the l^6 factor may be replaced by a small factor $e^{-\delta' l |\gamma|}$ if $| \gamma^* |$ or $d(j, \gamma)$ are nonzero, by a decrease in δ. Now suppose $| \gamma^* | = 0 = d(j, \gamma)$. Then $| \gamma | \leqslant 4$. Also $\Delta_j \subset X$ and so a derivative of e^{-Q} localized in Δ_j has only contributions from V(F_3 and F_4 do not contribute). For $\Delta_j \subset X$, the coefficient of V is $O(\lambda^{1/2})$ by Lemma 3.2.2 and since

$$l^6 \lambda^{1/2} \leqslant e^{-4\delta' l} \leqslant e^{-\delta' l |\gamma|}$$

for small λ, we obtain the desired factor $e^{-\delta' l |\gamma|}$ in this case. If the contraction is to $\chi_{\sigma(\Delta)}$, for Δ a unit lattice square in Δ_j, then $\Delta \subset Y \cap X'$ and we use the inequality

$$l^6 \leqslant e^{-\delta l |\gamma|} e^{l |\Delta|},$$

i.e., we use the contribution of $\Delta \subset Y \cap X'$ to the factor $\exp(l \mid Y \cap X' \mid)$ in (3.3.9).

For differentiation of a linear factor of A, we use the factor e^l from $e^{KlN(\Delta)}$ in (I.6.2) so that

$$l^6 \leqslant e^{-\delta' l|\gamma|} e^l.$$

This shows that l^6 is offset a factor by $e^{-\delta l|\gamma|}$ in all cases.

Next we consider the $O(\lambda^{-1})$ coefficient resulting from differentiating e^{-Q} in a square $\Delta \subset X^0$. In this case $d(j, \gamma) \geqslant L = |\log \lambda|^2$ and

$$O(\lambda^{-1}) e^{-\delta d(j,\gamma)} \leqslant {}^{-\delta d(j,\gamma)/2}.$$

This controls the modifications (a)–(c) mentioned above.

Let $M(\Delta)$ be the number of contractions in a unit lattice square Δ. As in [16],

$$\begin{aligned} \| A' W(\pi, J) \|_{p'} &\leqslant \prod_i \| f_i \|_{L_2} \prod_\Delta [p'(N(\Delta) + 4M(\Delta))]!^{1/p'} e^{O(1)(N(\Delta)+M(\Delta))} \\ &\quad \times \prod_{\gamma \in \pi} (K_6 e^{-\delta'(d(j,\gamma)+|\gamma|)}) e^{l|Y \cap X'|} e^{N(\Delta)}. \end{aligned} \tag{3.3.10}$$

To complete the proof, we must bound the factorials. We use the inequality,

$$(p'(N + 4M))!^{1/p'} \leqslant e^{\ln p'(N+4M)} (N + 4M)! \leqslant e^{\ln p'(N+4M)} N! (4M)^{4M} N^{4M}.$$

The factor $e^{\ln p' N}$ is bounded by $e^{K(p')lN}$ of (I.6.2). By the proof of [16, Lemma 10.2],

$$\exp(M^{3/2}) \leqslant e^{O|\Gamma|} e^{O\Sigma_\gamma d(j,\gamma)}.$$

Thus the factor $(4M)^{4M}$ is harmless. The $N!$ reappears in $|||\, A \,|||$, and so we need only analyze N^{4M}. First suppose $N \leqslant M^{4/3}$. Then

$$N^{4M} \leqslant \exp(4M \ln(M^{4/3})) = \exp((16/3)\, M \ln M) \leqslant O(1) \exp(\delta' M^{3/2}).$$

Next suppose $N \geqslant M^{4/3}$. Then $M \leqslant N^{3/4}$ and

$$N^{4M} \leqslant \exp(4N^{3/4} \ln N) \leqslant O(1)\, e^N.$$

Thus, the factorials in (3.3.10) can be bounded in all cases and the proof of Lemma 3.3.3 is complete.

4. Convergence of the Expansion

4.1. *The "Inside" Sum*

In this section we prove Proposition 1.2.1. Write

$$\phi - \xi_+ = (\phi - g) + (g - \xi_+) = \psi + (g - \xi_+). \tag{4.1.1}$$

Since

$$|g(x) - \xi_+| \leqslant O(\lambda^{-1/2}),$$

we can express $A(\phi - \xi_+) = A(\psi + (g - \xi_+))$ as a finite sum of terms C_j, each of the form (3.3.1), with

$$\sum |||\, C_j \,||| \leqslant \lambda^{-N/2} \,|||\, A \,|||.$$

Apply Theorem 3.3.1 in each connected component Y_i of Y to give

$$|R''(A, Y, \Sigma, \Gamma)| \leqslant |||\, A \,|||\, \lambda^{-N/2} e^{-dl|\Gamma^c|} e^{-b\lambda^{-1/2}|X^0 \cap Y|} e^{c\lambda^{1/2}|Y \cap \Lambda|}. \tag{4.1.2}$$

Each l-lattice square of $X \cap Y$ which does not meet $\mathscr{L}(A)$ must contain at least one derivative line from Γ^c, since otherwise the square is a component which contributes to the ratio of partition functions in (I.7.11) and not to Y. Thus we have a "density of derivative lines" in $X \cap Y$, as reflected in the inequality

$$2l^2 |\Gamma^c| \geqslant |X \cap Y| - Nl^2. \tag{4.1.3}$$

Hence

$$|Y \cap \Lambda| = |Y \cap X^0| + |Y \cap X| \leqslant |Y \cap X^0| + 2l^2 |\Gamma^c| + Nl^2. \tag{4.1.4}$$

We combine (4.1.2) and (4.1.4) to yield

$$|R(A, Y, \partial Y^\pm)| \leqslant \lambda^{-N/2} \,|||\, A \,|||\, \sum_{\{\Sigma \cap Y, \Gamma\}} \exp[-2\delta l^{-1} |Y \cap \Lambda| - b(2\lambda)^{-1/2} |X^0 \cap Y|], \tag{4.1.5}$$

where $\delta > 0$ is independent of λ. In (4.1.5) we remove the constraint on the Γ sum, and we increase K in (I.6.2) in order to bound the Nl^2 term from (4.1.4).

Since there are at most $4|Y \cap \Lambda|/l^2$ l-lattice lines in $\mathscr{B}(\Sigma, Y)$ and hence at most $2^{4|Y \cap \Lambda|/l^2}$ terms in the sum over Γ,

$$|R(A, Y, \partial Y^\pm)| \leqslant \lambda^{-N/2} \,|||\, A \,|||\, \sum_{\{\Sigma \cap Y\}} \exp[-\delta l^{-1} |Y \cap \Lambda| - b(2\lambda)^{-1/2} |X^0 \cap Y|]. \tag{4.1.6}$$

The sum over $\Sigma \cap Y$ is bounded by the observation that $\Sigma \upharpoonright X^\pm = \pm$, and since the same is true for any l-lattice square bordering $X^\pm$, thus $\Sigma \cap Y$ is uniquely specified by $X^0 \cap Y$ and $\Sigma \cap (X^0 \cap Y)$. With $|X^0 \cap Y| = r$ fixed, there are 2^r choices of $\Sigma \cap X^0 \cap Y$ and at most

$$\binom{|Y \cap \Lambda|}{r}$$

choices of $X_0 \cap Y$. Thus

$$\sum_{\Sigma \cap Y} \exp[-b(2\lambda)^{-1/2} \mid X^0 \cap Y \mid] \leqslant \sum_{r=0}^{|Y \cap \Lambda|} \binom{| Y \cap \Lambda |}{r} (2e^{-b(2\lambda)^{-1/2}})^r$$

$$= (1 + 2e^{-b(2\lambda)^{-1/2}})^{|Y \cap \Lambda|} \leqslant \exp(\lambda \mid Y \cap \Lambda \mid).$$

Substitution in (4.1.6) completes the proof in the case $\partial Y^- \neq \varnothing$.

Suppose $\partial Y^- = \varnothing$. The unique $\Sigma \cap Y$ for which $X^0 \cap Y = \varnothing$ is $\sigma \equiv 1$ (and hence $g \equiv \xi_+$). Thus $g - \xi_+$ does not appear in (4.1.1) and $\lambda^{-N/2}$ does not appear in (4.1.2). For the remaining $\Sigma \cap Y$'s, $| X^0 \cap Y | \geqslant 1$ and an extra convergence factor $\exp[-b(2\lambda)^{-1/2}]$ results.

4.2. *Ratios of Partition Functions*

In this section we prove a generalization of Proposition 1.2.2 and extend this bound to complex λ in a sector $| \operatorname{Im} \lambda | \leqslant \epsilon \operatorname{Re} \lambda$. For sets A and B, let $A \Delta B = (A \cup B) \sim (A \cap B)$ be the symmetric difference. For pairs $N_i = \{N_i^+, N_i^-\}$ let

$$N_{\text{plus}} \equiv \{N^+ \cup N^-, \varnothing\}. \tag{4.2.1}$$

It is always assumed that $N(\Gamma^\pm) \subset N^\pm$ and $\partial Y \subset \Gamma$ in the expression $Z(Y, N, \Gamma)$.

THEOREM 4.2.1. *Suppose $N_2^- = \varnothing$. There are constants c and ϵ such that for*

$$| \operatorname{Im} \lambda | \leqslant \epsilon \operatorname{Re} \lambda \leqslant \epsilon^2, \qquad \mu = 0, \tag{4.2.2}$$

as well as for

$$0 \leqslant \mu \leqslant \lambda^2 \leqslant \epsilon^2,$$

we have

$$| Z(Y, N_1, \Gamma_1) Z(Y, N_2, \Gamma_2)^{-1} | \leqslant \exp[c \mid \lambda \mid^{1/2} (| N_{1\text{plus}} \Delta N_2 | + l^2 \mid \Gamma_1 \Delta \Gamma_2 \mid)]. \tag{4.2.3}$$

If $N_1^- \neq \varnothing \neq N_1^+$, the right side of (4.2.3) *can be multiplied by* $\exp(-b \mid \lambda \mid^{-1/2})$, *for some $b > 0$.*

Remark. For $N_2 = \Gamma_2 = \Gamma_1 = \varnothing$ and $N_1^- = \Delta$, the theorem gives $\langle \chi_-(\Delta) \rangle \leqslant \exp(O(- \mid \lambda \mid^{-1/2}))$, which implies symmetry breaking.

LEMMA 4.2.2. *Let Δ^l be an l-lattice square. Let $N^l = \{N(\partial \Delta^l), \varnothing\}$ and $\Gamma^l = \partial \Delta^l$. There is a constant c such that for λ as in Theorem* 4.2.1

$$| Z(\Delta^l, N^l, \Gamma^l)^{\pm 1} | \leqslant \exp[c\lambda^{1/2} l^2].$$

Proof. By definition

$$Z(\Delta^l, N^l, \Gamma^l) = \int e^{-V(\Delta^l)} \prod_{\Delta \subset \Delta^l} \chi_+(\Delta)\, d\psi_\Gamma$$

$$= 1 + \int (e^{-V(\Delta^l)} - 1)\Big(\prod_{\Delta \subset \Delta^l} \chi_+(\Delta)\Big)\, d\psi_\Gamma + \int \Big(\prod_{\Delta \subset \Delta^l} \chi_+(\Delta) - 1\Big)\, d\psi_\Gamma$$

where V is given by (1.1.2). Here we have used the fact that $g \equiv h \equiv \xi_+$ and $F = 0$. By the Hölder inequality, used as in (3.2.9), the second term is bounded by $O(|\lambda|^{1/2} l^2)$. To bound the third term, we note that

$$0 \leqslant 1 - \prod_{\Delta \subset \Delta^l} \chi_+(\Delta) \leqslant \sum_{\Delta \subset \Delta^l} \chi_-(\Delta).$$

Since

$$\chi_-(\Delta) = -\pi^{-1/2} \int_{-\infty}^{-\xi_+} e^{-(\psi(\Delta)-a)^2}\, da,$$

a Schwarz inequality gives

$$\int \chi_-(\Delta)\, d\psi_\Gamma \leqslant \Big(\int e^{-2\psi(\Delta)^2}\, d\psi_\Gamma\Big)^{1/2} e^{-O(|\lambda|^{-1})} = e^{-O(|\lambda|^{-1})},$$

to complete the proof.

Proof of Proposition 1.2.2. We assume Theorem 4.2.1 and use it together with Lemma 4.2.2 to bound each factor on the right side of the following inequalities:

$$\frac{Z(\sim Y, N^+, N^-, \partial Y)}{Z(R^2, \varnothing, \varnothing, \varnothing)} = \frac{Z(\sim Y, N^+, N^-, \partial Y)}{Z(\sim Y, N_{\text{plus}}, \partial Y)} \frac{Z(\sim Y, N_{\text{plus}}, \partial Y)}{Z(R^2, \varnothing, \varnothing, \varnothing)},$$

$$\leqslant \exp(2c\,|\lambda|^{1/2}\,|N^-| + |\Gamma|) \frac{Z(\sim Y, N_{\text{plus}}, \partial Y)}{Z(R^2, \varnothing, \varnothing, \varnothing)}.$$

Since

$$Z(\sim Y, N_{\text{plus}}, \partial Y) = \frac{Z(R^2, N_{\text{plus}} \cup (Y \cap \Lambda), (\Gamma \cup \mathscr{B}(\Lambda \cap Y)))}{\prod_{\Delta^l \subset Y \cap \Lambda} Z(\Delta^l, N^l, \Gamma^l)},$$

$$\leqslant Z(R^2, \varnothing, \varnothing, \varnothing) \exp 2c\,|\lambda|^{1/2} (|\Lambda \cap Y| + |N| + l^2 |\Gamma|),$$

the desired inequality (1.2.2) follows (with a new choice of c). If $N^- \neq \varnothing$, then by Theorem 4.2.1, the right side can be multiplied by $\exp(-b\,|\lambda|^{-1/2})$. Because $\sim Y$ is unbounded, and thus meets the + region $\sim\Lambda$, the restriction $N_1^+ \neq \varnothing$ is not required, as we will see below.

Next we establish a special case of Theorem 4.2.1.

LEMMA 4.2.3. *For $0 \leqslant \mu \leqslant \lambda^2$ and λ sufficiently small positive,*

$$Z(Y, N, \Gamma)Z(Y, N_{\text{plus}}, \Gamma)^{-1} \leqslant 1.$$

Proof. Since Z factors across ∂Y, we assume without loss of generality that Y is connected. As in Section I.7, we introduce the approximate Dirichlet measures

$$d\phi_{Y\cap\Lambda,g} = \frac{\exp(-\int \delta m^2(x) :(\phi(x) - g(x))^2 dx)\, d\phi_{Y\cap\Lambda}}{\int \exp(-\int \delta m^2(x) :\psi(x)^2: dx)\, d\psi_1}, \tag{4.2.4}$$

where $g = g_\Sigma$ with Σ compatible with N and where $\delta m^2(\cdot)$ has support near Γ. Let

$$Z(Y, g, \mu) = \int d\phi_{Y\cap\Lambda,g}$$

and

$$E_{g,\mu}(A(\phi)) = Z(Y, g, \mu)^{-1} \int A\, d\phi_{Y\cap\Lambda,g}.$$

Let $g' = -g$, $g'' \equiv \xi_+$. Note that

$$|g(x)| = |g'(x)| \leqslant g''(x) \tag{4.2.5}$$

for all x, with equality for $x \in \text{suppt}\ \delta m^2$, by Lemma I.4.1.

We assert that

$$Z(Y, g, \mu) \leqslant Z(Y, g'', \mu). \tag{4.2.6}$$

In fact the change $g \to g''$ in the exponent in (4.2.4) affects only the linear term in g, $\int \delta m^2(x)\phi(x)g(x)\, dx$, by (4.2.5). The exponent of this linear term can be expanded as a power series. Each term in the corresponding expansion for $Z(Y, g, \mu)$ is a Schwinger function for the measure $d\phi_{Y\cap\Lambda}$. By Griffiths' first inequality, these Schwinger functions increase if the test functions are increased pointwise. By (4.2.5), $g \to g''$ increases the test functions, hence (4.2.6).

Because $\prod \chi_+$ is monotone increasing and $\prod \chi_-$ is monotone decreasing, the FKG inequalities imply

$$\begin{aligned} &E_{g,\mu}\left(\left[\prod_{\Delta \subset N^+} \chi_+(\phi(\Delta))\right]\left[\prod_{\Delta \subset N^-} \chi_-(\phi(\Delta))\right]\right) \\ &\qquad \leqslant E_{g,\mu}\left(\prod_{\Delta \subset N^+} \chi_+(\phi(\Delta))\right) E_{g,\mu}\left(\prod_{\Delta \subset N^-} \chi_-(\phi(\Delta))\right). \end{aligned} \tag{4.2.7}$$

Differentiation with respect to μ and another application of the FKG inequalities shows that $E_{g,\mu}(\prod \chi_+)$ is monotone increasing in μ. This statement holds for

$\mu < 0$ as well as $\mu > 0$, and as above, the same reasoning shows that $E_{g,\mu}(\prod \chi_+)$ is increasing in g, as long as $|g|$ is held fixed on suppt δm^2. Combining these inequalities with the $\phi \to -\phi$, $\mu \to -\mu$ symmetry allows us to bound (4.2.7) as follows:

$$\begin{aligned}(4.2.7) &\leqslant E_{g'',\mu}\Big(\prod_{\Delta\subset N^+} \chi_+(\phi(\Delta))\Big)\, E_{g,\mu}\Big(\prod_{\Delta\subset N^-} \chi_-(\phi(\Delta))\Big)\\ &= E_{g'',\mu}\Big(\prod_{\Delta\subset N^+} \chi_+(\phi(\Delta))\Big)\, E_{g',-\mu}\Big(\prod_{\Delta\subset N^-} \chi_+(\phi(\Delta))\Big)\\ &\leqslant E_{g'',\mu}\Big(\prod_{\Delta\subset N^+} \chi_+(\phi(\Delta))\Big)\, E_{g'',\mu}\Big(\prod_{\Delta\subset N^-} \chi_+(\phi(\Delta))\Big).\end{aligned}$$

Again by FKG inequalities, the above quantity is bounded by

$$E_{g'',\mu}\Big(\prod_{\Delta\subset N_{\text{plus}}} \chi_+(\phi(\Delta))\Big).$$

We multiply the resulting inequality by (4.2.6), and take a limit as $\delta m^2 \to \infty$, suppt $\delta m^2 \to \Gamma$ and the proof is complete.

Proof of Theorem 4.2.1. The proof could be based on a Neumann series to invert a contraction operator defined by Kirkwood–Salsburg equations. Instead, we take advantage of the triangular nature of the equations to give an inductive proof.

As a simple reduction, we begin by imposing Dirichlet conditions on the border of a large square S of diameter

$$r \geqslant 8(|\Lambda|^2 + |\Gamma_1| + |\Gamma_2|)$$

centered at the origin and containing Γ. Let us assume the theorem in case $\partial S \subset \Gamma$, and establish the general case. Using (I.5.6) to insert Dirichlet data on ∂S and Theorem 3.3.1 to estimate error terms, one can show that

$$|Z(Y, N, \Gamma_i \cup \partial S) - Z(Y, N, \Gamma_i)| \leqslant e^{-2(|\Lambda^2|+|\Gamma_1|+|\Gamma_2|)}. \tag{4.2.8}$$

Here we use convergence from a long contraction distance $O(r)$ to dominate the volume divergence $\exp(O|\lambda|^{1/2}|\Lambda|)$. Furthermore, Lemma 4.2.2 combined with the special case of the theorem we are assuming implies

$$|Z(Y, N_2, \Gamma_2 \cup \partial S)^{-1}| \leqslant e^{O(|\lambda|^{1/2})|\Lambda|}.$$

Combined with (4.2.8), this yields

$$|Z(Y, N_2, \Gamma_2 \cup \partial S)^{-1} - Z(Y, N_2, \Gamma_2)^{-1}| \leqslant e^{-(|\Lambda|^2+|\Gamma_1|+|\Gamma_2|)}$$

for λ small. The general case follows from these bounds and the identity

$$a^{-1}b = a^{-1}(b - B) + (a^{-1} - A^{-1})B + A^{-1}B,$$

where $a = Z(Y, N_2, \Gamma_2)$, $A = (Y, N_2, \Gamma_2 \cup \partial S)$, etc.

We now assume $\partial S \subset \Gamma_1 \cap \Gamma_2$, $\Gamma_1 \cup \Gamma_2 \subset S$, $Y \subset S$. We use two induction variables. The first induction is an increasing induction on $| Y |$; if $| Y | = \Delta^l$ there is nothing to prove. Thus we assume the theorem for all $Y_0 \subset Y$. The second induction is a decreasing induction on the variable

$$J = \min\{| \Gamma_1 |, | \Gamma_2 |\} + \min\{ |N_1 |, | N_2 |\}. \tag{4.2.9}$$

If J assumes its maximum value then the ratio Z/Z in (4.2.3) equals or is bounded by one, proving the theorem in this case.

Before proceeding, we reduce to three separate cases:

Case 1. $\Gamma_1 = \Gamma_2$, $N_1^- = \varnothing = N_2^-$

Case 2. $N_1^+ = N_2^+$, $N_1^- = \varnothing = N_2^-$

Case 3. $\Gamma_1 = \Gamma_2$, $N_2 = N_{1\text{plus}}$.

The general case follows from these three cases by the factorization

$$\frac{Z(Y, N_1, \Gamma_1)}{Z(Y, N_2, \Gamma_2)} = \frac{Z(Y, N, \Gamma_1)}{Z(Y, N_{1\text{plus}}, \Gamma_1)} \frac{Z(Y, N_{1\text{plus}}, \Gamma_1)}{Z(Y, N_{1\text{plus}} \cup N_2, \Gamma_1)} \times \frac{Z(Y, N_{1\text{plus}} \cup N_2, \Gamma_1)}{Z(Y, N_{1\text{plus}} \cup N_2, \Gamma_2)} \frac{Z(Y, N_{1\text{plus}} \cup N_2, \Gamma_2)}{Z(Y, N_2, \Gamma_2)}.$$

Consider first Case 1, and assume the theorem has been proved in Cases 1–3 for smaller Y and for larger J. Suppose $| N_1 | \leqslant | N_2 |$. (The case $| N_1 | \geqslant | N_2 |$ is similar.) Also assume $N_1 \neq N_2$, for otherwise (4.2.3) equals one. Choose $\Delta \in N_2 \sim N_1$ and define

$$N_3 = \{N_3^+, N_3^-\} = \{N_1, \Delta\}.$$

Then

$$\left| 1 - \frac{Z(Y, N_1, \Gamma_1)}{Z(Y, N_{3\text{plus}}, \Gamma_1)} \right| = \left| \frac{Z(Y, N_{3\text{plus}}, \Gamma_1) - Z(Y, N_1, \Gamma_1)}{Z(Y, N_{3\text{plus}}, \Gamma_1)} \right| = \left| \frac{Z(Y, N_3, \Gamma_1)}{Z(Y, N_{3\text{plus}}, \Gamma_1)} \right|. \tag{4.2.10}$$

This ratio is in Case 3, and since

$$| N_1 | = \min\{| N_1 |, | N_2 |\} < \min\{| N_3 |, | N_{3\text{plus}} |\} = | N_1 | + 1,$$

the inductive hypothesis applies, and (4.2.10) is bounded by $\exp(-b\,|\lambda|^{-1/2}/2)$. Here we assume $N_1 \neq \varnothing$. However, if $N_1 = \varnothing$, then $\Gamma_1 \cap \Lambda = \varnothing$ also and so $Y \cap (R^2 \backslash \Lambda) \neq 0$. In this case, the restriction $N_1^+ \neq \varnothing$ is not required for the insertion of the factor $\exp(-b\,|\,\lambda\,|^{-1/2})$. Thus without restriction on N_1,

$$[Z(Y, N_1, \Gamma_1) Z(Y, N_1 \cup \Delta, \Gamma_1)^{-1}]^{\pm 1} \leqslant \exp(c\,|\,\lambda\,|^{1/2})$$

for some constant c. Using this bound and the induction hypothesis again, we have

$$\left|\frac{Z(Y, N_1, \Gamma_1)}{Z(Y, N_2, \Gamma_2)}\right| \leqslant \left|\frac{Z(Y, N_1, \Gamma_1)}{Z(Y, N_1 \cup \Delta, \Gamma_1)}\right| \left|\frac{Z(Y, N_1 \cup \Delta, \Gamma_1)}{Z(Y, N_2, \Gamma_2)}\right|$$
$$\leqslant \exp(c\,|\,\lambda\,|^{1/2}\,|\,N_1 \Delta N_2\,|),$$

completing the proof in this case.

Next we consider Case 2. This case involves the insertion of Dirichlet data, and the methods are related to those of [16]. For $b \subset \Gamma_2 \sim \Gamma_1$, we show that

$$\left|1 - \frac{Z(\Gamma_1)}{Z(\Gamma_1 \cup b)}\right| = \left|\frac{Z(\Gamma_1 \cup b) - Z(\Gamma_1)}{Z(\Gamma_1 \cup b)}\right| \leqslant O(|\,\lambda\,|^{1/2}\, l^2) \qquad (4.2.11)$$

so that

$$|\,Z(\Gamma_1) Z(\Gamma_2)^{-1}\,|^{\pm 1} \leqslant (1 + O\,|\,\lambda\,|^{1/2} l^2)\,|\,Z(\Gamma_1 \cup b) Z(\Gamma_2)^{-1}|^{\pm 1}$$
$$\leqslant \exp(c\,|\,\lambda\,|^{1/2} l^2\,|\,\Gamma_1 \Delta \Gamma_2\,|),$$

where we have used the induction hypothesis for the last inequality and the notation $Z(\Gamma) = Z(Y, N_1, \Gamma)$. We expand $Z(\Gamma_1 \cup b) - Z(\Gamma_1)$ as in (I.5.5). For each term $Z''(Y, \Sigma, \Gamma)$ of the expansion, let $W(\Gamma)$ be the connected component of $Y \sim \Gamma$ which contains b. Each term in this construction factors,

$$Z'' = Z''(Y \sim W, \Sigma \sim W, \Gamma \sim W)\, Z''(W, \Sigma \cap W, \Gamma \cap W).$$

As in Section I.7, the Γ and Σ sums also factor (with ∂W fixed). The partial sums are subject to constraints and for the outside region $Y \sim W$, the partial sum can be resummed to yield

$$Z(\Gamma_1 \cup b) - Z(\Gamma_1) = \sum_{\partial W^\pm} Z(Y \sim W, N_3, \Gamma_3) \sum{}' Z''(W, \Sigma \cap W, \Gamma \cap W) \qquad (4.2.12)$$

where

$$N_3^\pm = [N_1^\pm \cup N(\partial W^\pm)] \sim W \qquad \Gamma_3 = (\Gamma_1 \sim W) \cup \partial W.$$

Also the sum $\sum'$ over $\Sigma \cap W$ and $\Gamma \cap W$ is subject to the constraint $(\Sigma \cap W)^\pm \supset N_1^\pm \cap W$, $\Gamma \supset (\Gamma_1 \cap W) \cup \partial W$, and $W = W(\Gamma)$.

As in the proof of Proposition 1.2.1 given in Section 4.1, we have

$$\sum{}' \mid Z''(W, \Sigma \cap W, \Gamma \cap W)| \leqslant c \mid \lambda \mid^{1/2} l^2 e^{-2\delta |W|/l}.$$

In fact, there is a density of derivative lines in $X \cap W$, as in (4.1.4) (with $N = N(A) = 0$), and further convergence is produced by the existence of at least one derivative $\partial/\partial s_b$ which produces either $\chi_\Delta{}'$ and the convergence $\exp(-O \mid \lambda \mid^{-1/2})$ or a vertex and the convergence $O(\mid \lambda \mid^{1/2})$. By Lemma 4.2.2, the same bound holds if we multiply the right side by

$$\mid Z^*(W)| = \prod_{\Delta^l \subset W} \mid Z(\Delta^l, N^l, \partial\Delta^l)|.$$

Substitution in (4.2.12) gives

$$\left| 1 - \frac{Z(\Gamma_1)}{Z(\Gamma_1 \cup b)} \right|$$
$$\leqslant c \mid \lambda \mid^{1/2} l^2 \sum_{\partial W^\pm} e^{-2\delta |W|/l} \mid Z(Y \sim W, N_3, \Gamma_3)\, Z^*(W)\, Z(\Gamma_1 \cup b)^{-1} \mid.$$

We express $Z(Y \sim W, N_3, \Gamma_3)\, Z^*(W)$ as a single partition function $Z(Y, N_3 \cup W, \Gamma_3 \cup \mathscr{B}(W^*))$. Because $\mid \Gamma_1 \cup b \mid > \mid \Gamma_1 \mid$ and $\mid N_3 \cup W \mid \geqslant \mid N_1 \mid$, the induction hypothesis applies and bounds the ratio of Z's above by $\exp(O(\mid \lambda \mid^{1/2} \mid W \mid))$.

To control the combinatorics in the above sum over $\partial W^\pm$, we observe that since W is a connected union of l-lattice squares containing b, the choice of W with $\mid W \mid$ fixed can be made in at most $\exp(O(\mid W \mid / l^2))$ ways, while the choice of $\pm$ for the contours of ∂W can also be made in at most $\exp(O(\mid W \mid / l^2))$ ways. Substitution in the above inequality yields (4.2.11) and completes the proof of Case 2.

Finally we consider Case 3. If $N_1^- = \varnothing$ there is nothing to prove. If $N_1^+ = \varnothing$, the theorem follows from Lemma 4.2.3 for real couplings and from the $\phi \to -\phi$ symmetry for complex coupling (with $\mu = 0$). Thus we suppose $N_1^- \neq \varnothing \neq N_1^+$. (To establish the mass gap, this remaining portion of Case 3 is not required.) In addition we suppose without loss of generality that Y is connected, and since $Y \subset S$, that Y is bounded. It follows that ∂Y has unique outer contour consisting of a simple closed curve. This curve is either $+$ or $-$.

As in I.5.5, we expand $Z(Y, N_1, \Gamma_1)$ as a sum of terms $Z''(Y, \Sigma, \Gamma)$. Choose a distinguished lattice square $\Delta \subset Y$, and with Δ fixed we define

$$Z^{(\pm)} = \Sigma^{(\pm)} Z'',$$

where $\Sigma^{(\pm)}$ denotes the sum over all terms Z'' which have a closed $+$ (resp. $-$) contour enclosing Δ. Observe that if the outer contour of Y is $+$ (resp. $-$) then $Z = Z^{(+)}$ (resp. $Z^{(-)}$). For each term Z'' of $Z^{(\pm)}$, we define $\gamma^\pm = \gamma^\pm(\Sigma, \Gamma, \Delta) = \gamma$ to be the closed $+$ (resp. $-$) contour of smallest area, enclosing Δ. Let W be the

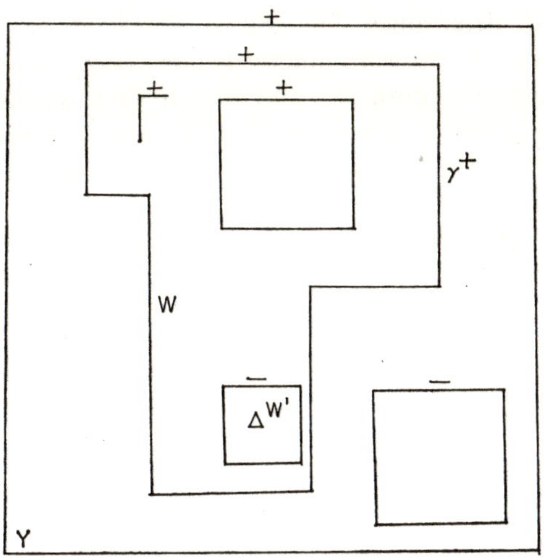

FIG. 1. The regions W and W' are indicated. The lines denote $\pm$Dirichlet lines which determine $\Gamma^{\pm}$.

connected component of $S \sim \Gamma$ which borders on γ and lies inside γ. There is only one such component, as we see from the minimality of γ, and $W \subset Y$. Let W' be the component of $Y \sim W$ which contains Δ_x if $\Delta \subset W$, let $W' = \varnothing$. Also set $U = W \cup W'$. See Fig. 1.

As before,

$$Z'' = Z''(Y \sim U, \Sigma \sim U, \Gamma \sim U)\, Z''(W, \Sigma \cap W, \Gamma \cap W)\, Z''(W', \Sigma \cap W', \Gamma \cap W'),$$

and as before, a partial resummation yields

$$Z^{(\pm)}(Y, N_1, \Gamma_1) = \sum_{\partial W^{\pm}} Z(Y \sim U, N_3, \Gamma_3)\, R(W)(Z - Z^{(\pm)})(W', N_4, \Gamma_4) \tag{4.2.13}$$

where

$$\begin{aligned} Z - Z^{\pm} &= 1, & \text{when } W' &= \varnothing, \\ N_3^{\pm} &= [N_1^{\pm} \cup N(\partial U^{\pm})] \sim U, & \Gamma_3 &= (\Gamma_1 \sim U) \cup \partial U, \\ N_4^{\pm} &= [N_1^{\pm} \cup N(\partial W'^{\pm})] \cap W', & \Gamma_4 &= (\Gamma_1 \cap W') \cup \partial W' \end{aligned}$$

and $R(W)$ denotes a constrained sum over the region W. In the region W', the absence of $+$ (resp. $-$) contours enclosing Δ is a constraint, leading to the difference $Z - Z^{(\pm)}$ above. (Note that all contours, by definition are connected.)

The occurrence of $Z^{(\pm)}$ on the right-hand side of (4.2.13) requires an augmentation of our inductive hypothesis. We assume inductively that

$$|\, Z^{(\pm)}(Y, N_1, \Gamma_1)\, Z(Y, N_{1\text{plus}}, \Gamma_1)^{-1} \,| \leqslant e^{-b|\lambda|^{-1/2}} \tag{4.2.14}$$

where $\Delta = \Delta^{\mp} \subset N_1^{\mp}$ determines $Z^{(\pm)}$. Since $Z^{(+)} = Z$ or $Z^{(-)} = Z$ and since $N_1^+ \neq \varnothing \neq N_1^-$, (4.2.14), proved inductively, also completes the proof of Case 3.

The region W must meet both N_1^- and N_1^+. In fact, one sign is achieved on γ and the other on $\partial W'$ if $W' \neq \varnothing$ and on $\Delta^{\mp}$ if $W' = \varnothing$. Here we use minimality of γ to infer that γ and $\partial W'$ have opposite signs. Thus for each term Z'' in the $R(W)$ sum, the convergence region $X^0 \cap W$ is not empty. Further convergence results from the density of derivative lines in $X \cap W$, leading to the estimate

$$| R(W)| \leqslant e^{-b|\lambda|^{-1/2}} e^{-2\delta|W|/l},$$

and as in Case 2, an estimate of the same form holds for $| R(W) \, Z^*(W)^{-1} |$. Substitution in (4.2.13) gives

$$| Z^{(\pm)}(Y, N_1, \Gamma_1)| \leqslant e^{-b|\lambda|^{-1/2}} \sum_{\partial W^{\pm}} e^{-2\delta|W|/l} \, | Z(Y \sim U, N_{3\text{plus}}, \Gamma_3) \times Z^*(W) \, Z(W', N_{4\text{plus}}, \Gamma_4)|. \tag{4.2.15}$$

Here we have applied the inductive hypothesis in the regions $Y \sim U$ and W'. Because $W \neq \varnothing$ by construction, these regions are necessarily smaller than Y. However, the inductive hypothesis in Y also applies to the product of Z's in (4.2.15). In fact, since $X^0 \cap W \neq \varnothing$, there are more Dirichlet lines in $Z^*(W)$ than in $\Gamma_1 \cap W$, resulting in an equal J-value for previously considered cases 1 and 2, and the inequality

$$| Z^{\pm}(Y, N_1, \Gamma_1) \, Z(Y, N_{1\text{plus}}, \Gamma_1)^{-1} | \leqslant e^{-b|\lambda|^{-1/2}} \sum_{\partial W^{\pm}} e^{[O(|\lambda|^{1/2}) - 2\delta/l]|W|}.$$

Substitution into (4.2.15) combined with combinatorics as in Case 2 control the sum over $\partial W^{\pm}$. In this argument there is one new issue which did not arise in Case 2, namely, the choice of a distinguished point in W. We choose this point to be the first intersection of γ with some ray starting at $\Delta^{\mp}$. That choice can be made in at most $|\gamma| \leqslant |W|$ ways. This completes the proof of (4.2.14), Case 3, and Theorem 4.2.1.

4.3. *Exponential Clustering*

Let A and B be Wick polynomials of the form (I.6.1) and localized in regions $\mathscr{L}(A)$ and $\mathscr{L}(B)$, respectively. Let

$$D = \text{dist}(\mathscr{L}(A), \mathscr{L}(B)).$$

THEOREM 4.3.1. *There is $\delta > 0$ and a constant c_{AB} such that for λ sufficiently small,*

$$|\langle AB\rangle - \langle A\rangle\langle B\rangle| \leqslant c_{AB} e^{-\delta D} \tag{4.3.1}$$

uniformly in Λ.

Remark. An immediate corollary to this theorem is the unique vacuum and positive mass of Theorem I.2.1.

Proof. As in [16], we introduce duplicate variables ϕ, $\phi^{\sim}$ so that

$$\langle AB\rangle - \langle A\rangle\langle B\rangle = \tfrac{1}{2}\langle (A - A^{\sim})(B - B^{\sim})\rangle. \tag{4.3.2}$$

With duplicate variables, there are independent Σ, $\Sigma^{\sim}$ expansions, and the terms in the double expansion are labeled by a pair $\{\Sigma, \Sigma^{\sim}\}$. Given $\{\Sigma, \Sigma^{\sim}\}$ we define the regions X, $X^{\pm}$ and $X^{\sim}$, $X^{\sim\pm}$ as before (see Section I.5) and set $\mathscr{B}(\Sigma, \Sigma^{\sim})$ to be the set of l-lattice lines in $(X \cap X^{\sim}) \sim (\mathscr{L}(A) \cup \mathscr{L}(B))$. As in [16], the cluster expansion treats ϕ and $\phi^{\sim}$ symmetrically, so that there is a single parameter s_b for both ϕ and $\phi^{\sim}$ variables, and a single label $\Gamma \subset \mathscr{B}(\Sigma, \Sigma^{\sim})$ for the terms of the expansion. Because the Σ and $\Sigma^{\sim}$ expansions are independent, the individual terms in this expansion are not invariant under the $\phi \leftrightarrow \phi^{\sim}$ symmetry. In particular, we consider four types of contours: $++$, $+-$, $-+$ and $--$, and in general

$$\Gamma = \Gamma^{++} \cup \Gamma^{+-} \cup \Gamma^{-+} \cup \Gamma^{--}.$$

With this notation the Γ^{++}, Γ^{--} contours are preserved by the $\phi \leftrightarrow \phi^{\sim}$ symmetry, while the Γ^{+-} and Γ^{-+} contours are interchanged.

For fixed Σ, $\Sigma^{\sim}$ and Γ, we define $\gamma = \gamma(\Sigma, \Sigma^{\sim}, \Gamma)$ to be the smallest $++$ or $--$ connected closed contour enclosing $\mathscr{L}(A)$. Because of $+$ boundary conditions, γ exists for nonzero terms in the expansion. Let Int γ denote the region enclosed by γ, and let W be the connected component of (Int γ) $\sim \Gamma$ which borders on γ. Also let W' be the component of $R^2 \sim W$ which contains $\mathscr{L}(A)$, if $\mathscr{L}(A) \not\subset W$. Otherwise let $W' = \varnothing$.

As a first step, we hold W fixed and resum the rest of the expansion. The result is a term (i) which depends only on ∂W, (ii) factors across γ, (iii) is symmetric under $\phi \leftrightarrow \phi^{\sim}$, and (iv) is symmetric under $\phi \leftarrow \phi^{\sim}$ in Int γ while ϕ and $\phi^{\sim}$ are fixed in $R^2 \sim$ Int γ. Because of (iv) and the subtractions in (4.3.2), this term is zero unless $\mathscr{L}(B) \subset$ Int γ, which we now assume. Consequently $|W|/l \geqslant D$.

We now proceed as in [16] and Section 4.2. The resummation in $R^2 \sim (W \cup W')$ is free from constraints and yields a partition function, while the resummation in W' involves contributions from A and a difference of partition functions, as in (4.2.13). Thus we obtain

$$|\langle AB\rangle - \langle A\rangle\langle B\rangle| \leqslant \sum_W |R(W)| \frac{Z(\sim W, N(\partial W), \partial W)}{Z(\Lambda, \varnothing, \varnothing)}$$

where $R(W)$ is the constrained sum associated with the region W. The theorem now follows as in Section 1.2 and [16], using a bound on $R(W)$, given below:

LEMMA 4.2.2. *There is a* $\delta > 0$ *such that for* λ *sufficiently small*

$$| R(W)| \leqslant |||\, A \,|||\ |||\, B \,|||\, e^{-\delta |W|/l}.$$

The summation over W is as in the proof of Theorem 4.2.1.

4.4. *Analyticity and the Infinite Volume Limit*

By the Lee–Yang theorem and results of Fröhlich and Simon [11b], the infinite volume limit of the Schwinger functions exists and is Euclidean invariant provided $\mu > 0$. Moreover, if A is a product of field operators, then in each finite volume Λ we have

$$|\langle A\rangle_{\mu_1} - \langle A\rangle_{\mu=0}| = \left| \int_0^{\mu_1} \frac{d}{d\mu} \langle A\rangle(\mu)\, d\mu \right| \leqslant \text{const}\ \mu_1$$

with the constant independent of Λ provided μ and λ are real. The bound on the above derivative is an immediate consequence of the mass gap as formulated in Theorem 4.3.1.

Thus the $\mu \searrow 0$ limit converges uniformly in Λ and so the Schwinger functions converge for $\mu = 0$ and λ real as $\Lambda \nearrow R^2$. To obtain analyticity in λ (for $\mu = 0$) we note that the finite volume Schwinger functions are analytic and uniformly bounded in Λ in the sector $0 < \operatorname{Re} \lambda \leqslant \epsilon | \lambda |$.

An application of Vitali's theorem yields analyticity for infinite volume. This completes the proof of Theorems I.2.1 and I.2.5.

Convergence of the $\Lambda \nearrow R^2$ limit also follows from the expansion methods of this paper and since the expansion method established the stronger result that the series expansion (I.7.15) converges term by term and is valid in the infinite volume limit, we sketch the method. Let Δ be a lattice square not in Λ, and let Δ_t be a strip of Δ of width t, $0 \leqslant t \leqslant 1$. Let Z_9 denote the Λ dependence of Z, and let $Z_t = Z_{\Lambda \cup \Delta_t}$. Then we must show that

$$\left| \frac{Z_0(R^2, N(\Gamma), \Gamma)}{Z_0(R^2, \varnothing, \varnothing)} - \frac{Z_1(R^2, N(\Gamma), \Gamma)}{Z_1(R^2, \varnothing, \varnothing)} \right|$$

is small exponentially in $D = \operatorname{dist}(\Gamma, \Delta)$. However,

$$\frac{d}{dt} \frac{Z_t(R_2, N(\Gamma), \Gamma)}{Z_t(R_2, \varnothing, \varnothing)}$$

is a truncated vacuum expectation value (Γ and $N(\Gamma)$ being regarded as observables) and the methods of Section 4.3 apply.

IV Phase Transitions and Critical Behavior

Introduction

The model with potential $\phi^4 + \sigma\phi^2$ can be studied as a function of the real parameter σ. For σ large in magnitude, expansion methods can be used to study the spectrum of the Hamiltonian H in detail. However the difficult region to study corresponds to intermediate values of σ, close to the critical value σ_c dividing the single phase region from the two phase region. The gap in the spectrum of H vanishes at the critical point, and the spectrum of H becomes a half-line. One of the most difficult problems of physics is to obtain detailed information about models in the neighborhood of the critical point. Some mathematically complete results and some conjectures in this direction are contained in these papers.

Bibliography

Critical point dominance in quantum field models, Ann. l'Inst. H. Poincaré A21, 27–41 (1974).

ϕ_2^4 quantum field model in the single phase region: differentiability of the mass and bounds on critical exponents, Phys. Rev. 10, 536–539 (1974).

A remark on the existence of ϕ_4^4, Phys. Rev. Lett. 33, 440–442 (1974).

On the approach to the critical point, Ann. l'Inst. H. Poincaré A22, 13–26 (1975).

Critical exponents and elementary particles, Commun. Math. Phys. 52, 203–209 (1977).

Critical point dominance in quantum field models

by

James GLIMM (*)
Courant Institute. New York University.
New York 10012

and

Arthur JAFFE (**)
Lyman Laboratory. Harvard University
Cambridge, Massachusetts

ABSTRACT. — We study the generating functional $\Gamma\{A\}$ for vertex functions in $\mathscr{P}(\varphi)_2$ models and prove the Callan-Symanzik equations for the weakly coupled $\lambda_0\varphi_2^4$ model. Under reasonable (but unproved) assumptions, we show that the dimensionless physical coupling constant g in $\varphi_{2,3}^4$ models achieves its maximum value at the critical point. We discuss a more general picture of « critical point dominance ».

1. THE CALLAN-SYMANZIK EQUATIONS

The Euclidean $\mathscr{P}(\varphi)_2$ quantum field models are parameterized by (m_0, g_0). Here m_0 is the bare mass and $g_0 = \lambda_0/m_0^2$ is the dimensionless bare charge. The bare charge λ_0 is the vector of coefficients of the interaction polynomial

$$\mathscr{P}(\varphi) = \sum_{j=1}^{2N} \lambda_0^{(j)} \frac{1}{j!} : \varphi^j : . \tag{1.1}$$

(*) Supported in part by the National Science Foundation, Grant NSF-GP-24003.
(**) Supported in part by the National Science Foundation, Grant NSF-GP-40354X.

We discuss the variation of the solution with respect to these bare parameters, and the transformation from bare parameters to a physical parameter space (m, g). Here m is the physical mass, inf (spectrum (M) $\sim \{0\}$) where the mass operator $M = (H^2 - P^2)^{\frac{1}{2}}$. A physical coupling constant g may be defined by the momentum space vertex function (for $l \geqq 3$) as

$$g^{(l)} = -\Gamma^{(l)}(p)/m^2 \,|_{p=0} \tag{1.2}$$

where

$$\Gamma^{(l)}(p)\delta\Big(\sum_i p_i\Big) = \int e^{i\Sigma p_j x_j}\tilde{\Gamma}^{(l)}(x)dx\,.$$

For $j = 1$, we take $g^{(1)} = \langle \varphi \rangle$. The field strength renormalization constant Z is

$$Z = (p^2 + m^2)\langle \Phi^{\sim}(p)\Phi(0)\rangle \,|_{p^2 = -m^2} \tag{1.3}$$

where $\langle\ \ \rangle$ denotes expectation in the Euclidean ground state measure. For increased flexibility we consider theories without the field strength renormalization, i. e. with $Z \neq 1$. Then the Green's functions and vertex functions can be considered as functions of Z as well as (m_0^2, g_0). We note that

$$Z^{l/2}\Gamma^{(l)}(p, m_0, g_0, Z) = \Gamma^{(l)}(p, m_0, g_0, 1)\,, \tag{1.3a}$$

or in terms of the generating functional

$$\Gamma\{Z^{\frac{1}{2}}A, m_0, g_0, Z\} = \Gamma\{A, m_0, g_0, 1\}\,, \tag{1.3b}$$

and the field strength renormalization amounts to choosing the theory with $Z = 1$.

Under scaling, one can check that Z is invariant (dimensionless) and that

$$\Gamma^{(l)}(\alpha p, \alpha m_0, g_0, Z)(\alpha m)^{-2} = \Gamma^{(l)}(p, m_0, g_0, Z)m^{-2}\,, \tag{1.4}$$

see the appendix. Thus the definition of $g^{(j)}$ in (1.2) is scale invariant.

The Callan-Symanzik equations [*11*] describe a one parameter family of perturbations of the form $\mathscr{P}(\varphi)_2 + \sigma\int :\varphi^2: d\bar{x}$. After a new choice of bare mass $m_0(\sigma)$ (see the appendix), this family defines a smooth curve $(m_0(\sigma), g_0(\sigma), Z(\sigma))$. The Callan-Symanzik equations express the variation of the theory with respect to σ. One side of the equation is an expression for $(d/d\sigma)\Gamma\{A, \sigma\}$, which can be recognized as a prescription for φ^2 vertex insertions into the lines of the vertex functions. The other side of the equation is the chain rule,

$$\frac{d}{d\sigma}\Gamma\{A, \sigma, Z\} = \frac{dm(\sigma)^2}{d\sigma}\frac{\partial\Gamma}{\partial m(\sigma)^2} + \frac{dg(\sigma)}{d\sigma}\frac{\partial\Gamma}{\partial g(\sigma)} + \frac{dZ(\sigma)}{d\sigma}\frac{\partial\Gamma}{\partial Z(\sigma)} \tag{1.5}$$

Here Γ is the generating function for the (amputated, one particle irreducible) vertex parts.

We now study (1.5) for weak coupling. In [5] we prove the existence of Γ, for weak coupling, and its analyticity in A and σ. Let $J_A(\sigma)$ be the solution of the equation

$$G_x\{J, \sigma\} - G_x\{0, \sigma\} = A(x)$$

where $G\{J, \sigma\}$ is the generating function for the Euclidean Green's functions and $G_x = \delta G/\delta J(x)$. From the Legendre transformation

$$\Gamma\{A, \sigma\} = G\{J_A, \sigma\} - A(J_A(\sigma)) - G^{(1)}(J_A(\sigma)),$$

we find

$$\frac{d}{d\sigma}\Gamma\{A, \sigma\} = \frac{d}{d\sigma}G\{J, \sigma\}|_{J=J_A(\sigma)},$$

for $A \in H_1$, $J \in H_{-1}$ and $\|A\|_1$ and $\|J\|_{-1}$ small. We use the notation

$$\mathscr{Z}\{J, \sigma, Z_0\} = \langle e^{Z_0^{-\frac{1}{2}}\Phi(J)}\rangle_\sigma$$
$$\mathscr{Z}_{:xx:}\{J, \sigma, Z_0\} = Z_0^{-1}\langle :\Phi(x)^2: e^{Z_0^{-\frac{1}{2}}\Phi(J)}\rangle_\sigma$$

(with $m_0(\sigma)$-Wick ordering). The field strength renormalization then amounts to choosing Z_0 to be the Z of (1.3), but for the time being we regard Z_0 as an independent variable.

We have

$$G\{J, \sigma, Z/Z_0\} = \ln \mathscr{Z}\{J, \sigma, Z_0\}$$

and

$$\frac{d}{d\sigma}\Gamma = \frac{d}{d\sigma}G = \frac{d}{d\sigma}\ln \mathscr{Z} = \mathscr{Z}^{-1}\frac{d}{d\sigma}\mathscr{Z}.$$

In order to simplify subsequent formulas, we assume that $\mathscr{P}$ is even, so $G^{(1)}(J) = 0$, and then

$$\text{(1.6)}\quad \frac{d}{d\sigma}\Gamma\{A, \sigma\} = \mathscr{Z}^{-1}Z_0\int[-\mathscr{Z}_{:xx:}\{J_A, \sigma\} + \mathscr{Z}\{J_A, \sigma\}\mathscr{Z}_{:xx:}\{0, \sigma\}]dx.$$

For J regular, we establish (1.6) by Dimock's asymptotic calculations [3]. First we keep m_0 fixed, and consider the interaction $\mathscr{P}(\varphi) + \sigma :\varphi^2:$. Then the mass renormalization transformation (see appendix) is used to obtain (1.6) with $m_0(\sigma)$ Wick order. For the techniques to prove convergence of the integral in (1.6) with $J \in H_{-1}$, see [5, Chapter 2]. The key point is that since H_{-1} is an L_2 norm, we must be careful to eliminate terms linear in J, or in localized portions J_α of J. Since $\mathscr{P}$ is even, these terms do not contribute to the cluster expansion of (1.6).

We note that the integrand

$$F(x) = -\langle :\Phi(x)^2: e^{Z_0^{-\frac{1}{2}}\Phi(J)}\rangle_\sigma + \langle :\Phi(x)^2:\rangle_\sigma\langle e^{Z_0^{-\frac{1}{2}}\Phi(J)}\rangle_\sigma,$$

which clusters can be written

$$F(x) = \lim_{a \to 0} F(x, x + a)$$

where

(1.7) $$F(x, y) = - \langle \Phi(x)\Phi(y)e^{Z_0^{-\frac{1}{2}}\Phi(J)} \rangle_\sigma + \langle \Phi(x)\Phi(y) \rangle_\sigma \langle e^{Z_0^{-\frac{1}{2}}\Phi(J)} \rangle_\sigma ,$$

so

$$\frac{1}{Z_0} \frac{d\mathscr{Z}\{J, \sigma\}}{d\sigma} = \lim_{a \to 0} \int F(x, x + a)dx .$$

Recall that $G\{J, \sigma\} = \ln \mathscr{Z}\{J, \sigma\}$, the generating function for connected parts. Then regarding Z_0, σ and λ_0 as independent variables,

(1.8) $$\frac{d}{d\sigma}\Gamma\{A, \sigma\} = \frac{d}{d\sigma}G\{J, \sigma\} = \mathscr{Z}\{J, \sigma\}^{-1}\frac{d}{d\sigma}\mathscr{Z}\{J, \sigma\}$$
$$= - \lim_{a \to 0} Z_0 \int [G_{x,x+a}\{J, \sigma\} + G_x\{J, \sigma\}^2 - G_{x,x+a}\{0, \sigma\}]dx$$

We eliminate the J dependence in (1.8) by using $G_x\{J, \sigma\} = A(x)$ and the fact that $-\Gamma_{xy}\{A, \sigma\}$ is the kernel of the operator inverse to $G_{xy}\{J, \sigma\}$. Thus by (1.8),

(1.9) $$\frac{d}{d\sigma}\Gamma\{A, \sigma\} = \lim_{a \to 0} Z_0 \int [\Gamma^{-1}_{x,x+a}\{A, \sigma\} - \Gamma^{-1}_{x,x+a}\{0, \sigma\} - A(x)^2]dx$$

In order to equate (1.9) to the right side of (1.5), the only problem is to show that the indicated derivatives exist. We specialize further to the case of a pure $\lambda_0\varphi^4$ interaction. For small coupling, there are no multiple phases, and so m and g are functions of m_0 and g_0. The existence of derivatives $\partial\Gamma/\partial m_0^2$ and $\partial\Gamma/\partial g_0$ follows from the analyticity of $\Gamma\{A, \sigma\}$ in the bare parameters [5], while $\partial\Gamma/\partial Z_0$ exists by (1.3*b*). The fact that $m_0(\sigma)^2$ is differentiable with a bounded inverse is the second transformation of the appendix. We regard Z as a function of m_0 and g_0, defined by (1.3), and we take $Z_0 = Z$. By scaling $\partial g/\partial m_0 = \partial Z/\partial m_0 = 0$ and $\partial m/\partial m_0 = 1$, while $\partial Z/\partial\lambda_0$ and $\partial m/\partial\lambda_0$ exist by [9]. Thus the derivatives $dZ(\sigma)/d\sigma$ and $dm(\sigma)/d\sigma$ taken at constant λ_0 exist also. We show below for weak coupling that $\partial g/\partial\lambda_0$ (m_0 constant) and $\partial g/\partial m_0$ (λ_0 constant) (i. e. $dg/d\sigma$) exist and are not zero. Furthermore, it follows that the Jacobian $\partial(m, g)/\partial(m_0, g_0)$ is not zero. The proof of the Callan-Symanzik equations for weak $\lambda_0\varphi_2^4$ models, namely

(1.10) $$\frac{dm(\sigma)^2}{d\sigma}\frac{\partial\Gamma}{\partial m(\sigma)^2}\{A, \sigma\} + \frac{dg(\sigma)}{d\sigma}\frac{\partial\Gamma}{\partial g(\sigma)}\{A, \sigma\} + \frac{dZ(\sigma)}{d\sigma}\frac{\partial\Gamma}{\partial Z(\sigma)}\{A, \sigma\}$$
$$= \lim_{a \to 0} Z_0 \int [\Gamma^{-1}_{x,x+a}\{A, \sigma\} - \Gamma^{-1}_{x,x+a}\{0, \sigma\} - A(x)^2]dx$$

is reduced to the following result:

THEOREM 1.1. — For a pure φ^4 interaction with weak coupling, the coordinate transformation $(m, g) \leftrightarrow (m_0, g_0)$ is differentiable with a nonvanishing Jacobian, for real bare parameters and for g defined by (1.4).

Proof. — We have already seen that it is sufficient to show $\partial g/\partial\lambda_0$ exists and is nonzero. With m_0 fixed, $\partial g/\partial\lambda_0$ is proportional to

$$-\left.\frac{\partial\Gamma^{(4)}(p)}{\partial\lambda_0}\right|_{p=0} = -\frac{\partial}{\partial\lambda_0}\int \left. G_a^{(4)}(x)\right|_{x_1=0} dx_2 dx_3 dx_4 , \tag{1.11}$$

where the amputated Green's function G_a equals

$$G_a(x) = \int dx' G(x') \prod_{i=1}^{4} \tilde{\Gamma}^{(2)}(x_i - x_i') .$$

Using Dimock's proofs [3] of the asymptotic formulas for the Euclidean Green's functions, (1.11) can be evaluated as

$$-\frac{\partial\Gamma^{(4)}(0)}{\partial\lambda_0} = 1 + 0(\lambda_0)$$

which is $\neq 0$ for small coupling.

2. CRITICAL POINT DOMINANCE

In this section we discuss the global nature of the transformation $(m, g) \leftrightarrow (m_0, g_0)$ for the φ^4 model. Unlike the mathematical proofs of Section 1, our discussion here is a heuristic picture, based on reasonable assumptions about the physical parameters (m, g). In a multiple phase region we consider a pure phase. We conclude that the charge g achieves its maximum value g_c for the φ_2^4 or φ_3^4 model at the critical point $(m = 0)$, which we presume exists. Here g_c may be infinite, but corresponds to finite values in the bare parameter space (m_0, g_0). We relate this behavior to $\beta(g)$, one of the coefficients in the Callan-Symanzik equation, and are led in the superrenormalizable cases $(d = 2, 3)$ to:

THE PICTURE OF CRITICAL POINT DOMINANCE. — The curve $m(\sigma)^2$ has a unique zero at $\sigma = \sigma_c$, and increases monotonically to $+\infty$ on each side of this zero. The curve $g(\sigma)$ decreases monotonically to zero as $|\sigma - \sigma_c| \to \infty$, with a (possibly infinite) maximum g_c at $\sigma = \sigma_c$, see Figure 1.

We now discuss our assumptions and conclusions. For $d = 2$, the cluster expansion deals with the $\sigma \to +\infty$ limit [6]. After scaling (to ensure that the mass remains bounded) the model converges as $\sigma \to \infty$ to a free theory. In this limit $\lambda \approx \lambda_0$, $m \approx m_0$, and $g \approx g_0$; the monotonicity of $g(\sigma)$ or $m(\sigma)$ follows by perturbation theory, see [3]. The more general result that $m(\sigma)$ is monotonic for $\sigma \geq \sigma_c$ was established in [7]. We remark that in the Ising model, a stronger result holds for the inverse correlation length ξ^{-1}. For Ising_2, ξ^{-1} is monotonic in both the single phase region and the two phase region, see, e. g. [8], and ξ^{-1} vanishes at the critical point, near which there is a linear power law behavior $\xi^{-1} = a\,|\,\sigma - \sigma_c\,|$.

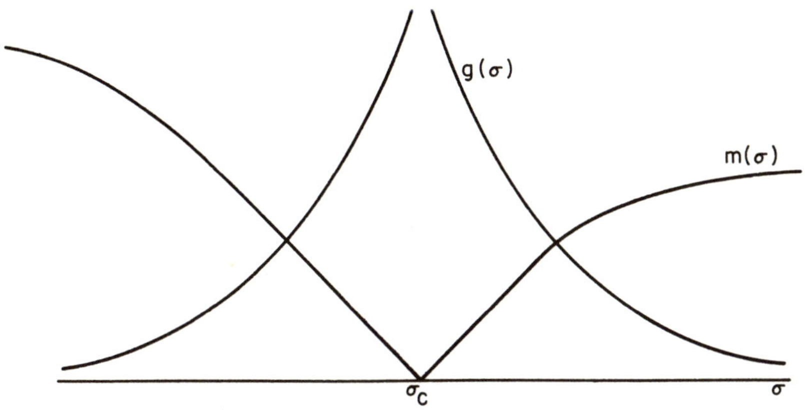

FIG. 1. — Critical Point Dominance.

For $d = 3$, the σ-perturbation produces no new ultraviolet divergences in a finite volume [4], and we expect the same behavior in an infinite volume. Thus for $d = 3$ the same picture should hold for $m(\sigma)$ as in $d = 2$. The dimensionless coupling constant in three dimensions is $g = -\,\Gamma^{(4)}(0)/m$.

This picture of monotonicity of $m(\sigma)$ is related to a choice of variables made by Symanzik in analyzing the characteristics of the Callan-Symanzik equations [11]. Symanzik chooses a dimensionless parameter s to reparameterize the $m(\sigma)$ curve, $\sigma = \sigma(s)$, starting from a point $m(0)$, and such that

$$\frac{d}{ds} m(s)^2 = m(s)^2 . \tag{2.1}$$

Since (2.1) can be integrated to yield $m(s)^2 = m(0)^2 e^s$, it is a valid assumption only on a branch of the σ curve for which $m(\sigma)^2$ varies monotonically between 0 and ∞. Since $m(\sigma)^2$ is finite for σ finite, and (assuming continuity) is bounded for σ bounded, $m(\sigma)^2 \to \infty$ can occur at most for $\sigma \to \pm\infty$. In terms of the new parameter, an s-curve corresponds to a monotone branch $\sigma > \sigma_c$ or $\sigma < \sigma_c$ of a σ-curve.

Let us assume that on each s-curve the use of physical coordinates $m(s)$ and $g(s)$ is justified. We now argue that g is monotone in s on each side of the critical point. Let $\beta = dg/ds$. By scaling, β is a function $\beta(g)$ of g alone. (Here we use the fact that the σ-curves and s-curves are mapped into one another by scaling. Then the tangent lines

$$dg(s)/dm(s)^2 = (dg(s)/ds) \times (dm(s)^2/ds)^{-1}$$

transform as m^{-2} and β is scale invariant.) Thus the equation

$$\frac{dg}{ds} = \beta(g) \tag{2.2}$$

is a first order ordinary differential equation.

PROPOSITION. — Assume g and $d\beta(g)/dg$ are bounded for $\sigma-\sigma_c \geqq \varepsilon > 0$, for all $\varepsilon > 0$. Then $g(s)$ is monotone in s and $\beta(g) < 0$.

Proof. — By the assumed bound, $\beta(g)$ satisfies

$$|\beta(g_1) - \beta(g_2)| \leq M|g_1 - g_2|,$$

giving existence and uniqueness for the solution of the equation (2.2). Let $(\bar{s}, \infty)$ be the maximum interval on which $g(s)$ is strictly monotonic in s. Then $\beta(g(\bar{s})) = 0$ and by the uniqueness of the solution to (2.2), $g(s) = g(\bar{s}) =$ const. for $s \in [\bar{s}, \infty)$. However, the $d = 2$ cluster expansion [6] concerns the $\sigma \to \infty$ $(s \to \infty)$ limit. In this limit $g \approx g_0$, $m \approx m_0$ and $g_0(s) \to 0$ as $s \to \infty$. By (2.1) and perturbation theory (which is asymptotic [3])

$$\beta(g) = -g + 0(g^2) \tag{2.3}$$

as $g \to 0$. We conclude for small g that $\beta(g) < 0$, a contradiction to $\beta(g) = \beta(g(s)) = 0$, so $s = -\infty$, completing the proof for $d = 2$. For $d=3$, we expect a similar result with $\beta = -$ const $g + 0(g^2)$, but the required estimates have not yet been proved.

We now argue that for $\sigma < \sigma_c$, $g(s) \to 0$ as $s \to \infty$ $(\sigma \to -\infty)$. In that case, the proposition also applies for $\sigma < \sigma_c$, i. e. in the two phase region. We appeal to the Goldstone picture of the ground state, in which the ground state is determined by the minimum of the Euclidean action function V, and where the curvature at the minimum approximates the mass. For the φ^4 model with $\sigma \ll 0$, the minimum of $V(\cdot)$ occurs at $\pm(6|\sigma|/\lambda_0)^{\frac{1}{2}} \equiv \pm\bar{x}$. At the minimum, the curvature if $V^{(2)}(\pm\bar{x}) = 2|\sigma|$, while $V^{(3)}(\pm\bar{x}) = \pm(6\lambda_0|\sigma|)^{\frac{1}{2}}$ and $V^{(4)}(\pm\bar{x}) = \lambda_0$. Hence the dimensionless couplings at the minimum $\pm\bar{x}$ are for $d = 2$,

$$g^{(j)} = V^{(j)}(\pm\bar{x})/V^{(2)}(\pm\bar{x}) = \begin{cases} \pm(3\lambda_0/2|\sigma|)^{\frac{1}{2}}, & j = 3 \\ \lambda_0/2|\sigma|, & j = 4 \end{cases} \tag{2.4}$$

We conclude that as $\sigma \to -\infty$, and after scaling so that the mass $(2|\sigma|)^{\frac{1}{2}}$ remains bounded, the model converges to a free field (or a sum of two free fields, depending on the boundary conditions). By (2.4), we also see that the

Goldstone picture suggests $g(s) \approx \lambda_0/m(s)^2$ in this limit, and by (2.1), $\beta(g) \approx -g$ in this limit. In other words (2.3) holds for $\sigma \to -\infty$ as well as for $\sigma \to +\infty$. A similar picture holds for the quartic coupling $g^{(4)}$ (but not $g^{(3)}$) in three dimensions.

With this picture of critical point dominance, the function $\beta(g)$ has a zero at $g = 0$, and a second zero at $g = g_c$ (possibly infinity). The coupling constant is bounded by g_c, and β is represented in Figure 2. The zero at $g = 0$ governs the canonical (short distance, ultraviolet) behavior of the theory, while the presumed zero at $g = g_c$ governs the infrared (long distance) behavior.

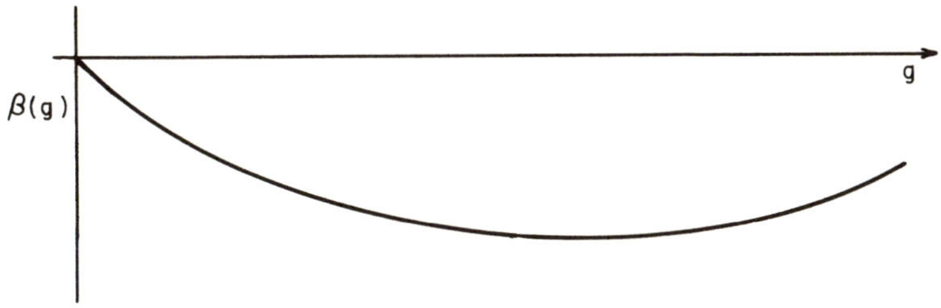

FIG. 2. — The curve $\beta(g)$ for $d = 2$.

This question remains open whether $g_c = \infty$, in which case the function $\beta(g) < 0$ for $g > 0$. In order for g_c to be finite, it is necessary that $\lambda_c = 0$ and that λ vanish as $\sigma \to \sigma_c$ in such a way that $\lambda/m^2 \to g_c$. If $g_c < \infty$, then our picture of critical point dominance says that g_c is the maximum value of g, and in the two phase region $\beta(g)$ returns to zero along another curve $\beta < 0$, see Figure 3.

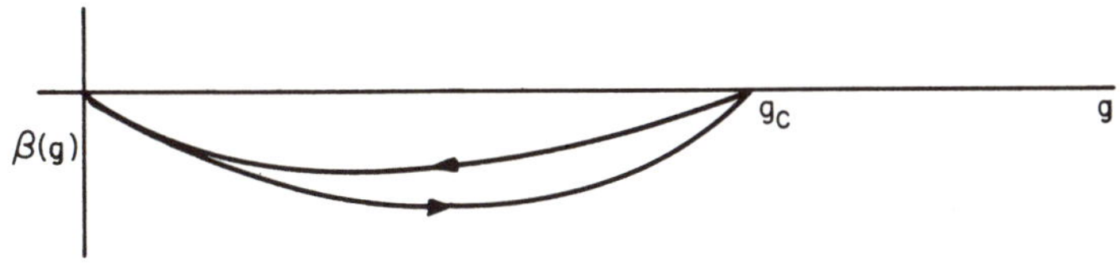

FIG. 3. — The trajectory of $\beta(g(\sigma))$, assuming $g_c < \infty$.

Critical theories ($g = g_c$) need not be scale invariant, but lie on the critical line of bare parameters $g_{0,c}$ (see the appendix). This line is transformed into itself by scale transformations. If a scale invariant φ_2^4 model exists, it is obtained by an infinite scaling of a critical (but not scale invariant) theory.

In four dimensions, as opposed to $d < 4$, the coupling constant λ is dimensionless, i. e. $g = \lambda$. Hence the first order contribution to $g(g_0)$ does not contribute to $\beta(g)$ and explicit calculation shows that the second order contribution is positive. Hence for weak coupling,

$$\beta(g) = \alpha g^2 + 0(g^3), \qquad \alpha > 0 .$$

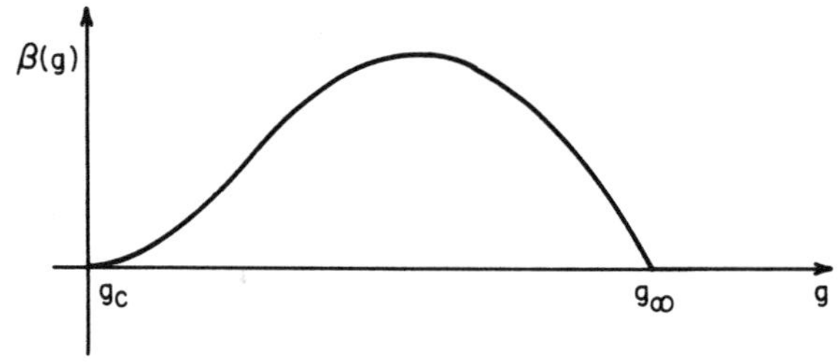

FIG. 4. — The curve $\beta(g)$ for $d = 4$.

The change in the sign of $\beta(g)$ from $d = 2$ or $d = 3$ reverses the roles of the zeros of $\beta(g)$. Thus the zero at $g = 0$ is the critical zero, $g_c = 0$, and governs the infrared (long distance) behavior of the theory. On the other hand, a zero at $g = g_\infty > 0$ would govern ultraviolet behavior (and is expected to be noncanonical). The arguments from the two dimensional case, applied to $\beta > 0$, yield characteristic curves drawn in Figure 5.

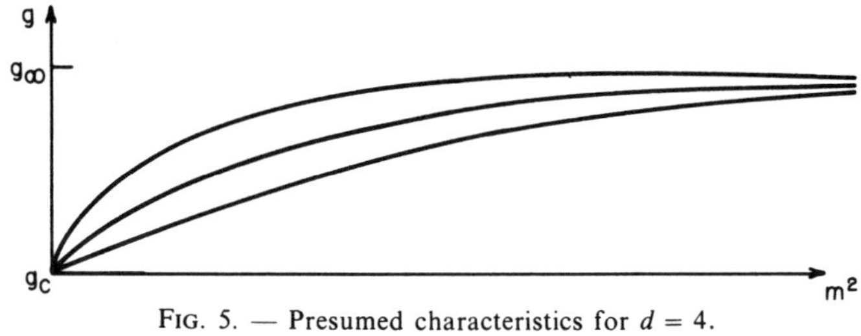

FIG. 5. — Presumed characteristics for $d = 4$.

As for $d < 4$, we expect in four dimensions that there are infinitely many critical theories, all interchanged by scaling. These theories presumably have canonical long distance behavior and infinite scaling of these theories in the long distance direction should produce a free theory of zero mass particles.

Finally we reconsider the behavior of

$$g = - Z^2 G^{(2)\sim}(0)^{-4} G^{(4)\sim}(0) m^{-4+d}$$

as we approach the critical point from the single phase region. By the definition of Z, $Z \leqq m^2 G^{(2)\sim}(0)$, so

$$g \leqq - G^{(2)\sim}(0)^{-2} G^{(4)\sim}(0) m^d .$$

Each of $G^{(2)\sim}$, $G^{(4)\sim}$ and m^2 is assumed to have behavior characterized by a critical exponent. We use the terminology of statistical mechanics, see [*10*]: the mass m is the inverse correlation length ξ^{-1}, and $G^{(2)\sim}(0)$ is proportional to susceptibility $(\partial M/\partial H)_T = \chi_T$. If $\varepsilon = (T/T_c - 1)$, the ε-dependence in the neighborhood of the critical point is given by

$$m = \xi^{-1} \sim \varepsilon^{\nu},$$
$$G^{(2)\sim}(0) \sim \chi_T \sim \varepsilon^{-\gamma}.$$

In field theory, we use $\varepsilon = \left(\frac{m_0^2}{m_{0,c}^2} - 1\right)$. The four point correlation function $G^{(4)}$ has the gap exponent Δ_4, which relates its critical behavior to the behavior of $G^{(2)}$,

$$-G^{(4)} \sim \varepsilon^{-2\Delta_4} G^{(2)} \sim \varepsilon^{-2\Delta_4 - \gamma}.$$

Thus

$$g \leqq 0(\varepsilon^{\gamma - 2\Delta_4 + d\nu}).$$

For the Ising model in the single phase region ($\varepsilon > 0$) the exponents are

$$d = 2: \quad \gamma = \frac{7}{4}, \quad \nu = 1, \quad \Delta_4 = \frac{15}{8},$$
$$d = 3: \quad \gamma = \frac{5}{4}, \quad \nu = \frac{5}{8}, \quad \Delta_4 = \frac{25}{16},$$

so

$$\gamma - 2\Delta_4 + d\nu = \begin{cases} 0, & d = 2 \\ 0, & d = 3 \end{cases}$$

Here we use exponents as in [*10*], and the $d = 3$ exponents are approximate. While there is no known relation between the Ising model exponents and the field theory exponents, Wilson and others have argued that they may coincide. These exponents do not distinguish for $d = 2, 3$ whether g has a finite limit at the critical point. Furthermore, these exponents indicate that the dimensional coupling constant $\lambda_c = 0$.

The critical exponent η describes the rate of decay of the pair correlation function at the critical point,

$$G(r) \sim r^{-(d-2+\eta)},$$

and $\eta \neq 0$ is the case of anomalous (noncanonical) dimensions. In the I_2 model, $\eta = \frac{1}{4}$. This anomalous dimension indicates that the propagator decays faster than it would in the presence of zero mass particles (i. e. $\eta = 0$). Thus we do not expect zero mass particles in the critical theory and we expect $Z_c = 0$. If $Z \sim \varepsilon^\zeta$ near the critical point, the bound $Z \leqq m^2 G^{(2)\sim}(0)$ shows

$$\zeta \geqq 2\nu - \gamma .$$

By the known inequality $(2 - \eta)\nu \geqq \gamma$, we have $\zeta \geqq \eta\nu$. Furthermore, $\zeta = \frac{1}{4}$ for I_2, using a recent calculation of Tracy and McCoy [*12*], and thus for I_2 the inequality is an identity. Thus anomalous dimensions, $\eta > 0$, indicate $Z_c = 0$.

Rigorous bounds on the two point function [*6*] exclude zero mass particles in the two point function, and assuming $m \to 0$ as the critical point is approached, these same bounds then prove $Z \to 0$. For I_3, $\eta \approx .04 \neq 0$, so again we expect $Z_c = 0$ and no zero mass particles in the φ_3^4 field theory. It is expected (up to logarithms) that the exponents are given by their canonical values in I_4.

APPENDIX

REVIEW OF SCALE AND MASS SHIFT TRANSFORMATIONS

We assume $d = 2$ in the appendix. The corresponding results for $d > 2$ are easily derived.

SCALE TRANSFORMATIONS. — On Fock space, the scale transformation is a unitary operator U defined on an n-particle state f by

$$(U^*f)(\vec{k}_1, \ldots, \vec{k}_n) = \alpha^{n/2} f(\alpha\vec{k}_1, \ldots, \alpha\vec{k}_n),$$

for $\alpha > 0$. The free field φ_{m_0} of mass m_0, and its Hamiltonian H_{0,m_0}, transform as

$$U\varphi_{m_0}(\vec{x})U^* = \varphi_{\alpha m_0}(\vec{x}/\alpha),$$
$$UH_{0,m_0}U^* = \alpha^{-1}H_{0,\alpha m_0}.$$

Also the no particle state Ω_0 is invariant.

$$U\Omega_0 = \Omega_0,$$

and $Ua(\vec{k})U^* = \alpha^{\frac{1}{2}}\alpha(k\alpha)$.

Furthermore, on Wick monomials, scale transformations map

$$U : \varphi_{m_0}(x)^j : U^* = : \varphi_{\alpha m_0}(\vec{x}/\alpha)^j : .$$

With an ultraviolet cutoff κ and box of length L,

$$UH_1(m_0, \lambda, \kappa, L)U^* = H_1(\alpha m_0, \alpha\lambda, \alpha\kappa, \alpha^{-1}L),$$

since the Wick ordering constant is invariant under scaling in two dimensions:

$$c(m_0, \kappa) = c(\alpha m_0, \alpha\kappa).$$

The vacuum energy for

$$H_{0,m_0} + H_1(m_0, \lambda_0, \kappa, L)$$

is the same as the vacuum energy for

$$U[H_{0,m_0} + H_1(m_0, \lambda_0, \kappa, L)]U^* = \alpha^{-1}H_{0,\alpha m_0} + H_1(\alpha m_0, \alpha\lambda_0, \alpha\kappa, \alpha^{-1}L)$$
$$= \alpha^{-1}[H_{0,\alpha m_0} + H_1(\alpha m_0, \alpha^2\lambda_0, \alpha\kappa, \alpha^{-1}L)].$$

Thus

$$E(m_0, \lambda_0, \kappa, L) = \alpha^{-1}E(\alpha m_0, \alpha^2\lambda_0, \alpha\kappa, \alpha^{-1}L)$$

and writing $H = H_0 + H_1 - E$,

$$UH(m_0, \lambda_0, \kappa, L)U^* = \alpha^{-1}H(\alpha m_0, \alpha^2\lambda_0, \alpha\kappa, \alpha^{-1}L).$$
$$U\Omega(m_0, \lambda_0, \kappa, L) = \Omega(\alpha m_0, \alpha^2\lambda_0, \alpha\kappa, \alpha^{-1}L).$$

Thus for a cutoff $\mathscr{P}(\varphi)_2$ field theory, scale transformations on Fock space provide a unitary map between Hamiltonians with different cutoffs, but the same g_0. Since the vacuum expectation values converge as κ, $L \to \infty$, we obtain in the limit an isomorphism of theories with constant g_0. We may define a unitary transformation

$$U : \mathscr{H}_1 \to \mathscr{H}_2$$

between the Hilbert spaces $\mathscr{H}_i$ of two such theories, where $U\Omega_1 = \Omega_2$,

$$UH_1U^* = \alpha^{-1}H_2,$$

and

$$U\varphi_1(x)U^* = \varphi_2(x/\alpha).$$

Theories described by different values of the parameter α are physically equivalent, and in the (m_0^2, λ_0) plane correspond to various points on lines of constant g_0, see Figure 6.

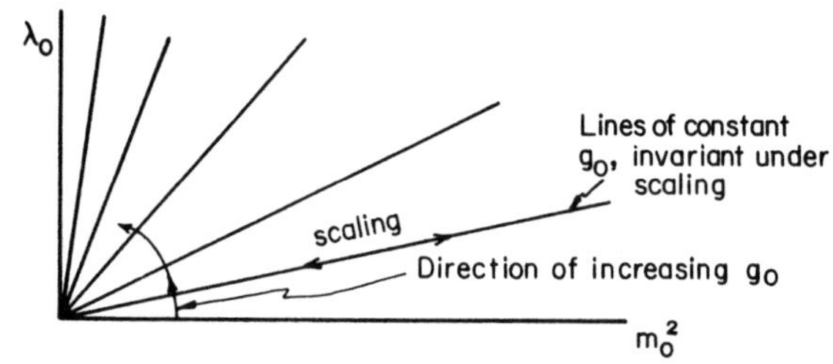

FIG. 6. — Scaling in the bare parameter space.

MASS SHIFT TRANSFORMATION. — The second transformation arises from a unitary operator on a Fock space over a finite volume (with e. g. periodic boundary conditions). In the infinite volume limit, we obtain an isomorphism of free fields of different mass, and of interacting fields as follows: we define in the Q-representation the scale (Bogoliubov) transformation,

$$U\varphi^{\sim}(\vec{k})U^* = \varphi^{\sim}(\vec{k})\alpha(\vec{k}),$$

where $\alpha(\vec{k}) = (\vec{k}^2 + m_0^2)^{\frac{1}{4}}(\vec{k}^2 + m_1^2)^{-\frac{1}{4}} = \mu_0^{\frac{1}{2}}\mu_1^{-\frac{1}{2}}$. Then by a standard calculation, see for example [1],

$$U\varphi_{m_0}(\vec{x})U^* = \varphi_{m_1}(\vec{x}),$$

$$U : \varphi_{m_0}(\vec{x})^j : U^* = \sum_{r=0}^{[j/2]} \frac{j!}{(j-2r)!r!2^r}(-\delta c)^r : \varphi_{m_1}(\vec{x})^{j-2r} :,$$

with

$$\delta c = \frac{1}{4\pi}\int\left[\frac{1}{k^2+m_0^2} - \frac{1}{k+m_1^2}\right]dk = \frac{1}{2}\ln\left(\frac{m_1}{m_0}\right). \tag{A1}$$

We note that δc varies over $(-\infty, \infty)$ as m_1/m_0 varies over $(0, \infty)$. Also

$$UH_{0,m_0}U^* = H_{0,m_1} + \frac{1}{2}(m_0^2 - m_1^2)\int : \varphi_{m_1}^2 : d\vec{x} - EL$$

where E is the (finite) vacuum energy density

$$E = -\int\left[\frac{1}{2}(\mu_0 - \mu_1) + \frac{1}{4}\frac{m_0^2 - m_1^2}{\mu_1}\right]d\vec{k}.$$

Let $\mathcal{P}$ be an interaction polynomial for mass m_0,

$$\mathcal{P}(\varphi_{m_0}) = \sum_{j=1}^{2N} \lambda_0^{(j)} : \varphi_{m_0}^j :.$$

Define $\hat{\mathcal{P}}$ as the transformed polynomial,

$$\hat{\mathcal{P}}(\varphi_{m_1}) = U\mathcal{P}(\varphi_{m_0})U^* = \sum_{j=0}^{2N} \hat{\lambda}_0^{(j)} : \varphi_{m_1}^j :,$$

and note $\lambda_0^{(j)} = \hat{\lambda}_0^{(j)}$ for $j = 2N, 2N-1$. Let H_1 and $\hat{H}_1$ be the respective interaction Hamiltonians. Then for $H(m_0, \lambda) = H_{0,m_0} + H_1(m_0, \lambda) - E(m_0, \lambda)$, where E is the vacuum energy, we have

$$UH(m_0, \lambda_0)U^* = H(m_1, \tilde{\lambda}_0). \tag{A2}$$

Here

(A3) $$\tilde{\lambda}_0^{(j)} = \hat{\lambda}_0^{(j)}, \qquad j \neq 2,$$
$$\tilde{\lambda}_0^{(2)} = \hat{\lambda}_0^{(2)} + \frac{1}{2}(m_0^2 - m_1^2).$$

We note in particular that $\lambda_0^{(2N)} > 0$ and

(A4) $$\tilde{\lambda}_0^{(2)} = \lambda_0^{(2)} + \frac{1}{2}(m_0^2 - m_1^2) + N(2N-1)!!\lambda_0^{(2N)}(-\delta c)^{N-1} + 0(|\delta c|^{N-2}).$$

By (A1), $-\delta c \to +\infty$ as $m_1 \to 0$. Also

$$|\delta c| \leqq 0\left(\left|\ln\frac{m_1}{m_0}\right|\right),$$

so $\tilde{\lambda}_0^{(2)} \to -\infty$ as $m_1 \to \infty$. Thus for m_0 fixed, $\tilde{\lambda}_0^{(2)}$ ranges over $(\infty, -\infty)$ as m_1 varies over $(0, \infty)$.

Remark 1. — Given $H(m_0, \lambda_0)$, there exists a bare mass m_1 for which

$$UH(m_0, \lambda_0)U^* = H(m_1, \tilde{\lambda}_0)$$

with $\tilde{\lambda}_0^{(2)} = 0$.

For the $\lambda_0 : \varphi_{m_0}^4 : + \sigma : \varphi_{m_0}^2 :$ interaction, $\tilde{\lambda}_0^{(2)}$ is monotonic in m_1, so the zero is unique. (More generally, $\mathscr{P}(\varphi_{m_0}) + \sigma\varphi_{m_0}^2$ has a unique zero for $|\sigma|$ large.) Hence with λ_0 fixed, we obtain a one parameter family of theories, parameterized by σ, with the interaction $\lambda_0 : \varphi_{m_1(\sigma)}^4 :$. In our diagram of couplings, this σ line is horizontal, with its height determined by λ_0, see Figure 7.

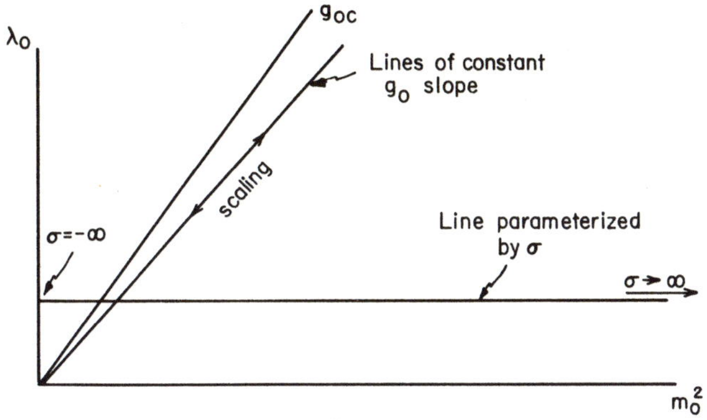

FIG. 7. — The $\lambda_0 : \varphi^4 : + \sigma : \varphi^2 :$ interaction.

For a pure φ^4 interaction, (A4) becomes

(A5) $$\tilde{\lambda}_0^2 = \lambda_0^{(2)} + \frac{1}{2}(m_0^2 - m_1^2) - 3\lambda \ln\left(\frac{m_1}{m_0}\right).$$

With the choice $\tilde{\lambda}_0^{(2)} = 0$, $\lambda_0^{(2)} = \sigma$, m_0^2 fixed, $m_1^2 = m_1^2(\sigma)$, we then have

(A6) $$\frac{dm_1^2(\sigma)}{d\sigma} = \left(\frac{1}{2} + \frac{3}{2}\frac{\lambda}{m_1^2(\sigma)}\right)^{-1}$$

Remark 2. — For the $\lambda\varphi_2^4$ interaction, $m_1^2(\sigma)$ is a differentiable function of σ, bounded away from zero for σ bounded away from $-\infty$.

Remark 3. — The finite volume may be introduced by periodic boundary conditions. Within the region of convergence of the cluster expansion, the infinite volume theory is independent of the choice of boundary conditions.

REFERENCES

[*1*] R. BAUMEL, Private communication.

[*2*] S. COLEMAN, Dilatations. *1971 Erice Lectures.*

[*3*] J. DIMOCK, Perturbation series asymptotic to Schwinger functions in $\mathscr{P}(\Phi)_2$. *Commun. Math. Phys.*, t. **35**, 1974, p. 347-356.

[*4*] J. FELDMAN, The Φ_3^4 field theory in a finite volume. *Commun. Math. Phys.*

[*5*] J. GLIMM and A. JAFFE, The entropy principle for vertex functions in quantum field models. *Ann. Institut Henri Poincaré*, t. **21**, 1974, p. 1-25.

[*6*] J. GLIMM, A. JAFFE and T. SPENCER, The particle structure of the weakly coupled $\mathscr{P}(\Phi)_2$ model and other applications of high temperature expansions. Part I. Physics of quantum field models. Part II. The cluster expansion. In: *Constructive quantum field theory*, Ed. by G. Velo and A. Wightman, Lecture notes in physics, Vol. 25, Springer Verlag, Berlin, 1973.

[*7*] F. GUERRA, L. ROSEN and B. SIMON, The $\mathscr{P}(\Phi)_2$ Euclidean quantum field theory as classical statistical mechanics. *Ann. Math.*

[*8*] B. MCCOY and T. T. WU, *The two dimensional ising model.* Harvard University Press, 1973.

[*9*] T. SPENCER, The mass gap for the $\mathscr{P}(\Phi)_2$ quantum field model with a strong external field. *Commun. Math. Phys.*

[*10*] H. E. STANLEY, *Introduction to phase transitions and critical phenomena.* Oxford University Press, 1971.

[*11*] K. SYMANZIK, Small distance behavior in field theory and power counting. *Commun. Math. Phys.*, t. **18**, 1970, p. 227-246.

[*12*] C. TRACY and B. MCCOY, Neutron scattering and the correlation functions of the Ising model near T_c. *Phys. Rev. Lett.*, t. **31**, 1973, p. 1500-1504.

(Manuscrit reçu le 11 février 1974)

$\phi_2{}^4$ quantum field model in the single-phase region: Differentiability of the mass and bounds on critical exponents

James Glimm*† and Arthur Jaffe‡§

Institut des Hautes Etudes Scientifiques, Bures-sur-Yvette, Essonne, France

(Received 25 March 1974)

We prove that the physical mass m is a differentiable function of the bare mass m_0 down to the critical point (defined by the occurrence of $m = 0$ or of the onset of symmetry breaking). We derive an upper bound on $m'(m_0)$. We show that the critical exponents η, ν, γ, ζ are bounded from below by their classical values.

The physical mass m in the $\phi_2{}^4$ quantum field model depends on the bare mass $m_0 > 0$, as well as on the other parameters of the theory. The cluster expansion converges for large m_0, and shows that $m > 0$ (Refs. 1 and 2) and that m is analytic in m_0.[3] As a consequence of correlation inequalities, m decreases as m_0 decreases for m_0 in an interval $(m_{0,c}, \infty)$.[4] According to the standard picture of symmetry breaking in the ground state (see Ref. 5), we expect that the lower end point $m_{0,c}$ of this interval is characterized by the condition $m = 0$. Two other alternatives, however, have not been excluded: $m_{0,c} = 0$ or $\langle \phi \rangle \neq 0$. In the Goldstone picture $m_0 > m_{0,c}$ corresponds to the single-phase region, and $m_{0,c} = 0$ would mean the absence of phase transitions. We expect that in a pure phase $\langle \phi \rangle$ is a continuous function of m_0, so we expect $\langle \phi \rangle = 0$ for $m_0 = m_{0,c}$. The two-dimensional Ising model is known to behave in this manner, while the open questions for the three-dimensional Ising model are similar to those discussed above for $\phi_2{}^4$.

One purpose of this note is to prove that m is differentiable in m_0, for $m_0 > m_{0,c}$. We define $m_{0,c}$ to be the supremum of the m_0 for which $m = 0$ or $\langle \phi \rangle \neq 0$, and we define the $\phi_2{}^4$ quantum field theory using Dirichlet boundary conditions. The convergence as $V \to \infty$ is based on monotonicity together with an upper bound on the Schwinger functions. For free boundary conditions, the upper bound follows from Ref. 6. The generalization of this upper bound to Dirichlet boundary conditions follows from the GKS inequality.[4,7] The monotonicity is established in Refs. 4 and 8. Our basic estimate is the correlation inequality that the four-point connected Euclidean Green's function is negative for $m_0 > m_{0,c}$.[5,9] We call this inequality the Lebowitz inequality.

To define m, choose $D > 0$ and consider the Euclidean expectation

$$F(D) = \int_{|\vec{x}|,\, |\vec{y}| < D} \langle \phi(x)\phi(y) \rangle_T \, d\vec{x}\, d\vec{y} \,\Big|_{x_0 = 0,\, y_0 = D}, \tag{1}$$

where $\langle\ \rangle_T$ denotes the truncated vacuum expectation value. For $m_0 > m_{0,c}$, $\langle\ \rangle_T = \langle\ \rangle$ in (1). Then

$$m = \lim_{D \to \infty} -\ln F(D)/D\,. \tag{2}$$

The validity of (2) follows from the spectral representation of the two-point function and from standard properties of the two-point function.[6,10] We fix the Wick-ordering constants and vary $m_0 = m_0(\sigma)$ by the addition of a quadratic (mass) perturbation, $\frac{1}{2}\sigma \int :\phi(x)^2: dx$.[11] We define

$$m' \equiv \frac{dm}{d\sigma}\,. \tag{3}$$

To bound the derivative m', we derive a bound on $-d[\ln F(D)/D]/d\sigma$ which is uniform in D and thus obtain the following:

Theorem 1. For each $\epsilon > 0$, $m(\sigma)$ is Lipschitz-continuous on the interval $\sigma \geq \sigma_c + \epsilon$. If $m(\sigma) \to 0$ as $\sigma \to \sigma_c$, then $m \leq (\sigma - \sigma_c)^{1/2}$.

Proof. Assuming $\sigma > \sigma_c$, we apply the GKS[4] inequality and the Lebowitz inequality[5] to obtain

$$\begin{aligned} 0 &\leq \frac{d}{d\sigma}[-\ln F(D)/D] \\ &= \frac{1}{2DF(D)} \int [\langle \phi(x) : \phi(z)^2 : \phi(y) \rangle \\ &\qquad - \langle \phi(x)\phi(y)\rangle\langle : \phi(z)^2 : \rangle]\, d\vec{x}\, d\vec{y}\, dz \\ &\leq D^{-1}F(D)^{-1} \int \langle \phi(x)\phi(z)\rangle\langle\phi(z)\phi(y)\rangle \, d\vec{x}\, d\vec{y}\, dz\,. \end{aligned} \tag{4}$$

To apply these inequalities we use a momentum cutoff as an intermediate step, in order to eliminate the Wick ordering.

The contribution of the region $z_0 \notin [0, D]$ or $|\vec{z}| \geq D$ to the upper bound in (4) is negligible, as $D \to \infty$ with σ fixed. As a consequence,

$$m' \leq \limsup_{D \to \infty} \sup_{0 \leq z_0 \leq D} \frac{\int \langle \phi(x)\phi(z)\rangle\langle\phi(z)\phi(y)\rangle \, d\vec{x}\, d\vec{y}\, d\vec{z}}{\int \langle \phi(x)\phi(y)\rangle \, d\vec{x}\, d\vec{y}}\,, \tag{5}$$

where the $\vec{z}$ integration is restricted to $|\vec{z}| \leq D$.

We now justify the exclusion of these regions. We first consider the region I defined by $0 \leq z_0 \leq D$, $|\vec{z}| \geq D + D^{3/4}$. The spectral representation for $\langle \phi(x)\phi(z)\rangle \langle \phi(z)\phi(y)\rangle$ is

$$\int da\, db\, \rho(a)\rho(b) C_a(x-z) C_b(z-y),$$

where $C_a(x-y)$ is the kernel of $(-\Delta + a)^{-1}$. We choose a point w on the line xy with $w_0 = z_0$. Then

$$\begin{aligned}|x-z| &\geq |x-w| + O(1)\min\{|z-w|^2/D, |z-w|\}\\ &\geq |x-w| + O(D^{1/2}) + O(|\vec{z}|^{1/2}),\end{aligned}$$

as $|\vec{z}| \to \infty$. For a given mass $a \geq m^2$,

$$C_a(x) \sim |x|^{-1/2} e^{-\sqrt{a}\,|x|} \quad \text{as } |x| \to \infty.$$

Thus,

$$C_a(x-z)C_b(z-y) \leq C_a(x-w)C_b(w-y)e^{-O(\delta)},$$

where $\delta = D^{1/2} + |\vec{z}|^{1/2}$. Thus we see that the integral over region I is bounded by $e^{-O(\delta)}$ times the integral over a strip of unit width about the line xy. Hence the integral over region I is negligible.

Furthermore, we see from the translation invariance of the integrands that the numerator and denominator both have magnitude $O(D^2 e^{-mD})$. Thus the region II defined by $0 \leq z_0 \leq D$, $D \leq |\vec{z}| \leq D + D^{3/4}$ contributes to the numerator at most the magnitude $O(D^{7/4} e^{-mD})$, and is negligible in the limit $D \to \infty$. Similarly, the region III defined by $z_0 \geq D + D^{3/4}$ and the region IV defined by $D + D^{3/4} \geq z_0 \geq D$ are negligible in (4) in the limit $D \to \infty$.

Let χ_D be the characteristic function of the interval $|\vec{z}| \leq D$ and let

$$f_D(z) = \int \langle \phi(x)\phi(z)\rangle \chi_D(\vec{z}) d\vec{x},$$

$$g_D(z) = \int \langle \phi(z)\phi(y)\rangle \chi_D(\vec{z}) d\vec{y},$$

so that the numerator in (5) is the L_2 inner product $\langle f, g\rangle$. Again, as $D \to \infty$, the angle between f_D and χ_D or between g_D and χ_D tends to zero. Thus as $D \to \infty$,

$$|\langle f, g\rangle| \leq \frac{\langle f, \chi\rangle\langle \chi, g\rangle}{\|\chi\|^2} + o(1),$$

or

$$m' \leq \limsup_{D\to\infty} \sup_{z_0 = w_0} \frac{\int \langle \phi(x)\phi(z)\rangle\, d\vec{x}\, d\vec{z} \int \langle \phi(w)\phi(z)\rangle\, d\vec{w}\, d\vec{z}}{2D \int \langle \phi(x)\phi(y)\rangle\, d\vec{x}\, d\vec{y}}, \tag{6}$$

where the integrals extend over $|\vec{x}|$, $|\vec{y}|$, or $|\vec{z}| \leq D$. We now show that we may replace the sup in (6) by the value at $z_0 = 0$ or D. Let

$$|t\rangle = e^{-tH/2} \int_{|\vec{x}| \leq D} \phi(\vec{x})\Omega, \quad F(t) = \langle t|t\rangle,$$

so the numerator in (6) is $G(t) = F(t)F(D-t)$, with $0 \leq t = z_0 = w_0 \leq D$. Note

$$(-1)^n F^{(n)}(t) = \langle t|H^n|t\rangle > 0,$$

and

$$\begin{aligned}G''(t) &= F''(t)F(D-t) + F(t)F''(D-t)\\ &\quad - 2F'(t)F'(D-t).\end{aligned} \tag{7}$$

By the Schwarz inequality, $-F'(t) \leq F(t)^{1/2} F''(t)^{1/2}$, so from (7) we have $G''(t) \geq 0$. Thus $G(t)$ has its maximum value at an end point $T = 0$ or D. It follows that (6) factors to give

$$m' \leq \int \langle \phi(\vec{x})\phi(\vec{0})\rangle\, d\vec{x} = \tfrac{1}{2}\int \rho(a) a^{-1/2} da \leq (2m)^{-1}. \tag{8}$$

Here $\rho(a)$ is the spectral measure for the two-point function. The existence of a Hilbert space ensures that $\rho(a)da$ is a positive measure, and the canonical commutation relations ensure $\int \rho da = 1$. Thus integration of (8) down to $\sigma = \sigma_c$ completes the proof of the theorem.

Furthermore, if the mass is defined for a limit of Dirichlet data in a strip of finite width L, the mass $m(\sigma)$ is upper semicontinuous. This follows since $m(\sigma, L)$ is continuous and decreasing to $m(\sigma)$ as $L \to \infty$. Thus

$$m(\sigma) = \lim_{\epsilon \to 0_+} m(\sigma_c + \epsilon).$$

We remark that our bound (8) also yields an inequality for the mass m as a function of the bare mass m_0.

Theorem 2. Assume a pure $\phi_2{}^4$ interaction which is Wick-ordered in the bare vacuum for a field of mass m_0, the bare mass.

(a) Let $\epsilon > 0$. The mass $m(m_0)$ is Lipshitz-continuous on the interval $m_0 > m_{0,c} + \epsilon$.

(b) Assume $m_{0,c} \neq 0$, and $m(m_0) \to 0$ as $m_0 \to m_{0,c}$. Then the mass m is bounded by its classical value, with a logarithmic correction,

$$m \leq \left(m_0^2 - m_{0,c}{}^2 + 3\lambda \ln \frac{m_0^2}{m_{0,c}{}^2}\right)^{1/2}. \tag{9}$$

This theorem concerns the change of variables $\sigma \to m_0^2$, since we now Wick-order in the bare mass m_0. In Ref. 11 we saw that the change in Wick ordering constants due to the change in bare mass yields

$$\frac{d\sigma}{dm_0^2} = 1 + 3\lambda m_0^{-2}. \tag{10}$$

From this combined with theorem 1, we infer theorem 2a. From (8), we then obtain

$$\frac{dm^2}{dm_0^2} = 2m \frac{dm}{d\sigma} \frac{d\sigma}{dm_0^2} \leq 1 + 3\lambda m_0^{-2},$$

which on integration yields (9).

The inequality (9) yields an inequality on the critical exponent ν governing the mass. Assuming that $m(m_{0,c}+\epsilon) \sim \epsilon^{\nu}$ as $\epsilon \to 0_+$, we see that $\nu \geq \frac{1}{2}$. (We give another elementary proof of this bound below.) The value $\nu = \frac{1}{2}$ is the classical value predicted by mean field theory or the Goldstone picture, while the value $\nu = 1$ occurs for the two-dimensional Ising model (see Ref. 12). We believe that $\nu = 1$ is more likely for ϕ_2^4 than $\nu = \nu_{cl} = \frac{1}{2}$, but we believe that $\nu \geq \frac{1}{2}$ is the best which can be obtained by methods which do not depend strongly on dimension, such as those used here. We note that Symanzik[13] has also used the Lebowitz inequality to bound an exponent.

We note that other exponents can also be bounded by their canonical values. The long-distance anomalous dimension η is greater than its classical value zero,

$$\eta \geq 0 = \eta_{cl},$$

as follows in two dimensions from the observation that the two-point function is a monotonic decreasing function of $|x-y|$ (see Ref. 10). In greater than two dimensions, the free field has an $m \to 0$ limit, so the proof that $\eta \geq 0$ is a straightforward consequence of the Källén-Lehmann representation.

If we assume the existence of an isolated single-particle state of mass m (this is proved for λ small[1]) then the field-strength renormalization constant Z is defined by

$$G^{(2)}(p) = \int \langle \Phi(x)\Phi(0)\rangle e^{ipx}\, dx$$

$$= \frac{Z}{p^2+m^2} + \int_{m^2+\delta}^{\infty} \frac{\rho(a)da}{p^2+a}.$$

Let ζ be the critical exponent for Z, $Z \sim \epsilon^{\zeta}$. Since $0 \leq Z \leq 1$, we have $\zeta \geq 0$. As discussed in Ref. 1, $\zeta = 0$ is the classical value and is compatible with the existence of zero-mass particles. Since ρ is positive with total integral 1,

$$Z/m^2 \leq G^{(2)}(0) \leq 1/m^2. \tag{11}$$

Let γ be the critical exponent for the susceptibility $\chi = G^{(2)}(0)$,

$$G^{(2)}(0) = \int \langle \Phi(x)\Phi(0)\rangle\, dx \sim \epsilon^{-\gamma}.$$

Thus from (11) we obtain

$$2\nu - \zeta \leq \gamma \leq 2\nu. \tag{12}$$

It is a reasonable hypothesis (apparently not accessible at present to proof) that $m' \leq O(1)Zm^{-1}$ and $\gamma = 2\nu - \zeta$, improving the bounds (8) and (12), respectively.

From the Lebowitz inequality we obtain, as in (4), (9),

$$0 \leq -\frac{dG^{(2)}(0)}{dm_0^2} = -(1+3\lambda m_0^{-2})\frac{dG^{(2)}(0)}{d\sigma}$$

$$\leq (1+3\lambda m_0^{-2})$$

$$\times \int \langle \Phi(x)\Phi(z)\rangle\langle\Phi(0)\Phi(z)\rangle\, dx\, dz$$

$$= (1+3\lambda m_0^{-2})G^{(2)}(0)^2. \tag{13}$$

Integrating (13) down to the critical point yields

$$G^{(2)}(0)^{-1} \leq m_0^2 - m_{0,c}^2 + 3\lambda \ln\frac{m_0^2}{m_{0,c}^2},$$

from which we obtain $\gamma \geq 1 = \gamma_{cl}$, and by the upper bound in (12), $\nu \geq \frac{1}{2} = \nu_{cl}$. We summarize these bounds as follows:

Theorem 3. Suppose $m_{0,c} > 0$ and $m \to 0$ as $m_0 \to m_{0,c}$. Then the exponents η, ν, γ, and ζ are bounded from below by their classical values,

$$\eta \geq 0, \quad \nu \geq \tfrac{1}{2}, \quad \gamma \geq 1, \quad \zeta \geq 0.$$

We raise the question whether other field-theory exponents are also bounded by their classical values.

Finally, we remark a simple inequality on α, the exponent for the specific heat C:

$$\epsilon^{-\alpha} \sim C \equiv \tfrac{1}{4}\int [\langle :\Phi(x)^2::\Phi(y)^2:\rangle$$

$$- \langle :\Phi(x)^2:\rangle\langle:\Phi(y)^2:\rangle]\, dy$$

$$\leq \tfrac{1}{2}\int \langle\Phi(x)\Phi(y)\rangle^2 dy = \tfrac{1}{2}\int G^{(2)}(p)^2 dp,$$

where again we use the Lebowitz inequality. From

$$G^{(2)}(p) = \int \frac{\rho(a)da}{p^2+a},$$

we find that

$$\int G^{(2)}(p)^2 dp = \pi \int \frac{\rho(a)\rho(b)\ln(b/a)}{b-a}\, da\, db$$

$$\leq O(m^{-2}).$$

Here we employ an elementary estimate on $(b-a)^{-1}\ln(b/a)$, and hence we infer

$$\alpha \leq 2\nu. \tag{14}$$

*Work supported in part by the National Science Foundation under Grant No. NSF-GP-24003.

†Permanent address: Rockefeller University, New York, New York 10021.

‡Work supported in part by the National Science Foundation under Grant No. NSF-GP-40354X.

§Permanent address: Lyman Laboratory of Physics, Harvard University, Cambridge, Massachusetts 02138.

[1]J. Glimm, A. Jaffe, and T. Spencer, Ann. Math. (to be published).

[2]J. Glimm, A. Jaffe, and T. Spencer, in *Constructive Quantum Field Theory*, edited by G. Velo and A. Wightman (Springer, Berlin, 1973).

[3]T. Spencer, Commun. Math. Phys. (to be published).

[4]F. Guerra, L. Rosen, and B. Simon, Ann. Math. (to be published).

[5]J. Glimm, A. Jaffe, and T. Spencer, in *Constructive Quantum Field Theory*, edited by G. Velo and A. Wightman (Ref. 2).

[6]J. Glimm and A. Jaffe, J. Math. Phys. 13, 1568 (1972).

[7]J. Fröhlich, Ann. Inst. Henri Poincaré (to be published).

[8]E. Nelson, in *Constructive Quantum Field Theory*, edited by G. Velo and A. Wightman (Ref. 2).

[9]J. Lebowitz, Commun. Math. Phys. 35, 87 (1974); B. Simon and R. Griffiths, *ibid.* 33, 145 (1973).

[10]B. Simon, Commun. Math. Phys. 31, 127 (1973); Ann. Math. (to be published).

[11]J. Glimm and A. Jaffe, Ann. Inst. Henri Poincaré 21, (1974).

[12]H. E. Stanley, *Introduction to Phase Transitions and Critical Phenomena* (Oxford Univ. Press, Oxford, England, 1971).

[13]K. Symanzik, Cargèse lectures, 1973 (unpublished).

Remark on the Existence of φ_4^4

James Glimm
Rockefeller University, New York, New York 10021

and

Arthur Jaffe*
Harvard University, Cambridge, Massachusetts 02138

(Received 10 June 1973)

The bound $|S_\epsilon^{(2)}(f\otimes g)| \leq \mathrm{const}\,|f|_{\mathcal{S}}|g|_{\mathcal{S}}$, uniform as the lattice spacing $\epsilon \to 0$, is sufficient to yield construction of a φ^4 field theory. In perturbation theory, this bound holds in space-time dimensions $d \leq 4$. The bounds on the n-point Schwinger functions follow from the given bound by the use of Lebowitz's inequalities. They yield the growth $|S^{(n)}| \leq O(n!)$, consistent with the Osterwalder-Schrader axioms, which allow reconstruction of a real-time theory.

We remark that a uniform bound on the two-point Schwinger function (with a lattice approximation) yields the existence of a $\varphi_4^4 - \mu\varphi$ model without cutoffs, including the case $\mu = 0$. We assume that the two-point functions are labeled by a cutoff parameter ϵ, the lattice spacing, and that for some $\mathcal{S}$ norm, independent of ϵ,

$$|S_\epsilon^{(2)}(f\otimes g)| \leq |f|_{\mathcal{S}}|g|_{\mathcal{S}}. \tag{1}$$

Theorem 1.—With the assumption of (1), there exists a set of Schwinger functions $S^{(n)}(x_1, \ldots, x_n) = \lim_{\epsilon_j} S_{\epsilon_j}^{(n)}(x_1, \ldots, x_n)$ which satisfy the Osterwalder-Schrader axioms,[1] with the possible exception of Euclidean covariance and clustering. Here ϵ_j denotes a subsequence of lattice cutoffs, and $\epsilon_j \to 0$.

Remark.—As Wilson observed,[2] the nontriviality of the φ_4^4 model may be more difficult to establish than its existence. In particular, the hypothesis of Theorem 1 does not require the correct charge or mass renormalization, even in perturbation theory. Thus the limiting theory could describe a free field, and, still worse, the possibility $S^{(n)} \equiv 0$ is not excluded. Correct mass (and field-strength) renormalization prevents the vanishing of $S^{(2)}$.

We obtain uniform bounds on the n-point function in terms of the two-point function, and convergence of the $S_{\epsilon_j}^{(n)}$ by compactness. The uniform bounds depend on correlation inequalities which were proved in statistical mechanics by Lebowitz,[3] and which follow in φ^4 field theory by the spin-$\frac{1}{2}$ approximation of Simon and Griffiths.[4] Our use of these inequalities in field theory was suggested by the work of Feldman[5] and especially the work of Spencer[6] on the elimination of even bound states in the φ^4 model. See also Glimm and Jaffe,[7] where the inequality on the connected four-point function, $S_T^{(4)} \leq 0$, yields bounds on critical exponents by their mean-field values.

Proposition 1.—Under the above hypothesis (1), with $f_i \geq 0$ and $\mu \geq 0$,

$$0 \leq S_\epsilon^{(n)}(f_1 \otimes \cdots \otimes f_n) \leq 2^{n-1}(n-1)!\prod_{i=1}^n |f_i|_{\mathcal{S}}, \tag{2}$$

where $|f_i|_{\mathcal{S}}$ is the same norm as in (1).

Proof.—The proof involves an induction hypothesis. We assume $\mu \geq 0$ and use the Lebowitz inequalities[3]

$$\langle qt\rangle - \langle q\rangle\langle t\rangle \leq 0. \tag{3}$$

Here

$$t(x) = \tfrac{1}{2}(\varphi + \tilde{\varphi}), \quad q(x) = \tfrac{1}{2}(\varphi - \tilde{\varphi}),$$

$$t = \Pi_i t(x_i), \quad q = \Pi_i q(x_i),$$

and $\tilde{\varphi}$ denotes a duplicate Euclidean field variable. The restriction $\mu \geq 0$ is no loss of generality, by the $\varphi \to -\varphi$ symmetry. We take

$$t = \prod_{i=1}^n t(x_i), \quad q = q(x_{n+1})q(x_{n+2}),$$

and we write (3) as

$$\textstyle\sum_X S(\sim X)[S(X, x_{n+1}, x_{n+2}) - S(X)S(x_{n+1}, x_{n+2}) + S(X)S(x_{n+1})S(x_{n+2})] \leq \sum_X S(X, x_{n+1})S(\sim X, x_{n+2}), \tag{4}$$

where X denotes a subset of $x_1, \ldots, x_n$, and $\sim X$ the complement of X. By the Griffiths-Kelly-Sherman inequalities,[8] each term on the left-hand side of (4) is positive. Thus, in particular,

$$0 \leq S^{(n+2)}(x_1, \ldots, x_{n+2}) \leq S^{(n)}(x_1, \ldots, x_n)S^{(2)}(x_{n+1}, x_{n+2}) + \textstyle\sum_X S(X, x_{n+1})S(\sim X, x_{n+2}), \tag{5}$$

and we have bounded $S^{(n+2)}$ by a sum of products of lower-order Schwinger functions.

The inductive hypothesis we use is (2) for $f_i \geq 0$ and $n \geq 1$. For $n=2$, the hypothesis follows from (1), while for $n=1$, it follows from the Schwarz inequality and from (1):

$$S^{(1)}(f) \leq S^{(2)}(f \otimes f)^{1/2} \leq |f|_{\mathcal{S}}.$$

Assuming the inductive hypothesis for $S^{(1)}, \ldots, S^{(n+1)}$, we prove it for $S^{(n+2)}$. By (5),

$$S^{(n+2)}(f_1 \otimes \cdots \otimes f_n) \leq 2^n[(n-1)! + \sum_X |X|!(n-|X|)!] \prod_{i=1}^{n+2} |f_i|_{\mathcal{S}} \leq 2^{n+1}(n+1)! \prod_{i=1}^{n+2} |f_i|_{\mathcal{S}}.$$

Here we use the fact that with $|X|=j$ there are $\binom{n}{j}$ choices for the subset. This completes the proof of the induction hypothesis.

We consider the norms on $\mathcal{S}(R^d)$,

$$|f|_r = \sup_x |(1+x^2)^r f(x)|.$$

Proposition 2.—With the assumption of (1), there exists $r<\infty$ such that

$$|S_\epsilon^{(2)}(f \otimes g)| \leq \text{const}\, |f|_r |g|_r,$$

with the constant uniform in ϵ.

Proof.—Because $S^{(2)}$ is positive, and defined on $\mathcal{S} \otimes \mathcal{S}$, it is a positive measure. Let $0 \leq \chi_j(x) = \chi(x-j)$ be a C_0^∞ partition of unity, with $j \in Z^d$. Then

$$|S^{(2)}(f \otimes g)| \leq \sum_{i,j} (\sup_{\text{supp}\chi_i \otimes \chi_j} |f \otimes g|) S^{(2)}(\chi_i \otimes \chi_j).$$

Since the χ_i are translates of a particular function, for some $m<\infty$,

$$S^{(2)}(\chi_i \otimes \chi_j) < |\chi_i|_{\mathcal{S}} |\chi_j|_{\mathcal{S}} \leq \text{const}(1+i^2)^m(1+j^2)^m,$$

with the constant uniform in ϵ. By choosing $r=m+d$, the proposition follows.

By linearity, we then obtain from Propositions 1 and 2 the following:

Corollary.—With the assumption of (1), there exists $r<\infty$, and $c<\infty$, such that

$$|S_\epsilon^{(n)}(f_1 \otimes \cdots \otimes f_n)| \leq c^n n! \prod_{i=1}^{n} |f_i|_r. \tag{6}$$

Here $f_i \in \mathcal{S}(R^d)$, and the constants are uniform in ϵ.

Proof of Theorem 1.—The existence of a convergent subsequence, as $\epsilon \to 0$, follows from the Corollary to Proposition 2 by compactness, and the same bounds hold in the limit. The symmetry and Osterwalder-Schrader positivity of the lattice Schwinger functions are known for a finite-volume theory[9] and easily extended to the infinite-volume and $\epsilon=0$ limits. This completes the proof.

Corollary.—With the assumption of (1), the generating function

$$Z\{f\} = \sum_{n=0}^{\infty} S^{(n)}(f \otimes \cdots \otimes f)/n!$$

is an analytic function for f in a neighborhood of the origin in $\mathcal{S}(R^d)$. It is also the characteristic function for a unique measure dq on $\mathcal{S}'(R^d)$, i.e.,

$$Z\{f\} = \int e^{\varphi(f)} dq, \quad S^{(n)}(f_1, \ldots, f_n) = \int \varphi(f_1) \cdots \varphi(f_n) dq. \tag{7}$$

Proof.—The existence and analyticity of Z follow immediately from (6), while the existence and uniqueness of dq follow by Minlos's theorem.

We remark that a number of other known properties of $(\varphi^4 - \mu\varphi)_2$ are also a consequence of hypothesis (1). In two and three dimensions, Theorem 1 may provide an alternative method for proof of existence of the φ^4 theory. In three dimensions, the finite-space-time-volume φ^4 theory exists[10-11] and is a limit as $\epsilon \to 0$ of a lattice theory.[12] It is sufficient to verify hypothesis (1) in the finite-volume, $\epsilon=0$ theory, since the Lebowitz inequality (3) and the proof of (2) extend to the $\epsilon=0$ limit. Then the norm $|f|_{\mathcal{S}}$ must be uniform in the volume.

*Work supported in part by the National Science Foundation under Grant No. NSF GP 40354X.

[1]K. Osterwalder, in *Constructive Quantum Field Theory*, edited by G. Velo and A. Wightman (Springer, Berlin, 1973). See also J. Fröhlich, to be published; K. Osterwalder and R. Schrader, to be published, Axiom (E0‴).

[2]K. Wilson and J. Kogut, to be published.

[3]J. Lebowitz, Commun. Math. Phys. 35, 87 (1974).

[4]B. Simon and R. Griffiths, Commun. Math. Phys. 33, 145 (1973).

[5]J. Feldman, "On the Absence of Bound States in the φ^4 Quantum Field Model without Symmetry Breaking" (to be published).

[6]T. Spencer, "The Absence of Even Bound States in φ_2^4" (to be published).

[7]J. Glimm and A. Jaffe, Phys. Rev. D (to be published).

[8]J. Ginibre, Commun. Math. Phys. 16, 310 (1970).

[9]F. Guerra, L. Rosen, and B. Simon, to be published; J. Fröhlich, to be published.

[10]J. Glimm and A. Jaffe, Fortschr. Phys. 21, 327 (1973).

[11]J. Feldman, to be published.

[12]Y. Park, "Lattice Approximation of the $(\varphi^4 - \mu\varphi)_3$ Field Theory in a Finite Volume" (to be published).

[13]We have been informed that J. Feldman has independently proved Proposition 1.

On the approach to the critical point [1]

by

James GLIMM [2]
Rockefeller University New York, New York 10021

and

Arthur JAFFE [3]
Harvard University Cambridge, Massachusetts 02138

ABSTRACT. — The mass perturbation yields a new construction for $\mathscr{P}(\varphi)_2$ = Even + Linear quantum field models. For φ^4 models, the Schwinger functions are differentiable in the parameter $\sigma = \delta m^2$, in the single phase region, and are continuous as $\sigma \searrow \sigma_c$. For the Ising_d model, $d \geq 2$ classical bounds on the critical exponents, *e. g.* $1 \leq \gamma \leq 2\nu$, are a consequence of an Ornstein-Zernike type upper bound on the two point function.

---

1. INTRODUCTION

For certain critical values of the parameters, the Schwinger functions of quantum field theory or correlation functions of statistical mechanics are expected to exhibit nonexponential decay,

$$S^{(2)}(x) \equiv \int \varphi(x)\varphi(0)dq \sim |x|^{-d+2-\eta}$$

---

(1) We thank the Aspen Center for Physics for hospitality.
(2) Supported in part by the National Science Foundation under Grant MPS 74-13252.
(3) Supported in part by the National Science Foundation under Grant MPS 73-05037.

where $0 \leq \eta \leq 2$ and with the result that the corresponding integrated quantities, for example the susceptibility,

$$\chi \equiv \int S^{(2)}(x)dx$$

become infinite. Let $S^{(n)}(\sigma; x)$ denote the dependence of the n-point Schwinger function in an even $\mathscr{P}(\varphi)_2$ model on the mass perturbation

$$\frac{1}{2}\sigma \int :\varphi(x)^2 : dx.$$

An elementary application of Griffiths' inequalities gives three ranges of parameter values as follows: $\sigma < \sigma_c^-$ defines a multiple phase theory, $\sigma_c^- < \sigma < \sigma_c^+$ defines a zero mass (critical) theory, and $\sigma_c^+ < \sigma$ defines a positive mass, single phase theory. It is expected that $\sigma_c^- = \sigma_c = \sigma_c^+$; however in some cases, σ_c may not define a zero mass theory, see chapter 2.

As $\sigma \to +\infty$, the limit $S^{(n)}(\sigma, x) \to 0$, $n \geq 1$, follows from the cluster expansion [*15*] [*16*]. Callan and Symanzik, see [*2*] [*29*], propose the mass perturbation, as in the formula

$$S^{(n)}(\sigma_c, x) = -\int_{\sigma_c}^{\infty} \partial_\sigma S^{(n)}(\sigma, x)d\sigma, \tag{1.1}$$

for the study of critical phenomena. In fact with g denoting the physical charge, hypotheses *e. g.* on

$$\beta(g) \equiv \frac{m}{2}\left(\frac{dg}{d\sigma}\frac{d\sigma}{dm}\right)_{\mathscr{P}=\text{const.}}$$

near $\sigma = \sigma_c$ lead to the scaling law and calculation of critical exponents in terms of *e. g.* $\beta'(g_c)$, see [*3*] [*24*].

In this paper, we show that the derivatives in (1.1) exist for φ^4 models in the single phase region, as a consequence of the Lebowitz inequalities [*12*]. The Lebowitz inequalities dominate $\partial S^{(n)}/\partial\sigma$ by a sum of products of two point functions,

$$0 \leq -\partial S^{(n)}/\partial\sigma \leq \sum \prod_{i,j \neq k,l} S^{(2)}(x_i - x_j) \times \int S^{(2)}(x_k - y)S^{(2)}(x_l - y)dy, \tag{1.2}$$

see chapter 3.

2. CONSTRUCTION OF THE $\mathscr{P}$ = EVEN $-\ \mu\varphi$ MODELS

We give a new construction of the $\mathscr{P}$ = Even + Linear model in two dimensions. We start from a polynomial $\mathscr{P}$ with a sufficiently large bare mass, so that it yields a Wightman theory as a consequence of a convergent cluster expansion [15] [16]. We study the associated Schwinger functions under a two parameter family of perturbations, of the form $\frac{1}{2}\sigma\varphi^2 - \mu\varphi$. Using a Griffiths inequality, we show that for $\sigma \leq 0$ and $\mu \geq 0$, the Schwinger functions converge monotonically to Schwinger functions for an Osterwalder-Schrader theory [23], with the possible exception of clustering. In the case of mixed phases, the decomposition into pure phases satisfying all the Osterwalder-Schrader axioms is given in [1] [5]. Each pure phase then gives rise to a $\mathscr{P}(\varphi)_2$ Wightman theory.

We consider the perturbation for the action,

$$\delta V = \frac{1}{2}\sigma \int_{\Lambda_1} :\varphi^2: dx - \mu \int_{\Lambda_2} \varphi dx, \tag{2.1}$$

where $\mu, -\sigma \geq 0$, and where Λ is a bounded space-time region with volume $|\Lambda|$. We suppose that the linear term in $\mathscr{P}$ has a nonpositive coefficient.

PROPOSITION 2.1. — Under the perturbation δV, the $\mathscr{P}(\varphi)_2$ Schwinger functions increase monotonically in $-\sigma, \mu, |\Lambda_i|$.

Proof [18]. — By the first Griffiths inequality, $S^{(n)}(X) \geq 0$, and by the second Griffiths inequality

$$-\frac{d}{d\sigma} S^{(n)}(X) = \int_{\Lambda_1} [S^{(n+2)}(Xzz) - S^{(n)}(X)S^{(2)}(zz)]dz \geq 0. \tag{2.2}$$

For the lattice approximation to field theory, the Griffiths inequalities follow from [6], for $\mathscr{P}$ = even $- \mu\varphi$, $\mu \geq 0$. These inequalities follow in field theory from convergence of the lattice approximation ([18] for $\mathscr{P}(\varphi)_2$, [25] for φ_3^4). Here we note the Wick ordering constant cancels between the two terms on the right of (2.2). This proves that $S^{(n)}$ increases monotonically in $-\sigma$, $|\Lambda_1|$. The proof of monotonicity in μ, $|\Lambda_2|$ is similar.

We remark that the Schwinger functions are uniformly bounded in Λ_i, σ, and μ, so long as σ and μ remain bounded (This is a consequence of Hamiltonian φ-bounds [8]. See Fröhlich [5] for the translation of these bounds to the Schwinger functions). Hence the Schwinger functions converge as $|\Lambda_i| \to \infty$, as $\sigma \searrow \bar{\sigma} > -\infty$, and as $\mu \nearrow \bar{\mu} < \infty$. Clearly the limiting Schwinger functions are Euclidean covariant, and the remaining Osterwalder-Schrader axioms (except clustering) follow from the properties of the approximate Schwinger functions.

THEOREM 2.2. — For each polynomial $\mathscr{P}(\varphi)$ with a convergent cluster expansion, we obtain a two parameter family of (possibly mixed phase) Osterwalder-Schrader theories, with Schwinger functions $S^{(n)}(\sigma, \mu; X)$.

Each theory $\{ S^{(n)}(\sigma, \mu; X) \}$ can be decomposed into a direct integral of pure phases, with each component an Osterwalder-Schrader theory.

REMARK 2.3. — We note that the theories $S^{(n)}(\sigma, \mu; X)$ are uniquely defined by the bare parameters, together with the bare mass. Thus with $\sigma_i \leq 0$, $\mu_i \geq 0$, we can add successively the perturbations

$$\delta V_i = \frac{1}{2}\sigma_i \int_{\Lambda_i} : \varphi^2 : dx - \mu_i \int_{\Lambda_i} \varphi dx,$$

taking the limit $\Lambda_i \nearrow R^2$ in any order, and obtain $S^{(n)}(\sigma, \mu; X)$, where $\sigma = \Sigma\sigma_i$, and $\mu = \Sigma_i\mu_i$. The fact that the limit $S^{(n)}(\sigma, \mu; X)$ is independent of the order in which the perturbations are added has the following interpretation. If a phase transition occurs at the parameter values σ, μ, we can regard $S^{(n)}(\sigma, \mu; X)$ as defined by a δV_i perturbation, starting from the boundary conditions $S^{(n)}(\sigma^{(i)}, \mu^{(i)}; X)$, where $\sigma^{(i)} = \Sigma_{j\neq i}\sigma_j$ and $\mu^{(i)} = \Sigma_{j\neq i}\mu_j$. Since $S^{(n)}(\sigma, \mu; X)$ is independent of the order in which the δV_i perturbations are added, the various boundary conditions $\sigma^{(i)}$, $\mu^{(i)}$ define the same theory. Allowing for an infinite series σ_j, μ_j, we see that

$$S^{(n)}(\sigma, \mu ; X) = \lim_{\sigma\searrow\sigma', \mu'\nearrow\mu} S^{(n)}(\sigma', \mu' ; X)$$

Because of the possibility of phase transitions, the iterated limits with some or all $\mu_i \leq 0$ may not coincide with the $S^{(n)}(\sigma, \mu; X)$. In fact the iterated limits with each $\sigma_i\mu_i \leq 0$ are always defined (in any order) by monotonicity, and should provide a general set of boundary conditions. In particular, for φ^4, both pure phases as well as the even mixture could be constructed by this method. From this point of view, the construction of Theorem 2.2 with a single δV perturbation could be called *weak coupling boundary conditions.*

REMARK 2.4. — For φ^6 and higher order $\mathscr{P}(\varphi)$ theories, it is possible that curves of phase transitions can occur in the σ, μ plane for $\mu \neq 0$. Because the above iterated limits coincide, any such curve in the $\mu > 0$ half plane must have a positive slope in the σ, μ plane, at least in the case of a first order phase transition with respect to order parameter $\langle \varphi \rangle$. Standard mean field approximations suggest that there should be a phase transition only at $\mu = 0$ if the φ^4, φ^6, ... coefficients in an even $\mathscr{P}$ are fixed and $\sigma \ll 0$. This picture also suggests that the curves for $\mu > 0$ or $\mu < 0$ move in toward and meet the $\mu = 0$ line as σ decreases. It is consistent with this picture that $\sigma_c^+ = \sigma_c^-$ (as defined in chapter 3) could give the value of σ for which two such curves meet the $\mu = 0$ line. In this case it is reasonable to suppose that $m(\sigma_c) > 0$. See figure 1. The variation of the φ^4 and higher coefficients is expected to lead to tricritical points, see [*26*]. There heuristic calculations indicate classical tricritical exponents for $d = 3$, in contrast to an ordinary critical point.

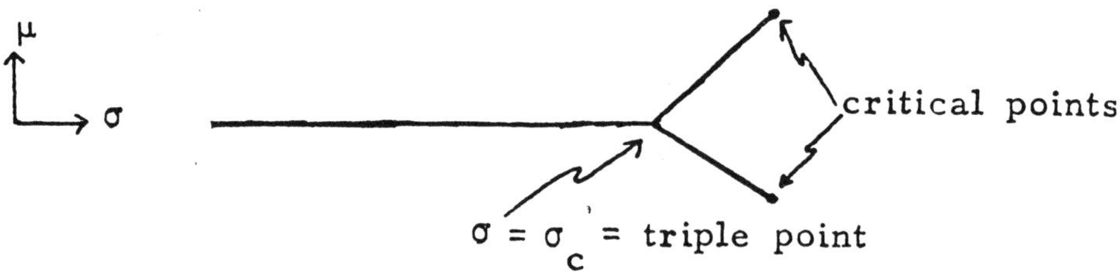

FIG. 1. — Postulated phase transition curves for $\varphi^6 + \lambda\varphi^4$ models, $\lambda \ll 0$.

REMARK 2.5. — The above construction (weak coupling boundary conditions) coincides with the theory defined by Dirichlet boundary data. In fact if the weakly coupled $\mathscr{P}(\varphi)_2$ theory defined above by a cluster expansion has Dirichlet data on the boundary of a region Λ_0, then $S(\sigma)$ is monotone increasing in Λ_0. The limits $\Lambda_0 \nearrow R^2$, $\Lambda_1 \nearrow R^2$, $\Lambda_2 \nearrow R^2$ may be taken in any order. With $\Lambda_0 \nearrow R^2$ first, we have weak $\mathscr{P}(\varphi)_2$ boundary conditions, since the cluster expansion defines a theory which is independent of boundary conditions. With $\Lambda_0 = \Lambda_1 = \Lambda_2 \nearrow R^2$, we have the conventional Dirichlet theory.

The weak coupling boundary conditions also coincide with free boundary data, provided only that $m_0 \gg 1$ and that the space cutoff in the mass term $\frac{1}{2}\sigma\int :\varphi^2: dx$ is removed after the space cutoff in the other terms of $H_I = \int \mathscr{P}(\varphi)dx$. It follows that the estimates of [*8*], established for the C*-algebra construction of $\mathscr{P}(\varphi)_2$ field theory [*7*], also have consequences for the theory constructed here; see section 5. An extensive analysis of even boundary conditions is given in [*19*]. It is an open question whether the periodic mass converges in the infinite volume limit to the mass of the infinite volume theory. A positive answer to this question would show that for φ^4, $m(\sigma) \searrow = 0$ as $\sigma \searrow \sigma_c$.

3. THE CRITICAL POINT IN EVEN $\mathscr{P}(\varphi)_2$ MODELS

The variation of the mass parameter σ in an even $\mathscr{P}(\varphi)_2$ model (see chapters 1, 2) yields three characteristic intervals:

a) a single phase interval (σ_c^+, ∞),

b) a critical interval (σ_c^-, σ_c^+),

c) a multiple phase interval $(-\infty, \sigma_c^-)$ with symmetry breaking.

The endpoints $\sigma_c^\pm$ may also be included in the intervals *a*) *c*).

We let $\mu = 0$, and let

$$M(\sigma)^2 = \lim_{|x|\to\infty} S^{(2)}(\sigma\,; x).$$

We define $m(\sigma)$ as the exponential decay rate of $S^{(2)}(\sigma; x) - M(\sigma)^2$. Consider the two conditions,

$$m(\sigma) > 0 \tag{3.1}$$

$$M(\sigma) = 0. \tag{3.2}$$

DEFINITION 3.1. — Let σ_c^+ be the infimum of the values of σ satisfying both (3.1) and (3.2). Let σ_c^- be the supremum of the values of σ which violate (3.2).

REMARK 3.2. — With the above definitions:

(*i*) $\sigma_c^- \leq \sigma_c^+ < \infty$.
(*ii*) On $(-\infty, \sigma_c^-)$, (3.2) is violated.
(*iii*) On (σ_c^-, σ_c^+), (3.2) holds, but $m = 0$.
(*iv*) On (σ_c^+, ∞), both (3.1) and (3.2) hold.

By scaling, $\sigma \to +\infty$ is the weak coupling limit. For weak coupling, the cluster expansion [*15*] [*16*] yields (3.1) and (3.2), so $\sigma_c^+ < \infty$. The remaining statements follow by the monotone decrease of $S^{(2)}(x; \sigma)$ in σ. The bound $|\sigma_c^-| < \infty$ is announced in [*4*]; see also [*5*].

THEOREM 3.3. — Let $\mu = 0$. Then $m(\sigma)$ is continuous on (σ_c^+, ∞) at all points of continuity of $S^{(2)}(\sigma; x)$.

THEOREM 3.4. — Let $\mu = 0$, and suppose (3.2) holds for $\sigma = \sigma_c^+$. Then $m(\sigma) \searrow m(\sigma_c^+)$ as $\sigma \searrow \sigma_c^+$.

THEOREM 3.5. — Let $\mu = 0$ and suppose $m(\sigma) \not\to 0$ as $\sigma \searrow \sigma_c^+$. Then (3.2) holds for $\sigma = \sigma_c^+$, and so $m(\sigma) \searrow m(\sigma_c^+)$ as $\sigma \searrow \sigma_c^+$.

REMARK 3.6. — Assuming φ_3^4 to be a Wightman theory, $M(\sigma_c) = 0$. This follows from the fact that $S^2(\sigma; x) \leq |x|^{-1}$. Thus $m(\sigma) \searrow m(\sigma_c^+)$.

REMARK 3.7. — See Figure 1. In that case, the limit $\sigma \searrow \sigma_c$ should yield the pure phase associated with $M(\sigma_c) = 0$, but with $m(\sigma_c) > 0$.

REMARK 3.8. — $\mathscr{P}(\varphi)_2$ critical theories, *i. e.* $m = 0 = M$, are not scale invariant. The continuum of such theories is parameterized by a scale length, see ([*16*], p. 155).

Proof of Theorems 3.3-3.5. — Since $S^{(2)}$ is monotone decreasing for $\sigma \in (-\infty, \infty)$, and continuous from above, it follows that $m(\sigma)$ is monotone increasing on (σ_c^-, ∞). If σ_0 satisfies (3.2), then $m(\sigma) \searrow m(\sigma_0)$ as $\sigma \searrow \sigma_0$, for the exponential decay rate of $S^{(2)}$ defines $m(\sigma_0)$. This proves Theorem 3.4. The same argument applied to convergence from below completes the proof of Theorem 3.3. Theorem 3.5 follows from the observation that if $m \not\to 0$, then $S^{(2)}$ has a uniform exponential decay as $\sigma \searrow \sigma_c$.

THEOREM 3.9. — For a φ_2^4 theory with $\mu = 0$, the derivatives $\partial S^{(n)}(\sigma; x)/\partial\sigma$ exist on (σ_c^+, ∞), and are bounded by (1.2). Furthermore $\partial Z(\sigma, f)/\partial\sigma$ exists, where $Z(\sigma, f) = \langle e^{\Phi(f)} \rangle_\sigma$, and $| f |_{L_p} \ll 1$.

Proof. — We use the Lebowitz inequalities

$$S^{(n+2)}(12U) - S^{(n)}(U)S^{(2)}(12) \leq \sum_{X \text{ odd}} S(1X)S(2 \sim X), \tag{3.3}$$

where X is a subset of U, see [*12*, eq. 5]. By induction, the right side of (3.3) can be bounded by a sum of products of two point functions of the stated form. The sum in (1.2) ranges over the $\leq (n+1)!$ terms which arise from this bound. The inductive hypothesis we use is the following: $S^{(2n)}$ is bounded by a sum of at most $(2n-1)!$ products of n two point functions.

We first differentiate with respect to a change in σ in a finite volume Λ. By Remark 2.3, with $\delta\sigma \leq 0$,

$$S^{(n)}(\sigma) - S^{(n)}(\sigma + \delta\sigma)$$
$$= \lim_{\Lambda\to\infty} \int_{\sigma+\delta\sigma}^{0} \int_\Lambda [S^{(n+2)}(s, \Lambda, z, z) - S^{(n)}(s, \Lambda, U)S^{(2)}(s, \Lambda, z, z)]dzds$$

where $S(s, \Lambda)$ has a mass perturbation σ in all of R^2, and a mass perturbation $s - \sigma$, $\delta\sigma \leq s - \sigma \leq 0$, in Λ. We substitute (3.3) and (1.2) as an upper bound for the integrand. By the Lebesgue monotone convergence theorem, we may take the limit $\Lambda \to \infty$ under the integral sign. Convergence of the z integration follows from the positive mass and exponential decay of $S^{(2)}$.

The derivative of $Z(\sigma, f)$ exists by summation of the exponential series.

REMARK 3.10. — The existence of the derivatives of the connected Euclidean Green's functions $S_T^{(n)}(x) \equiv G^{(n)}(x)$ follows from this result. The derivative $dG^{(n)}/d\sigma$ is bounded in the norms of [*9*].

THEOREM 3.11. — For a φ_2^4 theory with $\mu = 0$, the Euclidean vertex functions $\Gamma^{(n)}(x)$ have bounded derivatives $d\Gamma^{(n)}(x)/d\sigma$, on the interval (σ_c^+, ∞).

Proof. — By the definition of the $\Gamma^{(n)}(x)$ in terms of the $G^{(n)}(x)$ and $\Gamma^{(2)} = -G^{(2)-1}$, and Remark 3.10, it is sufficient to prove the boundedness of $d\Gamma^{(2)}/d\sigma$ as a map from H_1 to H_{-1}. By definition,

$$\frac{d\Gamma^{(2)}}{d\sigma} = \Gamma \frac{dG^{(2)}}{d\sigma} \Gamma.$$

Since Γ is a bounded transformation from H_1 to H_{-1}, we need only show that $dG^{(2)}/d\sigma$ is bounded from H_{-1} to H_1. However, as in (1.2) or [11],

$$0 \leq -\frac{dG^{(2)}(x)}{d\sigma} \leq \int G^{(2)}(x-z)G^{(2)}(z)dz. \tag{3.4}$$

For $\sigma \in (\sigma_c^+, \infty)$, we have $m(\sigma) > 0$, so the Fourier transform $G^{(2)\sim}(p)$ is bounded by $m(\sigma)^{-2}$. Since $G^{(2)}(x)$ is bounded fıom H_{-1} to H_1, it follows that $G^{(2)} {}_* G^{(2)}$ is also. Hence so is $dG^{(2)}/d\sigma$, by (3.4).

4. THE ISING MODEL CRITICAL POINT

In this section we obtain bounds on critical exponents in Ising models. The analysis of Section 3 may also be applied to the Ising model, with $\beta \nearrow \beta_c$ replacing $\sigma \searrow \sigma_c$. The bound on $dS^{(n)}/d\beta$ by a product of two point functions is known [20]. In our 'analysis of the Ising model we assume an Ornstein-Zernike upper bound on the two point function,

$$S^{(2)}(\beta\,;x) \leq \text{const.} \int \frac{e^{-ipx}}{p^2+m^2}\,dp = S_0^{(2)}(m^2\,;x) \tag{4.1}$$

where the constant is independent of β and where $m = \xi^{-1}$ is the inverse correlation length. In field theory, (4.1) follows from the Lehmann spectral formula. From this hypothesis, we have the bound $\eta \geq 0$ on the anomalous dimension. We now use (4.1) to derive other bounds on critical exponents.

THEOREM 4.1. — Assume the bound (4.1). Then:

(*a*) For the ferromagnetic Ising model with $\beta < \beta_c$, $d \geq 2$,

$$0 \leq \frac{d\chi}{d\beta} \leq 0(1)\chi^2. \tag{4.2}$$

(*b*) Assuming the existence of critical exponents, they satisfy

$$1 \leq \gamma \leq 2\nu \tag{4.3}$$

$$\Delta_4 \leq \frac{1}{2}(d\nu + \gamma) \tag{4.4}$$

$$\Delta_4 + \Delta_6 \leq d\nu + \gamma \tag{4.5}$$

REMARK 4.2. — The inequality

$$0 \leq -\frac{dm}{d\beta} \leq 0(m^{-1}) \tag{4.6}$$

also follows by slight modification of [11], using a transfer matrix formalism [22]. The bounds (4.1-6) hold for a φ^4 field theory.

Proof. — The inequality (4.2) follows as in [*11*], and yields $1 \leq \gamma$. Combined with (4.1) this yields (4.3). The inequalities (4.4) and (4.5) are a consequence of (4.1) and the bounds of [*13*].

5. PRESERVATION OF ESTIMATES

In this section, we sketch how to transfer uniform estimates, *e. g.* estimates of the form

$$e^{-(\mathrm{H} \pm \mathrm{A})} \leq \text{const.}$$

established for $\mathrm{V} < \infty$, to the infinite volume limit. This justifies the statement in Remark 2.5 that the Hamiltonian estimates [*8*] can be used in the infinite volume limit with weak coupling boundary conditions.

We are concerned with perturbations of the form

$$\mathrm{A}_j = \int : \varphi^j : (x) h_j x dx, \tag{5.1}$$

with $j \leq \deg \mathscr{P} \equiv d$, with suppt. $h_j \subset [-a, a]$, and with $h_j \in \mathrm{L}_{d/d-j}$, see [*5*]. In case $j = d$, we also require $\| h_j \|_\infty$ small. We also introduce the mass perturbation

$$\mathrm{B}_2 = \frac{1}{2}\sigma \int_{-b}^{b} : \varphi^2 : (x) dx. \tag{5.2}$$

The estimates will be formulated in semigroup language, because the properties [*5*] of

$$z \to e^{-t(\mathrm{H} + z_2 \mathrm{B}_2 + z_j \mathrm{A}_j)} \tag{5.3}$$

permit control of the semigroups without considering questions of operator domains. We use the path space notation implicit in the fact that the theories constructed here are Osterwalder-Schrader reconstructions [*23*].

THEOREM 5.1. — The Feynman-Kac density

$$e^{-\int_0^{\mathrm{T}} [\mathrm{B}_2(q(t)) + \mathrm{A}_j(q(t))] dt}$$

defines a semigroup on the physical Hilbert space $\mathscr{H}$.

Proof. — This statement follows by limits from the approximate theories, given the convergence of the weak coupling Schwinger functions with φ^j vertices, and the uniform bounds on the analytic functions (5.3).

In particular, with b and a fixed, the bounds of [*8*] are transfered into the weak coupling infinite volume limit by the argument of this section. See Remark 2.9.

The semigroup of Theorem 5.1 will be called $\exp[-t\mathrm{H}(\sigma, j)]$. We identify

$$\mathrm{H}(\sigma, j) = \mathrm{H} + \mathrm{B}_2 + \mathrm{A}_j \tag{5.4}$$

on the domain

$$\mathscr{D} = \{ e^{-sH}\mathscr{R}\Omega : s > 0 \}$$

where $\mathscr{R}$ is a polynomial in fields and Ω is the ground state of H. By the reconstruction theorem [23]. $\mathscr{D} \subset D(H)$ and by the cluster expansion $\mathscr{D} \subset D(B_2) \cap D(A_j)$. The limits of the difference quotients from finite volume then establish (5.4) on $\mathscr{D} \times \mathscr{D}$, and hence by the above domain inclusions (5.4) holds on $\mathscr{D}$. In the case of the mass perturbation, $H + B_2$, (5.4) extends to $D(H) \times D(H)$ by Spencer's local N_τ estimate [28]. The fact that $N_{\tau,loc} \geq 0$ allows the transfer of the first order estimate on $N_{\tau,loc}$ to the infinite volume limit [1]. Let

$$\delta E(\sigma, b, h_j) = - \ln \| \exp [- H(\sigma, j)] \|, \tag{5.5}$$

$$\delta E(\sigma, b) = \delta E(\sigma, b, \phi) = \inf \text{spect.} (H + B_2). \tag{5.6}$$

We normalize the Wick ordering so that

$$\langle : \varphi^2 : \rangle = 0, \tag{5.7}$$

with expectations taken in the weak coupling physical Hilbert space. Then $\langle : \varphi^j : \rangle = 0$, with the standard definition of Wick order. The absence of domain difficulties follows from the cluster expansion, see [27].

Let $\delta E^*(\sigma, b, h_j)$ be the lim sup of the finite volume (weak coupling) approximates to $\delta E(\sigma, b, h_j)$. Then

$$- \delta E \leq - \delta E^*, \tag{5.8}$$

as in the proof of Theorem 5.1. Our main result is

THEOREM 5.2. — With constants uniform in b,

$$- \delta E(\sigma, b, h_j) \leq - \delta E(\sigma, b) + 0\left(a + \| h_j \|_{d/d-j}^{d/d-j}\right). \tag{5.9}$$

LEMMA 5.3.

$$- \delta E(\sigma, b, h_j) = \lim_{T\to\infty} (2T)^{-1} \ln \int e^{-\int_{-T}^{T} [B_2 + A_j] dt} dq. \tag{5.10}$$

Proof. — Let ψ be a bounded function on path space, measurable with respect to the σ-algebra generated by $\phi(x, t)$ for $t \geq 0$, let Θ denote reflection about $t = 0$, and let $\psi \to \psi_T$ denote a time translation. Then

$$\begin{aligned} - \delta E(\sigma, b, h_j) &= \sup_{\psi} \lim_{T\to\infty} T^{-1} \ln \langle \psi, e^{-TH(\sigma,j)}\psi \rangle \\ &= \sup_{\psi} \lim_{T\to\infty} T^{-1} \ln \int \Theta\psi \exp \left[- \int_0^T (B_2 + A_j) dt \right] \psi_T dq \\ &\leq \sup_{\psi} \lim_{T\to\infty} \left\{ T^{-1} \ln \| \psi \|_\infty + T^{-1} \ln \int \exp \left[- \int_0^T (B_2 + A_j) dt \right] dq \right\}. \end{aligned}$$

This completes the proof, since the function $\Omega \equiv 1$ is an allowed choice of ψ.

LEMMA 5.4. — $-\ \delta E(\sigma, b, h_j)$ is non-negative and convex in σ, h_j.

Proof. — The convexity follows from a Schwarz inequality relative to the inner product defined by the Euclidean weak coupling measure dq, combined with Lemma 5.3 $-\ \delta E \geq 0$ by the normalization $\langle : \varphi^j : \rangle = 0$.

LEMMA 5.5. — $-\ \delta E(\sigma, b)$ is convex and monotone increasing, as a function of b.

Proof. — Since $-\ \delta E \geq 0$ and $-\ \delta E(\sigma, 0) = 0$, the derivative $d(-\ \delta E)/db$ is non-negative for $b = 0$. Assuming convexity, the derivative is increasing, and hence is always non-negative, so $-\ \delta E$ is monotonic increasing. To prove convexity, we use a Schwarz inequality relative to the Osterwalder-Schrader inner product defined by a line $x =$ const. Thus

$$-\ \delta E(\sigma, b) \leq -\frac{1}{2}(\delta E(\sigma, b') + \delta E(\sigma, b''))$$

where $b = \frac{1}{2}(b' + b'')$. Here $x =$ const. determines the choice of b' and b'', and by varying the constant we obtain an arbitrary division.

Remark. — The lemma applies to any localized perturbation, *e. g.* $B_2 + A_j$, so long as each h_j is the characteristic function of Λ_2.

Proof of Theorem 5.2. — We follow [*17*], which simplifies the original method of [*8*]. It is no loss of generality to assume $b \geq a$. As above, we use a Schwarz inequality relative to the lines $x = \pm a$. For $x \in [-a, a]$ we use the norm of the semigroup (with x and t interchanged). Since dq is rotation invariant, it is invariant under rotation by the angle $\theta = \pi/2$. The vacuum energy for fixed x is bounded by $0(1)[1 + |h(x)|^{d/d-j}]T$, where the second factor contributes for $|h|$ large. The factor $0(1)$ contains the σ-dependence. The bound is linear in T by the linear bound on the vacuum energy in the volume. Integrating over x, gives $0(T)\left(a + \|h\|_{d/d-j}^{d/d-j}\right)$ and $-\ \delta E(\sigma, b, h_j) \leq -\ \delta E(\sigma, b) + 0(1)\left(a + \|h_j\|_{d/d-j}^{d/d-j}\right)$.

In order to apply Theorem 5.2, we let $T_2 \nearrow \infty$, $b \nearrow \infty$, where $[-T_2, T_2]$ is the time interval in which B_2 is inserted in the path space integral. The $T_2 \nearrow \infty$ limit is taken first. In this limit, the effect of the ϕ^j perturbation is estimated using Theorem 5.2. We require,

LEMMA 5.6. — Let $\Omega \equiv 1$ and let

$$\widehat{H} = H + B_2 - \delta E(\sigma, b).$$

Then

$$\lim_{T\to\infty} \frac{\| e^{-(T-a)\widehat{H}}\Omega \|}{\| e^{-T\widehat{H}}\Omega \|} = 1. \tag{5.11}$$

Proof. — We let $\widehat{\mathrm{H}} = \int \lambda d\mathrm{E}_\lambda$. By Lemma 5.4, $0 = \inf \text{suppt.}\ d\langle \mathrm{E}_\lambda\Omega, \Omega\rangle \equiv \inf \text{suppt.}\ d\mu(\lambda)$.

We consider

$$dv_t(\lambda) = \frac{e^{-t\lambda}d\mu(\lambda)}{\int_0^\infty e^{-t\lambda}d\mu(\lambda)}.$$

Thus

$$\frac{\langle \Omega, e^{-(t-a)\widehat{\mathrm{H}}}\Omega\rangle}{\langle \Omega, e^{-t\widehat{\mathrm{H}}}\Omega\rangle} = \int_0^\infty e^{a\lambda}dv_t(\lambda).$$

Splitting the integral, $\int_0^\infty = \int_0^\varepsilon + \int_\varepsilon^\infty$, we find the contribution $\int_0^\varepsilon \to 1$, and $\int_\varepsilon^\infty \to 0$. which completes the proof.

By Theorem 5.1,

$$\int e^{-\int_{-t_j}^{t_j} \mathrm{A}_j(t)dt}\, e^{-\int_{-\mathrm{T}_2}^{\mathrm{T}_2} \mathrm{B}_2(t)dt}dq \Big/ \int e^{-\int_{-\mathrm{T}_2}^{\mathrm{T}_2} \mathrm{B}_2(t)dt}dq = \langle \theta_{\mathrm{T}_2}, e^{-2t_j\mathrm{H}_1}\theta_{\mathrm{T}_2}\rangle$$

where

$$\mathrm{H}_1 = \mathrm{H} + \mathrm{B}_2 - \delta\mathrm{E}(\sigma, b) + \mathrm{A}_j = \widehat{\mathrm{H}} + \mathrm{A}_j$$

and

$$\theta_{\mathrm{T}_2} = \frac{e^{-(\mathrm{T}_2 - t_j)\widehat{\mathrm{H}}}\Omega}{\| e^{-\mathrm{T}_2\widehat{\mathrm{H}}}\Omega \|}.$$

By Theorem 5.2, and Lemma 5.6,

$$\lim_{\mathrm{T}_2\to\infty} \langle \theta_{\mathrm{T}_2}, e^{-2t_j\mathrm{H}_1}\theta_{\mathrm{T}_2}\rangle/\| \theta_{\mathrm{T}_2}\|^2 \leq \exp t_j 0(1)\big(a + \| h_j \|_{d/d-j}^{d/d-j}\big).$$

Thus by Lemma 5.3,

$$\ln \| e^{-\mathrm{H}_1} \| \leq 0(1)\big(a + \| h_j \|_{d/d-j}^{d/d-j}\big)$$

in the limit $b = \infty$. Thus we have proved

THEOREM 5.7. — Let $\mathrm{H}(\sigma)$ be the even + linear $\mathscr{P}(\phi)_2$ Hamiltonian with weak coupling boundary conditions given by the mass perturbation

$$\frac{1}{2}\sigma\int :\phi^2 : dx,\ \sigma \leq 0.$$

Then,

$$e^{-(\mathrm{H}(\sigma)\pm \mathrm{A}_j)t} \leq e^{\mathrm{c}t} \tag{5.12}$$

where

$$c = 0(1)\big(a + \| h_j \|_{d/d-j}^{d/d-j}\big). \tag{5.13}$$

REMARK 5.8. — The same argument may be used with the Hamiltonian $H(\sigma, \mu)$, for a $\mathcal{P}(\phi)_2$ = even + linear Hamiltonian with weak coupling boundary conditions given by the perturbation $\frac{1}{2}\sigma\int :\phi^2: dx - \mu\int \phi dx$, with $-\sigma, \mu \geq 0$. For a general $\mathcal{P}(\phi)_2$ theory ($\mathcal{P} \neq$ even + linear), Theorem 5.7 remains valid, but in this case the $b = \infty$ infinite volume limit is defined by convergent subsequences.

REMARK 5.9. — The same argument can be used to obtain semigroup estimates of the type (5.12-13) for infinite volume, weakly coupled $\mathcal{P}(\phi)_2$ models. In this case, the convergence of the Schwinger functions as $T \to \infty$ follows from the cluster expansion, rather than monotonicity.

REMARK 5.10. — We improve on (5.13), in order to allow perturbations A_j and test functions h_j which do not have compact support. Using the bounds (5.12)-(5.13), valid for $b \leq \infty$, we have

$$\begin{aligned} -\delta\widehat{E}(\sigma, b, h_j) &\equiv \ln \| \exp(-\widehat{H} + A_j) \| \\ &\leq 0(1)\left(a + \| h_j \|_{d/d-j}^{d/d-j}\right). \end{aligned}$$

Since $-\delta\widehat{E}$ is convex in h_j and vanishes for $h_j = 0$, the above inequality implies

$$-\delta\widehat{E} \leq 0(1)\left(a^{j/d}\| h_j \|_{d/d-j} + \| h_j \|_{d/d-j}^{d/d-j}\right).$$

We write $h_j = \sum_{n=-\infty}^{\infty} \chi_{[n,n+1]} h_j$ and use convexity again to obtain the bound

$$-\delta\widehat{E} \leq 0(1)\left(\|(1 + |x|)^r h_j\|_{d/d-j} + \| h_j \|_{d/d-j}^{d/d-j}\right), \tag{5.14}$$

for some r. This bound holds uniformly for $b \leq \infty$ and for any h_j with $(1 + |x|)^r h_j \in L_{d/d-j}$.

REFERENCES

[1] O. BRATELLI, Conservation of estimates in quantum field theory. *Comm. Pure and Appl. Math.*, t. **25**, 1972, p. 759-779.

[2] C. CALLAN, Broken scale invariance in scalar field theory. *Phys. Rev.*, **D2**, 1970, p. 1541-1547.

[3] S. COLEMAN, Scaling Anomalies, in *Developments in High Energy Physics*, Academic Press, New York, 1972, p. 280-296.

[4] R. DOBRUSHIN and R. MINLOS, Construction of a one dimensional quantum field via a continuous Markov field. *Funct. Anal. and its Appl.*, t. 7, 1973, p. 324-325 (English transl.).

[5] J. FRÖHLICH, Schwinger functions and their generating functionals I, II. *Helv. Phys. Acta*, and *Adv. in Mathematics*.

[6] J. Ginibre, General formulation of Griffiths' inequalities. *Commun. Math. Phys.*, t. **16**, 1970, p. 310-328.

[7] J. Glimm and A. Jaffe, The $(\lambda\varphi^4)_2$ quantum field theory without cutoffs III. The physical vacuum. *Acta Math.*, t. **125**, 1970, p. 203-261.

[8] J. Glimm and A. Jaffe, The $\lambda(\varphi^4)_2$ quantum field theory IV. Perturbations of the Hamiltonian. *J. Math. Phys.*, t. **13**, 1972, p. 1558-1584.

[9] J. Glimm and A. Jaffe, Entropy principle for vertex functions in quantum field models. *Ann. Institut Henri Poincaré*, t. **21**, 1974, p. 1-26.

[10] J. Glimm and A. Jaffe, Critical point dominance in quantum field models. *Ann. Institut Henri Poincaré*, t. **21**, 1974, p. 27-41.

[11] J. Glimm and A. Jaffe, φ_2^4 quantum field model in the single-phase region: Differentiability of the mass and bounds on critical exponents, *Phys. Rev.*, **D15**, to appear.

[12] J. Glimm and A. Jaffe, A remark on the existence of φ_4^4. *Phys. Rev. Lett.*, t. **33**, 1974, p. 440-442.

[13] J. Glimm and A. Jaffe, Absolute bounds on vertices and couplings. *Ann. Institut Henri Poincaré*, t. **22**, 1975, p. 1-13.

[14] J. Glimm and A. Jaffe, Two and three body equations in quantum field models. *Commun. Math. Phys.*, to appear.

[15] J. Glimm, A. Jaffe and T. Spencer, The Wightman axioms and particle structure in the $P(\varphi)_2$ quantum field model. *Ann. Math.*, t. **100**, 1974, p. 585-632.

[16] J. Glimm, A. Jaffe and T. Spencer, The particle structure of the weakly coupled $P(\varphi)_2$ models and other applications of high temperature expansions, in: *Constructive Quantum Field Theory*, G. Velo and A. S. Wightman (eds.), Springer-Verlag, Berlin, 1973.

[17] F. Guerra, L. Rosen and B. Simon, Nelson's Symmetry and the infinite volume behavior of the vacuum in $P(\varphi)_2$. *Commun. Math. Phys.*, t. **27**, 1972, p. 10-22.

[18] F. Guerra, L. Rosen and B. Simon, The $P(\varphi)_2$ Euclidean quantum field theory as classical statistical mechanics. *Ann. Math.*, t. **101**, 1975, p. 111-259.

[19] F. Guerra, L. Rosen and B. Simon, *The pressure is independent of the boundary conditions*, preprint.

[20] J. Lebowitz, More inequalities for Ising ferromagnets. *Phys. Rev.*, **B5**, 1972, p. 2538-2541.

[21] J. Lebowitz, GHS and other inequalities. *Commun. Math. Phys.*, t. **35**, 1974, p. 87-92.

[22] R. Minlos and Ya. Sinai, Investigation of the spectra of stochastic operators arising in lattice models of a gas. *Theoretical and Math. Phys.*, t. **2**, 1970, p. 167-176 (English transl.).

[23] K. Osterwalder and R. Schrader, Axioms for Euclidean Green's functions I, II. *Commun. Math. Phys.*, t. **31**, 1973, p. 83-113 and *Commun. Math. Phys.*

[24] G. Parisi, Field theory approach to second order phase transitions in two and three dimensional systems. *1973 Cargèse lectures.*

[25] Y. Park, *Lattice approximation of the* $(\varphi^4 - \mu\varphi)_3$ *field theory in a finite volume*, preprint.

[26] E. Riedel and F. Wegner, Tricritical exponents and scaling fields. *Phys. Rev. Lett.*, t. **29**, 1972, p. 349-352.

[27] R. Schrader, *Local operator products and field equations in* $P(\varphi)_2$ *theories.*

[28] T. Spencer, Perturbation of the $P(\varphi)_2$ quantum field Hamiltonian. *J. Math. Phys.*, t. **14**, 1973, p. 823-828.

[29] K. Symanzik, Small distance behavior in field theory and power counting. *Commun. Math. Phys.*, t. **18**, 1970, p. 227-246.

(Manuscrit reçu le 2 septembre 1974)

Critical Exponents and Elementary Particles

J. Glimm*
The Rockefeller University, New York, NY 10021, USA

A. Jaffe**
Harvard University, Cambridge, MA 02138, USA

Abstract. Particles are shown to exist for a.e. value of the mass in single phase ϕ^4 lattice and continuum field theories and nearest neighbor Ising models. The particles occur in the form of poles at imaginary (Minkowski) momenta of the Fourier transformed two point function. The new inequality $dm^2/d\sigma \leqq Z$, where $\sigma = m_0^2$ is a bare mass^2 and Z is the strength of the particle pole, is basic to our method. This inequality implies inequalities for critical exponents.

1. Introduction

Euclidean ϕ^4 fields are believed to describe the asymptotic long distance behavior of certain lattice models of statistical mechanics at their critical points (e.g. the Ising model). It was proposed [9] that a construction of ϕ^4 fields could be based on this expectation, and partial results in this direction are given in [6–8, 10, 11, 1, 12]. In this construction, the field ϕ will be nontrivial (i.e. not a free field) only in the case in which the corresponding lattice model critical point is asymptotically nontrivial at long distances.

In order to better distinguish between the trivial and the nontrivial cases, we continue here our investigation [7, 10, 11] of critical exponents (see also § 5). In general, our results have the form

$$\text{canonical exponent} \leqq \text{general exponent}\,, \tag{1.1}$$

$$\phi\text{-exponent} \leqq \phi^2\text{-exponent}\,, \tag{1.2}$$

and in particular if the lattice ϕ^2 field (e.g. the Ising model energy-energy correlation) is canonical, then so is the corresponding lattice ϕ-field and also the resulting continuum ϕ-field. The converse to this statement seems to be false, and a counterexample may be found in the ϕ_4^4 lattice field at weak coupling. In this model, there is some evidence that ϕ^2 deviates from canonical by a logarithm.

* Supported in part by the National Science Foundation under grant PHY 76-17191
** Supported in part by the National Science Foundation under grant MPS 75-21212

The bounds (1.1) are given in [7] while the bound [10]

$$\eta \leqq \eta_E/2 \tag{1.3}$$

of the form (1.2) is an elementary consequence of Lebowitz' inequality. This inequality compares the anomalous dimension η for ϕ with the anomalous dimension η_E for ϕ^2. One result of this paper is an improvement on the bound [3] $1 \leqq (2-\eta)\nu$, namely

$$1 \leqq (2-\zeta/\nu)\nu \tag{1.4}$$

where ν is the exponent for the mass and ζ the exponent for field strength renormalization. (Note $\zeta/\nu \geqq \eta$ in the case of Euclidean covariance and $\zeta/\nu = \eta$, assuming scale relations.) The stronger upper bounds on η in [11] required stronger hypothesis.

Our proof of (1.4) follows from a stronger inequality for an Ising model, lattice ϕ^4 fields or $d=1,2,3$ continuum ϕ^4 fields. We show that

$$dm^2/d\sigma \leqq Z, \quad \text{assuming} \quad \sigma > \sigma_c \tag{1.5}$$

where Z is the strength of the one particle pole in the truncated two point function, m^{-1} is the correlation length and σ is the bare mass^2 (or inverse temperature). The equality (1.5) improves $dm^2/d\sigma \leqq 1$, established in [7], and integration of (1.5) yields the relation (1.4) for critical exponents. (Recall that $m \searrow 0$ as $\sigma \searrow \sigma_c$ [1, 11].)

The main result of the present paper is that $dm^2/d\sigma \neq 0$ for almost all masses m, so that by (1.5), $Z \neq 0$ for almost all masses. We infer that an elementary particle (a pole in the propagator) exists for almost all masses in the region $\sigma > \sigma_c$, i.e. the single phase region. Also as $\sigma \searrow \sigma_c$,

$$\chi = \int G^{(2)}(x)dx \to \infty, \tag{1.6}$$

and so

$$\Gamma^{(2)}(p=0) \to 0 \quad \text{as} \quad \sigma \searrow \sigma_c .$$

This convergence was assumed in [7] in the derivation of the inequality $\gamma \geqq 1$. We give proofs for continuum fields, and the same methods then carry over to lattice fields and nearest neighbor Ising models on a rectangular lattice. For the Ising models σ (the coefficient of ϕ^2 in the ϕ^4 interaction) is replaced by $-\beta$.

2. Critical Behavior of the Mass and the Field Strength Renormalization

For pure imaginary (Minkowski) momentum ip, p real, we define

$$\chi(p) = \int G(x) e^{-px} dx = G\tilde{}(ip)$$

with $\int \dots dx$ replaced by a summation in the lattice case. We choose ip inside the tube of analytically of $G\tilde{}$. Then

$$\begin{aligned} 0 \leqq -d\chi(p)/d\sigma &= \tfrac{1}{2} \int \langle \phi(x) : \phi(y)^2 : \phi(0) \rangle e^{-p(x-y)} e^{-py} dx dy \\ &\leqq \int \langle \phi(x)\phi(y) \rangle \langle \phi(y)\phi(0) \rangle e^{-p(x-y)} e^{-py} dx dy = \chi(p)^2 . \end{aligned}$$

Thus

$$0 \leqq -d\Gamma/d\sigma = -d(-\chi(p))^{-1}/d\sigma = -\chi^{-2}d\chi(p)/d\sigma \leqq 1 \tag{2.1}$$

for all such p. Since we have used translation invariance, the proof applies in the infinite volume case, and also in the finite volume periodic case.

To apply this inequality in a finite volume with periodic boundary conditions, we note that the Hamiltonian has discrete spectrum, and that its first excited state occurs in the spectral decomposition of the two point function. Furthermore, the first excited state occurs at zero momentum. Thus we take $p=(p_0, \boldsymbol{p})$, with $\boldsymbol{p}=0$, and $Z^{-1}<\infty$ for all volumes $V<\infty$. Let $m_{p,V}$ be the periodic mass. For $-p_0^2<m^2$, Γ is an analytic function of the independent variables σ and p_0^2 by a standard application of the Paley-Wiener theorem, cf. [11, §2]. The one particle curve $+p_0^2=m^2$ is analytic in any region free of level crossings in the first excited state, and thus is at least piecewise analytic. [15, VII, § 1] can be applied to the Hamiltonian $H(\sigma)$, which is analytic in σ by second order estimates [16]. For example the inequality $H(\mathrm{Re}\,\sigma)^2 \leqq \mathrm{const}\, H(\sigma^-)H(\sigma)+\mathrm{const}$ implies that $H(\sigma)$ is closed, on the domain $\mathscr{D}(H(\mathrm{Re}\,\sigma))$. With the same discrete exceptions at possible level crossings, Γ vanishes along the one particle curve $p^2=-m(\sigma)^2$. From these facts, it follows $d\Gamma/dp^2$ is a multiple of $d\Gamma/d\sigma$ on this one particle curve. In particular, in the $(\sigma, -p^2)$ plane, the vector $(1, dm^2/d\sigma)$ lies along the one particle curve, and so

$$0=\nabla\Gamma\cdot(dm^2/d\sigma, 1)=[\partial\Gamma/\partial(-p^2)]dm^2/d\sigma+\partial\Gamma/d\sigma \tag{2.2}$$

or

$$0 \leqq Z^{-1} = -\partial\Gamma/\partial p^2|_{p^2=-m^2} = -[\partial\Gamma/\partial\sigma]\,[dm^2/d\sigma]^{-1}\,.$$

Thus by (2.1),

$$0 \leqq dm^2/d\sigma = -Z\,\partial\Gamma/\partial\sigma \leqq Z\,, \tag{2.3}$$

proving (2.3). In the case of level crossings, one sided derivatives $dm^2/d\sigma^{\pm}$ satisfy (2.3), since Z is semicontinuous at a level crossing.

For later use, we note that

$$dm^2/d\sigma \leqq \mathrm{const.} \tag{2.4}$$

by [7] in the continuum case, or (2.3), since $Z \leqq 1$ for canonical continuum fields. For lattice fields and Ising models, the p-space canonical upper bound [4] on the two point function, and its Herglotz representation [11], show $Z \leqq$ const. Thus by (1.6), the bound (2.4) extends to ϕ^4 lattice fields and Ising models in a finite periodic volume, with a constant independent of the volume.

By passing to the subsequence $V_j \to \infty$, we have $m_{p,V} \to m_{p,\infty}$ but we do not know (yet) that $m_{p,\infty}$ is the mass m_p of the corresponding infinite volume theory. However, by semicontinuity of the spectrum.

$$m_{p,\infty} \leqq m_p\,.$$

By [5], $m_p=m_d$, where m_d is the mass in a theory with Dirichlet boundary conditions. We define the corresponding critical values $\sigma_{cd}=\sigma_{cp} \leqq \sigma_{c,p,\infty}$. By (2.4) and Ascoli's theorem, we have uniform convergence and uniform Lipschitz continuity of $m^2_{p,V}(\sigma)$ and the limit $m^2_{p,\infty}(\sigma)$. In particular, both m_d and $m_{p\infty}$ go continuously to zero at their critical points [7, 1, 12].

By definition of σ_c, the quantity $dm^2/d\sigma$ cannot vanish in a one-sided neighborhood $(\sigma_c, \sigma_c+\varepsilon]$ of σ_c, and by (1.5) the measure of the set on which $dm^2/d\sigma \neq 0$ is proportional to $m^2(\sigma_c+\varepsilon)$, uniformly in V. It follows that by the diagonal process, we can choose $V_j \to \infty$ so that there is a sequence $\sigma_i \searrow \sigma_{cp\infty}$ with the strict inequality

$$\lim_{V_j \to \infty} dm^2/d\sigma|_{\sigma=\sigma_i} > 0 .$$

For these values of σ_i, we also have Z bounded away from zero by (1.5), which implies $m_p = m_{p\infty}$ for $\sigma = \sigma_i$. Because the m's are monotone in σ, it follows that $\sigma_{cd} = \sigma_{cp} = \sigma_{cp\infty}$.

Extending the above argument, we can work with $\Sigma = \{\sigma : dm^2_{p\infty}/d\sigma > 0\}$, which is a set of positive measure in every interval $(\sigma_c, \sigma_c+\varepsilon)$. On this set $Z \neq 0$ and $m_{p\infty} = m_p$. However since $m_{p\infty}$ is constant on $\sim\Sigma$, while m_p is monotone (and hence monotone on Σ) we must have $m_{p\infty} = m_p$ everywhere. We summarize these results.

Theorem 1. *For* $\sigma > \sigma_c$, *we have*

$$m_{p\infty} = m_p = m_d$$

and

$$dm^2/d\sigma \leqq Z(\sigma) .$$

Furthermore $Z(\sigma)$ *uppersemicontinuous:*

$$Z(\sigma) \leqq \limsup_{\sigma' \to \sigma} Z(\sigma') .$$

We now argue that $Z \neq 0$ for all masses m not in a set of measure zero. Consider $m^2(\sigma)$ and $\sigma(m^2)$ as monotone functions. Thus they have bounded variation and are differentiable a.e. with derivatives $m^{2\prime}$, σ' which are positive and L_1. Thus

$$(\sigma')^{-1} = dm^2/d\sigma \neq 0 \text{ a.e.},$$

as a function of m, and by (2.3) $Z \neq 0$ a.e. as a function of m. The region of non uniqueness of mass renormalization corresponds to intervals in which $m^2(\sigma) = \text{const.}$, hence to δ-functions in the measures $d\sigma(m^2)$. These occur for a most countable number of values of m^2.

Remark. The above result completes the proof of the second of three steps of [9] for a possible construction[1] of Φ^4_d fields: that defined by the long distance behavior of the corresponding lattice field at its critical point. The first step was established in [7, 1, 12]. The third step seems to depend on showing scaling behavior as $\sigma \searrow \sigma_c$, i.e. the isolation of leading long distance behavior from next to leading, etc.

3. Divergence of χ at the Critical Point

In this section we show that $\chi_c = \chi(\sigma_c) = \infty$, which is related to the upper bound $\eta \leqq 2$. Again we work in the $(\sigma, -p^2)$ parameter space, but in the $V = \infty$ theory directly. Below or on the one particle curve $-p^2 = m^2$, the quantity $\partial\Gamma/\partial\sigma$ is bounded, while

[1] If the nontrivial part of the critical behavior occurs only in the next to leading terms for all values of the bare charge, then the construction of [9] is inappropriate (cf. § 5)

below this curve or on nonhorizontal portions of this curve, $\partial\Gamma/\partial(p^2)$ is bounded also, and Γ vanishes, at least on non-horizontal portions of the curve. Moreover, by the Herglotz representation for Γ, any bound on $\partial\Gamma/\partial(-p^2)$, with $p^2=-m^2$, extends to $0 \leqq -p^2 \leqq m^2$, since $\partial\Gamma/\partial(-p^2)$ is decreasing in $-p^2$ in this interval. Thus by (2.2) and integration,

$$0 \leqq -\Gamma(p^2=0, \sigma=\sigma_i) \leqq 0(m^2)(dm^2/d\sigma)^{-1}$$

where σ_i is chosen as in § 2, so that $Z \neq 0$, hence $G(p^2=-m^2)=\infty$ and $\Gamma(p^2=-m)=0$. Thus

$$0(1)m^{-2}dm^2/d\sigma \leqq \chi_{\sigma_i}.$$

Assuming χ_{σ_c} finite, we have

$$dm/d\sigma \leqq \text{const.}\, m.$$

Recall that $m(\sigma_c+0)=0$, so that after integration from σ_c to $\sigma>\sigma_c$,

$$0 \leqq m(\sigma) \leqq m(\sigma_c+0)\, e^{\text{const.}(\sigma-\sigma_c)}=0.$$

This is a contradiction, which shows that $\chi_c=\infty$. As a corollary of this result, we observe that σ_c is characterized as the largest value of σ for which the effective potential (see [2]) has zero curvature at the origin:

$$V''|_{\sigma=\sigma_c}=\Gamma^{(2)}(p=0)|_{\sigma=\sigma_c}=0.$$

4. The Ising Limit

We consider the interactions

$$\tfrac{1}{2}\textstyle\sum_{n.n}(\phi_i-\phi_{i'})^2+\sum_i[\lambda(\phi_i^2-1)^2+\tfrac{1}{2}\sigma\phi_i^2] \tag{4.1}$$

and

$$-\textstyle\sum_{n.n}\beta\psi_i\psi_{i'}+\sum_i\lambda_0(\psi_i^2-1)^2 \tag{4.2}$$

where $\sum_{n.n}$ denotes a sum over nearest neighbor pairs. With

$$\psi=(1-(\sigma+2d)/4\lambda)^{-1/2}\phi$$

and

$$\lambda_0=\lambda-(\sigma+2d)/4\lambda, \qquad \beta=1-(\sigma+2d)/4\lambda,$$

these two expressions differ by a constant, and thus define the same lattice field theory. For the purpose of studying the Ising limit (λ or $\lambda_0\to\infty$, β fixed), it is convenient to use the representation (4.2). Let β_{cI} be the critical value of β in the Ising model.

Theorem 2. *As $\lambda_0\to\infty$, $m(\beta,\lambda_0)\to m_I(\beta)$, the Ising model mass, for $\beta<\beta_{c,I}$, and $\beta_{c,I} \leqq \lim_{\lambda_0\to\infty}\beta_c(\lambda_0)$.*

Proof. As in (2.1), $-d\Gamma/d(-\beta)\leqq 2$ for imaginary momenta below the one particle curve, and as in (2.3), $0\leqq dm^2/d(-\beta)\leqq 2Z$. These bounds are uniform in $\lambda_0\leqq\infty$, in the one phase region. Thus they imply uniform convergence (through a subsequence

$\lambda_{0j} \to \infty$) and uniform Lipschitz continuity in β of $m^2(\beta, \lambda_0)$. Let $m^2(\beta, \infty)$ be the limit value of the mass^2, and let $m_I^2(\beta)$ be the Ising model mass^2. We have $m^2(\beta, \infty) \leqq m_I^2(\beta)$ for $\beta \geqq \beta_{c,I}$ by semicontinuity of the spectrum. Next we choose values β_i for which $Z(\beta_i, \lambda_0)$ is bounded away from zero as $\lambda_{0,j} \to \infty$. For these values of β_i, we have $m^2(\beta, \infty) = m_I^2(\beta)$. By monotonicity, $m^2(\beta, \infty) = m_I^2(\beta)$ throughout the single phase region. Convergence of the full sequence follows from uniqueness of the limit, in the single phase region. With convergence of the masses assured for $\beta \leqq \beta_{c,I}$, it follows that $\lim_{\lambda_0 \to \infty} \beta_c(\lambda_0) \geqq \beta_{c,I}$.

The same method applies to variation of the β's which couple between layers. Let $J = (J_1, \ldots, J_d)$ and let

$$H(\sigma) = -\tfrac{1}{2} \Sigma \beta J_j \sigma_i \sigma_{i+e_j} \tag{4.3}$$

where e_j is the unit vector in the jth coordinate direction. Then (4.3) defines a d-dimensional Ising model. For simplicity we consider only the case in which $J_d \to 0$ and set $J_1 = \ldots = J_{d-1} = 1$. Then $\beta_c = \beta_c(J_d)$ is a function of J_d. As above, we have

Theorem 3. *For $J_d \geqq 0$, m is continuous and β_c is semicontinuous as functions of J_d, in the single phase region.*

Remark. Continuity of β_c would follow from a uniform lower bound on $dm^2/d(-\beta)$, for example $\nu < \infty$.

5. Trivial vs. Nontrivial Fields

We give a simple sufficient condition for triviality of the ϕ^4 continuum limit. We parametrize the single phase even ϕ^4 lattice fields by $\lambda(=$bare charge$)$, $m(=$physical mass$)$, $\varepsilon(=$lattice spacing$)$ and $d(=$space time dimension$)$.

Theorem 4. *Suppose that $Z = Z(\lambda, m, \varepsilon, d)$ is continuous and bounded away from zero for $0 \leqq \lambda \leqq \infty$, $0 \leqq m \leqq \delta$, $\varepsilon = 1$, and for some $\delta > 0$, and some $d \geqq 3$. Then the continuum limit of an even single phase d-dimensional ϕ^4 lattice field with arbitrary charge renormalization has a free two point function.*

Proof. The continuum fields are constructed with the arbitrary renormalization $\lambda = \lambda(\varepsilon)$, $0 < m = m(\varepsilon) \leqq$ const, where $\varepsilon_j \to 0$ through a subsequence. Existence of the limit follows from the bounds of [6, 4], which establish compactness of the sequence of Schwinger functions. Properties of the $\varepsilon \to 0$ theory are reduced to properties of the $\varepsilon = 1$ theory by a (canonical) scale transformation. Z is invariant under scale transformations, so that

$$Z(\lambda(\varepsilon), m(\varepsilon), \varepsilon) = Z(\varepsilon^{4-d} \lambda(\varepsilon), \varepsilon\, m(\varepsilon), 1)\,.$$

There is a similar transformation law for the spectral weight of the two point function. The Fourier transformed two point function at zero momentum has the Herglotz representation [11]

$$G^{(2)\sim}(p_0, \boldsymbol{p} = 0) = Z(\cosh m = \cos p_0)^{-1} + \int_{m+0}^{\infty} (\cosh a - \cos p_0)^{-1} d\varrho(a, \lambda, m, \varepsilon, d)$$

for $\varepsilon=1$. Continuity of Z at $m=0$ can be expressed as the following property of the measure $d\varrho$:

$$\lim_{\gamma\to 0}\lim_{m\to 0}\int_{m^2+0}^{\gamma} d\varrho(a,\lambda_1(m),1,d)=0 \tag{5.1}$$

for any $\lambda=\lambda_1(m)$. We choose $m=\varepsilon m(\varepsilon)\to 0$ and

$$\lambda_1(\varepsilon m(\varepsilon))=\varepsilon^{4-d}\lambda(\varepsilon).$$

Under scaling, (5.1) transforms into

$$\lim_{\gamma\to 0}\lim_{\varepsilon\to 0}\int_{m(\varepsilon)^2+0}^{\varepsilon^{-2}\gamma} d\varrho(a,\lambda(\varepsilon),m(\varepsilon),\varepsilon,d)=0,$$

which implies that limit two point function is free.

References

1. Baker, G.: Self interacting boson quantum field theory and the thermodynamic limit in d dimensions. J. Math. Phys. **16**, 1324—1346 (1975)
2. Coleman, S.: Secret Symmetry: An introduction to spontaneous symmetry breakdown and guage fields. 1973 Erice Summer School of Physics
3. Fisher, M.: Rigorous Inequalities for critical-point correlation exponents. Phys. Rev. **180**, 594—600 (1969)
4. Fröhlich, J., Simon, B., Spencer, T.: A new method for the analysis of phase transitions and spontaneous breaking of discrete and continuous symmetries. Phys. Rev. Lett. **36**, 804—806 (1976)
5. Fröhlich, J., Simon, B.: Pure states for general $P(\phi)_2$ theories: Construction, regularity and variational equality. Ann. Math., to appear
6. Glimm, J., Jaffe, A.: A remark on the existence of ϕ_4^4. Phys. Rev. Lett. **33**, 440—442 (1974)
7. Glimm, J., Jaffe, A.: The ϕ_2^4 quantum field theory in the single phase region: Differentiability of the mass and bounds on critical exponents. Phys. Rev. D **10**, 536—539 (1974)
8. Glimm, J., Jaffe, A.: Three particle structure of ϕ^4 interactions and the scaling limit. Phys. Rev. D **11**, 2816—2827 (1975)
9. Glimm, J., Jaffe, A.: Critical problems in quantum fields, International Colloquium on Mathematical Methods of Quantum Field Theory, Marseille, June 1975
10. Glimm, J., Jaffe, A.: Critical exponents and renormalization in the ϕ^4 scaling limit, Conference on quantum dynamics: Models and mathematics. Acta Phys. Austriaca Suppl. XVI (1975)
11. Glimm, J., Jaffe, A.: Particles and scaling for lattice fields and Ising models. Commun. math. Phys. **51**, 1—14 (1976)
12. Rosen, J.: Mass renormalization for lattice $\lambda\phi_2^4$ fields. Adv. Math. (to appear)
13. Rosen, J.: The Ising limit of ϕ^4 lattice fields. The Rockefeller University (preprint)
14. Schrader, R.: A possible constructive approach to ϕ_4^4 I, III. Commun. math. Phys. **49**, 131—154 (1976); **50**, 97—102 (1976)
15. Kato, J.: Perturbation theory for linear operators. Berlin-Heidelberg-New York: Springer 1966
16. Glimm, J., Jaffe, A.: A $\lambda\phi^4$ Quantum field theory without cutoffs. I. Phys. Rev. **176**, 1945—1951 (1968) and Singular perturbations of self adjoint operators. Comm. Pure Appl. Math. **22**, 401—414 (1969)

Communicated by R. Haag

Received June 30, 1976

V Particle Structure

Introduction

The papers in this section study the particle spectrum of H, using a number of techniques. The one-particle irreducible functions (vertex functions) are obtained from the Green's functions by a Legendre transform. These functions can be used to study the two or three particle spectrum of H, because of their rapid decay rate. One conjectured property of the single phase ϕ^4 model is the absence of three-particle bound states. This would follow from the conjectured negativity of the six point vertex function, $\Gamma^6 \leq 0$. Other ideas are investigated in a paper on the resummation of one particle lines; these may be useful in a future analysis of the problem of asymptotic completeness.

Bibliography

The entropy principle for vertex functions in quantum field models, Ann. l'Inst. H. Poincaré A21. 1–26 (1974).

Three-particle structure of ϕ^4 interactions and the scaling limit, Phys. Rev. D11, 2816–2827 (1975).

Two and three body equations in quantum field models, Commun. Math. Phys. 44, 293–320 (1975).

Particles and scaling for lattice fields and Ising models, Commun. Math. Phys. 51, 1–13 (1976).

The resummation of one particle lines, Commun. Math. Phys. 67, 267–293 (1979).

The entropy principle for vertex functions in quantum field models

by

James GLIMM (*)
Courant Institute. New York University. New York 10012

and

Arthur JAFFE (**)
Harvard University. Cambridge, Massachusetts 02138

ABSTRACT. — We obtain the (amputated, one particle irreducible) vertex functions for weakly coupled Euclidean $\mathscr{P}(\Phi)_2$ models. The generating function $\Gamma_2\{ A \}$ for the vertex functions is jointly analytic in A and in the bare coupling constants λ_j. The generating function $G\{ J \}$ for connected Green's functions has a convergent tree-graph expansion in terms of vertex functions and propagators.

CONTENTS

(*) Supported in part by the National Science Foundation under Grant NSF-GP-24003.
(**) Supported in part by the National Science Foundation under Grant NSF-GP-40354X.

1. THE VERTEX FUNCTIONS

1.1. The partition function

The partition function in Euclidean $\mathscr{P}(\Phi)_2$ models,

$$Z\{J\} = \int e^{\hat{\Phi}(J)}dq = \int e^{\int(\Phi(x) - \langle\Phi(x)\rangle)J(x)dx}dq = \int e^{\Phi(J)}dq/e^{\langle\Phi\rangle\int J(x)dx}$$

is the generating function for the Schwinger functions

$$S^{(n)}(x_1, \ldots, x_n) = \int \hat{\Phi}(x_1) \ldots \hat{\Phi}(x_n)dq .$$

of the translated field $\hat{\Phi}(x) = \Phi(x) - \langle \Phi(x) \rangle$.

Here dq is the Euclidean measure on $\mathscr{S}'(\mathrm{R}^2)$, associated with the $\mathscr{P}(\Phi)_2$ quantum field. Indicating derivatives by subscripts,

$$S^{(n)}(x_1, \ldots, x_n) = Z_{x_1,\ldots,x_n}\{0\} .$$

The connected parts of the Schwinger functions $G^{(n)}(x_1, \ldots, x_n)$ have the effects of disconnected processes removed. (They have no zero particle intermediate states.) The $G^{(n)}$ functions cluster exponentially, with the exponential decay rate equal to the physical mass m. The generating function

$$G\{J\} = \ln Z\{J\}$$

yields these (connected) Green's functions,

$$G^{(n)}(x_1, \ldots, x_n) = G_{x_1,\ldots,x_n}\{0\} .$$

The vertex functions $\Gamma^{(n)}(x_1, \ldots, x_n)$ are the amputated, one particle irreducible parts of the $G^{(n)}(x_1, \ldots, x_n)$. They have the effects of one particle intermediate states removed, and are believed to cluster with a decay rate $> m$. Thus the vertex functions display the « upper mass gap » in the energy momentum spectrum. Furthermore, the physical charge is defined in terms of a vertex function, so understanding the vertex functions is a step toward understanding charge renormalization.

Construction of the vertex functions is the beginning of Symanzik's program of structure analysis for Green's functions [9]; he proposed construction of n-particle irreducible parts. See also [*11*]. Related to Symanzik's proposed analysis of Green's functions, is our analysis of the n-particle structure of the energy spectrum [3]. Consider the subspace $\mathscr{H}_E \subset \mathscr{H}$ for the spectral interval [0, E] of the Hamiltonian $H = H^*$. In weak $\mathscr{P}(\Phi)_2$ models, cluster properties yield an explicit construction of $\mathscr{H}_E$. Furthermore, this analysis exhibits n-particle structure, since states of energy $\leq (n + 1)m_0(1 - \varepsilon)$ are spanned by polynomials of degree n in the Euclidean field Φ. This result, for $n = 0, 1$, is used to prove the existence of isolated one particle states [3].

In this paper we construct the vertex functions $\Gamma^{(n)}$ and their generating function $\Gamma\{A\}$,

$$\Gamma^{(n)}(x_1, \ldots, x_n) = \Gamma_{x_1,\ldots,x_n}\{A = 0\},$$

for $\mathscr{P}(\Phi)_2$ models with weak coupling. We establish analyticity of $\Gamma\{A\}$ in A, as well as in the bare parameters. In a subsequent paper we establish the Callan-Symanzik equations, giving the change $d\Gamma/d\sigma$ for a mass perturbation $\sigma \int :\Phi^2: d\vec{x}$.

In the definition of the vertex functions, amputation means that in momentum space $G^{(n)}$ is divided by the product

$$\prod_{j=1}^{n} \int G^{(2)}(x, 0) e^{ip_j x} dx$$

of propagators in each momentum variable. In perturbation theory language, amputation removes all mass subdiagrams attached to external legs of diagrams contributing to $G^{(n)}$.

One particle irreducible (1PI) graphs are those which cannot be disconnected by removal of a single line. We use a Legendre transformation (variational or entropy principle) to obtain $\Gamma\{A\}$ from $G\{J\}$. The thermodynamic parameters are identified as follows. J is a chemical potential or magnetic field, and so its conjugate variable has the role of N (= number of particles) or magnetization, M. With $\ln Z = PV$, we see that Γ defined by (1.2.1) is essentially the negative of the Helmholtz free energy $N - PV$. Entropy is the variable conjugate to temperature, and $kT = \beta^{-1}$ is the inverse coefficient of the energy in dq. Below we introduce a coefficient λ_j for each homogeneous contribution to the energy. Conjugate to each λ_j is an « entropy » S_j associated with the jth Legendre transform. The case $j = 2$ is discussed in [11] and leads to two particle irreducible graphs. Suppression of the factor $e^{-\langle \mathscr{S}(J)\rangle}$ in Z leads to graphs which are one particle irreducible in a stronger sense than the graphs considered here, in that tadpoles (i. e. one particle reducible parts with no external legs) are also removed, see [11].

This method goes back to De Dominicis and Martin in statistical mechanics [2] and to Jona-Lasinio in quantum field theory [7]. We follow work of Symanzik [10]. Our contribution is to obtain bounds on $Z\{J\}$, $G\{J\}$ and $G^{(2)}$ which permit the analysis to be carried out rigorously. We also obtain bounds on the vertex functions $\Gamma^{(n)}$. The n-dependence of these bounds ensures convergence of the expansion of $G\{J\}$ (and also $Z\{J\}$ in terms of « tree-like graphs » with vertex kernels $\Gamma^{(r)}$ and propagators $G^{(2)}$. In this manner we exhibit graphically the 1PI property of the vertex functions.

Let H_p denote the complex Sobolev space $\mathscr{D}((-\Delta+1)^{p/2})$, with norm

$$\| f \|_p^2 = \int_{x\in\mathbb{R}^2} |((-\Delta+1)^{p/2}f)(x)|^2 dx .$$

Also let

$$H_{p,\varepsilon} = \{ f : f \in H_p, \| f \|_p < \varepsilon \} .$$

We define analytic functions and multilinear forms $\mathscr{ML}_n$ in the Appendix.

The partition function $Z\{J\}$ is a limit of finite volume partition functions, at least for $J \in C_0^\infty$. The finite volume partition functions are jointly analytic in $J \in H_{-1}$ and in the bare parameters of the theory in any « positive but weak coupling » region of the bare parameter space (See Chapter 2). By definition, in such a region each coupling constant λ_j, the coefficient of $:\Phi^j:$ in the interaction Hamiltonian, is sufficiently small with respect to the mass m_0 in the free Hamiltonian. We also require $|\lambda_j| \le O(1)|\operatorname{Re}\lambda_l|$, and $\operatorname{Re}\lambda_l > 0$ where λ_l is the coupling constant of the highest degree term, and of course l is even. In this region, with $\| J \|_{-1}$ small the finite volume partition functions are uniformly bounded. Hence by Proposition A3, the infinite volume limit $Z\{J\}$ is analytic for $\| J \|_{-1}$ small and for λ_j in the positive but weak coupling region. { We remark that this argument does not yield derivatives of $Z\{J\}$ with respect to λ_l at $\lambda_l = 0$, but Dimock [*1*] has shown that the Schwinger functions are C^∞ in the λ's in a region including $\lambda_l = 0^+$. } We establish our basic estimates for $Z\{J\}$ in Chapter 2. Fröhlich [*6*] previously used properties of $Z\{J\}$ to study the Schwinger functions in $\mathscr{P}(\Phi)_2$ models. Feldman [*5*] has results for the finite volume Φ_3^4 model.

We assume throughout this paper that the coupling constants λ_j lie in a weak coupling region and that $J \in H_{-1,\varepsilon}$.

PROPOSITION 1.1.1. — The partition function $Z\{J\}$ is jointly analytic in λ_j and $J \in H_{-1,\varepsilon}$, for $\varepsilon > 0$ sufficiently small.

We postpone the proof to Chapter 2. By Proposition A5 we have.

COROLLARY 1.1.2. — Let $J \in H_{-1,\varepsilon}$, for ε sufficiently small. Then

$$Z_{x_1,\dots,x_n}\{J\} \in \mathscr{ML}_n(H_{-1})$$

with norms bounded by $O(1)L^n n!$

Remark. — Fröhlich obtained a bound $(n!)^{1/2}$ by a more careful analysis [*6*], but with a different norm.

PROPOSITION 1.1.3. — $Z\{J\} \neq 0$ and $Z\{J\}$ is bounded away from zero for $\| J \|_{-1}$ small.

Proof. — For real λ_j and J, the Schwarz inequality yields

$$1 = Z\{0\} = \int dq \leq \left[\int e^{\Phi(J)}dq \int e^{-\Phi(J)}dq\right]^{1/2} = Z\{J\}^{1/2} Z\{-J\}^{1/2}.$$

By Proposition 1.1.1, $Z\{-J\}$ is bounded, so $Z\{J\}^{-1}$ is also bounded. For the general (complex) case, $Z\{0\} = 1$ and $Z\{J\} \neq 0$ by continuity.

Recall $G\{J\} = \ln Z\{J\}$. By Proposition 1.1.3, $G\{J\}$ exists.

PROPOSITION 1.1.4 [10]. — For λ_j, J real, $G\{J\}$ is a convex function of J.

Proof. — Let $J = \alpha_1 J_1 + \alpha_2 J_2$ be a convex sum with J_i real. By Hölder's inequality with $p_i = \alpha_i^{-1}$,

$$\begin{aligned} G\{J\} &= \ln Z\{\alpha_1 J_1 + \alpha_2 J_2\} \\ &\leq \ln Z\{J_1\}^{\alpha_1} Z\{J_2\}^{\alpha_2} = \alpha_1 G\{J_1\} + \alpha_2 G\{J_2\}. \end{aligned}$$

By Definition A1, we have

PROPOSITION 1.1.5. — $G\{J\}$ is analytic in H_{-1}, and in the λ_j. The Taylor series coefficients

$$G_{x_1,\dots,x_n}\{J\} \in \mathcal{M}\mathcal{L}_n(H_{-1})$$

have norms $O(1)L^n n!$. Here L is uniformly bounded for λ_j and J as above.

PROPOSITION 1.1.6. — For complex coupling constants in a weak coupling region, or for real coupling with positive physical mass, $G^{(2)}(x, y)$ is the integral kernel of a bounded operator $g^{(2)}$ from H_{-1} to H_1 with a bounded inverse from H_1 to H_{-1}.

We prove this result in Chapter 2.

By analyticity in λ_j, $dG^{(2)}(x, y)/d\lambda_j$ is also bounded, from H_{-1} to H_1.

We define the self energy part Π of $g^{(2)}$ by the resolvent equation,

$$g^{(2)} = c + g^{(2)}\Pi c, \tag{1.1.1}$$

or

$$\Pi = c^{-1} - \Gamma = -\Delta + m_0^2 - \Gamma.$$

Here $c = (-\Delta + m_0^2)^{-1}$ Let $c' = (-\Delta + m_0^2 + 2\lambda_2)^{-1}$.

PROPOSITION 1.1.7. — With λ_j in a weak coupling region, the self energy part has the form

$$\Pi = -2\lambda_2 + \Pi', \tag{1.1.2}$$

where Π' is a bounded transformation from H_1 to H_{-1} with norm $\sum_{j=3}^{l} \sum_{k=1}^{l} O(|\lambda_j \lambda_k|)$. Thus for sufficiently weak coupling,

$$
\begin{aligned}
g^{(2)} &= c + c \sum_{n=1}^{\infty} (\Pi c)^n \\
&= (-\Delta + m_0^2 - \Pi)^{-1} \\
&= (-\Delta + m_0^2 + 2\lambda_2 - \Pi')^{-1} \\
&= c' + c' \sum_{n=1}^{\infty} (\Pi' c')^n .
\end{aligned}
\tag{1.1.3}
$$

Remarks. — We give the proof in Chapter 2. The equation (1.1.4) shows how Π or Π' determines the mass renormalization in the propagator. For a pure quadratic interaction, by (1.1.3), $\Pi' = 0$, and

$$g^{(2)} = (-\Delta + m_0^2 + 2\lambda_2)^{-1} .$$

1.2. **The Legendre transformation** $G\{J\} \leftrightarrow \Gamma\{A\}$

The generating function $\Gamma\{A\}$ for the vertex functions is related to the generating function $G\{J\}$ for Green's functions by a Legendre transformation. We give two formulations [2] [7] [*10*]: the first formulation involves a variational principle, while the second formulation relies on solving a nonlinear functional equation. Both formulations are defined and agree for real λ_j, A, J, while the second formulation generalizes to the complex case, see Corollaries 1.2.4-1.2.5. Throughout this section we assume that $A \in H_{1,\delta}$. We obtain $\Gamma\{A\}$ and find that it is analytic in A, for δ sufficiently small.

DEFINITION 1.2.1. — Let A, λ_j be real. Define

$$\Gamma\{A\} = \inf_J [-A(J) + G\{J\}], \tag{1.2.1}$$

where the infimum runs over real $J \in H_{-1,\varepsilon}$. We take δ small enough so that the infimum is not attained on the boundary.

Given $J \in H_{-1,\varepsilon}$, the expression

$$A(x) = G_x\{J\} \tag{1.2.2}$$

defines a bounded transformation $J \to A$ from $H_{-1,\varepsilon}$ into H_1. The inverse mapping $A \to J$ from $H_{1,\delta}$ to H_{-1} is obtained by solving (1.2.2) as an equation for J.

PROPOSITION 1.2.2. — Given A, λ_j real, the following conditions are equivalent, for $J \in H_{-1,\varepsilon}$:

(*i*) J minimizes (1.2.1),

(*ii*) J satisfies equation (1.2.2).

Proof. — Since (*ii*) is the variational equation associated with (1.2.1), it follows that (*i*) $\Rightarrow$ (*ii*). We assume (*ii*), and let $M \in H_{-1}$. The function

$$f(\alpha) = -A(J + \alpha M) + G\{J + \alpha M\}$$

is convex (Proposition 1.1.4) and has derivative 0 at $\alpha = 0$, by (*ii*). Since convexity implies f' is monotone, $f' \geq 0$ for $\alpha > 0$ and $f' \leq 0$ for $\alpha < 0$. Since M is arbitrary, J is a minimum, implying (*i*).

A direct proof of the existence and uniqueness of a minimizing J in (1.2.1) can be given. However, we study here the solution to (1.2.2). In both approaches, our proofs require $\| A \|_1$ to be small.

THEOREM 1.2.3. — Consider the complex case and let $\Gamma A \in H_{-1,\varepsilon}$ for ε sufficiently small. Then the equation (1.2.2) has a unique solution $J_A \in H_{-1,2\varepsilon}$. The solution J_A is jointly analytic in A and in λ_j.

Remark. — As before, Γ denotes the linear operator inverse to $g^{(2)}$, the operator with kernel $G^{(2)}(x, y) = G_{x,y}\{0\}$.

Two immediate corollaries are:

COROLLARY 1.2.4. — Let A, λ_j be real, $A \in H_{1,\delta}$. Then (1.2.1) has a unique minimizing function $J = J_A$, and

$$\Gamma\{A\} = -A(J_A) + G\{J_A\}. \tag{1.2.3}$$

COROLLARY 1.2.5. — The generating function $\Gamma\{A\}$ is analytic in $A \in H_{1,\delta}$. In other words, for A, $A + B \in H_{1,\delta}$,

$$\Gamma\{A + B\} = \sum_{n=0}^{\infty} \frac{1}{n!} \Gamma^{(n)}_{B,B,\ldots,B}\{A\}.$$

The Taylor series coefficients $\Gamma^{(n)}_{x_1,\ldots,x_n}\{A\}$ are elements of $\mathcal{ML}_n(H_1)$, with norms $O(L^n n!)$, and are analytic functions of A and the λ_j.

PROPOSITION 1.2.6 [*10*]. — In the real case, $\Gamma\{A\}$ is concave: for real A, $B \in H_{1,\delta}$ and $0 \leq \alpha \leq 1$,

$$\Gamma\{\alpha A + (1 - \alpha)B\} \geq \alpha\Gamma\{A\} + (1 - \alpha)\Gamma\{B\}.$$

Proof. — By Definition 1.2.1,

$$\begin{aligned}\Gamma\{\alpha A + (1 - \alpha)B\} &= \inf_J [\alpha(-A(J) + G\{J\}) + (1 - \alpha)(-B(J) + G\{J\})] \\ &\geq \alpha \inf_J [-A(J) + G\{J\}] + (1 - \alpha) \inf_J [-B(J) + G\{J\}] \\ &= \alpha\Gamma\{A\} + (1 - \alpha)\Gamma\{B\}.\end{aligned}$$

Remark. — Corollary 1.2.5 holds, in particular, for A = 0. Then $\Gamma^{(n)}(x_1, \ldots, x_n) = \Gamma_{x_1,\ldots,x_n}\{0\}$ and we obtain bounds on the vertex functions for weak $\mathscr{P}(\Phi)_2$ models. We note $\Gamma^{(1)}(x) = \Gamma^{(1)}_x\{0\} = 0$, and $-\Gamma^{(2)}(x, y) = -\Gamma^{(2)}_{x,y}\{0\}$ is the kernel of the linear operator $\Gamma = [g^{(2)}]^{-1}$.

Remark. — By Proposition 1.1.6, the operators $\Gamma : H_1 \to H_{-1}$ and $\Gamma^{-1} : H_{-1} \to H_1$ are both bounded. Thus equation (1.2.2) is equivalent to the equation

$$J = \Gamma A + \Gamma K\{J\}, \tag{1.2.4}$$

where K is the nonlinear operator defined by

$$K\{J\}(x) = -G_x\{J\} + G^{(2)}(x, J).$$

Proof of Theorem 1.2.3. — Explicitly,

$$K\{J\}(x_1) = -\sum_{n=3}^{\infty} \frac{1}{(n-1)!} \int dx_2 \ldots dx_n G^{(n)}(x_1, \ldots, x_n) \cdot J(x_2) \ldots J(x_n). \tag{1.2.5}$$

By Propositions 1.1.5 and 1.1.6, the operator ΓK is a strict contraction on $H_{-1,2\varepsilon}$, for ε sufficiently small. In fact,

$$\begin{aligned} \|\Gamma K\{f\}\|_{-1} &\leq O(1) \sum_{n=3}^{\infty} \frac{n!L^n}{(n-1)!} \|f\|_{-1}^{n-1} \\ &\leq O(1)\|f\|_{-1}^2 \leq \frac{1}{2}\|f\|_{-1}. \end{aligned} \tag{1.2.6}$$

Thus for $\Gamma A \in H_{-1,\varepsilon}$, the operator

$$T_A : f \mapsto \Gamma A + \Gamma K\{f\}$$

maps into itself the sphere in H_{-1} centered at ΓA, and with radius ε. $T_A f$ is analytic in A. Likewise

$$\|T_A^n f\|_{-1} < 2\varepsilon, \tag{1.2.7}$$

and $T_A^n f$ is analytic in A.

Define $h_n(A) = T_A^n f$, with $\Gamma A \in H_{-1,\varepsilon}$. By (1.2.7), the norms $\|h_n(A)\|_{-1}$ are uniformly bounded. Also

$$\|h_{n+1}(A) - h_n(A)\|_{-1} = \|\Gamma K\{h_n(A)\} - \Gamma K\{h_{n-1}(A)\}\|_{-1} \tag{1.2.8}$$

We bound (1.2.8) by using (1.2.5) and Proposition 1.1.5, in a fashion similar to the derivation of (1.2.6). We obtain for $f \in H_{-1,\varepsilon_1}$ (with $\varepsilon_1 \leq \varepsilon$ sufficiently small),

$$\|h_{n+1}(A) - h_n(A)\|_{-1} \leq \frac{1}{2}\|h_n(A) - h_{n-1}(A)\|_{-1}.$$

By iteration,

$$\| h_n(A) - h_m(A) \|_{-1} \leq \sum_{j=m}^{n-1} 2^{-j+1} \leq O(2^{-m}).$$

By Proposition A3, there exists an analytic function $h : H_{1,\varepsilon} \to H_{-1,2\varepsilon}$ such that

$$\| h_n(A) - h(A) \|_{-1} \to 0.$$

In particular,

$$J_A \equiv h(A) = \lim_{n\to\infty} T_A^n f \in H_{-1,2\varepsilon}$$

exists and is a fixed point of T_A, i. e. a solution to (1.2.4) or (1.2.2). Uniqueness of the solution J_A follows. For two solutions $J_1, J_2 \in H_{-1,2\varepsilon}$, we obtain as above, for ε sufficiently small,

$$\begin{aligned} \| J_1 - J_2 \|_{-1} &= \| \Gamma K\{J_1\} - \Gamma K\{J_2\} \|_{-1} \\ &\leq O(1) \| K\{J_1\} - K\{J_2\} \|_{-1} \\ &\leq O(\varepsilon) \| J_1 - J_2 \|_{-1}. \end{aligned}$$

Choosing ε sufficiently small, $O(\varepsilon) < 1$, so $J_1 = J_2$.

The solution J_A is analytic in A, as follows by the analyticity of $h(A)$. Analyticity in λ_j follows similarly.

We now study the inverse Legendre transformation $\Gamma\{A\} \to G\{J\}$. Our treatment follows that for $G\{J\} \to \Gamma\{A\}$ above.

PROPOSITION 1.2.7. — In the real case, with $\| J \|_{-1}$ small,

$$G\{J\} = \sup_A [J(A) + \Gamma\{A\}], \tag{1.2.9}$$

where the supremum runs over $A \in H_{1,\delta}$. A maximizing A always exists and is given by (1.2.2).

Proof. — By Definition 1.2.1, $G\{J\} \geq \Gamma\{A\} + J(A)$, with equality for A defined by (1.2.2). This completes the proof.

By Corollary 1.2.5, the expression

$$J(x) = -\Gamma_x\{A\} \tag{1.2.10}$$

defines a bounded transformation $A \mapsto J$, from $H_{1,\varepsilon}$ to H_{-1}. The inverse mapping $J \mapsto A$ is obtained by solving (1.2.10). From the concavity of $\Gamma\{A\}$, Proposition 1.2.6, we obtain analogously to Proposition 1.2.2:

PROPOSITION 1.2.8. — Consider the real case. Given J, the following conditions are equivalent, for $A \in H_{1,\varepsilon}$:

(*i*) A maximizes (1.2.9),
(*ii*) A satisfies (1.2.10).

THEOREM 1.2.9. — Consider the complex case and let $\Gamma^{-1} J \in H_{1,\varepsilon}$, with ε sufficiently small. Then the equation (1.2.10) has a unique solution $A_J \in H_{1,2\varepsilon}$. The function A_J is jointly analytic in J and in λ_j, and is the inverse function to J_A of Theorem 1.2.3, i. e., $J_{A_J} = J$.

Proof. — Recall that the operator Γ has the kernel $-\Gamma^{(2)}(x, y)$. We then rewrite (1.2.10) as

$$A = \Gamma^{-1}J + \Gamma^{-1}\mathscr{F}\{A\} \equiv \mathscr{M}_J\{A\}, \tag{1.2.11}$$

where

$$\mathscr{F}\{A\}(x) = \Gamma_x\{A\} - \Gamma A = \sum_{n=3}^{\infty} \frac{1}{(n-1)!}\Gamma^{(n)}(x, A, \ldots, A). \tag{1.2.12}$$

We follow the proof of Theorem 1.2.3 to obtain the existence, uniqueness and analyticity of $A_J = \lim_n \mathscr{M}_J^n \Gamma^{-1}J$. Differentiating (1.2.3) and using (1.2.2) yields (1.2.10). Thus A_J and J_A are inverse functions.

COROLLARY 1.2.10. — In the real case, (1.2.9) has a unique maximizing function $A = A_J$, for $J \in H_{-1,\varepsilon}$ and

$$G\{J\} = J(A_J) + \Gamma\{A_J\}. \tag{1.2.13}$$

Legendre transforms have been considered from a general point of view, with $H_{\pm 1}$ replaced by topological vector spaces X and Y put in duality by a bilinear form, see [*8*]. The formula $J_A = J$ and Propositions 1.2.2 and 1.2.8 are valid in this general context.

1.3. Tree graphs

The Legendre transformation

$$G\{J\} \to \Gamma\{A\} \tag{1.3.1}$$

and its inverse

$$\Gamma\{A\} \to G\{J\} \tag{1.3.2}$$

generate convergent expansions for $G\{J\}$ and $\Gamma\{A\}$ in terms of tree-like graphs. For instance, in the series for $G\{J\}$, each vertex in a tree graph corresponds to a kernel $\Gamma^{(r)}(x_1, \ldots, x_r)$ and each line corresponds to a kernel $G^{(2)}(x, y)$ (propagator). In perturbation theory, such graph expansions are used to define the vertex functions.

We start from the identities

$$G_x\{J\} = +A_J(x) \tag{1.3.3}$$

and

$$\Gamma_x\{A\} = -J_A(x) \tag{1.3.4}$$

obtained from (1.2.2) and (1.2.10), along with Propositions 1.2.2 and

1.2.8. These identities relate $G\{J\}$ and $\Gamma\{A\}$ directly to A_J and J_A, and provide our tree graph expansions. For instance, the power series for (1.3.3),

$$A_J(x) = \sum_{n=2}^{\infty} \frac{1}{(n-1)!} G^{(n)}(x, J, \ldots, J) \tag{1.3.5}$$

has an alternative expression from iterating (1.2.11) to obtain A_J. We identify the coefficients of order $n-1$ in J in these two expressions, and thereby obtain an expression for $G^{(n)}(x_1, \ldots, x_n)$. This expression is a finite sum, and each term is an integral that we represent by a graph. The graphs contributing to $G^{(n)}(x_1, \ldots, x_n)$ have n external lines labelled by $x_1, \ldots, x_n$. The graphs have a « root », the external line labelled by x_1, and branch in one direction away from the root. Each vertex with r legs has the kernel $\Gamma^{(r)}(y_1, \ldots, y_r)$ and each line has the kernel $G^{(2)}(x, y)$, the kernel of Γ^{-1}. For instance, the tree-like graph of Figure 1 contributes to $G^{(5)}$ and corresponds (up to a combinatoric factor) to the integral

$$\int G^{(2)}(x_1, y_1)\, G^{(2)}(x_2, y_2)\, G^{(2)}(y_3, z_1)\, G^{(2)}(x_3, z_2)\, G^{(2)}(x_4, z_3) \\ G^{(2)}(x_5, z_4)\, \Gamma^{(3)}(y_1, y_2, y_3)\, \Gamma^{(4)}(z_1, z_2, z_3, z_4)\, dy\, dz\,. \tag{1.3.6}$$

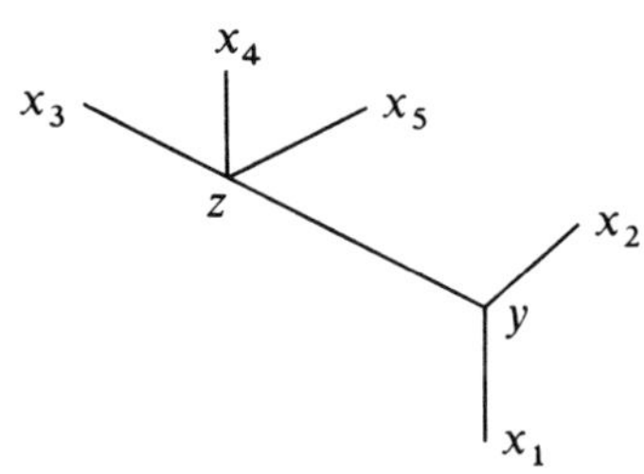

FIG. 1. — A tree graph contributing to $G^{(5)}$.

Retaining only the linear term $A = \Gamma^{-1}J$ in (1.2.11), we obtain the single vertex contribution to $G^{(n)}(x_1, \ldots, x_n)$, namely its one particle irreducible part. Each external leg has the operator Γ^{-1} (propagator) which replaces the mass subdiagrams removed by amputation. For $n = 3$, this is the only contribution to $G^{(3)}$ so

$$G^{(3)}(x_1, x_2, x_3) = \int dy\, G^{(2)}(x_1, y_1)\, G^{(2)}(x_2, y_2)\, G^{(2)}(x_3, y_3) \\ \Gamma^{(3)}(y_1, y_2, y_3)\,. \tag{1.3.7}$$

Substituting at least one nonlinear term from $\mathscr{F}\{A\}$ in (1.2.12) into (1.2.11), we obtain the one particle reducible part of $G^{(n)}(x_1, \ldots, x_n)$,

$G^{(4)}(1, \ldots, 4) =$ 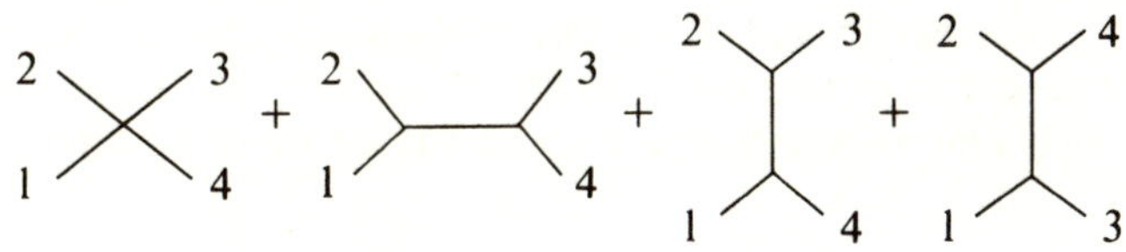

FIG. 2. — In the tree graph expansion of $G^{(n)}$, vertices have kernels $\Gamma^{(r)}$ and lines are propagators $G^{(2)}(x, y)$.

represented by tree graphs with more than one vertex. For example, $G^{(4)}(x_1, \ldots, x_4)$ has the expansion given in Figure 2. Here we abbreviate x_j by j.

For fixed n, the number of tree graphs contributing to $G^{(n)}$ is finite. This number diverges as $n \to \infty$, but nevertheless the power series expansion of $G\{J\}$ in powers of J is convergent, since $G_x\{J\} = A_J(x)$ and A_J is analytic in J. Thus we have proved,

PROPOSITION 1.3.1. — $G\{J\}$ has a convergent tree graph expansion in terms of vertices $\Gamma^{(r)}$ and propagators $G^{(2)}$.

Likewise, we use (1.3.4) and (1.2.4) to generate a tree graph expansion for $\Gamma\{A\}$. In this case, the vertices have kernels equal to the amputated Green's functions,

$$G_a^{(r)}(x_1, \ldots, x_r) = \left(\prod_{i=1}^{r} \Gamma(i)\right) G^{(r)}(x_1, \ldots, x_r),$$

where $\Gamma(i)$ denotes the inverse propagator acting on the variable x_i. Also $r \geq 3$. The lines have kernels given by the propagator $G^{(2)}(x, y)$, as before. In addition, because of the -1 in (1.2.5), there is a factor -1 for each vertex as well as an overall factor -1 from (1.3.4). The sign of a term is therefore $+1$ for an odd number of vertices and -1 for an even number.

Retaining the linear factor $J = \Gamma A$ in (1.2.4), we obtain the single vertex contribution $G_a^{(n)}(x_1, \ldots, x_n)$ to $\Gamma^{(n)}(x_1, \ldots, x_n)$. For $n = 3$, this is the only contribution, and $\Gamma^{(3)}(x_1, x_2, x_3) = G_a^{(3)}(x_1, x_2, x_3)$, which is another form of (1.3.7).

Substituting at least one nonlinear term from (1.2.5) into (1.2.4) we obtain the remaining contributions to $\Gamma^{(n)}$, represented by tree graphs as before. For example, $\Gamma^{(4)}(x_1, \ldots, x_4)$ has the expansion given in Figure 3.

$\Gamma^{(4)}(1, 2, 3, 4) =$ 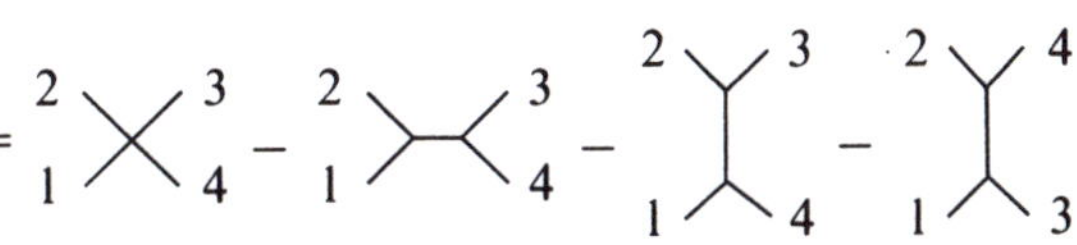

FIG. 3. — In the tree graph expansion of $\Gamma^{(n)}$, vertices have kernels $G_a^{(r)}$ and lines are propagators $G^{(2)}(x, y)$. A minus sign occurs for terms with an even number of vertices.

PROPOSITION 1.3.2. — The analytic function $\Gamma\{A\}$ has a convergent tree graph expansion in terms of vertices $G_a^{(r)}$, of propagators $G^{(2)}$, and with a minus sign for each graph having an even number of vertices.

2. ESTIMATES

In this chapter we establish the estimates stated in Chapter 1. One new feature of our estimate on $Z\{J\}$ is the global bound for $J \in H_{-1}$, as opposed to the local bounds for $J \in L_2$, suppt. J compact, established previously. It is clear that if $\langle \Phi \rangle \neq 0$,

$$\exp \langle \Phi(J) \rangle = \exp \left[\langle \Phi(\cdot) \rangle \int J(x)\, dx \right] \tag{2.1}$$

is bounded in L_1, but not in H_{-1}. This is the reason we substract $\langle \Phi \rangle$ from Φ in the definition of Z. It ensures that no contribution in ln Z is linear in J, and it eliminates the parts of Z which are unbounded in H_{-1}.

In order to estimate $\Phi - \langle \Phi \rangle$, it is convenient to introduce a new symmetry by a tensor product (doubling) procedure, introduced by Ginibre in statistical mechanics. Let $\mathscr{S}'(R^2)^\sim$, $dq^\sim$ be an isomorphic copy of $\mathscr{S}'(R^2)$, dq and let

$$\hat{\mathscr{S}}' = \mathscr{S}'(R^2) \otimes \mathscr{S}'(R^2)^\sim$$
$$d\hat{q} = dq \otimes dq^\sim .$$

For any function A on $\mathscr{S}'(R^2)$, let $A^\sim$ be its isomorphic image, a function on $\mathscr{S}'(R^2)^\sim$, and define an odd function $A_0 = A - A^\sim$ or in tensor product notation

$$A_0 = A \otimes I - I \otimes A^\sim .$$

Then

$$\langle AB \rangle - \langle A \rangle \langle B \rangle = \frac{1}{2} \langle A_0 B_0 \rangle , \tag{2.2}$$

which gives a natural way to perform the subtraction of $\langle \Phi \rangle$ as it occurs in the expansion.

Before getting into the detailed proof of Proposition 1.1.1, we give the proof of Propositions 1.1.6 and 1.1.7. In the case of real coupling constants, let

$$G^{(2)}(p)^\sim = \int_0^\infty \frac{d\rho(a)}{p^2 + a} \tag{2.3}$$

be the spectral representation for the Fourier transform of the connected two point Euclidean Green's function (We assume here Lorentz invariance together with the bound $\int (1 + a)^{-1} d\rho(a) < \infty$). The positivity condition ensures $d\rho(a)$ is a positive measure.

THEOREM 2.1 (*a*). — For real coupling constants $G^{(2)} : H_{-1} \to H_1$ is continuous if and only if

$$\int_0^\infty d\rho(a) + \int_0^\infty a^{-1} d\rho(a) < \infty . \tag{2.4}$$

(*b*) $\Gamma = G^{(2)^{-1}} : H_1 \to H_{-1}$ is continuous.

Proof. — (*a*) Note that $G^{(2)\sim}$ is the Fourier transform of $G^{(2)}(x, 0)$. Then $G^{(2)} : H_{-1} \to H_1$ is a bounded operator, if and only if $(1 + p^2)G^{(2)\sim}(p)$ is a bounded function of p. Assume $(1 + p^2)G^{(2)\sim}(p)$ is bounded. Since $d\rho(a)$ is positive,

$$\int (a^{-1} + 1) d\rho(a) = G^{(2)\sim}(0) + \sup p^2 G^{(2)\sim}(p) < \infty .$$

Conversely, assuming (2.4), we have

$$\sup_p (1 + p^2)G^{(2)\sim}(p) \le \sup \int (p^2 + a)^{-1} d\rho(a) + \sup \int p^2(p^2 + a^2)^{-1} d\rho(a)$$
$$= \int a^{-1} d\rho(a) + \int d\rho(a) < \infty .$$

(*b*) $\Gamma : H_1 \to H_{-1}$ is a bounded operator if and only if $[G^{(2)\sim}(p)(p^2 + 1)]^{-1}$ is a bounded function of p. There is an interval $[0, \beta]$, such that $\int_0^\beta d\rho(a)$ is nonzero. Then

$$\inf_p (p^2 + 1)G^{(2)\sim}(p) \ge \inf_p \int_0^\beta d\rho(a)(p^2 + 1)(p^2 + a)^{-1}$$
$$\ge \min \{ 1, \beta^{-1} \} \int_0^\beta d\rho(a) > 0 ,$$

and Γ is bounded.

Remarks. — 1. The hypotheses hold in weakly coupled $\mathscr{P}(\Phi)_2$ models. Hence both $G^{(2)} : H_{-1} \to H_1$ and its inverse $\Gamma : H_1 \to H_{-1}$ are bounded operators. In fact, all $\mathscr{P}(\Phi)_2$ quantum field models obey the canonical commutation relations (before field strength renormalization) so for these models

$$\int d\rho(a) = 1 .$$

Weakly coupled $\mathscr{P}(\Phi)_2$ models have isolated one particle states, so $\int a^{-1} d\rho(a) < \infty$ and Proposition 1.1.6 follows for λ_j real.

2. More generally, (*a*) holds for $\mathscr{P}(\Phi)_2$ models except at the critical points (points of zero mass). It should also hold for Φ_3^4, or for other models with a strictly positive mass and a finite field strength renormalization

constant $Z^{-1} = \int d\rho(a)$. Furthermore, (*b*) should hold in any Wightman field theory.

Proof of Proposition 1.1.7. — Let $c = (-\Delta + 1)^{-1}$ have the kernel $C(x - y)$. We integrate by parts on q-space to obtain

$$S^{(2)}(x - y) = C(x - y) - \int dx' C(x - x') \int V'(\Phi(x'))\Phi(y) dq ,$$

where V' is the derivative of the interaction polynomial V. Similarly,

$$G^{(2)}(\bar{f}, f) - \| f \|_{-1}^2 = - \langle V'(c\bar{f})\Phi(f) \rangle_T ,$$

where T denotes the connected (truncated) part,

$$\langle V'\Phi \rangle_T = \langle V'\Phi \rangle - \langle V' \rangle \langle \Phi \rangle .$$

Let $K(x, y) = \langle V'(\Phi(x))\Phi(y) \rangle_T$. By translation invariance of the measure dq, $K(x, y) = \zeta(x - y)$. Hence the operator k with kernel $K(x, y)$ commutes with c, and k is bounded on H_p if and only if k is bounded on H_0, as we also see by

$$(2.5) \qquad | K(f, g) | = | \langle f, kg \rangle | = | \langle c^{p/2} f, kc^{-p/2} g \rangle | \le \| k \| \ \| f \|_{-p} \| g \|_p .$$

Here $\| k \|$ denotes the operator norm on H_0 (hence on H_p).

We bound $\| k \|$ using estimates from the cluster expansion [3]. Locally, $K(x, y)$ is L_2 and hence L_1. It follows that $\zeta(x)$ is locally L_1. As a consequence of the spectral condition and the mass gap for the relativistic weak coupling $\mathcal{P}(\Phi)_2$ model with real coupling constants [3], $\zeta(x)$ is analytic for $x \neq 0$ and for $|x|$ bounded away from zero,

$$(2.6) \qquad | \zeta(x) | \le O(\sup_j | \lambda_j |) e^{-m|x|} .$$

Thus $\zeta(x)$ is L_1 and

$$(2.6') \qquad \| k \| \le \| \zeta \|_{L_1} < O(\sup_j | \lambda_j |) .$$

It follows by (2.5) that k is a bounded operator on H_1, and that

$$(2.7) \qquad \Pi = - \Gamma k$$

is a bounded operator from H_1 to H_{-1}.

We next isolate the first order part of Π, by writing

$$V'(\Phi(x)) = \lambda_1 + 2\lambda_2 \Phi(x) + W(x) ,$$

with

$$W(x) = \sum_{n=3}^{l} n\lambda_n : \Phi(x)^{n-1} : .$$

By definition,

$$K(x, y) = 2\lambda_2 G^{(2)}(x, y) + K_1(x, y),$$

where $K_1(x, y) = \langle W(x)\Phi(y) \rangle_T$ is the integral kernel of the operator k_1. Thus

$$\Pi = -2\lambda_2 - \Gamma k_1, \tag{2.8}$$

and as above, Γk_1 is bounded from H_1 to H_{-1}. Since deg $W \geq 2$, integrating by parts produces at least one more V' vertex in the evaluation of $\langle W\Phi \rangle_T$. This proves that $\| k_1 \|$ is at least second order in the λ_j, and at least first order in some λ_j, $j \geq 3$. The remaining assertions immediately follow for real coupling.

For complex coupling, we use the cluster expansion [4] to replace (2.6) by the estimate

$$\int_{|x| \leq 1} |\zeta(x + y)| \, dx \leq O(1)e^{-(m-\varepsilon)|y|}.$$

From (2.6′) we deduce Proposition 1.1.6 in the case of complex coupling constants, and then the remainder of the proof follows as before.

We begin the proof of Proposition 1.1.1 with two preliminary estimates. We use the notation of [4].

LEMMA 2.2. — The operator $(-\Delta + 1)^{1/2}$ maps $L_2(R^2)$ boundedly into $L_p(R^2)$ for all $p \in [2, \infty)$.

Proof. — This follows by Hausdorff-Young and Hölder inequalities. We consider the operators

$$\partial^b C_\gamma = \partial^b(-\Delta_\gamma + m_0^2)^{-1} = (-\Delta_\gamma + m_0^2)^{-1} - (\Delta_{\gamma \cup b} + m_0^2)^{-1}$$

and

$$\partial^\Gamma C_\gamma = \partial^{b_1} \ldots \partial^{b_m} C_\gamma$$

where b and b_i are lattice segments and $\Gamma = \{ b_1, \ldots, b_m \}$. Let

$$d(\alpha, \beta, \Gamma) = \sup_{\substack{i=\alpha,\beta \\ b\in\Gamma}} \text{dist}\,(\Delta_i, b).$$

Recall that Δ_i, Δ_α and Δ_β are lattice squares, subsets of R^2, while Δ_γ, $\Delta_{\gamma \cup b}$, are Laplace operators with zero Dirichlet boundary data on lattice curves γ, $\gamma \cup b$, ...

LEMMA 2.3. — Let q be given, $1 \leq q < \infty$, and let m_0 be sufficiently large. The operator

$$(-\Delta + 1)^{1/2} \partial^\Gamma C_\gamma (-\Delta + 1)^{1/2} : L_2(\Delta_\alpha) \to L_2(\Delta_\beta)$$

has a norm bounded by

$$m_0^{O(1)-|\Gamma|/2q} K(\Gamma) \exp\,[-m_0 d(\alpha, \beta, \Gamma)/2]$$

where the constant $K(\Gamma)$ is independent of α, β, γ, Γ, m_0. Furthermore, the combinatoric estimate of [*4*], Proposition 8.2 holds.

The proof of the combinatoric estimates follows as in [*4*]. The analytic estimate is based on

LEMMA 2.4. — Let $(\Delta_\gamma - m_0^2)u = 0$ on some subdomain $\mathscr{D}$, and let $u \in L_2$. Then $\nabla u \in L_2$ on any interior subdomain $\mathscr{E}$ which is bounded and bounded away from $\partial\mathscr{D}$. Also

$$\| \nabla u \|_{L_2(\mathscr{E})} \leq \text{const.} \| u \|_{L_2(\mathscr{D})} .$$

Proof. — Away from γ, standard regularity estimates for elliptic operators apply. Thus $u \in C^\infty$ and the Sobolev norms are dominated by $\| u \|_{L_2(\mathscr{D})}$. Thus we consider u up to γ, which consists of lattice line segments. Along the interior of a line segment, u is regular by a reflection principle. For instance, consider the data u given on a semicircle. Extend the data to the opposite semicircle by (-1) times the reflection about the diameter. This determines a harmonic function, zero along the diameter, and hence regular up to the diameter. In the case of an interior corner, a similar reflection argument gives regularity up to γ.

It remains to consider u in a neighborhood of an exterior corner or endpoint ξ of γ. Let C be a circle about ξ, and let $v = \zeta u$, with ζ a smooth function, defined on the interior of C, equal to zero near ξ and equal to one near C. Then v, ∇v and $f = \Delta_\gamma v$ are bounded by const. $\| u \|_{L_2(\mathscr{D})}$. However, $\Delta_{\gamma \cup C}(v - u) = f$, so

$$\nabla u = \nabla v - \nabla(\Delta_{\gamma \cup C})^{-1} f .$$

Since $\nabla(\Delta_{\gamma \cup C})^{-1}$ is a bounded operator, ∇u is bounded in L_2 norm on a neighborhood of ξ.

This completes the proof of Lemma 2.4.

Away from Γ, the kernel of $\partial^\Gamma C_\gamma$ is a solution of the equations

$$(-\Delta_{\gamma,x} + m_0^2)\partial^\Gamma C_\gamma(x, y) = 0 = (-\Delta_{\gamma,y} + m_0^2)\partial^\Gamma C_\gamma(x, y) .$$

Thus by Lemma 2.4, away from Γ, L_2 bounds on $\partial^\Gamma C_\gamma$ [4, Proposition 8.1] imply corresponding L_2 bounds on $\nabla_x \nabla_y \partial^\Gamma C_\gamma(x, y)$. For x or y near Γ, we consider several cases. If either x or y is near $\Gamma_0 \subset \Gamma$, but bounded away from $\Gamma \sim \Gamma_0$, then L_2 bounds on

$$\partial^{\Gamma \sim \Gamma_0} C_{\gamma \cup \Gamma_{00}}(x, y) , \qquad \Gamma_{00} \subset \Gamma_0$$

imply L_2 bounds on

$$\nabla_x \nabla_y \partial^{\Gamma \sim \Gamma_0} C_{\gamma \cup \Gamma_{00}}(x, y) , \qquad \Gamma_{00} \subset \Gamma_0 .$$

The latter give (by linear combination and the definition $\partial^\Gamma = \partial^{\Gamma_0}\partial^{\Gamma \sim \Gamma_0}$ above) L_2 bounds on

$$\nabla_x \nabla_y \partial^\Gamma C_\gamma(x, y) ,$$

with the same decay as the bounds on $\partial^\Gamma C_\gamma$ of [*4*, Proposition 8.1]. Here we assume $\Gamma_0 \neq \Gamma$ and use the distance $d(\alpha, \beta, \Gamma \sim \Gamma_0)$ to give the decay. If $\Gamma_0 = \Gamma$, then the decay comes only from the separation between x and y. For x bounded away from y

$$(-\Delta_y + m_0^2)C_\gamma(x, y) = 0 = (-\Delta_x + m_0^2)C_\gamma(x, y)$$

and as above we find that

$$\nabla_x\nabla_y C_{\gamma\cup\Gamma_{00}}(x, y), \qquad \Gamma_{00} \subset \Gamma_0 = \Gamma$$

and

$$\nabla_x\nabla_y\partial^\Gamma C_\gamma(x, y)$$

have L_2 bounds with the same decay as $\partial^\Gamma C_\gamma$.

Finally for x, y and Γ all near each other, so $|\Gamma| \leq O(1)$, we use the fact that

$$C_\gamma \leq C_\phi = (-\Delta + m_0^2)^{-1}$$

to see that

$$\| \nabla_x\nabla_y C_\gamma \|$$

is bounded, where $\| \cdot \|$ denotes the L_2 operator norm. In this case, we do not obtain the factor $m_0^{-(2/q)|\Gamma|} = m_0^{-O(1)}$. Combining all cases gives the proof of Lemma 2.3.

The cluster expansion of [*4*] is based on localizations of the interaction occurring in exponents such as exp $(\Phi(J))$. With $J \in H_{-1}$, we cannot localize by multiplying J by a characteristic function χ_α of a unit lattice square Δ_α. Instead we define

$$J_\alpha = (-\Delta + 1)^{1/2}\chi_\alpha(-\Delta + 1)^{-1/2}J .$$

Thus $J_\alpha \in H_{-1}$ and

$$J = \sum_\alpha J_\alpha, \qquad \| J \|_{-1}^2 = \sum_\alpha \| J_\alpha \|_{-1}^2 .$$

We let γ denote a union of lattice segments in R^2, and let Δ_γ be the Laplace operator with Dirichlet boundary conditions on γ. We let $d\Phi_\gamma$ denote the Gaussian measure on $\mathscr{S}'$ with mean zero and covariance $(-\Delta_\gamma + m_0^2)^{-1} = C_\gamma$. We may take the infinite volume limit for the interacting theory with the Dirichlet contour γ and define for $h \in C_0^\infty$,

$$dq_\gamma = e^{-V(h)}d\Phi_\gamma .$$

We note that dq_γ is not normalized, and refers to a finite volume approximation to the infinite volume measure. We let $J \in C_0^\infty$ and define

$$Z\{J, \gamma\} = \int e^{\Phi(J)}dq_\gamma / e^{\int\langle\Phi(x)\rangle J(x)dx}$$

$$Z_0\{J, \gamma\} = Z\{J, \gamma\}\, e^{\int\langle\Phi(x)\rangle J(x)dx} = \int e^{\Phi(J)}dq_\gamma .$$

Our estimates will be uniform for $\|J\|_{-1}$ small. The contour γ does not occur in the expectation $\langle \Phi \rangle$ in the exponent. In case $R^2 \sim \gamma$ has two components—Ext γ and Int γ—the measure dq_γ factors into an interior and an exterior measure. In an obvious notation

$$Z_0\{J, \text{Int } \gamma\} = \int \exp\left[\int_{\text{Int } \gamma} \Phi(x)J(x)dx\right] dq_\gamma ,$$
$$Z\{J, \gamma\} = Z\{J, \text{Ext } \gamma\} Z_0\{J, \text{Int } \gamma\}.$$

Observe that the normalization differs $\left(\text{by the factor } \int dq\right)$ from that of § 1.1. The statement of the following proposition is independent of the normalization, since it concerns a ratio of partition functions. In the proof we use the above unnormalized measures.

PROPOSITION 2.5. — Let $R^2 \sim \gamma$ have two components and let J belong to $H_{-1,\varepsilon}$. In a weak coupling region,

$$|Z\{J, \text{Ext } \gamma\}/Z\{J\}| \leq e^{O(|\text{Int } \gamma| + \|J\|_{-1}^2)} .$$

Proof. — The normalizing factor $\exp\left[\int \langle \Phi(x) \rangle J(x)\, dx\right]$ cancels between the numerator and denominator and plays no role in the proof. Let (Int γ)* be the set of lattice lines in Int γ. We write

$$(2.9) \quad \left|\frac{Z\{J, \text{Ext } \gamma\}}{Z\{J\}}\right| = \left|\frac{Z\{J, \gamma\}}{Z\{J\} Z_0\{J, \text{Int } \gamma\}}\right|$$
$$= \left|\frac{Z_0\{J, (\text{Int } \gamma)^*\}}{Z_0\{J, \text{Int } \gamma\}} \times \frac{1}{Z_0\{J, (\text{Int } \gamma)^*\}} \times \frac{Z\{J, \gamma\}}{Z\{J\}}\right|$$

and we bound each of these factors by $\exp\left[O(|\text{Int } \gamma| + \|J\|_{-1}^2)\right]$. We first bound $|Z\{J, \gamma\}/Z\{J\}|$. This bound is proved in the finite volume approximation, uniformly in the volume, using the Kirkwood-Salsburg equations [*4*, Proposition 5.2]. The same argument, applied within Int γ, bounds the ratio $|Z_0\{J, (\text{Int } \gamma)^*\}/Z_0\{J, \text{Int } \gamma\}|$. We have only to indicate how the proof of [*4*] is modified to allow estimates depending on $J \in H_{-1}$.

A typical term in the expansion has the form

$$(2.10) \quad \int K(\Phi, J, \gamma')e^{\int(\Phi - \langle\Phi\rangle)J(x)dx} dq_{\gamma'}/Z\{J\}$$
$$= \int K(\Phi, J, \gamma')e^{\int_{\text{Int}\gamma'}\Phi(x)J(x)dx} dq_{\text{Int}\gamma'} Z\{J, \text{Ext } \gamma\}/Z\{J\}$$

where $K(\Phi, J, \gamma')$ denotes some function on path space. The Kirkwood-Salsburg equations for $\rho(\gamma) = Z\{J, \gamma\}/Z\{J\}$ have the form

$$\rho = 1 + \mathscr{K}\rho,$$

see [4, Chapters 3, 6]. These equations have a unique solution $(I-\mathscr{K})^{-1}1=\rho$ satisfying

$$|\rho(\gamma)| \leq \exp(|\operatorname{Int}\gamma|)$$

if the operator $\mathscr{K}$ is a contraction. Using the Banach space as in [4], we require bounds of order $\exp(-K|\operatorname{Int}\gamma'|)$ on the kernel in (2.10),

$$\int K(\Phi, J, \gamma') e^{\int_{\operatorname{Int}\gamma'}\Phi(x)J(x)dx} dq_{\operatorname{Int}\gamma'} Z_0\{J, (\operatorname{Int}\gamma')^*\}^{-1}. \tag{2.11}$$

By the Schwarz inequality, (2.11) is dominated by

$$\begin{aligned}\left[\int |K(\Phi, J, \gamma')|^2 d\Phi_{\operatorname{Int}\gamma'}\right]^{1/2} \left[\int dq'_{\operatorname{Int}\gamma'}\right]^{1/4} \\ \times \left[\int \exp\left[4 \operatorname{Re}\int_{\operatorname{Int}\gamma'} \Phi(x)J(x)dx\right] d\Phi_{\operatorname{Int}\gamma'}\right]^{1/4} Z_0\{J, (\operatorname{Int}\gamma')^*\}^{-1}.\end{aligned} \tag{2.12}$$

Here the measure dq' has all the parameters λ_j replaced by $4 \operatorname{Re}\lambda_j$. The integral over dq' is independent of J, and by standard estimates is dominated by $\exp(O|\operatorname{Int}\gamma'|)$. The last factor in (2.12) can be calculated explicitly as

$$\exp\left[2\int_{\operatorname{Int}\gamma'} \operatorname{Re} J(x)C_\gamma(x, y) \operatorname{Re} J(y)dx\,dy\right] \leq \exp\left[2\|J\|_{-1}^2\right].$$

In the first integral, only the J-vertices require special treatment. We use the H_{-1} localization $J = \Sigma J_\alpha$ as above. The case of a contraction $\partial^\Gamma C_\gamma$ joining J_α and J_β contributes to K the numerical factor

$$|\langle J_\alpha \partial^\Gamma C_\gamma J_\beta\rangle| \leq \|\chi_{\Delta_\alpha}(-\Delta+1)^{1/2}\partial^\Gamma C_\gamma(-\Delta+1)^{1/2}\chi_{\Delta_\beta}\| \quad \|J_\alpha\|_{-1}\|J_\beta\|_{-1}.$$

The sum over α, β of these factors is bounded by Lemma 2.3. In the case of a contraction $\partial^\Gamma C_\gamma$ joining J_α to a vertex from dq, the vertex is multiplied by the function $\partial^\Gamma C_\gamma(x, J_\alpha)$. By Lemmas 2.2 and 2.3, this function is L_p for all $p<\infty$, with a norm bounded by the decay rate specified in Lemma 2.3. Hence for $p\in[2, \infty)$,

$$\|\partial^\Gamma C_\gamma(\cdot, J)\|_{L_p} \leq \text{const.}\ \|J\|_{-1}.$$

Multiplication of a vertex from dq by an L_p function does not affect the proofs of [4], and so with the above modifications the proof is that of [4].

Finally, to complete the bound on (2.12), we must show that

$$|Z_0\{J, (\text{Int}\,\gamma)^*\}|^{-1} \leq e^{O|\text{Int}\,\gamma|},$$

which then also yields the desired bound on (2.9). We note that

$$Z_0\{J, (\text{Int}\,\gamma)^*\} = \Pi Z_0\{J, \text{Int}\,\Delta_\alpha\}$$

where the product ranges over $\Delta_\alpha \subset \text{Int}\,\gamma$. Thus we must show that

$$|Z_0\{J, \text{Int}\,\Delta_\alpha\}|^{-1} \leq O(1). \tag{2.13}$$

Using a Duhamel formula and a Schwartz inequality with respect to the free measure, one shows that

$$|Z_0\{J, \text{Int}\,\Delta_\alpha\} - Z_0\{0, \text{Int}\,\Delta_\alpha\}| \leq O(\|J\|_{-1}).$$

Hence for small λ_j, (2.13) holds and the proof of the proposition is complete. In the foregoing, we have for notational convenience suppressed the dependence on the multiparameter s which interpolates between the various measures dq_γ. The factor $m_0^{O(1)}$ from Lemma 2.3 is compensated for by factors $O(|\lambda_j|) \ll 1$.

Proof of Proposition 1.1.1. — We obtain a uniform bound on the finite volume approximations to $Z\{J\}$. The desired result then follows from convergence for $J \in C_0^\infty$ and Proposition A3. We proceed by induction, adding one local piece J_α at each step. Let $\{\zeta_\alpha : \alpha \in Z^2\}$ be a C^∞ partition of unity, invariant under lattice translations, and let $J_\alpha = \zeta_\alpha J$. We will show later that $\Sigma \|J_\alpha\|_{-1}^2 \leq \text{const.} \|J\|_{-1}^2$. Let $B \subset Z^2$ be a finite set, let $\alpha \in Z^2 \sim B$, and define

$$J_B = \sum_{\beta \in B} J_\beta$$

$$J(s) = (1-s)J_B + s(J_\alpha + J_B) = J_B + sJ_\alpha .$$

To state the inductive hypothesis, we define

$$c_{\beta_1\beta_2} = O(1)e^{-(m_0-\varepsilon)|\beta_1-\beta_2|}$$

$$C_{B,\beta} = \sum_{\beta_1 \in B} c_{\beta_1,\beta} \leq C \equiv O(1) \sum_{\beta_1 \in Z^2} c_{\beta_1,\beta} .$$

The inductive hypothesis is

$$|Z\{J_B\}| \leq \exp\left[\sum_{\beta \in B} C_{B,\beta} \|J_\beta\|_{-1}^2\right], \tag{2.14}$$

which yields

$$|Z\{J\}| \leq \exp\left[C\Sigma \|J_\beta\|_{-1}^2\right] \leq \exp(C' \|J\|_{-1}^2).$$

We have, with $\hat{\Phi} = \Phi - \langle \Phi \rangle$,

$$\begin{aligned} Z\{J_B + J_\alpha\} - Z\{J_B\} &= \int_0^1 \frac{d}{ds} Z\{J(s)\}\, ds \\ &= \int_0^1 ds \int [(\Phi(x) - \langle \Phi \rangle) J_\alpha(x) dx] e^{\int \hat{\Phi}(J(s))} dq \\ &= \int_0^1 ds [\langle \Phi(J_\alpha) e^{\hat{\Phi}(J(s))} \rangle - \langle \Phi(J_\alpha) \rangle \langle e^{\hat{\Phi}(J(s))} \rangle]. \end{aligned}$$

Assuming the bound (2.14) on $Z\{J_B\}$,

$$|Z\{J_B + J_\alpha\}| \leq \exp\left[\sum_{\beta \in B} C_{B,\beta} \|J_\beta\|_{-1}^2\right] + \sup_s |\langle \Phi(J_\alpha) e^{\hat{\Phi}(J(s))} \rangle_T|. \tag{2.15}$$

We show that

$$|\langle \quad \rangle_T| \leq |Z\{J_B\}| \left[(C_{B,\alpha} + c_{\alpha\alpha}) \|J_\alpha\|_{-1}^2 + \sum_{\beta \in B} c_{\beta\alpha} \|J_\beta\|_{-1}^2\right]. \tag{2.16}$$

Substituting in (2.15) yields

$$|Z\{J_B + J_\alpha\}| \leq \exp\left[\sum_{\beta \in B \cup \alpha\}} C_{B \cup \alpha, \beta} \|J_\beta\|_{-1}^2\right]$$

and completes the induction.

To bound $\langle \quad \rangle_T$, we choose

$$A = \Phi(J_\alpha), \qquad B = e^{\hat{\Phi}(J(s))}$$

in (2.2). We apply the cluster expansion to $\langle A_0 B_0 \rangle$, and each expansion term is even under the symmetry $\Phi \leftrightarrow \Phi^\sim$, while A_0 and B_0 are individually odd. Using the fact that

$$B_0 = (e^{\hat{\Phi}(J(s))} - I)_0 = (e^{\hat{\Phi}(J(s))} - I) \otimes I - I \otimes (e^{\hat{\Phi}(J(s))} - I),$$

we see that every term in $\langle A_0 B_0 \rangle$ is at least quadratic in J:

$$(\text{linear in } J_\alpha) \times (\text{linear in } J(s)).$$

In the cluster expansion we choose contours γ which contain suppt J_α. Then the region exterior to the contour contributes the factor

$$O(1)|Z\{J_B, \text{Ext}\,\gamma\}| \leq e^{O|\text{Int}\,\gamma| + \|J_B\|_{-1}^2} |Z\{J_B\}|,$$

by Proposition 2.5. The region Int γ contributes $O(1) \exp[-K|\text{Int}\,\gamma|]$ as in [*4*, Proposition 5.2]. We choose K sufficiently large, and the remaining estimates follow as in [*4*].

To complete the proof, we only need to show that

$$\Sigma \|J_\alpha\|_{-1}^2 \leq \text{const.}\, \|J\|_{-1}^2.$$

In fact

$$\Sigma \| J_\alpha \|^2_{-1} = \Sigma \| \zeta_\alpha J \|^2_{-1} = \| J \|^2_{-1} + \Sigma \langle \zeta_\alpha J, [(-\Delta + 1)^{-1}, \zeta_\alpha] J \rangle .$$

The second term above is dominated by

$$\begin{aligned}(\Sigma \| J_\alpha \|^2_{-1})^{1/2} (\Sigma \| (-\Delta + 1)^{-1/2} [-\Delta, \zeta_\alpha] (-\Delta + 1)^{-1} J \|^2_0)^{1/2} \\ \leq (\Sigma \| J_\alpha \|^2_{-1})^{1/2} \| \Sigma - \Delta + 1)^{-1/2} [-\Delta, \zeta_\alpha] (-\Delta + 1)^{-1} \\ [-\Delta, \zeta_\alpha] (-\Delta + 1)^{-1/2} \|^{1/2} \| J \|_{-1} \\ \leq \text{const.} \| J \|_{-1} (\Sigma \| J_\alpha \|^2_{-1})^{1/2} ,\end{aligned}$$

from which the desired bound follows.

APPENDIX

ANALYTIC FUNCTIONS

We consider functions f which map a subset Ω of one Banach space E into a Banach space F. The case $E = C^n$ is the case of vector valued analytic functions. Consider two conditions:

(i) Ω is open in the norm topology.

(ii) Suppose that for each finite sequence $A_j \in E$, $z_j \in C$ such that $z_1 A_1 + \ldots + z_n A_n \in \Omega$, the function $f(z_1 A_1 + \ldots + z_n A_n)$ is analytic in $z = (z_1, \ldots, z_n)$.

DEFINITION A1. — If (i), (ii) are satisfied, the function $f : \Omega \to F$ is *analytic* in Ω. If $A \in \Omega$, we say that $f(\cdot)$ is analytic at A. If $\Omega = E$, then f is *entire*.

DEFINITION A2. — Let $\mathcal{ML}_n(E, F)$ be the space of bounded multilinear forms of n arguments in E, with values in F. Let $\mathcal{ML}_n(E, C) = \mathcal{ML}_n(E)$.

By Vitali's theorem and a 3ε-argument we obtain

PROPOSITION A3. — Let $f_n : \Omega \to F$ be a sequence of functions analytic in Ω, with $\| f_n \|_F$ uniformly bounded on Ω. If $f_n \to f$ pointwise on a dense subset of Ω, then f is analytic in Ω. Also both f_n and the derivatives of $f_n(\Sigma z_i B_i)$ converge pointwise in F norm to f or its derivatives.

PROPOSITION A4. — If $f : \Omega \to F$ is analytic at $A \in \Omega$, then the derivative

$$\frac{d}{dz_1} \cdots \frac{d}{dz_n} f\left(A + \sum_{i=1}^{n} z_i B_i\right)\Bigg|_{z=0} = h_A^{(n)}(B_1, \ldots, B_n)$$

is an element of $\mathcal{ML}_n(E, F)$ with norm less than $\alpha \beta^n n!$, for constants $\alpha = \alpha(A)$ and $\beta = \beta(A)$.

Proof. — Consider $n = 1$. By analyticity

$$\| f(A + z_1(B + C)) - f(A) - z_1 h_A^{(1)}(B + C) \|_F \leq O(|z_1|^2).$$

Also for $n = 2$,

$$\| f(A + z_1 B + z_2 C) - f(A) - z_1 h_A^{(1)}(B) - z_2 h_A^{(1)}(C) \|_F \leq O(|z_1|^2 + |z_1 z_2| + |z_2|^2).$$

Setting $z_1 = z_2$, we have

$$h_A^{(1)}(B + C) = h_A^{(1)}(B) + h_A^{(1)}(C).$$

The multilinearity of $h_A^{(n)}$ for $n > 1$ follows similarly by induction.

We choose $\| B_i \|_E < 1$. Then by the Cauchy integral formula (with integration over a product of circles of radius $O(n^{-1})$),

$$\| h_A^{(n)}(B_1, \ldots, B_n) \|_F \leq \alpha(A)(\beta n)^n .$$

By multilinearity, for arbitrary $B_i \in E$,

$$\| h_A^{(n)}(B_1, \ldots, B_n) \|_F \leq \alpha(A)(\beta n)^n \prod_{i=1}^{n} \| B_i \|_E ,$$

which completes the proof.

PROPOSITION A5. — If f is an entire function, and $\varepsilon > 0$,

$$\| h_{\mathrm{A}}^{(n)}(\mathrm{B}_1, \ldots, \mathrm{B}_n) \|_{\mathrm{F}} \leq \alpha(\mathrm{A})\varepsilon^n n! \prod_{i=1}^{n} \| \mathrm{B}_i \|_{\mathrm{E}} .$$

This proof follows the usual proof for $\mathrm{E} = \mathrm{C}^n$.

REFERENCES

[1] J. DIMOCK, Asymptotic perturbation expansion in the $\mathscr{P}(\Phi)_2$ quantum field theory. *Commun. Math. Phys.*, t. **35**, 1974, p. 347-356.

[2] C. DE DOMINICIS and P. MARTIN, Stationary entropy principle and renormalization in normal and superfluid systems I, II. *Jour. Math. Phys.*, t. **5**, 1964, p. 14-30, p. 31-59.

[3] J. GLIMM, A. JAFFE and T. SPENCER, The Wightman axioms and particle structure in the $\mathscr{P}(\Phi)_2$ quantum field model. *Ann. Math.*, to appear.

[4] J. GLIMM, A. JAFFE and T. SPENCER, The particle structure of the weakly coupled $\mathscr{P}(\Phi)_2$ models and other applications of high temperature expansions. Part II. The cluster expansion. In: *Constructive quantum field theory*, Ed. by G. Velo and A. Wightman, Lecture notes in physics, Vol. 25, Springer Verlag, Berlin, 1973.

[5] J. FELDMAN, *The $\lambda\Phi_3^4$ field theory in a finite volume.*

[6] J. FRÖHLICH, Schwinger functions and their generating functionals. *I. Helv. Phys. Acta.*

[7] G. JONA-LASINIO, Relativistic field theories with symmetry breaking solutions. *Nuovo Cimento* (L), t. **34**, 1964, p. 1790-1795.

[8] J. MOREAU, Sous-différentiabilité. In : *Colloquium on convexity*, ed. by Fenchel. Kobenhavns Universitet Matematisk Institut, Copenhagen, 1967.

[9] K. SYMANZIK, On the many-particle structure of Green's functions in quantum field theory. *J. Math. Phys.*, t. **1**, 1960, p. 249-273.

[10] K. SYMANZIK, Renormalizable models with simple symmetry breaking. I. Symmetry breaking by a source term. *Commun. Math. Phys.*, t. **16**, 1970, p. 48-80.

[11] A. N. VASILEV and A. K. KAZANSKII, Legendre transforms of the generating functionals in quantum field theory. *Teoret. i Mat. Fizika*, t. **12**, 1972, p. 352-369.

(Manuscrit reçu le 6 février 1974)

Three-particle structure of ϕ^4 interactions and the scaling limit

James Glimm*
Rockefeller University, New York, New York 10021

Arthur Jaffe†
Harvard University, Cambridge, Massachusetts 02138
(Received 13 January 1975)

We show that the inverse propagator $\Gamma(x)$ and the proper self-energy part $\Pi(x)$ are positive for all $x \neq 0$. These results combined with some postulates on the sign or the decay of the vertex functions, in particular, on the decay rate for $\Gamma(x)$, yield the absence of three-particle bound states and the existence of the scaling (Gell-Mann, Low, Wilson) limit for the ϕ^4 interaction.

I. INTRODUCTION

Weak interactions or repulsive interactions are often more tractable than general nonlinear interactions. In the study of particles in quantum field models, the most detailed information is known in these two cases. For weak interactions, the cluster expansions provide extensive information.[1,2] For even ϕ^4 interactions, i.e., ϕ^4 interactions without symmetry breaking, correlation inequalities exclude the even-bound-state spectrum;[3] for earlier partial results see Refs. 2 and 4. In both methods, the proofs rely on decay estimates for the Schwinger functions or Green's functions to establish spectral properties of the mass operator $M = (H^2 - P^2)^{1/2}$.

In this paper we call attention to the importance of correlation inequalities or decay estimates on the inverse propagator and the Euclidean vertex functions $\Gamma^{(n)}(x_1, \ldots, x_n)$ (one-particle irreducible Euclidean Green's functions). On the one hand, assuming certain decay estimates, we prove in Secs. V A and V B the absence of odd (three-particle) bound states. By "absence of odd bound states" we just mean the absence of spectrum for the mass operator $M \restriction \mathcal{H}_0$ in the spectral interval $(m, 3m)$. Here m is the physical mass and $\restriction \mathcal{H}_0$ denotes the restriction of M to the subspace $\mathcal{H}_0$ of vectors which are odd under the $\phi \to -\phi$ symmetry. In addition, we find that $\Gamma^{(6)} \leq 0$, or a decay estimate for $\Gamma^{(6)}$, yields a $3m$ exponential decay rate for the inverse propagator $\Gamma(x)$,

$$\Gamma(x) \leq \text{const} \times e^{-3m|x|} .$$

Furthermore, information on the spectral threshold in the inverse propagator, e.g.,

$$\Gamma(x) \leq \text{const} \times e^{-(m+\epsilon)|x|} ,$$

is sufficient to prove the existence of the scaling limit of Gell-Mann, Low, and Wilson. In this limit for $\lambda\phi_2^4$, λ (the bare coupling) $\to \infty$ while m (the physical mass) is held fixed. See Sec. V C. According to renormalization-group ideas, the control of this limit in two dimensions would be comparable (as far as renormalization difficulties go) to an existence proof for ϕ_4^4.

The remainder of this paper is concerned with the reduction of the problem to the decay estimates which we postulate in Sec. V. In Sec. II, we relate the propagator G and inverse propagator Γ to standard representations for Herglotz functions, and we show that the kernel of Γ (and of the proper self-energy part Π) satisfies

$$\Pi(x) = \Gamma(x) \geq 0, \quad x \neq 0 . \tag{1.1}$$

In Sec. III, we derive a functional-integration by parts formula,

$$\langle \phi(x)(1 - P_0)A \rangle = \int G(x-y)\left[\left\langle \frac{\delta A}{\delta\phi(y)} \right\rangle - \langle \mathcal{V}'(y)(1 - P_0 - P_1)A \rangle \right] dy . \tag{1.2}$$

Here $\mathcal{V}$ is the interaction polynomial, G is the propagator, and P_0, P_1 are the projections onto the Euclidean vacuum Ω and the states $(1 - P_0)\phi\Omega$, respectively. (See Sec. III for details.) In Sec. IV, we apply these ideas to derive the representation

$$\Gamma(x) = \langle\!\langle \mathcal{V}'(x)(1 - P_0 - P_1)\mathcal{V}'(0) \rangle\!\rangle , \quad |x| \neq 0 . \tag{1.3}$$

The bounds of Sec. V use this representation.

II. INTEGRAL REPRESENTATIONS FOR (THE PROPAGATOR AND) THE INVERSE PROPAGATOR

A. The propagator as a Herglotz function

Let $\phi(x)$ be a real, scalar field satisfying the Osterwalder-Schrader axioms.[5] [An Osterwalder-Schrader theory is the analytic continutation to imaginary time of a relativistic, Wightman theory. The Osterwalder-Schrader axioms give conditions on the imaginary-time (Euclidean) Schwinger functions which ensure the existence of an analytic continuation to a real-time theory satisfying the Wightman axioms. See Refs. 5 and 10 for details.] The two-point Green's function (propagator) is defined by

$$G(x-y)=\langle\phi(x)\phi(y)\rangle-\langle\phi(x)\rangle^2 . \tag{2.1}$$

By analytic continuation of the (real time) vacuum expectation value, it follows that for $x\neq 0$

$$G(x)=\int_{m^2}^{\infty} d\rho(a)C_a(x) , \tag{2.2}$$

where $C_a=(-\Delta+a)^{-1}$ and where $d\rho(a)$ is a positive, tempered measure.

In this section we assume an additional bound

$$\int_{m^2}^{\infty}\frac{d\rho(a)}{1+a^2}<\infty , \tag{2.3}$$

which does not follow from the axioms. The bound (2.3) ensures that the representation (2.2) determines a canonical extension of $G(x)$ to a tempered distribution. The assumption (2.3) is reasonable, since for all renormalizable interactions, perturbation theory suggests $\int(1+a)^{-1}d\rho<\infty$, which ensures (2.3). Furthermore, in canonical theories, $\int d\rho=1$. In fact, (2.3) is equivalent to the assumption that $G(x)$ has a Fourier transform $\tilde{G}(p)$ which as a function of $z=-p^2$ is a Herglotz function of class $P(-\infty,m^2)$, in case $m>0$.

By definition, a Herglotz function is an analytic function defined on the upper half plane and having positive imaginary part. A Herglotz function of class $P(a,b)$ is a Herglotz function which can be continued analytically into the lower half plane across the real interval (a,b) by the formula $f(z^*)=f^*(z)$. A Herglotz function $f(z)$ of class $P(-\infty,m^2)$ has a unique, canonical representation

$$f(-z)=-\alpha z+\beta+\int_{m^2}^{\infty}\left(\frac{1}{a+z}-\frac{a}{a^2+1}\right)d\mu(a) . \tag{2.4}$$

where $\alpha\geq 0$, β is real, and $(1+a^2)^{-1}d\mu(a)$ is a positive, finite measure supported on the interval $[m^2,\infty)$. The relation (2.4) is sometimes called a once-subtracted dispersion relation. An analysis of $\tilde{G}(p)$ in terms of representations such as (2.4) has also been given by Symanzik.[6] See Donoghue, Ref. 7, pp. 18–26, for general properties of Herglotz functions, including a proof of (2.4).

Proposition 2.1. Let $F(x)$ be a Euclidean invariant distribution with Fourier-Laplace transform $\tilde{F}(p)$ which as a function of $z=-p^2$ belongs to $P(-\infty,m^2)$. Then

$$F(x)=\int_{m^2}^{\infty} C_a(x)d\mu(a)>0 \text{ for } x\neq 0 , \tag{2.5}$$

where $(1+a^2)^{-1}d\mu(a)$ is a positive, finite measure. Furthermore, $\tilde{F}(p)=f(-p^2)$, where $f(-p^2)$ satisfies (2.4).

Proof. By general properties of Herglotz functions, $\tilde{F}(p)=f(-p^2)$, as described above. Taking the Fourier transform of (2.4) yields

$$F(x)=(\alpha\Delta+\beta)\delta(x)+\int_{m^2}^{\infty}C_a(x)\frac{d\mu(a)}{1+a^2}+\Delta\int_{m^2}^{\infty}C_a(x)\frac{a\,d\mu(a)}{1+a^2} , \tag{2.6}$$

so only the last two terms contribute for $x\neq 0$. For $a>0$ and $x\neq 0$, $C_a(x)$ is positive, real analytic, and exponentially decreasing in $\sqrt{a}|x|$, and Δ may be taken under the integral sign. Since $\Delta C_a(x)=aC_a(x)$ for $x\neq 0$, we have (2.5) for $x\neq 0$, assuming $m>0$. If $m=0$, we have

$$F_\epsilon(x)=(\alpha\Delta+\beta)\delta(x)+\int_{\epsilon^2}^{\infty}[C_a(x)+\Delta C_a(x)a]\frac{d\mu(a)}{1+a^2}=\int_{\epsilon^2}^{\infty}C_a(x)d\mu(a),\quad x\neq 0 , \tag{2.7}$$

and $0<F_\epsilon(x)\nearrow F(x)$ in $\mathcal{S}'$ as $\epsilon\searrow 0$ for $x\neq 0$. By positivity $F_\epsilon(x)\to F(x)$ pointwise and (2.5) is valid. We also obtain a regularity condition on $d\mu(a)$ at $a=0$ in $d=2$ dimensions. For $d>2$, we modify the reasoning to allow for a δ function in $d\mu$ at $a=0$.

B. Positivity of $\Gamma(x)$, the inverse propagator

Assuming (2.2) and (2.3), $G(x)$ is a tempered distribution. Thus, $f\mapsto G*f$ defines a linear operator on L_2 with domain $\mathcal{S}$. Define the inverse propagator Γ by

$$\Gamma\equiv -G^{-1} ,$$

which acts as convolution by $\Gamma(x)$. By definition, $\tilde{\Gamma}(p)=-1/\tilde{G}(p)$, and by positivity of the Euclidean Hilbert-space inner product, $\tilde{G}(p)>0$. Thus, $\tilde{\Gamma}(p)<0$ and Γ is a negative operator.

Theorem 2.2 Assume (2.2) and (2.3). Then $\Gamma(x)$ satisfies the conditions of Proposition 2.1. In particular,

$$\Gamma(x)=\int_{\overline{m}^2}^{\infty} d\nu(a)C_a(x),\quad x\neq 0, \tag{2.8}$$

where $(1+a^2)^{-1}d\nu(a)$ is a positive, finite measure. If in addition $\int(1+a)^{-1}d\rho<\infty$, then $\overline{m}\geq m$; thus for $x\neq 0$, $\Gamma(x)$ is real analytic, positive, monotonic decreasing in $|x|$, and has an exponential decay rate $\overline{m}\geq m$.

Remark. If the spectral measure $d\rho(a)$ for $\tilde{G}(p)$ has no mass in the (nonempty) interval (m^2,M^2), then $m<\overline{m}\leq M$. The Castillejo-Dalitz-Dyson (CDD) radius is the radius of analyticity of the inverse propagator. It equals $\overline{m}$, and $\overline{m}<M$ is the case of CDD poles in $\tilde{\Gamma}(p)$. See Ref. 8,

for example.

Proof. The function $f(z)\neq 0$ is a Herglotz function if and only if $-1/f(z)$ is a Herglotz function. If $f(z)\in P(-\infty, m^2)$ and $\int(1+a)^{-1}d\mu(a)<\infty$, then assuming (2.2),

$$f(-z)=\int_{m^2}^{\infty}\frac{1}{a+z}\,d\mu(a)\,.$$

Then $-1/f(z)\in P(-\infty, m^2)$.

Theorem 2.3. Assume (2.2) and $\int d\rho(a)=1$. Then

$$\tilde{\Gamma}(p)=-(p^2+\beta)+\int_{\overline{m}^2}^{\infty}\left(\frac{1}{p^2+a}-\frac{a}{a^2+1}\right)d\nu(a)\,, \tag{2.9}$$

where $(1+a^2)^{-1}d\nu(a)$ is a positive, finite measure and $\overline{m}\geq m$.

Proof. Since $\int d\rho(a)=1$, it follows that $p^2\tilde{G}(p)\to 1$ as $p^2\to\infty$, and hence $p^{-2}\tilde{\Gamma}(p)\to -1$. By (2.4),

$$\tilde{\Gamma}(p)=-(\alpha p^2+\beta)+\int_{\overline{m}^2}^{\infty}\left(\frac{1}{p^2+a}-\frac{a}{a^2+1}\right)d\nu(a)\,,$$

with $(1+a^2)^{-1}d\nu(a)$ a positive, finite measure. By the dominated convergence theorem, as $p^2\to\infty$,

$$p^{-2}\int\left(\frac{1}{p^2+a}-\frac{a}{a^2+1}\right)d\nu(a)$$

$$=p^{-2}\int\frac{(1-p^2a)}{(p^2+a)(a^2+1)}\,d\nu(a)\to 0\,.$$

Thus $p^{-2}\tilde{\Gamma}(p)\to-\alpha$ and $\alpha=1$. As above, $\overline{m}\geq m$.

We now assume a further regularity requirement on $d\rho(a)$, namely the existence of the first moment,

$$\int a\,d\rho(a)<\infty\,. \tag{2.10}$$

The bound (2.10) is known to hold for $\mathcal{P}(\phi)_2$ models. [This is, however, the only simple example with finite mass renormalization, and hence the only example where we expect (2.10).] In the case of (2.10), we may assume the canonical normalization of ϕ,

$$\int d\rho(a)=1\,. \tag{2.11}$$

Theorem 2.4. Assume (2.2), (2.10), and (2.11). Then

$$\tilde{\Gamma}(p)=-\left(p^2+\int a\,d\rho\right)+\int_{\overline{m}^2}^{\infty}\frac{d\nu(a)}{p^2+a}\,, \tag{2.12}$$

where $(1+a)^{-1}d\nu(a)$ is a positive, finite measure and $\overline{m}\geq m$.

Proof. We have

$$\begin{aligned}\tilde{\Gamma}(p)&=-(p^2+m^2)\left[\int\left(\frac{p^2+m^2}{p^2+a}\right)d\rho(a)\right]^{-1}\\&=-(p^2+m^2)\left[1-\int\left(\frac{a-m^2}{p^2+a}\right)d\rho(a)\right]^{-1}\\&=-p^2-m^2-\int(a-m^2)d\rho+o(1)\\&=-p^2-\int a\,d\rho+o(1).\end{aligned}$$

For a positive measure $d\mu(a)$ with $\int(1+a^2)^{-1}d\mu(a)<\infty$,

$$\int_{\overline{m}^2}^{\infty}\left(\frac{1}{p^2+a}-\frac{a}{a^2+1}\right)d\mu(a)=\text{const}+o(1)$$

if and only if $\int(1+a)^{-1}d\mu(a)<\infty$. Combining these two facts with (2.9) yields the theorem.

Remark. Assuming (2.2) and (2.10), but not (2.11),

$$\tilde{\Gamma}(p)=-\left(\int d\rho\right)^{-1}\left(p^2+\frac{\int a\,d\rho}{\int d\rho}\right)+\int_{\overline{m}^2}^{\infty}\frac{d\nu(a)}{p^2+a}\,. \tag{2.13}$$

C. Field-Strength renormalization

In this section we assume the existence of isolated one-particle spectrum in the propagator. We assume

$$\tilde{G}(p)=\int\frac{d\rho(a)}{p^2+a}=\frac{Z}{p^2+m^2}+\int_{M^2}^{\infty}\frac{d\sigma(a)}{p^2+a}\,, \tag{2.14}$$

where $m>0$ and where $d\sigma(a)$ is supported in $[M,\infty)$, with $m<M$. We also assume $\int d\rho<\infty$, so it is no loss of generality to assume a canonical theory, $\int d\rho=1$. A theory satisfying these assumptions is called a canonical, isolated-particle (CIP) theory.

Theorem 2.4. For a CIP theory,

$$Z^{-1}=1+\int_{\overline{m}^2}^{\infty}\frac{d\nu(a)}{(a-m^2)^2}\,. \tag{2.15}$$

Proof. Since $\tilde{\Gamma}(p)$ is analytic as a function of p^2 for $|p^2|<\overline{m}^2$, and since $m<\overline{m}$, it follows that $\tilde{\Gamma}(p)$ is analytic at $p^2=-m^2$, the pole of $\tilde{G}(p)$. Thus,

$$Z^{-1}=-\left(\frac{d\tilde{\Gamma}}{dp^2}\right)_{p^2=-m^2}\,,$$

and differentiating (2.9) yields (2.15).

Definition 2.5. The renormalized propagator $\tilde{G}_{\text{ren}}(p)$ for a CIP theory is

$$\tilde{G}_{\text{ren}}(p)=Z^{-1}\tilde{G}(p)=\int\frac{d\rho_{\text{ren}}(a)}{p^2+a}\,, \tag{2.16}$$

where

$$d\rho_{\rm ren}(a)=Z^{-1}d\rho(a)$$
$$=\left[1+\int_{\overline{m}^2}^{\infty}\frac{d\nu(b)}{(b-m^2)^2}\right]d\rho(a)\,. \qquad (2.17)$$

Theorem 2.6. For a CIP theory,

$$\frac{1}{p^2+m^2}\leq \tilde G_{\rm ren}(p)\leq\left(\frac{\overline{m}^2}{\overline{m}^2-m^2}\right)^2 m^{-2}\,, \qquad (2.18)$$

when p is real.

Comment. For $\overline{m}/m-1>\epsilon$, $m>\epsilon$, the bound (2.18) on $\tilde G_{\rm ren}(p)$ is independent of all parameters. Because $M\geq\overline{m}$, the bound $\overline{m}/m\geq 1+\epsilon$ also excludes level crossings in the state for the particle of lowest mass. For a canonical $\mathcal{P}(\phi)$ model without bound states or CDD zeros in $\tilde G(p)$, we have $\overline{m}/m\geq 2$. As a converse to Theorem 2.6 we have Theorem 2.7.

Theorem 2.7. For a CIP theory, $\overline{m}/m$ is bounded away from 1 if $M^2\tilde G_{\rm ren}(p=0)$ is bounded and if M/m is bounded away from one.

Proof of Theorem 2.6. The lower bound in (2.18) follows by definition, so we need only prove the upper bound. To begin, we note that

$$\int_{\overline{m}^2}^{\infty}\frac{d\nu(a)}{(a-m^2)^2}\leq\left(\frac{\overline{m}^2}{\overline{m}^2-m^2}\right)^2\int_{\overline{m}^2}^{\infty}\frac{d\nu(a)}{a^2}\,, \qquad (2.19)$$

since $a(a-m^2)^{-1}$ is decreasing in a for $a>m^2$.

We have two expressions for $d\tilde\Gamma(p)/dp^2$: One is obtained from (2.9), and the other is obtained from (2.14) and the relation $\tilde\Gamma'=\tilde G'G^{-2}$. Equating these we obtain

$$1+\int_{\overline{m}^2}^{\infty}\frac{d\nu(a)}{(p^2+a)^2}=\int_{m^2}^{\infty}\frac{d\rho(a)}{(p^2+a)^2}\left(\int_{m^2}^{\infty}\frac{d\rho(a)}{p^2+a}\right)^{-2}. \qquad (2.20)$$

Using (2.20) with $p^2=0$, we infer from (2.17)–(2.19)

$$\tilde G_{\rm ren}(p)\leq\left(\frac{\overline{m}^2}{\overline{m}^2-m^2}\right)^2\left(1+\int_{\overline{m}^2}^{\infty}\frac{d\nu(a)}{a^2}\right)\int_{m^2}^{\infty}\frac{d\rho(a)}{p^2+a}$$
$$\leq\left(\frac{\overline{m}^2}{\overline{m}^2-m^2}\right)^2\int_{m^2}^{\infty}\frac{d\rho(a)}{a^2}\left(\int_{m^2}^{\infty}\frac{d\rho(a)}{a}\right)^{-1}$$
$$\leq\left(\frac{\overline{m}^2}{\overline{m}^2-m^2}\right)^2 m^{-2}\,,$$

completing the proof.

Proof of Theorem 2.7. If $\overline{m}<M$, we refer to the definition [Ref. 8, formula (6.5)],

$$(\overline{m}/m)^2-1=\left[m^2\int_{M^2}^{\infty}\frac{d\rho_{\rm ren}(a)}{a-\overline{m}^2}\right]^{-1}$$
$$=[(M/m)^2-1]\left[(M^2-m^2)\int_{M^2}^{\infty}\frac{d\rho_{\rm ren}(a)}{a-\overline{m}^2}\right]^{-1}. \qquad (2.21)$$

Also, using (2.21) and the monotonicity of $a(a-\overline{m}^2)^{-1}$,

$$(M^2-m^2)\int_{M^2}^{\infty}\frac{d\rho_{\rm ren}(a)}{a-\overline{m}^2}=(M^2-\overline{m}^2)\int_{M^2}^{\infty}\frac{d\rho_{\rm ren}(a)}{a-\overline{m}^2}+1$$
$$\leq M^2\int_{M^2}^{\infty}\frac{d\rho_{\rm ren}(a)}{a}+1$$
$$\leq M^2\tilde G_{\rm ren}(0)+1\,.$$

So by (2.21),

$$(\overline{m}/m)^2-1\geq[(M/m)^2-1][M^2\tilde G_{\rm ren}(0)+1]^{-1}\,,$$

completing the proof.

III. INTEGRATION BY PARTS IN THE PHYSICAL MEASURE

We consider integration by parts in Euclidean $\mathcal{P}(\phi)$ boson models. The Euclidean Hilbert space is spanned by states $\phi(x_1)\cdots\phi(x_n)\Omega$, $n=0,1,\ldots$, where Ω is the Euclidean vacuum. See Refs. 5 and 10 for more details. Function-space integration by parts yields the formula

$$\langle\phi(x)(1-P_0)A\rangle$$
$$=\int C(x-y)\left[\left\langle\frac{\delta A}{\delta\phi(y)}\right\rangle-\langle\mathcal{V}'(y)(1-P_0)A\rangle\right]dy, \qquad (3.1)$$

where P_0 is the projection onto Ω and

$$C=(-\Delta+m_0^2)^{-1},\quad \mathcal{V}'(y)=:\mathcal{P}'(\phi(y)):\,.$$

In following the point of view that we should express our formulas in terms of physical (renormalized) parameters, we wish to replace C in (3.1) by the Green's function G. Let P_1 denote the projection onto the span of $(1-P_0)\phi\Omega$. Explicitly,

$$P_1\theta=-\int(1-P_0)\phi(x)\Omega\Gamma(x-y)\langle\phi(y)\Omega,1-P_0)\theta\rangle\,dx\,dy\,.$$

Theorem 3.1. In each pure phase of a noncritical $\mathcal{P}(\phi)_2$ model,

$$\langle\phi(x)(1-P_0)A\rangle$$
$$=\int G(x-y)\left[\left\langle\frac{\delta A}{\delta\phi(y)}\right\rangle-\langle\mathcal{V}'(y)(1-P_0-P_1)A\rangle\right]dy\,. \qquad (3.2)$$

Remark. We interpret subtraction of the one-particle contribution $\langle\mathcal{V}'(y)P_1A\rangle$ in (3.2) as providing the mass renormalization of (3.1). In this way we replace C (characterized by the bare mass m_0) by G (characterized by the physical mass m).

Proof. Using (3.1) with $A=\phi(y)$ gives

$$C(x-y)=G(x-y)+\langle\phi(x)(1-P_0)(C\mathcal{V}')(y)\rangle\,. \qquad (3.3)$$

The validity of (3.1) in each pure phase is proved in Ref. 8. Substituting (3.3) into (3.1) gives

$$\langle\phi(x)(1-P_0)A\rangle=\int G(x-y)\Big[\Big\langle\frac{\delta A}{\delta\phi(y)}\Big\rangle-\langle\mathcal{V}'(y)(1-P_0)A\rangle\Big]\,dy$$

$$+\int\langle\phi(x)(1-P_0)(C\mathcal{V}')(y)\rangle\Big[\Big\langle\frac{\delta A}{\delta\phi(y)}\Big\rangle-\langle\mathcal{V}'(y)(1-P_0)A\rangle\Big]\,dy$$

$$=\int G(x-y)\Big[\Big\langle\frac{\delta A}{\delta\phi(y)}\Big\rangle-\langle\mathcal{V}'(y)(1-P_0)A\rangle\Big]\,dy+\int\langle\phi(x)(1-P_0)\mathcal{V}'(y)\rangle\langle\phi(y)(1-P_0)A\rangle\,dy\,. \tag{3.4}$$

By Euclidean invariance, $\langle\phi(x)\mathcal{V}'(y)\rangle=\langle\mathcal{V}'(x)\phi(y)\rangle$, and by translation invariance the last term in (3.4) is

$$-\int G(x-y)\langle\mathcal{V}'(y)(1-P_0)\phi(z)\rangle\Gamma(z-t)\langle\phi(t)(1-P_0)A\rangle\,dy\,dz\,dt=\int G(x-y)\langle\mathcal{V}'(y)P_1A\rangle\,dy\,,$$

from which (3.2) follows.

Corollary 3.2. In each pure phase of a noncritical $\mathcal{P}(\phi)_2$ model,

$$\Gamma_x\langle\phi(x)(1-P_0)A\rangle=\int\Gamma(x-y)\langle\phi(y)(1-P_0)A\rangle\,dy$$
$$=-\Big\langle\frac{\delta A}{\delta\phi(x)}\Big\rangle+\langle\mathcal{V}'(x)(1-P_0-P_1)A\rangle\,. \tag{3.5}$$

Remark. In the case of an even theory, P_0 does not occur in the formulas above.

IV. THE PROPER SELF-ENERGY PART

The proper self-energy operator Π is defined by the equation

$$\Pi=C^{-1}+\Gamma\,, \tag{4.1}$$

where $C=(-\Delta+m_0^2)^{-1}$. Let $\Pi(x-y)$ be the kernel of Π. Since C^{-1} has the local kernel $(-\Delta+m_0^2)\times\delta(x-y)$, by Theorem 2.2 we have Theorem 4.1.

Theorem 4.1. Assuming (2.2)–(2.3),

$$\Pi(x)=\Gamma(x)\geq 0,\quad\text{for } x\neq 0\,. \tag{4.2}$$

Using the integration by parts formula of Sec. III, we obtain a representation for $\Pi(x)$.

Theorem 4.2. For a noncritical $\mathcal{P}(\phi)_2$ model,

$$\Pi(x-y)=-\langle\mathcal{V}''(x)\rangle\delta(x-y)+\langle\mathcal{V}'(x)(1-P_0-P_1)\mathcal{V}'(y)\rangle\,. \tag{4.3}$$

Proof. Applying the operators Γ and C^{-1} to (3.3) and using (4.1) gives

$$\Gamma(x-y)+C^{-1}(x-y)=\int\Gamma(x-z)\langle\phi(z)(1-P_0)\mathcal{V}'(y)\rangle\,dz$$
$$=\Pi(x-y)\,. \tag{4.4}$$

Thus by applying (3.5) to (4.4) we obtain (4.3).

Theorem 4.3. For a noncritical $\mathcal{P}(\phi)_2$ model with the canonical normalization (2.10),

$$m^2=m_0^2+\langle\mathcal{V}''(x)\rangle-\int_{m^2}^{\infty}(a-m^2)\,d\rho(a)\,. \tag{4.5}$$

Remark. If $\mathcal{V}$ is Wick ordered in the physical vacuum, then $\langle\mathcal{V}''(x)\rangle=0$ and

$$m^2-m_0^2=-\int_{m^2}^{\infty}(a-m^2)\,d\rho(a)\leq 0\,. \tag{4.6}$$

This is the old result of Lehmann that $m\leq m_0$ (see Ref. 9).

Proof. Let $h(p)$ be the Fourier transform of $\langle\mathcal{V}'(x)(1-P_0-P_1)\mathcal{V}'(y)\rangle$. We claim $|h(p)|\leq O(|p|^{-2(1-\epsilon)})$ as $|p|\to\infty$. This is the case for the Fourier transform of $\langle\mathcal{V}'(x)(1-P_0)\mathcal{V}'(y)\rangle$, by the local regularity estimates of Ref. 8. Let $g(p)$ be the Fourier transform of $\langle\mathcal{V}'(x)(1-P_0)\phi(y)\rangle$. Then the Fourier transform of $-\langle\mathcal{V}'(x)P_1\mathcal{V}'(y)\rangle$ is $g^2\tilde{\Gamma}$. For $\mathcal{P}(\phi)_2$, $|g(p)|\leq O(|p|^{-2+\epsilon})$ as $|p|\to\infty$, by a similar argument. Since $|\tilde{G}(p)|\geq O(|p|^{-2})$ by (2.2), $|g^2\tilde{\Gamma}|\leq O(|p|^{-2+2\epsilon})$, showing $|h|\leq O(|p|^{-2(1-\epsilon)})$.

Taking the Fourier transform of (4.3),

$$\tilde{\Gamma}(p)=-(p^2+m_0^2)+\tilde{\Pi}(p)$$
$$=-[p^2+m_0^2+\langle\mathcal{V}''(x)\rangle]+O(|p|^{-2(1-\epsilon)})\,.$$

Comparing with (2.12) yields

$$\int a\,d\rho(a)=m_0^2+\langle\mathcal{V}''(x)\rangle\,, \tag{4.7}$$

to complete the proof of (4.5).

Corollary 4.4. For a $\mathcal{P}(\phi)_2$ model,

$$\Gamma(x)+C_\alpha^{-1}(x)=\langle\mathcal{V}'(x)(1-P_0-P_1)\mathcal{V}'(0)\rangle\,, \tag{4.8}$$

where $\alpha=\int a\,d\rho$ and $\tilde{C}_\alpha(p)=(p^2+\alpha)^{-1}$. The Fourier transform of (4.8) is $O(|p|^{-2+\epsilon})$ as $|p|\to\infty$, and (4.8) is positive for all x.

V. ON THE ABSENCE OF THREE-PARTICLE BOUND STATES AND THE SCALING LIMIT FOR ϕ_2^4

A. $\Gamma^{(6)}$ and three-particle bound states

In an even $\mathcal{P}(\phi)$ model, the Hilbert space $\mathcal{H}$ decomposes into even and odd subspaces $\mathcal{H}_e$, $\mathcal{H}_o$ which are invariant under the symmetry $\phi\to-\phi$,

$$\mathcal{H}=\mathcal{H}_e\oplus\mathcal{H}_o\,. \tag{5.1}$$

The physical mass m is

$$m = \inf \sigma(M \upharpoonright \mathcal{H}_o), \tag{5.2}$$

where σ denotes the spectrum. The eigenfunctions of the mass operator M are the vacuum, the one-particle states, and bound states. We expect even-bound states only in the spectral interval $(m, 2m)$ and odd-bound states only in the spectral interval $(m, 3m)$.

In the even ϕ^4 model,

$$(0, 2m) \cap \sigma(M \upharpoonright \mathcal{H}_e) = \emptyset, \tag{5.3}$$

i.e., two-particle (even) bound states do not occur. This result was established by Spencer[3]; it extends the methods of Feldman[4] and weak-coupling results.[2] We consider here the odd subspace $\mathcal{H}_o$ in the even ϕ^4 model. We investigate the consequences of certain (as yet unproved) inequalities for vertex functions.

Let Γ be the inverse propagator, and let $\Gamma^{(n)}$ denote the *unamputated*, one-particle irreducible part of the connected n-point Green's function $G^{(n)}$. For instance,

$$\Gamma^{(4)}(x_1, \ldots, x_4) = G^{(4)}(x_1, \ldots, x_4)$$

and

$$\Gamma^{(6)}(x_1, \ldots, x_6) = G^{(6)}(x_1, \ldots, x_6) + \sum \int G^{(4)}(x_{i_1}, x_{i_2}, x_{i_3}, z)\Gamma(z - z') \times G^{(4)}(z', x_{i_4}, x_{i_5}, x_{i_6}) dz\, dz', \tag{5.4}$$

where the sum ranges over the ten choices of two three-element subsets $\{i_1, i_2, i_3\}, \{i_4, i_5, i_6\}$ of $\{1, \ldots, 6\}$.

For a ϕ^4 interaction, the lowest nonzero order of perturbation theory predicts that (at least for weak coupling)

$$\Gamma^{(4n+2)}(x_1, \ldots, x_{4n+2}) \le 0, \quad n = 1, 2, \ldots. \tag{5.5}$$

Here we assume a weaker form of (5.5) in the case $n = 1$, namely

$$\Gamma^{(6)}(x_1, x_2, x_3, y_1, y_2, y_3) \le O(1)e^{-3md}, \tag{5.6}$$

where

$$d = \text{dist}(\{x_1, x_2, x_3\}, \{y_1, y_2, y_3\}).$$

Let $\mathcal{E}_n^+$ be the subspace of the Euclidean Hilbert space $\mathcal{E}$ which is spanned by vectors of the form $\phi(f_1) \cdots \phi(f_n)\Omega$, with the f_i's supported at positive times. Let E be the canonical projection of $\mathcal{E}^+ = \{\text{span}\mathcal{E}_n^+, n = 0, 1, \cdots\}$ onto $\mathcal{H}$, and let $E\mathcal{E}_n^+ = \mathcal{H}_n$. See Ref. 5.

Theorem 5.1. Assume (5.6) in an even ϕ_2^4 model. Then

$$(m, 3m) \cap \sigma(M \upharpoonright \mathcal{H}_1) = \emptyset, \tag{5.7}$$

$$(m, 3m) \cap \sigma(M \upharpoonright \mathcal{H}_3) = \emptyset. \tag{5.8}$$

Here σ denotes the spectrum (of the mass operator).

Remark. We can interpret these statements as the absence of bound states in the two-point function (5.7) or the absence of bound states in the six-point function (5.8).

Theorem 5.2. Assume (5.6) in an even ϕ_2^4 model. Then $M \ge 3m$ where M is defined in (2.14) and

$$0 \le \Gamma(x) \le O(1)e^{-3m|x|}, \quad |x| \to \infty. \tag{5.9}$$

Thus $\tilde{G}(p)$ has no CDD zeros for $|p| < 3m$, and $\bar{m} \ge 3m$.

Theorem 5.3. Assume (5.6) in a weak-coupling $\lambda\phi_2^4$ model. Then for λ sufficiently small,

$$(m, 3m) \cap \sigma(M \upharpoonright \mathcal{H}_o) = \emptyset.$$

Remark. Theorem 5.3 is a corollary of Theorem 5.1 and the n-particle cluster expansion.[1] In particular, the states $\mathcal{H}_1 \oplus \mathcal{H}_3$ span $\mathcal{H}_o$ up to energy $4m - \epsilon$ for $\lambda \le \lambda(\epsilon)$. By the Haag-Ruelle scattering theory, we then have $\sigma(M \upharpoonright \mathcal{H}_o) = \{m\} \cup [3m, \infty)$.

Let $P_1^{\mathcal{H}}$ denote the projection of $\mathcal{H}$ onto the one-particle eigenspace of the mass operator M. We require a relation between $P_1^{\mathcal{H}}$ and the Euclidean projection P_1 of $\mathcal{E}$ onto $\mathcal{E}_1 = \text{span}(1 - P_0)\phi\Omega$. Let T_t denote Euclidean time translation and let Θ denote Euclidean time inversion. We show that on $\mathcal{E}^+ \times \mathcal{E}^+$ the operator $A_t \equiv \Theta(E^* P_1^{\mathcal{H}} E - P_1)T_t$ has an exponential decay rate $\bar{m}$, the decay rate of $\Gamma(x)$, i.e., the CDD radius.

Theorem 5.4. Consider a CIP theory. Let $\chi \in \mathcal{E}^+$ be supported in the Euclidean time interval $[0, T]$. Assume $E\chi \in C^\infty(H)$ and for $t \ge 0$ let

$$F(t) \equiv \langle \chi, A_t \chi \rangle_{\mathcal{E}} = \langle E\chi, P_1^{\mathcal{H}} e^{-tH} E\chi \rangle_{\mathcal{H}} - \langle \Theta\chi, P_1 T_t \chi \rangle_{\mathcal{E}}.$$

Then there exists $r < \infty$ such that for any $a < \bar{m}$,

$$\left(1 - \frac{d^2}{dt^2}\right)^{-r} e^{at}\theta(t)F(t) \in L_2, \tag{5.10}$$

where $\theta(t)$ is the Heaviside step function.

Proof of Theorem 5.2. We use Theorem 4.2 and Eq. (4.2). Thus for $x \ne y$,

$$\begin{aligned} 0 \le \Gamma(x - y) &= 16\lambda^2 \langle\!\langle :xxx: (1 - P_1) : yyy : \rangle\!\rangle \\ &= 16\lambda^2 \lim_{\substack{x_i \to x \\ y_i \to y}} \langle x_1 x_2 x_3 (1 - P_1) y_1 y_2 y_3 \rangle. \end{aligned} \tag{5.11}$$

Here $\langle :xy:z\rangle$ denotes $\langle\!\langle : \phi(x)\phi(y) : \phi(z) \rangle\!\rangle$, etc., and the Wick ordering is omitted since

$$(1 - P_1) : \phi(y_1)\phi(y_2)\phi(y_3) : \Omega = (1 - P_1)\phi(y_1)\phi(y_2)\phi(y_3)\Omega.$$

We expand (5.11) in terms of connected Green's functions, and we expand the term $G^{(6)}$ using (5.4). Thus for $x-y\neq 0$,

$$\begin{aligned}0 \leq (4\lambda)^{-2}\Gamma(x-y) &= 6G(x-y)^3+9G(x-y)G^{(4)}(xxyy)\\ &\quad -9\int G^{(4)}(xxyz)\Gamma(z-z')G^{(4)}(z'xyy)dz\,dz'\\ &\quad +\Gamma^{(6)}(xxxyyy)\,. \qquad (5.12)\end{aligned}$$

Diagrammatically, we write for $x\neq y$ the expansion in Fig. 1, where the $x-y$ channel is drawn from left to right. We show below that each term above is bounded above by $O(1)\exp(-3m|x-y|)$, and thus (5.9) follows. Thus, by the definition of $\overline{m}$, it follows that $\overline{m}\geq 3m$, and $\tilde{G}(p)$ has no CDD zeros for $|p|<3m$. Since $M\geq\overline{m}$, it follows that $M\geq 3m$, to complete the proof.

To establish the required estimates, we use

$$0\leq G(x)\leq O(1)|x|^{-1/2}e^{-m|x|}\,,\quad |x|\to\infty\,, \qquad (5.13)$$

which follows from the corresponding bound on $C(x)$ and $\int d\rho=1$. It follows that $G(x-y)^3$ has an exponential decay rate $3m$. Furthermore, by the Lebowitz inequality $G^{(4)}(xxyy)\leq 0$ (see Ref. 10), so the term linear in $G^{(4)}$ is bounded above by zero. The bound on $\Gamma^{(6)}$ follows by the assumption (5.6). Thus we obtain the inequality of Fig. 2.

In order to bound the integral in Fig. 2, we use the decay properties of $G^{(4)}$. In fact, by the second Griffiths inequality and the Lebowitz inequality (see Ref. 10)

$$0\leq -G^{(4)}(xyzt)\leq G(x-z)G(y-t)+G(x-t)G(y-z)\,. \qquad (5.14)$$

Furthermore, by Corollary 4.4,

$$\begin{aligned}-9\int & G^{(4)}(xxyz)\Gamma(z-z')G^{(4)}(z'xyy)dz\,dz'\\ &\leq 9\int G^{(4)}(xxyz)C_\alpha{}^{-1}(z-z')G^{(4)}(z'xyy)dz\,dz'\\ &= 9\int\langle xxyz\rangle_T C_{m_0{}^2}{}^{-1}(z-z')\langle z'xyy\rangle_T dz\,dz'\\ &\quad +9(\alpha-m_0{}^2)\int\langle xxyz\rangle_T\langle zxyy\rangle_T dz\,. \qquad (5.15)\end{aligned}$$

We use (5.12) and (5.14) to bound the last term in (5.15). Namely

$$\begin{aligned}\int\langle xxyz\rangle_T\langle zxyy\rangle_T dz &\leq 4\langle xy\rangle^2\int\langle xz\rangle\langle zy\rangle\,dz\\ &\leq O(1)e^{-3m|x-y|}\,. \qquad (5.16)\end{aligned}$$

Thus,

$$\begin{aligned}0 &\leq \Gamma(x-y)\\ &\leq O(1)e^{-3m|x-y|}\\ &\quad +144\lambda^2\int\langle xxyz\rangle_T C^{-1}(z-z')\langle z'xyy\rangle_T dz\,dz'\,.\end{aligned}$$

To bound the integral above, we use the integration by parts formula (3.1) to write

$$\begin{aligned}\int C^{-1}(z-z')\langle z'xyy\rangle_T dz' &= -\langle\!\langle \mathcal{V}'(z)\phi(x)\phi(y)\phi(y)\rangle\!\rangle_T\\ &\equiv -\langle\overline{z}xyy\rangle_T\,,\end{aligned}$$

where we let $\overline{z}$ denote $\mathcal{V}'(z)$. Thus,

$$\begin{aligned}0 &\leq \Gamma(x-y)\\ &\leq O(1)e^{-3m|x-y|}-144\lambda^2\int\langle xxyz\rangle_T\langle\overline{z}xyy\rangle_T dz\,. \qquad (5.17)\end{aligned}$$

In order to bound $|\langle\overline{z}xyy\rangle_T|$, we use Hamiltonian estimates. We introduce Euclidean coordinates $x=(x_0,x_1)$ in R^2 so that $x_1=y_1$ and $x_0<y_0$. We consider three cases: $z_0\leq x_0+1$, $x_0+1\leq z_0\leq y_0-1$, $y_0-1\leq z_0$. If $z_0\leq x_0+1$, we write

$$\langle\overline{z}xyy\rangle_T=\langle\overline{z}x(1-P_0)yy\rangle-2\langle\overline{z}y\rangle\langle xy\rangle\,.$$

Since the mass spectrum on $\mathcal{H}_e$ starts at $2m$, it follows that

$$|\langle\overline{z}xyy\rangle_T|\leq O(1)e^{-2m|x-y|}\,.$$

The absence of singularities at the coincidence $\overline{z}=x\neq y$ follows as in Ref. 8 from the fact that the local behavior is dominated by the low-order perturbation contribution. Thus, in this case by (5.14),

$$\begin{aligned}\left|\int_{z_0\leq x_0+1}\langle xxyz\rangle_T\langle\overline{z}xyy\rangle_T dz\right| &\leq O(1)e^{-3m|x-y|}\int\langle xz\rangle dz\\ &\leq O(1)e^{-3m|x-y|}\,.\end{aligned}$$

If $y_0-1\leq z_0$, then we use

$$|\langle\overline{z}xyy\rangle_T|\leq O(1)e^{-m|x-y|}\,,$$

and by (5.14),

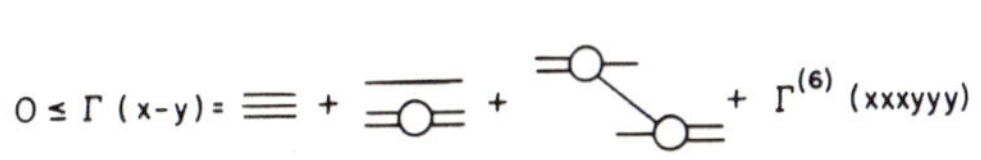

FIG. 1. Diagram expansion for Γ in a ϕ^4 theory according to Eq. (5.12).

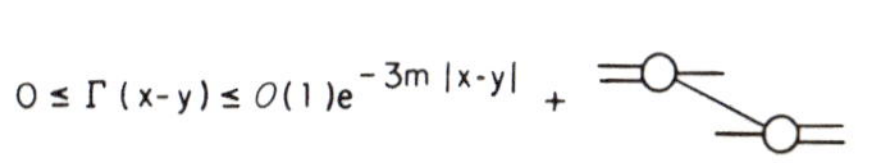

FIG. 2. Bound on $\Gamma(x-y)$.

$$\left|\int_{z_0\le y_0-1}\langle xxyz\rangle_T\langle \bar{z}xyy\rangle_T dz\right|$$
$$\le O(1)\int_{z_0\le y_0-1} e^{-m|x-z|-2m|x-y|}dz$$
$$\le O(1)e^{-3m|x-y|} .$$

Finally, if $x_0+1\le z_0\le y_0-1$, we have

$$|\langle x\bar{z}yy\rangle_T|\le O(1)e^{-m|x_0-z_0|-2m|z_0-y_0|} ,$$

and

$$\left|\int_{x_0+1\le z_0\le y_0-1}\langle xxyz\rangle_T\langle x\bar{z}yy\rangle_T dz\right|$$
$$\le O(1)\int\langle xz\rangle\langle xy\rangle e^{-m|x_0-z_0|}e^{-2m|z_0-y_0|}dz$$
$$\le O(1)e^{-3m|x-y|} .$$

Thus, we obtain from (5.17)

$$0\le\Gamma(x-y)\le O(1)e^{-3m|x-y|} \tag{5.18}$$

to complete the proof.

Proof of Theorem 5.1. We assume Theorem 5.4, and choose $\chi=\phi(f)^3\Omega$, where $f\in C_0^\infty$, $f\ge 0$, and support of f is contained in the strip $0<x_0\le T$. Linear combinations of such vectors (with different f, T) span $\mathcal{E}_3^+$. Thus, it is sufficient to show for $t\ge 0$ that

$$0\le W(t)\equiv\langle E\chi, e^{-tH}(1-P_1^{\mathcal{H}})E\chi\rangle_{\mathcal{H}}$$
$$\le O(1)e^{-at} , \tag{5.19}$$

for any $a<3m$. We establish (5.19) for $a<\overline{m}$, and appeal to Theorem 5.2 for the fact that $\overline{m}=3m$. The positivity of $W(t)$ follows by the positivity of the inner product in $\mathcal{H}$ and the spectral theorem. Thus, we need only establish the upper bound in (5.19).

We begin by repeating the proof of Theorem 5.2 with $\langle\Theta\chi, (1-P_1)T_t\chi\rangle_{\mathcal{E}}=\langle\phi(\vartheta f)^3(1-P_1)\phi(f_t)^3\rangle_{\mathcal{E}}$ replacing $\langle xxx(1-P_1)yyy\rangle_{\mathcal{E}}$. Here $f_t(x_0,x_1)=f(x_0-t,x_1)$. In this manner we obtain the upper bound

$$\langle\phi(\vartheta f)^3(1-P_1)\phi(f_t)^3\rangle_{\mathcal{E}}\le O(1)e^{-3m|t|} . \tag{5.20}$$

With $\chi=\phi(f)^3\Omega$ and $t\ge 0$, we see that the function $F(t)$ of Theorem 5.4 equals

$$F(t)=\langle\phi(\vartheta f)^3(1-P_1)\phi(f_t)^3\rangle_{\mathcal{E}}$$
$$-\langle E\phi(f)^3\Omega, e^{-tH}(1-P_1^{\mathcal{H}})E\phi(f)^3\Omega\rangle_{\mathcal{H}} .$$

Thus, by (5.20),

$$0\le\theta(t)W(t)=\theta(t)\langle\phi(\vartheta f)^3(1-P_1)\phi(f_t)^3\rangle_{\mathcal{E}}-\theta(t)F(t)$$
$$\le O(1)\theta(t)e^{-3mt}-\theta(t)F(t) , \tag{5.21}$$

where $F(t)$ satisfies (5.10), i.e., $\theta(t)F(t)$ belongs to a Sobolev-Paley-Wiener space.

Suppose the vector $(1-P_1^{\mathcal{H}})E\phi(f)^3\Omega$ has energy spectrum which intersects the interval $[0,\overline{m})$. Then by the spectral theorem, there exists $\epsilon>0$ such that for $a<\overline{m}$, but $\overline{m}-a$ sufficiently small,

$$e^{at}\theta(t)W(t)\ge\text{const}\times\theta(t)e^{\epsilon t} .$$

Since $(1-d^2/dt^2)^{-r}$ has a positive kernel, this lower bound contradicts (5.21) and (5.10) and completes the proof of Theorem 5.1.

Proof of Theorem 5.4. We study the functions

$$F_{\mathcal{H}}(t)=\theta(t)\langle E\chi, e^{-tH}P_1^{\mathcal{H}}E\chi\rangle_{\mathcal{H}}$$

and

$$F_{\mathcal{E}}(t)=\langle\Theta\chi, P_1T_t\chi\rangle_{\mathcal{E}} .$$

Furthermore, let

$$F_1(t)=F_{\mathcal{H}}(t)-F_{\mathcal{E}}(t) .$$

We note that $F_1(t)$ is continuous, except for a jump discontinuity at $t=0$, and by the spectral theorem

$$|F_1(t)|\le O(1)e^{-m|t|} .$$

We show below that F_1 can also be written

$$F_1(t)=F_A(t)+F_B(t) ,$$

where (i) F_A, F_B are each continuous, except for a jump discontinuity at $t=0$, (ii) $F_B(t)$ is supported in $t\le 0$, (iii) the Fourier transform $\tilde{F}_A(p_0)$ is an analytic function of p_0 in the strip $|\mathrm{Im}\,p_0|<\overline{m}$, and (iv) in each strip $|\mathrm{Im}\,p_0|\le a<\overline{m}$,

$$|\tilde{F}_A(p_0)|\le O(1)(1+|\mathrm{Re}\,p_0|)^s ,$$

where $O(1)$, but not s, depends on a. The following Paley-Wiener estimate then establishes the theorem: We let $1=\varphi_1(t)+\varphi_2(t)+\varphi_3(t)$ where each φ_j is C^∞, where $\varphi_j(t)$ is supported on the intervals $[-\infty,-1]$, $[-2,2]$, $[1,\infty]$ for $j=1,2,3$, respectively. We claim

$$\left(1-\frac{d^2}{dt^2}\right)^{-r}e^{at}\theta(t)\varphi_j(t)F_A(t)\in L_2 ,$$

for $j=1,2,3$. But by (i) and (ii),

$$\theta(t)F(t)=\theta(t)F_1(t)$$
$$=\theta(t)F_A(t)$$
$$=\theta(t)\sum_{j=1}^{3}\varphi_j(t)F_A(t) ,$$

so (5.10) follows.

Since $\varphi_1(t)\theta(t)=0$, the estimate for $j=1$ is trivial. For $j=2$, φ_j has compact support. Thus by (i),

$$\left\|\left(1-\frac{d^2}{dt^2}\right)^{-r}e^{at}\theta\varphi_2F_A\right\|_2\le\text{const}\times\|\varphi_2F_A\|_2\le\text{const}.$$

We thus are reduced to consider $\theta\varphi_3F_A=\varphi_3F_A$.

$$\left\|\left(1-\frac{d^2}{dt^2}\right)^{-r} e^{at}\varphi_3 F_A\right\|_2$$
$$\leq \sum_{n=0}^{\infty}\frac{a^n}{n!}\left\|\left(1-\frac{d^2}{dt^2}\right)^{-r} t^n\varphi_3 F_A\right\|_2$$
$$=\sum_{n=0}^{\infty}\frac{a^n}{n!}\|(1+p_0^2)^{-r}(\varphi_3 F_A)^{\sim(n)}(p_0)\|_2$$
$$\leq O(1)\sum_{n=0}^{\infty}\left(\frac{a}{a+\epsilon}\right)^n\|(1+p_0^2)^{-1}\|_2<\infty\,,$$

where the superscript tilde denotes Fourier transformation. Here we use

$$|(1+p_0^2)^{-r+1}(\varphi_3 F)^{\sim(n)}(p_0)|\leq \text{const}\times n!\,(a+\epsilon)^{-n}\,.$$

To establish this inequality we appeal to

$$|(\varphi_3 F_A)^{\sim(n)}(p_0)|$$
$$=\left|\int\tilde\varphi_3(p_0-q)\tilde F_A^{(n)}(q)dq\right|$$
$$\leq\|(1+r^2)^{-s}\tilde F_A^{(n)}(r)\|_\infty\int(1+q^2)^s|\tilde\varphi_3(p_0-q)|dq$$
$$\leq O(1)(1+p_0^2)^s\|(1+r^2)^{-s}\tilde F_A^{(n)}(r)\|_\infty$$
$$\leq O(1)(1+p_0^2)^s n!\,(a+\epsilon)^{-n}\,.$$

The bound on $\tilde F_A^{(n)}(r)$ follows by (iii), (iv), and the Cauchy integral formula. We choose $r=s+1$ to complete the proof.

Thus we are reduced to establishing the representation $F_1=F_A+F_B$, with the properties claimed above. We introduce some notation: Let $|p_1\rangle$ denote the one-particle state in $\mathcal{H}$ with momentum p_1, and normalized so that $\langle p|p'\rangle_{\mathcal{H}}=\delta(p-p')$. By Lorentz covariance

$$H|p_1\rangle=\mu(p_1)|p_1\rangle=(p_1^2+m^2)^{1/2}|p_1\rangle\,.$$

Since $E\chi\in C^\infty(H)$, the inner product $\langle E\chi|p_1\rangle_{\mathcal{H}}$ is a rapidly decreasing function of p_1.

Using the spectral theorem, we can compute $\tilde F_{\mathcal{H}}(p_0)$, namely

$$\tilde F_{\mathcal{H}}(p_0)=-i\int\frac{dp_1}{p_0-i\mu(p_1)}|\langle E\chi|p_1\rangle_{\mathcal{H}}|^2\,.$$

Next we study $\tilde F_{\mathcal{E}}(p_0)$. Define

$$G(z)\equiv\langle\Theta\chi,\,\phi(z)\rangle_{\mathcal{E}}\,.$$

By the definition of P_1,

$$F_{\mathcal{E}}(t)=-\int G(z)\Gamma(z-z')\bar G(-z_0'-t,\,-z_1')dz\,dz'$$

and

$$\tilde F_{\mathcal{E}}(p_0)=-\int\tilde G(p)\tilde G(-p)^-\tilde\Gamma(p)dp_1\,.$$

Let us assume the representation

$$\tilde G(p)=\frac{G_1(p_1)}{p_0-i\mu(p_1)}+\frac{G_2(p_1)}{p_0+i\mu(p_1)}+G_3(p_0,p_1)\,,\tag{5.22}$$

where

$$G_1(p_1)=-i\langle E\chi|p_1\rangle_{\mathcal{H}}\left(\frac{Z}{2\mu}\right)^{1/2}\,,$$

and where G_3 is analytic for $|\mathrm{Im}\,p_0|<\overline{m}$. Substituting the representation (5.22) into $\tilde F_{\mathcal{E}}$ gives nine terms. The G_1-G_1 term is

$$\int\left(\frac{Z}{2\mu}\right)|\langle E\chi|p_1\rangle_{\mathcal{H}}|^2\frac{1}{(p_0-i\mu)^2}\tilde\Gamma(p)dp_1$$
$$=-i\int\frac{1}{(p_0-i\mu)}|\langle E\chi|p_1\rangle_{\mathcal{H}}|^2\left(\frac{iZ\tilde\Gamma(p)}{2\mu(p_0-i\mu)}\right)dp_1$$
$$=\tilde F_{\mathcal{H}}(p_0)+R(p_0)\,,$$

where

$$R(p_0)=\int dp_1|\langle E\chi|p_1\rangle_{\mathcal{H}}|^2\frac{[Z/2\mu-i\int_{M^2}^\infty[(p_0+i\mu)/(p^2+a)]\,d\rho(a)]}{[Z+\int_{M^2}^\infty[(p^2+m^2)/(p^2+a)]\,d\rho(a)]}\,.$$

The properties of $R(p_0)$ depend on (v) the rapid decrease of $|\langle E\chi|p_1\rangle|$, (vi) the fact that $M\geq\overline{m}$, and (vii) the fact that

$$Z+\int_{M^2}^\infty\left(\frac{p^2+m^2}{p^2+a}\right)d\rho(a)$$

is analytic and nonzero for $|\mathrm{Im}\,p_0|<\overline{m}$ (see Ref. 8, Chapter 6). Thus, $R(p_0)$ is analytic if $|\mathrm{Im}\,p_0|<\overline{m}$. In each strip $|\mathrm{Im}\,p_0|\leq a<\overline{m}$, a bound $|R(p_0)|\leq O(1)[1+(\mathrm{Re}\,p_0)^2]^s$ follows from the representation for $R(p_0)$, and hence $R(p_0)$ is a contribution to $\tilde F_A(p_0)$.

Next we study the G_2-G_2 contribution to $\tilde F_{\mathcal{E}}$. As above, we find the sum of an integral over a pole at $p_0=-i\mu$ (instead of $p_0=i\mu$) and a piece $R_2(p_0)$ analytic in $|\mathrm{Im}\,p_0|<\overline{m}$. The inverse Fourier transform of the pole term has the form

$$F_B(t)=\theta(-t)\langle\psi,\,e^{tH}P_1^{\mathcal{H}}\psi\rangle_{\mathcal{H}}\,,$$

for some vector $\psi\in\mathcal{H}$. The term $R_2(p_0)$ is another contribution to $\tilde F_A$.

The remaining terms, G_i-G_j with $i\neq j$ or with $i=j=3$, are analytic for $|\mathrm{Im}\,p_0|<\overline{m}$, by similar calculation. They, too, obey a polynomial bound in $1+(\mathrm{Re}\,p_0)^2$, and thus contribute to $\tilde F_A$.

To complete the proof, we show that G_3 defined by (5.22) has the analytic properties that we assumed. We first study

$$G_+(p)\equiv\int_{z_0\ge 0} G(z)e^{-ipz}dz\ .$$

For $z_0\ge 0$,

$$G(z)=\langle E\chi,\, e^{-z_0H}[P_1^{\mathcal{K}}+(1-P_1^{\mathcal{K}})]E\phi(z_1,z_0=0)\Omega\rangle_{\mathcal{K}}\ .$$

The one-particle contribution gives the G_1 pole term in (5.22). We use

$$\langle p_1|\phi(z_1,z_0=0)\Omega\rangle_{\mathcal{K}}=\left(\frac{Z}{2\mu}\right)^{1/2}e^{-ip_1z_1}\ ,$$

and the spectral theorem to compute $G_1(p_1)$. The contribution from $(1-P_1^{\mathcal{K}})$ is part of $G_3(p)$, which follows by the spectral theorem and $M\ge\overline{m}$. Similarly, we analyze

$$G_-(p)=\int_{z_0\le -T} G(z)e^{-ipz}dz\ ,$$

and find the G_2 pole plus another contribution to G_3. Finally,

$$G_0(p)=\int_{-T}^{0}dz_0\int dz_1G(z)e^{-ipz}\ ,$$

is entire in p_0, since it has compact support in z_0.

B. $\Gamma^{(4n+2)}$ and three-particle bound states

In this section we assume that

$$\Gamma^{(4n+2)}(x_1,\ldots,x_{2n+1},y_1,\ldots,y_{2n+1})\le O(1)e^{-3md}\ ,\tag{5.23}$$

where $d=\text{dist}\{(x_1,\ldots,x_{2n+1}),(y_1,\ldots,y_{2n+1})\}$, and $n=1,2,\ldots$.

Theorem 5.5. Assume (5.23) in an even ϕ^4 model. Then

$$(m,3m)\cap\sigma(M\restriction\mathcal{K}_0)=\emptyset\ .\tag{5.24}$$

Proof. The basic idea of the proof is the same as the proof of Theorems 5.1 and 5.2 and as the proof of Feldman,[4] although the diagrams that occur and the estimates are slightly more complicated. We merely present the algebraic structure, without giving detailed estimates. As in the proof of (5.8), it is sufficient to establish the Euclidean estimate

$$\langle\hat\theta,\,T_t(1-P_1)\theta\rangle_{\mathcal{E}}\le O(1)e^{-3mt}\ ,\tag{5.25}$$

where

$$\theta=\phi(f_1)\cdots\phi(f_{2n+1})\Omega\in\mathcal{E}^+_{2n+1}\ ,\tag{5.26}$$

$f_i\ge 0$, $f_i\in C_0^\infty$, and $\hat\theta$ is the time inverted vector. The connection between the Euclidean estimate (5.25) on $\mathcal{E}_{2n+1}$ and the spectral estimate (5.24) on $\mathcal{K}_{2n+1}$ then follows by

$$|\langle\hat\theta,\,T_t(1-P_1)\theta\rangle_{\mathcal{E}}-\langle E\theta,\,e^{-tH}(1-P_1^{\mathcal{K}})E\theta\rangle_{\mathcal{K}}|\le O(1)e^{-at}\ ,\tag{5.27}$$

for all $a<3m$, as in the proof of (5.8). Since $\mathcal{K}_0=\oplus\mathcal{K}_{2n+1}$, the statement (5.24) follows.

To establish (5.25), we expand the inner product into connected functions. Further, expand the $(4n+2)$-point connected part $G^{(4n+2)}$ into $\Gamma^{(4n+2)}$ [bounded above by (5.23)] plus a sum of terms. Each term in the expansion of (5.25) is represented by a sum of (possibly disconnected) tree graphs. These graphs have $2(2n+1)$ external legs; of them $2n+1$ are labeled $x_1,\ldots,x_{2n+1}$ and called initial legs, while $2n+1$ are labeled $y_1,\ldots,y_{2n+1}$ and called final legs. The vertices of the graphs are connected Green's functions $G^{(r)}(x_1,\ldots,x_r)$, with $4\le r\le 4n+2$, while the lines of the graphs are either G (external lines) or Γ (internal lines). In addition there are combinatoric factors and a factor (-1) for each Γ (internal line). The subtraction $-\langle\hat\theta,\,T_tP_1\theta\rangle_{\mathcal{E}}$ removes graphs that are one-particle reducible in the channel $(\{x_1,\ldots,x_{2n+1}\},\{y_1,\ldots,y_{2n+1}\})$. Thus, each graph is formally one-particle irreducible in the $x-y$ channel, and hence they are formally two-particle irreducible by $\phi\to-\phi$ symmetry.

In order to establish the desired three-particle decay (5.25), we estimate each graph separately, as in the proof of Theorems 5.1 and 5.2. In performing these estimates we make use of the following facts:

(1) Connected graphs with an even number of initial legs also have an even number of final legs, by $\phi\to-\phi$ symmetry. These graphs decay as $O(e^{-2mt})$, since they reflect the absence of a two-particle bound-state spectrum on the even subspace $\mathcal{K}_e$ (see Ref. 3).

(2) In proving (5.25) for a given n, we can assume its validity for $\theta\in\mathcal{E}^+_{2r+1}$, $r<n$, i.e., we argue by induction on n.

As a consequence of (1) and (2), all disconnected graphs which contribute to $\langle\hat\theta,\,T_t(1-P_1)\theta\rangle_{\mathcal{E}}$ are bounded by $O(1)e^{-3mt}$. Thus,

$$\langle\hat\theta,\,T_t(1-P_1)\theta\rangle_{\mathcal{E}}=O(1)e^{-3mt}+C\ ,$$

where C is the sum of (connected) tree graphs generated by the expansion of $G^{(4n+2)}$ into $\Gamma^{(4n+2)}$ + tree graphs, and the graph in Fig. 3 does not occur. Each of these remaining graphs is estimated using the decay properties of the $G^{(r)}$ and integration by parts, as in the proof of Theorem 5.2.

FIG. 3. Connected graph excluded by the one-particle subtraction $(I-P_1)$.

C. The CDD radius and the scaling limit

Conventional ideas of the renormalization group place considerable importance in the scaling limit, namely the limit λ (bare coupling) $\to\infty$ with m (physical mass) fixed. If $d<4$, these theories are the limits ($\lambda=\infty, m_0=\infty$) of superrenormalizable models of mass m. Standard arguments indicate that the infinitely scaled theories (if they exist) are not renormalizable in dimension $d<4$. Thus, the construction of the infinitely scaled ϕ_2^4 models would establish the existence of ϕ_2^4 theories of greater singularity than ϕ_4^4. Furthermore, one expects that the $m=0$ limit of the infinitely scaled theories is scale invariant, and is the Gell-Mann–Low–Wilson fixed point. The $m=0$ limit of the infinitely scaled theories is also a hyperbolic fixed point of the scale transformation group. The infinitely scaled transformations form the unstable manifold associated with this fixed point, while the $m=0$, $\lambda<\infty$ critical theories form the stable manifold.

We find that $\bar{m}/m \geq 1+\epsilon$ is a sufficient condition to ensure the existence of the scaling limit. Combined with Ref. 11 and Ref. 5, the results of this paper are a step toward the construction of a ϕ_4^4 field theory. Indeed, the missing steps are (a) the proof that $\bar{m}/m$ is bounded away from 1, and (b) the adaptation of certain arguments depending on Lorentz invariance to the case of a cutoff field theory. Conventional ideas indicate that this construction for ϕ_d^4, with $d>4$, yields a free field. The nontriviality of this construction for ϕ_4^4 is related to the logarithmic divergence expected for the field-strength renormalization Z^{-1}.

We illustrate the scaling limit in Fig. 4, where we plot the physical mass m for theories with fixed bare mass m_0, as a function of the dimensionless bare coupling $g_0=\lambda m_0^{d-4}$. Scaling maps these curves onto one another, and in the figure we assume that $m\to 0$ continuously as we approach the critical coupling $g_{0,c}$. The infinite scaling (dotted) line lies over $g_{0,c}$, and theories on this line have $\lambda=\infty$, $m_0=\infty$. The infinite scaling line is parameterized by m. In Fig. 5 we draw our schematic picture of the presumed ϕ_2^4 scaling trajectories, as described above.

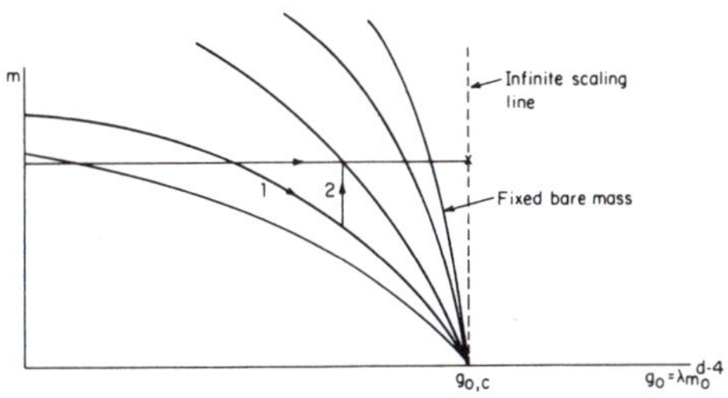

FIG. 4. The scaling limit for ϕ^4. Note movement along the m = const line can be achieved by (1) increasing λ with m_0 fixed, followed by (2) scaling to increase the mass back to m (thus increasing m_0 and λ with λm_0^{d-4} fixed).

Theorem 5.6. Assume $\bar{m}/m$ is bounded away from one for m fixed and $g_0<g_{0,c}$ in an even $\lambda\phi_2^4$ model. Then there exists an infinitely scaled, Osterwalder-Schrader theory with mass m. The Schwinger functions $S_m^{(n)}$ of this model are the limits (through subsequences) of the Schwinger functions $S_{m,\lambda}^{(n)}$ for the model with mass m and coupling λ,

$$S_m^{(n)}=\lim_{\lambda_j\to\infty} S_{m,\lambda_j}^{(n)}. \tag{5.28}$$

Proof. For m fixed and $\bar{m}/m$ bounded away from 1, we apply Theorem 2.6 to bound the two-point Schwinger function in momentum space,

$$(p^2+m^2)^{-1}\leq \tilde{S}_{\rm ren}(m,\lambda,p)=\tilde{G}_{\rm ren}(m,\lambda,p)\leq O(1)$$

with $O(1)$ uniform in λ. Thus,

$$\tilde{S}_{\rm ren}(m,p)=\lim_{\lambda_j\to\infty}\tilde{S}_{\rm ren}(m,\lambda_j,p)$$

exists and is bounded. The $\tilde{S}(m,\lambda,p)$ are Euclidean invariant and Osterwalder-Schrader positive, and hence so is $\tilde{S}_{\rm ren}(m,p)$. Thus,

$$\tilde{S}_{\rm ren}(m,p)=\int_{m^2}^{\infty}\frac{d\rho(a)}{p^2+a},$$

where $d\rho(a)$ is positive and

$$\int_{m^2}^{\infty}\frac{d\rho(a)}{a}<\infty.$$

As in Ref. 11 (see also Ref. 12 and 13,), the n-

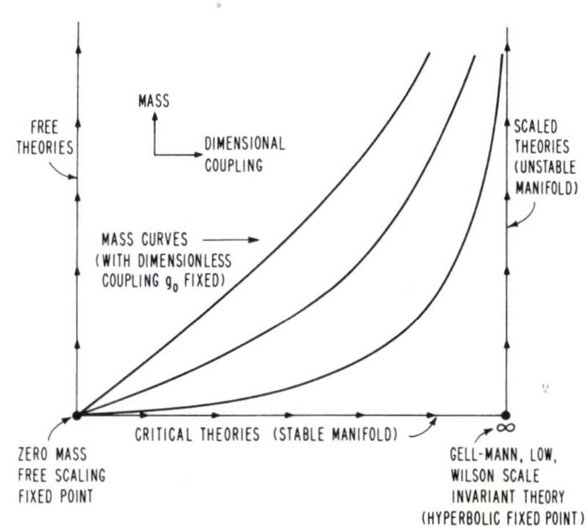

FIG. 5. Trajectories of the scaling group.

point Schwinger functions are bounded in terms of S_{ren}, namely

$$|S^{(n)}_{\text{ren}}(m,\lambda,f_1,\ldots,f_n)| = |Z^{-n/2}S^{(n)}(m,\lambda,f_1,\ldots,f_n)| \leq O(1)\prod_{i=1}^{n} S^{(2)}_{\text{ren}}(m,\lambda,f_i{}^*\otimes f_i)^{1/2}.$$

Hence, passing to subsequences, we obtain limits as $\lambda\to\infty$ for each $S^{(n)}_{\text{ren}}(m,\lambda)$. Since each approximate $S^{(n)}_{\text{ren}}$ satisfies the Osterwalder-Schrader axioms with a mass gap $m>0$, so do the limiting Schwinger functions, as covariance, positivity, symmetry, and absence of spectrum are all preserved under limits.

Remarks 5.7. While the hypotheses of Theorem 5.2 are sufficient to ensure $\overline{m}/m=3$, the weaker hypothesis

$$\Gamma^{(6)}(xxxyyy) \leq O(1)e^{-(m+\epsilon)|x-y|}, \quad |x-y|\to\infty, \tag{5.29}$$

uniformly for $g_0<g_{0,c}$ and m fixed, is sufficient to establish in a $\lambda\phi_2^4$ model that

$$\Gamma(x) \leq O(1)e^{-(m+\epsilon)|x|}, \quad |x|\to\infty. \tag{5.30}$$

Hence $\overline{m}/m-1\geq\epsilon/m$, and the scaling limit for mass m exists.

Remark 5.8. An alternative construction of an infinitely scaled ϕ_d^4 field theory starts from the d-dimensional Ising model I_d with lattice spacing ϵ, and takes $\epsilon\to0$ with $m=\xi^{-1}$ = inverse correlation length fixed. Since this construction differs from the previous one ($\lambda\to\infty$) only by a change in the order of the ϵ,λ limits, it is attractive to suppose that both constructions yield the same theory. For $d=2$, a detailed analysis by Wu and co-workers[14] on the closed-form solutions of I_2 suggests that the required bounds on the two-point function may be attainable in order to assert the existence of the infinitely scaled, n-point correlation functions of I_2, using compactness and subsequences as above.

---

*Work supported in part by the National Science Foundation under Grant No. MPS 74-13252.

†Work supported in part by the National Science Foundation under Grant No. MPS 73-05037.

[1]J. Glimm, A. Jaffe, and T. Spencer, Ann. Math. 100, 585 (1974).

[2]J. Glimm, A. Jaffe, and T. Spencer, in *Constructive Quantum Field Theory*, Springer Lecture Notes in Physics, edited by G. Velo and A. Wightman (Springer, Berlin, 1973), Vol. 25.

[3]T. Spencer, Commun. Math. Phys. 39, 77 (1974).

[4]J. Feldman, Can. J. Phys. 52, 1583 (1974).

[5]K. Osterwalder and R. Schrader, Commun. Math. Phys. 31, 83 (1973); *ibid.* (to be published).

[6]K. Symanzik, J. Math. Phys. 1, 249 (1960).

[7]W. Donoghue, Jr., *Monotone Matrix Functions and Analytic Continuation* (Springer, Berlin, 1974).

[8]J. Glimm and A. Jaffe, Commun. Math. Phys. (to be published.

[9]H. Lehmann, Nuovo Cimento 11, 342 (1954).

[10]*Constructive Quantum Field Theory*, Springer Lecture Notes in Physics, edited by G. Velo and A. Wightman (Springer, Berlin, 1973), Vol. 25.

[11]J. Glimm and A. Jaffe, Phys. Rev. Lett. 33, 440 (1974).

[12]J. Glimm and A. Jaffe, Ann. Inst. H. Poincaré 22, 1 (1975).

[13]C. Newman (unpublished).

[14]T. T. Wu, private communication. Also see C. Tracy and B. McCoy, Phys. Rev. Lett. 31, 1500 (1973).

Two and Three Body Equations in Quantum Field Models*

James Glimm**

Rockefeller University, New York, N.Y. 10021, USA

Arthur Jaffe***

Harvard University, Cambridge, Mass. 02138, USA

Received December 27, 1974

Abstract. In each pure phase of a $\mathscr{P}(\phi)_2$ quantum field model, we establish local regularity of the Green's functions and exponential decay for noncritical models. We establish the existence of two-particle and three-particle Bethe-Salpeter kernels in the Euclidean region.

Contents

1. Introduction

The study of particles in weakly coupled $\mathscr{P}(\phi)_2$ quantum field theories was begun in [25, 26]. The cluster expansions developed in these papers resulted in the construction of isolated one-particle states. According to the Haag-Ruelle theory, the existence of n-particle in and out states, and the existence of an isometric S-matrix, follow from the existence of isolated one particle states. The ϕ^4 model, in the single phase region, has been shown to be repulsive in the sense that no even mass spectrum occurs in the two particle bound state interval $(m, 2m)$ [46, 8, 25]. The presence of bound state mass spectrum was indicated for the $\phi^6-\phi^4$ interaction [25]. Spencer [47] and Spencer and Zirilli [48] have a more detailed analysis of the energy momentum spectrum for weak coupling, which goes up to the threshold $E \leqq 3m-\varepsilon$, and uses the Bethe-Salpeter equation. An early version of their work motivated the present paper.

* We thank the Institut des Hautes Etudes Scientifiques for hospitality.

** Supported in part by the National Science Foundation under Grant MPS 74-13252.

*** Supported in part by the National Science Foundation under Grant MPS 73-05037.

In this paper we study the Bethe-Salpeter kernel in the wider class of noncritical $\mathscr{P}(\phi)_2$ models, and we also study three body kernels. By definition, a quantum field is called critical if its mass is zero; otherwise it is noncritical. We expect particles and a complete set of scattering states for all noncritical $\mathscr{P}(\phi)_2$ theories.

There are at present five (partially overlapping) methods to construct the infinite volume $\mathscr{P}(\phi)_2$ model. The original method [14] of compactness and convergent subsequences uses operator algebra techniques and estimates based in part on Euclidean space time localization and decoupling in the Feyman-Kac formula. This construction applies to the widest class of interactions [14, 42, 43]. The second method is the cluster expansion [25, 26], which applies in the case of weak coupling or large external field [45], and which gives the most complete information about the resulting model.

The third method [39, 28, 9, 44] in based on monotone convergence and upper bounds. The monotonicity is based on Griffiths' inequalities [12, 28], while the upper bound originates in the estimate [15]

$$\pm\phi(h)\leqq C(h)(H+I), \tag{1.1}$$

see also [27, 44, 9, 10]. The passage from this ϕ-bound to estimates on Wightman functions is given in [17]. The same simple proof works for Schwinger functions. It is well known that bounds on Wightman functions imply bounds on Schwinger functions at noncoinciding points, see for example [40, 41]. In [28], a theory of conditional expectations based on the Markov property [49, 38] is developed to handle this point. A variant [24] of the monotonic construction replaces Dirichlet boundary conditions in (1.1) by weak coupling boundary conditions.

The main advantage of the third method is a proof of Lorentz invariance, of correlation inequalities and (for ϕ^4 only) of the Lee Yang theorem. See [8, 21–24, 28, 29, 44, 46] for applications.

In the fourth construction [5, 11], the compactness of method one is formulated in a probability framework. The fifth construction is to obtain the weakly coupled ϕ^4 model by Borel summation of the perturbation series [6]. The convergence of this method is based on new estimates which require cancellation of vacuum graphs in the cluster expansion. This method gives a type of uniqueness which is particularly satisfying in view of the importance of perturbation theory in calculations.

We study the $\mathscr{P}(\phi)_2$ model obtained by one of these procedures. Much of Chapter 2 is an assimilation of results known in the literature for some time. Our starting point is the ϕ^j and π bounds

$$\begin{aligned}\pm:\phi^j(h):&\leqq C(h,j)(H+I)\\ \pm\pi(h)&\leqq C_\pi(h)(H+I).\end{aligned}$$

In Chapter 3 we establish the integration by parts formula for the Schwinger functions $S^{(n)}$ in each pure phase. We use this result in Chapter 4 to analyze the connected parts of the Schwinger functions (Euclidean Green's functions $G^{(n)}$). Let $H_{\pm 1}$ be the Sobolev space defined by the inner product

$$\langle f,g\rangle_{\pm 1}=\langle f,(-\Delta+m_0^2)^{\pm 1}g\rangle_{L_2}. \tag{1.2}$$

Let $H_{\pm 1}^{\otimes j}$ be a j-fold tensor product of such spaces. For noncritical $\mathscr{P}(\phi)_2$ theories, we prove that $G^{(n)}(x_1, \ldots, x_n)$ is the kernel of a bounded operator

$$G^{(n)}: H_{-1}^{\otimes j} \to H_1^{\otimes n-j}, \tag{1.3}$$

for $0<j<n$. Here we use the ϕ^j bound, and exponential clustering, which follows from the existence of a mass gap. The stronger tree graph decay rate of the $G^{(n)}$ is known for weak coupling [6]. Since $G^{(n)}$ is translation invariant, it cannot be compact. However, we show that the restriction of $G^{(n)}$ to a subspace of definite total momentum is Hilbert-Schmidt. This is a crucial bound on the $G^{(n)}$.

An immediate corollary of this bound on $G^{(n)}$ is the fact that noncritical Schwinger functions $S^{(n)}$ for the field $\hat{\phi} = \phi - \langle \phi \rangle$ are continuous multilinear functionals on the n-fold product $H_{-1} \times \ldots \times H_{-1}$. This result was established for weak coupling in [20]. A second corollary is the existence of the vertex functions $\Gamma^{(n)}(x_1, \ldots, x_n)$ (direct correlation functions) for noncritical $\mathscr{P}(\phi)_2$ theories. The $\Gamma^{(n)}$ are bounded operators from $H_1^{\otimes j}$ to $H_{-1}^{\otimes n-j}$, $0<j<n$, improving slightly a bound from [20].

In Chapter 6 we show that the inverse propagator $\Gamma(x)$ has an exponential decay rate equal to the CDD radius $\bar{m}$. Here $m < \bar{m} \leqq M$ where m is the single particle mass and (m, M) is the upper gap in the spectrum of the two point function.

In Chapter 7 we construct the 2-particle Bethe-Salpeter kernel for non-critical $\mathscr{P}(\phi)_2$ models. We use the compactness of the $G^{(n)}$ on constant momentum subspaces, and apply a Fredholm alternative. The exclusion of the eigenvalue zero in the Fredholm alternative is related to the following property, which we establish in Chapter 5 for a $\mathscr{P}(\phi)_2$ measure: If a polynomial $\mathscr{R}(\phi)$ takes the value zero almost everywhere on $\mathscr{S}'(R^2)$, then $\mathscr{R}=0$, i.e. the coefficients of $\mathscr{R}$ vanish. This property can be rephrased by saying that the Euclidean $\mathscr{P}(\phi)_2$ measure is not supported on any algebraic variety in $\mathscr{S}'(R^2)$. We also require this property for polynomials $\mathscr{R}(\phi)$ with a definite momentum. It is a consequence of integration by parts, or of the existence [11] of Euclidean momentum operators π. We remark that some time ago Bros [3] constructed the two body Bethe-Salpeter kernel in the context of the Wightman axioms, plus additional assumptions. Finally in Chapter 8, we consider three particle equations.

The two body Bethe-Salpeter kernel is a bounded operator

$$K: H_1^{\otimes 2} \to H_{-1}^{\otimes 2} \tag{1.4}$$

satisfying the integral equation

$$R = R_0 + R_0 K R \tag{1.5}$$

in the case of even $\mathscr{P}(\phi)$ models. Here R and R_0 map $H_{-1}^{\otimes 2}$ into $H_1^{\otimes 2}$ and are defined by their kernels

$$\begin{aligned} R_0(x_1, x_2; y_1, y_2) &= G^{(2)}(x_1, y_1) G^{(2)}(x_2, y_2) + G^{(2)}(x_1, y_2) G^{(2)}(x_2, y_1), \\ R(x_1 x_2; y_1 y_2) &= G^{(4)}(x_1, x_2, y_1, y_2) + R_0(x_1, x_2; y_1, y_2). \end{aligned} \tag{1.6}$$

Thus

$$K = R_0^{-1} - R^{-1}, \tag{1.7}$$

and formally K is a sum of graphs two particle irreducible between initial and final states. The construction of K, however, does not have an immediate significance for the particle structure. In fact the estimates [47] which yield decay properties of $K(x_1, x_2; y_1, y_2)$ depend on the cluster expansion, and are known only for weak coupling. The methods of Lebowitz and Penrose [31, 32] from statistical mechanics given an analytic continuation of decay estimates. The Lebowitz-Penrose results pertain only to the gap between the ground state and the first excited state, and are not strong enough to give the existence of particles. The extension of the Lebowitz-Penrose result [32] to field theory is contained in [29]. It is an interesting speculation that nontrivial decay properties for $K(x_1, x_2, y_1, y_2)$ are a consequence of the existence of a mass gap. However analyticity properties in the coupling should not be a consequence of a positive mass, even in the case $\mathscr{P}=$even, cf. [24, Fig. 2.1].

2. ϕ^j Bounds and Schwinger Functions

Let $\mathfrak{A}$ be the C^*-algebra of the $V=\infty$ limit, acting on a Hilbert space $\mathscr{H}$. By [1], $\mathfrak{A}'=\mathscr{Z}$, the center of $\mathfrak{A}''$. The direct integral decomposition is defined with respect to $\mathscr{Z}$; we write it as

$$\mathscr{H}=\int \mathscr{H}(\xi)d\xi \tag{2.1}$$

so that $\xi \in \Xi$ is a label for the set Ξ of phases. Let $\Omega_\xi \in \mathscr{H}_\xi$ be the vacuum and let H_ξ be the Hamiltonian. For example, in the ϕ_2^4 model we expect a single phase for $\lambda<\lambda_c$, and two phases for $\lambda>\lambda_c$. We expect a continuum of phases in the case of phase transitions with a continuous group of internal symmetries, so this direct integral machinery may not be too elaborate for the use we make of it. See [25, 11] for a further discussion of phases.

We now state our basic ϕ and π bounds. Let

$$1 \leqq j \leqq d = \deg \mathscr{P}\,.$$

For general polynomials, all constructions [14, 11] assume free boundary conditions. For $\mathscr{P}=$even + linear, we allow either free or weak coupling boundary conditions [24]. We let V denote the volume of interaction; for weak coupling boundary conditions, V is the length of the interval $[-b, b]$ in which the mass perturbation $\sigma\int_{-b}^{b} :\phi^2(x): dx$ is inserted. Note that for $V=\infty$, weak coupling and Dirichlet boundary data define the same theory, for $\mathscr{P}=$even + linear [24].

Theorem 2.1 a *(ϕ^j and π bounds). For $V<\infty, j\leqq d$*

$$\pm\phi(h) \leqq \|h\|_1 C(H+I)\,, \tag{2.2a}$$

$$\pm\pi(h) \leqq \|h\|_2 C(H+I)\,, \tag{2.2b}$$

$$\pm:\phi^j:(h) \leqq H + O(\|(|x|^r+1)h\|_{d/d-j} + \|h\|_{d/d-j}^{d/d-j})\,, \tag{2.2c}$$

for h real and in the appropriate L_p space, and for $r>1+(j|d)$. Here C is a constant, independent of h and V. In the case $j=d$, we also require $\|h\|_\infty$ small.

Theorem 2.1 b (*ϕ^j bounds*). *For $j \leqq d$, $V=\infty$ and in a.e. pure phase, the bounds* (2.2a) *and* (2.2c) *are valid. Moreover for free boundary conditions*

$$\pm :\phi^j:(h) \leqq H + O(\|h\|_1 + \|h\|_{d/d-j}^{d/d-j}). \tag{2.2d}$$

Proof of Theorem 2.1a. (2.2a) was proved in [15]. The bound (2.2b) was established in [16, 17] for $V<\infty$, and the same methods can be used directly in the infinite volume limit. Bounds related to (2.2c) were first given in [15] and have been improved independently by Fröhlich [11] and the authors (unpublished). For weak coupling, Dirichlet, or free boundary conditions and $V<\infty$, we use the inequality

$$\pm :\phi^j:(h) \leqq H + O(a) + O(1)\|h\|_{d/d-j}^{d/d-j}$$

for support $h \subset [-a, a]$, cf. [24, 51]. Because the vacuum energy of the $:\phi^j:$ perturbation is convex, and is zero for $h=0$, this yields

$$\pm :\phi^j:(h) \leqq H + O(a)^{j/d}\|h\|_{d/d-j} + O(1)\|h\|_{d/d-j}^{d/d-j}.$$

We write $h=\sum h_n$, suppt $h_n \subset [n, n+1]$, and then by convexity again,

$$\pm :\phi^j:(h) \leqq H + O(1)\sum \|(|n|+1)h_n\|_{d/d-j} + O(1)\|h\|_{d/d-j}^{d/d-j}.$$

Finally,

$$\begin{aligned}\sum \|(|n|+1)h_n\|_{d/d-j} &= \sum (|n|+1)^{-\alpha}\|(|n|+1)^{1+\alpha}h_n\|_{d/d-j} \\ &\leqq \text{const. } \|(|x|+1)^{1+\alpha}h\|_{d/d-j}\end{aligned}$$

for $\alpha > j/d$.

Proof of Theorem 2.1b. The transfer of the ϕ bound to the infinite volume limit is given in [15]. The infinite volume limit of the $:\phi^j:$ bounds is established in [24].

There is also an abstract method for transferring an estimate of the form

$$\pm A \leqq H + I \tag{2.3}$$

to the infinite volume limit, and then into almost every phase [2]. This method requires an additional estimate, for example

$$\|A(H+I)^{-1}\| + \|[H, A](H+I)^{-1}\| < \text{const.} \tag{2.4}$$

uniform in the volume V. Such an estimate holds in the case $A=\phi(h)$, $h \in \mathscr{S}(R^1)$, using [15, Lemma 1.1] and (2.2a)–(2.2c). Hence we obtain (2.2a), with $h \in \mathscr{S}$, in the $V=\infty$ limit and in almost every pure phase. The role of the inequality (2.6) is to provide a domain of essential self-adjointness[1] for $A=\phi(h)$. Having established the bound and decomposition to pure phases for $h \in \mathscr{S}$, we pass to $h \in L_1$ using (2.2a). (2.2c) is a consequence of (2.2d), for $V=\infty$. The bound (2.2d) is given in [11].

[1] The general method of using commutation estimates to establish self adjointness was introduced in [13] to prove self adjointness of space-time averaged field operators $\phi(f)$. One abstraction of this method was presented in [18], and another abstraction was presented in [15]. These results established essential self adjointness of an operator or form A, satisfying relative bounds with respect to a positive self adjoint operator H. In our present applications, we use [15: Lemma 1.1, Theorem 1.2] to establish essential self adjointness of $\phi(h)$. Other abstractions and applications have also been given by Nelson [37], McBryan [33—35], Faris und Lavine [7], and McBryan and Park [36].

In the case of weak coupling boundary conditions, the basic method relies on uniform estimates [10, 24] for the analytic generating functionals of the ordinary and generalized Schwinger functions. By compactness, these functionals have a limit point, as $b\to\infty$, and the limit point satisfies the same estimate. However, convergence of the Schwinger functions as $b\to\infty$ [24; see also Theorem 2.2, 2.4 below] implies that the analytic functions actually have a limit as $b\to\infty$. The Schwinger functions are constructed from the generating functionals [10] and yield the infinite volume Hilbert space [10, 40, 41]. On this Hilbert space, the bound (2.2c) holds in the sense of a norm estimate on the perturbed semigroup [24]. Further details are omitted.

Theorem 2.2. *Let $h, \chi \in C_0^\infty$, $\int \chi dx = 1$, $0 \leqq \chi$, and $V < \infty$, and*

$$\delta H_\chi = \int \prod_{i=1}^{j} (\phi * \chi)(x) h(x) dx .$$

Then

$$\pm \delta H_\chi \leqq H + O(1) \tag{2.5}$$

uniformly as $V\to\infty$. Let $\chi_i = \kappa_i \chi(\kappa; x)$, $i=1,2$. Then

$$\pm(\delta H_{\chi_1} - \delta H_{\chi_2}) \leqq H + o(1) \tag{2.6}$$

uniformly in V, as $\kappa_1, \kappa_2 \to \infty$. Again for $j=d$, we require $\|h\|_\infty$ small.

Theorem 2.2 follows as an extension of the methods of [15] and [51] for free and Dirichlet boundary conditions, respectively. It is presumably also true for weak coupling boundary conditions, but we do not require this fact.

We need to introduce certain function spaces related to the estimates of Theorem 2.1 Let

$$\begin{cases} \mathscr{L}_{1,p} = \mathscr{L}_1(R^2) \cap \mathscr{L}_p(R^2), & p<\infty \\ \mathscr{L}_{1,\infty,\varepsilon} = \mathscr{L}_1(R^2) \cap \{f : \|f\|_\infty < \varepsilon\} . \end{cases} \tag{2.7}$$

Whenever $p=\infty$, the space $\mathscr{L}_{1,\infty,\varepsilon}$ will be understood, even if the ε is not explicitly included.

Before studying the quantum field on the space $\mathscr{H}(\xi)$, we require more information about its action on $\mathscr{H}$.

Theorem 2.3. *The quantum field on $\mathscr{H}$, specified by a C^* algebra construction, is unitarily equivalent to an Osterwalder-Schrader* [40, 41] *reconstructed field. The Schwinger functions are moments of a unique measure dq on $\mathscr{S}'(R^2)$.*

Remark. Since the covariance of the field under Lorentz transformations is not known, the corresponding imaginary time fields on $\mathscr{S}'(R^2)$ will be time translation covariant, but the Euclidean invariance is not known.

Proof. We verify the axioms (A1)–(A5), B, C of Fröhlich [9] in order to obtain the imaginary time fields and the reconstructed fields. Thus we want to study the generating functions

$$J(f) = \int e^{i\phi(f)} dq \tag{2.8}$$

of the imaginary time measure. For f piecewise constant in t, $J(f)$ has a Hamiltonian definition. If $h_1(x), \ldots, h_n(x)$ are piecewise constant values of $f(\cdot, t)$, then

$$J(if) = \langle \Omega, e^{-(t_2-t_1)(H+\phi(h_1))} \ldots e^{-(t_n-t_{n-1})(H+\phi(h_n))}\Omega \rangle .$$

For a finite space cutoff, we know that e^{-tH} is positivity preserving and that the imaginary time measure on $\mathscr{S}'(R^2)$ exists. Also $J(f)$ is complex analytic in $f \in \mathscr{S}(R^2)$, for example, and can be bounded using Theorem 2.1.

For any operator H, we define its vacuum energy to be $E(H) = \inf$ spectrum H. Let

$$\delta E(t, f) = E(H + \phi(f(\cdot, t))) - E(H) ,$$

for f real. Then by Theorem 2.1,

$$\begin{aligned} |J(f)| &\leqq \exp - \textstyle\int_{-\infty}^{\infty} \delta E(t, \operatorname{Im} f) dt \\ &\leqq \exp (\text{const.} (\|f\|_{d/d-1}^{d/d-1} + \textstyle\int \|(|x|^r + 1) f(\cdot, t)\|_{d/d-1} dt) \end{aligned} \tag{2.9}$$

with a constant uniform in the space cutoff. By Vitali's theorem, J is complex analytic for $f \in \mathscr{L}$ where $\mathscr{L}$ is defined by the norms in (2.9). Following Fröhlich [10], we conclude that the space cutoff Schwinger n-point function is a bounded multilinear functional on $\mathscr{L} \times \ldots \times \mathscr{L}$. By multilinearity, we can restrict each test function f in the n-point Schwinger function to be supported in a strip

$$\{x, t: |t - i| \leqq 1/2\} ,$$

$i \in Z$. Let n_i be the number of arguments of the Schwinger function restricted to this strip; thus $\sum n_i = n$. With this notation, the Cauchy integral formula combined with the bound (2.9) shows that the n-point Schwinger function is bounded by $O(\prod_i n_i!)K^n$, as a functional on $\prod^n \mathscr{L}$, $K = O(\int \|(|x|^2 + 1) \cdot \|_{d/d-1} dt + \| \cdot \|_{d/d-1}^{d/d-1})$. This bound is uniform in n and in the space cutoff region. We know that the space-cutoff Wightman functions converge [15] (after selection of a subsequence), and by analytic continuation the $V \to \infty$ limit of the Schwinger functions exists also. The $V = \infty$ Schwinger functions have the same $\mathscr{L}$ norm, $O(\prod n_i! K^n)$, which allows summation of the power series. The resulting functional $J(f)$ is complex analytic, satisfies (2.9), and is the limit of the finite volume J's. The axioms of Fröhlich are readily verified, because the estimates and positivity conditions are preserved in the $V = \infty$ limit. We only comment that his axiom CI, which yields sharp time fields, is proved using the second order estimate of [15, Lemma 1.1],

$$\pm \phi(f)^2 \leqq |f|_{\mathscr{S}}^2 (H + I)^2 , \tag{2.10}$$

where $|f|_{\mathscr{S}}$ is a Schwartz space norm on $f \in \mathscr{S}$, independent of $V \leqq \infty$.

From [10], we thus obtain the existence of an imaginary time field theory over $\mathscr{S}'(R^2)$ and an Osterwalder-Schrader reconstructed field theory whose Schwinger functions are just the functional derivatives of $J(f)$. Thus the reconstructed Schwinger functions (and Wightman functions) agree with those of the field theory on $\mathscr{H}$, and so the two field theories are unitarily equivalent [15].

Theorem 2.4. *Let* $1 \leqq j_\nu \leqq d = \deg \mathscr{P}$, $1 \leqq \nu \leqq n$. *The generalized Schwinger functions*

$$\mathscr{S}_j^{(n)}(x) = \int \prod_{\nu=1}^n :\phi^{j_\nu}(x_\nu): dq$$

are continuous as multilinear forms on $\prod_{\nu=1}^n \mathscr{L}_{1,d/d-j_\nu}$, *with norms bounded by* $(\prod n_i!)K^n$ *as above. They are the functional derivatives of*

$$Z(h_1, \ldots, h_d) = \int e^{-\Sigma_{j=1}^d :\phi^j(h_j):} dq .$$

Z is bounded and analytic in $h_j \in \mathscr{L}_{1,d/d-j}$.

Proof. The proof is the same as that of Theorem 2.3, once we know that the $V \to \infty$ limit of the generalized Schwinger functions exists. We insert a momentum cutoff κ and use the bounds which result from Theorem 2.2. We conclude that

$$\mathscr{S}_{j,\kappa}^{(n)} = \int \prod_{\nu=1}^n :\phi_\kappa^{j_\nu}(x): dq$$

is bounded uniformly in κ and V, and that

$$\mathscr{S}_{j,\kappa}^{(n)} \to \mathscr{S}_j^{(n)} \tag{2.11}$$

uniformly in V. Since $\mathscr{S}_{j,\kappa}^{(n)}$ can be expressed in terms of ordinary Schwinger functions by undoing the Wick order, the existence of the $V \to \infty$ limit in $\mathscr{S}_{j,\kappa}^{(n)}$ follows. By a 3ε argument, the $V \to \infty$ limit exists in $\mathscr{S}_n^{(n)}$ also, completing the proof.

Remark. The limit (2.11) holds for $V = \infty$ also and it holds for a.e. phase $\xi \in \Xi$ uniformly in ξ. We next show that the estimates of Theorem 2.4 on the generalized Schwinger functions are valid in a.e. phase.

Theorem 2.5. *For the* $\mathscr{P}(\phi)_2$ *field theory acting on* $\mathscr{H}(\xi)$, *the Schwinger functions satisfy the bounds of Theorem 2.4, uniformly in* ξ, *and are the functional derivatives of a complex analytic function* $Z_\xi(h_1, \ldots, h_d)$, *bounded as in Theorem 2.4.*

Proof. For test functions smooth in t, the Schwinger functions with $\phi(x)$, $:\phi_\kappa^j(x):$ and $:\phi_\kappa^j(x): - :\phi_{\kappa'}^j(x):$ vertices are defined, using the estimate of Theorem 2.1 in the argument of [13]. For E a projection in $\mathscr{Z}$, the same statement applies to the Schwinger functions $\mathscr{S}_E$ defined in the vacuum $\Omega_E = E\Omega$. Assuming that h has compact support in t, we bound $\mathscr{S}_E^{(N)}$ by $K^N(\prod_i N_i!)\|E\Omega\|^2$, using the Schwarz inequality:

$$\begin{aligned} |\langle \Omega_E A \Omega_E \rangle| &\leqq \|E\Omega\| \langle \Omega_E A^* A \Omega_E \rangle^{1/2} \\ &\leqq \ldots \leqq \|E\Omega\|^{2-2^n} \langle \Omega (A^*A)^{2^n} \Omega \rangle^{2^{-n-1}} . \end{aligned}$$

Here A is the integral of a time ordered product of $\phi(x), \ldots$ vertices as above and of e^{-tH} operators. We suppose that the test functions for A also have compact support, and we choose the strips in Theorem 2.4 to match the support of the test functions in A. Then $(A^*A)^{2^n}$ is also such a time ordered product, with test functions supported in 2^{n+1} times as many time strips. If A is defined by N functional derivatives of Z, $(A^*A)^n$ has $2^{n+1}N$ such functional derivatives; however, only N_i occur in a given time strip. Thus

$$\langle \Omega (A^*A)^{2^n} \Omega \rangle \leqq (K^N \textstyle\prod_i N_i!)^{2^{n+1}}$$

which yields

$$|\langle \Omega_E A \Omega_E \rangle| \leqq \|E\Omega\|^2 K^N \textstyle\prod_i N_i!\,.$$

Thus we see that $\mathscr{S}_E^{(n)}/\|E\Omega\|^2$ satisfies the bounds of Theorem 2.4, unformly in E. Passing to the pure phases, we have

$$\mathscr{S}_E^{(n)} = \int_{\Xi_E} \mathscr{S}_\xi^{(n)} d\xi$$

$$\|E\Omega\|^2 = \int_{\Xi_E} d\xi\,,$$

which Ξ_E is the subset of Ξ corresponding to $E \in \mathscr{Z}$. It follows that $\mathscr{S}_\xi^{(n)}$ is $L_\infty(\Xi, d\xi)$, as a functional on $\mathscr{L}_{1,d/d-j_1} \times \ldots \times \mathscr{L}_{1,d/d-j_n}$ depending on ξ. These bounds permit summation of the exponential series, yielding Z_ξ, for a.e. ξ. They also permit removal of the ultraviolet cutoff in the $\phi^j(x)$ vertices. This completes the proof.

Results similar to Theorems 2.3–2.5 were obtained independently by Fröhlich [11]. As a reformulation of results of [11], we have the following two theorems:

Theorem 2.6. *For the $\mathscr{P}(\phi)_2$ field theory acting on $\mathscr{H}(\xi)$, a.e. pure phase is unitarily equivalent to an Osterwalder-Schrader reconstructed theory. If dq_ξ denotes the associated path space measure on $\mathscr{S}'(R^2)$, then*

$$Z = \int Z_\xi dq\,.$$

Theorem 2.7. *Under the unitary equivalence of Theorem 2.3, $\mathscr{Z} = \mathfrak{A}'$ is identified with the algebra $\mathscr{M}_{\text{inv}}$ of time translation invariant bounded measurable function on $\mathscr{S}'(R^2)$.*

Proof. By [11, Lemma 2.17], $\mathscr{M}_{\text{inv}} \subset \mathscr{M}_\infty \cap \mathscr{M}_+ \cap \mathscr{M}_-$, where $\mathscr{M}_\infty$ is L_∞ of the σ-field $\sum_\infty$ of sets measurable at ∞, and $\mathscr{M}_\pm$ is L_∞ of the σ-field $\sum_\pm$ generated by functions $\phi(x, t)$, for $\pm t \geqq 0$. Let G be a positive element of $\mathscr{M}_{\text{inv}}$. Then G is invariant under the time reflection operator θ [11, Proof of Theorem 2.19]. Let E_{inv} be the conditional expectation projection onto $\mathscr{M}_{\text{inv}}$. Then for $f \in \mathscr{M}_+$,

$$\begin{aligned} 0 &\leqq \int (\theta G f)^- G f\, dq \\ &= \int (\theta f)^- f G^2 dq \end{aligned}$$

and since this is true for all $G \geqq 0$ in $\mathscr{M}_{\text{inv}}$,

$$0 \leqq E_{\text{inv}}(\theta f)^- f\,.$$

Hence

$$\begin{aligned} \int (\theta G f)^- G f\, dq &\leqq \|G^2\|_\infty \int E_{\text{inv}}(\theta f)^- f\, dq \\ &= \|G^2\|_\infty \int (\theta f)^- f\, dq = \|G^2\|_\infty \langle f, f\rangle\,. \end{aligned}$$

It follows that G defines a bounded operator on $\mathscr{H}$. G is seen to commute with $\mathscr{M}_0$, the functions of the time zero field, and G commutes with time translation, on $\mathscr{H}$. Thus $G \in \mathfrak{A}' = \mathscr{Z}$.

We have proved that $\mathscr{M}_{\text{inv}}$ is a subalgebra of $\mathscr{Z}$. Thus we can define a direct integral decomposition of $\mathfrak{A}$ with respect to $\mathscr{M}_{\text{inv}}$. Each component $\mathfrak{A}(\zeta)$ is an Osterwalder-Schrader reconstruction with path space measure dq_ζ equal to a component of dq under direct integral decomposition of dq with respect to $\mathscr{M}_{\text{inv}}$. Thus the measures dq_ζ are ergodic under time translation [11, Theorem 2.16].

Hence they have a cluster property, from which the irreducibility of $\mathfrak{A}(\zeta)$ follows. By general properties of C^*-algebra decompositions, this shows that $\mathscr{Z}=\mathscr{M}_{\text{inv}}$.

3.1. Integration by Parts

The Green's functions $\mathscr{G}^{(n)}$ are defined below as the connected (truncated) parts of the Schwinger functions $\mathscr{S}^{(n)}$, with the truncation defined in the standard fashion. For the case at hand, namely a direct integral of irreducible theories, the truncation is performed in each pure phase ξ, defining $\mathscr{G}^{(n)}_\xi$, and then by definition $\mathscr{G}^{(n)}=\int\mathscr{G}^{(n)}_\xi d\xi$. The ultraviolet behavior of $\mathscr{S}^{(n)}$ and $\mathscr{G}^{(n)}$ are determined by perturbation theory, and these singularities coincide with the local singularities of a massive free theory. In order to establish this fact, we integrate by parts on path space. Successive integration by parts yields a representation of $\mathscr{S}^{(n)}$ or $\mathscr{G}^{(n)}$ as a sum of low order perturbation contributions, plus a remainder which is regular in x. Such an expansion is possible because the $\mathscr{P}(\phi)_2$ model is superrenormalizable.

The formula for integration by parts in the $\mathscr{S}^{(n)}$ is

$$\int :\phi^j:(h)R(\phi)dq=\int dq\int A(x)\left[\frac{\delta R}{\delta\phi(x)}-R(\phi)\mathscr{P}'(\phi(x))\right]dx\,, \tag{3.1.1}$$

where

$$A(x)=\int dyC(x-y)h(y):\phi^{j-1}:(y)\,,$$
$$R(\phi)=\prod_{i=1}^{n}:\phi^{j_i}:(h_i)\,.$$
$$C=(-\Delta+m_0^2)^{-1}\,.$$

Theorem 3.1.1. *The partial integration formula* (3.1.1) *is valid for the* $V=\infty$ *field theory and in a.e. phase* $\xi\in\Xi$.

Remark. The left side of (3.1.1) is defined by Theorem 2.6 and 2.10, while the right side of (3.1.1) has kernels which do not factor. To deal with this situation, we integrate by parts in each linear factor in $:\phi^j:(h)R(\phi)$. This result produces a sum of many terms, of the form

$$W=\int\left[\prod_{\nu=1}^{r}:\mathscr{P}^{(\alpha_\nu)}(\phi(y_\nu)):\right]w(y)dy\,, \tag{3.1.2}$$

where

$$w(y)=\int v(x,y)\prod_{i=1}^{n}h_i(x_i)dx\,,$$

and where $v(x,y)$ is a constant times a product of factors $C(x_i-y_j)$ and $C(x_i-x_j)$.

Let $\mu^{-\varepsilon}$ be the operator $(-\Delta+m_0^2)^{-\varepsilon/2}$, and let $\mu^{-\varepsilon}_{\otimes j}$ be the product of $\mu^{-\varepsilon}$ acting on j variables.

Lemma 3.1.2. *In* (3.1.2), $\alpha_\nu\geqq 1$ *and* $C(x)\in\mathscr{L}_{d/\alpha_\nu}$.

Proof. By definition $\alpha_\nu\geqq 1$, so $d/\alpha_\nu<\infty$. Also $C(x)$ is L_p for all $p<\infty$.

Lemma 3.1.3. *For* $\varepsilon<0$ *and any choice of exponents* l_ν *in* Z^+, *the function*

$$\mu^{-\varepsilon}\prod_{\nu=1}^{r}C(x-y_\nu)^{l_\nu} \tag{3.1.3}$$

is bounded and continuous. Here $\mu^{-\varepsilon}$ acts on x.

Proof. $C(x-y_\nu)$ has a Fourier transform of magnitude $(p^2+m_0^2)^{-1}$. Thus the $l=\sum l_\nu$ fold convolution is bounded by $O(1)(p^2+m_0^2)^{-1+\varepsilon/2}$, by standard power counting estimates, see e.g. [19]. After multiplication by $\mu^{-\varepsilon}$, the Fourier transform is integrable, and hence (3.1.3) is bounded. Continuity follows similarly, since $|(e^{ipx}-1)|\leqq O(|px|^{\varepsilon/3})$.

Theorem 3.1.4. *For any* $\varepsilon>0$, $\mu_{\otimes n}^{-\varepsilon}\mathscr{S}^{(n)}$ *is a bounded continuous function, with bounds uniform in* $\xi\in\Xi$.

Theorem 3.1.5. *Let* $\mathscr{C}_{\{j\}}=\prod_{1\leqq i<i'\leqq n}(1+C(x_i,x_{i'})^{(j_i+j_{i'})})$. *Then* $\mathscr{C}_{\{j\}}^{-1}\mathscr{S}^{(n)}$ *is a bounded continuous function, with bounds uniform in* $\xi\in\Xi$.

Proof of Theorems 3.1.1, 3.1.4, 3.1.5. The integration by parts formula (3.1.1) is known for $V<\infty$. In that case by Theorem 2.6, the representation (3.1.2) and the Lemmas 3.1.2, 3.1.3 above,

$$\begin{aligned}|(\mu_{\otimes n}^{-\varepsilon}\mathscr{S}^{(n)})(h)| &= |\mathscr{S}^{(n)}(\mu_{\otimes n}^{-\varepsilon}h)| \\ &\leqq \textstyle\sum_{\text{finite sum}}\int dx\prod\|C(y_\nu-x_i)\|_{\mathscr{L}_{d/\alpha_\nu}}\prod C(x_i-x_j)|\mu_{\otimes n}^{-\varepsilon}h| \\ &\leqq \textstyle\sum_{\text{finite sum}} O(1)\int dx\prod C(x_i-x_j)\mu_{\otimes n}^{-\varepsilon}|h| \\ &\leqq \textstyle\sum_{\text{finite sum}} O(1)\int|h|\mu_{\otimes n}^{-\varepsilon}\prod C(x_i-x_j) \\ &\leqq O(1)\|h\|_{\mathscr{L}_1},\end{aligned}$$

where $h=\prod_{i=1}^n h_i(x_i)$. Since the dual of $\mathscr{L}^1$ is $\mathscr{L}^\infty$, in a finite volume $\mu_{\otimes n}^{-\varepsilon}\mathscr{S}^{(n)}$ is $\mathscr{L}^\infty$. Since $\mu_{\otimes n}^{-\varepsilon}$ maps $\mathscr{L}^\infty$ into bounded, continuous functions, Theorem 3.1.4 holds for $V<\infty$, with bounds uniform in V. Similarly, Theorem 3.1.5 holds for $V<\infty$ with bounds uniform in V. By limits, the same bounds are valid for $V=\infty$. As in Chapter 2, these estimates decompose into a.e. phase ξ. This establishes the existence of the right side of (3.1.1). The existence of the $V=\infty$ limit for (3.1.1) follows as above.

For a polynomial $Q(\phi)$, let $Q_T=T^{-1}\int_0^T Q(\phi(\cdot+t))dt$. Then Q_T is bounded in L_p, for all $T<\infty$ and $p<\infty$. Let Q_∞ be any w^*-L_p limit point. In the integration by parts formula, if we insert a factor Q_T in both sides, a new term is included due to contraction to Q_T. This term vanishes, however, as $T\to\infty$. Thus the formula remains valid even when the integrands on both sides are multiplied by Q_∞.

Now Q_∞ is time translation invariant, and thus bounded functions of Q_∞ belong to $\mathscr{Z}$, the center of $\mathfrak{A}$. Let $\mathscr{Z}_\infty$ denote the subalgebra of $\mathscr{Z}$, generated by all such Q_∞. Then integration by parts remains valid in the direct integral theory, decomposed with respect to $\mathscr{Z}_\infty$. We must only show that a theory for which $\mathscr{Z}_\infty=\{\text{constants}\}$ is irreducible.

We show reducibility implies $\mathscr{Z}\neq\{\text{constants}\}$. Consider a theory with a multiple vacuum and let P denote the projection onto the space of vacuum vectors. Suppose R is a polynomial such that $R\Omega\perp\Omega$, $P\Omega=\Omega$, and $\|PR\Omega\|=1$. Then

$$1=\lim_{T\to\infty}\frac{1}{T}\int_0^T dt\langle R\Omega, e^{-tH}R\Omega\rangle=\langle R\Omega, R_\infty\Omega\rangle$$

and

$$0=\langle\Omega, e^{-tH}R\Omega\rangle=\langle\Omega, R_\infty\Omega\rangle .$$

Thus const. $\neq R_\infty \in \mathscr{X}_\infty$, and the proof is complete.

3.2. Green's Functions and Graphs

The integration by parts formula will be used in our study of Green's functions, i.e. connected or truncated Schwinger functions. Here we define truncations for the generalized Green's functions and relate these to the integration by parts formula.

Let $A_1, \ldots, A_n$ be functions of ϕ, e.g. $A_i=:\phi^{j_i}:(h_i)$. Let P be the set of partitions of $\{1, \ldots, n\}$ and for $\varrho \in P$, let $\varrho=\{\sigma_1, \ldots, \sigma_j\}$, $|\varrho|=j$. Each σ_ν is a subset of $\{1, \ldots, n\}$.

Definition 3.2.1. $\langle\prod_i A_i\rangle_T$ is defined recursively by

$$\langle A_1 \ldots A_n\rangle=\sum_{\varrho\in P}\prod_{\sigma\in\varrho}\langle\prod_{i\in\sigma} A_i\rangle_T . \tag{3.2.1}$$

Remark. $\langle\prod A_i\rangle_T$ depends not only on $\prod A_i$, but also on its factorization into the A_i's.

Proposition 3.2.2. *The inversion of* (3.2.1) *is*

$$\langle A_1 \ldots A_n\rangle_T=\sum_{\varrho\in P}(-1)^{|\varrho|}(|\varrho|-1)!\prod_{\sigma\in\varrho}\langle\prod_{i\in\sigma} A_i\rangle . \tag{3.2.2}$$

Proof. This is standard.

Consider the case that each A_i is a monomial in the field. Successive integration by parts produces a sum of terms, and each term is labelled by a graph D. If $\mathscr{D}$ is the set of all such graphs, the identity has the form

$$\langle\prod_i A_i\rangle=\sum_{D\in\mathscr{D}}\langle\prod_\nu B_\nu\rangle \tag{3.2.3}$$

where B_ν are monomials in the field [formed from derivatives of the A_i or of $\mathscr{P}$, as in (3.1.1)] and determined by D. The vertices of the graph are the B_ν, while the lines arise from contractions $C(x-x')$.

Theorem 3.2.2. *With the above definitions of* $\mathscr{D}$ *and* B_ν

$$\langle\prod_{i=1}^n A_i\rangle_T=\sum_{D\in\mathscr{D}}\langle\prod_\nu B_\nu\rangle_T .$$

Proof. It is sufficient to treat the case of one integration by parts, in one linear factor from A_1. We use (3.2.2) to define $\langle\prod A_i\rangle_T$, and integrate by parts on the right side in the factor containing A_1. We group the result into $n+1$ terms. There are $n-1$ terms in which A_1 contracts to an A_i, $1\leqq i\leqq n$. There is one term with A_1 contracted to a new $:\mathscr{P}(\phi):$ vertex, and if A_1 is not Wick ordered, a contraction of A_1 to itself is also possible. Let B_1' be the function of ϕ which results from the contraction of A_1 to $:\mathscr{P}:$, and let B_1'' be the result of contracting A_1 to itself. Since these last two terms occur for each $\varrho\in P$, they contribute

$$\begin{aligned}&\sum_{\varrho\in P}(-1)^{|\varrho|}(|\varrho|-1)!\Big(\prod_{\substack{\sigma\in\varrho\\1\notin\sigma}}\langle\prod_{i\in\sigma} A_i\rangle\Big)\Big\langle\Big(\prod_{\substack{i\in\sigma_i\\i\neq 1}} A_i\Big)B_1'+B_1''\Big)\Big\rangle\\&\quad=\langle B_1'\prod_{i>1} A_i\rangle_T+\langle B_1''\prod_{i>1} A_i\rangle_T .\end{aligned}$$

Next consider a term with A_1 contracted to A_i, and let B_{1i} be the result of this contraction. Such a term arises only for certain $\varrho \in P$, namely ϱ which do not separate 1 and i. Let P_i be the set of all such partitions. This term is

$$\sum_{\varrho \in P_i} (-1)^{|\varrho|}(|\varrho|-1)! \left(\prod_{\substack{\sigma \in \varrho \\ 1 \notin \sigma}} \langle \prod_{i \in \sigma} A_i \rangle \right) \left\langle B_{1i} \prod_{\substack{v \in \sigma_1 \\ 1 \neq v \neq i}} A_v \right\rangle$$
$$= \langle B_{1i} \prod_{1 \neq v \neq i} A_v \rangle_T \, .$$

Combining these $n+1$ terms completes the proof of Theorem 3.2.2.

Remark 3.2.3. For any graph D which is not connected, but which has a vacuum component (so that $\deg B_v = 0$, or $B_v = \text{const.}$ for some v), the term $\langle \prod_v B_v \rangle_T$ vanishes.

The truncations defined by (3.2.2) are not linear in the state $\langle \ \rangle$. Thus there is a difference between truncating in the state defined by $\Omega \in \mathcal{H}$ and truncating in a.e. pure phase ξ. The latter is the correct definition, because it leads to Green's functions with nice cluster properties. Using the measure dq_ξ, we define the general Green's functions as

$$G_\xi^{(n)}(x_1, \ldots, x_n) = (\int :\phi(x_1)^{j_1}: \ldots :\phi(x_n)^{j_n}: dq_\xi)_T \tag{3.2.4}$$

and

$$G^{(n)}(x_1, \ldots, x_n) = \int_\Xi G_\xi^{(n)}(x_1, \ldots, x_n) d\xi \, . \tag{3.2.5}$$

It follows immediately that $G^{(n)}$ and $G_\xi^{(n)}$ satisfy the bounds of Theorems 3.1.4 and 3.1.5. We improve these bounds in Chapter 4. Using integration by parts, as in Theorem 3.1.1, we no longer need to restrict j_v to be less than $d = \deg \mathcal{P}$. However, in this case, we lose the nice $n!$ bound on the $\mathcal{S}^{(n)}$ from Chapter 2. In principle, the $G^{(n)}$ should have an $n!$ estimate (as derivatives of an analytic function) for factorizing test functions, and for $j_v \leqq d$. See [20, 11].

We call *ordinary* the Green's functions

$$G_\xi^{(n)}(x_1, \ldots, x_n) = (\int \phi(x_1) \ldots \phi(x_n) dq_\xi)_T \, , \tag{3.2.6}$$

or their integral, and *generalized* the Green's functions (3.2.4).

4. Regularity of the Green's Functions

There are two aspects of the regularity of the Green's functions: their short distance singularities (ultraviolet behavior) and their long distance decay (infrared behavior). The intermediate range behavior is settled by the general fact that the $\mathcal{S}^{(n)}$ are real analytic at noncoinciding points. The infrared behavior is controlled by the assumption that we are dealing with a noncritical theory. In the case of a physical mass $m>0$, the $G^{(n)}$ decay exponentially under large separation of points, $|G^{(n)}| \leqq e^{-mr}$. We assume that decay occurs in all Euclidean directions, as is the case for a Lorentz invariant theory with $m>0$. The following theorem bounds the ultraviolet behavior:

Theorem 4.1. *For $n>2$, the ordinary $\mathcal{P}(\phi)_2$ Green's functions* (3.2.6) *are bounded and continuous for a.e. $\xi \in \Xi$, with bounds uniform in ξ. For $n=2$,*

$G_\xi^{(2)}(x-x')-C(x-x')$ *has this property. For the generalized Green's functions of a* $\mathscr{P}(\phi)_2$ *theory,* $\mu_{\otimes n}^{-\varepsilon}G_\xi^{(n)}$ *and* $\mathscr{C}_{\{j\}}^{-1}G_\xi^{(n)}$ *are bounded and continuous for a.e.* $\xi\in\Xi$ *(cf. Theorems 3.1.4–3.1.5), with bounds uniform* ξ.

Proof. The bounds on $\mu_{\otimes n}^{-\varepsilon}G^{(n)}$ and $\mathscr{C}_{\{j\}}^{-1}G^{(n)}$ follow from Theorems 3.1.4–3.1.5 and the definition of $G^{(n)}$. The bound on the ordinary Green's functions (all $j_\nu=1$) follows from the proof of these theorems, once we observe that there are no $\phi(x_i)-\phi(x_j)$ contractions in $G^{(n)}$, because of the truncations. Here we require $n>2$, or if $n=2$, we exclude the one (zero order) term with such a contraction.

For the long distance behavior we use the comparison function

$$\mathscr{E}_\varepsilon=\prod\nolimits_{1\leqq i<l\leqq n}e^{\varepsilon|x_i-x_l|}\,.$$

Theorem 4.2. *For a noncritical* $\mathscr{P}(\phi)_2$ *model with exponential decay* $\geqq m$ *in all Euclidean directions, the functions*

$$\mathscr{E}_\varepsilon\mu_{\otimes n}^{-\delta}G^{(n)},\qquad \mathscr{E}_\varepsilon\mathscr{C}_{\{j\}}^{-1}G^{(n)}$$

are bounded and continuous for some $\varepsilon>0$ *and for any* $\delta>0$. *For the ordinary Green's functions, the same is true for*

$$\mathscr{E}_\varepsilon G^{(n)}\,.$$

Proof. We modify the proof of Theorems 3.1.4–3.1.5 to obtain the exponential decay. Let d be the diameter of $\{x_1,\dots,x_n\}$, and let (x_i,x_j) be the pair determining this diameter. We consider two parallel lines separated by distance r which are perpendicular to the line (x_i,x_j) and which separate $\{x_1,\dots,x_n\}$ into two nonempty clusters $\{I_1,I_2\}$. We assume each x_i has a distance at least r from the lines. We assert such lines always exist with $r\geqq d/3n$, see [30].

For a term labelled by a graph D, let $\varrho(D)$ be the partition of $\{1,\dots,n\}$ determined by the connected components of D. If ϱ does not refine $\{I_1,I_2\}$, then G contains a sequence of contractions of length r. In this case, the term is dominated by $O(e^{-m_0r})$, so the corresponding contributions to $G^{(n)}$ have exponential decay. Thus we consider partitions ϱ which refine $\{I_1,I_2\}$. In each integration dz over a generalized Green's function in a connected component, we restrict dz to lie on the same side of the parallel lines as the x_i variables of that component. The contribution from the complementary z region is again exponentially small. However, the contribution with z restricted in this fashion is also exponentially small, as one sees by using the Schwarz inequality and the mass $m>0$ in a standard fashion.

We now obtain expressions for the constant momentum subspace of $H_{-1}^{\otimes j}$ and for the restriction of $G^{(n)}$ to this subspace. In $L_2(R^{2j})=L_2(R^2)^{\otimes j}$, the functions $f(x_1,\dots,x_n)$ of zero total momentum are invariant under the transformation $x_i\to x_i+a$. We define a norm on such functions

$$|f|=[\textstyle\int|f(x_1,\dots,x_n)|^2dx_2\dots dx_n]^{1/2}\,. \tag{4.1}$$

namely the L_2 norm in $x_2,\dots,x_n$, and we let $\mathscr{H}(0)$ denote the zero momentum Hilbert space. Functions of total momentum P have the form

$$f_P=e^{iP\Sigma x_j/n}f\in\mathscr{H}(P)\,,\qquad f\in\mathscr{H}(0)\,,$$

and $|f_P|=|f|$. We also write $\mathscr{H}^{(n)}(P)$ to indicate the number of particles.

The norm in $H^{\otimes j}_{\pm 1}$ is determined by $\mu^{\pm 1}_{\otimes j}$ and the mappings

$$H^{\otimes j}_{1} \xrightarrow[\mu\otimes j]{} L_2(R^2)^{\otimes j} \xrightarrow[\mu\otimes j]{} H^{\otimes j}_{-1}$$

are unitary. We define the momentum P vectors $\mathscr{H}_{\pm}(P)$ in $H^{\otimes j}_{-1}$ by

$$\mathscr{H}_{\pm}(P) = \mu^{\pm 1}_{\otimes j}\mathscr{H}(P)\,, \tag{4.2}$$

with the norm $|\mu^{\pm 1}_{\otimes j} f_P| = |f_P|_{\pm}$.

We consider $G^{(n)}(x_1, \ldots, x_n)$ as the kernel of an integral operator $G^{(n)}$ from $H^{\otimes j}_{-1}$ to $H^{\otimes n-j}_{1}$. This is unitarily equivalent to the operator

$$\mu_{\otimes n} G^{(n)} \equiv \mu_{\otimes n-j} G^{(n)} \mu_{\otimes j}$$

as a transformation from $L_2^{\otimes j}$ to $L_2^{\otimes n-j}$. The unitary equivalence commutes with translations.

If we extend $G^{(n)}$ to a map from $\mathscr{H}^{(j)}_{-}(P)$ to $\mathscr{H}^{(nn-j)}_{+}(P)$, this is unitarily equivalent to $\mu_{\otimes n} G^{(n)}$ as a map from $\mathscr{H}^{(j)}(P)$ to $\mathscr{H}^{(n-j)}(P)$.

Theorem 4.3. (a) *For a noncritical $\mathscr{P}(\phi)_2$ model, and $0<j<n$, the ordinary Green's functions $G^{(n)}$ is a bounded operator from $H^{\otimes j}_{-1}$ to $H^{\otimes n-j}_{1}$.*

(b) *If in addition $n>2$, then $G^{(n)}$ defines a Hilbert-Schmidt map, $\mu_{\otimes n} G^{(n)}$, from $\mathscr{H}^{(j)}(P)$ to $\mathscr{H}^{(n-j)}(P)$.*

Remark. In essence, this theorem states that after differentiating each variable of $G^{(n)}$ once, the result $\mu_{\otimes n} G^{(n)}$ is an L_2 function of the difference variables. The proof also shows that $\mu_{\otimes n} G^{(n)}$ is continuous in P, and vanishes in Hilbert-Schmidt norm as $P \to \infty$.

Theorem 4.4. *For a noncritical $\mathscr{P}(\phi)_2$ model, and $0<j<n$, the generalized $G^{(n)}$ is a bounded operator from $H^{\otimes j}_{-1+\varepsilon}$ to $H^{\otimes n-j}_{1-\varepsilon}$, for $\varepsilon>0$.*

We prove Theorem 4.3; the proof of Theorem 4.4 is similar. We integrate by parts in each linear factor $\phi(x_i)$ in $G^{(n)}$. Because of truncation (see Theorem 3.2.2) there are no $\phi(x_i)-\phi(x_k)$ contractions. If $n \leqq \deg \mathscr{P}$, there is one factor proportional to

$$\int \prod_{i=1}^{n} C(x_i - z)dz = F(x_1, \ldots, x_n) \tag{4.2}$$

and this term fulfills the conclusions of the theorem. In fact, the Fourier transform of $\mu_{\otimes n} F$ is proportional to

$$\delta\left(\sum_{i=1}^{n} p_i\right) \prod_{i=1}^{n} \mu(p_i)^{-1}\,,$$

which is the kernel of a bounded operator from $L_2^{\otimes j}$ to $L_2^{\otimes n-j}$, verifying the conclusion of (a). Furthermore,

$$\begin{aligned} |\mu_{\otimes n} F|^2 &= \text{const.}\ \left\|\mu\left(\sum_{i=2}^{n} p_i\right)^{-1} \prod_{i=2}^{n} \mu(p_i)\right\|^2_{L_2} \\ &\leqq \text{const.}\,, \end{aligned}$$

proving that (4.2) also satisfies the conclusion of (b).

The remaining terms are all of the form

$$T = \int \prod C(x_i - y_{l_i}) G^{(r)}(y_1, \ldots, y_r) dy\,,$$

where $G^{(r)}$ is a generalized Green's function. By Theorem 4.2, $\mu_{\otimes n} T$ is continuous, and exponentially decreasing in its difference variables. The theorem follows from these facts.

5. The Support of dq

Following Fröhlich [11], we introduce the Euclidean π operators which generate the translation group $\phi \to \phi + f$.

$$e^{i\pi(f)} : F(\phi) \to F(\phi + f)\left(\frac{d(q+f)}{dq}\right)^{1/2} . \tag{5.1}$$

The Jacobian in (5.1) is

$$\frac{d(q+f)}{dq} = \exp - [\delta\mathscr{P}(f) + \phi((-\Delta + m_0^2)f) + \tfrac{1}{2}\langle f, (-\Delta + m_0^2)f\rangle] , \tag{5.2}$$

where

$$\delta\mathscr{P}(f) = \int [:\mathscr{P}(\phi + f): - :\mathscr{P}(\phi):] dx .$$

In view of the bounds of Chapter 2, we choose

$$f \in \mathscr{L}_{1,d} \cap (-\Delta + m_0^2)^{-1} \mathscr{L}_{1,d/d-1} \cap H_1 , \tag{5.3}$$

where H_1 is a Sobolev space. Then (5.2) belongs to $L_p(\mathscr{S}'(R^2), dq)$ for all $p < \infty$.

Theorem 5.1 [11]. *The Euclidean π operators are self adjoint operators on the Euclidean space $\mathscr{H}_{\mathscr{E}} = L_2(\mathscr{S}'(R^2), dq)$. The generating functional*

$$E(f, g) = \langle \Omega, e^{i\phi(g)} e^{i\pi(f)} \Omega \rangle \tag{5.4}$$

is entire analytic for $g \in L_{1,d/d-1}$ and f in (5.3).

Theorem 5.2. *The Euclidean operators exist as self adjoint operators on $\mathscr{H}_{\mathscr{E}}(\xi) = L_2(\mathscr{S}'(R^2), dq_\xi)$. The generating functional E_ξ, defined as in (5.4), characterizes the Weyl form of the commutation relations. (It satisfies positive definiteness, continuity and commutation relations.) Moreover $E = \int E_\xi d\xi$, and E_ξ satisfies the bounds of Theorem 5.1.*

This theorem is essentially contained in [11]. The key step is to show that (5.2) is the Jacobean of the translated measure on $\mathscr{H}_{\mathscr{E}}(\xi)$.

It is convenient to work with a domain of smooth vectors $\mathscr{D}$ which is dense in $\mathscr{H}_{\mathscr{E}}(\xi)$. Let $\mathscr{D}$ be the linear span of $\Omega \equiv 1$ and the vectors

$$A\Omega, \quad A = A_n = (\textstyle\prod_{i=1}^{n} \phi)(g_n) , \tag{5.5}$$

where g_n is a symmetric function in $\mathscr{S}(R^2)^{\otimes n}$. Then

$$\begin{aligned} &\phi(f)\mathscr{D} \subset \mathscr{D}, \quad f \in \mathscr{S}(R^2) , \\ &U(a)\mathscr{D} \subset \mathscr{D} , \end{aligned} \tag{5.6}$$

where we assume that Ω is invariant under the action of Euclidean translations $U(a)$. $\mathscr{D}$ is not invariant under $\pi(f)$, and so we introduce the domain $\mathscr{D}'$, spanned by the vector Ω and

$$A\Omega,\ A=\prod_{i=1}^{n}:\phi^{\nu_i}(g_n):,\qquad \nu_i\leqq d\,. \tag{5.5'}$$

Then

$$:\phi^{\nu}(f):\mathscr{D}'\subset\mathscr{D}'\,;\qquad U(a)\mathscr{D}'\subset\mathscr{D}'\,; \tag{5.6'}$$

for $\nu\leqq d$, and

$$\pi(f)\mathscr{D}'\subset\mathscr{D}',\ f\in\mathscr{S}(R^2)\,. \tag{5.7}$$

The relation (5.7) follows from the canonical commutation relations and the identity

$$\pi(f)\Omega=-\frac{i}{2}\{\mathscr{P}'(f)+\phi((-\Delta+m_0^2)f)\}\Omega\,, \tag{5.8}$$

which is the infinitesimal form of (5.2).

Theorem 5.3. *The L_2 inner product defined by dq_ξ is nondegenerate on $\mathscr{D}\times\mathscr{D}$.*

Remark. The equation $B=0$, where

$$B=\sum_{n=0}^{N}A_n(q) \tag{5.9}$$

defines a polynomial variety in $\mathscr{S}'(R^2)$. The theorem states that dq_ξ is not supported on any polynomial variety. Here A_n is defined by (5.5).

Proof. Assume that for all $C\in\mathscr{D}$, $\int CBdq_\xi=0$. Since $B\in L_2(dq_\xi)$ and since $\mathscr{D}$ is dense in $\mathscr{D}'$ in the $L_2(dq_\xi)$ norm, by the remark following Theorem 2.4, $B\perp\mathscr{D}'$ also. Thus

$$0=\int C_1BC_2dq_\xi=\langle C_1BC_2\rangle\,,\qquad C_i\in\mathscr{D}'\,,$$

and so for all $f_j\in\mathscr{S}(R^2)$, we have

$$\begin{aligned}0&=\langle\Omega[\pi(f_1),[\pi(f_2),[\ldots,[\pi(f_N),B]\ldots]]]\Omega\rangle\\&=N!(-i)^N g_N(f_1\otimes\ldots\otimes f_N)\,.\end{aligned} \tag{5.10}$$

It follows that $g_N\equiv 0$, and by induction $g_n\equiv 0$ for all n. This completes the proof.

We need to improve this result by allowing A_n to lie in a constant momentum subspace. We denote the pure phase vacuum $\int\cdot\,dq_\xi$ by $\langle\cdot\rangle=\langle\cdot\rangle_\xi$, as above, and we assume a nonzero mass gap.

Theorem 5.4. *Assume an exponential decay rate for correlations, uniformly in all Euclidean directions. Given $p\in R^2$, suppose*

$$0=\int[\langle C,U(a)e^{ipa}B\rangle-\langle C\rangle\langle e^{ipa}B\rangle]da \tag{5.11}$$

for all $C\in\mathscr{D}$. Then

$$0=\int g_n(x_1-a,\ldots,x_n-a)e^{ipa}da,\qquad n\geqq 1\,. \tag{5.12}$$

Proof. The integrand in (5.11) is rapidly decreasing in a, as a consequence of the nonzero mass gap and Theorem 4.2. In fact, the expansion into connected Green's functions yields a sum of products, and each product contains at least one factor with variables from both B and C. This factor decays rapidly in $|a|$ by Theorem 4.2 and the fact that $g_n \in \mathscr{S}$, while the other factors are bounded. We then follow the proof of Theorem 5.3.

For application in Chapters 7,8, we want to consider kernels $g_n \in H_{-1}^{\otimes n}$ rather than in $\mathscr{S}$. In this case, A_n given by (5.5) is not in general defined (even for the free field) but requires some subtraction, e.g. Wick ordering would be sufficient. We define the Wick ordering with respect to the measure dq_ξ, i.e. with respect to the function $\Omega \equiv 1$.

In order to analyze higher portions of the energy spectrum, it is also necessary to subtract states associated with the span of $n=1,2,\dots$ fields, $\phi(f)$, $\phi(f)\phi(g)$, etc. We consider here the cases $n=0,1$. We discuss briefly the $n=2$ subtraction in Chapter 8, when we introduce equations for three particle states.

Let P_j be the projection of $\mathscr{H}_{\mathscr{E}}(\xi)$ onto the closed subspace of degree j polynomials in ϕ which are orthogonal to all degree l polynomials, $l<j$.

Theorem 5.5. *With the hypotheses of Theorem 5.4,*

$$(1-P_0)B = B - \langle B\rangle \tag{5.13}$$

$$P_1 B = -\int \langle B\hat{\phi}(y)\rangle \Gamma^{(2)}(y-z)\hat{\phi}(z)dydz \tag{5.14}$$

where $\hat{\phi} = \phi - \langle\phi\rangle$, *$B$ is defined by* (5.9) *and each* $g_n \in \mathscr{S}(R^{2n})$.

Proof. Since (5.13) is obvious, we need only verify (5.14). By Theorem 4.2, $g(y) = \langle B\hat{\phi}(y)\rangle$ is a rapidly decreasing element of H_1. Since $\Gamma^{(2)}: H_1 \to H_{-1}$ [20], $h = \Gamma^{(2)} g \in H_{-1}$, and

$$\langle \hat{\phi}(\bar{h})\hat{\phi}(h)\rangle = G^{(2)}(\bar{h}\otimes h) \leqq O(1)\|h\|_{-1}^2\,, \tag{5.14a}$$

cf. [20]. Thus the right side of (5.14) is in $L_2(\mathscr{S}'(R^2), dq_\xi)$. Finally we note that the right side of (5.14) equals: 0 for $B=1$; $\hat{\phi}(x)$ for $B=\hat{\phi}(x)$; and 0 if $B\perp$ to 1 and $B\perp\phi(x)$, all x. These properties characterize P_1, hence (5.14).

Corollary 5.6. *Let*

$$B = \int \hat{\phi}(x)\hat{\phi}(y)g(x,y)dxdy\,. \tag{5.15}$$

Then (5.13) *and* (5.14) *admit extensions to* $g \in H_{-1}^{\otimes 2}$, *by continuity.*

Proof. We note

$$\langle B\hat{\phi}(z)\rangle = \int G^{(3)}(x,y,z)g(x,y)dxdy\,.$$

Thus by Theorem 4.3, we see that (5.14) extends by continuity to $g \in H_{-1}^{\otimes 2}$. To establish (5.13) on $H_{-1}^{\otimes 2}$, we note that

$$\|(1-P_0)B\|^2 = G^{(4)}(g^- \otimes g) + \langle g, G^{(2)}\otimes G^{(2)} g\rangle_{L_2(R^2)}\,,$$

and each term extends to $H_{-1}^{\otimes 2}$ by continuity, using Theorem 4.3 and (5.14a).

We now assume the decay properties of $\Gamma^{(2)}(x)$ established in Chapter 6, to sharpen our estimates.

Theorem 5.7. *Suppose g is symmetric and is exponentially decreasing in the variable $x+y$, so that for all β, with $|\beta|$ sufficiently small,*

$$e^{\beta\cdot(x+y)}g(x,y) \in H_{-1}^{\otimes 2}\,.$$

Then with B given by (5.15),

$$\Gamma^{(2)} * \langle B\hat{\phi}(\cdot)\rangle \in H_{-1} \tag{5.16}$$

is exponentially decreasing.

Proof. As above, $\langle B\hat{\phi}(\cdot)\rangle$ is exponentially decreasing, and hence an element of $H_{1,m}$, as defined in (6.9). By Theorem 6.4, and $\bar{m} \geqq m$, (5.16) is an element of $H_{-1,m}$.
We remark that in the hypothesis of Theorem 5.4, we could write (5.11) as $0 = \int \langle C, (1-P_0)U(a)e^{ipa}B\rangle da$.

Theorem 5.8. *With B, g as above, suppose that*

$$0 = \int \langle C, (1-P_0-P_1)U(a)e^{ipa}B\rangle da \tag{5.17}$$

for all $C \in \mathcal{D}$. *Then*

$$0 = \int g(x_1 - a, x_2 - a)e^{ipa} da\,. \tag{5.18}$$

Proof. We treat $(1-P_0)$ and P_1 separately, since each of these projections has rapid decay in a. We see this by expanding each term as a sum of products of connected Green's functions. Using the exponential decay of g in the variable $x+y$, the (mass gap) exponential decay of the connected Green's functions and Theorem 6.4, the decay in a follows. Two commutators with π then yield (5.18).

Remark 5.9. The decay of the integrand does not depend on the local regularity of $C \in \mathcal{D}$. In particular, we may take $C = B$.

6. Decay of $\Gamma^{(2)}(x)$ and CDD Zeros

We consider an Osterwalder-Schrader theory with the (connected) momentum space two point function

$$\chi(p) = G^{(2)}(P) = \frac{Z}{p^2+m^2} + \int_{a \geqq M^2} \frac{dv(a)}{p^2+a}\,. \tag{6.1}$$

We assume $0 < m \leqq M$. If, in addition, $m < M$ and $Z \neq 0$, the model has an isolated one particle state of mass m, and in that case Z is the field strength renormalization constant. By positivity, $Z \geqq 0$ and $dv \geqq 0$, and we assume $\int dv(a)a^{-1} < \infty$. If the theory is canonical, then in addition

$$Z + \int dv(a) = 1\,. \tag{6.2}$$

The inverse propagator $-\chi(p)^{-1}$ is also the one particle irreducible part of $\chi(p)$ The Fourier transform of $-\chi(p)^{-1}$ is $\Gamma^{(2)}(x)$, and convolution by $\Gamma^{(2)}$ is an isomorphism of H_1 onto H_{-1}, by [20, Theorem 2.1 and Remark 2].

We are concerned here with the exponential decay rate $\bar{m}$ of $\Gamma^{(2)}(x)$, since such decay is related to the spectral properties of the theory. We find that if $m < M$, then

$$m < \bar{m} \leqq M\,, \tag{6.3}$$

and $\bar{m}$ is related to the complex zeros of $\chi(p)$. In case $\bar{m} < M$, then the decay rate $\bar{m}$ is not associated with any physical particle. For instance, if M is the mass of an isolated bound state, we find

$$m < \bar{m} < M, \tag{6.4}$$

In case the strict inequality (6.4) holds, the propagator $\chi(p)$ has a CDD zero, see Castillejo, Dalitz, and Dyson [4].

Our results are nontrivial only if $m < M$, i.e. if an upper mass gap exists in the propagator.

Definition 6.1. We define the CDD root $\bar{m}$ as the root, in the interval $(m, M]$, of the equation

$$Z + \int_{a \geqq M^2} \frac{m^2 - \bar{m}^2}{a - \bar{m}^2} \, d\nu(a) = 0\,. \tag{6.5}$$

If no such root exists, or if $m = M$, define $\bar{m} = M$.

Remark 6.2. Assume a canonical theory (6.4). Then for $Z \neq 0$

$$\bar{m} = \left[\frac{\zeta}{\zeta + Z} m^2 + \frac{Z}{\zeta + Z} M^2 \right]^{1/2}, \tag{6.6}$$

where $\zeta \in [0, 1 - Z]$ is defined by the equation

$$\zeta = \int \frac{M^2 - \bar{m}^2}{a - \bar{m}^2} \, d\nu(a)\,. \tag{6.7}$$

For example, if $d\nu(a)$ is a point mass at $a = M^2$, then $\bar{m} = [(1 - Z)m^2 + ZM^2]^{1/2}$, with $\zeta = 1 - Z$ and (6.4) holds. CDD zeros are absent if and only if

$$\int \frac{M^2 - m^2}{a - M^2} \, d\nu(a) < Z\,. \tag{6.8}$$

We show below that $\Gamma^{(2)}(x)$ has the decay rate $\bar{m}$. In order to make precise the notion of decay rate we introduce the Sobolev-Paley-Wiener space $H_{p,\lambda}$. For $\lambda = 0$, we let $H_{p,\lambda} = H_p$, the usual Sobolev space. For $\lambda > 0$, we have

Definition 6.3. Let $H_{p,\lambda}$ be those distributions f for which

$$e^{\alpha x}(-\Delta + a)^{p/2} f \in L_2 \tag{6.9}$$

for all α with $|\alpha| < \lambda$. Here $a > \lambda^{1/2}$ and such a is sufficiently large not to affect the exponential decay rate of f.

Theorem 6.4. (*a*) *The convolution operator $\Gamma^{(2)}(x)$ is a continuous transformation from $H_{1,\lambda}$ to $H_{-1,\lambda}$ for any $\lambda \leqq \bar{m}$.*

(b) *If $\bar{m} < M$, then $\Gamma^{(2)}(x)$ is not densely defined, from $H_{1,\lambda}$ to $H_{-1,\lambda}$ for any $\lambda > \bar{m}$.*

Remark. This is the precise sense in which $\Gamma^{(2)}(x)$ decays at the rate $\bar{m}$. Thus the presence of CDD zeros is equivalent to a decay rate $\bar{m}$ not associated with a physical particle.

Lemma 6.5. *Let T be a convolution operator with kernel $T(x)$, and suppose that when a is sufficiently large,*

$$F(k) = (a + k^2)^{(q-p)/2} \tilde{T}(k)$$

is analytic for $k=\sigma+i\tau$, *with* $|\tau|<\lambda$, *and is bounded in* σ *for each fixed* $|\tau|<\lambda$. *Then the operator*

$$T: H_{p,\lambda} \rightarrow H_{q,\lambda}$$

is continuous. (Here $|\tau|$ *is the Euclidean norm.)*

Proof. By Fourier transformation, $H_{p,\lambda}$ contains those functions f such that for $|\alpha|<\lambda$ and k real,

$$e^{\alpha\cdot\nabla}(a+k^2)^{p/2}\tilde{f}(k)\in L_2 .$$

Thus a convolution operator T is bounded from $H_{p,\lambda}$ to $H_{q,\lambda}$ if for k real,

$$F_\alpha \equiv e^{\alpha\cdot\nabla}(a+k^2)^{(q-p)/2}\tilde{T}(k)e^{-\alpha\cdot\nabla} \tag{6.10}$$

is a bounded operator on L_2 for all real α, $|\alpha|<\lambda$. In case the series converges, F_α is given by the Baker-Campbell-Hausdorff formula as a sum of multiplication operators on L_2,

$$F_\alpha=\sum_{n=0}^{\infty}\frac{1}{n!}\left[(\alpha\cdot\nabla)^n(a+k^2)^{(q-p)/2}\tilde{T}(k)\right].$$

Thus the operator norm of F_α on L_2 is bounded by

$$\|F_\alpha\|\leqq\sum_{n=0}^{\infty}\frac{1}{n!}\|(\alpha\cdot\nabla)^n(a+k^2)^{(q-p)/2}\tilde{T}_{(k)}\|_{L_\infty}. \tag{6.11}$$

We estimate (6.11), using the Cauchy integral formula and the assumption of boundedness on $(a+k^2)^{(q-p)/2}\tilde{T}(k)$. By a rotation of coordinates, we may suppose that α is parallel to the x_1 coordinate axis. Thus we estimate the derivatives of $(a+k^2)^{(q-p)/2}\tilde{T}(k)$, for $k=\sigma$ real, by integrating over a disc $|\sigma_1-z_1|\leqq\beta$, centered at σ, and with $|\alpha|<\beta<\lambda$. Then

$$\left|\left(\frac{\partial}{\partial k_1}\right)^n(a+k^2)^{(q-p)/2}\tilde{T}(k)\right|\leqq A\beta^{-n},$$

where $A=A(\beta)$ is independent of n and σ. We conclude that

$$\|F_\alpha\|\leqq A\sum_{n=0}^{\infty}|\alpha|^n\beta^{-n}<\infty .$$

Proof of Theorem 6.4. We note

$$-\tilde{\Gamma}^{(2)}(k)(k^2+m^2)^{-1}=\left[Z+\int_{a\geqq M^2}\left(\frac{k^2+m^2}{k^2+a}\right)dv(a)\right]^{-1}, \tag{6.12}$$

and we first establish (*b*). If $\Gamma^{(2)}(x)$ has a decay rate β, the Fourier transform $\tilde{\Gamma}^{(2)}(k)$ has an analytic continuation defined for $|\tau|<\beta$. In particular, for $\sigma=0$,

$$0\neq Z+\int_{a\geqq M^2}\frac{m^2-\tau^2}{a-\tau^2}dv(a). \tag{6.13}$$

The integral in (6.13) decreases from 0 to

$$\int\frac{m^2-M^2}{a-M^2}dv(a)<0$$

as τ^2 increases from m^2 to M^2. If $\bar{m}<M$, then $\bar{m}$ is the first zero of (6.13). Thus we have $\beta\leqq\bar{m}$, proving (b).

To establish (a), we note that for $A\leqq B$, $B+x>0$, the function $(A+x)(B+x)^{-1}$ is monotonic increasing in x. Thus for $k^2=(\sigma+i\tau)^2$,

$$\begin{aligned}\operatorname{Re}\left(\frac{k^2+m^2}{k^2+a}\right)&=\left.\frac{(\sigma_1^2-\tau^2+m^2)(\sigma_1^2-\tau^2+a)+4\sigma_2^2\tau^2}{(\sigma_1^2-\tau^2+a)^2+4\sigma_2^2\tau^2}\right|_{\sigma_1^2=\sigma_2^2=\sigma^2}\\&\geqq\frac{(\sigma_1^2-\tau^2+m^2)(\sigma_1^2-\tau^2+a)}{(\sigma_1^2-\tau^2+a)^2}=\frac{\sigma_1^2-\tau^2+m^2}{\sigma_1^2-\tau^2+a}\\&\geqq\frac{-\tau^2+m^2}{-\tau^2+a}.\end{aligned}\tag{6.14}$$

For $|\tau|<\bar{m}$, (6.13) is strictly positive. Thus by (6.14),

$$0<Z+\int_{a\geqq M^2}\frac{m^2-\tau^2}{a-\tau^2}\,dv(a)\leqq\operatorname{Re}\left[Z+\int\frac{k^2+m^2}{k^2+a}\,dv(a)\right].$$

We conclude that the right side of (6.12) is nonsingular and bounded away from zero uniformly in σ, for fixed $|\tau|<\bar{m}$. Furthermore in (6.12).

$$\int_{a\geqq M^2}\frac{k^2+m^2}{k^2+a}\,dv(a)$$

is analytic in k^2 for $\tau^2<\bar{m}^2$, since $\operatorname{Re}(k^2+a)=\sigma^2-\tau^2+a\geqq\sigma^2+M^2-\bar{m}^2>0$. Thus $(k^2+m^2)^{-1}\tilde{\Gamma}(k)$ is analytic $|\tau|<\bar{m}$ and bounded uniformly in σ for fixed $|\tau|<\bar{m}$. We conclude that Lemma 6.5 applies to $\Gamma^{(2)}(x)$ in the case $p=-q=1$, $\lambda\leqq\bar{m}$, and the proof is complete.

7. The Two Particle Bethe-Salpeter Kernel

In this chapter, we prove the existence of the Bethe-Salpeter kernel K_ξ in the Euclidean region, for almost every pure phase ξ. Let R and R_0 be defined as in (1.6) so that

$$R_0(x_1,x_2,y_1,y_2)=G_\xi^{(2)}(x_1,y_1)G_\xi^{(2)}(x_2,y_2)+G_\xi^{(2)}(x_1,y_2)G_\xi^{(2)}(x_2,y_1)\tag{7.1}$$

and

$$\begin{aligned}R(x_1,x_2,y_1,y_2)&=R_0(x_1,x_2,y_1,y_2)+G_\xi^{(4)}(x_1,x_2,y_1,y_2)\\&\quad\int G_\xi^{(3)}(x_1,x_2,z)\Gamma_\xi^{(2)}(z,z')G_\xi^{(3)}(z',y_1,y_2)dzdz'\\&=\langle\phi(x_1)\phi(x_2)(I-P_0-P_1)\phi(y_1)\phi(y_2)\rangle\end{aligned}\tag{7.2}$$

Here $\Gamma_\xi^{(2)}=-G_\xi^{(2)-1}$, so that R is one particle irreducible between the initial variables $x=(x_1,x_2)$ and the final variables $y=(y_1,y_2)$. Then K_ξ is defined by

$$R=R_0+R_0K_\xi R\tag{7.3}$$

or

$$K_\xi=R_0^{-1}-R^{-1}.\tag{7.4}$$

Theorem 7.1. *Consider a $\mathscr{P}(\phi)_2$ model and suppose that* (a) *the $V=\infty$ limit is Lorentz invariant and* (b) *the mass $m=m_\xi$ in the pure phase ξ is strictly positive. Then K_ξ defined by* (7.3)–(7.4) *is a bounded operator from* Sym $H_1^{\otimes 2}$ *to* $H_{-1}^{\otimes 2}$.

Remark. We expect that the hypotheses are valid for all noncritical $\lambda\mathscr{P}(\phi)_2-\mu\phi$ models. They have been verified for weak coupling (i.e. $\lambda \ll m_0^2$ or $\mu \gg m_0^2$) [25, 26, 45] and for even $\mathscr{P}(\phi)_2$ models in the single phase region or $\lambda\phi^4-\mu\phi$ with $\mu \neq 0$ [39, 28, 9, 44, 29, 24].

The kernel K_ξ is formally two particle irreducible between initial and final states, so we expect a configuration exponential decay rate $3m$, or $4m$ for an even theory, assuming the absence of bound states and CDD zeros.

Proof. We know that $R_0^{-1}: H_1^{\otimes 2} \to H_{-1}^{\otimes 2}$ is bounded, from the spectral representation for the two point function [20]. To establish the same result for R^{-1}, we work in subspaces $\mathscr{H}_\pm(P)$ of $H_{\pm 1}^{\otimes 2}$ with definite total momentum, see (4.2). By Theorem 4.3 (b), $\mu_{\otimes n} G^{(4)}$ is Hilbert-Schmidt on $\mathscr{H}(P)$, so $G^{(4)}$ is Hilbert-Schmidt from $\mathscr{H}_-(P)$ to $\mathscr{H}_+(P)$. Since $\Gamma^{(2)}$ is an isomorphism of H_1 onto H_{-1} [20], the last term in (7.2) is also Hilbert-Schmidt from $\mathscr{H}_-(P)$ to $\mathscr{H}_+(P)$, with norm 0 as $|P| \to \infty$. By the Fredholm alternative, R^{-1} is defined if R has no null vectors on $\mathscr{H}_-(P)$.

For simplicity, we suppose $\langle\phi\rangle_\xi=0$. In other words, we replace ϕ by $\hat{\phi}$ of Chapter 5. With this assumption, $G^{(2)}$ and $G^{(3)}$ are also Schwinger functions, and (7.2) becomes

$$R(x, y)=S_\xi^{(4)}(x, y)-S_\xi^{(2)}(x_1, x_2)S_\xi^{(2)}(y_1, y_2) - S_\xi^{(3)}(x, z)[G^{(2)}]^{-1}(z, z')S_\xi^{(3)}(z', y)dzdz' . \tag{7.5}$$

The above subtractions remove intermediate vacuum and one particle states, and are related to the projections P_0 and P_1 of Chapter 5. Let

$$\chi(h)=\int \phi(x_1)\phi(x_2)h(x_1, x_2)dx_1dx_2 . \tag{7.6}$$

Then

$$R(g\otimes h)=\langle\chi(g)(1-P_0-P_1)\chi(h)\rangle_\xi . \tag{7.7}$$

Furthermore, representing $(1-P_0-P_1)\chi$ as a polynomial of degree 2, we find

$$(1-P_0-P_1)\chi(h)=\chi(h)-\langle\chi(h)\rangle_\xi + \int \langle\chi(h)\phi(z)\rangle_\xi \Gamma^{(2)}(z, z')\phi(z')dzdz' , \tag{7.8}$$

and

$$\langle\chi(h)\phi(z)\rangle_\xi=\int G^{(3)}(x_1, x_2, z)h(x_1, x_2)dx . \tag{7.9}$$

We note

$$\langle|\chi(h)-\langle\chi\rangle|^2\rangle_\xi=\int G^{(4)}-R_0)h^-(x)h(y)dxdy . \tag{7.10}$$

We now consider R acting from $\mathscr{H}_-(P)$ to $\mathscr{H}(P)$, and we suppose that R has a nonzero null space. For simplicity, we treat the case $P=0$. We suppose that there is a nonzero even function f in $L_2(R^2)$ such that

$$0=\int (\mu_{\otimes 4}R)(x)f(x_1-x_2)^- f(x_3-x_4)dx_2dx_3dx_4 .$$

We multiply this equation by

$$\int g(x_1+x_2-w)^- g(x_3+x_4+w)dw = g^- * g(x_1+x_2+x_3+x_4)$$

and integrate over x_1. For $g \in C_0^\infty$, we obtain

$$0 = \int (\mu_{\otimes 4} R)(x) h(x_1, x_2)^- h(x_3+w, x_4+w) dx dw\,,$$

where

$$h(y, z) = f(y-z) g(y+z)\,.$$

In terms of (7.6)–7.9),

$$0 = \int dw \langle \chi(h), (I - P_0 - P_1)\chi(h_w)\rangle_\xi\,, \tag{7.11}$$

where $h_w(y, z) = h(y+w, z+w)$. By Theorem 5.8 and Remark 5.9, with $X(h) = B = C$, we have

$$0 = \int h(y+a, z+a) da = \int f(y-z) g(y+z+2a) da = \text{const.}\ f(y-z)\,.$$

Thus $\chi(h) = 0$ and so R has no null vectors (except zero) in the space of definite momentum.

It follows that $R \restriction \mathscr{H}_-(P)$ has a bounded inverse. Since the inverse depends continuously on P, and remains bounded uniformly as $P \to \infty$, R, defined on $\mathscr{H}_{-1}$ has a bounded inverse. This completes the proof of Theorem 7.1.

8. Three Particle Equations

The three particle scattering equation is defined in analogy to Chapter 7, after subtraction of the two body rescattering processes. We give two procedures for these subtractions, one based on the two-body Bethe-Salpeter kernel, the other based on Faddeev-type equations for two-body rescattering processes. We expect that a quantitative analysis of the three-body equation will yield insight into such problems as: Do three-particle bound states occur? Is S unitary up to the four-particle threshold?

Let $\alpha = 12, 13, 23$ labeled pairs of initial or final particles. A two particle process occurs in a channel labeled by a pair α, β. The free Green's function for a two particle channel is

$$R_0^{\alpha\beta} = G^{(2)}(x_i, y_j)[G^{(2)}(x_{\alpha_1}, y_{\beta_1})G^{(2)}(x_{\alpha_2}, y_{\beta_2}) + G^{(2)}(x_{\alpha_1}, y_{\beta_2})G^{(2)}(x_{\alpha_2}, y_{\beta_1})] \tag{8.1}$$

where $i \notin \alpha$, $j \notin \beta$. The total R_0 is

$$R_0 = \sum_\beta R_0^{\alpha\beta} = \sum_\pi G^{(2)}(x_1 - y_i) G^{(2)}(x_2 - y_j) G^{(2)}(x_3 - y_k)\,, \tag{8.2}$$

where $\sum_\pi$ ranges over the six permutations (i, j, k) of $(1, 2, 3)$.

We also define the two particle scattering kernels

$$R_{\alpha\beta} = G^{(2)}(x_i, y_j) G^{(4)}(x_\alpha, y_\beta)\,, \tag{8.3}$$

where $i \notin \alpha, j \notin \beta$. Graphically, $R_{\alpha\beta}$ represents scattering processes with initial state (i, α), final state (j, β), and two particle scattering in the α, β channel. See Fig. 8.1.

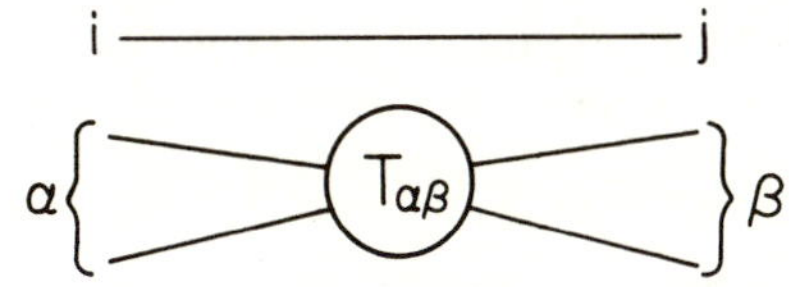

Fig. 8.1. A two particle process

Proposition 8.1. *The operators* $R_0^{\alpha\beta}$, R_0, *and* $R_{\alpha\beta}$ *are bounded transformations* $H_{-1}^{\otimes 3}$ *to* $H_1^{\otimes 3}$. *The norm of* $R_{\alpha\beta}$ *is* $O(\lambda)$ *as* $\lambda \equiv \sup|\lambda_j| \to 0$.

Proof. This follows immediately by the bounds of [20].

Definition 8.2. The two-body Bethe-Salpeter kernel $K_{\alpha\alpha}$ for the three body problem is

$$K_{\alpha\alpha} = R_0^{\alpha\alpha})^{-1} - (R_0^{\alpha\alpha} + R_{\alpha\alpha})^{-1}. \tag{8.4}$$

$K_{\alpha\beta}$ is defined by a permutation of the y variables of $K_{\alpha\alpha}$. The two body amplitude $R_{2\text{-body}}$ satisfies

$$R_{2\text{-body}} = R_0 + \tfrac{1}{9} R_0 \textstyle\sum_{\alpha,\beta} K_{\alpha\beta} R_{2\text{-body}}. \tag{8.5}$$

We remark that $K_{\alpha\beta}$ has the form

$$K_{\alpha\beta} = K \otimes (-\Gamma), \tag{8.6}$$

where K is the Bethe-Salpeter kernel (7.4) for the channel $(\alpha\beta)$ and $-\Gamma$ is the inverse propagator for the remaining particle.

Theorem 8.3. *Under the hypotheses of Theorem 7.1,* $K_{\alpha\beta}$ *and*

$$R_{2\text{-body}}^{-1} = R_0^{-1} - \tfrac{1}{9} \textstyle\sum_{\alpha,\beta} K_{\alpha\beta}$$

are bounded maps from Sym $H_1^{\otimes 3}$ *to* $H_{-1}^{\otimes 3}$, *with* $\|K_{\alpha\beta}\| \leqq O(\lambda)$ *as* $\lambda \to 0$. *Also for* λ *sufficiently small,* $R_{2\text{-body}}$ *is a bounded map from* Sym $H_{-1}^{\otimes 3}$ *to* $H_1^{\otimes 3}$.

Proof. The bound on $K_{\alpha\beta}$ follows by Theorem 7.1 which bounds K of (7.4) and the properties of Γ established in [20]. Since R_0^{-1} is bounded from Sym $H_1^{\otimes 3}$ to $H_{-1}^{\otimes 3}$, the existence and boundedness of $R_{2\text{-body}}^{-1}$ follows. We note by (8.5),

$$(I - \tfrac{1}{9} R_0 \textstyle\sum K_{\alpha\beta}) R_{2\text{-body}} = R_0,$$

and, by the above, the norm of $R_0 K_{\alpha\beta}$ from $H_1^{\otimes 3}$ to $H_1^{\otimes 3}$ is $O(\lambda)$. Thus for λ sufficiently small,

$$(I - \tfrac{1}{9} R_0 \textstyle\sum K_{\alpha\beta})^{-1} : H_1^{\otimes 3} \to H_1^{\otimes 3}$$

is bounded. It follows that

$$R_{2\text{-body}} : H_{-1}^{\otimes 3} \to H_1^{\otimes 3}$$

exists and is bounded.

Theorem 8.4. *Consider* $R_{2\text{-body}}^{-1}$ *restricted to a constant momentum subspace:*

$$R_{2\text{-body}}^{-1}(P) : \text{Sym}\,\mathscr{H}_+(P) \to \mathscr{H}_-(P).$$

Then for $|P|$ *sufficiently large,* $R_{2\text{-body}}(P)$ *is bounded from* Sym $\mathscr{H}_{-}(P)$ *to* $\mathscr{H}_{+}(P)$. *(Here the minimum* $|P|$ *may depend on* λ.*) Also as* $|P|\to\infty$,

$$\|R_{2\text{-body}}(P)-R_0(P)\|_{-,+}\to 0\,.$$

Proof. Each of the operators $R_0, R_0^{\alpha\beta}, R_{\alpha\beta}, K_{\alpha\beta}$ is translation invariant, and thus acts on constant momentum subspaces. By Theorem 4.3 (b), $R_{\alpha\beta}(R_0^{\alpha\beta})^{-1}$ is a bounded operator on Sym $\mathscr{H}_{+}(P)$, with norm $o(1)$ as $|P|\to\infty$. We choose $|P|$ sufficiently large so that

$$R_0K_{\alpha\beta}=-R_0(R_0^{\alpha\beta})^{-1}\textstyle\sum_{n=1}^{\infty}(-R_{\alpha\beta}(R_0^{\alpha\beta})^{-1})^n$$

has norm on sym $\mathscr{H}_{+}(P)$ less than 1. Thus $\|\sum_{\alpha\beta}\frac{1}{9}R_0K_{\alpha\beta}\|<1$ on sym $\mathscr{H}_{+}(P)$ and

$$R_{2\text{-body}}^{-1}=R_0^{-1}(1-\tfrac{1}{9}\textstyle\sum_{\alpha,\beta}R_0K_{\alpha\beta})$$

is invertible. The inverse $R_{2\text{-body}}(P)$ is bounded from Sym$\mathscr{H}_{-}(P)$ to $\mathscr{H}_{+}(P)$, since $R_0(P)$ is. Also $(R_0K_{\alpha\beta})(P)$ has norm $o(1)$ as $|P|\to\infty$, so $\|R_{2\text{-body}}(P)-R_0(P)\|_{-,-}$ is $o(1)$. Here $\|\cdot\|_{-+}$ denotes the norm from sym$\mathscr{H}_{-}(P)$ to $\mathscr{H}_{+}(P)$.

We remark that instead of using the direct definition (8.5) of $R_{2\text{-body}}$, we could also use Faddeev type equations to separate the two body forces. Let $R(\alpha)$ denote the sum of 2-body rescattering diagrams in which the initial two-body scattering involves the pair α. Then the Feddeev equations are

$$\begin{aligned}R(\alpha)&=R_0^{\alpha\alpha}+R_{\alpha\alpha}\\&\quad+\textstyle\sum_{\beta\neq\alpha}R_{\alpha\beta}(R_0^{\beta\beta}+R_{\beta\beta})^{-1}R(\beta)\\&=R_0^{\alpha\alpha}+R_{\alpha\alpha}\\&\quad\textstyle\sum_{\beta\neq\alpha}R_{\alpha\beta}[(R_0^{\beta\beta})^{-1}-K_{\beta\beta}]R(\beta)\,.\end{aligned}\tag{8.7}$$

As above, we find

Theorem 8.5. *For* λ *sufficiently small, the Eq.* (8.7) *have a unique solution* $R(\alpha):H_{-1}^{\otimes 3}\to H_1^{\otimes 3}$. *For constant total momentum P, the equations* (8.7) *have a unique solution* $R(\alpha, P):\mathscr{H}_{-}(P)\to\mathscr{H}_{+}(P)$, *for* $|P|$ *sufficiently large (depending on* λ*).*

For small λ, we define

$$\bar{R}_{2\text{-body}}=\textstyle\sum_\alpha(R(\alpha)-R_0^{\alpha\alpha})+R_0\,.\tag{8.8}$$

We expect that $R_{2\text{-body}}=\bar{R}_{2\text{-body}}$.

We now define the three body Bethe-Salpeter kernel $K^{(3)}$, by the equations

$$R^{(3)}(x_1,x_2,x_3,y_1,y_2,y_3)=\langle\textstyle\prod_{i=1}^3\phi(x_i)(I-\sum_{j=0}^2P_j)\prod_{k=1}^3\phi(x_6)\rangle\,,\tag{8.9}$$

and

$$R^{(3)^{-1}}=R_{2\text{-body}}^{-1}-K^{(3)}\,,\tag{8.10}$$

or

$$R^{(3)}=R_{2\text{-body}}+R_{2\text{-body}}K^{(3)}R^{(3)}\,.\tag{8.11}$$

Theorem 8.6. *Under the hypothesis of Theorem 7.1,* $R^{(3)^{-1}}$ *and* $K^{(3)}$ *exist as bounded operators from* Sym $H_1^{\otimes 3}$ *to* $H_{-1}^{\otimes 3}$, *and satisfy* (8.10). *For weak coupling,* (8.11) *also holds.*

Remark. N-body equations, as well as a further study of the three body equations, will be contained in a subsequent article. By definition $-K^{(n)}$ is the connected part of $R^{(n)^{-1}}$.

References

1. Araki, H.: On the algebra of all local observables. Prog. Theor. Phys. **32**, 844—854 (1964)
2. Bratteli, O.: Conservation of estimates in quantum field theory. Comm. Pure and Appl. Math. **25**, 759—779 (1972)
3. Bros, J.: Some analyticity properties implied by the two particle structure of Green's functions in general quantum field theory. In: Gilbert, R., Newton, R. (Eds.): Analytic methods in mathematical physics. New York: Gordon and Breach 1970
4. Castillejo, L., Dalitz, R., Dyson, F.: Low's scattering equation for the charged and neutral scalar theories. Phys. Rev. **101**, 453—458 (1956)
5. Dobrushyn, R., Minlos, R.: Construction of a one dimensional quantum field via a continuous Markov field. Funct. Anal. and its Appl. **7**, 324—325 (1973) (English transl.)
6. Eckmann, J.-P. Magnen, J., Seneor, R.: Decay properties and Borel summability for the Schwinger functions in $P(\phi)_2$ theories. Commun. math. Phys. **39**, 251—271 (1975)
7. Faris, W., Lavine, R.: Commutators and self-adjointness of Hamiltonian operators. Commun. math. Phys. **35**, 39—48 (1974)
8. Feldman, J.: On the absence of bound states in the ϕ^4 quantum field model without symmetry breaking. Canad. J. Phys. **52**, 1583—1587 (1974)
9. Fröhlich, J.: Verification of axioms for Euclidean and relativistic fields and Haag's theorem in a class of $P(\phi)_2$ models. Ann. Inst. H. Poincaré (to appear)
10. Fröhlich, J.: Schwinger functions and their generating functionals. I. Helv. Phys. Acta (to appear)
11. Fröhlich, J.: Schwinger functions and their generating functionals. II. Adv. Math. (to appear)
12. Ginibre, J.: General formulation of Griffiths' inequalities. Comm. Math. Phys. **16**, 310—328 (1970)
13. Glimm, J., Jaffe, A.: A $\lambda\phi^4$ quantum field theory without cutoffs II, the field operators and the approximate vacuum. Ann. of Math. **91**, 362—401 (1970)
14. Glimm, J., Jaffe, A.: The $(\lambda\phi^4)_2$ quantum field theory without cutoffs. III. The physical vacuum. Acta Math. **125**, 203—261 (1970)
15. Glimm, J., Jaffe, A.: The $(\lambda\phi^4)_2$ quantum field theory without cutoffs. IV. Perturbations of the Hamiltonian. J. Math. Phys. **13**, 1558—1584 (1972)
16. Glimm, J., Jaffe, A.: The energy momentum spectrum and vacuum espectation values in quantum field theory. J. Math. Phys. **11**, 3335—3338 (1970)
17. Glimm, J., Jaffe, A.: The energy momentum spectrum and vacuum expectation values in quantum field theory. II. Commun. math. Phys. **22**, 1—22 (1971)
18. Glimm, J., Jaffe, A.: Boson quantum field models. In: Streater, R. (Ed.): Mathematics of contemporary physics. New York: Academic Press 1972
19. Glimm, J., Jaffe, A.: Positivity of the ϕ_3^4 Hamiltonian. Fortschritte der Physik **21**, 327—376 (1973)
20. Glimm, J., Jaffe, A.: Entropy principle for vertex functions in quantum field models. Ann. Inst. H. Poincaré **21**, 1—26 (1974)
21. Glimm, J., Jaffe, A.: ϕ_2^4 quantum field model in the single phase region: Differentiability of the mass and bounds on critical exponents. Phys. Rev. D **10**, 536—539 (1974)
22. Glimm, J., Jaffe, A.: A remark on the existence of ϕ_4^4. Phys. Rev. Lett. **33**, 440—442 (1974)
23. Glimm, J., Jaffe, A.: Absolute bounds on vertices and couplings. Ann. l'Inst. H. Poincaré **22** (to appear)
24. Glimm, J., Jaffe, A.: On the approach to the critical point. Ann. l'Inst. H. Poincaré **22** (to appear)
25. Glimm, J., Jaffe, A., Spencer, T.: The particle structure of the weakly coupled $P(\phi)_2$ model and other applications of high temperature expansions. In: Velo, G., Wightman, A. (Eds.): Constructive quantum field theory. Lecture notes in Physics 25. Berlin Heidelberg New York: Springer 1973
26. Glimm, J., Jaffe, A., Spencer, T.: The Wightman axioms and particle structure in the $P(\phi)_2$ quantum field model. Ann. Math. **100**, 585—632 (1974)
27. Guerra, F., Rosen, L., Simon, B.: Nelson's symmetry and the infinite volume behavior of the vacuum in $P(\phi)_2$. Commun. math. Phys. **27**, 10—22 (1972)

28. Guerra, F., Rosen, L., Simon, B.: The $P(\phi)_2$ Euclidean quantum field theory as çlassical statistical mechanics. Ann. Math. **101**, 111—259 (1975)
29. Guerra, F., Rosen, L., Simon, B.: Correlation inequalities and the mass gap. III. Mass gap for a class of strongly coupled theories with nonzero external field. Commun. math. Phys., to appear
30. Jost, R.: The general theory of quantized fields. American Mathematical Society, Providence, 1965
31. Lebowitz, J., Penrose, O.: Analytic and clustering properties of the thermodynamic functions and distribution functions for classical lattice and continuum systems **11**, 99—124 (1968)
32. Lebowitz, J., Penrose, O.: Decay of correlations. Phys. Rev. Lett. **31**, 749—752 (1973)
33. McBryan, O.: Local generators for the Lorentz group in the $P(\phi)_2$ model. Nuovo Cimento, **18** A, 654—662 (1973)
34. McBryan, O.: Vector currents in the Yukawa_2 quantum field theory (preprint)
35. McBryan, O.: Self adjointness of relatively bounded quadratic forms and operators. J. Funct. Anal. (to appear)
36. McBryan, O., Park, Y.: Lorentz covariance of the Yukawa_2 quantum field theory. J. Math. Phys. **16**, 105—110 (1975)
37. Nelson, E.: Time ordered operator products of sharp time quadratic forms. J. Funct. Anal. **11**, 211—219 (1972)
38. Nelson, E.: The construction of quantum fields from Markov fields. J. Funct. Anal. **12**, 97—112 (1973)
39. Nelson, E.: Probability theory and Euclidean field theory. In: Velo, G., Wightman, A. (Eds.): Constructive quantum field theory. Lecture notes in Physics 25. Berlin Heidelberg New York: Springer 1973
40. Osterwalder, K., Schrader, R.: Axioms for Euclidean Green's functions. Commun. math. Phys. **31**, 83—112 (1974)
41. Osterwalder, K., Schrader, R.: Axioms for Euclidean Green's functions. II. Commun. math. Phys. **42**, 281—305 (1975)
42. Rosen, L.: A $\lambda\phi^{2n}$ field theory without cutoffs. Commun. math. Phys. **16**, 157—183 (1970)
43. Schrader, R.: Yukawa quantum field theory in two space-time dimensions without cutoff. Ann. Phys. **70**, 412—457 (1972)
44. Simon, B.: The $P(\phi)_2$ Euclidean quantum field theory. Princeton: Princeton University Press 1974
45. Spencer, T.: The mass gap for the $P(\phi)_2$ quantum field model with a strong external field. Commun. math. Phys. **39**, 63—76 (1974)
46. Spencer, T.: The absence of even bound states in ϕ_2^4. Commun. math. Phys. **39**, 77—79 (1974)
47. Spencer, T.: The decay of the Bethe Salpeter kernel in $P(\phi)_2$ quantum field models. Commun. math. Phys. (to appear)
48. Spencer, T., Zirilli, F.: Private communication
49. Symanzik, K.: A modified model for Euclidean quantum field theory. N.Y.U. (preprint) 1964
50. Tucciarone, A.: A relativistic treatment of the three body problem. Nuovo Cimento **41** A, 204—221 (1966)
51. Glimm, J., Jaffe, A.: ϕ^j bounds in $P(\phi)_2$ quantum field models. Proceedings of the International Colloquium on Mathematical Methods in Quantum Field Theory, Marseille, 1975

Communicated by A. S. Wightman

Particles and Scaling for Lattice Fields and Ising Models

James Glimm*
Rockefeller University, New York, N.Y. 10021, USA

Arthur Jaffe**
Harvard University, Cambridge, Mass. 02138, USA

Abstract. The conjectured inequality $\Gamma^{(6)} \leqq 0$ leads to the existence of φ_d^4 fields and the scaling (continuum) limit for d-dimensional Ising models. Assuming $\Gamma^{(6)} \leqq 0$ and Lorentz covariance of this construction, we show that for $d \geqq 6$ these φ_d^4 fields are free fields unless the field strength renormalization Z^{-1} diverges. Let λ be the bare charge and ε the lattice spacing. Under the same assumptions ($\Gamma^{(6)} \leqq 0$, Lorentz covariance and $d \geqq 6$) we show that if $\lambda\varepsilon^{4-d}$ is bounded as $\varepsilon \to 0$, then Z^{-1} is bounded and the limit field is free.

1. Introduction

Even φ^4 fields in the single phase region describe particles which interact through repulsive forces. The evidence in support of this statement includes the absence of even bound states [24, 3, 34] the canonical lower bound on critical exponents [16], the arguments of [19] concerning absence of three-particle bound states and *CDD* zeros (which depend on the conjectured inequality $\Gamma^{(6)} \leqq 0$) and the scaling and numerical arguments of [22, 29] in favor of this $\Gamma^{(6)}$ inequality for $d \leqq 3$.

The purpose of this note is to extend some of these ideas to lattice φ^4 field theories and Ising models. There are two reasons for studying lattice (as opposed to continuum) field theories. The first reason is that all dimensions, including $d \geqq 4$, are possible, and that the results obtained may be related to the existence [17, 16, 1, 30, 21] and triviality/nontriviality problems in these dimensions, see also [32, 33]. The second reason is that the lattice φ^4 field theory is an intermediate link between the continuum φ^4 field theory and the Ising model, and may serve to clarify the relation between Ising model and field theory critical behavior, see [20–22, 26–27, 31]. Beyond these two reasons, we note that lattice field theories are at the outset easier because of the absence of ultraviolet problems but ultimately more difficult because of the absence of the Euclidean rotation symmetry.

* Supported in part by the National Science Foundation under Grant MPS 74-13252
** Supported in part by the National Science Foundation under Grant MPS 75-21212

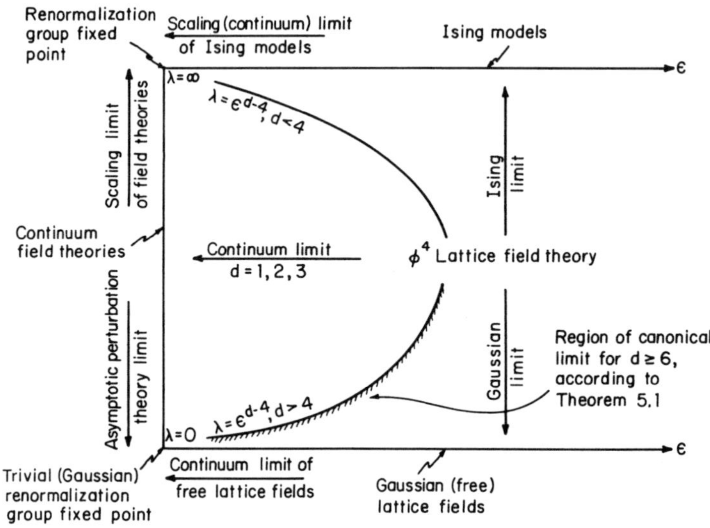

Fig. 1. The $\lambda\varphi^4+\sigma\varphi^2$ parameter plane. $\varepsilon=$ lattice spacing; $\sigma=\sigma(\lambda, \varepsilon, m)$ chosen to give mass m in single phase region; m held fixed (e.g. $m=1$). The λ axis represents (continuum) field theory. The ε axis at $\lambda=\infty$ represents Ising models [30]. The ε axis at $\lambda=0$ represents Gaussian lattice fields. The curve $\lambda=\varepsilon^{d-4}$ represents constant dimensionless bare charge. Charge renormalization for $d\geqq 4$ may lead to a reparametrization of the λ axis

As a compensation for the loss of Euclidean invariance and of the resulting Lehmann spectral formula, we use the momentum space canonical upper bound [9] on the two point function, as well as the $\Gamma^{(6)}\leqq 0$ inequality as used in [19].

Our first main result assumes only $\Gamma^{(6)}\leqq 0$. The conclusion is the existence of a continuum φ_d^4 field theory with an arbitrary choice of charge renormalization. Included is the existence of the scaling (continuum) limit for Ising models. The second main result assumes in addition Lorentz invariance of the continuum theory, as well as the (finite) charge renormalization, $\lambda\leqq\varepsilon^{d-4}$, ε being the lattice spacing. Note that $\varepsilon^{d-4}\to 0$ in the continuum limit $\varepsilon\to 0$, for $d\geqq 5$. In this case, the $\varepsilon=0$ theory is canonical for $d\geqq 6$ (§ 5).

The limits considered here are summarized in Figure 1 above. Coincidence of the two fixed points seems to correspond to the picture in which nontrivial φ^4 field theories cannot be constructed as a limit of lattice cutoff approximations. Similarly, it would seem that the Ising model would be governed by the trivial fixed point only in the case for which a nontrivial fixed point does not occur in Figure 1.

A renormalizability curve $\lambda=0(\varepsilon^{d-4})$ of constant dimensionless bare charge is indicated on this figure. A limit constructed along any such curve is renormalizable. Those limits constructed below this family of curves are superrenormalizable (e.g. the continuum limit with constant $\lambda, d=1,2,3$). Theories constructed above these curves are nonrenormalizable (i.e. $\lambda\varepsilon^{4-d}\to\infty$ as $\varepsilon\to 0$). The charge λ has dimension $(\text{length})^{d-4}$, so that $\lambda\varepsilon^{4-d}$ is a dimensionless bare charge.

2. The Wave Function Renormalization and Proper Self Energy for Noncritical Lattice Theories

In this chapter, we consider a $P(\varphi)$ lattice field or Ising model. We suppose that the infinite volume limit is defined and is invariant under lattice translations, rotations and reflections. We also assume *FKG* inequalities [6], a positive mass $m>0$ (i.e. an exponential decay rate for correlations in lattice directions) and a single phase. These restrictions should be viewed as restrictions on the boundary conditions and/or coupling constants. In particular they are satisfied for φ^4 interactions and the Ising model approaching the critical point from a single phase direction. Let

$$G^{(2)}(x)=\langle\varphi(x)\varphi(0)\rangle-\langle\varphi(x)\rangle\langle\varphi(0)\rangle\geqq 0 \tag{2.1}$$

be the truncated two point function. The Fourier transform

$$G^{(2)\sim}(p)=\sum e^{-ipx}G^{(2)}(x) \tag{2.2}$$

is nonnegative because $G^{(2)}(x)$ is positive definite. Using invariance of $G^{(2)}(x)$ under reflections and rotations, $G^{(2)\sim}(p)$ is a symmetric function of the variable $\{\cos p_\nu\}_{\nu=1}^{d}$. Using the selfadjoint transfer matrix, one shows that

$$G^{(2)}(x_1,\ldots,x_d)\leqq G^{(2)}(0,\ldots 0,x_\nu,\ldots 0) \tag{2.3}$$

for any ν [30]. In particular $G^{(2)}$ has an exponential decay rate in all directions and $G^{(2)\sim}$ is analytic for p in a tube τ. By Hartog's theorem, τ is convex and also using the $p_\nu\to -p_\nu$ symmetry, we see that

$$\left\{p:\sum_{\nu=1}^{d}|\mathrm{Im}\,p_\nu|\leqq m\right\}\subset\tau\subset\left\{p:\sup_\nu|\mathrm{Im}\,p_\nu|\leqq m\right\}. \tag{2.4}$$

The use of Hartog's theorem can be avoided by a Schwarz inequality. For $p_1,p_2\in\tau$,

$$\left|\sum G^{(2)}(x)e^{-i(p_1+p_2)x/2}\right|\leqq\left(\sum G^{(2)}(x)e^{\mathrm{Im}\,p_1x}\right)^{\frac12}\left(\sum G^{(2)}(x)e^{\mathrm{Im}\,p_2x}\right)^{\frac12}.$$

Thus not only is τ convex, but its border coincides with the singularities of $G^{(2)\sim}(p)$, for $\mathrm{Re}\,p=0$. We have proved the following:

Proposition 2.1. *For a lattice field or Ising model as above, $G^{(2)\sim}$ is analytic in a convex tube τ, satisfying* (2.4).

We distinguish a single lattice direction x, and write $\boldsymbol{x}=\{x_2,\ldots,x_d\}$, $\boldsymbol{p}=\{p_2,\ldots,p_d\}$. The existence of a self adjoint transfer matrix means that $G^{(2)\sim}$ has an analytic continuation for p_1 in the half plane $p_1^2\neq$ neg. real, for $\boldsymbol{p}$ real.

Proposition 2.2. *Consider a lattice field or Ising model as above with $\boldsymbol{p}$ real. Then $G^{(2)\sim}(p)$ is Herglotz as a function of the variable $1-\cos p_1$. For some positive measure $d\varrho(a,\boldsymbol{p})$ depending on $\boldsymbol{p}$,*

$$G^{(2)\sim}(p)=\int_{-1+\cosh m}^{\infty}\frac{d\varrho(a,\boldsymbol{p})}{1-\cos p_1+a}, \tag{2.5}$$

and

$$0<\int\frac{d\varrho}{a}<\infty\,. \tag{2.6}$$

Proof. An explicit calculation shows that the Fourier transform of $e^{-\mu|n|}$ is

$$\Sigma_n e^{-\mu|n|+ipn}=\sinh\mu/(\cosh\mu-\cos p)\,.$$

Substitution of this formula into the spectral representation for the transfer matrix establishes (2.5) after the change of variable $a=-1+\cosh\mu$, while the upper bound in (2.6) follows from

$$\sum_x G^{(2)}(x)e^{-i\mathbf{p}x}=G^{(2)\sim}(0,\mathbf{p})=\int\frac{d\varrho(a,\mathbf{p})}{a}\leqq\sum_x G^{(2)}(x)<\infty\,. \tag{2.7}$$

From (2.5) we see that $G^{(2)\sim}(p)$ is monotone decreasing in each argument p_ν for $0\leqq p_\nu\leqq\pi$. We show that $G^{(2)\sim}(\pi,\ldots,\pi)\neq0$ and then it follows from (2.5) and monotonicity that $d\varrho(a,\mathbf{p})$ is nonzero for any real p. If $G^{(2)\sim}(\pi,\ldots,\pi)=0$, then $d\varrho(a,\pi)=0$ by (2.5), and so $G^{(2)\sim}(p_1,\pi,\ldots,\pi)=0$ for all p_1. Interchanging first and second coordinates, $G^{(2)\sim}(\pi,p_2,\pi,\ldots,\pi)=0$, and so $G^{(2)\sim}(p_1,p_2,\pi,\ldots,\pi)=0$ for arbitrary p_1 and p_2. Continuing in this fashion, we conclude that $G^{(2)\sim}$ and $G^{(2)}$ both vanish, which is a contradiction.

Definition. $\Gamma^{(2)}(x)$ is the convolution inverse to $-G^{(2)}$, so that

$$\Gamma^{(2)\sim}(p)=-1/G^{(2)\sim}(p)\,.$$

The next result is elementary, cf. [19].

Proposition 2.3. *Consider a lattice field theory or Ising model as above. $\Gamma^{(2)\sim}$ is analytic in a convex tube τ_Γ satisfying*

$$\{p:\sum|\mathrm{Im}\,p_\nu|\leqq m\}\subset\tau_\Gamma\,. \tag{2.8}$$

For $\mathbf{p}$ *real, $\Gamma^{(2)\sim}$ is Herglotz as a function of $1-\cos p_1$, and has a unique representation*

$$\Gamma^{(2)\sim}(p)=-\alpha(\mathbf{p})(1-\cos p_1)+\beta(\mathbf{p})+\int_{-1+\cosh m}^{\infty}\left[\frac{1}{1-\cos p_1+a}-\frac{a}{a^2+1}\right]d\nu(a,\mathbf{p}) \tag{2.9}$$

with $\alpha\geqq0$, β real, $d\nu$ positive and

$$\int d\nu/(1+a^2)<\infty\,.$$

Remark. For $d\geqq3$, there is a constant K depending only on d and $G^{(2)}(0)$ such that

$$\alpha+|\beta|+\int(1+a^2)^{-1}d\nu\leqq K\,. \tag{2.10}$$

Proof. By [9],

$$G^{(2)\sim}(p)\leqq\text{const.}\,|p|^{-2}\,. \tag{2.11}$$

By the Herglotz property, used successively in each coordinate, as in the proof of Proposition 2.2,

$$G^{(2)\sim}(p) \leqq \text{const.} \prod_{\nu=1}^{d} |p_\nu|^{-2} G^{(2)\sim}(\pi, \ldots, \pi).$$

Thus for any $r>0$ and for $d>3$,

$$\begin{aligned} G^{(2)}(0) &= (2\pi)^{-d} \int G^{(2)\sim}(p)dp \\ &\leqq \text{const.} \Bigg(\int\limits_{\substack{\text{Inf}\,|p_\nu| \leqq r \\ \nu}} |p|^{-2} dp + r^{-2d} G^{(2)\sim}(\pi, \ldots, \pi) \Bigg) \\ &\leqq \text{const.} (r + r^{-2d} G^{(2)\sim}(\pi, \ldots, \pi)). \end{aligned}$$

Here the optimal choice of r is

$$r = 0(G^{(2)\sim}(\pi, \ldots, \pi)^{1/(2d+1)}$$

and so

$$G^{(2)}(0)^{2d+1} \leqq \text{const.}\, G^{(2)\sim}(\pi, \ldots, \pi) \leqq \text{const.}\, G^{(2)\sim}(p).$$

A uniform upper bound on (2.9) follows. For $d=3$, the exponents must be changed slightly to dominate $r \ln r$ in place of r, but a bound on (2.9) in still valid. Linear combinations of $\cos p_1 = 0, -1$ show that β and $\alpha + \int (a^2+1)^{-1} d\nu(a)$ are each bounded. Since α and $d\nu$ are each nonnegative, the individual terms α and $\int (a^2+1)^{-1} d\nu$ are also bounded, proving (2.10).

We now write $x = (x', x'')$, $p = (p', p'')$ for some decomposition

$$x' = (x_1, \ldots, x_j), \qquad x'' = (x_{j+1}, \ldots, x_d)$$

of the coordinates, and consider the partial Fourier transform

$$\Gamma^{(2)\sim}(x', p'') \equiv \sum_{x''} e^{-ip''x''} \Gamma^{(2)}(x', x'').$$

Proposition 2.4. *Consider a lattice field or Ising model as above. For* $x_1 \neq 0$, $x_2 = \ldots = x_j = 0$, p'' *real,* $\Gamma^{(2)\sim}(x', p'')$ *is positive and has an exponential decay rate* $\bar{m} \geqq m$. *(Note* $j \neq 0$, *by assumption.)*

Proof. First we consider the case $x' = x_1$ and $p'' = \mathbf{p}$. The Herglotz representation (2.9) in x_1-space has three terms. The first is nonnegative for $x_1 \neq 0$. The second, proportional to β, vanishes for $x_1 \neq 0$. The third is positive for all x_1, so the proof is complete in this case. The general case follows from the case just considered, because integration with respect to $p_2, \ldots, p_j$ is equivalent to evaluation at $x_2 = \ldots = x_j = 0$.

Definition. If the measure $d\varrho(a, \mathbf{p}=0)$ has a δ-function at $a = -1 + \cosh m$, then $Z>0$ is the strength of this δ-function. Otherwise by definition $Z=0$.

Proposition 2.5. *For a pure phase lattice* $P(\varphi)$ *field theory or Ising model with* $m>0$ *and* $Z>0$,

$$\begin{aligned} Z^{-1} &= -(\partial/\partial(1-\cos p_1))\Gamma^{(2)\sim}(p_1, \mathbf{p})|_{\cos p = \cosh m,\, \mathbf{p}=0} \\ &= \alpha(0) + \int_0^\infty (a - \cosh m + 1)^{-2} d\nu(a, 0) \end{aligned} \qquad (2.12)$$

Proof. We write

$$G^{(2)\sim}(p_1, 0) = Z(\cosh m - \cos p_1)^{-1} + \int \frac{d\hat{\varrho}}{1 - \cos p_1 + a}$$

where $\int_{-1+\cosh m}^{-1+\cosh m+\varepsilon} d\hat{\varrho} \to 0$ as $\varepsilon \to 0$. Thus as $\cos p_1 \to \cosh m$,

$$Z(\cosh m - \cos p_1)^{-1} \leqq G^{2\sim}(p_1, 0) \leqq (Z+\varepsilon)(\cosh m - \cos p_1)^{-1} \tag{2.12a}$$

and the derivative of $\Gamma^{(2)}$ can be evaluated as Z^{-1} from this inequality.

Remark. We have not assumed an upper mass gap, in contrast to [19].

Definition. C_{L,m_0} is the free (Gaussian) lattice covariance,

$$C_{L,m_0} = \left[-1 + \cosh m_0 + \sum_{\nu=1}^{d} (1 - \cos p_\nu) \right]^{-1} \tag{2.13a}$$

and

$$\Pi^\sim(p) = \Gamma^{(2)\sim}(p) + C_{L,m_0}^{-1} \,. \tag{2.13b}$$

Proposition 2.6. *Consider a lattice field theory with the hypotheses of § 2. Let P_0, P_1 be projections onto the vacuum $\Omega (\equiv 1)$ and the span of $(1-P_0)\varphi\Omega$ respectively. Then*

$$\Pi(x) = -\langle P'' \rangle \delta_0(x) + \langle P'(x), (1 - P_0 - P_1) P'(0) \rangle \tag{2.14}$$

if the Wick ordering covariance for P is C_{L,m_0}.

Proof. As in [19]. The Ising model is excluded because the δ-functions in the single spin measure prevent integration by parts.

Proposition 2.7. *Consider a lattice field theory as above.*

$$\Gamma^{(2)}(x_1, \ldots, x_j, p_{j+1}, \ldots, p_d)|_{x_1=0=\ldots=x_j} \geqq 1 - \cosh m_0 - \langle P'' \rangle - 2d$$

Proof. We use the integration by parts representation (2.13–14), and omit

$$\sum_{x''} e^{-ip''x''} \langle P'(x'=0, x'')(1 - P_0 - P_1) P'(0) \rangle \,.$$

Here $x' = x_1, \ldots, x_j$ and $x'' = x_{j+1}, \ldots, x_d$. The omitted quantity is positive because

$$x'' \mapsto \langle P'(x'=0, x'')(1 - P_0 - P_1) P'(0) \rangle$$

is positive definite and hence the Fourier transform of a positive measure.

3. Three Particle Decay for $\mathscr{P}(\varphi) = \varphi^4$

In this section we consider $\lambda\varphi^4 + \sigma\varphi^2$ lattice field theories with $\sigma > \sigma_c$. Then all hypotheses of §2 are satisfied. Ising models are excluded because we integrate by parts.

Proposition 3.1. *Assume* $\Gamma^{(6)}(xxxyyy) \leqq 0$ *for* $x-y$ *large. Then the CDD radius* $\bar{m}$ *satisfies* $\bar{m} \geqq 3m$. *The analyticity tube* τ_Γ *for* $\Gamma^{(6)}$ *satisfies*

$$\{p: \Sigma |\operatorname{Im} p_\nu| \leqq 3m\} \subset \tau_\Gamma$$

Proof. As in [19]. See also Proposition 3.2 below.

This result leads to the existence of φ_d^4 field theories and the scaling limit for Ising models, see §4. In order to establish the triviality of some of these field theories, we need to assume canonical upper bounds for the two point decay rate (as would follow from the Lehmann spectral formula in the continuum case). Explicitly, we assume

$$0 \leqq G^{(2)}(x) \leqq \begin{cases} K_1 |x|^{-d+2} e^{-m|x|}, & |x| \leqq m^{-1} \\ K_1 m^{d-2} e^{-m|x|}, & |x| \geqq m^{-1} \end{cases} \tag{3.1}$$

This bound is always satisfied for fixed σ, λ, but eventually we will consider the postulate that K_1 is independent of σ and/or λ, which is not known.

Proposition 3.2. *Assume* (3.1) *and* $\Gamma^{(6)}(xxxyyy) \leqq 0$ *for a* $\lambda\varphi^4 + \sigma\varphi^2$ *lattice field theory with* $\sigma > \sigma_c$. *Then for*

$$K = (K_1^3 + K_1^4) K_2 (|\cosh m_0 + \langle P'' \rangle| + 1)$$

where K_2 *is a universal constant and* K_1 *is defined by* (3.1),

$$\Pi(x_1, p=0) \leqq \begin{cases} \lambda^2 K |x_1|^{-2d+7} e^{-3m|x_1|} & \text{for} \quad |x_1| \leqq m^{-1} \\ \lambda^2 K m^{2d-7} e^{-3m|x_1|} & \text{for} \quad |x_1| \geqq m^{-1}, \end{cases} \tag{3.2}$$

Proof. As in [19], we expand Π in terms of connected and one particle irreducible subdiagrams. Negative terms, including $\Gamma^{(6)}$ can be omitted, yielding the bound

$$\Pi(x_1, p=0) \leqq \sum_x \lambda^2 \left\{ 6G^{(2)}(x)^3 - 36 \sum_{z,z'} G^{(4)}(xx0z) \Gamma^{(2)}(z-z') G^{(4)}(z'00x) \right\}. \tag{3.3}$$

By Lebowitz' inequality, $G^4 \leqq 0$. Let p_j be the momentum variable conjugate to $z_j - z_j'$. By Proposition 2.4, applied with $p'' = (p_2, \ldots, p_d) = 0$, the terms with $z_1 - z_1' \neq 0$ may be omitted. Having done so, we apply Proposition 2.4 again with $z_1 - z_1' = 0$, $p'' = (p_3, \ldots, p_d) = 0$; we omit terms with $z_2 \neq z_2'$. Continuing in this fashion we have

$$\Pi(x_1, p=0) \leqq \sum_x \lambda^2 \left\{ 6G^{(2)}(x)^3 + 36(2d+1+\langle P'' \rangle) \sum_x G^{(4)}(xx0z) G^{(4)}(z00x) \right\},$$

where Proposition 2.7 provides the lower bound on $\Gamma^{(2)}(0)$. Lebowitz' inequality bounds the $G^{(4)}$ factors above, so that

$$\begin{aligned} \Pi(x_1, p=0) \leqq \sum_x \lambda^2 \{ & 6G^{(2)}(x)^3 \\ & + 144(2d-1+\langle P'' \rangle) G^{(2)}(x)^2 (G^{(2)} * G^{(2)})(x) \}. \end{aligned} \tag{3.4}$$

Substitution of (3.1) yields (3.2).

4. Existence of Continuum φ_d^4 Fields and the Ising Scaling Limit

The original existence, regularity and Lorentz covariance theorems for φ_2^4 fields were given in [10–12, 15, 2]. A survey of this work, including the contributions of Nelson, Segal and others may be found in [13, 14]. The remaining two Wigthman axioms — uniqueness of the vacuum and its invariance under Lorentz rotations — were verified in [23] for weak coupling, and in [8] for arbitrary coupling. The completion of the φ_3^4 existence program has recently been achieved, see [4, 28] for weak coupling and [5, 7] for arbitrary coupling. The original nonuniqueness proof for relativistic fields was given in [25].

In this section we consider the continuum limit of lattice φ_d^4 fields.

Theorem 4.1. *Assume $\Gamma^{(6)}(xxxyyy) \leqq 0$ for all $\sigma > \sigma_c(\lambda)$ and all λ for $\lambda\varphi^4 + \sigma\varphi^2$ lattice fields. Choose a mass $m > 0$, and for each choice of lattice spacing $\varepsilon > 0$, choose a bare charge $\lambda(\varepsilon)$. Then the limit $\varepsilon \to 0$ defines (through convergent subsequences) a continuum field theory limit of the $\lambda(\varepsilon)\varphi^2 + \sigma\varphi^2$ ε-lattice theories. This limit has mass $m > 0$ and satisfies the Wightman-Osterwalder-Schrader axioms with the possible exception of Lorentz invariance.*

Proof. By [1, 30], there is a $\sigma = \sigma(\varepsilon) > \sigma_c(\lambda(\varepsilon))$ which gives mass $m > 0$ to the ε-lattice field theory. By [17], existence of the limit (through subsequences) follows from a uniform bound on the two point function. As in [19], the bound on the *CDD* radius gives the required bound on the wave function renormalized two point function. Because of the wave function renormalization, the limit theory has a mass m. This completes the proof.

A portion of this proof extends to Ising models.

Proposition 4.2. *Assume $\bar{m}$ (the CDD radius) $\geqq (1+\delta)m$, $\delta > 0$, for $T > T_c$ in the d-dimensional Ising model. Then the scaling limit exists and defines a field theory in the sense specified in Theorem 4.1.*

Theorem 4.3. *Assume $\Gamma^{(6)}(xxxyyy) \leqq 0$ for $\lambda\varphi^4 + \sigma\varphi^2$ fields with mass m fixed, positive and λ sufficiently large ($\sigma > \sigma_c$). Then the Ising model scale limit exists.*

Proof. By [30], the Schwinger functions of the lattice field theory converge, as $\lambda \to \infty$ to the Schwinger functions of the Ising model. Thus $G^{(2)\sim}(p)$ and $\Gamma^{(2)\sim}(p)$ converge for real p. In fact from (2.3), (2.7) and $m > 0$ we have a uniform bound on $\int d\varrho(a, \mathbf{p})/a$ which implies pointwise convergence of $G^{(2)\sim}(p)$ for real p. As in the proof of Proposition 2.2, $G^{(2)\sim}(p)$ is bounded away from zero, so $\Gamma^{(2)\sim}(p)$ converges pointwise and is bounded also. However the remark following Proposition 2.3 yields bounds on α, $|\beta|$ and $\int dv/(1+a^2)$, which by Propositions 2.3 and 3.1 imply that $\Gamma^{(2)\sim}(p_1, 0)$ is analytic up to $|\mathrm{Im}\, p_1| < 3m$, and bounded uniformly in λ. Hence the same is true for the $\Gamma^{(2)\sim}$ of the Ising model. Thus $\bar{m} \geqq 3m$. This statement is uniform in m, and existence of the scaling limit follows from Proposition 4.2.

5. Critical φ^4 Lattice Theories, $d \geqq 6$

Theorem 5.1. *Consider a $\lambda\varphi^4 + \sigma\varphi^2$ lattice field theory with $\sigma > \sigma_c(\lambda)$ and $\Gamma^6(xxxyyy) \leqq 0$. Then $Z \neq 0$. For $d \geqq 6$, the bound on Z^{-1} is independent of $\sigma > \sigma_c(\lambda)$ and λ for λ and m bounded.*

Proof. $Z>0$ because $\Gamma^{(6)}\leqq 0$ implies an upper mass gap. To establish the uniform bound on Z^{-1} for $d\geqq 6$, we use (2.12). By the remark following Proposition 2.3, it is sufficient to estimate the small a behavior of the dv integral, which we do using (2.11), [9] and (3.4). From (2,12),

$$Z^{-1}=\text{const.}+\int_{a=1-\cosh m}^{\infty}\frac{dv}{(a+1-\cosh m)^2}\leqq \text{const.}+\text{const.}\int a^{-2}dv$$
$$\leqq \text{const.}\left(1-\left(\frac{\partial}{\partial p_1}\right)^2\Gamma^{\frown}(p)|_{p=0}\right).$$

(Note in general $\partial/\partial p_1^2)\neq(\partial/\partial p_1)^2)$. By (2.13)–(2.14),

$$Z^{-1}\leqq \text{const.}\left(1-\left(\frac{\partial}{\partial p_1}\right)^2\Pi^{\frown}(p)|_{p=0}\right)$$

and by (3.4)

$$\begin{aligned} Z^{-1}&\leqq \text{const.}\left(1+\sum_x x_1^2\Pi(x)\right)\\ &\leqq \text{const.}\left[1+\sum_x x_1^2(G^{(2)}(x)^3+G^{(2)}(x)^2G^{(2)}*G^{(2)}(x))\right] \qquad (5.1)\\ &\leqq \text{const.}\left[1-\left(\frac{\partial}{\partial p_1}\right)^2(G^{(2)\frown}*G^{(2)\frown}*G^{(2)\frown}+G^{(2)\frown}*G^{(2)\frown}*(G^{(2)\frown})^2)\right]\Bigg|_{p=0}.\end{aligned}$$

The point to these manipulations is to replace a derivative of Γ by a derivative of the upper bound on Γ. By the Herglotz property, we are again able to transfer derivatives to the upper bound (2.11) as follows. For simplicity, we consider only the final (and worst) term above. It is

$$-\left(\frac{\partial}{\partial p_1}\right)^2\int(\cos(p-k)^2+a)^{-1}(\cos(k-l)^2+b)^{-1}(\cos(l-q)^2+c)^{-1}(\cos(l-q)^2+e)^{-1}$$
$$\times\, d\varrho(a)d\varrho(b)d\varrho(c)d\varrho(e)dkdldq|_{p=0}\,.$$

Without loss of generality, we restrict all $d\varrho(a)\ldots d\varrho(e)$ integrations to the interval [0, 1]. By (2.11), it follows that

$$\int_0^1 d\varrho(a,0)\leqq\text{const.}$$

and so we may replace the $d\varrho$ integrations by $\sup\limits_{a,b,c,e}$. This supremum is evaluated as $0=a=b=c=e$. Thus we have accomplished our goal. The derivative is placed on the upper bound, in this case the zero mass free lattice field. It is convenient to return to x-space, where we find

$$Z^{-1}\leqq\text{const.}\left[1+\sum_x x_1^2\left(x^{-3d+6}+\sum_y(x-y)^{-d+2}y^{-d+2}\right)x^{-2d+4}\right]$$

The theorem follows from this bound.

Theorem 5.2. *Under the hypothesis of Theorem 5.1, and the construction of Theorem 4.1, the two point function is free ($\Pi=0$) at momentum $\mathbf{p}=0$, in the limit $\varepsilon\to 0$.*

Remark. If $\lambda=\lambda(\varepsilon)$ is the bare charge (ε = lattice spacing) then $\lambda(\varepsilon)\leqq\varepsilon^{d-4}$ in the above case.

Corollary 5.3. *Suppose in addition that the limit $\varepsilon\to 0$ above is Euclidean invariant. Then it is a free scalar field of mass m.*

Proof. The two point function is free by the theorem, and the theory as a whole is free by [35, p. 163].

Proof of Theorem 5.2. Let $\Gamma^{\sim}_{\lambda,m,\varepsilon}(p)$ be the inverse propagator for the theory with bare charge λ, mass m and lattice spacing ε. By scaling,

$$\Gamma_{s^{d-4}\lambda,s^{-1}m,s\varepsilon}(p)=s^{-2}\Gamma_{\lambda,m,\varepsilon}(sp)\,. \tag{5.2}$$

Here we choose $\varepsilon=1$, $s\to 0$, and

$$s^{-1}m=\text{const.},\quad \text{i.e.}\quad m=\text{const.}\,s\to 0$$
$$s^{d-4}\lambda=\lambda(s)\leqq O(s^{d-4})$$

so that the λ on the right side of (5.2) is bounded, but possibly s-dependent as $s\to 0$. Now $\Gamma_{\lambda,m,\varepsilon}(p_1,0)$ has a zero at $\cos p=\cosh m$ and as we have just seen, bounded first and second derivatives with respect to p_1 evaluated at $\boldsymbol{p}=0$, $p_1=O(m)$. Furthermore the first derivative vanishes at $p_1=0$ because of the $p_1\to -p_1$ symmetry. The same method bounds one more p_1 derivative, and

$$s^{-2}\Gamma_{\lambda,m,\varepsilon}(sp)|_{\boldsymbol{p}=0}=s^{-2}(Z^{-1}(\cosh m-\cos sp_1)+O(s^3p_1^3))\to Z^{-1}(p_1^2+\text{const.})$$

as $s\to 0$. This completes the proof.

Remark. The vanishing of Γ on the mass shell gives a cancellation which formally allows $d\geqq 5$ in place of $d\geqq 6$ above.

6. Upper Bounds on Critical Exponents

We give upper bounds on certain critical exponents in a single phase $\varphi^4+\sigma\varphi^2$ (lattice) quantum field model. In particular we study the anomalous dimension η, the exponent ζ for the field strength renormalization constant Z and the exponent ν for the physical mass. We choose $\sigma>\sigma_c$ (the single phase region) and consider the limit $\sigma\searrow\sigma_c$.

In this section we assume $\Gamma^{(6)}\leqq 0$, as well as the x-space upper bound

$$G^{(2)}(x,\sigma)\leqq K(\min\{|x|,m^{-1}\})^{-(d-2+\eta)}e^{-m|x|} \tag{6.1}$$

with K uniform in the allowed x, σ. We expect (6.1) to hold for lattice field theories. We then show $\eta=0=\zeta$ for $d\geqq 5$, complementing the results of § 5.

Proposition 6.1. *For a pure phase $\lambda\varphi^4+\sigma\varphi^2$ (lattice) field theory with $m>0$, $Z>0$ and satisfying* (6.1), *we have*

$$0\leqq\eta\leqq\zeta/\nu\,. \tag{6.2}$$

If in addition $\Gamma^{(6)}\leqq 0$, then

$$\zeta/\nu\leqq 2\,. \tag{6.3}$$

Here ζ is the exponent for the field strength renormalization constant and ν is the exponent for the mass. Thus $Z \sim m^{\zeta/\nu}$ as $m \to 0$.

Proof. Using (2.12a) evaluated at $p_1 = 0$ gives

$$m^{-2} \leqq \text{const.}\, Z^{-1} G^{2\sim}(0, \sigma).$$

We bound $G^{2\sim}(0, \sigma)$ by integrating (6.1) to obtain

$$m^{-2} \leqq \text{const.}\, Z^{-1} m^{\eta-2}.$$

Thus

$$m^{-\eta} Z \leqq \text{const.},$$

or

$$-\eta + \zeta/\nu \geqq 0.$$

Since the inequality $0 \leqq \eta$ is known we have established (6.2).

Assuming $\Gamma^{(6)} \leqq 0$, we may use (5.1). Then

$$m^{-\zeta/\nu} \sim Z^{-1} \leqq \text{const.}(1 + \sum x_1^2 \Pi(x)).$$

We show $\Sigma \Pi(x) \leqq$ const., from which we infer

$$1 + \sum x_1^2 \Pi(x) \leqq 0(m^{-2}),$$

proving $\zeta/\nu \leqq 2$. The summability of $\Pi(x)$ follows from

$$\Sigma \Pi(x) = \Pi^{\sim}(0) = 0(1) + \Gamma^{\sim}(0). \tag{6.4}$$

Consider σ in the interval (σ_c, σ_1), for $\sigma_1 > \sigma_c$. Since $\Gamma^{\sim}(0) = -G^{2\sim}(0)^{-1} \leqq 0$, we need only bound $\Gamma^{\sim}(0)$ from below. However $G^2(x, \sigma)$ is decreasing in σ (a consequence of the second Griffiths inequality). Thus for $\sigma > \sigma_c$

$$\sum_x G^2(x, \sigma) = -\Gamma^{\sim}(0, \sigma)^{-1}$$

is decreasing in σ and $\Gamma^{\sim}(0, \sigma)$ is also decreasing in σ. Thus $\Gamma^{\sim}(0, \sigma_c) \geqq \Gamma^{\sim}(0, \sigma_1)$.

Theorem 6.2. *Assume $\Gamma^{(6)} \leqq 0$ and (6.1) in a $\lambda\varphi^4 + \sigma\varphi^2$ lattice theory. Then*

$$\eta = \zeta/\nu = 0 \quad \textit{for} \quad d \geqq 5 \tag{6.5}$$

and for $d \leqq 4$

$$0 \leqq \eta \leqq 2(1 - d/5) = \begin{cases} .4 & d = 4 \\ .8 & d = 3. \\ 1.2 & d = 2 \end{cases} \tag{6.6}$$

Proof. We use the bound (5.1) for

$$\begin{aligned} m^{-\zeta/\nu} \sim Z^{-1} &\leqq \text{const.} \left(1 + \sum_x x_1^2 \Pi(x)\right) \\ &\leqq \text{const.} \left(1 + \sum_x x_1^2 [G(x)^3 + G(x)^2 (G * G)(x)]\right). \end{aligned} \tag{6.7}$$

Substituting (6.1) we find the G^3 term in (6.7) sums to

$$O(m^{-8-2d-3\eta)}), \tag{6.8}$$

while the $G^2(G*G)$ term sums to

$$O(m^{-(10-2d-4\eta)}). \tag{6.9}$$

By Proposition 6.1, $2-\eta \geqq 0$ and

$$(10-2d-4\eta)-(8-2d-3\eta) \geqq 0.$$

Thus the $G^2(G*G)$ term in (6.7) dominates and by (6.2), (6.3), (6.7),

$$0 \leqq \eta \leqq \zeta/\nu \leqq 10-2d-4\eta. \tag{6.10}$$

For $d \geqq 5$,

$$0 \leqq \eta \leqq \zeta/\nu \leqq 10-2d \leqq 0,$$

proving (6.5). For $d \leqq 4$, (6.10) gives $5\eta \leqq 10-2d$, which is (6.6).

Remark. If the upper bound on $\Pi(x)$ in (3.3) could be improved to show that the $G^{(4)}\Gamma G^{(4)}$ term has the same long distance behavior as the $(G^{(2)})^3$ term, then the above bounds on η are improved to $\eta \leqq \zeta/\nu \leqq 8-2d-3\eta$. In this case $\eta=\zeta=0$ for $d \geqq 4$ (as expected), and for $d \leqq 4$, $\eta \leqq (2-d/2)$.

Acknowledgement. We thank T. Spencer and K. Symanzik for helpful conversations.

References

1. Baker, G.: Self-interacting Boson quantum field theory and the thermodynamic limit in d dimensions. J. Math. Phys. **16**, 1324—1346 (1975)
2. Cannon, J., Jaffe, A.: Lorentz covariance of the $\lambda(\varphi^4)_2$ quantum field theory. Commun. math. Phys. **17**, 261—321 (1970)
3. Feldman, J.: On the absence of bound states for $\lambda\varphi_2^4$ quantum field models without symmetry breaking. Canad. J. Phys. **52**, 1583—1587 (1974)
4. Feldman, J., Osterwalder, K.: The Wightman axioms and the mass gap for weakly coupled Φ_3^4 quantum field theories. Ann. Phys., to appear
5. Feldman, J., Osterwalder, K.: The construction of $\lambda(\varphi^4)_3$ quantum field models, presented at the International Colloquium on Mathematical Methods in Quantum Field Theory, Marseille, June 1975
6. Fortuin, C., Kastelyn, P., Ginibre, J.: Correlation inequalities on some partially ordered sets. Commun. math. Phys. **22**, 89—103 (1971)
7. Fröhlich, J.: To appear
8. Fröhlich, J., Simon, B.: Pure states for general $P(\varphi)_2$ theories: construction, regularity, and variational equality, preprint, 1976
9. Fröhlich, J., Simon, B., Spencer, T.: Infrared bounds, phase transitions and continuous symmetry breaking. Commun. math. Phys. **50**, 79—85 (1976)
10. Glimm, J., Jaffe, A.: A $\lambda\varphi^4$ field theory without cutoffs I. Phys. Rev. **176**, 1945—1951 (1968)
11. Glimm, J., Jaffe, A.: The $\lambda(\varphi^4)_2$ quantum field theory without cutoffs. II. The field operators and the approximate Hamiltonian. Ann. Math. **91**, 362—401 (1970)
12. Glimm, J., Jaffe, A.: The $\lambda(\varphi^4)_2$ quantum field theory without cutoffs. III. The physical vacuum. Acta Math. **125**, 203—267 (1970)
13. Glimm, J., Jaffe, A.: Quantum field models. In: Statistical mechanics (C. DeWitt, R. Stora, eds.), 1970 Les Houches Lectures. New York: Gordon and Breach Science Publishers 1971

14. Glimm, J., Jaffe, A.: Boson quantum field models. In: Mathematics of contemporary physics. London Mathematical Society (R. Streater ed.). New York-London: Academic Press 1972
15. Glimm, J., Jaffe, A.: The $\lambda\varphi_2^4$ quantum field theory without cutoffs. IV. Pérturbations of the Hamiltonian. J. Math. Phys. **13**, 1568—1584 (1972)
16. Glimm, J., Jaffe, A.: The φ_2^4 quantum field model in the single phase region: Differentiability of the mass and bounds on critical exponents. Phys. Rev. D **10**, 536—539 (1974)
17. Glimm, J., Jaffe, A.: A remark on the existence of φ_4^4. Phys. Rev. Letters **33**, 440—442 (1974)
18. Glimm, J., Jaffe, A.: On the approach to the critical point. Ann. de l'Inst. Henri Poincaré **22**, 13—26 (1975)
19. Glimm, J., Jaffe, A.: Three particle structure of φ^4 interactions and the infinite scaling limit. Phys. Rev. D **11**, 2816—2827 (1975)
20. Glimm, J., Jaffe, A.: Particles and bound states and progress toward unitarity and scaling. Presented at the International Symposium on Mathematical Problems in Theoretical Physics. Kyoto, January 23—29, 1975
21. Glimm, J., Jaffe, A.: Critical problems in quantum fields, presented at the International Colloquium on Mathematical Methods of Quantum Field Theory. Marseille, June 1975
22. Glimm, J., Jaffe, A.: Critical exponents and renormalization in the φ^4 scaling limit, presented at the Conference on Quantum Dynamics: Models and Mathematics. Bielefeld, September 1975
23. Glimm, J., Jaffe, A., Spencer, T.: The Wightman axioms and particle structure in the $P(\varphi)_2$ quantum field model. Ann. Math. **100**, 585—632 (1974)
24. Glimm, J., Jaffe, A., Spencer, T.: The particle structure of the weakly coupled $P(\varphi)_2$ model and other applications of high temperature expansions, Part I: Physics of quantum field models; Part II: The cluster expansion. In: Constructive quantum field theory (G. Velo, A. S. Wightman eds.), 1973 Erice Lectures, Lecture Notes in Physics, Vol. 25. Berlin-Heidelberg-New York: Springer 1973
25. Glimm, J., Jaffe, A., Spencer, T.: Phase transitions for φ_2^4 quantum fields. Commun. math. Phys. **45**, 293—320 (1975)
26. Isaacson, D.: The critical behavior of the anharmonic oscillator. Commun. math. Phys., to appear
27. McBryan, O., Rosen, J.: Existence of the critical point in φ^4 field theory. Commun. math. Phys., to appear
28. Magnen, J., Sénéor, R.: The infinite volume limit of the φ_3^4 model. Ann. l'Inst. H. Poincaré, to appear
29. Marcheson, D.: Private communication
30. Rosen, J.: Mass Renormalization for the $\lambda\varphi^4$ Euclidean lattice field theory, preprint, 1975
31. Rosen, J.: The Ising model limit of φ^4 lattice fields, preprint, 1976
32. Schrader, R.: A possible constructive approach to φ_4^4. Commun. math. Phys. **49**, 131—154 (1976)
33. Schrader, R.: A possible constructive approach to φ_4^4 II, preprint, 1976
34. Spencer, T.: The absence of even bound states for $\lambda(\varphi^4)_2$. Commun. math. Phys. **39**, 77—79 (1974)
35. Streater, R., Wightman, A. S.: PCT, spin and statistics and all that. New York: Benjamin 1964

Communicated by K. Hepp

Received May 11, 1976

The Resummation of One Particle Lines*

James Glimm[1] and Arthur Jaffe[2]

[1] The Rockefeller University, New York, New York 10021, USA
[2] Harvard University, Cambridge, Massachusetts 02138, USA

Abstract. We propose a partial resummation for a weak coupling cluster expansion. The resummation gives one particle lines with in/out field propagators. We give a Bethe-Salpeter equation in which one particle subtractions are defined using physical one particle states. By these methods, we show that $P(\phi)_2$ quantum fields in the weak coupling region have only isolated bound state spectrum below the $2m$ threshold. Here P is not restricted to be even.

1. Introduction

We present a new method for the study of mass spectrum, asymptotic completeness and related questions. The method uses exact subtractions of the physical one particle states; technically, it is based on the n-particle cluster expansion [8], and a (new)partial resummation which identifies the physical one particle lines in closed form.

Previous work [3, 4, 11, 12] on spectral properties of $P(\phi)_2$ quantum fields was based on the Euclidean program of defining n-particle irreducible amplitudes [13]. In this program, the vacuum subtractions are performed exactly, while one and higher particle subtractions are replaced by an orthogonalization of the polynomials in the Euclidean field ϕ. Thus in this program, a j-particle irreducible n-point function is defined by projecting $\phi(x_1)\phi(x_2)\ldots\phi(x_n)$ onto the orthogonal complement, in the Euclidean Hilbert space $\mathscr{E}\equiv L_2(\mathscr{S}', d\mu)$, of the subspace spanned by the vectors

$$\phi(y_1)\ldots\phi(y_l), l\leqq j, y_\nu\in\mathbb{R}^2. \tag{1.1}$$

While the resulting subtraction agrees with the subtraction of (physical) j-particle intermediate states for $j=0$, it does not agree in general, and in particular if bound

* Supported in part by the National Science Foundation under Grants PHY 78-08066 and PHY 77-18762. Both authors thank the I.H.E.S., Bures-sur-Yvette, and A. J. thanks C.E.N., Saclay, for their hospitality.

states occur. For $j \geqq 1$ spurious zeros and poles (CDD poles) can be introduced by these subtractions.

The Euclidean program has been successful when $j=0$ (the vacuum) was the only subtraction required. Within this limit, particles and bound states below the continuum threshold have been analyzed, and for even theories asymptotic completeness up to energies below the four particle threshold is known. Various extensions of this work have been given [1, 2, 7, 9]. This work makes it appears likely that the j particle Euclidean subtractions can be incorporated into the spectral analysis program.

We study here the simplest problem requiring a one particle subtraction: the analysis of the four point function up to energies below the three particle threshold, for a general (not necessarily even) interaction at weak coupling. These methods appear useful in the study of higher spectral properties in field theory, and presumably they have an analog in statistical mechanics.

In §2, we define normal ordering: $\phi(x_1,0)\phi(x_2,0)$: of the square of the time $t=0$ field by subtraction of the physical one particle state. In other words, if $|\Omega\rangle$ is the vacuum state in the physical Hilbert space $\mathscr{H}$, then

$$:\phi^2:|\Omega\rangle \perp |\text{one particle states}\rangle . \tag{1.2}$$

The central object of the study in this paper is then

$$S_{0,T}=\langle\Omega| :\phi(x_1,0)\phi(x_2,0): :\phi(y_1,T)\phi(y_2,T):|\Omega\rangle , \tag{1.3}$$

where $\Omega\equiv 1$ is the Euclidean vacuum state, and the inner product is in the Euclidean Hilbert space $\mathscr{E}$. We let $\langle\cdot\rangle$ now denote an expectation in the Euclidean vacuum Ω. Let

$$\begin{aligned} G_{0,T}=&\langle\phi(x_1,0)\phi(y_1,T)\rangle\langle\phi(x_2,0)\phi(y_2,T)\rangle \\ &+\langle\phi(x_1,0)\phi(y_2,T)\rangle\langle\phi(x_2,0)\phi(y_1,T)\rangle . \end{aligned} \tag{1.4}$$

We replace the Euclidean Bethe – Salpeter equation of the form (in momentum space)

$$S^{\sim}=G^{\sim}+G^{\sim}K^{\sim}S^{\sim}, \tag{1.5}$$

by an equation of the form

$$S_{0,T}=A_{0,T}+\sum_{0\leq j\leq T} B_{0,T}S_{j,T}. \tag{1.6}$$

By summing over T, this leads to an equation

$$\sum_{0\leqq T} S_{0,T}=\sum_{0\leqq T} A_{0,T}+\Big(\sum_{0\leqq j} B_{0,j}\Big)\sum_{0\leqq T} S_{0,T} \tag{1.7}$$

with the solution

$$\sum_{0\leqq T} S_{0,T}=\Big(1-\sum_{0\leqq j} B_{0,j}\Big)^{-1}\sum_{0\leqq T} A_{0,T}, \tag{1.8}$$

since $S_{0,T}=S_{j,T+j}$, etc. We see that $\sum_{0\leqq j} B_{0,j}$ plays the role of $G^{\sim}K^{\sim}$ in the Bethe-Salpeter equation. As with $G^{\sim}K^{\sim}$, we achieve a factorization of $\sum_{0\leqq j} B_{0,j}$. The first factor contains the free field behavior of the four point function, and has a square root singularity with a two-sheeted analytic continuation. The first factor arises from two physical one particle lines, up to the time of their first interaction. The second ("amputated") factor is analytic up to energies less than $3(m-\varepsilon)$, and contains the effects of interaction between the one particle lines. Moreover $\sum_{0\leqq j} B_{0,j}$ is compact on a suitable space of functions, analytic in p. The spectral properties of the field theory, for $E \leqq 3(m-\varepsilon)$, follow from these facts.

2. One Particle Subtractions

In this section we review properties of one particle subtractions and establish notation. To simplify the formulas, we assume

$$\langle \Omega|\phi(x,t)|\Omega\rangle = 0. \tag{2.1}$$

In other words, for a general theory, we make the substitution $\phi \to \hat{\phi} \equiv \phi - \langle\phi\rangle$. Let $|p\rangle$ be the one particle state of momentum p. with the normalization

$$\langle p|p'\rangle = (2\mu(p))^{-1}\delta(p-p'), \tag{2.2}$$

where $\mu(p) = (p^2+m^2)$ and m is the one particle mass. Then with

$$\begin{aligned} |x\rangle &= (2\pi)^{-1/2}\int e^{-ipx}|p\rangle dp \\ |f\rangle &= \int f(x)|x\rangle dx, \end{aligned} \tag{2.3}$$

we have

$$\langle x|y\rangle = (2\mu)^{-1}(x-y); \langle f|g\rangle = \langle f, (2\mu)^{-1} g\rangle_{L_2}, \tag{2.4}$$

where $\mu = (-d^2/dx^2 + m^2)^{1/2}$. The field strength renormalization constant Z is defined by

$$\langle y|Z^{-1/2}\phi(x,0)|\Omega\rangle = (2\mu)^{-1}(x-y), \quad \text{or equivalently} \tag{2.5}$$

$$Z^{1/2} = (2\pi)^{1/2} e^{ipx} 2\mu(p)\langle p|\phi(x)|\Omega\rangle. \tag{2.6}$$

By Lorentz invariance, Z is independent of x, y, p.

Let $M_0 = |\Omega\rangle\langle\Omega|$ and let

$$M_1 - M_0 = \int |p\rangle\langle p| 2\mu(p) dp \tag{2.7}$$

be the orthogonal projection onto one particle states. In general, M_a is the projection onto states of mass $M \equiv (H^2-P^2)^{1/2} \leqq am$. Then $M_0\phi\Omega = 0$,

$$M_1\phi(f,0)|\Omega\rangle = Z^{1/2}|f\rangle, \tag{2.8}$$

and

$$\begin{aligned} e^{-tH} M_1\phi(f,0)|\Omega\rangle &= Z^{1/2} e^{-tH}|f\rangle \\ &= Z^{1/2}|e^{-t\mu}f\rangle = M_1\phi(e^{-t\mu}f,0)|\Omega\rangle. \end{aligned} \tag{2.9}$$

In weakly coupled $P(\phi)_2$ theories, as considered in this paper, $I-M_1$ projects onto states with energy greater than $2(m_0-\varepsilon)$, where m_0 is the bare mass. Thus

$$\|e^{-TH}\{e^{-H}\phi(f,0)|\Omega\rangle-\phi(e^{-\mu}f,0)|\Omega\rangle\}\| \leq 0(1)e^{-2(m_0-\varepsilon)T}. \tag{2.10}$$

We use this formula in § 5 to identify resummed one particle lines.

We define the (physical) normal ordering, so that $:\phi^2:|\Omega\rangle \perp M_1$. Let

$$\alpha_0(x-y)=\langle\Omega|\phi(x,0)\phi(y,0)|\Omega\rangle \tag{2.11}$$

$$\alpha_1(z-x,x-y)=Z^{-1/2}(2\mu_z)\langle z|\phi(x,0)\phi(y,0)|\Omega\rangle. \tag{2.12}$$

Then

$$:\phi(x,0)\phi(y,0):=\phi(x,0)\phi(y,0)-\alpha_0(x-y) -\int\alpha_1(z-x,x-y)\phi(z,0)dz. \tag{2.13}$$

We require for later use regularity properties of α_0 and α_1. Note α_0, as a truncated Schwinger function, is C^∞ except at $x-y=0$, where it has a *ln* singularity, and α_0 and all derivatives are exponentially decreasing as $x-y\to\infty$. The estimates on α_1 in this section are elementary. They will be improved in § 5. We work mainly in a space of fixed total momentum p, and so we seek a bound on

$$\tilde{\alpha}_1(p,x-y)\equiv(2\pi)^{-1/2}\int e^{+ip(x+y-2z)/2}\alpha_1(z-x,x-y)dz = Z^{-1/2}2\mu(p)\langle p|\phi\left(\frac{x-y}{2},0\right)\phi\left(\frac{y-x}{2},0\right)|\Omega\rangle. \tag{2.14}$$

Let $\mathscr{H}(p)$ denote the Hilbert space of fixed total momentum p. with inner product $\langle,\rangle_p$. It is convenient to exclude Ω from $\mathscr{H}(0)$, so that

$$\mathscr{H}\oplus\{\lambda\Omega\}=\int_\oplus\mathscr{H}(p)dp. \tag{2.15}$$

In general, vectors in $\mathscr{H}(p)$ are continuum states in $\mathscr{H}$. We can, however, given Φ, $\Psi\in\mathscr{H}-\{\lambda\Omega\}$ write a formula for $\langle\Phi(p),\Psi(p)\rangle_p$, namely

$$\langle\Phi(p),\Psi(p)\rangle_p=\langle\Phi,\Psi\rangle_p=(2\pi)^{-1/2}\int\langle\Phi,e^{-ipa}T_a\Psi\rangle_{\mathscr{H}}da, \tag{2.16}$$

where $T_a\equiv e^{iPa}$ is translation by a.

From the exponential cluster properties of the truncated Schwinger functions, we see that

$$\{\phi(x,0)\phi(y,0)-\alpha_0(x-y)\}\Omega \tag{2.17}$$

has finite norm in $\mathscr{H}(p)$, with a bound uniform in x,y. Optimal estimates give

$$\int f(x-y)\{\phi(x,0)\phi(y,0)-\alpha_0(x-y)\}\Omega d(x-y)=\theta \tag{2.18}$$

a finite norm in $\mathscr{H}(p)$ when f is in the Sobolev space H_{-1}, and in fact

$$\|\theta(p)\|_p^2\leq\text{const.}\|\mu^{-1}f\|_{L_2}. \tag{2.19}$$

Such bounds are true more generally of vectors

$$\Psi=[\phi(x_1,t_1)\ldots\phi(x_n,t_n)-\langle\phi(x_1,t_1)\ldots\phi(x_n,t_n)\rangle]\Omega, \tag{2.20}$$

with all $t_j \geqq 0$ and no two (x_j, t_j)'s coinciding. Such estimates are proved by integration by parts, cf. [5], to obtain optimal momentum space behavior (i.e. free field singularities at coinciding points in x space). Such vectors Ψ (when smoothed with test functions to become elements of $\mathscr{H}$) are dense in $\mathscr{H}$. Thus the corresponding vectors (2.20) are dense in $\mathscr{H}(p)$ for a.e. p. Furthermore they are continuous in p in the sense that $\langle \psi_1(p), \psi_2(p)\rangle_p$ is continuous in p.

With $f \in \mathscr{S}$, let

$$\theta_t = \phi(e^{t\mu} f, t)\Omega = \int \phi(x,t)(e^{t\mu} f)(x)dx\,\Omega\,. \tag{2.21}$$

A calculation with the Lehmann spectral formula shows that as $t \to \infty$, the vectors θ_t converge in $\mathscr{H}$ at the rate $O(e^{-t(m-\varepsilon)})$. The same is true, furthermore for $\theta_t(p)$ in $\mathscr{H}(p)$, with convergence uniform in p. The limit is

$$\lim_{t\to\infty} \theta_t = Z^{1/2}\Phi_{ex}(f)\Omega\,, \tag{2.22}$$

where the time zero asymptotic field Φ_{ex} is given by standard Haag-Ruelle scattering theory. We now use Φ to denote the real time (Minkowski) field, and Φ_{ex} its asymptotic limit. At time zero we identify the Euclidean and Minkowski fields $\Phi(x) \equiv \Phi(x,0) = \phi(x,0) \equiv \phi(x)$.

We claim that after mapping $\Phi_{ex}(f)\Omega$ into $\mathscr{H}(p_0)$ by (2.16), we may take the limit $f(p) \to \delta(p - p_0)$ in the test function. Thus we identify $\tilde{\Phi}_{ex}(p)\Omega \tilde{f}(p)$ as the component of $\Phi_{ex}(f)\Omega$ with momentum p.

To justify this exchange of limits first appeal to the uniformity of the convergence in (2.22) as $t \to \infty$ in each $\mathscr{H}(p)$. It follows that the inner product $\langle \Psi, \Phi_{ex}(f)\Omega\rangle_p$ is a continuous function of p. Take $\tilde{f}(p) = \tilde{f}_1(p)\tilde{f}_2(p)$. By (2.16)

$$\langle \Psi, \Phi_{ex}(f)\Omega\rangle_p = \langle \Psi, \Phi_{ex}(f_1)\Omega\rangle_p \tilde{f}_2(p)\,. \tag{2.23}$$

Because of the continuity of the inner product in p, we can take the limit $\tilde{f}_2(p) \to \delta(p - p_0)$, where $\tilde{f}_1(p_0) = 1$.

Thus

$$\langle \Psi, \tilde{\Phi}_{ex}(p_0)\Omega\rangle_{\mathscr{H}} = \int \langle \Psi, \Phi_{ex}(f)\Omega\rangle_p dp \tag{2.24}$$

and we define $\tilde{\Phi}_{ex}(p_0)\Omega$ as an improper element of $\mathscr{H}$. The vector $\tilde{\Phi}_{ex}(p)\Omega$ has a finite inner product with all vectors Ψ of the form (2.20). Thus also by (2.23),

$$\langle \Psi, \tilde{\Phi}_{ex}(p_0)\Omega\rangle_p = \langle \Psi, \tilde{\Phi}_{ex}(p)\Omega\rangle_{\mathscr{H}}\,\delta(p - p_0)\,. \tag{2.25}$$

Since the definition, $|p_0\rangle = \tilde{\Phi}_{ex}(p_0)\Omega$, it follows that $\langle \Psi | p_0\rangle$ is also well defined and bounded uniformly in (x_j, t_j) for all $t_j \geqq 0$ and no coinciding points, unless $n=2$. In particular, since $\langle p_0 | \Omega\rangle = 0$, it follows that $\tilde{\alpha}_1(p, x-y)$ is bounded as a function of $x-y$. Furthermore, $\tilde{\alpha}_1(p, x-y)$ is Hölder continuous in $x-y$ with exponent less than 1/2.

In terms of the function spaces A_δ of § 6, $\tilde{\alpha}_1(p, \cdot) \in A_\delta$ for $\delta_1 \leqq 3/2 - \varepsilon$, $\delta_2 < 0$. To obtain this improvement on the Hölder continuity above, we write

$$\begin{aligned}\langle p | \phi(x_1)\phi(x_2)\Omega\rangle &= e^{\mu(p)}\langle p | e^{-H}\phi(x_1)\phi(x_2)\Omega\rangle \\ &= e^{\mu(p)}\langle p | e^{-H}(I - M_0)\phi(x_1)\phi(x_2)\Omega\rangle\,,\end{aligned}$$

and the bound the norm of $e^{-H}(I-M_0)\Phi(x_1)\phi(x_2)\Omega$ in $\mathscr{H}(p)$. The improved smoothing comes from e^{-H} and can be exhibited by integration by parts formulas as above. Then (2.19) is replaced by $\|\mu^{-3/2+\varepsilon}f\|_{L_2}$.

3. Span of $:\Phi^n:$

Let

$$:\Phi^j(f,t): \equiv e^{itH}:\Phi(x_1)\dots\Phi(x_j):e^{-itH} \tag{3.1}$$

where $:\Phi(x_1)\dots\Phi(x_j):$ denotes normal ordering defined so that $:\Phi:=\Phi$, $M_0\Phi\Omega=0$, and

$$M_1:\Phi(x_1)\dots\Phi(x_j):\Omega=0, j\geqq 2. \tag{3.2}$$

Theorem 3.1. *Choose n integer $\geqq 2$, $\varepsilon>0$, and $\lambda<\lambda(n,\varepsilon)$. Then states of the form*

$$M_{n+1-\varepsilon}:\Phi^j(f,t):\Omega, \quad 2\leqq j\leqq n, \tag{3.3}$$

span $(M_{n+1-\varepsilon}-M_1)\mathscr{H}$. Here f ranges over $\mathscr{S}(\mathbb{R}^n)$, and t belongs to an open interval (a,b).

Remark. Taking $n=2$, the vectors $M_{3-\varepsilon}:\Phi^2:(f,t)\Omega$ span all bound states and two particle states with mass in the interval $(m,(3-\varepsilon)m)$. In other words, all states with mass in this interval occur in the four point function $S_{0,T}$.

Proof. Let E_a denote the projection onto states with energy $\leqq am$. Let $|f,t\rangle_j \equiv :\Phi^j:(f,t)\Omega$. We show that $E_{n+1-\varepsilon}|f,t\rangle_j, j=2,\dots,n$, span $E_{n+1-\varepsilon}(I-M_1)\mathscr{H}$. By Lorentz invariance, the states $M_{n+1-\varepsilon}|f,t\rangle_j$ then span $(M_{n+1-\varepsilon}-M_1)\mathscr{H}$. For states $E_a|f,t\rangle_j$ we note that the total momentum P satisfies

$$P^2\leqq(a^2-1)m^2. \tag{3.4}$$

Secondly we remark that $E_a|f,t\rangle_j$ is an entire function of t. Thus if $\langle\chi|E_a|f,t\rangle_j=0$, for $t\in(a,b)$, it follows by analytic continuation that $\langle\chi|E_a|f,it\rangle_j=0$. Thus it is sufficient to show that $E_{n+1-\varepsilon}|f,it\rangle_j, j=2,3,\dots,n, t\in\mathbb{Z}_+$ span $E_{n+1-\varepsilon}(I-E_1)\mathscr{H}$.

It has already been proved by the cluster expansion [8] that the states

$$E_{n+1-\varepsilon}\{\Omega, \Phi(f,t)\Omega, |f,it\rangle_j, 2\leqq j\leqq n, t\in\mathbb{Z}_+\}, \tag{3.5}$$

span $E_{n+1-\varepsilon}\mathscr{H}$. In fact the same method [8], taking (3.4) into account, shows that $\Phi(f,0)\Omega$ can replace $\Phi(f,t)\Omega$ in (3.5). Thus a dense set of states χ in $E_{n+1-\varepsilon}\mathscr{H}$ have the form

$$\chi=E_{n+1-\varepsilon}\{\alpha\Omega+\phi(f)\Omega+\psi\} \tag{3.6}$$

where ψ is in the span of $|f,it\rangle$, $t\in\mathbb{Z}_+$ and $2\leqq j\leqq n$. In (3.6), we restrict f to have Fourier transform $f^\sim(p)$ vanishing unless $p^2\leqq[(n+1-\varepsilon)-1]m^2$ c.f. (3.4).

We take a sequence χ_n of such vectors, converging to a $\chi\in(1-M_1)\mathscr{H}$. Then

$$M_0\chi_n=\alpha_n\Omega\to M_0\chi=0$$

so that $\alpha_n \to 0$. Since the norms $f \to \|\phi(f)\Omega\|$ and $f \to \|M_1 \phi(f)\Omega\|$ are equivalent, and since

$$(M_1 - M_0)\chi_n = M_1 \phi(f_n)\Omega \to M_1 \chi = 0.$$

$\phi(f_n)\Omega \to 0$ also. Thus $\psi_n \to \chi$ and the proof is complete.

4. Integration by Parts

The basic expansion steps depend on integration by parts. Let

$$C_0 = (-\Delta + M_0^2)^{-1}(x, y) = (2\mu)^{-1}[\exp(-|x_0 - y_0|\mu)](\mathbf{x}, \mathbf{y}), \tag{4.1}$$

and let C be a covariance of the class considered in [8]. This class is generated from C_0 by two operations: (a) the block diagonalization

$$C \to \chi_X C \chi_X + \chi_{\mathbb{R}^2 \setminus X} C \chi_{\mathbb{R}^2 \setminus X} \tag{4.2}$$

for X a rectangle in $\mathbb{R}^2$, and (b) the convex sum

$$\{C_j\} \to \sum \beta_j C_j, \sum \beta_j = 1, \beta_j \geqq 0. \tag{4.3}$$

For this class of covariance operators, we define the normalized measure on $\mathscr{S}'(\mathbb{R}^2)$,

$$d\mu_C = \lim_{\Lambda \uparrow \mathbb{R}^2} Z(\Lambda)^{-1} \exp(-\textstyle\int_\Lambda \mathscr{P}(\phi(x))dx) d\mu_C, \tag{4.4}$$

with existence of the limit following from [8].

With $R(\phi)$ a polynomial in ϕ, the formula

$$\frac{DR(\phi)}{D\phi(y)} = \left[\frac{\partial}{\partial \phi(y)} - \mathscr{P}'(\phi(y))\right] R(\phi) \tag{4.5}$$

defines $D/D\phi(y)$ as a densely defined linear operator on $\mathscr{E}$. Integration by parts is the identity

$$\phi(x) = \int_{R^2} C(x, y) \frac{D}{D\phi(y)} dy, \tag{4.6}$$

valid when substituted in the integral

$$\int \phi R(\phi) d\mu_C. \tag{4.7}$$

We use (4.6) to decompose ϕ into a sum of "annihilation" and "creation" operators:

$$\phi(x) = a(x) + a^*(x), \tag{4.8}$$

$$a(x) = \int_{x_0 \leqq y_0} C(x, y) \frac{D}{D\phi(y)} dy; \; a(x)^* = \int_{x_0 \geqq y_0} C(x, y) \frac{D}{D\phi(y)} dy. \tag{4.9}$$

Next consider a time interval (s, t) without contours. This means

$$C(x, y) = C_0(x, y), s \leqq x_0, y_0 \leqq t.$$

Then for $x_0 \leqq z_0 \leqq y_0$

$$C(x,y) = \int_{z_0=\text{const}} C_0(x,z)(2\mu C_0)(z,y)d\mathbf{z},$$

which has the consequence

$$\int_{x_0 \leqq y_0 \leqq z_0} C(x,y)\frac{D}{D\phi(y)}dy = a(x) - (e^{-(z_0-x_0)\mu}a)(z_0,\mathbf{x})$$
$$= a(z_0,\mathbf{x})^* - (e^{-(z_0-x_0)\mu}a^*)(x). \tag{4.10}$$

These identities are used to write $\phi = a + a^*$ and to move a, a^* forward and backward in time. Normally a^* is moved to time $t = -\infty$, i.e. eliminated. Also $a = \phi - a^*$ can be rewritten in terms of ϕ.

Remaining expansion steps concern the insertion and removal of contours, achieved by variation of the parameters β_j in (4.3).

5. The One Particle Line

The main technical tool of this paper is the use of virtual contraction terms to allow the resummation of one particle lines. We explain these ideas here, and for clarity we give a very simple example, namely the evaluation of

$$\int \phi(f,0)R_T(\phi)d\mu = \langle \phi(f,0)R_T\rangle \tag{5.1}$$

with R_T localized at times $t \geqq T$. As an application, we obtain improved bounds on α_1.

The goal of our expansion is to write

$$\langle \phi(f,0)R_T\rangle = \langle \{\alpha + \phi(w,1)\}R_T\rangle + O(e^{-2(m_0-\varepsilon)T}), \tag{5.2}$$

where the right hand side is a partial resummation of the n-particle cluster expansion given in [8]. Having identified the resummation leading to (5.2), we can use (5.2) as an elementary expansion step in the definition of our modified Bethe-Salpeter equation (1.6). In terms of vectors in the physical Hilbert space $\mathscr{H}$, we identify (5.2). Let

$$|\chi\rangle \equiv \phi(f,0)|\Omega\rangle - \langle\Omega|\phi(f,0)|\Omega\rangle|\Omega\rangle - e^{H}\phi(e^{-\mu}f,0)|\Omega\rangle. \tag{5.3}$$

Then (2.10) says that for any $\theta = R_0|\Omega\rangle$,

$$|\langle\chi, e^{-TH}\theta\rangle| \leqq O(1)e^{-2(m_0-\varepsilon)T}, \tag{5.4}$$

and comparison with (5.2) shows that $w = e^{-\mu}f$, $\alpha = \langle\phi(f)\rangle$.

In order to evaluate (5.1), we use the 1-particle expansion in [8]. Writing $\phi = a + a^*$, we use (4.8), (4.10) to advance the annihilator in time to become a field ϕ at $t=1$. We use (4.9) to contract a^*. New vertices from the $D/D\phi$ terms are removed by insertion and removal of contours, as in [8]. We divide the expansion terms into two sets as follows: Terms in the first set have $y_0 < T$ (where y_0 is defined in [8, pp. 609–610]). These terms have the form $\langle QR\rangle$, where

$$Q = \sum_j Q_j, \quad Q_j = \alpha(j) + \phi(w(j), t_j), \tag{5.5}$$

where t_j is an integer time, $t_j < T$. The second set of terms has $y_0 = +\infty$, and contains contradictions to times $\geqq T$. These are all remainder terms, and the sum of the norms of remainder terms is bounded by $O(1)\|\phi(f)\| \|R\| e^{-2(m_0-\varepsilon)T}$ with norms defined in [8]. Thus we write

$$\langle \phi R \rangle = \langle QR \rangle + \text{Remainder}. \tag{5.6}$$

Furthermore, we choose $f(x)$ with $f^{\sim}$ supported in

$$\{p : |p| \leqq \sqrt{3}(m-\varepsilon)\} = \{p : \mu(p) \leqq 2(m-\varepsilon')\}. \tag{5.7}$$

Then the terms $\phi(w(j), t_j)$ in (5.5) can be returned to time $t=1$.

In summary, we have for $n=1$

$$\langle \phi(f,0)R \rangle = \alpha_T \langle R \rangle + \langle \phi(w_T, 1)R \rangle + O(e^{-2(m_0-\varepsilon)T}). \tag{5.8}$$

In (5.8), α_T and $\phi(w_T, 1)$ are the linear and constant terms in Q. Each term in Q is independent of T. However writing Q as a sum of terms $Q = \sum_i Q_i$, the index set $\{i\}$ depends on T. Thus α_T and w_T also depend on T. In order to cure this defect, we add and subtract certain terms on the right side of (5.8). To identify the missing terms we consider the case $T = \infty$, when (5.8) becomes

$$\langle \phi(f,0)R \rangle = \alpha \langle R \rangle + \langle \phi(w, 1)R \rangle. \tag{5.9}$$

For $T = \infty$, there are no error terms, but there are terms in α resulting from contraction of $\phi(f, 0)$ to the vacuum in the region $t \geqq T$. We can think of these terms as contractions to the vacuum in the region blocked by R, and we call these terms, virtual contractions of ϕ to R.

The sum of the virtual contraction terms is absolutely convergent and bounded by $O(\exp - 2(m-\varepsilon)T)$, because this sum is contained in the expansion of (5.9) to which the estimates of [8] also apply. Thus

$$\alpha pR \rangle - \alpha_T \langle R \rangle = \sum \text{virtual terms} \tag{5.10}$$

$$\langle \phi(w, 1)R \rangle - \langle \phi(w_T)R \rangle = \sum \text{virtual terms} \tag{5.11}$$

where in (5.10) we sum over all virtual terms which are independent of ϕ, and in (5.11) the sums runs over all terms linear in ϕ. Note that the estimates of [6] apply to operator kernels, such as w_T, $w_{T=\infty}$ and $w_{T=\infty} - w_T$, as well as to expectation values such as $\langle \phi(w_{T=\infty}, 1) \rangle$. In the norm of [8],

$$\|w_{T=\infty} - w_T\| \leqq O(e^{-2(m-\varepsilon)T}). \tag{5.12}$$

We add and subtract (5.10) and (5.11) to the right side of (5.5). The addition modifies α and w, while the subtraction is absorbed into the error term $O(e^{-2(m_0-\varepsilon)T})$. This yields

$$\langle r(f,0)R \rangle = \alpha \langle R \rangle + \langle \phi(w, 1)R \rangle + O(e^{-2(m-\varepsilon)T}), \tag{5.13}$$

where now $\alpha = \alpha_{T-\infty}$ and $w = w_{T=\infty}$ are independent of T.

Now we identify $\alpha = \langle \phi(f, 0) \rangle$, and with the convention (2.1), $\alpha = 0$. Comparison of (5.2) and (5.4) shows that $w = e^{-\mu} f$.

Thus in (5.13), we have achieved an explicit resummation of the one particle line, with the exact physical mass. The error term is an absolutely convergent series of direct and virtual contraction terms, with two particle decay $\sim e^{-2(m-\varepsilon)T}$.

To conclude the description of the expansion, we correct an error in [8]. In order to return field operator $\phi(f,t)$ to earlier times $s<t$, we must be dealing with translation invariant, i.e., momentum conserving expansion operations. The expansion is based on an integer lattice, artificially introduced in position space. Because of the translation invariance of the original theory, the expansion is term by term invariant under integer (lattice preserving) translations, but not under arbitrary space translations. To recover full translation invariance, we integrate each term over relative lattice positions, generated by the translations

$$\mathbb{Z}\to\mathbb{Z}+\theta,\theta\in(0,1]. \tag{5.14}$$

This integration is done at the end of each expansion step containing no contours, i.e. before moving $\phi(f,t)$ backward in time.

The norms of [8] are not invariant under translation by non-integer θ, as in (5.14), but they are easily replaced by equivalent norms, which are θ-invariant. This is done explicitly in § 6. The expansion remains norm decreasing in these equivalent norms. The extra expansion step,

$$F(\theta,\ldots)\to\int_0^1 F(\theta,\ldots)d\theta$$

is norm preserving, in the new translation invariant, norms. Thus the modified expansion is norm decreasing, and convergent, as before.

The improved estimates on $\tilde{\alpha}_1$ give exponential decrease, in addition to the local regularity established in § 2. In terms of the space A_δ of § 6, we state our result as

Theorem 5.1. *As a function of* $x_{\rm rel}=(x-y)/2$, $\tilde{\alpha}_1(p,x_{\rm rel})\in A_\delta$ *for* $\delta_1<2-\varepsilon$, $\delta_2>0$ *sufficiently small. The norm of* $\tilde{\alpha}_1$ *goes to zero as* $\lambda\to 0$.

We give part of the proof here, and complete the argument in § 9, 10. For $x_{\rm rel}$ in a bounded set, the analysis of § 2 is sufficient. Thus we let $j\leqq x_{\rm rel}\leqq j+1$ and consider j large. As in § 2,

$$\| |(\Phi_{ex}(f)-Z^{-1/2}e^{-jH}(\phi(e^{j\mu}f)))\Omega> \| \leqq O(e^{-(m-\varepsilon)j})$$

by a calculation with the Lehmann spectral formula, and the same bound holds in each space $\mathscr{H}(p)$ of total momentum p. If χ_j represents a smoothed version of the characteristic function of $[j,j+1]$, then we see as in § 2 that for $\chi_j\equiv\chi_j(x_{\rm rel})$,

$$\chi_j\{\tilde{\alpha}_1(p_{\rm tot},x_{\rm rel})-\langle Z^{-1/2}e^{-jH}\phi(e^{j\mu}f)\Omega,\phi^2(x_{\rm rel},p_{\rm tot})\Omega\rangle_p\} \tag{5.15}$$

has an $H_{3/2-\varepsilon}$ norm bounded by $e^{-(m-\varepsilon)j}$, and thus finite norm in A_δ, which is moreover summable in j. By (5.2), we can write

$$\begin{aligned}&\langle\phi(e^{l\mu}f,0),\int\phi(x_1,l)\,\phi(x_2,l)g(x)dx\rangle\\&\quad=\langle\phi(e^{(l-1)\mu}f,0),\int\phi(x_1,l-1)\,\phi(x_2,l-1)g\,dx\rangle\\&\qquad+O(e^{-2(m_0-\varepsilon)l})e^{-\delta j}\|e^{l\mu}f\|\,\|g\|,\end{aligned} \tag{5.16}$$

for $1 \leqq l \leqq j$. The exponent proportional to j will be explained below. The norms in the error term are from [8], and the error term can be bounded by $e^{-\varepsilon' j}\|f\|\,\|g\|$, provided p_{tot}, i.e. the support of f, is restricted to $|p_{\text{tot}}| \leqq \sqrt{3}m_0(1-\varepsilon'')$. We now iterate this identity, and sum over $0 \leqq l \leqq j$ to obtain

$$\begin{aligned}&\langle \phi(e^{j\mu}f, 0), \textstyle\int \phi(x_1, j)\,\phi(x_2, j)g(x)dx\rangle \\ &\quad = \langle \phi(f) \textstyle\int \phi(x_1)\,\phi(x_2)g(x)dx\rangle + O(e^{-\varepsilon' j})\,\|f\|\,\|g\|\,. \end{aligned} \tag{5.17}$$

The first term has exponential decrease in x_{rel}. In fact because of the normalization $\langle\phi\rangle = 0$, the ordinary and truncated three point functions coincide. Thus the first term, at constant momentum p, belongs to A_δ in the variable x_{rel}.

The error terms in (5.16) result from direct and virtual contraction to the observables $\phi(x_1, l-1)$ and $\phi(x_2, l-1)$. However terms with contraction to only one of these two observables can be resummed, and yield zero, again because of the normalization $\langle\phi\rangle = 0$. The terms with direct or virtual contraction to both ϕ's at time $l-1$ have extra decay, determined by $|x_{\text{rel}}|$, i.e., j. The relation of the norms of [8], as in (5.17), to the norms in A_δ is analyzed in §9, 10, and will complete the proof.

6. Function Spaces

In this section we define function spaces suitable for the analysis of our expansion. First we introduce spaces necessary to obtain a translation invariant expansion. Let $\{\Delta_i\}$, $\{\Delta_i'\}$ each be disjoint covers of $\mathbb{R}^n$, and define

$$\|f\| = \sum_i \|f\|_{L_2(\Delta_i)} \tag{6.1}$$

$$\|f\|' = \sum_i \|f\|_{L_2(\Delta_i')}\,. \tag{6.2}$$

We assume that each Δ_i is contained in the union of at most K of the Δ_i''s, and conversely.

Proposition 6.1. *The norms* (6.1) *and* (6.2) *are equivalent, and*

$$\frac{1}{K}\|f\|' \leqq \|f\| \leqq K\|f\|'\,. \tag{6.3}$$

In addition, if $f_a(x) = f(x+a)$, then for all $a \in \mathbb{R}^n$, let

$$|||f||| = \int_\Delta \|f_a\|\,da\,, \tag{6.4}$$

where Δ is a unit n-cube. We assume $(\Delta_i + a) \cap \Delta_j \neq \emptyset$ for at most K j's, for each fixed choice of i and a.

Proposition 6.2. *The norms* (6.1) *and* (6.4) *are equivalent.*

As an application, we change norms in the expansion of [8]. In place of the norms (2.43) of [8], we replace $\|w(\alpha, u, s, y_0, \cdot\,)\|_{L_2}$ by

$$\int_0^1 \|w(\alpha, u, s, y_0, \cdot\, + a\theta)\|_{L_2}d\theta, \qquad a = (1, 1, \ldots, 1)\,. \tag{6.5}$$

for a localized kernel w. As described in § 5, the integration over θ arises in the expansion from integration over choices of the position of the space lattice. The step of averaging the expansion over θ preserves the sum of the (averaged) norms of elementary expansion terms.

A second application involves separation of center of mass coordinates. For a function $f(x_1, \ldots, x_j)$ of j variables, we introduce

$$x_{\text{tot}} = \frac{1}{j} \sum_{k=1}^{j} x_k \tag{6.6}$$

$$x_{\text{rel},k} = x_k - x_{\text{tot}} .$$

For $i \in \mathbb{Z}^j$ we define for $I_k = [k, k+1]$,

$$\Delta_i = I_{i_1}(x_{\text{rel},i_1}) \ldots I_{i_{j-1}}(r_{\text{rel},j-1}) I_{i_j(x_{\text{tot}})} . \tag{6.7}$$

With these new Δ's in (6.7), we obtain a new set of equivalent norms. A translation invariant kernel $k(x; y)$, acts on spaces of definite total momentum as follows:

$$k_{\text{rel}}(x_{\text{rel}}; y_{\text{rel}}; p_{\text{tot}}) = (2\pi)^{-1/2} \int e^{ip_{\text{tot}}(y_{\text{tot}} - x_{\text{tot}})} k(x; y) dy_{\text{tot}} . \tag{6.8}$$

In the special case $j=2$, we deal with kernels for two particle scattering. These are kernels of integral operators from functions of a single relative variable $\frac{1}{2}(y_1 - y_2)$ to functions of a single relative variable $\frac{1}{2}(x_1 - x_2)$. Thus we introduce Hilbert spaces of functions of one variable on which these operators act. We use the space A_δ of symmetric functions with δ_1 fractional L_2 derivatives and with Fourier transforms analytic in a strip $|\text{Im}\, p| < \delta_2 \leqq 4m$, Our choice of the space A_δ is motivated by [12]. Related spaces are considered in [10].

We let $\delta = (\delta_1, \delta_2)$, $|\delta_2| < 4\text{m}$, and let

$$K_\delta = e^{\delta_2 |x|} \mu^{\delta_1} , \tag{6.9}$$

where $\mu = (-d^2/dx^2 + (4m)^2)^{1/2}$. Let us define the inner product

$$\langle f, g \rangle_\delta = \langle K_\delta f, K_\delta g \rangle_{L_2} ; \ \| f \|_\delta = \| K_\delta f \| . \tag{6.10}$$

Then A_δ is the Hilbert space C_0^∞ completed in the norm $\| \cdot \|_\delta$. In § 7 we show that K_δ^* may be used in place of K_δ in (6.10). Likewise, since

$$K_\delta^{-1} = K_{-\delta}^* , \tag{6.11}$$

we can define A_δ as the unitary image under $K_{-\delta}^*$ of $L_2(R)$,

$$K_{-\delta}^* : L_2(R) \to A_\delta . \tag{6.12}$$

By (6.11), it follows that the dual space of A_δ is

$$A_\delta^* = A_{-\delta} . \tag{6.13}$$

We say that $\delta > \delta'$ if $\delta > \delta_1'$ and $\delta_2 > \delta_2'$. We state properties of A_δ in three propositions.

Proposition 6.3. *Let $\delta \geqq \delta'$. Then*

$$A_\delta \subset A_{\delta'} \tag{6.14}$$

Let $I(\delta', \delta)$ denote the injection of A_δ into $A_{\delta'}$, for $\delta \geqq \delta'$,

$$I(\delta', \delta) : A_{\delta'} \to A_{\delta'} . \tag{6.15}$$

Proposition 6.4. *Let $\delta > \delta'$.*
(i) *$I(\delta', \delta)$ is compact.*
(ii) *If $\delta > \delta' + (\frac{1}{2}, 0)$, then* (6.15) *is Hilbert Schmidt*
(iii) *If $\delta > \delta' + (1, 0)$, then* (6.15) *is trace class.*

We let A^0 denote the Banach space of functions f with norm

$$\|f\| = \sum_n \|f\|_{L_2(n, n+1)} < \infty , \tag{6.16}$$

Proposition 6.5. *Let $\delta_1 \geqq 0$, $\delta_2 > 0$. Then*

$$A_\delta \subset A^0 \subset A_{-\delta} . \tag{6.17}$$

Furthermore the injections $A_\delta \to A^0 \to A_{-\delta}$ are bounded.

Proofs. In the proofs of these three propositions, we refer ahead to § 7. The proof of Propositions 6.3–6.4 involve the study of the operator $K_{\delta'} K_\delta^{-1}$. In fact let $\{f_n\}$ denote a sequence in A_δ. Then

$$K_{\delta'} f_n = (K_{\delta'} K_\delta^{-1})(K_\delta f_n) . \tag{6.18}$$

Thus if $\{f_n\}$ is bounded in A_δ, i.e. $\|K_\delta f_n\|_{L_2} \leqq$ const., it is bounded (compact) in $A_{\delta'}$ if $K_{\delta'} K_\delta^{-1}$ is bounded (compact). By Theorem 7.6,

$$K_{\delta'} K_\delta^{-1} = e^{\delta_2' |x|} \mu^{-(\delta_1 - \delta_1')} e^{-\delta_2 |x|} \tag{6.19}$$

is bounded for $\delta \geqq \delta'$ and compact for $\delta > \delta'$. Furthermore, (6.19) is Hilbert-Schmidt under condition (ii) and trace class under condition (iii).

We now proceed to Proposition 6.5. Using Proposition 6.3,

$$A_\delta \subset A_{0, \delta_2} \subset A_{0, -\delta_2} \subset A_{-\delta} .$$

Thus it is sufficient to establish (6.17) for the case $\delta_1 = 0$. Let δ now denote $\delta_2 > 0$, and let

$$M(\delta) = \sum_n e^{-\delta |n|} .$$

Then

$$\|f\|_{-\delta} = \|e^{-\delta |x|} f\|_{L_2} \leqq M(\delta) \sup_n \|f\|_{L_2(n, n+1)} \leqq M(\delta) \|f\| . \tag{6.20}$$

Also

$$\|f\|_{L_2(n, n+1)} \leqq e^{-\delta(|n| - 1)} \|e^{\delta |x|} f\|_{L_2(n, n+1)} .$$

Thus by the Schwarz inequality

$$\|f\| \leqq e^{\delta} M(2\delta)^{1/2} \Big(\sum_n \|e^{\delta|x|} f\|^2_{L_2(n,n+1)}\Big)^{1/2}$$
$$= e^{\delta} M(2\delta)^{1/2} \|f\|_{\delta}.$$

This completes the proof.

We end this section with a remark generalizing the consequences of (6.12) (6.18) for the injection $A_\delta \to A_{\delta'}$.

Proposition 6.6. *The transformation* $R: A_\delta \to A_{\delta'}$ *is bounded, compact, Hilbert Schmidt or trace class if and only if the operator*

$$K_{\delta'} R K_\delta^{-1}$$

is bounded, compact, Hilbert Schmidt or trace class on $L_2(\mathbb{R})$.

7. L_2 Estimates

In this section let $\mu = (-d^2/dx^2 + a^2)^{1/2}$, and let $k(\alpha, a; x-y)$ denote the kernel of $\mu^{-\alpha}$. Note that $k(\alpha, a; x) = a^{1-\alpha} k(\alpha, 1, ax)$, a scaling relation.

Proposition 7.1. *Let* $0 < \alpha < 1$. *Then*

$$0 < k(\alpha, a, x) < w(\alpha, a; x) \equiv \text{const.} |x|^{-1+\alpha} e^{-a|x|}. \tag{7.1}$$

If $\alpha > 1$,

$$0 < k(\alpha, a, x) < w(\alpha, a; x) \equiv \text{const.} a^{1-\alpha} e^{-a|x|}. \tag{7.2}$$

Proof. Let A be a strictly positive operator with bounded inverse. By the Cauchy integral formula, for $0 < \alpha < 2$,

$$A^{-\alpha/2} = \frac{\sin(\alpha\pi/2)}{\pi} \int_0^\infty (A+b)^{-1} b^{-\alpha/2} db. \tag{7.3}$$

Take $A = \mu^2$. The kernel of $(A+b)^{-1}$ is

$$(A+b)^{-1}(x,y) = (2\pi)^{-1/2} \int (p^2 + a^2 + b)^{-1} e^{-ip(x-y)} dp$$
$$= (2\pi)^{1/2} (2m(b))^{-1} e^{-m(b)|x-y|}, \tag{7.4}$$

where $m(b) = (a^2 + b)^{1/2}$. The positivity of k then follows by the positivity of (7.3), (7.4). We now specialize to the case $a = 1$, $\alpha < 1$ from which (7.1) follows by scaling. Inserting (7.4) in (7.3) and dividing the b integration into $\int_0^1 + \int_1^\infty$, we first bound the $\int_0^1$ integral. For $0 \leqq b \leqq 1$, $m(b) \geqq 1 + a/4$. Thus

$$\int_0^1 (A+b)^{-1}(x,0) b^{-\alpha/2} db \leqq \text{const.} e^{-|x|} \int_0^1 e^{-b|x|/4} b^{-\alpha/2} db$$
$$\leqq \text{const.} e^{-|x|} \min\{|x|^{-1+\alpha/2}, 1\}$$
$$\leqq \text{const.} |x|^{-1+\alpha} e^{-|x|}. \tag{7.5}$$

For the $\int_1^\infty$ integral,

$$\int_1^\infty m(b)^{-1} e^{-m(b)|x|} b^{-\alpha/2} db$$

$$=2\int_x^\infty \frac{x}{(t^2+x^2)^{1/2}} t^{1-\alpha}|x|^{-2+\alpha} e^{-(t^2+x^2)^{1/2}} dt$$

$$\leq 2|x|^{-1+\alpha} \int_x^\infty t^{-\alpha} e^{-(t^2+x^2)^{1/2}} dt. \tag{7.6}$$

For $t \geqq x$, it follows that for $\varepsilon > 0$ there exists $\beta < \infty$ satisfying

$$(t^2+x^2)^{1/2} \geqq |x| + \varepsilon t - \beta .$$

Thus (7.6) is bounded by

$$\text{const}\,|x|^{-1+\alpha} e^{-|x|}. \tag{7.7}$$

Combining (7.5), (7.7) yields (7.1).

If $\alpha > 1$, we again obtain (7.2) by scaling the $m=1$ bound. In this case we use the n-fold convolution

$$k(\alpha, 1, x) = w * \ldots * w, \qquad w = k(\alpha/n, 1, x) \qquad 0 < \alpha/n < 1, \tag{7.8}$$

and the bound (7.1) to establish $k(\alpha, 1, x) \leqq \text{const}\, e^{-|x|}$. Similarly

Proposition 7.2. *Let* $|x| \geqq \varepsilon > 0$, $-1 < \alpha < \infty$. *Then*

$$\left| \left(\frac{d}{dx}\right)^n k(\alpha, a, x) \right| \leqq C_{n,\varepsilon,a} e^{-a|x|}.$$

Theorem 7.3. *Let* $-1 < \alpha < 1$, $-a < \beta < a$. *Then*

$$\mu^\alpha e^{\beta|x|} \mu^{-\alpha} e^{-\beta|x|} \quad \textit{and} \quad e^{\beta|x|} \mu^\alpha e^{-\beta|x|} \mu^{-\alpha} \tag{7.9}$$

are bounded, as operators on L_2

Proof. For simplicity consider $\alpha \geqq 0$, $\beta \geqq 0$ in the first operator. The other cases are similar. Let $\zeta \in C_0^\infty$, with $\zeta(x) = 1$ in a neighborhood of $x = 0$. We write

$$\mu^{\pm\alpha} = (\mu^{\pm\alpha})_\zeta + (\mu^{\pm\alpha})_{1-\zeta}$$

where the subscript ζ or $1-\zeta$ replaces an operator with another whose kernel is multiplied by $\zeta(x-y)$ or $1-\zeta(x-y)$ respectively. Now $(\mu^\alpha)_{1-\zeta} e^{\beta|x|} \mu^{-\alpha} e^{-\beta|x|}$ is bounded by Propositions 7.1 and 7.2 and

$$e^{\beta|x|} (\mu^\alpha)_\zeta \mu^{-\alpha} e^{-\beta|x|} = I + e^{\beta|x|} (\mu^\alpha)_{1-\zeta} \mu^{-\alpha} e^{-\beta|x|}$$

is also bounded, for the same reason, as is $e^{\beta|x|} \mu^{-\alpha} e^{-\beta|x|}$. Thus our problem is reduced to bounding

$$T = [(\mu^\alpha)_\zeta, e^{\beta|x|}] e^{-\beta|x|}.$$

Because of the ζ in T, the unboundedness of $e^{\beta|x|}$ plays no role. In fact we choose ζ so that $\sum_j \zeta(x+j)=1$. In momentum space, the kernel of

$$T_j=[(\mu^\alpha)_\zeta, \zeta(x+j)e^{\beta|x|}]e^{-\beta j}$$

is bounded by a multiple of

$$f(p,q)=(\mu(p)^\alpha-\mu(q)^\alpha)(|p-q|+1)^{-2}.$$

Since

$$\sup_p \int |f(p,q)|dq<\infty,$$

T_j is bounded. Moreover $\|T\| \leqq \text{const} \sup_j \|T_j\|$ and the proof is complete.

Corollary 7.4. *The space A_δ is the closure of $C_0^\infty(R)$ in the norm*

$$\|f\|_\delta=\|K_\delta f\|_{L_2} \tag{7.10}$$

or in the norm

$$\|f\|'_\delta=\|K_\delta^* f\|_{L_2}. \tag{7.11}$$

The norms $\|\cdot\|$ and $\|\cdot\|'$ are equivalent.

Proof. We estimate

$$\begin{aligned}\|f\|_\delta&=\|K_\delta f\|=\|K_\delta(K_\delta^*)^{-1}K_\delta^* f\|\\ &\leqq \|K_\delta(K_\delta^*)^{-1}\|\,\|f\|'_\delta \leqq \|K_\delta(K_\delta^*)^{-1}\|\,\|K_\delta^* K_\delta^{-1}\|\,\|f\|_\delta.\end{aligned}$$

However

$$K_\delta(K_\delta^*)^{-1}=e^{\delta_2|x|}\mu^{\delta_1}e^{-\delta_2|x|}\mu^{-\delta_1}, \tag{7.12}$$

and

$$K_\delta^* K_\delta^{-1}=\mu^{\delta_1}e^{\delta_2|x|}\mu^{-\delta_1}e^{-\delta_2|x|}. \tag{7.13}$$

Applying the theorem to (7.12)–(7.13) shows that they are bounded, so $\|\cdot\|$ and $\|\cdot\|'$ are equivalent.

Lemma 7.5. *Let $\delta<1$ and let $v(x)$ be bounded, positive and square integrable. Let $T(x,y)$ be the kernel of an integral operator satisfying*

$$|T(x,y)|\leqq v(x)v(y)|x-y|^{-\delta}. \tag{7.14}$$

Then T is compact.

Proof. Let $\chi_n(x,y)$ be the characteristic function of $|x-y|>n^{-1}$, and let $T_n(x,y) = T(x,y)\chi_n(x,y)$. Then T_n is Hilbert Schmidt, and hence compact. Also

$$\|T-T_n\| \leqq \sup_x \int_{|x-y|<n^{-1}} v(x)v(y)|x-y|^{-\delta}dy \leqq O(n^{-1+\delta}).$$

Hence T, as the norm limit of compact operators, is compact.

Theorem 7.6. *Let* $\alpha<\beta$, $\gamma<\varepsilon$, *with* $\alpha;\beta\in(-a,a),\gamma,\varepsilon\in(-1,1)$. *Let*

$$T\equiv e^{+\alpha|x|}\mu^{+\gamma}e^{-\beta|x|}\mu^{-\varepsilon}. \tag{7.15}$$

(i) *T is compact.*
(ii) *If* $\varepsilon-\gamma>\frac{1}{2}$, *T is Hilbert Schmidt.*
(iii) *If* $\varepsilon-\gamma>1$, *T is trace class.*

Proof. We write

$$T=(e^{\alpha|x|}\mu^{-(\varepsilon-\gamma)}e^{-\beta|x|})(e^{\beta|x|}\mu^{\varepsilon}e^{-\beta|x|}\mu^{-\varepsilon}). \tag{7.16}$$

The second factor in (7.16) is bounded by Theorem 7.3. We now show that the first factor is compact, bounded or Hilbert Schmidt. Let S denote this first factor. Then by (7.1), for any $\delta>0$, and assuming $0<\varepsilon-\gamma<1$,

$$\begin{aligned}|S(x,y)|&\leqq \text{const}\, e^{\alpha|x|}|x-y|^{-1+(\varepsilon-\gamma)}e^{-a|x-y|}e^{-\beta|y|}\\ &\leqq \text{const}\, e^{-(\beta-\alpha)|y|}|x-y|^{-1+(\varepsilon-\gamma)}e^{-(a-\alpha)|x-y|}\\ &\leqq \text{const}\, e^{-\delta|x|}e^{-(\beta-\alpha-\delta)|y|}e^{-(a-\alpha-\delta)|x-y|}|x-y|^{-1+\varepsilon-\gamma}\end{aligned} \tag{7.17}$$

By Lemma 7.5, S is compact. Furthermore, if $\varepsilon-\gamma>\frac{1}{2}$, then S is Hilbert Schmidt. If $\varepsilon-\gamma>1$, $0<\delta<\beta-\alpha$,

$$S=(e^{\alpha|x|}\mu^{-(\varepsilon-\gamma)/2}e^{-(\beta-\delta)|x|})(e^{(\beta-\delta)|x|}\mu^{-(\varepsilon-\gamma)/2}e^{-\beta|x|}). \tag{7.18}$$

Each factor in parentheses in (7.18) is Hilbert Schmidt, by (ii). Thus S, as the product of two Hilbert Schmidt operators, is trace class.

Proposition 7.7. *Let* $\delta_2>0$. *The norm*

$$|||f|||_\delta\equiv\max_{\pm}\|e^{\pm\delta_2 x}\mu^{\delta_1}f\|_{L_2} \tag{7.19}$$

is equivalent to the norm $\|f\|_\delta$.

Proof. For $\alpha\geqq 0$, we have the pointwise bound

$$\exp(\pm\alpha x)\leqq\exp\alpha|x|\leqq\exp(\alpha x)+\exp(-\alpha x).$$

Thus

$$|||f|||_\delta\leqq\|f\|_\delta\leqq 2|||f|||_\delta. \tag{7.20}$$

Proposition 7.8. *Let* $|\delta_2|<a$. *Then as an operator from* A_δ *to* A_δ,

$$|||e^{-t\mu}|||_{A_\delta}=\sup_f|||e^{-t\mu}f|||_\delta/|||f|||_\delta\leqq\exp(-t\{a^2-\delta_2^2\}^{1/2}). \tag{7.21}$$

Proof. First we note that the result for $\delta_2<0$ follows from the result for $\delta_2\geqq 0$. In fact if $T:A_\delta\to A_\delta$ is bounded with norm $\|T\|_{A_\delta}$, then $T^*:A_\delta^*\to A_\delta^*$ is bounded with the same norm. By (6.13), $A_\delta^*=A_{-\delta}$, so we take $\delta_2\geqq 0$; let $T=\exp(-t\mu)$. We show that $|||T|||_{A_\delta}$ is bounded as in (7.21). By Proposition 7.7, this shows that T is bounded on A_δ. For convenience, we choose f in the dense set C_0^∞ in A_δ, so $f(p)$ is entire. Then

$$\|e^{\pm\alpha x}\mu^\beta Tf\|_{L_2}=\|\mu(p\pm i\alpha)^\beta\exp(-t\mu(p\pm i\alpha))\tilde{f}(p\pm i\alpha)\|_{L_2}. \tag{7.22}$$

We note that

$$\operatorname{Re}\mu(p\pm i\alpha)=\operatorname{Re}(p^2+a^2-\alpha^2\pm 2ip\alpha)^{1/2}$$
$$\geqq (p^2+a^2-\alpha^2)^{1/2}\geqq (a^2-\alpha^2)^{1/2}\,. \tag{7.23}$$

Thus with $\alpha=\delta_2$, $\beta=\delta_1$,

$$|||Tf|||_\delta \leqq \exp(-t(a^2-\alpha^2)^{1/2}) \max_{\pm} \|\mu(p\pm i\alpha)^\beta \tilde{f}(p\pm i\alpha)\|_{L_2}$$
$$=\exp(-t(a^2-\alpha^2)^{1/2}) \max_{\pm} \|e^{\pm\alpha x}\mu^\beta f\|_{L_2}$$
$$=\exp(-t(a^2-\alpha^2)^{1/2})|||f|||_\delta\,. \tag{7.24}$$

Proposition 7.9. *Let* $T=(e^{-\mu}\otimes e^{-\mu})$. *Then acting on* A_δ *as functions of* p_{rel} *with* p_{tot} *fixed,*

$$|||T^t|||_{A_\delta}\leqq \exp(-2t\sqrt{a^2+p_{\mathrm{tot}}^2-\delta_2^2}) \tag{7.25}$$

Proof. As above, we assume $\delta_2\geqq 0$. Then

$$\|e^{\pm\alpha x_{\mathrm{rel}}}\mu^\beta Tf\|_{L_2}$$
$$=\|\mu(p_{\mathrm{rel}}\pm i\alpha)^\beta \exp\{-t\mu(p_{\mathrm{tot}}+p_{\mathrm{rel}}\pm i\alpha)-t\mu(p_{\mathrm{tot}}-p_{\mathrm{rel}}\pm i\alpha)\}\tilde{f}(p_{\mathrm{rel}}\pm i\alpha)\|_{L_2} \tag{7.26}$$

using $p_{\mathrm{tot}}=\frac{1}{2}(p_1+p_2)$, $p_{\mathrm{rel}}=\frac{1}{2}(p_1-p_2)$. Then as in (7.23),

$$\textstyle\sum_\pm \operatorname{Re}\mu(p_{\mathrm{tot}}\pm p_{\mathrm{rel}}\pm i\alpha)\geqq 2(a^2+p_{\mathrm{tot}}^2-\alpha^2)^{1/}$$

Thus

$$\|e^{\pm\alpha x_{\mathrm{rel}}}\mu^\beta Tf\|_{L_2}\leqq e^{-2t(a^2+p_{\mathrm{tot}}^2-\alpha^2)1/2}\|\mu(p_{\mathrm{rel}}\pm i\alpha)^\beta\tilde{f}\|_{L_2}\,.$$

Corollary 7.10. *For* $\operatorname{Re}E<2(a^2+p_{\mathrm{tot}}^2-\delta_2^2)^{1/2}$,

$$|||(1-e^E e^{-\mu}\otimes e^{-\mu})^{-1}|||_{A_\delta\to A_\delta}\leqq (1-e^{-[2(a^2+p_{\mathrm{tot}}^2-\delta_2^2)^{1/2}-\operatorname{Re}E]})^{-1}$$

Remark. With T as in Proposition 7.9, the Corollary bounds

$$\tilde{T}(E)\equiv\sum_{j\geqq 0} T^j e^{jE} \tag{7.27}$$

We now show that $\tilde{T}(E)$ has a two sheeted analytic continuation around a square root branch point at threshold. For simplicity, we take $p_{\mathrm{tot}}=0$ and we do not analyze the singularity at threshold.

Proposition 7.11. *For*

$$\zeta=(4a^2-E^2)^{1/2}\neq 0\,,$$

the operator

$$\tilde{T}(E(\zeta))=\frac{1}{1-\exp[(4a^2-\zeta^2)^{1/2}-(4a^2+4p_{\mathrm{rel}}^2)^{1/2}]} \tag{7.28}$$

acting from $A_{(\delta_1,\delta_2)}$ *to* $A_{(\delta_1,-\delta_2)}$, $\delta_2>0$, *has a bounded analytic continuation in* ζ *from* $\operatorname{Re}\zeta>2\delta_2$ *(i.e.* $\operatorname{Re}E<2(a^2-\delta_2^2)^{1/2}$*) to the region* $\operatorname{Re}\zeta>-2\delta_2$.

Proof. We see that $\tilde{T}(E(\zeta))$ has poles in p_{rel} at

$$p_{\mathrm{rel}} = \pm i\zeta/2. \tag{7.29}$$

As long as the poles do not cross the strip $|\mathrm{Im}\, p_{\mathrm{rel}}| \leqq \delta_2$, i.e. as long as $\mathrm{Re}\,\zeta > -2\delta_2$, the contour integral defining the inner product

$$\langle f, \tilde{T}(E) f \rangle_{L_2}, \quad f \in A_\delta, \tag{7.30}$$

as written in the momentum space variable p_{rel}, can be deformed to avoid the poles (7.29). This procedure defines the analytic continuation of (7.28) to $\mathrm{Re}\,\zeta > -2\delta_2$. Mapping back to the variable E gives the two sheeted analytic continuation of (7.27) around the square root branch point $E = 2a$. For $\zeta \neq 0$, the contour can be chosen so that (7.28) is bounded along the contours. As $\zeta \to 0$, the poles come together, and the contour, which must go above one pole and below the other, cannot be bounded away from the poles. See also [3, 4, 12].

It is only in the analytic continuation argument that we use the fact that the functions $f \in A_\delta$ are symmetric in x-space. Because of this assumption, (7.30) can be written in momentum space as

$$\langle f, \tilde{T}(E) f \rangle_{L_2} = \int_{-\infty}^{\infty} \tilde{\tilde{f}}(p) (\tilde{T}(E) \tilde{f})(p) dp. \tag{7.31}$$

Thus no absolute values occur in (7.31) and the analytic continuation in E can be accompanied by the deformation of the p integration contour (inside the strip $|\mathrm{Im}\, p| < \delta_2$) to avoid the poles (7.29) of $\tilde{T}(E(\zeta))$. By the above argument, if $C(\zeta)$ denotes the deformed contour,

$$|\langle f, \tilde{T}(E) f \rangle_{L_2}| \leqq \sup_{C(\zeta)} |\tilde{T}(E(\zeta))| \int_{C(\zeta)} |\tilde{\tilde{f}}(p) \tilde{f}(p)| dp. \tag{7.32}$$

since the sup is dominated by $O(1)$. For $|\mathrm{Im}\, p|$ bounded away from δ_2, $|\tilde{f}(p)| \leqq \mathrm{const}\, \|f\|_{A_\delta}$. Thus in (7.31),

$$\left| \int_{-L}^{L} \tilde{\tilde{f}}(p) \tilde{f}(p) \tilde{T}(E) dp \right| \leqq \mathrm{const}\, \|f\|_{A_\delta}^2.$$

The tails, $\int_{-\infty}^{-L} + \int_{L}^{\infty}$ are bounded directly, and can be continued analytically to $\mathrm{Re}\, E \leqq L$ without deformation of contour, as in Proposition 7.9, and Corollary 7.10. Since L is arbitrary, we get the desired analytic continuation.

8. One Particle Lines in a Two Particle Channel

We modify the two particle cluster expansion [8] applied to $S_{0,T}$ in order to define our integral equation. The first modification is inclusion of virtual contraction terms to the $:\phi(T)^2:$ observable, to allow resummation of one particle lines, as explained in § 5. The second modification, which is new to this section, is the inclusion of virtual interaction terms between the independent one particle lines.

The operator

$$e^{-t\mu}\otimes e^{-t\mu} \tag{8.1}$$

generates dynamics in two independent one particle lines, and we want to find this operator by resumming certain terms from the expansion of $:\phi(x_1)\phi(x_2):$. It is easy to find a candidate for this resummation, namely all terms in which there is no contraction between x_1 and x_2 vertices, nor any overlap between the x_1 and x_2 contours. Each term of this nature factors, and thus expresses some type of independent dynamics for x_1 and x_2. However, the sum of these terms does not factor, and does not define independent dynamics for x_1 and x_2. In fact, some terms, expected as part of the $e^{-t\mu}$ dynamics for x_1 do not occur, because the region of space required to insert this term in the expansion is already occupied by other vertices and/or contours belonging to x_2 or to x_1 and x_2 jointly. To compensate, we add and subtract these missing terms. With the addition, the sum of the independent terms also factors, and can be identified with (8.1). When subtracted, the terms are called virtual interaction terms; they contribute along with the direct interaction terms to the Bethe-Salpeter kernel.

In more detail, we restrict the momenta of each x_j test function as in (5.7). Using a smooth momentum space cutoff function, we write the x_j test function as a sum of two test functions. The first has high momentum and the second has purely low momentum. The term with the high momentum has an enhanced decay rate. This fact can be obtained from the convergence proof of [8]. Because of the high decay rate ($\geqq 2m_0-\varepsilon$ for the single particle) a simplier expansion, not requiring a backward movement in time is possible for the high momentum term. This expansion will be explained later.

The more difficult term involves low momentum. We now explain the expansion for these terms. Let $:\phi^2:_W$ denote normal ordering relative to the bare vacuum. This is the normal ordering used in [8]. By adding and subtracting a constant term and a term linear in ϕ, we can pass from $:\phi^2:$ to $:\phi^2:_W$ or back, as in (2.13). We use this identity to start the expansion with $:\phi^2:_W$, but to yield $:\phi^2:$ terms ($1<t=i\leqq T$) as a result of the expansion. It is also used to write the time T: $:\phi^2:$ observable as $:\phi^2:_W+\phi+\text{const}$. The linear term $\phi|_{t=0}$ is expanded as in § 5, as will be explained in more detail below.

The $:\phi^2:_W|_{t=0}$ is expanded according to the 2-particle expansion of [8]. Virtual terms are added and subtracted, for interaction with the time T observables and also for interaction between the two one particle lines. For terms without interaction between the two ϕ^2 lines, each ϕ is returned independently to $t=1$. The expansion terminates with any of these terms:

(i) a constant (i.e. ϕ-independent) term,

(ii) a term linear in ϕ, returned to $t=1$,

(iii) two independent ϕ's in the form $:\phi(y_1)\phi(y_2):$, each returned to $t=1$,

(iv) a $:\phi^2:$ at integer time $t=i$, $1\leqq i<T$ with direct or virtual interaction between the two variables y_1 and y_2

(v) direct or virtual contraction to $:\phi^2:(T)$, the time T observable.

Because of the virtual terms in (v), the terms (i) and (ii) are T independent; because of normal ordering in $:\phi^2:$, they sum to zero. The terms (iii) are exactly the tensor product of the one particle line expansion of § 5, and thus can be identified with the dynamics (8.1). These terms are then $e^{-\mu}\otimes e^{-\mu}S_{1,T}$. The terms (iv) are proportional to $S_{i,T}$, and give $B_{0,i}^{\mathrm{amp}}S_{i,T}$, where $B_{0,i}^{\mathrm{amp}}$ denotes the sum of the proportionality operators for fixed i. Finally the terms (v) have three particle decay, and are bounded by $O(\exp(-3(m_0-\varepsilon)T))$. We denote the terms (v) by $A_{0,T}^{\mathrm{amp}}$.

The expansion of $\phi|_{t=0}$ follows § 5. All terms linear in ϕ are returned to $t=1$. The expansion terminates with terms of type (i), (ii), (iv) or (v), but gives no contribution to the (iii) terms.

In the high momentum term, the two ϕ's in $:\phi^2:_W$ are moved forward in time independently until they interact with each other (direct or virtual interaction or until they contract (direct or virtual contraction) to a time T observable. This produces terms of the type (i), (ii), (iv), (v), as above, but not of type (iii). The (i) and (ii) terms sum to zero, and the (iv) and (v) terms are estimated as below. We note that as long as the two lines do not interact, momentum is conserved within each line. In particular, the high momentum property is retained until interaction.

In both high and low momentum terms, we require an independent average over the lattice displacement $\theta\in(0,1]$ for each of the independent one particle lines. This means that the contours for the distinct lines are inserted along distinct, incompatible lattices.

The estimates on the (iv) and (v) terms of the expansion follow from [8], because all terms, including virtual terms, arise from a finite number of sums, differences or products of convergent expansions. Thus the (iv) and (v) terms are absolutely convergent in the norms of [8]. In view of regularization (see § 9, 10), they are also bounded and absolutely convergent in the operator norms on A_δ, on the space of equal time functions.

In summary, we have

$$S_{0,T}=e^{-\mu}\otimes e^{-\mu}S_{1,T}+\sum_{i=1} B_{0,i}^{\mathrm{amp}}S_{iT}+A_{0,T}^{\mathrm{amp}}. \tag{8.2}$$

Iterating the expansion, we obtain (1.6) with

$$A_{0,T}=\sum_{0\leqq j\leqq T}(e^{-\mu}\otimes e^{-\mu})^j A_{j,T}^{\mathrm{amp}} \tag{8.3}$$

$$B_{0,T}=\sum_{0\leqq j\leqq T-1}(e^{-\mu}\otimes e^{-\mu})^j B_{j,T}^{\mathrm{amp}}. \tag{8.4}$$

Thus

$$\tilde{A}(E)=\sum_{0\leqq T} A_{0,T}e^{ET}=(I-e^E e^{-\mu}\otimes e^{-\mu})^{-1}\sum_{0\leqq T} A_{0,T}^{\mathrm{amp}}e^{ET}=\tilde{T}(E)\tilde{A}^{\mathrm{amp}}(E), \tag{8.5}$$

$$\tilde{B}(E)=\sum_{0\leqq T} B_{0,T}e^{ET}=(I-e^E e^{-\mu}\otimes e^{-\mu})^{-1}\sum_{1<T} B_{0,T}^{\mathrm{amp}}e^{ET}=\tilde{T}(E)\tilde{B}^{\mathrm{amp}}(E). \tag{8.6}$$

Thus we write (1.6) as

$$\tilde{S}(E)=\tilde{A}(E)+\tilde{B}(E)\tilde{S}(E). \tag{8.7}$$

9. Decay Estimates in x-Spaces

The two particle expansion of § 8 introduces seven operators which map kernels of const, $\phi(x)$ or $:\phi(x_1)\phi(x_2):$ into kernels of const, ϕ or $:\phi^2:$ at possibly later times.

The simplest of these operators is $e^{-\mu}\otimes e^{-\mu}$. This operator is described in § 7, and it acts on the kernel of $:\phi^2:$ in the mapping

$$:\phi(x_1)\phi(x_2):|_{t=0}\to :\phi(y_1)\phi(y_2):|_{t=1}\,. \tag{9.1}$$

Another simple operation is re-Wick ordering $\delta\alpha_0$. Since integration by parts yields

$$\delta\alpha_0\in A_{\frac{3}{2}-\varepsilon,m-\varepsilon}=A^*_{-\frac{3}{2}+\varepsilon,-m+\varepsilon} \tag{9.2}$$

$\delta\alpha_0$ is a bounded linear functional on all A_δ spaces we consider. Also $\tilde{\alpha}_1(p,x_{\text{rel}})$ for fixed p satisfies

$$\alpha_1(p,\cdot)\in A_{\frac{3}{2}-\varepsilon,\delta_2}=A^*_{-\frac{3}{2}+\varepsilon,-\delta_2},\qquad \delta_2>0\,. \tag{9.3}$$

and thus is bounded on all A_δ spaces we consider.

Let $\mathscr{K}_j^{(\text{iv})}$ denote the operator which is the sum of all (iv) terms associated with the map

$$:\phi(x_1)\phi(x_2):_W|_{t=0}\to :\phi(y_1)\phi(y_2):|_{t=j}\,. \tag{9.4}$$

The operator $\mathscr{K}^{(\text{iv})}$ has a kernel depending on x_{rel}, y_{rel}, $x_{\text{tot}}-y_{\text{tot}}$.

We similarly define an operator $\mathscr{K}_T^{(\text{v})}$ by the sum of type (v) terms. Then $\mathscr{K}_T^{(\text{v})}$ is determined as a bilinear form on the kernels $f(x_{\text{rel}},p_{\text{rel}})$ and $g(y_{\text{rel}},p_{\text{rel}})$ of the time $t=0$ and time $t=T$ observables.

Namely at time $t=0$, the observable is

$$\int f(x_{\text{rel}},p_{\text{tot}}):\phi^2(x_{\text{rel}},p_{\text{tot}}):_W dx_{\text{rel}}$$

and at Euclidean time $t=T$, it is

$$\begin{aligned}
&\int g(y_{\text{rel}},p_{\text{tot}}):\phi^2(y_{\text{rel}},p_{\text{tot}}):dy_{\text{rel}}\\
&\quad=\int g(y_{\text{rel}},p_{\text{tot}}):\phi^2(y_{\text{rel}},p_{\text{tot}}):_W dy_{\text{rel}}\\
&\qquad-\int g(y_{\text{rel}},p_{\text{tot}})\tilde{\alpha}_1(p_{\text{tot}},y_{\text{rel}})dy_{\text{rel}}\tilde{\phi}(p_{\text{tot}})\\
&\qquad+\int g(y_{\text{rel}},p_{\text{tot}})\delta\alpha_0(y_{\text{rel}})dy_{\text{rel}}\,.
\end{aligned} \tag{9.5}$$

With $f\in A_\delta$, as a function of x_{rel}, $g\in A_{\delta 1}$, then $\mathscr{K}_T^{(\text{v})}$ is an operator from A_δ to $A^*_{\delta 1}=A_{-\delta 1}$.

Also let $\mathscr{L}_j^{(\text{iv})}$ and $\mathscr{L}_j^{(\text{v})}$ be associated with the map

$$\phi(x)|_{t=0}\to :\phi(y_1)\phi(y_2):|_{t=j} \tag{9.6}$$

and with $\phi(x)$ contractions at time T. Then $\mathscr{L}^{(\text{iv})}$ has a kernel depending on y_{rel}, $x-y_{\text{tot}}$.

For comparison with [8], we restrict $\mathscr{K}$, $\mathscr{L}$ to functions supported in products of unit intervals. The $\mathscr{K}$, $\mathscr{L}$ are bounded from L_2 to L_2. According to § 6, the unit lattice squares in x_1, x_2 space and in y_1, y_2 space can be rotated by $\pi/4$ so that they are alligned with x_{rel}, x_{tot}, y_{rel}, y_{tot}. Let Δ_x, etc. denote the characteristic function of such a square. A minor improvement in the result of [8] includes some x as well as

t exponential decay. With $:\phi^2:$ as the time T observable this yields

$$\begin{aligned}\|\Delta_x \mathscr{K}_T^{(\mathrm{v})} \Delta_y\|_{L_2 \to L_2^*} &\leqq e^{-3(m-\varepsilon)T} e^{-\varepsilon'|x_{\mathrm{tot}}-y_{\mathrm{tot}}|} \\ &\leqq [e^{-\varepsilon'|x_{\mathrm{rel}}-y_{\mathrm{rel}}|} + e^{-\varepsilon'|x_{\mathrm{rel}}+y_{\mathrm{rel}}|}]. \end{aligned} \tag{9.7}$$

The type (iv) terms have an improved decay because the interaction between the two one particle lines give graphs which are connected, and hence decay in $|x_{\mathrm{rel}}| + |y_{\mathrm{rel}}|$. Thus

$$\|\Delta_x \mathscr{K}_j^{(\mathrm{iv})} \Delta_j\|_{L_2 \to L_2} \leqq e^{-3(m-\varepsilon)j} e^{-\varepsilon'|x_{\mathrm{tot}}-y_{\mathrm{tot}}|} e^{-\varepsilon'|x_{\mathrm{rel}}|-\varepsilon'|y_{\mathrm{rel}}|}. \tag{9.8}$$

The graphs in $\mathscr{L}$ are all connected.

$$\|\Delta_x \mathscr{L}_j^{(\mathrm{iv})} \Delta_y\|_{L_2 \to L_2} \leqq e^{-3(m-\varepsilon)j} e^{-\varepsilon'|x-y_{\mathrm{tot}}|} e^{-\varepsilon'|y_{\mathrm{rel}}|} \tag{9.9}$$

and

$$\|\Delta_x \mathscr{L}_j^{(\mathrm{v})} \Delta_j\|_{L_2 \to L_2^*} \leqq e^{-3(m-\varepsilon)T} e^{-\varepsilon'|x-y_{\mathrm{tot}}|} e^{-\varepsilon'|y_{\mathrm{rel}}|}. \tag{9.10}$$

Next we consider the operators $\mathscr{K}$, $\mathscr{L}$ at fixed total momentum p, as mappings on the function of relative variables:

$$\begin{aligned} &\mathscr{K}_j^{(\mathrm{iv})} : A_\delta^* = A_{-\delta} \to A_\delta, \qquad \mathscr{K}_T^{(\mathrm{v})} : A_\delta \to A_\delta \\ &\mathscr{L}_j^{(\mathrm{iv})} \in A_\delta, \qquad \mathscr{L}_T^{(\mathrm{v})} \in A_\delta. \end{aligned} \tag{9.11}$$

For $\delta_1 = 0$ these four norms are bounded as follows:

Theorem 9.1. *For δ_2 sufficiently small and greater than zero, the four operators (9.11) are bounded. The norms are $O(e^{-3(m-\varepsilon)j})$ for $\mathscr{K}_j$ and $\mathscr{L}_j$. Moreover $\mathscr{K}$ and $\mathscr{L}$ are complex analytic in p_{tot}, for $|\mathrm{Im}\, p_{\mathrm{tot}}|$ small.*

Proof. For $\mathscr{K}_j^{(\mathrm{iv})}$ we have

$$\begin{aligned}\|\mathscr{K}_j^{(\mathrm{iv})}\|_{A_\delta \to A_\delta} &\leqq \sum e^{\delta_2|x_{\mathrm{rel}}| + \delta_2|y_{\mathrm{rel}}|} \|\Delta_x \mathscr{K}_j^{(\mathrm{iv})} \Delta_y\|_{L_2 \to L_2} \\ &\leqq O(\sum e^{-3(m-3)j - |x_{\mathrm{tot}}-y_{\mathrm{tot}}|(\varepsilon' - |\mathrm{Im}\, p_{\mathrm{tot}}|)} \leqq O(e^{-3(m-\varepsilon)j}). \end{aligned}$$

Similarly

$$\|\mathscr{K}_T^{(\mathrm{v})}\|_{A_\delta \to A_\delta} \leqq \sum e^{-\delta_2|x_{\mathrm{rel}}| + \delta_2|y_{\mathrm{rel}}|} \|\Delta_x \mathscr{K}_T^{(\mathrm{v})} \Delta_y\|_{L_2 \to L_2^*} \leqq O(e^{-3(m-\varepsilon)T}).$$

The remaining bounds on

$$\begin{aligned} &\|\mathscr{L}_j^{(\mathrm{iv})}\|_{A_\delta} \leqq \sum e^{\delta|y_{\mathrm{rel}}|} \|\Delta_x \mathscr{L}_j^{(\mathrm{iv})} \Delta_y\|_{L_2 \to L_2} \leqq e^{-3(m-\varepsilon)j} \\ &\|\mathscr{L}_T^{(\mathrm{v})}\|_{A_\delta} \leqq \sum e^{\delta|y_{\mathrm{rel}}|} \|\Delta_x \mathscr{L}_T^{(\mathrm{v})} \Delta_y\|_{L_2 \to L_2^*} \leqq e^{-3(m-\varepsilon)T} \end{aligned} \tag{9.12}$$

are similar, to complete the proof.

The proof of (9.12) applies to the error terms in (5.17) and shows their contribution to $\tilde{\alpha}_1$ lies in A_{0,δ_2}, $\delta_2 > 0$.

10. Smoothing and Compactness in x-Space

In this section we exhibit the smoothing property of the operators $\mathscr{K}$ and $\mathscr{L}$ defined in § 8 and § 9. In particular, this gives the properties of A^{amp} and B^{amp} on the spaces A_δ. Our main technical result, which leads to an analysis of $S_{0,T}$ is the following:

Theorem 10.1. *The connected expansion terms (§8, type* iv*) define bounded operators* $B_{0,j}^{\text{amp}}$ *such that*

$$B_{0,j}^{\text{amp}} : A_{-\delta} \to A_\delta \tag{10.1}$$

with

$$\|B_{0,j}^{\text{amp}}\|_{A_{-\delta}\to A_\delta} \leqq O(\lambda)e^{-3(m-\varepsilon)j} \tag{10.2}$$

when

$$\delta_1 \leqq 2-\varepsilon, \quad \delta_2 < \varepsilon, \tag{10.3}$$

for some $\varepsilon > 0$ *sufficiently small. The terms of §8, type* v *define the bounded operator* $A_{0,T}^{\text{amp}}$, *where*

$$A_{0,T}^{\text{amp}} : A_{-1+\varepsilon,\delta_2} \to A_{1-\varepsilon,\delta_2} \tag{10.4}$$

with

$$\|A_{0,T}^{\text{amp}}\| \leqq O(1)e^{-3(m-\varepsilon)T}. \tag{10.5}$$

Moreover, $B_{0,j}^{\text{amp}}$ *is compact on the indicated spaces, and trace class if* $\delta_1 \leqq \frac{3}{2} - \varepsilon$.

Remark. It follows that the Laplace transformed operators

$$\begin{aligned} A^{\text{amp}}(E) &= \sum_{0\leqq T} A_{0,T}^{\text{amp}} e^{ET} \\ B^{\text{amp}}(E) &= \sum_{0\leqq T} B_{0,T}^{\text{amp}} e^{ET} \end{aligned} \tag{10.6}$$

are complex analytic in E for $\operatorname{Re} E \leqq 3(m_0 - \varepsilon)$. This statement is true for $\tilde{A}^{\text{amp}}$ and $\tilde{B}^{\text{amp}}$ as operators on functions of the relative position variable. In particular, they are bounded as operators from A_δ to $A_{\delta'}$ where δ, δ' are given by Theorem 10.1, and $\tilde{B}^{\text{amp}}$ is compact, and has small norm as $\lambda \to 0$.

Proof. The improvement of Theorem 10.1 over the §9 involves a careful examination of the initial (and final) expansion steps. Each term at the end of the expansion has the structure given in §8. The sum of all expansion terms converge in the norms of [8]. We thus inspect individual terms. The $\mathscr{K}_j^{(\text{iv})}$ terms, for example, have the structure

$:\phi^2:(w)$ 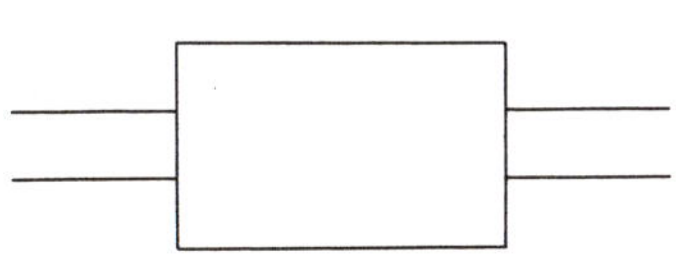 $:\phi^2:(v)$ (10.7)

where w is the kernel of $:\phi^2:(t=0)$ and v is the kernel of $:\phi^2:(t=j)$. The initial expansion step inserts a contour around the $:\phi^2:(t=0)$ vertices; this contour may contain other vertices and partial contours. Likewise the expansion ends by removing the $:\phi^l:(t<j)$ vertices by a contour. In [8] estimates on the inside vertices $R(\phi)$ and function space integration was obtained by Schwarz inequality in path space

$$\left|\int_{\text{inside}} e^{-P(\phi)}R(\phi)d\phi_{\text{inside}}\right| \leqq |(\int R^2 d\phi \int e^{-2P(\phi)}d\phi). \tag{10.8}$$

However, optimal bounds on the Schwinger functions, e.g. $\int e^{-P(\phi)}R(\phi)d\phi_{\text{inside}}$ improve on this estimate, see [5]. These optimal bounds are obtained by integration by parts of each field in $R(\phi)$, before applying the Schwarz inequality.

In expressions such as for $\mathscr{K}^{(\text{iv})}$ in (10.7), this integration by parts produces "initial" and "final" vertices which contract with the $:\phi^2:(w)$ or $:\phi^2:(v)$. An analysis of these diagrams shows that their continuity in v and w lies not only in L_2, but in the appropriate Sobolev space for Theorem 10.1. For example, the dominant diagrams in $\mathscr{K}^{(\text{iv})}$ have the form

(10.8)

with a single ϕ^4 vertex. Their momentum space decay is estimated for fixed p_{tot} by $O(\lambda)\|\mu^{-2}w\|_{L_2}\|\mu^{-2}v\|_{L_2}$ in the variable p_{rel}. Here each line carries a factor μ^{-1}, and the integration over the time of the intermediate vertex gives an additional factor μ^{-1} which can be divided between w and v. More generally, a large $\mathscr{K}^{(\text{iv})}$ diagram can be estimated by isolating a vertex on either end, and using and L_2 operator norm on the intermediate diagram:

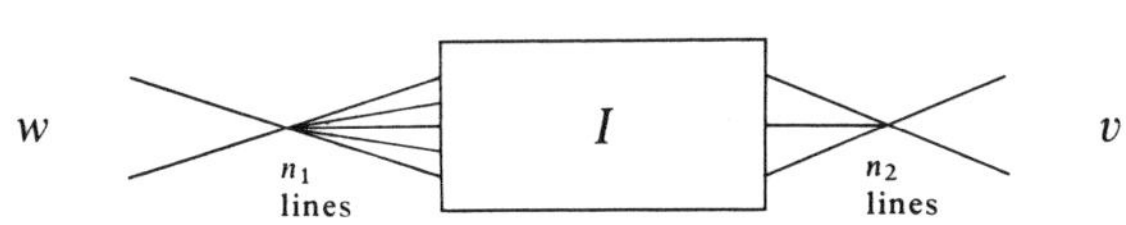

Then I as an operator from $L_2(R^{n_2})$ to $L_2(R^{n_1})$ has an operator norm $\|I\|_{n_1,n_2}$ which can be estimated by the cluster expansion and Schwarz inequality as mentioned above. The single vertex diagrams have Hilbert-Schmidt norms $\leqq O(1)\|\mu^{-2}w\|\ \|\mu^{-2}v\|$ as above. The reduction to these norms is systematized in [6, §6].

Other typical terms which arise involve α_1. (The analysis of such terms completes the proof of Theorem 5.1.) These terms will have a dominant diagram

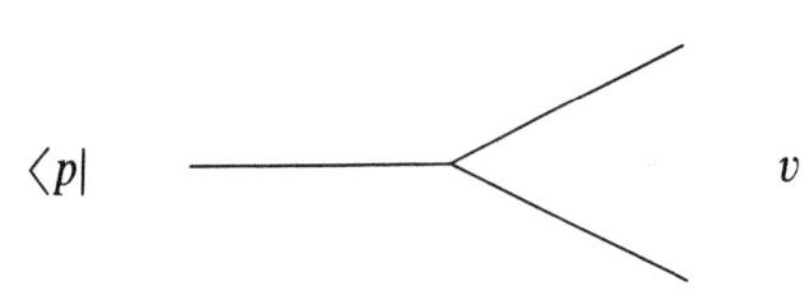

which as above is estimated by

$$O(1)\|\mu^{-2}v\|.$$

The dominant diagram for $\tilde{A}^{\mathrm{amp}}(E)$ is

arising from the $T=0$ contribution. Here we obtain the estimate $O(1)\|\mu^{-1}v\|_{L_2}\|\mu^{-1}w\|_{L_2}$ arising from a factor μ^{-1} on each line.

We now return to the Eqs. (1.6), (8.3)–(8.7), (10.6)–(10.7). In Laplace transform space the solution for $\tilde{S}(E)$ is

$$\tilde{S}(E)=[1-\tilde{B}(E)]^{-1}\tilde{A}(E). \tag{10.9}$$

To analyze bound states, it is sufficient to regard $\tilde{S}(E)$ as a bilinear form between a dense set of states. We use states $\theta_p=\int f(x_{\mathrm{rel}},p_{\mathrm{tot}}):\phi(x_1)\phi(x_2):dx_{\mathrm{rel}}\Omega$, dense in $\mathscr{H}(p)$ by §3. Thus we study the bilinear form $\langle g,\tilde{S}(E)f\rangle$, or equivalently, we realize (10.9) as an operator equation from $A_{-1+\varepsilon,\delta_2}$ to $A_{1-\varepsilon,\delta_2}=A^*_{-1+\varepsilon,\delta_2}$.

By Proposition 7.11 and Theorem 10.1,

$$\tilde{A}(E):A_{-1+\varepsilon,\delta_2}\to A_{1-\varepsilon,-\delta_2}, \tag{10.10}$$

for $\operatorname{Re}E\leqq 3(m_0-\varepsilon)$, with an analytic continuation onto the second sheet, for E not at threshold. Similarly

$$\tilde{B}(E):A_{-2+\varepsilon,-\delta_2}\to A_{2-\varepsilon,-\delta_2} \tag{10.11}$$

is bounded, and

$$\tilde{B}(E):A_{-\frac{3}{2}+\varepsilon,-\delta_2}\to A_{\frac{3}{2}-\varepsilon,-\delta_2} \tag{10.12}$$

is trace class, with norm (away from threshold)

$$\|\tilde{B}(E)\|_{A_\delta\to A_\delta}\leqq O(\lambda)<1, \quad \text{for} \quad \delta_1>0, \quad \delta_2<0. \tag{10.13}$$

Thus $(I-\tilde{B}(E))^{-1}$ is bounded on the same spaces, and

$$\tilde{S}(E):A_{-1+\varepsilon,\delta_2}\to A_{1-\varepsilon,-\delta_2}$$

is bounded.

Near threshold, since $\tilde{B}(E)$ is compact, $(I-\tilde{B}(E))^{-1}$ and $\tilde{S}(E)$ are meromorphic, and the only possible point of accumulation of poles is at threshold.

The asserted properties of bound states follow. See [3,4,12] for a more complete analysis of a similar situation.

References

1. Bros, J., Lasalle, M.: Analyticity properties and many body structure in general quantum field theory. III. Commun. Math. Phys. **54**, 33–62 (1977)
2. Combescure, M., Dunlop, F.: N-particle irreducible functions in Euclidean quantum field theory. Ann. Phys. (N.Y.) to appear

3. Dimock, J., Eckmann, J.P.: On the bound state in weakly coupled $\lambda(\phi^6 - \phi^4)_2$. Commun. Math. Phys. **51**, 41–54 (1976)
4. Dimock, J., Eckmann, J.P.: Spectral properties and bound state scattering for weakly coupled $P(\phi)_2$ models. Ann. Phys. **103**, 289–314 (1977)
5. Glimm, J., Jaffe, A.: Two and three body equations. Commun. Math. Phys. **44**, 293–320 (1975)
6. Glimm, J., Jaffe, A.: Positivity of the ϕ_3^4 Hamiltonian. Fortschr. Physik **21**, 327–376 (1973)
7. Glimm, J., Jaffe, A.: In: International symposium on mathematical problems in theoretical physics, Araki, H. (ed.). Lecture notes in physics, Vol. 39. Berlin, Heidelberg, New York: Springer 1975
8. Glimm, J., Jaffe, A., Spencer, T.: The Wightman axioms and particle structure in the $P(\phi)_2$ quantum field model. Ann. Math. **100**, 585–632 (1974)
9. Koch, H.: Private communication
10. Gelfand, I., Shilov, G.: Generalized functions, Vol. II, III. New York: Academic Press 1967
11. Spencer, T.: The decay of the Bethe-Salpeter kernel in $P(\phi)_2$ quantum field. Commun. Math. Phys. **44**, 143–164 (1975)
12. Spencer, T., Zirilli, F.: Scattering states and bound states in $\lambda P(\phi)_2$. Commun. Math. Phys. **49**, 1–16 (1976)
13. Symanzik, K.: On the many particle structure of Green's functions in quantum field theory. J. Math. Phys. **1**, 249–273 (1960)

Communicated by A. Jaffe

Received January 9, 1979

VI Bounds on Coupling Constants

Introduction

The ϕ^4 interaction in the single phase region is repulsive, as can be demonstrated either using perturbation theory for weak coupling or using correlation inequalities for all couplings. Furthermore a remarkably simple argument of the authors based on Griffiths' and Lebowitz' inequalities shows that the dimensional physical coupling constant g, defined in terms of the expectation of four fields, is bounded. Thus $0 \leq g \leq$ const. The universal constant depends only on the dimension d of space-time. In other words the nonlinearity in the ϕ^4 equation of motion forces ϕ itself to be small. Using K. Symanzik's Brownian motion representation of correlations, improved correlation inequalities have been derived by Aizenman, Brydges, Fröhlich and Spencer which show that the constant vanishes for $d \leq 5$.

Bibliography

Absolute bounds on vertices and couplings, Ann. l'Inst. H. Poincaré A22, 97–107 (1975).

The coupling constant in a ϕ^4 quantum field theory, in Recent Developments in Gauge Theories, G.'t Hooft et al. Editors, Plenum, New York, 1980.

Absolute bounds on vertices and couplings

by

James GLIMM (1)
Rockefeller University, New York, New York 10021

and

Arthur JAFFE (2)
Harvard University Cambridge, Mass. 02138

ABSTRACT. — We prove absolute upper and/or lower bounds on φ^4 and φ^6 dimensionless vertices and physical coupling constants.

---

1. DEFINITIONS AND ASSUMPTIONS

We derive absolute bounds on φ^4 and φ^6 dimensionless coupling constants g. For instance, in a pure φ^4 model we prove

$$0 \leq g \leq \text{const.}, \tag{1.1}$$

in the single phase region (no symmetry breaking). We also obtain bounds on associated vertex functions and on connected parts. Our general methods presumably have other consequences for n-particle amplitudes, $n \geq 8$, which we do not pursue here.

DEFINITIONS. — Let

$$\langle 1 \ldots n \rangle \equiv \langle \Phi(x_1) \ldots \Phi(x_n) \rangle \tag{1.2}$$

---

(1) Supported in part by the National Science Foundation under Grant MPS 74-13252.
(2) Supported in part by the National Science Foundation under Grant MPS 73-05037.

denote the n-point (Euclidean) Schwinger function, and let $\langle 1 \dots n \rangle_T$ denote its connected part. We let

$$(1.3)\quad S_T^{(n)}(p) = S_T^{(n)}(p_1, \dots, p_{n-1}) = \int \langle 1 \dots n \rangle_T \exp\left[i \sum_{j=1}^{n=1} p_j x_j\right] dx_1 \dots dx_{n-1}\Big|_{x_n=0}$$

denote the Fourier transform of $\langle 1 \dots n \rangle_T$. We encounter $S_T^{(2)}(p)$ often, and denote it $\chi(p)$. We use the standard spectral representation

$$(1.4)\qquad \chi(p) = S_T^{(2)}(p) = \int \frac{d\rho(a)}{p^2 + a},$$

where $d\rho(a)$ is the spectral measure. Furthermore, we let

$$(1.5)\qquad \Gamma^{(n)}(p) = \Gamma^{(n)}(p_1, \dots, p_{n-1})$$

denote the n-point Euclidean vertex function (thus $\Gamma^{(n)}$ is the amputated, one particle irreducible part of $S_T^{(n)}(p)$). In an even theory, we have for example,

$$(1.6)\qquad \Gamma^{(4)}(p) = S_T^{(4)}(p) \prod_{i=1}^{4} \chi(p_i)^{-1},$$

where $p_4 = -(p_1 + p_2 + p_3)$. Also

$$(1.7)\quad \Gamma^{(6)}(p) = \left[S_T^{(6)}(p) - \sum_X S_T^{(4)}(X)\chi(P_X)^{-1} S_T^{(4)}(\sim X)\right] \prod_{i=1}^{6} \chi(p_i)^{-1},$$

where $p_1 + \dots + p_6 = 0$, where $(X, \sim X)$ is a partition of $(p_1, \dots, p_6)$ into two subsets of three elements, and where P_X is the sum of the momenta in X.

We let $d \le 4$ denote the space-time dimension and let m denote the mass gap. Note that $\Gamma^{(n)}$ has the dimension of mass to the power $d - \frac{1}{2}n(d-2)$. We define the dimensionless amplitude $g^{(n)}(p)$ by

$$(1.8)\qquad g^{(n)}(p) = -m^{-d-n+\frac{1}{2}nd}\Gamma^{(n)}(p),$$

and the associated coupling constant $g^{(n)}$ by

$$(1.9)\qquad g^{(n)} = g^{(n)}(0).$$

For example,

$$(1.10)\qquad g^{(4)} = -m^{d-4}\chi^{-4} \int \langle 1234 \rangle_T dx_1 dx_2 dx_3,$$

where $\chi \equiv \chi(0)$.

ASSUMPTIONS. — We make some or all of the following assumptions. In the case of weakly coupled $\mathscr{P}(\varphi)_2$, assumptions (*a*)-(*f*) have been proved, while for single phase, even φ_2^4 models (*h*)-(*i*) have been proved. We expect (*a*)-(*f*) to hold for single phase φ_3^4, φ_3^6, and φ_4^4 models, while (*h*)-(*i*) are expected for all single phase even φ_d^4 models, $d \leq 4$. The evidence for the validity of (1.4) in φ_4^4 comes from perturbation theory. The proof of (*i*) in even, single phase φ_2^4 has been established by Cartier (private communication).

(*a*) The Wightman axioms for a scalar boson field, or the Osterwalder-Schrader axioms for the corresponding Euclidean theory. The two point function $\chi(p)$ satisfies (1.4). In particular we are assuming that $\int \frac{d\rho(a)}{a}$ is finite.

(*b*) A positive mass gap, $m > 0$.

(*c*) $\langle \varphi \rangle = 0$.

(*d*) The second Griffiths inequality,

$$\langle \mathrm{A}_1\mathrm{A}_2 \rangle \geq \langle \mathrm{A}_1 \rangle \langle \mathrm{A}_2 \rangle,$$

where $(\mathrm{A}_1, \mathrm{A}_2)$ is a partition of $(1, \ldots, n)$. For instance,

$$-\langle 1234 \rangle_{\mathrm{T}} = -\{\langle 1234 \rangle - \langle 14 \rangle\langle 23 \rangle - \langle 12 \rangle\langle 34 \rangle - \langle 13 \rangle\langle 24 \rangle\} \leq \langle 12 \rangle\langle 34 \rangle + \langle 13 \rangle\langle 24 \rangle. \tag{1.11}$$

(*f*) The measure $d\rho(a)$ contains a delta function at $a = \mathrm{m}^2$, of unit strength (This is a renormalization hypothesis).

(*h*) In the case of a φ^4 interaction, it is appropriate to assume the Lebowitz inequality $\langle 1234 \rangle_{\mathrm{T}} \leq 0$, or the set of Lebowitz inequalities,

$$\langle 12\mathrm{U} \rangle \leq \langle 12 \rangle \langle \mathrm{U} \rangle + \sum_{\mathrm{W}} \langle \mathrm{W}, 1 \rangle \langle \sim \mathrm{W}, 2 \rangle \tag{1.12}$$

where W is an (odd) subset of U.

(*i*) In the case of a pure φ^4 model, we assume

$$\langle 123456 \rangle_{\mathrm{T}} \geq 0. \tag{1.13}$$

2. BOUNDS ON QUARTIC COUPLINGS

THEOREM 2.1.

(*i*) Assume (*a*)-(*d*) above. Then

$$g^{(4)} \leq \text{const.}\, \chi^{-2} m^{-4}. \tag{2.1}$$

(*ii*) Assume (*f*) also, then $\chi \geq m^{-2}$, so

$$g^{(4)} \leq \text{const.} \tag{2.2}$$

(*iii*) Assume (*h*) also, then

$$0 \leq g^{(4)} \leq \text{const.}\, \chi^{-2} m^{-4} \leq \text{const.} \tag{2.3}$$

Remark 1. — The constants are pure numbers, independent of all parameters. They can be determined explicitly by the proof.

Remark 2. — The upper bound on $g^{(4)}$ is related to the picture of « critical point dominance » discussed in [*1*]. According to this picture, for $d < 4$, $g^{(4)}$ assumes its maximum value at the critical point, *i. e.* at the onset of phase transitions. According to standard ideas, $m = 0$ at the critical point.

Remark 3. — For $d < 4$, the limit λ_0 (= bare φ^4 coupling constant) $\to \infty$ with m fixed plays a role in the application of the Callan-Symanzik equations to the study of critical phenomena [*2*]. We use Theorem 2.1 below, combined with bounds from [*3*], to show that either this limit exists, or that in this limit $g^{(4)} \to 0$ (see Corollary 2.4). If $g^{(4)}$ is monotone in the dimensionless charge $\lambda_0 m_0^{d-4}$, then the alternative $g^{(4)} \to 0$ is excluded. The possibility that $g^{(4)}$ is monotone is discussed in [*1*], and is a consequence of conventional assumptions made in the study of the Callan-Symanzik equations.

Remark 4. — Other bounds on quartic couplings have been established by Lukaszuk and Martin [*4*], see also Healy [*5*]. Their bounds require different assumptions and methods, and yield different conclusions.

In the case of a pure φ^4 model we also obtain bounds on $g^{(4)}(p)$ for $p \neq 0$.

THEOREM 2.2. — Assume (*a*)-(*h*) above. There exists $\delta > 0$, such that if $|p_i| < \delta m$, then $g^{(4)}(p)$ is analytic in p and bounded uniformly by

$$|g^{(4)}(p)| \leq \text{const.}\, \chi^{-2} m^{-4} \leq \text{const.} \tag{2.4}$$

Proof of Theorem 2.1. — By the positivity of $d\rho$ (Wightman positivity condition)

$$\langle 12 \rangle \geq 0, \tag{2.5}$$

so the right side of (1.11) is positive. By symmetry we then have

$$\begin{aligned} -\langle 1234 \rangle_T &\leq (\langle 12 \rangle \langle 34 \rangle + \langle 13 \rangle \langle 24 \rangle)^{\frac{1}{3}} (\langle 14 \rangle \langle 23 \rangle \\ &\quad + \langle 12 \rangle \langle 34 \rangle)^{\frac{1}{3}} (\langle 13 \rangle \langle 24 \rangle + \langle 14 \rangle \langle 23 \rangle)^{\frac{1}{3}} \\ &= \Big(\sum_{i=1}^{8} A_i\Big)^{\frac{1}{3}} \leq 2 \sup A_i^{\frac{1}{3}}. \end{aligned} \tag{2.6}$$

Here each A_i is either

$$\langle 12 \rangle \langle 13 \rangle \langle 14 \rangle \langle 23 \rangle \langle 24 \rangle \langle 34 \rangle = \prod_{1 \le i < j \le 4} (ij)$$

or else a permutation of

$$\langle 12 \rangle^2 \langle 34 \rangle^2 \langle 13 \rangle \langle 24 \rangle .$$

We bound each A_i using Proposition A2.1, of Appendix 2,

$$\langle x0 \rangle \le O(1) \mid x \mid^{-d_e - m(1-\varepsilon)|x|} \chi .$$

An elementary estimate shows that each $A_i^{\frac{1}{3}}$ is integrable over $dx_1 dx_2 dx_3$. By homogeneity, and (A2.1),

$$\int A_i^{\frac{1}{3}} dx_1 dx_2 dx_3 \le O(1) m^{-d} \chi^2 ,$$

from which we conclude

$$- S_T^{(4)}(0) \le O(1) m^{-d} \chi^2 . \tag{2.7}$$

By (1.8), with $n = 4$, and (1.6) we have (2.1). The theorem then follows.

Proof of Theorem 2.2. — Assuming (*h*) and (2.6),

$$\mid \langle 1234 \rangle_T \mid \le 2 \sup_i A_i^{\frac{1}{3}} . \tag{2.8}$$

We bound the power series coefficients of $S_T^{(4)}(p)$, expanding about $p = 0$. The derivatives $\partial/\partial p_j$ become multiplication by ix_j in the Fourier transform, and are dominated by the exponential decrease of (2.8), estimated by (A2.1). Thus the power series in momenta converges for $|p_i| < m/3$, and establishes the analyticity and boundedness of $S_T^{(4)}(p)$,

$$\left| S_T^{(4)}(p) \right| \le O(1) m^{-d} \chi^2 . \tag{2.9}$$

Finally, the analyticity and boundedness of $\chi(p)^{-1}$ follows from (1.4), for $|p| < m$, yielding

$$\mid \chi(p) \mid^{-1} \le 2\chi(0)^{-1} .$$

Thus $g^{(4)}(p)$ is analytic in p and

$$\mid g^{(4)}(p) \mid \le O(1) m^{-d} \chi^{-2} \le O(1) m^{-d+4} ,$$

with the constant independent of p, m, λ for $|p_i| \le \delta m$. This completes the proof. We have also proved.

COROLLARY 2.3. — Assuming (*a*)-(*h*), $S_T^{(4)}(p)$ is analytic for $|p_i| \le \delta m$ and bounded uniformly by (2.9).

We now let $S_j^{(n)}$ be the n-point Schwinger function for a model satisfying (*a*)-(*h*) and labelled by the index j (for instance, $S^{(n)}_{\lambda_0,m}$ may be the φ^4_2 model, labelled by the bare coupling λ_0 and mass m).

COROLLARY 2.4. — Assume m_j is bounded away from zero. Then either there exists a convergent subsequence of Schwinger functions

$$S^{(n)} = \lim_i S^{(n)}_j, \tag{2.10}$$

with $S^{(n)}$ satisfying (*a*)-(*h*), or else $g^{(4)}_T(p) \to 0$.

Proof. — Assume for some p, with $|p_i| \leq \delta m$, there is a subsequence of $g_r(p)$ bounded away from zero. Then by (2.4)

$$\chi^2_r \leq \text{const. } m_r^{-4}(g^{(4)}_r(p))^{-1} \leq \text{const.}$$

This bound (uniform in r) on

$$\int \langle 12 \rangle_{T,r} dx_1 = \int \langle 12 \rangle_r dx_1 \leq \text{const.}$$

proves a uniform bound

$$|\langle fg \rangle| \leq |f|_{\mathcal{S}} |g|_{\mathcal{S}} \tag{2.11}$$

for a suitable Schwartz space norm $|\,.\,|_{\mathcal{S}}$. Thus by Theorem 1 of [*3*], a convergent subsequence of Schwinger functions $S^{(n)}_{j_i}$ exists, and $S^{(n)}$ is a tempered distribution. The properties (*a*)-(*h*) follow from the corresponding properties in the approximate theories.

Conversely, if $g_j(p) \to 0$ for each p with $|p_i| < \delta m$, the uniform bound (2.4) ensures that the family of analytic functions $g_j(p)$ converges to the analytic function zero.

3. BOUNDS ON $S^{(6)}_T(p)$

In this section we prove preliminary bounds on S^6_T.

THEOREM 3.1.

(*i*) Assume (*a*)-(*h*). Then there exists $\delta > 0$ such that if $|p_i| \leq m\delta$, then $S^{(6)}_T(p)$ is analytic and bounded by

$$m^{2d}\chi^{-3} |S^{(6)}_T(p)| \leq \text{const.} \tag{3.1}$$

(*ii*) Assume (*a*)-(*i*). Then

$$0 \leq m^{2d}\chi^{-3} S^{(6)}_T(0) \leq \text{const.} \tag{3.2}$$

The constants are pure numbers, independent of the parameters.

Proof. — We consider $p = 0$. The extension to $p \neq 0$ follows the proof of Theorem 2.2. Using (f), we have

$$\langle 123456 \rangle - \langle 12 \rangle \langle 3456 \rangle \geq 0. \tag{3.3}$$

The definition of the six point connected part is

$$\langle 123456 \rangle_T = \langle 123456 \rangle - \sum_Y \langle Y_1 \rangle \langle Y_2 \rangle + 2 \sum_Z \langle Z_1 \rangle \langle Z_2 \rangle \langle Z_3 \rangle. \tag{3.4}$$

Here $Y = (Y_1, Y_2)$ ranges over the fifteen partitions of $(1, 2, \ldots, 6)$ into two and four element subsets, while $Z = (Z_1, Z_2, Z_3)$ ranges over the fifteen partitions of $(1, \ldots, 6)$ into three pairs. Thus

$$-\langle 123456 \rangle_T \leq \sum_Y{}' \langle Y_1 \rangle \langle Y_2 \rangle - 2 \sum_Z \langle Z_1 \rangle \langle Z_2 \rangle \langle Z_3 \rangle. \tag{3.5}$$

where Σ' omits the partition $\langle 12 \rangle \langle 3456 \rangle$. We use ($h$) and the definition of $\langle 1234 \rangle_T$ to obtain

$$-\langle 123456 \rangle_T \leq \sum_Z{}' \langle Z_1 \rangle \langle Z_2 \rangle \langle Z_3 \rangle, \tag{3.6}$$

where Σ' omits the three partitions which contain the factor $\langle Z_1 \rangle = \langle 12 \rangle$. By symmetry,

$$-\langle 123456 \rangle_T \leq 12 \sup \prod_{j=1}^{15} \prod_{i=1}^{3} \langle Z_i^{(j)} \rangle^{1/15}, \tag{3.7}$$

where $j = 1, 2, \ldots, 15$ labels the two element subsets of $(1, \ldots, 6)$, and $(Z_1^{(j)}, Z_2^{(j)}, Z_3^{(j)})$ one of the twelve partitions of $(1, \ldots, 6)$ into three pairs, no one of which is j.

We now follow the proof of Theorem 2.1 to obtain

$$-S_T^{(6)}(0) \leq \text{const.}\ m^{-2d} \chi^3.$$

proving one side of (3.1).

We next use (1.12) (see [*3*]) with $U = (3456)$ to obtain

$$\langle 123456 \rangle \leq \langle 12 \rangle \langle 3456 \rangle + \sum_W \langle W, 1 \rangle \langle \sim W, 2 \rangle,$$

where W ranges over the eight odd subsets of U. Thus by (3.4),

$$\langle 123456 \rangle_T \leq -\sum_Y{}'' \langle Y_1 \rangle \langle Y_2 \rangle + 2 \sum_Z \langle Z_1 \rangle \langle Z_2 \rangle \langle Z_3 \rangle$$

where Σ'' ranges over the six partitions containing both 1 and 2 in the four element subset. By the definition of $\langle 1234 \rangle_T$,

$$\langle 123456 \rangle_T \leq - \sum_Y{}'' \langle Y_1 \rangle_T \langle Y_2 \rangle_T + \sum_Z{}' \langle Z_1 \rangle \langle Z_2 \rangle \langle Z_3 \rangle.$$

By (1.11) we note that

$$\langle 123456 \rangle_T \leq 2 \sum_Z{}' \langle Z_1 \rangle \langle Z_2 \rangle \langle Z_3 \rangle,$$

which is just a bound by twice the right hand side of (3.6). We now proceed as above to obtain the bound (3.2) for $p = 0$.

Finally, we remark that assuming (*i*) yields $0 \leq S_T^{(6)}(0)$, to complete the proof.

4. BOUNDS ON $g^{(6)}$

We use the bounds of Chapter 3 to establish bounds on the six point vertex function in the φ^4 model.

THEOREM 4.1. — Assuming (*a*)-(*h*), for $| p_i | \leq m\delta$, $g^{(6)}(p)$ is analytic and bounded by

$$| g^{(6)}(p) | \leq \text{const.}$$

where δ is the smaller of the constants given by Theorems 2.2 and 3.1.

Proof. — By Theorem 3.1, $m^{2d}\chi^{-3}S_T^{(6)}(p)$ is analytic and bounded. By Theorem 2.2 $g^{(4)}(p)$ is analytic and bounded. Note

$$g^{(4)}(p) = m^{d-4} \prod_{i=1}^{4} \chi(p_i)^{-1} S_T^{(4)}(p).$$

The desired bound follows by

$$g^{(6)}(p) = - m^{2d-6}\Gamma^{(6)}(p)$$

and the representation (1.7).

APPENDIX 1

THE FREE EUCLIDEAN PROPAGATOR

We establish the elementary bounds on

$$c(m^2 ; x) = \frac{1}{\pi} \int \frac{e^{-ipx} dp}{p^2 + m^2} = m^{d-2} c(1 ; mx)$$

used below. By Euclidean invariance, we may evaluate c for $x = x_t = (t, \vec{0}), t > 0$, and

$$f(t) = c(1 ; x_t) = \int \frac{e^{-t\mu}}{\mu} d\vec{p}:$$

where $\mu = \mu(\vec{p}) = (\vec{p}^2 + 1)^{\frac{1}{2}}$.

PROPOSITION A1.1. — There is a constant O(1) depending only on d, such that

$$f(t) \leq \begin{cases} O(1) t^{(1-d)/2} e^{-t} & t \geq \frac{1}{2} \\ O(1) t^{-(d-2)} & t \leq \frac{1}{2}, d \geq 3 \\ O(1) \mid \ln t \mid & t \leq \frac{1}{2}, d = 2 \end{cases}$$

REMARK. — The long and short range exponents of t differ.

Proof. — Let $\mu_t(\vec{p}) = (\vec{p}^2 + t^2)^{\frac{1}{2}}$, so

$$f(t) = t^{-(d-2)} e^{-t} \mathrm{I}(t)$$

where

$$\mathrm{I}(t) = \int \mu_1^{-1} \exp\left[-(\mu_t - t)\right] d\vec{p}.$$

For $t \geq \frac{1}{2}$, and $t \geq |\vec{p}|$, we use

$$\mu_t - t \leq \text{const.}\ \vec{p}^2 t^{-1}$$

to establish

$$\mathrm{I}(t) \leq O(t^{(d-3)/2}).$$

Furthermore, for $\frac{1}{2} \leq t \leq |\vec{p}|$, we use $\mu_t - t \geq \text{const.}\ |\vec{p}|$ to give

$$\mathrm{I}(t) \leq O(t^{-1}).$$

Thus for $t \geq \frac{1}{2}$, we have established the proposition.

For $t \leq \frac{1}{2}$, and $d \geq 3$, we have

$$\mathrm{I}(t) \leq \int |\vec{p}|^{-1} \exp\left[-|\vec{p}| + \frac{1}{2}\right] d\vec{p} \leq O(1).$$

To bound the case $d = 2$, with $t \leq \frac{1}{2}$, we use

$$\mathrm{I}(t) \leq \text{const.} \int_0^\infty e^{-u} \frac{du}{(t^2 + u^2)^{\frac{1}{2}}} \leq O(|\ln t|).$$

APPENDIX 2

THE EUCLIDEAN PROPAGATOR

The Euclidean propagator $\langle xy \rangle$ has the spectral representation given by (1.4),

$$\langle xy \rangle = \langle \Phi(x)\Phi(y) \rangle = \int d\rho(a)c(a\,;\,x-y).$$

PROPOSITION A2.1. — Assume (*a*)-(*c*) above, and let $\varepsilon > 0$. Then

$$\langle x0 \rangle \leq O(1) \mid x \mid^{-d} e^{-m(1-\varepsilon)|x|}\chi, \tag{A2.1}$$

where O(1) depends only on the dimension d and on ε.

PROOF. — We use proposition A1.1. Let $d \geq 3$. Then

$$\begin{aligned}\langle x0 \rangle &= \int d\rho(a)a^{(d-2)/2}c(1\,;\,\sqrt{a}x).\\ &\leq O(1)\int \frac{d\rho(a)}{a}\,a \mid x \mid^{-d+2} e^{-\sqrt{a}|x|(1-\varepsilon/2)}\\ &\leq O(1)e^{-m(1-\varepsilon)|x|}\left(\int \frac{d\rho(a)}{a}\right) \mid x \mid^{-d+2} \left(\sup_a ae^{-\sqrt{a}|x|\varepsilon/2}\right).\end{aligned}$$

The sup occurs for $a = 4(\varepsilon x)^{-2}$. Thus (A2.1) follows. For $d = 2$,

$$\begin{aligned}\langle x0 \rangle &= \int d\rho(a)c(1\,;\,\sqrt{a}x)\\ &\leq O(1)\int_{\mathrm{I}} \frac{d\rho(a)}{a}\,a \mid \ln\sqrt{a}x \mid + O(1)\int_{\mathrm{II}} \frac{d\rho(a)}{a}\,ae^{-\sqrt{a}|x|},\end{aligned}$$

where

$$\mathrm{I} = \left\{ a : m \leq \sqrt{a} \leq \mid 2x \mid^{-1} \right\} \quad \text{and} \quad \mathrm{II} = \left\{ a : \sqrt{a} \geq \mid 2x \mid^{-1} \right\}.$$

The integral over II is bounded as above, by (A2.1). The integral over I is bounded using

$$\sup_{a\in \mathrm{I}} a \left| \ln \sqrt{a} \mid x \mid \right| \leq \mid x \mid^{-2} \sup_{0\leq u \leq 1} u^2 \mid \ln u \mid \leq O(1) \mid x \mid^{-2}.$$

This yields the bound $O(1) \mid x \mid^{-2}\chi$. Since $\mid mx \mid \leq \frac{1}{2}$, for $a \in \mathrm{I}$, the integral over I is bounded by (A2.1).

REMARK. — The increase in the bound on the short range singularities (from $x^{(2-d)}$ to x^{-d}) is associated with an allowed divergence of the field strength renormalization integral

$$Z^{-1} = \int d\rho(a).$$

REFERENCES

[1] J. Glimm and A. Jaffe, Critical point dominance in quantum field models. *Ann. Institut Henri Poincaré*, t. **21**, 1974, p. 27-41.

[2] G. Parisi, Field theory approach to second order phase transitions in two and three dimensional systems, *1973 Cargèse lectures.*

[3] J. Glimm and A. Jaffe, Remark on the existence of φ^4. *Phys. Rev. Lett.*, t. **33**, 1974, p. 440-442.

[4] L. Lukaszuk and A. Martin, Absolute upper bounds for $\pi\pi$ scattering. *Nuovo Cimento*, t. **52** *A*, 1967, p. 122-145.

[5] J. Healy, New rigorous bounds on coupling constants in field theory. *Phys. Rev.*, **D8**, 1973, p. 1904-1914.

(Manuscrit reçu le 9 août 1974)

THE COUPLING CONSTANT IN A ϕ^4 FIELD THEORY

James Glimm*[1] and Arthur Jaffe+[2]

*The Rockefeller University, New York, NY 10021

+Harvard University, Cambridge, MA 02138

Abstract

Under simple hypothesis, the following statements are proved in a ϕ^4 field theory: (1) The coupling constant is continuous in the parameters, e.g. as the scaling limit is approached or as ultraviolet cutoffs are removed. (2) Zero coupling corresponds to a free field or a generalized free field. (3) Different types of amputation in the definition of the coupling constant differ only by a finite nonzero factor.

* * * * *

In this note, we show by elementary methods that the ϕ^4 coupling constant is a continuous function of the parameters under assumptions which should be appropriate for the critical point scaling limit or removal of the ultraviolet cutoff. In particular the limit theory is nontrivial (not a generalized free field) if and only if the coupling constants of the approximating theories are bounded away from zero. Our analysis applies also to the Ising model, at its critical point.

The motivation for this paper is the statement of Baker [1] that high temperature series expansions for the d=3 Ising model

---

1. Supported in part by the National Science Foundation under Grant PHY 78-08066.

2. Supported in part by the National Science Foundation under Grant PHY 79-16812.

give small but significant violations of hyperscaling: $d\nu + \gamma \neq 2\Delta$. If these numbers are significant, we agree with Baker that the critical point defines a trivial field theory.

We work with a coupling constant defined off mass shell (at p=0), in order to be able to perform the analysis. Particle amputation multiplies an external leg of a n-point function by

$$m^2 - p^2 \Big|_{p=0} = m^{+2}$$

while propagator amputation multiplies by

$$\left(Z^{-1} G^{(2)\sim}(p)\right)^{-1} \Big|_{p=0} = Z / \int G^{(2)}(x-y)\, d(x-y)$$

where

$$G^{(2)} = \langle\phi(x)\,\phi(y)\rangle - \langle\phi\rangle^2$$

is the truncated two point function of the ϕ^4 field theory. Also let $\chi = \int G^{(2)}\, d(x-y)$. The the coupling constant is

$$0 \le g_\ell = -Z^{-2} m^{d+4} \int G^{(4)}(x_1, \ldots, x_4)\, dx_1 \ldots dx_3 \left(\frac{Z}{m^2\chi}\right)^\ell \tag{1}$$

In g_ℓ, ℓ legs have propagator amputation and $4 - \ell$ legs have particle amputation. The bound

$$Z/(m^2\chi) \le 1 \tag{2}$$

follows from the definition of field strength renormalization, so that

$$g \equiv g_0 \ge g_1 \ge \cdots \ge g_4 \quad . \tag{3}$$

It was first shown in [2] that

$$g_2 = -m^d \int G^{(4)}/\chi^2 \le \text{const.} \tag{4}$$

Since g_2 has the critical exponents,

(5) $$g_2 \sim \tau^{-2\Delta+\gamma+d\nu} \quad ,$$

this leads to the inequality

(6) $$-2\Delta + \gamma + d\nu \geq 0 \quad .$$

The subsequent consideration of (4)-(6) by other authors is summarized in [1].

In this paper we work only with g_0 . The finiteness of

(7) $$g_\ell / g_{\ell+1} = Z^{-1} m^2 \chi$$

follows from conventional scaling ideas, which we now assume. In particular we assume

$$0 \leq Z^{-1} m^{-d+2} G^{(2)}(x/m) \leq \begin{cases} 0(1)\ x^{-(d-1)/2} e^{-|x|}, & |x| \geq 1 \\ 0(1)\ x^{-d+2-\eta}, & |x| \leq 1 \end{cases}$$

We take $\eta < 2$ and the constants 0(1) above to be universal constants. Then (7) is bounded uniformly in all parameters. This gives us

<u>Theorem 1</u>. Assume (8) with universal constants. Then with the hypothesis of [2], g and all g_ℓ are bounded, by a universal constant. Either the g_ℓ are all nonzero or they are all zero. In the latter case, and only then, the theory is a generalized free field.

The final statement follows from [3].

Now we turn to the continuity of g under a limit process. We assume that m is bounded away from zero in the limit process. This allows a (positive mass) scaling limit at the critical point, and removal of ultraviolet cutoffs, for noncritical couplings. By scaling, we assume m is fixed at 1.

To illustrate the idea, we see that $g_\ell / g_{\ell+1}$ is convergent by the Lebesgue bounded convergence theorem. Thus is is sufficient to consider g_0 .

<u>Theorem 2</u>. Assume (8) with universal constants and the hypothesis of [2]. The each g_ℓ is continuous in a limit process for which the mass m is fixed and not zero, and for which the two and four point functions $G^{(2)}(x)$ and $G^{(4)}(x_1,\ldots,x_4)$ are pointwise convergent a. e.

Proof. By Lebowitz' and Griffiths' inequalities,

$$0 \leq -G^{(4)}(x_1,\ldots,x_4) \leq G^{(2)}(x_{i_1} - x_{i_2})G^{(2)}(x_{i_3} - x_{i_4})$$

for any permutation $\{i_1,\ldots,i_4\}$ of $\{1,\ldots,4\}$. After a permutation and translation of variables, we can suppose that $i_\nu = \nu$, $x_1 = 0$, and $|x_2 - x_3| \leq |x_3 - x_4|$. Also we may take $m = 1$. Then with $F(x)$ denoting the right hand member of (8),

$$\int F(x_1 - x_2)\ F(x_3 - x_4)\ dx_2\, dx_3\, dx_4$$

$$\leq \int_{|x_2-x_3| \leq |x_3-x_4|} F(x_3 - x_4)F(x_2)\ d(x_2 - x_3)\ d(x_3 - x_4)\ dx_2$$

$$\leq \int x^d\ F(x)\ dx \int F(x)\ dx < \infty\ .$$

Thus continuity follows from the Lebesgue bounded convergence theorem.

References

1. G. Baker, Phys. Rev. B, 15 p.1553 (1977).

2. J. Glimm and A. Jaffe, Ann. Inst. H. Poincaré 22 p.97 (1975).

3. C. Newman, Commun. Math. Phys. 41 p.1 (1975).

VII Confinement and Instantons

Introduction

These papers are two in a series of articles in which the authors investigate the physics of instanton gases. The abelian four dimensional lattice gauge theory yields a Coulomb dipole gas. The second paper discusses confinement in terms of vortex configurations and their relation to the center of the gauge group.

Bibliography

Instantons in a U(1) lattice gauge theory, Commun. Math. Phys. 56, 195–212 (1977).

Charges, vortices, and confinement, Nuclear Physics B149, 49–60 (1979).

Instantons in a $U(1)$ Lattice Gauge Theory: A Coulomb Dipole Gas

James Glimm*
Rockefeller University, New York, NY 10021, USA

Arthur Jaffe**
Harvard University, Cambridge, MA 02138, USA

Abstract. We study the decomposition $A=A_I+A_{SW}$ of a $U(1)$ lattice gauge field into instanton and spin wave parts. The action also decomposes, $\mathscr{A}=\mathscr{A}_I+\mathscr{A}_{SW}+R$. Here $\mathscr{A}_I$ is a Coulomb dipole gas, $\mathscr{A}_{SW}$ is a zero mass free field, and R is a higher order remainder. We study $\mathscr{A}_I$ in detail, for $d\geqq 4$, in the dilute gas case (which corresponds to the low temperature limit of the gauge field theory). We establish the leading behavior of the free energy: $f\sim\varepsilon^{-d}a\zeta$. Here ε is the lattice spacing, a is a geometrical constant and ζ is an activity defined in terms of a small number of instanton configurations. Our methods suggest the absence of screening in the dilute dipole gas, $d\geqq 4$, in contrast to Debye screening for the dilute monopole gas.

1. Introduction

It has been proposed by Gell-Mann and others that a gauge field coupled to a fermion (quark) field may describe the internal structure of protons, neutrons, etc., as quark triplets, etc. In order to account for the strong binding of individual quarks, as well as the observed property of asymptotic freedom of the physical particles, a qualitative understanding of the phase transitions and critical points of the gauge-quark system appears necessary [11]. The important properties of the critical points are inferred from renormalization group and semiclassical methods.

In this paper we study some steps in this program on a rigorous mathematical level. See [8] for an introduction to the mathematics of lattice gauge fields. Following Polyakov [9], we formally derive the instanton interaction in terms of a dilute Coulomb gas for a $U(1)$ lattice gauge theory. For $d\geqq 4$, the gas consists of vector charges, and a conservation law restricts the charges to occur only in dipoles. Since the dipole-dipole interaction is not integrable at long distances, standard methods do not apply. Nevertheless, we prove an upper and lower bound on the free

* Supported in part by the National Science Foundation under Grant PHY 76-17191
** Supported in part by the National Science Foundation under Grant PHY 75-21212

energy of the dipole gas. The essential step is to use a cancellation between the forces produced on a test dipole by oppositely oriented dipoles occurring in a block. Our conclusions suggest that higher order corrections to the dipole forces do not generate a mass (finite correlation length at sufficiently low temperature. This is in contrast to the $d=3$ dilute Coulomb gas of scalar charges for which Debye suggested a finite correlation length, see [1].

We study in detail the action

$$\mathscr{A} = -\beta \sum \cos(\varepsilon^2 g F) . \tag{1.1}$$

We first separate $F = F_I + F_{SW}$ into two parts: The spin wave part F_{SW} characterizes the oscillations withing one cosine well, while the instanton part F_I characterizes the transitions between different wells of the cosine function. The splitting gives rise to a magnetic charge density Q with the geometric interpretation

$$Q = \partial S ; \tag{1.2}$$

namely Q is the boundary ($d-3$ dimensions) of a Dirac surface S ($d-2$ dimensions). We call Q magnetic (rather than electric) since S is the geometric dual of $[F]$ (a discretization of F). The divergence of the dual of F is generally called a magnetic charge.

For $d=3$ dimensions, the monopole charge Q occurs as the boundary of the Dirac string S. For $d \geqq 4$, Q satisfies the conservation law

$$\partial Q = 0 . \tag{1.3}$$

An elementary charge configuration for $d=4$ is illustrated in Figure 1. In that case the vector charges (arrows) bound an elementary plaquette in the dual lattice.

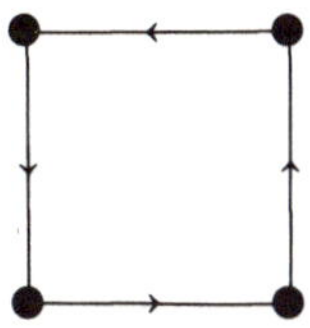

Fig. 1. Elementary dipoles, $d=4$

Because of the conservation law (1.3), the elementary charge configurations in $d \geqq 4$ are dipolar, in fact two orthogonal dipoles for $d=4$.

With the above splitting of F we obtain for the action the splitting

$$\mathscr{A} = \mathscr{A}_I + \mathscr{A}_{SW} + R . \tag{1.4}$$

Each of the three parts in (1.4) is individually gauge invariant. Here $\mathscr{A}_I$ is a Coulomb interaction for the charges Q, while $\mathscr{A}_{SW}$ is the action for a zero mass free field. The remainder R is of order three or more in Q or F_{SW}, and formally vanishes at zero lattice spacing. In the detailed analysis which follows, we neglect R and consider the approximate action $\mathscr{A}_I$ and $\mathscr{A}_{SW}$ composed of independent parts. The free field part $\mathscr{A}_{SW}$ can be solved exactly, while the instanton part $\mathscr{A}_I$ requires a detailed analysis. $\mathscr{A}_I$ also vanishes formally at zero lattice spacing.

In this manner we arrive at the analysis of a Coulomb gas of monopoles ($d=3$) and of dipoles ($d \geqq 4$). The remainder of the paper is devoted to the detailed study of the statistical mechanics of a dipole gas, see also [12]. The basic difficulties arise from the long range nature of the Coulomb action. We develop an expansion method to study the free energy of the dipoles at low density, i.e. $\beta \gg 1$.

Our detailed analysis involves several steps. First we define dipole clusters Q_L (i.e. average or block charges) to replace the elementary dipole excitation Q of Figure 1. We choose the dipole clusters so that the long range interaction between two clusters is the same as the long range interaction between two elementary dipoles, up to a quadrapole error. Here we have the lengths scales

$$\varepsilon \ll R_L \ll r\,; \tag{1.5}$$

ε is the lattice spacing (diameter of the elementary excitation), $R_L = \varepsilon O(\beta^{1/d-2})$ is the diameter of the dipole cluster and r is the inter-cluster separation. The quadrapole-quadrapole interaction is $O(r^{-(d+2)})$, which is integrable at infinity and thus can be treated by standard methods. (The possibility of charge-quadrapole interactions actually requires a second averaging and the replacement of quadrapole by octapole error terms.)

Each dipole cluster has a much reduced self action. The elementary dipoles have self action $O(\beta)$, while the dipole clusters have self action $O(\beta^{-\varepsilon})$. This requires the introduction of an activity

$$\zeta = e^{-O(\beta)} \tag{1.6}$$

which characterizes each averaged dipole cluster. The small activity results in a dilute gas of dipoles in typical configurations, i.e. in an expected density $O(\zeta) \ll 1$.

Using this typical density, we can sum the action up to a statistical length scale $R_{\text{stat}} = \varepsilon \exp(\zeta^{-\delta})$. Thus

$$\varepsilon \ll R_L \ll R_{\text{stat}}\,. \tag{1.7}$$

For $r > R_{\text{stat}}$, we take advantage of the cancellation due to charge symmetry of the dipole gas. For every dipole of positive orientation in Figure 1, there is an elementary dipole with opposite orientation. Thus we can define the total charge δ of a large number ν of dipoles. Configurations which have a large dipole charge are accompanied by a small statistical weight

$$P \leqq e^{-\delta^2/2\nu}\,. \tag{1.8}$$

Here $\nu = \nu_+ + \nu_-$ and $\delta = \nu_+ - \nu_-$ is the deviation from dipole neutrality. The entropy factor (1.8) is just the binomial distribution

$$P = 2^{-\nu} \binom{\nu}{\nu_+}\,. \tag{1.9}$$

We use this small factor to produce conditional convergence on the distance scale $r > R_{\text{stat}}$.

2. The Definition of Instantons and Spin Waves

We consider the simple cubic lattice Z_ε^d with lattice spacing ε. The $U(1)$ gauge field is a Lie algebra valued 1-cochain defined on lattice bonds (i.e. simplices connecting nearest neighbor lattice sites). An element γ of the $U(1)$ group can be written

$$\gamma = \exp(i\varepsilon g A), \qquad -\pi \leqq \varepsilon g A \leqq \pi, \tag{2.1}$$

and we define A to be the gauge field. The 2-cochain defined as the coboundary of A is

$$F = \partial^* A = \varepsilon^{-1} \sum_{i \in \partial p} A(i). \tag{2.2}$$

Thus F is defined on lattice squares (plaquettes p). Since four links bound any plaquette,

$$|\varepsilon^2 g F| \leqq 4\pi. \tag{2.3}$$

We have introduced the factor ε^{-1} in the definition of ∂^* so both A and F formally have limits as $\varepsilon \to 0$. In this limit A is a 1-form, and $F = dA$ is a 2-form. The de Rham cohomology of differential forms on a compact manifold is isomorphic to the simplicial cohomology of a smooth triangulation of the manifold, see [10]. See [7] for the geometry of gauge fields.

The action $\mathscr{A}$ associated with F is defined by

$$\mathscr{A} = -\varepsilon^{d-4} g^{-2} \sum_p \cos(\varepsilon^2 g F) = -\varepsilon^{d-4} g^{-2} \sum_p \cos(\varepsilon^2 g \partial^* A), \tag{2.4}$$

where the sum ranges over plaquettes. We note that

$$\beta = \varepsilon^{d-4} g^{-2} \tag{2.5}$$

is a dimensionless coupling constant which is inversely related to the strength of the interaction. Thus

$$\mathscr{A} = -\beta \sum_p \cos(\varepsilon^2 g F). \tag{2.6}$$

We let $[F]$ denote the discretization of F. By definition

$$\varepsilon^2 g [F] = 2\pi j, \quad \text{when} \quad j - \tfrac{1}{2} < (2\pi)^{-1} \varepsilon^2 g F \leqq j + \tfrac{1}{2}, \tag{2.7}$$

and where j is an integer. According to (2.3), the possible values of j are 0, ± 1, ± 2.

We note that $\partial^* F = (\partial^*)^2 A = 0$, which we can also write $\delta^* * F = 0$. Here we introduce the duality map $*$. For the purpose of geometric interpretation we let $*$ be the composition of Hodge duality with the metric. Hodge duality defines an isomorphism between r-cochains on Z_ε^d to $(d-r)$-cochains on the dual lattice $*Z_\varepsilon^d$. The metric defines an isomorphism between $(d-r)$ cochains and $(d-r)$ chains (raising and lowering indices). Thus

$$\begin{aligned} * : r \text{ cochains} &\to d-r \text{ chains} \\ r \text{ chains} &\to d-r \text{ cochains} \end{aligned}$$

and satisfies

$$** = (-1)^{r(d-r)}.$$

Furthermore, under the $*$ isomorphism,

$$*\partial^* * = \partial, \qquad *\delta^* * = \delta,$$

where ∂ is the boundary operator, and δ is its adjoint.

In contrast to the above, $[F]$ does not have vanishing coderivative, and we define the instanton charge density $*Q$ by

$$\partial^*[F]=*Q\,, \tag{2.8}$$

or with $S=[*F]$, $\partial S=\partial[*F]=Q$, cf. (1, 2). Maxwell's equations have the form

$$\partial^* F=0, \quad \delta^* F=-\omega_{d-1} J\,, \tag{2.9}$$

which says that J is the source for F. Similarly Q is the source for $*[F]$. Of course $\partial Q=0$, or equivalently, $\partial^* * Q=0$, which is a conservation law for $d>3$. For $d=4$, the Dirac surface has its boundary Q defined on links, i.e. Q is a 1-cochain. Then $\partial Q=0$ is a Kirkoff law: $\sum_i Q(i)=0$, where the sum runs over $2d=8$ oriented links meeting at a given lattice site. Thus Q is concentrated on closed $d-3$ surfaces (loops, for $d=4$) and it follows that the charge distribution consists of dipoles. For $d=4$, the corresponding quantum mechanical soliton ("boxiton") has also been studied [6]. For $d=3$, however, Q is a 0-cochain, concentrated at points, which corresponds to isolated charges.

This distinction of loop charges (or closed $d-3$ surfaces, $d>3$) vs. point charges ($d=3$) is basic to the study of phase transition in $U(1)$ gauge theories, see also [9]. In particular the $d=4$ loop charge explain the expected occurrence of an ordered phase at sufficiently low temperatures $T>0$ for $d\geqq 4$, and the absence of this phase for $d=3$. The mechanism, that an appropriate boundary first has dimension greater than zero, and that its entropy is thus dominated at low temperature by its energy, is the same as the mechanism governing phase transitions in the Ising model. Thus we expect that the critical dimension for phase transitions in $U(1)$ gauge theories is $d_c=3$, in contrast to the critical dimension $d_c=1$ for the Ising model. The inequality $d_c\geqq 3$ was established in [3].

Now consider the transformation $A(x)\to A(x)+2\pi n_x(\varepsilon g)^{-1}$, with n_x an integer. Under this change of variables,

$$F(p)=(\partial^* A)(p)\to F(p)+2\pi(\varepsilon g)^{-1}(\partial^* n)(p) \tag{2.10}$$

and

$$[F]\to[F]+2\pi\sum_{\partial p} n_x\,.$$

Since $\partial^*\partial^*=0$, Q is unaffected, as is the action $\mathscr{A}$. It follows that S, regarded as a $d-2$ surface, can be deformed by such a change of variables, to an arbitrary $d-2$ surface with the same boundary Q. Unfortunately, there does not seem to be a natural determination of a unique $*[F]$.

The lattice Laplace operator is

$$\Delta=-\delta^*\partial^*-\partial^*\delta^*\,, \tag{2.11}$$

and its dual is

$$\Delta=-\delta\partial-\partial\delta\,.$$

Because the topology of R^d is trivial, we can form Δ^{-1}. The same is true for a finite lattice with Dirichlet boundary conditions or with periodic boundary conditions.

(In the latter case, the topology is trivial except in dimension one.) Consider

$$F_I = [F] + \delta^* \Delta^{-1} * Q . \tag{2.12}$$

Then

$$\begin{aligned}\partial^* F_I &= \partial^* F + \partial^* \delta^* \Delta^{-1} * Q \\ &= *Q + (\partial^* \delta^* + \delta^* \partial^*) \Delta^{-1} * Q = *Q - *Q = 0 .\end{aligned} \tag{2.13}$$

Since the topology is trivial and f is a 2-cochain, we can define A_I by the equation

$$\partial^* A_I = F_I = [F] + \delta^* \Delta^{-1} * Q . \tag{2.14}$$

This is the vector potential for instantons. Finally we define the spinwave potential $A_{SW} \equiv A - A_I$ and $F_{SW} = F - F_I$.

For the action we now have

$$\begin{aligned}\mathscr{A} &= -\beta \sum \cos(\varepsilon^2 g \partial^* A) \\ &= -\beta \sum \cos(\varepsilon^2 g \partial^* A_I) \cos(\varepsilon^2 g \partial^* A_{SW}) + \beta \sum \sin(\varepsilon^2 g \partial^* A_I) \sin(\varepsilon^2 g \partial^* A_{SW}).\end{aligned} \tag{2.15}$$

Observe that

$$\begin{aligned}\cos(\varepsilon^2 g \partial^* A_I) &= \cos(\varepsilon^2 g \partial^* (-\Delta)^{-1} * Q) \\ \sin(\varepsilon^2 g \partial^* A_I) &= \sin(\varepsilon^2 g \partial^* (-\Delta)^{-1} * Q) .\end{aligned} \tag{2.16}$$

so the Dirac surface S or $*[F]$ does not contribute to the action. Also note

$$\langle \delta^* (-\Delta)^{-1} * Q, \partial^* A_{SW} \rangle = \langle (-\Delta)^{-1} * Q, \partial^* \partial^* A_{SW} \rangle = 0 .$$

Based on the φ^4 analysis [4], we expect that the dominant effect on long range order will be given by the decoupled instanton and spin wave fields, i.e. from the terms in (2.15) which are lowest order in εg (and formally ε, g independent as $\varepsilon \to 0$). Thus following Polyakov [9] we introduce the approximate action

$$\begin{aligned}\mathscr{A}_0 &= \text{const} + \tfrac{1}{2} \sum \varepsilon^d (\delta(-\Delta)^{-1} Q)^2 + \tfrac{1}{2} \sum \varepsilon^d (\partial^* A_{SW})^2 \\ &= \text{const} + \mathscr{A}_I + \mathscr{A}_{SW} .\end{aligned} \tag{2.17}$$

After decoupling, we will have long range order only if it occurs in each of the fields A_I and A_{SW}. Since the approximate action for A_{SW} is Gaussian, it can be explicitly computed that long range order occurs in A_{SW} for $d \geqq 4$, but not for $d \leqq 3$ (cf. [3]). Thus for the remainder of the paper we consider only the instanton field generated by the charges Q. Since

$$\begin{aligned}\langle \delta(-\Delta)^{-1} Q, \delta(-\Delta)^{-1} Q \rangle &= \langle Q, (-\Delta)^{-1} \partial \delta (-\Delta)^{-1} Q \rangle \\ &= \langle Q, (-\Delta)^{-2} (\partial \delta + \delta \partial) Q \rangle = \langle Q, (-\Delta)^{-1} Q \rangle ,\end{aligned} \tag{2.18}$$

we see that the instanton action is

$$\begin{aligned}\mathscr{A}_I &= \tfrac{1}{2} \langle Q, (-\Delta)^{-1} Q \rangle = \tfrac{1}{2} \sum \varepsilon^d q_i (-\Delta)^{-1} q_j \\ &\sim \tfrac{1}{2} \omega_{d-1} \sum_{i,j} \varepsilon^{2d} q_i r_{ij}^{-d+2} q_j .\end{aligned} \tag{2.19}$$

Thus the $U(1)$ instantons interact with Coulomb forces in the approximation (2.17). Returning to the exact action (2.15), we expect that the Coulomb forces are the leading long distance parts of the instanton action.

3. The Landau Length Scale and the Activity

In the approximation of § 2, the pure instanton action is

$$\mathscr{A}_I = \tfrac{1}{2} \sum \varepsilon^d q_i (-\Delta)^{-1} q_j \, . \tag{3.1}$$

We recognize q_i as the charge density and $\varepsilon^d q_i$ as a charge. Each q_i takes values which are integer multiples of $2\pi\varepsilon^{-3}g^{-1}$, by virtue of (2.7). We let $\beta \equiv \varepsilon^{d-4} g^{-2}$ define the Landau length scale R_L by the formula

$$\beta = (R_L/\varepsilon)^{d-2} \, . \tag{3.2}$$

R_L characterizes the size of individual contributions to (3.1). In other words,

$$R_L = (\varepsilon^{2d-6} g^{-2})1/^{(d-2)} = \varepsilon \beta^{1/(d-2)} \, , \tag{3.3}$$

and the interactions in (3.1) on a length scale less than R_L are strong, while those on a length scale greater than R_L are weak. In particular, if $R_L \gg \varepsilon$, then the *self interaction* is large compared to the *interaction between distinct charges.* The self interaction of a single dipole loop is of the order

$$\varepsilon^{2d} q^2 \varepsilon^{-d+2} = O(\varepsilon^{2d-6} g^{-2} \varepsilon^{-d+2}) = O(\beta) \, . \tag{3.4}$$

It is convenient to modify the definition (3.3) in order to make R_L/ε an integer. We set

$$R_L = \varepsilon [\beta^{1/(d-2)}] \, , \tag{3.3$'$}$$

where here $[\,\cdot\,]$ denotes the integer part. With this definition, the Landau lattice $Z^d_{R_L}$ is a sublattice of Z^d_ε.

Given a charge configuration Q, we will define a decomposition into a typical part Q_t and an atypical part Q_a,

$$Q = Q_a + Q_t \, , \tag{3.5}$$

see Chapter 4. The main property of Q_t is that it is a union of connected components of Q, and that Q_t has a low density of $d-3$ lattice elements. (Here we consider Q as a $d-3$ chain, and hence as a $d-3$ surface in R^d.) Since ∂ is a local operator, we have

$$\partial Q_t = 0 \, . \tag{3.6}$$

We now show how to replace the charge distribution Q_t by an averaged charge distribution $Q_{t,L}$ on the Landau lattice. Each elementary dipole of Q_t in the dual Z^d_ε lattice will be replaced by a large number of weaker dipoles in the dual Landau lattice. In making this replacement, we define the error term δQ_t by the equation

$$\delta Q_t = Q_t - Q_{t,L} \, . \tag{3.7}$$

The physical interpretation of δQ_t is a quadrapole interaction to compensate for the shorter distance (quadrapolar) forces which differ in Q_t and $Q_{t,L}$. The long range (dipolar) forces arising from Q_t to $Q_{t,L}$ will agree, since we define them to have equal (average) dipole densities. We also define Q' by

$$Q' = Q_a + Q_{t,L} = Q - \delta Q_t \,. \tag{3.8}$$

Note

$$Q = Q_a + Q_{t,L} + \delta Q_t \,. \tag{3.9}$$

Since $\mathscr{A}_I(Q)$ is quadratic in Q, for two charge configurations Q_i, Q_j it is convenient to define their mutual action $\mathscr{A}_I(Q_i, Q_j)$ by

$$\mathscr{A}_I(Q_i, Q_j) = \tfrac{1}{2}(\mathscr{A}_I(Q_i + Q_j) - \mathscr{A}_I(Q_i) - \mathscr{A}_I(Q_j)) \,. \tag{3.10}$$

Proposition 3.1. *For any charge distributions* Q, Q_j,

$$\mathscr{A}_I(Q, Q) = \mathscr{A}_I(Q) \,, \tag{3.11}$$

$$\mathscr{A}_I\left(Q, \sum_{j=1}^{n} Q_j\right) = \sum_{j=1}^{n} \mathscr{A}_I(Q, Q_j) \tag{3.12}$$

and

$$\mathscr{A}_I\left(\sum_{j=1}^{n} Q_j\right) = \sum_{i,j=1}^{n} \mathscr{A}_I(Q_i, Q_j) \,. \tag{3.13}$$

Proof. These identities follow by expanding the quadratic function $\mathscr{A}_I$, for example

$$\mathscr{A}_I(Q, Q) = \tfrac{1}{2}(\mathscr{A}_I(2Q) - 2\mathscr{A}_I(Q)) = \tfrac{1}{2}(4-2)\mathscr{A}_I(Q) \,.$$

Corollary. *For* $Q = Q_a + Q_t = Q_a + Q_{t,L} + \delta Q_t = Q' + \delta Q_t$,

$$\mathscr{A}_I(Q) - \mathscr{A}_I(Q') = \mathscr{A}_I(Q_t) - \mathscr{A}_I(Q_{t,L}) + 2\mathscr{A}_I(Q_a, \delta Q_t) \,. \tag{3.14}$$

Thus the change in action due to replacing typical ε-dipoles by averaged R_L-dipoles is the sum of a quadrapole remainder term, $2\mathscr{A}_I(Q_a, \delta Q_t)$, and the change due to the typical configurations. We estimate (3.14) in Chapter 4.

We now define $Q_{t,L}$. Using (3.6), we write

$$Q_t = \partial S_t \tag{3.15}$$

for some $d-2$ lattice surface S_t. (Henceforth we assume $d \geqq 4$.) This surface S_t is the Dirac surface for the typical charges. We choose S_t to have minimal $d-2$ area. It follows that if $\{Q_t^{(j)}\}$ is a collection of charge configurations supported in lattice volumes of bounded diameter, then the area of $\{S_t^{(j)}\}$ is also bounded. Define

$$S_a = S - S_t \tag{3.16}$$

so

$$\partial S_t = Q_t \,, \quad \partial S_a = Q_a \,. \tag{3.17}$$

We first assume that S_t is a single $d-2$ cell in the dual ε lattice, and we define $S_{t,L}$. In the general case where S_t is the sum of a bounded number of elementary $d-2$ cells, we define $S_{t,L}$ by linearity.

Let S_t be a single $d-2$ cell in $*Z_\varepsilon^d$, and consider the collection of $(R_L/\varepsilon)^{d-2}$ dipoles (i.e. $d-2$ surfaces) in $*Z_\varepsilon^d$ which both (i) lie in a single $d-2$ cell in $Z_{R_L}^d$ parallel to S_t and (ii) are closest to S_t. We define $S_{t,L}$ to be this $d=2$ Landau lattice cell, and we assign a charge strength as follows: Since S_t has strength $2\pi j\varepsilon^{d-2}g^{-1}$, we take the strength of each weak dipole (ε lattice dipole making up $S_{t,L}$) to be

$$2\pi j\varepsilon^{d-2}g^{-1}(\varepsilon/R_L)^{d-2}=2\pi^j\varepsilon^{d-2}g^{-1}\beta^{-1} . \tag{3.18}$$

Thus the dipole charge is scaled down by β^{-1}.

We justify this choice by calculating the long range part of the interaction between two pairs of dipole charges, oriented as in Figure 2.

Fig. 2. Two dipole pairs on a line, separated by distance r, illustrated for $d=4$

Before averaging, the mutual action, up to quadrapole corrections is

$$\omega qq'\left(\frac{d^2}{dr^2}r^{-d+2}\right)\varepsilon^2 ,$$

where q, q' are the charges, ω is a constant and ε is the charge separation within a dipole pair. After averaging, there are $(R_L/\varepsilon)^{d-3}$ charges on each face of the boundary of S_L, and each charge is scaled down by β^{-1}. The dipole charge separation is R_L in place of ε. Thus for $r \gg R_L$, the mutual action is

$$\begin{aligned}&\omega\beta^{-2}qq'\left(\frac{d^2}{dr^2}r^{-d+2}\right)R_L^2(R_L/\varepsilon)^{2d-6}\\&=\omega qq'\left(\frac{d^2}{dr^2}r^{-d+2}\right)\varepsilon^2(R_L/\varepsilon)^{2(d-2)}\beta^{-2}\\&=\omega qq'\left(\frac{d^2}{dr^2}r^{-d+2}\right)\varepsilon^2 ,\end{aligned}$$

up to quadrapole corrections; namely the action is unchanged up to a quadrapole error.

Proposition 3.3. *Let Q_t contain at most a bounded number N of charges. Then the self action $\mathscr{A}(Q_{t,L})$ is of magnitude $O(\beta^{-1/(d-2)})$.*

Proof. The charge strength per elementary charge in $Q_{t,L}$ is of the order $\varepsilon^{d-3}g^{-1}\beta^{-1}$, since $Q_{t,L}=\partial S_{t,L}$, and the dipoles have strength $\varepsilon^{d-2}g^{-1}\beta^{-1}$ as explained above. Summation over relative separation of the charge pairs yields the factor ε^{-d+2}, while for fixed relative separation there are $(R_L/\varepsilon)^{d-3}$ charge pairs. Thus the self action has magnitude

$$(\varepsilon^{d-3}g^{-1}\beta^{-1})^2(R_L/\varepsilon)^{d-3}\varepsilon^{-d+2}=O(\varepsilon/R_L)=O(\beta^{-1/(d-2)}) . \tag{3.19}$$

Note the bound (3.19) uses the fact that $S_{t,L}$ contains a bounded number N of dipoles, since N^2 terms occur as a multiplicative factor in the bound.

As a consequence of this argument, each small charge cluster Q_t can be replaced by weaker averaged charge clusters $Q_{t,L}$ without changing the long range interaction. Aside from the necessity of introducing a quadrapole interaction (which we denote by $\mathcal{Q}$), it is also necessary to introduce an activity ζ to compensate for the large change in self action. For a single ε lattice dipole, the self action is $c_0\beta$, cf. (3.4). Since the averaged dipole has self action $O(\beta^{-1/(d-2)})$, the activity $\zeta = e^{-O(\beta)}$ is associated with a single charge. Likewise, each small dipole cluster $S_t^{(j)}$ will have an activity $\zeta_j = e^{-O(\beta)}$ to compensate for the large change in self action. We discuss this issue in detail in Chapter 4. In terms of the above equations, the change in self action (yielding the activity) arises from the term $\mathcal{A}_I(Q_t) - \mathcal{A}_I(Q_{t,L}')$ in (3.14).

We give a lower bound on $\mathcal{A}_I(Q)$ in terms of the charges in Q which have density $2\pi j\varepsilon^{-3}g^{-1}$, i.e. ε lattice charges. We can apply this bound to Q' defined by (3.8), where Q_a are the ε lattice charges.

Proposition 3.4. *Let $Q = Q_a + Q_t$ where Q_a are ε lattice charges. There is a constant $c_1 > 0$ such that*

$$\mathcal{A}_I(Q) \geqq c_1 \beta \sum_{Q_a} j^2 , \tag{3.20}$$

where $\sum_{Q_a}$ denotes the sum over all $d-3$ cells with nonvanishing charge Q_a, and where the charge density q_i in Q_a takes on values $2\pi\varepsilon^{-3}g^{-1}j$, for $|j| \leqq 12$.

Proof. Let $B = (-\Delta)^{-1}Q$ and observe that

$$\mathcal{A}_I(Q) = \tfrac{1}{2}\langle \delta B, \delta B \rangle .$$

For a given charge q in Q, let $\sum^{(q)}$ denote the sum over all $d-2$ cells of the dual lattice for which q lies on the boundary. By definition

$$\delta = \varepsilon^{-1} \sum{}^{(q)} .$$

Note that

$$\delta B = \partial\delta(-\Delta)^{-1} S = S + \partial\delta\Delta^{-1} S$$

and hence

$$\partial\delta B = \partial S = Q . \tag{3.21}$$

In other words, δB satisfies a conservation law, $\delta\delta B = 0$, and in particular for a charge $q \in Q_a$,

$$\sum{}^{(q)} \delta B = 2\pi\varepsilon^{-2} g^{-1} j . \tag{3.22}$$

However, each $d-2$ cell occurs at most $2(d-2)$ times in the iterated sum $\sum_{Q_a}\sum^{(q)}$, and so omitting terms from $Q_{t,L}$,

$$\mathcal{A}_I(Q) = \tfrac{1}{2}\langle \delta B, \delta B \rangle \geqq \varepsilon^d 2^{-2}(d-2) \sum_{Q_a} \sum{}^{(q)} (\delta B)^2 . \tag{3.23}$$

Each sum $\sum^{(q)}$ contains $6 = 2(d-(d-3))$ terms, and so by the Schwarz inequality

$$|\sum{}^{(q)} \delta B|^2 \leqq 36 \sum{}^{(q)} (\delta B)^2 .$$

Thus by (3.22)–(3.23),

$$\mathscr{A}_I Q') \geqq [\pi^2/36(d-2)]\varepsilon^{d-4} g^{-2} \sum_{Q_a} j^2 = c_1 \beta \sum_{Q_a} j^2 ,$$

where $c_1 = \pi^2/36(d-2)$.

We end this section by remarking that a further change in the charge configuration eliminates the long distance part of the quadrapolar interaction $\mathscr{Q}$. In fact, replacing each array of $(R_L/\varepsilon)^{d-2}$ piecewise constant dipoles in $S_{t,L}$ by a linear combination of 2^d translates, we eliminate the quadrapole interaction up to an octapole correction $\mathscr{O}$. So that all error terms arising from linear contributions to charge minus average charge can be estimated with octapole remainders, we perform this one additional averaging in the definition of $Q_{t,L}$.

4. The Dilute Dipole Gas for $d \geqq 4$: Upper Bound on Z

We study $\mathscr{A}_I$ defined in (2.19) for a dipole (instanton) gas and establish bounds on the partition function

$$Z = \sum_Q e^{\mathscr{A}_I(Q)} . \tag{4.1}$$

We obtain the leading contribution to the free energy

$$f_V = V^{-1} \ln Z , \quad \beta \to \infty . \tag{4.2}$$

To any finite configuration of charges Q, we associate the action

$$\zeta_Q = e^{-\mathscr{A}_I(Q)} \tag{4.3}$$

defined by (2.19). As in (3.4), we see that

$$\mathscr{A}_I(Q) = c_Q \beta , \tag{4.4}$$

for a constant c_Q which can be expressed explicitly in terms of $(-\Delta)^{-1}$ via (2.19). The constant c_Q is invariant under scale transformations, and thus (if we use scale transformations to identify charge configurations Q on different lattice spacings) the constant c_Q is independent of ε.

Our main assertion is that it is sufficient to examine a finite number of Q's to determine f_V to leading order. Let $\mathscr{Q}$ be the set of all finite charge configurations, and let

$$\mathscr{Q}_M = \{Q \in \mathscr{Q} : \text{diam } Q \leqq M\varepsilon\}$$
$$\zeta_0 = \sup_Q \{\zeta_Q : Q \in \mathscr{Q}_M\} . \tag{4.5}$$

Since ζ_Q is translation invariant, the sup above is taken over the finite set $\mathscr{Q}_M/Z^d$. Here M is a constant to be chosen later. Let a_0 denote the multiplicity of the maximum, i.e. the number of Q's for which $\zeta_Q = \zeta_0$. In counting a_0, we identify two charge configurations if they are related one to another by a lattice translation. We state our main result.

Theorem 4.1. *There is a $\gamma>0$ and a choice of M in (4.5) such that*

$$f_V = a_0 \varepsilon^{-d} \zeta_0 (1 + O(\beta^{-\gamma})) , \tag{4.6}$$

for $V > V(\beta)$ sufficiently large and for β sufficiently large.

Remark. The choice $V > V(\beta)$ eliminates surface boundary effects in leading order. Fabrizio De Angelis, de Falco and Guerra [2] have shown that $\lim_V f_V$ exists for the lattice gauge theory, and Griffiths [5] show this for $d=3$ magnetic dipoles. For the instanton gas considered here, we believe that $\lim f_V = f$ also exists and that $f = \varepsilon^{-d} \sum \zeta_Q + O(\zeta_0^{1+\theta})$, for $\theta < 1$. In this section we prove a lower bound on f_V of the form (4.6).

First let us explain the significance of Proposition 3.4. Let c_0 be the coefficient c_Q in (4.4) for Q consisting of a single ε lattice charge loop (charge with density of order $\varepsilon^{-3} g^{-1}$). Let $Q = Q_a + Q_t$ be a decomposition into ε lattice charges Q_a and disjoint charges Q_t. If Q_a contains $L \geqq c_0/c_1$ charges, then Q does not contribute to the maximum in (4.5). Thus only a bounded number of charges will contribute to the maximum in (4.5), independent of M.

We now specify precisely the typical configurations and the application of the construction of Section 3. Decompose the lattice into hypercubic blocks Δ with side length $2R_L$. Let Q be a charge configuration, and let $N(Q, \Delta)$ be the total number of charges in Δ and in $\bar{\Delta}$, the union of Δ and the $3^d - 1$ blocks adjacent to Δ. If

$$N(Q, \Delta) \leqq 4 \cdot 3^d c_0 / c_1 \equiv N , \tag{4.7}$$

then $Q \restriction \Delta$ is called typical; otherwise $Q \restriction \Delta$ is called atypical. The typical charge configurations are more complicated than the bounded charge configurations (4.5). In fact a typical configuration may contain several widely separated clusters.

Proposition 4.2. *Let $\delta > 0$ be given and let $Q = \sum Q_i$ be a typical configuration, composed of clusters Q_i having minimal separation $D\varepsilon$. Then for $D \geqq D(N, \delta)$ sufficiently large*

$$|\mathscr{A}(Q_i, Q_j)| \leqq \delta \beta N^{-2} , \tag{4.8}$$

for $i \neq j$.

Proof. From the definition (3.10), we have

$$\mathscr{A}_I(Q_i, Q_j) \sim \tfrac{1}{2} \omega_{d-1} \sum_{\substack{q_i \in Q_i \\ q_j \in Q_j}} \varepsilon^{2d} q_i r_{ij}^{-d+2} q_j .$$

Summing over charges $2\pi j \varepsilon^{-3} g^{-1}$ and $r_{ij} \geqq D\varepsilon$ gives

$$|\mathscr{A}_I(Q_i, Q_j)| \leqq c_{Q_i, Q_j} \beta D^{-d+2} , \tag{4.9}$$

where c_{Q_i, Q_j} is a constant. In fact $c_{Q_i, Q_j} \leqq O(N^2)$. Thus (4.8) holds for $D(N, \delta)$ sufficiently large.

With the choice of $D = D(N, \delta)$ as required by Proposition 4.2, we divide $Q \restriction \bar{\Delta}$ into clusters separated by at least $D\varepsilon$. Each cluster Q_j (as $Q \restriction \bar{\Delta}$ itself) has at most N charges, so each cluster has diameter at most $ND\varepsilon$. Choose $M = ND$ in (4.5). Thus if Q is typical in Δ,

$$Q \restriction \bar{\Delta} = \sum Q_j, \quad Q_j \in \mathscr{Q}_M , \tag{4.10}$$

and (4.8) holds. In case the cluster Q_j lies in two or more of the 3^d blocks of $\bar{\Delta}$, we assign Q_j to the block containing its "lower corner" in some lexicographic order of lattice sites in Z^d. For β sufficiently large ($R_L \gg \varepsilon$) each Q_j associated with Δ satisfies

$$\partial Q_j = 0\,. \tag{4.11}$$

Let $J(\Delta)$ index clusters associated with Δ. For typical blocks Δ let

$$Q_\Delta = \sum_{j \in J(\Delta)} Q_j\,. \tag{4.12}$$

Then Q_Δ is the configuration to which we apply the construction of Chapter 3, giving $Q_{\Delta,L}$, which agrees with Q_Δ up to the octapole remainder. Let

$$\begin{aligned} &Q_t = \sum_\Delta Q_\Delta\,, \qquad Q_{t,L} = \sum_\Delta Q_{\Delta,L}\,, \\ &\partial Q_\Delta = Q_\Delta - Q_{\Delta,L}\,. \end{aligned} \tag{4.13}$$

Lemma 4.3. *For β sufficiently large,*

$$\mathscr{A}_I(Q_\Delta) - \mathscr{A}_I(Q_{\Delta,L}) \geqq \sum_{j \in J_\Delta} \mathscr{A}_I(Q_j)(1 - \delta - \beta^{-1})\,. \tag{4.14}$$

Moreover we may take $\delta = 0$ if J_Δ contains a single element.

Proof. Using Proposition 3.1,

$$\mathscr{A}_I(Q_\Delta) = \sum_{i,j \in J_\Delta} \mathscr{A}_I(Q_i, Q_j) = \sum_{i \in J_\Delta} \mathscr{A}_I(Q_i) + \sum_{\substack{i \neq j \\ i,j \in J_\Delta}} \mathscr{A}_I(Q_i, Q_j)\,.$$

By Proposition 4.2, we bound the intercluster action by

$$\sum_{i \neq j} \delta \beta N^{-2} \leqq \delta \beta\,.$$

Since Q_j belongs to the finite set up to translations, of typical charge clusters,

$$\beta \leqq \text{const} \sum_{j \in J_\Delta} \mathscr{A}_I(Q_j)\,, \tag{4.15}$$

using the constants of (4.4). Thus with a new δ,

$$\mathscr{A}_I(Q_\Delta) \geqq (1 - \delta) \sum_{j \in J_\Delta} \mathscr{A}_I(Q_j)\,. \tag{4.16}$$

Clearly we may take $\delta = 0$ if J_Δ contains a single element.

We bound $\mathscr{A}_I(Q_{\Delta,L})$ using Proposition 3.3 Thus

$$|\mathscr{A}_I(Q_{\Delta,L})| \leqq O(1) \beta^{-1/(d-2)}\,,$$

and by (4.15)

$$\begin{aligned} |\mathscr{A}_I(Q_{\Delta,L})| &\leqq O(1) \beta^{-1/(d-2)} \beta^{-1} \sum_{j \in J_\Delta} \mathscr{A}_I(Q_j) \\ &\leqq \beta^{-1} \sum_{j \in J_\Delta} \mathscr{A}_I(Q_j) \end{aligned} \tag{4.17}$$

for β large. The desired bound follows from (4.16–(4.17).

Lemma 4.4. *There exists $\gamma>0$ such that for β sufficiently large,*

$$\mathscr{A}_I(Q_t)-\mathscr{A}_I(Q_{t,L})\geqq\sum_{\Delta}\sum_{j\in J_\Delta}\mathscr{A}_I(Q_j)(1-\delta_\Delta-\beta^{-\gamma}) \tag{4.18}$$

where $\delta_\Delta=0$ if only one cluster is associated with Δ.

Proof. We expand using Proposition 3.1,

$$\mathscr{A}_I(Q_t)-\mathscr{A}_I(Q_{t,L}) = \sum_{\Delta}(\mathscr{A}_I(Q_\Delta)-\mathscr{A}_I(Q_{\Delta,L}))+\sum_{\Delta\neq\Delta'}(\mathscr{A}_I(Q_\Delta,Q_{\Delta'})-\mathscr{A}_I(Q_{\Delta,L},Q_{\Delta',L})) . \tag{4.19}$$

The first terms in (4.19) are bounded below by Lemma 4.3. The sum over $\Delta\neq\Delta'$ is split into two parts: the sum over adjacent pairs and the sum over nonadjacent pairs. The sum over adjacent pairs is bounded using Proposition 4.2 and the argument in the proof of Lemma 4.3. For nonadjacent pairs we write the terms as

$$\sum_{\Delta,\Delta'}(\mathscr{A}_I(Q_\Delta-Q_{\Delta,L},Q_{\Delta'})+\mathscr{A}_I(Q_{\Delta,L},Q_{\Delta'}-Q_{\Delta',L})) = \sum_{\Delta,\Delta'}\mathscr{A}_I(\delta Q_\Delta,Q_{\Delta'}+Q_{\Delta',L}) . \tag{4.20}$$

The sum (4.20) contributes to the octapole remainder. Due to the subtractions in the definition of δQ_Δ,

$$|\mathscr{A}_I(\delta Q_\Delta,Q_{\Delta'}+Q_{\Delta',L})|\leqq O(1)\beta D^{-d-1} ,$$

where

$$D=\varepsilon^{-1}\operatorname{dist}(\Delta,\Delta')\geqq(R_L/\varepsilon)=\beta^{1/(d-2)} .$$

This bound follows as in the proof of Proposition 4.2, cf. (4.9). The additional powers of D in the denominator arise from the multiple differences which occur in the definition of δQ_Δ.

The summation over Δ and Δ' is achieved by taking $\operatorname{dist}(\Delta,\Delta')=2\boldsymbol{n}R_L$ where $\boldsymbol{n}$ is a vector with integer components, and summing $\boldsymbol{n}$ over Z^d. Thus, using (4.15) and $\sum D^{-d-1}\leqq O(R_L/\varepsilon)^{-d-1}\leqq O(\beta^{-1-\gamma})$,

$$\sum_{\Delta,\Delta'}|\mathscr{A}_I(\delta Q_\Delta,Q_{\Delta'}+Q_{\Delta',L})|\leqq O(\beta^{-\gamma})\sum_{\Delta}\sum_{j\in J_\Delta}\mathscr{A}_I(Q_j) .$$

Lemma 4.5. *There exists $\gamma>0$ such that for $\delta>0$ and β sufficiently large*

$$\mathscr{A}_I(Q)\geqq(c_1-\delta)\beta\sum_{Q_a}j^2+\sum_{\Delta}\sum_{j\in J_\Delta}\mathscr{A}_I(Q_j)(1-\delta_\Delta-\beta^{-\gamma}) . \tag{4.21}$$

Proof. We write

$$\mathscr{A}_I(Q)=\mathscr{A}_I(Q')+\mathscr{A}_I(Q_t)-\mathscr{A}_I(Q_{t,L})+2\mathscr{A}_I(Q_a,\delta Q_t) . \tag{4.22}$$

The action $\mathscr{A}_I(Q')$ is bounded below using Proposition 3.4, with $Q'=Q_a+Q_{t,L}$. Thus

$$\mathscr{A}_I(Q')\geqq c_1\beta\sum_{Q_a}j^2 .$$

The difference $\mathcal{A}_I(Q_t)-\mathcal{A}_I(Q_{t,L})$ is bounded below by Lemma 4.4. The remaining term $\mathcal{A}_I(Q_a,\delta Q_t)$ contributes to the octapole remainder, and it is bounded by

$$|\mathcal{A}_I(Q_a,\delta Q_t)|\leqq O(\delta\beta)\sum_{Q_a} j^2 .$$

Proof of Theorem 4.1 (upper bound on Z). By Lemma 4.5,

$$e^{-\mathcal{A}_I(Q)}\leqq e^{-(c_1-\delta)\beta\sum_{Q_a} j^2}\prod_{\Delta}\prod_{J_\Delta} e^{-\mathcal{A}_I(Q_j)(1-\delta_\Delta-\beta^{-\gamma})} .$$

We substitute $\zeta_j=e^{-\mathcal{A}_I(Q_j)}$. In order to simplify the sum over configurations, and to make the Q_a, Q_t sums factorize, we introduce some spurious terms. We allow overlapping Q_a, Q_t configurations, and we regard two or more elementary clusters in the same or adjacent blocks as separate events with activity

$$(\zeta_{j_1}\dots\zeta_{j_l})^{1-\delta_\Delta}=\zeta_{j_1\dots j_l} . \tag{4.23}$$

We then allow arbitrary multiplicities of these $\zeta_{j_1},\zeta_{j_1j_2}\dots$ events in any block. This allows us to bound the summation over charge configurations, $\sum_{\{Q\}}$, by independent sums $\sum_{\{Q_a\}}\sum_{\{Q_t\}}$ over atypical and typical charge configurations. Thus

$$Z=\sum_{\{Q\}}e^{-\mathcal{A}_I(Q)}\leqq\left(\sum_{\{Q_a\}}e^{-\frac{1}{2}c_1\beta\Sigma Q_a j^2}\right)\left(\sum_{\{Q_t\}}\prod_{\Delta}\prod_{J_\Delta}\zeta_{j_1}\dots\zeta_{j_1\dots j_2}\right)$$

$$\leqq\left(\sum_{\{Q_a\}}e^{-\frac{1}{2}c_1\beta\Sigma Q_a j^2}\right)\left(1+\sum_j\left(\frac{R_L}{\varepsilon}\right)^d\zeta_j+\sum_{j_1j_2}\left(\frac{R_L}{\varepsilon}\right)^{2d}\zeta_{j_1j_2}+\dots\right)^{VR_L^{-d}} .$$

Here we remark there are a finite number of terms for the typical summation in a given block Δ. The factors $(R_L/\varepsilon)^d$, $(R_L/\varepsilon)^{2d}$, etc. arise from possible locations of the clusters in Δ. Since $(R_L/\varepsilon)^d=\beta^{d/d-2}$, and $\zeta_j\leqq\zeta_0$, $\zeta_{j_1j_2}\leqq\zeta_0^{2(1-\delta_\Delta)},\dots,$

$$Z\leqq\left(\sum_{\{Q_a\}}e^{-\frac{1}{2}c_1\beta\Sigma Q_a j^2}\right)\left(1+\left(\frac{R_L}{\varepsilon}\right)^d a_0\zeta_0(1+O(\beta^{-\gamma}))\right)^{VR_L^{-d}}$$

$$\leqq\left(\sum_{\{Q_a\}}e^{-\frac{1}{2}c_1\beta\Sigma Q_a j^2}\right)e^{a_0\zeta_0(1+O(\beta^{-\gamma}))V\varepsilon^{-d}} .$$

Here we have used the fact that δ_Δ does not appear for a single cluster, to identify the leading term in the exponent as $a_0\zeta_0$.

The sum over atypical configurations can be carried out explicitly. Since Q_a has at least $4\cdot 3^d c_0/c_1$ charges on the ε lattice, the activity of each atypical configuration in Δ is at least ζ_0^2. Thus the atypical terms can be included in the error $O(\beta^{-\delta})$.

5. The Lower Bound on Z

In this section we prove the lower bound on $Z=\sum_{\{Q\}}e^{-\mathcal{A}_I}$, which completes the proof of Theorem 4.1. Since each term in the sum over configurations is nonnegative, we achieve a lower bound by omitting certain configurations Q. The remaining configurations, called typical, differ from the typical configurations of Section 4 (and in any case are a set of measure zero in the infinite volume measure).

We introduce an infinite sequence $R_1=\varepsilon$, $R_2=R_L$,

$$\begin{aligned} R_3 &= \zeta_0^{-2\delta} R_L \\ R_4 &= \exp(\zeta_0^{-\delta}) R_L \\ R_n &= 2^n R_4\,, \quad n \geqq 4 \end{aligned} \tag{5.1}$$

of length scales. The first restriction defining a typical configuration Q is that Q must be a sum of translates of the a_0 distinct $Q_j \in \mathcal{Q}_M$ with activity $\zeta_0=\zeta_{Q_j}$, as in (4.5). Furthermore we allow only one such elementary cluster in each block of length R_3; within this block, the cluster must have a distance at least $\zeta_0^{-\delta} R_L = \zeta_0^{\delta} R_3$ from the boundary of the block.

Let

$$\mathscr{A}_I(Q_i. R, R') = \sum \mathscr{A}_I(Q_i, Q_j) \tag{5.2}$$

where the sum runs over all elementary clusters $Q_j \neq Q_i$ in Q at a distance between R and R' from the cluster Q_i. For typical configurations,

$$\mathscr{A}_I(Q_i, 0, \zeta_0^{\delta} R_3) = 0 \tag{5.3}$$

because the minumum intercluster distance is $2\zeta_0^{\delta} R_3$.

Lemma 5.1. *For β sufficiently large and for a typical configuration Q,*

$$|\mathscr{A}_I(Q_i, 0, R_4)| \leqq \tfrac{1}{2}\zeta_0^{\delta}\,. \tag{5.4}$$

Proof. At separation r,

$$|\mathscr{A}_I(Q_i, Q_j)| \leqq O(r/R_L)^{-d} \beta^{-1/(d-2)}\,.$$

At minimum separation $r = 2\zeta_0^{\delta} R_3 = 2\zeta_0^{-\delta} R_L$

$$|\mathscr{A}_I(Q_i, Q_j)| \leqq 2^{-d} \zeta_0^{\delta d} \beta^{-1/(d-2)}\,.$$

For Q_j's separated by one or more intermediate block of size R_3, the discrete sum can be estimated by an integral, giving

$$\begin{aligned} |\mathscr{A}_I(Q_i, 0, R_4)| &\leqq O(\zeta_0^{\delta d}) \beta^{-1/(d-2)} + O(R_3/R_L)^{-d} \beta^{-1/(d-2)} \ln(R_4/R_L) \\ &\leqq \tfrac{1}{2}\zeta_0^{\delta}\,. \end{aligned}$$

Here we have substituted R_3^{-d} as an upper bound on the density of charge clusters.

For distances $R > R_4$, the bound on $\mathscr{A}_I(Q_i, 0, R)$ depends on cancellations between oppositely oriented dipoles. To achieve this cancellation, we impose further restrictions on the typical configurations. The map $q \to -q$ is a symmetry of $\mathscr{A}_I$ which reverses the orientation of the elementary dipole clusters. Thus the number a_0 of these clusters is even, and we can label them as $Q_{\pm j}, j=1, \ldots, a_0/2$, with $Q_{-j} = -Q_j$ the orientation reversed image of Q_j. We divide R^d into d-cells (block spins) of length R_n, for each length scale R_n, $n \geqq 5$. In any block spin of length R_n, let v_j be the number of times the elementary dipole cluster Q_j occurs. Our final restriction defining a typical configuration is that

$$|v_j - v_{-j}| \leqq (R_n/R_3)^{3d/4} \tag{5.5}$$

for each $j = 1, \ldots, a_0/2$, each R_n-block and each $n = 5, 6, \ldots$.

Lemma 5.2. *For β sufficiently large and for typical configurations,*

$$|\mathscr{A}_I(Q_i, 0, \infty)| \leqq \zeta_0^\delta . \tag{5.6}$$

Proof. Write

$$\mathscr{A}_I(Q_i, 0, \infty) = \mathscr{A}_I(Q_i, 0, R_4) + \sum_{n=4}^{\infty} \mathscr{A}_I(Q_i, R_n, R_{n+1}) .$$

We divide the spherical shell contributing to $\mathscr{A}_I(Q_i, R_n, R_{n+1})$ into $O(R_n/R_{4[n/5]})^d$ blocks of length $R_{4[n/5]}$. In each block

$$|v_j - v_{-j}| \leqq (R_{4[n/5]}/R_3)^{3d/4}$$

is the deviation from neutrality. The dipole interaction strength for a dipole with separation R is bounded by $O(\beta^{-1/(d-2)})(R/R_L)^{-d}$. Thus

$$\begin{aligned} |\mathscr{A}_I(Q_i, R_n, R_{n+1})| &\geqq \sum_j O(\beta^{-1/(d-2)})(R_n/R_L)^{-d}|v_j - v_{-j}|(R_n/R_{4[n/5]})^d \\ &\leqq O(\beta^{-1/(d-2)}) R_L^d R_3^{-3d/4} R_{4[n/5]}^{-d/4} \\ &\leqq O(\beta^{-1/(d-2)}) 2^{-[n/5]d} \zeta_0^{3d/2} e^{-\zeta_0^\delta/4} . \end{aligned}$$

Summing over n, and choosing β large,

$$\sum_{n=4}^{\infty} |\mathscr{A}_I(Q_I, R_n, R_{n+1})| \leqq \tfrac{1}{2}\zeta_0^d$$

so that Lemma 5.1 completes the proof.

Proof of Theorem 4.1 (completed). We now let $\sum'$ denote the sum over typical configurations. In a bounded region Λ of volume V, a typical configuration Q is specified by the number $N = N(Q)$ of dipole clusters, the choice of $\pm$ orientation for each cluster, the $(a_0/2)^N$ choices of $|j|$ and the

$$\binom{V/R_3^d}{N} (R_3/\varepsilon)^{dN} (1 - O(\zeta_0^\delta)^N \tag{5.7}$$

possible positions of the clusters. Here the factor $(1 - O(\zeta_0^\delta)^N$ results from the excluded boundary region for the location of each cluster in its block of length R_3. Thus

$$Z = \sum_Q e^{-\mathscr{A}_I} \geqq \sum_Q{}' e^{-\mathscr{A}_I} \leqq \sum_N \sum_{\{Q:N(Q)=N\}} \sum_{\pm} e^{-\zeta_0^\delta N} (a_0/2)^N \zeta_0^N . \tag{5.8}$$

Substitution of (5.7) and $\sum_{\pm} = 2^N$ in (5.8) would give

$$\begin{aligned} Z &\geqq (1 + a_0(R_3/\varepsilon)^d e^{-\zeta_0} \zeta_0 (1 - O(\zeta_0^\delta))^{V/R_3^d} \\ &\geqq [1 + a_0 \varepsilon^{-d} \zeta_0 (1 - O(\beta^{-\delta}))]^V \end{aligned} \tag{5.9}$$

and complete the proof. Thus it remains to analyze the restriction placed on $\sum'$ by the restriction to typical configurations.

For a particular R_n-length block, the number of choices of $\pm$ which violate (5.5) is bounded by

$$e^{-O(|\nu_j-\nu_{-j}|^2/|\nu_j+\nu_{-j}|)}2^{|\nu_j+\nu_{-j}|}\leqq e^{-O(R_n/R_3)^{6d/4}/(R_n/R_3)^d}2^{|\nu_j+\nu_{-j}}$$
$$\leqq e^{-O(2^{n/2}\zeta_0^\delta e^{\frac{1}{2}\zeta_0^{-\delta}})}2^{|\nu_j+\nu_{-j}|}\,.$$

Thus the complementary set occurs for at least

$$[1-e^{-O(^{n/2}\zeta_0^\delta e^{\frac{1}{2}\zeta_0^{-\delta}})}]2^{|\nu_j+\nu_{-j}|}$$
$$\cong\left[1-\left(\frac{R_3}{R_n}\right)^d e^{-O(2^{n/2}\zeta_0^\delta e^{\frac{1}{2}\zeta_0^{-\delta}})}\right]^{(R_n/R_3)^d}2^{|\nu_j+\nu_{-j}|}$$

choices. Multiplying these values, we see that (5.5) is satisfied on all R_n-length blocks at once for

$$\left[1-\left(\frac{R_3}{R_n}\right)^d e^{-O(2^{n/2}\zeta_0^\delta e^{\frac{1}{2}\zeta_0^{-\delta}})}\right]^{V/R_3^d}2^N$$

choices of $\pm$. Multiplying over n shows that

$$\sum_\pm\geqq\left[1-O(1)\sum_n\left(\frac{R_3}{R_n}\right)^d e^{-O(2^{n/2}\zeta_0^\delta e^{\frac{1}{2}\zeta_0^{-\delta}}}\right]^{V/R_3^d}2^N\,.$$

Substitution in (5.8) now yields (5.9) and completes the proof.

Acknowledgement. We wish to thank O. McBryan and T. Spencer for helpful discussions.

References

1. Brydges, D.: A rigorous approach to Debye screening in dilute classical Coulomb systems. Commun. math. Phys., to appear
2. Fabrizio De Angelis, G., de Falco, D., Guerra, F.: Lattice gauge models in the strong coupling regime. Preprint
3. Glimm, J., Jaffe, A.: Quark trapping for $U(1)$ lattice gauge fields. Phys. Letters **66** B, 67—69 (1977)
4. Glimm, J., Jaffe, A. Spencer, T.: Convergent expansion about mean field theory, Parts I and II. Ann. Phys. **101**, 610—630; **101**, 631—669 (1976)
5. Griffiths, R. B.: Free energy of interacting magnetic dipoles. Phys. Rev. **176**, 655—659 (1968)
6. Kogut, J., Susskind, L.: Hamiltonian formulation of Wilson's lattice gauge theories. Phys. Rev. D **11**, 395—408 (1975)
7. Mayer, M.: Gauge fields as quantized connection forms. In: Conference on differential geometric methods in mathematical physics (eds. K. Bleuler, A. Reetz). Lecture notes in mathematics, Vol. 570. Berlin-Heidelberg-New York: Springer 1977
8. Osterwalder, K., Seiler, E.: Gauge field theories on the lattice. Ann. Phys. to appear
9. Polyakov, A. M.: Quark confinement and topology of gauge theories. Nucl. Phys. B **120**, 429—458 (1977)
10. Singer, I., Thorpe, J. A.: Lecture notes on elementary topology and geometry. Berlin-Heidelberg-New York: Springer 1976
11. Wilson, K.: Confinement of quarks. Phys. Rev. D **10**, 2245—2259 (1974)
12. Høye, J. S., Stell, G.: Ionic solution theory for non-ideal solvents—potential of mean force between ions. Preprint

Communicated by A. Jaffe

Received May 16, 1977

CHARGES, VORTICES AND CONFINEMENT *

James GLIMM
The Rockefeller University, New York, NY 10021, USA

Arthur JAFFE **
Harvard University, Cambridge, MA 02138, USA

Received 28 August 1978
(Revised 20 November 1978)

We study the confinement phase transition. We point out the relation between the decay rates of the Wilson loop integral, Osterwalder and Seiler's n-ality restriction, and Polyakov and 't Hooft's analysis of the center of the gauge group. Vortices and merons are analyzed as possible mechanisms to produce disorder in the center. We show how to realize the 't Hooft commutation relations, and we estimate the critical coupling for vortices to be in approximate agreement with the critical coupling determined by Callan, Dashen and Gross from the analysis of instantons.

1. Introduction

In SU(n) gauge theories, the area decay of the Wilson functional

$$W_\sigma(\mathrm{C}) = \langle \mathrm{Tr}_\sigma \mathrm{P e}^{\oint_\mathrm{C} A_\mu \mathrm{d}x^\mu} \rangle \sim \mathrm{e}^{-\mathrm{O}\,|\mathrm{area\ C}|} \tag{1.1}$$

is taken as an indication of quark confinement. Although the sign of the β-function in the renormalization group may be the underlying cause of the disorder of the gauge field which leads to area decay, there have been efforts to describe the dominant modes of disorder directly, e.g., in terms of point charges (merons) [1,2] and vortices (disorder operators) [3]. We examine these two approaches.

2. Particle and quark representations

To distinguish between confined and observable particles, we call the former quarks and the latter particles. The distinction between these possibilities depends

* Supported in part by the National Science Foundation under Grants PHY-78-08066 and PHY-77-18762.
** John S. Guggenheim Fellow.

both on the representation σ of the gauge group G in (1.1) and on the vacuum functional. We use the Wilson functional as the criterion for confinement.

In various examples, one finds that fractional charges are confined, integer charges are not. We make this statement precise by proposing that there is a subgroup $\mathcal{Z}_0$ of the center $\mathcal{Z}$ of G with the property that

$$W_\sigma \sim e^{-O(\text{area})}, \qquad \text{if } \sigma(\mathcal{Z}_0) \neq I\,, \tag{2.1a}$$

$$W_\sigma \sim e^{-O(\text{length})}, \qquad \text{if } \sigma(\mathcal{Z}_0) = I\,. \tag{2.1b}$$

With charges labeled by the dual group $\hat{\mathcal{Z}}$, $\mathcal{Z}_0^\perp$ is the subgroup of $\hat{\mathcal{Z}}$ which occurs in (2.1b), and labels the "integer" charges, while $\hat{\mathcal{Z}} \setminus \mathcal{Z}_0^\perp)$ gives the representation in (2.1a) and labels fractional charges. The subgroup $\mathcal{Z}_0$ depends on the vacuum functional, and thus labels phase transitions in the transverse modes of the gauge theory.

According to perturbation theory, in the asymptotically free phase, all particles are observable and (2.1) holds with $\mathcal{Z}_0 = I$, $\mathcal{Z}_0^\perp = \hat{\mathcal{Z}}$. In this case all charges are integral. At high temperature, on a lattice, we extend [4] to show that (2.1) holds with $\mathcal{Z}_0 = \mathcal{Z}$, $\mathcal{Z}_0^\perp = 0 \in \hat{\mathcal{Z}}$, at least for G = SU(2), or SU(3). In this case all non-zero charges labeled by $\hat{\mathcal{Z}}$ are fractional. To establish (2.1) in general, we require a further hypothesis: that bound states formed from observable particles are also observable.

For pure gauge theories, in $d = 2$, a transfer matrix argument shows all representations $\sigma \neq$ identity are quark representations (have an area decay). Thus $d = 2$, for pure gauge theories does not follow (2.1), but realizes the exceptional case Σ = identity of theorem 2.1 below.

We begin with an abstract characterization of particle states. On physical grounds, we require that particle states yield antiparticle states and also multiparticle states. Expressed in mathematical language, we postulate that if σ and τ are particle representations, then so are the irreducible representations contained in $\sigma \otimes \tau$. Because $W_{\bar{\sigma}} = W_{\sigma^*}$, σ^* is a particle representation if and only if σ is. This class of representations is determined by the following elementary theorem.

Theorem 2.1. Let G be a connected, compact simple Lie group. Let Σ denote a class of irreducible representations closed under complex conjugation and decomposition of tensor products. Then either Σ equals the identity, or Σ is the set of all representations of G which annihilate some subgroup $\mathcal{Z}_0$ of the center $\mathcal{Z}$ of G, i.e., $\Sigma(\mathcal{Z}_0) = I$.

Proof. cf. Chevalley [6]. Let K = K(Σ) denote the kernel of the class Σ. The representative functions separate points on G/K, so by the Stone-Weierstrass theorem the representative functions contain all functions on G/K. Thus Σ is the class of all representations on G/K. For a connected simple Lie group G, and a normal subgroup K, either K = G or else K is discrete. In the latter case K is pointwise invariant under inner automorphisms, and hence contained in $\mathcal{Z}$.

As an illustration, let G = U(1), so that the dual group $\hat{G}$ is the integers. Then Σ above is closed under products and inverses, and thus is a subgroup of $\hat{G}$. This agrees with the known [17,18] analysis of the Abelian Higgs model in two dimensions in which confined (quark) representations are labeled by elements of $\hat{G}/\Sigma$. Apparently the methods of [18] would also show that the elements of Σ label particle representations.

We now introduce an action $\mathcal{L}$. For mathematical definiteness we consider a lattice gauge theory, though our results can also be formulated for the continuum limit. We study the plaquette action

$$\mathcal{L}(P)_\tau = \text{Re Tr}_\tau\, U(P)_\tau\,, \qquad U(P)_\tau = \prod_b \tau(U(b))\,, \tag{2.2}$$

where $U(b) \in G$ is the gauge group variable for a bond b, τ is a representation of G and the product $\prod$ ranges over the ordered bonds $b \in \partial P$. The total action is summed over plaquettes, and may also be summed over certain representations,

$$\mathcal{L} = \sum_\tau a_\tau \sum_P \mathcal{L}(P)_\tau\,. \tag{2.3}$$

The action defines an expectation $\langle \cdot \rangle$ and the Wilson functional

$$W_\sigma(C) = \langle \text{Tr}_\sigma(\text{P} \exp \int_C A)\rangle\,, \tag{2.4}$$

namely the expectation of the trace of the parallel transport about the closed curve C. Here $\text{P}\exp(\int_C A) = U(C)_\sigma$.

We now investigate this mathematical structure in more detail. Since $e^{\mathcal{L}}$ is a function on G, the Peter-Weyl theorem tells us it has a Fourier expansion in terms of representations of G. Let $\Sigma_{\mathcal{L}}$ be the class of representations thus obtained. Since $\Sigma_{\mathcal{L}}$ satisfies the hypotheses of theorem 2.1, $K(\Sigma_{\mathcal{L}})$ equals either G or a subgroup of $\mathcal{Z}$. On physical grounds we eliminate K = G as being trivial (i.e., $\mathcal{L}$ = const.). Furthermore, we also eliminate $K = \mathcal{Z}$ since in this case W_σ would vanish identically for all quark representations, as we show in theorem 2.2 below. In case G = SU(2) or SU(3), $\mathcal{Z}$ has no proper subgroups and therefore we require $K(\Sigma_{\mathcal{L}}) = I$ in that case. In the case of G = SU(n), the center is $\mathcal{Z} = \{z^j, j = 0, 1, ..., n-1\}$, where $z = \exp(2\pi i/n)I$. For the representation σ, the "n-ality" n_σ is defined by the equation:

$$\sigma(z) = \exp(2\pi\, i n_\sigma/n) I\,. \tag{2.5}$$

Abstractly, $n_\sigma : \hat{G} \to \hat{\mathcal{Z}}$ is defined by $n_\sigma = \sigma \,|\, z \in \hat{\mathcal{Z}}$. Hence the requirement $n_\sigma \neq 0$ is equivalent to the requirement $\sigma(\mathcal{Z}) \not\equiv I$.

Theorem 2.2. Let $K(\Sigma_{\mathcal{L}}) = \mathcal{Z}_0 \subset \mathcal{Z}$, $\sigma(\mathcal{Z}_0) \neq I$. Then $W_\sigma = 0$.

Proof.

$$W_\sigma(C) = \langle \text{Tr}_\sigma\, U(C)_\sigma\rangle = \int_{G/\mathcal{Z}_0} e^{\mathcal{L}}\left(\int \text{Tr}_\sigma\, U(C)_\sigma\right) d\mu/\text{normalization},$$

where $d\mu$ is the product of Haar measures $dU(b)$ on bonds b. Fixing all variables but a single bond b, we integrate a single factor $\sigma(U(\mathrm{b}))$. By the orthogonality of group representations, this integral vanishes for $\sigma \not\equiv \mathrm{I}$.

The high temperature result of Osterwalder and Seiler is:

Theorem 2.3. For any choice of representation τ in the Lagrangian of an SU(n) lattice gauge theory, for sufficiently high temperature ($a_\tau << 1$), and for $n_\sigma \neq 0$, it follows that W_σ has the area decay (2.5). In other words, if $\sigma(\mathcal{Z}) \not\equiv \mathrm{I}$, then at high temperature σ is a quark representation. Theorem 2.3 establishes the quark criterion (2.1) at high temperature.

The next three theorems establish the particle criterion (2.1) at high temperatures, at least for G = SU(2) and SU(3). Consider the set P of irreducible representations σ such that

$$\mathrm{P} = \{\sigma : \sigma \in \Sigma\tau \otimes \tau^*, \text{ where } \tau(\mathcal{Z}) = \mathrm{I}\} . \tag{2.6}$$

At high temperatures, for G = SU(2), SU(3) we show that P is exactly the set of particle representations. This may also hold for SU(n).

Theorem 2.4. For any compact group G, the $\sigma \in \mathrm{P}$ are representations of $\mathrm{G}/\mathcal{Z}$, and $\oplus_{\sigma\in\mathrm{P}}$ is a faithful representation of $\mathrm{G}/\mathcal{Z}$.

Proof. Since $\sigma(\mathcal{Z}) = \mathrm{I}$, σ is a representation of $\mathrm{G}/\mathcal{Z}$. To prove the second statement, let $\gamma \in \mathrm{G}, \gamma \notin \mathcal{Z}$. Then there exists $\chi \in \mathrm{G}$ with $\gamma^{-1}\chi\gamma \neq \chi$ and a function f with $f(\gamma^{-1}\chi\gamma) \neq f(\chi)$. But P results from the decomposition of $U_\gamma: f \to f(\gamma^{-1}\chi\gamma)$ on $L_2(\mathrm{G}/\mathcal{Z})$.

Theorem 2.5. For G = SU(2) or SU(3), P is the set of all representations σ for which $\sigma(\mathcal{Z}) = \mathrm{I}$. In other words, P is the set of n-ality zero representations.

Proof. This follows from a calculation with Clebsch-Gordon coefficients [8,9,15].

Theorem 2.6. Let dimension $d \geqslant 3$, G = SU(n), $\sigma \in \mathrm{P}$. Then W_σ has length decay at high temperatures. In other words, at high temperature, all the representations P are particle representations.

Proof. Expand $\exp(\mathcal{L}(\mathrm{P}))$ to obtain $\tau(U(\mathrm{P}))$ for selected plaquettes P. For an oriented bond variable b belonging to exactly two of the selected plaquettes (say $\mathrm{b} \in \mathrm{P}_1, -\mathrm{b} \in \mathrm{P}_2$) it follows from the orthogonality of group representations that

$$\int \mathrm{d}U(\mathrm{b}) \, \mathrm{Tr}_{\tau_1} U(\mathrm{P}_1) \, \mathrm{Tr}_{\tau_2} U(\mathrm{P}_2) = \delta_{\tau_1,\tau_2} \mathrm{Tr}_{\tau_1} U(\mathrm{P}_1 \cup \mathrm{P}_2) . \tag{2.7}$$

We choose plaquettes forming a tube which contains the curve C. Then all integrations except those in C can be performed by the convolution formula (2.7). The result is

$$\int \mathrm{d}U(\mathrm{b}) \, \mathrm{Tr}_\sigma U(\mathrm{C}) \prod_{\mathrm{b}\in\mathrm{C}} \mathrm{Tr}_\tau (U(\mathrm{b}) \, \mathrm{Tr}_\tau (U(\mathrm{b})^*) . \tag{2.8}$$

For $\sigma \in \mathrm{P}$, it follows that $\sigma^* \in \mathrm{P}$. For each of these remaining integrals, the orthogonality of group representations selects from (2.8) the σ^* component in $\tau \otimes \tau^*$. This then yields a contribution $a_\tau^{\mathrm{O(length\,C)}}$ to $W_\sigma(\mathrm{C})$. For a_τ sufficiently small (high temperature) the cluster expansion [4,7] shows that this term dominates the expectation $\langle \cdot \rangle$.

3. Vortices

In this and remaining sections of the paper we consider mechanisms which produce disorder and confinement. Here we are concerned with singular gauge transformations which produce vortex configurations in the gauge field. Since vortex configurations reflect geometrical properties of the gauge group G and space-time R^d, we consider a continuum field A.

The singular gauge transformations considered here are multivalued, and not defined on a manifold $S \subset R^d$. On $R^d \setminus S$ we consider a smooth function $x \to U(x)$ with values in G. Along each path in $R^d \setminus S$ there will be a unique continuation of a single valued branch of $U(x)$. Thus $\Pi_1(R^d \setminus S)$ is the obstruction to a single valued choice of U. Since $\Pi_1(R^d \setminus S)$ is trivial if S has codimension three or more, we suppose that codim S = 2.

With U given, the transformed gauge potential is

$$A' = U^{-1}AU + U^{-1}dU\,. \tag{3.1}$$

We require that A is single values. This is the case for the group G = SU(n) when the ambiguity of $U(x)$ lies in the center $\mathcal{Z}$ of G. In other words, suppose that $x \to U(x)/\mathcal{Z}$ is a single valued gauge transformation defined on $R^d \setminus S$, with gauge group $G/\mathcal{Z}$. In that case, two different values $U_1(x)$, $U_2(x)$ of $U(x)$ give in (3.1),

$$A_1' - A_2' = U^{-1}z^{-1}dzU \tag{3.2}$$

for an element $z \in \mathcal{Z}$. Since $\mathcal{Z}$ is discrete for SU(n), $dz = 0$ and A' is single valued. A is also single valued for $U(1)$.

On $R^d \setminus S$, the gauge transformation (3.1) gives $F' = U^{-1}FU$. On S, F' acquires a Dirac delta contribution. In particular, if μ, ν are orthogonal directions, normal to S, then

$$F'_{\mu,\nu} = U^{-1}F_{\mu\nu}U + 2\pi\alpha\delta_{\mu\nu}\,. \tag{3.3}$$

Here α is a function on S with values in the Lie algebra of G and $\delta_{\mu\nu}$ is the δ-function in the two directions normal to S. This formula follows by considering the line integral of A on a curve C, and using the identity for path ordered integrals.

$$\begin{aligned} R(C) &= \mathrm{Tr}\, \mathrm{P}e^{\oint_C A_\mu dx^\mu} \\ &= \mathrm{Tr}\, e^{\int_{\mathrm{Int}\, C} F_{\mu\nu} d\sigma^{\mu\nu}}\,, \end{aligned} \tag{3.4}$$

where C is an infinitesimal circuit around S (not homotopic to zero). Note that $R(C)$ is just the character of the parallel transport of a fibre around C. Similarly we have (using the assumption that the ambiguity of U lies in $\mathcal{Z}$)

$$R'(C) = \mathrm{Tr}\, \mathrm{P}e^{\oint_C A'}$$

$$= \mathrm{Tr}\, \mathrm{Pe}^{\oint_C U^{-1}(A + dUU^{-1})U}$$

$$= \mathrm{Tr}\, \mathrm{Pe}^{\oint_C (A + dUU^{-1})}$$

$$= \mathrm{Tr}(\mathrm{Pe}^{\oint_C A}\, \mathrm{e}^{\oint_C dUU^{-1}})$$

$$= \mathrm{Tr}(\mathrm{Pe}^{\oint_C A}\, \mathrm{e}^{\oint_C d \ln U}) = \mathrm{Tr}(z(\mathrm{C})\, \mathrm{Pe}^{\oint_C A}) \tag{3.5}$$

If G is simple,

$$R'(\mathrm{C}) = z(\mathrm{C})\, R(\mathrm{C})\,. \tag{3.6}$$

Here $z(\mathrm{C}) \in \mathcal{Z}$ is the element of the center given by the homotopy class of C in $\Pi_1(\mathrm{R}^d \setminus \mathrm{S})$. Rewriting (3.5) in infinitesimal form, and using (3.4) yields (3.3).

If Z is discrete, α in (3.3) is normalized in the sense that $z = \mathrm{e}^{2\pi\alpha} \in \mathcal{Z}$ is constant on connected components of S. The singular gauge transformation U introduces vorticity into A. The gauge transformation satisfies the 't Hooft commutation relations [3] with the Wilson function $R(\mathrm{C})$. Then (3.6) is the commutation rule

$$R'(\mathrm{C}) = z^j R(\mathrm{C}) \tag{3.7}$$

where j is the winding number of the curve C, and $z = \exp(2\pi i/n)$ for G = SU(n).

Because of the δ function in F', the action corresponding to F' is infinite (quadratically divergent). Thus an ultraviolet cutoff is required. We introduce the cutoff in the form of a lattice, and in the lattice theory we define the effective coupling constant g_{eff} by the lattice length and the renormalization group. Then g_{eff} contains all effects of fluctuations on scales shorter than the lattice spacing, and these fluctuations have been integrated out of the theory.

We introduce S as a surface in the dual lattice of codimension 2. Suppose S is the boundary of an oriented hypersurface Σ,

$$\mathrm{S} = \partial\Sigma\,.$$

This assumption leads to $\partial\mathrm{S} = 0$ and the Bianchi identity for F', and will be justified statistically at high temperature in sect. 7. Given $z \in \mathcal{Z}$, we define a multivalued gauge transformation $U(x)$ on sites of the original lattice, to be constant on each side of Σ, and to change by a factor of z on crossing Σ. The resulting $\gamma(\mathrm{b})$ for a bond b transforms as

$$\gamma(\mathrm{b})' = \begin{cases} \gamma(\mathrm{b})z, & \text{if b crosses } \Sigma \text{ in a + direction,} \\ \gamma(\mathrm{b}), & \text{otherwise,} \end{cases}$$

while the plaquette $U(\mathrm{P})$'s have the transformation law

$$U(\mathrm{P})' = \begin{cases} U(\mathrm{P})z, & \text{if P winds S in a + direction,} \\ U(\mathrm{P}), & \text{otherwise.} \end{cases}$$

It follows that the action is modified by an amount proportional to the $d-2$ dimensional area $|S|$ of S. These singular gauge transformations form an Abelian group; they are Euclidean regularizations of 't Hooft's disorder operators. See also [14].

4. The vortex phase transition

The action of the singular gauge transformation is proportional to $g_{\text{eff}}^{-2}|S|$, while its entropy is proportional to $|S|$. For large g_{eff}, the entropy dominates the action and large sized vortex sheets are favored.

One can see that this leads to an area decay in $W_\sigma(\mathrm{C}) = \langle R_\sigma(\mathrm{C})\rangle$ since the number of independent degrees of freedom in the vortex fluctuation is proportional to the area of C, when we are restricted to vortex sheets S passing through the surface spanned by C. With S unbounded, all such vortex sheets (not only those near C) may pass through this surface once, or a non-zero number (mod $n-1$) of times. Then [3,14]

$$R(\mathrm{C}) = z^{\Sigma\,(\mathrm{S,\,C\ winding\ numbers})}\,,$$

and

$$W_\sigma(\mathrm{C}) = \langle R_\sigma(\mathrm{C})\rangle = \mathrm{e}^{-\mathrm{O(no.\ S\ degrees\ of\ freedom)}}$$

$$= \mathrm{e}^{-\mathrm{O(area)}}\,.$$

We now argue that the effects of interaction is to lower the critical value of g_{eff}, at least for G = SU(2). For SU(2), singular sets cancel on overlap. Thus the composition of singular gauge transformations with singular sets S_i, $i = 1, 2$ is equivalent to a single singular gauge transformation with singular set

$$\mathrm{S} = \mathrm{S}_1 \Delta \mathrm{S}_2 = \mathrm{S}_1 \cup \mathrm{S}_2 \setminus (\mathrm{S}_1 \cap \mathrm{S}_2)\,.$$

Thus overlap reduces the action, favoring more singular sets S_i at any fixed value of g_{eff}. Overlap does not change the (mod 2) winding number

$$\text{winding number } (\mathrm{C}, \mathrm{S}_1 \Delta \mathrm{S}_2) = \sum_{i=1}^{2} \text{winding number } (\mathrm{C}, \mathrm{S}_i), \qquad (\mathrm{mod}\ 2).$$

One can similarly show that the interaction of S with small, but arbitrary fluction $U(\mathrm{P})$ is attractive. In fact, if the plaquette P winds S, then

$$\begin{aligned} \mathcal{L}'(\mathrm{P}) - \mathcal{L}(\mathrm{P})|_{U=U(\mathrm{P})} &\equiv -\mathrm{Tr}(zU(\mathrm{P})) + \mathrm{Tr}\, U(\mathrm{P}) \\ &= 2\, \mathrm{Tr}\, U(\mathrm{P}) \\ &\leqslant \mathcal{L}'(\mathrm{P}) - \mathcal{L}(\mathrm{P})|_{U=1} \\ &\equiv -\mathrm{Tr}(z) + \mathrm{Tr}\, I = 2\, \mathrm{Tr}\, I\,. \end{aligned}$$

The right-hand side is the incremental action of a singular gauge at P, in the presence of a small fluctuation $U(\mathrm{P})$, and the left-hand side is the incremental action of a singular gauge at P, in the presence of the vacuum at P.

For SU(3), the interaction of singular sets S_i again lowers the critical coupling, and presumably the effect of the interaction of S with small fluctuations also lowers the critical coupling.

5. Merons with O(2) symmetry

The influence of merons on confinement has been analyzed previously, using classical meron configurations with O(4) and O(3) symmetry, the suppressed degrees of freedom being reintroduced statistically. Here we use an O(2) symmetric ansatz, *cf.* Taubes [10], which, because of its lower symmetry, allows for arbitrary meron pair locations in a plane to be treated classically. We define the plane by

$$\rho \equiv (x^2 + y^2)^{1/2} = 0 \,.$$

In a suitable gauge (for which the volume and surface charge densities vanish), the merons satisfy boundary conditions

$$F_{z,t}(\rho = 0) = \pi i \sum_j \pm\delta(z - z_j)\delta(t - t_j) \,.$$

For a loop C lying in the place $\rho = 0$, the integrand $A_\mu \, \mathrm{d}x^\mu$ lies in an Abelian subalgebra, and so we can evaluate

$$\oint_{\mathrm{C}} A_\mu \, \mathrm{d}x^\mu = \int_{\mathrm{Int\,C}} F_{zt} \, \mathrm{d}z \, \mathrm{d}t$$

$$= i\pi(n_+ - n_-) \,,$$

where n_+ (n_-) is the number of merons (antimerons) within C. Thus

$$R(\mathrm{C}) \equiv \mathrm{e}^{\oint A_\mu \, \mathrm{d}x^\mu} = (-1)^{n_+ - n_-} \,.$$

This formula leads to

$$\langle R(\mathrm{C}) \rangle = \mathrm{e}^{-\mathrm{O}(N)} \,,$$

where N is the number of effective meron degrees of freedom (the number of independent choices of -1) in C.

At low densities, where the separation between meron pairs is large relative to the separation within a given pair, the pairing of the merons at given locations is essentially unique. Pairs far away from the boundary, C, do not contribute to N, because there is no plausible rearrangement of pairing. Stated differently, a single meron could not be introduced into the interior, because there is no flexibility to allow a sequence of repairing, which would bring in a partner from outside C.

At high densities, the situation is different. When the distance between distinct pairs is similar to the distance within pairs, the choice of pairings is not uniquely determined by the meron locations. In this sense the meron pairs at high density resemble a fluid. A single meron can consistently be introduced into or removed from the interior of C, because the required sequence of repairings of other merons is allowed. Thus in this case all merons within C contribute to N.

Our dense fluid picture results from the idea that meron pairs do not dissociate, either classically or quantum mechanically [2]. Our picture is distinct from the picture of Callan, Dashen and Gross, in which the merons dissociate into a plasma [1], but was inspired by their picture.

Various evidence supports our picture of non-dissociation. A classical analysis by Gidas [19] of infinitesimal deformations about the two-meron solution indicates that there is no relative group freedom, i.e., an 8-parameter family of gauge-inequivalent solutions. This is consistent with the pairings of classical merons on a line [20]. The classical instability of regularized merons by half instantons was known to Callan, Dashen and Gross, *cf.* also [19,21]. Furthermore, it was shown in [22] that dissociated merons are unstable quantum mechanically.

In summary we have $N = \mathrm{O(length)}$ at low density (small g_{eff}) and $N = \mathrm{O(area)}$ at high density (large g_{eff}). Thus merons are capable of causing the confinement transition.

The SU(3) merons which have been analyzed to date [16] seem to be totally unrelated to triality and to the center $\mathcal{Z}$ of SU(3) (*cf.* sect. 2).

6. Comparison of critical couplings

With an action

$$A = \tfrac{1}{8} g_{\text{eff}}^{-2}\, \mathrm{Tr} \int F_{\mu\nu}^2(x)\, \mathrm{d}x$$

$$= \tfrac{1}{4} g_{\text{eff}}^{-2} \int F_{\mu\nu}^a(x)\, F_a^{\mu\nu}(x)\, \mathrm{d}x\,, \tag{6.1}$$

Callen, Dashen and Gross [1] have estimated the critical value of g_{eff}^2 at

$$g_{\text{c, meron}}^2 = 5.8\,. \tag{6.2}$$

Here g_{eff}^2 is defined in terms of a renormalization group transformation, and is a function of the length scale.

Because interactions lower $g_{\text{c, vortex}}^2$, we get an upper bound on the critical temperature from the Z_2 lattice gauge theory. The latter theory is self-dual, so its critical temperature is known in closed form [11]. With the action

$$A = \tfrac{1}{4}g^{-2}\,\mathrm{Tr}\sum_{p} U_p\,,$$

we have

$$g^2_{\mathrm{c,\ vortex,\ lattice}} \leqslant g^2_{\mathrm{c,\ Z_2,\ lattice}} = 1.14\,. \tag{6.3}$$

To find the equivalent critical coupling $g^2_{\mathrm{c,\ vortex,\ YM}}$ for the Yang-Mills interaction (6.1), we follow the method of Foerster [14] (omitting the factor of 2 for a Higgs field). However in place of $g^2_{\mathrm{c,\ Z_2}}$ from (6.3), we begin this calculation with $g^2_{\mathrm{c,\ vortex,\ lattice}} = 0.85$ (see below). This gives

$$g^2_{\mathrm{c,\ vortex,\ YM}} = \pi^2 g^2_{\mathrm{c,\ vortex,\ lattice}} = 8.4\,, \tag{6.4}$$

in order of magnitude agreement with the meron critical value (6.2). We now discuss the correction to (6.3) due to the effect of the SU(2) fluctuations on the Z_2 critical point.

It is convenient to expand $e^{\mathcal{L}}$ in terms of SU(2) group characters. Let Tr_j be the trace of the spin-j representation of SU(2). Then

$$e^{g^{-2}\mathrm{Tr}_{1/2}} = \sum_{j\ \text{half integer}} \alpha_j(g)\,\mathrm{Tr}_j\,. \tag{6.5}$$

Theorem 6.1. In the expansion (6.5),

$$\alpha_j(g) = (2j+1)\,g^2 I_{2j+1}(2g^{-2})\,, \tag{6.6}$$

where $I_\nu(z)$ is the modified Bessel function of the first kind,

$$I_\nu(z) = \sum_{k=0}^{\infty} \frac{(\frac{1}{2}z)^{\nu+2k}}{k!(\nu+k)!}\,.$$

Proof. The effect of multiplication, $\otimes^n\delta \to \otimes^{n+1}\delta$, where δ is the spin-$\frac{1}{2}$ representation of SU(2), generates a random walk on the Clebsch-Gordon coefficients, with an absorbing barrier at $j = -\frac{1}{2}$. This random walk can be solved explicitly in terms of binomial coefficients, see [12], and yields

$$\alpha_j(g) = \sum_{k=0}^{\infty} g^{-4(k+j)}\,\frac{2j+1}{k!(k+2j+1)!}\,,$$

to complete the proof.

We define the single-site expectation

$$\langle \mathrm{Tr}_j\, U\rangle \equiv \int \mathrm{Tr}_j\, U e^{\mathcal{L}}\, \mathrm{d}U/\text{normalization} = \alpha_j/\alpha_0\,,$$

and so

$$\lim_{g\to 0} \alpha_j/\alpha_0 = 2j + 1 \ , \tag{6.7}$$

$$\lim_{g\to\infty} \alpha_j/\alpha_0 = \delta_{j,0} \ . \tag{6.8}$$

We use the first term in the series, to determine the factor

$$|1 - \tfrac{1}{4}(2 - \langle \mathrm{Tr}_{1/2}\, U\rangle)| = \tfrac{1}{2} + \tfrac{1}{4}(\alpha_{1/2}/a_0) \ ,$$

which we regard as a measure of the change in length of the Z_2 spins due to interaction with the SU(2) fluctuations. This change in spin length can be compensated for by a change in coupling constant renormalization, so that

$$g^2_{\mathrm{c,\,vortex}} \simeq (\tfrac{1}{2} + \alpha_{1/2}/4\alpha_0)\ 1.14 \ ,$$

or

$$g^2_{\mathrm{c,\,vortex}} \simeq 0.85 \ .$$

7. The high-temperature limit

At high temperature, we attempt a coherent description of the large-scale fluctuations in the gauge field. Vortices play a role, as do magnetic (dipole) charge loops and topological point charges (i.e., merons and instantons). These fluctuations occur on singular sets of dimension 2, 1, 0, respectively.

For a general configuration γ of an SU(2) lattice gauge field, define the set of plaquettes $X = X_\gamma$ as those with $\mathrm{Tr}\ U(\mathrm{P}) < 0$. We wish to approximate γ by a vortex configuration γ_S defined by the singular set S, so that

$$U_\mathrm{S}(\mathrm{P}) = \mathrm{sgn}\ \mathrm{Tr}\ U(\mathrm{P}) \ , \tag{7.1}$$

for most plaquettes P and most configurations γ. The problem is that the dual surface X^* is not in general closed, and not in general the boundary of a hypersurface Σ. However, without loss of generality, we may add or subtract from X^* arbitrary closed 2-surfaces S, since these are boundaries of hypersurfaces Σ. Thus, without loss of generality, we may suppose that X^* is the minimal 2-surface with boundary ∂X^*. Stated differently, minimal surfaces with given boundaries ∂X^* span the degrees of freedom for sgn Tr U(P) which are not already included in the vortex degrees of freedom. Because $U(\mathrm{P}) = -1$ is unstable relative to $U(\mathrm{P}) = +1$, the action of this degree of freedom is proportional to the area $|X^*|$. However, the entropy of this degree of freedom is given by a length $|\partial X^*|$. Thus entropy never dominates over action for large connected portions of X^*. Hence the minimal surfaces X^* remain small (or at least narrow) deviations in sgn Tr U(P), and the dominant large scale fluctuations in sgn Tr U(P) are vortices.

We remark that ∂X^* are magnetic charge loops, and were analyzed at low density in the Abelian case [13]. Because of the cancellation from opposite sides of the loops, the magnetic charge loops are presumably dipoles, hence have forces which are too short range to influence confinement.

References

[1] C. Callan, R. Dashen and D. Gross, Phys. Rev. D17 (1978) 2717.
[2] J. Glimm and A. Jaffe, Phys. Rev. D18 (1978) 463.
[3] G. 't Hooft, Nucl. Phys. B138 (1978) 1.
[4] K. Osterwalder and E. Seiler, Ann. of Phys. 110 (1978) 440.
[5] A.M. Polyakov, Thermal properties of gauge fields and quark liberation, preprint.
[6] C. Chevalley, The theory of Lie groups (Princeton Press, 1946).
[7] J. Glimm, A. Jaffe and T. Spencer, *in* Constructive field theory, Springer Lecture Notes in Physics, vol. 25 (1974).
[8] S. Coleman, J. Math. Phys. 5 (1964) 1343; 6 (1965) 2014.
[9] W. Miller, Symmetry groups and their applications (Academic press, New York, 1972).
[10] C. Taubes, O(2) symmetric solutions to an SU(2) Yang-Mills theory.
[11] R. Balian, J. Drouffe and C. Itzykson, Phys. Rev. D11 (1975) 2098.
[12] W. Feller, Introduction to probability (Wiley, New York, 1957).
[13] J. Glimm and A. Jaffe, Comm. Math. Phys. 56 (1977) 195.
[14] D. Foerster, On quark confinement in a condensate of non-Abelian vortices; An estimate of the coupling needed for vortices to condense in Yang-Mills theory, Cornell preprints.
[15] G. de B. Robinson, Representation of symmetric group (Univ. of Toronto Press, 1961).
[16] J.Z. Imbrie, O(3) symmetric merons in an SU(3) Yang-Mills theory, Lett. Math. Phys. 2 (1978) 483.
[17] C. Callan, R. Dashen and D. Gross, Phys. Lett. 66B (1977) 345.
[18] R. Israel and C. Nappi, Comm. Math. Phys., to appear.
[19] B. Gidas, Deformations and spectral properties of merons, J. Math. Phys., to appear.
[20] J. Glimm and A. Jaffe, Phys. Lett. 73B (1978) 167.
[21] D. Williams, Mathematical remarks on positivity and self-adjointness of the Euclidean, Yang-Mills stability operator, preprint.
[22] J. Glimm and A. Jaffe, Phys. Rev. Lett. 40 (1978) 277.
[23] M. Lüscher, Absence of spontaneous gauge symmetry breaking in Hamiltonian lattice gauge theories, DESY 77/16.

VIII Reflection Positivity

Introduction

Reflection positivity of a Gaussian measure is shown to be equivalent to monotonicity of the Green's function for the Laplace operator with respect to classical Dirichlet or Neumann boundary conditions.

Bibliography

A note on reflection positivity, Lett. Math. Phys. 3, 377–378 (1979).

A NOTE ON REFLECTION POSITIVITY

JAMES GLIMM[1]
The Rockefeller University, New York, NY 10021, U.S.A.

and

ARTHUR JAFFE[2]
Harvard University, Cambridge, MA 02138, U.S.A.

ABSTRACT. We establish reflection positivity of covariance operators, using properties of Dirichlet or Neumann boundary data.

Osterwalder and Schrader isolated a basic positivity property in classical statistical mechanics which leads to the construction of a self adjoint contraction semigroup $\exp(-tH)$ acting on a Hilbert space $\mathcal{H}$. This yields the standard connection between quantum mechanics and path integrals. Verification of their condition in specific examples ultimately relies on a particular positivity condition for the inverse $(-\Delta_B + I)^{-1}$ of a Laplace operator Δ_B acting on $L_2(R^d)$. Here B denotes classical boundary conditions on the finite union of piecewise smooth hypersurfaces Γ in R^d. We show here that the positivity condition can be derived from monotonicity $C_D \leqslant C_N$ of Dirichlet and of Neumann Laplace operators.

Let Π denote a hyperplane in R^d and let $\theta = \theta_\Pi$ denote reflection in Π. Let $R^d_\pm$ denote the two connected components of $R^d \backslash \Pi$. Note that on $L_2(R^d)$, θ is self adjoint and $\theta^2 = I$. Let $\Pi_\pm$ denote the orthogonal projection onto $L_2(R^d_\pm)$.

DEFINITION 1. *An operator C is θ invariant if* $[\theta, C] = 0$.

DEFINITION 2. *A θ invariant operator C is OS positive with respect to θ if*:

$$\Pi_+ \theta C \Pi_+ \geqslant 0. \tag{1}$$

Remark. If $[\theta, C] = 0$, then (1) is equivalent to $\Pi_- \theta C \Pi_- \geqslant 0$.

THEOREM 1. *Let $C = (-\Delta_B + I)^{-1}$ be θ-invariant and let B denote Dirichlet and/or Neumann data on Γ. Then C is OS positive.*

---

[1] Supported in part by National Science Foundation Grant No. PHY 78-08066.
[2] Supported in part by National Science Foundation Grant No. PHY 77-18762.

Proof. Let $P_\pm = \frac{1}{2}(I \pm \theta)$ be the projections onto the even and odd eigenspaces of θ. Note $(I - \theta)C = 2P_-CP_-$, so the positivity condition (1) can be written:

$$\Pi_+ [C - 2P_-CP_-] \Pi_+ \geqslant 0. \qquad (2)$$

We assert first, that $2P_-CP_-$, restricted to $\Pi_\pm$, equals C_D restricted to $\Pi_\pm$. Here $C_D = (-\Delta'_B + I)^{-1}$ where Δ'_B is the Laplace operator with Dirichlet data on Π and B data on $\Gamma\backslash\Pi$. Secondly, we assert $C_D \leqslant C$. Hence (2) is a consequence of these assertions.

The first assertion follows from the observation that $(I - \theta)C(x, y)$ vanishes on Π, and that $(I - \theta)C$ satisfies $(-\Delta + I)(I - \theta)C(x, y) = \delta(x - y)$. The second assertion follows from the characterization of the bilinear form $\Pi_\pm C_D^{-1} \Pi_\pm$ as a restriction of the bilinear form C^{-1} to functions vanishing on Π and supported in $\Pi_\pm$.

We remark that on $\Pi_\pm$, $2P_+CP_+ = (I + \theta)C = C_N \geqslant C$ is the Neumann covariance, and a proof could be based on these facts. Note also that on $\Pi_\pm$, $\theta C = \frac{1}{2}(C_N - C_D)$.

We now allow periodic boundary conditions. Consider R^d replaced by the torus T^d; replace Π by a cross-section

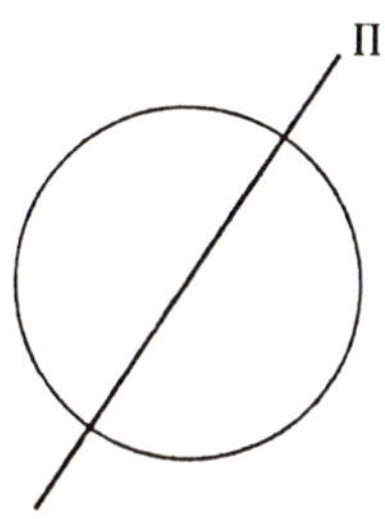

Fig. 1. Hyperplane reflection for a torus.

dividing T^d in two equal parts, cf. Fig. 1. Now $\Pi_\pm$ are the projections onto L_2 functions supported on either side of Π. We allow Dirichlet and/or Neumann data B on Γ as well. With these replacements, Definitions 1 and 2 are made. As above, we show:

THEOREM 2. *Let* $C = (-\Delta_B + I)^{-1}$ *be θ-invariant on* $L_2(T^d)$. *Then C is OS positive.*

Finally we remark that R^d or T^d can be replaced by a lattice R^d_δ or T^d_δ with spacing δ. In this case we obtain *OS* positivity for classical, θ-invariant boundary conditions on a lattice. Again we use the inequality $C_D \leqslant C$, which can also be established by introducing Dirichlet data with a mass perturbation of C^{-1} on the lattice.

THEOREM 3. *The lattice versions of Theorems* 1 *and* 2 *hold.*

REFERENCE

Osterwalder, K., and Schrader, R., *Commun. Math. Phys.* **31**, 83–112 (1973); **42**, 281–305 (1975).

(Received July 2, 1979)

378